TANGENTS AND SECANTS

30. $\displaystyle\int \tan u \, du = \ln|\sec u| + C$

31. $\displaystyle\int \sec u \, du = \ln|\sec u + \tan u| + C$

32. $\displaystyle\int \tan^2 u \, du = \tan u - u + C$

33. $\displaystyle\int \sec^2 u \, du = \tan u + C$

34. $\displaystyle\int \sec u \tan u \, du = \sec u + C$

35. $\displaystyle\int \tan^3 u \, du = \tfrac{1}{2}\tan^2 u + \ln|\cos u| + C$

36. $\displaystyle\int \sec^3 u \, du = \tfrac{1}{2}\sec u \tan u + \tfrac{1}{2}\ln|\sec u + \tan u| + C$

37. $\displaystyle\int \tan^n u \, du = \frac{\tan^{n-1} u}{n-1} - \int \tan^{n-2} u \, du$

38. $\displaystyle\int \sec^n u \, du = \frac{\sec^{n-2} u \tan u}{n-1} + \frac{n-2}{n-1}\int \sec^{n-2} u \, du$

CONTANGENTS AND COSECANTS

39. $\displaystyle\int \cot u \, du = \ln|\sin u| + C$

40. $\displaystyle\int \csc u \, du = \ln|\csc u - \cot u| + C$

41. $\displaystyle\int \cot^2 u \, du = -\cot u - u + C$

42. $\displaystyle\int \csc^2 u \, du = -\cot u + C$

43. $\displaystyle\int \csc u \cot u \, du = -\csc u + C$

44. $\displaystyle\int \cot^3 u \, du = -\tfrac{1}{2}\cot^2 u - \ln|\sin u| + C$

45. $\displaystyle\int \csc^3 u \, du = -\tfrac{1}{2}\csc u \cot u + \tfrac{1}{2}\ln|\csc u - \cot u| + C$

46. $\displaystyle\int \cot^n u \, du = -\frac{\cot^{n-1} u}{n-1} - \int \cot^{n-2} u \, du$

47. $\displaystyle\int \csc^n u \, du = -\frac{\csc^{n-2} u \cot u}{n-1} + \frac{n-2}{n-1}\int \csc^{n-2} u \, du$

HYPERBOLIC FUNCTIONS

48. $\displaystyle\int \sinh u \, du = \cosh u + C$

49. $\displaystyle\int \cosh u \, du = \sinh u + C$

50. $\displaystyle\int \tanh u \, du = \ln(\cosh u) + C$

51. $\displaystyle\int \coth u \, du = \ln|\sinh u| + C$

52. $\displaystyle\int \operatorname{sech} u \, du = \arctan(\sinh u) + C$

53. $\displaystyle\int \operatorname{csch} u \, du = \ln|\tanh \tfrac{1}{2}u| + C$

54. $\displaystyle\int \operatorname{sech}^2 u \, du = \tanh u + C$

55. $\displaystyle\int \operatorname{csch}^2 u \, du = -\coth u + C$

56. $\displaystyle\int \operatorname{sech} u \tanh u \, du = -\operatorname{sech} u + C$

57. $\displaystyle\int \operatorname{csch} u \coth u \, du = -\operatorname{csch} u + C$

58. $\displaystyle\int \sinh^2 u \, du = \tfrac{1}{4}\sinh 2u - \tfrac{1}{2}u + C$

59. $\displaystyle\int \cosh^2 u \, du = \tfrac{1}{4}\sinh 2u + \tfrac{1}{2}u + C$

60. $\displaystyle\int \tanh^2 u \, du = u - \tanh u + C$

61. $\displaystyle\int \coth^2 u \, du = u - \coth u + C$

62. $\displaystyle\int u \sinh u \, du = u \cosh u - \sinh u + C$

63. $\displaystyle\int u \cosh u \, du = u \sinh u - \cosh u + C$

(table continued at the back)

10 NBC

THE GREEK ALPHABET

A	α	alpha
B	β	beta
Γ	γ	gamma
Δ	δ	delta
E	ϵ	epsilon
Z	ζ	zeta
H	η	eta
Θ	θ	theta
I	ι	iota
K	κ	kappa
Λ	λ	lambda
M	μ	mu
N	ν	nu
Ξ	ξ	xi
O	o	omicron
Π	π	pi
P	ρ	rho
Σ	σ	sigma
T	τ	tau
Υ	υ	upsilon
Φ	ϕ	phi
X	χ	chi
Ψ	ψ	psi
Ω	ω	omega

FOR INSTRUCTORS

Wiley**PLUS** is built around the activities you perform in your class each day. With Wiley**PLUS** you can:

Prepare & Present
Create outstanding class presentations using a wealth of resources such as PowerPoint™ slides, image galleries, interactive simulations, and more. You can even add materials you have created yourself.

Create Assignments
Automate the assigning and grading of homework or quizzes by using the provided question banks, or by writing your own.

Track Student Progress
Keep track of your students' progress and analyze individual and overall class results.

Now Available with WebCT and Blackboard!

"It has been a great help, and I believe it has helped me to achieve a better grade."

Michael Morris,
Columbia Basin College

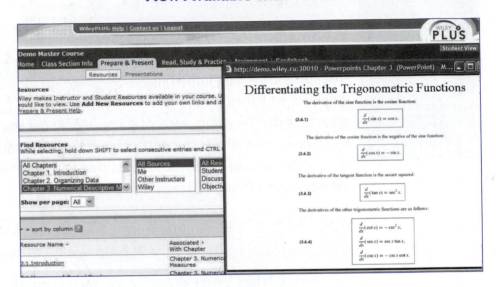

FOR STUDENTS

You have the potential to make a difference!

WileyPLUS is a powerful online system packed with features to help you make the most of your potential and get the best grade you can!

With Wiley**PLUS** you get:

• A complete online version of your text and other study resources.

• Problem-solving help, instant grading, and feedback on your homework and quizzes.

• The ability to track your progress and grades throughout the term.

For more information on what *WileyPLUS* can do to help you and your students reach their potential, please visit www.wiley.com/college/wileyplus.

76% of students surveyed said it made them better prepared for tests. *

*Based on a survey of 972 student users of *WileyPLUS*

THE WILEY BICENTENNIAL—KNOWLEDGE FOR GENERATIONS

Each generation has its unique needs and aspirations. When Charles Wiley first opened his small printing shop in lower Manhattan in 1807, it was a generation of boundless potential searching for an identity. And we were there, helping to define a new American literary tradition. Over half a century later, in the midst of the Second Industrial Revolution, it was a generation focused on building the future. Once again, we were there, supplying the critical scientific, technical, and engineering knowledge that helped frame the world. Throughout the 20th Century, and into the new millennium, nations began to reach out beyond their own borders and a new international community was born. Wiley was there, expanding its operations around the world to enable a global exchange of ideas, opinions, and know-how.

For 200 years, Wiley has been an integral part of each generation's journey, enabling the flow of information and understanding necessary to meet their needs and fulfill their aspirations. Today, bold new technologies are changing the way we live and learn. Wiley will be there, providing you the must-have knowledge you need to imagine new worlds, new possibilities, and new opportunities.

Generations come and go, but you can always count on Wiley to provide you the knowledge you need, when and where you need it!

WILLIAM J. PESCE
PRESIDENT AND CHIEF EXECUTIVE OFFICER

PETER BOOTH WILEY
CHAIRMAN OF THE BOARD

TENTH EDITION

SALAS

HILLE

ETGEN

CALCULUS

ONE VARIABLE

JOHN WILEY & SONS, INC.

In fond remembrance of
EINAR HILLE

ACQUISITIONS EDITOR	Mary Kittell
PUBLISHER	Laurie Rosatone
MARKETING MANAGER	Amy Sell
EDITORIAL ASSISTANT	Danielle Amico
MARKETING ASSISTANT	Tara Martinho
SENIOR PRODUCTION EDITOR	Sandra Dumas
DESIGNER	Hope Miller
SENIOR MEDIA EDITOR	Stefanie Liebman
PRODUCTION MANAGEMENT	Suzanne Ingrao
FREELANCE DEVELOPMENTAL EDITOR	Anne Scanlan-Rohrer
COVER IMAGE	Steven Puetzer/Masterfile

This book was set in New Times Roman by Techbooks, Inc. and printed and bound by Courier(Westford). The cover was printed by Courier(Westford).

This book is printed on acid-free paper. ∞

To order books or for customer service please, call 1-800-CALL WILEY (225-5945).

Library of Congress Cataloging-in-Publication Data

Salas, Saturnino L.
    Calculus—10th ed/Saturnino Salas, Einar Hille, Garrett Etgen.

    ISBN 978-0-470-07333-9

Printed in the United States of America
10 9 8 7 6 5 4 3

This text is devoted to the study of single variable calculus. While applications from the sciences, engineering, and economics are often used to motivate or illustrate mathematical ideas, the emphasis is on the three basic concepts of calculus: limit, derivative, and integral.

This edition is the result of a collaborative effort with S.L. Salas, who scrutinized every single sentence in the text for possible improvement in precision and readability. His gift for writing and his uncompromising standards of mathematical accuracy and clarity illuminate the beauty of the subject while increasing its accessibility to students. It has been a pleasure for me to work with him.

FEATURES OF THE TENTH EDITION

Precision and Clarity

The emphasis is on mathematical exposition; the topics are treated in a clear and understandable manner. Mathematical statements are careful and precise; the basic concepts and important points are not obscured by excess verbiage.

Balance of Theory and Applications

Problems drawn from the physical sciences are often used to introduce basic concepts in calculus. In turn, the concepts and methods of calculus are applied to a variety of problems in the sciences, engineering, business, and the social sciences through text examples and exercises. Because the presentation is flexible, instructors can vary the balance of theory and applications according to the needs of their students.

Accessibility

This text is designed to be completely accessible to the beginning calculus student without sacrificing appropriate mathematical rigor. The important theorems are explained and proved, and the mathematical techniques are justified. These may be covered or omitted according to the theoretical level desired in the course.

Visualization

The importance of visualization cannot be over-emphasized in developing students' understanding of mathematical concepts. For that reason, over 1200 illustrations accompany the text examples and exercise sets.

Technology

The technology component of the text has been strengthened by revising existing exercises and by developing new exercises. Well over half of the exercise sets have problems requiring either a graphing utility or a computer algebra system (CAS). Technology exercises (designated by ▷) are designed to illustrate or expand material developed within the sections.

Projects

Projects with an emphasis on problem solving offer students the opportunity to investigate a variety of special topics that supplement the text material. The projects typically require an approach that involves both theory and applications, including the use of technology. Many of the projects are suitable for group-learning activities.

Early Coverage of Differential Equations

Differential equations are formally introduced in Chapter 7 in connection with applications to exponential growth and decay. First-order linear equations, separable equations, and second linear equations with constant coefficients, plus a variety of applications, are treated in a separate chapter immediately following the techniques of integration material in Chapter 8.

CHANGES IN CONTENT AND ORGANIZATION

In our effort to produce an even more effective text, we consulted with the users of the Ninth Edition and with other calculus instructors. Our primary goals in preparing the Tenth Edition were the following:

1. **Improve the exposition.** As noted above, every topic has been examined for possible improvement in the clarity and accuracy of its presentation. Essentially every section in the text underwent some revision; a number of sections and subsections were completely rewritten.

2. **Improve the illustrative examples.** Many of the existing examples have been modified to enhance students' understanding of the material. New examples have been added to sections that were rewritten or substantially revised.

3. **Revise the exercise sets.** Every exercise set was examined for balance between drill problems, midlevel problems, and more challenging applications and conceptual problems. In many instances, the number of routine problems was reduced and new midlevel to challenging problems were added.

Specific changes made to achieve these goals and meet the needs of today's students and instructors include:

Comprehensive Chapter-End Review Exercise Sets

The Skill Mastery Review Exercise Sets introduced in the Ninth Edition have been expanded into chapter-end exercise sets. Each chapter concludes with a comprehensive set of problems designed to test and to re-enforce students' understanding of basic concepts and methods developed within the chapter. These review exercise sets average over 50 problems per set.

Precalculus Review (Chapter 1)

The subject matter of this chapter—inequalities, basic analytic geometry, the function concept and the elementary functions—is unchanged. However, much of the material has been rewritten and simplified.

Limits (Chapter 2)

The approach to limits is unchanged, but many of the explanations have been revised. The illustrative examples throughout the chapter have been modified, and new examples have been added.

Differentiation and Applications (Chapters 3 and 4)

There are some significant changes in the organization of this material. Realizing that our treatments of linear motion, rates of change per unit time, and the Newton-Raphson method depended on an understanding of increasing/decreasing functions and the concavity of graphs, we moved these topics from Chapter 3 (the derivative) to Chapter 4 (applications of the derivative). Thus, Chapter 3 is now a shorter chapter which focuses solely on the derivative and the processes of differentiation, and Chapter 4 is expanded to encompass all of the standard applications of the derivative—curve-sketching, optimization, linear motion, rates of change, and approximation. As in all previous editions, Chapter 4 begins with the mean-value theorem as the theoretical basis for all the applications.

Integration and Applications (Chapters 5 and 6)

In a brief introductory section, area and distance are used to motivate the definite integral in Chapter 5. While the definition of the definite integral is based on upper and lower sums, the connection with Riemann sums is also given. Explanations, examples, and exercises throughout Chapters 5 and 6 have been modified, but the topics and organization remain as in the Ninth Edition.

The Transcendental Functions, Techniques of Integration (Chapters 7 and 8)

The coverage of the inverse trigonometric functions (Chapter 7) has been reduced slightly. The treatment of powers of the trigonometric functions (Chapter 8) has been completely rewritten. The optional sections on first-order linear differential equations and separable differential equations have been moved to Chapter 9, the new chapter on differential equations.

Some Differential Equations (Chapter 9)

This new chapter is a brief introduction to differential equations and their applications. In addition to the coverage of first-order linear equations and separable equations noted above, we have moved the section on second-order linear homogeneous equations with constant coefficients from the Ninth Edition's Chapter 18 to this chapter.

Sequences and Series (Chapters 11 and 12)

Efforts were made to reduce the overall length of these chapters through rewriting and eliminating peripheral material. Eliminating extraneous problems reduced several exercise sets. Some notations and terminology have been modified to be consistent with common usage.

SUPPLEMENTS

An Instructor's Solutions Manual, ISBN 0470127309, includes solutions for all problems in the text.

A Student Solutions Manual, ISBN 0470105534, includes solutions for selected problems in the text.

A Companion Web site, www.wiley.com/college/salas, provides a wealth of resources for students and instructors, including:

- **PowerPoint Slides** for important ideas and graphics for study and note taking.
- **Online Review Quizzes** to enable students to test their knowledge of key concepts. For further review diagnostic feedback is provided that refers to pertinent sections of the text.
- **Animations** comprise a series of interactive Java applets that allow students to explore the geometric significance of many major concepts of Calculus.
- **Algebra and Trigonometry Refreshers** is a self-paced, guided review of key algebra and trigonometry topics that are essential for mastering calculus.
- **Personal Response System Questions** provide a convenient source of questions to use with a variety of personal response systems.
- **Printed Test Bank** contains static tests which can be printed for quick tests.
- **Computerized Test Bank** includes questions from the printed test bank with algorithmically generated problems.

WILEYPLUS

Expect More from Your Classroom Technology

This text is supported by *WileyPLUS*—a powerful and highly integrated suite of teaching and learning resources designed to bridge the gap between what happens in the classroom and what happens at home. *WileyPLUS* includes a complete online version of the text, algorithmically generated exercises, all of the text supplements, plus course and homework management tools, in one easy-to-use website.

Organized Around the Everyday Activities You Perform in Class, *WileyPLUS* Helps You:

Prepare and present: *WileyPLUS* lets you create class presentations quickly and easily using a wealth of Wiley-provided resources, including an online version of the textbook, PowerPoint slides, and more. You can adapt this content to meet the needs of your course.

Create assignments: *WileyPLUS* enables you to automate the process of assigning and grading homework or quizzes. You can use algorithmically generated problems from the text's accompanying test bank, or write your own.

Track student progress: An instructor's grade book allows you to analyze individual and overall class results to determine students' progress and level of understanding.

Promote strong problem-solving skills: *WileyPLUS* can link homework problems to the relevant section of the online text, providing students with context-sensitive help. *WileyPLUS* also features mastery problems that promote conceptual understanding of key topics and video walkthroughs of example problems.

Provide numerous practice opportunities: Algorithmically generated problems provide unlimited self-practice opportunities for students, as well as problems for homework and testing.

Support varied learning styles: *WileyPLUS* includes the entire text in digital format, enhanced with varied problem types to support the array of different student learning styles in today's classroom.

Administer your course: You can easily integrate *WileyPLUS* with another course management system, grade books, or other resources you are using in your class, enabling you to build your course your way.

WileyPLUS Includes A Wealth of Instructor and Student Resources:

Student Solutions Manual: Includes worked-out solutions for all odd-numbered problems and study tips.

Instructor's Solutions Manual: Presents worked out solutions to all problems.

PowerPoint Lecture Notes: In each section of the book a corresponding set of lecture notes and worked out examples are presented as PowerPoint slides that are tied to the examples in the text.

View an online demo at www.wiley.com/college/wileyplus or contact your local Wiley representative for more details.

The Wiley Faculty Network—Where Faculty Connect

The Wiley Faculty Network is a faculty-to-faculty network promoting the effective use of technology to enrich the teaching experience. The Wiley Faculty Network facilitates the exchange of best practices, connects teachers with technology, and helps to enhance instructional efficiency and effectiveness. The network provides technology training and tutorials, including *WileyPLUS* training, online seminars, peer-to-peer exchanges of experiences and ideas, personalized consulting, and sharing of resources.

Connect with a Colleague

Wiley Faculty Network mentors are faculty like you, from educational institutions around the country, who are passionate about enhancing instructional efficiency and effectiveness through best practices. You can engage a faculty mentor in an online conversation at www.wherefacultyconnect.com.

Connect with the Wiley Faculty Network

Web: www.wherefacultyconnect.com
Phone: 1-866-FACULTY

ACKNOWLEDGMENTS

The revision of a text of this magnitude and stature requires a lot of encouragement and help. I was fortunate to have an ample supply of both from many sources. The present book owes much to the people who contributed to the first nine editions, most recently: Omar Adawi, Parkland College; Mihaly Bakonyi, Georgia State University; Edward B. Curtis, University of Washington; Boris A. Datskovsky, Temple University; Kathy Davis University of Texas-Austin; Dennis DeTurck, University of Pennsylvania; John R. Durbin, University of Texas-Austin; Ronald Gentle, Eastern Washington University; Robert W. Ghrist, Georgia Institute of Technology; Charles H. Giffen, University of Virginia–Charlottesville; Michael Kinyon, Indiana University-South Bend; Susan J. Lamon, Marquette University; Peter A. Lappan, Michigan State University; Nicholas Macri, Temple University; James Martino, Johns Hopkins University; James R. McKinney, California State Polytechnic University-Pomona; Jeff Morgan, Texas A & M University; Peter J. Mucha, Georgia Institute of Technology; Elvira Munoz-Garcia, University of California, Los Angeles; Ralph W. Oberste-Vorth, University of South Florida; Charles Odion, Houston Community College; Charles Peters, University of Houston; Clifford S. Queen, Lehigh University; J. Terry Wilson, San Jacinto College Central and Yang Wang, Georgia Institute of Technology. I am deeply indebted to all of them.

The reviewers and contributors to the Tenth Edition supplied detailed criticisms and valuable suggestions. I offer my sincere appreciation to the following individuals:

Omar Adawi,	Parkland College
Ulrich Albrecht,	Auburn University
Joseph Borzellino,	California Polytechnic State University, San Luis Obispo
Michael R. Colvin,	The University of St. Thomas
James Dare,	Indiana-Purdue University at Fort Wayne
Nasser Dastrange,	Buena Vista University
David Dorman,	Middlebury College
Martin E. Flashman,	Humboldt State University/Occidental College
David Frank,	University of Minnesota
Melanie Fulton,	High Point University
Isobel Gaensler,	Georgia State University
Frieda Ganter,	West Hills College, Lemoore
Murli M. Gupta,	George Washington University
Aida Kadic-Galeb,	University of Tampa
Mohammad Ghomi,	Georgia Institute of Technology
Robert W. Ghrist,	University of Illinois, Urbana-Champaign
Semion Gutman,	University of Oklahoma
Rahim G. Karimpour,	Southern Illinois University- Edwardsville
Robert Keller,	Loras College
Kevin P. Knudson,	Mississippi State University
Ashok Kumar,	Valdosta State University
Jeff Leader,	Rose-Hulman Institute of Technology
Xin Li,	University of Central Florida
Doron Lubinsky,	Georgia Institute of Technology

Edward T. Migliore,	Monterey Peninsula College/University of California, Santa Cruz
Maya Mukherjee,	Morehouse College
Sanjay Mundkur,	Kennesaw State University
Michael M. Neumann,	Mississippi State University
Charles Odion,	Houston Community College
Dan Ostrov,	Santa Clara University
Shahrokh Parvini,	San Diego Mesa College
Chuang Peng,	Morehouse College
Kanishka Perera,	Florida Institute of Technology
Denise Reid,	Valdosta State University
Paul Seeburger,	Monroe Community College
Constance Schober,	University of Central Florida
Peter Schumer,	Middlebury College
Kimberly Shockey	
Henry Smith,	River Parishes Community College
James Thomas,	Colorado State University
James L. Wang,	University of Alabama
Ying Wang,	Augusta State University
Charles Waters,	Minnesota State University, Mankato
Mark Woodard,	Furman University
Yan Wu,	Georgia Southern University
Dekang Xu,	University of Houston

I am especially grateful to Paul Lorczak and Neil Wigley, who carefully read the revised material. They provided many corrections and helpful comments. I would like to thank Bill Ardis for his advice and guidance in the creation of the new technology exercises in the text.

I am deeply indebted to the editorial staff at John Wiley & Sons. Everyone involved in this project has been encouraging, helpful, and thoroughly professional at every stage. In particular, Laurie Rosatone, Publisher, Mary Kittell, Acquisitions Editor, Anne Scanlan-Rohrer, Freelance Developmental Editor and Danielle Amico, Editorial Assistant, provided organization and support when I needed it and prodding when prodding was required. Special thanks go to Sandra Dumas, Production Editor, and Suzanne Ingrao, Production Coordinator, who were patient and understanding as they guided the project through the production stages; Hope Miller, Designer, whose creativity produced the attractive interior design as well as the cover and Stefanie Liebman, Senior Media Editor, who has produced valuable media resources to support my text.

Finally, I want to acknowledge the contributions of my wife, Charlotte; without her continued support, I could not have completed this work.

Garret J. Etgen

CONTENTS

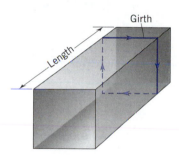

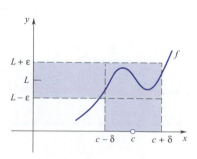

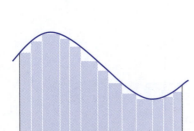

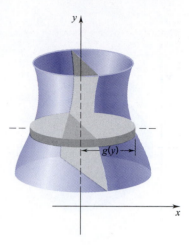

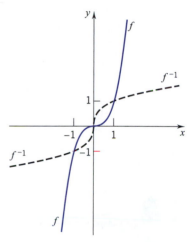

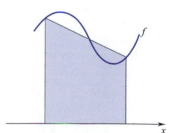

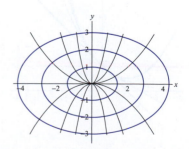

*Denotes optional section.

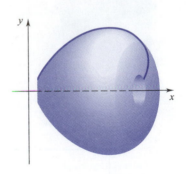
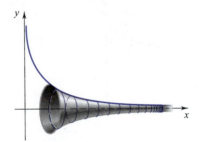
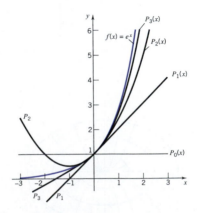

TENTH EDITION

SALAS

HILLE

ETGEN

CALCULUS

ONE VARIABLE

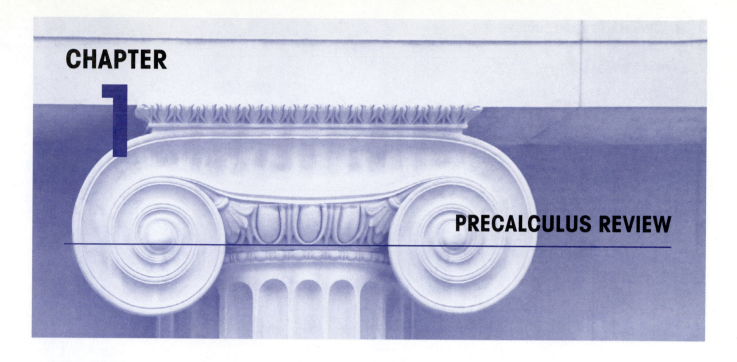

CHAPTER 1

PRECALCULUS REVIEW

In this chapter we gather together for reference and review those parts of elementary mathematics that are necessary for the study of calculus. We assume that you are familiar with most of this material and that you don't require detailed explanations. But first a few words about the nature of calculus and a brief outline of the history of the subject.

■ 1.1 WHAT IS CALCULUS?

To a Roman in the days of the empire, a "calculus" was a pebble used in counting and gambling. Centuries later, "calculare" came to mean "to calculate," "to compute," "to figure out." For our purposes, calculus is elementary mathematics (algebra, geometry, trigonometry) enhanced by *the limit process*.

Calculus takes ideas from elementary mathematics and extends them to a more general situation. Some examples are on pages 2 and 3. On the left-hand side you will find an idea from elementary mathematics; on the right, this same idea as extended by calculus.

It is fitting to say something about the history of calculus. The origins can be traced back to ancient Greece. The ancient Greeks raised many questions (often paradoxical) about tangents, motion, area, the infinitely small, the infinitely large—questions that today are clarified and answered by calculus. Here and there the Greeks themselves provided answers (some very elegant), but mostly they provided only questions.

Elementary Mathematics	Calculus
slope of a line $y = mx + b$	slope of a curve $y = f(x)$

(Table continues)

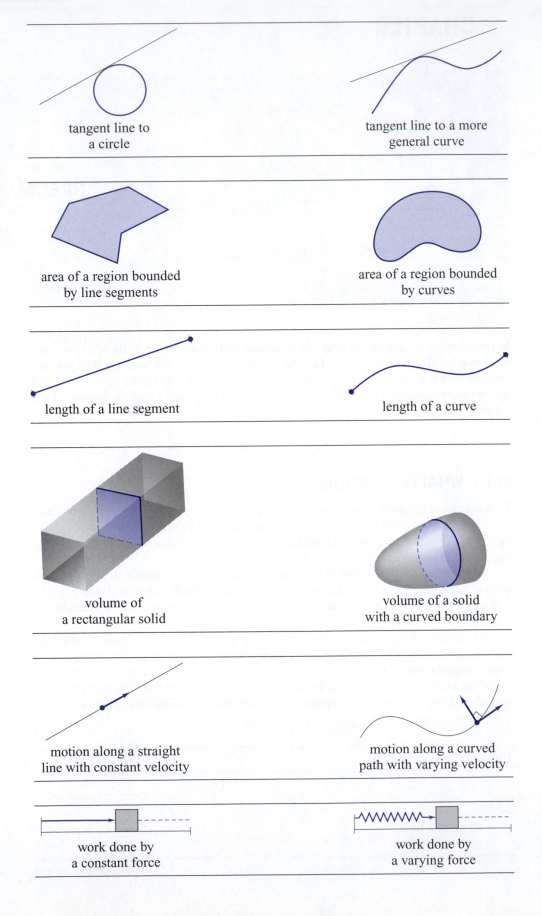

tangent line to
a circle

tangent line to a more
general curve

area of a region bounded
by line segments

area of a region bounded
by curves

length of a line segment

length of a curve

volume of
a rectangular solid

volume of a solid
with a curved boundary

motion along a straight
line with constant velocity

motion along a curved
path with varying velocity

work done by
a constant force

work done by
a varying force

mass of an object of constant density	mass of an object of varying density

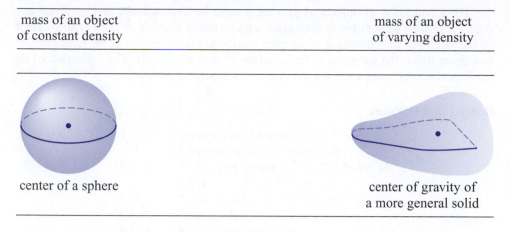

center of a sphere	center of gravity of a more general solid

After the Greeks, progress was slow. Communication was limited, and each scholar was obliged to start almost from scratch. Over the centuries, some ingenious solutions to calculus-type problems were devised, but no general techniques were put forth. Progress was impeded by the lack of a convenient notation. Algebra, founded in the ninth century by Arab scholars, was not fully systematized until the sixteenth century. Then, in the seventeenth century, Descartes established analytic geometry, and the stage was set.

The actual invention of calculus is credited to Sir Isaac Newton (1642–1727), an Englishman, and to Gottfried Wilhelm Leibniz (1646–1716), a German. Newton's invention is one of the few good turns that the great plague did mankind. The plague forced the closing of Cambridge University in 1665, and young Isaac Newton of Trinity College returned to his home in Lincolnshire for eighteen months of meditation, out of which grew his *method of fluxions*, his *theory of gravitation*, and *his theory of light*. The method of fluxions is what concerns us here. A treatise with this title was written by Newton in 1672, but it remained unpublished until 1736, nine years after his death. The new method (calculus to us) was first announced in 1687, but in vague general terms without symbolism, formulas, or applications. Newton himself seemed reluctant to publish anything tangible about his new method, and it is not surprising that its development on the Continent, in spite of a late start, soon overtook Newton and went beyond him.

Leibniz started his work in 1673, eight years after Newton. In 1675 he initiated the basic modern notation: dx and $\int$. His first publications appeared in 1684 and 1686. These made little stir in Germany, but the two brothers Bernoulli of Basel (Switzerland) took up the ideas and added profusely to them. From 1690 onward, calculus grew rapidly and reached roughly its present state in about a hundred years. Certain theoretical subtleties were not fully resolved until the twentieth century.

■ 1.2 REVIEW OF ELEMENTARY MATHEMATICS

In this section we review the terminology, notation, and formulas of elementary mathematics.

Sets

A *set* is a collection of distinct objects. The objects in a set are called the *elements* or *members* of the set. We will denote sets by capital letters A, B, C, ... and use lowercase letters a, b, c, ... to denote the elements.

For a collection of objects to be a set it must be *well-defined*; that is, given any object x, it must be possible to determine with certainty whether or not x is an element of the set. Thus the collection of all even numbers, the collection of all lines parallel to a given line l, the solutions of the equation $x^2 = 9$ are all sets. The collection of all intelligent adults is not a set. It's not clear who should be included.

Notions and Notation

the object x is in the set A	$x \in A$
the object x is not in the set A	$x \notin A$
the set of all x which satisfy property P	$\{x : P\}$
$(\{x : x^2 = 9\} = \{-3, 3\})$	
A is a subset of B, A is contained in B	$A \subseteq B$
B contains A	$B \supseteq A$
the union of A and B	$A \cup B$
$(A \cup B = \{x : x \in A \text{ or } x \in B\})$	
the intersection of A and B	$A \cap B$
$(A \cap B = \{x : x \in A \text{ and } x \in B\})$	
the empty set	$\emptyset$

These are the only notions from set theory that you will need at this point.

Real Numbers

Classification

positive integers[†]	$1, 2, 3, \ldots$
integers	$0, 1, -1, 2, -2, 3, -3, \ldots$
rational numbers	p/q, with p, q integers, $q \neq 0$;
	for example, $5/2, -19/7, -4/1 = -4$
irrational numbers	real numbers which are not rational;
	for example $\sqrt{2}, \sqrt[3]{7}, \pi$

Decimal Representation

Each real number can be expressed as a decimal. To express a rational number p/q as a decimal, we divide the denominator q into the numerator p. The resulting decimal either *terminates* or *repeats*:

$$\frac{3}{5} = 0.6, \qquad \frac{27}{20} = 1.35, \qquad \frac{43}{8} = 5.375$$

are terminating decimals;

$$\frac{2}{3} = 0.6666\cdots = 0.\overline{6}, \qquad \frac{15}{11} = 1.363636\cdots = 1.\overline{36}, \qquad \text{and}$$

$$\frac{116}{37} = 3.135135\cdots = 3.\overline{135}$$

are repeating decimals. (The bar over the sequence of digits indicates that the sequence repeats indefinitely.) The converse is also true; namely, every terminating or repeating decimal represents a rational number.

[†] Also called *natural numbers*.

The decimal expansion of an irrational number can neither terminate nor repeat. The expansions

$$\sqrt{2} = 1.414213562\cdots \qquad \text{and} \qquad \pi = 3.141592653\cdots$$

do not terminate and do not develop any repeating pattern.

If we stop the decimal expansion of a given number at a certain decimal place, then the result is a rational number that approximates the given number. For instance, $1.414 = 1414/1000$ is a rational number approximation to $\sqrt{2}$ and $3.14 = 314/100$ is a rational number approximation to π. More accurate approximations can be obtained by using more decimal places from the expansions.

The Number Line (Coordinate Line, Real Line)

On a horizontal line we choose a point O. We call this point the *origin* and assign to it *coordinate* 0. Now we choose a point U to the right of O and assign to it *coordinate* 1. See Figure 1.2.1. The distance between O and U determines a scale (a unit length). We go on as follows: the point a units to the right of O is assigned coordinate a; the point a units to the left of O is assigned coordinate $-a$.

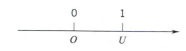

Figure 1.2.1

In this manner we establish a one-to-one correspondence between the points of a line and the numbers of the real number system. Figure 1.2.2 shows some real numbers represented as points on the number line. Positive numbers appear to the right of 0, negative numbers to the left of 0.

Figure 1.2.2

Order Properties

  (i) Either $a < b$, $b < a$, or $a = b$.   (trichotomy)

 (ii) If $a < b$ and $b < c$, then $a < c$.

(iii) If $a < b$, then $a + c < b + c$ for all real numbers c.   (transitivity)

 (iv) If $a < b$ and $c > 0$, then $ac < bc$.

  (v) If $a < b$ and $c < 0$, then $ac > bc$.

(Techniques for solving inequalities are reviewed in Section 1.3.)

Density

Between any two real numbers there are infinitely many rational numbers and infinitely many irrational numbers. In particular, *there is no smallest positive real number.*

Absolute Value

$$|a| = \begin{cases} a, & \text{if } a \geq 0 \\ -a, & \text{if } a < 0. \end{cases}$$

other characterizations $|a| = \max\{a, -a\}$; $|a| = \sqrt{a^2}$.

geometric interpretation $|a| =$ distance between a and 0;

$|a - c| =$ distance between a and c.

properties   **(i)** $|a| = 0$ iff $a = 0.$[†]
          (ii) $|-a| = |a|.$
          (iii) $|ab| = |a||b|.$
          (iv) $|a + b| \leq |a| + |b|.$ (the triangle inequality)[††]
          (v) $||a| - |b|| \leq |a - b|.$ (a variant of the triangle inequality)
          (vi) $|a|^2 = |a^2| = a^2.$

Techniques for solving inequalities that feature absolute value are reviewed in Section 1.3.

Intervals

Suppose that $a < b$. The *open interval* (a, b) is the set of all numbers between a and b:

$$(a, b) = \{x : a < x < b\}.$$

The *closed interval* $[a, b]$ is the open interval (a, b) together with the endpoints a and b:

$$[a, b] = \{x : a \leq x \leq b\}.$$

There are seven other types of intervals:

$$(a, b] = \{x : a < x \leq b\},$$

$$[a, b) = \{x : a \leq x \leq b\},$$

$$(a, \infty) = \{x : a < x\},$$

$$[a, \infty) = \{x : a \leq x\},$$

$$(-\infty, b) = \{x : x < b\},$$

$$(-\infty, b] = \{x : x \leq b\},$$

$$(-\infty, \infty) = \text{the set of real numbers.}$$

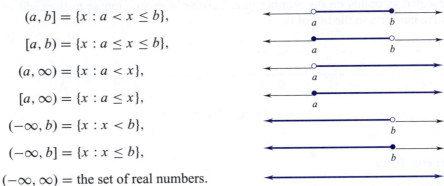

Interval notation is easy to remember: we use a square bracket to include an endpoint and a parenthesis to exclude it. On a number line, inclusion is indicated by a solid dot, exclusion by an open dot. The symbols ∞ and $-\infty$, read "infinity" and "negative infinity" (or "minus infinity"), do not represent real numbers. In the intervals listed above, the symbol ∞ is used to indicate that the interval extends indefinitely in the positive direction; the symbol $-\infty$ is used to indicate that the interval extends indefinitely in the negative direction.

Open and Closed

Any interval that contains no endpoints is called *open*: $(a, b), (a, \infty), (-\infty, b),$ $(-\infty, \infty)$ are open. Any interval that contains each of its endpoints (there may be one or two) is called *closed*: $[a, b], [a, \infty), (-\infty, b]$ are closed. The intervals $(a, b]$ and $[a, b)$ are called *half-open (half-closed)*: $(a, b]$ is open on the left and closed on the right; $[a, b)$ is closed on the left and open on the right. Points of an interval which are not endpoints are called *interior* points of the interval.

[†]By "iff" we mean "if and only if." This expression is used so often in mathematics that it's convenient to have an abbreviation for it.

[††]The absolute value of the sum of two numbers cannot exceed the sum of their absolute values. This is analogous to the fact that in a triangle the length of one side cannot exceed the sum of the lengths of the other two sides.

Boundedness

A set S of real numbers is said to be:

(i) *Bounded above* if there exists a real number M such that

$$x \leq M \qquad \text{for all} \qquad x \in S;$$

such a number M is called an *upper bound* for S.

(ii) *Bounded below* if there exists a real number m such that

$$m \leq x \qquad \text{for all} \qquad x \in S;$$

such a number m is called a *lower bound* for S.

(iii) *Bounded* if it is bounded above and below.[†]

Note that if M is an upper bound for S, then any number greater than M is also an upper bound for S, and if m is a lower bound for S, than any number less than m is also a lower bound for S.

Examples The intervals $(-\infty, 2]$ and $(-\infty, 2)$ are both bounded above by 2 (and by every number greater than 2), but these sets are not bounded below. The set of positive integers $\{1, 2, 3, \ldots\}$ is bounded below by 1 (and by every number less than 1), but the set is not bounded above; there being no number M greater than or equal to all positive integers, the set has no upper bound. All finite sets of numbers are bounded— (bounded below by the least element and bounded above by the greatest). Finally, the set of all integers, $\{\cdots, -3, -2, -1, 0, 1, 2, 3, \cdots\}$, is unbounded in both directions; it is unbounded above and unbounded below. ❏

Factorials

Let n be a positive integer. By *n factorial*, denoted $n!$, we mean the product of the integers from n down to 1:

$$n! = n(n-1)(n-2)\cdots 3 \cdot 2 \cdot 1.$$

In particular

$$1! = 1, \ 2! = 2 \cdot 1 = 2, \ 3! = 3 \cdot 2 \cdot 1 = 6, \ 4! = 4 \cdot 3 \cdot 2 \cdot 1 = 24, \ \text{and so on.}$$

For convenience we define $0! = 1$.

Algebra

Powers and Roots

a real, p a positive integer	$a^1 = a, \quad a^p = \overbrace{a \cdot a \cdots\cdot a}^{p \text{ factors}}$
	$a \neq 0: \quad a^0 = 1, \quad a^{-p} = 1/a^p$
laws of exponents	$a^{p+q} = a^p a^q, \quad a^{p-q} = a^p a^{-q}, \quad (a^q)^p = a^{pq}$
a real, q odd	$a^{1/q}$, called the qth root of a, is the number b such that $b^q = a$
a nonnegative, q even	$a^{1/q}$ is the nonnegative number b such that $b^q = a$
notation	$a^{1/q}$ can be written $\sqrt[q]{a}$ ($a^{1/2}$ is written $\sqrt{a}$)
rational exponents	$a^{p/q} = (a^{1/q})^p$

[†]In defining *bounded above*, *bounded below*, and *bounded* we used the conditional "if," not "iff." We could have used "iff," but that would have been unnecessary. Definitions are by their very nature "iff" statements.

Examples

$$2^0 = 1, \ 2^1 = 1, \ 2^2 = 2 \cdot 2 = 4, \ 2^3 = 2 \cdot 2 \cdot 2 = 8, \ \text{and so on}$$

$$2^{5+3} = 2^5 \cdot 2^3 = 32 \cdot 8 = 256, \ 2^{3-5} = 2^{-2} = 1/2^2 = 1/4$$

$$(2^2)^3 = 2^{3 \cdot 2} = 2^6 = 64, \ (2^3)^2 = 2^{2 \cdot 3} = 2^6 = 64$$

$$8^{1/3} = 2, \ (-8)^{1/3} = -2, \ 16^{1/2} = \sqrt{16} = 4, \ 16^{1/4} = 2$$

$$8^{5/3} = (8^{1/3})^5 = 2^5 = 32, \ 8^{-5/3} = (8^{1/3})^{-5} = 2^{-5} = 1/2^5 = 1/32 \quad \square$$

Basic Formulas

$$(a + b)^2 = a^2 + 2ab + b^2$$

$$(a - b)^2 = a^2 - 2ab + b^2$$

$$(a + b)^3 = a^3 + 3a^2b + 3ab^2 + b^3$$

$$(a - b)^3 = a^3 - 3a^2b + 3ab^2 - b^3$$

$$a^2 - b^2 = (a - b)(a + b)$$

$$a^3 - b^3 = (a - b)(a^2 + ab + b^2)$$

$$a^4 - b^4 = (a - b)(a^3 + a^2b + ab^2 + b^3)$$

More generally:

$$a^n - b^n = (a - b)(a^{n-1} + a^{n-2}b + \cdots + ab^{n-2} + b^{n-1})$$

Quadratic Equations

The roots of a quadratic equation

$$ax^2 + bx + c = 0 \qquad \text{with } a \neq 0$$

are given by the general quadratic formula

$$r = \frac{-b \pm \sqrt{b^2 - 4ac}}{2a}.$$

If $b^2 - 4ac > 0$, the equation has two real roots; if $b^2 - 4ac = 0$, the equation has one real root; if $b^2 - 4ac < 0$, the equation has no real roots. (*It has two complex roots.*)

Geometry

Elementary Figures

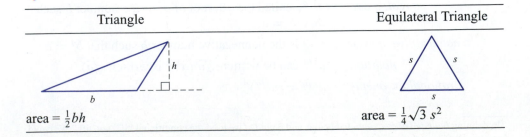

Triangle	Equilateral Triangle
area $= \frac{1}{2}bh$	area $= \frac{1}{4}\sqrt{3}\, s^2$

Rectangle	Rectangular Solid

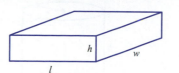

area = lw
perimeter = $2l + 2w$
diagonal = $\sqrt{l^2 + w^2}$

volume = lwh
surface area = $2lw + 2lh + 2wh$

Square	Cube

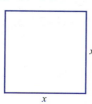

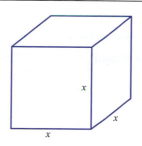

area = x^2
perimeter = $4x$
diagonal = $x\sqrt{2}$

volume = x^3
surface area = $6x^2$

Circle	Sphere

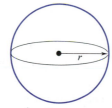

area = πr^2
circumference = $2\pi r$

volume = $\frac{4}{3}\pi r^3$
surface area = $4\pi r^2$

Sector of a Circle: radius r, central angle θ measured in radians (see Section 1.6).

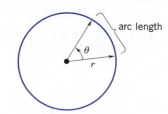

arc length

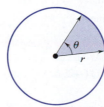

arc length = $r\theta$

area = $\frac{1}{2}r^2\theta$

(*Table continues*)

Right Circular Cylinder	Right Circular Cone
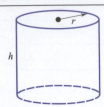	

volume $= \pi r^2 h$

lateral area $= 2\pi rh$

total surface area $= 2\pi r^2 + 2\pi rh$

volume $= \frac{1}{3}\pi r^2 h$

slant height $= \sqrt{r^2 + h^2}$

lateral area $= \pi r \sqrt{r^2 + h^2}$

total surface area $= \pi r^2 + \pi r \sqrt{r^2 + h^2}$

EXERCISES 1.2

Exercises 1–10. Is the number rational or irrational?

1. $\frac{17}{7}$.

2. -6.

3. $2.131313\ldots = 2.\overline{13}$.

4. $\sqrt{2} - 3$.

5. 0.

6. $\pi - 2$.

7. $\sqrt[3]{8}$.

8. 0.125.

9. $-\sqrt{9}$.

10. $(\sqrt{2} - \sqrt{3})(\sqrt{2} + \sqrt{3})$

Exercises 11–16. Replace the symbol $*$ by $<$, $>$, or $=$ to make the statement true.

11. $\frac{3}{4} * 0.75$.

12. $0.33 * \frac{1}{3}$.

13. $\sqrt{2} * 1.414$.

14. $4 * \sqrt{16}$.

15. $-\frac{2}{7} * -0.285714$.

16. $\pi * \frac{22}{7}$.

Exercises 17–23. Evaluate

17. $|6|$.

18. $|-4|$.

19. $|-3-7|$.

20. $|-5| - |8|$.

21. $|-5| + |-8|$.

22. $|2 - \pi|$.

23. $|5 - \sqrt{5}|$.

Exercises 24–33. Indicate on a number line the numbers x that satisfy the condition.

24. $x \geq 3$

25. $x \leq -\frac{3}{2}$.

26. $-2 \leq x \leq 3$.

27. $x^2 < 16$.

28. $x^2 \geq 16$.

29. $|x| \leq 0$.

30. $x^2 \geq 0$.

31. $|x - 4| \leq 2$.

32. $|x + 1| > 3$.

33. $|x + 3| \leq 0$.

Exercises 34–40. Sketch the set on a number line.

34. $[3, \infty)$.

35. $(-\infty, 2)$.

36. $(-4, 3]$.

37. $[-2, 3] \cup [1, 5]$.

38. $\left[-3, \frac{3}{2}\right) \cap \left(\frac{3}{2}, \frac{5}{2}\right]$.

39. $(-\infty, -1) \cup (-2, \infty)$.

40. $(-\infty, 2) \cap [3, \infty)$.

Exercises 41–47. State whether the set is bounded above, bounded below, bounded. If a set is bounded above, give an upper bound; if it is bounded below, give a lower bound; if it is bounded, give an upper bound and a lower bound.

41. $\{0, 1, 2, 3, 4\}$.

42. $\{0, -1, -2, -3, \ldots\}$.

43. The set of even integers.

44. $\{x : x \leq 4\}$.

45. $\{x : x^2 > 3\}$.

46. $\left\{\frac{n-1}{n} : n = 1, 2, 3 \ldots\right\}$.

47. The set of rational numbers less than $\sqrt{2}$.

Exercises 48–50.

48. Order the following numbers and place them on a number line: $\sqrt[3]{\pi}$, $2^{\sqrt{\pi}}$, $\sqrt{2}$, 3^π, π^3.

49. Let $x_0 = 2$ and define $x_n = \dfrac{17 + 2x_{n-1}^3}{3x_{n-1}^2}$ for $n = 1, 2, 3, 4, \ldots$ Find at least five values for x_n. Is the set $S = \{x_0, x_1, x_2, \ldots, x_n, \ldots\}$ bounded above, bounded below, bounded? If so, give a lower bound and/or an upper bound for S. If n is a large positive integer, what is the approximate value of x_n?

50. Rework Exercise 49 with $x_0 = 3$ and $x_n = \dfrac{231 + 4x_{n-1}^5}{5x_{n-1}^4}$

Exercises 51–56. Write the expression in factored form.

51. $x^2 - 10x + 25$.

52. $9x^2 - 4$.

53. $8x^6 + 64$.

54. $27x^3 - 8$.

55. $4x^2 + 12x + 9$.

56. $4x^4 + 4x^2 + 1$.

Exercises 57–64. Find the real roots of the equation.

57. $x^2 - x - 2 = 0$.

58. $x^2 - 9 = 0$.

59. $x^2 - 6x + 9 = 0$.

60. $2x^2 - 5x - 3 = 0$.

61. $x^2 - 2x + 2 = 0$.

62. $x^2 + 8x + 16 = 0$.

63. $x^2 + 4x + 13 = 0$.

64. $x^2 - 2x + 5 = 0$.

Exercises 65–69. Evaluate.

65. $5!$.

66. $\dfrac{5!}{8!}$.

67. $\dfrac{8!}{3!5!}$.

68. $\dfrac{9!}{3!6!}$.

69. $\dfrac{7!}{0!7!}$.

70. Show that the sum of two rational numbers is a rational number.

71. Show that the sum of a rational number and an irrational number is irrational.

72. Show that the product of two rational numbers is a rational number.

73. Is the product of a rational number and an irrational number necessarily rational? necessarily irrational?

74. Show by example that the sum of two irrational numbers (a) can be rational; (b) can be irrational. Do the same for the product of two irrational numbers.

75. Prove that $\sqrt{2}$ is irrational. HINT: Assume that $\sqrt{2} = p/q$ with the fraction written in lowest terms. Square both sides of this equation and argue that both p and q must be divisible by 2.

76. Prove that $\sqrt{3}$ is irrational.

77. Let S be the set of all rectangles with perimeter P. Show that the square is the element of S with largest area.

78. Show that if a circle and a square have the same perimeter, then the circle has the larger area. Given that a circle and a rectangle have the same perimeter, which has the larger area?

The following mathematical tidbit was first seen by one of the authors many years ago in Granville, Longley, and Smith, *Elements of Calculus*, now a Wiley book.

79. *Theorem* (a phony one): $1 = 2$.

PROOF (a phony one): Let a and b be real numbers, both different from 0. Suppose now that $a = b$. Then

$$ab = b^2$$
$$ab - a^2 = b^2 - a^2$$
$$a(b - a) = (b + a)(b - a)$$
$$a = b + a.$$

Since $a = b$, we have

$$a = 2a.$$

Division by a, which by assumption is not 0, gives

$$1 = 2. \quad \square$$

What is wrong with this argument?

■ 1.3 REVIEW OF INEQUALITIES

All our work with inequalities is based on the order properties of the real numbers given in Section 1.2. In this section we work with the type of inequalities that arise frequently in calculus, inequalities that involve a variable.

To solve an inequality in x is to find the numbers x that satisfy the inequality. These numbers constitute a set, called the *solution set* of the inequality.

We solve inequalities much as we solve an equation, but there is one important difference. We can maintain an inequality by adding the same number to both sides, or by subtracting the same number from both sides, or by multiplying or dividing both sides by the same *positive* number. But if we multiply or divide by a *negative* number, then the inequality is *reversed*:

$$x - 2 < 4 \quad \text{gives} \quad x < 6, \qquad x + 2 < 4 \quad \text{gives} \quad x < 2,$$

$$\frac{1}{2}x < 4 \quad \text{gives} \quad x < 8,$$

but

$$-\frac{1}{2}x < 4 \quad \text{gives} \quad x > -8.$$
$$\text{\quad\quad\quad\quad\quad\quad\quad\quad}\uparrow\text{—note, the inequality is reversed}$$

Example 1 Solve the inequality

$$-3(4 - x) \leq 12.$$

SOLUTION Multiplying both sides of the inequality by $-\frac{1}{3}$, we have

$$4 - x \geq -4. \qquad \text{(the inequality has been reversed)}$$

Subtracting 4, we get

$$-x \geq -8.$$

To isolate x, we multiply by -1. This gives

$$x \leq 8. \qquad \text{(the inequality has been reversed again)}$$

The solution set is the interval $(-\infty, 8]$. ❑

There are generally several ways to solve a given inequality. For example, the last inequality could have been solved as follows:

$$-3(4 - x) \leq 12,$$

$$-12 + 3x \leq 12,$$

$$3x \leq 24, \qquad \text{(we added 12)}$$

$$x \leq 8. \qquad \text{(we divided by 3)}$$

To solve a quadratic inequality, we try to factor the quadratic. Failing that, we can complete the square and go on from there. This second method always works.

Example 2 Solve the inequality

$$x^2 - 4x + 3 > 0.$$

SOLUTION Factoring the quadratic, we obtain

$$(x - 1)(x - 3) > 0.$$

The product $(x - 1)(x - 3)$ is zero at 1 and 3. Mark these points on a number line (Figure 1.3.1). The points 1 and 3 separate three intervals:

$$(-\infty, 1), \qquad (1, 3), \qquad (3, \infty).$$

Figure 1.3.1

On each of these intervals the product $(x - 1)(x - 3)$ keeps a constant sign:

$$\begin{aligned}
&\text{on } (-\infty, 1) \quad \text{[to the left of 1]} \quad &&\text{sign of } (x - 1)(x - 3) = (-)(-) = +; \\
&\text{on } (1, 3) \quad \text{[between 1 and 3]} \quad &&\text{sign of } (x - 1)(x - 3) = (+)(-) = -; \\
&\text{on } (3, \infty) \quad \text{[to the right of 3]} \quad &&\text{sign of } (x - 1)(x - 3) = (+)(+) = +.
\end{aligned}$$

The product $(x - 1)(x - 3)$ is positive on the open intervals $(-\infty, 1)$ and $(3, \infty)$. The solution set is the union $(-\infty, 1) \cup (3, \infty)$. ❑

Example 3 Solve the inequality

$$x^2 - 2x + 5 \leq 0.$$

SOLUTION Not seeing immediately how to factor the quadratic, we use the method that always works: completing the square. Note that

$$x^2 - 2x + 5 = (x^2 - 2x + 1) + 4 = (x - 1)^2 + 4.$$

This tells us that

$$x^2 - 2x + 5 \geq 4 \qquad \text{for all real } x,$$

and thus there are no numbers that satisfy the inequality we are trying to solve. To put it in terms of sets, the solution set is the empty set $\emptyset$. ❑

In practice we frequently come to expressions of the form

$$(x - a_1)^{k_1}(x - a_2)^{k_2} \ldots (x - a_n)^{k_n}$$

$k_1, k_2, \ldots, k_n$ positive integers, $a_1 < a_2 < \cdots < a_n$. Such an expression is zero at $a_1, a_2, \ldots, a_n$. It is positive on those intervals where the number of negative factors is even and negative on those intervals where the number of negative factors is odd.

Take, for instance,

$$(x + 2)(x - 1)(x - 3).$$

This product is zero at $-2, 1, 3$. It is

negative on	$(-\infty, -2)$,	(3 negative terms)
positive on	$(-2, 1)$,	(2 negative terms)
negative on	$(1, 3)$,	(1 negative term)
positive on	$(3, \infty)$.	(0 negative terms)

See Figure 1.3.2

Figure 1.3.2

Example 4 Solve the inequality

$$(x + 3)^5(x - 1)(x - 4)^2 < 0.$$

SOLUTION We view $(x + 3)^5(x - 1)(x - 4)^2$ as the product of three factors: $(x + 3)^5$, $(x - 1)$, $(x - 4)^2$. The product is zero at $-3, 1, 4$. These points separate the intervals

$$(-\infty, -3), \qquad (-3, 1), \qquad (1, 4), \qquad (4, \infty).$$

On each of these intervals the product keeps a constant sign:

positive on	$(-\infty, -3)$,	(2 negative factors)
negative on	$(-3, 1)$,	(1 negative factor)
positive on	$(1, 4)$,	(0 negative factors)
positive on	$(4, \infty)$.	(0 negative factors)

See Figure 1.3.3.

Figure 1.3.3

The solution set is the open interval $(-3, 1)$. ❑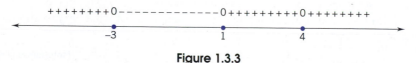

This approach to solving inequalities will be justified in Section 2.6

Inequalities and Absolute Value

Now we take up inequalities that involve absolute values. With an eye toward developing the concept of limits (Chapter 2), we introduce two Greek letters: δ (delta) and ϵ (epsilon).

As you know, for each real number a

(1.3.1)
$$|a| = \begin{cases} a & \text{if } a \geq 0. \\ -a, & \text{if } a < 0, \end{cases} \qquad |a| = \max\{a, -a\}, \qquad |a| = \sqrt{a^2}.$$

We begin with the inequality

$$|x| < \delta$$

where δ is some positive number. To say that $|x| < \delta$ is to say that x lies within δ units of 0 or, equivalently, that x lies between $-\delta$ and δ. Thus

(1.3.2)
$$|x| < \delta \qquad \text{iff} \qquad -\delta < x < \delta.$$

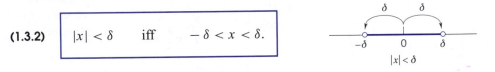

The solution set is the open interval $(-\delta, \delta)$.

To say that $|x - c| < \delta$ is to say that x lies within δ units of c or, equivalently, that x lies between $c - \delta$ and $c + \delta$. Thus

(1.3.3)
$$|x - c| < \delta \qquad \text{iff} \qquad c - \delta < x < c + \delta.$$

The solution set is the open interval $(c - \delta, c + \delta)$.

Somewhat more delicate is the inequality

$$0 < |x - c| < \delta.$$

Here we have $|x - c| < \delta$ with the additional requirement that $x \neq c$. Consequently,

(1.3.4)
$$0 < |x - c| < \delta \qquad \text{iff} \qquad c - \delta < x < c \qquad \text{or} \qquad c < x < c + \delta.$$

The solution set is the union of two open intervals: $(c - \delta, c) \cup (c, c + \delta)$.

The following results are an immediate consequence of what we just showed.

$|x| < \frac{1}{2}$ iff $-\frac{1}{2} < x < \frac{1}{2}$; $\qquad\qquad$ [solution set: $(-\frac{1}{2}, \frac{1}{2})$]

$|x - 5| < 1$ iff $4 < x < 6$; $\qquad\qquad$ [solution set: $(4, 6)$]

$0 < |x - 5| < 1$ iff $4 < x < 5$ or $5 < x < 6$; $\qquad$ [solution set: $(4, 5) \cup (5, 6)$]

Example 5 Solve the inequality

$$|x + 2| < 3.$$

SOLUTION Once we recognize that $|x + 2| = |x - (-2)|$, we are in familiar territory.

$$|x - (-2)| < 3 \qquad \text{iff} \qquad -2 - 3 < x < -2 + 3 \qquad \text{iff} \qquad -5 < x < 1.$$

The solution set is the open interval $(-5, 1)$. ❑

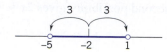

Example 6 Solve the inequality

$$|3x - 4| < 2.$$

SOLUTION Since

$$|3x - 4| = \left|3\left(x - \tfrac{4}{3}\right)\right| = |3|\left|x - \tfrac{4}{3}\right| = 3\left|x - \tfrac{4}{3}\right|,$$

the inequality can be written

$$3\left|x - \tfrac{4}{3}\right| < 2.$$

This gives

$$\left|x - \tfrac{4}{3}\right| < \tfrac{2}{3}, \qquad \tfrac{4}{3} - \tfrac{2}{3} < x < \tfrac{4}{3} + \tfrac{2}{3}, \qquad \tfrac{2}{3} < x < 2.$$

The solution set is the open interval $(\tfrac{2}{3}, 2)$.

ALTERNATIVE SOLUTION There is usually more than one way to solve an inequality. In this case, for example, we can write

$$|3x - 4| < 2$$

as

$$-2 < 3x - 4 < 2$$

and proceed from there. Adding 4 to the inequality, we get

$$2 < 3x < 6.$$

Division by 3 gives the result we had before:

$$\tfrac{2}{3} < x < 2. \quad ❑$$

Let $\epsilon > 0$. If you think of $|a|$ as the distance between a and 0, then you see that

(1.3.5) $\quad \boxed{|a| > \epsilon \quad \text{iff} \quad a > \epsilon \quad \text{or} \quad a < -\epsilon.}$

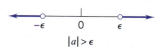

$|a| > \epsilon$

Example 7 Solve the inequality

$$|2x + 3| > 5.$$

SOLUTION In general

$$|a| > \epsilon \quad \text{iff} \quad a > \epsilon \quad \text{or} \quad a < -\epsilon.$$

So here

$$2x + 3 > 5 \quad \text{or} \quad 2x + 3 < -5.$$

The first possibility gives $2x > 2$ and thus

$$x > 1.$$

The second possibility gives $2x < -8$ and thus

$$x < -4$$

The total solution is therefore the union

$$(-\infty, -4) \cup (1, \infty). \quad \square$$

We come now to one of the fundamental inequalities of calculus: for all real numbers a and b,

(1.3.6)

$$|a + b| \leq |a| + |b|.$$

This is called the *triangle inequality* in analogy with the geometric observation that "in any triangle the length of each side is less than or equal to the sum of the lengths of the other two sides."

PROOF OF THE TRIANGLE INEQUALITY The key here is to think of $|x|$ as $\sqrt{x^2}$. Note first that

$$(a + b)^2 = a^2 + 2ab + b^2 \leq |a|^2 + 2|a||b| + |b|^2 = (|a| + |b|)^2.$$

Comparing the extremes of the inequality and taking square roots, we have

$$\sqrt{(a + b)^2} \leq |a| + |b|. \qquad \text{(Exercise 51)}$$

The result follows from observing that

$$\sqrt{(a + b)^2} = |a + b|. \quad \square$$

Here is a variant of the triangle inequality that also comes up in calculus: for all real numbers a and b,

(1.3.7)

$$\bigl||a| - |b|\bigr| \leq |a - b|.$$

The proof is left to you as an exercise.

EXERCISES 1.3

Exercises 1–20. Solve the inequality and mark the solution set on a number line.

1. $2 + 3x < 5$.

2. $\frac{1}{2}(2x + 3) < 6$.

3. $16x + 64 \leq 16$.

4. $3x + 5 > \frac{1}{4}(x - 2)$.

5. $\frac{1}{2}(1 + x) < \frac{1}{3}(1 - x)$.

6. $3x - 2 \leq 1 + 6x$.

7. $x^2 - 1 < 0$.

8. $x^2 + 9x + 20 < 0$.

9. $x^2 - x - 6 \geq 0$.

10. $x^2 - 4x - 5 > 0$.

11. $2x^2 + x - 1 \leq 0$.

12. $3x^2 + 4x - 4 \geq 0$.

13. $x(x - 1)(x - 2) > 0$.

14. $x(2x - 1)(3x - 5) \leq 0$.

15. $x^3 - 2x^2 + x \geq 0$.

16. $x^2 - 4x + 4 \leq 0$.

17. $x^3(x - 2)(x + 3)^2 < 0$.

18. $x^2(x - 3)(x + 4)^2 > 0$.

19. $x^2(x - 2)(x + 6) > 0$.

20. $7x(x - 4)^2 < 0$.

Exercises 21–36. Solve the inequality and express the solution set as an interval or as the union of intervals.

21. $|x| < 2$.

22. $|x| \geq 1$.

23. $|x| > 3$.

24. $|x - 1| < 1$.

25. $|x - 2| < \frac{1}{2}$.

26. $|x - \frac{1}{2}| < 2$.

27. $0 < |x| < 1$.

28. $0 < |x| < \frac{1}{2}$.

29. $0 < |x - 2| < \frac{1}{2}$.

30. $0 < |x - \frac{1}{2}| < 2$.

31. $0 < |x - 3| < 8$.

32. $|3x - 5| < 3$.

33. $|2x + 1| < \frac{1}{4}$.

34. $|5x - 3| < \frac{1}{2}$.

35. $|2x + 5| > 3$.

36. $|3x + 1| > 5$.

Exercises 37–42. Each of the following sets is the solution of an inequality of the form $|x - c| < \delta$. Find c and δ.

37. $(-3, 3)$.

38. $(-2, 2)$.

39. $(-3, 7)$.

40. $(0, 4)$.

41. $(-7, 3)$.

42. (a, b).

Exercises 43–46. Determine all numbers $A > 0$ for which the statement is true.

43. If $|x - 2| < 1$, then $|2x - 4| < A$.

44. If $|x - 2| < A$, then $|2x - 4| < 3$.

45. If $|x + 1| < A$, then $|3x + 3| < 4$.

46. If $|x + 1| < 2$, then $|3x + 3| < A$.

47. Arrange the following in order :$1, x, \sqrt{x}, 1/x, 1/\sqrt{x}$, given that: $(a)\ x > 1; (b)\ 0 < x < 1$.

48. Given that $x > 0$, compare

$$\sqrt{\frac{x}{x + 1}} \quad \text{and} \quad \sqrt{\frac{x + 1}{x + 2}}.$$

49. Suppose that $ab > 0$. Show that if $a < b$, then $1/b < 1/a$.

50. Given that $a > 0$ and $b > 0$, show that if $a^2 \le b^2$, then $a \le b$.

51. Show that if $0 \le a \le b$, then $\sqrt{a} \le \sqrt{b}$.

52. Show that $|a - b| \le |a| + |b|$ for all real numbers a and b.

53. Show that $\big||a| - |b|\big| \le |a - b|$ for all real numbers a and b. HINT: Calculate $\big||a| - |b|\big|^2$.

54. Show that $|a + b| = |a| + |b|$ iff $ab \ge 0$.

55. Show that

$$\text{if} \quad 0 \le a \le b, \quad \text{then} \quad \frac{a}{1 + a} \le \frac{b}{1 + b}.$$

56. Let a, b, c be nonnegative numbers. Show that

$$\text{if} \quad a \le b + c, \quad \text{then} \quad \frac{a}{1 + a} \le \frac{b}{1 + b} + \frac{c}{1 + c}.$$

57. Show that if a and b are real numbers and $a < b$, then $a < (a + b)/2 < b$. The number $(a + b)/2$ is called the *arithmetic mean* of a and b.

58. Given that $0 \le a \le b$, show that

$$a \le \sqrt{ab} \le \frac{a + b}{2} \le b.$$

The number $\sqrt{ab}$ is called the *geometric mean* of a and b.

■ 1.4 COORDINATE PLANE; ANALYTIC GEOMETRY

Rectangular Coordinates

The one-to-one correspondence between real numbers and points on a line can be used to construct a coordinate system for the plane. In the plane, we draw two number lines that are mutually perpendicular and intersect at their origins. Let O be the point of intersection. We set one of the lines horizontally with the positive numbers to the right of O and the other vertically with the positive numbers above O. The point O is called the *origin*, and the number lines are called the *coordinate axes*. The horizontal axis is usually labeled the *x-axis* and the vertical axis is usually labeled the *y-axis*. The coordinate axes separate four regions, which are called *quadrants*. The quadrants are numbered *I, II, III, IV* in the counterclockwise direction starting with the upper right quadrant. See Figure 1.4.1.

Rectangular coordinates are assigned to points of the plane as follows (see Figure 1.4.2.). The point on the *x*-axis with line coordinate a is assigned rectangular coordinates $(a, 0)$. The point on the *y*-axis with line coordinate b is assigned rectangular coordinates $(0, b)$. Thus the origin is assigned coordinates $(0, 0)$. A point P not on one of the coordinate axes is assigned coordinates (a, b) provided that the line l_1 that passes through P and is parallel to the *y*-axis intersects the *x*-axis at the point with coordinates $(a, 0)$, and the line l_2 that passes through P and is parallel to the *x*-axis intersects the *y*-axis at the point with coordinates $(0, b)$.

This procedure assigns an ordered pair of real numbers to each point of the plane. Moreover, the procedure is reversible. Given any ordered pair (a, b) of real numbers, there is a unique point P in the plane with coordinates (a, b).

To indicate P with coordinates (a, b) we write $P(a, b)$. The number a is called the *x-coordinate* (the *abscissa*); the number b is called the *y-coordinate* (the *ordinate*). The coordinate system that we have defined is called a *rectangular coordinate system*. It is often referred to as a *Cartesian coordinate system* after the French mathematician René Descartes (1596–1650).

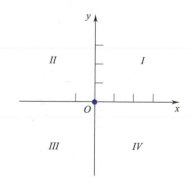

Figure 1.4.1

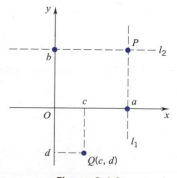

Figure 1.4.2

Distance and Midpoint Formulas

Let $P_0(x_0, y_0)$ and $P_1(x_1, y_1)$ be points in the plane. The formula for the distance $d(P_0, P_1)$ between P_0 and P_1 follows from the *Pythagorean theorem*:

$$d(P_0, P_1) = \sqrt{|x_1 - x_0|^2 + |y_1 - y_0|^2} = \sqrt{(x_1 - x_0)^2 + (y_1 - y_0)^2}.$$

$$\underset{\longleftarrow |a|^2 = a^2}{}$$

(Figure 1.4.3)

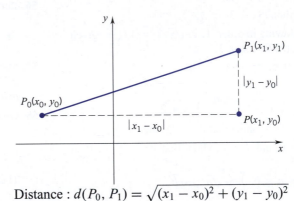

Distance : $d(P_0, P_1) = \sqrt{(x_1 - x_0)^2 + (y_1 - y_0)^2}$

Figure 1.4.3

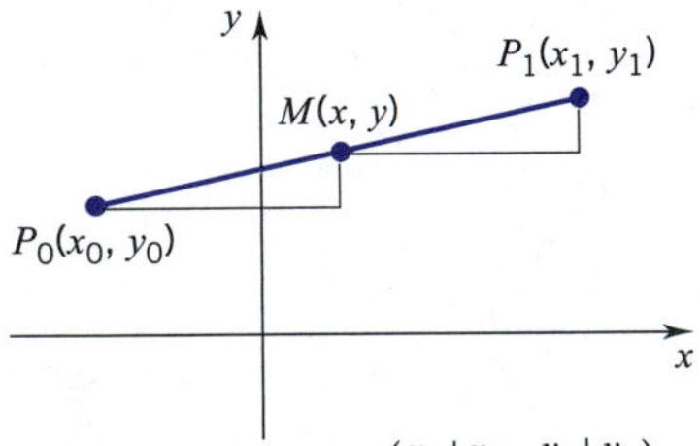

Let $M(x, y)$ be the midpoint of the line segment $\overline{P_0 P_1}$. That

$$x = \frac{x_0 + x_1}{2} \quad \text{and} \quad y = \frac{y_0 + y_1}{2}$$

follows from the congruence of the triangles shown in Figure 1.4.4

Midpoint: $M = \left(\frac{x_0 + x_1}{2}, \frac{y_0 + y_1}{2} \right)$

Figure 1.4.4

Lines

(i) Slope Let l be the line determined by $P_0(x_0, y_0)$ and $P_1(x_1, y_1)$. If l is not vertical, then $x_1 \neq x_0$ and the slope of l is given by the formula

$$m = \frac{y_1 - y_0}{x_1 - x_0}.$$

(Figure 1.4.5)

With θ (as indicated in the figure) measured counterclockwise from the x-axis,

$$m = \tan \theta.^\dagger$$

The angle θ is called the *inclination* of l. If l is vertical, then $\theta = \pi/2$ and the slope of l is not defined.

(ii) Intercepts If a line intersects the x-axis, it does so at some point $(a, 0)$. We call a the x-intercept. If a line intersects the y-axis, it does so at some point $(0, b)$. We call b the y-intercept. Intercepts are shown in Figure 1.4.6.

---

†The trigonometric functions are reviewed in Section 1.6.

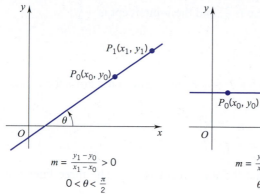

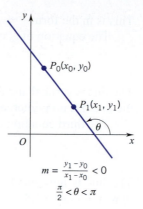

$$m = \frac{y_1 - y_0}{x_1 - x_0} > 0$$

$$0 < \theta < \frac{\pi}{2}$$

$$m = \frac{y_1 - y_0}{x_1 - x_0} = 0$$

$$\theta = 0$$

$$m = \frac{y_1 - y_0}{x_1 - x_0} < 0$$

$$\frac{\pi}{2} < \theta < \pi$$

Figure 1.4.5

(iii) Equations

vertical line	$x = a.$
horizontal line	$y = b.$
point-slope form	$y - y_0 = m(x - x_0).$
slope-intercept form	$y = mx + b.$
two-intercept form	$\dfrac{x}{a} + \dfrac{y}{b} = 1.$
general form	$Ax + By + C = 0.$

$(y = b$ at $x = 0)$

$(x$-intercept a; y-intercept $b)$

$(A$ and B not both 0)

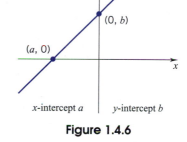

Figure 1.4.6

(iv) Parallel and Perpendicular Nonvertical Lines

$$\begin{aligned} \textit{parallel} \quad & m_1 = m_2. \\ \textit{perpendicular} \quad & m_1 m_2 = -1. \end{aligned}$$

(v) The Angle Between Two Lines The angle between two lines that meet at right angles is $\pi/2$. Figure 1.4.7 shows two lines (l_1, l_2 with inclinations θ_1, θ_2) that intersect but not at right angles. These lines form two angles, marked α and $\pi - \alpha$ in the figure. The smaller of these angles, the one between 0 and $\pi/2$, is called *the angle* between l_1 and l_2. This angle, marked α in the figure, is readily obtained from θ_1 and θ_2.

If neither l_1 nor l_2 is vertical, the angle α between l_1 and l_2 can also be obtained from the slopes of the lines:

$$\tan \alpha = \frac{m_1 - m_2}{1 + m_1 m_2}.$$

The derivation of this formula is outlined in Exercise 75 of Section 1.6.

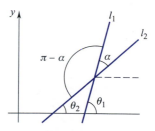

Figure 1.4.7

Example 1 Find the slope and the y-intercept of each of the following lines:

$$l_1 : 20x - 24y - 30 = 0, \qquad l_2 : 2x - 3 = 0, \qquad l_3 : 4y + 5 = 0.$$

SOLUTION The equation of l_1 can be written

$$y = \tfrac{5}{6}x - \tfrac{5}{4}.$$

This is in the form $y = mx + b$. The slope is $\frac{5}{6}$, and the y-intercept is $-\frac{5}{4}$.

The equation of l_2 can be written

$$x = \tfrac{3}{2}.$$

The line is vertical and the slope is not defined. Since the line does not cross the y-axis, the line has no y-intercept.

The third equation can be written

$$y = -\tfrac{5}{4}.$$

The line is horizontal. The slope is 0 and the y-intercept is $-\frac{5}{4}$. The three lines are drawn in Figure 1.4.8. ❏

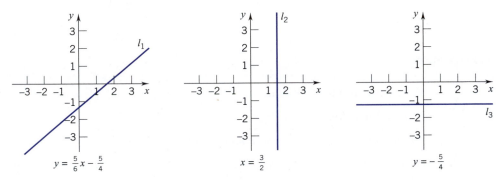

Figure 1.4.8

Example 2   Write an equation for the line l_2 that is parallel to

$$l_1 : 3x - 5y + 8 = 0$$

and passes through the point $P(-3, 2)$.

SOLUTION   The equation for l_1 can be written

$$y = \tfrac{3}{5}x + \tfrac{8}{5}.$$

The slope of l_1 is $\frac{3}{5}$. The slope of l_2 must also be $\frac{3}{5}$. (For nonvertical parallel lines, $m_1 = m_2$.)

Since l_2 passes through $(-3, 2)$ with slope $\frac{3}{5}$, we can use the point-slope formula and write the equation as

$$y - 2 = \tfrac{3}{5}(x + 3). \quad ❏$$

Example 3   Write an equation for the line that is perpendicular to

$$l_1 : x - 4y + 8 = 0$$

and passes through the point $P(2, -4)$.

SOLUTION   The equation for l_1 can be written

$$y = \tfrac{1}{4}x + 2.$$

The slope of l_1 is $\frac{1}{4}$. The slope of l_2 is therefore -4. (For nonvertical perpendicular lines, $m_1 m_2 = -1$.)

Since l_2 passes through $(2, -4)$ with slope -4, we can use the point-slope formula and write the equation as

$$y + 4 = -4(x - 2). \quad ❏$$

Example 4 Show that the lines

$$l_1 : 3x - 4y + 8 = 0 \quad \text{and} \quad l_2 : 12x - 5y - 12 = 0$$

intersect and find their point of intersection.

SOLUTION The slope of l_1 is $\frac{3}{4}$ and the slope of l_2 is $\frac{12}{5}$. Since l_1 and l_2 have different slopes, they intersect at a point.

To find the point of intersection, we solve the two equations simultaneously:

$$3x - 4y + 8 = 0$$
$$12x - 5y - 12 = 0.$$

Multiplying the first equation by -4 and adding it to the second equation, we obtain

$$11y - 44 = 0$$
$$y = 4.$$

Substituting $y = 4$ into either of the two given equations, we find that $x = \frac{8}{3}$. The lines intersect at the point $\left(\frac{8}{3}, 4\right)$. ❏

Circle, Ellipse, Parabola, Hyperbola

These curves and their remarkable properties are thoroughly discussed in Section 10.1. The information we give here suffices for our present purposes.

Circle

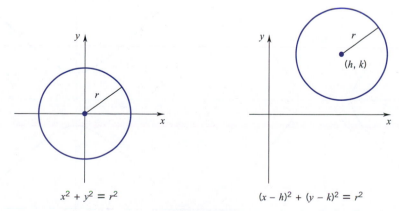

$$x^2 + y^2 = r^2 \qquad\qquad (x - h)^2 + (y - k)^2 = r^2$$

Figure 1.4.9

Ellipse

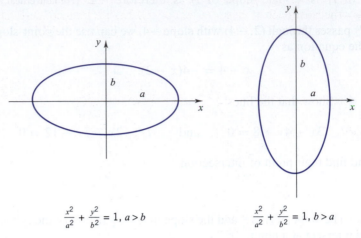

$$\frac{x^2}{a^2} + \frac{y^2}{b^2} = 1, \, a > b \qquad\qquad \frac{x^2}{a^2} + \frac{y^2}{b^2} = 1, \, b > a$$

Figure 1.4.10

Parabola

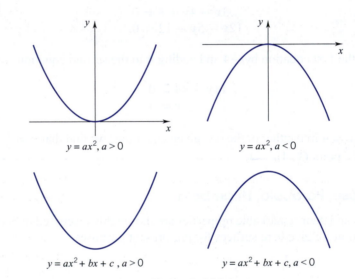

$$y = ax^2, \, a > 0 \qquad\qquad y = ax^2, \, a < 0$$

$$y = ax^2 + bx + c \,, \, a > 0 \qquad\qquad y = ax^2 + bx + c, \, a < 0$$

Figure 1.4.11

Hyperbola

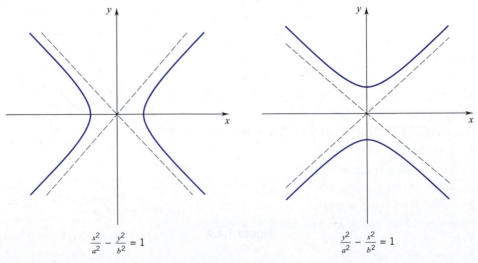

$$\frac{x^2}{a^2} - \frac{y^2}{b^2} = 1 \qquad\qquad\qquad \frac{y^2}{a^2} - \frac{x^2}{b^2} = 1$$

Figure 1.4.12

Remark   The circle, the ellipse, the parabola, and the hyperbola are known as the *conic sections* because each of these configurations can be obtained by slicing a "double right circular cone" by a suitably inclined plane. (See Figure 1.4.13.)   ❏

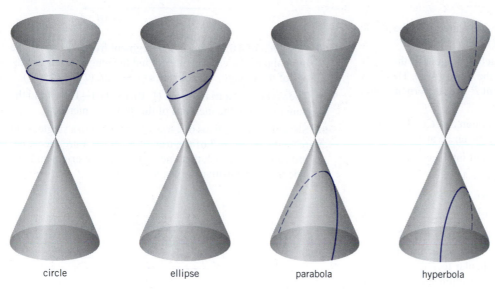

| circle | ellipse | parabola | hyperbola |

Figure 1.4.13

EXERCISES 1.4

Exercises 1–4. Find the distance between the points.

1. $P_0(0, 5)$, $P_1(6, -3)$. **2.** $P_0(2, 2)$, $P_1(5, 5)$.

3. $P_0(5, -2)$, $P_1(-3, 2)$. **4.** $P_0(2, 7)$, $P_1(-4, 7)$.

Exercises 5–8. Find the midpoint of the line segment $\overline{P_0 P_1}$.

5. $P_0(2, 4)$, $P_1(6, 8)$. **6.** $P_0(3, -1)$, $P_1(-1, 5)$.

7. $P_0(2, -3)$, $P_1(7, -3)$. **8.** $P_0(a, 3)$, $P_1(3, a)$.

Exercises 9–14. Find the slope of the line through the points.

9. $P_0(-2, 5)$, $P_1(4, 1)$. **10.** $P_0(4, -3)$, $P_1(-2, -7)$.

11. $P(a, b)$, $Q(b, a)$. **12.** $P(4, -1)$, $Q(-3, -1)$.

13. $P(x_0, 0)$, $Q(0, y_0)$. **14.** $O(0, 0)$, $P(x_0, y_0)$.

Exercises 15–20. Find the slope and y-intercept.

15. $y = 2x - 4$. **16.** $6 - 5x = 0$.

17. $3y = x + 6$. **18.** $6y - 3x + 8 = 0$.

19. $7x - 3y + 4 = 0$. **20.** $y = 3$.

Exercises 21–24. Write an equation for the line with

21. slope 5 and y-intercept 2.

22. slope 5 and y-intercept -2.

23. slope -5 and y-intercept 2.

24. slope -5 and y-intercept -2.

Exercises 25–26. Write an equation for the horizontal line 3 units

25. above the x-axis.

26. below the x-axis.

Exercises 27–28. Write an equation for the vertical line 3 units

27. to the left of the y-axis.

28. to the right of the y-axis.

Exercises 29–34. Find an equation for the line that passes through the point $P(2, 7)$ and is

29. parallel to the x-axis.

30. parallel to the y-axis.

31. parallel to the line $3y - 2x + 6 = 0$.

32. perpendicular to the line $y - 2x + 5 = 0$.

33. perpendicular to the line $3y - 2x + 6 = 0$.

34. parallel to the line $y - 2x + 5 = 0$.

Exercises 35–38. Determine the point(s) where the line intersects the circle.

35. $y = x$, $x^2 + y^2 = 1$.

36. $y = mx$, $x^2 + y^2 = 4$.

37. $4x + 3y = 24$, $x^2 + y^2 = 25$.

38. $y = mx + b$, $x^2 + y^2 = b^2$.

Exercises 39–42. Find the point where the lines intersect.

39. $l_1 : 4x - y - 3 = 0$, $l_2 : 3x - 4y + 1 = 0$.

40. $l_1 : 3x + y - 5 = 0$, $l_2 : 7x - 10y + 27 = 0$.

41. $l_1 : 4x - y + 2 = 0$, $l_2 : 19x + y = 0$.

42. $l_1 : 5x - 6y + 1 = 0, \quad l_2 : 8x + 5y + 2 = 0.$

43. Find the area of the triangle with vertices $(1, -2), (-1, 3),$ $(2, 4).$

44. Find the area of the triangle with vertices $(-1, 1), (3, \sqrt{2}),$ $(\sqrt{2}, -1).$

45. Determine the slope of the line that intersects the circle $x^2 + y^2 = 169$ only at the point $(5, 12).$

46. Find an equation for the line which is tangent to the circle $x^2 + y^2 - 2x + 6y - 15 = 0$ at the point $(4, 1).$ HINT: A line is tangent to a circle at a point P iff it is perpendicular to the radius at $P.$

47. The point $P(1, -1)$ is on a circle centered at $C(-1, 3).$ Find an equation for the line tangent to the circle at $P.$

Exercises 48–51. Estimate the point(s) of intersection.

48. $l_1 : 3x - 4y = 7, \quad l_2 : -5x + 2y = 11.$

49. $l_1 : 2.41x + 3.29y = 5, \quad l_2 : 5.13x - 4.27y = 13.$

50. $l_1 : 2x - 3y = 5, \quad$ circle $: x^2 + y^2 = 4.$

51. circle $: x^2 + y^2 = 9, \quad$ parabola $: y = x^2 - 4x + 5.$

Exercises 52–53. The *perpendicular bisector* of the line segment $\overline{PQ}$ is the line which is perpendicular to $\overline{PQ}$ and passes through the midpoint of $\overline{PQ}.$ Find an equation for the perpendicular bisector of the line segment that joins the two points.

52. $P(-1, 3), \quad Q(3, -4).$

53. $P(1, -4), \quad Q(4, 9).$

Exercises 54–56. The points are the vertices of a triangle. State whether the triangle is *isosceles* (two sides of equal length), a right triangle, both of these, or neither of these.

54. $P_0(-4, 3), \quad P_1(-4, -1), \quad P_2(2, 1).$

55. $P_0(-2, 5), \quad P_1(1, 3), \quad P_2(-1, 0).$

56. $P_0(-1, 2), \quad P_1(1, 3), \quad P_2(4, 1).$

57. Show that the distance from the origin to the line $Ax + By + C = 0$ is given by the formula

$$d(0, l) = \frac{|C|}{\sqrt{A^2 + B^2}}.$$

58. An *equilateral triangle* is a triangle the three sides of which have the same length. Given that two of the vertices of an equilateral triangle are $(0, 0)$ and $(4, 3),$ find all possible locations for a third vertex. How many such triangles are there?

59. Show that the midpoint M of the hypotenuse of a right triangle is equidistant from the three vertices of the triangle. HINT: Introduce a coordinate system in which the sides of the triangle are on the coordinate axes; see the figure.

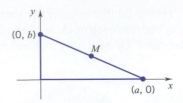

60. A *median* of a triangle is a line segment from a vertex to the midpoint of the opposite side. Find the lengths of the medians of the triangle with vertices $(-1, -2), (2, 1), (4, -3).$

61. The vertices of a triangle are $(1, 0), (3, 4), (-1, 6).$ Find the point(s) where the medians of this triangle intersect.

62. Show that the medians of a triangle intersect in a single point (called the *centroid* of the triangle). HINT: Introduce a coordinate system such that one vertex is at the origin and one side is on the positive x-axis; see the figure.

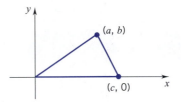

63. Prove that each diagonal of a parallelogram bisects the other. HINT: Introduce a coordinate system with one vertex at the origin and one side on the positive x-axis.

64. $P_1(x_1, y_1), P_2(x_2, y_2), P_3(x_3, y_3), P_4(x_4, y_4)$ are the vertices of a quadrilateral. Show that the quadrilateral formed by joining the midpoints of adjacent sides is a parallelogram.

65. Except in scientific work, temperature is usually measured in degrees Fahrenheit (F) or in degrees Celsius (C). The relation between F and C is linear. (In the equation that relates F to C, both F and C appear to the first degree.) The freezing point of water in the Fahrenheit scale is $32°$F; in the Celsius scale it is $0°$C. The boiling point of water in the Fahrenheit scale is $212°$F; in the Celsius scale it is $100°$C. Find an equation that gives the Fahrenheit temperature F in terms of the Celsius temperature C. Is there a temperature at which the Fahrenheit and Celsius readings are equal? If so, find it.

66. In scientific work, temperature is measured on an absolute scale, called the Kelvin scale (after Lord Kelvin, who initiated this mode of temperature measurement). The relation between Fahrenheit temperature F and absolute temperature K is linear. Given that $K = 273°$ when $F = 32°,$ and $K = 373°$ when $F = 212°,$ express K in terms of F. Then use your result in Exercise 65 to determine the connection between Celsius temperature and absolute temperature.

■ **1.5 FUNCTIONS**

The fundamental processes of calculus (called *differentiation* and *integration*) are processes applied to functions. To understand these processes and to be able to carry them out, you have to be comfortable working with functions. Here we review some of the basic ideas and the nomenclature. We assume that you are familiar with all of this.

Functions can be applied in a very general setting. At this stage, and throughout the first thirteen chapters of this text, we will be working with what are called *real-valued functions of a real variable*, functions that assign real numbers to real numbers.

Domain and Range

Let's suppose that D is some set of real numbers and that f is a function defined on D. Then f assigns a unique number $f(x)$ to each number x in D. The number $f(x)$ is called the *value* of f at x, or the *image* of x under f. The set D, the set on which the function is defined, is called the *domain* of f, and the set of values taken on by f is called the *range* of f. In set notation

$$\text{dom}\,(f) = D, \qquad \text{range}\,(f) = \{f(x) : x \in D\}.$$

We can specify the function f by indicating exactly what $f(x)$ is for each x in D.

Some examples. We begin with the squaring function

$$f(x) = x^2, \qquad \text{for all real numbers } x.$$

The domain of f is explicitly given as the set of real numbers. Particular values taken on by f can be found by assigning particular values to x. In this case, for example,

$$f(4) = 4^2 = 16, \qquad f(-3) = (-3)^2 = 9, \qquad f(0) = 0^2 = 0.$$

As x runs through the real numbers, x^2 runs through all the nonnegative numbers. Thus the range of f is $[0, \infty)$. In abbreviated form, we can write

$$\text{dom}\,(f) = (-\infty, \infty), \qquad \text{range}\,(f) = [0, \infty)$$

and we can say that f *maps* $(-\infty, \infty)$ onto $[0, \infty)$.

Now let's look at the function g defined by

$$g(x) = \sqrt{2x + 4}, \qquad x \in [0, 6].$$

The domain of g is given as the closed interval $[0, 6]$. At $x = 0$, g takes on the value 2:

$$g(0) = \sqrt{2 \cdot 0 + 4} = \sqrt{4} = 2;$$

at $x = 6$, g has the value 4:

$$g(6) = \sqrt{2 \cdot 6 + 4} = \sqrt{16} = 4.$$

As x runs through the numbers in $[0, 6]$, $g(x)$ runs through the numbers from 2 to 4. Therefore, the range of g is the closed interval $[2, 4]$. The function g maps $[0, 6]$ onto $[2, 4]$.

Some functions are defined *piecewise*. As an example, take the function h, defined by setting

$$h(x) = \begin{cases} 2x + 1, & \text{if } x < 0 \\ x^2, & \text{if } x \geq 0. \end{cases}$$

As explicitly stated, the domain of h is the set of real numbers. As you can verify, the range of h is also the set of real numbers. Thus the function h maps $(-\infty, \infty)$ onto $(-\infty, \infty)$. A more familiar example is the *absolute value function* $f(x) = |x|$. Here

$$f(x) = \begin{cases} x, & \text{if } x \geq 0 \\ -x, & \text{if } x < 0. \end{cases}$$

The domain of this function is $(-\infty, \infty)$ and the range is $[0, \infty)$.

Remark Functions are often given by equations of the form $y = f(x)$ with x restricted to some set D, the domain of f. In this setup x is called the *independent variable* (or the *argument* of the function) and y, except in special cases, clearly dependent on x, is called the *dependent variable*. ❑

The Graph of a Function

If f is a function with domain D, then the *graph of f* is the set of all points $P(x, f(x))$ with x in D. Thus the graph of f is the graph of the equation $y = f(x)$ with x restricted to D; namely

the graph of $f = \{(x, y) : x \in D, y = f(x)\}$.

The most elementary way to sketch the graph of a function is to plot points. We plot enough points so that we can "see" what the graph may look like and then connect the points with a "curve." Of course, if we can identify the curve in advance (for example, if we know that the graph is a straight line, a parabola, or some other familiar curve), then it is much easier to draw the graph.

The graph of the squaring function

$$f(x) = x^2, \qquad x \in (-\infty, \infty)$$

is the parabola shown in Figure 1.5.1. The points that we plotted are indicated in the table and marked on the graph. The graph of the function

$$g(x) = \sqrt{2x + 4}, \qquad x \in [0, 6]$$

is the arc shown in Figure 1.5.2

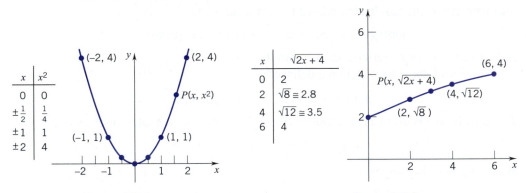

Figure 1.5.1

Figure 1.5.2

The graph of the function

$$h(x) = \begin{cases} 2x + 1, & \text{if } x < 0 \\ x^2, & \text{if } x \geq 0 \end{cases}$$

and the graph of the absolute value function are shown in Figures 1.5.3 and 1.5.4.

Although the graph of a function is a "curve" in the plane, not every curve in the plane is the graph of a function. This raises a question: How can we tell whether a curve is the graph of a function?

A curve C which intersects each vertical line at most once is the graph of a function: for each $P(x, y) \in C$, define $f(x) = y$. A curve C which intersects some vertical line more than once is not the graph of a function: If $P(x, y_1)$ and $P(x, y_2)$ are both on C, then how can we decide what $f(x)$ is? Is it y_1; or is it y_2?

These observations lead to what is called the *vertical line test*: a curve C in the plane is the graph of a function iff no vertical line intersects C at more than one point. Thus circles, ellipses, hyperbolas are not the graphs of functions. The curve shown in Figure 1.5.5 is the graph of a function, but the curve shown in Figure 1.5.6 is not the graph of a function.

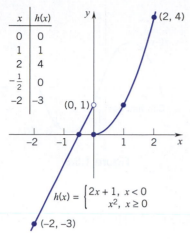

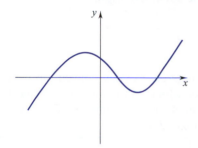

Figure 1.5.3

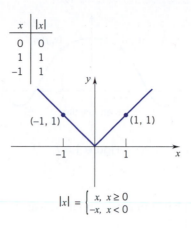

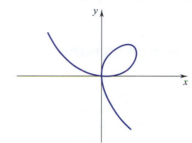

Figure 1.5.4

Figure 1.5.5

Figure 1.5.6

Graphing calculators and computer algebra systems (CAS) are valuable aids to graphing, but, used mindlessly, they can detract from the understanding necessary for more advanced work. We will not attempt to teach the use of graphing calculators or the ins and outs of computer software, but technology-oriented exercises will appear throughout the text.

Even Functions, Odd Functions; Symmetry

For even integers n, $(-x)^n = x^n$; for odd integers n, $(-x)^n = -x^n$. These simple observations prompt the following definitions:

A function f is said to be *even* if

$$f(-x) = f(x) \qquad \text{for all} \qquad x \in \text{dom}\,(f);$$

a function f is said to be *odd* if

$$f(-x) = -f(x) \qquad \text{for all} \qquad x \in \text{dom}\,(f).$$

The graph of an even function is *symmetric about the y-axis*, and the graph of an odd function is *symmetric about the origin*. (Figures 1.5.7 and 1.5.8.)

The absolute value function is even:

$$f(-x) = |-x| = |x| = f(x).$$

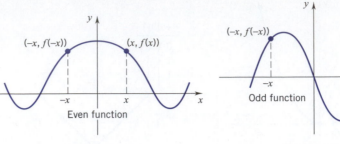

Figure 1.5.7   **Figure 1.5.8**

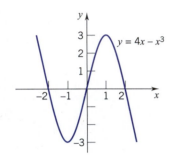

Figure 1.5.9

Its graph is symmetric about the y-axis. (See Figure 1.5.4.) The function $f(x) = 4x - x^3$ is odd:

$$f(-x) = 4(-x) - (-x^3) = -4x + x^3 = -(4x - x^3) = -f(x).$$

The graph, shown in Figure 1.5.9, is symmetric about the origin.

Convention on Domains

If the domain of a function f is not explicitly given, then by convention we take as domain the maximal set of real numbers x for which $f(x)$ is a real number. For the function $f(x) = x^3 + 1$, we take as domain the set of real numbers. For $g(x) = \sqrt{x}$, we take as domain the set of nonnegative numbers. For

$$h(x) = \frac{1}{x - 2}$$

we take as domain the set of all real numbers $x \neq 2$. In interval notation.

$$\text{dom}(f) = (-\infty, \infty), \quad \text{dom}(g) = [0, \infty), \quad \text{and} \quad \text{dom}(h) = (-\infty, 2) \cup (2, \infty).$$

The graphs of the three functions are shown in Figure 1.5.10.

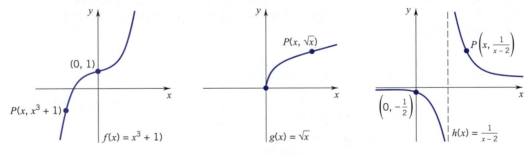

Figure 1.5.10

Example 1   Give the domain of each function:

(a) $f(x) = \dfrac{x + 1}{x^2 + x - 6}$,   **(b)** $g(x) = \dfrac{\sqrt{4 - x^2}}{x - 1}$.

SOLUTION   **(a)** You can see that $f(x)$ is a real number iff $x^2 + x - 6 \neq 0$. Since

$$x^2 + x - 6 = (x + 3)(x - 2),$$

the domain of f is the set of real numbers other than -3 and 2. This set can be expressed as

$$(-\infty, -3) \cup (-3, 2) \cup (2, \infty).$$

(b) For $g(x)$ to be a real number, we need

$$4 - x^2 \geq 0 \qquad \text{and} \qquad x \neq 1.$$

Since $4 - x^2 \geq 0$ iff $x^2 \leq 4$ iff $-2 \leq x \leq 2$, the domain of g is the set of all numbers x in the closed interval $[-2, 2]$ other than $x = 1$. This set can be expressed as the union of two half-open intervals:

$$[-2, 1) \cup (1, 2]. \quad \square$$

Example 2 Give the domain and range of the function:

$$f(x) = \frac{1}{\sqrt{2 - x}} + 5.$$

SOLUTION First we look for the domain. Since $\sqrt{2 - x}$ is a real number iff $2 - x \geq 0$, we need $x \leq 2$. But at $x = 2$, $\sqrt{2 - x} = 0$ and its reciprocal is not defined. We must therefore restrict x to $x < 2$. The domain is $(-\infty, 2)$.

Now we look for the range. As x runs through $(-\infty, 2)$, $\sqrt{2 - x}$ takes on all positive values and so does its reciprocal. The range of f is therefore $(5, \infty)$. The function f maps $(-\infty, 2)$ onto $(5, \infty)$. $\quad \square$

Functions are used in applications to show how variable quantities are related. The domain of a function that appears in an application is dictated by the requirements of the application.

Example 3 U.S. Postal Service regulations require that the length plus the girth (the perimeter of a cross section) of a package for mailing cannot exceed 108 inches. A rectangular box with a square end is designed to meet the regulation exactly (see Figure 1.5.11). Express the volume V of the box as a function of the edge length of the square ends and give the domain of the function.

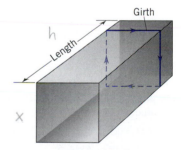

Figure 1.5.11

SOLUTION Let x denote the edge length of the square ends and let h denote the length of the box. The girth is the perimeter of the square, or $4x$. Since the box meets the regulations exactly,

$$4x + h = 108 \qquad \text{and therefore} \qquad h = 108 - 4x.$$

The volume of the box is given by $V = x^2 h$ and so it follows that

$$V(x) = x^2(108 - 4x) = 108x^2 - 4x^3.$$

Since neither the edge length of the square end nor the length of the box can be negative, we have

$$x \geq 0 \qquad \text{and} \qquad h = 108 - 4x \geq 0.$$

The second condition requires $x \leq 27$. The full requirement on x, $0 \leq x \leq 27$, gives $\text{dom}(V) = [0, 27]$. $\quad \square$

Example 4 A soft-drink manufacturer wants to fabricate cylindrical cans. (See Figure 1.5.12.) The can is to have a volume of 12 fluid ounces, which we take to be approximately 22 cubic inches. Express the total surface area S of the can as a function of the radius and give the domain of the function.

Figure 1.5.12

SOLUTION Let r be the radius of the can and h the height. The total surface area (top, bottom, and lateral area) of a right circular cylinder is given by the formula

$$S = 2\pi r^2 + 2\pi rh.$$

Since the volume $V = \pi r^2 h$ is to be 22 cubic inches, we have

$$\pi r^2 h = 22 \quad \text{and} \quad h = \frac{22}{\pi r^2}$$

and therefore

$$S(r) = 2\pi r^2 + 2\pi r \left(\frac{22}{\pi r^2}\right) = 2\pi r^2 + \frac{44}{r}. \qquad \text{(square inches)}$$

Since r can take on any positive value, dom $(S) = (0, \infty)$. ❏

EXERCISES 1.5

Exercises 1–6. Calculate (a) $f(0)$, (b) $f(1)$, (c) $f(-2)$, (d) $f(3/2)$.

1. $f(x) = 2x^2 - 3x + 2.$

2. $f(x) = \dfrac{2x - 1}{x^2 + 4}.$

3. $f(x) = \sqrt{x^2 + 2x}.$

4. $f(x) = |x + 3| - 5x.$

5. $f(x) = \dfrac{2x}{|x + 2| + x^2}.$

6. $f(x) = 1 - \dfrac{1}{(x + 1)^2}.$

Exercises 7–10. Calculate (a) $f(-x)$, (b) $f(1/x)$, (c) $f(a + b)$.

7. $f(x) = x^2 - 2x.$

8. $f(x) = \dfrac{x}{x^2 + 1}.$

9. $f(x) = \sqrt{1 + x^2}.$

10. $f(x) = \dfrac{x}{|x^2 - 1|}.$

Exercises 11 and 12. Calculate $f(a + h)$ and $[f(a + h) - f(a)]/h$ for $h \neq 0$.

11. $f(x) = 2x^2 - 3x.$

12. $f(x) = \dfrac{1}{x - 2}.$

Exercises 13–18. Find the number(s) x, if any, where f takes on the value 1.

13. $f(x) = |2 - x|.$

14. $f(x) = \sqrt{1 + x}.$

15. $f(x) = x^2 + 4x + 5.$

16. $f(x) = 4 + 10x - x^2.$

17. $f(x) = \dfrac{2}{\sqrt{x^2 - 5}}.$

18. $f(x) = \dfrac{x}{|x|}.$

Exercises 19–30. Give the domain and range of the function.

19. $f(x) = |x|.$

20. $g(x) = x^2 - 1.$

21. $f(x) = 2x - 3.$

22. $g(x) = \sqrt{x} + 5.$

23. $f(x) = \dfrac{1}{x^2}.$

24. $g(x) = \dfrac{4}{x}.$

25. $f(x) = \sqrt{1 - x}.$

26. $g(x) = \sqrt{x - 3}.$

27. $f(x) = \sqrt{7 - x} - 1.$

28. $g(x) = \sqrt{x - 1} - 1.$

29. $f(x) = \dfrac{1}{\sqrt{2 - x}}.$

30. $g(x) = \dfrac{1}{\sqrt{4 - x^2}}.$

Exercises 31–40. Give the domain of the function and sketch the graph.

31. $f(x) = 1.$

32. $f(x) = -1.$

33. $f(x) = 2x.$

34. $f(x) = 2x + 1.$

35. $f(x) = \frac{1}{2}x + 2.$

36. $f(x) = -\frac{1}{2}x - 3.$

37. $f(x) = \sqrt{4 - x^2}.$

38. $f(x) = \sqrt{9 - x^2}.$

39. $f(x) = x^2 - x - 6.$

40. $f(x) = |x - 1|.$

Exercises 41–44. Sketch the graph and give the domain and range of the function.

41. $f(x) = \begin{cases} -1, & x < 0 \\ 1, & x > 0. \end{cases}$

42. $f(x) = \begin{cases} x^2, & x \leq 0 \\ 1 - x, & x > 0. \end{cases}$

43. $f(x) = \begin{cases} 1 + x, & 0 \leq x \leq 1 \\ x, & 1 < x < 2 \\ \frac{1}{2}x + 1, & 2 \leq x. \end{cases}$

44. $f(x) = \begin{cases} x^2, & x < 0 \\ -1, & 0 < x < 2 \\ x, & 2 < x. \end{cases}$

Exercises 45–48. State whether the curve is the graph of a function. If it is, give the domain and the range.

45.

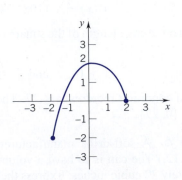

46.

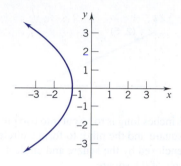

47.

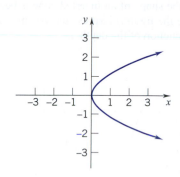

48.

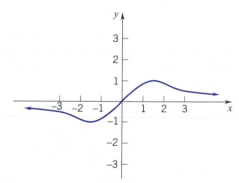

Exercises 49–56. State whether the function is odd, even, or neither.

49. $f(x) = x^3$.

50. $f(x) = x^2 + 1$.

51. $g(x) = x(x - 1)$.

52. $g(x) = x(x^2 + 1)$.

53. $f(x) = \dfrac{x^2}{1 - |x|}$.

54. $F(x) = x + \dfrac{1}{x}$.

55. $f(x) = \dfrac{x}{x^2 - 9}$.

56. $f(x) = \sqrt[5]{x - x^3}$.

C▶ 57. The graph of $f(x) = \frac{1}{3}x^3 + \frac{1}{2}x^2 - 12x - 6$ looks something like this:

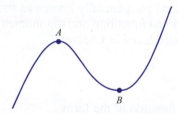

(a) Use a graphing utility to sketch an accurate graph of f.

(b) Find the zero(s) of f (the values of x for which $f(x) = 0$) accurate to three decimal places.

(c) Find the coordinates of the points marked A and B, accurate to three decimal places.

C▶ 58. The graph of $f(x) = -x^4 + 8x^2 + x - 1$ looks something like this:

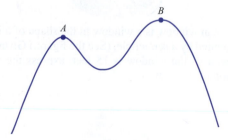

(a) Use a graphing utility to sketch an accurate graph of f.

(b) Find the zero(s) of f, if any. Use three decimal place accuracy.

(c) Find the coordinates of the points marked A and B, accurate to three decimal places.

C▶ Exercises 59 and 60. Use a graphing utility to draw several views of the graph of the function. Select the one that most accurately shows the important features of the graph. Give the domain and range of the function.

59. $f(x) = |x^3 - 3x^2 - 24x + 4|$.

60. $f(x) = \sqrt{x^3 - 8}$.

61. Determine the range of $y = x^2 - 4x - 5$:

(a) by writing y in the form $(x - a)^2 + b$.

(b) by first solving the equation for x.

62. Determine the range of $y = \dfrac{2x}{4 - x}$:

(a) by writing y in the form $a + \dfrac{b}{4 - x}$.

(b) by first solving the equation for x.

63. Express the area of a circle as a function of the circumference.

64. Express the volume of a sphere as a function of the surface area.

65. Express the volume of a cube as a function of the area of one of the faces.

66. Express the volume of a cube as a function of the total surface area.

67. Express the surface area of a cube as a function of the length of the diagonal of a face.

68. Express the volume of a cube as a function of one of the diagonals.

69. Express the area of an equilateral triangle as a function of the length of a side.

70. A right triangle with hypotenuse c is revolved about one of its legs to form a cone. (See the figure.) Given that x is the length of the other leg, express the volume of the cone as a function of x.

71. A Norman window is a window in the shape of a rectangle surmounted by a semicircle. (See the figure.) Given that the perimeter of the window is 15 feet, express the area as a function of the width x.

72. A window has the shape of a rectangle surmounted by an equilateral triangle. Given that the perimeter of the window is 15 feet, express the area as a function of the length of one side of the equilateral triangle.

73. Express the area of the rectangle shown in the accompanying figure as a function of the x-coordinate of the point P.

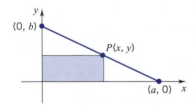

74. A right triangle is formed by the coordinate axes and a line through the point (2,5). (See the figure.) Express the area of the triangle as a function of the x-intercept.

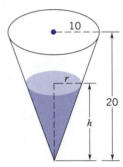

75. A string 28 inches long is to be cut into two pieces, one piece to form a square and the other to form a circle. Express the total area enclosed by the square and circle as a function of the perimeter of the square.

76. A tank in the shape of an inverted cone is being filled with water. (See the figure.) Express the volume of water in the tank as a function of the depth h.

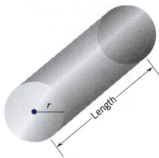

77. Suppose that a cylindrical mailing container exactly meets the U.S. Postal Service regulations given in Example 3. (See the figure.) Express the volume of the container as a function of the radius of an end.

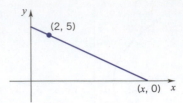

■ 1.6 THE ELEMENTARY FUNCTIONS

The functions that figure most prominently in single-variable calculus are the polynomials, the rational functions, the trigonometric functions, the exponential functions, and the logarithm functions. These functions are generally known as the *elementary functions*. Here we review polynomials, rational functions, and trigonometric functions. Exponential and logarithm functions are introduced in Chapter 7.

Polynomials

We begin with a nonnegative integer n. A function of the form

$$P(x) = a_n x^n + a_{n-1} x^{n-1} + \cdots + a_1 x + a_0 \qquad \text{for all real } x,$$

where the *coefficients* $a_n, a_{n-1}, \ldots, a_1, a_0$ are real numbers and $a_n \neq 0$ is called a (*real*) *polynomial* of *degree n*.

If $n = 0$, the polynomial is simply a constant function:

$$P(x) = a_0 \qquad \text{for all real } x.$$

Nonzero constant functions are polynomials of degree 0. The function $P(x) = 0$ for all real x is also a polynomial, but we assign no degree to it.

Polynomials satisfy a condition known as the *factor theorem*: if P is a polynomial and r is a real number, then

$$P(r) = 0 \qquad \text{iff} \qquad (x - r) \text{ is a factor of } P(x).$$

The real numbers r at which $P(x) = 0$ are called the *zeros* of the polynomial.

The linear functions

$$P(x) = ax + b, \qquad a \neq 0$$

are the polynomials of degree 1. Such a polynomial has only one zero: $r = -b/a$. The graph is the straight line $y = ax + b$.

The quadratic functions

$$P(x) = ax^2 + bx + c, \qquad a \neq 0$$

are the polynomials of degree 2. The graph of such a polynomial is the parabola $y = ax^2 + bx + c$. If $a > 0$, the vertex is the lowest point on the curve; the curve opens up. If $a < 0$, the vertex is the highest point on the curve. (See Figure 1.6.1.)

vertex

vertex

$a > 0$ $\qquad\qquad$ $a < 0$

Figure 1.6.1

The zeros of the quadratic function $P(x) = ax^2 + bx + c$ are the roots of the quadratic equation

$$ax^2 + bx + c = 0.$$

The three possibilities are depicted in Figure 1.6.2. Here we are taking $a > 0$.

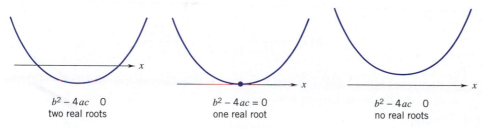

$b^2 - 4ac$ 0 $\qquad\qquad$ $b^2 - 4ac = 0$ $\qquad\qquad$ $b^2 - 4ac$ 0
two real roots $\qquad\qquad$ one real root $\qquad\qquad$ no real roots

Figure 1.6.2

Polynomials of degree 3 have the form $P(x) = ax^3 + bx^2 + cx + d, a \neq 0$. These functions are called *cubics*. In general, the graph of a cubic has one of the two following shapes, again determined by the sign of a (Figure 1.6.3). Note that we have not tried to locate these graphs with respect to the coordinate axes. Our purpose here is simply to indicate the two typical shapes. You can see, however, that for a cubic there are three possibilities: three real roots, two real roots, one real root. (Each cubic has at least one real root.)

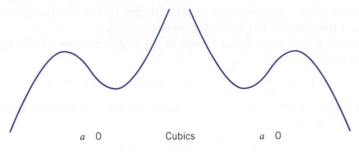

Figure 1.6.3

Polynomials become more complicated as the degree increases. In Chapter 4 we use calculus to analyze polynomials of higher degree.

Rational Functions

A *rational function* is a function of the form

$$R(x) = \frac{P(x)}{Q(x)}$$

where P and Q are polynomials. Note that every polynomial P is a rational function: $P(x) = P(x)/1$ is the quotient of two polynomials. Since division by 0 is meaningless, a rational function $R = P/Q$ is not defined at those points x (if any) where $Q(x) = 0$; R is defined at all other points. Thus, dom $(R) = \{x : Q(x) \neq 0\}$.

Rational functions $R = P/Q$ are more difficult to analyze than polynomials and more difficult to graph. In particular, we have to examine the behavior of R near the zeros of the denominator and the behavior of R for large values of x, both positive and negative. If, for example, the denominator Q is zero at $x = a$ but the numerator P is not zero at $x = a$, then the graph of R tends to the vertical as x tends to a and the line $x = a$ is called a *vertical asymptote*. If as x becomes very large positive or very large negative the values of R tend to some number b, then the line $y = b$ is called a *horizontal asymptote*. Vertical and horizontal asymptotes are mentioned here only in passing. They will be studied in detail in Chapter 4. Below are two simple examples.

(i) The graph of

$$R(x) = \frac{1}{x^2 - 4x + 4} = \frac{1}{(x - 2)^2}$$

is shown in Figure 1.6.4. The line $x = 2$ is a vertical asymptote; the line $y = 0$ (the x-axis) is a horizontal asymptote.

(ii) The graph of

$$R(x) = \frac{x^2}{x^2 - 1} = \frac{x^2}{(x - 1)(x + 1)}$$

is shown in Figure 1.6.5. The lines $x = 1$ and $x = -1$ are vertical asymptotes; the line $y = 1$ is a horizontal asymptote.

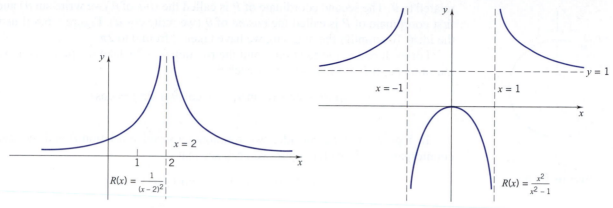

$R(x) = \dfrac{1}{(x-2)^2}$

Figure 1.6.4

$R(x) = \dfrac{x^2}{x^2-1}$

Figure 1.6.5

The Trigonometric Functions

Radian Measure   Degree measure, traditionally used to measure angles, has a serious drawback. It is artificial; there is no intrinsic connection between a degree and the geometry of a rotation. Why choose 360° for one complete revolution? Why not 100°? or 400°?

There is another way of measuring angles that is more natural and lends itself better to the methods of calculus: measuring angles in *radians*.

Angles arise from rotations. We will measure angles by measuring rotations. Suppose that the points of the plane are rotated about some point 0. The point 0 remains fixed, but all other points P trace out circular arcs on circles centered at 0. The farther P is from 0, the longer the circular arc (Figure 1.6.6). The *magnitude* of a rotation about 0 is by definition the length of the arc generated by the rotation as measured on a circle at a unit distance from 0.

Now let θ be any real number. The rotation of *radian measure θ* (we shall simply call it the *rotation θ*) is by definition the rotation of magnitude $|\theta|$ in the counterclockwise direction if $\theta > 0$, in the clockwise direction if $\theta < 0$. If $\theta = 0$, there is no movement; every point remains in place.

In degree measure a full turn is effected over the course of 360°. In radian measure, a full turn is effected during the course of 2π radians. (The circumference of a circle of radius 1 is 2π.) Thus

$$2\pi \text{ radians} = 360 \text{ degrees}$$

$$\text{one radian} = 360/2\pi \text{ degrees} \cong 57.30°$$

$$\text{one degree} = 2\pi/360 \text{ radians} \cong 0.0175 \text{ radians.}$$

The following table gives some common angles (rotations) measured both in degrees and in radians.

Figure 1.6.6

degrees	0°	30°	45°	60°	90°	120°	135°	150°	180°	270°	360°
radians	0	$\frac{1}{6}\pi$	$\frac{1}{4}\pi$	$\frac{1}{3}\pi$	$\frac{1}{2}\pi$	$\frac{2}{3}\pi$	$\frac{3}{4}\pi$	$\frac{5}{6}\pi$	π	$\frac{3}{2}\pi$	2π

Cosine and Sine   In Figure 1.6.7 you can see a circle of radius 1 centered at the origin of a coordinate plane. We call this the *unit circle*. On the circle we have marked the point A (1, 0).

Now let θ be any real number. The rotation θ takes A (1, 0) to some point P, also on the unit circle. The coordinates of P are completely determined by θ and have names

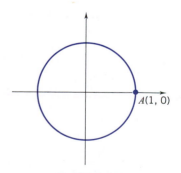

$A(1, 0)$

Figure 1.6.7

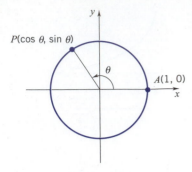

Figure 1.6.8

related to θ. The second coordinate of P is called the *sine* of θ (we write $\sin\theta$) and the first coordinate of P is called the *cosine* of θ (we write $\cos\theta$). Figure 1.6.8 illustrates the idea. To simplify the diagram, we have taken θ from 0 to 2π.

For each real θ, the rotation θ and the rotation $\theta + 2\pi$ take the point A to exactly the same point P. It follows that for each θ,

$$\sin(\theta + 2\pi) = \sin\theta, \qquad \cos(\theta + 2\pi) = \cos\theta.$$

In Figure 1.6.9 we consider two rotations: a positive rotation θ and its negative counterpart $-\theta$. From the figure, you can see that

$$\sin(-\theta) = -\sin\theta, \qquad \cos(-\theta) = \cos\theta.$$

The sine function is an odd function and the cosine function is an even function.

In Figure 1.6.10 we have marked the effect of consecutive rotations of $\frac{1}{2}\pi$ radians:

$$(a, b) \to (-b, a) \to (-a, -b) \to (b, -a).$$

In each case, $(x, y) \to (-y, x)$. Thus,

$$\sin(\theta + \tfrac{1}{2}\pi) = \cos\theta, \qquad \cos(\theta + \tfrac{1}{2}\pi) = -\sin\theta.$$

A rotation of π radians takes each point to the point antipodal to it: $(x, y) \to (-x, -y)$. Thus

$$\sin(\theta + \pi) = -\sin\theta, \qquad \cos(\theta + \pi) = -\cos\theta.$$

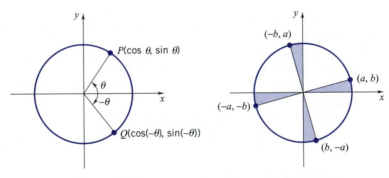

Figure 1.6.9   **Figure 1.6.10**

Tangent, Cotangent, Secant, Cosecant   There are four other trigonometric functions: the tangent, the cotangent, the secant, the cosecant. These are obtained as follows:

$$\tan\theta = \frac{\sin\theta}{\cos\theta}, \qquad \cot\theta = \frac{\cos\theta}{\sin\theta}, \qquad \sec\theta = \frac{1}{\cos\theta}, \qquad \csc\theta = \frac{1}{\sin\theta}.$$

The most important of these functions is the tangent. Note that the tangent function is an odd function

$$\tan(-\theta) = \frac{\sin(-\theta)}{\cos(-\theta)} = \frac{-\sin\theta}{\cos\theta} = -\tan\theta$$

and repeats itself every π radians:

$$\tan(\theta + \pi) = \frac{\sin(\theta + \pi)}{\cos(\theta + \pi)} = \frac{-\sin\theta}{-\cos\theta} = \tan\theta.$$

Particular Values The values of the sine, cosine, and tangent at angles (rotations) frequently encountered are given in the following table.

	0	$\frac{1}{6}\pi$	$\frac{1}{4}\pi$	$\frac{1}{3}\pi$	$\frac{1}{2}\pi$	$\frac{2}{3}\pi$	$\frac{3}{4}\pi$	$\frac{5}{6}\pi$	π	$\frac{3}{2}\pi$	2π
$\sin\theta$	0	$\frac{1}{2}$	$\frac{1}{2}\sqrt{2}$	$\frac{1}{2}\sqrt{3}$	1	$\frac{1}{2}\sqrt{3}$	$\frac{1}{2}\sqrt{2}$	$\frac{1}{2}$	0	-1	0
$\cos\theta$	1	$\frac{1}{2}\sqrt{3}$	$\frac{1}{2}\sqrt{2}$	$\frac{1}{2}$	0	$-\frac{1}{2}$	$-\frac{1}{2}\sqrt{2}$	$-\frac{1}{2}\sqrt{3}$	-1	0	1
$\tan\theta$	0	$\frac{1}{3}\sqrt{3}$	1	$\sqrt{3}$	$--$	$-\sqrt{3}$	-1	$-\frac{1}{3}\sqrt{3}$	0	$--$	0

The (approximate) values of the trigonometric functions for any angle θ can be obtained with a hand calculator or from a table of values.

Identities Below we list the basic trigonometric identities. Some are obvious; some have just been verified; the rest are derived in the exercises.

(i) *unit circle*

$$\sin^2\theta + \cos^2\theta = 1, \qquad \tan^2\theta + 1 = \sec^2\theta, \qquad 1 + \cot^2\theta = \csc^2\theta.$$

(the first identity is obvious; the other two follow from the first)

(ii) *periodicity*[†]

$$\sin(\theta + 2\pi) = \sin\theta, \qquad \cos(\theta + 2\pi) = \cos\theta, \qquad \tan(\theta + \pi) = \tan\theta$$

(iii) *odd and even*

$$\sin(-\theta) = -\sin\theta, \qquad \cos(-\theta) = \cos\theta, \qquad \tan(-\theta) = -\tan\theta.$$

(the sine and tangent are odd functions; the cosine is even)

(iv) *sines and cosines*

$$\sin(\theta + \pi) = -\sin\theta, \quad \cos(\theta + \pi) = -\cos\theta,$$
$$\sin(\theta + \tfrac{1}{2}\pi) = \cos\theta, \quad \cos(\theta + \tfrac{1}{2}\pi) = -\sin\theta,$$
$$\sin(\tfrac{1}{2}\pi - \theta) = \cos\theta, \quad \cos(\tfrac{1}{2}\pi - \theta) = \sin\theta.$$

(only the third pair of identities still has to be verified)

(v) *addition formulas*

$$\sin(\alpha + \beta) = \sin\alpha\cos\beta + \cos\alpha\sin\beta,$$
$$\sin(\alpha - \beta) = \sin\alpha\cos\beta - \cos\alpha\sin\beta,$$
$$\cos(\alpha + \beta) = \cos\alpha\cos\beta - \sin\alpha\sin\beta,$$
$$\cos(\alpha - \beta) = \cos\alpha\cos\beta + \sin\alpha\sin\beta.$$

(taken up in the exercises)

(vi) *double-angle formulas*

$$\sin 2\theta = 2\sin\theta\cos\theta, \qquad \cos 2\theta = \cos^2\theta - \sin^2\theta = 2\cos^2\theta - 1 = 1 - 2\sin^2\theta.$$

(follow from the addition formulas)

[†]A function f with an unbounded domain is said to be *periodic* if there exists a number $p > 0$ such that, if θ is in the domain of f, then $\theta + p$ is in the domain and $f(\theta + p) = f(\theta)$. The least number p with this property (if there is a least one) is called the *period* of the function. The sine and cosine have period 2π. Their reciprocals, the cosecant and secant, also have period 2π. The tangent and cotangent have period π.

(vii) *half-angle formulas*

$$\sin^2 \theta = \tfrac{1}{2}(1 - \cos 2\theta), \quad \cos^2 \theta = \tfrac{1}{2}(1 + \cos 2\theta)$$

(follow from the double-angle formulas)

In Terms of a Right Triangle For angles θ between 0 and $\pi/2$, the trigonometric functions can also be defined as ratios of the sides of a right triangle. (See Figure 1.6.11.)

$$\sin \theta = \frac{\text{opposite side}}{\text{hypotenuse}}, \qquad \csc \theta = \frac{\text{hypotenuse}}{\text{opposite side}},$$

$$\cos \theta = \frac{\text{adjacent side}}{\text{hypotenuse}}, \qquad \sec \theta = \frac{\text{hypotenuse}}{\text{adjacent side}}, \qquad \text{(Exercise 81)}$$

$$\tan \theta = \frac{\text{opposite side}}{\text{adjacent side}}, \qquad \cot \theta = \frac{\text{adjacent side}}{\text{opposite side}}.$$

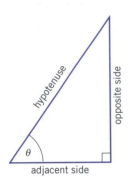

Figure 1.6.11

Arbitrary Triangles Let a, b, c be the sides of a triangle and let A, B, C be the opposite angles. (See Figure 1.6.12.)

$$area \quad \tfrac{1}{2}ab \, \sin C = \tfrac{1}{2}ac \, \sin B = \tfrac{1}{2}bc \, \sin A.$$

$$law \ of \ sines \quad \frac{\sin A}{a} = \frac{\sin B}{b} = \frac{\sin C}{c}. \qquad \text{(taken up in the exercises)}$$

$$law \ of \ cosines \quad a^2 = b^2 + c^2 - 2bc \cos A,$$
$$b^2 = a^2 + c^2 - 2ac \cos B,$$
$$c^2 = a^2 + b^2 - 2ab \cos C.$$

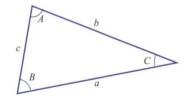

Figure 1.6.12

Graphs Usually we work with functions $y = f(x)$ and graph them in the xy-plane. To bring the graphs of the trigonometric functions into harmony with this convention, we replace θ by x and write $y = \sin x$, $y = \cos x$, $y = \tan x$. (These are the only functions that we are going to graph here.) The functions have not changed, only the symbols: x is a rotation that takes $A(1, 0)$ to the point $P(\cos x, \sin x)$. The graphs of the sine, cosine, and tangent appear in Figure 1.6.13.

The graphs of sine and cosine are waves that repeat themselves on every interval of length 2π. These waves appear to chase each other. They do chase each other. In the chase the cosine wave remains $\tfrac{1}{2}\pi$ units behind the sine wave:

$$\cos x = \sin(x + \tfrac{1}{2}\pi).$$

Changing perspective, we see that the sine wave remains $\tfrac{3}{2}\pi$ units behind the cosine wave:

$$\sin x = \cos(x + \tfrac{3}{2}\pi).$$

All these waves crest at $y = 1$, drop down to $y = -1$, and then head up again.

The graph of the tangent function consists of identical pieces separated every π units by asymptotes that mark the points x where $\cos x = 0$.

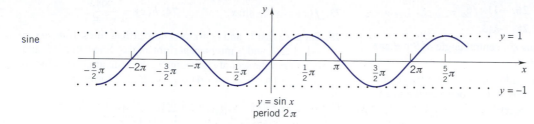

$y = \sin x$
period 2π

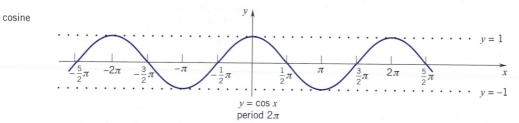

$y = \cos x$
period 2π

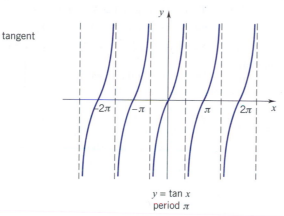

$y = \tan x$
period π

vertical asymptotes $x = (n + \frac{1}{2})\pi$, n an integer

Figure 1.6.13

EXERCISES 1.6

Exercises 1–10. State whether the function is a polynomial, a rational function (but not a polynomial), or neither a polynomial nor a rational function. If the function is a polynomial, give the degree.

1. $f(x) = 3$.

2. $f(x) = 1 + \frac{1}{2}x$.

3. $g(x) = \dfrac{1}{x}$.

4. $h(x) = \dfrac{x^2 - 4}{\sqrt{2}}$.

5. $F(x) = \dfrac{x^3 - 3x^{3/2} + 2x}{x^2 - 1}$.

6. $f(x) = 5x^4 - \pi x^2 + \frac{1}{2}$.

7. $f(x) = \sqrt{x}(\sqrt{x} + 1)$.

8. $g(x) = \dfrac{x^2 - 2x - 8}{x + 2}$.

9. $f(x) = \dfrac{\sqrt{x^2 + 1}}{x^2 - 1}$.

10. $h(x) = \dfrac{(\sqrt{x} + 2)(\sqrt{x} - 2)}{x^2 + 4}$.

Exercises 11–16. Determine the domain of the function and sketch the graph.

11. $f(x) = 3x - \frac{1}{2}$.

12. $f(x) = \dfrac{1}{x + 1}$.

13. $g(x) = x^2 - x - 6$.

14. $F(x) = x^3 - x$.

15. $f(x) = \dfrac{1}{x^2 - 4}$.

16. $g(x) = x + \dfrac{1}{x}$.

Exercises 17–22. Convert the degree measure into radian measure.

17. $225°$.

18. $-210°$.

19. $-300°$.

20. $450°$.

21. $15°$.

22. $3°$.

Exercises 23–28. Convert the radian measure into degree measure.

23. $-3\pi/2$.

24. $5\pi/4$.

25. $5\pi/3$.　　　　　　**26.** $-11\pi/6$.

27. 2.　　　　　　　　**28.** $-\sqrt{3}$.

29. Show that in a circle of radius r, a central angle of θ radians subtends an arc of length $r\theta$.

30. Show that in a circular disk of radius r, a sector with a central angle of θ radians has area $\frac{1}{2}r^2\theta$. Take θ between 0 and 2π. HINT: The area of the circle is πr^2.

Exercises 31–38. Find the number(s) x in the interval $[0, 2\pi]$ which satisfy the equation.

31. $\sin x = 1/2$.　　　　**32.** $\cos x = -1/2$.

33. $\tan x/2 = 1$.　　　　**34.** $\sqrt{\sin^2 x} = 1$.

35. $\cos x = \sqrt{2}/2$.　　　**36.** $\sin 2x = -\sqrt{3}/2$.

37. $\cos 2x = 0$.　　　　**38.** $\tan x = -\sqrt{3}$.

Exercises 39–44. Evaluate to four decimal place accuracy.

39. $\sin 51°$.　　　　　　**40.** $\cos 17°$.

41. $\sin(2.352)$.　　　　**42.** $\cos(-13.461)$.

43. $\tan 72.4°$.　　　　　**44.** $\cot(7.311)$.

Exercises 45–52. Find the solutions x that are in the interval $[0, 2\pi]$. Express your answers in radians and use four decimal place accuracy.

45. $\sin x = 0.5231$.　　　**46.** $\cos x = -0.8243$.

47. $\tan x = 6.7192$.　　　**48.** $\cot x = -3.0649$.

49. $\sec x = -4.4073$.　　　**50.** $\csc x = 10.260$.

Exercises 51–52. Solve the equation $f(x) = y_0$ for x in $[0, 2\pi]$ by using a graphing utility. Display the graph of f and the line $y = y_0$ in one figure; then use the trace function to find the point(s) of intersection.

51. $f(x) = \sin 3x$;　　$y_0 = -1/\sqrt{2}$.

52. $f(x) = \cos \frac{1}{2}x$;　　$y_0 = \frac{3}{4}$

Exercises 53–58. Give the domain and range of the function.

53. $f(x) = |\sin x|$　　　**54.** $g(x) = \sin^2 x + \cos^2 x$.

55. $f(x) = 2\cos 3x$.　　　**56.** $F(x) = 1 + \sin x$.

57. $f(x) = 1 + \tan^2 x$.　　**58.** $h(x) = \sqrt{\cos^2 x}$.

Exercises 59–62. Determine the period. (The least positive number p for which $f(x + p) = f(x)$ for all x.)

59. $f(x) = \sin \pi x$.　　　**60.** $f(x) = \cos 2x$.

61. $f(x) = \cos \frac{1}{3}x$.　　**62.** $f(x) = \sin \frac{1}{2}x$.

Exercises 63–68. Sketch the graph of the function.

63. $f(x) = 3\sin 2x$.　　　**64.** $f(x) = 1 + \sin x$.

65. $g(x) = 1 - \cos x$.　　　**66.** $F(x) = \tan \frac{1}{2}x$.

67. $f(x) = \sqrt{\sin^2 x}$.　　**68.** $g(x) = -2\cos x$.

Exercises 69–74. State whether the function is odd, even, or neither.

69. $f(x) = \sin 3x$.　　　**70.** $g(x) = \tan x$.

71. $f(x) = 1 + \cos 2x$.　　**72.** $g(x) = \sec x$.

73. $f(x) = x^3 + \sin x$.　　**74.** $h(x) = \dfrac{\cos x}{x^2 + 1}$.

75. Suppose that l_1 and l_2 are two nonvertical lines. If $m_1 m_2 = -1$, then l_1 and l_2 intersect at right angles. Show that if l_1 and l_2 do not intersect at right angles, then the angle α between l_1 and l_2 (see Section 1.4) is given by the formula

$$\tan \alpha = \left| \frac{m_1 - m_2}{1 + m_1 m_2} \right|.$$

HINT: Derive the identity

$$\tan(\theta_1 - \theta_2) = \frac{\tan \theta_1 - \tan \theta_2}{1 + \tan \theta_1 \tan \theta_2}$$

by expressing the right side in terms of sines and cosines.

Exercises 76–79. Find the point where the lines intersect and determine the angle between the lines.

76. $l_1: 4x - y - 3 = 0$,　　$l_2: 3x - 4y + 1 = 0$.

77. $l_1: 3x + y - 5 = 0$,　　$l_2: 7x - 10y + 27 = 0$.

78. $l_1: 4x - y + 2 = 0$,　　$l_2: 19x + y = 0$.

79. $l_1: 5x - 6y + 1 = 0$,　　$l_2: 8x + 5y + 2 = 0$.

80. Show that the function $f(x) = \begin{cases} 1, & x \text{ rational} \\ 0, & x \text{ irrational} \end{cases}$ is periodic but has no period.

81. Verify that, for angles θ between 0 and $\pi/2$, the definition of the trigonometric functions in terms of the unit circle and the definitions in terms of a right triangle are in agreement. HINT: Set the triangle as in the figure.

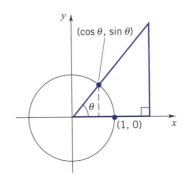

The setting for Exercises 82, 83, 84 is a triangle with sides a, b, c and opposite angles A, B, C.

82. Show that the area of the triangle is given by the formula $A = \frac{1}{2}ab \sin C$.

83. Confirm the law of sines:

$$\frac{\sin A}{a} = \frac{\sin B}{b} = \frac{\sin C}{c}.$$

HINT: Drop a perpendicular from one vertex to the opposite side and use the two right triangles formed.

84. Confirm the law of cosines:

$$a^2 = b^2 + c^2 - 2bc \cos A.$$

HINT: Drop a perpendicular from angle B to side b and use the two right triangles formed.

85. Verify the identity

$$\cos(\alpha - \beta) = \cos \alpha \cos \beta + \sin \alpha \sin \beta.$$

HINT: With P and Q as in the accompanying figure, calculate the length of $\overline{PQ}$ by applying the law of cosines.

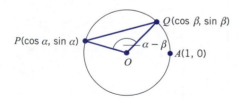

86. Use Exercise 85 to show that

$$\cos(\alpha + \beta) = \cos \alpha \cos \beta - \sin \alpha \sin \beta.$$

87. Verify the following identities:

$$\sin(\tfrac{1}{2}\pi - \theta) = \cos \theta, \qquad \cos(\tfrac{1}{2}\pi - \theta) = \sin \theta.$$

88. Verify that

$$\sin(\alpha + \beta) = \sin \alpha \cos \beta + \cos \alpha \sin \beta.$$

HINT: $\sin(\alpha + \beta) = \cos[(\tfrac{1}{2}\pi - \alpha) - \beta]$.

89. Use Exercise 88 to show that

$$\sin(\alpha - \beta) = \sin \alpha \cos \beta - \cos \alpha \sin \beta.$$

90. It has been said that "all of trigonometry lies in the undulations of the sine wave." Explain.

91. (a) Use a graphing utility to graph the polynomials

$$f(x) = x^4 + 2x^3 - 5x^2 - 3x + 1,$$

$$g(x) = -x^4 + x^3 + 4x^2 - 3x + 2.$$

(b) Based on your graphs in part (a), make a conjecture about the general shape of the graphs of polynomials of degree 4.

(c) Test your conjecture by graphing

$$f(x) = x^4 - 4x^2 + 4x + 2 \quad \text{and} \quad g(x) = -x^4.$$

Conjecture a property shared by the graphs of all polynomials of the form

$$P(x) = x^4 + ax^3 + bx^2 + cx + d.$$

Make an analogous conjecture for polynomials of the form.

$$Q(x) = -x^4 + ax^3 + bx^2 + cx + d.$$

92. (a) Use a graphing utility to graph the polynomials.

$$f(x) = x^5 - 7x^3 + 6x + 2,$$

$$g(x) = -x^5 + 5x^3 - 3x - 3.$$

(b) Based on your graphs in part (a), make a conjecture about the general shape of the graph of a polynomial of degree 5.

(c) Now graph

$$P(x) = x^5 + ax^4 + bx^3 + cx^2 + dx + e$$

for several choices of a, b, c, d, e. (For example, try $a = b = c = d = e = 0$.) How do these graphs compare with your graph of f from part (a)?

93. (a) Use a graphing utility to graph $f_A(x) = A \cos x$ for several values of A; use both positive and negative values. Compare your graphs with the graph of $f(x) = \cos x$.

(b) Now graph $f_B(x) = \cos Bx$ for several values of B. Since the cosine function is even, it is sufficient to use only positive values for B. Use some values between 0 and 1 and some values greater than 1. Again, compare your graphs with the graph of $f(x) = \cos x$.

(c) Describe the effects that the coefficients A and B have on the graph of the cosine function.

94. Let $f_n(x) = x^n$, $n = 1, 2, 3 \dots$.

(a) Using a graphing utility, draw the graphs of f_n for $n = 2, 4, 6$ in one figure, and in another figure draw the graphs of f_n for $n = 1, 3, 5$.

(b) Based on your results in part (a), make a general sketch of the graph of f_n for even n and for odd n.

(c) Given a positive integer k, compare the graphs of f_k and f_{k+1} on $[0, 1]$ and on $(1, \infty)$.

■ 1.7 COMBINATIONS OF FUNCTIONS

In this section we review the elementary ways of combining functions.

Algebraic Combinations of Functions

Here we discuss with some precision ideas that were used earlier without comment.

On the intersection of their domains, functions can be added and subtracted:

$$(f + g)(x) = f(x) + g(x), \qquad (f - g)(x) = f(x) - g(x);$$

they can be multiplied:

$$(fg)(x) = f(x)g(x);$$

and, at the points where $g(x) \neq 0$, we can form the quotient:

$$\left(\frac{f}{g}\right)(x) = \frac{f(x)}{g(x)},$$

a special case of which is the reciprocal:

$$\left(\frac{1}{g}\right)(x) = \frac{1}{g(x)}.$$

Example 1   Let

$$f(x) = \sqrt{x+3} \qquad \text{and} \qquad g(x) = \sqrt{5-x} - 2.$$

(a) Give the domain of f and of g.

(b) Determine the domain of $f + g$ and specify $(f + g)(x)$.

(c) Determine the domain of f/g and specify $(f/g)(x)$.

SOLUTION

(a) We can form $\sqrt{x+3}$ iff $x + 3 \geq 0$, which holds iff $x \geq -3$. Thus $\text{dom}(f) = [-3, \infty)$. We can form $\sqrt{5-x} - 2$ iff $5 - x \geq 0$, which holds iff $x \leq 5$. Thus $\text{dom}(g) = (-\infty, 5]$.

(b) $\text{dom}(f + g) = \text{dom}(f) \cap \text{dom}(g) = [-3, \infty) \cap (-\infty, 5] = [-3, 5]$,

$$(f + g)(x) = f(x) + g(x) = \sqrt{x+3} + \sqrt{5-x} - 2.$$

(c) To obtain the domain of the quotient, we must exclude from $[-3, 5]$ the numbers x at which $g(x) = 0$. There is only one such number: $x = 1$. Therefore

$$\text{dom}\left(\frac{f}{g}\right) = \{x \in [-3, 5] : x \neq 1\} = [-3, 1) \cup (1, 5],$$

$$\left(\frac{f}{g}\right)(x) = \frac{f(x)}{g(x)} = \frac{\sqrt{x+3}}{\sqrt{5-x} - 2}. \qquad \square$$

We can multiply functions f by real numbers α and form what are called *scalar multiples* of f:

$$(\alpha f)(x) = \alpha f(x).$$

With functions f and g and real numbers α and β, we can form *linear combinations*:

$$(\alpha f + \beta g)(x) = \alpha f(x) + \beta g(x).$$

These are just specific instances of the products and sums that we defined at the beginning of the section.

You have seen all these algebraic operations many times before:

(i) The polynomials are simply finite linear combinations of powers x^n, each of which is a finite product of identity functions $f(x) = x$. (Here we are taking the point of view that $x^0 = 1$.)

(ii) The rational functions are quotients of polynomials.

(iii) The secant and cosecant are reciprocals of the cosine and the sine.

(iv) The tangent and cotangent are quotients of sine and cosine.

Vertical Translations (Vertical Shifts)   Adding a positive constant c to a function raises the graph by c units. Subtracting a positive constant c from a function lowers the graph by c units. (Figure 1.7.1.)

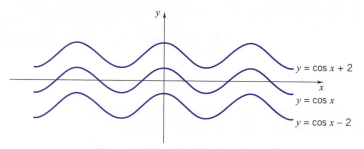

Figure 1.7.1

Composition of Functions

You have seen how to combine functions algebraically. There is another (probably less familiar) way to combine functions, called *composition*. To describe it, we begin with two functions, f and g, and a number x in the domain of g. By applying g to x, we get the number $g(x)$. If $g(x)$ is in the domain of f, then we can apply f to $g(x)$ and thereby obtain the number $f(g(x))$.

What is $f(g(x))$? It is the result of first applying g to x and then applying f to $g(x)$. The idea is illustrated in Figure 1.7.2. This new function—it takes x in the domain of g to $g(x)$ in the domain of f, and assigns to it the value $f(g(x))$—is called the *composition of f with g* and is denoted by $f \circ g$. (See Figure 1.7.3.) The symbol $f \circ g$ is read "f circle g."

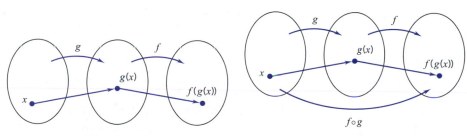

Figure 1.7.2   Figure 1.7.3

DEFINITION 1.7.1   COMPOSITION

Let f and g be functions. For those x in the domain of g for which $g(x)$ is in the domain of f, we define the *composition of f with g*, denoted $f \circ g$, by setting

$$(f \circ g)(x) = f(g(x)).$$

In set notation,

$$\operatorname{dom}(f \circ g) = \{x \in \operatorname{dom}(g) : g(x) \in \operatorname{dom}(f)\}$$

Example 2   Suppose that

$$g(x) = x^2 \qquad \text{(the squaring function)}$$

and

$$f(x) = x + 3. \qquad \text{(the function that adds 3)}$$

Then

$$(f \circ g)(x) = f(g(x)) = g(x) + 3 = x^2 + 3.$$

Thus, $f \circ g$ is the function that *first* squares and *then* adds 3.

On the other hand, the composition of g with f gives

$$(g \circ f)(x) = g(f(x)) = (x + 3)^2.$$

Thus, $g \circ f$ is the function that *first* adds 3 and *then* squares.

Since f and g are everywhere defined, both $f \circ g$ and $g \circ f$ are also everywhere defined. Note that $g \circ f$ is *not* the same as $f \circ g$. ❏

Example 3 Let $f(x) = x^2 - 1$ and $g(x) = \sqrt{3 - x}$.

The domain of g is $(-\infty, 3]$. Since f is everywhere defined, the domain of $f \circ g$ is also $(-\infty, 3]$. On that interval

$$(f \circ g)(x) = f(g(x)) = \left(\sqrt{3 - x}\right)^2 - 1 = (3 - x) - 1 = 2 - x.$$

Since $g(f(x)) = \sqrt{3 - f(x)}$, we can form $g(f(x))$ only for those x in the domain of f for which $f(x) \leq 3$. As you can verify, this is the set $[-2, 2]$. On $[-2, 2]$

$$(g \circ f)(x) = g(f(x)) = \sqrt{3 - (x^2 - 1)} = \sqrt{4 - x^2}. \; ❏$$

Horizontal Translations (Horizontal Shifts)

Adding a positive constant c to the argument of a function shifts the graph c units left: the function $g(x) = f(x + c)$ takes on at x the value that f takes on at $x + c$. Subtracting a positive constant c from the argument of a function shifts the graph c units to the right: the function $h(x) = f(x - c)$ takes on at x the value that f takes on at $x - c$. (See Figure 1.7.4.)

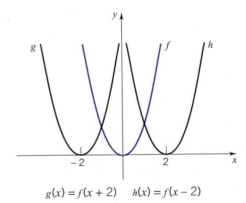

$$g(x) = f(x + 2) \qquad h(x) = f(x - 2)$$

Figure 1.7.4

We can form the composition of more than two functions. For example, the triple composition $f \circ g \circ h$ consists of first h, then g, and then f:

$$(f \circ g \circ h)(x) = f[g(h(x))].$$

We can go on in this manner with as many functions as we like.

Example 4 If $f(x) = \dfrac{1}{x}$, $\quad g(x) = x^2 + 1$, $\quad h(x) = \cos x$,

then
$$(f \circ g \circ h)(x) = f[g(h(x))] = \frac{1}{g(h(x))} = \frac{1}{[h(x)]^2 + 1}$$
$$= \frac{1}{\cos^2 x + 1}. \quad ❑$$

Example 5 Find functions f and g such that $f \circ g = F$ given that
$$F(x) = (x + 1)^5.$$

A SOLUTION The function consists of first adding 1 and then taking the fifth power. We can therefore set

$$g(x) = x + 1 \qquad\qquad \text{(adding 1)}$$

and

$$f(x) = x^5. \qquad\qquad \text{(taking the fifth power)}$$

As you can see,
$$(f \circ g)(x) = f(g(x)) = [g(x)]^5 = (x + 1)^5. \quad ❑$$

Example 6 Find three functions f, g, h such that $f \circ g \circ h = F$ given that
$$F(x) = \frac{1}{|x| + 3}.$$

A SOLUTION F takes the absolute value, adds 3, and then inverts. Let h take the absolute value:

$$\text{set} \quad h(x) = |x|.$$

Let g add 3:

$$\text{set} \quad g(x) = x + 3.$$

Let f do the inverting:

$$\text{set} \quad f(x) = \frac{1}{x}.$$

With this choice of f, g, h, we have

$$(f \circ g \circ h)(x) = f[g(h(x))] = \frac{1}{g(h(x))} = \frac{1}{h(x) + 3} = \frac{1}{|x| + 3}. \quad ❑$$

EXERCISES 1.7

Exercises 1–8. Set $f(x) = 2x^2 - 3x + 1$ and $g(x) = x^2 + 1/x$. Calculate the indicated value.

1. $(f + g)(2)$.

2. $(f - g)(-1)$.

3. $(f \cdot g)(-2)$.

4. $\left(\dfrac{f}{g}\right)(1)$.

5. $(2f - 3g)(\tfrac{1}{2})$.

6. $\left(\dfrac{f + 2g}{f}\right)(-1)$.

7. $(f \circ g)(1)$.

8. $(g \circ f)(1)$.

Exercises 9–12. Determine $f + g, f - g, f \cdot g, f/g$, and give the domain of each

9. $f(x) = 2x - 3$, $g(x) = 2 - x$.

10. $f(x) = x^2 - 1$, $g(x) = x + 1/x$.

11. $f(x) = \sqrt{x - 1}$, $g(x) = x - \sqrt{x + 1}$.

12. $f(x) = \sin^2 x$, $g(x) = \cos 2x$.

13. Given that $f(x) = x + 1/\sqrt{x}$ and $g(x) = \sqrt{x} - 2\sqrt{x}$, find
(a) $6f + 3g$, (b) $f - g$, (c) f/g.

14. Given that

$$f(x) = \begin{cases} 1 - x, & x \le 1 \\ 2x - 1, & x > 1, \end{cases} \quad \text{and } g(x) = \begin{cases} 0, & x < 2 \\ -1, & x \ge 2, \end{cases}$$

find $f + g$, $f - g$, $f \cdot g$. HINT: Break up the domains of the two functions in the same manner.

Exercises 15–22. Sketch the graph with f and g as shown in the figure.

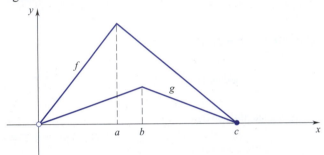

15. $2g$.

16. $\frac{1}{2}f$.

17. $-f$.

18. $0 \cdot g$.

19. $-2g$.

20. $f + g$.

21. $f - g$.

22. $f + 2g$.

Exercises 23–30. Form the composition $f \circ g$ and give the domain.

23. $f(x) = 2x + 5$, $g(x) = x^2$.

24. $f(x) = x^2$, $g(x) = 2x + 5$.

25. $f(x) = \sqrt{x}$, $g(x) = x^2 + 5$.

26. $f(x) = x^2 + x$, $g(x) = \sqrt{x}$.

27. $f(x) = 1/x$, $g(x) = (x - 2)/x$.

28. $f(x) = 1/(x - 1)$, $g(x) = x^2$.

29. $f(x) = \sqrt{1 - x^2}$, $g(x) = \cos 2x$.

30. $f(x) = \sqrt{1 - x}$, $g(x) = 2\cos x$ for $x \in [0, 2\pi]$.

Exercises 31–34. Form the composition $f \circ g \circ h$ and give the domain.

31. $f(x) = 4x$, $g(x) = x - 1$, $h(x) = x^2$.

32. $f(x) = x - 1$, $g(x) = 4x$, $h(x) = x^2$.

33. $f(x) = \dfrac{1}{x}$, $g(x) = \dfrac{1}{2x + 1}$, $h(x) = x^2$.

34. $f(x) = \dfrac{x + 1}{x}$, $g(x) = \dfrac{1}{2x + 1}$, $h(x) = x^2$.

Exercises 35–38. Find f such that $f \circ g = F$ given that

35. $g(x) = \dfrac{1 + x^2}{1 + x^4}$, $F(x) = \dfrac{1 + x^4}{1 + x^2}$.

36. $g(x) = x^2$, $F(x) = ax^2 + b$.

37. $g(x) = 3x$, $F(x) = 2\sin 3x$.

38. $g(x) = -x^2$, $F(x) = \sqrt{a^2 + x^2}$.

Exercises 39–42. Find g such that $f \circ g = F$ given that

39. $f(x) = x^3$, $F(x) = (1 - 1/x^4)^2$.

40. $f(x) = x + \dfrac{1}{x}$, $F(x) = a^2x^2 + \dfrac{1}{a^2x^2}$.

41. $f(x) = x^2 + 1$, $F(x) = (2x^3 - 1)^2 + 1$.

42. $f(x) = \sin x$, $F(x) = \sin 1/x$.

Exercises 43–46. Find $f \circ g$ and $g \circ f$.

43. $f(x) = \sqrt{x}$, $g(x) = x^2$.

44. $f(x) = 3x + 1$, $g(x) = x^2$.

45. $f(x) = 1 - x^2$, $g(x) = \sin x$.

46. $f(x) = x^3 + 1$, $g(x) = \sqrt[3]{x - 1}$.

47. Find g given that $(f + g)(x) = f(x) + c$.

48. Find f given that $(f \circ g)(x) = g(x) + c$.

49. Find g given that $(fg)(x) = cf(x)$.

50. Find f given that $(f \circ g)(x) = cg(x)$.

51. Take f as a function on $[0, a]$ with range $[0, b]$ and take g as defined below. Compare the graph of g with the graph of f; give the domain of g and the range of g.
(a) $g(x) = f(x - 3)$.
(b) $g(x) = 3f(x + 4)$.
(c) $g(x) = f(2x)$.
(d) $g(x) = f(\frac{1}{2}x)$.

52. Suppose that f and g are odd functions. What can you conclude about $f \cdot g$?

53. Suppose that f and g are even functions. What can you conclude about $f \cdot g$?

54. Suppose that f is an even function and g is an odd function. What can you conclude about $f \cdot g$?

55. For $x \ge 0$, f is defined as follows:

$$f(x) = \begin{cases} x, & 0 \le x \le 1 \\ 1, & x > 1. \end{cases}$$

How is f defined for $x < 0$ if (a) f is even? (b) f is odd?

56. For $x \ge 0$, $f(x) = x^2 - x$. How is f defined for $x < 0$ if (a) f is even? (b) f is odd?

57. Given that f is defined for all real numbers, show that the function $g(x) = f(x) + f(-x)$ is an even function.

58. Given that f is defined for all real numbers, show that the function $h(x) = f(x) - f(-x)$ is an odd function.

59. Show that every function defined for all real numbers can be written as the sum of an even function and an odd function.

60. For $x \ne 0, 1$, define

$$f_1(x) = x, \quad f_2(x) = \frac{1}{x}, \quad f_3(x) = 1 - x,$$

$$f_4(x) = \frac{1}{1 - x}, \quad f_5(x) = \frac{x - 1}{x}, \quad f_6(x) = \frac{x}{x - 1}.$$

This family of functions is *closed* under composition; that is, the composition of any two of these functions is again one of these functions. Tabulate the results of composing these functions one with the other by filling in the table shown in the figure. To indicate that $f_i \circ f_j = f_k$, write "f_k" in the ith row, jth column. We have already made two entries in the table. Check out these two entries and then fill in the rest of the table.

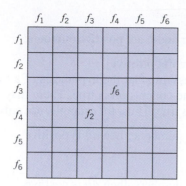

both positive and negative values. Compare your graphs with the graph of f, and describe the effect that varying b has on the graph of F.

(b) Now fix a value of b and graph F for several values of a; again, use both positive and negative values. Compare your graphs with the graph of f, and describe the effect that varying a has on the graph of F.

(c) Choose values for a and b, and graph $-F$. What effect does changing the sign of F have on the graph?

64. For all values of a and b, the graph of F is a parabola which opens upward. Find values for a and b such that the parabola will have x-intercepts at $-\frac{3}{2}$ and 2. Verify your result algebraically.

▶**Exercises 61–62.** Set $f(x) = x^2 - 4$, $g(x) = \dfrac{3x}{2-x}$, $h(x) = \sqrt{x+4}$, and $k(x) = \dfrac{2x}{3+x}$. Use a CAS to find the indicated composition.

61. (a) $f \circ g$; (b) $g \circ k$; (c) $f \circ k \circ g$.

62. (a) $g \circ f$; (b) $k \circ g$; (c) $g \circ f \circ k$.

▶**Exercises 63 and 64.** Set $f(x) = x^2$ and $F(x) = (x-a)^2 + b$.

63. (a) Choose a value for a and, using a graphing utility, graph F for several different values of b. Be sure to choose

▶**Exercises 65–66.** Set $f(x) = \sin x$.

65. (a) Using a graphing utility, graph cf for $c = -3, -2, -1, 2, 3$. Compare your graphs with the graph of f.

(b) Now graph $g(x) = f(cx)$ for $c = -3, -2, -\frac{1}{2}, \frac{1}{3}, \frac{1}{2}, 2, 3$. Compare your graphs with the graph of f.

66. (a) Using a graphing utility, graph $g(x) = f(x-c)$ for $c = -\frac{1}{2}\pi, -\frac{1}{4}\pi, \frac{1}{3}\pi, \frac{1}{2}\pi, \pi, 2\pi$. Compare your graphs with the graph of f.

(b) Now graph $g(x) = af(bx - c)$ for several values of a, b, c. Describe the effect of a, the effect of b, the effect of c.

■ 1.8 A NOTE ON MATHEMATICAL PROOF; MATHEMATICAL INDUCTION

Mathematical Proof

The notion of proof goes back to Euclid's *Elements*, and the rules of proof have changed little since they were formulated by Aristotle. We work in a deductive system where truth is argued on the basis of assumptions, definitions, and previously proved results. We cannot claim that such and such is true without clearly stating the basis on which we make that claim.

A theorem is an implication; it consists of a hypothesis and a conclusion:

$$\text{if (hypothesis)} \ldots, \text{then (conclusion)} \ldots.$$

Here is an example:

$$\text{If } a \text{ and } b \text{ are positive numbers, then } ab \text{ is positive.}$$

A common mistake is to ignore the hypothesis and persist with the conclusion: to insist, for example, that $ab > 0$ just because a and b are numbers.

Another common mistake is to confuse a theorem

$$\text{if } A, \text{ then } B$$

with its converse

$$\text{if } B, \text{ then } A.$$

The fact that a theorem is true does not mean that its converse is true: While it is true that

$$\text{if } a \text{ and } b \text{ are positive numbers, then } ab \text{ is positive,}$$

it is *not* true that

> if ab is positive, then a and b are positive numbers;

$[(-2)(-3)$ is positive but -2 and -3 are not positive].

A third, more subtle mistake is to assume that the hypothesis of a theorem represents the only condition under which the conclusion is true. There may well be other conditions under which the conclusion is true. Thus, for example, not only is it true that

> if a and b are positive numbers, then ab is positive

but it is also true that

> if a and b are negative numbers, then ab is positive.

In the event that a theorem

> if A, then B

and its converse

> if B, then A

are both true, then we can write

> A if and only if B or more briefly A iff B.

We know, for example, that

> if $x \geq 0$,   then $|x| = x$;

we also know that

> if $|x| = x$,   then $x \geq 0$.

We can summarize this by writing

> $x \geq 0$ iff $|x| = x$.

Remark We'll use "iff" frequently in this text but not in definitions. As stated earlier in a footnote, definitions are by their very nature *iff statements*. For example, we can say that "a number r is called a *zero* of P if $P(r) = 0$;" we don't have to say "a number r is called a *zero* of P iff $P(r) = 0$." In this situation, the "only if" part is taken for granted. ❏

A final point. One way of proving

> if A, then B

is to assume that

(1) A holds and B does not hold

and then arrive at a contradiction. The contradiction is taken to indicate that (1) is a false statement and therefore

> if A holds, then B must hold.

Some of the theorems of calculus are proved by this method.

Calculus provides procedures for solving a wide range of problems in the physical and social sciences. The fact that these procedures give us answers that seem to make sense is comforting, but it is only because we can prove our theorems that we can have confidence in the mathematics that is being applied. Accordingly, the study of calculus should include the study of some proofs.

Mathematical Induction

Mathematical induction is a method of proof which can be used to show that certain propositions are true for all positive integers n. The method is based on the following axiom:

1.8.1 AXIOM OF INDUCTION

Let S be a set of positive integers. If

(A) $1 \in S$, and
(B) $k \in S$ implies that $k + 1 \in S$,

then all the positive integers are in S.

You can think of the axiom of induction as a kind of "domino theory." If the first domino falls (Figure 1.8.1), and if each domino that falls causes the next one to fall, then, according to the axiom of induction, each domino will fall.

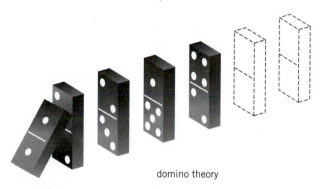

domino theory

Figure 1.8.1

While we cannot prove that this axiom is valid (axioms are by their very nature assumptions and therefore not subject to proof), we can argue that it is *plausible*.

Let's assume that we have a set S that satisfies conditions (A) and (B). Now let's choose a positive integer m and "argue" that $m \in S$.

From (A) we know that $1 \in S$. Since $1 \in S$, we know that $1 + 1 \in S$, and thus that $(1 + 1) + 1 \in S$, and so on. Since m can be obtained from 1 by adding 1 successively $(m - 1)$ times, it *seems clear* that $m \in S$.

To prove that a given proposition is true for *all* positive integers n, we let S be the set of positive integers for which the proposition is true. We prove first that $1 \in S$; that is, that the proposition is true for $n = 1$. Next we assume that the proposition is true for some positive integer k, and show that it is true for $k + 1$; that is, we show that $k \in S$ implies that $k + 1 \in S$. Then by the axiom of induction, we conclude that S contains the set of positive integers and therefore the proposition is true for all positive integers.

Example 1 We'll show that

$$1 + 2 + 3 + \cdots + n = \frac{n(n + 1)}{2} \quad \text{for all positive integers } n.$$

SOLUTION   Let S be the set of positive integers n for which

$$1 + 2 + 3 + \cdots + n = \frac{n(n+1)}{2}.$$

Then $1 \in S$ since

$$1 = \frac{1(1+1)}{2}.$$

Next, we assume that $k \in S$; that is, we assume that

$$1 + 2 + 3 \cdots + k = \frac{k(k+1)}{2}.$$

Adding up the first $k + 1$ integers, we have

$$1 + 2 + 3 + \cdots + k + (k+1) = [1 + 2 + 3 + \cdots + k] + (k+1)$$

$$= \frac{k(k+1)}{2} + (k+1) \qquad \text{(by the induction hypothesis)}$$

$$= \frac{k(k+1) + 2(k+1)}{2}$$

$$= \frac{(k+1)(k+2)}{2},$$

and so $k + 1 \in S$. Thus, by the axiom of induction, we can conclude that all positive integers are in S; that is, we can conclude that

$$1 + 2 + 3 + \cdots + n = \frac{n(n+1)}{2} \qquad \text{for all positive integers } n. \quad ❑$$

Example 2   We'll show that, if $x \geq -1$, then

$$(1 + x)^n \geq 1 + nx \quad \text{for all positive integers } n.$$

SOLUTION   We take $x \geq -1$ and let S be the set of positive integers n for which

$$(1 + x)^n \geq 1 + nx.$$

Since

$$(1 + x)^1 = 1 + 1 \cdot x,$$

we have $1 \in S$.

We now assume that $k \in S$. By the definition of S,

$$(1 + x)^k \geq 1 + kx.$$

Since

$$(1 + x)^{k+1} = (1 + x)^k (1 + x) \geq (1 + kx)(1 + x) \qquad \text{(explain)}$$

and

$$(1 + kx)(1 + x) = 1 + (k+1)x + kx^2 \geq 1 + (k+1)x,$$

we can conclude that

$$(1 + x)^{k+1} \geq 1 + (k+1)x$$

and thus that $k + 1 \in S$.

We have shown that

$$1 \in S \qquad \text{and that} \qquad k \in S \qquad \text{implies} \qquad k + 1 \in S.$$

By the axiom of induction, all positive integers are in S. ❑

Remark An induction does not have to begin with the integer 1. If, for example, you want to show that some proposition is true for all integers $n \geq 3$, all you have to do is show that it is true for $n = 3$, and that, if it is true for $n = k$, then it is true for $n = k + 1$. (Now you are starting the chain reaction by pushing on the third domino.) ❑

EXERCISES 1.8

Exercises 1–10. Show that the statement holds for all positive integers n.

1. $2n \leq 2^n$.

2. $1 + 2n \leq 3^n$.

3. $2^0 + 2^1 + 2^2 + 2^3 + \cdots + 2^{n-1} = 2^n - 1$.

4. $1 + 3 + 5 + \cdots + (2n - 1) = n^2$.

5. $1^2 + 2^2 + 3^2 + \cdots + n^2 = \frac{1}{6}n(n + 1)(2n + 1)$.

6. $1^3 + 2^3 + 3^3 + \cdots + n^3 = (1 + 2 + 3 + \cdots + n)^2$.
HINT: Use Example 1.

7. $1^3 + 2^3 + \cdots + (n - 1)^3 < \frac{1}{4}n^4 < 1^3 + 2^3 + \cdots + n^3$.

8. $1^2 + 2^2 + \cdots + (n - 1)^2 < \frac{1}{3}n^3 < 1^2 + 2^2 + \cdots + n^2$.

9. $\dfrac{1}{\sqrt{1}} + \dfrac{1}{\sqrt{2}} + \dfrac{1}{\sqrt{3}} + \cdots + \dfrac{1}{\sqrt{n}} \geq \sqrt{n}$.

10. $\dfrac{1}{1 \cdot 2} + \dfrac{1}{2 \cdot 3} + \dfrac{1}{3 \cdot 4} + \cdots + \dfrac{1}{n(n + 1)} = \dfrac{n}{n + 1}$.

11. For what integers n is $3^{2n+1} + 2^{n+2}$ divisible by 7? Prove that your answer is correct.

12. For what integers n is $9^n - 8n - 1$ divisible by 64? Prove that your answer is correct.

13. Find a simplifying expression for the product

$$\left(1 - \frac{1}{2}\right)\left(1 - \frac{1}{3}\right) \cdots \left(1 - \frac{1}{n}\right)$$

and verify its validity for all integers $n \geq 2$.

14. Find a simplifying expression for the product

$$\left(1 - \frac{1}{2^2}\right)\left(1 - \frac{1}{3^2}\right) \cdots \left(1 - \frac{1}{n^2}\right)$$

and verify its validity for all integers $n \geq 2$.

15. Prove that an N-sided convex polygon has $\frac{1}{2}N(N - 3)$ diagonals. Take $N > 3$.

16. Prove that the sum of the interior angles in an N-sided convex polygon is $(N - 2)180°$. Take $N > 2$.

17. Prove that all sets with n elements have 2^n subsets. Count the empty set ∅ and the whole set as subsets.

18. Show that, given a unit length, for each positive integer n, a line segment of length $\sqrt{n}$ can be constructed by straight edge and compass.

19. Find the first integer n for which $n^2 - n + 41$ is *not* a prime number.

■ CHAPTER 1. REVIEW EXERCISES

Exercises 1–4. Is the number rational or irrational?

1. 1.25.

2. $\sqrt{16/9}$.

3. $\sqrt{5} + 1$.

4. $1.001001001\ldots$.

Exercises 5–8. State whether the set is bounded above, bounded below, bounded. If the set is bounded above, give an upper bound; if it is bounded below, give a lower bound; if it is bounded, give an upper bound and a lower bound.

5. $S = \{1, 3, 5, 7, \cdots\}$.

6. $S = \{x : x \leq 1\}$.

7. $S = \{x : |x + 2| < 3\}$.

8. $S = \{(-1/n)^n : n = 1, 2, 3, \cdots\}$.

Exercises 9–12. Find the real roots of the equation.

9. $2x^2 + x - 1 = 0$.

10. $x^2 + 2x + 5 = 0$.

11. $x^2 - 10x + 25 = 0$.

12. $9x^3 - x = 0$.

Exercises 13–22. Solve the inequality. Express the solution as an interval or as the union of intervals. Mark the solution on a number line.

13. $5x - 2 < 0$.

14. $3x + 5 < \frac{1}{2}(4 - x)$.

15. $x^2 - x - 6 \geq 0$.

16. $x(x^2 - 3x + 2) \leq 0$.

17. $\dfrac{x + 1}{(x + 2)(x - 2)} > 0$.

18. $\dfrac{x^2 - 4x + 4}{x^2 - 2x - 3} \leq 0$.

19. $|x - 2| < 1$.

20. $|3x - 2| \geq 4$.

21. $\left|\dfrac{2}{x + 4}\right| > 2$.

22. $\left|\dfrac{5}{x + 1}\right| < 1$.

Exercises 23–24. (a) Find the distance between the points P, Q. (b) Find the midpoint of the line segment $\overline{PQ}$.

23. $P(2, -3), Q(1, 4)$.

24. $P(-3, -4), Q(-1, 6)$.

Exercises 25–28. Find an equation for the line that passes through the point $(2, -3)$ and is

25. parallel to the y-axis.

26. parallel to the line $y = 1$.

27. perpendicular to the line $2x - 3y = 6$.

28. parallel to the line $3x + 4y = 12$.

Exercises 29–30. Find the point where the lines intersect.

29. $l_1 : x - 2y = -4, \qquad l_2 : 3x + 4y = 3$.

30. $l_1 : 4x - y = -2, \qquad l_2 : 3x + 2y = 0$.

31. Find the point(s) where the line $y = 8x - 6$ intersects the parabola $y = 2x^2$.

32. Find an equation for the line tangent to the circle

$$x^2 + y^2 + 2x - 6y - 3 = 0$$

at the point $(2, 1)$.

Exercises 33–38. Give the domain and range of the function.

33. $f(x) = 4 - x^2$.

34. $f(x) = 3x - 2$.

35. $f(x) = \sqrt{x - 4}$.

36. $f(x) = \frac{1}{2}\sqrt{1 - 4x^2}$.

37. $f(x) = \sqrt{1 + 4x^2}$.

38. $f(x) = |2x + 1|$.

Exercises 39–40. Sketch the graph and give the domain and range of the function.

39. $f(x) = \begin{cases} 4 - 2x, & x \le 2 \\ x - 2, & x > 2 \end{cases}$.

40. $f(x) = \begin{cases} x^2 + 2, & x \le 0 \\ 2 - x^2, & x > 0 \end{cases}$.

Exercises 41–44. Find the number(s) x in the interval $[0, 2\pi]$ which satisfy the equation.

41. $\sin x = -\frac{1}{2}$.

42. $\cos 2x = -\frac{1}{2}$.

43. $\tan(x/2) = -1$.

44. $\sin 3x = 0$.

Exercises 45–48. Sketch the graph of the function.

45. $f(x) = \cos 2x$.

46. $f(x) = -\cos 2x$.

47. $f(x) = 3 \cos 2x$.

48. $f(x) = \frac{1}{3} \cos 2x$.

Exercises 49–51. Form the combinations $f + g$, $f - g$, $f \cdot g$, f/g and specify the domain of combination.

49. $f(x) = 3x + 2, \quad g(x) = x^2 - 1$.

50. $f(x) = x^2 - 4, \quad g(x) = x + 1/x$.

51. $f(x) = \cos^2 x, \quad g(x) = \sin 2x, \quad$ for $x \in [0, 2\pi]$.

Exercises 52–54. Form the compositions $f \circ g$ and $g \circ f$, and specify the domain of each of these combinations.

52. $f(x) = x^2 - 2x, \quad g(x) = x + 1$.

53. $f(x) = \sqrt{x + 1}, \quad g(x) = x^2 - 5$.

54. $f(x) = \sqrt{1 - x^2}, \quad g(x) = \sin 2x$.

55. (a) Write an equation in x and y for an arbitrary line l that passes through the origin.

(b) Verify that if $P(a, b)$ lies on l and α is a real number, then the point $Q(\alpha a, \alpha b)$ also lies on l.

(c) What additional conclusion can you draw if $\alpha > 0$? if $\alpha < 0$?

56. *The roots of a quadratic equation.* You can find the roots of a quadratic equation by resorting to the quadratic formula. The approach outlined below is more illuminating. Since division by the leading coefficient does not alter the roots of the equation, we can make the coefficient 1 and work with the equation

$$x^2 + ax + b = 0.$$

(a) Show that the equation $x^2 + ax + b = 0$ can be written as

$$(x - \alpha)^2 - \beta^2 = 0, \quad \text{or}$$
$$(x - \alpha)^2 = 0, \quad \text{or}$$
$$(x - \alpha)^2 + \beta^2 = 0.$$

HINT: Set $\alpha = -a/2$, complete the square, and go on from there.

(b) What are the roots of the equation $(x - \alpha)^2 - \beta^2 = 0$?

(c) What are the roots of the equation $(x - \alpha)^2 = 0$?

(d) Show that the equation $(x - \alpha)^2 + \beta^2 = 0$ has no real roots.

57. Knowing that

$$|a + b| \le |a| + |b| \quad \text{for all real } a, b$$

show that

$$|a| - |b| \le |a - b| \quad \text{for all real } a, b.$$

58. (a) Express the perimeter of a semicircle as a function of the diameter.

(b) Express the area of a semicircle as a function of the diameter.

CHAPTER

2

LIMITS

AND

CONTINUITY

■ 2.1 THE LIMIT PROCESS (AN INTUITIVE INTRODUCTION)

We could begin by saying that limits are important in calculus, but that would be a major understatement. *Without limits, calculus would not exist. Every single notion of calculus is a limit in one sense or another.* For example,

What is the slope of a curve? It is the limit of slopes of secant lines. (Figure 2.1.1.)

What is the length of a curve? It is the limit of the lengths of polygonal paths inscribed in the curve. (Figure 2.1.2)

What is the area of a region bounded by a curve? It is the limit of the sum of areas of approximating rectangles. (Figure 2.1.3)

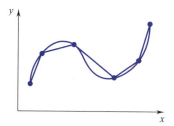

Figure 2.1.1

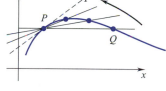

Figure 2.1.2

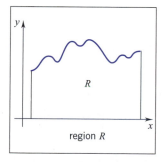

region R

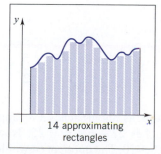

1 approximating rectangle

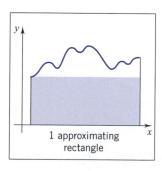

3 approximating rectangles

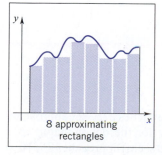

8 approximating rectangles

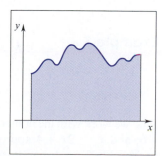

14 approximating rectangles

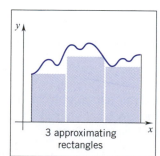

Figure 2.1.3

The Idea of Limit

Technically there are several limit processes, but they are all very similar. Once you master one of them, the others will pose few difficulties. The limit process that we start with is the one that leads to the notion of *continuity* and the notion of *differentiability*. At this stage our approach is completely informal. All we are trying to do here is lay an intuitive foundation for the mathematics that begins in Section 2.2

We start with a number c and a function f defined at all numbers x near c but not necessarily at c itself. In any case, whether or not f is defined at c and, if so, how is totally irrelevant.

Now let L be some real number. We say that *the limit of $f(x)$ as x tends to c is L* and write

$$\lim_{x \to c} f(x) = L$$

provided that (roughly speaking)

as x approaches c, f(x) approaches L

or (somewhat more precisely) provided that

f(x) is close to L for all $x \neq c$ which are close to c.

Let's look at a few functions and try to apply this limit idea. Remember, our work at this stage is entirely intuitive.

Example 1 Set $f(x) = 4x + 5$ and take $c = 2$. As x approaches 2, $4x$ approaches 8 and $4x + 5$ approaches $8 + 5 = 13$. We conclude that

$$\lim_{x \to 2} f(x) = 13. \quad ❏$$

Example 2 Set $f(x) = \sqrt{1 - x}$ and take $c = -8$. As x approaches -8, $1 - x$ approaches 9 and $\sqrt{1 - x}$ approaches 3. We conclude that

$$\lim_{x \to -8} f(x) = 3.$$

If for that same function we try to calculate

$$\lim_{x \to 2} f(x),$$

we run into a problem. The function $f(x) = \sqrt{1 - x}$ is defined only for $x \leq 1$. It is therefore not defined for x near 2, and the idea of taking the limit as x approaches 2 makes no sense at all:

$$\lim_{x \to 2} f(x) \qquad \text{does not exist.} \quad ❏$$

Example 3

$$\lim_{x \to 3} \frac{x^3 - 2x + 4}{x^2 + 1} = \frac{5}{2}.$$

First we work with the numerator: as x approaches 3, x^3 approaches 27, $-2x$ approaches -6, and $x^3 - 2x + 4$ approaches $27 - 6 + 4 = 25$. Now for the denominator: as x approaches 3, $x^2 + 1$ approaches 10. The quotient (it would seem) approaches $25/10 = 5/2$. ❏

The curve in Figure 2.1.4 represents the graph of a function f. The number c is on the x-axis and the limit L is on the y-axis. As x approaches c along the x-axis, $f(x)$ approaches L along the y-axis.

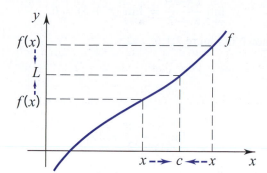

Figure 2.1.4

As we have tried to emphasize, in taking the limit of a function f as x tends to c, it does not matter whether f is defined at c and, if so, how it is defined there. The only thing that matters is the values taken on by f at numbers x near c. Take a look at the three cases depicted in Figure 2.1.5. In the first case, $f(c) = L$. In the second case, f is not defined at c. In the third case, f is defined at c, but $f(c) \neq L$. However, in each case

$$\lim_{x \to c} f(x) = L$$

because, as suggested in the figures,

as x approaches c, $f(x)$ approaches L.

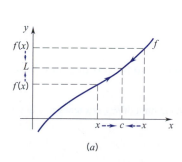

(a)

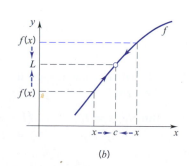

(b)

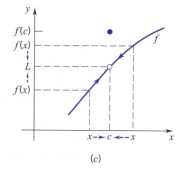

(c)

Figure 2.1.5

Example 4 Set $f(x) = \dfrac{x^2 - 9}{x - 3}$ and let $c = 3$. Note that the function f is not defined at 3: at 3, both numerator and denominator are 0. But that doesn't matter. For $x \neq 3$, and therefore *for all x near* 3,

$$\frac{x^2 - 9}{x - 3} = \frac{(x - 3)(x + 3)}{x - 3} = x + 3.$$

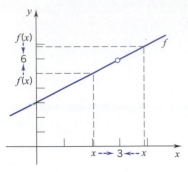

Figure 2.1.6

Therefore, if x is close to 3, then $\dfrac{x^2 - 9}{x - 3} = x + 3$ is close to $3 + 3 = 6$. We conclude that

$$\lim_{x \to 3} \frac{x^2 - 9}{x - 3} = \lim_{x \to 3} (x + 3) = 6.$$

The graph of f is shown in Figure 2.1.6. ❑

Example 5

$$\lim_{x \to 2} \frac{x^3 - 8}{x - 2} = 12.$$

The function $f(x) = \dfrac{x^3 - 8}{x - 2}$ is undefined at $x = 2$. But, as we said before, that doesn't matter. For all $x \neq 2$,

$$\frac{x^3 - 8}{x - 2} = \frac{(x - 2)(x^2 + 2x + 4)}{x - 2} = x^2 + 2x + 4.$$

Therefore,

$$\lim_{x \to 2} \frac{x^3 - 8}{x - 2} = \lim_{x \to 2} (x^2 + 2x + 4) = 12. ❑$$

Example 6 If $f(x) = \begin{cases} 3x - 4, & x \neq 0 \\ 10, & x = 0, \end{cases}$ then $\lim\limits_{x \to 0} f(x) = -4$.

It does not matter that $f(0) = 10$. For $x \neq 0$, and thus for all x near 0,

$$f(x) = 3x - 4 \quad \text{and therefore } \lim_{x \to 0} f(x) = \lim_{x \to 0} (3x - 4) = -4. ❑$$

One-Sided Limits

Numbers x near c fall into two natural categories: those that lie to the left of c and those that lie to the right of c. We write

$$\lim_{x \to c^-} f(x) = L \qquad \text{[The left-hand limit of } f(x) \text{ as } x \text{ tends to } c \text{ is } L.]$$

to indicate that

as x approaches c from the left, f(x) approaches L.

We write

$$\lim_{x \to c^+} f(x) = L \qquad \text{[The right-hand limit of } f(x) \text{ as } x \text{ tends to } c \text{ is } L.]$$

to indicate that

as x approaches c from the right, f(x) approaches L.[†]

[†]The left-hand limit is sometimes written $\lim\limits_{x \uparrow c} f(x)$ and the right-hand limit, $\lim\limits_{x \downarrow c} f(x)$.

As an example, take the function indicated in Figure 2.1.7. As x approaches 5 from the left, $f(x)$ approaches 2; therefore

$$\lim_{x \to 5^-} f(x) = 2.$$

As x approaches 5 from the right, $f(x)$ approaches 4; therefore

$$\lim_{x \to 5^+} f(x) = 4.$$

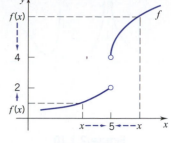

Figure 2.1.7

The full limit, $\lim_{x \to 5} f(x)$, does not exist: consideration of $x < 5$ would force the limit to be 2, but consideration of $x > 5$ would force the limit to be 4.

For a full limit to exist, both one-sided limits have to exist and they have to be equal.

Example 7 For the function f indicated in Figure 2.1.8,

$$\lim_{x \to (-2)^-} f(x) = 5 \qquad \text{and} \qquad \lim_{x \to (-2)^+} f(x) = 5.$$

In this case

$$\lim_{x \to -2} f(x) = 5.$$

It does not matter that $f(-2) = 3$.

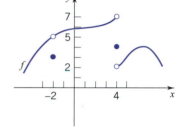

Figure 2.1.8

Examining the graph of f near $x = 4$, we find that

$$\lim_{x \to 4^-} f(x) = 7 \qquad \text{whereas} \qquad \lim_{x \to 4^+} f(x) = 2.$$

Since these one-sided limits are different,

$$\lim_{x \to 4} f(x) \qquad \text{does not exist.} \quad ❏$$

Example 8 Set $f(x) = x/|x|$. Note that $f(x) = 1$ for $x > 0$, and $f(x) = -1$ for $x < 0$:

$$f(x) = \begin{cases} 1, & \text{if } x > 0 \\ -1, & \text{if } x < 0. \end{cases} \qquad \text{(Figure 2.1.9)}$$

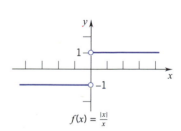

$f(x) = \frac{|x|}{x}$

Figure 2.1.9

Let's try to apply the limit process at different numbers c.

If $c < 0$, then for all x sufficiently close to c, $x < 0$ and $f(x) = -1$. It follows that for $c < 0$

$$\lim_{x \to c} f(x) = \lim_{x \to c} (-1) = -1.$$

If $c > 0$, then for all x sufficiently close to c, $x > 0$ and $f(x) = 1$. It follows that for $c < 0$

$$\lim_{x \to c} f(x) = \lim_{x \to c} (1) = 1.$$

However, the function has no limit as x tends to 0:

$$\lim_{x \to 0^-} f(x) = -1 \qquad \text{but} \qquad \lim_{x \to 0^+} f(x) = 1. \quad ❏$$

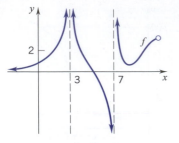

Figure 2.1.10

Example 9   We refer to the function indicated in Figure 2.1.10 and examine the behavior of $f(x)$ for x close to 3 and x close to 7.

As x approaches 3 from the left or from the right, $f(x)$ becomes arbitrarily large and cannot stay close to any number L. Therefore

$$\lim_{x \to 3} f(x) \qquad \text{does not exist.}$$

As x approaches 7 from the left, $f(x)$ becomes arbitrarily large negative and cannot stay close to any number L. Therefore

$$\lim_{x \to 7} f(x) \qquad \text{does not exist.}$$

The same conclusion can be reached by noting that as x approaches 7 from the right, $f(x)$ becomes arbitrarily large. ❑

Remark   To indicate that $f(x)$ becomes arbitrarily large, we can write $f(x) \to \infty$. To indicate that $f(x)$ becomes arbitrarily large negative, we can write $f(x) \to -\infty$.

Go back to Figure 2.1.10, and note that for the function depicted there the following statements hold:

$$\text{as } x \to 3^-, \quad f(x) \to \infty \qquad \text{and} \qquad \text{as } x \to 3^+, \quad f(x) \to \infty.$$

Consequently,

$$\text{as } x \to 3, \qquad f(x) \to \infty.$$

Also,

$$\text{as } x \to 7^-, \quad f(x) \to -\infty \qquad \text{and} \qquad \text{as } x \to 7^+, \quad f(x) \to \infty.$$

We can therefore write

$$\text{as } x \to 7, \qquad |f(x)| \to \infty. ❑$$

Example 10   We set

$$f(x) = \frac{1}{x - 2}$$

and examine the behavior of $f(x)$ (a) as x tends to 4 and then (b) as x tends to 2.

(a) As x tends to 4, $x - 2$ tends to 2 and the quotient tends to 1/2. Thus

$$\lim_{x \to 4} f(x) = \frac{1}{2}.$$

(b) As x tends to 2 from the left, $f(x) \to -\infty$. (See Figure 2.1.11.) As x tends to 2 from the right, $f(x) \to \infty$. The function can have no numerical limit as x tends to 2. Thus

$$\lim_{x \to 2} f(x) \qquad \text{does not exist.}$$

However, it is true that

$$\text{as } x \to 2, \qquad |f(x)| \to \infty. ❑$$

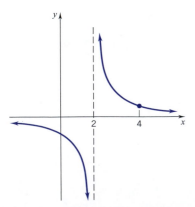

Figure 2.1.11

Example 11 Set $f(x) = \begin{cases} 1 - x^2, & x < 1 \\ 1/(x - 1), & x > 1. \end{cases}$

For $x < 1$, $f(x) = 1 - x^2$. Thus

$$\lim_{x \to 1^-} f(x) = 0.$$

For $x > 1$, $f(x) = 1/(x - 1)$. Therefore, as $x \to 1^+$, $f(x) \to \infty$. The function has no numerical limit as $x \to 1$:

$$\lim_{x \to 1} f(x) \qquad \text{does not exist.}$$

We now assert that

$$\lim_{x \to 1.5} f(x) = 2.$$

To see this, note that for x close to 1.5, $x > 1$ and therefore $f(x) = 1/(x - 1)$. It follows that

$$\lim_{x \to 1.5} f(x) = \lim_{x \to 1.5} \frac{1}{x - 1} = \frac{1}{0.5} = 2.$$

See Figure 2.1.12. ❏

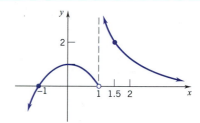

Figure 2.1.12

Example 12 Here we set $f(x) = \sin(\pi/x)$ and show that the function can have no limit as $x \to 0$.

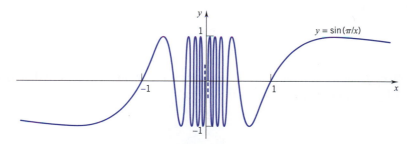

Figure 2.1.13

The function is not defined at $x = 0$, but, as you know, that's irrelevant. What keeps f from having a limit as $x \to 0$ is indicated in Figure 2.1.13. As $x \to 0$, $f(x)$ keeps oscillating between $y = 1$ and $y = -1$ and therefore cannot remain close to any one number L.[†] ❏

In our final example we rely on a calculator and deduce a limit from numerical calculation.

[†]We can approach $x = 0$

$$\text{by numbers } a_n = \frac{2}{4n + 1} \qquad \text{and} \qquad \text{by numbers } b_n = -\frac{2}{4n + 1},$$

$n = 0, 1, 2, 3, \ldots$. As you can check, $f(a_n) = 1$ and $f(b_n) = -1$. This confirms the oscillatory behavior of f near $x = 0$.

Example 13 Let $f(x) = (\sin x)/x$. If we try to evaluate f at 0, we get the meaningless ratio $0/0$; f is not defined at $x = 0$. However, f is defined for all $x \neq 0$, and so we can consider

$$\lim_{x \to 0} \frac{\sin x}{x}.$$

We select numbers that approach 0 closely from the left and numbers that approach 0 closely from the right. Using a calculator, we evaluate f at these numbers. The results are tabulated in Table 2.1.1.

■ **Table 2.1.1**

(Left side)		(Right side)	
x (radians)	$\frac{\sin x}{x}$	x (radians)	$\frac{\sin x}{x}$
-1	0.84147	1	0.84147
-0.5	0.95885	0.5	0.95885
-0.1	0.99833	0.1	0.99833
-0.01	0.99998	0.01	0.99998
-0.001	0.99999	0.001	0.99999

These calculations suggest that

$$\lim_{x \to 0^-} \frac{\sin x}{x} = 1 \qquad \text{and} \qquad \lim_{x \to 0^+} \frac{\sin x}{x} = 1$$

and therefore that

$$\lim_{x \to 0} \frac{\sin x}{x} = 1.$$

The graph of f, shown in Figure 2.1.14, supports this conclusion. A proof that this limit is indeed 1 is given in Section 2.5. ❏

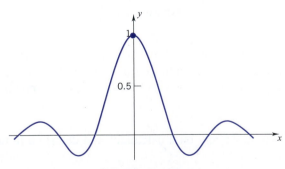

Figure 2.1.14

If you have found all this to be imprecise, you are absolutely right. Our work so far has been imprecise. In Section 2.2 we will work with limits in a more coherent manner.

EXERCISES 2.1

Exercises 1–10. You are given a number c and the graph of a function f. Use the graph to find

(a) $\lim\limits_{x \to c^-} f(x)$ (b) $\lim\limits_{x \to c^+} f(x)$ (c) $\lim\limits_{x \to c} f(x)$ (d) $f(c)$

1. $c = 2$.

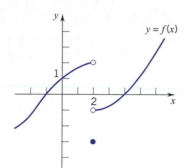

2. $c = 3$.

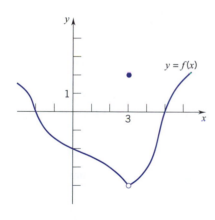

3. $c = 3$.

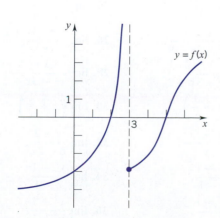

4. $c = 4$.

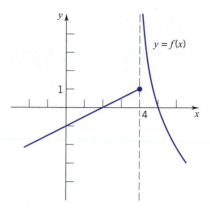

5. $c = -2$.

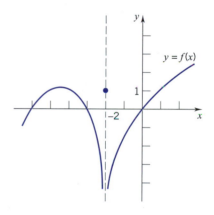

6. $c = 1$.

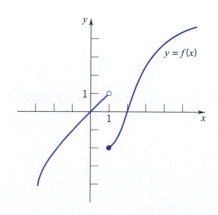

7. $c = 1$.

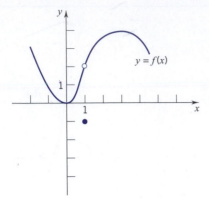

8. $c = -1$.

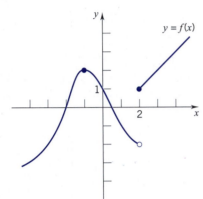

9. $c = 2$.

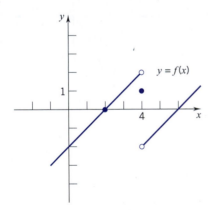

10. $c = 3$.

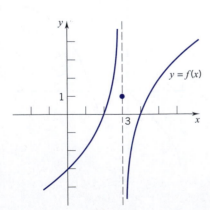

Exercises 11–12. Give the values of c for which $\lim\limits_{x \to c} f(x)$ does not exist.

11.

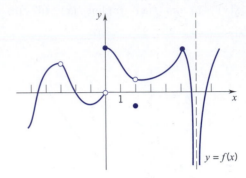

12.

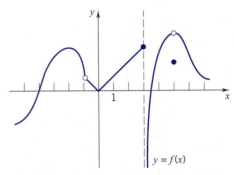

Exercises 13–49. Decide on intuitive grounds whether or not the indicated limit exists; evaluate the limit if it does exist.

13. $\lim\limits_{x \to 0} (2x - 1)$.

14. $\lim\limits_{x \to 1} (2 - 5x)$.

15. $\lim\limits_{x \to -2} (x^2 - 2x + 4)$.

16. $\lim\limits_{x \to 4} \sqrt{x^2 + 2x + 1}$.

17. $\lim\limits_{x \to -3} (|x| - 2)$.

18. $\lim\limits_{x \to 0} \dfrac{1}{|x|}$.

19. $\lim\limits_{x \to 1} \dfrac{3}{x + 1}$.

20. $\lim\limits_{x \to -1} \dfrac{4}{x + 1}$.

21. $\lim\limits_{x \to -1} \dfrac{-2}{x + 1}$.

22. $\lim\limits_{x \to 2} \dfrac{1}{3x - 6}$.

23. $\lim\limits_{x \to 3} \dfrac{2x - 6}{x - 3}$.

24. $\lim\limits_{x \to 3} \dfrac{x^2 - 6x + 9}{x - 3}$.

25. $\lim\limits_{x \to 3} \dfrac{x - 3}{x^2 - 6x + 9}$.

26. $\lim\limits_{x \to 2} \dfrac{x^2 - 3x + 2}{x - 2}$.

27. $\lim\limits_{x \to 2} \dfrac{x - 2}{x^2 - 3x + 2}$.

28. $\lim\limits_{x \to 1} \dfrac{x - 2}{x^2 - 3x + 2}$.

29. $\lim\limits_{x \to 0} \left(x + \dfrac{1}{x} \right)$.

30. $\lim\limits_{x \to 1} \left(x + \dfrac{1}{x} \right)$.

31. $\lim\limits_{x \to 0} \dfrac{2x - 5x^2}{x}$.

32. $\lim\limits_{x \to 3} \dfrac{x - 3}{6 - 2x}$.

33. $\lim\limits_{x \to 1} \dfrac{x^2 - 1}{x - 1}$.

34. $\lim\limits_{x \to 1} \dfrac{x^3 - 1}{x - 1}$.

35. $\lim\limits_{x \to 1} \dfrac{x^3 - 1}{x + 1}$.

36. $\lim\limits_{x \to 1} \dfrac{x^2 + 1}{x^2 - 1}$.

37. $\lim\limits_{x \to 0} f(x); \quad f(x) = \begin{cases} 1, & x \neq 0 \\ 3, & x = 0. \end{cases}$

38. $\lim\limits_{x \to 1} f(x); \quad f(x) = \begin{cases} 3x, & x < 1 \\ 3, & x > 1. \end{cases}$

39. $\lim\limits_{x \to 4} f(x); \quad f(x) = \begin{cases} x^2, & x \neq 4 \\ 0, & x = 4. \end{cases}$

40. $\lim\limits_{x \to 0} f(x); \quad f(x) = \begin{cases} -x^2, & x < 0 \\ x^2, & x > 0. \end{cases}$

41. $\lim\limits_{x \to 0} f(x); \quad f(x) = \begin{cases} x^2, & x < 0 \\ 1 + x, & x > 0. \end{cases}$

42. $\lim\limits_{x \to 1} f(x); \quad f(x) = \begin{cases} 2x, & x < 1 \\ x^2 + 1, & x > 1. \end{cases}$

43. $\lim\limits_{x \to 2} f(x); \quad f(x) = \begin{cases} 3x, & x < 1 \\ x + 2, & x \geq 1. \end{cases}$

44. $\lim\limits_{x \to 0} f(x); \quad f(x) = \begin{cases} 2x, & x \leq 1 \\ x + 1, & x > 1. \end{cases}$

45. $\lim\limits_{x \to 0} f(x); \quad f(x) = \begin{cases} 2, & x \text{ rational} \\ -2, & x \text{ irrational.} \end{cases}$

46. $\lim\limits_{x \to 1} f(x); \quad f(x) = \begin{cases} 2x, & x \text{ rational} \\ 2, & x \text{ irrational.} \end{cases}$

47. $\lim\limits_{x \to 1} \dfrac{\sqrt{x^2 + 1} - \sqrt{2}}{x - 1}.$

48. $\lim\limits_{x \to 5} \dfrac{\sqrt{x^2 + 5} - \sqrt{30}}{x - 5}.$

49. $\lim\limits_{x \to 1} \dfrac{x^2 + 1}{\sqrt{2x + 2} - 2}.$

Exercises 50–54. After estimating the limit using the prescribed values of x, validate or improve your estimate by using a graphing utility.

50. Estimate

$$\lim\limits_{x \to 0} \frac{1 - \cos x}{x} \qquad \text{(radian measure)}$$

by evaluating the quotient at $x = \pm 1, \pm 0.1, \pm 0.01, \pm 0.001$.

51. Estimate

$$\lim\limits_{x \to 0} \frac{\tan 2x}{x} \qquad \text{(radian measure)}$$

by evaluating the quotient at $x = \pm 1, \pm 0.1, \pm 0.01, \pm 0.001$.

52. Estimate

$$\lim\limits_{x \to 0} \frac{x - \sin x}{x^3} \qquad \text{(radian measure)}$$

after evaluating the quotient at $x = \pm 1, \pm 0.1, \pm 0.01, \pm 0.001, \pm 0.0001$.

53. Estimate

$$\lim\limits_{x \to 1} \frac{x^{3/2} - 1}{x - 1}$$

by evaluating the quotient at $x = 0.9, 0.99, 0.99, 0.9999$ and at $x = 1.1, 1.01, 1.001, 1.0001$.

54. Estimate

$$\lim\limits_{x \to 0} \frac{2\cos x - 2 + x^2}{x^4} \qquad \text{(radian measure)}$$

by evaluating the quotient at $x = \pm 1, \pm 0.1, \pm 0.01, \pm 0.0001, \pm 0.0001$.

55. (a) Use a graphing utility to estimate $\lim\limits_{x \to 4} f(x)$:

  (i) $f(x) = \dfrac{2x^2 - 11x + 12}{x - 4};$

  (ii) $f(x) = \dfrac{2x^2 - 11x + 12}{x^2 - 8x + 16}.$

  (b) Use a CAS to find each of the limits in part (a).

56. (a) Use a graphing utility to estimate $\lim\limits_{x \to 4} f(x)$:

  (i) $f(x) = \dfrac{3x^2 - 10x - 8}{5x^2 + 16x - 16};$

  (ii) $f(x) = \dfrac{5x^2 - 26x + 24}{4x^2 - 11x - 20}.$

  (b) Use a CAS to find each of the limits in part (a).

57. (a) Use a graphing utility to estimate $\lim\limits_{x \to 2} f(x)$:

  (i) $f(x) = \dfrac{\sqrt{6 - x} - x}{x - 2};$ (ii) $f(x) = \dfrac{x^2 - 4x + 4}{x - \sqrt{6 - x}}.$

  (b) Use a CAS to find each of the limits in part (a).

58. (a) Use a graphing utility to estimate $\lim\limits_{x \to 2} f(x)$

  (i) $f(x) = \dfrac{2x - \sqrt{18 - x}}{4 - x^2};$ (ii) $f(x) = \dfrac{2 - \sqrt{2x}}{\sqrt{8x} - 4}.$

  (b) Use a CAS to find each of the limits in part (a).

Exercises 59–62. Use a graphing utility to find at least one number c at which $\lim\limits_{x \to c} f(x)$ does not exist.

59. $f(x) = \dfrac{x + 1}{|x^3 + 1|}.$

60. $f(x) = \dfrac{|6x^2 - x - 35|}{2x - 5}.$

61. $f(x) = \dfrac{|x|}{x^5 + 2x^4 + 13x^3 + 26x^2 + 36x + 72}.$

62. $f(x) = \dfrac{5x^3 - 22x^2 + 15x + 18}{x^3 - 9x^2 + 27x - 27}.$

63. Use a graphing utility to draw the graphs of

$$f(x) = \frac{1}{x}\sin x \qquad \text{and} \qquad g(x) = x\sin\left(\frac{1}{x}\right)$$

for $x \neq 0$ between $-\pi/2$ and $\pi/2$. Describe the behavior of $f(x)$ and $g(x)$ for x close to 0.

64. Use a graphing utility to draw the graphs of

$$f(x) = \frac{1}{x}\tan x \qquad \text{and} \qquad g(x) = x\tan\left(\frac{1}{x}\right)$$

for $x \neq 0$ between $-\pi/2$ and $\pi/2$. Describe the behavior of $f(x)$ and $g(x)$ for x close to 0.

■ 2.2 DEFINITION OF LIMIT

In Section 2.1 we tried to give you an intuitive feeling for the limit process. However, our description was too vague to be called "mathematics." We relied on statements such as

"as x approaches c, $f(x)$ approaches L"

and

"$f(x)$ is close to L for all $x \neq c$ which are close to c."

But what exactly do these statements mean? What are we saying by stating that "$f(x)$ approaches L"? How close is close?

In this section we formulate the limit process in a coherent manner and, by so doing, establish a foundation for more advanced work.

As before, in taking the limit of $f(x)$ as x approaches c, we don't require that f be defined at c, but we do require that f be defined at least on an open interval $(c - p, c + p)$ except possibly at c itself.

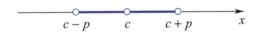

To say that

$$\lim_{x \to c} f(x) = L$$

is to say that $|f(x) - L|$ can be made as small as we choose, *less than any $\epsilon > 0$ we choose*, by restricting x to a sufficiently small set of the form $(c - \delta, c) \cup (c, c + \delta)$, *by restricting x by an inequality of the form $0 < |x - c| < \delta$ with $\delta > 0$ sufficiently small.*

Phrasing this idea precisely, we have the following definition.

DEFINITION 2.2.1   THE LIMIT OF A FUNCTION

Let f be a function defined at least on an open interval $(c - p, c + p)$ except possibly at c itself. We say that

$$\lim_{x \to c} f(x) = L$$

if for each $\epsilon > 0$, there exists a $\delta > 0$ such that

$$\text{if} \quad 0 < |x - c| < \delta, \quad \text{then} \quad |f(x) - L| < \epsilon.$$

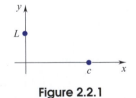

Figure 2.2.1

Figures 2.2.1 and 2.2.2 illustrate this definition.

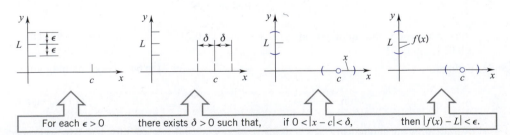

Figure 2.2.2

Except in the case of a constant function, the choice of δ depends on the previous choice of ϵ. We do not require that there exists a number δ which "works" for *all* ϵ, but rather, that for each ϵ there exists a δ which "works" for that particular ϵ.

In Figure 2.2.3, we give two choices of ϵ and for each we display a suitable δ. For a δ to be suitable, all points within δ of c (with the possible exception of c itself) must be taken by the function f to within ϵ of L. In part (b) of the figure, we began with a smaller ϵ and had to use a smaller δ.

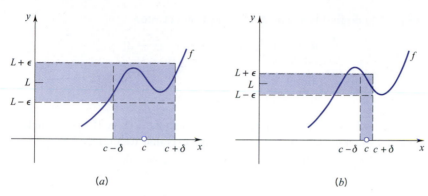

(a)   (b)

Figure 2.2.3

The δ of Figure 2.2.4 is too large for the given ϵ. In particular, the points marked x_1 and x_2 in the figure are not taken by f to within ϵ of L.

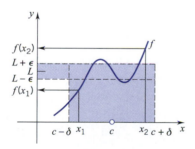

Figure 2.2.4

As these illustrations suggest, the limit process can be described entirely in terms of open intervals. (See Figure 2.2.5.)

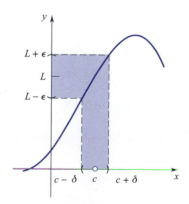

Figure 2.2.5

(2.2.2)

> Let f be defined at least on an open interval $(c - p, c + p)$ except possibly at c itself. We say that
>
> $$\lim_{x \to c} f(x) = L$$
>
> if for each open interval $(L - \epsilon, L + \epsilon)$ there is an open interval $(c - \delta, c + \delta)$ such that all the numbers in $(c - \delta, c + \delta)$, with the possible exception of c itself, are mapped by f into $(L - \epsilon, L + \epsilon)$.

Next we apply the ϵ, δ definition of limit to a variety of functions. At first you may find the ϵ, δ arguments confusing. It usually takes a little while for the ϵ, δ idea to take hold.

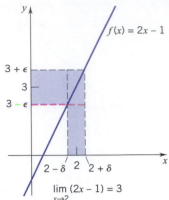

$$\lim_{x \to 2}(2x - 1) = 3$$

Figure 2.2.6

Example 1 Show that

$$\lim_{x \to 2}(2x - 1) = 3. \qquad \text{(Figure 2.2.6)}$$

Finding a δ. Let $\epsilon > 0$. We seek a number $\delta > 0$ such that

$$\text{if} \qquad 0 < |x - 2| < \delta, \qquad \text{then} \qquad |(2x - 1) - 3| < \epsilon.$$

What we have to do first is establish a connection between

$$|(2x - 1) - 3| \qquad \text{and} \qquad |x - 2|.$$

The connection is evident:

$$(*) \qquad\qquad |(2x - 1) - 3| = |2x - 4| = 2|x - 2|.$$

To make $|(2x - 1) - 3|$ less than ϵ, we need to make $2|x - 2| < \epsilon$, which we can accomplish by making $|x - 2| < \epsilon/2$. This suggests that we choose $\delta = \frac{1}{2}\epsilon$.

Showing that the δ "works." If $0 < |x - 2| < \frac{1}{2}\epsilon$, then $2|x - 2| < \epsilon$ and, by (*), $|(2x - 1) - 3| < \epsilon$. ❑

Remark In Example 1 we chose $\delta = \frac{1}{2}\epsilon$, but we could have chosen *any* positive number δ less than $\frac{1}{2}\epsilon$. In general, if a certain δ^* "works" for a given ϵ, then any δ less than δ^* will also work. ❑

Example 2 Show that

$$\lim_{x \to -1}(2 - 3x) = 5. \qquad \text{(Figure 2.2.7)}$$

Finding a δ. Let $\epsilon > 0$. We seek a number $\delta > 0$ such that

$$\text{if} \qquad 0 < |x - (-1)| < \delta, \qquad \text{then} \qquad |(2 - 3x) - 5| < \epsilon.$$

To find a connection between

$$|x - (-1)| \qquad \text{and} \qquad |(2 - 3x) - 5|,$$

we simplify both expressions:

$$|x - (-1)| = |x + 1|$$

and

$$|(2 - 3x) - 5| = |-3x - 3| = |-3||x + 1| = 3|x + 1|.$$

We can conclude that

$$(**) \qquad\qquad |(2 - 3x) - 5| = 3|x - (-1)|.$$

We can make the expression on the left less than ϵ by making $|x - (-1)|$ less than $\epsilon/3$. This suggests that we set $\delta = \frac{1}{3}\epsilon$.

Showing that the δ "works." If $0 < |x - (-1)| < \frac{1}{3}\epsilon$, then $3|x - (-1)| < \epsilon$ and, by (**), $|(2 - 3x) - 5| < \epsilon$. ❑

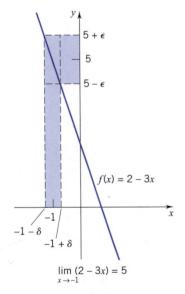

$$\lim_{x \to -1}(2 - 3x) = 5$$

Figure 2.2.7

Three Basic Limits

Here we apply the ϵ, δ method to confirm three basic limits that are intuitively obvious. (If the ϵ, δ method did not confirm these limits, then the method would have been thrown out a long time ago.)

Example 3 For each number c,

(2.2.3)

$$\lim_{x \to c} x = c.$$

(Figure 2.2.8)

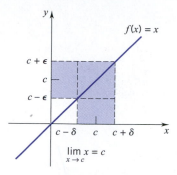

Figure 2.2.8

PROOF Let c be a real number and let $\epsilon > 0$. We must find a $\delta > 0$ such that

$$\text{if} \quad 0 < |x - c| < \delta, \quad \text{then} \quad |x - c| < \epsilon.$$

Obviously we can choose $\delta = \epsilon$. ❏

Example 4 For each real number c

(2.2.4)

$$\lim_{x \to c} |x| = |c|.$$

(Figure 2.2.9)

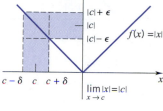

Figure 2.2.9

PROOF Let c be a real number and let $\epsilon > 0$. We seek a $\delta > 0$ such that

$$\text{if} \quad 0 < |x - c| < \delta, \quad \text{then} \quad \big||x| - |c|\big| < \epsilon.$$

Since

$$\big||x| - |c|\big| \le |x - c|,$$

[(1.3.7)]

we can choose $\delta = \epsilon$, for

$$\text{if} \quad 0 < |x - c| < \epsilon, \quad \text{then} \quad \big||x| - |c|\big| < \epsilon. \quad ❏$$

Example 5 For each constant k

(2.2.5)

$$\lim_{x \to c} k = k.$$

(Figure 2.2.10)

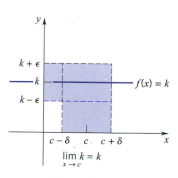

Figure 2.2.10

PROOF Here we are dealing with the constant function

$$f(x) = k.$$

Let $\epsilon > 0$. We must find a $\delta > 0$ such that

$$\text{if} \quad 0 < |x - c| < \delta, \quad \text{then} \quad |k - k| < \epsilon.$$

Since $|k - k| = 0$, we always have

$$|k - k| < \epsilon$$

no matter how δ is chosen; in short, any positive number will do for δ. ❏

Usually ϵ, δ arguments are carried out in two stages. First we do a little scratch work, labeled "finding a δ" in Examples 1 and 2. This scratch work involves working backward from $|f(x) - L| < \epsilon$ to find a $\delta > 0$ sufficiently small so that we can begin with the inequality $0 < |x - c| < \delta$ and arrive at $|f(x) - L| < \epsilon$. This first stage is just preliminary, but it shows us how to proceed in the second stage. The second stage consists of showing that the δ "works" by verifying that, for our choice of δ, it is true that

$$\text{if} \quad 0 < |x - c| < \delta, \quad \text{then} \quad |f(x) - L| < \epsilon.$$

The next two examples will give you a better feeling for this idea of working backward to find a δ.

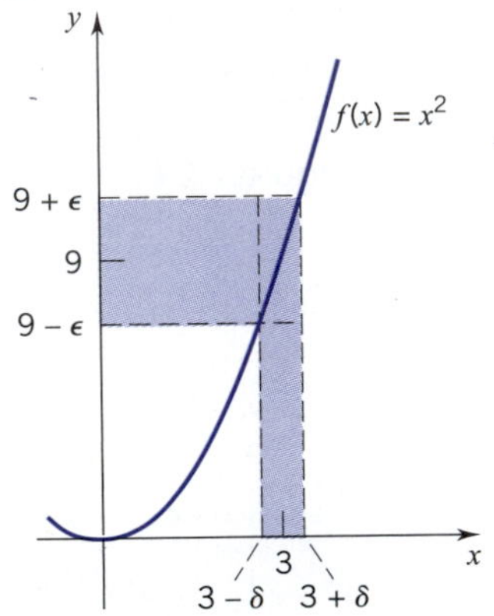

Example 6

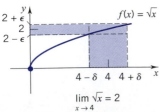

$$\lim_{x \to 3} x^2 = 9 \qquad \text{(Figure 2.2.11)}$$

Figure 2.2.11

Finding a δ. Let $\epsilon > 0$. We seek a $\delta > 0$ such that

$$\text{if} \quad 0 < |x - 3| < \delta, \quad \text{then} \quad |x^2 - 9| < \epsilon.$$

The connection between $|x - 3|$ and $|x^2 - 9|$ can be found by factoring:

$$x^2 - 9 = (x + 3)(x - 3),$$

and thus,

$$(*) \qquad\qquad |x^2 - 9| = |x + 3||x - 3|.$$

At this point, we need to get an estimate for the size of $|x + 3|$ for x close to 3. For convenience, we'll take x within one unit of 3.

If $|x - 3| < 1$, then $2 < x < 4$ and

$$|x + 3| \leq |x| + |3| = x + 3 < 7.$$

Therefore, by $(*)$,

$$(**) \qquad\qquad \text{if} \quad |x - 3| < 1, \quad \text{then} \quad |x^2 - 9| < 7|x - 3|.$$

If, in addition, $|x - 3| < \epsilon/7$, then it will follow that

$$|x^2 - 9| < 7(\epsilon/7) = \epsilon.$$

This means that we can let $\delta = $ the minimum of 1 and $\epsilon/7$.

Showing that the δ "works." Let $\epsilon > 0$. Choose $\delta = \min\{1, \epsilon/7\}$ and assume that

$$0 < |x - 3| < \delta.$$

Then

$$|x - 3| < 1 \quad \text{and} \quad |x - 3| < \epsilon/7.$$

By $(**)$,

$$|x^2 - 9| < 7|x - 3|,$$

and since $|x - 3| < \epsilon/7$, we have

$$|x^2 - 9| < 7(\epsilon/7) = \epsilon. \qquad \square$$

Example 7

$$\lim_{x \to 4} \sqrt{x} = 2. \qquad \text{(Figure 2.2.12)}$$

Figure 2.2.12

Finding a δ. Let $\epsilon > 0$. We seek a $\delta > 0$ such that

$$\text{if} \quad 0 < |x - 4| < \delta, \quad \text{then} \quad |\sqrt{x} - 2| < \epsilon.$$

To be able to form $\sqrt{x}$, we need to have $x \geq 0$. To ensure this, we must have $\delta \leq 4$. (Explain.)

Remembering that we must have $\delta \leq 4$, let's move on to find a connection between $|x - 4|$ and $|\sqrt{x} - 2|$. With $x \geq 0$, we can form $\sqrt{x}$ and write

$$x - 4 = (\sqrt{x})^2 - 2^2 = (\sqrt{x} + 2)(\sqrt{x} - 2).$$

Taking absolute values, we have

$$|x - 4| = |\sqrt{x} + 2||\sqrt{x} - 2|.$$

Since $|\sqrt{x} + 2| \geq 2 > 1$, it follows that

$$|\sqrt{x} - 2| < |x - 4|.$$

This last inequality suggests that we can simply set $\delta \leq \epsilon$. But remember the requirement $\delta \leq 4$. We can meet both requirements on δ by setting $\delta =$ the minimum of 4 and ϵ.

Showing that the δ "works." Let $\epsilon > 0$. Choose $\delta = \min\{4, \epsilon\}$ and assume that

$$0 < |x - 4| < \delta.$$

Since $\delta \leq 4$, we have $x \geq 0$, and so $\sqrt{x}$ is defined. Now, as shown above,

$$|x - 4| = |\sqrt{x} + 2||\sqrt{x} - 2|.$$

Since $|\sqrt{x} + 2| \geq 2 > 1$, we can conclude that

$$|\sqrt{x} - 2| < |x - 4|.$$

Since $|x - 4| < \delta$ and $\delta \leq \epsilon$, it does follow that $|x - 2| < \epsilon$. ❏

There are several different ways of formulating the same limit statement. Sometimes one formulation is more convenient, sometimes another. In particular, it is useful to recognize that the following four statements are equivalent:

(2.2.6)

$$\text{(i) } \lim_{x \to c} f(x) = L \qquad \text{(ii) } \lim_{h \to 0} f(c + h) = L$$

$$\text{(iii) } \lim_{x \to c}(f(x) - L) = 0 \qquad \text{(iv) } \lim_{x \to c} |f(x) - L| = 0.$$

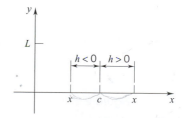

Figure 2.2.13

The equivalence of (i) and (ii) is illustrated in Figure 2.2.13: simply think of h as being the signed distance from c to x. Then $x = c + h$, and x approaches c iff h approaches 0. It is a good exercise in ϵ, δ technique to prove that (i) is equivalent to (ii).

Example 8 For $f(x) = x^2$, we have

$$\lim_{x \to 3} x^2 = 9 \qquad \lim_{h \to 0}(3 + h)^2 = 9$$

$$\lim_{x \to 3} (x^2 - 9) = 0 \qquad \lim_{x \to 3} |x^2 - 9| = 0. \quad ❏$$

We come now to the ϵ, δ definitions of one-sided limits. These are just the usual ϵ, δ statements, except that for a left-hand limit, the δ has to "work" only for x to the left of c, and for a right-hand limit, the δ has to "work" only for x to the right of c.

> **DEFINITION 2.2.7 LEFT-HAND LIMIT**
>
> Let f be a function defined at least on an open interval of the form $(c - p, c)$. We say that
>
> $$\lim_{x \to c^-} f(x) = L$$
>
> if for each $\epsilon > 0$ there exists a $\delta > 0$ such that
>
> $$\text{if} \quad c - \delta < x < c, \quad \text{then} \quad |f(x) - L| < \epsilon.$$

> **DEFINITION 2.2.8 RIGHT-HAND LIMIT**
>
> Let f be a function defined at least on an open interval of the form $(c, c + p)$. We say that
>
> $$\lim_{x \to c^+} f(x) = L$$
>
> if for each $\epsilon > 0$ there exists a $\delta > 0$ such that
>
> $$\text{if} \quad c < x < c + \delta \quad \text{then} \quad |f(x) - L| < \epsilon.$$

As our intuitive approach in Section 2.1 suggested,

$$(2.2.9) \quad \boxed{\lim_{x \to c} f(x) = L \quad \text{iff} \quad \lim_{x \to c^-} f(x) = L \quad \text{and} \quad \lim_{x \to c^+} f(x) = L.}$$

The result follows from the fact that any δ that "works" for the limit will work for both one-sided limits, and any δ that "works" for both one-sided limits will work for the limit.

Example 9 For the function defined by setting

$$f(x) = \begin{cases} 2x + 1, & x \le 0 \\ x^2 - x, & x > 0, \end{cases} \quad \text{(Figure 2.2.14)}$$

$\lim_{x \to 0} f(x)$ does not exist.

PROOF The left- and right-hand limits at 0 are as follows:

$$\lim_{x \to 0^-} f(x) = \lim_{x \to 0^-} (2x + 1) = 1, \quad \lim_{x \to 0^+} f(x) = \lim_{x \to 0^+} (x^2 - x) = 0.$$

Since these one-sided limits are different, $\lim_{x \to 0} f(x)$ does not exist. ❑

Example 10 For the function defined by setting

$$g(x) = \begin{cases} 1 + x^2, & x < 1 \\ 3, & x = 1 \\ 4 - 2x, & x > 1, \end{cases} \quad \text{(Figure 2.2.15)}$$

$\lim_{x \to 1} g(x) = 2.$

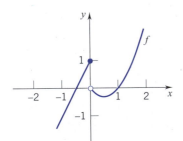

Figure 2.2.14

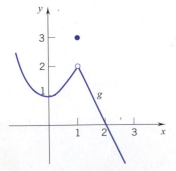

Figure 2.2.15

PROOF The left- and right-hand limits at 1 are as follows:

$$\lim_{x\to 1^-} g(x) = \lim_{x\to 1^-} (1 + x^2) = 2, \qquad \lim_{x\to 1^+} g(x) = \lim_{x\to 1^+} (4 - 2x) = 2.$$

Thus, $\lim_{x\to 1} g(x) = 2$. NOTE: It does not matter that $g(1) \neq 2$. ❑

At an endpoint of the domain of a function we can't take a (full) limit and we can't take a one-sided limit from the side on which the function is not defined, but we can try to take a limit from the side on which the function is defined. For example, it makes no sense to write

$$\lim_{x\to 0} \sqrt{x} \quad \text{or} \quad \lim_{x\to 0^-} \sqrt{x}.$$

But it does make sense to try to find

$$\lim_{x\to 0^+} \sqrt{x}. \qquad \text{(Figure 2.2.16)}$$

As you probably suspect, this one-sided limit exists and is 0.

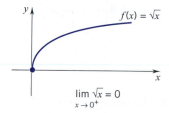

Figure 2.2.16

EXERCISES 2.2

Exercises 1–20. Decide in the manner of Section 2.1 whether or not the indicated limit exists. Evaluate the limits that do exist.

1. $\lim_{x\to 1} \dfrac{x}{x + 1}$.

2. $\lim_{x\to 0} \dfrac{x^2(1 + x)}{2x}$.

3. $\lim_{x\to 0} \dfrac{x(1 + x)}{2x^2}$.

4. $\lim_{x\to 4} \dfrac{x}{\sqrt{x + 1}}$.

5. $\lim_{x\to 1} \dfrac{x^4 - 1}{x - 1}$.

6. $\lim_{x\to -1} \dfrac{1 - x}{x + 1}$.

7. $\lim_{x\to 0} \dfrac{x}{|x|}$.

8. $\lim_{x\to 1} \dfrac{x^2 - 1}{x^2 - 2x + 1}$.

9. $\lim_{x\to -2} \dfrac{|x|}{x}$.

10. $\lim_{x\to 9} \dfrac{x - 3}{\sqrt{x} - 3}$.

11. $\lim_{x\to 3^+} \dfrac{x + 3}{x^2 - 7x + 12}$.

12. $\lim_{x\to 0^-} \dfrac{x}{|x|}$.

13. $\lim_{x\to 1^+} \dfrac{\sqrt{x} - 1}{x}$.

14. $\lim_{x\to 3^-} \sqrt{9 - x^2}$.

15. $\lim_{x\to 2^+} f(x)$ if $f(x) = \begin{cases} 2x - 1, & x \leq 2 \\ x^2 - x, & x > 2. \end{cases}$

16. $\lim_{x\to -1^-} f(x)$ if $f(x) = \begin{cases} 1, & x \leq -1 \\ x + 2, & x > -1. \end{cases}$

17. $\lim_{x\to 2} f(x)$ if $f(x) = \begin{cases} 3, & x \text{ an integer} \\ 1, & \text{otherwise.} \end{cases}$

18. $\lim_{x\to 3} f(x)$ if $f(x) = \begin{cases} x^2, & x < 3 \\ 7, & x = 3 \\ 2x + 3, & x > 3. \end{cases}$

19. $\lim_{x\to 2} f(x)$ if $f(x) = \begin{cases} 3, & x \text{ an integer} \\ 1, & \text{otherwise.} \end{cases}$

20. $\lim_{x\to 2} f(x)$ if $f(x) = \begin{cases} x^2, & x \leq 1 \\ 5x, & x > 1. \end{cases}$

21. Which of the δ's displayed in the figure "works" for the given ϵ?

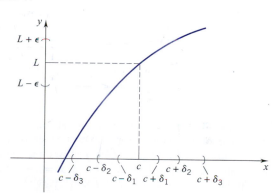

22. For which of the ϵ's given in the figure does the specified δ work?

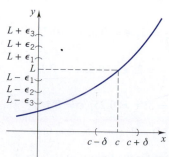

Exercises 23–26. Find the largest δ that "works" for the given ϵ.

23. $\lim_{x\to 1} 2x = 2$; $\epsilon = 0.1$.

24. $\lim_{x\to 4} 5x = 20$; $\epsilon = 0.5$.

25. $\lim_{x\to 2} \frac{1}{2}x = 1$; $\epsilon = 0.01$.

26. $\lim_{x\to 2} \frac{1}{5}x = \frac{2}{5}$; $\epsilon = 0.1$.

27. The graphs of $f(x) = \sqrt{x}$ and the horizontal lines $y = 1.5$ and $y = 2.5$ are shown in the figure. Use a graphing utility to find a $\delta > 0$ which is such that

if $\quad 0 < |x - 4| < \delta, \quad$ then $\quad |\sqrt{x} - 2| < 0.5.$

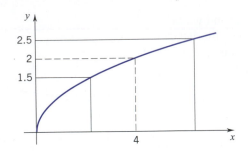

28. The graphs of $f(x) = 2x^2$ and the horizontal lines $y = 1$ and $y = 3$ are shown in the figure. Use a graphing utility to find a $\delta > 0$ which is such that

if $\quad 0 < |x + 1| < \delta, \quad$ then $\quad |2x^2 - 2| < 1.$

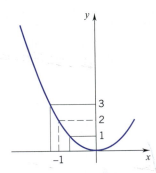

Exercises 29–34. For each of the limits stated and the ϵ's given, use a graphing utility to find a $\delta > 0$ which is such that if $0 < |x - c| < \delta$, then $|f(x) - L| < \epsilon$. Draw the graph of f together with the vertical lines $x = c - \delta, x = c + \delta$ and the horizontal lines $y = L - \epsilon, y = L + \epsilon$.

29. $\lim_{x \to 2} \left(\frac{1}{4}x^2 + x + 1\right) = 4$; $\epsilon = 0.5$, $\epsilon = 0.25$.

30. $\lim_{x \to -2} (x^3 + 4x + 2) = 2$; $\epsilon = 0.5$, $\epsilon = 0.25$.

31. $\lim_{x \to 1} \dfrac{x - 1}{\sqrt{x} - 1} = 2$; $\epsilon = 0.5$, $\epsilon = 0.25$.

32. $\lim_{x \to -1} \dfrac{1 - 3x}{2x + 4} = 2$; $\epsilon = 0.5$, $\epsilon = 0.1$.

33. $\lim_{x \to 0} \dfrac{\sin 3x}{x} = 3$; $\epsilon = 0.25$, $\epsilon = 0.1$.

34. $\lim_{x \to 1} \tan(\pi x/4) = 1$; $\epsilon = 0.5$, $\epsilon = 0.1$.

Give an ϵ, δ proof for the following statements.

35. $\lim_{x \to 4} (2x - 5) = 3$.

36. $\lim_{x \to 2} (3x - 1) = 5$.

37. $\lim_{x \to 3} (6x - 7) = 11$.

38. $\lim_{x \to 0} (2 - 5x) = 2$.

39. $\lim_{x \to 2} |1 - 3x| = 5$.

40. $\lim_{x \to 2} |x - 2| = 0$.

41. Let f be some function for which you know only that

if $\quad 0 < |x - 3| < 1, \quad$ then $\quad |f(x) - 5| < 0.1.$

Which of the following statements are necessarily true?

(a) If $|x - 3| < 1$, then $|f(x) - 5| < 0.1$.

(b) If $|x - 2.5| < 0.3$, then $|f(x) - 5| < 0.1$.

(c) $\lim_{x \to 3} f(x) = 5$.

(d) If $0 < |x - 3| < 2$, then $|f(x) - 5| < 0.1$.

(e) If $0 < |x - 3| < 0.5$, then $|f(x) - 5| < 0.1$.

(f) If $0 < |x - 3| < \frac{1}{4}$, then $|f(x) - 5| < \frac{1}{4}(0.1)$.

(g) If $0 < |x - 3| < 1$, then $|f(x) - 5| < 0.2$.

(h) If $0 < |x - 3| < 1$, then $|f(x) - 4.95| < 0.05$.

(i) If $\lim_{x \to 3} f(x) = L$, then $4.9 \le L \le 5.1$.

42. Suppose that $|A - B| < \epsilon$ for each $\epsilon > 0$. Prove that $A = B$.
HINT: Suppose that $A \ne B$ and set $\epsilon = \frac{1}{2}|A - B|$.

Exercises 43–44. Give the four limit statements displayed in (2.2.6), taking

43. $f(x) = \dfrac{1}{x - 1}$, $c = 3$

44. $f(x) = \dfrac{x}{x^2 + 2}$, $c = 1$.

45. Prove that

$$\textbf{(2.2.10)} \qquad \lim_{x \to c} f(x) = 0, \quad \text{iff} \quad \lim_{x \to c} |f(x)| = 0.$$

46. (a) Prove that

if $\quad \lim_{x \to c} f(x) = L, \quad$ then $\quad \lim_{x \to c} |f(x)| = |L|.$

(b) Show that the converse is false. Give an example where

$$\lim_{x \to c} |f(x)| = |L| \qquad \text{and} \qquad \lim_{x \to c} f(x) = M \ne L,$$

and then give an example where

$$\lim_{x \to c} |f(x)| \quad \text{exists but} \quad \lim_{x \to c} f(x) \quad \text{does not exist.}$$

47. Give an ϵ, δ proof that statement (i) in (2.2.6) is equivalent to (ii).

48. Give an ϵ, δ proof of (2.2.9).

49. (a) Show that $\lim_{x \to c} \sqrt{x} = \sqrt{c}$ for each $c > 0$.
HINT: If x and c are positive, then

$$0 \le |\sqrt{x} - \sqrt{c}| = \frac{|x - c|}{\sqrt{x} + \sqrt{c}} < \frac{1}{\sqrt{c}}|x - c|.$$

(b) Show that $\lim_{x \to 0^+} \sqrt{x} = 0$.

Give an ϵ, δ proof for the following statements.

50. $\lim_{x \to 2} x^2 = 4$.

51. $\lim_{x \to 1} x^3 = 1$.

52. $\lim_{x \to 3} \sqrt{x + 1} = 2$.

53. $\lim_{x \to 3^-} \sqrt{3 - x} = 0$.

54. Prove that, for the function

$$g(x) = \begin{cases} x, & x \text{ rational} \\ 0, & x \text{ irrational,} \end{cases}$$

$$\lim_{x \to 0} g(x) = 0.$$

55. The function

$$f(x) = \begin{cases} 1, & x \text{ rational} \\ 0, & x \text{ irrational} \end{cases}$$

is called the *Dirichlet function*. Prove that for no number c does $\lim_{x \to c} f(x)$ exist.

Prove the limit statement.

56. $\lim_{x \to c^-} f(x) = L$ iff $\lim_{h \to 0} f(c - |h|) = L$.

57. $\lim_{x \to c^+} f(x) = L$ iff $\lim_{h \to 0} f(c + |h|) = L$.

58. $\lim_{x \to c} f(x) = L$ iff $\lim_{x \to c}[f(x) - L] = 0$.

59. Suppose that $\lim_{x \to c} f(x) = L$.

(a) Prove that if $L > 0$, then $f(x) > 0$ for all $x \neq c$ in an interval of the form $(c - \gamma, c + \gamma)$.
HINT: Use an ϵ, δ argument, setting $\epsilon = L$.

(b) Prove that if $L < 0$, then $f(x) < 0$ for all $x \neq c$ in an interval of the form $(c - \gamma, c + \gamma)$.

60. Prove or give a counterexample: if $f(c) > 0$ and $\lim_{x \to c} f(x)$ exists, then $f(x) > 0$ for all x in an interval of the form $(c - \gamma, c + \gamma)$.

61. Suppose that $f(x) \leq g(x)$ for all $x \in (c - p, c + p)$, except possibly at c itself.

(a) Prove that $\lim_{x \to c} f(x) \leq \lim_{x \to c} g(x)$, provided each of these limits exist.

(b) Suppose that $f(x) < g(x)$ for all $x \in (c - p, c + p)$, except possibly at c itself. Does it follow that $\lim_{x \to c} f(x) < \lim_{x \to c} g(x)$?

62. Prove that if $\lim_{x \to c} f(x) = L$, then there are positive numbers δ and B such that if $0 < |x - c| < \delta$, then $|f(x)| < B$.

■ 2.3 SOME LIMIT THEOREMS

As you probably gathered by working through the previous section, it can become rather tedious to apply the ϵ, δ definition of limit time and time again. By proving some general theorems, we can avoid some of this repetitive work. Of course, the theorems themselves (at least the first ones) will have to be proved by ϵ, δ methods.

We begin by showing that if a limit exists, it is unique.

THEOREM 2.3.1   THE UNIQUENESS OF A LIMIT

If $\quad \lim_{x \to c} f(x) = L \quad$ and $\quad \lim_{x \to c} f(x) = M, \quad$ then $\quad L = M.$

PROOF We show $L = M$ by proving that the assumption $L \neq M$ leads to the false conclusion that

$$|L - M| < |L - M|.$$

Assume that $L \neq M$. Then $|L - M|/2 > 0$. Since $\lim_{x \to c} f(x) = L$, we know that there exists a $\delta_1 > 0$ such that

(1)   if $0 < |x - c| < \delta_1,$ then $|f(x) - L| < |L - M|/2.$

(Here we are using $|L - M|/2$ as ϵ.)

Since $\lim_{x \to c} f(x) = M$, we know that there exists a $\delta_2 > 0$ such that

(2)   if $0 < |x - c| < \delta_2,$ then $|f(x) - L| < |L - M|/2.$

(Again, we are using $|L - M|/2$ as ϵ.)

Now let x_1 be a number that satisfies the inequality

$$0 < |x_1 - c| < \text{minimum of } \delta_1 \text{ and } \delta_2.$$

Then, by (1) and (2),

$$|f(x_1) - L| < \frac{|L - M|}{2} \quad \text{and} \quad |f(x_1) - M| < \frac{|L - M|}{2}.$$

It follows that

$$
\begin{aligned}
|L - M| \; &= \; |[L - f(x_1)] + [f(x_1) - M]| \\
&\leq \; |L - f(x_1)| + |f(x_1) - M|
\end{aligned}
$$

by the triangle inequality ⟶↑

$$
= \; |f(x_1) - L| + |f(x_1) - M| < \frac{|L - M|}{2} + \frac{|L - M|}{2} = |L - M|.
$$

$|a| = |-a|$ ⟶↑

❑

THEOREM 2.3.2

If $\lim_{x \to c} f(x) = L$ and $\lim_{x \to c} g(x) = M$, then

(i) $\lim_{x \to c} [f(x) + g(x)] = L + M$,

(ii) $\lim_{x \to c} [\alpha f(x)] = \alpha L$ $|\alpha|$ a real number

(iii) $\lim_{x \to c} [f(x) g(x)] = LM$.

PROOF Let $\epsilon > 0$. To prove (i), we must show that there exists a $\delta > 0$ such that

$$
\text{if} \quad 0 < |x - c| < \delta, \quad \text{then} \quad |[f(x) + g(x)] - [L + M]| < \epsilon.
$$

Note that

$$
(*) \qquad \begin{aligned}
|[f(x) + g(x)] - [L + M]| &= |[f(x) - L] + [g(x) - M]| \\
&\leq |f(x) - L| + |g(x) - M|.
\end{aligned}
$$

We can make $|[f(x) + g(x)] - [L + M]|$ less than ϵ by making $|f(x) - L|$ and $|g(x) - M|$ each less than $\frac{1}{2}\epsilon$. Since $\epsilon > 0$, we know that $\frac{1}{2}\epsilon > 0$. Since

$$
\lim_{x \to c} f(x) = L \quad \text{and} \quad \lim_{x \to c} g(x) = M,
$$

we know that there exist positive numbers δ_1 and δ_2 such that

$$
\text{if} \quad 0 < |x - c| < \delta_1, \quad \text{then} \quad |f(x) - L| < \tfrac{1}{2}\epsilon
$$

and

$$
\text{if} \quad 0 < |x - c| < \delta_2, \quad \text{then} \quad |g(x) - M| < \tfrac{1}{2}\epsilon.
$$

Now we set $\delta = $ the minimum of δ_1 and δ_2 and note that, if $0 < |x - c| < \delta$, then

$$
|f(x) - L| < \tfrac{1}{2}\epsilon \quad \text{and} \quad |g(x) - M| < \tfrac{1}{2}\epsilon.
$$

Thus, by $(*)$,

$$
|[f(x) + g(x)] - [L + M]| < \epsilon.
$$

In summary, by setting $\delta = \min\{\delta_1, \delta_2\}$, we find that

$$
\text{if} \quad 0 < |x - c| < \delta \quad \text{then} \quad |[f(x) + g(x)] - [L + M]| < \epsilon.
$$

This completes the proof of (i). For proofs of (ii) and (iii), see the supplement to this section. ❑

If you are wondering about $\lim_{x \to c}[f(x) - g(x)]$, note that

$$f(x) - g(x) = f(x) + (-1)g(x),$$

and so the result

(2.3.3)
$$\lim_{x \to c}[f(x) - g(x)] = L - M$$

follows from (i) and (ii).

Theorem 2.3.2 can be extended (by mathematical induction) to any finite collection of functions; in particular, if

$$\lim_{x \to c} f_1(x) = L_1, \qquad \lim_{x \to c} f_2(x) = L_2, \qquad \ldots, \qquad \lim_{x \to c} f_n(x) = L_n,$$

and $\alpha_1, \alpha_2, \ldots, \alpha_n$ are real numbers, then

(2.3.4)
$$\lim_{x \to c}[\alpha_1 f_1(x) + \alpha_2 f_2(x) + \cdots + \alpha_n f_n(x)]$$
$$= \alpha_1 L_1 + \alpha_2 L_2 + \cdots + \alpha_n L_n.$$

Also,

(2.3.5)
$$\lim_{x \to c}[f_1(x) f_2(x) \cdots f_n(x)] = L_1 L_2 \cdots L_n.$$

For each polynomial $P(x) = a_n x^n + \cdots + a_1 x + a_0$ and each real number c

(2.3.6)
$$\lim_{x \to c} P(x) = P(c).$$

PROOF We already know that

$$\lim_{x \to c} x = c.$$

From (2.3.5) we know that

$$\lim_{x \to c} x^k = c^k \qquad \text{for each positive integer } k.$$

We also know that $\lim_{x \to c} a_0 = a_0$. It follows from (2.3.4) that

$$\lim_{x \to c}[a_n x^n + \cdots + a_1 x + a_0] = a_n c^n + \cdots + a_1 c + a_0,$$

which says that

$$\lim_{x \to c} P(x) = P(c).$$

A function f for which $\lim_{x \to c} f(x) = f(c)$ is said to be *continuous* at c. What we just showed is that polynomials are continuous at each number c. Continuous functions, our focus in Section 2.4, have a regularity and a predictability not shared by other functions.

Examples

$$\lim_{x \to 1} (5x^2 - 12x + 2) = 5(1)^2 - 12(1) + 2 = -5,$$

$$\lim_{x \to 0} (14x^5 - 7x^2 + 2x + 8) = 14(0)^5 - 7(0)^2 + 2(0) + 8 = 8,$$

$$\lim_{x \to -1} (2x^3 + x^2 - 2x - 3) = 2(-1)^3 + (-1)^2 - 2(-1) - 3 = -2. \quad \square$$

We come now to reciprocals and quotients.

THEOREM 2.3.7

If $\quad \lim_{x \to c} g(x) = M \quad$ with $\quad M \neq 0, \quad$ then $\quad \lim_{x \to c} \dfrac{1}{g(x)} = \dfrac{1}{M}.$

PROOF Given in the supplement to this section. $\quad \square$

Examples

$$\lim_{x \to 4} \frac{1}{x^2} = \frac{1}{16}, \qquad \lim_{x \to 2} \frac{1}{x^3 - 1} = \frac{1}{7}, \qquad \lim_{x \to -3} \frac{1}{|x|} = \frac{1}{|-3|} = \frac{1}{3}. \quad \square$$

Once you know that reciprocals present no trouble, quotients become easy to handle.

THEOREM 2.3.8

If $\lim_{x \to c} f(x) = L \quad$ and $\quad \lim_{x \to c} g(x) = M \quad$ with $\quad M \neq 0, \quad$ then $\quad \lim_{x \to c} \dfrac{f(x)}{g(x)} = \dfrac{L}{M}.$

PROOF The key here is to observe that the quotient can be written as a product:

$$\frac{f(x)}{g(x)} = f(x) \frac{1}{g(x)}.$$

With $$\lim_{x \to c} f(x) = L \qquad \text{and} \qquad \lim_{x \to c} \frac{1}{g(x)} = \frac{1}{M},$$

the product rule [part (iii) of Theorem 2.3.2] gives

$$\lim_{x \to c} \frac{f(x)}{g(x)} = L \frac{1}{M} = \frac{L}{M}. \quad \square$$

This theorem on quotients applied to the quotient of two polynomials gives us the limit of a rational function. If $R = P/Q$ where P and Q are polynomials and c is a real number, then

---

(2.3.9) $\qquad \lim_{x \to c} R(x) = \lim_{x \to c} \dfrac{P(x)}{Q(X)} = \dfrac{P(c)}{Q(c)} = R(c), \qquad$ provided $Q(c) \neq 0.$

This says that a rational function is *continuous* at all numbers c where the denominator is different from zero.

Examples

$$\lim_{x \to 2} \frac{3x - 5}{x^2 + 1} = \frac{6 - 5}{4 + 1} = \frac{1}{5}, \qquad \lim_{x \to 3} \frac{x^3 - 3x^2}{1 - x^2} = \frac{27 - 27}{1 - 9} = 0. \quad \square$$

There is no point looking for a limit that does not exist. The next theorem gives a condition under which a quotient does not have a limit.

THEOREM 2.3.10

If $\lim\limits_{x \to c} f(x) = L$ with $L \neq 0$ and $\lim\limits_{x \to c} g(x) = 0,$ then $\lim\limits_{x \to c} \dfrac{f(x)}{g(x)}$ does not exist.

PROOF Suppose, on the contrary, that there exists a real number K such that

$$\lim_{x \to c} \frac{f(x)}{g(x)} = K.$$

Then

$$L = \lim_{x \to c} f(x) = \lim_{x \to c} \left[g(x) \cdot \frac{f(x)}{g(x)} \right] = \lim_{x \to c} g(x) \cdot \lim_{x \to c} \frac{f(x)}{g(x)} = 0 \cdot K = 0.$$

This contradicts our assumption that $L \neq 0$. $\quad \square$

Examples
From Theorem 2.3.10 you can see that

$$\lim_{x \to 1} \frac{x^2}{x - 1}, \qquad \lim_{x \to 2} \frac{3x - 7}{x^2 - 4}, \qquad \text{and} \qquad \lim_{x \to 0} \frac{5}{x}$$

all fail to exist. $\quad \square$

Now we come to quotients where both the numerator and denominator tend to zero. Such quotients will be particularly important to us as we go on.

Example 1 Evaluate the limits that exist:

(a) $\lim\limits_{x \to 3} \dfrac{x^2 - x - 6}{x - 3},$ **(b)** $\lim\limits_{x \to 4} \dfrac{(x^2 - 3x - 4)^2}{x - 4},$ **(c)** $\lim\limits_{x \to -1} \dfrac{x + 1}{(2x^2 + 7x + 5)^2}.$

SOLUTION

(a) First we factor the numerator:

$$\frac{x^2 - x - 6}{x - 3} = \frac{(x + 2)(x - 3)}{x - 3}.$$

For $x \neq 3$,

$$\frac{x^2 - x - 6}{x - 3} = x + 2.$$

Therefore

$$\lim_{x \to 3} \frac{x^2 - x - 6}{x - 3} = \lim_{x \to 3} (x + 2) = 5.$$

(b) Note that

$$\frac{(x^2 - 3x - 4)^2}{x - 4} = \frac{[(x + 1)(x - 4)]^2}{x - 4} = \frac{(x + 1)^2(x - 4)^2}{x - 4}.$$

Thus for $x \neq 4$,

$$\frac{(x^2 - 3x - 4)^2}{x - 4} = (x + 1)^2(x - 4).$$

It follows that

$$\lim_{x \to 4} \frac{(x^2 - 3x - 4)^2}{x - 4} = \lim_{x \to 4} (x + 1)^2(x - 4) = 0.$$

(c) Since

$$\frac{x + 1}{(2x^2 + 7x + 5)^2} = \frac{x + 1}{[(2x + 5)(x + 1)]^2} = \frac{x + 1}{(2x + 5)^2(x + 1)^2},$$

for $x \neq -1$,

$$\frac{x + 1}{(2x^2 + 7x + 5)^2} = \frac{1}{(2x + 5)^2(x + 1)}.$$

As $x \to -1$, the denominator tends to 0 but the numerator tends to 1. It follows from Theorem 2.3.10 that

$$\lim_{x \to -1} \frac{1}{(2x + 5)^2(x + 1)} \qquad \text{does not exist.}$$

Therefore

$$\lim_{x \to -1} \frac{x + 1}{(2x^2 + 7x + 5)^2} \qquad \text{does not exist.} \quad \square$$

Example 2 Justify the following assertions.

(a) $\displaystyle \lim_{x \to 2} \frac{1/x - 1/2}{x - 2} = -\frac{1}{4}$ **(b)** $\displaystyle \lim_{x \to 9} \frac{x - 9}{\sqrt{x} - 3} = 6.$

SOLUTION

(a) For $x \neq 2$,

$$\frac{1/x - 1/2}{x - 2} = \frac{\dfrac{2 - x}{2x}}{x - 2} = \frac{-(x - 2)}{2x(x - 2)} = \frac{-1}{2x}.$$

Thus

$$\lim_{x \to 2} \frac{1/x - 1/2}{x - 2} = \lim_{x \to 2} \left[\frac{-1}{2x} \right] = -\frac{1}{4}.$$

(b) Before working with the fraction, we remind you that for each positive number c

$$\lim_{x \to c} \sqrt{x} = \sqrt{c}. \qquad \text{(Exercise 49, Section 2.2)}$$

Now to the fraction. First we "rationalize" the denominator:

$$\frac{x-9}{\sqrt{x}-3} = \frac{x-9}{\sqrt{x}-3} \cdot \frac{\sqrt{x}+3}{\sqrt{x}+3} = \frac{(x-9)(\sqrt{x}+3)}{x-9} = \sqrt{x}+3 \quad (x \neq 9).$$

It follows that

$$\lim_{x \to 9} \frac{x-9}{\sqrt{x}-3} = \lim_{x \to 9}[\sqrt{x}+3] = 6. \quad ❑$$

Remark In this section we phrased everything in terms of two-sided limits. Although we won't stop to prove it, *analogous results carry over to one-sided limits.* ❑

EXERCISES 2.3

1. Given that

$$\lim_{x \to c} f(x) = 2, \qquad \lim_{x \to c} g(x) = -1, \qquad \lim_{x \to c} h(x) = 0,$$

evaluate the limits that exist. If the limit does not exist, state how you know that.

(a) $\lim_{x \to c}[f(x) - g(x)]$. (b) $\lim_{x \to c}[f(x)]^2$.

(c) $\lim_{x \to c} \dfrac{f(x)}{g(x)}$. (d) $\lim_{x \to c} \dfrac{h(x)}{f(x)}$.

(e) $\lim_{x \to c} \dfrac{f(x)}{h(x)}$. (f) $\lim_{x \to c} \dfrac{1}{f(x) - g(x)}$.

2. Given that

$$\lim_{x \to c} f(x) = 3, \qquad \lim_{x \to c} g(x) = 0, \qquad \lim_{x \to c} h(x) = -2,$$

evaluate the limits that exist. If the limit does not exist, state how you know that.

(a) $\lim_{x \to c}[3f(x) - 2h(x)]$. (b) $\lim_{x \to c}[h(x)]^3$.

(c) $\lim_{x \to c} \dfrac{h(x)}{x - c}$. (d) $\lim_{x \to c} \dfrac{g(x)}{h(x)}$.

(e) $\lim_{x \to c} \dfrac{4}{f(x) - h(x)}$. (f) $\lim_{x \to c}[3 + g(x)]^2$.

3. When asked to evaluate

$$\lim_{x \to 4}\left(\frac{1}{x} - \frac{1}{4}\right)\left(\frac{1}{x-4}\right),$$

Moe replies that the limit is zero since $\lim\limits_{x \to 4}\left[\dfrac{1}{x} - \dfrac{1}{4}\right] = 0$ and cites Theorem 2.3.2 as justification. Verify that the limit is actually $-\frac{1}{16}$ and identify Moe's error.

4. When asked to evaluate

$$\lim_{x \to 3} \frac{x^2 + x - 12}{x - 3},$$

Moe says that the limit does not exist since $\lim\limits_{x \to 3}(x - 3) = 0$ and cites Theorem 2.3.10 (limit of a quotient) as justification. Verify that the limit is actually 7 and identify Moe's error.

Exercises 5–38. Evaluate the limits that exist.

5. $\lim\limits_{x \to 2} 3$.

6. $\lim\limits_{x \to 3}(5 - 4x)^2$.

7. $\lim\limits_{x \to -4}(x^2 + 3x - 7)$.

8. $\lim\limits_{x \to -2} 3|x - 1|$.

9. $\lim\limits_{x \to \sqrt{3}}|x^2 - 8|$.

10. $\lim\limits_{x \to -1} \dfrac{x^2 + 1}{3x^5 + 4}$.

11. $\lim\limits_{x \to 0}\left(x - \dfrac{4}{x}\right)$.

12. $\lim\limits_{x \to 5} \dfrac{2 - x^2}{4x}$.

13. $\lim\limits_{x \to 0} \dfrac{x^2 + 1}{x - 1}$.

14. $\lim\limits_{x \to 0} \dfrac{x^2}{x^2 + 1}$.

15. $\lim\limits_{x \to 2} \dfrac{x}{x^2 - 4}$.

16. $\lim\limits_{h \to 0} h\left(1 - \dfrac{1}{h}\right)$.

17. $\lim\limits_{h \to 0} h\left(1 + \dfrac{1}{h}\right)$.

18. $\lim\limits_{x \to 2} \dfrac{x - 2}{x^2 - 4}$.

19. $\lim\limits_{x \to 2} \dfrac{x^2 - 4}{x - 2}$.

20. $\lim\limits_{x \to -2} \dfrac{(x^2 - x - 6)^2}{x + 2}$.

21. $\lim\limits_{x \to 4} \dfrac{\sqrt{x} - 2}{x - 4}$.

22. $\lim\limits_{x \to 1} \dfrac{x - 1}{\sqrt{x} - 1}$.

23. $\lim\limits_{x \to 1} \dfrac{x^2 - x - 6}{(x + 2)^2}$.

24. $\lim\limits_{x \to -2} \dfrac{x^2 - x - 6}{(x + 2)^2}$.

25. $\lim\limits_{h \to 0} \dfrac{1 - 1/h^2}{1 - 1/h}$.

26. $\lim\limits_{h \to 0} \dfrac{1 - 1/h^2}{1 + 1/h^2}$.

27. $\lim\limits_{h \to 0} \dfrac{1 - 1/h}{1 + 1/h}$.

28. $\lim\limits_{h \to 0} \dfrac{1 + 1/h}{1 + 1/h^2}$.

29. $\lim\limits_{t \to -1} \dfrac{t^2 + 6t + 5}{t^2 + 3t + 2}$.

30. $\lim\limits_{x \to 2^+} \dfrac{\sqrt{x^2 - 4}}{x - 2}$.

31. $\lim\limits_{t \to 0} \dfrac{t + a/t}{t + b/t}$.

32. $\lim\limits_{x \to 1} \dfrac{x^2 - 1}{x^3 - 1}$.

33. $\lim\limits_{x \to 1} \dfrac{x^5 - 1}{x^4 - 1}$.

34. $\lim\limits_{h \to 0} h^2\left(1 + \dfrac{1}{h}\right)$.

35. $\lim\limits_{h \to 0} h\left(1 + \dfrac{1}{h^2}\right)$.

36. $\lim\limits_{x \to -4}\left(\dfrac{3x}{x + 4} + \dfrac{8}{x + 4}\right)$.

37. $\lim\limits_{x \to -4}\left(\dfrac{2x}{x + 4} + \dfrac{8}{x + 4}\right)$.

38. $\lim\limits_{x \to -4}\left(\dfrac{2x}{x + 4} - \dfrac{8}{x + 4}\right)$.

39. Evaluate the limits that exist.

(a) $\lim\limits_{x \to 4} \left(\dfrac{1}{x} - \dfrac{1}{4} \right)$.

(b) $\lim\limits_{x \to 4} \left[\left(\dfrac{1}{x} - \dfrac{1}{4} \right) \left(\dfrac{1}{x-4} \right) \right]$.

(c) $\lim\limits_{x \to 4} \left[\left(\dfrac{1}{x} - \dfrac{1}{4} \right) (x-2) \right]$.

(d) $\lim\limits_{x \to 4} \left[\left(\dfrac{1}{x} - \dfrac{1}{4} \right) \left(\dfrac{1}{x-4} \right)^2 \right]$.

40. Evaluate the limits that exist.

(a) $\lim\limits_{x \to 3} \dfrac{x^2 + x + 12}{x - 3}$.

(b) $\lim\limits_{x \to 3} \dfrac{x^2 + x - 12}{x - 3}$.

(c) $\lim\limits_{x \to 3} \dfrac{(x^2 + x - 12)^2}{x - 3}$.

(d) $\lim\limits_{x \to 3} \dfrac{x^2 + x - 12}{(x - 3)^2}$.

41. Given that $f(x) = x^2 - 4x$, evaluate the limits that exist.

(a) $\lim\limits_{x \to 4} \dfrac{f(x) - f(4)}{x - 4}$.

(b) $\lim\limits_{x \to 1} \dfrac{f(x) - f(1)}{x - 1}$.

(c) $\lim\limits_{x \to 3} \dfrac{f(x) - f(1)}{x - 3}$.

(d) $\lim\limits_{x \to 3} \dfrac{f(x) - f(2)}{x - 3}$.

42. Given that $f(x) = x^3$, evaluate the limits that exist.

(a) $\lim\limits_{x \to 3} \dfrac{f(x) - f(3)}{x - 3}$.

(b) $\lim\limits_{x \to 3} \dfrac{f(x) - f(2)}{x - 3}$.

(c) $\lim\limits_{x \to 3} \dfrac{f(x) - f(3)}{x - 2}$.

(d) $\lim\limits_{x \to 1} \dfrac{f(x) - f(1)}{x - 1}$.

43. Show by example that $\lim\limits_{x \to c} [f(x) + g(x)]$ can exist even if $\lim\limits_{x \to c} f(x)$ and $\lim\limits_{x \to c} g(x)$ do not exist.

44. Show by example that $\lim\limits_{x \to c} [f(x)g(x)]$ can exist even if $\lim\limits_{x \to c} f(x)$ and $\lim\limits_{x \to c} g(x)$ do not exist.

Exercises 45–51. True or false? Justify your answers.

45. If $\lim\limits_{x \to c} [f(x) + g(x)]$ exists but $\lim\limits_{x \to c} f(x)$ does not exist, then $\lim\limits_{x \to c} g(x)$ does not exist.

46. If $\lim\limits_{x \to c} [f(x) + g(x)]$ and $\lim\limits_{x \to c} f(x)$ exist, then it can happen that $\lim\limits_{x \to c} g(x)$ does not exist.

47. If $\lim\limits_{x \to c} \sqrt{f(x)}$ exists, then $\lim\limits_{x \to c} f(x)$ exists.

48. If $\lim\limits_{x \to c} f(x)$ exists, then $\lim\limits_{x \to c} \sqrt{f(x)}$ exists.

49. If $\lim\limits_{x \to c} f(x)$ exists, then $\lim\limits_{x \to c} \dfrac{1}{f(x)}$ exists.

50. If $f(x) \le g(x)$ for all $x \ne c$, then $\lim\limits_{x \to c} f(x) \le \lim\limits_{x \to c} g(x)$.

51. If $f(x) < g(x)$ for all $x \ne c$, then $\lim\limits_{x \to c} f(x) < \lim\limits_{x \to c} g(x)$.

52. (a) Verify that
$$\max\{f(x), g(x)\} = \tfrac{1}{2}\{[f(x) + g(x)] + |f(x) - g(x)|\}.$$

(b) Find a similar expression for $\min \{f(x), g(x)\}$.

53. Let $h(x) = \min\{f(x), g(x)\}$ and $H(x) = \max\{f(x), g(x)\}$. Show that

if $\lim\limits_{x \to c} f(x) = L$ and $\lim\limits_{x \to c} g(x) = L$,

then $\lim\limits_{x \to c} h(x) = L$ and $\lim\limits_{x \to c} H(x) = L$.

HINT: Use Exercise 52.

54. (*Stability of limit*) Let f be a function defined on some interval $(c - p, c + p)$. Now change the value of f at a finite number of points $x_1, x_2, \ldots, x_n$ and call the resulting function g.

(a) Show that if $\lim\limits_{x \to c} f(x) = L$, then $\lim\limits_{x \to c} g(x) = L$.

(b) Show that if $\lim\limits_{x \to c}$ does not exist, then $\lim\limits_{x \to c} g(x)$ does not exist.

55. (a) Suppose that $\lim\limits_{x \to c} f(x) = 0$ and $\lim\limits_{x \to c} [f(x)g(x)] = 1$. Prove that $\lim\limits_{x \to c} g(x)$ does not exist.

(b) Suppose that $\lim\limits_{x \to c} f(x) = L \ne 0$ and $\lim\limits_{x \to c} [f(x)g(x)] = 1$. Does $\lim\limits_{x \to c} g(x)$ exist, and if so, what is it?

56. Let f be a function defined at least on an interval $(c - p, c + p)$. Suppose that for each function g

$$\lim\limits_{x \to c} [f(x) + g(x)] \quad \text{does not exist if} \quad \lim\limits_{x \to c} g(x)$$

does not exist.

Show that $\lim\limits_{x \to c} f(x)$ does exist.

(*Difference quotients*) Let f be a function and let c and $c + h$ be numbers in an interval on which f is defined. The expression

$$\dfrac{f(c + h) - f(c)}{h}$$

is called a *difference quotient* for f. (Limits of difference quotients as $h \to 0$ are at the core of Chapter 3.) In Exercises 57–60, calculate

$$\lim\limits_{h \to 0} \dfrac{f(c + h) - f(c)}{h}$$

for the function f and the number c.

57. $f(x) = 2x^2 - 3x; \quad c = 2$.

58. $f(x) = x^3 + 1; \quad c = -1$.

59. $f(x) = \sqrt{x}; \quad c = 4$.

60. $f(x) = 1/(x + 1); \quad c = 1$.

61. Calculate

$$\lim\limits_{h \to 0} \dfrac{f(x + h) - f(x)}{h}$$

for each of the following functions:

(a) $f(x) = x$.

(b) $f(x) = x^2$.

(c) $f(x) = x^3$.

(d) $f(x) = x^4$.

(e) $f(x) = x^n$, n an arbitrary positive integer.
Make a guess and confirm your guess by induction.

*SUPPLEMENT TO SECTION 2.3

PROOF OF THEOREM 2.3.2 (II)

We consider two cases: $\alpha \neq 0$ and $\alpha = 0$. If $\alpha \neq 0$, then $\epsilon/|\alpha| > 0$ and, since

$$\lim_{x \to c} f(x) = L,$$

we know that there exists $\delta > 0$ such that,

$$\text{if} \quad 0 < |x - c| < \delta, \quad \text{then} \quad |f(x) - L| < \frac{\epsilon}{|\alpha|}.$$

From the last inequality, we obtain

$$|\alpha| |f(x) - L| < \epsilon \quad \text{and thus} \quad |\alpha f(x) - \alpha L| < \epsilon.$$

The case $\alpha = 0$ was treated before. (2.2.5) ❑

PROOF OF THEOREM 2.3.2 (III)

We begin with a little algebra:

$$\begin{aligned}
|f(x)g(x) - LM| &= |[f(x)g(x) - f(x)M] + [f(x)M - LM]| \\
&\leq |f(x)g(x) - f(x)M| + |f(x)M - LM| \\
&= |f(x)||g(x) - M| + |M||f(x) - L| \\
&\leq |f(x)||g(x) - M| + (1 + |M|)|f(x) - L|.
\end{aligned}$$

Now let $\epsilon > 0$. Since $\lim_{x \to c} f(x) = L$ and $\lim_{x \to c} g(x) = M$, we know the following:

1. There exists $\delta_1 > 0$ such that, if $0 < |x - c| < \delta_1$, then

$$|f(x) - L| < 1 \quad \text{and thus} \quad |f(x)| < 1 + |L|.$$

2. There exists $\delta_2 > 0$ such that

$$\text{if} \quad 0 < |x - c| < \delta_2, \quad \text{then} \quad |g(x) - M| < \left(\frac{\frac{1}{2}\epsilon}{1 + |L|} \right).$$

3. There exists $\delta_3 > 0$ such that

$$\text{if} \quad 0 < |x - c| < \delta_3, \quad \text{then} \quad |f(x) - L| < \left(\frac{\frac{1}{2}\epsilon}{1 + |M|} \right).$$

We now set $\delta = \min\{\delta_1, \delta_2, \delta_3\}$ and observe that, if $0 < |x - c| < \delta$, then

$$\begin{aligned}
|f(x) - LM| &\leq |f(x)||g(x) - M| + (1 + |M|)|f(x) - L| \\
&< (1 + |L|) \left(\frac{\frac{1}{2}\epsilon}{1 + |L|} \right) + (1 + |M|) \left(\frac{\frac{1}{2}\epsilon}{1 + |M|} \right) = \epsilon. \quad ❑
\end{aligned}$$

$$\underset{\text{by (1)}}{\uparrow} \qquad \underset{\text{by (2)}}{\uparrow} \qquad \underset{\text{by (3)}}{\uparrow}$$

PROOF OF THEOREM 2.3.7

For $g(x) \neq 0$,

$$\left| \frac{1}{g(x)} - \frac{1}{M} \right| = \frac{|g(x) - M|}{|g(x)||M|}.$$

Choose $\delta_1 > 0$ such that

$$\text{if} \quad 0 < |x - c| < \delta_1, \quad \text{then} \quad |g(x) - M| < \frac{|M|}{2}.$$

For such x,

$$|g(x)| > \frac{|M|}{2} \qquad \text{so that} \qquad \frac{1}{|g(x)|} < \frac{2}{|M|}$$

and thus

$$\left| \frac{1}{g(x)} - \frac{1}{M} \right| = \frac{|g(x) - M|}{|g(x)||M|} \le \frac{2}{|M|^2} |g(x) - M| = \frac{2}{M^2} |g(x) - M|.$$

Now let $\epsilon > 0$ and choose $\delta_2 > 0$ such that

$$\text{if} \qquad 0 < |x - c| < \delta_2, \qquad \text{then} \qquad |g(x) - M| < \frac{M^2}{2} \epsilon.$$

Setting $\delta = \min\{\delta_1, \delta_2\}$, we find that

$$\text{if} \qquad 0 < |x - c| < \delta, \qquad \text{then} \qquad \left| \frac{1}{g(x)} - \frac{1}{M} \right| < \epsilon. \qquad \square$$

■ 2.4 CONTINUITY

In ordinary language, to say that a certain process is "continuous" is to say that it goes on without interruption and without abrupt changes. In mathematics the word "continuous" has much the same meaning.

The concept of continuity is so important in calculus and its applications that we discuss it with some care. First we treat *continuity at a point c* (a number c), and then we discuss *continuity on an interval*.

Continuity at a Point

The basic idea is as follows: We are given a function f and a number c. We calculate (if we can) both $\lim_{x \to c} f(x)$ and $f(c)$. If these two numbers are equal, we say that f is *continuous* at c. Here is the definition formally stated.

DEFINITION 2.4.1

Let f be a function defined at least on an open interval $(c - p, c + p)$. We say that f is *continuous at c* if

$$\lim_{x \to c} f(x) = f(c).$$

If the domain of f contains an interval $(c - p, c + p)$, then f can fail to be continuous at c for only one of two reasons: either

(i) f has a limit as x tends to c, but $\lim_{x \to c} f(x) \ne f(c)$, or

(ii) f has no limit as x tends to c.

In case (i) the number c is called a *removable* discontinuity. The discontinuity can be removed by redefining f at c. If the limit is L, redefine f at c to be L.

In case (ii) the number c is called an *essential* discontinuity. You can change the value of f at a billion points in any way you like. The discontinuity will remain. (Exercise 51.)

The function depicted in Figure 2.4.1 has a removable discontinuity at c. The discontinuity can be removed by lowering the dot into place (i.e., by redefining f at c to be L).

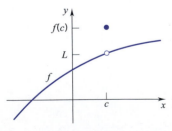

Figure 2.4.1

The functions depicted in Figures 2.4.2, 2.4.3, and 2.4.4 have essential discontinuities at c. The discontinuity in Figure 2.4.2 is, for obvious reasons, called a *jump* discontinuity. The functions of Figure 2.4.3 have *infinite* discontinuities.

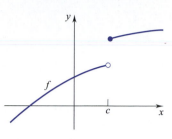

Figure 2.4.2

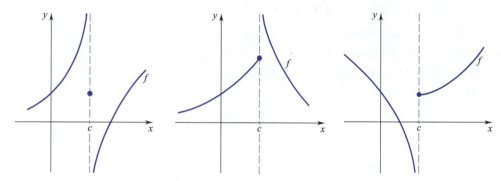

Figure 2.4.3

In Figure 2.4.4, we have tried to portray the Dirichlet function

$$f(x) = \begin{cases} 1, & x \text{ rational} \\ -1, & x \text{ irrational}. \end{cases}$$

At no point c does f have a limit. Each point is an essential discontinuity. The function is everywhere discontinuous.

Most of the functions that you have encountered so far are continuous at each point of their domains. In particular, this is true for polynomials P,

$$\lim_{x \to c} P(x) = P(c), \qquad [(2.3.6)]$$

for rational functions (quotients of polynomials) $R = P/Q$,

$$\lim_{x \to c} R(x) = \lim_{x \to c} \frac{P(x)}{Q(x)} = \frac{P(c)}{Q(c)} = R(c) \qquad \text{provided} \qquad Q(c) \neq 0, \quad [(2.3.9)]$$

and for the absolute value function,

$$\lim_{x \to c} |x| = |c|. \qquad [(2.2.4)]$$

As you were asked to show earlier (Exercise 49, Section 2.2),

$$\lim_{x \to c} \sqrt{x} = \sqrt{c} \qquad \text{for each } c > 0.$$

This makes the square-root function continuous at each positive number. What happens at $c = 0$, we discuss later.

With f and g continuous at c, we have

$$\lim_{x \to c} f(x) = f(c) \qquad \lim_{x \to c} g(x) = g(c)$$

and thus, by the limit theorems,

$$\lim_{x \to c} [f(x) + g(x)] = f(c) + g(c), \qquad \lim_{x \to c} [f(x) - g(x)] = f(c) - g(c)$$

$$\lim_{x \to c} [\alpha f(x)] = \alpha f(c) \quad \text{for each real } \alpha \qquad \lim_{x \to c} [f(x)g(x)] = f(c)g(c)$$

and, if $g(c) \neq 0$, $\qquad \lim_{x \to c} [f(x)/g(x)] = f(c)/g(c).$

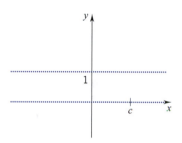

Figure 2.4.4

We summarize all this in a theorem.

THEOREM 2.4.2

If f and g are continuous at c, then

(i) $f + g$ is continuous at c;

(ii) $f - g$ is continuous at c;

(iii) αf is continuous at c for each real α;

(iv) $f \cdot g$ is continuous at c;

(v) f/g is continuous at c provided $g(c) \neq 0$.

These results can be combined and extended to any finite number of functions.

Example 1   The function $F(x) = 3|x| + \dfrac{x^3 - x}{x^2 - 5x + 6} + 4$ is continuous at all real numbers other than 2 and 3. You can see this by noting that

$$F = 3f + g/h + k$$

where

$$f(x) = |x|, \qquad g(x) = x^3 - x, \qquad h(x) = x^2 - 5x + 6, \qquad k(x) = 4.$$

Since f, g, h, k are everywhere continuous, F is continuous except at 2 and 3, the numbers at which h takes on the value 0. (At those numbers F is not defined.) ❏

Our next topic is the continuity of composite functions. Before getting into this, however, let's take a look at continuity in terms of ϵ, δ. A direct translation of

$$\lim_{x \to c} f(x) = f(c)$$

into ϵ, δ terms reads like this: for each $\epsilon > 0$, there exists a $\delta > 0$ such that

$$\text{if} \quad 0 < |x - c| < \delta, \quad \text{then} \quad |f(x) - f(c)| < \epsilon.$$

Here the restriction $0 < |x - c|$ is unnecessary. We can allow $|x - c| = 0$ because then $x = c$, $f(x) = f(c)$, and thus $|f(x) - f(c)| = 0$. Being 0, $|f(x) - f(c)|$ is certainly less than ϵ.

Thus, an ϵ, δ characterization of continuity at c reads as follows:

(2.4.3) $\quad f$ is continuous at c if $\begin{cases} \text{for each } \epsilon > 0 \text{ there exists a } \delta > 0 \text{ such that} \\ \text{if} \quad |x - c| < \delta, \quad \text{then} \quad |f(x) - f(c)| < \epsilon. \end{cases}$

In intuitive terms

$$f \text{ is continuous at } c \quad \text{if} \quad \text{for } x \text{ close to } c, \quad f(x) \text{ is close to } f(c).$$

We are now ready to take up the continuity of composite functions. Remember the defining formula: $(f \circ g)(x) = f(g(x))$. (You may wish to review Section 1.7.)

THEOREM 2.4.4

If g is continuous at c and f is continuous at $g(c)$, then the composition $f \circ g$ is continuous at c.

The idea here is as follows: with g continuous at c, we know that

for x close to c, $g(x)$ is close to $g(c)$;

from the continuity of f at $g(c)$, we know that

with $g(x)$ close to $g(c)$, $f(g(x))$ is close to $f(g(c))$.

In summary,

with x close to c, $f(g(x))$ is close to $f(g(c))$.

The argument we just gave is too vague to be a proof. Here, in contrast, is a proof. We begin with $\epsilon > 0$. We must show that there exists a number $\delta > 0$ such that

if $\quad |x - c| < \delta, \quad$ then $\quad |f(g(x)) - f(g(c))| < \epsilon$.

In the first place, we observe that, since f is continuous at $g(c)$, there does exist a number $\delta_1 > 0$ such that

(1) $\qquad$ if $\quad |t - g(c)| < \delta_1, \quad$ then $\quad |f(t) - f(g(c))| < \epsilon$.

With $\delta_1 > 0$, we know from the continuity of g at c that there exists a number $\delta > 0$ such that

(2) $\qquad$ if $\quad |x - c| < \delta, \quad$ then $\quad |g(x) - g(c)| < \delta_1$.

Combining (2) and (1), we have what we want: by (2),

if $\quad |x - c| < \delta, \quad$ then $\quad |g(x) - g(c)| < \delta_1$

so that by (1)

$$|f(g(x)) - f(g(c))| < \epsilon.$$

This proof is illustrated in Figure 2.4.5. The numbers within δ of c are taken by g to within δ_1 of $g(c)$, and then by f to within ϵ of $f(g(c))$.

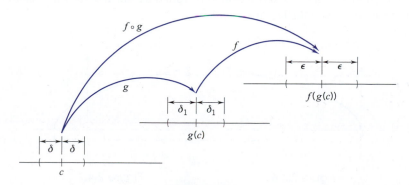

Figure 2.4.5

It's time to look at some examples.

Example 2 The function $F(x) = \sqrt{\dfrac{x^2 + 1}{x - 3}}$ is continuous at all numbers greater than 3. To see this, note that $F = f \circ g$, where

$$f(x) = \sqrt{x} \qquad \text{and} \qquad g(x) = \frac{x^2 + 1}{x - 3}.$$

Now, take any $c > 3$. Since g is a rational function and g is defined at c, g is continuous at c. Also, since $g(c)$ is positive and f is continuous at each positive number, f is continuous at $g(c)$. By Theorem 2.4.4, F is continuous at c. ❑

The continuity of composites holds for any finite number of functions. The only requirement is that each function be continuous *where it is applied*.

Example 3 The function $F(x) = \dfrac{1}{5 - \sqrt{x^2 + 16}}$ is continuous everywhere except at $x = \pm 3$, where it is not defined. To see this, note that $F = f \circ g \circ k \circ h$, where

$$f(x) = \frac{1}{x}, \qquad g(x) = 5 - x, \qquad k(x) = \sqrt{x}, \qquad h(x) = x^2 + 16,$$

and observe that each of these functions is being evaluated only where it is continuous. In particular, g and h are continuous everywhere, f is being evaluated only at nonzero numbers, and k is being evaluated only at positive numbers. ❑

Just as we considered one-sided limits, we can consider one-sided continuity.

DEFINITION 2.4.5   ONE-SIDED CONTINUITY

A function f is called

$\qquad$ *continuous from the left at c* $\qquad$ if $\qquad \displaystyle\lim_{x \to c^-} f(x) = f(c)$.

It is called

$\qquad$ *continuous from the right at c* $\qquad$ if $\qquad \displaystyle\lim_{x \to c^+} f(x) = f(c)$.

The function of Figure 2.4.6 is continuous from the right at 0; the function of Figure 2.4.7 is continuous from the left at 1.

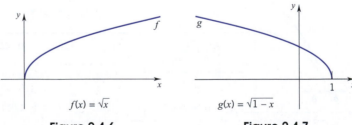

$f(x) = \sqrt{x}$

$g(x) = \sqrt{1 - x}$

Figure 2.4.6 $\qquad\qquad\qquad$ **Figure 2.4.7**

It follows from (2.2.9) that a function is continuous at c iff it is continuous from both sides at c. Thus

(2.4.6)

f is continuous at c iff $f(c)$, $\displaystyle\lim_{x \to c^-} f(x)$, $\displaystyle\lim_{x \to c^+} f(x)$

all exist and are equal.

Example 4   Determine the discontinuities, if any, of the following function:

$$f(x) = \begin{cases} 2x + 1, & x \le 0 \\ 1, & 0 < x \le 1 \\ x^2 + 1, & x > 1. \end{cases} \qquad \text{(Figure 2.4.8)}$$

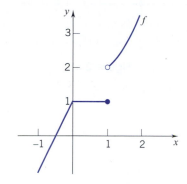

Figure 2.4.8

SOLUTION Clearly f is continuous at each point in the open intervals $(-\infty, 0), (0, 1), (1, \infty)$. (On each of these intervals f is a polynomial.) Thus, we have to check the behavior of f at $x = 0$ and $x = 1$. The figure suggests that f is continuous at 0 and discontinuous at 1. Indeed, that is the case:

$$f(0) = 1, \qquad \lim_{x \to 0^-} f(x) = \lim_{x \to 0^-} (2x + 1) = 1, \qquad \lim_{x \to 0^+} f(x) = \lim_{x \to 0^+} (1) = 1.$$

This makes f continuous at 0. The situation is different at $x = 1$:

$$\lim_{x \to 1^-} f(x) = \lim_{x \to 1^-} (1) = 1 \qquad \text{and} \qquad \lim_{x \to 1^+} f(x) = \lim_{x \to 1^+} (x^2 + 1) = 2.$$

Thus f has an essential discontinuity at 1, a jump discontinuity. ❑

Example 5   Determine the discontinuities, if any, of the following function:

$$f(x) = \begin{cases} x^3, & x \le -1 \\ x^2 - 2, & -1 < x < 1 \\ 6 - x, & 1 \le x < 4 \\ \dfrac{6}{7 - x}, & 4 < x < 7 \\ 5x + 2, & x \ge 7. \end{cases}$$

SOLUTION   It should be clear that f is continuous at each point of the open intervals $(-\infty, -1), (-1, 1), (1, 4), (4, 7), (7, \infty)$. All we have to check is the behavior of f at $x = -1, 1, 4, 7$. To do so, we apply (2.4.6).

The function is continuous at $x = -1$ since $f(-1) = (-1)^3 = -1$,

$$\lim_{x \to -1^-} f(x) = \lim_{x \to -1^-} (x^3) = -1, \qquad \text{and} \qquad \lim_{x \to -1^+} f(x) = \lim_{x \to -1^+} (x^2 - 2) = -1.$$

Our findings at the other three points are displayed in the following chart. Try to verify each entry.

c	$f(c)$	$\displaystyle\lim_{x \to c^-} f(x)$	$\displaystyle\lim_{x \to c^+} f(x)$	Conclusion
1	5	-1	5	discontinuous
4	not defined	2	2	discontinuous
7	37	does not exist	37	discontinuous

The discontinuity at $x = 4$ is removable: if we redefine f at 4 to be 2, then f becomes continuous at 4. The numbers 1 and 7 are essential discontinuities. The discontinuity at 1 is a jump discontinuity; the discontinuity at 7 is an infinite discontinuity: $f(x) \to \infty$ as $x \to 7^-$. ❑

Continuity on Intervals

A function f is said to be *continuous on an interval* if it is continuous at each interior point of the interval and one-sidedly continuous at whatever endpoints the interval may contain.

For example:

(i) The function

$$f(x) = \sqrt{1 - x^2}$$

is continuous on $[-1, 1]$ because it is continuous at each point of $(-1, 1)$, continuous from the right at -1, and continuous from the left at 1. The graph of the function is the semicircle shown in Figure 2.4.9.

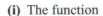

Figure 2.4.9

(ii) The function

$$f(x) = \frac{1}{\sqrt{1 - x^2}}$$

is continuous on $(-1, 1)$ because it is continuous at each point of $(-1, 1)$. It is not continuous on $[-1, 1)$ because it is not continuous from the right at -1. It is not continuous on $(-1, 1]$ because it is not continuous from the left at 1.

(iii) The function graphed in Figure 2.4.8 is continuous on $(-\infty, 1]$ and continuous on $(1, \infty)$. It is not continuous on $[1, \infty)$ because it is not continuous from the right at 1.

(iv) Polynomials, being everywhere continuous, are continuous on $(-\infty, \infty)$.

Continuous functions have special properties not shared by other functions. Two of these properties are featured in Section 2.6. Before we get to these properties, we prove a very useful theorem and revisit the trigonometric functions.

EXERCISES 2.4

1. The graph of f is given in the figure.
   (a) At which points is f discontinuous?
   (b) For each point of discontinuity found in (a), determine whether f is continuous from the right, from the left, or neither.
   (c) Which, if any, of the points of discontinuity found in (a) is removable? Which, if any, is a jump discontinuity?

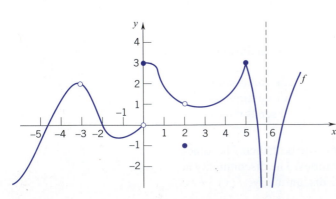

2. The graph of g is given in the figure. Determine the intervals on which g is continuous.

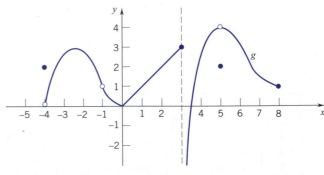

Exercises 3–16. Determine whether or not the function is continuous at the indicated point. If not, determine whether the discontinuity is a removable discontinuity or an essential discontinuity. If the latter, state whether it is a jump discontinuity, an infinite discontinuity, or neither.

3. $f(x) = x^3 - 5x + 1$; $x = 2$.

4. $g(x) = \sqrt{(x-1)^2 + 5}$; $x = 1$.

5. $f(x) = \sqrt{x^2 + 9}$; $x = 3$.

6. $f(x) = |4 - x^2|$; $x = 2$.

7. $f(x) = \begin{cases} x^2 + 4, & x < 2 \\ x^3, & x \geq 2; \end{cases}$ $x = 2$.

8. $h(x) = \begin{cases} x^2 + 5, & x < 2 \\ x^3, & x \geq 2; \end{cases}$ $x = 2$.

9. $g(x) = \begin{cases} x^2 + 4, & x < 2 \\ 5, & x = 2 \\ x^3, & x > 2; \end{cases}$ $x = 2$.

10. $g(x) = \begin{cases} x^2 + 5, & x < 2 \\ 10, & x = 2 \\ 1 + x^3, & x > 2; \end{cases}$ $x = 2$.

11. $f(x) = \begin{cases} \dfrac{|x-1|}{x-1}, & x \neq 1 \\ 0, & x = 1; \end{cases}$ $x = 1$.

12. $f(x) = \begin{cases} 1 - x, & x < 1 \\ 1, & x = 1; \\ x^2 - 1, & x > 1; \end{cases}$ $x = 1$.

13. $h(x) = \begin{cases} \dfrac{x^2 - 1}{x + 1}, & x \neq -1 \\ -2, & x = -1; \end{cases}$ $x = -1$.

14. $g(x) = \begin{cases} \dfrac{1}{x+1}, & x \neq -1 \\ 0, & x = -1; \end{cases}$ $x = -1$.

15. $f(x) = \begin{cases} \dfrac{x+2}{x^2-4}, & x \neq 2 \\ 4, & x = 2; \end{cases}$ $x = 2$

16. $f(x) = \begin{cases} -x^2, & x < 0 \\ 0, & x = 0 \\ 1/x^2, & x > 0; \end{cases}$ $x = 0$

Exercises 17–28. Sketch the graph and classify the discontinuities (if any) as being removable or essential. If the latter, is it a jump discontinuity, an infinite discontinuity, or neither.

17. $f(x) = |x - 1|$.

18. $h(x) = |x^2 - 1|$.

19. $f(x) = \begin{cases} \dfrac{x^2 - 4}{x - 2}, & x \neq 2 \\ 4, & x = 2. \end{cases}$

20. $f(x) = \begin{cases} \dfrac{x - 3}{x^2 - 9}, & x \neq 3, -3 \\ \frac{1}{6}, & x = 3, -3 \end{cases}$

21. $f(x) = \begin{cases} \dfrac{x + 2}{x^2 - x - 6}, & x \neq -2, 3 \\ -\frac{1}{5}, & x = -2, 3. \end{cases}$

22. $g(x) = \begin{cases} 2x - 1, & x < 1 \\ 0, & x = 1 \\ 1/x^2, & x > 1. \end{cases}$

23. $f(x) = \begin{cases} -1, & x < -1 \\ x^3, & -1 \leq x \leq 1 \\ 1, & 1 < x. \end{cases}$

24. $g(x) = \begin{cases} 1, & x \leq -2 \\ \frac{1}{2}x, & -2 < x < 4 \\ \sqrt{x}, & 4 \leq x. \end{cases}$

25. $h(x) = \begin{cases} 1, & x \leq 0 \\ x^2, & 0 < x < 1 \\ 1, & 1 \leq x < 2 \\ x, & 2 \leq x. \end{cases}$

26. $g(x) = \begin{cases} -x^2, & x < -1 \\ 3, & x = -1 \\ 2 - x, & -1 < x \leq 1 \\ 1/x^2, & 1 < x. \end{cases}$

27. $f(x) = \begin{cases} 2x + 9, & x < -2 \\ x^2 + 1, & -2 < x \leq 1, \\ 3x - 1, & 1 < x < 3 \\ x + 6, & 3 < x. \end{cases}$

28. $g(x) = \begin{cases} x + 7, & x < -3 \\ |x - 2|, & -3 < x < -1 \\ x^2 - 2x, & -1 < x < 3 \\ 2x - 3, & 3 \leq x. \end{cases}$

29. Sketch a graph of a function f that satisfies the following conditions:
  1. $\text{dom}(f) = [-3, 3]$.
  2. $f(-3) = f(-1) = 1$; $f(2) = f(3) = 2$.
  3. f has an infinite discontinuity at -1 and a jump discontinuity at 2.
  4. f is right continuous at -1 and left continuous at 2.

30. Sketch a graph of a function f that satisfies the following conditions:
  1. $\text{dom}(f) = [-2, 2]$.
  2. $f(-2) = f(-1) = f(1) = f(2) = 0$.
  3. f has an infinite discontinuity at -2, a jump discontinuity at -1, a jump discontinuity at 1, and an infinite discontinuity at 2.
  4. f is continuous from the right at -1 and continuous from the left at 1.

Exercises 31–34. If possible, define the function at 1 so that it becomes continuous at 1.

31. $f(x) = \dfrac{x^2 - 1}{x - 1}$.

32. $f(x) = \dfrac{1}{x - 1}$.

33. $f(x) = \dfrac{x - 1}{|x - 1|}$.

34. $f(x) = \dfrac{(x - 1)^2}{|x - 1|}$.

35. Let $f(x) = \begin{cases} x^2, & x < 1 \\ Ax - 3, & x \geq 1. \end{cases}$ Find A given that f is continuous at 1.

36. Let $f(x) = \begin{cases} A^2 x^2, & x \leq 2 \\ (1 - A)x, & x > 2. \end{cases}$ For what values of A is f continuous at 2?

37. Give necessary and sufficient conditions on A and B for the function

$$f(x) = \begin{cases} Ax - B, & x \leq 1 \\ 3x, & 1 < x < 2 \\ Bx^2 - A, & 2 \leq x \end{cases}$$

to be continuous at $x = 1$ but discontinuous at $x = 2$.

38. Give necessary and sufficient conditions on A and B for the function in Exercise 37 to be continuous at $x = 2$ but discontinuous at $x = 1$.

39. Set $f(x) = \begin{cases} 1 + cx, & x < 2 \\ c - x, & x \geq 2. \end{cases}$ Find a value of c that makes f continuous on $(-\infty, \infty)$. Use a graphing utility to verify your result.

40. Set $f(x) = \begin{cases} 1 - cx + dx^2, & x \leq -1 \\ x^2 + x, & -1 < x < 2 \\ cx^2 + dx + 4, & x \geq 2. \end{cases}$ Find values of c and d that make f continuous on $(-\infty, \infty)$. Use a graphing utility to verify your result.

Exercises 41–44. Define the function at 5 so that it becomes continuous at 5.

41. $f(x) = \dfrac{\sqrt{x+4}-3}{x-5}$.

42. $f(x) = \dfrac{\sqrt{x+4}-3}{\sqrt{x-5}}$.

43. $f(x) = \dfrac{\sqrt{2x-1}-3}{x-5}$.

44. $f(x) = \dfrac{\sqrt{x^2-7x+16}-\sqrt{6}}{(x-5)\sqrt{x+1}}$.

Exercises 45–47. At what points (if any) is the function continuous?

45. $f(x) = \begin{cases} 1, & x \text{ rational} \\ 0, & x \text{ irrational.} \end{cases}$

46. $g(x) = \begin{cases} x, & x \text{ rational} \\ 0, & x \text{ irrational.} \end{cases}$

47. $h(x) = \begin{cases} 2x, & x \text{ an integer} \\ x^2, & \text{otherwise.} \end{cases}$

48. The following functions are important in science and engineering:

1. The *Heaviside function* $H_c(x) = \begin{cases} 0, & x < c \\ 1, & x \geq c. \end{cases}$

2. The *unit pulse function*

$$P_{\epsilon,c}(x) = \frac{1}{\epsilon}[H_c(x) - H_{c+\epsilon}(x)].$$

(a) Graph H_c and $P_{\epsilon,c}$.
(b) Determine where each of the functions is continuous.
(c) Find $\lim_{x\to c^-} H_c(x)$ and $\lim_{x\to c^+} H_c(x)$. What can you say about $\lim_{x\to c} H(x)$?

49. (*Important*) Prove that

$$f \text{ is continuous at } c \quad \text{iff} \quad \lim_{h\to 0} f(c+h) = f(c).$$

50. (*Important*) Let f and g be continuous at c. Prove that if:
(a) $f(c) > 0$, then there exists $\delta > 0$ such that $f(x) > 0$ for all $x \in (c - \delta, c + \delta)$.
(b) $f(c) < 0$, then there exists $\delta > 0$ such that $f(x) < 0$ for all $x \in (c - \delta, c + \delta)$.
(c) $f(c) < g(c)$, then there exists $\delta > 0$ such that $f(x) < g(x)$ for all $x \in (c - \delta, c + \delta)$.

51. Suppose that f has an essential discontinuity at c. Change the value of f as you choose at any finite number of points $x_1, x_2, \ldots, x_n$ and call the resulting function g. Show that g also has an essential discontinuity at c.

52. (a) Prove that if f is continuous everywhere, then $|f|$ is continuous everywhere.
(b) Give an example to show that the continuity of $|f|$ does not imply the continuity of f.
(c) Give an example of a function f such that f is continuous nowhere, but $|f|$ is continuous everywhere.

53. Suppose the function f has the property that there exists a number B such that

$$|f(x) - f(c)| \leq B|x - c|$$

for all x in the interval $(c - p, c + p)$. Prove that f is continuous at c.

54. Suppose the function f has the property that

$$|f(x) - f(t)| \leq |x - t|$$

for each pair of points x, t in the interval (a, b). Prove that f is continuous on (a, b).

55. Prove that if

$$\lim_{h\to 0} \frac{f(c+h) - f(c)}{h}$$

exists, then f is continuous at c.

56. Suppose that the function f is continuous on $(-\infty, \infty)$. Show that f can be written

$$f = f_e + f_o,$$

where f_e is an even function which is continuous on $(-\infty, \infty)$ and f_o is an odd function which is continuous on $(-\infty, \infty)$.

Exercises 57–60. The function f is not defined at $x = 0$. Use a graphing utility to graph f. Zoom in to determine whether there is a number k such that the function

$$F(x) = \begin{cases} f(x), & x \neq 0 \\ k, & x = 0 \end{cases}$$

is continuous at $x = 0$. If so, what is k? Support your conclusion by calculating the limit using a CAS.

57. $f(x) = \dfrac{\sin 5x}{\sin 2x}$.

58. $f(x) = \dfrac{x^2}{1 - \cos 2x}$.

59. $f(x) = \dfrac{\sin x}{|x|}$.

60. $f(x) = \dfrac{x \sin 2x}{\sin x^2}$.

■ 2.5 THE PINCHING THEOREM; TRIGONOMETRIC LIMITS

Figure 2.5.1 shows the graphs of three functions f, g, h. Suppose that, as suggested by the figure, for x close to c, f is trapped between g and h. (The values of these functions at c itself are irrelevant.) If, as x tends to c, both $g(x)$ and $h(x)$ tend to the same limit L, then $f(x)$ also tends to L. This idea is made precise in what we call *the pinching theorem*.

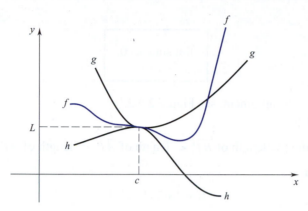

Figure 2.5.1

THEOREM 2.5.1   THE PINCHING THEOREM

Let $p > 0$. Suppose that, for all x such that $0 < |x - c| < p$,

$$h(x) \leq f(x) \leq g(x).$$

If

$$\lim_{x \to c} h(x) = L \qquad \text{and} \qquad \lim_{x \to c} g(x) = L,$$

then

$$\lim_{x \to c} f(x) = L.$$

PROOF Let $\epsilon > 0$. Let $p > 0$ be such that

$$\text{if} \quad 0 < |x - c| < p, \quad \text{then} \quad h(x) \leq f(x) \leq g(x).$$

Choose $\delta_1 > 0$ such that

$$\text{if} \quad 0 < |x - c| < \delta_1, \quad \text{then} \quad L - \epsilon < h(x) < L + \epsilon.$$

Choose $\delta_2 > 0$ such that

$$\text{if} \quad 0 < |x - c| < \delta_2, \quad \text{then} \quad L - \epsilon < g(x) < L + \epsilon.$$

Let $\delta = \min\{p, \delta_1, \delta_2\}$. For x satisfying $0 < |x - c| < \delta$, we have

$$L - \epsilon < h(x) \leq f(x) \leq g(x) < L + \epsilon,$$

and thus

$$|f(x) - L| < \epsilon. \quad \square$$

Remark   With straightforward modifications, the pinching theorem holds for one-sided limits. We do not spell out the details here because throughout this section we will be working with two-sided limits.   ❑

We come now to some trigonometric limits. All calculations are based on radian measure.

As our first application of the pinching theorem, we prove that

(2.5.2)

$$\lim_{x \to 0} \sin x = 0.$$

PROOF   To follow the argument, see Figure 2.5.2.[†]

For small $x \neq 0$

$$0 < |\sin x| = \text{length of } \overline{BP} < \text{ length of } \overline{AP} < \text{ length of } \overset{\frown}{AP} = |x|.$$

Thus, for such x

$$0 < |\sin x| < |x|.$$

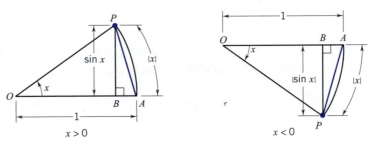

$$x > 0 \qquad\qquad\qquad\qquad x < 0$$

Figure 2.5.2

Since

$$\lim_{x \to 0} 0 = 0 \qquad \text{and} \qquad \lim_{x \to 0} |x| = 0,$$

we know from the pinching theorem that

$$\lim_{x \to 0} |\sin x| = 0 \qquad \text{and therefore} \qquad \lim_{x \to 0} \sin x = 0. \quad ❑$$

From this it follows readily that

(2.5.3)

$$\lim_{x \to 0} \cos x = 1.$$

PROOF   In general, $\cos^2 x + \sin^2 x = 1$. For x close to 0, the cosine is positive and we have

$$\cos x = \sqrt{1 - \sin^2 x}.$$

As x tends to 0, $\sin x$ tends to 0, $\sin^2 x$ tends to 0, and therefore $\cos x$ tends to 1.   ❑

[†]Recall that in a circle of radius 1, a central angle of x radians subtends an arc of length $|x|$.

Next we show that the sine and cosine functions are everywhere continuous; which is to say, for all real numbers c,

(2.5.4)

$$\lim_{x \to c} \sin x = \sin c \quad \text{and} \quad \lim_{x \to c} \cos x = \cos c.$$

PROOF Take any real number c. By (2.2.6) we can write

$$\lim_{x \to c} \sin x \quad \text{as} \quad \lim_{h \to 0} \sin(c + h).$$

This form of the limit suggests that we use the addition formula

$$\sin(c + h) = \sin c \, \cos h + \cos c \, \sin h.$$

Since $\sin c$ and $\cos c$ are constants, we have

$$\lim_{h \to 0} \sin(c + h) = (\sin c)(\lim_{h \to 0} \cos h) + (\cos c)(\lim_{h \to 0} \sin h)$$

$$= (\sin c)(1) + (\cos c)(0) = \sin c.$$

The proof that $\lim_{x \to c} \cos x = \cos c$ is left to you. ❏

The remaining trigonometric functions

$$\tan x = \frac{\sin x}{\cos x}, \qquad \cot x = \frac{\cos x}{\sin x}, \qquad \sec x = \frac{1}{\cos x}, \qquad \csc x = \frac{1}{\sin x}$$

are all continuous where defined. Justification? They are all quotients of continuous functions.

We turn now to two limits, the importance of which will become clear in Chapter 3:

(2.5.5)

$$\lim_{x \to 0} \frac{\sin x}{x} = 1 \quad \text{and} \quad \lim_{x \to 0} \frac{1 - \cos x}{x} = 0.$$

Remark These limits were investigated by numerical methods in Section 2.1, the first in the text, the second in the exercises. ❏

PROOF We show that

$$\lim_{x \to 0} \frac{\sin x}{x} = 1$$

by using some simple geometry and the pinching theorem. For any x satisfying $0 < x \leq \pi/2$ (see Figure 2.5.3), length of $\overline{PB} = \sin x$, length of $\overline{OB} = \cos x$, and length $\overline{OA} = 1$. Since triangle OAQ is a right triangle, $\tan x = \overline{QA}/1 = \overline{QA}$. Now

$$\text{area of triangle } OAP = \frac{1}{2}(1)\sin x = \frac{1}{2}\sin x$$

$$\text{area of sector } OAP = \frac{1}{2}(1)^2 x = \frac{1}{2}x$$

$$\text{area of triangle } OAQ = \frac{1}{2}(1)\tan x = \frac{1}{2}\frac{\sin x}{\cos x}.$$

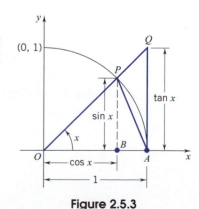

Figure 2.5.3

Since triangle $OAP \subseteq$ sector $OAP \subseteq$ triangle OAQ (and these are all proper containments), we have

$$\frac{1}{2}\sin x < \frac{1}{2}x < \frac{1}{2}\frac{\sin x}{\cos x}$$

$$1 < \frac{x}{\sin x} < \frac{1}{\cos x}$$

$$\cos x < \frac{\sin x}{x} < 1.$$

This inequality was derived for $x > 0$, but since

$$\cos(-x) = \cos x \qquad \text{and} \qquad \frac{\sin(-x)}{-x} = \frac{-\sin x}{-x} = \frac{\sin x}{x},$$

this inequality also holds for $x < 0$.

We can now apply the pinching theorem. Since

$$\lim_{x \to 0} \cos x = 1 \qquad \text{and} \qquad \lim_{x \to 0} 1 = 1,$$

we can conclude that

$$\lim_{x \to 0} \frac{\sin x}{x} = 1.$$

Now let's show that

$$\lim_{x \to 0} \frac{1 - \cos x}{x} = 0.$$

For small $x \neq 0$, $\cos x$ is close to 1 and so $\cos x \neq -1$. Therefore, we can write

$$\frac{1 - \cos x}{x} = \left(\frac{1 - \cos x}{x}\right)\left(\frac{1 + \cos x}{1 + \cos x}\right)^{\dagger}$$

$$= \frac{1 - \cos^2 x}{x(1 + \cos x)}$$

$$= \frac{\sin^2 x}{x(1 + \cos x)}$$

$$= \left(\frac{\sin x}{x}\right)\left(\frac{\sin x}{1 + \cos x}\right).$$

Since

$$\lim_{x \to 0} \frac{\sin x}{x} = 1 \qquad \text{and} \qquad \lim_{x \to 0} \frac{\sin x}{1 + \cos x} = \frac{0}{2} = 0,$$

it follows that

$$\lim_{x \to 0} \frac{1 - \cos x}{x} = 0. \quad \square$$

---

†This "trick" is a fairly common procedure with trigonometric expressions. It is much like using "conjugates" to revise algebraic expressions:

$$\frac{3}{4 + \sqrt{2}} = \frac{3}{4 + \sqrt{2}} \cdot \frac{4 - \sqrt{2}}{4 - \sqrt{2}} = \frac{3(4 - \sqrt{2})}{14}.$$

Remark   The limit statements in (2.5.5) can be generalized as follows:

(2.5.6)

> For each number $a \neq 0$
>
> $$\lim_{x \to 0} \frac{\sin ax}{ax} = 1 \qquad \lim_{x \to 0} \frac{1 - \cos ax}{ax} = 0.$$

Exercise 38.   ❏

We are now in a position to evaluate a variety of trigonometric limits.

Example 1   Find $\displaystyle\lim_{x \to 0} \frac{\sin 4x}{3x}$ and $\displaystyle\lim_{x \to 0} \frac{1 - \cos 2x}{5x}$.

SOLUTION   To calculate the first limit, we "pair off" $\sin 4x$ with $4x$ and use (2.5.6):

$$\frac{\sin 4x}{3x} = \frac{4}{4} \cdot \frac{\sin 4x}{3x} = \frac{4}{3} \cdot \frac{\sin 4x}{4x}.$$

Therefore,

$$\lim_{x \to 0} \frac{\sin 4x}{3x} = \lim_{x \to 0} \left[\frac{4}{3} \cdot \frac{\sin 4x}{4x} \right] = \frac{4}{3} \lim_{x \to 0} \frac{\sin 4x}{4x} = \frac{4}{3}(1) = \frac{4}{3}.$$

The second limit can be obtained the same way:

$$\lim_{x \to 0} \frac{1 - \cos 2x}{5x} = \lim_{x \to 0} \frac{2}{5} \cdot \frac{1 - \cos 2x}{2x} = \frac{2}{5} \lim_{x \to 0} \frac{1 - \cos 2x}{2x} = \frac{2}{5}(0) = 0. \quad ❏$$

Example 2   Find $\displaystyle\lim_{x \to 0} x \cot 3x$.

SOLUTION   We begin by writing

$$x \cot 3x = x \frac{\cos 3x}{\sin 3x} = \frac{1}{3} \left(\frac{3x}{\sin 3x} \right) (\cos \ 3x).$$

Since

$$\lim_{x \to 0} \frac{\sin 3x}{3x} = 1 \qquad \text{gives} \qquad \lim_{x \to 0} \frac{3x}{\sin 3x} = 1,$$

and $\displaystyle\lim_{x \to 0} \cos \ 3x = \cos \ 0 = 1$, we see that

$$\lim_{x \to 0} x \ \cot 3x = \frac{1}{3} \lim_{x \to 0} \left(\frac{3x}{\sin 3x} \right) \lim_{x \to 0} (\cos 3x) = \frac{1}{3}(1)(1) = \frac{1}{3}. \quad ❏$$

Example 3   Find $\displaystyle\lim_{x \to \pi/4} \frac{\sin \left(x - \frac{1}{4}\pi \right)}{\left(x - \frac{1}{4}\pi \right)^2}$.

SOLUTION $\displaystyle\frac{\sin \left(x - \frac{1}{4}\pi \right)}{(x - \frac{1}{4}\pi)^2} = \left[\frac{\sin \left(x - \frac{1}{4}\pi \right)}{(x - \frac{1}{4}\pi)} \right] \cdot \frac{1}{x - \frac{1}{4}\pi}.$

We know that

$$\lim_{x \to \pi/4} \frac{\sin \left(x - \frac{1}{4}\pi \right)}{x - \frac{1}{4}\pi} = 1.$$

Since $\lim\limits_{x\to\pi/4}\left(x-\tfrac{1}{4}\pi\right)=0$, you can see by Theorem 2.3.10 that

$$\lim_{x\to\pi/4}\frac{\sin\left(x-\tfrac{1}{4}\pi\right)}{\left(x-\tfrac{1}{4}\pi\right)^2}\qquad\text{does not exist.}\quad\square$$

Example 4 Find $\lim\limits_{x\to0}\dfrac{x^2}{\sec x-1}$.

SOLUTION The evaluation of this limit requires a little imagination. Since both the numerator and denominator tend to zero as x tends to zero, it is not clear what happens to the fraction. However, we can rewrite the fraction in a more amenable form by multiplying both numerator and denominator by $\sec x+1$.

$$\frac{x^2}{\sec x-1}=\frac{x^2}{\sec x-1}\left(\frac{\sec x+1}{\sec x+1}\right)$$

$$=\frac{x^2(\sec x+1)}{\sec^2 x-1}=\frac{x^2(\sec x+1)}{\tan^2 x}$$

$$=\frac{x^2\cos^2 x(\sec x+1)}{\sin^2 x}$$

$$=\left(\frac{x}{\sin x}\right)^2(\cos^2 x)(\sec x+1).$$

Since each of these factors has a limit as x tends to 0, the fraction we began with has a limit:

$$\lim_{x\to0}\frac{x^2}{\sec x-1}=\lim_{x\to0}\left(\frac{x}{\sin x}\right)^2\cdot\lim_{x\to0}\cos^2 x\cdot\lim_{x\to0}(\sec x+1)=(1)(1)(2)=2.\quad\square$$

EXERCISES 2.5

Exercises 1–32. Evaluate the limits that exist.

1. $\lim\limits_{x\to0}\dfrac{\sin 3x}{x}$.

2. $\lim\limits_{x\to0}\dfrac{2x}{\sin x}$.

3. $\lim\limits_{x\to0}\dfrac{3x}{\sin 5x}$.

4. $\lim\limits_{x\to0}\dfrac{\sin 3x}{2x}$.

5. $\lim\limits_{x\to0}\dfrac{\sin 4x}{\sin 2x}$.

6. $\lim\limits_{x\to0}\dfrac{\sin 3x}{5x}$.

7. $\lim\limits_{x\to0}\dfrac{\sin x^2}{x}$.

8. $\lim\limits_{x\to0}\dfrac{\sin x^2}{x^2}$.

9. $\lim\limits_{x\to0}\dfrac{\sin x}{x^2}$.

10. $\lim\limits_{x\to0}\dfrac{\sin^2 x^2}{x^2}$.

11. $\lim\limits_{x\to0}\dfrac{\sin^2 3x}{5x^2}$.

12. $\lim\limits_{x\to0}\dfrac{\tan^2 3x}{4x^2}$.

13. $\lim\limits_{x\to0}\dfrac{2x}{\tan 3x}$.

14. $\lim\limits_{x\to0}\dfrac{4x}{\cot 3x}$.

15. $\lim\limits_{x\to0}x\csc x$.

16. $\lim\limits_{x\to0}\dfrac{\cos x-1}{2x}$.

17. $\lim\limits_{x\to0}\dfrac{x^2}{1-\cos 2x}$.

18. $\lim\limits_{x\to0}\dfrac{x^2-2x}{\sin 3x}$.

19. $\lim\limits_{x\to0}\dfrac{1-\sec^2 2x}{x^2}$.

20. $\lim\limits_{x\to0}\dfrac{1}{2x\csc x}$.

21. $\lim\limits_{x\to0}\dfrac{2x^2+x}{\sin x}$.

22. $\lim\limits_{x\to0}\dfrac{1-\cos 4x}{9x^2}$.

23. $\lim\limits_{x\to0}\dfrac{\tan 3x}{2x^2+5x}$.

24. $\lim\limits_{x\to0}x^2(1+\cot^2 3x)$.

25. $\lim\limits_{x\to0}\dfrac{\sec x-1}{x\sec x}$.

26. $\lim\limits_{x\to\pi/4}\dfrac{1-\cos x}{x}$.

27. $\lim\limits_{x\to\pi/4}\dfrac{\sin x}{x}$.

28. $\lim\limits_{x\to0}\dfrac{\sin^2 x}{x(1-\cos x)}$.

29. $\lim\limits_{x\to\pi/2}\dfrac{\cos x}{x-\tfrac{1}{2}\pi}$.

30. $\lim\limits_{x\to\pi}\dfrac{\sin x}{x-\pi}$.

31. $\lim\limits_{x\to\pi/4}\dfrac{\sin\left(x+\tfrac{1}{4}\pi\right)-1}{x-\tfrac{1}{4}\pi}$. HINT: $x+\tfrac{1}{4}\pi=x-\tfrac{1}{4}\pi+\tfrac{1}{2}\pi$.

32. $\lim\limits_{x\to\pi/6}\dfrac{\sin\left(x+\tfrac{1}{3}\pi\right)-1}{x-\tfrac{1}{6}\pi}$.

33. Show that $\lim\limits_{x\to c}\cos x=\cos c$ for all real numbers c.

Exercises 34–37. Evaluate the limit, taking a and b as nonzero constants.

34. $\lim\limits_{x \to 0} \dfrac{\sin ax}{bx}$.

35. $\lim\limits_{x \to 0} \dfrac{1 - \cos ax}{bx}$.

36. $\lim\limits_{x \to 0} \dfrac{\sin ax}{\sin bx}$.

37. $\lim\limits_{x \to 0} \dfrac{\cos ax}{\cos bx}$.

38. Show that

if $\lim\limits_{x \to 0} f(x) = L$, then $\lim\limits_{x \to 0} f(ax) = L$ for each $a \neq 0$.

HINT: Let $\epsilon > 0$. If $\delta_1 > 0$ "works" for the first limit, then $\delta = \delta_1 / |a|$ "works" for the second limit.

Exercises 39–42. Evaluate $\lim\limits_{h \to 0} [f(c + h) - f(c)]/h$.

39. $f(x) = \sin x$, $c = \pi/4$. HINT: Use the addition formula for the sine function.

40. $f(x) = \cos x$, $c = \pi/3$.

41. $f(x) = \cos 2x$, $c = \pi/6$.

42. $f(x) = \sin 3x$, $c = \pi/2$.

43. Show that $\lim\limits_{x \to 0} x \sin(1/x) = 0$. HINT: Use the pinching theorem.

44. Show that $\lim\limits_{x \to \pi} (x - \pi) \cos^2 [1/(x - \pi)] = 0$.

45. Show that $\lim\limits_{x \to 1} |x - 1| \sin x = 0$.

46. Let f be the Dirichlet function

$$f(x) = \begin{cases} 1, & x \text{ rational} \\ 0, & x \text{ irrational.} \end{cases}$$

Show that $\lim\limits_{x \to 0} x f(x) = 0$.

47. Prove that if there is a number B such that $|f(x)| \leq B$ for all $x \neq 0$, then $\lim\limits_{x \to 0} x f(x) = 0$. NOTE: Exercises 43–46 are special cases of this general result.

48. Prove that if there is a number B such that $|f(x)/x| \leq B$ for all $x \neq 0$, then $\lim\limits_{x \to 0} f(x) = 0$.

49. Prove that if there is a number B such that $|f(x) - L|/|x - c| \leq B$ for all $x \neq c$, then $\lim\limits_{x \to c} f(x) = L$.

50. Given that $\lim\limits_{x \to c} f(x) = 0$ and $|g(x)| \leq B$ for all x in an interval $(c - p, c + p)$, prove that

$$\lim\limits_{x \to c} f(x) g(x) = 0.$$

Exercises 51–52. Use the limit utility in a CAS to evaluate the limit.

51. $\lim\limits_{x \to 0} \dfrac{20x - 15x^2}{\sin 2x}$.

52. $\lim\limits_{x \to 0} \dfrac{\tan x}{x^2}$.

53. Use a graphing utility to plot $f(x) = \dfrac{x}{\tan 3x}$ on $[-0.2, 0.2]$. Estimate $\lim\limits_{x \to 0} f(x)$; use the zoom function if necessary. Verify your result analytically.

54. Use a graphing utility to plot $f(x) = \dfrac{\tan x}{\tan x + x}$ on $[-0.2, 0.2]$. Estimate $\lim\limits_{x \to 0} f(x)$; use the zoom function if necessary. Verify your result analytically.

■ 2.6 TWO BASIC THEOREMS

A function which is continuous on an interval does not "skip" any values, and thus its graph is an "unbroken curve." There are no "holes" in it and no "jumps." This idea is expressed coherently by the *intermediate-value theorem*.

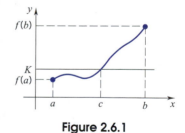

Figure 2.6.1

THEOREM 2.6.1 THE INTERMEDIATE-VALUE THEOREM
If f is continuous on $[a, b]$ and K is any number between $f(a)$ and $f(b)$, then there is at least one number c in the interval (a, b) such that $f(c) = K$.

We illustrate the theorem in Figure 2.6.1. What can happen in the discontinuous case is illustrated in Figure 2.6.2. There the number K has been "skipped."

It's a small step from the intermediate-value theorem to the following observation:

"continuous functions map intervals onto intervals."

A proof of the intermediate-value theorem is given in Appendix B. We will assume the result and illustrate its usefulness.

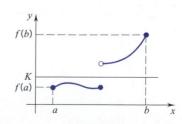

Figure 2.6.2

Here we apply the theorem to the problem of locating the zeros of a function. In particular, suppose that the function f is continuous on $[a, b]$ and that either

$$f(a) < 0 < f(b) \qquad \text{or} \qquad f(b) < 0 < f(a) \qquad \text{(Figure 2.6.3)}$$

Then, by the intermediate-value theorem, we know that the equation $f(x) = 0$ has at least one root between a and b.

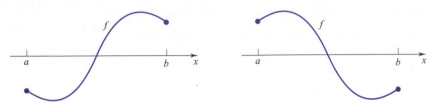

Figure 2.6.3

Example 1   We set $f(x) = x^2 - 2$. Since $f(1) = -1 < 0$ and $f(2) = 2 > 0$, there exists a number c between 1 and 2 such that $f(c) = 0$. Since f increases on $[1, 2]$, there is only one such number. This is the number we call $\sqrt{2}$.

So far we have shown only that $\sqrt{2}$ lies between 1 and 2. We can locate $\sqrt{2}$ more precisely by evaluating f at 1.5, the midpoint of the interval $[1, 2]$. Since $f(1.5) = 0.25 > 0$ and $f(1) < 0$, we know that $\sqrt{2}$ lies between 1 and 1.5. We now evaluate f at 1.25, the midpoint of $[1, 1.5]$. Since $f(1.25) \cong -0.438 < 0$ and $f(1.5) > 0$, we know that $\sqrt{2}$ lies between 1.25 and 1.5. Our next step is to evaluate f at 1.375, the midpoint of $[1.25, 1.5]$. Since $f(1.375) \cong -0.109 < 0$ and $f(1.5) > 0$, we know that $\sqrt{2}$ lies between 1.375 and 1.5. We now evaluate f at 1.4375, the midpoint of $[1.375, 1.5]$. Since $f(1.4375) \cong 0.066 > 0$ and $f(1.375) < 0$, we know that $\sqrt{2}$ lies between 1.375 and 1.4375. The average of these two numbers, rounded off to two decimal places, is 1.41. A calculator gives $\sqrt{2} \cong 1.4142$. So we are not far off.   ❏

The method used in Example 1 is called the *bisection method*. It can be used to locate the roots of a wide variety of equations. The more bisections we carry out, the more accuracy we obtain.

As you will see in the exercise section, the intermediate-value theorem gives us results that are otherwise elusive, but, as our next example makes clear, the theorem has to be applied with some care.

Example 2   The function $f(x) = 2/x$ takes on the value -2 at $x = -1$ and it takes on the value 2 at $x = 1$. Certainly 0 lies between -2 and 2. Does it follow that f takes on the value 0 somewhere between -1 and 1? No: the function is not continuous on $[-1, 1]$, and therefore it can and does skip the number 0.   ❏

Boundedness; Extreme Values

A function f is said to be *bounded* or *unbounded* on a set I in the sense in which the set of values taken on by f on the set I is bounded or unbounded.

For example, the sine and cosine functions are bounded on $(-\infty, \infty)$:

$$-1 \le \sin x \le 1 \qquad \text{and} \qquad -1 \le \cos x \le 1 \qquad \text{for all } x \in (-\infty, \infty).$$

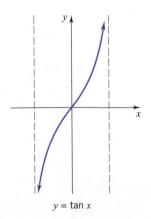

$y = \tan x$

Figure 2.6.4

Both functions map $(-\infty, \infty)$ onto $[-1, 1]$.

The situation is markedly different in the case of the tangent. (See Figure 2.6.4.) The tangent function is bounded on $[0, \pi/4]$; on $[0, \pi/2]$ it is bounded below but not bounded above; on $(-\pi/2, 0]$ it is bounded above but not bounded below; on $(-\pi/2, \pi/2)$ it is unbounded both below and above.

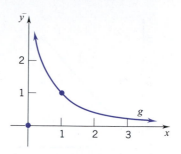

Example 3 Let

$$g(x) = \begin{cases} 1/x^2, & x > 0 \\ 0, & x = 0. \end{cases}$$

(Figure 2.6.5)

It is clear that g is unbounded on $[0, \infty)$. (It is unbounded above.) However, it is bounded on $[1, \infty)$. The function maps $[0, \infty)$ onto $[0, \infty)$, and it maps $[1, \infty)$ onto $(0, 1]$. ❏

Figure 2.6.5

A function may take on a maximum value; it may take on a minimum value; it may take on both a maximum value and a minimum value; it may take on neither.

Here are some simple examples:

$$f(x) = \begin{cases} 1, & x \text{ rational} \\ 0, & x \text{ irrational} \end{cases}$$

takes on both a maximum value (the number 1) and a minimum value (the number 0) on every interval of the real line.

The function

$$f(x) = x^2 \qquad x \in (0, 5]$$

takes on a maximum value (the number 25), but it has no minimum value.

The function

$$f(x) = \frac{1}{x}, \qquad x \in (0, \infty)$$

has no maximum value and no minimum value.

For a function continuous on a bounded closed interval, the existence of both a maximum value and a minimum value is guaranteed. The following theorem is fundamental.

THEOREM 2.6.2 THE EXTREME-VALUE THEOREM

If f is continuous on a bounded closed interval $[a, b]$, then on that interval f takes on both a maximum value M and a minimum value m.

For obvious reasons, M and m are called the *extreme values* of the function.

The result is illustrated in Figure 2.6.6. The maximum value M is taken on at the point marked d, and the minimum value m is taken on at the point marked c.

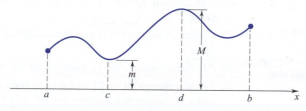

Figure 2.6.6

In Theorem 2.6.2, the full hypothesis is needed. If the interval is not bounded, the result need not hold: the cubing function $f(x) = x^3$ has no maximum on the interval $[0, \infty)$. If the interval is not closed, the result need not hold: the identity function

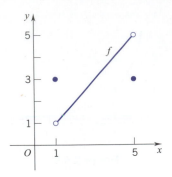

Figure 2.6.7

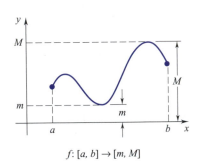

$f: [a, b] \to [m, M]$

Figure 2.6.8

$f(x) = x$ has no maximum and no minimum on $(0, 2)$. If the function is not continuous, the result need not hold. As an example, take the function

$$f(x) = \begin{cases} 3, & x = 1 \\ x, & 1 < x < 5 \\ 3, & x = 5. \end{cases}$$

The graph is shown in Figure 2.6.7. The function is defined on $[1, 5]$, but it takes on neither a maximum value nor a minimum value. The function maps the closed interval $[1, 5]$ onto the open interval $(1, 5)$.

One final observation. From the intermediate-value theorem we know that

"continuous functions map intervals onto intervals."

Now that we have the extreme-value theorem, we know that

"continuous functions map bounded closed intervals $[a, b]$ onto bounded closed intervals $[m, M]$." (See Figure 2.6.8.)

Of course, if f is constant, then $M = m$ and the interval $[m, M]$ collapses to a point.

A proof of the extreme-value theorem is given in Appendix B. Techniques for determining the maximum and minimum values of functions are developed in Chapter 4. These techniques require an understanding of "differentiation," the subject to which we devote Chapter 3.

EXERCISES 2.6

Exercises 1–8. Use the intermediate-value theorem to show that there is a solution of the given equation in the indicated interval.

1. $2x^3 - 4x^2 + 5x - 4 = 0$; $[1, 2]$.
2. $x^4 - x - 1 = 0$; $[-1, 1]$.
3. $\sin x + 2 \cos x - x^2 = 0$; $[0, \pi/2]$.
4. $2 \tan x - x = 1$; $[0, \pi/4]$.
5. $x^2 - 2 + \dfrac{1}{2x} = 0$; $[\tfrac{1}{4}, 1]$.
6. $x^{5/3} + x^{1/3} = 1$; $[-1, 1]$.
7. $x^3 = \sqrt{x + 2}$; $[1, 2]$.
8. $\sqrt{x^2 - 3x - 2} = 0$; $[3, 5]$.
9. Let $f(x) = x^5 - 2x^2 + 5x$. Show that there is a number c such that $f(c) = 1$.
10. Let $f(x) = \dfrac{1}{x - 1} + \dfrac{1}{x - 4}$. Show that there is a number $c \in (1, 4)$ such that $f(c) = 0$.
11. Show that the equation $x^3 - 4x + 2 = 0$ has three distinct roots in $[-3, 3]$ and locate the roots between consecutive integers.
12. Use the intermediate-value theorem to prove that there exists a positive number c such that $c^3 = 2$.

Exercises 13–24. Sketch the graph of a function f that is defined on $[0, 1]$ and meets the given conditions (if possible).

13. f is continuous on $[0, 1]$, minimum value 0, maximum value $\tfrac{1}{2}$.
14. f is continuous on $[0, 1)$, minimum value 0, no maximum value.
15. f is continuous on $(0, 1)$, takes on the values 0 and 1, but does not take on the value $\tfrac{1}{2}$.
16. f is continuous on $[0, 1]$, takes on the values -1 and 1, but does not take on the value 0.
17. f is continuous on $[0, 1]$, maximum value 1, minimum value 1.
18. f is continuous on $[0, 1]$ and nonconstant, takes on no integer values.
19. f is continuous on $[0, 1]$, takes on no rational values.
20. f is not continuous on $[0, 1]$, takes on both a maximum value and a minimum value and every value in between.
21. f is continuous on $(0, 1)$, takes on only two distinct values.
22. f is continuous on $(0, 1)$, takes on only three distinct values.
23. f is continuous on $(0, 1)$, and the range of f is an unbounded interval.

24. f is continuous on $[0, 1]$, and the range of f is an unbounded interval.

25. (*Fixed-point property*) Show that if f is continuous on $[0, 1]$ and $0 \le f(x) \le 1$ for all $x \in [0, 1]$, then there exists at least one point c in $[0, 1]$ at which $f(c) = c$. HINT: Apply the intermediate-value theorem to the function $g(x) = x - f(x)$.

26. Given that f and g are continuous on $[a, b]$, that $f(a) < g(a)$, and $g(b) < f(b)$, show that there exists at least one number c in (a, b) such that $f(c) = g(c)$. HINT: Consider $f(x) - g(x)$.

27. From Exercise 25 we know that if f is continuous on $[0, 1]$ and $0 \le f(x) \le 1$ for all $x \in [0, 1]$, then the graph of f intersects the diagonal of the unit square that joins the vertices $(0, 0)$ and $(1, 1)$. (See the figure.) Show that under these conditions

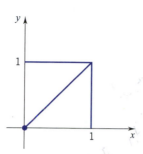

(a) the graph of f also intersects the other diagonal of the unit square
(b) and, more generally, if g is continuous on $[0, 1]$ with $g(0) = 0$ and $g(1) = 1$, or with $g(0) = 1$ and $g(1) = 0$, then the graph of f intersects the graph of g.

28. Use the intermediate-value theorem to prove that every real number has a cube root. That is, prove that for any real number a there exists a number c such that $c^3 = a$.

29. The intermediate-value theorem can be used to prove that each polynomial equation of odd degree

$$x^n + a_{n-1}x^{n-1} + \cdots + a_1x + a_0 = 0 \quad \text{with } n \text{ odd}$$

has at least one real root. Show that the cubic equation

$$x^3 + ax^2 + bx + c = 0$$

has at least one real root.

30. Let n be a positive integer.
(a) Prove that if $0 \le a < b$, then $a^n < b^n$.
HINT: Use mathematical induction.
(b) Prove that every nonnegative real number x has a unique nonnegative nth root $x^{1/n}$.
HINT: The existence of $x^{1/n}$ can be seen by applying the intermediate-value theorem to the function $f(t) = t^n$ for $t \ge 0$. The uniqueness follows from part (a).

31. The temperature T (in °C) at which water boils depends on the elevation above sea level. The formula

$$T(h) = 100.862 - 0.0415\sqrt{h + 431.03}$$

gives the approximate value of T as a function of the elevation h measured in meters. Use the intermediate-value theorem to show that water boils at about 98°C at an elevation of between 4000 and 4500 meters.

32. Assume that at any given instant, the temperature on the earth's surface varies continuously with position. Prove that there is at least one pair of points diametrically opposite each other on the equator where the temperature is the same. HINT: Form a function that relates the temperature at diametrically opposite points of the equator.

33. Let $\mathcal{C}$ denote the set of all circles with radius less than or equal to 10 inches. Prove that there is at least one member of $\mathcal{C}$ with an area of exactly 250 square inches.

34. Fix a positive number P. Let $\mathcal{R}$ denote the set of all rectangles with perimeter P. Prove that there is a member of $\mathcal{R}$ that has maximum area. What are the dimensions of the rectangle of maximum area? HINT: Express the area of an arbitrary element of $\mathcal{R}$ as a function of the length of one of the sides.

35. Given a circle C of radius R. Let $\mathcal{F}$ denote the set of all rectangles that can be inscribed in C. Prove that there is a member of $\mathcal{F}$ that has maximum area.

▷ **Exercises 36–39.** Use the intermediate-value theorem to estimate the location of the zeros of the function. Then use a graphing utility to approximate these zeros to within 0.001.

36. $f(x) = 2x^3 + 4x - 4$.

37. $f(x) = x^3 - 5x + 3$.

38. $f(x) = x^5 - 3x + 1$.

39. $f(x) = x^3 - 2\sin x + \frac{1}{2}$.

▷ **Exercises 40–43.** Determine whether the function f satisfies the hypothesis of the intermediate-value theorem on the interval $[a, b]$. If it does, use a graphing utility or a CAS to find a number c in (a, b) such that $f(c) = \frac{1}{2}[f(a) + f(b)]$.

40. $f(x) = \dfrac{x+1}{x^2+1}; \quad [-2, 3]$.

41. $f(x) = \dfrac{4x+3}{(x-1)}; \quad [-3, 2]$.

42. $f(x) = \sec x; \quad [-\pi, 2\pi]$.

43. $f(x) = \sin x - 3\cos 2x; \quad [\pi/2, 2\pi]$

▷ **Exercises 44–47.** Use a graphing utility to graph f on the given interval. Is f bounded? Does it have extreme values? If so, what are these extreme values?

44. $f(x) = \dfrac{x^3 - 8x + 6}{4x + 1}; \quad [0, 5]$.

45. $f(x) = \dfrac{2x}{1 + x^2}; \quad [-2, 2]$.

46. $f(x) = \dfrac{\sin 2x}{x^2}; \quad [-\pi/2, \pi/2]$.

47. $f(x) = \dfrac{1 - \cos x}{x^2}; \quad [-2, 2]$.

■ PROJECT 2.6 The Bisection Method for Finding the Roots of $f(x) = 0$

If the function f is continuous on $[a, b]$, and if $f(a)$ and $f(b)$ have opposite signs, then, by the intermediate-value theorem, the equation $f(x) = 0$ has at least one root in (a, b). For simplicity, let's assume that there is only one such root and call it c. How can we estimate the location of c? The intermediate-value theorem itself gives us no clue. The simplest method for approximating c is called the *bisection method*. It is an iterative process—a basic step is iterated (carried out repeatedly) until c is approximated with as much accuracy as we wish.

It is standard practice to label the elements of successive approximations by using subscripts $n = 1, 2, 3$, and so forth. We begin by setting $u_1 = a$ and $v_1 = b$. Now bisect $[u_1, v_1]$. If c is the midpoint of $[u_1, v_1]$, then we are done. If not, then c lies in one of the halves of $[u_1, v_1]$. Call it $[u_2, v_2]$. If c is the midpoint

of $[u_2, v_2]$, then we are done. If not, then c lies in one of the halves of $[u_2, v_2]$. Call that half $[u_3, v_3]$ and continue. The first three iterations for a particular function are shown in the figure.

After n bisections, we are examining the midpoint m_n of the interval $[u_n, v_n]$. Therefore, we can be certain that

$$|c - m_n| \leq \frac{1}{2}(v_n - u_n) = \frac{1}{2}\left(\frac{v_{n-1} - u_{n-1}}{2}\right) = \cdots = \frac{b-a}{2^n}.$$

Thus, m_n approximates c to within $(b-a)/2^n$. If we want m_n to approximate c to within a given number ϵ, then we must carry out the iteration to the point where

$$\frac{b-a}{2^n} < \epsilon.$$

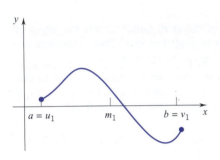
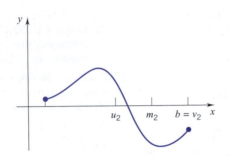
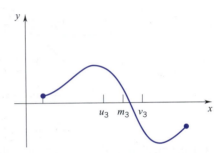

Problem 1. In Example 1 we used the function $f(x) = x^2 - 2$ and the bisection method to obtain an estimate of $\sqrt{2}$ accurate to within two decimal places.

(a) Suppose we want a numerical estimate accurate to within 0.001. How many iterations would be required to achieve this accuracy?

(b) How many iterations would be required to obtain a numerical estimate accurate to within 0.0001? 0.00001?

Problem 2. The function $f(x) = x^3 + x - 9$ has one zero c. Locate c between two consecutive integers.

(a) How many iterations of the bisection method would be required to approximate c to within 0.01? Use the bisection method to approximate c to within 0.01.

(b) How many iterations would be required to approximate c to within 0.001? 0.0001?

Problem 3. The function $f(x) = \sin x + x + 3$ has one zero c. Locate c between two consecutive integers.

(a) How many iterations of the bisection method would be required to approximate c to within 0.01? Use the bisection method to approximate c to within 0.01.

(b) How many iterations would be required to approximate c to within 0.00001? 0.000001?

The following modification of the bisection method is sometimes used. Suppose that the function f, continuous on $[a, b]$, has exactly one zero c in the interval (a, b). The line connecting $(a, f(a))$ and $(b, f(b))$ is drawn and the x-intercept is used as the first approximation for c instead of the midpoint of the interval. The process of bisection is then applied and continued until the desired degree of accuracy is obtained.

Problem 4. Carry out three iterations of the modified bisection method for the functions given in Problems 1, 2, and 3. How does this method compare with the bisection method in terms of the rate at which the approximations converge to the zero c?

■ **CHAPTER 2. REVIEW EXERCISES**

Exercises 1–30. State whether the limit exists; evaluate the limit if it does exist.

1. $\lim\limits_{x\to 3} \dfrac{x^2-3}{x+3}$.

2. $\lim\limits_{x\to 2} \dfrac{x^2+4}{x^2+2x+1}$.

3. $\lim\limits_{x\to 3} \dfrac{(x-3)^2}{x+3}$.

4. $\lim\limits_{x\to 3} \dfrac{x^2-9}{x^2-5x+6}$.

5. $\lim\limits_{x\to 2^+} \dfrac{x-2}{|x-2|}$.

6. $\lim\limits_{x\to -2} \dfrac{|x|}{x-2}$.

7. $\lim\limits_{x\to 0}\left(\dfrac{1}{x}-\dfrac{1-x}{x}\right)$.

8. $\lim\limits_{x\to 3^+} \dfrac{\sqrt{x-3}}{|x-3|}$.

9. $\lim\limits_{x\to 1} \dfrac{|x-1|}{x}$.

10. $\lim\limits_{x\to 1} \dfrac{x^3-1}{|x^3-1|}$.

11. $\lim\limits_{x\to 1} \dfrac{\sqrt{x}-1}{x-1}$.

12. $\lim\limits_{x\to 3^+} \dfrac{x^2-2x-3}{\sqrt{x-3}}$.

13. $\lim\limits_{x\to 3^+} \dfrac{\sqrt{x^2-2x-3}}{x-3}$.

14. $\lim\limits_{x\to 4} \dfrac{\sqrt{x+5}-3}{x-4}$.

15. $\lim\limits_{x\to 2} \dfrac{x^3-8}{x^4-3x^2-4}$.

16. $\lim\limits_{x\to 0} \dfrac{5x}{\sin 2x}$.

17. $\lim\limits_{x\to 0} \dfrac{\tan^2 2x}{3x^2}$.

18. $\lim\limits_{x\to 0} x\csc 4x$.

19. $\lim\limits_{x\to 0} \dfrac{x^2-3x}{\tan x}$.

20. $\lim\limits_{x\to \pi/2} \dfrac{\cos x}{2x-\pi}$.

21. $\lim\limits_{x\to 0} \dfrac{\sin 3x}{5x^2-4x}$.

22. $\lim\limits_{x\to 0} \dfrac{5x^2}{1-\cos 2x}$.

23. $\lim\limits_{x\to -\pi} \dfrac{x+\pi}{\sin x}$.

24. $\lim\limits_{x\to 2} \dfrac{x^2-4}{x^3-8}$.

25. $\lim\limits_{x\to 2^-} \dfrac{x-2}{|x^2-4|}$.

26. $\lim\limits_{x\to 2} \dfrac{1-2/x}{1-4/x^2}$.

27. $\lim\limits_{x\to 1^+} \dfrac{x^2-3x+2}{\sqrt{x-1}}$.

28. $\lim\limits_{x\to 3} \dfrac{1-9/x^2}{1+3x}$.

29. $\lim\limits_{x\to 2} f(x)$ if $f(x)=\begin{cases} x+1, & x<1 \\ 3x-x^2, & x>1. \end{cases}$

30. $\lim\limits_{x\to -2} f(x)$ if $f(x)=\begin{cases} 3+x, & x<-2 \\ 5, & x=-2 \\ x^2-3, & x>-2. \end{cases}$

31. Let f be some function for which you know only that

if $0<|x-2|<1$, then $|f(x)-4|<0.1$.

Which of the following statements are necessarily true?

(a) If $0\le|x-2|<1$, then $|f(x)-4|\le 0.1$.

(b) If $0<|x-2|<\frac{1}{2}$, then $|f(x)-4|<0.05$.

(c) If $0\le|x-2.5|<0.2$, then $|f(x)-4|<0.1$.

(d) If $0<|x-1.5|<1$, then $|f(x)-4|<0.1$.

(e) If $\lim\limits_{x\to 2} f(x)=L$, then $3.9\le L\le 4.1$.

32. Find a number a for which $\lim\limits_{x\to 3} \dfrac{2x^2-3ax+x-a-1}{x^2-2x-3}$ exists and then evaluate the limit.

33. (a) Sketch the graph of

$$f(x)=\begin{cases} 3x+4, & x<-1 \\ -2x-2, & -1<x<2 \\ 2x, & x>2 \\ x^2, & x=-1,2. \end{cases}$$

(b) Evaluate the limits that exist.

(i) $\lim\limits_{x\to -1^-} f(x)$.

(ii) $\lim\limits_{x\to -1^+} f(x)$.

(iii) $\lim\limits_{x\to -1} f(x)$.

(iv) $\lim\limits_{x\to 2^-} f(x)$.

(v) $\lim\limits_{x\to 2^+} f(x)$.

(vi) $\lim\limits_{x\to 2} f(x)$.

(c) (i) Is f continuous from the left at -1? Is f continuous from the right at -1?

(ii) Is f continuous from the left at 2? Is f continuous from the right at 2?

34. (a) Does $\lim\limits_{x\to 0}\cos\left(\dfrac{1-\cos x}{2x}\right)$ exist? If so, what is the limit?

(b) Does $\lim\limits_{x\to 0}\cos\left(\dfrac{\pi\sin x}{2x}\right)$ exist? If so, what is the limit?

35. Set $f(x)=\begin{cases} 2x^2-1, & x<2 \\ A, & x=2 \\ x^3-2Bx, & x>2. \end{cases}$ For what values of A and B is f continuous at 2?

36. Give necessary and sufficient conditions on A and B for the function

$$f(x)=\begin{cases} Ax+B, & x<-1 \\ 2x, & -1\le x\le 2 \\ 2Bx-A, & 2<x \end{cases}$$

to be continuous at $x=-1$ but discontinuous at $x=2$.

Exercises 37–40. The function f is continuous everywhere except at a. If possible, define f at a so that it becomes continuous at a.

37. $f(x)=\dfrac{x^3-2x-15}{x+3}$, $a=-3$.

38. $f(x)=\dfrac{\sqrt{x+1}-2}{x-3}$, $a=3$.

39. $f(x)=\dfrac{\sin\pi x}{x}$, $a=0$.

40. $f(x)=\dfrac{1-\cos x}{x^2}$, $a=0$.

41. A function f is defined on the interval $[a,b]$. Which of the following statements are necessarily true?

(a) If $f(a)>0$ and $f(b)<0$, then there must exist at least one number c in (a,b) for which $f(c)=0$.

(b) If f is continuous on $[a, b]$ with $f(a) < 0$ and $f(b) > 0$, then there must exist at least one number c in (a, b) for which $f(c) = 0$.

(c) If f is continuous on (a, b) with $f(a) > 0$ and $f(b) < 0$, then there must exist at least one number c in (a, b), for which $f(c) = 0$.

(d) If f is continuous on $[a, b]$ with $f(c) = 0$ for some number c in (a, b), then $f(a)$ and $f(b)$ have opposite signs.

Exercises 42–43. Use the intermediate-value theorem to show that the equation has a solution in the interval specified.

42. $x^3 - 3x - 4 = 0$, $[2, 3]$.

43. $2\cos x - x + 1$, $[1, 2]$.

Exercises 44–46. Give an ϵ, δ proof for each statement.

44. $\lim_{x \to 2}(5x - 4) = 6$.

45. $\lim_{x \to -4} |2x + 5| = 3$.

46. $\lim_{x \to 9} \sqrt{x - 5} = 2$.

47. Prove that if $\lim_{x \to 0}[f(x)/x]$ exists, then $\lim_{x \to 0} f(x) = 0$.

48. Prove that if $\lim_{x \to c} g(x) = l$ and if f is continuous at l, then $\lim_{x \to c} f(g(x)) = f(l)$. HINT: See Theorem 2.4.4.

49. A function f is continuous at all $x \geq 0$. Can f take on the value zero at and only

(a) at the positive integers?

(b) at the reciprocals of the positive integers?

If the answer is yes, sketch a figure that supports your answer; if the answer is no, prove it.

50. Two functions f and g are everywhere defined. Can they both be everywhere continuous

(a) if they differ only at a finite number of points?

(b) if they differ only on a bounded closed interval $[a, b]$?

(c) if they differ only on a bounded open interval (a, b)?

Justify your answers.

CHAPTER 3

THE DERIVATIVE;
THE PROCESS OF
DIFFERENTIATION

Introduction

We begin with a function f. On the graph of f we choose a point $(x, f(x))$ and a nearby point $(x + h, f(x + h))$. (See Figure 3.1.1.) Through these two points we draw a line. We call this line a *secant line* because it cuts through the graph of f.[†] The figure shows this secant line first with $h > 0$ and then with $h < 0$.

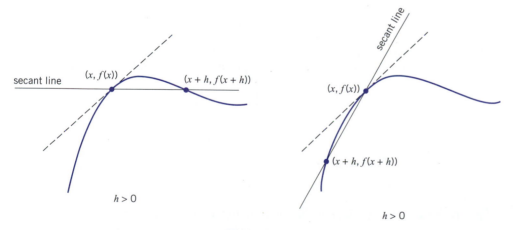

Figure 3.1.1

Whether h is positive or negative, the slope of the secant line is the *difference quotient*

$$\frac{f(x + h) - f(x)}{h}.$$
(Check this out.)

[†] The word "secant" comes from the Latin "secare," to cut.

If we let h tend to zero (from one side or the other), then ideally the point $(x + h, f(x + h))$ slides along the curve toward $(x, f(x))$, $x + h$ tends to x, $f(x + h)$ tends to $f(x)$, and the slope of the secant

$$(*) \qquad \frac{f(x + h) - f(x)}{h}$$

tends to a limit that we denote by $f'(x)^\dagger$. While $(*)$ represents the slope of the approaching secant, the number $f'(x)$ represents the slope of the graph at the point $(x, f(x))$.

What we call "differential calculus" is the implementation of this idea.

Derivatives and Differentiation

> **DEFINITION 3.1.1**
>
> A function f is said to be differentiable at x if
>
> $$\lim_{h \to 0} \frac{f(x + h) - f(x)}{h} \qquad \text{exists.}$$
>
> If this limit exists, it is called the *derivative of f at x* and is denoted by $f'(x)$.

As indicated in the introduction, the derivative

$$f'(x) = \lim_{h \to 0} \frac{f(x + h) - f(x)}{h}$$

represents *slope of the graph of f at the point* $(x, f(x))$. The line that passes through the point $(x, f(x))$ with slope $f'(x)$ is called the *tangent line* at the point $(x, f(x))$. (This line is marked by dashes in Figure 3.1.1.)

Example 1   We begin with a linear function

$$f(x) = mx + b .$$

The graph of this function is the line $y = mx + b$, a line with constant slope m. We therefore expect $f'(x)$ to be constantly m. Indeed it is: for $h \neq 0$,

$$\frac{f(x + h) - f(x)}{h} = \frac{[m(x + h) + b] - [mx + b]}{h} = \frac{mh}{h} = m$$

and therefore

$$f'(x) = \lim_{h \to 0} \frac{f(x + h) - f(x)}{h} = \lim_{h \to 0} m = m. \quad \square$$

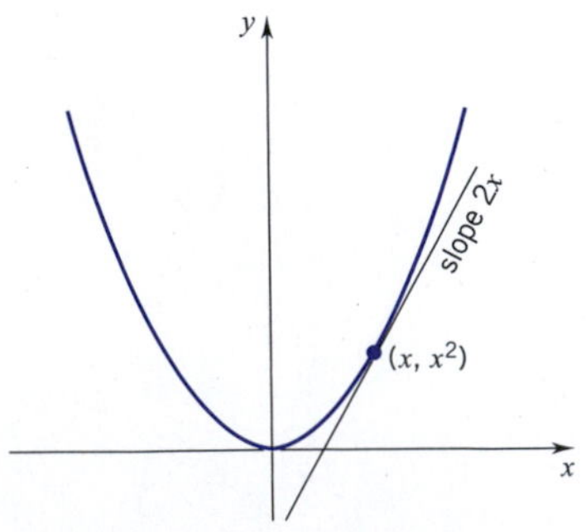

Example 2   Now we look at the squaring function

$$f(x) = x^2. \qquad \text{(Figure 3.1.2)}$$

To find $f'(x)$, we form the difference quotient

$$\frac{f(x + h) - f(x)}{h} = \frac{(x + h)^2 - x^2}{h}$$

Figure 3.1.2

$\dagger$ This prime notation goes back to the French mathematician Joseph-Louis Lagrange (1736–1813). Other notations are introduced later.

and take the limit as $h \to 0$. Since

$$\frac{(x+h)^2 - x^2}{h} = \frac{(x^2 + 2xh + h^2) - x^2}{h} = \frac{2xh + h^2}{h} = 2x + h,$$

we have

$$\frac{f(x+h) - f(x)}{h} = 2x + h.$$

Therefore

$$f'(x) = \lim_{h \to 0} \frac{f(x+h) - f(x)}{h} = \lim_{h \to 0} (2x + h) = 2x.$$

The slope of the graph changes with x. For $x < 0$, the slope is negative and the curve tends down; at $x = 0$, the slope is 0 and the tangent line is horizontal; for $x > 0$, the slope is positive and the curve tends up. ❏

Example 3   Here we look for $f'(x)$ for the square-root function

$$f(x) = \sqrt{x}, \qquad x \geq 0. \qquad \text{(Figure 3.1.3)}$$

Since $f'(x)$ is a two-sided limit, we can expect a derivative at most for $x > 0$.

We take $x > 0$ and form the difference quotient

$$\frac{f(x+h) - f(x)}{h} = \frac{\sqrt{x+h} - \sqrt{x}}{h}.$$

We simplify this expression by multiplying both numerator and denominator by $\sqrt{x+h} + \sqrt{x}$. This gives us

$$\frac{f(x+h) - f(x)}{h} = \left(\frac{\sqrt{x+h} - \sqrt{x}}{h} \right) \left(\frac{\sqrt{x+h} + \sqrt{x}}{\sqrt{x+h} + \sqrt{x}} \right)$$

$$= \frac{(x+h) - x}{h(\sqrt{x+h} + \sqrt{x})} = \frac{1}{\sqrt{x+h} + \sqrt{x}}.$$

It follows that

$$f'(x) = \lim_{h \to 0} \frac{f(x+h) - f(x)}{h} = \lim_{h \to 0} \frac{1}{\sqrt{x+h} + \sqrt{x}} = \frac{1}{2\sqrt{x}}.$$

At each point of the graph to the right of the origin the slope is positive. As x increases, the slope diminishes and the graph flattens out. ❏

The *derivative* f' is a function, its value at x being $f'(x)$. However, this function f' is defined only at those numbers x where f is differentiable. As you just saw in Example 3, while the square-root function is defined on $[0, \infty)$, its derivative f' is defined only on $(0, \infty)$:

$$f(x) = \sqrt{x} \quad \text{for all } x \geq 0; \qquad f'(x) = \frac{1}{2\sqrt{x}} \quad \text{only for } x > 0.$$

To *differentiate* a function f is to find its derivative f'.

Example 4   Let's differentiate the reciprocal function

$$f(x) = \frac{1}{x}. \qquad \text{(Figure 3.1.4)}$$

We begin by forming the difference quotient

$$\frac{f(x+h) - f(x)}{h} = \frac{\dfrac{1}{x+h} - \dfrac{1}{x}}{h}.$$

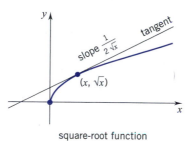

square-root function

Figure 3.1.3

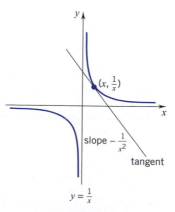

$y = \frac{1}{x}$

Figure 3.1.4

Now we simplify:

$$\frac{\frac{1}{x+h} - \frac{1}{x}}{h} = \frac{\frac{x}{x(x+h)} - \frac{x+h}{x(x+h)}}{h} = \frac{\frac{-h}{x(x+h)}}{h} = \frac{-1}{x(x+h)}.$$

It follows that

$$f'(x) = \lim_{h \to 0} \frac{f(x+h) - f(x)}{h} = \lim_{h \to 0} \frac{-1}{x(x+h)} = -\frac{1}{x^2}.$$

The graph of the function consists of two curves. On each curve the slope, $-1/x^2$, is negative: large negative for x close to 0 (each curve steepens as x approaches 0 and tends toward the vertical) and small negative for x far from 0 (each curve flattens out as x moves away from 0 and tends toward the horizontal). ❏

Evaluating Derivatives

Example 5   We take $f(x) = 1 - x^2$ and calculate $f'(-2)$.
  We can first find $f'(x)$:

$$f'(x) = \lim_{h \to 0} \frac{f(x+h) - f(x)}{h}$$

$$= \lim_{h \to 0} \frac{[1 - (x+h)^2] - [1 - x^2]}{h} = \lim_{h \to 0} \frac{-2xh - h^2}{h} = \lim_{h \to 0}(-2x - h) = -2x$$

and then substitute -2 for x:

$$f'(-2) = -2(-2) = 4.$$

We can also evaluate $f'(-2)$ directly:

$$f'(-2) = \lim_{h \to 0} \frac{f(-2+h) - f(-2)}{h}$$

$$= \lim_{h \to 0} \frac{[1 - (-2+h)^2] - [1 - (-2)^2]}{h} = \lim_{h \to 0} \frac{4h - h^2}{h} = \lim_{h \to 0}(4 - h) = 4. \quad ❏$$

Example 6   Let's find $f'(-3)$ and $f'(1)$ given that

$$f(x) = \begin{cases} x^2, & x \le 1 \\ 2x - 1, & x > 1. \end{cases}$$

By definition,

$$f'(-3) = \lim_{h \to 0} \frac{f(-3+h) - f(-3)}{h}.$$

For all x sufficiently close to -3, $f(x) = x^2$. Thus

$$f'(-3) = \lim_{h \to 0} \frac{(-3+h)^2 - (-3)^2}{h} = \lim_{h \to 0} \frac{(9 - 6h + h^2) - 9}{h} = \lim_{h \to 0}(-6 + h) = -6.$$

Now let's find

$$f'(1) = \lim_{h \to 0} \frac{f(1+h) - f(1)}{h}.$$

Since f is not defined by the same formula on both sides of 1, we will evaluate this limit by taking one-sided limits. Note that $f(1) = 1^2 = 1$.

To the left of 1, $f(x) = x^2$. Thus

$$\lim_{h \to 0^-} \frac{f(1+h) - f(1)}{h} = \lim_{h \to 0^-} \frac{(1+h)^2 - 1}{h}$$

$$= \lim_{h \to 0^-} \frac{(1 + 2h + h^2) - 1}{h} = \lim_{h \to 0^-} (2 + h) = 2.$$

To the right of 1, $f(x) = 2x - 1$. Thus

$$\lim_{h \to 0^+} \frac{f(1+h) - f(1)}{h} = \lim_{h \to 0^+} \frac{[2(1+h) - 1] - 1}{h} = \lim_{h \to 0^+} 2 = 2.$$

The limit of the difference quotient exists and is 2:

$$f'(1) = \lim_{h \to 0} \frac{f(1+h) - f(1)}{h} = 2. \quad \square$$

Tangent Lines

If f is differentiable at c, the line that passes through the point $(c, f(c))$ with slope $f'(c)$ is the tangent line at that point. As an equation for this line we can write

(3.1.2)

$$\boxed{y - f(c) = f'(c)(x - c).}$$ (point-slope form)

This is the line through $(c, f(c))$ that best approximates the graph of f near the point $(c, f(c))$.

Example 7   We go back to the square-root function

$$f(x) = \sqrt{x}$$

and write an equation for the tangent line at the point $(4, 2)$.

As we showed in Example 3, for $x > 0$

$$f'(x) = \frac{1}{2\sqrt{x}}.$$

Thus $f'(4) = \frac{1}{4}$. The equation for the tangent line at the point $(4, 2)$ can be written

$$y - 2 = \tfrac{1}{4}(x - 4). \quad \square$$

Example 8   We differentiate the function

$$f(x) = x^3 - 12x$$

and seek the points of the graph where the tangent line is horizontal. Then we write an equation for the tangent line at the point of the graph where $x = 3$.

First we calculate the difference quotient:

$$\frac{f(x+h) - f(x)}{h} = \frac{[(x+h)^3 - 12(x+h)] - [x^3 - 12x]}{h}$$

$$= \frac{x^3 + 3x^2h + 3xh^2 + h^3 - 12x - 12h - x^3 + 12x}{h}$$

$$= \frac{3x^2h + 3xh^2 + h^3 - 12h}{h} = 3x^2 + 3xh + h^2 - 12.$$

Now we take the limit as $h \to 0$:

$$f'(x) = \lim_{h \to 0} \frac{f(x+h) - f(x)}{h} = \lim_{h \to 0}(3x^2 + 3xh + h^2 - 12) = 3x^2 - 12.$$

The function has a horizontal tangent at the points $(x, f(x))$ where $f'(x) = 0$. In this case

$$f'(x) = 0 \qquad \text{iff} \qquad 3x^2 - 12 = 0 \qquad \text{iff} \qquad x = \pm 2.$$

The graph has a horizontal tangent at the points

$$(-2, f(-2)) = (-2, 16) \qquad \text{and} \qquad (2, f(2)) = (2, -16).$$

The graph of f and the horizontal tangents are shown in Figure 3.1.5.

The point on the graph where $x = 3$ is the point $(3, f(3)) = (3, -9)$. The slope at this point is $f'(3) = 15$, and the equation of the tangent line at this point can be written

$$y + 9 = 15(x - 3). \quad \square$$

A Note on Vertical Tangents

The cube-root function

$$f(x) = x^{1/3}$$

is everywhere continuous, but as we show below, it is not differentiable at $x = 0$. The difference quotient at $x = 0$,

$$\frac{f(0+h) - f(0)}{h} = \frac{h^{1/3} - 0}{h} = \frac{1}{h^{2/3}},$$

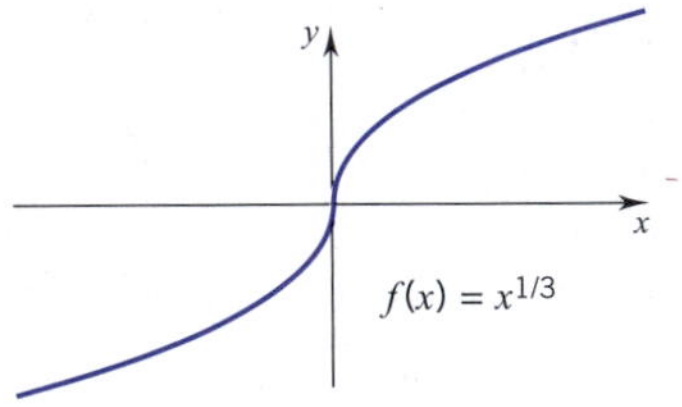

increases without bound as $h \to 0$. In the notation established in Section 2.1,

$$\text{as } h \to 0, \qquad \frac{f(0+h) - f(0)}{h} \to \infty.$$

Thus f is not differentiable at $x = 0$.

The behavior of f at $x = 0$ is depicted in Figure 3.1.6. For reasons geometrically evident, we say that the graph of f has a *vertical tangent* at the origin.[†]

Differentiability and Continuity

A function can be continuous at a number x without being differentiable there. Viewed geometrically, this can happen for only one of two reasons: either the tangent line at $(x, f(x))$ is vertical (you just saw an example of this), or there is no tangent line at $(x, f(x))$. The lack of a tangent line at a point of continuity is illustrated below.

The graph of the absolute value function

$$f(x) = |x|$$

is shown in Figure 3.1.7. The function is continuous at 0 (it is continuous everywhere), but it is not differentiable at 0:

$$\frac{f(0+h) - f(0)}{h} = \frac{|0 + h| - |0|}{h} = \frac{|h|}{h} = \begin{cases} -1, & h < 0 \\ 1, & h > 0 \end{cases}$$

so that

$$\lim_{h \to 0^-} \frac{f(0+h) - f(0)}{h} = -1, \qquad \lim_{h \to 0^+} \frac{f(0+h) - f(0)}{h} = 1$$

and thus

$$\lim_{h \to 0} \frac{f(0+h) - f(0)}{h} \qquad \text{does not exist.}$$

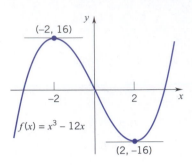

(−2, 16)

−2   2

$f(x) = x^3 - 12x$

(2, −16)

Figure 3.1.5

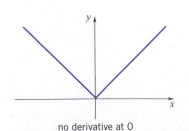

$f(x) = x^{1/3}$

Figure 3.1.6

no derivative at 0

Figure 3.1.7

[†] Vertical tangents will be considered in detail in Section 4.7.

The lack of differentiability at 0 is evident geometrically. At $x = 0$ the graph changes direction abruptly and there is no tangent line.

Another example of this sort of behavior is offered by the function

$$f(x) = \begin{cases} x^2, & x \le 1 \\ \frac{1}{2}x + \frac{1}{2}, & x > 1. \end{cases}$$

(Figure 3.1.8)

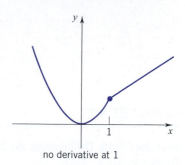

no derivative at 1

Figure 3.1.8

As you can check, the function is everywhere continuous, but at the point $(1, 1)$ the graph has an abrupt change of direction. The calculation below confirms that f is not differentiable at $x = 1$:

$$\lim_{h \to 0^-} \frac{f(1 + h) - f(1)}{h} = \lim_{h \to 0^-} \frac{(1 + h)^2 - 1}{h} = \lim_{h \to 0^-} \frac{h^2 + 2h}{h} = \lim_{h \to 0^-} (h + 2) = 2,$$

$$\lim_{h \to 0^+} \frac{f(1 + h) - f(1)}{h} = \lim_{h \to 0^+} \frac{\frac{1}{2}(1 + h) + \frac{1}{2} - 1}{h} = \lim_{h \to 0^+} \left(\frac{1}{2}\right) = \frac{1}{2}.$$

Since these one-sided limits are different,

$$\lim_{h \to 0} \frac{f(1 + h) - f(1)}{h} \qquad \text{does not exist.}$$

For our last example,

$$f(x) = |x^3 - 6x^2 + 8x| + 3,$$

(Figure 3.1.9)

we used a graphing utility.[†] So doing, it appeared that f is differentiable except, possibly, at $x = 0$, at $x = 2$, and at $x = 4$. There abrupt changes in direction seem to occur. By zooming in near the point $(2, f(2))$, we confirmed that the left-hand limits and right-hand limits of the difference quotient both exist at $x = 2$ but are not equal. See Figure 3.1.10. A similar situation was seen at $x = 0$ and $x = 4$. From the look of it, f fails to be differentiable at $x = 0$, at $x = 2$, and at $x = 4$.

Although not every continuous function is differentiable, every differentiable function is continuous.

$-1 \le x \le 5, \ 0 \le y \le 10$

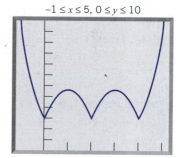

Figure 3.1.9

$1.997 \le x \le 2.003, \ 2.996 \le y \le 3.006$

THEOREM 3.1.3

If f is differentiable at x, then f is continuous at x.

PROOF For $h \ne 0$ and $x + h$ in the domain of f,

$$f(x + h) - f(x) = \frac{f(x + h) - f(x)}{h} \cdot h \ .$$

With f differentiable at x,

$$\lim_{h \to 0} \frac{f(x + h) - f(x)}{h} = f'(x).$$

Since $\lim_{h \to 0} h = 0$, we have

$$\lim_{h \to 0} [f(x + h) - f(x)] = \left[\lim_{h \to 0} \frac{f(x + h) - f(x)}{h}\right] \cdot \left[\lim_{h \to 0} h\right] = f'(x) \cdot 0 = 0.$$

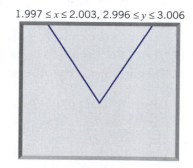

Figure 3.1.10

[†] It wasn't necessary to use a graphing utility here, but we figured that the use of it might make for a pleasant change of pace

It follows that

$$\lim_{h \to 0} f(x + h) = f(x),$$ (2.2.6)

and thus f is continuous at x. ❑

EXERCISES 3.1

Exercises 1–10. Differentiate the function by forming the difference quotient

$$\frac{f(x + h) - f(x)}{h}$$

and taking the limit as h tends to 0.

1. $f(x) = 2 - 3x$.　　　**2.** $f(x) = k$, k constant.

3. $f(x) = 5x - x^2$.　　**4.** $f(x) = 2x^3 + 1$.

5. $f(x) = x^4$.　　　　**6.** $f(x) = 1/(x + 3)$.

7. $f(x) = \sqrt{x - 1}$.　　**8.** $f(x) = x^3 - 4x$.

9. $f(x) = 1/x^2$.　　　**10.** $f(x) = 1/\sqrt{x}$.

Exercises 11–16. Find $f'(c)$ by forming the difference quotient

$$\frac{f(c + h) - f(c)}{h}$$

and taking the limit as $h \to 0$.

11. $f(x) = x^2 - 4x$; $c = 3$.　**12.** $f(x) = 7x - x^2$; $c = 2$

13. $f(x) = 2x^3 + 1$; $c = 1$.　**14.** $f(x) = 5 - x^4$; $c = -1$.

15. $f(x) = \dfrac{8}{x + 4}$; $c = -2$.　**16.** $f(x) = \sqrt{6 - x}$; $c = 2$.

Exercises 17–20. Write an equation for the tangent line at $(c, f(c))$.

17. $f(x) = 5x - x^2$; $c = 4$.　**18.** $f(x) = \sqrt{x}$; $c = 4$.

19. $f(x) = 1/x^2$; $c = -2$.　**20.** $f(x) = 5 - x^3$; $c = 2$.

21. The graph of f is shown below.

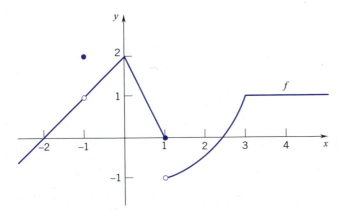

(a) At which numbers c is f discontinuous? Which of the discontinuities is removable?

(b) At which numbers c is f continuous but not differentiable?

22. Exercise 21 for the function f graphed below.

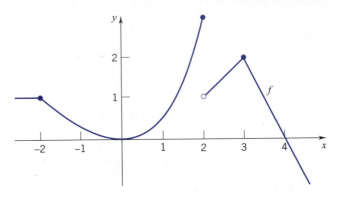

Exercises 23–28. Draw the graph of f; indicate where f is not differentiable.

23. $f(x) = |x + 1|$.　　**24.** $f(x) = |2x - 5|$.

25. $f(x) = \sqrt{|x|}$.　　**26.** $f(x) = |x^2 - 4|$.

27. $f(x) = \begin{cases} x^2, & x \le 1 \\ 2 - x, & x > 1. \end{cases}$　**28.** $f(x) = \begin{cases} x^2 - 1, & x \le 2 \\ 3, & x > 2. \end{cases}$

Exercises 29–32. Find $f'(c)$ if it exists.

29. $f(x) = \begin{cases} 4x, & x < 1 \\ 2x^2 + 2, & x \ge 1; \end{cases}$　$c = 1$.

30. $f(x) = \begin{cases} 3x^2, & x \le 1 \\ 2x^3 + 1, & x > 1; \end{cases}$　$c = 1$.

31. $f(x) = \begin{cases} x + 1, & x \le -1 \\ (x + 1)^2, & x > -1; \end{cases}$　$c = -1$.

32. $f(x) = \begin{cases} -\frac{1}{2}x^2, & x < 3 \\ -3x, & x \ge 3; \end{cases}$　$c = 3$.

Exercises 33–38. Sketch the graph of the derivative of the function indicated.

33.

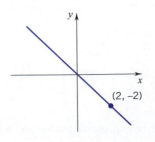

34.

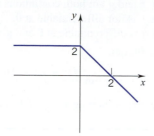

35.

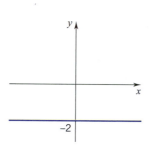

36.

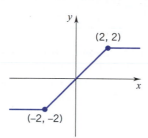

37.

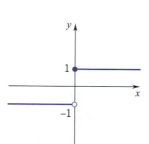

38.

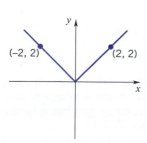

39. Show that

$$f(x) = \begin{cases} x^2, & x \le 1 \\ 2x, & x > 1 \end{cases}$$

is not differentiable at $x = 1$.

40. Set

$$f(x) = \begin{cases} (x + 1)^2, & x \le 0 \\ (x - 1)^2, & x > 0. \end{cases}$$

(a) Determine $f'(x)$ for $x \ne 0$.

(b) Show that f is not differentiable at $x = 0$.

41. Find A and B given that the function

$$f(x) = \begin{cases} x^3, & x \le 1 \\ Ax + B, & x > 1. \end{cases}$$

is differentiable at $x = 1$.

42. Find A and B given that the function

$$f(x) = \begin{cases} x^2 - 2, & x \le 2 \\ Bx^2 + Ax, & x > 2 \end{cases}$$

is differentiable at $x = 2$.

Exercises 43–48. Give an example of a function f that is defined for all real numbers and satisfies the given conditions.

43. $f'(x) = 0$ for all real x.

44. $f'(x) = 0$ for all $x \ne 0$; $f'(0)$ does not exist.

45. $f'(x)$ exists for all $x \ne -1$; $f'(-1)$ does not exist.

46. $f'(x)$ exists for all $x \ne \pm 1$; neither $f'(1)$ nor $f'(-1)$ exists.

47. $f'(1) = 2$ and $f(1) = 7$.

48. $f'(x) = 1$ for $x < 0$ and $f'(x) = -1$ for $x > 0$.

49. Set $f(x) = \begin{cases} x^2 - x, & x \le 2 \\ 2x - 2, & x > 2. \end{cases}$

(a) Show that f is continuous at 2.

(b) Is f differentiable at 2?

50. Let $f(x) = x\sqrt{x}$, $x \ge 0$. Calculate $f'(x)$ for each $x > 0$.

51. Set $f(x) = \begin{cases} 1 - x^2, & x \le 0 \\ x^2, & x > 0. \end{cases}$

(a) Is f differentiable at 0?

(b) Sketch the graph of f.

52. Set

$$f(x) = \begin{cases} x, & x \text{ rational} \\ 0, & x \text{ irrational,} \end{cases} \qquad g(x) = \begin{cases} x^2, & x \text{ rational} \\ 0, & x \text{ irrational.} \end{cases}$$

(a) Show that f is not differentiable at 0.

(b) Show that g is differentiable at 0 and give $g'(0)$.

(Normal lines) If the graph of f has a tangent line at $(c, f(c))$, then

(3.1.4) the line through $(c, f(c))$ that is perpendicular to the tangent line is called the *normal line.*

53. Write an equation for the normal line at $(c, f(c))$ given that the tangent line at this point

(a) is horizontal;

(b) has slope $f'(c) \ne 0$;

(c) is vertical.

54. All the normals through a circular arc pass through one point. What is this point?

55. As you saw in Example 7, the line $y - 2 = \frac{1}{4}(x - 4)$ is tangent to the graph of the square-root function at the point $(4, 2)$. Write an equation for the normal line through this point.

56. (*A follow-up to Exercise 55*) Sketch the graph of the square-root function displaying both the tangent and the normal at the point (4, 2).

57. The lines tangent and normal to the graph of the squaring function at the point (3, 9) intersect the x-axis at points s units apart. What is s?

58. The graph of the function $f(x) = \sqrt{1 - x^2}$ is the upper half of the unit circle. On that curve (see the figure below) we have marked a point $P(x, y)$.

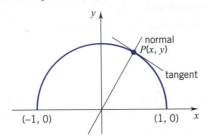

(a) What is the slope of the normal at P? Express your answer in terms of x and y.

(b) Deduce from (a) the slope of the tangent at P. Express your answer in terms of x and y.

(c) Confirm your answer in (b) by calculating

$$f'(x) = \lim_{h \to 0} \frac{\sqrt{1 - (x + h)^2} - \sqrt{1 - x^2}}{h}.$$

HINT: First rationalize the numerator of the difference quotient by multiplying both numerator and denominator by $\sqrt{1 - (x + h)^2} + \sqrt{1 - x^2}$.

59. Let $f(x) = \begin{cases} x \sin(1/x), & x \neq 0 \\ 0, & x = 0 \end{cases}$ and $g(x) = xf(x)$. The graphs of f and g are indicated in the figures below.

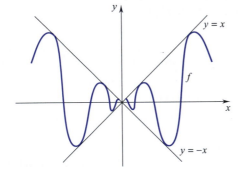

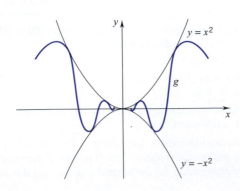

(a) Show that f and g are both continuous at 0.

(b) Show that f is not differentiable at 0.

(c) Show that g is differentiable at 0 and give $g'(0)$.

(*Important*). By definition

$$f'(c) = \lim_{h \to 0} \frac{f(c + h) - f(c)}{h}$$

provided this limit exists. Setting $x = c + h$, we can write

(3.1.5) $$f'(c) = \lim_{x \to c} \frac{f(x) - f(c)}{x - c}$$

This is an alternative definition of derivative which has advantages in certain situations. Convince yourself of the equivalence of both definitions by calculating $f'(c)$ by both methods.

60. $f(x) = x^3 + 1$; $c = 2$. **61.** $f(x) = x^2 - 3x$; $c = 1$.

62. $f(x) = \sqrt{1 + x}$; $c = 3$. **63.** $f(x) = x^{1/3}$; $c = -1$.

64. $f(x) = \dfrac{1}{x + 2}$; $c = 0$.

65. Set $f(x) = x^{5/2}$ and consider the difference quotient

$$D(h) = \frac{f(2 + h) - f(2)}{h}.$$

(a) Use a graphing utility to graph D for $h \neq 0$. Estimate $f'(2)$ to three decimal places from the graph.

(b) Create a table of values to estimate $\lim_{h \to 0} D(h)$. Estimate $f'(2)$ to three decimal places from the table.

(c) Compare your results from (a) and (b).

66. Exercise 65 with $f(x) = x^{2/3}$.

67. Use the definition of the derivative with a CAS to find $f'(x)$ in general and $f'(c)$ in particular.

(a) $f(x) = \sqrt{5x - 4}$; $c = 3$.

(b) $f(x) = 2 - x^2 + 4x^4 - x^6$; $c = -2$.

(c) $f(x) = \dfrac{3 - 2x}{2 + 3x}$; $c = -1$.

68. Use a CAS to evaluate, if possible,

$$f'(c) = \lim_{h \to 0} \frac{f(c + h) - f(c)}{h}.$$

(a) $f(x) = |x - 1| + 2$; $c = 1$.

(b) $f(x) = (x + 2)^{5/3} - 1$; $c = -2$.

(c) $f(x) = (x - 3)^{2/3} + 3$; $c = 3$.

69. Let $f(x) = 5x^2 - 7x^3$ on $[-1, 1]$.

(a) Use a graphing utility to draw the graph of f.

(b) Use the trace function to approximate the points on the graph where the tangent line is horizontal.

(c) Use a CAS to find $f'(x)$.

(d) Use a solver to solve the equation $f'(x) = 0$ and compare what you find to what you found in (b).

70. Exercise 69 with $f(x) = x^3 + x^2 - 4x + 3$ on $[-2, 2]$.

71. Set $f(x) = 4x - x^3$.

(a) Use a CAS to find $f'(\frac{3}{2})$. Then find equations for the tangent T and the normal N at the point $(\frac{3}{2}, f(\frac{3}{2}))$.

(b) Use a graphing utility to display N, T, and the graph of f in one figure.

(c) Note that T is a good approximation to the graph of f for x close to $\frac{3}{2}$. Determine the interval on which the vertical separation between T and the graph of f is of absolute value less than 0.01.

72. If $f(x) = x$, then $f'(x) = 1 \cdot x^0 = 1$.
If $f(x) = x^2$, then $f'(x) = 2x^1 = 2x$.

(a) Show that
$$\text{if} \quad f(x) = x^3, \quad \text{then} \quad f'(x) = 3x^2.$$

(b) Prove by induction that for each positive integer n,
$$f(x) = x^n \quad \text{has derivative} \quad f'(x) = nx^{n-1}.$$

HINT:
$$(x+h)^{k+1} - x^{k+1} = x(x+h)^k - x \cdot x^k + h(x+h)^k.$$

■ 3.2 SOME DIFFERENTIATION FORMULAS

Calculating the derivative of

$$f(x) = (x^3 + 2x - 3)(4x^2 + 1) \quad \text{or} \quad f(x) = \frac{6x^2 - 1}{x^4 + 5x + 1}$$

by forming the appropriate difference quotient

$$\frac{f(x+h) - f(x)}{h}$$

and then taking the limit as h tends to 0 is somewhat laborious. Here we derive some general formulas that enable us to calculate such derivatives quite quickly and easily.

We begin by pointing out that constant functions have derivative identically 0:

(3.2.1)
$$\text{if} \quad f(x) = \alpha, \quad \alpha \text{ any constant,} \quad \text{then} \quad f'(x) = 0 \quad \text{for all } x,$$

and the identity function $f(x) = x$ has constant derivative 1:

(3.2.2)
$$\text{if} \quad f(x) = x, \quad \text{then} \quad f'(x) = 1 \quad \text{for all } x.$$

PROOF For $f(x) = \alpha$,
$$f'(x) = \lim_{h \to 0} \frac{f(x+h) - f(x)}{h} = \lim_{h \to 0} \frac{\alpha - \alpha}{h} = \lim_{h \to 0} 0 = 0.$$

For $f(x) = x$,
$$f'(x) = \lim_{h \to 0} \frac{f(x+h) - f(x)}{h} = \lim_{h \to 0} \frac{(x+h) - x}{h} = \lim_{h \to 0} \frac{h}{h} = \lim_{h \to 0} 1 = 1. \quad \square$$

Remark These results can be verified geometrically. The graph of a constant function $f(x) = \alpha$ is a horizontal line, and the slope of a horizontal line is 0. The graph of the identity function $f(x) = x$ is the graph of the line $y = x$. The line has slope 1. $\square$

THEOREM 3.2.3 DERIVATIVES OF SUMS AND SCALAR MULTIPLES

Let α be a real number. If f and g are differentiable at x, then $f + g$ and αf are differentiable at x. Moreover,
$$(f + g)'(x) = f'(x) + g'(x) \quad \text{and} \quad (\alpha f)'(x) = \alpha f'(x).$$

PROOF To verify the first formula, note that

$$\frac{(f+g)(x+h)-(f+g)(x)}{h} = \frac{[f(x+h)+g(x+h)]-[f(x)+g(x)]}{h}$$

$$= \frac{f(x+h)-f(x)}{h} + \frac{g(x+h)-g(x)}{h}.$$

By definition,

$$\lim_{h\to 0}\frac{f(x+h)-f(x)}{h} = f'(x) \quad \text{and} \quad \lim_{h\to 0}\frac{g(x+h)-g(x)}{h} = g'(x).$$

Thus

$$\lim_{h\to 0}\frac{(f+g)(x+h)-(f+g)(x)}{h} = f'(x)+g'(x),$$

which means that

$$(f+g)'(x) = f'(x)+g'(x).$$

To verify the second formula, we must show that

$$\lim_{h\to 0}\frac{(\alpha f)(x+h)-(\alpha f)(x)}{h} = \alpha f'(x).$$

This follows directly from the fact that

$$\frac{(\alpha f)(x+h)-(\alpha f)(x)}{h} = \frac{\alpha f(x+h)-\alpha f(x)}{h} = \alpha\left[\frac{f(x+h)-f(x)}{h}\right]. \quad ❑$$

Remark In this section and in the next few sections we will derive formulas for calculating derivatives. It will be to your advantage to commit these formulas to memory. You may find it useful to put these formulas into words. According to Theorem 3.2.3,

"the derivative of a sum is the sum of the derivatives" and

"the derivative of a scalar multiple is the scalar multiple of the derivative." ❑

Since $f - g = f + (-g)$, it follows that if f and g are differentiable at x, then $f - g$ is differentiable at x, and

(3.2.4)
$$(f-g)'(x) = f'(x) - g'(x).$$

"The derivative of a difference is the difference of the derivatives."

These results can be extended by induction to any finite collection of functions: if $f_1, f_2, \ldots, f_n$ are differentiable at x, and $\alpha_1, \alpha_2, \ldots, \alpha_n$ are numbers, then the linear combination $\alpha_1 f_1 + \alpha_2 f_2 + \ldots + \alpha_n f_n$ is differentiable at x and

(3.2.5)
$$(\alpha_1 f_1 + \alpha_2 f_2 + \cdots + \alpha_n f_n)'(x) = \alpha_1 f_1'(x) + \alpha_2 f_2'(x) + \cdots + \alpha_n f_n'(x).$$

"The derivative of a linear combination is the linear combination of the derivatives."

> **THEOREM 3.2.6   THE PRODUCT RULE**
>
> If f and g are differentiable at x, then so is their product, and
>
> $$(f \cdot g)'(x) = f(x)g'(x) + g(x)f'(x).$$

"The derivative of a product is the first function times the derivative of the second plus the second function times the derivative of the first."

PROOF   We form the difference quotient

$$\frac{(f \cdot g)(x + h) - (f \cdot g)(x)}{h} = \frac{f(x + h)g(x + h) - f(x)g(x)}{h}$$

$$= \frac{f(x + h)g(x + h) - f(x + h)g(x) + f(x + h)g(x) - f(x)g(x)}{h}$$

and rewrite it as

$$f(x + h)\left[\frac{g(x + h) - g(x)}{h}\right] + g(x)\left[\frac{f(x - h) - f(x)}{h}\right].$$

[Here we have added and subtracted $f(x + h)g(x)$ in the numerator and then regrouped the terms so as to display the difference quotients for f and g.] Since f is differentiable at x, we know that f is continuous at x (Theorem 3.1.5) and thus

$$\lim_{h \to 0} f(x + h) = f(x). \qquad \text{(Exercise 49, Section 2.4)}$$

Since

$$\lim_{h \to 0} \frac{g(x + h) - g(x)}{h} = g'(x) \qquad \text{and} \qquad \lim_{h \to 0} \frac{f(x + h) - f(x)}{h} = f'(x),$$

we obtain

$$\lim_{h \to 0} \frac{(f \cdot g)(x + h) - (f \cdot g)(x)}{h} =$$

$$\lim_{h \to 0} f(x + h) \lim_{h \to 0}\left[\frac{g(x + h) - g(x)}{h}\right] + g(x) \lim_{h \to 0}\left[\frac{f(x + h) - f(x)}{h}\right]$$

$$= f(x)g'(x) + g(x)f'(x). \quad \square$$

Using the product rule, it is not hard to show that

(3.2.7)

> for each positive integer n
>
> $p(x) = x^n$ has derivative $p'(x) = nx^{n-1}$.

In particular,

$$p(x) = x \qquad \text{has derivative} \qquad p'(x) = 1 = 1 \cdot x^{0},{}^{\dagger}$$
$$p(x) = x^2 \qquad \text{has derivative} \qquad p'(x) = 2x,$$
$$p(x) = x^3 \qquad \text{has derivative} \qquad p'(x) = 3x^2,$$
$$p(x) = x^4 \qquad \text{has derivative} \qquad p'(x) = 4x^3,$$

and so on.

---

†In this setting we are following the convention that x^0 is identically 1 even though in itself 0^0 is meaningless.

PROOF OF (3.2.7)   We proceed by induction on n. If $n = 1$, then we have the identity function

$$p(x) = x,$$

which we know has derivative

$$p'(x) = 1 = 1 \cdot x^0.$$

This means that the formula holds for $n = 1$.

We assume now that the result holds for $n = k$; that is, we assume that if $p(x) = x^k$, then $p'(x) = kx^{k-1}$, and go on to show that it holds for $n = k + 1$. We let

$$p(x) = x^{k+1}$$

and note that

$$p(x) = x \cdot x^k.$$

Applying the product rule (Theorem 3.2.6) and our induction hypothesis, we obtain

$$p'(x) = x \cdot kx^{k-1} + x^k \cdot 1 = (k+1)x^k.$$

This shows that the formula holds for $n = k + 1$.

By the axiom of induction, the formula holds for all positive integers n. ❏

Remark   Formula (3.2.7) can be obtained without induction. From the difference quotient

$$\frac{p(x+h) - p(x)}{h} = \frac{(x+h)^n - x^n}{h},$$

apply the formula

$$a^n - b^n = (a-b)(a^{n-1} + a^{n-2}b + \cdots + ab^{n-2} + b^{n-1}), \quad \text{(Section 1.2)}$$

and you'll see that the difference quotient becomes the sum of n terms, each of which tends to x^{n-1} as h tends to zero. ❏

The formula for differentiating polynomials follows from (3.2.5) and (3.2.7):

(3.2.8)

> If $\quad P(x) = a_n x^n + a_{n-1} x^{n-1} + \cdots + a_2 x^2 + a_1 x + a_0,$
>
> then $\quad P'(x) = na_n x^{n-1} + (n-1)a_{n-1} x^{n-2} + \cdots + 2a_2 x + a_1.$

For example,

$$P(x) = 12x^3 - 6x - 2 \quad \text{has derivative} \quad P'(x) = 36x^2 - 6$$

and

$$Q(x) = \tfrac{1}{4}x^4 - 2x^2 + x + 5 \quad \text{has derivative} \quad Q'(x) = x^3 - 4x + 1.$$

Example 1   Differentiate $F(x) = (x^3 - 2x + 3)(4x^2 + 1)$ and find $F'(-1)$.

SOLUTION   We have a product $F(x) = f(x)g(x)$ with

$$f(x) = x^3 - 2x + 3 \quad \text{and} \quad g(x) = 4x^2 + 1.$$

The product rule gives

$$F'(x) = f(x)g'(x) + g(x)f'(x)$$
$$= (x^3 - 2x + 3)(8x) + (4x^2 + 1)(3x^2 - 2)$$
$$= 8x^4 - 16x^2 + 24x + 12x^4 - 5x^2 - 2$$
$$= 20x^4 - 21x^2 + 24x - 2.$$

Setting $x = -1$, we have

$$F'(-1) = 20(-1)^4 - 21(-1)^2 + 24(-1) - 2 = 20 - 21 - 24 - 2 = -27. \quad \square$$

Example 2 Differentiate $F(x) = (ax + b)(cx + d)$, where a, b, c, d are constants.

SOLUTION We have a product $F(x) = f(x)g(x)$ with

$$f(x) = ax + b \quad \text{and} \quad g(x) = cx + d.$$

Again we use the product rule

$$F'(x) = f(x)g'(x) + g(x)f'(x).$$

In this case

$$F'(x) = (ax + b)c + (cx + d)a = 2acx + bc + ad.$$

We can also do this problem without using the product rule by first carrying out the multiplication

$$F(x) = acx^2 + bcx + adx + bd$$

and then differentiating

$$F'(x) = 2acx + bc + ad. \quad \square$$

Example 3 Suppose that g is differentiable at each x and that $F(x) = (x^3 - 5x)g(x)$. Find $F'(2)$ given that $g(2) = 3$ and $g'(2) = -1$.

SOLUTION Applying the product rule, we have

$$F'(x) = [(x^3 - 5x)g(x)]' = (x^3 - 5x)g'(x) + g(x)(3x^2 - 5).$$

Therefore,

$$F'(2) = (-2)g'(2) + (7)g(2) = (-2)(-1) + (7)(3) = 23. \quad \square$$

We come now to reciprocals.

THEOREM 3.2.9 THE RECIPROCAL RULE

If g is differentiable at x and $g(x) \neq 0$, then $1/g$ is differentiable at x and

$$\left(\frac{1}{g}\right)'(x) = -\frac{g'(x)}{[g(x)]^2}.$$

PROOF Since g is differentiable at x, g is continuous at x. (Theorem 3.1.5) Since $g(x) \neq 0$, we know that $1/g$ is continuous at x and thus that

$$\lim_{h \to 0} \frac{1}{g(x + h)} = \frac{1}{g(x)}.$$

For h different from 0 and sufficiently small, $g(x + h) \neq 0$. The continuity of g at x and the fact that $g(x) \neq 0$ guarantee this. (Exercise 50, Section 2.4) The difference quotient for $1/g$ can be written

$$\frac{1}{h}\left[\frac{1}{g(x+h)} - \frac{1}{g(x)}\right] = \frac{1}{h}\left[\frac{g(x) - g(x+h)}{g(x+h)g(x)}\right]$$

$$= -\left[\frac{g(x+h) - g(x)}{h}\right]\frac{1}{g(x+h)g(x)}.$$

As h tends to zero, the right-hand side (and thus the left) tends to

$$-\frac{g'(x)}{[g(x)]^2}. \quad \square$$

Using the reciprocal rule, we can show that Formula (3.2.7) also holds for negative integers:

(3.2.10)
> for each negative integer n,
> $p(x) = x^n$ has derivative $p'(x) = nx^{n-1}$.

This formula holds at all x except, of course, at $x = 0$, where no negative power is even defined. In particular, for $x \neq 0$,

$p(x) = x^{-1}$ has derivative $p'(x) = (-1)x^{-2} = -x^{-2}$,

$p(x) = x^{-2}$ has derivative $p'(x) = -2x^{-3}$,

$p(x) = x^{-3}$ has derivative $p'(x) = -3x^{-4}$,

and so on.

PROOF OF (3.2.10) Note that

$$p(x) = \frac{1}{g(x)} \quad \text{where} \quad g(x) = x^{-n} \text{ and } -n \text{ is a positive integer.}$$

The rule for reciprocals gives

$$p'(x) = -\frac{g'(x)}{[g(x)]^2} = -\frac{(-nx^{-n-1})}{x^{-2n}} = (nx^{-n-1})x^{2n} = nx^{n-1}. \quad \square$$

Example 4 Differentiate $f(x) = \dfrac{5}{x^2} - \dfrac{6}{x}$ and find $f'(\tfrac{1}{2})$.

SOLUTION To apply (3.2.10), we write

$$f(x) = 5x^{-2} - 6x^{-1}.$$

Differentiation gives

$$f'(x) = -10x^{-3} + 6x^{-2}.$$

Back in fractional notation,

$$f'(x) = -\frac{10}{x^3} + \frac{6}{x^2}.$$

Setting $x = \tfrac{1}{2}$, we have

$$f'\left(\tfrac{1}{2}\right) = -\frac{10}{\left(\tfrac{1}{2}\right)^3} + \frac{6}{\left(\tfrac{1}{2}\right)^2} = -80 + 24 = -56. \quad \square$$

Example 5 Differentiate $f(x) = \dfrac{1}{ax^2 + bx + c}$, where a, b, c are constants.

SOLUTION Here we have a reciprocal $f(x) = 1/g(x)$ with

$$g(x) = ax^2 + bx + c.$$

The reciprocal rule (Theorem 3.2.9) gives

$$f'(x) = -\frac{g'(x)}{[g(x)]^2} = -\frac{2ax + b}{[ax^2 + bx + c]^2}. \quad ❏$$

Finally we come to quotients in general.

THEOREM 3.2.11   THE QUOTIENT RULE

If f and g are differentiable at x and $g(x) \neq 0$, then the quotient f/g is differentiable at x and

$$\left(\frac{f}{g}\right)'(x) = \frac{g(x)f'(x) - f(x)g'(x)}{[g(x)]^2}.$$

"The derivative of a quotient is the denominator times the derivative of the numerator minus the numerator times the derivative of the denominator, all divided by the square of the denominator."

Since $f/g = f(1/g)$, the quotient rule can be obtained from the product and reciprocal rules. The proof of the quotient rule is left to you as an exercise. Finally, note that the reciprocal rule is just a special case of the quotient rule. [Take $f(x) = 1$.]

From the quotient rule you can see that all rational functions (quotients of polynomials) are differentiable wherever they are defined.

Example 6 Differentiate $F(x) = \dfrac{6x^2 - 1}{x^4 + 5x + 1}$.

SOLUTION Here we are dealing with a quotient $F(x) = f(x)/g(x)$. The quotient rule,

$$F'(x) = \frac{g(x)f'(x) - f(x)g'(x)}{[g(x)]^2},$$

gives

$$F'(x) = \frac{(x^4 + 5x + 1)(12x) - (6x^2 - 1)(4x^3 + 5)}{(x^4 + 5x + 1)^2}$$

$$= \frac{-12x^5 + 4x^3 + 30x^2 + 12x + 5}{(x^4 + 5x + 1)^2}. \quad ❏$$

Example 7 Find equations for the tangent and normal lines to the graph of

$$f(x) = \frac{3x}{1 - 2x}$$

at the point $(2, f(2)) = (2, -2)$.

SOLUTION We need to find $f'(2)$. Using the quotient rule, we get

$$f'(x) = \frac{(1-2x)(3) - 3x(-2)}{(1-2x)^2} = \frac{3}{(1-2x)^2}.$$

This gives

$$f'(2) = \frac{3}{[1-2(2)]^2} = \frac{3}{(-3)^2} = \frac{1}{3}.$$

As an equation for the tangent, we can write

$$y - (-2) = \tfrac{1}{3}(x - 2), \qquad \text{which simplifies to} \qquad y + 2 = \tfrac{1}{3}(x - 2).$$

The equation for the normal line can be written $y + 2 = -3(x - 2)$. ❏

Example 8 Find the points on the graph of

$$f(x) = \frac{4x}{x^2 + 4}$$

where the tangent line is horizontal.

SOLUTION The quotient rule gives

$$f'(x) = \frac{(x^2 + 4)(4) - 4x(2x)}{(x^2 + 4)^2} = \frac{16 - 4x^2}{(x^2 + 4)^2}.$$

The tangent line is horizontal only at the points $(x, f(x))$ where $f'(x) = 0$. Therefore, we set $f'(x) = 0$ and solve for x:

$$\frac{16 - 4x^2}{(x^2 + 4)^2} = 0 \qquad \text{iff} \qquad 16 - 4x^2 = 0 \qquad \text{iff} \qquad x = \pm 2.$$

The tangent line is horizontal at the points where $x = -2$ or $x = 2$. These are the points $(-2, f(-2)) = (-2, -1)$ and $(2, f(2)) = (2, 1)$. See Figure 3.2.1. ❏

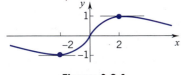

Figure 3.2.1

Remark Some expressions are easier to differentiate if we rewrite them in more convenient form. For example, we can differentiate

$$f(x) = \frac{x^5 - 2x}{x^2} = \frac{x^4 - 2}{x}$$

by the quotient rule, or we can write

$$f(x) = (x^4 - 2)x^{-1}$$

and use the product rule; even better, we can write

$$f(x) = x^3 - 2x^{-1}$$

and proceed from there:

$$f'(x) = 3x^2 + 2x^{-2}. ❏$$

EXERCISES 3.2

Exercises 1–20. Differentiate.

1. $F(x) = 1 - x$.

2. $F(x) = 2(1 + x)$.

3. $F(x) = 11x^5 - 6x^3 + 8$.

4. $F(x) = \dfrac{3}{x^2}$.

5. $F(x) = ax^2 + bx + c$; a, b, c constant.

6. $F(x) = \dfrac{x^4}{4} - \dfrac{x^3}{3} + \dfrac{x^2}{2} - \dfrac{x}{1}$.

7. $F(x) = -\dfrac{1}{x^2}$.

8. $F(x) = \dfrac{(x^2 + 2)}{x^3}$.

9. $G(x) = (x^2 - 1)(x - 3)$.

10. $F(x) = x - \dfrac{1}{x}.$

11. $G(x) = \dfrac{x^3}{1-x}.$

12. $F(x) = \dfrac{ax-b}{cx-d};$ a, b, c, d constant.

13. $G(x) = \dfrac{x^2-1}{2x+3}.$

14. $G(x) = \dfrac{7x^4+11}{x+1}.$

15. $G(x) = (x^3-2x)(2x+5).$

16. $G(x) = \dfrac{x^3+3x}{x^2-1}.$

17. $G(x) = \dfrac{6-1/x}{x-2}.$

18. $G(x) = \dfrac{1+x^4}{x^2}.$

19. $G(x) = (9x^8 - 8x^9)\left(x + \dfrac{1}{x}\right).$

20. $G(x) = \left(1 + \dfrac{1}{x}\right)\left(1 + \dfrac{1}{x^2}\right).$

Exercises 21–26. Find $f'(0)$ and $f'(1)$.

21. $f(x) = \dfrac{1}{x-2}.$

22. $f(x) = x^2(x+1).$

23. $f(x) = \dfrac{1-x^2}{1+x^2}.$

24. $f(x) = \dfrac{2x^2+x+1}{x^2+2x+1}.$

25. $f(x) = \dfrac{ax+b}{cx+d};$ a, b, c, d constant.

26. $f(x) = \dfrac{ax^2+bx+c}{cx^2+bx+a};$ a, b, c constant.

Exercises 27–30. Find $f'(0)$ given that $h(0) = 3$ and $h'(0) = 2$.

27. $f(x) = xh(x).$

28. $f(x) = 3x^2 h(x) - 5x.$

29. $f(x) = h(x) - \dfrac{1}{h(x)}.$

30. $f(x) = h(x) + \dfrac{x}{h(x)}.$

Exercises 31–34. Find an equation for the tangent line at the point $(c, f(c))$.

31. $f(x) = \dfrac{x}{x+2};$ $c = -4.$

32. $f(x) = (x^3 - 2x + 1)(4x - 5);$ $c = 2.$

33. $f(x) = (x^2 - 3)(5x - x^3);$ $c = 1.$

34. $f(x) = x^2 - \dfrac{10}{x};$ $c = -2.$

Exercises 35–38. Find the point(s) where the tangent line is horizontal.

35. $f(x) = (x-2)(x^2 - x - 11).$

36. $f(x) = x^2 - \dfrac{16}{x}.$

37. $f(x) = \dfrac{5x}{x^2+1}.$

38. $f(x) = (x+2)(x^2 - 2x - 8).$

Exercises 39–42. Find all x at which (a) $f'(x) = 0$; (b) $f'(x) > 0$; (c) $f'(x) < 0$.

39. $f(x) = x^4 - 8x^2 + 3.$

40. $f(x) = 3x^4 - 4x^3 - 2.$

41. $f(x) = x + \dfrac{4}{x^2}.$

42. $f(x) = \dfrac{x^2 - 2x + 4}{x^2 + 4}.$

Exercises 43–44. Find the points where the tangent to the graph of

43. $f(x) = -x^2 - 6$ is parallel to the line $y = 4x - 1.$

44. $f(x) = x^3 - 3x$ is perpendicular to the line $5y - 3x = 8.$

Exercises 45–48. Find a function f with the given derivative.

45. $f'(x) = 3x^2 + 2x + 1.$

46. $f'(x) = 4x^3 - 2x + 4.$

47. $f'(x) = 2x^2 - 3x - \dfrac{1}{x^2}.$

48. $f'(x) = x^4 + 2x^3 + \dfrac{1}{2\sqrt{x}}.$

49. Find A and B given that the derivative of

$$f(x) = \begin{cases} Ax^3 + Bx + 2, & x \le 2 \\ Bx^2 - A, & x > 2 \end{cases}$$

is everywhere continuous. HINT: First of all, f must be continuous.

50. Find A and B given that the derivative of

$$f(x) = \begin{cases} Ax^2 + B, & x < -1 \\ Bx^5 + Ax + 4, & x \ge -1 \end{cases}$$

is everywhere continuous.

51. Find the area of the triangle formed by the x-axis, the tangent to the graph of $f(x) = 6x - x^2$ at the point $(5, 5)$, and the normal through this point (the line through this point that is perpendicular to the tangent).

52. Find the area of the triangle formed by the x-axis and the lines tangent and normal to the graph of $f(x) = 9 - x^2$ at the point $(2, 5)$.

53. Find A, B, C such that the graph of $f(x) = Ax^2 + Bx + C$ passes through the point $(1, 3)$ and is tangent to the line $4x + y = 8$ at the point $(2, 0)$.

54. Find A, B, C, D such that the graph of $f(x) = Ax^3 + Bx^2 + Cx + D$ is tangent to the line $y = 3x - 3$ at the point $(1, 0)$ and is tangent to the line $y = 18x - 27$ at the point $(2, 9)$.

55. Find the point where the line tangent to the graph of the quadratic function $f(x) = ax^2 + bx + c$ is horizontal. NOTE: This gives a way to find the vertex of the parabola $y = ax^2 + bx + c$.

56. Find conditions on a, b, c, d which guarantee that the graph of the cubic $p(x) = ax^3 + bx^2 + cx + d$ has:

(a) exactly two horizontal tangents.
(b) exactly one horizontal tangent.
(c) no horizontal tangents.

57. Find the points $(c, f(c))$ where the line tangent to the graph of $f(x) = x^3 - x$ is parallel to the secant line that passes through the points $(-1, f(-1))$ and $(2, f(2))$.

58. Find the points $(c, f(c))$ where the line tangent to the graph of $f(x) = x/(x+1)$ is parallel to the secant line that passes through the points $(1, f(1))$ and $(3, f(3))$.

59. Let $f(x) = 1/x, x > 0.$ Show that the triangle that is formed by each line tangent to the graph of f and the coordinate axes has an area of 2 square units.

60. Find two lines through the point $(2, 8)$ that are tangent to the graph of $f(x) = x^3.$

61. Find equations for all the lines tangent to the graph of $f(x) = x^3 - x$ that pass through the point $(-2, 2)$.

62. Set $f(x) = x^3$.

(a) Find an equation for the line tangent to the graph of f at $(c, f(c))$, $c \neq 0$.

(b) Determine whether the tangent line found in (a) intersects the graph of f at a point other than (c, c^3).

If it does, find the x-coordinate of the second point of intersection.

63. Given two functions f and g, show that if f and $f + g$ are differentiable, then g is differentiable. Give an example to show that the differentiability of $f + g$ does not imply that f and g are each differentiable.

64. We are given two functions f and g, with f and $f \cdot g$ differentiable. Does it follow that g is differentiable? If not, find a condition that guarantees that g is differentiable if both f and $f \cdot g$ are differentiable.

65. Prove the validity of the quotient rule.

66. Verify that, if f, g, h are differentiable, then

$$(fgh)'(x) = f'(x)g(x)h(x) + f(x)g'(x)h(x) + f(x)g(x)h'(x).$$

HINT: Apply the product rule to $[f(x)g(x)]h(x)$.

67. Use the result in Exercise 66 to find the derivative of $F(x) = (x^2 + 1)[1 + (1/x)](2x^3 - x + 1)$.

68. Use the result in Exercise 66 to find the derivative of $G(x) = \sqrt{x}[1/(1 + 2x)](x^2 + x - 1)$.

69. Use the product rule to show that if f is differentiable, then

$$g(x) = [f(x)]^2 \quad \text{has derivative} \quad g'(x) = 2f(x)f'(x).$$

70. Use the result in Exercise 69 to find the derivative of $g(x) = (x^3 - 2x^2 + x + 2)^2$.

▶ **Exercises 71–74.** Use a CAS to find where $f'(x) = 0$, $f'(x) > 0$, $f'(x) < 0$. Verify your results with a graphing utility.

71. $f(x) = \dfrac{x^2}{x + 1}$.

72. $f(x) = 8x^5 - 60x^4 + 150x^3 - 125x^2$.

73. $f(x) = \dfrac{x^4 - 16}{x^2}$. **74.** $f(x) = \dfrac{x^3 + 1}{x^4}$.

▶ **75.** Set $f(x) = \sin x$.

(a) Estimate $f'(x)$ at $x = 0$, $x = \pi/6$, $x = \pi/4$, $x = \pi/3$, and $x = \pi/2$ using the difference quotient

$$\frac{f(x + h) - f(x)}{h}$$

taking $h = \pm 0.001$.

(b) Compare the estimated values of $f'(x)$ found in (a) with the values of $\cos x$ at each of these points.

(c) Use your results in (b) to guess the derivative of the sine function.

▶ **76.** Let $f(x) = x^4 + x^3 - 5x^2 + 2$.

(a) Use a graphing utility to graph f on the interval $[-4, 4]$ and estimate the x-coordinates of the points where the tangent line to the graph of f is horizontal.

(b) Use a graphing utility to graph $|f|$. Are there any points where f is not differentiable? If so, estimate the numbers where f fails to be differentiable.

■ **3.3 THE d/dx NOTATION; DERIVATIVES OF HIGHER ORDER**

The d/dx Notation

So far we have indicated the derivative by a prime. There are, however, other notations that are widely used, particularly in science and in engineering. The most popular of these is the "double-d" notation of Leibniz.[†] In the Leibniz notation, the derivative of a function y is indicated by writing

$$\frac{dy}{dx} \quad \text{if} \quad y \quad \text{is a function of } x,$$

$$\frac{dy}{dt} \quad \text{if} \quad y \quad \text{is a function of } t,$$

$$\frac{dy}{dz} \quad \text{if} \quad y \quad \text{is a function of } z,$$

and so on. Thus,

$$\text{if } y = x^3, \frac{dy}{dx} = 3x^2; \quad \text{if } y = \frac{1}{t^2}, \frac{dy}{dt} = -\frac{2}{t^3}; \quad \text{if } y = \sqrt{z}, \frac{dy}{dz} = \frac{1}{2\sqrt{z}}$$

[†] Gottfried Wilhelm Leibniz (1646–1716), the German mathematician whose role in the creation of calculus was outlined on page 3.

The symbols

$$\frac{d}{dx}, \quad \frac{d}{dt}, \quad \frac{d}{dz}, \quad \text{and so forth}$$

are also used as prefixes before expressions to be differentiated. For example,

$$\frac{d}{dx}(x^3 - 4x) = 3x^2 - 4, \quad \frac{d}{dt}(t^2 + 3t + 1) = 2t + 3, \quad \frac{d}{dz}(z^5 - 1) = 5z^4.$$

In the Leibniz notation the differentiation formulas read:

$$\frac{d}{dx}[f(x) + g(x)] = \frac{d}{dx}[f(x)] + \frac{d}{dx}[g(x)], \quad \frac{d}{dx}[\alpha f(x)] = \alpha \frac{d}{dx}[f(x)],$$

$$\frac{d}{dx}[f(x)g(x)] = f(x)\frac{d}{dx}[g(x)] + g(x)\frac{d}{dx}[f(x)],$$

$$\frac{d}{dx}\left[\frac{1}{g(x)}\right] = -\frac{1}{[g(x)]^2}\frac{d}{dx}[g(x)],$$

$$\frac{d}{dx}\left[\frac{f(x)}{g(x)}\right] = \frac{g(x)\frac{d}{dx}[f(x)] - f(x)\frac{d}{dx}[g(x)]}{[g(x)]^2}.$$

Often functions f and g are replaced by u and v and the x is left out altogether. Then the formulas look like this:

$$\frac{d}{dx}(u + v) = \frac{du}{dx} + \frac{dv}{dx}, \quad \frac{d}{dx}(\alpha u) = \alpha \frac{du}{dx},$$

$$\frac{d}{dx}(uv) = u\frac{du}{dx} + v\frac{du}{dx},$$

$$\frac{d}{dx}\left(\frac{1}{v}\right) = -\frac{1}{v^2}\frac{dv}{dx}, \quad \frac{d}{dx}\left(\frac{u}{v}\right) = \frac{v\frac{du}{dx} - u\frac{dv}{dx}}{v^2}.$$

The only way to develop a feeling for this notation is to use it. Below we work out some examples.

Example 1 Find $\dfrac{dy}{dx}$ for $y = \dfrac{3x - 1}{5x + 2}$.

SOLUTION We use the quotient rule:

$$\frac{dy}{dx} = \frac{(5x + 2)\frac{d}{dx}(3x - 1) - (3x - 1)\frac{d}{dx}(5x + 2)}{(5x + 2)^2}$$

$$= \frac{(5x + 2)(3) - (3x - 1)(5)}{(5x + 2)^2} = \frac{11}{(5x + 2)^2}. \quad ❑$$

Example 2 Find $\dfrac{dy}{dx}$ for $y = (x^3 + 1)(3x^5 + 2x - 1)$.

SOLUTION Here we use the product rule:

$$\frac{dy}{dx} = (x^3 + 1)\frac{d}{dx}(3x^5 + 2x - 1) + (3x^5 + 2x - 1)\frac{d}{dx}(x^3 + 1)$$

$$= (x^3 + 1)(15x^4 + 2) + (3x^5 + 2x - 1)(3x^2)$$

$$= (15x^7 + 15x^4 + 2x^3 + 2) + (9x^7 + 6x^3 - 3x^2)$$

$$= 24x^7 + 15x^4 + 8x^3 - 3x^2 + 2. \quad ❑$$

Example 3 Find $\dfrac{d}{dt}\left(t^3 - \dfrac{t}{t^2 - 1}\right)$.

SOLUTION

$$\frac{d}{dt}\left(t^3 - \frac{t}{t^2 - 1}\right) = \frac{d}{dt}(t^3) - \frac{d}{dt}\left(\frac{t}{t^2 - 1}\right)$$

$$= 3t^2 - \left[\frac{(t^2 - 1)(1) - t(2t)}{(t^2 - 1)^2}\right] = 3t^2 + \frac{t^2 + 1}{(t^2 - 1)^2}. \quad \square$$

Example 4 Find $\dfrac{du}{dx}$ for $u = x(x + 1)(x + 2)$.

SOLUTION You can think of u as

$$[x(x + 1)](x + 2) \quad \text{or as} \quad x[(x + 1)(x + 2)].$$

From the first point of view,

$$\frac{du}{dx} = [x(x + 1)](1) + (x + 2)\frac{d}{dx}[x(x + 1)]$$

$$= x(x + 1) + (x + 2)[x(1) + (x + 1)(1)]$$

$(*)$
$$= x(x + 1) + (x + 2)(2x + 1).$$

From the second point of view,

$$\frac{du}{dx} = x\frac{d}{dx}[(x + 1)(x + 2)] + (x + 1)(x + 2)(1)$$

$$= x[(x + 1)(1) + (x + 2)(1)] + (x + 1)(x + 2)$$

$(**)$
$$= x(2x + 3) + (x + 1)(x + 2).$$

Both $(*)$ and $(**)$ can be multiplied out to give

$$\frac{du}{dx} = 3x^2 + 6x + 2.$$

Alternatively, this same result can be obtained by first carrying out the multiplication and then differentiating

$$u = x(x + 1)(x + 2) = x(x^2 + 3x + 2) = x^3 + 3x^2 + 2x$$

so that

$$\frac{du}{dx} = 3x^2 + 6x + 2. \quad \square$$

Example 5 Evaluate dy/dx at $x = 0$ and $x = 1$ given that $y = \dfrac{x^2}{x^2 - 4}$.

SOLUTION

$$\frac{dy}{dx} = \frac{(x^2 - 4)2x - x^2(2x)}{(x^2 - 4)^2} = -\frac{8x}{(x^2 - 4)^2}.$$

At $x = 0$, $\dfrac{dy}{dx} = -\dfrac{8 \cdot 0}{(0^2 - 4)^2} = 0$; at $x = 1$, $\dfrac{dy}{dx} = -\dfrac{8 \cdot 1}{(1^2 - 4)^2} = -\dfrac{8}{9}. \quad \square$

Remark The notation

$$\frac{dy}{dx}\bigg|_{x=a}$$

is sometimes used to emphasize the fact that we are evaluating the derivative dy/dx at $x = a$. Thus, in Example 5, we have

$$\left.\frac{dy}{dx}\right|_{x=0} = 0 \quad \text{and} \quad \left.\frac{dy}{dx}\right|_{x=1} = -\frac{8}{9} \quad \square$$

Derivatives of Higher Order

As we noted in Section 3.1, when we differentiate a function f we get a new function f', the derivative of f. Now suppose that f' can be differentiated. If we calculate $(f')'$, we get the *second derivative of f*. This is denoted f''. So long as we have differentiability, we can continue in this manner, forming the *third derivative of f*, written f''', and so on. The prime notation is not used beyond the third derivative. For the *fourth derivative of* f, we write $f^{(4)}$ and more generally, for the *n*th derivative we write $f^{(n)}$. The functions $f', f'', f''', f^{(4)}, \ldots, f^{(n)}$ are called the derivatives of f of *orders* $1, 2, 3, 4, \ldots, n$, respectively. For example, if $f(x) = x^5$, then

$$f'(x) = 5x^4, \quad f''(x) = 20x^3, \quad f'''(x) = 60x^2, \quad f^{(4)}(x) = 120x, \quad f^{(5)}(x) = 120.$$

In this case, all derivatives of orders higher than five are identically zero. As a variant of this notation, you can write $y = x^5$ and then

$$y' = 5x^4, \quad y'' = 20x^3, \quad y''' = 60x^2, \quad \text{and so on.}$$

Since each polynomial P has a derivative P' that is in turn a polynomial, and each rational function Q has a derivative Q' that is in turn a rational function, polynomials and rational functions have derivatives of all orders. In the case of a polynomial of degree n, derivatives of order greater than n are all identically zero. (Explain.)

In the Leibniz notation the derivatives of higher order are written

$$\frac{d^2 y}{dx^2} = \frac{d}{dx}\left(\frac{dy}{dx}\right), \quad \frac{d^3 y}{dx^3} = \frac{d}{dx}\left(\frac{d^2 y}{dx^2}\right), \ldots, \text{and so on.}$$

or

$$\frac{d^2}{dx^2}[f(x)] = \frac{d}{dx}\left[\frac{d}{dx}[f(x)]\right], \quad \frac{d^3}{dx^3}[f(x)] = \frac{d}{dx}\left[\frac{d^2}{dx^2}[f(x)]\right], \ldots, \text{and so on.}$$

Below we work out some examples.

Example 6 If $f(x) = x^4 - 3x^{-1} + 5$, then

$$f'(x) = 4x^3 + 3x^{-2} \quad \text{and} \quad f''(x) = 12x^2 - 6x^{-3}. \quad \square$$

Example 7

$$\frac{d}{dx}(x^5 - 4x^3 + 7x) = 5x^4 - 12x^2 + 7,$$

$$\frac{d^2}{dx^2}(x^5 - 4x^3 + 7x) = \frac{d}{dx}(5x^4 - 12x^2 + 7) = 20x^3 - 24x,$$

$$\frac{d^3}{dx^3}(x^5 - 4x^3 + 7x) = \frac{d}{dx}(20x^3 - 24x) = 60x^2 - 24. \quad \square$$

Example 8 Finally, we consider $y = x^{-1}$. In the Leibniz notation

$$\frac{dy}{dx} = -x^{-2}, \quad \frac{d^2 y}{dx^2} = 2x^{-3}, \quad \frac{d^3 y}{dx^3} = -6x^{-4}, \quad \frac{d^4 y}{dx^4} = 24x^{-5}, \ldots.$$

On the basis of these calculations, we are led to the general result

$$\frac{d^n y}{dx^n} = (-1)^n n! x^{-n-1}. \quad [\text{Recall that } n! = n(n-1)(n-2)\cdots 3 \cdot 2 \cdot 1.]$$

In Exercise 61 you are asked to prove this result. In the prime notation we have

$$y' = -x^{-2}, \qquad y'' = 2x^{-3}, \qquad y''' = -6x^{-4}, \qquad y^{(4)} = 24x^{-5}, \ldots.$$

In general

$$y^{(n)} = (-1)^n n! x^{-n-1}. \quad \square$$

EXERCISES 3.3

Exercises 1–10. Find dy/dx.

1. $y = 3x^4 - x^2 + 1$.

2. $y = x^2 + 2x^{-4}$.

3. $y = x - \dfrac{1}{x}$.

4. $y = \dfrac{2x}{1-x}$.

5. $y = \dfrac{x}{1+x^2}$.

6. $y = x(x-2)(x+1)$.

7. $y = \dfrac{x^2}{1-x}$.

8. $y = \left(\dfrac{x}{1+x}\right)\left(\dfrac{2-x}{3}\right)$.

9. $y = \dfrac{x^3 + 1}{x^3 - 1}$.

10. $y = \dfrac{x^2}{(1+x)}$.

Exercises 11–22. Find the indicated derivative.

11. $\dfrac{d}{dx}(2x - 5)$.

12. $\dfrac{d}{dx}(5x + 2)$.

13. $\dfrac{d}{dx}[(3x^2 - x^{-1})(2x + 5)]$.

14. $\dfrac{d}{dx}[(2x^2 + 3x^{-1})(2x - 3x^{-2})]$.

15. $\dfrac{d}{dt}\left(\dfrac{t^4}{2t^3 - 1}\right)$.

16. $\dfrac{d}{dt}\left(\dfrac{2t^3 + 1}{t^4}\right)$.

17. $\dfrac{d}{du}\left(\dfrac{2u}{1 - 2u}\right)$.

18. $\dfrac{d}{du}\left(\dfrac{u^2}{u^3 + 1}\right)$.

19. $\dfrac{d}{du}\left(\dfrac{u}{u-1} - \dfrac{u}{u+1}\right)$.

20. $\dfrac{d}{du}[u^2(1 - u^2)(1 - u^3)]$.

21. $\dfrac{d}{dx}\left(\dfrac{x^3 + x^2 + x + 1}{x^3 - x^2 + x - 1}\right)$.

22. $\dfrac{d}{dx}\left(\dfrac{x^3 + x^2 + x - 1}{x^3 - x^2 + x + 1}\right)$.

Exercises 23–26. Evaluate dy/dx at $x = 2$.

23. $y = (x+1)(x+2)(x+3)$.

24. $y = (x+1)(x^2+2)(x^3+3)$.

25. $y = \dfrac{(x-1)(x-2)}{(x+2)}$.

26. $y = \dfrac{(x^2+1)(x^2-2)}{x^2+2}$.

Exercises 27–32. Find the second derivative.

27. $f(x) = 7x^3 - 6x^5$.

28. $f(x) = 2x^5 - 6x^4 + 2x - 1$.

29. $f(x) = \dfrac{x^2 - 3}{x}$.

30. $f(x) = x^2 - \dfrac{1}{x^2}$.

31. $f(x) = (x^2 - 2)(x^{-2} + 2)$.

32. $f(x) = (2x - 3)\left(\dfrac{2x + 3}{x}\right)$.

Exercises 33–38. Find d^3y/dx^3.

33. $y = \frac{1}{3}x^3 + \frac{1}{2}x^2 + x + 1$.

34. $y = (1 + 5x)^2$.

35. $y = (2x - 5)^2$.

36. $y = \frac{1}{6}x^3 - \frac{1}{4}x^2 + x - 3$.

37. $y = x^3 - \dfrac{1}{x^3}$.

38. $y = \dfrac{x^4 + 2}{x}$.

Exercises 39–44. Find the indicated derivative.

39. $\dfrac{d}{dx}\left[x\dfrac{d}{dx}(x - x^2)\right]$.

40. $\dfrac{d^2}{dx^2}\left[(x^2 - 3x)\dfrac{d}{dx}(x + x^{-1})\right]$.

41. $\dfrac{d^4}{dx^4}[3x - x^4]$.

42. $\dfrac{d^5}{dx^5}[ax^4 + bx^3 + cx^2 + dx + e]$.

43. $\dfrac{d^2}{dx^2}\left[(1 + 2x)\dfrac{d^2}{dx^2}(5 - x^3)\right]$.

44. $\dfrac{d^3}{dx^3}\left[\dfrac{1}{x}\dfrac{d^2}{dx^2}(x^4 - 5x^2)\right]$.

Exercises 45–48. Find a function $y = f(x)$ for which:

45. $y' = 4x^3 - x^2 + 4x$.

46. $y' = x - \dfrac{2}{x^3} + 3$.

47. $\dfrac{dy}{dx} = 5x^4 + \dfrac{4}{x^5}$.

48. $\dfrac{dy}{dx} = 4x^5 - \dfrac{5}{x^4} - 2$.

49. Find a quadratic polynomial p with $p(1) = 3$, $p'(1) = -2$, and $p''(1) = 4$.

50. Find a cubic polynomial p with $p(-1) = 0$, $p'(-1) = 3$, $p''(-1) = -2$, and $p'''(-1) = 6$.

51. Set $f(x) = x^n$, n a positive integer.

(a) Find $f^{(k)}(x)$ for $k = n$.

(b) Find $f^{(k)}(x)$ for $k > n$.

(c) Find $f^{(k)}(x)$ for $k < n$.

52. Let p be an arbitrary polynomial

$$p(x) = a_n x^n + a_{n-1} x^{n-1} + \cdots + a_1 x + a_0, \quad a_n \neq 0.$$

(a) Find $(d^n/dx^n)[p(x)]$.

(b) What is $(d^k/dx^k)[p(x)]$ for $k > n$?

53. Set $f(x) = \begin{cases} x^2, & x \geq 0 \\ 0, & x < 0 \end{cases}$.

(a) Show that f is differentiable at 0 and give $f'(0)$.

(b) Determine $f'(x)$ for all x.

(c) Show that $f''(0)$ does not exist.

(d) Sketch the graph of f and f'.

54. Set $g(x) = \begin{cases} x^3, & x \geq 0 \\ 0, & x < 0 \end{cases}$.

(a) Find $g'(0)$ and $g''(0)$.

(b) Determine $g'(x)$ and $g''(x)$ for all other x.

(c) Show that $g'''(0)$ does not exist.

(d) Sketch the graphs of g, g', g''.

55. Show that in general

$$(f \cdot g)''(x) \neq f(x)g''(x) + f''(x)g(x).$$

56. Verify the identity

$$f(x)g''(x) - f''(x)g(x) = \frac{d}{dx}[f(x)g'(x) - f'(x)g(x)].$$

Exercises 57–60. Find the numbers x for which (a) $f''(x) = 0$, (b) $f''(x) > 0$, (c) $f''(x) < 0$.

57. $f(x) = x^3$.

58. $f(x) = x^4$.

59. $f(x) = x^4 + 2x^3 - 12x^2$.

60. $f(x) = x^4 + 3x^3 - 6x^2 - x$.

61. Prove by induction that

$$\text{if} \quad y = x^{-1}, \quad \text{then} \quad \frac{d^n y}{dx^n} = (-1)^n n! x^{-n-1}.$$

62. Calculate y', y'', y''' for $y = 1/x^2$. Use these results to guess a formula for $y^{(n)}$ for each positive integer n, and then prove the validity of your conjecture by induction.

63. Let u, v, w be differentiable functions of x. Express the derivative of the product uvw in terms of the functions u, v, w, and their derivatives.

64. (a) Find $\dfrac{d^n}{dx^n}(x^n)$ for $n = 1, 2, 3, 4, 5$. Give the general formula.

(b) Give the general formula for $\dfrac{d^{n+1}}{dx^{n+1}}(x^n)$.

65. Set $f(x) = \dfrac{1}{1-x}$. Find a formula for $\dfrac{d^n}{dx^n}[f(x)]$.

66. Set $f(x) = \dfrac{1-x}{1+x}$. Use a CAS to find a formula for $\dfrac{d^n}{dx^n}[f(x)]$.

67. Set $f(x) = x^3 - x$.

(a) Use a graphing utility to display in one figure the graph of f and the line $l : x - 2y + 12 = 0$.

(b) Find the points on the graph of f where the tangent is parallel to l.

(c) Verify the results you obtained in (b) by adding these tangents to your previous drawing.

68. Set $f(x) = x^4 - x^2$.

(a) Use a graphing utility to display in one figure the graph of f and the line $l : x - 2y - 4 = 0$.

(b) Find the points on the graph of f where the normal is perpendicular to l.

(c) Verify the results you obtained in (b) by adding these normals to your previous drawing.

69. Set $f(x) = x^3 + x^2 - 4x + 1$.

(a) Calculate $f'(x)$.

(b) Use a graphing utility to display in one figure the graphs of f and f'. If possible, graph f and f' in different colors.

(c) What can you say about the graph of f where $f'(x) < 0$? What can you say about the graph of f where $f'(x) > 0$?

70. Set $f(x) = x^4 - x^3 - 5x^2 - x - 2$.

(a) Calculate $f'(x)$.

(b) Use a graphing utility to display in one figure the graphs of f and f'. If possible, graph f and f' in different colors.

(c) What can you say about the graph of f where $f'(x) < 0$? What can you say about the graph of f where $f'(x) > 0$?

71. Set $f(x) = \frac{1}{2}x^3 - 3x^2 + 3x + 3$.

(a) Calculate $f'(x)$.

(b) Use a graphing utility to display in one figure the graphs of f and f'. If possible, graph f and f' in different colors.

(c) What can you say about the graph of f where $f'(x) = 0$?

(d) Find the x-coordinate of each point where the tangent to the graph of f is horizontal by finding the zeros of f' to three decimal places.

72. Set $f(x) = \frac{1}{2}x^3 - 3x^2 + 4x + 1$.

(a) Calculate $f'(x)$.

(b) Use a graphing utility to display in one figure the graphs of f and f'. If possible, graph f and f' in different colors.

(c) What can you say about the graph of f where $f'(x) = 0$?

(d) Find the x-coordinate of each point where the tangent to the graph of f is horizontal by finding the zeros of f' to three decimal places.

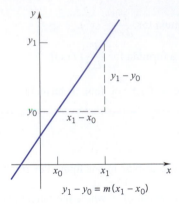

Figure 3.4.1

■ 3.4 THE DERIVATIVE AS A RATE OF CHANGE

In the case of a linear function $y = mx + b$, the graph is a straight line and the slope m measures the steepness of the line by giving the rate of climb of the line, *the rate of change of y with respect to x.*

As x changes from x_0 to x_1, y changes m times as much:

$$y_1 - y_0 = m(x_1 - x_0) \qquad \text{(Figure 3.4.1)}$$

Thus the slope m gives the change in y per unit change in x.

In the more general case of a differentiable function

$$y = f(x)$$

the graph is a curve. The slope

$$\frac{dy}{dx} = f'(x)$$

still gives *the rate of change of y with respect to x*, but this rate of change can vary from point to point. At $x = x_1$ (see Figure 3.4.2) the rate of change of y with respect to x is $f'(x_1)$; the steepness of the graph is that of a line of slope $f'(x_1)$. At $x = x_2$, the rate of change of y with respect to x is $f'(x_2)$; the steepness of the graph is that of a line of slope $f'(x_2)$. At $x = x_3$, the rate of change of y with respect to x is $f'(x_3)$; the steepness of the graph is that of a line of slope $f'(x_3)$.

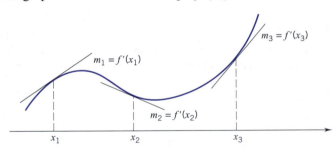

Figure 3.4.2

The derivative as a rate of change is one of the fundamental ideas of calculus. Keep it in mind whenever you see a derivative. This section is only introductory. We'll develop the idea further as we go on.

Example 1 The area of a square is given by the formula $A = x^2$ where x is the length of a side. As x changes, A changes. The rate of change of A with respect to x is the derivative

$$\frac{dA}{dx} = \frac{d}{dx}(x^2) = 2x.$$

When $x = \frac{1}{4}$, this rate of change is $\frac{1}{2}$: the area is changing at half the rate of x. When $x = \frac{1}{2}$, the rate of change of A with respect to x is 1: the area is changing at the same rate as x. When $x = 1$, the rate of change of A with respect to x is 2 : the area is changing at twice the rate of x.

In Figure 3.4.3 we have plotted A against x. The rate of change of A with respect to x at each of the indicated points appears as the slope of the tangent line. ❑

Example 2 An equilateral triangle of side x has area

$$A = \tfrac{1}{4}\sqrt{3}x^2. \qquad \text{(Check this out.)}$$

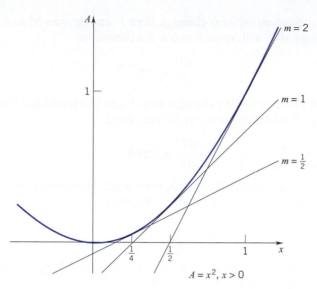

Figure 3.4.3

The rate of change of A with respect to x is the derivative

$$\frac{dA}{dx} = \tfrac{1}{2}\sqrt{3}x.$$

When $x = 2\sqrt{3}$, the rate of change of A with respect to x is 3. In other words, when the side has length $2\sqrt{3}$, the area is changing three times as fast as the length of the side. ❏

Example 3 Set $y = \dfrac{x-2}{x^2}$.

(a) Find the rate of change of y with respect to x at $x = 2$.
(b) Find the value(s) of x at which the rate of change of y with respect to x is 0.

SOLUTION The rate of change of y with respect to x is given by the derivative, dy/dx:

$$\frac{dy}{dx} = \frac{x^2(1) - (x-2)(2x)}{x^4} = \frac{-x^2 + 4x}{x^4} = \frac{4-x}{x^3}.$$

(a) At $x = 2$,

$$\frac{dy}{dx} = \frac{4-2}{2^3} = \frac{1}{4}.$$

(b) Setting $\dfrac{dy}{dx} = 0$, we have $\dfrac{4-x}{x^3} = 0$, and therefore $x = 4$. The rate of change of y with respect to x at $x = 4$ is 0. ❏

Example 4 Suppose that we have a right circular cylinder of changing dimensions. (Figure 3.4.4.) When the base radius is r and the height is h, the cylinder has volume

$$V = \pi r^2 h.$$

Figure 3.4.4

If r remains constant while h changes, then V can be viewed as a function of h. The rate of change of V with respect to h is the derivative

$$\frac{dV}{dh} = \pi r^2.$$

If h remains constant while r changes, then V can be viewed as a function of r. The rate of change of V with respect to r is the derivative

$$\frac{dV}{dh} = 2\pi rh.$$

Suppose now that r changes but V is kept constant. How does h change with respect to r? To answer this, we express h in terms of r and V:

$$h = \frac{V}{\pi r^2} = \frac{V}{\pi} r^{-2}.$$

Since V is held constant, h is now a function of r. The rate of change of h with respect to r is the derivative

$$\frac{dh}{dr} = -\frac{2V}{\pi} r^{-3} = -\frac{2(\pi r^2 h)}{\pi} r^{-3} = -\frac{2h}{r}. \quad \square$$

EXERCISES 3.4

1. Find the rate of change of the area of a circle with respect to the radius r. What is the rate when $r = 2$?

2. Find the rate of change of the volume of a cube with respect to the length s of a side. What is the rate when $s = 4$?

3. Find the rate of change of the area of a square with respect to the length z of a diagonal. What is the rate when $z = 4$?

4. Find the rate of change of $y = 1/x$ with respect to x at $x = -1$.

5. Find the rate of change of $y = [x(x + 1)]^{-1}$ with respect to x at $x = 2$.

6. Find the values of x at which the rate of change of $y = x^3 - 12x^2 + 45x - 1$ with respect to x is zero.

7. Find the rate of change of the volume of a sphere with respect to the radius r.

8. Find the rate of change of the surface area of a sphere with respect to the radius r. What is this rate of change when $r = r_0$? How must r_0 be chosen so that the rate of change is 1?

9. Find x_0 given that the rate of change of $y = 2x^2 + x - 1$ with respect to x at $x = x_0$ is 4.

10. Find the rate of change of the area A of a circle with respect to (a) the diameter d; (b) the circumference C.

11. Find the rate of change of the volume V of a cube with respect to

(a) the length w of a diagonal on one of the faces.
(b) the length z of one of the diagonals of the cube.

12. The dimensions of a rectangle are changing in such a way that the area of the rectangle remains constant. Find the rate of change of the height h with respect to the base b.

13. The area of a sector in a circle is given by the formula $A = \frac{1}{2}r^2\theta$ where r is the radius and θ is the central angle measured in radians.

(a) Find the rate of change of A with respect to θ if r remains constant.
(b) Find the rate of change of A with respect to r if θ remains constant.
(c) Find the rate of change of θ with respect to r if A remains constant.

14. The total surface area of a right circular cylinder is given by the formula $A = 2\pi r(r + h)$ where r is the radius and h is the height.

(a) Find the rate of change of A with respect to h if r remains constant.
(b) Find the rate of change of A with respect to r if h remains constant.
(c) Find the rate of change of h with respect to r if A remains constant.

15. For what value of x is the rate of change of

$$y = ax^2 + bx + c \text{ with respect to } x$$

the same as the rate of change of

$$z = bx^2 + ax + c \text{ with respect to } x?$$

Assume that a, b, c are constant with $a \neq b$.

16. Find the rate of change of the product $f(x)g(x)h(x)$ with respect to x at $x = 1$ given that

$$f(1) = 0, \quad g(1) = 2, \quad h(1) = -2,$$
$$f'(1) = 1, \quad g'(1) = -1, \quad h'(1) = 0$$

■ 3.5 THE CHAIN RULE

In this section we take up the differentiation of composite functions. Until we get to Theorem 3.5.6, our approach is completely intuitive—no real definitions, no proofs, just informal discussion. Our purpose is to give you some experience with the standard computational procedures and some insight into why these procedures work. Theorem 3.5.6 puts this all on a sound footing.

Suppose that y is a differentiable function of u and u in turn is a differentiable function of x. Then y is a composite function of x. Does y have a derivative with respect to x? Yes it does, and dy/dx is given by a formula that is easy to remember:

(3.5.1)
$$\frac{dy}{dx} = \frac{dy}{du}\frac{du}{dx}.$$

This formula, known as the *chain rule*, says that

> *"the rate of change of y with respect to x is the rate of change of y with respect to u times the rate of change of u with respect to x."*

Plausible as all this sounds, remember that we have proved nothing. All we have done is assert that the composition of differentiable functions is differentiable and given you a formula—a formula that needs justification and is justified at the end of this section.

Before using the chain rule in elaborate computations, let's confirm its validity in some simple instances.

If $y = 2u$ and $u = 3x$, then $y = 6x$. Clearly

$$\frac{dy}{dx} = 6 = 2 \cdot 3 = \frac{dy}{du}\frac{du}{dx},$$

and so, in this case, the chain rule is confirmed:

$$\frac{dy}{dx} = \frac{dy}{du}\frac{du}{dx}.$$

If $y = u^3$ and $u = x^2$, then $y = (x^2)^3 = x^6$. This time

$$\frac{dy}{dx} = 6x^5, \quad \frac{dy}{du} = 3u^2 = 3(x^2)^2 = 3x^4, \quad \frac{du}{dx} = 2x$$

and once again

$$\frac{dy}{dx} = 6x^5 = 3x^4 \cdot 2x = \frac{dy}{du}\frac{du}{dx}.$$

Example 1 Find dy/dx by the chain rule given that

$$y = \frac{u-1}{u+1} \quad \text{and} \quad u = x^2.$$

SOLUTION

$$\frac{dy}{du} = \frac{(u+1)(1) - (u-1)(1)}{(u+1)^2} = \frac{2}{(u+1)^2} \quad \text{and} \quad \frac{du}{dx} = 2x$$

so that

$$\frac{dy}{dx} = \frac{dy}{du}\frac{dy}{dx} = \left[\frac{2}{(u+1)^2}\right]2x = \frac{4x}{(x^2+1)^2}. \quad \square$$

Remark We would have obtained the same result without the chain rule by first writing y as a function of x and then differentiating:

$$\text{with} \quad y = \frac{u-1}{u+1} \quad \text{and} \quad u = x^2, \quad \text{we have} \quad y = \frac{x^2-1}{x^2+1}$$

and

$$\frac{dy}{dx} = \frac{(x^2+1)2x - (x^2-1)2x}{(x^2+1)^2} = \frac{4x}{(x^2+1)^2}. \quad \square$$

Suppose now that you were asked to calculate

$$\frac{d}{dx}[(x^2-1)^{100}].$$

You could expand $(x^2-1)^{100}$ into a polynomial by using the binomial theorem (that's assuming that you are familiar with the theorem and are adept at applying it) or you could try repeated multiplication, but in either case you would have a terrible mess on your hands: $(x^2-1)^{100}$ has 101 terms. Using the chain rule, we can derive a formula that will render such calculations almost trivial.

By the chain rule, we can show that, if u is a differentiable function of x and n is a positive or negative integer, then

(3.5.2)
$$\boxed{\frac{d}{dx}(u^n) = nu^{n-1}\frac{du}{dx}.}$$

If n is a positive integer, the formula holds without restriction. If n is negative, the formula is valid except at those numbers where $u(x) = 0$.

PROOF Set $y = u^n$. In this case,

$$\frac{dy}{dx} = \frac{dy}{du}\frac{du}{dx}$$

gives

$$\frac{d}{dx}(u^n) = \frac{d}{du}(u^n)\frac{du}{dx} = nu^{n-1}\frac{du}{dx}. \quad \square$$

To calculate

$$\frac{d}{dx}[(x^2-1)^{100}],$$

we set $u = x^2 - 1$. Then by our formula

$$\frac{d}{dx}[(x^2-1)^{100}] = 100(x^2-1)^{99}\frac{d}{dx}(x^2-1) = 100(x^2-1)^{99}2x = 200x(x^2-1)^{99}.$$

Remark While it is clear that (3.5.2) is the only practical way to calculate the derivative of $y = (x^2-1)^{100}$, you do have a choice when differentiating a similar, but simpler, function such as $y = (x^2-1)^4$. By (3.5.2)

$$\frac{d}{dx}[(x^2-1)^4] = 4(x^2-1)^3\frac{d}{dx}(x^2-1) = 4(x^2-1)^3 2x = 8x(x^2-1)^3.$$

On the other hand, if we were to first expand the expression $(x^2-1)^4$, we would get

$$y = x^8 - 4x^6 + 6x^4 - 4x^2 + 1$$

and then

$$\frac{dy}{dx} = 8x^7 - 24x^5 + 24x^3 - 8x.$$

As a final answer, this is correct but somewhat unwieldy. To reconcile the two results, note that $8x$ is a factor of dy/dx:

$$\frac{dy}{dx} = 8x(x^6 - 3x^4 + 3x^2 - 1),$$

and the expression in parentheses is $(x^2 - 1)^3$ multiplied out. Thus,

$$\frac{dy}{dx} = 8x(x^2 - 1)^3,$$

as we saw above. However, (3.5.2) gave us this neat, compact result much more efficiently. ❑

Here are additional examples of a similar sort.

Example 2

$$\frac{d}{dx}\left[\left(x + \frac{1}{x}\right)^{-3}\right] = -3\left(x + \frac{1}{x}\right)^{-4}\frac{d}{dx}\left(x + \frac{1}{x}\right) = -3\left(x + \frac{1}{x}\right)^{-4}\left(1 - \frac{1}{x^2}\right). \quad ❑$$

Example 3

$$\frac{d}{dx}[1 + (2 + 3x)^5]^3 = 3[1 + (2 + 3x)^5]^2\frac{d}{dx}[1 + (2 + 3x)^5].$$

Since

$$\frac{d}{dx}[1 + (2 + 3x)^5] = 5(2 + 3x)^4\frac{d}{dx}(2 + 3x) = 5(2 + 3x)^4(3) = 15(2 + 3x)^4,$$

we have

$$\frac{d}{dx}[1 + (2 + 3x)^5]^3 = 3[1 + (2 + 3x)^5]^2[15(2 + 3x)^4]$$

$$= 45(2 + 3x)^4[1 + (2 + 3x)^5]^2. \quad ❑$$

Example 4 Calculate the derivative of $f(x) = 2x^3(x^2 - 3)^4$.

SOLUTION Here we need to use the product rule and the chain rule:

$$\frac{d}{dx}[2x^3(x^2 - 3)^4] = 2x^3\frac{d}{dx}[(x^2 - 3)^4] + (x^2 - 3)^4\frac{d}{dx}(2x^3)$$

$$= 2x^3[4(x^2 - 3)^3(2x)] + (x^2 - 3)^4(6x^2)$$

$$= 16x^4(x^2 - 3)^3 + 6x^2(x^2 - 3)^4 = 2x^2(x^2 - 3)^3(11x^2 - 9). \quad ❑$$

The formula

$$\frac{dy}{dx} = \frac{dy}{du}\frac{du}{dx}$$

can be extended to more variables. For example, if x itself depends on s, then we have

(3.5.3)

$$\boxed{\frac{dy}{ds} = \frac{dy}{du}\frac{du}{dx}\frac{dx}{ds}.}$$

If, in addition, s depends on t, then

(3.5.4)

$$\frac{dy}{dt} = \frac{dy}{du}\frac{du}{dx}\frac{dx}{ds}\frac{ds}{dt},$$

and so on. Each new dependence adds a new link to the chain.

Example 5 Find dy/ds given that $y = 3u + 1$, $u = x^{-2}$, $x = 1 - s$.

SOLUTION

$$\frac{dy}{du} = 3, \quad \frac{du}{dx} = -2x^{-3}, \quad \frac{dx}{ds} = -1.$$

Therefore

$$\frac{dy}{ds} = \frac{dy}{du}\frac{du}{dx}\frac{dx}{ds} = (3)(-2x^{-3})(-1) = 6x^{-3} = 6(1-s)^{-3}. \quad \square$$

Example 6 Find dy/dt at $t = 9$ given that

$$y = \frac{u+2}{u-1}, \quad u = (3s - 7)^2, \quad s = \sqrt{t}.$$

SOLUTION As you can check,

$$\frac{dy}{du} = -\frac{3}{(u-1)^2}, \quad \frac{du}{ds} = 6(3s - 7), \quad \frac{ds}{dt} = \frac{1}{2\sqrt{t}}.$$

At $t = 9$, we have $s = 3$ and $u = 4$, so that

$$\frac{dy}{du} = -\frac{3}{(4-1)^2} = -\frac{1}{3}, \quad \frac{du}{ds} = 6(9 - 7) = 12, \quad \frac{ds}{dt} = \frac{1}{2\sqrt{9}} = \frac{1}{6}.$$

Thus, at $t = 9$,

$$\frac{dy}{dt} = \frac{dy}{du}\frac{du}{ds}\frac{ds}{dt} = \left(-\frac{1}{3}\right)(12)\left(\frac{1}{6}\right) = -\frac{2}{3}. \quad \square$$

Example 7 Gravel is being poured by a conveyor onto a conical pile at the constant rate of 60π cubic feet per minute. Frictional forces within the pile are such that the height is always two-thirds of the radius. How fast is the radius of the pile changing at the instant the radius is 5 feet?

SOLUTION The formula for the volume V of a right circular cone of radius r and height h is

$$V = \tfrac{1}{3}\pi r^2 h.$$

However, in this case we are told that $h = \tfrac{2}{3}r$, and so we have

(∗)

$$V = \tfrac{2}{9}\pi r^3.$$

Since gravel is being poured onto the pile, the volume, and hence the radius, are functions of time t. We are given that $dV/dt = 60\pi$ and we want to find dr/dt at the instant $r = 5$. Differentiating (∗) with respect to t by the chain rule, we get

$$\frac{dV}{dt} = \frac{dV}{dr}\frac{dr}{dt} = \left(\tfrac{2}{3}\pi r^2\right)\frac{dr}{dt}.$$

Solving for dr/dt and using the fact that $dV/dt = 60\pi$, we find that

$$\frac{dr}{dt} = \frac{180\pi}{2\pi r^2} = \frac{90}{r^2}.$$

When $r = 5$,

$$\frac{dr}{dt} = \frac{90}{(5)^2} = \frac{90}{25} = 3.6.$$

Thus, the radius is increasing at the rate of 3.6 feet per minute at the instant the radius is 5 feet. ❑

So far we have worked entirely in Leibniz's notation. What does the chain rule look like in prime notation? Let's go back to the beginning. Once again, let y be a differentiable function of u: say

$$y = f(u).$$

Let u be a differentiable function of x: say

$$u = g(x).$$

Then

$$y = f(u) = f(g(x)) = (f \circ g)(x)$$

and, according to the chain rule (as yet unproved),

$$\frac{dy}{dx} = \frac{dy}{du}\frac{du}{dx}.$$

Since

$$\frac{dy}{dx} = \frac{d}{dx}[(f \circ g)(x)] = (f \circ g)'(x), \quad \frac{dy}{du} = f'(u) = f'(g(x)), \quad \frac{du}{dx} = g'(x),$$

the chain rule can be written

(3.5.5)

$$\boxed{(f \circ g)'(x) = f'(g(x))\,g'(x).}$$

The chain rule in prime notation says that

"the derivative of a composition $f \circ g$ at x is the derivative of f at $g(x)$ times the derivative of g at x."

In Leibniz's notation the chain rule *appears* seductively simple, to some even obvious. "After all, to prove it, all you have to do is cancel the du's":

$$\frac{dy}{dx} = \frac{dy}{d\!\!\!/u}\frac{d\!\!\!/u}{dx}.$$

Of course, this is just nonsense. What would one cancel from

$$(f \circ g)'(x) = f'(g(x))g'(x)?$$

Although Leibniz's notation is useful for routine calculations, mathematicians generally turn to prime notation where precision is required.

It is time for us to be precise. How do we know that the composition of differentiable functions is differentiable? What assumptions do we need? Under what circumstances is it true that

$$(f \circ g)'(x) = f'(g(x))g'(x)?$$

The following theorem provides the definitive answer.

THEOREM 3.5.6   THE CHAIN-RULE THEOREM

If g is differentiable at x and f is differentiable at $g(x)$, then the composition $f \circ g$ is differentiable at x and

$$(f \circ g)'(x) = f'(g(x))g'(x).$$

A proof of this theorem appears in the supplement to this section. The argument is not as easy as "canceling" the du's.

One final point. The statement

$$(f \circ g)'(x) = f'(g(x))g'(x)$$

is often written

$$\frac{d}{dx}[f(g(x))] = f'(g(x))g'(x).$$

EXERCISES 3.5

Exercises 1–6. Differentiate the function: (a) by expanding before differentiation, (b) by using the chain rule. Then reconcile your results.

1. $y = (x^2 + 1)^2$.

2. $y = (x^3 - 1)^2$.

3. $y = (2x + 1)^3$.

4. $y = (x^2 + 1)^3$.

5. $y = (x + x^{-1})^2$.

6. $y = (3x^2 - 2x)^2$.

Exercises 7–20. Differentiate the function.

7. $f(x) = (1 - 2x)^{-1}$.

8. $f(x) = (1 + 2x)^5$.

9. $f(x) = (x^5 - x^{10})^{20}$.

10. $f(x) = \left(x^2 + \dfrac{1}{x^2}\right)^3$.

11. $f(x) = \left(x - \dfrac{1}{x}\right)^4$.

12. $f(t) = \left(\dfrac{1}{1+t}\right)^4$.

13. $f(x) = (x - x^3 + x^5)^4$.

14. $f(t) = (t - t^2)^3$.

15. $f(t) = (t^{-1} + t^{-2})^4$.

16. $f(x) = \left(\dfrac{4x + 3}{5x - 2}\right)^3$.

17. $f(x) = \left(\dfrac{3x}{x^2 + 1}\right)^4$.

18. $f(x) = [(2x + 1)^2 + (x + 1)^2]^3$.

19. $f(x) = \left(\dfrac{x^3}{3} + \dfrac{x^2}{2} + \dfrac{x}{1}\right)^{-1}$.

20. $f(x) = [(6x + x^5)^{-1} + x]^2$.

Exercises 21–24. Find dy/dx at $x = 0$.

21. $y = \dfrac{1}{1 + u^2}$, $u = 2x + 1$.

22. $y = u + \dfrac{1}{u}$, $u = (3x + 1)^4$.

23. $y = \dfrac{2u}{1 - 4u}$, $u = (5x^2 + 1)^4$.

24. $y = u^3 - u + 1$, $u = \dfrac{1 - x}{1 + x}$.

Exercises 25–26. Find dy/dt.

25. $y = \dfrac{1 - 7u}{1 + u^2}$, $u = 1 + x^2$, $x = 2t - 5$.

26. $y = 1 + u^2$, $u = \dfrac{1 - 7x}{1 + x^2}$, $x = 5t + 2$.

Exercises 27–28. Find dy/dx at $x = 2$.

27. $y = (s + 3)^2$, $s = \sqrt{t - 3}$, $t = x^2$.

28. $y = \dfrac{1 + s}{1 - s}$, $s = t - \dfrac{1}{t}$, $t = \sqrt{x}$.

Exercises 29–38. Evaluate the following, given that

$$f(0) = 1, \quad f'(0) = 2, \quad f(1) = 0, \quad f'(1) = 1,$$
$$f(2) = 1, \quad f'(2) = 1,$$
$$g(0) = 2, \quad g'(0) = 1, \quad g(1) = 1, \quad g'(1) = 0,$$
$$g(2) = 1, \quad g'(2) = 1,$$
$$h(0) = 1, \quad h'(0) = 2, \quad h(1) = 2, \quad h'(1) = 1,$$
$$h(2) = 0, \quad h'(2) = 2,$$

29. $(f \circ g)'(0).$ **30.** $(f \circ g)'(1).$

31. $(f \circ g)'(2).$ **32.** $(g \circ f)'(0).$

33. $(g \circ f)'(1).$ **34.** $(g \circ f)'(2).$

35. $(f \circ h)'(0).$ **36.** $(f \circ h \circ g)'(1).$

37. $(g \circ f \circ h)'(2).$ **38.** $(g \circ h \circ f)'(0).$

Exercises 39–42. Find $f''(x)$.

39. $f(x) = (x^3 + x)^4.$

40. $f(x) = (x^2 - 5x + 2)^{10}.$

41. $f(x) = \left(\dfrac{x}{1 - x}\right)^3.$

42. $f(x) = \sqrt{x^2 + 1}$ $\left(\text{recall that} \dfrac{d}{dx}[\sqrt{x}] = \dfrac{1}{2\sqrt{x}}\right).$

Exercises 43–46. Express the derivative in prime notation.

43. $\dfrac{d}{dx}[f(x^2 + 1)].$ **44.** $\dfrac{d}{dx}\left[f\left(\dfrac{x - 1}{x + 1}\right)\right].$

45. $\dfrac{d}{dx}[[f(x)]^2 + 1].$ **46.** $\dfrac{d}{dx}\left[\dfrac{f(x) - 1}{f(x) + 1}\right].$

Exercises 47–50. Determine the values of x for which
(a) $f'(x) = 0$; (b) $f'(x) > 0$; (c) $f'(x) < 0$.

47. $f(x) = (1 + x^2)^{-2}.$ **48.** $f(x) = (1 - x^2)^2.$

49. $f(x) = x(1 + x^2)^{-1}.$ **50.** $f(x) = x(1 - x^2)^3.$

Exercises 51–53. Find a formula for the nth derivative.

51. $y = \dfrac{1}{1 - x}.$ **52.** $y = \dfrac{x}{1 + x}.$

53. $y = (a + bx)^n;$ n a positive integer, a, b constants.

54. $y = \dfrac{a}{bx + c},$ a, b, c constants.

Exercises 55–58. Find a function $y = f(x)$ with the given derivative. Check your answer by differentiation.

55. $y' = 3(x^2 + 1)^2(2x).$ **56.** $y' = 2x(x^2 - 1).$

57. $\dfrac{dy}{dx} = 2(x^3 - 2)(3x^2).$ **58.** $\dfrac{dy}{dx} = 3x^2(x^3 + 2)^2.$

59. A function L has the property that $L'(x) = 1/x$ for $x \neq 0$. Determine the derivative with respect to x of $L(x^2 + 1)$.

60. Let f and g be differentiable functions such that $f'(x) = g(x)$ and $g'(x) = f(x)$, and let

$$H(x) = [f(x)]^2 - [g(x)]^2.$$

Find $H'(x)$.

61. Let f and g be differentiable functions such that $f'(x) = g(x)$ and $g'(x) = -f(x)$, and let

$$T(x) = [f(x)]^2 + [g(x)]^2.$$

Find $T'(x)$.

62. Let f be a differentiable function. Use the chain rule to show that:

(a) if f is even, then f' is odd.
(b) if f is odd, then f' is even.

63. The number a is called a *double zero* (or a zero of *multiplicity* 2) of the polynomial P if

$$P(x) = (x - a)^2 q(x) \qquad \text{and} \qquad q(a) \neq 0.$$

Prove that if a is a double zero of P, then a is a zero of both P and P', and $P''(a) \neq 0$.

64. The number a is called a *triple zero* (or a zero of *multiplicity* 3) of the polynomial P if

$$P(x) = (x - a)^3 q(x) \qquad \text{and} \qquad q(a) \neq 0.$$

Prove that if a is a triple zero of P, then a is a zero of P, P', and P'', and $P'''(a) \neq 0$.

65. The number a is called a *zero of multiplicity k* of the polynomial P if

$$P(x) = (x - a)^k q(x) \qquad \text{and} \qquad q(a) \neq 0.$$

Use the results in Exercises 63 and 64 to state a theorem about a zero of multiplicity k.

66. An equilateral triangle of side length x and altitude h has area A given by

$$A = \frac{\sqrt{3}}{4}x^2 \qquad \text{where} \qquad x = \frac{2\sqrt{3}}{3}h.$$

Find the rate of change of A with respect to h and determine this rate of change when $h = 2\sqrt{3}$.

67. As air is pumped into a spherical balloon, the radius increases at the constant rate of 2 centimeters per second. What is the rate of change of the balloon's volume when the radius is 10 centimeters? (The volume V of a sphere of radius r is $\frac{4}{3}\pi r^3$.)

68. Air is pumped into a spherical balloon at the constant rate of 200 cubic centimeters per second. How fast is the surface area of the balloon changing when the radius is 5 centimeters? (The surface area S of a sphere of radius r is $4\pi r^2$.)

69. Newton's law of gravitational attraction states that if two bodies are at a distance r apart, then the force F exerted by one body on the other is given by

$$F(r) = -\frac{k}{r^2}$$

where k is a positive constant. Suppose that, as a function of time, the distance between the two bodies is given by

$$r(t) = 49t - 4.9t^2, \qquad 0 \leq t \leq 10.$$

(a) Find the rate of change of F with respect to t.
(b) Show that $(F \circ r)'(3) = -(F \circ r)'(7)$.

70. Set $f(x) = \sqrt[3]{1 - x}$.

(a) Use a CAS to find $f'(9)$. Then find an equation for the line l tangent to the graph of f at the point $(9, f(9))$.
(b) Use a graphing utility to display l and the graph of f in one figure.
(c) Note that l is a good approximation to the graph of f for x close to 9. Determine the interval on which the vertical separation between l and the graph of f is of absolute value less than 0.01.

▷**71.** Set $f(x) = \dfrac{1}{1 + x^2}$.

(a) Use a CAS to find $f'(1)$. Then find an equation for the line l tangent to the graph of f at the point $(1, f(1))$.

(b) Use a graphing utility to display l and the graph of f in one figure.

(c) Note that l is a good approximation to the graph of f for x close to 1. Determine the interval on which the vertical separation between l and the graph of f is of absolute value less than 0.01.

▷**72.** Use a CAS to find $\dfrac{d}{dx}\left[x^2 \dfrac{d^4}{dx^4}(x^2 + 1)^4 \right]$.

▷**73.** Use a CAS to express the following derivatives in f' notation.

(a) $\dfrac{d}{dx}\left[f\left(\dfrac{1}{x} \right) \right]$, (b) $\dfrac{d}{dx}\left[f\left(\dfrac{x^2 - 1}{x^2 + 1} \right) \right]$,

(c) $\dfrac{d}{dx}\left[\dfrac{f(x)}{1 + f(x)} \right]$.

▷**74.** Use a CAS to find the following derivatives:

(a) $\dfrac{d}{dx}[u_1(u_2(x))]$, (b) $\dfrac{d}{dx}[u_1(u_2(u_3(x)))]$,

(c) $\dfrac{d}{dx}[u_1(u_2(u_3(u_4(x))))]$.

▷**75.** Use a CAS to find a formula for $\dfrac{d^2}{dx^2}[f(g(x))]$.

*SUPPLEMENT TO SECTION 3.5

To prove Theorem 3.5.6, it is convenient to use a slightly different formulation of derivative.

THEOREM 3.5.7

The function f is differentiable at x iff

$$\lim_{t \to x} \frac{f(t) - f(x)}{t - x} \text{ exists.}$$

If this limit exists, it is $f'(x)$.

PROOF Fix x. For each $t \neq x$ in the domain of f, define

$$G(t) = \frac{f(t) - f(x)}{t - x}.$$

Note that

$$G(x + h) = \frac{f(x + h) - f(x)}{h}$$

and therefore

$$f \text{ is differentiable at } x \quad \text{iff} \quad \lim_{h \to 0} G(x + h) \text{ exists.}$$

The result follows from observing that

$$\lim_{h \to 0} G(x + h) = L \quad \text{iff} \quad \lim_{t \to x} G(t) = L.$$

For the equivalence of these two limits we refer you to (2.2.6). ❏

PROOF OF THEOREM 3.5.6 By Theorem 3.5.7 it is enough to show that

$$\lim_{t \to x} \frac{f(g(t)) - f(g(x))}{t - x} = f'(g(x))g'(x).$$

We begin by defining an auxiliary function F on the domain of f by setting

$$F(y) = \begin{cases} \dfrac{f(y) - f(g(x))}{y - g(x)}, & y \neq g(x) \\ f'(g(x)), & y = g(x) \end{cases}$$

F is continuous at $g(x)$ since

$$\lim_{y \to g(x)} F(y) = \lim_{y \to g(x)} \frac{f(y) - f(g(x))}{y - g(x)},$$

and the right-hand side is (by Theorem 3.5.7) $f'(g(x))$, which is the value of F at $g(x)$. For $t \neq x$,

(1)
$$\frac{f(g(t)) - f(g(x))}{t - x} = F(g(t)) \left[\frac{g(t) - g(x)}{t - x} \right].$$

To see this we note that, if $g(t) = g(x)$, then both sides are 0. If $g(t) \neq g(x)$, then

$$F(g(t)) = \frac{f(g(t)) - f(g(x))}{g(t) - g(x)},$$

so that again we have equality.

Since g, being differentiable at x, is continuous at x and since F is continuous at $g(x)$, we know that the composition $F \circ g$ is continuous at x. Thus

$$\lim_{t \to x} F(g(t)) = F(g(x)) = f'(g(x)).$$
$$\uparrow\!\!\!\rule{0.5cm}{0.4pt}\ \text{by our definition of } F$$

This, together with (1), gives

$$\lim_{t \to x} \frac{f(g(t)) - f(g(x))}{t - x} = f'(g(x))g'(x). \quad \square$$

■ PROJECT 3.5 ON THE DERIVATIVE OF u^n

If n is a positive or negative integer and the function u is differentiable at x, then by the chain rule

$$\frac{d}{dx}[u(x)]^n = n[u(x)]^{n-1}\frac{d}{dx}[u(x)],$$

except that, if n is negative, the formula fails at those numbers x where $u(x) = 0$.

We can obtain this result without appealing to the chain rule by using the product rule and carrying out an induction on n.

Let u be a differentiable function of x. Then

$$\frac{d}{dx}[u(x)]^2 = \frac{d}{dx}[u(x) \cdot u(x)]$$

$$= u(x)\frac{d}{dx}[u(x)] + u(x)\frac{d}{dx}[u(x)]$$

$$= 2u(x)\frac{d}{dx}[u(x)];$$

$$\frac{d}{dx}[u(x)]^3 = \frac{d}{dx}[u(x) \cdot [u(x)]^2]$$

$$= u(x)\frac{d}{dx}[u(x)]^2 + [u(x)]^2\frac{d}{dx}[u(x)]$$

$$= 2[u(x)]^2\frac{d}{dx}[u(x)] + [u(x)]^2\frac{d}{dx}[u(x)]$$

$$= 3[u(x)]^2\frac{d}{dx}[u(x)].$$

Problem 1. Show that

$$\frac{d}{dx}[u(x)]^4 = 4[u(x)]^3\frac{d}{dx}[u(x)].$$

Problem 2. Show by induction that

$$\frac{d}{dx}[u(x)]^n = n[u(x)]^{n-1}\frac{d}{dx}[u(x)] \quad \text{for all positive integers } n.$$

Problem 3. Show that if n is a negative integer, then

$$\frac{d}{dx}[u(x)]^n = n[u(x)]^{n-1}\frac{d}{dx}[u(x)]$$

except at those numbers x where $u(x) = 0$. HINT: Problem 2 and the reciprocal rule.

■ 3.6 DIFFERENTIATING THE TRIGONOMETRIC FUNCTIONS

An outline review of trigonometry—definitions, identities, and graphs—appears in Chapter 1. As indicated there, the calculus of the trigonometric functions is simplified by the use of radian measure. We will use radian measure throughout our work and refer to degree measure only in passing.

The derivative of the sine function is the cosine function:

(3.6.1)
$$\frac{d}{dx}(\sin x) = \cos x.$$

PROOF Fix any number x. For $h \neq 0$,

$$\frac{\sin(x+h) - \sin x}{h} = \frac{[\sin x \cos h + \cos x \sin h] - [\sin x]}{h}$$
$$= \sin x \frac{\cos h - 1}{h} + \cos x \frac{\sin h}{h}.$$

Now, as shown in Section 2.5

$$\lim_{h \to 0} \frac{\cos h - 1}{h} = 0 \quad \text{and} \quad \lim_{h \to 0} \frac{\sin h}{h} = 1.$$

Since x is fixed, $\sin x$ and $\cos x$ remain constant as h approaches zero. It follows that

$$\lim_{h \to 0} \frac{\sin(x+h) - \sin x}{h} = \lim_{h \to 0} \left(\sin x \frac{\cos h - 1}{h} + \cos x \frac{\sin h}{h} \right)$$
$$= \sin x \left(\lim_{h \to 0} \frac{\cos h - 1}{h} \right) + \cos x \left(\lim_{h \to 0} \frac{\sin h}{h} \right).$$

Thus

$$\lim_{h \to 0} \frac{\sin(x+h) - \sin x}{h} = (\sin x)(0) + (\cos x)(1) = \cos x. \quad ❏$$

The derivative of the cosine function is the negative of the sine function:

(3.6.2)
$$\frac{d}{dx}(\cos x) = -\sin x.$$

PROOF Fix any number x. For $h \neq 0$,

$$\cos(x+h) = \cos x \cos h - \sin x \sin h.$$

Therefore

$$\lim_{h \to 0} \frac{\cos(x+h) - \cos x}{h} = \lim_{h \to 0} \frac{[\cos x \cos h - \sin x \sin h] - [\cos x]}{h}$$
$$= \cos x \left(\lim_{h \to 0} \frac{\cos h - 1}{h} \right) - \sin x \left(\lim_{h \to 0} \frac{\sin h}{h} \right)$$
$$= -\sin x. \quad ❏$$

Example 1 To differentiate $f(x) = \cos x \sin x$, we use the product rule:

$$f'(x) = \cos x \frac{d}{dx}(\sin x) + \sin x \frac{d}{dx}(\cos x)$$
$$= \cos x (\cos x) + \sin x (-\sin x) = \cos^2 x - \sin^2 x. \quad ❏$$

We come now to the tangent function. Since $\tan x = \sin x / \cos x$, we have

$$\frac{d}{dx}(\tan x) = \frac{\cos x \dfrac{d}{dx}(\sin x) - \sin x \dfrac{d}{dx}(\cos x)}{\cos^2 x} = \frac{\cos^2 x + \sin^2 x}{\cos^2 x} = \frac{1}{\cos^2 x} = \sec^2 x.$$

The derivative of the tangent function is the secant squared:

(3.6.3)
$$\frac{d}{dx}(\tan x) = \sec^2 x.$$

The derivatives of the other trigonometric functions are as follows:

(3.6.4)
$$\frac{d}{dx}(\cot x) = -\csc^2 x,$$
$$\frac{d}{dx}(\sec x) = \sec x \tan x,$$
$$\frac{d}{dx}(\csc x) = -\csc x \cot x.$$

The verification of these formulas is left as an exercise.

It is time for some sample problems.

Example 2 Find $f'(\pi/4)$ for $f(x) = x \cot x$.

SOLUTION We first find $f'(x)$. By the product rule,

$$f'(x) = x\frac{d}{dx}(\cot x) + \cot x\frac{d}{dx}(x) = -x\csc^2 x + \cot x.$$

Now we evaluate f' at $\pi/4$:

$$f'(\pi/4) = -\frac{\pi}{4}(\sqrt{2})^2 + 1 = 1 - \frac{\pi}{2}. \quad ❑$$

Example 3 Find $\dfrac{d}{dx}\left[\dfrac{1 - \sec x}{\tan x}\right]$.

SOLUTION By the quotient rule,

$$\frac{d}{dx}\left[\frac{1 - \sec x}{\tan x}\right] = \frac{\tan x\dfrac{d}{dx}(1 - \sec x) - (1 - \sec x)\dfrac{d}{dx}(\tan x)}{\tan^2 x}$$

$$= \frac{\tan x(-\sec x \tan x) - (1 - \sec x)(\sec^2 x)}{\tan^2 x}$$

$$= \frac{\sec x(\sec^2 x - \tan^2 x) - \sec^2 x}{\tan^2 x}$$

$$= \frac{\sec x - \sec^2 x}{\tan^2 x} = \frac{\sec x(1 - \sec x)}{\tan^2 x}.$$

$$(\sec^2 x - \tan^2 x = 1) \longrightarrow \uparrow$$

Example 4 Find an equation for the line tangent to the curve $y = \cos x$ at the point where $x = \pi/3$.

SOLUTION Since $\cos \pi/3 = 1/2$, the point of tangency is $(\pi/3, 1/2)$. To find the slope of the tangent line, we evaluate the derivative

$$\frac{dy}{dx} = -\sin x$$

at $x = \pi/3$. This gives $m = -\sqrt{3}/2$. The equation for the tangent line can be written

$$y - \frac{1}{2} = -\frac{\sqrt{3}}{2}\left(x - \frac{\pi}{3}\right). \quad \square$$

Example 5 Set $f(x) = x + 2\sin x$. Find the numbers x in the open interval $(0, 2\pi)$ at which (a) $f'(x) = 0$, (b) $f'(x) > 0$, (c) $f'(x) < 0$.

SOLUTION The derivative of f is the function

$$f'(x) = 1 + 2\cos x.$$

The only numbers in $(0, 2\pi)$ at which $f'(x) = 0$ are the numbers at which $\cos x = -\frac{1}{2}$: $x = 2\pi/3$ and $x = 4\pi/3$. These numbers separate the interval $(0, 2\pi)$ into three open subintervals $(0, 2\pi/3), (2\pi/3, 4\pi/3), (4\pi/3, 2\pi)$. On each of these subintervals f' keeps a constant sign. The sign of f' is recorded below:

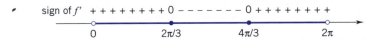

sign of f' $+ + + + + + + + 0 - - - - - - - 0 + + + + + + + +$

$\quad\quad\quad\quad 0 \quad\quad\quad\quad 2\pi/3 \quad\quad\quad 4\pi/3 \quad\quad\quad 2\pi$

Answers:

(a) $f'(x) = 0$ at $x = 2\pi/3$ and $x = 4\pi/3$.

(b) $f'(x) > 0$ on $(0, 2\pi/3) \cup (4\pi/3, 2\pi)$.

(c) $f'(x) < 0$ on $(2\pi/3, 4\pi/3)$. $\square$

The Chain Rule Applied to the Trigonometric Functions

If f is a differentiable function of u and u is a differentiable function of x, then, as you saw in Section 3.5,

$$\frac{d}{dx}[f(x)] = \frac{d}{du}[f(u)]\frac{du}{dx} = f'(u)\frac{du}{dx}.$$

Written in this form, the derivatives of the six trigonometric functions appear as follows:

(3.6.5)

$$\frac{d}{dx}(\sin u) = \cos u \frac{du}{dx}, \quad\quad \frac{d}{dx}(\cos u) = -\sin u \frac{du}{dx},$$

$$\frac{d}{dx}(\tan u) = \sec^2 u \frac{du}{dx}, \quad\quad \frac{d}{dx}(\cot u) = -\csc^2 u \frac{du}{dx},$$

$$\frac{d}{dx}(\sec u) = \sec u \tan u \frac{du}{dx}, \quad\quad \frac{d}{dx}(\csc u) = -\csc u \cot u \frac{du}{dx}.$$

Example 6

$$\frac{d}{dx}(\cos 2x) = -\sin 2x \frac{d}{dx}(2x) = -2\sin 2x. \quad \square$$

Example 7

$$\frac{d}{dx}[\sec(x^2 + 1)] = \sec(x^2 + 1)\tan(x^2 + 1)\frac{d}{dx}(x^2 + 1)$$

$$= 2x \sec(x^2 + 1)\tan(x^2 + 1). \quad \square$$

Example 8

$$\frac{d}{dx}(\sin^3 \pi x) = \frac{d}{dx}(\sin \pi x)^3$$

$$= 3(\sin \pi x)^2 \frac{d}{dx}(\sin \pi x)$$

$$= 3(\sin \pi x)^2 \cos \pi x \frac{d}{dx}(\pi x)$$

$$= 3(\sin \pi x)^2 \cos \pi x(\pi) = 3\pi \sin^2 \pi x \cos \pi x. \quad ❑$$

Our treatment of the trigonometric functions has been based entirely on radian measure. When degrees are used, the derivatives of the trigonometric functions contain the extra factor $\frac{1}{180}\pi \cong 0.0175$.

Example 9 Find $\frac{d}{dx}(\sin x°)$.

SOLUTION Since $x° = \frac{1}{180}\pi x$ radians,

$$\frac{d}{dx}(\sin x°) = \frac{d}{dx}(\sin \frac{1}{180}\pi x) = \frac{1}{180}\pi \cos \frac{1}{180}\pi x = \frac{1}{180}\pi \cos x°. \quad ❑$$

The extra factor $\frac{1}{180}\pi$ is a disadvantage, particularly in problems where it occurs repeatedly. This tends to discourage the use of degree measure in theoretical work.

EXERCISES 3.6

Exercises 1–12. Differentiate the function.

1. $y = 3\cos x - 4\sec x$.
2. $y = x^2 \sec x$.
3. $y = x^3 \csc x$.
4. $y = \sin^2 x$.
5. $y = \cos^2 t$.
6. $y = 3t^2 \tan t$.
7. $y = \sin^4 \sqrt{u}$.
8. $y = u \csc u^2$.
9. $y = \tan x^2$.
10. $y = \cos \sqrt{x}$.
11. $y = [x + \cot \pi x]^4$.
12. $y = [x^2 - \sec 2x]^3$.

Exercises 13–24. Find the second derivative.

13. $y = \sin x$.
14. $y = \cos x$.
15. $y = \dfrac{\cos x}{1 + \sin x}$.
16. $y = \tan^3 2\pi x$.
17. $y = \cos^3 2u$.
18. $y = \sin^5 3t$.
19. $y = \tan 2t$.
20. $y = \cot 4u$.
21. $y = x^2 \sin 3x$.
22. $y = \dfrac{\sin x}{1 - \cos x}$.
23. $y = \sin^2 x + \cos^2 x$.
24. $y = \sec^2 x - \tan^2 x$.

Exercises 25–30. Find the indicated derivative.

25. $\dfrac{d^4}{dx^4}(\sin x)$.
26. $\dfrac{d^4}{dx^4}(\cos x)$.
27. $\dfrac{d}{dt}\left[t^2 \dfrac{d^2}{dt^2}(t \cos 3t)\right]$.
28. $\dfrac{d}{dt}\left[t \dfrac{d}{dt}(\cos t^2)\right]$.
29. $\dfrac{d}{dx}[f(\sin 3x)]$.
30. $\dfrac{d}{dx}[\sin(f(3x))]$.

Exercises 31–36. Find an equation for the line tangent to the curve at the point with x coordinate a.

31. $y = \sin x$; $a = 0$.
32. $y = \tan x$; $a = \pi/6$.
33. $y = \cot x$; $a = \pi/6$.
34. $y = \cos x$; $a = 0$.
35. $y = \sec x$; $a = \pi/4$.
36. $y = \csc x$; $a = \pi/3$.

Exercises 37–46. Determine the numbers x between 0 and 2π where the line tangent to the curve is horizontal.

37. $y = \cos x$.
38. $y = \sin x$.
39. $y = \sin x + \sqrt{3}\cos x$.
40. $y = \cos x - \sqrt{3}\sin x$.
41. $y = \sin^2 x$.
42. $y = \cos^2 x$.
43. $y = \tan x - 2x$.
44. $y = 3\cot x + 4x$.
45. $y = 2\sec x + \tan x$.
46. $y = \cot x - 2\csc x$.

Exercises 47–50. Find all x in $(0, 2\pi)$ at which (a) $f'(x) = 0$; (b) $f'(x) > 0$; (c) $f'(x) < 0$.

47. $f(x) = x + 2\cos x$.
48. $f(x) = x - \sqrt{2}\sin x$.
49. $f(x) = \sin x + \cos x$.
50. $f(x) = \sin x - \cos x$.

Exercises 51–54. Find dy/dt (a) by the chain rule and (b) by writing y as a function of t and then differentiating.

51. $y = u^2 - 1$, $u = \sec x$, $x = \pi t$.
52. $y = [\frac{1}{2}(1 + u)]^3$, $u = \cos x$, $x = 2t$.
53. $y = [\frac{1}{2}(1 - u)]^4$, $u = \cos x$, $x = 2t$.
54. $y = 1 - u^2$, $u = \csc x$, $x = 3t$.

55. It can be shown by induction that the nth derivative of the sine function is given by the formula

$$\frac{d^n}{dx^n}(\sin x) = \begin{cases} (-1)^{(n-1)/2}\cos x, & n \text{ odd} \\ (-1)^{n/2}\sin x, & n \text{ even.} \end{cases}$$

Persuade yourself that this formula is correct and obtain a similar formula for the nth derivative of the cosine function.

56. Verify the following differentiation formulas:

(a) $\dfrac{d}{dx}(\cot x) = -\csc^2 x.$

(b) $\dfrac{d}{dx}(\sec x) = \sec x \tan x.$

(c) $\dfrac{d}{dx}(\csc x) = -\csc x \cot x.$

57. Use the identities

$$\cos x = \sin\left(\frac{\pi}{2} - x\right) \quad \text{and} \quad \sin x = \cos\left(\frac{\pi}{2} - x\right)$$

to give an alternative proof of (3.6.2).

58. The double-angle formula for the sine function takes the form: $\sin 2x = 2\sin x\cos x$. Differentiate this formula to obtain a double-angle formula for the cosine function.

59. Set $f(x) = \sin x$. Show that finding $f'(0)$ from the definition of derivative amounts to finding

$$\lim_{x\to 0}\frac{\sin x}{x}. \qquad \text{(see Section 2.5)}$$

60. Set $f(x) = \cos x$. Show that finding $f'(0)$ from the definition of derivative amounts to finding

$$\lim_{x\to 0}\frac{\cos x - 1}{x}.$$

Exercises 61–66. Find a function f with the given derivative. Check your answer by differentiation.

61. $f'(x) = 2\cos x - 3\sin x.$

62. $f'(x) = \sec^2 x - \csc^2 x.$

63. $f'(x) = 2\cos 2x + \sec x \tan x.$

64. $f'(x) = \sin 3x - \csc 2x \cot 2x.$

65. $f'(x) = 2x\cos(x^2) - 2\sin 2x.$

66. $f'(x) = x^2 \sec^2(x^3) + 2\sec 2x \tan 2x.$

67. Set $f(x) = \begin{cases} x\sin(1/x), & x \ne 0 \\ 0, & x = 0. \end{cases}$ and $g(x) = xf(x).$

In Exercise 62, Section 3.1, you were asked to show that f is continuous at 0 but not differentiable there, and that g is differentiable at 0. Both f and g are differentiable at each $x \ne 0$.

(a) Find $f'(x)$ and $g'(x)$ for $x \ne 0$.
(b) Show that g' is not continuous at 0.

68. Set $f(x) = \begin{cases} \cos x, & x \ge 0 \\ ax + b, & x < 0. \end{cases}$

(a) For what values of a and b is f differentiable at 0?
(b) Using the values of a and b you found in part (a), sketch the graph of f.

69. Set $g(x) = \begin{cases} \sin x, & 0 \le x \le 2\pi/3 \\ ax + b, & 2\pi/3 < x \le 2\pi. \end{cases}$

(a) For what values of a and b is g differentiable at $2\pi/3$?
(b) Using the values of a and b you found in part (a), sketch the graph of g.

70. Set $f(x) = \begin{cases} 1 + a\cos x, & x \le \pi/3 \\ b + \sin(x/2), & x > \pi/3. \end{cases}$

(a) For what values of a and b is f differentiable at $\pi/3$?
(b) Using the values of a and b you found in part (a), sketch the graph of f.

71. Let $y = A\sin\omega t + B\cos\omega t$ where A, B, ω are constants. Show that y satisfies the equation

$$\frac{d^2y}{dt^2} + \omega^2 y = 0.$$

72. A simple pendulum consists of a mass m swinging at the end of a rod or wire of negligible mass. The figure shows a simple pendulum of length L. The angular displacement θ at time t is given by a trigonometric expression:

$$\theta(t) = A\sin(\omega t + \phi)$$

where A, ω, ϕ are constants.

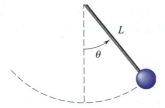

(a) Show that the function θ satisfies the equation

$$\frac{d^2\theta}{dt^2} + \omega^2\theta = 0.$$

(Except for notation, this is the equation of Exercise 71.)
(b) Show that θ can be written in the form

$$\theta(t) = A\sin\omega t + B\cos\omega t$$

where A, B, ω are constants.

73. An isosceles triangle has two sides of length c. The angle between them is x radians. Express the area A of the triangle as a function of x and find the rate of change of A with respect to x.

74. A triangle has sides of length a and b, and the angle between them is x radians. Given that a and b are kept constant, find the rate of change of the third side c with respect to x. HINT: Use the law of cosines.

▷ 75. Let $f(x) = \cos kx$, k a positive integer. Use a CAS to find

(a) $\dfrac{d^n}{dx^n}[f(x)],$

(b) all positive integers m for which $y = f(x)$ is a solution of the equation $y'' + my = 0.$

▷ 76. Use a CAS to show that $y = A\cos\sqrt{2}x + B\sin\sqrt{2}x$ is a solution of the equation $y'' + 2y = 0$. Find A and B given that $y(0) = 2$ and $y'(0) = -3$. Verify your results analytically.

77. Let $f(x) = \sin x - \cos 2x$ for $0 \le x \le 2\pi$.

(a) Use a graphing utility to estimate the points on the graph where the tangent is horizontal.

(b) Use a CAS to estimate the numbers x at which $f'(x) = 0$.

(c) Reconcile your results in (a) and (b).

78. Exercise 77 with $f(x) = \sin x - \sin^2 x$ for $0 \le x \le 2\pi$.

Exercises 79–80. Find an equation for the line l tangent to the graph of f at the point with x-coordinate c. Use a graphing utility to display l and the graph of f in one figure. Note that l is a good approximation to the graph of f for x close to c. Determine the interval on which the vertical separation between l and the graph of f is of absolute value less than 0.01.

79. $f(x) = \sin x$; $c = 0$.   **80.** $f(x) = \tan x$; $c = \pi/4$.

■ 3.7 IMPLICIT DIFFERENTIATION; RATIONAL POWERS

Up to this point we have been differentiating functions defined *explicitly* in terms of an independent variable. We can also differentiate functions not explicitly given in terms of an independent variable.

Suppose we know that y is a differentiable function of x and satisfies a particular equation in x and y. If we find it difficult to obtain the derivative of y, either because the calculations are burdensome or because we are unable to express y *explicitly* in terms of x, we may still be able to obtain dy/dx by a process called *implicit differentiation*. This process is based on differentiating both sides of the equation satisfied by x and y.

Example 1 We know that the function $y = \sqrt{1 - x^2}$ (Figure 3.7.1) satisfies the equation

$$x^2 + y^2 = 1. \qquad \text{(Figure 3.7.2)}$$

We can obtain dy/dx by carrying out the differentiation in the usual manner, or we can do it more simply by working with the equation $x^2 + y^2 = 1$.

Differentiating both sides of the equation with respect to x (remembering that y is a differentiable function of x), we have

$$\frac{d}{dx}(x^2) + \frac{d}{dx}(y^2) = \frac{d}{dx}(1)$$

$$2x + \underbrace{2y\frac{dy}{dx}}_{} = 0$$

$$\uparrow \text{———— (by the chain rule)}$$

$$\frac{dy}{dx} = -\frac{x}{y}.$$

We have obtained dy/dx in terms of x and y. Usually this is as far as we can go. Here we can go further since we have y explicitly in terms of x. The relation $y = \sqrt{1 - x^2}$ gives

$$\frac{dy}{dx} = -\frac{x}{\sqrt{1 - x^2}}.$$

Verify this result by differentiating $y = \sqrt{1 - x^2}$ in the usual manner. ❑

Example 2 Assume that y is a differentiable function of x which satisfies the given equation. Use implicit differentiation to express dy/dx in terms of x and y.

(a) $2x^2 y - y^3 + 1 = x + 2y$.   **(b)** $\cos(x - y) = (2x + 1)^3 y$.

SOLUTION

(a) Differentiating both sides of the equation with respect to x, we have

$$\underbrace{2x^2\frac{dy}{dx} + 4xy}_{} - \underbrace{3y^2\frac{dy}{dx}}_{} = 1 + 2\frac{dy}{dx}$$

(by the product rule) ———↑   ↑———(by the chain rule)

$$(2x^2 - 3y^2 - 2)\frac{dy}{dx} = 1 - 4xy.$$

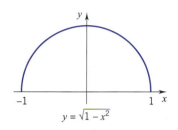

Figure 3.7.1

$y = \sqrt{1 - x^2}$

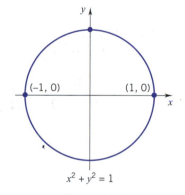

$x^2 + y^2 = 1$

Figure 3.7.2

Therefore

$$\frac{dy}{dx} = \frac{1 - 4xy}{2x^2 - 3y^2 - 2}.$$

(b) We differentiate both sides of the equation with respect to x:

$$\underbrace{-\sin(x - y)\left[1 - \frac{dy}{dx}\right]}_{\text{(by the chain rule)}} = (2x + 1)^3 \frac{dy}{dx} + 3(2x + 1)^2(2)y$$

$$[\sin(x - y) - (2x + 1)^3]\frac{dy}{dx} = 6(2x + 1)^2 y + \sin(x - y).$$

Thus

$$\frac{dy}{dx} = \frac{6(2x + 1)^2 y + \sin(x - y)}{\sin(x - y) - (2x + 1)^3}. \quad \square$$

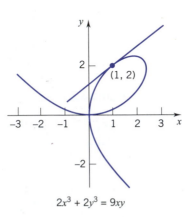

$2x^3 + 2y^3 = 9xy$

Figure 3.7.3

Example 3 Figure 3.7.3 shows the curve $2x^3 + 2y^3 = 9xy$ and the tangent line at the point $(1, 2)$. What is the slope of the tangent line at that point?

SOLUTION We want dy/dx where $x = 1$ and $y = 2$. We proceed by implicit differentiation:

$$6x^2 + 6y^2 \frac{dy}{dx} = 9x \frac{dy}{dx} + 9y$$

$$2x^2 + 2y^2 \frac{dy}{dx} = 3x \frac{dy}{dx} + 3y.$$

Setting $x = 1$ and $y = 2$, we have

$$2 + 8\frac{dy}{dx} = 3\frac{dy}{dx} + 6, \qquad 5\frac{dy}{dx} = 4, \qquad \frac{dy}{dx} = \frac{4}{5}.$$

The slope of the tangent line at the point $(1, 2)$ is $4/5$. $\quad \square$

We can also find higher derivatives by implicit differentiation.

Example 4 The function $y = (4 + x^2)^{1/3}$ satisfies the equation

$$y^3 - x^2 = 4.$$

Use implicit differentiation to express d^2y/dx^2 in terms of x and y.

SOLUTION Differentiation with respect to x gives

(∗) $$3y^2 \frac{dy}{dx} - 2x = 0.$$

Differentiating again, we have

$$\underbrace{3y^2 \frac{d}{dx}\left(\frac{dy}{dx}\right) + \left(\frac{dy}{dx}\right) \frac{d}{dx}(3y^2) - 2 = 0}_{\text{(by the product rule)}}$$

$$3y^2 \frac{d^2y}{dx^2} + 6y\left(\frac{dy}{dx}\right)^2 - 2 = 0.$$

Since (∗) gives

$$\frac{dy}{dx} = \frac{2x}{3y^2},$$

we have

$$3y^2\frac{d^2y}{dx^2} + 6y\left(\frac{2x}{3y^2}\right)^2 - 2 = 0.$$

As you can check, this gives

$$\frac{d^2y}{dx^2} = \frac{6y^3 - 8x^2}{9y^5}. \quad \square$$

Remark If we differentiate $x^2 + y^2 = -1$ implicitly, we find that

$$2x + 2y\frac{dy}{dx} = 0 \quad \text{and therefore} \quad \frac{dy}{dx} = -\frac{x}{y}.$$

However, the result is meaningless. It is meaningless because there is no real-valued function y of x that satisfies the equation $x^2 + y^2 = -1$. Implicit differentiation can be applied meaningfully to an equation in x and y only if there is a differentiable function y of x that satisfies the equation. $\square$

Rational Powers

You have seen that the formula

$$\frac{d}{dx}(x^n) = nx^{n-1}$$

holds for all real x if n is a positive integer and for all $x \neq 0$ if n is a negative integer. For $x \neq 0$, we can stretch the formula to $n = 0$ (and it is a bit of a stretch) by writing

$$\frac{d}{dx}(x^0) = \frac{d}{dx}(1) = 0 = 0x^{-1}.$$

The formula can then be extended to all rational exponents p/q:

(3.7.1)
$$\boxed{\frac{d}{dx}(x^{p/q}) = \frac{p}{q}x^{(p/q)-1}.}$$

The formula applies to all $x \neq 0$ where $x^{p/q}$ is defined.

DERIVATION OF (3.7.1) We operate under the assumption that the function $y = x^{1/q}$ is differentiable at all x where $x^{1/q}$ is defined. (This assumption is readily verified from considerations explained in Section 7.1.)

From $y = x^{1/q}$ we get

$$y^q = x.$$

Implicit differentiation with respect to x gives

$$qy^{q-1}\frac{dy}{dx} = 1$$

and therefore

$$\frac{dy}{dx} = \frac{1}{q}y^{1-q} = \frac{1}{q}x^{(1-q)/q} = \frac{1}{q}x^{(1/q)-1}.$$

So far we have shown that

$$\frac{d}{dx}(x^{1/q}) = \frac{1}{q}x^{(1/q)-1}.$$

The function $y = x^{p/q}$ is a composite function:

$$y = x^{p/q} = (x^{1/q})^p.$$

Applying the chain rule, we have

$$\frac{dy}{dx} = p(x^{1/q})^{p-1}\frac{d}{dx}(x^{1/q}) = px^{(p-1)/q}\frac{1}{q}x^{(1/q)-1} = \frac{p}{q}x^{(p/q)-1}$$

as asserted. ❏

Here are some simple examples:

$$\frac{d}{dx}(x^{2/3}) = \tfrac{2}{3}x^{-1/3}, \qquad \frac{d}{dx}(x^{5/2}) = \tfrac{5}{2}x^{3/2}, \qquad \frac{d}{dx}(x^{-7/9}) = -\tfrac{7}{9}x^{-16/9}.$$

If u is a differentiable function of x, then, by the chain rule

(3.7.2)
$$\boxed{\frac{d}{dx}(u^{p/q}) = \frac{p}{q}u^{(p/q)-1}\frac{du}{dx}.}$$

The verification of this is left to you. The result holds on every open x-interval where $u^{(p/q)-1}$ is defined.

Example 5

(a) $\dfrac{d}{dx}[(1+x^2)^{1/5}] = \tfrac{1}{5}(1+x^2)^{-4/5}(2x) = \tfrac{2}{5}x(1+x^2)^{-4/5}.$

(b) $\dfrac{d}{dx}[(1-x^2)^{2/3}] = \tfrac{2}{3}(1-x^2)^{-1/3}(-2x) = -\tfrac{4}{3}x(1-x^2)^{-1/3}.$

(c) $\dfrac{d}{dx}[(1-x^2)^{1/4}] = \tfrac{1}{4}(1-x^2)^{-3/4}(-2x) = -\tfrac{1}{2}x(1-x^2)^{-3/4}.$

The first statement holds for all real x, the second for all $x \neq \pm 1$, and the third only for $x \in (-1, 1)$. ❏

Example 6

$$\frac{d}{dx}\left[\left(\frac{x}{1+x^2}\right)^{1/2}\right] = \frac{1}{2}\left(\frac{x}{1+x^2}\right)^{-1/2}\frac{d}{dx}\left(\frac{x}{1+x^2}\right)$$

$$= \frac{1}{2}\left(\frac{x}{1+x^2}\right)^{-1/2}\frac{(1+x^2)(1) - x(2x)}{(1+x^2)^2}$$

$$= \frac{1}{2}\left(\frac{1+x^2}{x}\right)^{1/2}\frac{1-x^2}{(1+x^2)^2}$$

$$= \frac{1-x^2}{2x^{1/2}(1+x^2)^{3/2}}.$$

The result holds for all $x > 0$. ❏

EXERCISES 3.7

Preliminary note. In many of the exercises below you are asked to use implicit differentiation. We assure you that in each case there is a function $y = y(x)$ that satisfies the indicated equation and has the requisite derivative.

Exercises 1–10. Use implicit differentiation to express dy/dx in terms of x and y.

1. $x^2 + y^2 = 4.$

2. $x^3 + y^3 - 3xy = 0.$

3. $4x^2 + 9y^2 = 36.$

4. $\sqrt{x} + \sqrt{y} = 4.$

5. $x^4 + 4x^3y + y^4 = 1.$

6. $x^2 - x^2y + xy^2 + y^2 = 1.$

7. $(x - y)^2 - y = 0.$

8. $(y + 3x)^2 - 4x = 0.$

9. $\sin(x + y) = xy.$

10. $\tan xy = xy.$

Exercises 11–16. Express d^2y/dx^2 in terms of x and y.

11. $y^2 + 2xy = 16$. **12.** $x^2 - 2xy + 4y^2 = 3$.

13. $y^2 + xy - x^2 = 9$. **14.** $x^2 - 3xy = 18$.

15. $4 \tan y = x^3$. **16.** $\sin^2 x + \cos^2 y = 1$.

Exercises 17–20. Evaluate dy/dx and d^2y/dx^2 at the point indicated.

17. $x^2 - 4y^2 = 9$; $(5, 2)$.

18. $x^2 + 4xy + y^3 + 5 = 0$; $(2, -1)$.

19. $\cos(x + 2y) = 0$; $(\pi/6, \pi/6)$.

20. $x = \sin^2 y$; $(\frac{1}{2}, \pi/4)$.

Exercises 21–26. Find equations for the tangent and normal lines at the point indicated.

21. $2x + 3y = 5$; $(-2, 3)$.

22. $9x^2 + 4y^2 = 72$; $(2, 3)$.

23. $x^2 + xy + 2y^2 = 28$; $(-2, -3)$.

24. $x^3 - axy + 3ay^2 = 3a^3$; (a, a).

25. $x = \cos y$; $(\frac{1}{2}, \frac{\pi}{3})$.

26. $\tan xy = x$; $(1, \frac{\pi}{4})$.

Exercises 27–32. Find dy/dx.

27. $y = (x^3 + 1)^{1/2}$. **28.** $y = (x + 1)^{1/3}$.

29. $y = \sqrt[4]{2x^2 + 1}$. **30.** $y = (x + 1)^{1/3}(x + 2)^{2/3}$.

31. $y = \sqrt{2 - x^2}\sqrt{3 - x^2}$. **32.** $y = \sqrt{(x^4 - x + 1)^3}$.

Exercises 33–36. Carry out the differentiation.

33. $\dfrac{d}{dx}\left(\sqrt{x} + \dfrac{1}{\sqrt{x}}\right)$. **34.** $\dfrac{d}{dx}\left(\sqrt{\dfrac{3x + 1}{2x + 5}}\right)$.

35. $\dfrac{d}{dx}\left(\dfrac{x}{\sqrt{x^2 + 1}}\right)$. **36.** $\dfrac{d}{dx}\left(\dfrac{\sqrt{x^2 + 1}}{x}\right)$.

37. (*Important*) Show the general form of the graph.

    (a) $f(x) = x^{1/n}$, n a positive even integer.

    (b) $f(x) = x^{1/n}$, n a positive odd integer.

    (c) $f(x) = x^{2/n}$, n an odd integer greater than 1.

Exercises 38–42. Find the second derivative.

38. $y = \sqrt{a^2 + x^2}$. **39.** $y = \sqrt[3]{a + bx}$.

40. $y = x\sqrt{a^2 - x^2}$. **41.** $y = \sqrt{x} \tan \sqrt{x}$.

42. $y = \sqrt{x} \sin \sqrt{x}$.

43. Show that all normals to the circle $x^2 + y^2 = r^2$ pass through the center of the circle.

44. Determine the x-intercept of the tangent to the parabola $y^2 = x$ at the point where $x = a$.

The angle between two curves is the angle between their tangent lines at the point of intersection. If the slopes are m_1 and m_2, then the angle of intersection α can be obtained from the formula

$$\tan \alpha = \left| \frac{m_2 - m_1}{1 + m_1 m_2} \right|.$$

45. At what angles do the parabolas $y^2 = 2px + p^2$ and $y^2 = p^2 - 2px$ intersect?

46. At what angles does the line $y = 2x$ intersect the curve $x^2 - xy + 2y^2 = 28$?

47. The curves $y = x^2$ and $x = y^3$ intersect at the points $(1, 1)$ and $(0, 0)$. Find the angle between the curves at each of these points.

48. Find the angles at which the circles $(x - 1)^2 + y^2 = 10$ and $x^2 + (y - 2)^2 = 5$ intersect.

Two curves are said to be *orthogonal* iff, at each point of intersection, the angle between them is a right angle. Show that the curves given in Exercises 49 and 50 are orthogonal.

49. The hyperbola $x^2 - y^2 = 5$ and the ellipse $4x^2 + 9y^2 = 72$.

50. The ellipse $3x^2 + 2y^2 = 5$ and $y^3 = x^2$.
    HINT: The curves intersect at $(1, 1)$ and $(-1, 1)$

Two families of curves are said to be *orthogonal trajectories* (of each other) if each member of one family is orthogonal to each member of the other family. Show that the families of curves given in Exercises 51 and 52 are orthogonal trajectories.

51. The family of circles $x^2 + y^2 = r^2$ and the family of lines $y = mx$.

52. The family of parabolas $x = ay^2$ and the family of ellipses $x^2 + \frac{1}{2}y^2 = b$.

53. Find equations for the lines tangent to the ellipse $4x^2 + y^2 = 72$ that are perpendicular to the line $x + 2y + 3 = 0$.

54. Find equations for the lines normal to the hyperbola $4x^2 - y^2 = 36$ that are parallel to the line $2x + 5y - 4 = 0$.

55. The curve $(x^2 + y^2)^2 = x^2 - y^2$ is called a *lemniscate*. The curve is shown in the figure. Find the four points of the curve at which the tangent line is horizontal.

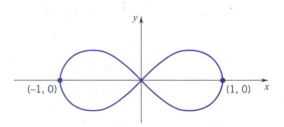

56. The curve $x^{2/3} + y^{2/3} = a^{2/3}$ is called an *astroid*. The curve is shown in the figure.

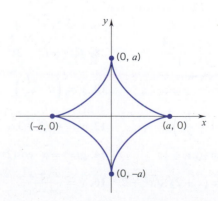

(a) Find the slope of the graph at an arbitrary point (x_1, y_1), which is not a vertex.

(b) At what points of the curve is the slope of the tangent line 0, 1, −1?

57. Show that the sum of the x- and y-intercepts of any line tangent to the graph of

$$x^{1/2} + y^{1/2} = c^{1/2}$$

is constant and equal to c.

58. A circle of radius 1 with center on the y-axis is inscribed in the parabola $y = 2x^2$. See the figure. Find the points of contact.

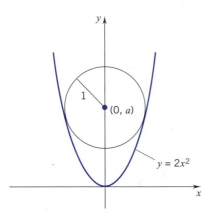

59. Set $f(x) = 3\sqrt[3]{x}$. Use a CAS to

(a) Find $d(h) = \dfrac{f(h) - f(0)}{h}$.

(b) Find $\lim\limits_{h \to 0^-} d(h)$ and $\lim\limits_{h \to 0^+} d(h)$.

(c) Is there a tangent line at $(0, 0)$? Explain.

(d) Use a graphing utility to draw the graph of f on $[-2, 2]$.

60. Exercise 59 with $f(x) = 3\sqrt[3]{x^2}$.

Exercises 61 and 62. Use a graphing utility to determine where

$$\text{(a) } f'(x) = 0; \quad \text{(b) } f'(x) > 0; \quad \text{(c) } f'(x) < 0,$$

61. $f(x) = x\sqrt[3]{x^2 + 1}$.

62. $f(x) = \dfrac{x^2 + 1}{\sqrt{x}}$.

63. A graphing utility in parametric mode can be used to graph some equations in x and y. Draw the graph of the equation $x^2 + y^2 = 4$ first by setting $x = t$, $y = \sqrt{4 - t^2}$ and then by setting $x = t$, $y = -\sqrt{4 - t^2}$.

Exercises 64–67. Use a CAS to find the slope of the line tangent to the curve at the given point. Use a graphing utility to draw the curve and the tangent line together in one figure.

64. $3x^2 + 4y^2 = 16$; $P(2, 1)$.

65. $4x^2 - y^2 = 20$; $P(3, 4)$.

66. $2 \sin y - \cos x = 0$; $P(0, \pi/6)$.

67. $\sqrt[3]{x^2} + \sqrt[3]{y^2} = 4$; $P(1, 3\sqrt{3})$.

68. (a) Use a graphing utility to draw the graph of the equation $x^3 + y^3 = 6xy$.

(b) Use a CAS to find equations for the lines tangent to the curve at the points where $x = 3$.

(c) Draw the graph of the equation and the tangent lines in one figure.

69. (a) Use a graphing utility to draw the *figure-eight* curve

$$x^4 = x^2 - y^2.$$

(b) Find the x-coordinates of the points of the graph where the tangent line is horizontal.

70. Use a graphing utility to draw the curve $(2 - x)y^2 = x^3$. Such a curve is called a *cissoid*.

■ CHAPTER 3. REVIEW EXERCISES

Exercises 1–4. Differentiate by taking the limit of the appropriate difference quotient.

1. $f(x) = x^3 - 4x + 3$.

2. $f(x) = \sqrt{1 + 2x}$.

3. $g(x) = \dfrac{1}{x - 2}$.

4. $F(x) = x \sin x$.

Exercises 5–22. Find the derivative.

5. $y = x^{2/3} - 7^{2/3}$.

6. $y = 2x^{3/4} - 4x^{-1/4}$.

7. $y = \dfrac{1 + 2x + x^2}{x^3 - 1}$.

8. $f(t) = (2 - 3t^2)^3$.

9. $f(x) = \dfrac{1}{\sqrt{a^2 - x^2}}$.

10. $y = \left(a - \dfrac{b}{x}\right)^2$.

11. $y = \left(a + \dfrac{b}{x^2}\right)^3$.

12. $y = x\sqrt{2 + 3x}$.

13. $y = \tan\sqrt{2x + 1}$.

14. $g(x) = x^2 \cos(2x - 1)$.

15. $F(x) = (x + 2)^2\sqrt{x^2 + 2}$.

16. $y = \dfrac{a^2 + x^2}{a^2 - x^2}$.

17. $h(t) = t \sec t^2 + 2t^3$.

18. $y = \dfrac{\sin 2x}{1 + \cos x}$.

19. $s = \sqrt[3]{\dfrac{2 - 3t}{2 + 3t}}$.

20. $r = \theta^2\sqrt{3 - 4\theta}$.

21. $f(\theta) = \cot(3\theta + \pi)$.

22. $y = \dfrac{x \sin 2x}{1 + x^2}$.

Exercises 23–26. Find $f'(c)$.

23. $f(x) = \sqrt[3]{x} + \sqrt{x}$; $c = 64$.

24. $f(x) = x\sqrt{8 - x^2}$; $c = 2$.

25. $f(x) = x^2 \sin^2 \pi x$; $c = \dfrac{1}{6}$.

26. $f(x) = \cot 3x$; $c = \dfrac{1}{9}\pi$.

Exercises 27–30. Find equations for the lines tangent and normal to the graph of f at the point indicated.

27. $f(x) = 2x^3 - x^2 + 3$; $(1, 4)$.

28. $f(x) = \dfrac{2x - 3}{3x + 4}$; $(-1, -5)$.

29. $f(x) = (x + 1) \sin 2x$; $(0, 0)$.

30. $f(x) = x\sqrt{1 + x^2}$; $(1, \sqrt{2})$.

Exercises 31–34. Find the second derivative.

31. $f(x) = \cos(2 - x)$. **32.** $f(x) = (x^2 + 4)^{3/2}$.

33. $y = x \sin x$. **34.** $g(u) = \tan^2 u$.

Exercises 35–36. Find a formula for the nth derivative.

35. $y = (a - bx)^n$. **36.** $y = \dfrac{a}{bx + c}$.

Exercises 37–40. Use implicit differentiation to express dy/dx in terms of x and y.

37. $x^3 y + xy^3 = 2$. **38.** $\tan(x + 2y) = x^2 y$.

39. $2x^3 + 3x \cos y = 2xy$. **40.** $x^2 + 3x\sqrt{y} = 1 + x/y$.

Exercises 41–42. Find equations for the lines tangent and normal to the curve at the point indicated.

41. $x^2 + 2xy - 3y^2 = 9$; $(3, 2)$.

42. $y \sin 2x - x \sin y = \frac{1}{4}\pi$; $(\frac{1}{4}\pi, \frac{1}{2}\pi)$.

Exercises 43–44. Find all x at which (a) $f'(x) = 0$; (b) $f'(x) > 0$; (c) $f'(x) < 0$.

43. $f(x) = x^3 - 9x^2 + 24x + 3$.

44. $f(x) = \dfrac{2x}{1 + 2x^2}$.

Exercises 45–46. Find all x in $(0, 2\pi)$ at which (a) $f'(x) = 0$; (b) $f'(x) > 0$; (c) $f'(x) < 0$.

45. $f(x) = x + \sin 2x$ **46.** $f(x) = \sqrt{3}x - 2\cos x$.

47. Find the points on the curve $y = \frac{2}{3}x^{3/2}$ where the inclination of the tangent line is (a) $\pi/4$, (b) $60°$, (c) $\pi/6$.

48. Find equations for all tangents to the curve $y = x^3$ that pass through the point $(0, 2)$.

49. Find equations for all tangents to the curve $y = x^3 - x$ that pass through the point $(-2, 2)$.

50. Find A, B, C given that the curve $y = Ax^2 + Bx + C$ passes through the point $(1, 3)$ and is tangent to the line $x - y + 1 = 0$ at the point $(2, 3)$.

51. Find A, B, C, D given that the curve $y = Ax^3 + Bx^2 + Cx + D$ is tangent to the line $y = 5x - 4$ at the point $(1, 1)$ and is tangent to the line $y = 9x$ at the point $(-1, -9)$.

52. Show that $d/dx(x^{-n}) = -n/x^{n+1}$ for all positive integers n by showing that

$$\lim_{h \to 0} \frac{1}{h}\left[\frac{1}{(x + h)^n} - \frac{1}{x^n}\right] = -\frac{n}{x^{n+1}}.$$

Exercises 53–57. Evaluate the following limits. HINT: Apply either Definition 3.1.1 or (3.1.5).

53. $\displaystyle\lim_{h \to 0} \frac{(1 + h)^2 - 2(1 + h) + 1}{h}$.

54. $\displaystyle\lim_{h \to 0} \frac{\sqrt{9 + h} - 3}{h}$. **55.** $\displaystyle\lim_{h \to 0} \frac{\sin(\frac{1}{6}\pi + h) - \frac{1}{2}}{h}$.

56. $\displaystyle\lim_{x \to 2} \frac{x^5 - 32}{x - 2}$. **57.** $\displaystyle\lim_{x \to \pi} \frac{\sin x}{x - \pi}$.

58. The figure is intended to depict a function f which is continuous on $[x_0, \infty)$ and differentiable on (x_0, ∞).

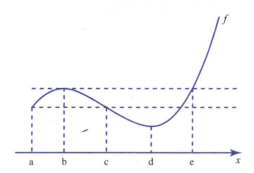

For each $x \in (x_0, \infty)$ define

$$M(x) = \text{maximum value of } f \text{ on } [x_0, x].$$

$$m(x) = \text{minimum value of } f \text{ on } [x_0, x].$$

a. Sketch the graph of M and specify the number(s) at which M fails to be differentiable.

b. Sketch the graph of m and specify the number(s) at which m fails to be differentiable.

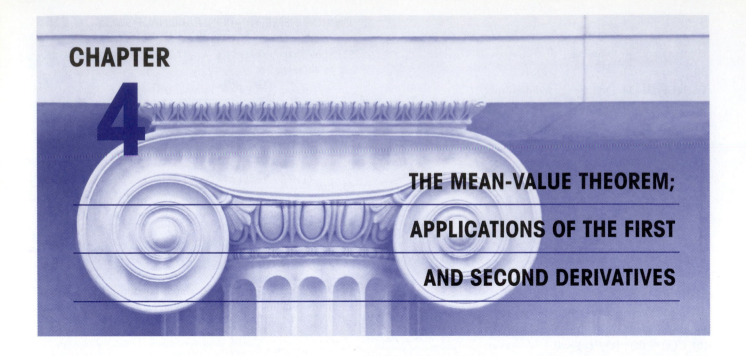

CHAPTER 4

THE MEAN-VALUE THEOREM; APPLICATIONS OF THE FIRST AND SECOND DERIVATIVES

■ 4.1 THE MEAN-VALUE THEOREM

We come now to the *mean-value theorem*. From this theorem flow most of the results that give power to the process of differentiation.[†]

THEOREM 4.1.1   THE MEAN-VALUE THEOREM

If f is differentiable on the open interval (a, b) and continuous on the closed interval $[a, b]$, then there is at least one number c in (a, b) for which

$$f'(c) = \frac{f(b) - f(a)}{b - a}.$$

Note that for this number

$$f(b) - f(a) = f'(c)(b - a).$$

The quotient

$$\frac{f(b) - f(a)}{b - a}$$

is the slope of the line l that passes through the points $(a, f(a))$ and $(b, f(b))$. To say that there is at least one number c for which

$$f'(c) = \frac{f(b) - f(a)}{b - a}$$

is to say that the graph of f has at least one point $(c, f(c))$ at which the tangent line is parallel to the line l. See Figure 4.1.1.

[†]The theorem was first stated and proved by the French mathematician Joseph-Louis Lagrange (1736–1813).

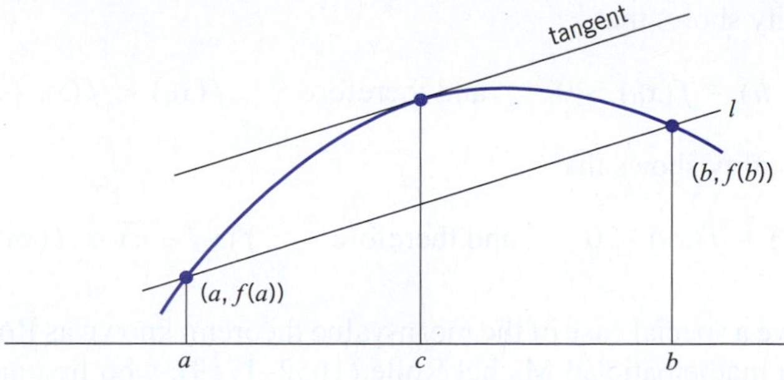

Figure 4.1.1

We will prove the mean-value theorem in steps. First we will show that if a function f has a nonzero derivative at some point x_0, then, for x close to x_0, $f(x)$ is greater than $f(x_0)$ on one side of x_0 and less than $f(x_0)$ on the other side of x_0.

THEOREM 4.1.2

Suppose that f is differentiable at x_0. If $f'(x_0) > 0$, then

$$f(x_0 - h) < f(x_0) < f(x_0 + h)$$

for all positive h sufficiently small. If $f'(x_0) < 0$, then

$$f(x_0 - h) > f(x_0) > f(x_0 + h)$$

for all positive h sufficiently small.

PROOF We take the case $f'(x_0) > 0$ and leave the other case to you. By the definition of the derivative,

$$\lim_{k \to 0} \frac{f(x_0 + k) - f(x_0)}{k} = f'(x_0).$$

With $f'(x_0) > 0$ we can use $f'(x_0)$ itself as ϵ and conclude that there exists $\delta > 0$ such that

$$\text{if} \quad 0 < |k| < \delta, \quad \text{then} \quad \left| \frac{f(x_0 + k) - f(x_0)}{k} - f'(x_0) \right| < f'(x_0).$$

For such k we have

$$-f'(x_0) < \frac{f(x_0 + k) - f(x_0)}{k} - f'(x_0) < f'(x_0)$$

and thus

$$0 < \frac{f(x_o + k) - f(x_0)}{k} < 2f'(x_0). \qquad \text{(Why?)}$$

In particular,

$$(*) \qquad \frac{f(x_0 + k) - f(x_0)}{k} > 0.$$

We have shown that $(*)$ holds for all numbers k which satisfy the condition $0 < |k| < \delta$. If $0 < h < \delta$, then $0 < |h| < \delta$ and $0 < |-h| < \delta$. Consequently,

$$\frac{f(x_0 + h) - f(x_0)}{h} > 0 \quad \text{and} \quad \frac{f(x_0 - h) - f(x_0)}{-h} > 0.$$

The first inequality shows that

$$f(x_0 + h) - f(x_0) > 0 \qquad \text{and therefore} \qquad f(x_0) < f(x_0 + h).$$

The second inequality shows that

$$f(x_0 - h) - f(x_0) < 0 \qquad \text{and therefore} \qquad f(x_0 - h) < f(x_0). \quad \square$$

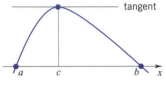

Figure 4.1.2

Next we prove a special case of the mean-value theorem, known as Rolle's theorem [after the French mathematician Michel Rolle (1652–1719), who first announced the result in 1691]. In Rolle's theorem we make the additional assumption that $f(a)$ and $f(b)$ are both 0. (See Figure 4.1.2.) In this case the line through $(a, f(a))$ and $(b, f(b))$ is horizontal. (It is the x-axis.) The conclusion is that there is a point $(c, f(c))$ at which the tangent line is horizontal.

THEOREM 4.1.3   ROLLE'S THEOREM

Suppose that f is differentiable on the open interval (a, b) and continuous on the closed interval $[a, b]$. If $f(a)$ and $f(b)$ are both 0, then there is at least one number c in (a, b) for which

$$f'(c) = 0.$$

PROOF   If f is constantly 0 on $[a, b]$, then $f'(c) = 0$ for all c in (a, b). If f is not constantly 0 on $[a, b]$, then f takes on either some positive values or some negative values. We assume the former and leave the other case to you.

Since f is continuous on $[a, b]$, f must take on a maximum value at some point c of $[a, b]$ (Theorem 2.6). This maximum value, $f(c)$, must be positive. Since $f(a)$ and $f(b)$ are both 0, c cannot be a and it cannot be b. This means that c must lie in the open interval (a, b) and therefore $f'(c)$ exists. Now $f'(c)$ cannot be greater than 0 and it cannot be less than 0 because in either case f would have to take on values greater than $f(c)$. (This follows from Theorem 4.1.2.) We can conclude therefore that $f'(c) = 0$. $\quad \square$

Remark   Rolle's theorem is sometimes formulated as follows:

Suppose that g is differentiable on the open interval (a, b) and continuous on the closed interval $[a, b]$. If $g(a) = g(b)$, then there is at least one number c in (a, b) for which

$$g'(c) = 0.$$

That these two formulations are equivalent is readily seen by setting $f(x) = g(x) - g(a)$ (Exercise 44). $\quad \square$

Rolle's theorem is not just a stepping stone toward the mean-value theorem. It is in itself a useful tool.

Example 1   We use Rolle's theorem to show that $p(x) = 2x^3 + 5x - 1$ has exactly one real zero.

SOLUTION Since p is a cubic, we know that p has at least one real zero (Exercise 29, Section 2.6). Suppose that p has more than one real zero. In particular, suppose that $p(a) = p(b) = 0$ where a and b are real numbers and $a \neq b$. Without loss of generality, we can assume that $a < b$. Since every polynomial is everywhere differentiable, p is differentiable on (a, b) and continuous on $[a, b]$. Thus, by Rolle's theorem, there is a number c in (a, b) for which $p'(c) = 0$. But

$$p'(x) = 6x^2 + 5 \geq 5 \qquad \text{for all } x,$$

and $p'(c)$ cannot be 0. The assumption that p has more than one real zero has led to a contradiction. We can conclude therefore that p has only one real zero. ❏

We are now ready to give a proof of the mean-value theorem.

PROOF OF THE MEAN-VALUE THEOREM We create a function g that satisfies the conditions of Rolle's theorem and is so related to f that the conclusion $g'(c) = 0$ leads to the conclusion

$$f'(c) = \frac{f(b) - f(a)}{b - a}.$$

The function

$$g(x) = f(x) - \left[\frac{f(b) - f(a)}{b - a}(x - a) + f(a) \right]$$

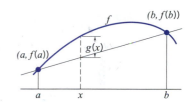

Figure 4.1.3

is exactly such a function. A geometric view of $g(x)$ is given in Figure 4.1.3. The line that passes through $(a, f(a))$ and $(b, f(b))$ has equation

$$y = \frac{f(b) - f(a)}{b - a}(x - a) + f(a).$$

[This is not hard to verify. The slope is right, and, at $x = a$, $y = f(a)$.] The difference

$$g(x) = f(x) - \left[\frac{f(b) - f(a)}{b - a}(x - a) + f(a) \right]$$

is simply the vertical separation between the graph of f and the line featured in the figure.

If f is differentiable on (a, b) and continuous on $[a, b]$, then so is g. As you can check, $g(a)$ and $g(b)$ are both 0. Therefore, by Rolle's theorem, there is at least one number c in (a, b) for which $g'(c) = 0$. Since

$$g'(x) = f'(x) - \frac{f(b) - f(a)}{b - a},$$

we have

$$g'(c) = f'(c) - \frac{f(b) - f(a)}{b - a}.$$

Since $g'(c) = 0$,

$$f'(c) = \frac{f(b) - f(a)}{b - a}. \quad ❏$$

Example 2 The function

$$f(x) = \sqrt{1 - x}, \qquad -1 \leq x \leq 1$$

satisfies the conditions of the mean-value theorem: it is differentiable on $(-1, 1)$ and continuous on $[-1, 1]$. Thus, we know that there exists a number c between -1 and 1 at which

$$f'(c) = \frac{f(1) - f(-1)}{1 - (-1)} = -\tfrac{1}{2}\sqrt{2}.$$

$$\underset{\raisebox{2pt}{$f(1) = 0,\ f(-1) = \sqrt{2}$}}{}$$

What is c in this case? To answer this, we differentiate f. By the chain rule,

$$f'(x) = -\frac{1}{2\sqrt{1-x}}.$$

The condition $f'(c) = -\tfrac{1}{2}\sqrt{2}$ gives

$$-\frac{1}{2\sqrt{1-c}} = -\frac{1}{2}\sqrt{2}.$$

Solve this equation for c and you'll find that $c = \tfrac{1}{2}$.

The tangent line at $(\tfrac{1}{2}, f(\tfrac{1}{2})) = (\tfrac{1}{2}, \tfrac{1}{2}\sqrt{2})$ is parallel to the secant line that passes through the endpoints of the graph. (Figure 4.1.4) ❏

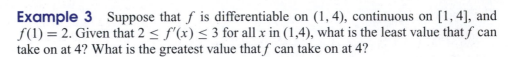

Figure 4.1.4

Example 3   Suppose that f is differentiable on $(1, 4)$, continuous on $[1, 4]$, and $f(1) = 2$. Given that $2 \le f'(x) \le 3$ for all x in $(1,4)$, what is the least value that f can take on at 4? What is the greatest value that f can take on at 4?

SOLUTION   By the mean-value theorem, there is at least one number c between 1 and 4 at which

$$f(4) - f(1) = f'(c)(4 - 1) = 3f'(c).$$

Solving this equation for $f(4)$, we have

$$f(4) = f(1) + 3f'(c).$$

Since $f'(x) \ge 2$ for every x in $(1, 4)$, we know that $f'(c) \ge 2$. It follows that

$$f(4) \ge 2 + 3(2) = 8.$$

Similarly, since $f'(x) \le 3$ for every x in $(1, 4)$, we know that $f'(c) \le 3$, and therefore

$$f(4) \le 2 + 3(3) = 11.$$

We have shown that $f(4)$ is at least 8 and no more than 11. ❏

Functions which do not satisfy the hypotheses of the mean-value theorem (differentiability on (a, b), continuity on $[a, b]$) may fail to satisfy the conclusion of the theorem. This is demonstrated in the Exercises.

EXERCISES 4.1

Exercises 1–4. Show that f satisfies the conditions of Rolle's theorem on the indicated interval and find all numbers c on the interval for which $f'(c) = 0$.

1. $f(x) = x^3 - x$; $[0, 1]$.

2. $f(x) = x^4 - 2x^2 - 8$; $[-2, 2]$.

3. $f(x) = \sin 2x$; $[0, 2\pi]$.

4. $f(x) = x^{2/3} - 2x^{1/3}$; $[0, 8]$.

Exercises 5–10. Verify that f satisfies the conditions of the mean-value theorem on the indicated interval and find all numbers c that satisfy the conclusion of the theorem.

5. $f(x) = x^2$; $[1, 2]$.

6. $f(x) = 3\sqrt{x} - 4x$; $[1, 4]$.

7. $f(x) = x^3$; $[1, 3]$.

8. $f(x) = x^{2/3}$; $[1, 8]$.

9. $f(x) = \sqrt{1 - x^2}$; $[0, 1]$.

10. $f(x) = x^3 - 3x$; $[-1, 1]$.

11. Determine whether the function $f(x) = \sqrt{1 - x^2}/(3 + x^2)$ satisfies the conditions of Rolle's theorem on the interval $[-1, 1]$. If so, find the numbers c for which $f'(c) = 0$.

12. The function $f(x) = x^{2/3} - 1$ has zeros at $x = -1$ and at $x = 1$.

    (a) Show that f' has no zeros in $(-1, 1)$.
    (b) Show that this does not contradict Rolle's theorem.

13. Does there exist a differentiable function f with $f(0) = 2$, $f(2) = 5$, and $f'(x) \leq 1$ for all x in $(0, 2)$? If not, why not?

14. Does there exist a differentiable function f with $f(x) = 1$ only at $x = 0, 2, 3$, and $f'(x) = 0$ only at $x = -1, 3/4, 3/2$? If not, why not?

15. Suppose that f is differentiable on $(2, 6)$ and continuous on $[2, 6]$. Given that $1 \leq f'(x) \leq 3$ for all x in $(2, 6)$, show that

$$4 \leq f(6) - f(2) \leq 12.$$

16. Find a point on the graph of $f(x) = x^2 + x + 3$, x between -1 and 2, where the tangent line is parallel to the line through $(-1, 3)$ and $(2, 9)$.

17. Sketch the graph of

$$f(x) = \begin{cases} 2x + 2, & x \leq -1 \\ x^3 - x, & x > -1 \end{cases}$$

and find the derivative. Determine whether f satisfies the conditions of the mean-value theorem on the interval $[-3, 2]$ and, if so, find the numbers c that satisfy the conclusion of the theorem.

18. Sketch the graph of

$$f(x) = \begin{cases} 2 + x^3, & x \leq 1 \\ 3x, & x > 1 \end{cases}$$

and find the derivative. Determine whether f satisfies the conditions of the mean-value theorem on the interval $[-1, 2]$ and, if so, find the numbers c that satisfy the conclusion of the theorem.

19. Set $f(x) = Ax^2 + Bx + C$. Show that, for any interval $[a, b]$, the number c that satisfies the conclusion of the mean-value theorem is $(a + b)/2$, the midpoint of the interval.

20. Set $f(x) = x^{-1}$, $a = -1$, $b = 1$. Verify that there is no number c for which

$$f'(c) = \frac{f(b) - f(a)}{b - a}.$$

Explain how this does not violate the mean-value theorem.

21. Exercise 20 with $f(x) = |x|$.

22. Graph the function $f(x) = |2x - 1| - 3$. Verify that $f(-1) = 0 = f(2)$ and yet $f'(x)$ is never 0. Explain how this does not violate Rolle's theorem.

23. Show that the equation $6x^4 - 7x + 1 = 0$ does not have more than two distinct real roots.

24. Show that the equation $6x^5 + 13x + 1 = 0$ has exactly one real root.

25. Show that the equation $x^3 + 9x^2 + 33x - 8 = 0$ has exactly one real root.

26. (a) Let f be differentiable on (a, b). Prove that if $f'(x) \neq 0$ for each $x \in (a, b)$, then f has at most one zero in (a, b).
    (b) Let f be twice differentiable on (a, b). Prove that if $f''(x) \neq 0$ for each $x \in (a, b)$, then f has at most two zeros in (a, b).

27. Let $P(x) = a_n x^n + \cdots + a_1 x + a_0$ be a nonconstant polynomial. Show that between any two consecutive roots of the equation $P'(x) = 0$ there is at most one root of the equation $P(x) = 0$.

28. Let f be twice differentiable. Show that, if the equation $f(x) = 0$ has n distinct real roots, then the equation $f'(x) = 0$ has at least $n - 1$ distinct real roots and the equation $f''(x) = 0$ has at least $n - 2$ distinct real roots.

29. A number c is called a *fixed point* of f if $f(c) = c$. Prove that if f is differentiable on an interval I and $f'(x) < 1$ for all $x \in I$, then f has at most one fixed point in I. HINT: Form $g(x) = f(x) - x$.

30. Show that the equation $x^3 + ax + b = 0$ has exactly one real root if $a \geq 0$ and at most one real root between $-\frac{1}{3}\sqrt{3}|a|$ and $\frac{1}{3}\sqrt{3}|a|$ if $a < 0$.

31. Set $f(x) = x^3 - 3x + b$.
    (a) Show that $f(x) = 0$ for at most one number x in $[-1, 1]$.
    (b) Determine the values of b which guarantee that $f(x) = 0$ for some number x in $[-1, 1]$.

32. Set $f(x) = x^3 - 3a^2 x + b$, $a > 0$. Show that $f(x) = 0$ for at most one number x in $[-a, a]$.

33. Show that the equation $x^n + ax + b = 0$, n an even positive integer, has at most two distinct real roots.

34. Show that the equation $x^n + ax + b = 0$, n an odd positive integer, has at most three distinct real roots.

35. Given that $|f'(x)| \leq 1$ for all real numbers x, show that $|f(x_1) - f(x_2)| \leq |x_1 - x_2|$ for all real numbers x_1 and x_2.

36. Let f be differentiable on an open interval I. Prove that, if $f'(x) = 0$ for all x in I, then f is constant on I.

37. Let f be differentiable on (a, b) with $f(a) = f(b) = 0$ and $f'(c) = 0$ for some c in (a, b). Show by example that f need not be continuous on $[a, b]$.

38. Prove that for all real x and y
    (a) $|\cos x - \cos y| \leq |x - y|$.
    (b) $|\sin x - \sin y| \leq |x - y|$.

39. Let f be differentiable on (a, b) and continuous on $[a, b]$.
    (a) Prove that if there is a constant M such that $f'(x) \leq M$ for all $x \in (a, b)$, then

$$f(b) \leq f(a) + M(b - a).$$

(b) Prove that if there is a constant m such that $f'(x) \geq m$ for all $x \in (a, b)$, then

$$f(b) \geq f(a) + m(b - a).$$

(c) Parts (a) and (b) together imply that if there exists a constant K such that $|f'(x)| \leq K$ on (a, b), then

$$f(a) - K(b - a) \leq f(b) \leq f(a) + K(b - a).$$

Show that this is the case.

40. Suppose that f and g are differentiable functions and $f(x)g'(x) - g(x)f'(x)$ has no zeros on some interval I. Assume that there are numbers a, b in I with $a < b$ for which $f(a) = f(b) = 0$, and that f has no zeros in (a, b). Prove that if $g(a) \neq 0$ and $g(b) \neq 0$, then g has exactly one zero in (a, b). HINT: Suppose that g has no zeros in (a, b) and consider $h = f/g$. Then consider $k = g/f$.

41. Suppose that f and g are nonconstant, everywhere differentiable functions and that $f' = g$ and $g' = -f$. Show that between any two consecutive zeros of f there is exactly one zero of g and between any two consecutive zeros of g there is exactly one zero of f.

42. (*Important*) Use the mean-value theorem to show that if f is continuous at x and at $x + h$ and is differentiable between these two numbers, then

$$f(x + h) - f(x) = f'(x + \theta h)h$$

for some number θ between 0 and 1. (In some texts this is how the mean-value theorem is stated.)

43. Let $h > 0$. Suppose f is continuous on $[x_0 - h, x_0 + h]$ and differentiable on $(x_0 - h, x_0 + h)$. Show that if

$$\lim_{x \to x_0} f'(x) = L,$$

then f is differentiable at x_0 and $f'(x_0) = L$. HINT: Exercise 42.

44. Suppose that g is differentiable on (a, b) and continuous on $[a, b]$. Without appealing to the mean-value theorem, show that if $g(a) = g(b)$, then there is at least one number c in (a, b) for which $g'(c) = 0$. HINT: Figure out a way to use Rolle's theorem.

45. (*Generalization of the mean-value theorem*) Suppose that f and g both satisfy the hypotheses of the mean-value theorem. Prove that if g' has no zeros in (a, b), then there is at least one number c in (a, b) for which

$$\frac{f(b) - f(a)}{g(b) - g(a)} = \frac{f'(c)}{g'(c)}.$$

This result is known as the *Cauchy mean-value theorem*. It reduces to the mean-value theorem if $g(x) = x$. HINT: To prove the result, set

$$F(x) = [f(b) - f(a)]g(x) - [g(b) - g(a)]f(x).$$

Exercises 46–47. Show that the given function satisfies the hypotheses of Rolle's theorem on the indicated interval. Use a graphing utility to graph f' and estimate the number(s) c where $f'(c) = 0$. Round off your estimates to three decimal places.

46. $f(x) = 2x^3 + 3x^2 - 3x - 2$; $[-2, 1]$.

47. $f(x) = 1 - x^3 - \cos(\pi x/2)$; $[0, 1]$.

48. Set $f(x) = x^4 - x^3 + x^2 - x$. Find a number b, if possible, such that Rolle's theorem is satisfied on $[0, b]$. If such a number b exists, find a number c that confirms Rolle's theorem on $[0, b]$ and use a graphing utility to draw the graph of f together with the line $y = f(c)$.

49. Exercise 49 with $f(x) = x^4 + x^3 + x^2 - x$.

Exercises 50–52. Use a CAS. Find the x-intercepts of the graph. Between each pair of intercepts, find, if possible, a number c that confirms Rolle's theorem.

50. $f(x) = \dfrac{x^2 - x}{x^2 + 2x + 2}$.

51. $f(x) = \dfrac{x^4 - 16}{x^2 + 4}$.

52. $f(x) = 125x^7 - 300x^6 - 760x^5 + 2336x^4 + 80x^3 - 4288x^2 + 3840x - 1024$.

Suppose that the function f satisfies the hypotheses of the mean-value theorem on an interval $[a, b]$. We can find the numbers c that satisfy the conclusion of the mean-value theorem by finding the zeros of the function

$$g(x) = f'(x) - \frac{f(b) - f(a)}{b - a}.$$

Exercises 53–54. Use a graphing utility to graph the function g that corresponds to the given f on the indicated interval. Estimate the zeros of g to three decimal places. For each zero c in the interval, graph the line tangent to the graph of f at $(c, f(c))$, and graph the line through $(a, f(a))$ and $(b, f(b))$. Verify that these lines are parallel.

53. $f(x) = x^4 - 7x^2 + 2$; $[1, 3]$.

54. $f(x) = x \cos x + 4 \sin x$; $[-\pi/2, \pi/2]$.

Exercises 55–56. The function f satisfies the hypotheses of the mean-value theorem on the given interval $[a, b]$. Use a CAS to find the number(s) c that satisfy the conclusion of the theorem. Then graph the function, the line through the endpoints $(a, f(a))$ and $(b, f(b))$, and the tangent line(s) at $(c, f(c))$.

55. $f(x) = x^3 - x^2 + x - 1$; $[1, 4]$.

56. $f(x) = x^4 - 2x^3 - x^2 - x + 1$; $[-2, 3]$.

■ 4.2 INCREASING AND DECREASING FUNCTIONS

We are going to talk about functions "increasing" or "decreasing" on an interval. To place our discussion on a solid footing, we will define these terms.

DEFINITION 4.2.1

A function f is said to

(i) *increase* on the interval I if for every two numbers x_1, x_2 in I,

$$x_1 < x_2 \qquad \text{implies that} \qquad f(x_1) < f(x_2);$$

(ii) *decrease* on the interval I if for every two numbers x_1, x_2 in I,

$$x_1 < x_2 \qquad \text{implies that} \qquad f(x_1) > f(x_2).$$

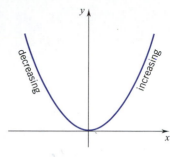

Figure 4.2.1

Preliminary Examples

(a) The squaring function

$$f(x) = x^2 \qquad \text{(Figure 4.2.1)}$$

decreases on $(-\infty, 0]$ and increases on $[0, \infty)$.

(b) The function

$$f(x) = \begin{cases} 1, & x < 0 \\ x, & x \geq 0 \end{cases} \qquad \text{(Figure 4.2.2)}$$

is constant on $(-\infty, 0)$; there it neither increases nor decreases. On $[0, \infty)$ the function increases.

(c) The cubing function

$$f(x) = x^3 \qquad \text{(Figure 4.2.3)}$$

is everywhere increasing.

(d) In the case of the Dirichlet function,

$$g(x) = \begin{cases} 1, & x \text{ rational} \\ 0, & x \text{ irrational,} \end{cases} \qquad \text{(Figure 4.2.4)}$$

there is no interval on which the function increases and no interval on which the function decreases. On every interval the function jumps back and forth between 0 and 1 an infinite number of times. ❏

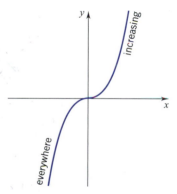

Figure 4.2.2

If f is a differentiable function, then we can determine the intervals on which f increases and the intervals on which f decreases by examining the sign of the first derivative.

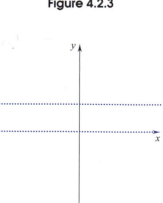

Figure 4.2.3

THEOREM 4.2.2

Suppose that f is differentiable on an open interval I.

(i) If $f'(x) > 0$ for all x in I, then f increases on I.
(ii) If $f'(x) < 0$ for all x in I, then f decreases on I.
(iii) If $f'(x) = 0$ for all x in I, then f is constant on I.

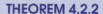
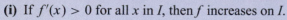

Dirichlet function

Figure 4.2.4

PROOF Choose any two numbers x_1 and x_2 in I with $x_1 < x_2$. Since f is differentiable on I, it is continuous on I. Therefore we know that f is differentiable on (x_1, x_2) and continuous on $[x_1, x_2]$. By the mean-value theorem there is a number c in (x_1, x_2) for which

$$f'(c) = \frac{f(x_2) - f(x_1)}{x_2 - x_1}.$$

In (i), $f'(x) > 0$ for all x. Therefore, $f'(c) > 0$ and we have

$$\frac{f(x_2) - f(x_1)}{x_2 - x_1} > 0, \qquad \text{which implies that} \qquad f(x_1) < f(x_2).$$

In (ii), $f'(x) < 0$ for all x. Therefore, $f'(c) < 0$ and we have

$$\frac{f(x_2) - f(x_1)}{x_2 - x_1} < 0, \qquad \text{which implies that} \qquad f(x_1) > f(x_2).$$

In (iii), $f'(x) = 0$ for all x. Therefore, $f'(c) = 0$ and we have

$$\frac{f(x_2) - f(x_1)}{x_2 - x_1} = 0, \qquad \text{which implies that} \qquad f(x_1) = f(x_2). \quad ❏$$

Remark In Section 3.2 we showed that if f is constant on an open interval I, then $f'(x) = 0$ for all $x \in I$. Part (iii) of Theorem 4.2.2 gives the converse: if $f'(x) = 0$ for all x in an open interval I, then f is constant on I. Combining these two statements, we can assert that

> if I is an open interval, then
>
> $\qquad f$ is constant on I $\qquad$ iff $\qquad$ $f'(x) = 0$ for all $x \in I$.

❏

Theorem 4.2.2 is useful but doesn't tell the complete story. Look, for example, at the function $f(x) = x^2$. The derivative $f'(x) = 2x$ is negative for x in $(-\infty, 0)$, zero at $x = 0$, and positive for x in $(0, \infty)$. Theorem 4.2.2 assures us that

$$f \text{ decreases on } (-\infty, 0) \text{ and increases on } (0, \infty),$$

but actually

$$f \text{ decreases on } (-\infty, 0] \text{ and increases on } [0, \infty).$$

To get these stronger results, we need a theorem that applies to closed intervals.

To extend Theorem 4.2.2 so that it works for an arbitrary interval I, the only additional condition we need is continuity at the endpoint(s).

> **THEOREM 4.2.3**
>
> Suppose that f is differentiable on the interior of an interval I and continuous on all of I.
>
> **(i)** If $f'(x) > 0$ for all x in the interior of I, then f increases on all of I.
> **(ii)** If $f'(x) < 0$ for all x in the interior of I, then f decreases on all of I.
> **(iii)** If $f'(x) = 0$ for all x in the interior of I, then f is constant on all of I.

The proof of this theorem is a simple modification of the proof of Theorem 4.2.2. It is time for examples.

Example 1 The function $f(x) = \sqrt{1 - x^2}$ has derivative $f'(x) = -x/\sqrt{1 - x^2}$. Since $f'(x) > 0$ for all x in $(-1, 0)$ and f is continuous on $[-1, 0]$, f increases on $[-1, 0]$. Since $f'(x) < 0$ for all x in $(0, 1)$ and f is continuous on $[0, 1]$, f decreases on $[0, 1]$. The graph of f is the semicircle shown in Figure 4.2.5. ❑

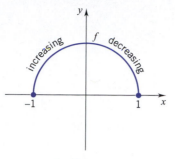

Figure 4.2.5

Example 2 The function $f(x) = 1/x$ is defined for all $x \neq 0$. The derivative $f'(x) = -1/x^2$ is negative for all $x \neq 0$. Thus the function f decreases on $(-\infty, 0)$ and on $(0, \infty)$. (See Figure 4.2.6.) Note that we did not say that f decreases on $(-\infty, 0) \cup (0, \infty)$; it does not. If $x_1 < 0 < x_2$, then $f(x_1) < f(x_2)$. ❑

Example 3 The function $g(x) = \frac{4}{5}x^5 - 3x^4 - 4x^3 + 22x^2 - 24x + 6$ is a polynomial. It is therefore everywhere continuous and everywhere differentiable.
Differentiation gives

$$g'(x) = 4x^4 - 12x^3 - 12x^2 + 44x - 24$$
$$= 4(x^4 - 3x^3 - 3x^2 + 11x - 6)$$
$$= 4(x + 2)(x - 1)^2(x - 3).$$

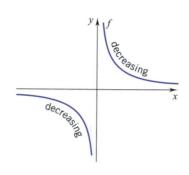

Figure 4.2.6

The derivative g' takes on the value 0 at -2, at 1, and at 3. These numbers determine four intervals on which g' keeps a constant sign:

$$(-\infty, -2), \quad (-2, 1), \quad (1, 3), \quad (3, \infty).$$

The sign of g' on these intervals and the consequences for g are as follows:

Since g is everywhere continuous, g increases on $(-\infty, -2]$, decreases on $[-2, 3]$, and increases on $[3, \infty)$. (See Figure 4.2.7.) ❑

Example 4 Let $f(x) = x - 2\sin x$, $0 \leq x \leq 2\pi$. Find the intervals on which f increases and the intervals on which f decreases.

SOLUTION In this case $f'(x) = 1 - 2\cos x$. Setting $f'(x) = 0$, we have

$$1 - 2\cos x = 0 \quad \text{and therefore} \quad \cos x = \tfrac{1}{2}.$$

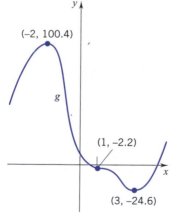

Figure 4.2.7

The only numbers in $[0, 2\pi]$ at which the cosine takes on the value $1/2$ are $x = \pi/3$ and $x = 5\pi/3$. It follows that on the intervals $(0, \pi/3)$, $(\pi/3, 5\pi/3)$, $(5\pi/3, 2\pi)$, the derivative f' keeps a constant sign. The sign of f' and the behavior of f are recorded below.

Since f is continuous throughout, f decreases on $[0, \pi/3]$, increases on $[\pi/3, 5\pi/3]$, and decreases on $[5\pi/3, 2\pi]$. (See Figure 4.2.8.) ❑

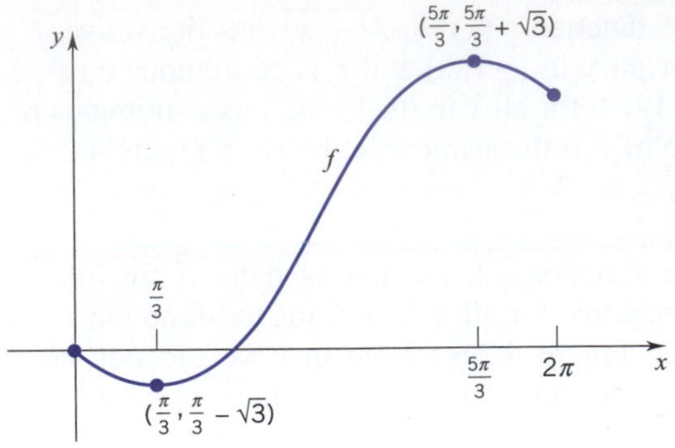

Figure 4.2.8

While the theorems we have proven have wide applicability, they do not tell the whole story.

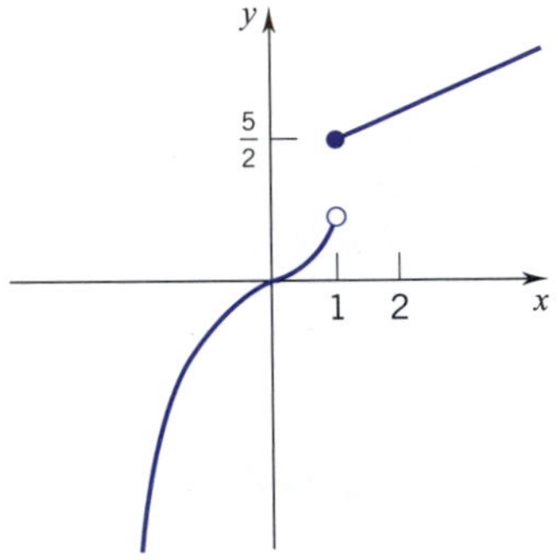

Example 5   The function

$$f(x) = \begin{cases} x^3, & x < 1 \\ \frac{1}{2}x + 2, & x \geq 1 \end{cases}$$

is graphed in Figure 4.2.9. Obviously there is a discontinuity at $x = 1$. The derivative $f'(x)$ is

$$3x^2 \text{ on } (-\infty, 1), \qquad \text{nonexistent at } x = 1, \qquad \tfrac{1}{2} \text{ on } (1, \infty).$$

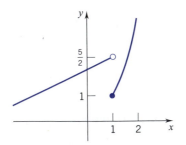

Figure 4.2.9

Since $f'(x) > 0$ on $(-\infty, 0)$ and f is continuous on $(-\infty, 0]$, f increases on $(-\infty, 0]$ (Theorem 4.2.3). Since $f'(x) > 0$ on $(0, 1)$ and is continuous on $[0, 1)$, f increases on $[0, 1)$ (Theorem 4.2.3). Since f increases on $(-\infty, 0]$ and on $[0, 1)$, f increases on $(-\infty, 1)$. (We don't need a theorem to tell us that.) Since $f'(x) > 0$ on $(1, \infty)$ and f is continuous on $[1, \infty)$, f increases on $[1, \infty)$. (Theorem 4.2.3) That f increases on $(-\infty, \infty)$ is not derivable from the theorems we've stated but is obvious by inspection. ❑

Example 6   The function

$$g(x) = \begin{cases} \frac{1}{2}x + 2, & x < 1 \\ x^3, & x \geq 1 \end{cases}$$

is graphed in Figure 4.2.10. Again, there is a discontinuity at $x = 1$. Note that $g'(x)$ is

$$\tfrac{1}{2} \text{ on } (-\infty, 1), \qquad \text{nonexistent at } x = 1, \qquad 3x^2 \text{ on } (1, \infty).$$

The function g increases on $(-\infty, 1)$ and on $[1, \infty)$ but does not increase on $(-\infty, \infty)$. The figure makes this clear. ❑

Figure 4.2.10

Equality of Derivatives

If two differentiable functions differ by a constant,

$$f(x) = g(x) + C,$$

then their derivatives are equal:

$$f'(x) = g'(x).$$

The converse is also true. In fact, we have the following theorem.

THEOREM 4.2.4

(i) Let I be an open interval. If $f'(x) = g'(x)$ for all x in I, then f and g differ by a constant on I.

(ii) Let I be an arbitrary interval. If $f'(x) = g'(x)$ for all x in the interior of I, and f and g are continuous on I, then f and g differ by a constant on I.

PROOF Set $H = f - g$. For the first assertion apply (iii) of Theorem 4.2.2 to H. For the second assertion apply (iii) of Theorem 4.2.3 to H. We leave the details as an exercise. ❏

We illustrate the theorem in Figure 4.2.11. At points with the same x-coordinate the slopes are equal, and thus the curves have the same steepness. The separation between the curves remains constant; the curves are "parallel."

Example 7 Find f given that $f'(x) = 6x^2 - 7x - 5$ for all real x and $f(2) = 1$.

SOLUTION It is not hard to find a function with the required derivative:

$$\frac{d}{dx}\left(2x^3 - \frac{7}{2}x^2 - 5x\right) = 6x^2 - 7x - 5.$$

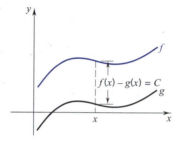

Figure 4.2.11

By Theorem 4.2.4 we know that $f(x)$ differs from $g(x) = 2x^3 - \frac{7}{2}x^2 - 5x$ only by some constant C. Thus we can write

$$f(x) = 2x^3 - \frac{7}{2}x^2 - 5x + C.$$

To evaluate C we use the fact that $f(2) = 1$. Since $f(2) = 1$ and

$$f(2) = 2(2)^3 - \frac{7}{2}(2)^2 - 5(2) + C = 16 - 14 - 10 + C = -8 + C,$$

we have $-8 + C = 1$. Therefore, $C = 9$. The function

$$f(x) = 2x^3 - \frac{7}{2}x^2 - 5x + 9$$

is the function with the specified properties. ❏

EXERCISES 4.2

Exercises 1–24. Find the intervals on which f increases and the intervals on which f decreases.

1. $f(x) = x^3 - 3x + 2.$

2. $f(x) = x^3 - 3x^2 + 6.$

3. $f(x) = x + \dfrac{1}{x}.$

4. $f(x) = (x - 3)^3.$

5. $f(x) = x^3(1 + x).$

6. $f(x) = x(x + 1)(x + 2).$

7. $f(x) = (x + 1)^4.$

8. $f(x) = 2x - \dfrac{1}{x^2}.$

9. $f(x) = \dfrac{1}{|x - 2|}.$

10. $f(x) = \dfrac{x}{1 + x^2}.$

11. $f(x) = \dfrac{x^2 + 1}{x^2 - 1}.$

12. $f(x) = \dfrac{x^2}{x^2 + 1}.$

13. $f(x) = |x^2 - 5|.$

14. $f(x) = x^2(1 + x)^2.$

15. $f(x) = \dfrac{x - 1}{x + 1}.$

16. $f(x) = x^2 + \dfrac{16}{x^2}.$

17. $f(x) = \sqrt{\dfrac{1 + x^2}{2 + x^2}}.$

18. $f(x) = |x + 1||x - 2|.$

19. $f(x) = x - \cos x, \quad 0 \le x \le 2\pi.$

20. $f(x) = x + \sin x, \quad 0 \le x \le 2\pi.$

21. $f(x) = \cos 2x + 2 \cos x, \quad 0 \le x \le \pi.$

22. $f(x) = \cos^2 x, \quad 0 \le x \le \pi.$

23. $f(x) = \sqrt{3}x - \cos 2x, \quad 0 \le x \le \pi.$

24. $f(x) = \sin^2 x - \sqrt{3} \sin x, \quad 0 \le x \le \pi.$

Exercises 25–32. Define f on the domain indicated given the following information.

25. $(-\infty, \infty)$; $\quad f'(x) = x^2 - 1$; $\quad f(1) = 2$.

26. $(-\infty, \infty)$; $\quad f'(x) = 2x - 5$; $\quad f(2) = 4$.

27. $(-\infty, \infty)$; $\quad f'(x) = 5x^4 + 4x^3 + 3x^2 + 2x + 1$; $\quad f(0) = 5$.

28. $(0, \infty)$; $\quad f'(x) = 4x^{-3}$; $\quad f(1) = 0$.

29. $(0, \infty)$; $\quad f'(x) = x^{1/3} - x^{1/2}$; $\quad f(0) = 1$.

30. $(0, \infty)$; $\quad f'(x) = x^{-5} - 5x^{-1/5}$; $\quad f(1) = 0$.

31. $(-\infty, \infty)$; $\quad f'(x) = 2 + \sin x$; $\quad f(0) = 3$.

32. $(-\infty, \infty)$; $\quad f'(x) = 4x + \cos x$; $\quad f(0) = 1$.

Exercises 33–36. Find the intervals on which f increases and the intervals on which f decreases.

33. $f(x) = \begin{cases} x + 7, & x < -3 \\ |x + 1|, & -3 \le x < 1 \\ 5 - 2x, & 1 \le x. \end{cases}$

34. $f(x) = \begin{cases} (x - 1)^2 & x < 1 \\ 5 - x, & 1 \le x < 3 \\ 7 - 2x, & 3 \le x. \end{cases}$

35. $f(x) = \begin{cases} 4 - x^2, & x < 1 \\ 7 - 2x, & 1 \le x < 3 \\ 3x - 10, & 3 \le x. \end{cases}$

36. $f(x) = \begin{cases} x + 2, & x < 0 \\ (x - 1)^2, & 0 < x < 3 \\ 8 - x, & 3 < x < 7 \\ 2x - 9, & 7 < x \\ 6, & x = 0, 3, 7. \end{cases}$

Exercises 37–40. The graph of f' is given. Draw a rough sketch of the graph of f given that $f(0) = 1$.

37.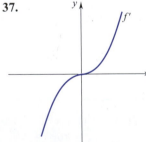

38.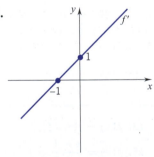

39.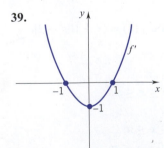

40.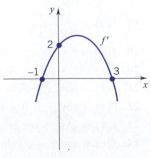

Exercises 41–42. The graph of a function f is given. Sketch the graph of f'. Give the intervals on which $f'(x) > 0$ and the intervals on which $f'(x) < 0$.

41.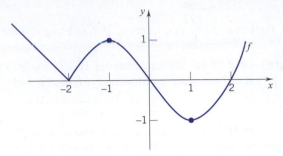

42.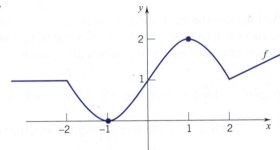

Exercises 43–46. Sketch the graph of a differentiable function f that satisfies the given conditions, if possible. If it's not possible, explain how you know it's not possible.

43. $f(x) > 0$ for all x, $f(0) = 1$, and $f'(x) < 0$ for all x.

44. $f(1) = -1$, $f'(x) < 0$ for all $x \ne 1$, and $f'(1) = 0$.

45. $f(-1) = 4$, $f(2) = 2$, and $f'(x) > 0$ for all x.

46. $f(x) = 0$ only at $x = 1$ and at $x = 2$, $f(3) = 4$, $f(5) = -1$.

Exercises 47–50. Either prove the assertion or show that the assertion is not valid by giving a counterexample. A pictorial counterexample suffices.

47. (a) If f increases on $[a, b]$ and increases on $[b, c]$, then f increases on $[a, c]$.

  (b) If f increases on $[a, b]$ and increases on $(b, c]$, then f increases on $[a, c]$.

48. (a) If f decreases on $[a, b]$ and decreases on $[b, c]$, then f decreases on $[a, c]$.

  (b) If f decreases on $[a, b)$ and decreases on $[b, c]$, then f decreases on $[a, c]$.

49. (a) If f increases on (a, b), then there is no number x in (a, b) at which $f'(x) < 0$.

  (b) If f increases on (a, b), then there is no number x in (a, b) at which $f'(x) = 0$.

50. If $f'(x) = 0$ at $x = 1, x = 2, x = 3$, then f cannot possibly increase on $[0, 4]$.

51. Set $f(x) = x - \sin x$.

  (a) Show that f increases on $(-\infty, \infty)$.

  (b) Use the result in part (a) to show that $\sin x < x$ on $(0, \infty)$ and $\sin x > x$ on $(-\infty, 0)$.

52. Prove Theorem 4.2.4.

53. Set $f(x) = \sec^2 x$ and $g(x) = \tan^2 x$ on the interval $(-\frac{\pi}{2}, \frac{\pi}{2})$. Show that $f'(x) = g'(x)$ for all x in $(-\frac{\pi}{2}, \frac{\pi}{2})$.

54. Having carried out Exercise 53, you know from Theorem 4.2.4 that there exists a constant C such that $f(x) - g(x) = C$ for all x in $(-\pi/2, \pi/2)$. What is C?

55. Suppose that for all real x

$$f'(x) = -g(x) \quad \text{and} \quad g'(x) = f(x).$$

(a) Show that $f^2(x) + g^2(x) = C$ for some constant C.
(b) Suppose that $f(0) = 0$ and $g(0) = 1$. What is C?
(c) Give an example of a pair of functions that satisfy parts (a) and (b).

56. Assume that f and g are differentiable on the interval $(-c, c)$ and $f(0) = g(0)$.

(a) Show that if $f'(x) > g'(x)$ for all $x \in (0, c)$, then $f(x) > g(x)$ for all $x \in (0, c)$.
(b) Show that if $f'(x) > g'(x)$ for all $x \in (-c, 0)$, then $f(x) < g(x)$ for all $x \in (-c, 0)$.

57. Show that $\tan x > x$ for all $x \in (0, \pi/2)$.

58. Show that $1 - x^2/2 < \cos x$ for all $x \in (0, \infty)$.

59. Let n be an integer greater than 1. Show that $(1 + x)^n > 1 + nx$ for all $x < 0$.

60. Show that $x - x^3/6 < \sin x$ for all $x > 0$.

61. It follows from Exercises 51 and 60 that

$$x - \tfrac{1}{6}x^3 < \sin x < x \quad \text{for all } x > 0.$$

Use this result to estimate $\sin 4°$. (The x above is in radians).

62. (a) Show that $\cos x < 1 - \frac{1}{2}x^2 + \frac{1}{24}x^4$ for all $x > 0$.
(b) It follows from part (a) and Exercise 58 that

$$1 - \tfrac{1}{2}x^2 < \cos x < 1 - \tfrac{1}{2}x^2 + \tfrac{1}{24}x^4 \quad \text{for all } x > 0.$$

Use this result to estimate $\cos 6°$. (The x above is in radians.)

▷**Exercises 63–66.** Use a graphing utility to graph f and its derivative f' on the indicated interval. Estimate the zeros of f' to three decimal places. Estimate the subintervals on which f increases and the subintervals on which f decreases.

63. $f(x) = 3x^4 - 10x^3 - 4x^2 + 10x + 9;$ $[-2, 5]$.

64. $f(x) = 2x^3 - x^2 - 13x - 6;$ $[-3, 4]$.

65. $f(x) = x \cos x - 3 \sin 2x;$ $[0, 6]$.

66. $f(x) = x^4 + 3x^3 - 2x^2 + 4x + 4;$ $[-5, 3]$.

▷**Exercises 67–70.** Use a CAS to find the numbers x at which
(a) $f'(x) = 0$, (b) $f'(x) > 0$, (c) $f'(x) < 0$.

67. $f(x) = \cos^3 x$, $0 \le x \le 2\pi$.

68. $f(x) = \dfrac{x}{\sqrt{x^2 + 4}}$. **69.** $f(x) = \dfrac{x^2 - 1}{x^2 + 1}$.

70. $f(x) = 8x^5 - 36x^4 + 6x^3 + 73x^2 + 48x + 9$.

▷**71.** Use a graphing utility to draw the graph of

$$f(x) = \sin x \sin (x + 2) - \sin^2(x + 1).$$

From the graph, what do you conclude about f and f'? Confirm your conclusions by calculating f'.

■ 4.3 LOCAL EXTREME VALUES

In many problems in economics, engineering, and physics it is important to determine how large or how small a certain quantity can be. If the problem admits a mathematical formulation, it is often reducible to the problem of finding the maximum or minimum value of some function.

Suppose that f is a function defined at some number c. We call c an *interior point* of the domain of f provided f is defined not only at c but at all numbers within an open interval $(c - \delta, c + \delta)$. This being the case, f is defined at all numbers x within δ of c.

DEFINITION 4.3.1 LOCAL EXTREME VALUES

Suppose that f is a function and c is an interior point of the domain. The function f is said to have a *local maximum at c* provided that

$$f(c) \ge f(x) \quad \text{for all } x \text{ sufficiently close to } c.$$

The function f is said to have a local *minimum at c* provided that

$$f(c) \le f(x) \quad \text{for all } x \text{ sufficiently close to } c.$$

The local maxima and minima of f comprise the *local extreme values of f*.

We illustrate these notions in Figure 4.3.1. A careful look at the figure suggests that local maxima and minima occur only at points where the tangent is horizontal $[f'(c) = 0]$ or where there is no tangent line $[f'(c)$ does not exist$]$. This is indeed the case.

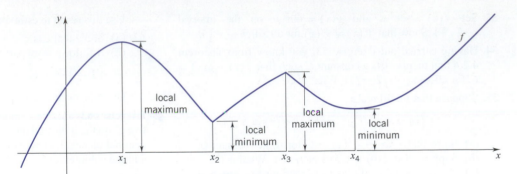

Figure 4.3.1

THEOREM 4.3.2

Suppose that c is an interior point of the domain of f. If f has a local maximum or local minimum at c, then

$$f'(c) = 0 \qquad \text{or} \qquad f'(c) \text{ does not exist.}$$

PROOF Let's suppose that f has a local extreme value at c, and let's suppose that $f'(c)$ exists. If $f'(c) > 0$ or $f'(c) < 0$, then, by Theorem 4.1.2, there must be points x_1 and x_2 arbitrarily close to c for which

$$f(x_1) < f(c) < f(x_2).$$

This makes it impossible for a local maximum or a local minimum to occur at c. Therefore, if $f'(c)$ exists, it must have the value 0. The only other possibility is that $f'(c)$ does not exist. ❏

On the basis of this result, we make the following definition (an important one):

DEFINITION 4.3.3 CRITICAL POINT

The interior points c of the domain of f for which

$$f'(c) = 0 \qquad \text{or} \qquad f'(c) \text{ does not exist}$$

are called the *critical points* for f.[†]

As a consequence of Theorem 4.3.2, in searching for local maxima and local minima, the only points we need to consider are the critical points.

We illustrate the technique for finding local maxima and minima by some examples. In each case the first step is to find the critical points.

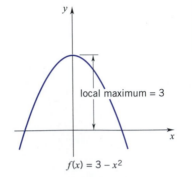

$f(x) = 3 - x^2$

Figure 4.3.2

Example 1 For

$$f(x) = 3 - x^2, \tag{Figure 4.3.2}$$

[†] Also called the *critical numbers* for f. We prefer the term "critical point" because it is more in consonance with the term used in the study of functions of several variables.

the derivative

$$f'(x) = -2x$$

exists everywhere. Since $f'(x) = 0$ only at $x = 0$, the number 0 is the only critical point. The number $f(0) = 3$ is a local maximum. ❏

Example 2 In the case of

$$f(x) = |x + 1| + 2 = \begin{cases} -x + 1, & x < -1 \\ x + 3, & x \geq -1, \end{cases} \quad \text{(Figure 4.3.3)}$$

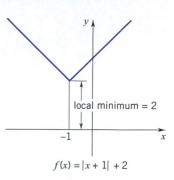

$f(x) = |x + 1| + 2$

Figure 4.3.3

differentiation gives

$$f'(x) = \begin{cases} -1, & x < -1 \\ \text{does not exist}, & x = -1 \\ 1, & x > -1. \end{cases}$$

This derivative is never 0. It fails to exist only at -1. The number -1 is the only critical point. The value $f(-1) = 2$ is a local minimum. ❏

Example 3 Figure 4.3.4 shows the graph of the function $f(x) = \dfrac{1}{x - 1}$. The domain is $(-\infty, 1) \cup (1, \infty)$. The derivative

$$f'(x) = -\frac{1}{(x - 1)^2}$$

exists throughout the domain of f and is never 0. Thus there are no critical points. In particular, 1 is not a critical point for f because 1 is not in the domain of f. Since f has no critical points, there are no local extreme values. ❏

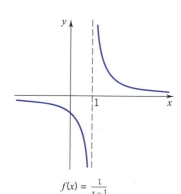

$f(x) = \frac{1}{x - 1}$

Figure 4.3.4

CAUTION The fact that c is a critical point for f does not gurantee that $f(c)$ is a local extreme value. This is made clear by the next two examples. ❏

Example 4 In the case of the function

$$f(x) = x^3, \quad \text{(Figure 4.3.5)}$$

the derivative $f'(x) = 3x^2$ is 0 at 0, but $f(0) = 0$ is not a local extreme value. The function is everywhere increasing. ❏

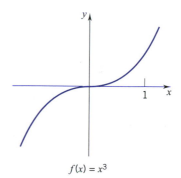

$f(x) = x^3$

Figure 4.3.5

Example 5 The function

$$f(x) = \begin{cases} -2x + 5, & x < 2 \\ -\frac{1}{2}x + 2, & x \geq 2 \end{cases} \quad \text{(Figure 4.3.6)}$$

is everywhere decreasing. Although 2 is a critical point [$f'(2)$ does not exist], $f(2) = 1$ is not a local extreme value. ❏

There are two widely used tests for determining the behavior of a function at a critical point. The first test (given in Theorem 4.3.4) requires that we examine the sign of the first derivative on both sides of the critical point. The second test (given in Theorem 4.3.5) requires that we examine the sign of the second derivative at the critical point itself.

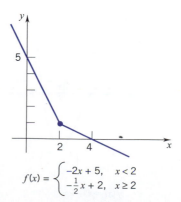

$f(x) = \begin{cases} -2x + 5, & x < 2 \\ -\frac{1}{2}x + 2, & x \geq 2 \end{cases}$

Figure 4.3.6

THEOREM 4.3.4   THE FIRST-DERIVATIVE TEST

Suppose that c is a critical point for f and f is continuous at c. If there is a positive number δ such that:

(i) $f'(x) > 0$ for all x in $(c - \delta, c)$ and $f'(x) < 0$ for all x in $(c, c + \delta)$, then $f(c)$ is a local maximum.   (Figures 4.3.7 and 4.3.8)

(ii) $f'(x) < 0$ for all x in $(c - \delta, c)$ and $f'(x) > 0$ for all x in $(c, c + \delta)$, then $f(c)$ is a local minimum.   (Figures 4.3.9 and 4.3.10)

(iii) $f'(x)$ keeps constant sign on $(c - \delta, c) \cup (c, c + \delta)$, then $f(c)$ is not a local extreme value.

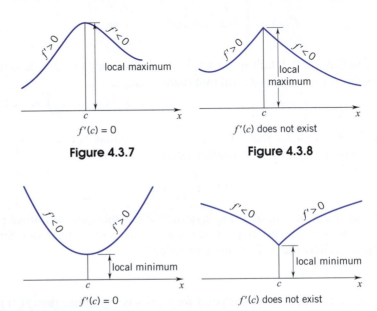

Figure 4.3.7   Figure 4.3.8

Figure 4.3.9   Figure 4.3.10

PROOF The result is a direct consequence of Theorem 4.2.3. ❏

Example 6   The function $f(x) = x^4 - 2x^3$ has derivative

$$f'(x) = 4x^3 - 6x^2 = 2x^2(2x - 3).$$

The only critical points are 0 and $\frac{3}{2}$. The sign of f' is recorded below.

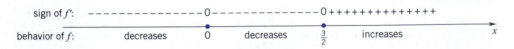

Since f' keeps the same sign on both sides of 0, $f(0) = 0$ is not a local extreme value. However, $f(\frac{3}{2}) = -\frac{27}{16}$ is a local minimum. The graph of f appears in Figure 4.3.11. ❏

Example 7   The function $f(x) = 2x^{5/3} + 5x^{2/3}$ is defined for all real x. The derivative of f is given by

$$f'(x) = \tfrac{10}{3}x^{2/3} + \tfrac{10}{3}x^{-1/3} = \tfrac{10}{3}x^{-1/3}(x + 1), \qquad x \neq 0.$$

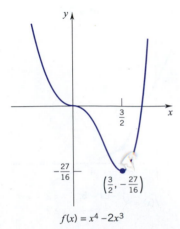

$f(x) = x^4 - 2x^3$

Figure 4.3.11

Since $f'(-1) = 0$ and $f'(0)$ does not exist, the critical points are -1 and 0. The sign of f' is recorded below. (To save space in the diagram, we write "dne" for "does not exist.")

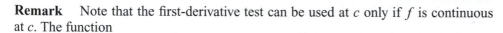

In this case $f(-1) = 3$ is a local maximum and $f(0) = 0$ is a local minimum. The graph appears in Figure 4.3.12. ❏

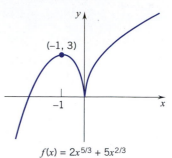

$f(x) = 2x^{5/3} + 5x^{2/3}$

Figure 4.3.12

Remark Note that the first-derivative test can be used at c only if f is continuous at c. The function

$$f(x) = \begin{cases} 1 + 2x, & x \leq 1 \\ 5 - x, & x > 1 \end{cases} \qquad \text{(Figure 4.3.13)}$$

has no derivative at $x = 1$. Therefore 1 is a critical point. While it is true that $f'(x) > 0$ for $x < 1$ and $f'(x) < 0$ for $x > 1$, it does not follow that $f(1)$ is a local maximum. The function is discontinuous at $x = 1$ and the first-derivative test does not apply. ❏

There are cases where it is difficult to determine the sign of f' on both sides of a critical point. If f is twice differentiable, then the following test may be easier to apply.

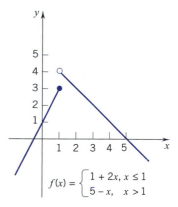

$f(x) = \begin{cases} 1 + 2x, x \leq 1 \\ 5 - x, \quad x > 1 \end{cases}$

Figure 4.3.13

THEOREM 4.3.5   THE SECOND-DERIVATIVE TEST

Suppose that $f'(c) = 0$ and $f''(c)$ exists.

 (i) If $f''(c) > 0$, then $f(c)$ is a local minimum.
 (ii) If $f''(c) < 0$, then $f(c)$ is a local maximum.

(Note that no conclusion is drawn if $f''(c) = 0$.)

PROOF We handle the case $f''(c) > 0$. The other is left as an exercise. (Exercise 32) Since f'' is the derivative of f', we see from Theorem 4.1.2 that there exists a $\delta > 0$ such that, if

$$c - \delta < x_1 < c < x_2 < c + \delta,$$

then

$$f'(x_1) < f'(c) < f'(x_2).$$

Since $f'(c) = 0$, we have

$$f'(x) < 0 \qquad \text{for } x \text{ in } (c - \delta, c) \qquad \text{and} \qquad f'(x) > 0 \qquad \text{for } x \text{ in } (c, c + \delta).$$

By the first-derivative test, $f(c)$ is a local minimum. ❏

Example 8   For $f(x) = 2x^3 - 3x^2 - 12x + 5$ we have
$$f'(x) = 6x^2 - 6x - 12 = 6(x^2 - x - 2) = 6(x - 2)(x + 1)$$
and
$$f''(x) = 12x - 6.$$

The critical points are 2 and -1; the first derivative is 0 at each of these points. Since $f''(2) = 18 > 0$ and $f''(-1) = -18 < 0$, we can conclude from the second-derivative test that $f(2) = -15$ is a local minimum and $f(-1) = 12$ is a local maximum. ❏

Comparing the First- and Second-Derivative Tests

The first-derivative test is more general than the second-derivative test. The first-derivative test can be applied at a critical point c even if f is not differentiable at c (provided of course that f is continuous at c). In contrast, the second-derivative test can be applied at c only if f is twice differentiable at c, and, even then, the test gives us information only if $f''(c) \neq 0$.

Example 9 Set $f(x) = x^{4/3}$. Here $f'(x) = \frac{4}{3}x^{1/3}$ so that

$$f'(0) = 0, \qquad f'(x) < 0 \quad \text{for} \quad x < 0, \qquad f'(x) > 0 \quad \text{for} \quad x > 0.$$

By the first-derivative test, $f(0) = 0$ is a local minimum. We cannot get this information from the second-derivative test because $f''(x) = \frac{4}{9}x^{-2/3}$ is not defined at $x = 0$. ❏

Example 10 To show what can happen if the second derivative is zero at a critical point c, we examine the functions

$$f(x) = x^3, \qquad g(x) = x^4, \qquad h(x) = -x^4. \qquad \text{(Figure 4.3.14)}$$

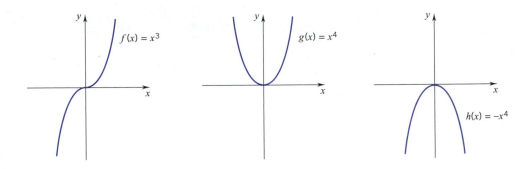

Figure 4.3.14

In each case $x = 0$ is a critical point:

$$f'(x) = 3x^2, \qquad g'(x) = 4x^3, \qquad h'(x) = -4x^3,$$
$$f'(0) = 0, \qquad g'(0) = 0, \qquad h'(0) = 0.$$

In each case the second derivative is zero at $x = 0$:

$$f''(x) = 6x, \qquad g''(x) = 12x^2, \qquad h''(0) = -12x^2,$$
$$f''(0) = 0, \qquad g''(0) = 0, \qquad h''(0) = 0.$$

The first function, $f(x) = x^3$, has neither a local maximum nor a local minimum at $x = 0$. The second function, $g(x) = x^4$, has derivative $g'(x) = 4x^3$. Since

$$g'(x) < 0 \quad \text{for} \quad x < 0, \qquad g'(x) > 0 \quad \text{for} \quad x > 0,$$

$g(0)$ is a local minimum. (The first-derivative test.) The last function, being the negative of g, has a local maximum at $x = 0$. ❏

EXERCISES 4.3

Exercises 1–28. Find the critical points and the local extreme values.

1. $f(x) = x^3 + 3x - 2$.

2. $f(x) = 2x^4 - 4x^2 + 6$.

3. $f(x) = x + \dfrac{1}{x}$.

4. $f(x) = x^2 - \dfrac{3}{x^2}$.

5. $f(x) = x^2(1 - x)$.

6. $f(x) = (1 - x)^2(1 + x)$.

7. $f(x) = \dfrac{1 + x}{1 - x}$.

8. $f(x) = \dfrac{2 - 3x}{2 + x}$.

9. $f(x) = \dfrac{2}{x(x + 1)}$.

10. $f(x) = |x^2 - 16|$.

11. $f(x) = x^3(1 - x)^2$.

12. $f(x) = \left(\dfrac{x - 2}{x + 2}\right)^3$.

13. $f(x) = (1 - 2x)(x - 1)^3$.

14. $f(x) = (1 - x)(1 + x)^3$.

15. $f(x) = \dfrac{x^2}{1 + x}$.

16. $f(x) = x\sqrt[3]{1 - x}$.

17. $f(x) = x^2\sqrt[3]{2 + x}$.

18. $f(x) = \dfrac{1}{x + 1} - \dfrac{1}{x - 2}$.

19. $f(x) = |x - 3| + |2x + 1|$.

20. $f(x) = x^{7/3} - 7x^{1/3}$.

21. $f(x) = x^{2/3} + 2x^{-1/3}$.

22. $f(x) = \dfrac{x^3}{x + 1}$.

23. $f(x) = \sin x + \cos x$, $\quad 0 < x < 2\pi$.

24. $f(x) = x + \cos 2x$, $\quad 0 < x < \pi$.

25. $f(x) = \sin^2 x - \sqrt{3} \sin x$, $\quad 0 < x < \pi$.

26. $f(x) = \sin^2 x$, $\quad 0 < x < 2\pi$.

27. $f(x) = \sin x \cos x - 3 \sin x + 2x$, $\quad 0 < x < 2\pi$.

28. $f(x) = 2\sin^3 x - 3 \sin x$, $\quad 0 < x < \pi$.

Exercises 29–30. The graph of f' is given. (a) Find the intervals on which f increases and the intervals on which f decreases. (b) Find the local maximum(s) and the local minimum(s) of f. Sketch the graph of f given that $f(0) = 1$.

29.

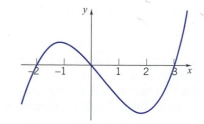

30.

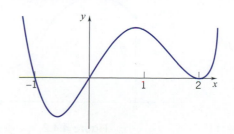

31. Let f and g be the differentiable functions, with graphs shown below. The point c is the point in the interval $[a, b]$ where the vertical separation between the two curves is greatest. Show that the line tangent to the graph of f at $x = c$ is parallel to the line tangent to the graph of g at $x = c$.

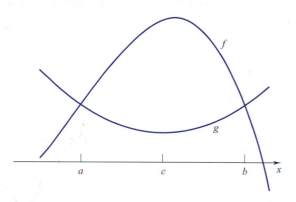

32. Prove the validity of the second-derivative test in the case that $f''(c) < 0$.

33. Let $f(x) = ax^2 + bx + c$, $a \neq 0$. Show that f has a local maximum at $x = -b/(2a)$ if $a < 0$ and a local minimum there if $a > 0$.

34. Let $f(x) = ax^3 + bx^2 + cx + d$, $a \neq 0$. Under what conditions on a, b, c will f have: (1) two local extreme values, (2) only one local extreme value, (3) no local extreme values?

35. Find the critical points and the local extreme values of the polynomial

$$P(x) = x^4 - 8x^3 + 22x^2 - 24x + 4.$$

Show that the equation $P(x) = 0$ has exactly two real roots, both positive.

36. A function f has derivative

$$f'(x) = x^3(x - 1)^2(x + 1)(x - 2).$$

At what numbers x, if any, does f have a local maximum? A local minimum?

37. Suppose that $p(x) = a_n x^n + a_{n-1}x^{n-1} + \cdots + a_1 x + a_0$ has critical points $-1, 1, 2, 3$, and corresponding values $p(-1) = 6$, $p(1) = 1$, $p(2) = 3$, $p(3) = 1$. Sketch a possible graph for p if: (a) n is odd, (b) n is even.

38. Suppose that $f(x) = Ax^2 + Bx + C$ has a local minimum at $x = 2$ and the graph passes through the points $(-1, 3)$ and $(3, -1)$. Find A, B, C.

39. Find a and b given that $f(x) = ax/(x^2 + b^2)$ has a local minimum at $x = -2$ and $f'(0) = 1$.

40. Let $f(x) = x^p(1 - x)^q$, p and q integers greater than or equal to 2.

(a) Show that the critical points of f are 0, $p/(p + q)$, 1.

(b) Show that if p is even, then f has a local minimum at 0.

(c) Show that if q is even, then f has a local minimum at 1.

(d) Show that f has a local maximum at $p/(p + q)$ for all p and q under consideration.

41. Prove that a polynomial of degree n has at most $n - 1$ local extreme values.

42. Let $y = f(x)$ be differentiable and suppose that the graph of f does not pass through the origin. The distance D from the origin to a point $P(x, f(x))$ of the graph is given by

$$D = \sqrt{x^2 + [f(x)]^2}.$$

Show that if D has a local extreme value at c, then the line through $(0, 0)$ and $(c, f(c))$ is perpendicular to the line tangent to the graph of f at $(c, f(c))$.

43. Show that $f(x) = x^4 - 7x^2 - 8x - 3$ has exactly one critical point c in the interval $(2, 3)$.

44. Show that $f(x) = \sin x + \frac{1}{2}x^2 - 2x$ has exactly one critical point c in the interval $(2, 3)$.

▶45. Set $f(x) = \dfrac{ax^2 + b}{cx^2 + d}, d \neq 0$. Use a CAS to show that f has a local minimum at $x = 0$ if $ad - bc > 0$ and a local maximum at $x = 0$ if $ad - bc < 0$. Confirm this by calculating $ad - bc$ for each of the functions given below and using a graphing utility to draw the graph.

(a) $f(x) = \dfrac{2x^2 + 3}{4 - x^2}.$ (b) $f(x) = \dfrac{3 - 2x^2}{x^2 + 2}.$

46. Set

$$f(x) = \begin{cases} x^2 \sin(1/x), & x \neq 0. \\ 0, & x = 0. \end{cases}$$

Earlier we stated that f is differentiable at 0 and that $f'(0) = 0$. Show that f has neither a local maximum nor a local minimum at $x = 0$.

▶Exercises 47–49. Use a graphing utility to graph the function on the indicated interval. (a) Use the graph to estimate the critical points and local extreme values. (b) Estimate the intervals on which the function increases and the intervals on which the function decreases. Round off your estimates to three decimal places.

47. $f(x) = 3x^3 - 7x^2 - 14x + 24; \; [-3, 4].$

48. $f(x) = |3x^3 + x^2 - 10x + 2| + 3x; \; [-4, 4].$

49. $f(x) = \dfrac{8 \sin 2x}{1 + \frac{1}{2}x^2}; \; [-3, 3].$

▶Exercises 50–52. Find the local extreme values of f by using a graphing utility to draw the graph of f and noting the numbers x at which $f'(x) = 0$.

50. $f(x) = -x^5 + 13x^4 - 67x^3 + 171x^2 - 216x + 108.$

51. $f(x) = x^2\sqrt{3x - 2}.$

52. $f(x) = \cos^2 2x.$

▶Exercises 53–54. The derivative f' of a function f is given. Use a graphing utility to graph f' on the indicated interval. Estimate the critical points of f and determine at each such point whether f has a local maximum, a local minimum, or neither. Round off your estimates to three decimal places.

53. $f'(x) = \sin^2 x + 2 \sin 2x; \; [-2, 2].$

54. $f'(x) = 2x^3 + x^2 - 4x + 3; \; [-4, 4].$

■ 4.4 ENDPOINT EXTREME VALUES; ABSOLUTE EXTREME VALUES

We will work with functions defined on an interval or on an interval with a finite number of points removed.

A number c is called the *left endpoint* of the domain of f if f is defined at c but undefined to the left of c. We call c the *right endpoint* of the domain of f if f is defined at c but undefined to the right of c.

The assumptions made on the structure of the domain guarantee that if c is the left endpoint of the domain, then f is defined at least on an interval $[c, c + \delta)$, and if c is the right endpoint, then f is defined at least on an interval $(c - \delta, c]$.

Endpoints of the domain can give rise to what are called *endpoint extreme values*. Endpoint extreme values (illustrated in Figures 4.4.1–4.4.4) are defined below.

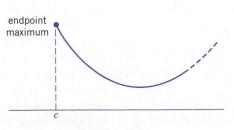

Figure 4.4.1

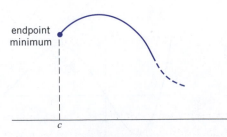

Figure 4.4.2

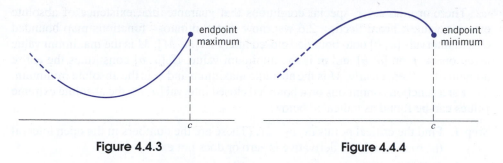

Figure 4.4.3   **Figure 4.4.4**

DEFINITION 4.4.1   ENDPOINT EXTREME VALUES

If c is an endpoint of the domain of f, then f is said to have an *endpoint maximum at c* provided that

$$f(c) \geq f(x) \qquad \text{for all } x \text{ in the domain of } f \text{ sufficiently close to } c.$$

It is said to have an *endpoint minimum at c* provided that

$$f(c) \leq f(x) \qquad \text{for all } x \text{ in the domain of } f \text{ sufficiently close to } c.$$

Endpoints in the domain of a continuous function which is differentiable at all points of the domain near that endpoint can be tested by examining the sign of the first derivative at nearby points and then reasoning as we did in Section 4.3. Suppose, for example, that c is a left endpoint and that f is continuous from the right at c. If $f'(x) < 0$ at all nearby $x > c$, then f decreases on an interval $[c, c + \delta)$ and $f(c)$ is an endpoint maximum. (Figure 4.4.1) If, on the other hand, $f'(x) > 0$ at all nearby $x > c$, then f increases on an interval $[c, c + \delta)$ and $f(c)$ is an endpoint minimum. (Figure 4.4.2) Similar reasoning can be applied to right endpoints.

Absolute Maxima and Absolute Minima

Whether or not a function f has a local extreme value or an endpoint extreme value at some point c depends entirely on the behavior of f at c and at points close to c. Absolute extreme values, which we define below, depend on the behavior of the function on its entire domain.

We begin with a number d in the domain of f. Here d can be an interior point or an endpoint.

DEFINITION 4.4.2   ABSOLUTE EXTREME VALUES

The function f is said to have an *absolute maximum at d* provided that

$$f(d) \geq f(x) \qquad \text{for all } x \text{ in the domain of } f;$$

f is said to have an *absolute minimum at d* provided that

$$f(d) \leq f(x) \qquad \text{for all } x \text{ in the domain of } f.$$

A function can be continuous on an interval (or even differentiable there) without taking on an absolute maximum or an absolute minimum. All we can say in general is that if f takes on an absolute extreme value, then it does so at a critical point or at an endpoint.

There are, however, special conditions that guarantee the existence of absolute extreme values. From Section 2.6 we know that continuous functions map bounded closed intervals $[a, b]$ onto bounded closed intervals $[m, M]$; M is the maximum value taken on by f on $[a, b]$ and m is the minimum value. If $[a, b]$ constitutes the entire domain of f, then, clearly, M is the absolute maximum and m is the absolute minimum.

For a function continuous on a bounded closed interval $[a, b]$, the absolute extreme values can be found as indicated below.

Step 1. Find the critical points $c_1, c_2, \ldots$. (These are the numbers in the open interval (a, b) at which the derivative is zero or does not exist.)

Step 2. Calculate $f(a), f(c_1), f(c_2), \ldots, f(b)$.

Step 3. The greatest of these numbers is the absolute maximum value of f and the least of these numbers is the absolute minimum.

Example 1 Find the critical points of the function

$$f(x) = 1 + 4x^2 - \tfrac{1}{2}x^4, \quad x \in [-1, 3].$$

Then find and classify all the extreme values.

SOLUTION Since f is continuous and the entire domain is the bounded closed interval $[-1, 3]$, we know that f has an absolute maximum and an absolute minimum. To find the critical points of f, we differentiate:

$$f'(x) = 8x - 2x^3 = 2x(4 - x^2) = 2x(2 - x)(2 + x).$$

The numbers x in $(-1, 3)$ at which $f'(x) = 0$ are $x = 0$ and $x = 2$. Thus, 0 and 2 are the critical points.

The sign of f' and the behavior of f are as follows:

sign of f' $- - - - - - 0 + + + + + + + + 0 - - - - - -$

behavior of f -1 decreases 0 increases 2 decreases 3 x

Taking the endpoints into consideration, we have:

$$f(-1) = 1 + 4(-1)^2 - \tfrac{1}{2}(-1)^4 = \tfrac{9}{2} \qquad \text{is an endpoint maximum;}$$
$$f(0) = 1 \qquad \text{is a local minimum;}$$
$$f(2) = 1 + 4(2)^2 - \tfrac{1}{2}(2^4) = 9 \qquad \text{is a local maximum;}$$
$$f(3) = 1 + 4(3)^2 - \tfrac{1}{2}(3)^4 = -\tfrac{7}{2} \qquad \text{is an endpoint minimum.}$$

The least of these extremes, $f(3) = -\tfrac{7}{2}$, is the absolute minimum; the greatest of these extremes, $f(2) = 9$, is the absolute maximum. The graph of the function is shown in Figure 4.4.5. ❑

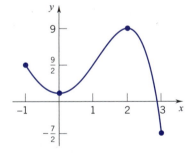

Figure 4.4.5

Example 2 Find the critical points of the function

$$f(x) = \begin{cases} x^2 + 2x + 2, & -\tfrac{1}{2} \le x < 0 \\ x^2 - 2x + 2, & 0 \le x \le 2. \end{cases}$$

Then find and classify all the extreme values.

SOLUTION Since f is continuous on its entire domain, which is the bounded closed interval $[-\tfrac{1}{2}, 2]$, we know that f has an absolute maximum and an absolute minimum. Differentiating f, we see that $f'(x)$ is

$$2x + 2 \text{ on } (-\tfrac{1}{2}, 0), \qquad \text{nonexistent at } x = 0, \qquad 2x - 2 \text{ on } (0, 2).$$

This makes $x = 0$ a critical point. Since $f'(x) = 0$ at $x = 1$, 1 is a critical point.
The sign of f' and the behavior of f are as follows:

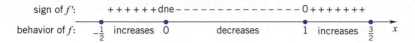

Therefore

$$f(-\tfrac{1}{2}) = \tfrac{1}{4} - 1 + 2 = \tfrac{5}{4} \qquad \text{is an endpoint minimum;}$$

$$f(0) = 2 \qquad \text{is a local maximum;}$$

$$f(1) = 1 - 2 + 2 = 1 \qquad \text{is a local minimum;}$$

$$f(2) = 2 \qquad \text{is an endpoint maximum.}$$

The least of these extremes, $f(1) = 1$, is the absolute minimum; the greatest of these extremes, $f(0) = f(2) = 2$, is the absolute maximum. The graph of the function is shown in Figure 4.4.6. ❑

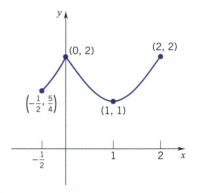

Figure 4.4.6

Behavior of $f(x)$ as $x \to \infty$ and as $x \to -\infty$.

We now state four definitions. Once you grasp the first one, the others become transparent.

To say that

$$\text{as} \qquad x \to \infty, \qquad f(x) \to \infty$$

is to say that, *as x increases without bound, $f(x)$ becomes arbitrarily large.* More precisely, given any positive number M, there exists a positive number K such that

$$\text{if} \qquad x \geq K, \qquad \text{then} \qquad f(x) \geq M.$$

For example, as $x \to \infty$,

$$x^2 \to \infty, \qquad \sqrt{1 + x^2} \to \infty, \qquad \tan\left(\frac{\pi}{2} - \frac{1}{x^2}\right) \to \infty$$

To say that

$$\text{as} \qquad x \to \infty, \qquad f(x) \to -\infty$$

is to say that, *as x increases without bound, $f(x)$ becomes arbitrarily large negative:* given any negative number M, there exists a positive number K such that

$$\text{if} \qquad x \geq K, \qquad \text{then} \qquad f(x) \leq M.$$

For example, as $x \to \infty$,

$$-x^4 \to -\infty, \qquad 1 - \sqrt{x} \to -\infty, \qquad \tan\left(\frac{1}{x^2} - \frac{\pi}{2}\right) \to -\infty.$$

To say that

$$\text{as} \qquad x \to -\infty, \qquad f(x) \to \infty$$

is to say that, *as x decreases without bound, $f(x)$ becomes arbitrarily large:* given any positive number M, there exists a negative number K such that

$$\text{if} \qquad x \leq K, \qquad \text{then} \qquad f(x) \geq M.$$

For example, as $x \to -\infty$,

$$x^2 \to \infty, \qquad \sqrt{1 - x} \to \infty, \qquad \tan\left(\frac{\pi}{2} - \frac{1}{x^2}\right) \to \infty.$$

Finally, to say that

$$\text{as} \qquad x \to -\infty, \qquad f(x) \to -\infty$$

is to say that, *as x decreases without bound, f(x) becomes arbitrarily large negative*: given any negative number M, there exists a negative number K such that,

$$\text{if} \quad x \leq K, \quad \text{then} \quad f(x) \leq M.$$

For example, as $x \to -\infty$,

$$x^3 \to -\infty, \qquad -\sqrt{1-x} \to -\infty, \qquad \tan\left(\frac{1}{x^2} - \frac{\pi}{2}\right) \to -\infty.$$

Remark   As you can readily see, $f(x) \to -\infty$ iff $-f(x) \to \infty$.   ❏

Suppose now that P is a nonconstant polynomial:

$$P(x) = a_n x^n + a_{n-1} x^{n-1} + \cdots + a_1 x + a_0 \quad (a_n \neq 0, \ n \geq 1).$$

For large $|x|$ — that is, for large positive x and for large negative x — the leading term $a_n x^n$ dominates. Thus, what happens to $P(x)$ as $x \to \pm\infty$ depends entirely on what happens to $a_n x^n$. (You are asked to confirm this in Exercise 43.)

Example 3

(a) As $x \to \infty$, $3x^4 - 100x^3 + 2x - 5 \to \infty$ since $3x^4 \to \infty$.

(b) As $x \to -\infty$, $5x^3 + 12x^2 + 80 \to -\infty$ since $5x^3 \to -\infty$.   ❏

Finally, we point out that if $f(x) \to \infty$, then f cannot have an absolute maximum value, and if $f(x) \to -\infty$, then f cannot have an absolute minimum value.

Example 4   Find the critical points of the function

$$f(x) = 6\sqrt{x} - x\sqrt{x}.$$

Then find and classify all the extreme values.

SOLUTION   The domain is $[0, \infty)$. To simplify the differentiation, we use fractional exponents:

$$f(x) = 6x^{1/2} - x^{3/2}.$$

On $(0, \infty)$

$$f'(x) = 3x^{-1/2} - \frac{3}{2}x^{1/2} = \frac{3(2-x)}{2\sqrt{x}}. \qquad \text{(Verify this.)}$$

Since $f'(x) = 0$ at $x = 2$, we see that 2 is a critical point.
The sign of f' and the behavior of f are as follows:

sign of f': $+ + + + + + + \; 0 \; - - - - - - - -$

behavior of f: 0   increases   1   decreases x

Therefore,

$$f(0) = 0 \text{ is an endpoint minimum;}$$

$$f(2) = 6\sqrt{2} - 2\sqrt{2} = 4\sqrt{2} \text{ is a local maximum.}$$

Since $f(x) = \sqrt{x}(6 - x) \to -\infty$ as $x \to \infty$, the function has no absolute minimum value. Since f increases on $[0, 2]$ and decreases on $[2, \infty)$, the local maximum is the absolute maximum. The graph of f appears in Figure 4.4.7.   ❏

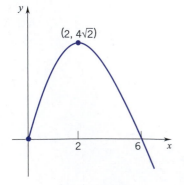

Figure 4.4.7

A Summary for Finding All the Extreme Values (Local, Endpoint, and Absolute) of a Continuous Function f

Step 1. Find the critical points — the interior points c at which $f'(c) = 0$ or $f'(c)$ does not exist.

Step 2. Test each endpoint of the domain by examining the sign of the first derivative at nearby points.

Step 3. Test each critical point c by examining the sign of the first derivative on both sides of c (the first-derivative test) or by checking the sign of the second derivative at c itself (second-derivative test).

Step 4. If the domain is unbounded on the right, determine the behavior of $f(x)$ as $x \to \infty$; if unbounded on the left, check the behavior of $f(x)$ as $x \to -\infty$.

Step 5. Determine whether any of the endpoint extremes and local extremes are absolute extremes.

Example 5 Find the critical points of the function

$$f(x) = \tfrac{1}{4}(x^3 - \tfrac{3}{2}x^2 - 6x + 2), \qquad x \in [-2, \infty).$$

The find and classify all the extreme values.

SOLUTION To find the critical points, we differentiate:

$$f'(x) = \tfrac{1}{4}(3x^2 - 3x - 6) = \tfrac{3}{4}(x + 1)(x - 2).$$

Since $f'(x) = 0$ at $x = -1$ and $x = 2$, the numbers -1 and 2 are critical points. The sign of f' and the behavior of f are as follows:

sign of f': $+ + + + + + + 0 - - - - - - - - - - - - - - 0 + + + + + + +$

behavior of f: -2 increases -1 decreases 2 increases x

We can see from the sign of f' that

$$f(-2) = \tfrac{1}{4}(-8 - 6 + 12 + 2) = 0 \qquad \text{is an endpoint minimum;}$$

$$f(-1) = \tfrac{1}{4}(-1 - \tfrac{3}{2} + 6 + 2) = \tfrac{11}{8} \qquad \text{is a local maximum;}$$

$$f(2) = \tfrac{1}{4}(8 - 6 - 12 + 2) = -2 \qquad \text{is a local minimum.}$$

The function takes on no absolute maximum value since $f(x) \to \infty$ as $x \to \infty$; $f(2) = -2$ is the absolute minimum value. The graph of f is shown in Figure 4.4.8. ❏

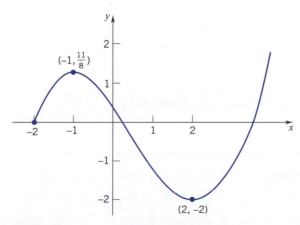

Figure 4.4.8

Example 6 Find the critical points of the function
$$f(x) = \sin x - \sin^2 x, \qquad x \in [0, 2\pi].$$
Then find and classify all the extreme values.

SOLUTION On the interval $(0, 2\pi)$
$$f'(x) = \cos x - 2 \sin x \cos x = \cos x (1 - 2 \sin x).$$
Setting $f'(x) = 0$, we have
$$\cos x (1 - 2 \sin x) = 0.$$

This equation is satisfied by the numbers x at which $\cos x = 0$ and the numbers x at which $\sin x = \frac{1}{2}$. On $(0, 2\pi)$, the cosine is 0 only at $x = \pi/2$ and $x = 3\pi/2$, and the sine is $\frac{1}{2}$ only at $x = \pi/6$ and $x = 5\pi/6$. The critical points, listed in order, are
$$\pi/6, \qquad \pi/2, \qquad 5\pi/6, \qquad 3\pi/2.$$

The sign of f' and the behavior of f are as follows:

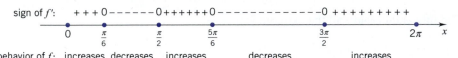

Therefore

$f(0) = 0$ is an endpoint minimum; $f(\pi/6) = \frac{1}{4}$ is a local maximum;

$f(\pi/2) = 0$ is a local minimum; $f(5\pi/6) = \frac{1}{4}$ is a local maximum;

$f(3\pi/2) = -2$ is a local minimum; $f(2\pi) = 0$ is an endpoint maximum.

Note that $f(\pi/6) = f(5\pi/6) = \frac{1}{4}$ is the absolute maximum and $f(3\pi/2) = -2$ is the absolute minimum. The graph of the function is shown in Figure 4.4.9. ❏

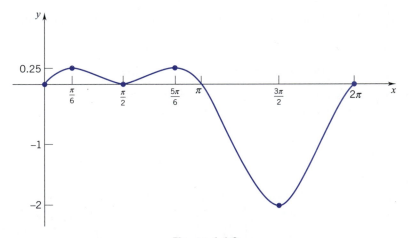

Figure 4.4.9

EXERCISES 4.4

Exercises 1–30. Find the critical points. Then find and classify all the extreme values.

* **1.** $f(x) = \sqrt{x + 2}$.

2. $f(x) = (x - 1)(x - 2)$.

3. $f(x) = x^2 - 4x + 1$, $x \in [0, 3]$.

∘ **4.** $f(x) = 2x^2 + 5x - 1$, $x \in [-2, 0]$.

5. $f(x) = x^2 + \dfrac{1}{x}$.   **6.** $f(x) = x + \dfrac{1}{x^2}$.

7. $f(x) = x^2 + \dfrac{1}{x}$, $\quad x \in \left[\frac{1}{10}, 2\right]$.

8. $f(x) = x + \dfrac{1}{x^2}$, $\quad x[1, \sqrt{2}]$.

9. $f(x) = (x - 1)(x - 2)$, $\quad x \in [0, 2]$.

10. $f(x) = (x - 1)^2(x - 2)^2$, $\quad x \in [0, 4]$.

11. $f(x) = \dfrac{x}{4 + x^2}$, $\quad x \in [-3, 1]$.

12. $f(x) = \dfrac{x^2}{1 + x^2}$, $\quad x \in [-1, 2]$.

13. $f(x) = (x - \sqrt{x})^2$. $\qquad$ **14.** $f(x) = x\sqrt{4 - x^2}$.

15. $f(x) = x\sqrt{3 - x}$. $\qquad$ **16.** $f(x) = \sqrt{x} - \dfrac{1}{\sqrt{x}}$.

17. $f(x) = 1 - \sqrt[3]{x - 1}$.

18. $f(x) = (4x - 1)^{1/3}(2x - 1)^{2/3}$.

19. $f(x) = \sin^2 x - \sqrt{3} \cos x$, $\quad 0 \le x \le \pi$.

20. $f(x) = \cot x + x$, $\quad 0 \le x \le \frac{2}{3}\pi$.

21. $f(x) = 2\cos^3 x + 3\cos x$, $\quad 0 \le x \le \pi$.

22. $f(x) = \sin 2x - x$, $\quad 0 \le x \le \pi$.

23. $f(x) = \tan x - x$, $\quad -\frac{1}{3}\pi \le x \le \frac{1}{2}\pi$.

24. $f(x) = \sin^4 x - \sin^2 x$, $\quad 0 \le x \le \frac{2}{3}\pi$.

25. $f(x) = \begin{cases} -2x, & 0 \le x < 1 \\ x - 3, & 1 \le x \le 4 \\ 5 - x, & 4 < x \le 7. \end{cases}$

26. $f(x) = \begin{cases} x + 9, & -8 \le x < -3 \\ x^2 + x, & -3 \le x \le 2 \\ 5x - 4, & 2 < x < 5. \end{cases}$

27. $f(x) = \begin{cases} x^2 + 1, & -2 \le x < -1 \\ 5 + 2x - x^2, & -1 \le x \le 3 \\ x - 1, & 3 < x < 6. \end{cases}$

28. $f(x) = \begin{cases} 2 - 2x - x^2, & -2 \le x \le 0 \\ |x - 2|, & 0 < x < 3 \\ \frac{1}{3}(x - 2)^3, & 3 \le x \le 4. \end{cases}$

29. $f(x) = \begin{cases} |x + 1|, & -3 \le x < 0 \\ x^2 - 4x + 2, & 0 \le x < 3 \\ 2x - 7, & 3 \le x < 4. \end{cases}$

30. $f(x) = \begin{cases} -x^2, & 0 \le x < 1 \\ -2x, & 1 < x < 2 \\ -\frac{1}{2}x^2, & 2 \le x \le 3. \end{cases}$

Exercises 31–34. Sketch the graph of an everywhere differentiable function that satisfies the given conditions. If you find that the conditions are contradictory and therefore no such function exists, explain your reasoning.

31. Local maximum at -1, local minimum at 1, $f(3) = 6$ the absolute maximum, no absolute minimum.

32. $f(0) = 1$ the absolute minimum, local maximum at 4, local minimum at 7, no absolute maximum.

33. $f(1) = f(3) = 0$, $f'(x) > 0$ for all x.

34. $f'(x) = 0$ at each integer x; f has no extreme values.

35. Show that the cubic $p(x) = x^3 + ax^2 + bx + c$ has extreme values iff $a^2 > 3b$.

36. Let r be a rational number, $r > 1$, and set

$$f(x) = (1 + x)^r - (1 + rx) \qquad \text{for} \qquad x \ge -1.$$

Show that 0 is a critical point for f and show that $f(0) = 0$ is the absolute minimum value.

37. Suppose that c is a critical point for f and $f'(x) > 0$ for $x \ne c$. Show that if $f(c)$ is a local maximum, then f is not continuous at c.

38. What can you conclude about a function f continuous on $[a, b]$, if for some c in (a, b), $f(c)$ is both a local maximum and a local minimum?

39. Suppose that f is continuous on $[a, b]$ and $f(a) = f(b)$. Show that f has at least one critical point in (a, b).

40. Suppose that $c_1 < c_2$ and that f takes on local maxima at c_1 and c_2. Prove that if f is continuous on $[c_1, c_2]$, then there is at least one point c in (c_1, c_2) at which f takes on a local minimum.

41. Give an example of a nonconstant function that takes on both its absolute maximum and absolute minimum on every interval.

42. Give an example of a nonconstant function that has an infinite number of distinct local maxima and an infinite number of distinct local minima.

43. Let P be a polynomial with positive leading coefficient:

$$P(x) = a_n x^n + a_{n-1}x^{n-1} + \cdots + a_1 x + a_0, \qquad n \ge 1.$$

Clearly, as $x \to \infty$, $a_n x^n \to \infty$. Show that, as $x \to \infty$, $P(x) \to \infty$ by showing that, given any positive number M, there exists a positive number K such that, if $x \ge K$, then $P(x) \ge M$.

44. Show that of all rectangles with diagonal of length c, the square has the largest area.

45. Let p and q be positive rational numbers and set $f(x) = x^p(1 - x)^q$, $0 \le x \le 1$. Find the absolute maximum value of f.

46. The sum of two numbers is 16. Find the numbers given that the sum of their cubes is an absolute minimum.

47. If the angle of elevation of a cannon is θ and a projectile is fired with muzzle velocity v ft/sec, then the range of the projectile is given by the formula

$$R = \frac{v^2 \sin 2\theta}{32} \text{ feet.}$$

What angle of elevation maximizes the range?

48. A piece of wire of length L is to be cut into two pieces, one piece to form a square and the other piece to form an equilateral triangle. How should the wire be cut so as to

(a) maximize the sum of the areas of the square and the triangle?

(b) minimize the sum of the areas of the square and the triangle?

▶**Exercises 49–52.** Use a graphing utility to graph the function on the indicated interval. Estimate the critical points of the function and classify the extreme values. Round off your estimates to three decimal places.

49. $f(x) = x^3 - 4x + 2x \sin x$; $[-2.5, 3]$.

50. $f(x) = x^4 - 7x^2 + 10x + 3$; $[-3, 3]$.

51. $f(x) = x \cos 2x - \cos^2 x$; $[-\pi, \pi]$.

52. $f(x) = 5x^{2/3} + 3x^{5/3} + 1$; $[-3, 1]$.

▶**Exercises 53–55.** Use a graphing utility to determine whether the function satisfies the hypothesis of the extreme-value theo-

rem on $[a, b]$ (Theorem 2.6.2). If the hypothesis is satisfied, find the absolute maximum value M and the absolute minimum value m. If the hypothesis is not satisfied, find M and m if they exist.

53. $f(x) = \begin{cases} 1 - \sqrt{2 - x}, & \text{if } 1 \le x \le 2 \\ 1 - \sqrt{x - 2}, & \text{if } 2 < x \le 3; \end{cases}$ $[a, b] = [1, 3]$.

54. $f(x) = \begin{cases} \frac{11}{4}x - \frac{19}{4}, & \text{if } 0 \le x \le 3 \\ \sqrt{x - 3} + 2, & \text{if } 3 < x \le 4; \end{cases}$ $[a, b] = [0, 4]$.

55. $f(x) = \begin{cases} \frac{1}{2}x + 1, & \text{if } 1 \le x < 4 \\ \sqrt{x - 3} + 2, & \text{if } 4 \le x \le 6; \end{cases}$ $[a, b] = [1, 6]$.

■ 4.5 SOME MAX-MIN PROBLEMS

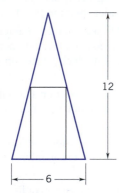

Figure 4.5.1

The techniques of the preceding two sections can be brought to bear on a wide variety of max-min problems. The *key* idea is to express the quantity to be maximized or minimized as a function of one variable. If the function is differentiable, we can differentiate and analyze the results. We begin with a geometric example.

Example 1   An isosceles triangle has a base of 6 units and a height of 12 units. Find the maximum possible area for a rectangle that is inscribed in the triangle and has one side resting on the base of the triangle. What are the dimensions of the rectangle(s) of maximum area?

SOLUTION   Figure 4.5.1 shows the isosceles triangle and a rectangle inscribed in the specified manner. In Figure 4.5.2 we have introduced a rectangular coordinate system. With x and y as in the figure, the area of the rectangle is given by the product

$$A = 2xy.$$

This is the quantity we want to maximize. To do this we have to express A as a function of only one variable.

Since the point (x, y) lies on the line that passes through $(0, 12)$ and $(3, 0)$,

$$y = 12 - 4x.$$ (Verify this.)

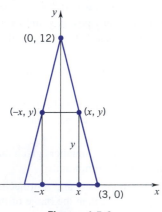

Figure 4.5.2

The area of the rectangle can now be expressed entirely in terms of x:

$$A(x) = 2x(12 - 4x) = 24x - 8x^2.$$

Since x and y represent lengths, x and y cannot be negative. As you can check, this restricts x to the closed interval $[0, 3]$.

Our problem can now be formulated as follows: find the absolute maximum of the function

$$A(x) = 24x - 8x^2, \qquad x \in [0, 3].$$

The derivative

$$A'(x) = 24 - 16x$$

is defined for all $x \in (0, 3)$. Setting $A'(x) = 0$, we have

$$24 - 16x = 0 \qquad \text{which implies} \qquad x = \tfrac{3}{2}.$$

The only critical point is $x = \frac{3}{2}$. Evaluating A at the endpoints of the interval and at the critical point, we have:

$$A(0) = 24(0) - 8(0)^2 = 0,$$

$$A\left(\tfrac{3}{2}\right) = 24\left(\tfrac{3}{2}\right) - 8\left(\tfrac{3}{2}\right)^2 = 18,$$

$$A(3) = 24(3) - 8(3)^2 = 0.$$

The function has an absolute maximum of 18, and this value is taken on at $x = \frac{3}{2}$. At $x = \frac{3}{2}$, the base $2x = 3$ and the height $y = 12 - 4x = 6$.

The maximum possible area is 18 square units. The rectangle that produces this area has a base of 3 units and a height of 6 units. ❑

The example we just considered suggests a basic strategy for solving max-min problems.

Strategy

Step 1. Draw a representative figure and assign labels to the relevant quantities.

Step 2. Identify the quantity to maximized or minimized and find a formula for it.

Step 3. Express the quantity to be maximized or minimized in terms of a single variable; use the conditions given in the problem to eliminate the other variable(s).

Step 4. Determine the domain of the function generated by Step 3.

Step 5. Apply the techniques of the preceding sections to find the extreme value(s).

Figure 4.5.3

Example 2 A paint manufacturer wants cylindrical cans for its specialty enamels. The can is to have a volume of 12 fluid ounces, which is approximately 22 cubic inches. Find the dimensions of the can that will require the least amount of material. See Figure 4.5.3.

SOLUTION Let r be the radius of the can and h the height. The total surface area (top, bottom, lateral area) of a circular cylinder of radius r and height h is given by the formula

$$S = 2\pi r^2 + 2\pi r h.$$

This is the quantity that we want to minimize.

Since the volume $V = \pi r^2 h$ is to be 22 cubic inches, we require that

$$\pi r^2 h = 22 \qquad \text{and thus} \qquad h = \frac{22}{\pi r^2}.$$

It follows from these equations that r and h must both be positive. Thus, we want to minimize the function

$$S(r) = 2\pi r^2 + 2\pi r \left(\frac{22}{\pi r^2}\right) = 2\pi r^2 + \frac{44}{r}, \qquad r \in (0, \infty).$$

Differentiation gives

$$\frac{dS}{dr} = 4\pi r - \frac{44}{r^2} = \frac{4\pi r^3 - 44}{r^2} = \frac{4(\pi r^3 - 11)}{r^2}.$$

The derivative is 0 where $\pi r^3 - 11 = 0$, which is the point $r_0 = (11/\pi)^{1/3}$. Since

$$\frac{dS}{dr} \text{ is } \begin{cases} \text{negative} & \text{for} & r < r_0 \\ \quad\quad 0 & \text{at} & r = r_0 \\ \text{positive} & \text{for} & r > r_0, \end{cases}$$

S decreases on $(0, r_0]$ and increases on $[r_0, \infty)$. Therefore, the function S is minimized by setting $r = r_0 = (11/\pi)^{1/3}$.

The dimensions of the can that will require the least amount of material are as follows:

$$\text{radius } r = (11/\pi)^{1/3} \cong 1.5 \text{ inches,} \quad \text{height } h = \frac{22}{\pi(11/\pi)^{2/3}} = 2(11/\pi)^{1/3}$$

$$\cong 3 \text{ inches.}$$

The can should be as wide as it is tall. ❏

Example 3 A window in the shape of rectangle capped by a semicircle is to have perimeter p. Choose the radius of the semicircular part so that the window admits the most light.

Figure 4.5.4

SOLUTION We take the point of view that the window which admits the most light is the one with maximum area. As in Figure 4.5.4, we let x be the radius of the semicircular part and y be the height of the rectangular part. We want to express the area

$$A = \tfrac{1}{2}\pi x^2 + 2xy$$

as a function of x alone. To do this, we must express y in terms of x.

Since the perimeter is p, we have

$$p = 2x + 2y + \pi x$$

and thus

$$y = \tfrac{1}{2}[p - (2 + \pi)x].$$

Since x and y represent lengths, these variables must be nonnegative. For both x and y to be nonnegative, we must have $0 \le x \le p/(2 + \pi)$.

The area can now be expressed in terms of x alone:

$$A(x) = \tfrac{1}{2}\pi x^2 + 2xy$$

$$= \tfrac{1}{2}\pi x^2 + 2x\left\{\tfrac{1}{2}[p - (2 + \pi)x]\right\}$$

$$= \tfrac{1}{2}\pi x^2 + px - (2 + \pi)x^2 = px - \left(2 + \tfrac{1}{2}\pi\right)x^2.$$

We want to maximize the function

$$A(x) = px - \left(2 + \tfrac{1}{2}\pi\right)x^2, \qquad 0 \le x \le p/(2 + \pi).$$

The derivative

$$A'(x) = p - (4 + \pi)x$$

is 0 only at $x = p/(4 + \pi)$. Since $A(0) = A[p/(2 + \pi)] = 0$, and since $A'(x) > 0$ for $0 < x < p/(4 + \pi)$ and $A'(x) < 0$ for $p/(4 + \pi) < x < p/(2 + \pi)$, the function A is maximized by setting $x = p/(4 + \pi)$. For the window to have maximum area, the radius of the semicircular part must be $p/(4 + \pi)$. ❏

Example 4 The highway department is asked to construct a road between point A and point B. Point A lies on an abandoned road that runs east-west. Point B is 3 miles north of the point of the old road that is 5 miles east of A. The engineering division proposes that the road be constructed by restoring a section of the old road from A to some point P and constructing a new road from P to B. Given that the cost of restoring the old road is $2,000,000 per mile and the cost of a new road is $4,000,000 per mile, how much of the old road should be restored so as to minimize the cost of the project?

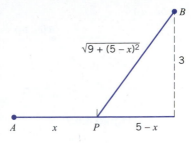

Figure 4.5.5

SOLUTION Figure 4.5.5 shows the geometry of the problem. Notice that we have chosen a straight line joining P and B rather than some curved path. (The shortest connection between two points is provided by the straight-line path.) We let x be the amount of old road that will be restored. Then

$$\sqrt{9+(5-x)^2} = \sqrt{34-10x+x^2}$$

is the length of the new part. The total cost of constructing the two sections of road is

$$C(x) = 2 \cdot 10^6 x + 4 \cdot 10^6 [34 - 10x + x^2]^{1/2}, \qquad 0 \le x \le 5.$$

We want to find the value of x that minimizes this function.
     Differentiation gives

$$C'(x) = 2 \cdot 10^6 + 4 \cdot 10^6 \left(\tfrac{1}{2}\right) [34 - 10x + x^2]^{-1/2}(2x - 10)$$

$$= 2 \cdot 10^6 + \frac{4 \cdot 10^6 (x - 5)}{[34 - 10x + x^2]^{1/2}}, \qquad 0 < x < 5.$$

Setting $C'(x) = 0$, we find that

$$1 + \frac{2(x - 5)}{[34 - 10x + x^2]^{1/2}} = 0$$

$$2(x - 5) = -[34 - 10x + x^2]^{1/2}$$

$$4(x^2 - 10x + 25) = 34 - 10x + x^2 \qquad 4x^2 - 10x + 100$$

$$3x^2 - 30x + 66 = 0 \qquad 3x^2 - 30x + 66$$

$$x^2 - 10x + 22 = 0.$$

By the general quadratic formula, we have

$$x = \frac{10 \pm \sqrt{100 - 4(22)}}{2} = 5 \pm \sqrt{3}.$$

The value $x = 5 + \sqrt{3}$ is not in the domain of our function; the value we want is $x = 5 - \sqrt{3}$. We analyze the sign of C':

sign of C': $- - - - - - - - 0 + + + + +$

behavior of C: 0 $5 - \sqrt{3}$ 5 x

            decreases   increases

Since the function is continuous on $[0, 5]$, it decreases on $[0, 5 - \sqrt{3}]$ and increases on $[5 - \sqrt{3}, 5]$. The number $x = 5 - \sqrt{3} \cong 3.27$ gives the minimum value of C. The highway department will minimize its costs by restoring 3.27 miles of the old road.   ❏

Example 5 (*The angle of incidence equals the angle of reflection.*) Figure 4.5.6 depicts light from point A reflected by a mirror to point B. Two angles have been marked: the *angle of incidence*, θ_i, and the *angle of reflection*, θ_r. Experiment shows

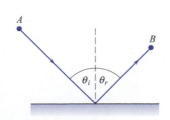

Figure 4.5.6

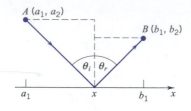

Figure 4.5.7

that $\theta_i = \theta_r$. Derive this result by postulating that the light travels from A to the mirror and then on to B by the shortest possible path.[†]

SOLUTION We write the length of the path as a function of x. In the setup of Figure 4.5.7,

$$L(x) = \sqrt{(x - a_1)^2 + a_2^2} + \sqrt{(x - b_1)^2 + b_2^2}, \quad x \in [a_1, b_1].$$

Differentiation gives

$$L'(x) = \frac{x - a_1}{\sqrt{(x - a_1)^2 + a_2^2}} + \frac{x - b_1}{\sqrt{(x - b_1)^2 + b_2^2}}.$$

Therefore

$$L'(x) = 0 \quad \text{iff} \quad \frac{x - a_1}{\sqrt{(x - a_1)^2 + a_2^2}} = \frac{b_1 - x}{\sqrt{(x - b_1)^2 + b_2^2}}$$

$$\text{iff} \; \sin \theta_i = \sin \theta_r \qquad \qquad \text{(see the figure)}$$

$$\text{iff} \; \theta_i = \theta_r.$$

That $L(x)$ is minimal when $\theta_i = \theta_r$ can be seen by noting that $L''(x)$ is always positive,

$$L''(x) = \frac{a_2^2}{[(x - a_1)^2 + a_2^2]^{3/2}} + \frac{b_2^2}{[(x - b_1)^2 + b_2^2]^{3/2}} > 0,$$

and applying the second-derivative test. ❑

(We must admit that there is a much simpler way to do Example 5, a way that requires no calculus at all. Can you find it?)

Now we will work out a simple problem in which the function to be maximized is defined not on an interval or on a union of intervals, but on a discrete set of points, in this case a finite collection of integers.

Example 6 A small manufacturer of fine rugs has the capacity to produce 25 rugs per week. Assume (for the sake of this example) that the production of the rugs per week leads to an annual profit which, measured in thousands of dollars, is given by the function $P = 100n - 600 - 3n^2$. Find the level of weekly production that maximizes P.

SOLUTION Since n is an integer, it makes no sense to differentiate

$$P = 100n - 600 - 3n^2$$

with respect to n.

Table 4.5.1, compiled by direct calculation, shows the profit P that corresponds to each production level n from 8 to 25. (For $n < 8$, P is negative; 25 is full capacity.) The table shows that the largest profit comes from setting production at 17 units per week.

We can avoid the arithmetic required to construct the table by considering the function

$$f(x) = 100x - 600 - 3x^2, \qquad 8 \le x \le 25.$$

[†]This is a special case of Fermat's *principle of least time*, which says that, of all (neighboring) paths, light chooses the one that requires the least time. If light passes from one medium to another, the geometrically shortest path is not necessarily the path of least time.

■ **Table 4.5.1**

n	P	n	P	n	P
8	8	14	212	20	200
9	57	15	225	21	177
10	100	16	232	22	148
11	137	17	233	23	113
12	168	18	228	24	72
13	193	19	217	25	25

For integral values of x, the function agrees with P. It is continuous on $[8, 25]$ and differentiable on $(8, 25)$ with derivative

$$f'(x) = 100 - 6x.$$

Obviously, $f'(x) = 0$ at $x = \frac{100}{6} = 16\frac{2}{3}$. Since $f'(x) > 0$ on $(8, 16\frac{2}{3})$ and is continuous at the endpoints, f increases on $[8, 16\frac{2}{3}]$. Since $f'(x) < 0$ on $(16\frac{2}{3}, 25)$ and is continuous at the endpoints, f decreases on $[16\frac{2}{3}, 25]$. The largest value of f corresponding to an integer value of x will therefore occur at $x = 16$ or at $x = 17$. Direct calculation of $f(16)$ and $f(17)$ shows that the choice $x = 17$ is correct. ❏

EXERCISES 4.5

1. Find the greatest possible value of xy given that x and y are both positive and $x + y = 40$.

2. Find the dimensions of the rectangle of perimeter 24 that has the largest area.

3. A rectangular garden 200 square feet in area is to be fenced off against rabbits. Find the dimensions that will require the least amount of fencing given that one side of the garden is already protected by a barn.

4. Find the largest possible area for a rectangle with base on the x-axis and upper vertices on the curve $y = 4 - x^2$.

5. Find the largest possible area for a rectangle inscribed in a circle of radius 4.

6. Find the dimensions of the rectangle of area A that has the smallest perimeter.

7. How much fencing is needed to define two adjacent rectangular playgrounds of the same width and total area 15,000 square feet?

8. A rectangular warehouse will have 5000 square feet of floor space and will be separated into two rectangular rooms by an interior wall. The cost of the exterior walls is $150 per linear foot and the cost of the interior wall is $100 per linear foot. Find the dimensions that will minimize the cost of building the warehouse.

9. Rework Example 3; this time assume that the semicircular portion of the window admits only one-third as much light per square foot as does the rectangular portion.

10. A rectangular plot of land is to be defined on one side by a straight river and on three sides by post-and-rail fencing. Eight hundred feet of fencing are available. How should the fencing be deployed so as to maximize the area of the plot?

11. Find the coordinates of P that maximize the area of the rectangle shown in the figure.

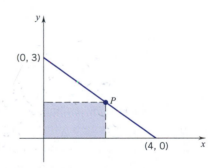

12. A triangle is to be formed as follows: the base of the triangle is to lie on the x-axis, one side is to lie on the line $y = 3x$, and the third side is to pass through the point $(1, 1)$. Assign a slope to the third side that maximizes the area of the triangle.

13. A triangle is to be formed as follows: two sides are to lie on the coordinate axes and the third side is to pass through the point $(2, 5)$. Assign a slope to the third side that minimizes the area of the triangle.

14. Show that, for the triangle of Exercise 13, it is impossible to assign a slope to the third side that maximizes the area of the triangle.

15. What are the dimensions of the base of the rectangular box of greatest volume that can be constructed from 100 square inches of cardboard if the base is to be twice as long as it is wide? Assume that the box has a top.

16. Exercise 15 under the assumption that the box has no top.

17. Find the dimensions of the isosceles triangle of largest area with perimeter 12.

18. Find the point(s) on the parabola $y = \frac{1}{8}x^2$ closest to the point $(0, 6)$.

19. Find the point(s) on the parabola $x = y^2$ closest to the point $(0, 3)$.

20. Find A and B given that the function $y = Ax^{-1/2} + Bx^{1/2}$ has a minimum of 6 at $x = 9$.

21. Find the maximal possible area for a rectangle inscribed in the ellipse $16x^2 + 9y^2 = 144$.

22. Find the maximal possible area for a rectangle inscribed in the ellipse $b^2x^2 + a^2y^2 = a^2b^2$.

23. A pentagon with a perimeter of 30 inches is to be constructed by adjoining an equilateral triangle to a rectangle. Find the dimensions of the rectangle and triangle that will maximize the area of the pentagon.

24. A 10-foot section of gutter is made from a 12-inch-wide strip of sheet metal by folding up 4-inch strips on each side so that they make the same angle with the bottom of the gutter. Determine the depth of the gutter that has the greatest carrying capacity. *Caution*: There are two ways to sketch the trapezoidal cross section. (See the figure.)

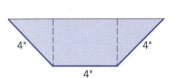

25. From a 15×8 rectangular piece of cardboard four congruent squares are to be cut out, one at each corner. (See the figure.) The remaining crosslike piece is then to be folded into an open box. What size squares should be cut out so as to maximize the volume of the resulting box?

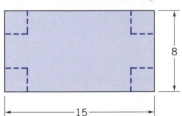

26. A page is to contain 81 square centimeters of print. The margins at the top and bottom are to be 3 centimeters each and, at the sides, 2 centimeters each. Find the most economical dimensions given that the cost of a page varies directly with the perimeter of the page.

27. Let ABC be a triangle with vertices $A = (-3, 0)$, $B = (0, 6)$, $C = (3, 0)$. Let P be a point on the line segment that joins B to the origin. Find the position of P that minimizes the sum of the distances between P and the vertices.

28. Solve Exercise 27 with $A = (-6, 0)$, $B = (0, 3)$, $C = (6, 0)$.

29. An 8-foot-high fence is located 1 foot from a building. Determine the length of the shortest ladder that can be leaned against the building and touch the top of the fence.

30. Two hallways, one 8 feet wide and the other 6 feet wide, meet at right angles. Determine the length of the longest ladder that can be carried horizontally from one hallway into the other.

31. A rectangular banner is to have a red border and a rectangular white center. The width of the border at top and bottom is to be 8 inches, and along the sides 6 inches. The total area is to be 27 square feet. Find the dimensions of the banner that maximize the area of the white center.

32. Conical paper cups are usually made so that the depth is $\sqrt{2}$ times the radius of the rim. Show that this design requires the least amount of paper per unit volume.

33. A string 28 inches long is to be cut into two pieces, one piece to form a square and the other to form a circle. How should the string be cut so as to (a) maximize the sum of the two areas? (b) minimize the sum of the two areas?

34. What is the maximum volume for a rectangular box (square base, no top) made from 12 square feet of cardboard?

35. The figure shows a cylinder inscribed in a right circular cone of height 8 and base radius 5. Find the dimensions of the cylinder that maximize its volume.

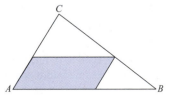

36. As a variant of Exercise 35, find the dimensions of the cylinder that maximize the area of its curved surface.

37. A rectangular box with square base and top is to be made to contain 1250 cubic feet. The material for the base costs 35 cents per square foot, for the top 15 cents per square foot, and for the sides 20 cents per square foot. Find the dimensions that will minimize the cost of the box.

38. What is the largest possible area for a parallelogram inscribed in a triangle ABC in the manner of the figure?

39. Find the dimensions of the isosceles triangle of least area that circumscribes a circle of radius r.

40. What is the maximum possible area for a triangle inscribed in a circle of radius r?

41. The figure shows a right circular cylinder inscribed in a sphere of radius r. Find the dimensions of the cylinder that maximize the volume of the cylinder.

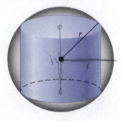

42. As a variant of Exercise 41, find the dimensions of the right circular cylinder that maximize the lateral area of the cylinder.

43. A right circular cone is inscribed in a sphere of radius r as in the figure. Find the dimensions of the cone that maximize the volume of the cone.

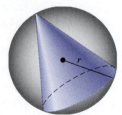

44. What is the largest possible volume for a right circular cone of slant height a?

45. A power line is needed to connect a power station on the shore of a river to an island 4 kilometers downstream and 1 kilometer offshore. Find the minimum cost for such a line given that it costs $50,000 per kilometer to lay wire under water and $30,000 per kilometer to lay wire under ground.

46. A tapestry 7 feet high hangs on a wall. The lower edge is 9 feet above an observer's eye. How far from the wall should the observer stand to obtain the most favorable view? Namely, what distance from the wall maximizes the visual angle of the observer? HINT: Use the formula for $\tan(A - B)$.

47. An object of weight W is dragged along a horizontal plane by means of a force P whose line of action makes an angle θ with the plane. The magnitude of the force is given by the formula

$$P = \frac{\mu W}{\mu \sin\theta + \cos\theta}$$

where μ denotes the coefficient of friction. Find the value of θ that minimizes P.

48. The range of a projectile fired with elevation angle θ at an inclined plane is given by the formula

$$R = \frac{2v^2 \cos\theta \sin(\theta - \alpha)}{g \cos^2 \alpha}$$

where α is the inclination of the target plane, and v and g are constants. Calculate θ for maximum range.

49. Two sources of heat are placed s meters apart—a source of intensity a at A and a source of intensity b at B. The intensity of heat at a point P on the line segment between A and B is given by the formula

$$I = \frac{a}{x^2} + \frac{b}{(s - x)^2},$$

where x is the distance between P and A measured in meters. At what point between A and B will the temperature be lowest?

50. The distance from a point to a line is the distance from that point to the closest point of the line. What point of the line $Ax + By + C = 0$ is closest to the point (x_1, y_1)? What is the distance from (x_1, y_1) to the line?

51. Let f be a differentiable function defined on an open interval I. Let $P(a, b)$ be a point not on the graph of f. Show that if $\overline{PQ}$ is the longest or shortest line segment that joins P to the graph of f, then $\overline{PQ}$ is perpendicular to the graph of f.

52. Draw the parabola $y = x^2$. On the parabola mark a point $P \neq O$. Through P draw the normal line. The normal line intersects the parabola at another point Q. Show that the distance between P and Q is minimized by setting $P = \left(\pm\frac{\sqrt{2}}{2}, \frac{1}{2}\right)$.

53. For each integer n, set $f(n) = 6n^4 - 16n^3 + 9n^2$. Find the integer n that minimizes $f(n)$.

54. A local bus company offers charter trips to Blue Mountain Museum at a fare of $37 per person if 16 to 35 passengers sign up for the trip. The company does not charter trips for fewer than 16 passengers. The bus has 48 seats. If more than 35 passengers sign up, then the fare for every passenger is reduced by 50 cents for each passenger in excess of 35 that signs up. Determine the number of passengers that generates the greatest revenue for the bus company.

55. The Hotwheels Rent-A-Car Company derives an average net profit of $12 per customer if it services 50 customers or fewer. If it services more than 50 customers, then the average net profit is decreased by 6 cents for each customer over 50. What number of customers produces the greatest total net profit for the company?

56. A steel plant has the capacity to produce x tons per day of low-grade steel and y tons per day of high-grade steel where

$$y = \frac{40 - 5x}{10 - x}.$$

Given that the market price of low-grade steel is half that of high-grade steel, show that about $5\frac{1}{2}$ tons of low-grade steel should be produced per day for maximum revenue.

57. The path of a ball is the curve $y = mx - \frac{1}{400}(m^2 + 1)x^2$. Here the origin is taken as the point from which the ball is thrown and m is the initial slope of the trajectory. At a distance which depends on m, the ball returns to the height from which it was thrown. What value of m maximizes this distance?

58. Given the trajectory of Exercise 57, what value of m maximizes the height at which the ball strikes a vertical wall 300 feet away?

59. A truck is to be driven 300 miles on a freeway at a constant speed of v miles per hour. Speed laws require that $35 \leq v \leq 70$. Assume that the fuel costs $2.60 per gallon and is consumed at the rate of $1 + (\frac{1}{400})v^2$ gallons per hour. Given that the driver's wages are $20 per hour, at what speed should the truck be driven to minimize the truck owner's expenses?

60. A tour boat heads out on a 100-kilometer sight-seeing trip. Given that the fixed costs of operating the boat total $2500 per hour, that the cost of fuel varies directly with the square of the speed of the boat, and at 10 kilometers per hour the cost of the fuel is $400 per hour, find the speed that minimizes the boat owner's expenses. Is the speed that minimizes the owner's expenses dependent on the length of the trip?

61. An oil drum is to be made in the form of a right circular cylinder to contain 16π cubic feet. The upright drum is to be taller than it is wide, but not more than 6 feet tall. Determine the dimensions of the drum that minimize surface area.

62. The cost of erecting a small office building is $1,000,000 for the first story, $1,100,000 for the second, $1,200,000 for the third, and so on. Other expenses (lot, basement, etc.) are $5,000,000. Assume that the annual rent is $200,000 per story. How many stories will provide the greatest return on investment?[†]

63. Points A and B are opposite points on the shore of a circular lake of radius 1 mile. Maggie, now at point A, wants to reach point B. She can swim directly across the lake, she can walk along the shore, or she can swim part way and walk part way. Given that Maggie can swim at the rate of 2 miles per hour and walks at the rate of 5 miles per hour, what route should she take to reach point B as quickly as possible? (No running allowed.)

64. Our friend Maggie of Exercise 63 finds a row boat. Given that she can row at the rate of 3 miles per hour, what route should she take now? Row directly across, walk all the way, or row part way and walk part way?

_____

[†]Here by "return on investment" we mean the ratio of income to cost.

65. Set $f(x) = x^2 - x$ and let P be the point $(4, 3)$.
  (a) Use a graphing utility to draw f and mark P.
  (b) Use a CAS to find the point(s) on the graph of f that are closest to P.
  (c) Let Q be a point which satisfies part (b). Determine the equation for the line l_{PQ} through P and Q; then display in one figure the graph of f, the point P, and the line l_{PQ}.
  (d) Determine the equation of the line l_N normal to the graph of f at $(Q, f(Q))$.
  (e) Compare l_{PQ} and l_N.

66. Exercise 65 with $f(x) = x - x^3$ and $P(1, 8)$.

67. Find the distance $D(x)$ from a point (x, y) on the line $y + 3x = 7$ to the origin. Use a graphing utility to draw the graph of D and then use the trace function to estimate the point on the line closest to the origin.

68. Find the distance $D(x)$ from a point (x, y) on the graph of $f(x) = 4 - x^2$ to the point $P(4, 3)$. Use a graphing utility to draw the graph of D and then use the trace function to estimate the point on the graph of f closest to P.

■ PROJECT 4.5 Flight Paths of Birds

Ornithologists studying the flight of birds have determined that certain species tend to avoid flying over large bodies of water during the daylight hours of summer. A possible explanation for this is that it takes more energy to fly over water than land because on a summer day air typically rises over land and falls over water. Suppose that a bird with this tendency is released from an island that is 6 miles from the nearest point A of a straight shoreline. It flies to a point B on the shore and then flies along the shore to its nesting area C, which is 12 miles from A. (See the figure.)

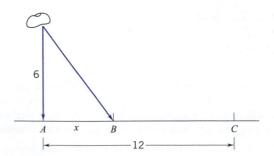

Let W denote the energy per mile required to fly over water, and let L denote the energy per mile required to fly over land.

Problem 1. Show that the total energy E expended by the bird in flying from the island to its nesting area is given by

$$E(x) = W\sqrt{36 + x^2} + L(12 - x), \qquad 0 \le x \le 12$$

where x is the distance from A to B measured in miles.

Problem 2. Suppose that $W = 1.5L$; that is, suppose it takes 50% more energy to fly over water than over land.

  (a) Use the methods of Section 4.5 to find the point B to which the bird should fly to minimize the total energy expended.

  (b) Use a graphing utility to graph E, and then find the minimum value to confirm your result in part (a). Take $L = 1$.

Problem 3. In general, suppose $W = kL, k > 1$.

  (a) Find the point B (as a function of k) to which the bird should fly to minimize the total energy expended.

  (b) Use a graphing utility to experiment with different values of k to find out how the point B moves as k increases/decreases. Take $L = 1$.

  (c) Find the value(s) of k such that the bird will minimize the total energy expended by flying directly to its nest.

  (d) Are there any values of k such that the bird will minimize the total energy expended by flying directly to the point A and then along the shore to C?

■ 4.6 CONCAVITY AND POINTS OF INFLECTION

We begin with a sketch of the graph of a function f, Figure 4.6.1. To the left of c_1 and between c_2 and c_3, the graph "curves up" (we call it *concave up*); between c_1 and c_2, and to the right of c_3, the graph "curves down" (we call it *concave down*). These terms deserve a precise definition.

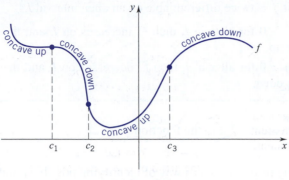

Figure 4.6.1

DEFINITION 4.6.1   CONCAVITY

Let f be a function differentiable on an open interval I. The graph of f is said to be *concave up* on I if f' increases on I; it is said to be *concave down* on I if f' decreases on I.

Stated more geometrically, the graph is concave up on an open interval where the slope increases and concave down on an open interval where the slope decreases.

One more observation: where concave up, the tangent line lies below the graph; where concave down, the tangent line lies above the graph. (Convince yourself of this by adding some tangent lines to the curve shown in Figure 4.6.1.)

Points that join arcs of opposite concavity are called *points of inflection*. The graph in Figure 4.6.1 has three of them: $(c_1, f(c_1)), (c_2, f(c_2)), (c_3, f(c_3))$. Here is the formal definition:

DEFINITION 4.6.2   POINT OF INFLECTION

Let f be a function continuous at c and differentiable near c. The point $(c, f(c))$ is called a *point of inflection* if there exists a $\delta > 0$ such that the graph of f is concave in one sense on $(c - \delta, c)$ and concave in the opposite sense on $(c, c + \delta)$.

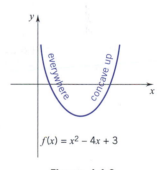

Figure 4.6.2

Example 1   The graph of the quadratic function $f(x) = x^2 - 4x + 3$ is concave up everywhere since the derivative $f'(x) = 2x - 4$ is everywhere increasing. (See Figure 4.6.2.) The graph has no point of inflection.   ❏

Example 2   For the cubing function $f(x) = x^3$, the derivative

$$f'(x) = 3x^2 \qquad \text{decreases on } (-\infty, 0] \text{ and increases on } [0, \infty).$$

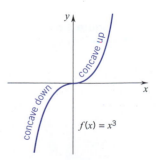

Figure 4.6.3

Thus, the graph of f is concave down on $(-\infty, 0)$ and concave up on $(0, \infty)$. The origin, $(0, 0) = (0, f(0))$, is a point of inflection, the only point of inflection. (See Figure 4.6.3.)   ❏

If f is twice differentiable, we can determine the concavity of the graph from the sign of the second derivative.

> **THEOREM 4.6.3**
>
> Suppose that f is twice differentiable on an open interval I.
>
> **(i)** If $f''(x) > 0$ for all x in I, then f' increases on I, and the graph of f is concave up.
>
> **(ii)** If $f''(x) < 0$ for all x in I, then f' decreases on I, and the graph of f is concave down.

PROOF Apply Theorem 4.2.2 to the function f'. ❏

The following result gives us a way of identifying possible points of inflection.

> **THEOREM 4.6.4**
>
> If the point $(c, f(c))$ is a point of inflection, then
>
> $$f''(c) = 0 \quad \text{or} \quad f''(c) \quad \text{does not exist.}$$

PROOF Suppose that $(c, f(c))$ is a point of inflection. Let's assume that the graph of f is concave up to the left of c and concave down to the right of c. The other case can be handled in a similar manner.

In this situation f' increases on an interval $(c - \delta, c)$ and decreases on an interval $(c, c + \delta)$.

Suppose now that $f''(c)$ exists. Then f' is continuous at c. It follows that f' increases on the half-open interval $(c - \delta, c]$ and decreases on the half-open interval $[c, c + \delta)$. This says that f' has a local maximum at c. Since by assumption $f''(c)$ exists, $f''(c) = 0$. (Theorem 4.3.2 applied to f'.)

We have shown that if $f''(c)$ exists, then $f''(c) = 0$. The only other possibility is that $f''(c)$ does not exist. (Such is the case for the function examined in Example 4 below.) ❏

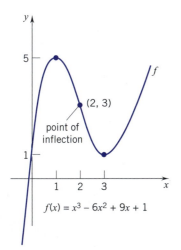

$f(x) = x^3 - 6x^2 + 9x + 1$

Figure 4.6.4

Example 3 For the function

$$f(x) = x^3 - 6x^2 + 9x + 1 \qquad \text{(Figure 4.6.4)}$$

we have

$$f'(x) = 3x^2 - 12x + 9 = 3(x^2 - 4x + 3)$$

and

$$f''(x) = 6x - 12 = 6(x - 2).$$

Note that $f''(x) = 0$ only at $x = 2$, and f'' keeps a constant sign on $(-\infty, 2)$ and on $(2, \infty)$. The sign of f'' on these intervals and the consequences for the graph of f are as follows:

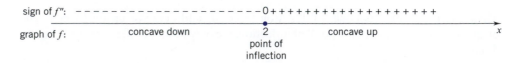

The point $(2, f(2)) = (2, 3)$ is a point of inflection. ❏

Example 4 For

$$f(x) = 3x^{5/3} - 5x \qquad \text{(Figure 4.6.5)}$$

we have

$$f'(x) = 5x^{2/3} - 5 \quad \text{and} \quad f''(x) = \tfrac{10}{3}x^{-1/3}.$$

The second derivative does not exist at $x = 0$. Since

$$f''(x) \text{ is } \begin{cases} \text{negative,} & \text{for } x < 0 \\ \text{positive,} & \text{for } x > 0, \end{cases}$$

the graph is concave down on $(-\infty, 0)$ and concave up on $(0, \infty)$. Since f is continuous at 0, the point $(0, f(0)) = (0, 0)$ is a point of inflection. ❏

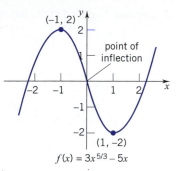

$f(x) = 3x^{5/3} - 5x$

Figure 4.6.5

CAUTION The fact that $f''(c) = 0$ or $f''(c)$ does not exist does not guarantee that $(c, f(c))$ is a point of inflection. (The statement that constitutes Theorem 4.6.4 is not an iff statement.) As you can verify, the function $f(x) = x^4$ satisfies the condition $f''(0) = 0$, but the graph is always concave up and there are no points of inflection. If f is discontinuous at c, then $f''(c)$ does not exist, but $(c, f(c))$ cannot be a point of inflection. A point of inflection occurs at c iff f is continuous at c and the point $(c, f(c))$ joins arcs of opposite concavity. ❏

Example 5 Determine the concavity and find the points of inflection (if any) of the graph of

$$f(x) = x + \cos x, \qquad x \in [0, 2\pi].$$

SOLUTION For $x \in [0, 2\pi]$, we have

$$f'(x) = 1 - \sin x \quad \text{and} \quad f''(x) = -\cos x.$$

On the interval under consideration $f''(x) = 0$ only at $x = \pi/2$ and $x = 3\pi/2$, and f'' keeps constant sign on $(0, \pi/2)$, on $(\pi/2, 3\pi/2)$, and on $(3\pi/2, 2\pi)$. The sign of f'' on these intervals and the consequences for the graph of f are as follows:

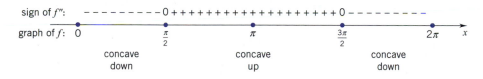

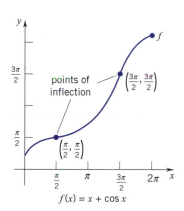

points of inflection $\left(\frac{3\pi}{2}, \frac{3\pi}{2}\right)$

$\left(\frac{\pi}{2}, \frac{\pi}{2}\right)$

$f(x) = x + \cos x$

Figure 4.6.6

The points $(\pi/2, f(\pi/2)) = (\pi/2, \pi/2)$, and $(3\pi/2, f(3\pi/2)) = (3\pi/2, 3\pi/2)$ are points of inflection. The graph of f is shown in Figure 4.6.6. ❏

EXERCISES 4.6

1. The graph of a function f is given in the figure. (a) Determine the intervals on which f increases and the intervals on which f decreases; (b) determine the intervals on which the graph of f is concave up, the intervals on which the graph is concave down, and give the x-coordinate of each point of inflection.

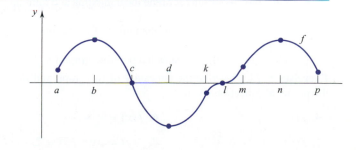

2. Exercise 1 applied to the function f graphed below.

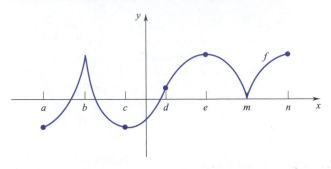

3. The figure below gives the graph of a function f, the graph of its first derivative f', and the graph of its second derivative f'', but not in the correct order. Which curve is the graph of which function?

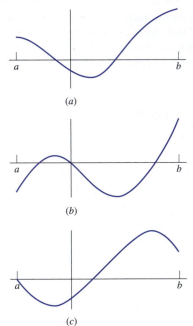

(a)

(b)

(c)

4. A function f is continuous on $[-4, 4]$ and twice differentiable on $(-4, 4)$. Some information on f, f', and f'' is tabulated below:

x	$(-4, -2)$	-2	$(-2, 0)$	0	$(0, 2)$	2	$(2, 4)$
$f'(x)$	positive	0	negative	negative	negative	0	negative
$f''(x)$	negative	negative	negative	0	positive	0	negative

(a) Give the x-coordinates of the local maxima and minima of f.

(b) Give the x-coordinates of the points of inflection of the graph of f.

(c) Given that $f(0) = 0$, sketch a possible graph for f.

Exercises 5–22. Describe the concavity of the graph and find the points of inflection (if any).

5. $f(x) = \dfrac{1}{x}$.

6. $f(x) = x + \dfrac{1}{x}$.

7. $f(x) = x^3 - 3x + 2$.

8. $f(x) = 2x^2 - 5x + 2$.

9. $f(x) = \frac{1}{4}x^4 - \frac{1}{2}x^2$.

10. $f(x) = x^3(1 - x)$.

11. $f(x) = \dfrac{x}{x^2 - 1}$.

12. $f(x) = \dfrac{x + 2}{x - 2}$.

13. $f(x) = (1 - x)^2(1 + x)^2$.

14. $f(x) = \dfrac{6x}{x^2 + 1}$.

15. $f(x) = \dfrac{1 - \sqrt{x}}{1 + \sqrt{x}}$.

16. $f(x) = (x - 3)^{1/5}$.

17. $f(x) = (x + 2)^{5/3}$.

18. $f(x) = x\sqrt{4 - x^2}$.

19. $f(x) = \sin^2 x$, $x \in [0, \pi]$.

20. $f(x) = 2\cos^2 x - x^2$, $x \in [0, \pi]$.

21. $f(x) = x^2 + \sin 2x$, $x \in [0, \pi]$.

22. $f(x) = \sin^4 x$, $x \in [0, \pi]$.

▶ **Exercises 23–26.** Find the points of inflection of the graph of f by using a graphing utility.

23. $f(x) = \dfrac{x^4 - 81}{x^2}$.

24. $f(x) = \sin^2 x - \cos x$, $-2\pi \leq x \leq 2\pi$.

25. $f(x) = x^5 + 9x^4 + 26x^3 + 18x^2 - 27x - 27$.

26. $f(x) = \dfrac{x}{\sqrt{3 - x}}$.

Exercises 27–34. Find: (a) the intervals on which f increases and the intervals on which f decreases; (b) the local maxima and the local minima; (c) the intervals on which the graph is concave up and the intervals on which the graph is concave down; (d) the points of inflection. Use this information to sketch the graph of f.

27. $f(x) = x^3 - 9x$.

28. $f(x) = 3x^4 + 4x^3 + 1$.

29. $f(x) = \dfrac{2x}{x^2 + 1}$.

30. $f(x) = x^{1/3}(x - 6)^{2/3}$.

31. $f(x) = x + \sin x$, $x \in [-\pi, \pi]$.

32. $f(x) = \sin x + \cos x$, $x \in [0, 2\pi]$.

33. $f(x) = \begin{cases} x^3, & x < 1 \\ 3x - 2, & x \geq 1. \end{cases}$

34. $f(x) = \begin{cases} 2x + 4, & x \leq -1 \\ 3 - x^2, & x > -1. \end{cases}$

Exercises 35–38. Sketch the graph of a continuous function f that satisfies the given conditions.

35. $f(0) = 1$, $f(2) = -1$; $f'(0) = f'(2) = 0$, $f'(x) > 0$ for $|x - 1| > 1$, $f'(x) < 0$ for $|x - 1| < 1$; $f''(x) < 0$ for $x < 1$, $f''(x) > 0$ for $x > 1$.

36. $f''(x) > 0$ if $|x| > 2$, $f''(x) < 0$ if $|x| < 2$; $f'(0) = 0$, $f'(x) > 0$ if $x < 0$, $f'(x) < 0$ if $x > 0$; $f(0) = 1$, $f(-2) = f(2) = \frac{1}{2}$, $f(x) > 0$ for all x, f is an even function.

37. $f''(x) < 0$ if $x < 0$, $f''(x) > 0$ if $x > 0$; $f'(-1) = f'(1) = 0$, $f'(0)$ does not exist, $f'(x) > 0$ if $|x| > 1$ $f'(x) < 0$

if $|x| < 1$ $(x \neq 0)$; $f(-1) = 2$, $f(1) = -2$; f is an odd function.

38. $f(-2) = 6$, $f(1) = 2$, $f(3) = 4$; $f'(1) = f'(3) = 0$, $f'(x) < 0$ if $|x - 2| > 1$, $f'(x) > 0$ if $|x - 2| < 1$; $f''(x) < 0$ if $|x + 1| < 1$ or $x > 2$, $f''(x) > 0$ if $|x - 1| < 1$ or $x < -2$.

39. Find d given that $(d, f(d))$ is a point of inflection of the graph of

$$f(x) = (x - a)(x - b)(x - c).$$

40. Find c given that the graph of $f(x) = cx^2 + x^{-2}$ has a point of inflection at $(1, f(1))$.

41. Find a and b given that the graph of $f(x) = ax^3 + bx^2$ passes through the point $(-1, 1)$ and has a point of inflection where $x = \frac{1}{3}$.

42. Determine A and B so that the curve

$$y = Ax^{1/2} + Bx^{-1/2}$$

has a point of inflection at $(1, 4)$.

43. Determine A and B so that the curve

$$y = A \cos 2x + B \sin 3x$$

has a point of inflection at $(\pi/6, 5)$.

44. Find necessary and sufficient conditions on A and B for $f(x) = Ax^2 + Bx + C$

(a) to decrease between A and B with graph concave up.
(b) to increase between A and B with graph concave down.

45. Find a function f with $f'(x) = 3x^2 - 6x + 3$ for all real x and $(1, -2)$ a point of inflection. How many such functions are there?

46. Set $f(x) = \sin x$. Show that the graph of f is concave down above the x-axis and concave up below the x-axis. Does $g(x) = \cos x$ have the same property?

47. Set $p(x) = x^3 + ax^2 + bx + c$.

(a) Show that the graph of p has exactly one point of inflection. What is x at that point?
(b) Show that p has two local extreme values iff $a^2 > 3b$.
(c) Show that p cannot have only one local extreme value.

48. Show that if a cubic polynomial $p(x) = x^3 + ax^2 + bx + c$ has a local maximum and a local minimum, then the midpoint of the line segment that connects the local high point to the local low point is a point of inflection.

49. (a) Sketch the graph of a function that satisfies the following conditions: for all real x, $f(x) > 0$, $f'(x) > 0$, $f''(x) > 0$; $f(0) = 1$.
(b) Does there exist a function which satisfies the conditions: $f(x) > 0$, $f'(x) < 0$, $f''(x) < 0$ for all real x? Explain.

50. Prove that a polynomial of degree n can have at most $n - 2$ points of inflection.

▷ **Exercises 51–54.** Use a graphing utility to graph the function on the indicated interval. (a) Estimate the intervals where the graph is concave up and the intervals where it is concave down. (b) Estimate the x-coordinate of each point of inflection. Round off your estimates to three decimal places.

51. $f(x) = x^4 - 5x^2 + 3$; $[-4, 4]$.

52. $f(x) = x \sin x$; $[-2\pi, 2\pi]$.

53. $f(x) = 1 + x^2 - 2x \cos x$; $[-\pi, \pi]$.

54. $f(x) = x^{2/3}(x^2 - 4)$; $[-5, 5]$.

▷ **Exercises 55–58.** Use a CAS to determine where:
(a) $f''(x) = 0$,   (b) $f''(x) > 0$,
(c) $f''(x) < 0$.

55. $f(x) = 2 \cos^2 x - \cos x$, $0 \leq x \leq 2\pi$.

56. $f(x) = \dfrac{x^2}{x^4 - 1}$.

57. $f(x) = x^{11} - 4x^9 + 6x^7 - 4x^5 + x^3$.

58. $f(x) = x\sqrt{16 - x^2}$.

■ 4.7 VERTICAL AND HORIZONTAL ASYMPTOTES; VERTICAL TANGENTS AND CUSPS

Vertical and Horizontal Asymptotes

In Figure 4.7.1 you can see the graph of

$$f(x) = \frac{1}{|x - c|} \qquad \text{for } x \text{ close to } c.$$

As $x \to c$, $f(x) \to \infty$; that is, given any positive number M, there exists a positive number δ such that

$$\text{if} \qquad 0 < |x - c| < \delta, \qquad \text{then} \qquad f(x) \geq M.$$

The line $x = c$ is called a *vertical asymptote*. Figure 4.7.2 shows the graph of

$$g(x) = -\frac{1}{|x - c|} \qquad \text{for } x \text{ close to } c.$$

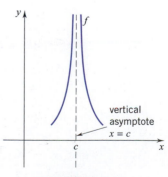

Figure 4.7.1

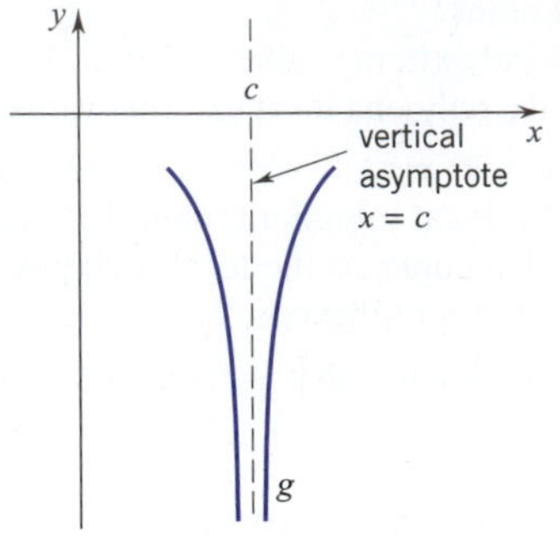

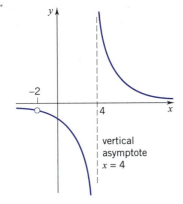

Figure 4.7.2

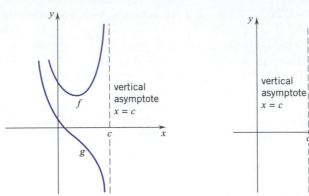

Figure 4.7.3

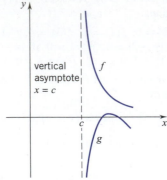

Figure 4.7.4

In this case, as $x \to c$, $g(x) \to -\infty$. Again, the line $x = c$ is called a *vertical asymptote*.

Vertical asymptotes can arise from one-sided behavior. With f and g as in Figure 4.7.3, we write

$$\text{as } x \to c^-, \qquad f(x) \to \infty \qquad \text{and} \qquad g(x) \to -\infty.$$

With f and g as in Figure 4.7.4, we write

$$\text{as } x \to c^+, \qquad f(x) \to \infty \qquad \text{and} \qquad g(x) \to -\infty.$$

In each case the line $x = c$ is a vertical asymptote for both functions.

Example 1   The graph of

$$f(x) = \frac{3x + 6}{x^2 - 2x - 8} = \frac{3(x + 2)}{(x + 2)(x - 4)}$$

has a vertical asymptote at $x = 4$: as $x \to 4^+$, $f(x) \to \infty$ and as $x \to 4^-$, $f(x) \to -\infty$. The vertical line $x = -2$ is not a vertical asymptote since as $x \to -2$, $f(x)$ tends to a finite limit: $\lim_{x \to -2} f(x) = \lim_{x \to -2} 3/(x - 4) = -\frac{1}{2}$.   (Figure 4.7.5)   ❏

From your knowledge of trigonometry you know that as $x \to \pi/2^-$, $\tan x \to \infty$ and as $x \to \pi/2^+$, $\tan x \to -\infty$. Therefore the line $x = \pi/2$ is a vertical asymptote. In fact, the lines $x = (2n + 1)\pi/2$, $n = 0, \pm 1, \pm 2, \ldots$, are all vertical asymptotes for the tangent function. (Figure 4.7.6)

Figure 4.7.5

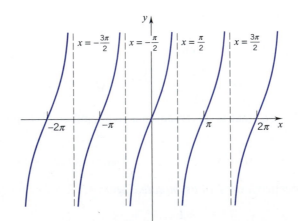

Figure 4.7.6

The graph of a function can have a *horizontal asymptote*. Such is the case (see Figures 4.7.7 and 4.7.8) if, as $x \to \infty$ or as $x \to -\infty$, $f(x)$ tends to a finite limit.

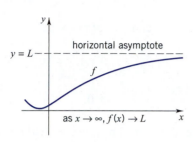

Figure 4.7.7

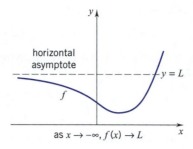

Figure 4.7.8

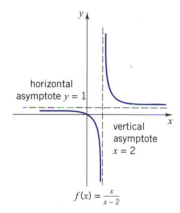

Figure 4.7.9

Example 2 Figure 4.7.9 shows the graph of the function

$$f(x) = \frac{x}{x - 2}.$$

As $x \to 2^-$, $f(x) \to -\infty$; as $x \to 2^+$, $f(x) \to \infty$. The line $x = 2$ is a vertical asymptote.

As $x \to \infty$,

$$f(x) = \frac{x}{x - 2} = \frac{1}{1 - 2/x} \to 1.$$

The same holds true as $x \to -\infty$. The line $y = 1$ is a horizontal asymptote. ❏

Example 3 Figure 4.7.10 shows the graph of the function

$$f(x) = \frac{\cos x}{x}, \quad x > 0.$$

As $x \to 0^+$, $\cos x \to 1$, $1/x \to \infty$, and

$$f(x) = \frac{\cos x}{x} = (\cos x)\left(\frac{1}{x}\right) \to \infty.$$

The line $x = 0$ (the y-axis) is a vertical asymptote.

As $x \to \infty$,

$$|f(x)| = \frac{|\cos x|}{|x|} \leq \frac{1}{|x|} \to 0$$

and therefore

$$f(x) = \frac{\cos x}{x} \to 0.$$

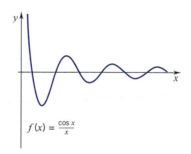

Figure 4.7.10

The line $y = 0$ (the x-axis) is a horizontal asymptote. In this case the graph does not stay to one side of the asymptote. Instead, it wiggles about it with oscillations of ever decreasing amplitude. ❏

Example 4 Here we examine the behavior of

$$g(x) = \frac{x + 1 - \sqrt{x}}{x^2 - 2x + 1} = \frac{x + 1 - \sqrt{x}}{(x - 1)^2}.$$

First, two observations: (a) Because of the presence of $\sqrt{x}$, g is not defined for negative numbers. The domain of g is $[0, 1) \cup (1, \infty)$. (b) On its domain, g remains positive.

As $x \to 1$,

$$x + 1 - \sqrt{x} \to 1, \qquad (x-1)^2 \to 0, \qquad \text{and} \qquad g(x) \to \infty.$$

Thus, the line $x = 1$ is a vertical asymptote.

As $x \to \infty$,

$$g(x) = \frac{x + 1 - \sqrt{x}}{x^2 - 2x + 1} = \frac{1 + 1/x - 1/\sqrt{x}}{x - 2 + 1/x} \to 0.$$

(The numerator tends to 1 and the denominator tends to ∞.) The line $y = 0$ (the x-axis) is a horizontal asymptote. ❏

The behavior of a rational function

$$R(x) = \frac{a_n x^n + \cdots + a_1 x + a_0}{b_k x^k + \cdots + b_1 x + b_0} \qquad (a_n \neq 0, b_k \neq 0)$$

as $x \to \infty$ and as $x \to -\infty$ is readily understood after division of numerator and denominator by the highest power of x that appears in the configuration.

Examples

(a) For $x \neq 0$, set

$$f(x) = \frac{x^4 - 4x^3 - 1}{2x^5 - x} = \frac{1/x - 4/x^2 - 1/x^5}{2 - 1/x^4}.$$

Both as $x \to \infty$ and as $x \to -\infty$,

$$1/x - 4/x^2 - 1/x^5 \to 0, \qquad 2 - 1/x^4 \to 2, \qquad \text{and} \qquad f(x) \to 0.$$

(b) For $x \neq 0$, set

$$f(x) = \frac{x^2 - 3x + 1}{4x^2 - 1} = \frac{1 - 3/x + 1/x^2}{4 - 1/x^2}.$$

Both as $x \to \infty$ and as $x \to -\infty$,

$$1 - 3/x + 1/x^2 \to 1, \qquad 4 - 1/x^2 \to 4, \qquad \text{and} \qquad f(x) \to 1/4.$$

(c) For $x \neq 0$, set

$$f(x) = \frac{3x^3 - 7x^2 + 1}{x^2 - 9} = \frac{3 - 7/x + 1/x^3}{1/x - 9/x^3}.$$

Note that for large positive x, $f(x)$ is positive, but for large negative x, $f(x)$ is negative. As $x \to \infty$, the numerator tends to 3, the denominator tends to 0, and the quotient, being positive, tends to ∞; as $x \to -\infty$, the numerator still tends to 3, the denominator still tends to 0, and the quotient, being negative this time, tends to $-\infty$. ❏

Vertical Tangents; Vertical Cusps

Suppose that f is a function continuous at $x = c$. We say that the graph of f has a *vertical tangent* at the point $(c, f(c))$ if

$$\text{as} \quad x \to c, \quad f'(x) \to \infty \quad \text{or} \quad f'(x) \to -\infty.$$

Examples (Figure 4.7.11)

(a) The graph of the cube-root function $f(x) = x^{1/3}$ has a vertical tangent at the point $(0, 0)$:

$$\text{as} \quad x \to 0, \qquad f'(x) = \tfrac{1}{3}x^{-2/3} \to \infty.$$

The vertical tangent is the line $x = 0$ (the y-axis).

(b) The graph of the function $f(x) = (2 - x)^{1/5}$ has a vertical tangent at the point $(2, 0)$:

$$\text{as} \quad x \to 2, \qquad f'(x) = -\tfrac{1}{5}(2 - x)^{-4/5} \to -\infty.$$

The vertical tangent is the line $x = 2$. ❑

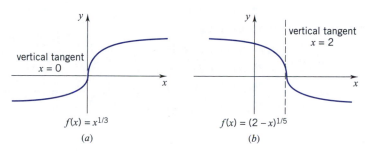

$f(x) = x^{1/3}$ $f(x) = (2 - x)^{1/5}$

       (a)   (b)

Figure 4.7.11

Occasionally you will see a graph tend to the vertical from one side, come to a sharp point, and then virtually double back on itself on the other side. Such a pattern signals the presence of a "vertical cusp."

Suppose that f is continuous at $x = c$. We say that the graph of f has a *vertical cusp* at the point $(c, f(c))$ if

$$\text{as } x \text{ tends to } c \text{ from one side, } f'(x) \to \infty$$

and

$$\text{as } x \text{ tends to } c \text{ from the other side, } f'(x) \to -\infty.$$

Examples (Figure 4.7.12)

(a) The function $f(x) = x^{2/3}$ is continuous at $x = 0$ and has derivative $f'(x) = \tfrac{2}{3}x^{-1/3}$. As $x \to 0^+$, $f'(x) \to \infty$; as $x \to 0^-$, $f'(x) \to -\infty$. This tells us that the graph of f has a vertical cusp at the point $(0, 0)$.

(b) The function $f(x) = 2 - (x - 1)^{2/5}$ is continuous at $x = 1$ and has derivative $f'(x) = -\tfrac{2}{5}(x - 1)^{-3/5}$. As $x \to 1^-$, $f'(x) \to \infty$; as $x \to 1^+$, $f'(x) \to -\infty$. The graph has a vertical cusp at the point $(1, 2)$.

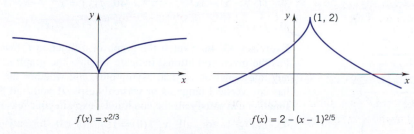

$f(x) = x^{2/3}$ $f(x) = 2 - (x - 1)^{2/5}$

Figure 4.7.12

EXERCISES 4.7

1. The graph of a function f is given in the figure.

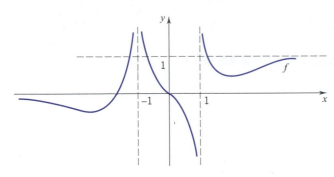

(a) As $x \to -1$, $f(x) \to$?
(b) As $x \to 1^-$, $f(x) \to$?
(c) As $x \to 1^+$, $f(x) \to$?
(d) As $x \to \infty$, $f(x) \to$?
(e) As $x \to -\infty$, $f(x) \to$?
(f) What are the vertical asymptotes?
(g) What are the horizontal asymptotes?

2. The graph of a function f is given in the figure.

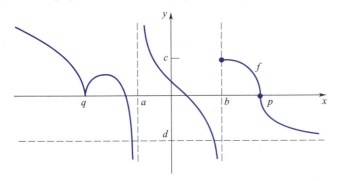

(a) As $x \to \infty$, $f(x) \to$?
(b) As $x \to b^+$, $f(x) \to$?
(c) What are the vertical asymptotes?
(d) What are the horizontal asymptotes?
(e) Give the numbers c, if any, at which the graph of f has a vertical tangent.
(f) Give the numbers c, if any, at which the graph of f has a vertical cusp.

Exercises 3–20. Find the vertical and horizontal asymptotes

3. $f(x) = \dfrac{x}{3x - 1}$.

4. $f(x) = \dfrac{x^3}{x + 2}$.

5. $f(x) = \dfrac{x^2}{x - 2}$.

6. $f(x) = \dfrac{4x}{x^2 + 1}$.

7. $f(x) = \dfrac{2x}{x^2 - 9}$.

8. $f(x) = \dfrac{\sqrt{x}}{4\sqrt{x} - x}$.

9. $f(x) = \left(\dfrac{2x - 1}{4 + 3x}\right)^2$.

10. $f(x) = \dfrac{4x^2}{(3x - 1)^2}$.

11. $f(x) = \dfrac{3x}{(2x - 5)^2}$.

12. $f(x) = \left(\dfrac{x}{1 - 2x}\right)^3$.

13. $f(x) = \dfrac{3x}{\sqrt{4x^2 + 1}}$.

14. $f(x) = \dfrac{x^{1/3}}{x^{2/3} - 4}$.

15. $f(x) = \dfrac{\sqrt{x}}{2\sqrt{x} - x - 1}$.

16. $f(x) = \dfrac{2x}{\sqrt{x^2 - 1}}$.

17. $f(x) = \sqrt{x + 4} - \sqrt{x}$.

18. $f(x) = \sqrt{x} - \sqrt{x - 2}$.

19. $f(x) = \dfrac{\sin x}{\sin x - 1}$.

20. $f(x) = \dfrac{1}{\sec x - 1}$.

Exercises 21–34. Determine whether or not the graph of f has a vertical tangent or a vertical cusp at c.

21. $f(x) = (x + 3)^{4/3}$; $c = -3$.

22. $f(x) = 3 + x^{2/5}$; $c = 0$.

23. $f(x) = (2 - x)^{4/5}$; $c = 2$.

24. $f(x) = (x + 1)^{-1/3}$; $c = -1$.

25. $f(x) = 2x^{3/5} - x^{6/5}$; $c = 0$.

26. $f(x) = (x - 5)^{7/5}$; $c = 5$.

27. $f(x) = (x + 2)^{-2/3}$; $c = -2$.

28. $f(x) = 4 - (2 - x)^{3/7}$; $c = 2$.

29. $f(x) = \sqrt{|x - 1|}$; $c = 1$.

30. $f(x) = x(x - 1)^{1/3}$; $c = 1$.

31. $f(x) = |(x + 8)^{1/3}|$; $c = -8$.

32. $f(x) = \sqrt{4 - x^2}$; $c = 2$.

33. $f(x) = \begin{cases} x^{1/3} + 2, & x \le 0 \\ 1 - x^{1/5}, & x > 0; \end{cases}$ $c = 0$.

34. $f(x) = \begin{cases} 1 + \sqrt{-x}, & x \le 0 \\ (4x - x^2)^{1/3}, & x > 0; \end{cases}$ $c = 0$.

Exercises 35–38. Sketch the graph of the function, showing all asymptotes.

35. $f(x) = \dfrac{x + 1}{x - 2}$.

36. $f(x) = \dfrac{1}{(x + 1)^2}$.

37. $f(x) = \dfrac{x}{1 + x^2}$.

38. $f(x) = \dfrac{x - 2}{x^2 - 5x + 6}$.

▶**Exercises 39–42.** Find (a) the intervals on which f increases and the intervals on which f decreases, and (b) the intervals on which the graph of f is concave up and the intervals on which it is concave down. Also, determine whether the graph of f has any vertical tangents or vertical cusps. Confirm your results with a graphing utility.

39. $f(x) = x - 3x^{1/3}$.

40. $f(x) = x^{2/3} - x^{1/3}$.

41. $f(x) = \frac{3}{5}x^{5/3} - 3x^{2/3}$.

42. $f(x) = \sqrt{|x|}$.

Exercises 43–46. Sketch the graph of a function f that satisfies the given conditions. Indicate whether the graph of f has any horizontal or vertical asymptotes, and whether the graph has any vertical tangents or vertical cusps. If you find that no function can satisfy all the conditions, explain your reasoning.

43. $f(x) \ge 1$ for all x, $f(0) = 1$; $f''(x) < 0$ for all $x \ne 0$; $f'(x) \to \infty$ as $x \to 0^+$, $f'(x) \to -\infty$ as $x \to 0^-$.

44. $f(0) = 0$, $f(3) = f(-3) = 0$; $f(x) \to -\infty$ as $x \to 1$, $f(x) \to -\infty$ as $x \to -1$, $f(x) \to 1$ as $x \to \infty$, $f(x) \to 1$ as $x \to -\infty$; $f''(x) < 0$ for all $x \neq \pm 1$.

45. $f(0) = 0$; $f(x) \to -1$ as $x \to \infty$, $f(x) \to 1$ as $x \to -\infty$; $f'(x) \to -\infty$ as $x \to 0$; $f''(x) < 0$ for $x < 0$, $f''(x) > 0$ for $x > 0$; f is an odd function.

46. $f(0) = 1$; $f(x) \to 4$ as $x \to \infty$, $f(x) \to -\infty$ as $x \to -\infty$; $f'(x) \to \infty$ as $x \to 0$; $f''(x) > 0$ for $x < 0$, $f''(x) < 0$ for $x > 0$.

47. Let p and q be positive integers, q odd, $p < q$. Let $f(x) = x^{p/q}$. Specify conditions on p and q so that

    (a) the graph of f has a vertical tangent at $(0, 0)$.
    (b) the graph of f has a vertical cusp at $(0, 0)$.

48. (*Oblique asymptotes*) Let $r(x) = p(x)/q(x)$ be a rational function. If (degree of p) = (degree of q) + 1, then r can be written in the form

$$r(x) = ax + b + \frac{Q(x)}{q(x)} \quad \text{with} \quad (\text{degree } Q) < (\text{degree } q).$$

Show that $[r(x) - (ax + b)] \to 0$ both as $x \to \infty$ and as $x \to -\infty$. Thus the graph of f "approaches the line $y = ax + b$" both as $x \to \infty$ and as $x \to -\infty$. The line $y = ax + b$ is called an *oblique asymptote*.

Exercises 49–52. Sketch the graph of the function showing all vertical and oblique asymptotes.

49. $f(x) = \dfrac{x^2 - 4}{x}$.

50. $f(x) = \dfrac{2x^2 + 3x - 2}{x + 1}$.

51. $f(x) = \dfrac{x^3}{(x - 1)^2}$.

52. $f(x) = \dfrac{1 + x - 3x^2}{x}$.

▸**Exercises 53–54.** Use a CAS to find the oblique asymptotes. Then use a graphing utility to draw the graph of f and its asymptotes, and thereby confirm your findings.

53. $f(x) = \dfrac{3x^4 - 4x^3 - 2x^2 + 2x + 2}{x^3 - x}$.

54. $f(x) = \dfrac{5x^3 - 3x^2 + 4x - 4}{x^2 + 1}$.

▸**Exercises 55–56.** Use a graphing utility to determine whether or not the graph of f has a horizontal asymptote. Confirm your findings analytically.

55. $f(x) = \sqrt{x^2 + 2x} - x$.

56. $f(x) = \sqrt{x^4 - x^2} - x^2$.

■ 4.8 SOME CURVE SKETCHING

During the course of the last few sections you have seen how to find the extreme values of a function, the intervals on which a function increases, and the intervals on which it decreases; how to determine the concavity of a graph and how to find the points of inflection; and, finally, how to determine the asymptotic properties of a graph. This information enables us to sketch a pretty accurate graph without having to plot point after point after point.

Before attempting to sketch the graph of a function, we try to gather together the information available to us and record it in an organized form. Here is an outline of the procedure we will follow to sketch the graph of a function f.

(1) *Domain*   Determine the domain of f; identify endpoints; find the vertical asymptotes; determine the behavior of f as $x \to \infty$ and as $x \to -\infty$.

(2) *Intercepts*   Determine the x- and y-intercepts of the graph. [The y-intercept is the value $f(0)$; the x-intercepts are the solutions of the equation $f(x) = 0$.]

(3) *Symmetry/periodicity*   If f is an even function [$f(-x) = f(x)$], then the graph of f is symmetric about the y-axis; if f is an odd function [$f(-x) = -f(x)$], then the graph of f is symmetric about the origin. If f is periodic with period p, then the graph of f replicates itself on intervals of length p.

(4) *First derivative*   Calculate f'. Determine the critical points; examine the sign of f' to determine the intervals on which f increases and the intervals on which f decreases; determine the vertical tangents and cusps.

(5) *Second derivative*   Calculate f''. Examine the sign of f'' to determine the intervals on which the graph is concave up and the intervals on which the graph is concave down; determine the points of inflection.

(6) *Points of interest and preliminary sketch*   Plot the points of interest in a preliminary sketch: intercept points, extreme points (local extreme points, absolute extreme points, endpoint extreme points), and points of inflection.

(7) *The graph*   Sketch the graph of f by connecting the points in a preliminary sketch, making sure that the curve "rises," "falls," and "bends" in the proper way. You may wish to verify your sketch by using a graphing utility.

Figure 4.8.1 gives some examples of elements to be included in a preliminary sketch

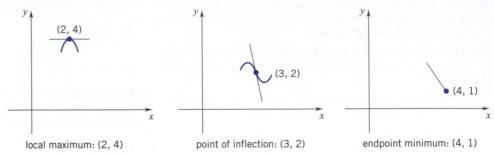

| local maximum: (2, 4) | point of inflection: (3, 2) | endpoint minimum: (4, 1) |

Figure 4.8.1

Example 1   Sketch the graph of $f(x) = \frac{1}{4}x^4 - 2x^2 + \frac{7}{4}$.

SOLUTION

(1) *Domain*   This is a polynomial function; so its domain is the set of all real numbers. Since the leading term is $\frac{1}{4}x^4$, $f(x) \to \infty$ both as $x \to \infty$ and as $x \to -\infty$. There are no asymptotes.

(2) *Intercepts*   The y-intercept is $f(0) = \frac{7}{4}$. To find the x-intercepts, we solve the equation $f(x) = 0$:

$$\frac{1}{4}x^4 - 2x^2 + \frac{7}{4} = 0.$$

$$x^4 - 8x^2 + 7 = 0,$$

$$(x^2 - 1)(x^2 - 7) = 0,$$

$$(x + 1)(x - 1)(x + \sqrt{7})(x - \sqrt{7}) = 0,$$

The x-intercepts are $x = \pm 1$ and $x = \pm\sqrt{7}$.

(3) *Symmetry/periodicity*   Since

$$f(-x) = \frac{1}{4}(-x)^4 - 2(-x^2) + \frac{7}{4} = \frac{1}{4}x^4 - 2x^2 + \frac{7}{4} = f(x),$$

f is an even function, and its graph is symmetric about the y-axis; f is not a periodic function.

(4) *First derivative*

$$f'(x) = x^3 - 4x = x(x^2 - 4) = x(x + 2)(x - 2).$$

The critical points are $x = 0$, $x = \pm 2$. The sign of f' and behavior of f:

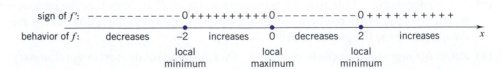

(5) *Second derivative*

$$f''(x) = 3x^2 - 4 = 3\left(x - \frac{2}{\sqrt{3}}\right)\left(x + \frac{2}{\sqrt{3}}\right).$$

The sign of f'' and the concavity of the graph of f:

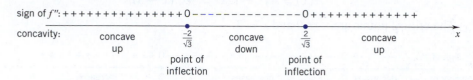

(6) *Points of interest and preliminary sketch* (Figure 4.8.2)

$(0, \frac{7}{4})$: y-intercept point.

$(-1, 0), (1, 0), (-\sqrt{7}, 0), (\sqrt{7}, 0)$: x-intercept points.

$(0, \frac{7}{4})$:   local maximum point.

$(-2, -\frac{9}{4}), (2, -\frac{9}{4})$:   local and absolute minimum points.

$(-2/\sqrt{3}, -17/36), (2/\sqrt{3}, -17/36)$:   points of inflection.

(7) *The graph* Since the graph is symmetric about the y-axis, we can sketch the graph for $x \geq 0$, and then obtain the graph for $x \leq 0$ by a reflection in the y-axis. See Figure 4.8.3. ❑

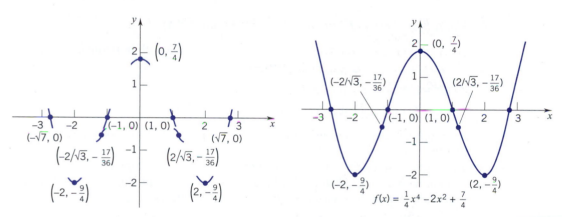

Figure 4.8.2   **Figure 4.8.3**

Example 2   Sketch the graph of $f(x) = x^4 - 4x^3 + 1$, $-1 \leq x < 5$.

SOLUTION

(1) *Domain* The domain is $[-1, 5)$; -1 is the left endpoint, and 5 is a "missing" right endpoint. There are no asymptotes. We do not consider the behavior of f as $x \to \pm\infty$ since f is defined only on $[-1, 5)$.

(2) *Intercepts* The y-intercept is $f(0) = 1$. To find the x-intercepts, we must solve the equation

$$x^4 - 4x^3 + 1 = 0.$$

We cannot do this exactly, but we can verify that $f(0) > 0$ and $f(1) < 0$, and that $f(3) < 0$ and $f(4) > 0$. Thus there are x-intercepts in the interval $(0, 1)$ and in the interval $(3, 4)$. We could find approximate values for these intercepts, but we won't stop to do this since our aim here is a sketch of the graph, not a detailed drawing.

(3) *Symmetry/periodicity* The graph is not symmetric about the y-axis: $f(-x) \neq f(x)$. It is not symmetric about the origin: $f(-x) \neq -f(x)$. The function is not periodic.

(4) *First derivative*   For $x \in (-1, 5)$

$$f'(x) = 4x^3 - 12x^2 = 4x^2(x - 3).$$

The critical points are $x = 0$ and $x = 3$.

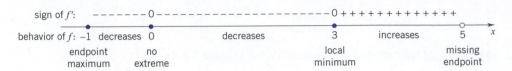

(5) *Second derivative.*

$$f''(x) = 12x^2 - 24x = 12x(x - 2).$$

The sign of f'' and the concavity of the graph of f:

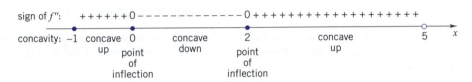

(6) *Points of interest and preliminary sketch* (Figure 4.8.4)

$(0, 1):$	y-intercept point; point of inflection with horizontal tangent.
$(-1, 6):$	endpoint maximum point.
$(2, -15):$	point of inflection.
$(3, -26):$	local and absolute minimum point.

As x approaches the missing endpoint 5 from the left, $f(x)$ increases toward a value of 126.

(7) *The graph*   Since the range of f makes a scale drawing impractical, we must be content with a rough sketch as in Figure 4.8.5. In cases like this, it is particularly important to give the coordinates of the points of interest. ❏

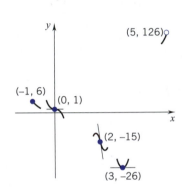

Figure 4.8.4

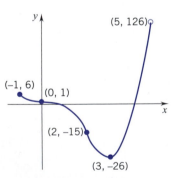

Figure 4.8.5

Example 3   Sketch the graph of $f(x) = \dfrac{x^2 - 3}{x^3}$.

SOLUTION

(1) *Domain*   The domain of f consists of all $x \neq 0$, the set $(-\infty, 0) \cup (0, \infty)$. The y-axis (the line $x = 0$) is a vertical asymptote: $f(x) \to \infty$ as $x \to 0^-$ and $f(x) \to -\infty$ as $x \to 0^+$. The x-axis (the line $y = 0$) is a horizontal asymptote: $f(x) \to 0$ both as $x \to \infty$ and as $x \to -\infty$.

(2) *Intercepts.*   There is no y-intercept since f is not defined at $x = 0$. The x-intercepts are $x = \pm\sqrt{3}$.

(3) *Symmetry*   Since

$$f(-x) = \frac{(-x)^2 - 3}{(-x)^3} = -\frac{x^2 - 3}{x^3} = -f(x),$$

the graph is symmetric about the origin; f is not periodic.

(4) *First derivative.*   It is easier to calculate f' if we first rewrite $f(x)$ using negative exponents:

$$f(x) = \frac{x^2 - 3}{x^3} = x^{-1} - 3x^{-3}$$

gives

$$f'(x) = -x^{-2} + 9x^{-4} = \frac{9 - x^2}{x^4}.$$

The critical points of f are $x = \pm 3$. NOTE: $x = 0$ is not a critical point since 0 is not in the domain of f.

The sign of f' and the behavior of f:

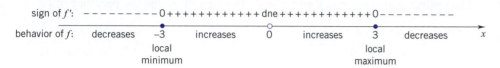

(5) *Second derivative*

$$f''(x) = 2x^{-3} - 36x^{-5} = \frac{2(x^2 - 18)}{x^5} = \frac{2(x - 3\sqrt{2})(x + 3\sqrt{2})}{x^5}.$$

The sign of f'' and the concavity of the graph of f:

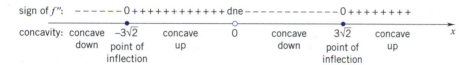

(6) *Points of interest and preliminary sketch* (Figure 4.8.6)

$$(-\sqrt{3}, 0), (\sqrt{3}, 0): \qquad x\text{-intercept points.}$$
$$(-3, -2/9): \qquad \text{local minimum point.}$$
$$(3, 2/9): \qquad \text{local maximum point.}$$
$$(-3\sqrt{2}, -5\sqrt{2}/36), (3\sqrt{2}, 5\sqrt{2}/36): \qquad \text{points of inflection.}$$

(7) *The graph*   See Figure 4.8.7.   ❏

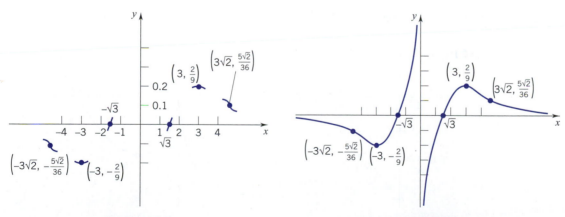

Figure 4.8.6   **Figure 4.8.7**

Example 4   Sketch the graph of $f(x) = \frac{3}{5}x^{5/3} - 3x^{2/3}$.

SOLUTION

1. *Domain*   The domain of f is the set of real numbers. Since we can express $f(x)$ as $\frac{3}{5}x^{2/3}(x - 5)$, we see that, as $x \to \infty$, $f(x) \to \infty$, and as $x \to -\infty$, $f(x) \to -\infty$. There are no asymptotes.

2. *Intercepts* Since $f(0) = 0$, the graph passes through the origin. Thus $x = 0$ is an x-intercept and $y = 0$ is the y-intercept; $x = 5$ is also an x-intercept.

3. *Symmetry/periodicity* There is no symmetry; f is not periodic.

4. *First derivative*

$$f'(x) = x^{2/3} - 2x^{-1/3} = \frac{x-2}{x^{1/3}}.$$

The critical points are $x = 0$ and $x = 2$. The sign of f' and the behavior of f:

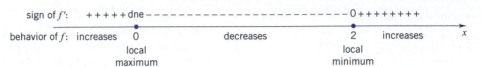

Note that, as $x \to 0^-$, $f'(x) \to \infty$, and as $x \to 0^+$, $f'(x) \to -\infty$. Since f is continuous at $x = 0$, and $f(0) = 0$, the graph of f has a vertical cusp at $(0, 0)$.

5. *Second derivative*

$$f''(x) = \tfrac{2}{3}x^{-1/3} + \tfrac{2}{3}x^{-4/3} = \tfrac{2}{3}x^{-4/3}(x + 1).$$

The sign of f'' and the concavity of the graph of f:

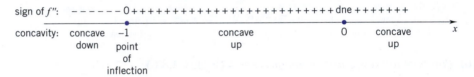

6. *Points of interest and preliminary sketch* (Figure 4.8.8)

$(0, 0)$: y-intercept point, local maximum point; vertical cusp.

$(0, 0), (5, 0)$: x-intercepts points.

$(2, -9\sqrt[3]{4}/5)$:   local minimum point, $f(2) \cong -2.9$.

$(-1, -18/5)$:   point of inflection.

7. *The graph* See Figure 4.8.9. ❏

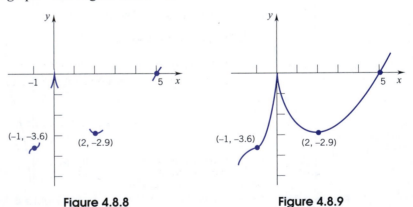

Figure 4.8.8   **Figure 4.8.9**

Example 5 Sketch the graph of $f(x) = \sin 2x - 2\sin x$.

SOLUTION

(1) *Domain* The domain of f is the set of all real numbers. There are no asymptotes and, as you can verify, the graph of f oscillates between $\tfrac{3}{2}\sqrt{3}$ and $-\tfrac{3}{2}\sqrt{3}$ both as $x \to \infty$ and as $x \to -\infty$.

(2) *Intercepts*   The *y*-intercept is $f(0) = 0$. To find the *x*-intercepts, we set $f(x) = 0$:

$$\sin 2x - 2\sin x = 2\sin x \cos x - 2\sin x$$

$$= 2\sin x(\cos x - 1) = 0.$$

Since $\sin x = 0$ at all integral multiples of π and $\cos x = 1$ at all integral multiples of 2π, the *x*-intercepts are the integral multiples of π: all $x = \pm n\pi$.

(3) *Symmetry/periodicity*   Since the sine is an odd function,

$$f(-x) = \sin(-2x) - 2\sin(-x) = -\sin 2x + 2\sin x = -f(x).$$

Thus, *f* is an odd function and the graph is symmetric about the origin. Also, *f* is periodic with period 2π. On the basis of these two properties, it would be sufficient to sketch the graph of *f* on the interval $[0, \pi]$. The result could then be extended to the interval $[-\pi, 0]$ using the symmetry, and then to $(-\infty, \infty)$ using the periodicity. However, for purposes of illustration here, we will consider *f* on the interval $[-\pi, \pi]$.

(4) *First derivative*

$$f'(x) = 2\cos 2x - 2\cos x$$

$$= 2(2\cos^2 x - 1) - 2\cos x$$

$$= 4\cos^2 x - 2\cos x - 2$$

$$= 2(2\cos x + 1)(\cos x - 1).$$

The critical points in $[-\pi, \pi]$ are $x = -2\pi/3, x = 0, x = 2\pi/3$.

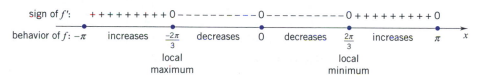

(5) *Second derivative*

$$f''(x) = -4\sin 2x + 2\sin x$$

$$= -8\sin x \cos x + 2\sin x$$

$$= 2\sin x(-4\cos x + 1).$$

$f''(x) = 0$ at $x = -\pi, 0, \pi$, and at the numbers x in $[-\pi, \pi]$ where $\cos x = \frac{1}{4}$, which are approximately ± 1.3. The sign of f'' and the concavity of the graph on $[-\pi, \pi]$:

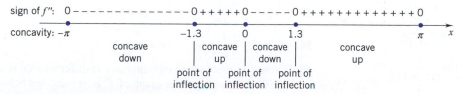

(6) *Points of interest and preliminary sketch* (Figure 4.8.10)

$(0, 0)$:   *y*-intercept point.

$(-\pi, 0), (0, 0), (\pi, 0)$:   *x*-intercept points; $(0, 0)$ is also a point of inflection.

$\left(-\frac{2}{3}\pi, \frac{3}{2}\sqrt{3}\right)$:   local and absolute maximum point; $\frac{3}{2}\sqrt{3} \cong 2.6$.

$\left(\frac{2}{3}\pi, -\frac{3}{2}\sqrt{3}\right)$:   local and absolute minimum point; $-\frac{3}{2}\sqrt{3} \cong -2.6$.

$(-1.3, 1.4), (1.3, -1.4)$:   points of inflection (approximately).

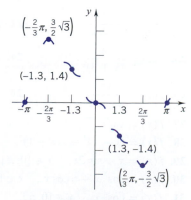

Figure 4.8.10

Figure 4.8.11

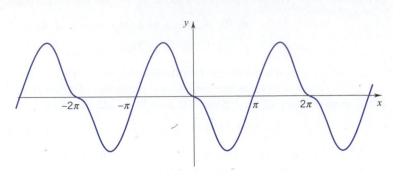

Figure 4.8.12

(7) *The graph*   The graph of f on the interval $[-\pi, \pi]$ is shown in Figure 4.8.11. An indication of the complete graph is given in Figure 4.8.12.   ❑

EXERCISES 4.8

Exercises 1–54. Sketch the graph of the function using the approach presented in this section.

1. $f(x) = (x - 2)^2$.

2. $f(x) = 1 - (x - 2)^2$.

3. $f(x) = x^3 - 2x^2 + x + 1$.

4. $f(x) = x^3 - 9x^2 + 24x - 7$.

5. $f(x) = x^3 + 6x^2$, $x \in [-4, 4]$.

6. $f(x) = x^4 - 8x^2$, $x \in (0, \infty)$.

7. $f(x) = \frac{2}{3}x^3 - \frac{1}{2}x^2 - 10x - 1$.

8. $f(x) = x(x^2 + 4)^2$.

9. $f(x) = x^2 + \dfrac{2}{x}$.

10. $f(x) = x - \dfrac{1}{x}$.

11. $f(x) = \dfrac{x - 4}{x^2}$.

12. $f(x) = \dfrac{x + 2}{x^3}$.

13. $f(x) = 2\sqrt{x} - x$, $x \in [0, 4]$.

14. $f(x) = \frac{1}{4}x - \sqrt{x}$, $x \in [0, 9]$.

15. $f(x) = 2 + (x + 1)^{6/5}$.

16. $f(x) = 2 + (x + 1)^{7/5}$.

17. $f(x) = 3x^5 + 5x^3$.

18. $f(x) = 3x^4 + 4x^3$.

19. $f(x) = 1 + (x - 2)^{5/3}$.

20. $f(x) = 1 + (x - 2)^{4/3}$.

21. $f(x) = \dfrac{x^2}{x^2 + 4}$.

22. $f(x) = \dfrac{2x^2}{x + 1}$.

23. $f(x) = \dfrac{x}{(x + 3)^2}$.

24. $f(x) = \dfrac{x}{x^2 + 1}$.

25. $f(x) = \dfrac{x^2}{x^2 - 4}$.

26. $f(x) = \dfrac{1}{x^3 - x}$.

27. $f(x) = x\sqrt{1 - x}$.

28. $f(x) = (x - 1)^4 - 2(x - 1)^2$.

29. $f(x) = x + \sin 2x$, $x \in [0, \pi]$.

30. $f(x) = \cos^3 x + 6\cos x$, $x \in [0, \pi]$.

31. $f(x) = \cos^4 x$, $x \in [0, \pi]$.

32. $f(x) = \sqrt{3}x - \cos 2x$, $x \in [0, \pi]$.

33. $f(x) = 2\sin^3 x + 3\sin x$, $x \in [0, \pi]$.

34. $f(x) = \sin^4 x$, $x \in [0, \pi]$.

35. $f(x) = (x + 1)^3 - 3(x + 1)^2 + 3(x + 1)$.

36. $f(x) = x^3(x + 5)^2$.

37. $f(x) = x^2(5 - x)^3$.

38. $f(x) = 4 - |2x - x^2|$.

39. $f(x) = 3 - |x^2 - 1|$.

40. $f(x) = x - x^{1/3}$.

41. $f(x) = x(x - 1)^{1/5}$.

42. $f(x) = x^2(x - 7)^{1/3}$.

43. $f(x) = x^2 - 6x^{1/3}$.

44. $f(x) = \dfrac{2x}{\sqrt{x^2 + 1}}$.

45. $f(x) = \sqrt{\dfrac{x}{x - 2}}$.

46. $f(x) = \sqrt{\dfrac{x}{x + 4}}$.

47. $f(x) = \dfrac{x^2}{\sqrt{x^2 - 2}}$.

48. $f(x) = 3\cos 4x$, $x \in [0, \pi]$.

49. $f(x) = 2\sin 3x$, $x \in [0, \pi]$.

50. $f(x) = 3 + 2\cot x + \csc^2 x$, $x \in (0, \frac{1}{2}\pi)$.

51. $f(x) = 2\tan x - \sec^2 x$, $x \in (0, \frac{1}{2}\pi)$.

52. $f(x) = 2\cos x + \sin^2 x$.

53. $f(x) = \dfrac{\sin x}{1 - \sin x}$, $x \in (-\pi, \pi)$.

54. $f(x) = \dfrac{1}{1 - \cos x}$, $x \in (-\pi, \pi)$.

55. Given: f is everywhere continuous, f is differentiable at all $x \neq 0$, $f(0) = 0$, and the graph of f' is as indicated below.

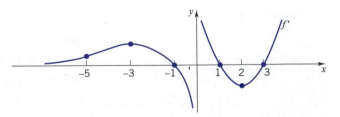

(a) Determine the intervals on which f increases and the intervals on which it decreases; find the critical points of f.
(b) Sketch the graph of f''; determine the intervals on which the graph of f is concave up and those on which it is concave down.
(c) Sketch the graph of f.

56. Set

$$F(x) = \begin{cases} \sin(1/x), & x \neq 0 \\ 0, & x = 0, \end{cases}$$

$$G(x) = \begin{cases} x\sin(1/x), & x \neq 0 \\ 0, & x = 0, \end{cases}$$

$$H(x) = \begin{cases} x^2\sin(1/x), & x \neq 0 \\ 0, & x = 0. \end{cases}$$

(a) Sketch a figure that shows the general nature of the graph of F.
(b) Sketch a figure that shows the general nature of the graph of G.
(c) Sketch a figure that shows the general nature of the graph of H.
(d) Which of these functions is continuous at 0?
(e) Which of these functions is differentiable at 0?

57. Set $f(x) = \dfrac{x^3 - x^{1/3}}{x}$. Show that $f(x) - x^2 \to 0$ as $x \to \pm\infty$. This says that the graph of f is *asymptotic* to the parabola $y = x^2$. Sketch the graph of f and feature this asymptotic behavior.

58. The lines $y = (b/a)x$ and $y = -(b/a)x$ are called *asymptotes* of the hyperbola

$$\frac{x^2}{a^2} - \frac{y^2}{b^2} = 1.$$

(a) Draw a figure that illustrates this asymptotic behavior.
(b) Show that the first-quadrant arc of the hyperbola, the curve

$$y = \frac{b}{a}\sqrt{x^2 - a^2},$$

is indeed asymptotic to the line $y = (b/a)x$ by showing that

$$\frac{b}{a}\sqrt{x^2 - a^2} - \frac{b}{a}x \to 0 \quad \text{as} \quad x \to \infty.$$

(c) Proceeding as in part (b), show that the second-quadrant arc of the hyperbola is asymptotic to the line $y = -(b/a)x$ by taking a suitable limit as $x \to -\infty$. (The asymptotic behavior in the other quadrants can be verified in an analogous manner, or by appealing to symmetry.)

■ 4.9 VELOCITY AND ACCELERATION; SPEED

Suppose that an object (some solid object) moves along a straight line. On the line of motion we choose a point of reference, a positive direction, a negative direction, and a unit distance. This gives us a coordinate system by which we can indicate the position of the object at any given time. Using this coordinate system, we denote by $x(t)$ the position of the object at time t.[†] There is no loss in generality in taking the line of motion as the x-axis. We can arrange this by choosing a suitable frame of reference.

You have seen that the derivative of a function gives the rate of change of that function at the point of evaluation. Thus, if $x(t)$ gives the position of the object at time t and the position function is differentiable, then the derivative $x'(t)$ gives *the rate of change of the position function at time t*. We call this the *velocity at time t and denote it by* $v(t)$. In symbols,

(4.9.1)

$$\boxed{v(t) = x'(t).}$$

Velocity at a particular time t (some call it "instantaneous velocity" at time t) can be obtained as the limit of average velocities. At time t the object is at $x(t)$ and at time $t + h$ it is at $x(t + h)$. If $h > 0$, then $[t, t + h]$ is a time interval and the quotient

$$\frac{x(t + h) - x(t)}{(t + h) - t} = \frac{x(t + h) - x(t)}{h}$$

[†]If the object is larger than a point mass, we can choose a spot on the object and view the location of that spot as the position of the object. In a course in physics an object is usually located by the position of its center of mass. (Section 17.6.)

gives the *average velocity* during this time interval. If $h < 0$, then $[t + h, t]$ is a time interval and the quotient

$$\frac{x(t) - x(t + h)}{t - (t + h)} = \frac{x(t) - x(t + h)}{-h},$$

which also can be written

$$\frac{x(t + h) - x(t)}{h},$$

gives the *average velocity* during this time interval. Thus, whether h is positive or negative, the difference quotient

$$\frac{x(t + h) - x(t)}{h}$$

gives the average velocity of the object during the time interval of length $|h|$ that begins or ends at t. The statement

$$v(t) = x'(t) = \lim_{h \to 0} \frac{x(t + h) - x(t)}{h}$$

expresses the velocity at time t as the limit as $h \to 0$ of these average velocities.

If the velocity function is itself differentiable, then its rate of change with respect to time is called the *acceleration*; in symbols,

(4.9.2)

$$\boxed{a(t) = v'(t) = x''(t).}$$

In the Leibniz notation,

(4.9.3)

$$\boxed{v = \frac{dx}{dt} \quad \text{and} \quad a = \frac{dv}{dt} = \frac{d^2 x}{dt^2}.}$$

The magnitude of the velocity, by which we mean the absolute value of the velocity, is called the *speed* of the object:

(4.9.4)

$$\boxed{\text{speed at time } t = v(t) = |v(t)|.}$$

The four notions that we have just introduced — position, velocity, acceleration, speed — provide the framework for the description of all straight-line motion.[†] The following observations exploit the connections that exist between these fundamental notions:

(1) Positive velocity indicates motion in the positive direction (x is increasing). Negative velocity indicates motion in the negative direction (x is decreasing).

(2) Positive acceleration indicates increasing velocity (increasing speed in the positive direction, decreasing speed in the negative direction). Negative acceleration indicates decreasing velocity (decreasing speed in the positive direction, increasing speed in the negative direction).

[†]Extended by vector methods (Chapter 14), these four notions provide the framework for the description of all motion.

(3) If the velocity and acceleration have the same sign, the object is speeding up, but if the velocity and acceleration have opposite signs, the object is slowing down.

PROOF OF (1) Note that $v = x'$. If $v > 0$, then $x' > 0$ and x increases. If $v < 0$, then $x' < 0$ and x decreases. ❑

PROOF OF (2) Note that

$$v = \begin{cases} v, & \text{in the positive direction} \\ -v, & \text{in the negative direction.} \end{cases}$$

Suppose that $a > 0$. Then v increases. In the positive direction, $\nu = v$ and therefore ν increases; in the negative direction, $\nu = -v$ and therefore ν decreases.

Suppose that $a < 0$. Then v decreases. In the positive direction, $\nu = v$ and therefore ν decreases; in the negative direction, $\nu = -v$ and therefore ν increases. ❑

PROOF OF (3) Note that

$$\nu^2 = v^2 \quad \text{and} \quad \frac{d}{dt}(v^2) = 2vv' = 2va.$$

If v and a have the same sign, then $va > 0$ and $\nu^2 = v^2$ increases. Therefore ν increases, which means the object is speeding up. If v and a have opposite sign, then $va < 0$ and $\nu^2 = v^2$ decreases. Therefore ν decreases, which means the object is slowing down. ❑

Example 1 An object moves along the x-axis; its position at each time t given by the function

$$x(t) = t^3 - 12t^2 + 36t - 27.$$

Let's study the motion from time $t = 0$ to time $t = 9$.

The object starts out at 27 units to the left of the origin:

$$x(0) = 0^3 - 12(0)^2 + 36(0) - 27 = -27$$

and ends up 54 units to the right of the origin:

$$x(9) = 9^3 - 12(9)^2 + 36(9) - 27 = 54.$$

We find the velocity function by differentiating the position function:

$$v(t) = x'(t) = 3t^2 - 24t + 36 = 3(t - 2)(t - 6).$$

We leave it to you to verify that

$$v(t) \text{ is } \begin{cases} \text{positive} & \text{for } 0 \leq t < 2 \\ 0, & \text{at } t = 2 \\ \text{negative,} & \text{for } 2 < t < 6 \\ 0, & \text{at } t = 6 \\ \text{positive,} & \text{for } 6 < t \leq 9. \end{cases}$$

We can interpret all this as follows: the object begins by moving to the right [$v(t)$ is positive for $0 \leq t < 2$]; it comes to a stop at time $t = 2 [v(2) = 0]$; it then moves left [$v(t)$ is negative for $2 < t < 6$]; it stops at time $t = 6 [v(6) = 0]$; it then moves right and keeps going right [$v(t) > 0$ for $6 < t \leq 9$].

We find the acceleration by differentiating the velocity:

$$a(t) = v'(t) = 6t - 24 = 6(t - 4).$$

We note that

$$a(t) \text{ is } \begin{cases} \text{negative,} & \text{for } 0 \le t < 4 \\ 0, & \text{at } t = 4 \\ \text{positive,} & \text{for } 4 < t \le 9. \end{cases}$$

At the beginning the velocity decreases, reaching a minimum at time $t = 4$. Then the velocity starts to increase and continues to increase.

Figure 4.9.1 shows a diagram for the sign of the velocity and a corresponding diagram for the sign of the acceleration. Combining the two diagrams, we have a brief description of the motion in convenient form. The direction of the motion at each time $t \in [0, 9]$ is represented schematically in Figure 4.9.2.

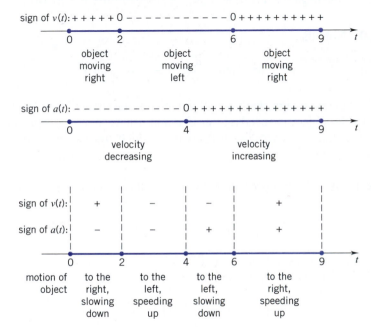

Figure 4.9.1

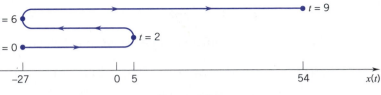

Figure 4.9.2

Another way to represent the motion is to graph x as a function of t, as we do in Figure 4.9.3. The velocity $v(t) = x'(t)$ then appears as the slope of the curve. From the figure, we see that we have positive velocity from $t = 0$ up to $t = 2$, zero velocity at time $t = 2$, then negative velocity up to $t = 6$, zero velocity at $t = 6$, then positive velocity to $t = 9$. The acceleration $a(t) = v'(t)$ can be read from the concavity of the curve. Where the graph is concave down (from $t = 0$ to $t = 4$), the velocity decreases; where the graph is concave up (from $t = 4$ to $t = 9$), the velocity increases. The speed is reflected by the steepness of the curve. The speed decreases from $t = 0$ to $t = 2$, increases from $t = 2$ to $t = 4$, decreases from $t = 4$ to $t = 6$, increases from $t = 6$ to $t = 9$. ❑

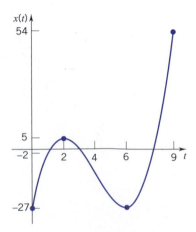

Figure 4.9.3

A few words about units. The units of velocity, speed, and acceleration depend on the units used to measure distance and the units used to measure time. The units of velocity are units of distance per unit time:

feet per second,   meters per second,   miles per hour,   and so forth.

The units of acceleration are units of distance per unit time per unit time:

feet per second per second,   meters per second per second,

miles per hour per hour,   and so forth.

Free Fall Near the Surface of the Earth

(In what follows, the line of motion is clearly vertical. So, instead of writing $x(t)$ to indicate position, we'll follow custom and write $y(t)$. Velocity is then $y'(t)$, acceleration is $y''(t)$, and speed is $|y'(t)|$.)

Imagine an object (for example, a rock or an apple) falling to the ground. (Figure 4.9.4.) We will assume that the object is in *free fall*; namely, that the gravitational pull on the object is constant throughout the fall and that there is no air resistance.[†]

Galileo's formula for the free fall gives the height of the object at each time t of the fall:

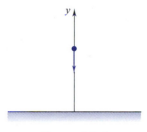

Figure 4.9.4

(4.9.5)
$$y(t) = -\tfrac{1}{2}gt^2 + v_0 t + y_0$$

where g is a positive constant the value of which depends on the units used to measure time and the units used to measure distance.[††]

Let's examine this formula. First, the point of reference is at ground level and the positive y direction is up. Next, since $y(0) = y_0$, the constant y_0 represents the height of the object at time $t = 0$. This is called the *initial position*. Differentiation gives

$$y'(t) = -gt + v_0.$$

Since $y'(0) = v_0$, the constant v_0 gives the velocity of the object at time $t = 0$. This is called the *initial velocity*. A second differentiation gives

$$y''(t) = -g.$$

This indicates that the object falls with constant negative acceleration $-g$.

The constant g is a *gravitational constant*. If time is measured in seconds and distance in feet, then g is approximately 32 feet per second per second[§]; if time is measured in seconds and distance in meters, then g is approximately 9.8 meters per

---

[†]In practice, neither of these conditions is ever fully met. Gravitational attraction near the surface of the earth does vary somewhat with altitude, and there is always some air resistance. Nevertheless, in the setting in which we will be working, the results that we obtain are good approximations of the actual motion.

[††]Galileo Galilei (1564–1642), a great Italian astronomer and mathematician, is popularly known today for his early experiments with falling objects. His astronomical observations led him to support the Copernican view of the solar system. For this he was brought before the Inquisition.

[§]The value of this constant varies slightly with latitude and elevation. It is approximately 32 feet per second per second at the equator, elevation zero. In Greenland it is about 32.23.

second per second. In making numerical calculations, we will take g as 32 feet per second per second or as 9.8 meters per second per second. Equation 4.9.5 then reads

$$y(t) = -16t^2 + v_0 t + y_0 \qquad \text{(distance in feet)}$$

or

$$y(t) = 4.9t^2 + v_0 t + y_0. \qquad \text{(distance in meters)}$$

Example 2   A stone is dropped from a height of 98 meters. In how many seconds does it hit the ground? What is the speed at impact?

SOLUTION   Here $y_0 = 98$ and $v_0 = 0$. Consequently, we have

$$y(t) = -4.9t^2 + 98.$$

To find the time t at impact, we set $y(t) = 0$. This gives

$$-4.9t^2 + 98 = 0, \qquad t^2 = 20, \qquad t = \pm\sqrt{20} = \pm 2\sqrt{5}.$$

We disregard the negative value and conclude that it takes $2\sqrt{5} \cong 4.47$ seconds for the stone to hit the ground.

The velocity at impact is the velocity at time $t = 2\sqrt{5}$. Since

$$v(t) = y'(t) = -9.8t,$$

we have

$$v(2\sqrt{5}) = -(19.6)\sqrt{5} \cong -43.83.$$

The speed at impact is about 43.83 meters per second.   ❏

Example 3   An explosion causes some debris to rise vertically with an initial velocity of 72 feet per second.

(a) In how many seconds does this debris attain maximum height?

(b) What is this maximum height?

(c) What is the speed of the debris as it reaches a height of 32 feet (i) going up? (ii) coming back down?

SOLUTION   Since we are measuring distances in feet, the basic equation reads

$$y(t) = -16t^2 + v_0 t + y_0.$$

Here $y_0 = 0$ (it starts at ground level) and $v_0 = 72$ (the initial velocity is 72 feet per second). The equation of motion is therefore

$$y(t) = -16t^2 + 72t.$$

Differentiation gives

$$v(t) = y'(t) = -32t + 72.$$

The maximum height is attained when the velocity is 0. This occurs at time $t = \frac{72}{32} = \frac{9}{4}$. Since $y(\frac{9}{4}) = 81$, the maximum height attained is 81 feet.

To answer part (c), we must find those times t for which $y(t) = 32$. Since

$$y(t) = -16t^2 + 72t,$$

the condition $y(t) = 32$ yields $-16t^2 + 72t = 32$, which simplifies to

$$16t^2 - 72t + 32 = 0.$$

This quadratic has two solutions, $t = \frac{1}{2}$ and $t = 4$. Since $v(\frac{1}{2}) = 56$ and $v(4) = -56$, the velocity going up is 56 feet per second and the velocity coming down is -56 feet per second. In each case the speed is 56 feet per second. ❏

EXERCISES 4.9

Exercises 1–6. An object moves along a coordinate line, its position at each time $t \geq 0$ given by $x(t)$. Find the position, velocity, and acceleration at time t_0. What is the speed at time t_0?

1. $x(t) = 4 + 3t - t^2$; $\quad t_0 = 5$

2. $x(t) = 5t - t^3$; $\quad t_0 = 3$.

3. $x(t) = \dfrac{18}{t+2}$; $\quad t_0 = 1$. $\qquad$ 4. $x(t) = \dfrac{2t}{t+3}$; $\quad t_0 = 3$.

5. $x(t) = (t^2 + 5t)(t^2 + t - 2)$; $\quad t_0 = 1$.

6. $x(t) = (t^2 - 3t)(t^2 + 3t)$; $\quad t_0 = 2$.

Exercises 7–10. An object moves along the x-axis, its position at each time $t \geq 0$ given by $x(t)$. Determine the times, if any, at which (a) the velocity is zero, (b) the acceleration is zero.

7. $x(t) = 5t + 1$. $\qquad\qquad$ 8. $x(t) = 4t^2 - t + 3$.

9. $x(t) = t^3 - 6t^2 + 9t - 1$. $\quad$ 10. $x(t) = t^4 - 4t^3 + 4t^2 + 2$.

Exercises 11–20. Objects A, B, C move along the x-axis. Their positions $x(t)$ from time $t = 0$ to time $t = t_3$ have been graphed in the figure as functions of t.

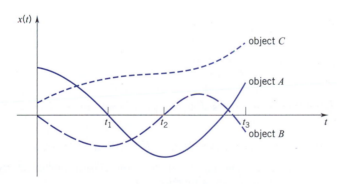

11. Which object begins farthest to the right?

12. Which object finishes farthest to the right?

13. Which object has the greatest speed at time t_1?

14. Which object maintains the same direction during the time interval $[t_1, t_3]$?

15. Which object begins moving left?

16. Which object finishes moving left?

17. Which object changes direction at time t_2?

18. Which object speeds up throughout the time interval $[0, t_1]$?

19. Which objects slow down during the time interval $[t_1, t_2]$?

20. Which object changes direction during the time interval $[t_2, t_3]$.

Exercises 21–28. An object moves along the x-axis, its position at each time $t \geq 0$ given by $x(t)$. Determine the time interval(s), if any, during which the object satisfies the given condition.

21. $x(t) = t^4 - 12t^3 + 28t^2$; moves right.

22. $x(t) = t^3 - 12t^2 + 21t$; moves left.

23. $x(t) = 5t^4 - t^5$; speeds up.

24. $x(t) = 6t^2 - t^4$; slows down.

25. $x(t) = t^3 - 6t^2 - 15t$; moves left slowing down.

26. $x(t) = t^3 - 6t^2 - 15t$; moves right slowing down.

27. $x(t) = t^4 - 8t^3 + 16t^2$; moves right speeding up.

28. $x(t) = t^4 - 8t^3 + 16t^2$; moves left speeding up.

Exercises 29–32. An object moves along a coordinate line, its position at each time $t \geq 0$ being given by $x(t)$. Find the times t at which the object changes direction.

29. $x(t) = (t + 1)^2(t - 9)^3$. $\qquad$ 30. $x(t) = t(t - 8)^3$.

31. $x(t) = (t^3 - 12t)^4$. $\qquad\qquad$ 32. $x(t) = (t^2 - 8t + 15)^3$.

Exercises 33–38. An object moves along the x-axis, its position at each time t given by $x(t)$. Determine those times from $t = 0$ to $t = 2\pi$ at which the object is moving to the right with increasing speed.

33. $x(t) = \sin 3t$. $\qquad\qquad$ 34. $x(t) = \cos 2t$.

35. $x(t) = \sin t - \cos t$. $\qquad$ 36. $x(t) = \sin t + \cos t$.

37. $x(t) = t + 2\cos t$. $\qquad\quad$ 38. $x(t) = t - \sqrt{2}\sin t$.

In Exercises 39–52, neglect air resistance. For the numerical calculations, take g as 32 feet per second per second or as 9.8 meters per second per second.

39. An object is dropped and hits the ground 6 seconds later. From what height, in feet, was it dropped?

40. Supplies are dropped from a stationary helicopter and seconds later hit the ground at 98 meters per second. How high was the helicopter?

41. An object is projected vertically upward from ground level with velocity v. Find the height in meters attained by the object.

42. An object projected vertically upward from ground level returns to earth in 8 seconds. Give the initial velocity in feet per second.

43. An object projected vertically upward passes every height less than the maximum twice, once on the way up and once on the way down. Show that the speed is the same in each direction. Measure height in feet.

44. An object is projected vertically upward from the ground. Show that it takes the object the same amount of time to reach its maximum height as it takes for it to drop from that height back to the ground. Measure height in meters.

45. A rubber ball is thrown straight down from a height of 224 feet at a speed of 80 feet per second. If the ball always rebounds with one-fourth of its impact speed, what will be the speed of the ball the third time it hits the ground?

46. A ball is thrown straight up from ground level. How high will the ball go if it reaches a height of 64 feet in 2 seconds?

47. A stone is thrown upward from ground level. The initial speed is 32 feet per second. (a) In how many seconds will the stone hit the ground? (b) How high will it go? (c) With what minimum speed should the stone be thrown so as to reach a height of at least 36 feet?

48. To estimate the height of a bridge, a man drops a stone into the water below. How high is the bridge (a) if the stone hits the water 3 seconds later? (b) if the man hears the splash 3 seconds later? (Use 1080 feet per second as the speed of sound.)

49. A falling stone is at a certain instant 100 feet above the ground. Two seconds later it is only 16 feet above the ground. (a) From what height was it dropped? (b) If it was thrown down with an initial speed of 5 feet per second, from what height was it thrown? (c) If it was thrown upward with an initial speed of 10 feet per second, from what height was it thrown?

50. A rubber ball is thrown straight down from a height of 4 feet. If the ball rebounds with one-half of its impact speed and returns exactly to its original height before falling again, how fast was it thrown originally?

51. Ballast dropped from a balloon that was rising at the rate of 5 feet per second reached the ground in 8 seconds. How high was the balloon when the ballast was dropped?

52. Had the balloon of Exercise 51 been falling at the rate of 5 feet per second, how long would it have taken for the ballast to reach the ground?

53. Two race horses start a race at the same time and finish in a tie. Prove that there must have been at least one time t *during* the race at which the two horses had exactly the same speed.

54. Suppose that the two horses of Exercise 53 cross the finish line together at the same speed. Show that they had the same acceleration at some instant during the race.

55. A certain tollroad is 120 miles long and the speed limit is 65 miles per hour. If a driver's entry ticket at one end of the tollroad is stamped 12 noon and she exits at the other end at 1:40 P.M., should she be given a speeding ticket? Explain.

56. At 1:00 P.M. a car's speedometer reads 30 miles per hour and at 1:15 P.M. it reads 60 miles per hour. Prove that the car's acceleration was exactly 120 miles per hour per hour at least once between 1:00 and 1:15.

57. A car is stationary at a toll booth. Twenty minutes later, at a point 20 miles down the road, the car is clocked at 60 mph. Explain how you know that the car must have exceeded the 60-mph speed limit some time before being clocked at 60 mph.

58. The results of an investigation of a car accident showed that the driver applied his brakes and skidded 280 feet in 6 seconds. If the speed limit on the street where the accident occurred was 30 miles per hour, was the driver exceeding the speed limit at the instant he applied his brakes? Explain. HINT: 30 miles per hour = 44 feet per second.

59. (*Simple harmonic motion*) A bob suspended from a spring oscillates up and down about an equilibrium point, its vertical position at time t given by

$$y(t) = A \sin(\omega t + \varphi_0)$$

where A, ω, φ_0 are positive constants. (This is an idealization in which we are disregarding friction.)

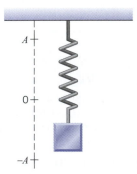

(a) Show that at all times t the acceleration of the bob $y''(t)$ is related to the position of the bob by the equation

$$y''(t) + \omega^2 y(t) = 0.$$

(b) It is clear that the bob oscillates from $-A$ to A, and the speed of the bob is zero at these points. At what position does the bob attain maximum speed? What is this maximum speed?

(c) What are the extreme values of the acceleration function? Where does the bob attain these extreme values?

60. An object moves along the x-axis, its position from $t = 0$ to $t = 5$ given by

$$x(t) = t^3 - 7t^2 + 10t + 5.$$

(a) Determine the velocity function v. Use a graphing utility to graph v as a function of t.

(b) Use the graph to estimate the times when the object is moving right and the times when it is moving left.

(c) Use the graphing utility to graph the speed v of the object as a function of t. Estimate the time(s) when the object stops. Estimate the maximum speed from $t = 1$ to $t = 4$.

(d) Determine the acceleration function a and use the graphing utility to graph it as a function of t. Estimate the times when the object is speeding up and the times when it is slowing down.

(e) Graph the velocity and acceleration functions on the same set of axes and use the graphs to estimate the times when the object is speeding up and the times when it is slowing down.

■ PROJECT 4.9A Angular Velocity; Uniform Circular Motion

As a particle moves along a circle of radius r, it effects a change in the central angle, marked θ in Figure A. We measure θ in radians. The *angular velocity*, ω,[†] of the particle is the time rate of change of θ; that is, $\omega = d\theta/dt$[†]. Circular motion with constant, positive angular velocity is called *uniform circular motion*.

Problem 1. A particle in uniform circular motion traces out a circular arc. The time rate of change of the length of that arc is called the *speed* of the particle. What is the speed of a particle that moves around a circle of radius r with constant, positive angular velocity ω?

Problem 2. The *kinetic energy*, KE, of a particle of mass m is given by the formula

$$\text{KE} = \tfrac{1}{2}mv^2$$

where v is the speed of the particle. Suppose the particle in Problem 1 has mass m. What is the kinetic energy of the particle?

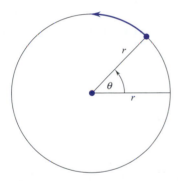

Figure A

Problem 3. A point P moves uniformly along the circle $x^2 + y^2 = r^2$ with constant angular velocity ω. Find the x- and y-coordinates of P at time t given that the motion starts at time $t = 0$ with $\theta = \theta_0$. Then find the velocity and acceleration of the projection of P onto the x-axis and onto the y-axis. [The projection of P onto the x-axis is the point $(x, 0)$; the projection of P onto the y-axis is the point $(0, y)$.]

Problem 4. Figure B shows a sector in a circle of radius r. The sector is the union of the triangle T and the segment S. Suppose that the radius vector rotates counterclockwise with a constant angular velocity of ω radians per second. Show that the area of the sector changes at a constant rate but that the area of T and the area of S do not change at a constant rate.

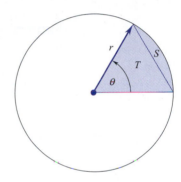

Figure B

Problem 5. Take S and T as in Problem 4. While the area of S and the area of T change at different rates, there is one value of θ between 0 and π at which both areas change at the same rate. Find this value of θ.

■ PROJECT 4.9B Energy of a Falling Body (Near the Surface of the Earth)

If we lift an object, we counteract the force of gravity. In so doing, we increase what physicists call the *gravitational potential energy* of the object. The gravitational potential energy of an object is defined by the formula

$$\text{GPE} = \text{weight} \times \text{height}.$$

Since the weight of an object of mass m is mg where g is the gravitational constant (we take this from physics), we can write

$$\text{GPE} = mgy$$

where y is the height of the object.

If we lift an object and release it, the object drops. As it drops, it loses height and therefore loses gravitational potential energy, but its speed increases. The speed with which the object

falls gives the object a form of energy called *kinetic energy*, the energy of motion. The kinetic energy of an object in motion is given by the formula

$$\text{KE} = \tfrac{1}{2}mv^2$$

where v is the speed of the object. For straight-line motion with velocity v we have $v^2 = v^2$ and therefore

$$\text{KE} = \tfrac{1}{2}mv^2.$$

Problem 1. Prove the *law of conservation of energy*:

$$\text{GPE} + \text{KE} = C, \text{ constant.}$$

HINT: Differentiate the expression $\text{GPE} + \text{KE}$ and use the fact that $dv/dt = -g$.

Problem 2. An object initially at rest falls freely from height y_0. Show that the speed of the object at height y is given by

$$v = \sqrt{2g(y_0 - y)}.$$

Problem 3. According to the results in Section 4.9, the position of an object that falls from rest from a height y_0 is given by

$$y(t) = -\tfrac{1}{2}gt^2 + y_o.$$

Calculate the speed of the object from this equation and show that the result obtained is equivalent to the result obtained in Problem 2.

■ 4.10 RELATED RATES OF CHANGE PER UNIT TIME

In Section 4.9 we studied straight-line motion and defined velocity as the rate of change of position with respect to time and acceleration as the rate of change of velocity with respect to time. In this section we work with other quantities that vary with time. The fundamental point is this: *if Q is any quantity that varies with time, then the derivative dQ/dt gives the rate of change of that quantity with respect to time.*

Example 1 A spherical balloon is expanding. Given that the radius is increasing at the rate of 2 inches per minute, at what rate is the volume increasing when the radius is 5 inches?

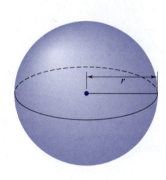

SOLUTION Find dV/dt when $r = 5$ inches, given that $dr/dt = 2$ in./min and

$$V = \tfrac{4}{3}\pi r^3. \qquad \text{(volume of a sphere of radius } r)$$

Both r and V are functions of t. Differentiating $V = \tfrac{4}{3}\pi r^3$ with respect to t, we have

$$\frac{dV}{dt} = 4\pi r^2 \frac{dr}{dt}.$$

Setting $r = 5$ and $dr/dt = 2$, we find that

$$\frac{dV}{dt} = 4\pi(5^2)2 = 200\pi.$$

When the radius is 5 inches, the volume is increasing at the rate of 200π cubic inches per minute. ❑

Example 2 A particle moves clockwise along the unit circle $x^2 + y^2 = 1$. As it passes through the point $(1/2, \sqrt{3}/2)$, its y-coordinate decreases at the rate of 3 units per second. At what rate does the x-coordinate change at this point?

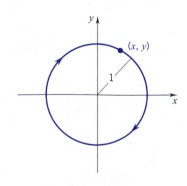

SOLUTION Find dx/dt when $x = 1/2$ and $y = \sqrt{3}/2$, given that $dy/dt = -3$ units/sec and

$$x^2 + y^2 = 1. \qquad \text{(equation of circle)}$$

Differentiating $x^2 + y^2 = 1$ with respect to t, we have

$$2x\frac{dx}{dt} + 2y\frac{dy}{dt} = 0 \qquad \text{and thus} \qquad x\frac{dx}{dt} + y\frac{dy}{dt} = 0.$$

Setting $x = 1/2$, $y = \sqrt{3}/2$, and $dy/dt = -3$, we find that

$$\frac{1}{2}\frac{dx}{dt} + \frac{\sqrt{3}}{2}(-3) = 0 \qquad \text{and therefore} \qquad \frac{dx}{dt} = 3\sqrt{3}.$$

As the object passes through the point $(1/2, \sqrt{3}/2)$, the x-coordinate increases at the rate $3\sqrt{3}$ units per second. ❑

Example 3 A 13-foot ladder leans against the side of a building, forming an angle θ with the ground. Given that the foot of the ladder is being pulled away from the building at the rate of 0.1 feet per second, what is the rate of change of θ when the top of the ladder is 12 feet above the ground?

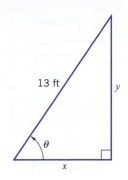

SOLUTION Find $d\theta/dt$ when $y = 12$ feet, given that $dx/dt = 0.1$ ft/sec and

$$\cos\theta = \frac{x}{13}.$$

Differentiation with respect to t gives

$$-\sin\theta\,\frac{d\theta}{dt} = \frac{1}{13}\frac{dx}{dt}.$$

When $y = 12$, $\sin\theta = \frac{12}{13}$. Setting $\sin\theta = \frac{12}{13}$ and $dx/dt = 0.1$, we have

$$-\left(\frac{12}{13}\right)\frac{d\theta}{dt} = \frac{1}{13}(0.1) \qquad \text{and thus} \qquad \frac{d\theta}{dt} = -\frac{1}{120}.$$

When the top of the ladder is 12 feet above the ground, θ decreases at the rate of $\frac{1}{120}$ radians per second (about half a degree per second). ❑

Example 4 Two ships, one heading west and the other east, approach each other on parallel courses 8 nautical miles apart.[†] Given that each ship is cruising at 20 nautical miles per hour (knots), at what rate is the distance between them diminishing when the ships are 10 nautical miles apart?

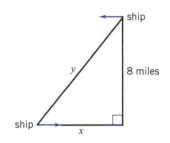

SOLUTION Let y be the distance between the ships measured in nautical miles. Since the ships are moving in opposite directions at the rate of 20 knots each, their horizontal separation (see the figure) is decreasing at the rate of 40 knots. Thus, we want to find dy/dt when $y = 10$, given that $dx/dt = -40$ knots. (We take dx/dt as negative since x is decreasing.) The variables x and y are related by the equation

$$x^2 + 8^2 = y^2. \qquad \text{(Pythagorean theorem)}$$

Differentiating $x^2 + 8^2 = y^2$ with respect to t, we find that

$$2x\frac{dx}{dt} + 0 = 2y\frac{dy}{dt} \qquad \text{and consequently} \qquad x\frac{dx}{dt} = y\frac{dy}{dt}.$$

When $y = 10$, $x = 6$. (Explain.) Setting $x = 6$, $y = 10$, and $dx/dt = -40$, we have

$$6(-40) = 10\frac{dy}{dt} \qquad \text{so that} \qquad \frac{dy}{dt} = -24.$$

(Note that dy/dt is negative since y is decreasing.) When the two ships are 10 miles apart, the distance between them is diminishing at the rate of 24 knots. ❑

The preceding examples were solved by the same general method, a method that we recommend to you for solving problems of this type.

Step 1. Draw a suitable diagram, and indicate the quantities that vary.

Step 2. Specify in mathematical form the rate of change you are looking for, and record all relevant information.

Step 3. Find an equation that relates the relevant variables.

Step 4. Differentiate with respect to time t the equation found in Step 3.

Step 5. State the final answer in coherent form, specifying the units that you are using.

[†]The international nautical mile measures 6080 feet.

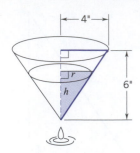

Figure 4.10.1

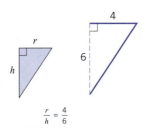

Figure 4.10.2

Example 5 A conical paper cup 8 inches across the top and 6 inches deep is full of water. The cup springs a leak at the bottom and loses water at the rate of 2 cubic inches per minute. How fast is the water level dropping when the water is exactly 3 inches deep?

SOLUTION We begin with a diagram that represents the situation after the cup has been leaking for a while. (Figure 4.10.1.) We label the radius and height of the remaining "cone of water" r and h. We can relate r and h by similar triangles. (Figure 4.10.2.) We measure r and h in inches. Now we seek dh/dt when $h = 3$, given that $dV/dt = -2$ in^3/min,

$$V = \frac{1}{3}\pi r^2 h \qquad \text{(volume of cone)} \qquad \text{and} \qquad \frac{r}{h} = \frac{4}{6} = \frac{2}{3}. \qquad \text{(similar triangles)}$$

Using the second equation to eliminate r from the first equation, we have

$$V = \frac{1}{3}\pi \left(\frac{2h}{3}\right)^2 h = \frac{4}{27}\pi h^3.$$

Differentiation with respect to t gives

$$\frac{dV}{dt} = \frac{4}{9}\pi h^2 \frac{dh}{dt}.$$

Setting $h = 3$ and $dV/dt = -2$, we have

$$-2 = \frac{4}{9}\pi (3)^2 \frac{dh}{dt} \qquad \text{and thus} \qquad \frac{dh}{dt} = -\frac{1}{2\pi}.$$

When the water is exactly 3 inches deep, the water level is dropping at the rate of $1/2\pi$ inches per minute (about 0.16 inches per minute). ❏

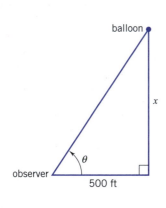

Example 6 A balloon leaves the ground 500 feet away from an observer and rises vertically at the rate of 140 feet per minute. At what rate is the inclination of the observer's line of sight increasing when the balloon is exactly 500 feet above the ground?

SOLUTION Let x be the altitude of the balloon and θ the inclination of the observer's line of sight. Find $d\theta/dt$ when $x = 500$, given that $dx/dt = 140$ ft/min and

$$\tan\theta = \frac{x}{500}.$$

Differentiation with respect to t gives

$$\sec^2\theta \frac{d\theta}{dt} = \frac{1}{500}\frac{dx}{dt}.$$

When $x = 500$, the triangle is isosceles. This implies that $\theta = \pi/4$ and $\sec\theta = \sqrt{2}$. Setting $\sec\theta = \sqrt{2}$ and $dx/dt = 140$, we have

$$(\sqrt{2})^2 \frac{d\theta}{dt} = \frac{1}{500}(140) \qquad \text{and therefore} \qquad \frac{d\theta}{dt} = 0.14.$$

When the balloon is exactly 500 feet above the ground, the inclination of the observer's line of sight is increasing at the rate of 0.14 radians per minute (about 8 degrees per minute). ❏

Example 7 A water trough with vertical cross section in the form of an equilateral triangle is being filled at a rate of 4 cubic feet per minute. Given that the trough is 12 feet long, how fast is the level of the water rising when the water reaches a depth of $1\frac{1}{2}$ feet?

SOLUTION Let x be the depth of the water measured in feet and V the volume of water measured in cubic feet. Find dx/dt when $x = 3/2$, given that $dV/dt = 4 \text{ ft}^3/\text{min}$.

$$\text{area of cross section} = \frac{1}{2}\left(\frac{2x}{\sqrt{3}}\right)x = \frac{\sqrt{3}}{3}x^2.$$

$$\text{volume of water} = 12\left(\frac{\sqrt{3}}{3}x^2\right) = 4\sqrt{3}x^2.$$

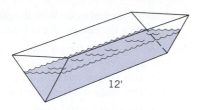

Differentiation of $V = 4\sqrt{3}x^2$ with respect to t gives

$$\frac{dV}{dt} = 8\sqrt{3}x\frac{dx}{dt}.$$

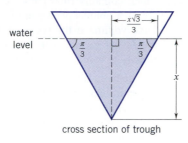

water level

cross section of trough

Setting $x = 3/2$ and $dV/dt = 4$, we have

$$4 = 8\sqrt{3}\left(\frac{3}{2}\right)\frac{dx}{dt} \qquad \text{and thus} \qquad \frac{dx}{dt} = \frac{1}{3\sqrt{3}} = \frac{1}{9}\sqrt{3}.$$

When the water reaches a depth of $1\frac{1}{2}$ feet, the water level is rising at the rate of $\frac{1}{9}\sqrt{3}$ feet per minute (about 0.19 feet per minute). ❏

EXERCISES 4.10

1. A point moves along the line $x + 2y = 2$. Find (a) the rate of change of the y-coordinate, given that the x-coordinate is increasing at the rate of 4 units per second; (b) the rate of change of the x-coordinate, given that the y-coordinate is decreasing at the rate of 2 units per second.

2. A particle is moving in the circular orbit $x^2 + y^2 = 25$. As it passes through the point $(3, 4)$, its y-coordinate is decreasing at the rate of 2 units per second. At what rate is the x-coordinate changing?

3. A particle is moving along the parabola $y^2 = 4(x + 2)$. As it passes through the point $(7, 6)$, its y-coordinate is increasing at the rate of 3 units per second. How fast is the x-coordinate changing at this instant?

4. A particle is moving along the parabola $4y = (x + 2)^2$ in such a way that its x-coordinate is increasing at the constant rate of 2 units per second. How fast is the particle's distance from the point $(-2, 0)$ changing as it passes through the point $(2, 4)$?

5. A particle is moving along the ellipse $x^2/16 + y^2/4 = 1$. At each time t its x- and y-coordinates are given by $x = 4\cos t$, $y = 2\sin t$. At what rate is the particle's distance from the origin changing at time t? At what rate is this distance from the origin changing when $t = \pi/4$?

6. A particle is moving along the curve $y = x\sqrt{x}$, $x \geq 0$. Find the points on the curve, if any, at which both coordinates are changing at the same rate.

7. A heap of rubbish in the shape of a cube is being compacted into a smaller cube. Given that the volume decreases at the rate of 2 cubic meters per minute, find the rate of change of an edge of the cube when the volume is exactly 27 cubic meters. What is the rate of change of the surface area of the cube at that instant?

8. The volume of a spherical balloon is increasing at the constant rate of 8 cubic feet per minute. How fast is the radius increasing when the radius is exactly 10 feet? How fast is the surface area increasing at that time?

9. At a certain instant the side of an equilateral triangle is α centimeters long and increasing at the rate of k centimeters per minute. How fast is the area increasing?

10. The dimensions of a rectangle are changing in such a way that the perimeter remains 24 inches. Show that when the area is 32 square inches, the area is either increasing or decreasing 4 times as fast as the length is increasing.

11. A rectangle is inscribed in a circle of radius 5 inches. If the length of the rectangle is decreasing at the rate of 2 inches per second, how fast is the area changing when the length is 6 inches?

12. A boat is held by a bow line that is wound about a bollard 6 feet higher than the bow of the boat. If the boat is drifting away at the rate of 8 feet per minute, how fast is the line unwinding when the bow is 30 feet from the bollard?

13. Two boats are racing with constant speed toward a finish marker, boat A sailing from the south at 13 mph and boat B approaching from the east. When equidistant from the marker, the boats are 16 miles apart and the distance between them is decreasing at the rate of 17 mph. Which boat will win the race?

14. A spherical snowball is melting in such a manner that its radius is changing at a constant rate, decreasing from 16 cm to 10 cm in 30 minutes. How fast is the volume of the snowball changing when the radius is 12 cm?

15. A 13-foot ladder is leaning against a vertical wall. If the bottom of the ladder is being pulled away from the wall at the rate of 2 feet per second, how fast is the area of the

triangle formed by the wall, the ground, and the ladder changing when the bottom of the ladder is 12 feet from the wall?

16. A ladder 13 feet long is leaning against a wall. If the foot of the ladder is pulled away from the wall at the rate of 0.5 feet per second, how fast will the top of the ladder be dropping when the base is 5 feet from the wall?

17. A tank contains 1000 cubic feet of natural gas at a pressure of 5 pounds per square inch. Find the rate of change of the volume if the pressure decreases at a rate of 0.05 pounds per square inch per hour. (Assume Boyle's law: *pressure × volume = constant*.)

18. The adiabatic law for the expansion of air is $PV^{1.4} = C$. At a given instant the volume is 10 cubic feet and the pressure is 50 pounds per square inch. At what rate is the pressure changing if the volume is decreasing at a rate of 1 cubic foot per second?

19. A man standing 3 feet from the base of a lamppost casts a shadow 4 feet long. If the man is 6 feet tall and walks away from the lamppost at a speed of 400 feet per minute, at what rate will his shadow lengthen? How fast is the tip of his shadow moving?

20. A light is attached to the wall of a building 64 feet above the ground. A ball is dropped from that height, but 20 feet away from the side of the building. The height y of the ball at time t is given by $y(t) = 64 - 16t^2$. Here we are measuring y in feet and t in seconds. How fast is the shadow of the ball moving along the ground 1 second after the ball is dropped?

21. An object that weighs 150 pounds on the surface of the earth will weigh $150(1 + \frac{1}{4000}r)^{-2}$ pounds when it is r miles above the earth. Given that the altitude of the object is increasing at the rate of 10 miles per minute, how fast is the weight decreasing when the object is 400 miles above the surface?

22. In the special theory of relativity the mass of a particle moving at speed v is given by the expression

$$\frac{m}{\sqrt{1 - v^2/c^2}}$$

where m is the mass at rest and c is the speed of light. At what rate is the mass of the particle changing when the speed of the particle is $\frac{1}{2}c$ and is increasing at the rate of $0.01c$ per second?

23. Water is dripping through the bottom of a conical cup 4 inches across and 6 inches deep. Given that the cup loses half a cubic inch of water per minute, how fast is the water level dropping when the water is 3 inches deep?

24. Water is poured into a conical container, vertex down, at the rate of 2 cubic feet per minute. The container is 6 feet deep and the open end is 8 feet across. How fast is the level of the water rising when the container is half full?

25. At what rate is the volume of a sphere changing at the instant when the surface area is increasing at the rate of 4 square centimeters per minute and the radius is increasing at the rate of 0.1 centimeter per minute?

26. Water flows from a faucet into a hemispherical basin 14 inches in diameter at the rate of 2 cubic inches per second.

How fast does the water rise (a) when the water is exactly halfway to the top? (b) just as it runs over? (The volume of a spherical segment is given by $\pi r h^2 - \frac{1}{3}\pi h^3$ where r is the radius of the sphere and h is the depth of the segment.)

27. The base of an isosceles triangle is 6 feet. Given that the altitude is 4 feet and increasing at the rate of 2 inches per minute, at what rate is the vertex angle changing?

28. As a boy winds up the cord, his kite is moving horizontally at a height of 60 feet with a speed of 10 feet per minute. How fast is the inclination of the cord changing when the cord is 100 feet long?

29. A revolving searchlight $\frac{1}{2}$ mile from a straight shoreline makes 1 revolution per minute. How fast is the light moving along the shore as it passes over a shore point 1 mile from the shore point nearest to the searchlight?

30. A revolving searchlight 1 mile from a straight shoreline turns at the rate of 2 revolutions per minute in the counterclockwise direction.
    (a) How fast is the light moving along the shore when it makes an angle of 45° with the shore?
    (b) How fast is the light moving when the angle is 90°?

31. A man starts at a point A and walks 40 feet north. He then turns and walks due east at 4 feet per second. A searchlight placed at A follows him. At what rate is the light turning 15 seconds after the man started walking east?

32. The diameter and height of a right circular cylinder are found at a certain instant to be 10 centimeters and 20 centimeters, respectively. If the diameter is increasing at the rate of 1 centimeter per second, what change in height will keep the volume constant?

33. A horizontal trough 12 feet long has a vertical cross section in the form of a trapezoid. The bottom is 3 feet wide, and the sides are inclined to the vertical at an angle with sine $\frac{4}{5}$. Given that water is poured into the trough at the rate of 10 cubic feet per minute, how fast is the water level rising when the water is exactly 2 feet deep?

34. Two cars, car A traveling east at 30 mph and car B traveling north at 22.5 mph, are heading toward an intersection I. At what rate is the angle IAB changing when cars A and B are 300 feet and 400 feet, respectively, from the intersection?

35. A rope 32 feet long is attached to a weight and passed over a pulley 16 feet above the ground. The other end of the rope is pulled away along the ground at the rate of 3 feet per second. At what rate is the angle between the rope and the ground changing when the weight is exactly 4 feet off the ground?

36. A slingshot is made by fastening the two ends of a 10-inch rubber strip 6 inches apart. If the midpoint of the strip is drawn back at the rate of 1 inch per second, at what rate is the angle between the segments of the strip changing 8 seconds later?

37. A balloon is released 500 feet away from an observer. If the balloon rises vertically at the rate of 100 feet per minute and at the same time the wind is carrying it away horizontally at the rate of 75 feet per minute, at what rate is the inclination of the observer's line of sight changing 6 minutes after the balloon has been released?

38. A searchlight is continually trained on a plane that flies directly above it at an altitude of 2 miles at a speed of 400 miles per hour. How fast does the light turn 2 seconds after the plane passes directly overhead?

39. A baseball diamond is a square 90 feet on a side. A player is running from second base to third base at the rate of 15 feet per second. Find the rate of change of the distance from the player to home plate at the instant the player is 10 feet from third base. (If you are not familiar with baseball, skip this problem.)

40. An airplane is flying at constant speed and altitude on a line that will take it directly over a radar station on the ground. At the instant the plane is 12 miles from the station, it is noted that the plane's angle of elevation is 30° and is increasing at the rate of 0.5° per second. Give the speed of the plane in miles per hour.

41. An athlete is running around a circular track of radius 50 meters at the rate of 5 meters per second. A spectator is 200 meters from the center of the track. How fast is the distance between the two changing when the runner is approaching the spectator and the distance between them is 200 meters?

Exercises 42–44. Here x and y are functions of t and are related as indicated. Obtain the desired derivative from the information given.

42. $2xy^2 - y = 22$. Given that $\dfrac{dy}{dt} = -2$ when $x = 3$ and $y = 2$, find $\dfrac{dx}{dt}$.

43. $x - \sqrt{xy} = 4$. Given that $\dfrac{dy}{dt} = 3$ when $x = 8$ and $y = 2$, find $\dfrac{dx}{dt}$.

44. $\sin x = 4 \cos y - 1$. Given that $\dfrac{dx}{dt} = -1$ when $x = \pi$ and $y = \dfrac{\pi}{3}$, find $\dfrac{dy}{dt}$.

■ 4.11 DIFFERENTIALS

In Figure 4.11.1 we have sketched the graph of a differentiable function f and below it the tangent line at the point $(x, f(x))$.

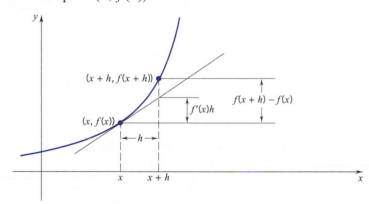

Figure 4.11.1

As the figure suggests, for small $h \neq 0$, $f(x + h) - f(x)$, the change in f from x to $x + h$ can be approximated by the product $f'(x)h$:

(4.11.1)
$$f(x + h) - f(x) \cong f'(x)h.$$

How good is this approximation? It is good in the sense that, for small h the difference between the two quantities,

$$[f(x + h) - f(x)] - f'(x)h,$$

is small compared to h. How small compared to h? Small enough compared to h that its ratio to h, the quotient

$$\frac{[f(x + h) - f(x)] - f'(x)h}{h},$$

tends to 0 as h tends to 0:

$$\lim_{h \to 0} \frac{[f(x+h) - f(x)] - f'(x)h}{h} = \lim_{h \to 0} \frac{f(x+h) - f(x)}{h} - \lim_{h \to 0} \frac{f'(x)h}{h}$$

$$= f'(x) - f'(x) = 0.$$

The quantities $f(x+h) - f(x)$ and $f'(x)h$ have names:

DEFINITION 4.11.2

For $h \neq 0$ the difference $f(x+h) - f(x)$ is called the *increment of f from x to $x + h$* and is denoted by Δf:

$$\Delta f = f(x+h) - f(x).^{\dagger}$$

The product $f'(x)h$ is called the *differential of f at x with increment h* and is denoted by df:

$$df = f'(x)h.$$

Display 4.11.1 says that, for small h, Δf and df are approximately equal:

$$\Delta f \cong df.$$

How close is the approximation? Close enough (as we just showed) that the quotient

$$\frac{\Delta f - df}{h}$$

tends to 0 as h tends to 0.

Let's see what all this amounts to in a very simple case. The area of a square of side x is given by the function

$$f(x) = x^2, \qquad x > 0.$$

If the length of each side increases from x to $x + h$, the area increases from $f(x)$ to $f(x+h)$. The change in area is the increment Δf:

$$\Delta f = f(x+h) - f(x)$$
$$= (x+h)^2 - x^2$$
$$= (x^2 + 2xh + h^2) - x^2$$
$$= 2xh + h^2.$$

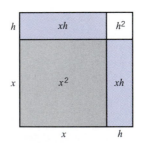

Figure 4.11.2

As an estimate for this change, we can use the differential

$$df = f'(x)h = 2xh. \qquad \text{(Figure 4.11.2)}$$

The error of this estimate, the difference between the actual change Δf and the estimated change df, is the difference

$$\Delta f - df = h^2.$$

As promised, the error is small compared to h in the sense that

$$\frac{\Delta f - df}{h} = \frac{h^2}{h} = h$$

tends to 0 as h tends to 0.

$^{\dagger}\Delta$ is a Greek letter, the capital of δ. Δf is read "delta f."

Example 1 Use a differential to estimate the change in $f(x) = x^{2/5}$
(a) as x increases from 32 to 34, **(b)** as x decreases from 1 to $\frac{9}{10}$.

SOLUTION Since $f'(x) = \frac{2}{5}x^{-3/5} = 2/(5x^{3/5})$, we have

$$df = f'(x)h = \frac{2}{5x^{3/5}}h.$$

(a) We set $x = 32$ and $h = 2$. The differential then becomes

$$df = \frac{2}{5(32)^{3/5}}(2) = \frac{4}{40} = 0.1.$$

A change in x from 32 to 34 increases the value of f by approximately 0.1. For comparison, our hand calculator gives

$$\Delta f = f(34) - f(32) \cong 4.0982 - 4 = 0.0982.$$

(b) We set $x = 1$ and $h = -\frac{1}{10}$. In this case, the differential is

$$df = \frac{2}{5(1)^{3/5}}\left(-\frac{1}{10}\right) = -\frac{2}{50} = -0.04.$$

A change in x from 1 to $\frac{9}{10}$ decreases the value of f by approximately 0.04. For comparison, our hand calculator gives

$$\Delta f = f(0.9) - f(1) = (0.9)^{2/5} - (1)^{2/5} \cong 0.9587 - 1 = -0.0413. \quad \square$$

Example 2 Use a differential to estimate: **(a)** $\sqrt{104}$, **(b)** $\cos 40°$.

SOLUTION
(a) We know that $\sqrt{100} = 10$. We need an estimate for the increase of

$$f(x) = \sqrt{x}$$

as x increases from 100 to 104. Here

$$f'(x) = \frac{1}{2\sqrt{x}} \quad \text{and} \quad df = f'(x)h = \frac{h}{2\sqrt{x}}.$$

With $x = 100$ and $h = 4$, df becomes

$$\frac{4}{2\sqrt{100}} = \frac{1}{5} = 0.2.$$

A change in x from 100 to 104 increases the value of the square root by approximately 0.2. It follows that

$$\sqrt{104} \cong \sqrt{100} + 0.2 = 10 + 0.2 = 10.2.$$

As you can check, $(10.2)^2 = 104.04$. Our estimate is not far off.

(b) Let $f(x) = \cos x$, where as usual x is given in radians. We know that $\cos 45° = \cos(\pi/4) = \sqrt{2}/2$. Converting 40° to radians, we have

$$40° = 45° - 5° = \frac{\pi}{4} - \left(\frac{\pi}{180}\right)5 = \frac{\pi}{4} - \frac{\pi}{36} \text{ radians.}$$

We use a differential to estimate the change in $\cos x$ as x decreases from $\pi/4$ to $(\pi/4) - (\pi/36)$:

$$f'(x) = -\sin x \quad \text{and} \quad df = f'(x)h = -h\sin x.$$

With $x = \pi/4$ and $h = -\pi/36$, df is given by

$$df = -\left(-\frac{\pi}{36}\right)\sin\left(\frac{\pi}{4}\right) = \frac{\pi}{36}\frac{\sqrt{2}}{2} = \frac{\pi\sqrt{2}}{72} \cong 0.0617.$$

A decrease in x from $\pi/4$ to $(\pi/4) - (\pi/36)$ increases the value of the cosine by approximately 0.0617. Therefore,

$$\cos 40° \cong \cos 45° + 0.0617 \cong 0.7071 + 0.0616 = 0.7688.$$

Our hand calculator gives $\cos 40° \cong 0.7660$. ❏

Example 3   A metal sphere with a radius of 10 cm is to be covered by a 0.02 cm coating of silver. Approximately how much silver will be required?

SOLUTION   We will use a differential to estimate the increase in the volume of a sphere if the radius is increased from 10 cm to 10.02 cm. The volume of a sphere of radius r is given by the formula $V = \frac{4}{3}\pi r^3$. Therefore

$$dV = 4\pi r^2 h.$$

Taking $r = 10$ and $h = 0.02$, we have

$$dV = 4\pi(10)^2(0.02) = 8\pi \cong 25.133.$$

It will take approximately 25.133 cubic cm of silver to coat the sphere.   ❏

Example 4   A metal cube is heated and the length of each edge is thereby increased by 0.1%. Use a differential to show that the surface area of the cube is then increased by about 0.2%.

SOLUTION   Let x be the initial length of an edge. The initial surface area is then $S(x) = 6x^2$. As the length increases from x to $x + h$, the surface area increases from $S(x)$ to $S(x + h)$. We will estimate the ratio

$$\frac{\Delta S}{S} = \frac{S(x + h) - S(x)}{S(x)}$$

by

$$\frac{dS}{S} \qquad \text{taking} \qquad h = 0.001x.$$

Here

$$S(x) = 6x^2, \qquad dS = 12xh = 12x(0.001x),$$

and therefore

$$\frac{dS}{S} = \frac{12x(0.001x)}{6x^2} = 0.002.$$

If the length of each edge is increased by 0.1%, the surface area is increased by about 0.2%.   ❏

EXERCISES 4.11

1. Use a differential to estimate the change in the volume of a cube caused by an increase h in the length of each side. Interpret geometrically the error of your estimate $\Delta V - dV$.

2. Use a differential to estimate the area of a ring of inner radius r and width h. What is the exact area?

▷**Exercises 3–8.** Use a differential to estimate the value of the indicated expression. Then compare your estimate with the result given by a calculator.

3. $\sqrt[3]{1002}$.

4. $1/\sqrt{24.5.}$.

5. $\sqrt[4]{15.5}$.

6. $(26)^{2/3}$.

7. $(33)^{3/5}$

8. $(33)^{-1/5}$.

Exercises 9–12. Use a differential to estimate the value of the expression. (Remember to convert to radian measure.) Then compare your estimate with the result given by a calculator.

9. $\sin 46°$.

10. $\cos 62°$.

11. $\tan 28°$

12. $\sin 43°$.

13. Estimate $f(2.8)$ given that $f(3) = 2$ and $f'(x) = (x^3 + 5)^{1/5}$.

14. Estimate $f(5.4)$ given that $f(5) = 1$ and $f'(x) = \sqrt[3]{x^2 + 2}$.

15. Find the approximate volume of a thin cylindrical shell with open ends given that the inner radius is r, the height is h, and the thickness is t.

16. The diameter of a steel ball is measured to be 16 centimeters, with a maximum error of 0.3 centimeters. Estimate by differentials the maximum error (a) in the surface area as calculated from the formula $S = 4\pi r^2$; (b) in the volume as calculated from the formula $V = \frac{4}{3}\pi r^3$.

17. A box is to be constructed in the form of a cube to hold 1000 cubic feet. Use a differential to estimate how accurately the inner edge must be made so that the volume will be correct to within 3 cubic feet.

18. Use differentials to estimate the values of x for which

(a) $\sqrt{x+1} - \sqrt{x} < 0.01$.

(b) $\sqrt[4]{x+1} - \sqrt[4]{x} < 0.002$.

19. A hemispherical dome with a 50-foot radius will be given a coat of paint 0.01 inch thick. The contractor for the job wants to estimate the number of gallons of paint that will be needed. Use a differential to obtain an estimate. (There are 231 cubic inches in a gallon.)

20. View the earth as a sphere of radius 4000 miles. The volume of ice that covers the north and south poles is estimated to be 8 million cubic miles. Suppose that this ice melts and the water produced distributes itself uniformly over the surface of the earth. Estimate the depth of this water.

21. The period P of the small oscillations of a simple pendulum is related to the length L of the pendulum by the equation

$$P = 2\pi \sqrt{\frac{L}{g}}$$

where g is the (constant) acceleration of gravity. Show that a small change dL in the length of a pendulum produces a change dP in the period that satisfies the equation

$$\frac{dP}{P} = \frac{1}{2}\frac{dL}{L}.$$

22. Suppose that the pendulum of a clock is 90 centimeters long. Use the result in Exercise 21 to determine how the length of the pendulum should be adjusted if the clock is losing 15 seconds per hour.

23. A pendulum of length 3.26 feet goes through one complete oscillation in 2 seconds. Use Exercise 21 to find the approximate change in P if the pendulum is lengthened by 0.01 feet.

24. A metal cube is heated and the length of each edge is thereby increased by 0.1%. Use a differential to show that the volume of the cube is then increased by about 0.3%.

25. We want to determine the area of a circle by measuring the diameter x and then applying the formula $A = \frac{1}{4}\pi x^2$. Use a differential to estimate how accurately we must measure the diameter for our area formula to yield a result that is accurate within 1%.

26. Estimate by differentials how precisely x must be determined (a) for our calculation of x^n to be accurate within 1%; (b) for our estimate of $x^{1/n}$ to be accurate within 1%. (Here n is a positive integer.)

Little-o(h) Let g be a function defined at least on some open interval that contains the number 0. We say that $g(h)$ is *little-o(h)* and write $g(h) = o(h)$ to indicate that, for small h, $g(h)$ is so small compared to h that

$$\lim_{h \to 0} \frac{g(h)}{h} = 0.$$

27. Determine whether the statement is true.

(a) $h^3 = o(h)$

(b) $\dfrac{h^2}{h-1} = o(h)$.

(c) $h^{1/3} = o(h)$.

28. Show that if $g(h) = o(h)$, then $\lim_{h \to 0} g(h) = 0$.

29. Show that if $g_1(h) = o(h)$ and $g_2(h) = o(h)$, then

$$g_1(h) + g_2(h) = o(h) \qquad \text{and} \qquad g_1(h)g_2(h) = o(h).$$

30. The figure shows the graph of a differentiable function f and a line with slope m that passes through the point $(x, f(x))$. The vertical separation at $x + h$ between the line with slope m and the graph of f has been labeled $g(h)$.

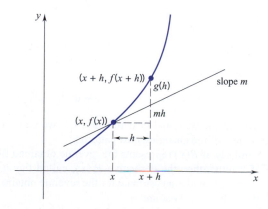

(a) Calculate $g(h)$.

(b) Show that, of all lines that pass through $(x, f(x))$, the tangent line is the line that best approximates the graph of f near the point $(x, f(x))$ by showing that

$$g(h) = o(h) \qquad \text{iff} \qquad m = f'(x).$$

Here we dispense with $g(h)$ and call $o(h)$ any expression in h which, for small h, is so small compared to h that

$$\lim_{h \to 0} \frac{o(h)}{h} = 0.$$

31. (*A differentiable function is locally almost linear.*) If f is a linear function,

$$f(x) = mx + b,$$

then

$$f(x + h) - f(x) = mh.$$

Show that

(4.11.3)

> a function f is differentiable at x iff there exists a number m such that
>
> $$f(x + h) - f(x) = mh + o(h).$$

What is m here?

■ PROJECT 4.11 Marginal Cost, Marginal Revenue, Marginal Profit

In business and economics we recognize that costs, revenues, and profits depend on many factors. Of special interest to us here is the study of how costs, revenues, and profits are affected by changes in production and sales. In this brief section we make the simplifying assumption that all production is sold and therefore units sold = units produced.

Suppose that $C(x)$ represents the cost of producing x units. Although x is usually a nonnegative integer, in theory and practice it is convenient to assume that $C(x)$ is defined for x in some interval and that the function C is differentiable. In this context, the derivative $C'(x)$ is called the *marginal cost at x*.

This terminology deserves some explanation. The difference $C(x + 1) - C(x)$ represents the cost of increasing production from x units to $x + 1$ units and, as such, gives the cost of producing the $(x + 1)$-st unit. *The derivative $C'(x)$ is called the marginal cost at x because it provides an estimate for the cost of the $(x + 1)$-st unit*: in general,

$$C(x + h) - C(x) \cong C'(x)h. \qquad \text{(differential estimate)}$$

At $h = 1$ this estimate reads

$$C(x + 1) - C(x) \cong C'(x).$$

Thus, as asserted,

$$C'(x) \cong \text{cost of } (x + 1)\text{-st unit.} \quad \square$$

By studying the marginal cost function C', we can obtain an overall view of the changing cost patterns.

Similarly, if $R(x)$ represents the revenue obtained from the sale of x units, then the derivative $R'(x)$, called the *marginal revenue at x*, provides an estimate for the revenue obtained from the sale of the $(x + 1)$-st unit.

If $C = C(x)$ and $R = R(x)$ are the cost and revenue functions associated with the production and sale of x units, then the function

$$P(x) = R(x) - C(x)$$

is called the *profit function*. The points x (if any) at which $C(x) = R(x)$ — that is, the values at which "cost" = "revenue"—

are called *break-even points*. The derivative $P'(x)$ is called the *marginal profit at x*. By Theorem 4.3.2, maximum profit occurs at a point where $P'(x) = 0$, a point where the marginal profit is zero, which, since

$$P'(x) = R'(x) - C'(x),$$

is a point where the marginal revenue $R'(x)$ equals the marginal cost $C'(x)$.

A word about revenue functions. The revenue obtained from the sale of x units at an average price $p(x)$ is the product of x and $p(x)$:

$$R(x) = xp(x).$$

In classical supply-demand theory, if too many units are sold, the price at which they can be sold comes down. It may come down so much that the product $xp(x)$ starts to decrease. If the market is flooded with units, $p(x)$ tends to zero and revenues are severely impaired. Thus it is that the revenue $R(x)$ increases with x up to a point and then decreases. The figure gives a graphical representation of a pair of cost and revenue functions, shows the break-even points, and indicates the regions of profit and loss.

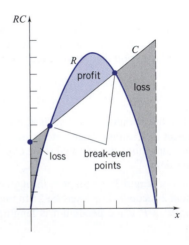

Problem 1. A manufacturer determines that the total cost of producing x units per hour is given by the function

$$C(x) = 2000 + 50x - \frac{x^2}{20}. \qquad \text{(dollars)}$$

What is the marginal cost at production level 20 units per hour? What is the exact cost of producing the 21st component?

Problem 2. A manufacturer determines that the costs and revenues that result from the production and sale of x units per day are given by the functions

$$C(x) = 12{,}000 + 30x \qquad \text{and} \qquad R(x) = 650x - 5x^2.$$

Find the profit function and determine the break-even points. Find the marginal profit function and determine the production/sales level for maximum profit.

Problem 3. The cost and revenue functions for the production and sale of x units are

$$C(x) = 4x + 1400 \qquad \text{and} \qquad R(x) = 20x - \frac{x^2}{50}.$$

(a) Find the profit function and determine the break-even points.

(b) Find the marginal profit function and determine the production level at which the marginal profit is zero.

(c) Sketch the cost and revenue functions in the same coordinate system and indicate the regions of profit and loss. Estimate the production level that produces maximum profit.

Problem 4. The cost and revenue functions are

$$C(x) = 3000 + 5x \qquad \text{and} \qquad R(x) = 60x - 2x\sqrt{x},$$

with x measured in thousands of units.

(a) Use a graphing utility to graph the cost function together with the revenue function. Estimate the break-even points.

(b) Graph the profit function and estimate the production level that yields the maximum profit.

Problem 5. The cost and revenue functions are

$$C(x) = 4 + 0.75x \qquad \text{and} \qquad R(x) = \frac{10x}{1 + 0.25x^2},$$

with x measured in hundreds of units.

(a) Use a graphing utility to graph the cost function together with the revenue function. Estimate the break-even points.

(b) Graph the profit function and estimate the production level that yields the maximum profit. Exactly how many units should be produced to maximize profit?

Problem 6. Let $C(x)$ be the cost of producing x units. The *average cost per unit* is $A(x) = C(x)/x$. Show that if $C''(x) > 0$, then the average cost per unit is a minimum at the production levels x where the marginal cost equals the average cost per unit.

Problem 7. Let $R(x)$ be the revenue that results from the sale of x units. The *average revenue per unit* is $B(x) = R(x)/x$. Show that if $R''(x) < 0$, then the average revenue per unit is a maximum at the values x where the marginal revenue equals the average revenue per unit.

■ 4.12 NEWTON-RAPHSON APPROXIMATIONS

Figure 4.12.1 shows the graph of a function f. Since the graph of f crosses the x-axis at $x = c$, the number c is a solution (root) of the equation $f(x) = 0$. In the setup of Figure 4.12.1, we can approximate c as follows: Start at x_1 (see the figure). The tangent line at $(x_1, f(x_1))$ intersects the x-axis at a point x_2 which is closer to c than x_1. The tangent line at $(x_2, f(x_2))$ intersects the x-axis at a point x_3, which in turn is closer to c than x_2. In this manner, we obtain numbers $x_1, x_2, x_3, \ldots, x_n, x_{n+1}$, which more and more closely approximate c.

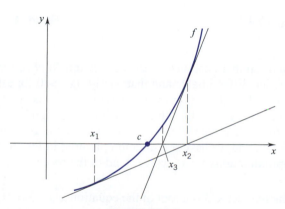

Figure 4.12.1

There is an algebraic connection between x_n and x_{n+1} that we now develop. The tangent line at $(x_n, f(x_n))$ has the equation

$$y - f(x_n) = f'(x_n)(x - x_n).$$

The x-intercept of this line, x_{n+1}, can be found by setting $y = 0$:

$$0 - f(x_n) = f'(x_n)(x_{n+1} - x_n).$$

Solving this equation for x_{n+1}, we have

(4.12.1)
$$x_{n+1} = x_n - \frac{f(x_n)}{f'(x_n)}.$$

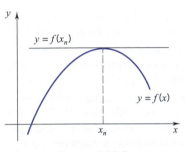

Figure 4.12.2

This method of locating a root of an equation $f(x) = 0$ is called the *Newton-Raphson method*. The method does not work in all cases. First, there are some conditions that must be placed on the function f. Clearly, f must be differentiable at points near the root c. Also, if $f'(x_n) = 0$ for some n, then the tangent line at $(x_n, f(x_n))$ is horizontal and the next approximation x_{n+1} cannot be calculated. See Figure 4.12.2. Thus, we will assume that $f'(x) \neq 0$ at points near c.

The method can also fail for other reasons. For example, it can happen that the first approximation x_1 produces a second approximation x_2, which in turn takes us back to x_1. In this case the approximations simply alternate between x_1 and x_2. See Figure 4.12.3. Another type of difficulty can arise if $f'(x_1)$ is close to zero. In this case the second approximation x_2 can be worse than x_1, the third approximation x_3 can be worse than x_2, and so forth. See Figure 4.12.4.

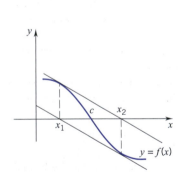

Figure 4.12.3

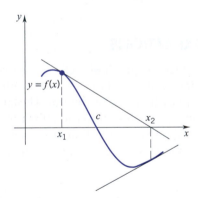

Figure 4.12.4

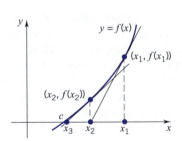

Figure 4.12.5

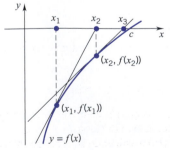

Figure 4.12.6

There is a condition that guarantees that the Newton-Raphson method will work. Suppose that f is twice differentiable and that $f(x)f''(x) > 0$ for all x between c and x_1. If $f(x) > 0$ for such x, then $f''(x) > 0$ for such x and (as shown in Section 4.6) the graph bends up, and we have the situation pictured in Figure 4.12.5. On the other hand, if $f(x) < 0$ for such x, then $f''(x) < 0$ for such x and (as shown in Section 4.6) the graph bends down, and we have the situation pictured in Figure 4.12.6. In each of these cases the approximations $x_1, x_2, x_3, \ldots$ tend to the root c.

Example 1 The number $\sqrt{3}$ is a root of the equation $x^2 - 3 = 0$. We will estimate $\sqrt{3}$ by applying the Newton-Raphson method to the function $f(x) = x^2 - 3$ starting at $x_1 = 2$. [As you can check, $f(x)f''(x) > 0$ on $(\sqrt{3}, 2)$ and therefore we can be sure that the method applies.] Since $f'(x) = 2x$, the Newton-Raphson formula gives

$$x_{n+1} = x_n - \left(\frac{x_n^2 - 3}{2x_n}\right) = \frac{x_n^2 + 3}{2x_n}.$$

Successive calculations with this formula (using a calculator) are given in the following table:

n	x_n	$x_{n+1} = \frac{x_n^2+3}{2x_n}$
1	2	1.75000
2	1.75000	1.73214
3	1.73214	1.73205

Since $(1.73205)^2 \cong 2.999997$, the method has generated a very accurate estimate of $\sqrt{3}$ in only three steps. ❑

EXERCISES 4.12

▷ **Exercises 1–8.** Use the Newton-Raphson method to estimate a root of the equation $f(x) = 0$ starting at the indicated value of x: (a) Express x_{n+1} in terms of x_n. (b) Give x_4 rounded off to five decimal places and evaluate f at that approximation.

1. $f(x) = x^2 - 24$; $x_1 = 5$.

2. $f(x) = x^3 - 4x + 1$; $x_1 = 2$.

3. $f(x) = x^3 - 25$; $x_1 = 3$.   4. $f(x) = x^5 - 30$; $x_1 = 2$.

5. $f(x) = \cos x - x$; $x_1 = 1$.

6. $f(x) = \sin x - x^2$; $x_1 = 1$.

7. $f(x) = \sqrt{x + 3} - x$; $x_1 = 1$.

8. $f(x) = x + \tan x$; $x_1 = 2$.

9. The function $f(x) = x^{1/3}$ is 0 at $x = 0$. Verify that the condition $f(x)f''(x) > 0$ fails everywhere. Show that the Newton-Raphson method starting at *any* number $x_1 \neq 0$ fails to generate numbers that approach the solution $x = 0$. Describe the numbers $x_1, x_2, x_3, \ldots$ that the method generates.

10. What results from the application of the Newton-Raphson method to a function f if the starting approximation x_1 is precisely the desired zero of f?

▷ 11. Set $f(x) = 2x^3 - 3x^2 - 1$.
   (a) Show that the equation $f(x) = 0$ has a root between 1 and 2.
   (b) Show that the Newton-Raphson method process started at $x_1 = 1$ fails to generate numbers that approach the root that lies between 1 and 2.
   (c) Estimate this root by starting at $x_1 = 2$. Determine x_4 rounded off to four decimal places and evaluate $f(x_4)$.

▷ 12. The function $f(x) = x^4 - 2x^2 - \frac{17}{16}$ has two zeros, one at a point a between 0 and 2, and the other at $-a$. (f is an even function.)
   (a) Show that the Newton-Raphson method fails in the search for a if we start at $x = \frac{1}{2}$. What are the outputs $x_1, x_2, x_3, \ldots$ in this case?

   (b) Estimate a by starting at $x_1 = 2$. Determine x_4 rounded off to five decimal places and evaluate $f(x_4)$.

13. Set $f(x) = x^2 - a$, $a > 0$. The roots of the equation $f(x) = 0$ are $\pm\sqrt{a}$.
   (a) Show that if x_1 is any initial estimate for $\sqrt{a}$, then the Newton-Raphson method gives the iteration formula

   $$x_{n+1} = \frac{1}{2}\left(x_n + \frac{a}{x_n}\right), \qquad n \geq 1.$$

   (b) Take $a = 5$. Starting at $x_1 = 2$, use the formula in part (a) to calculate x_4 to five decimal places and evaluate $f(x_4)$.

14. Set $f(x) = x^k - a$, k a positive integer, $a > 0$. The number $a^{1/k}$ is a root of the equation $f(x) = 0$.
   (a) Show that if x_1 is any initial estimate for $a^{1/k}$, then the Newton-Raphson method gives the iteration formula

   $$x_{n+1} = \frac{1}{k}\left[(k-1)x_n + \frac{a}{x_n^{k-1}}\right].$$

   Note that for $k = 2$ this formula reduces to the formula given in Exercise 13.
   (b) Use the formula in part (a) to approximate $\sqrt[3]{23}$. Begin at $x_1 = 3$ and calculate x_4 rounded off to five decimal places. Evaluate $f(x_4)$.

15. Set $f(x) = \frac{1}{x} - a$, $a \neq 0$.
   (a) Apply the Newton-Raphson method to derive the iteration formula

   $$x_{n+1} = 2x_n - ax_n^2, \qquad n \geq 1.$$

   Note that this formula provides a method for calculating reciprocals without recourse to division.
   (b) Use the formula in part (a) to calculate $1/2.7153$ rounded off to five decimal places.

16. Set $f(x) = x^4 - 7x^2 - 8x - 3$.

   (a) Show that f has exactly one critical point c in the interval $(2, 3)$.

   (b) Use the Newton-Raphson method to estimate c by calculating x_3. Round off your answer to four decimal places. Does f have a local maximum at c, a local minimum, or neither?

17. Set $f(x) = \sin x + \frac{1}{2}x^2 - 2x$.

   (a) Show that f has exactly one critical point c in the interval $(2, 3)$.

   (b) Use the Newton-Raphson method to estimate c by calculating x_3. Round off your answer to four decimal places.

Does f have a local maximum at c, a local minimum, or neither?

18. Approximations to π can be obtained by applying the Newton-Raphson method to $f(x) = \sin x$ starting at $x_1 = 3$.

   (a) Find x_4 rounded off to four decimal places.

   (b) What are the approximations if we start at $x_1 = 6$?

19. The equation $x + \tan x = 0$ has an infinite number of positive roots $r_1, r_2, r_3, \ldots, r_n$ slightly larger than $(n - \frac{1}{2})\pi$. Use the Newton-Raphson method to find r_1 and r_2 to three decimal place accuracy.

■ CHAPTER 4. REVIEW EXERCISES

Exercises 1–2. Show that f satisfies the conditions of Rolle's theorem on the indicated interval and find all the numbers c on the interval for which $f'(c) = 0$.

1. $f(x) = x^3 - x$; $[-1, 1]$.

2. $f(x) = \sin x + \cos x - 1$; $[0, 2\pi]$.

Exercises 3–6. Verify that f satisfies the conditions of the mean-value theorem on the indicated interval and find all the numbers c that satisfy the conclusion of the theorem.

3. $f(x) = x^3 - 2x + 1$; $[-2, 3]$.

4. $f(x) = \sqrt{x - 1}$; $[2, 5]$.

5. $f(x) = \frac{x + 1}{x - 1}$; $[2, 4]$.   **6.** $f(x) = x^{3/4}$; $[0, 16]$.

7. Set $f(x) = x^{2/3} - 1$. Note that $f(-1) = f(1) = 0$. Verify that there does not exist a number c in $(-1, 1)$ for which $f'(c) = 0$. Explain how this does not violate Rolle's theorem.

8. Set $f(x) = (x + 1)/(x - 2)$. Show that there does not exist a number c in $(1, 4)$ for which $f(4) - f(1) = f'(c)(4 - 1)$. Explain how this does not violate the mean-value theorem.

9. Does there exist a differentiable function f with $f(1) = 5$, $f(4) = 1$, and $f'(x) \geq -1$ for all x in $(1, 4)$? If not, how do you know?

10. Let $f(x) = x^3 - 3x + k$, k constant.

   a. Show that $f(x) = 0$ for at most one x in $[-1, 1]$.

   b. For what values of k does $f(x) = 0$ for some x in $[-1, 1]$?

Exercises 11–16. Find the intervals on which f increases and the intervals on which f decreases; find the critical points and the local extreme values.

11. $f(x) = 2x^3 + 3x^2 + 1$.   **12.** $f(x) = x^4 - 4x + 3$.

13. $f(x) = (x + 2)^2(x - 1)^3$.   **14.** $f(x) = x + \frac{4}{x^2}$.

15. $f(x) = \frac{x}{1 + x^2}$.

16. $f(x) = \sin x - \cos x$, $0 \leq x \leq 2\pi$.

Exercises 17–22. Find the critical points. Then find and classify all the extreme values.

17. $f(x) = x^3 + 2x^2 + x + 1$; $x \in [-2, 1]$.

18. $f(x) = x^4 - 8x^2 + 2$; $x \in [-1, 3]$.

19. $f(x) = x^2 + \frac{4}{x^2}$; $x \in [1, 4]$.

20. $f(x) = \cos^2 x + \sin x$; $x \in [0, 2\pi]$.

21. $f(x) = x\sqrt{1 - x}$; $x \in (-\infty, 1]$.

22. $f(x) = \frac{x^2}{x - 2}$; $x \in (2, \infty)$.

Exercises 23–25. Find all vertical, horizontal, and oblique (see Exercises 4.7) asymptotes.

23. $f(x) = \frac{3x^2 - 9x}{x^2 - x - 12}$.   **24.** $f(x) = \frac{x^2 - 4}{x^2 - 5x + 6}$.

25. $f(x) = \frac{x^4}{x^3 - 1}$.

Exercises 26–28. Determine whether or not the graph of f has a vertical tangent or a vertical cusp at c.

26. $f(x) = (x - 1)^{3/5}$; $c = 1$.

27. $f(x) = \frac{5}{7}x^{7/5} - 5x^{2/5}$; $c = 0$.

28. $f(x) = 3x^{1/3}(2 + x)$; $c = 0$.

Exercises 29–36. Sketch the graph of the function using the approach outlined in Section 4.8.

29. $f(x) = 6 + 4x^3 - 3x^4$.   **30.** $f(x) = 3x^5 - 5x^3 + 1$.

31. $f(x) = \frac{2x}{x^2 + 4}$.   **32.** $f(x) = x^{2/3}(x - 10)$.

33. $f(x) = x\sqrt{4 - x}$.   **34.** $f(x) = x^4 - 2x^2 + 3$.

35. $f(x) = \sin x + \sqrt{3}\cos x$, $x \in [0, 2\pi]$.

36. $f(x) = \sin^2 x - \cos x$, $x \in [0, 2\pi]$.

37. Sketch the graph of a function f that satisfies the following conditions:

$$f(-1) = 3, \quad f(0) = 0, \quad f(2) = -4;$$

$$f'(-1) = f'(2) = 0;$$

$$f'(x) > 0 \text{ for } x < -1 \text{ and for } x > 2, \quad f'(x) < 0$$

$$\text{if } -1 < x < 2;$$

$$f''(x) < 0 \text{ for } x < \tfrac{1}{2}, \quad f''(x) > 0 \text{ for } x > \tfrac{1}{2}.$$

38. Given that the surface area of a sphere plus the surface area of a cube is constant, show that the sum of the volumes is minimized by letting the diameter of the sphere equal the length of a side of the cube. What dimensions maximize the sum of the volumes?

39. A closed rectangular box with a square base is to be built subject to the following conditions: the volume is to be 27 cubic feet, the area of the base may not exceed 18 square feet, the height of the box may not exceed 4 feet. Determine the dimensions of the box (a) for minimal surface area; (b) for maximal surface area.

40. The line through $P(1, 2)$ intersects the positive x-axis at $A(a, 0)$ and the positive y-axis at $B(0, b)$. Determine the values of a and b that minimize the area of the triangle OAB.

41. A right circular cylinder is generated by revolving a rectangle of given perimeter P about one of its sides. What dimensions of the rectangle will generate the cylinder of maximum volume?

42. A printed page is to have a total area of 80 square inches. The margins at the top and on the sides are to be 1 inch each; the bottom margin is to be 1.5 inches. Determine the dimensions of the page that maximize the area available for print.

43. An object moves along a coordinate line, its position at time t given by the function $x(t) = t + 2\cos t$. Find those times from $t = 0$ to $t = 2\pi$ when the object is slowing down.

44. An object moves along a coordinate line, its position at time t given by the function $x(t) = (4t - 1)(t - 1)^2$, $t \geq 0$. (a) When is the object moving to the right? When to the left? When does it change direction? (b) What is the maximum speed of the object when moving left?

45. An object moves along a coordinate line, its position at time t given by the function $x(t) = \sqrt{t + 1}$, $t \geq 0$. (a) Show that the acceleration is negative and proportional to the cube of the velocity. (b) Use differentials to obtain numerical estimates for the position, velocity, and acceleration at time $t = 17$. Base your estimate on $t = 15$.

46. A rocket is fired from the ground straight up with an initial velocity of 128 feet per second. (a) When does the rocket reach maximum height? What is maximum height? (b) When does the rocket hit the ground and at what speed?

47. Ballast dropped from a balloon that was rising at the rate of 8 feet per second reached the ground in 10 seconds. How high was the balloon when the ballast was released?

48. A ball thrown straight up from the ground reaches a height of 24 feet in 1 second. How high will the ball go?

49. A boy walks on a straight, horizontal path away from a light that hangs 12 feet above the path. How fast does his shadow lengthen if he is 5 feet tall and walks at the rate of 168 feet per minute?

50. The radius of a cone increases at the rate of 0.3 inches per minute, but the volume remains constant. At what rate does the height of the cone change when the radius is 4 inches and the height is 15 inches?

51. A railroad track crosses a highway at an angle of 60°. A locomotive is 500 feet from the intersection and moving away from it at the rate of 60 miles per hour. A car is 500 feet from the intersection and moving toward it at the rate of 30 miles per hour. What is the rate of change of the distance between them?

52. A square is inscribed in a circle. Given that the radius of the circle is increasing at the rate of 5 centimeters per minute, at what rate is the area of the square changing when the radius is 10 centimeters?

53. A horizontal water trough 12 feet long has a vertical cross section in the form of an isosceles triangle (vertex down). The base and height of the triangle are each 2 feet. Given that water is being drained out of the trough at the rate of 3 cubic feet per minute, how fast is the water level falling when the water is 1.5 feet deep?

54. Use a differential to estimate $f(3.8)$ given that $f(4) = 2$ and $f'(x) = \sqrt[3]{3x - 4}$.

55. Use a differential to estimate $f(4.2)$ given that $f(x) = \sqrt{x} + 1/\sqrt{x}$.

Exercises 56–57. Use a differential to estimate the value of the expression.

56. $\sqrt[4]{83}$.   **57.** $\tan 43°$.

58. A spherical tank with a diameter of 20 feet will be given a coat of paint 0.05 inches thick. Estimate by a differential the amount of paint needed. (Assume that there are 231 cubic inches in a gallon.)

▷ **Exercises 59–60.** Use the Newton-Raphson method to estimate a root of $f(x) = 0$ starting at the indicated value: (a) Express x_{n+1} in terms of x_n. (b) Give x_4 rounded off to five decimal places and evaluate f at that approximation.

59. $f(x) = x^3 - 10$; $x_1 = 2$.

60. $f(x) = x\sin x - \cos x$; $x_1 = 1$.

CHAPTER 5

INTEGRATION

■ 5.1 AN AREA PROBLEM; A SPEED-DISTANCE PROBLEM

An Area Problem

In Figure 5.1.1 you can see a region Ω bounded above by the graph of a continuous function f, bounded below by the x-axis, bounded on the left by the line $x = a$, and bounded on the right by the line $x = b$. The question before us is this: What number, if any, should be called the area of Ω?

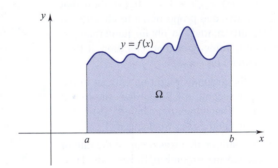

Figure 5.1.1

To begin to answer this question, we split up the interval $[a, b]$ into a finite number of subintervals

$$[x_0, x_1], [x_1, x_2], \ldots, [x_{n-1}, x_n] \qquad \text{with} \qquad a = x_0 < x_1 < \cdots < x_n = b.$$

This breaks up the region Ω into n subregions:

$$\Omega_1, \Omega_2, \ldots, \Omega_n. \qquad \text{(Figure 5.1.2)}$$

We can estimate the total area of Ω by estimating the area of each subregion Ω_i and adding up the results. Let's denote by M_i the maximum value of f on $[x_{i-1}, x_i]$ and by

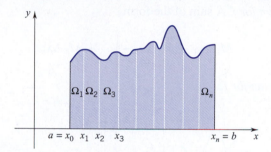

Figure 5.1.2

m_i the minimum value. (We know that there are such numbers because f is continuous.) Consider now the rectangles r_i and R_i of Figure 5.1.3. Since

$$r_i \subseteq \Omega_i \subseteq R_i,$$

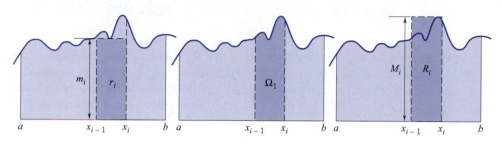

Figure 5.1.3

we must have

$$\text{area of } r_i \leq \text{area of } \Omega_i \leq \text{ area of } R_i.$$

Since the area of a rectangle is the length times the width,

$$m_i(x_i - x_{i-1}) \leq \text{area of } \Omega_i \leq M_i(x_i - x_{i-1}).$$

Setting $\Delta x_i = x_i - x_{i-1}$, we have

$$m_i \Delta x_i \leq \text{area of } \Omega_i \leq M_i \Delta x_i.$$

This inequality holds for $i = 1, i = 2, \ldots, i = n$. Adding up these inequalities, we get on the one hand

(5.1.1)
$$m_1 \Delta x_1 + m_2 \Delta x_2 + \cdots + m_n \Delta x_n \leq \text{area of } \Omega,$$

and on the other hand

(5.1.2)
$$\text{area of } \Omega \leq M_1 \Delta x_1 + M_2 \Delta x_2 + \cdots + M_n \Delta x_n.$$

A sum of the form

$$m_1 \Delta x_1 + m_2 \Delta x_2 + \cdots + m_n \Delta x_n \qquad \text{(Figure 5.1.4)}$$

is called a *lower sum for f.* A sum of the form

$$M_1 \Delta x_1 + M_2 \Delta x_2 + \cdots + M_n \Delta x_n \qquad \text{(Figure 5.1.5)}$$

is called an *upper sum for f.*

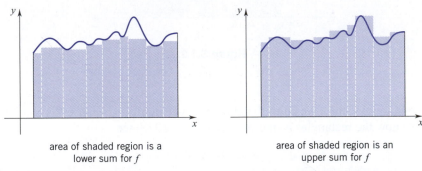

area of shaded region is a
lower sum for f

Figure 5.1.4

area of shaded region is an
upper sum for f

Figure 5.1.5

Inequalities 5.1.1 and 5.1.2 together tell us that for a number to be a candidate for the title "area of Ω," it must be greater than or equal to every lower sum for f and it must be less than or equal to every upper sum. It can be proven that with f continuous on $[a, b]$ there is one and only one such number. This number we call *the area of* Ω.

A Speed-Distance Problem

If an object moves at a constant speed for a given period of time, then the total distance traveled is given by the familiar formula

$$\text{distance} = \text{speed} \times \text{time.}$$

Suppose now that during the course of the motion the speed v does not remain constant; suppose that it varies continuously. How can we calculate the distance traveled in that case?

To answer this question, we suppose that the motion begins at time a, ends at time b, and during the time interval $[a, b]$ the speed varies continuously.

As in the case of the area problem, we begin by breaking up the interval $[a, b]$ into a finite number of subintervals:

$$[t_0, t_1], [t_1, t_2], \ldots, [t_{n-1}, t_n] \qquad \text{with} \qquad a = t_0 < t_1 < \cdots < t_n = b.$$

On each subinterval $[t_{i-1}, t_i]$ the object attains a certain maximum speed M_i and a certain minimum speed m_i. (How do we know this?) If throughout the time interval $[t_{i-1}, t_i]$ the object were to move constantly at its minimum speed, m_i, then it would cover a distance of $m_i \Delta t_i$ units. If instead it were to move constantly at its maximum speed, M_i, then it would cover a distance of $M_i \Delta t_i$ units. As it is, the actual distance traveled, call it s_i, must lie somewhere in between; namely, we must have

$$m_i \Delta t_i \leq s_i \leq M_i \Delta t_i.$$

The total distance traveled during full the time interval $[a, b]$, call it s, must be the sum of the distances traveled during the subintervals $[t_{i-1}, t_i]$; thus we must have

$$s = s_1 + s_2 + \cdots + s_n.$$

Since

$$m_1 \Delta t_1 \le s_1 \le M_1 \Delta t_1$$
$$m_2 \Delta t_2 \le s_2 \le M_2 \Delta t_2,$$
$$\vdots$$
$$m_n \Delta t_n \le s_n \le M_n \Delta t_n,$$

it follows by the addition of these inequalities that

$$m_1 \Delta t_1 + m_2 \Delta t_2 + \cdots + m_n \Delta t_n \le s \le M_1 \Delta t_1 + M_2 \Delta t_2 + \cdots + M_n \Delta t_n.$$

A sum of the form

$$m_1 \Delta t_1 + m_2 \Delta t_2 + \cdots + m_n \Delta t_n$$

is called a *lower sum* for the speed function. A sum of the form

$$M_1 \Delta t_1 + M_2 \Delta t_2 + \cdots + M_n \Delta t_n$$

is called an *upper sum* for the speed function. The inequality we just obtained for s tells us that s must be greater than or equal to every lower sum for the speed function, and it must be less than or equal to every upper sum. As in the case of the area problem, it turns out that there is one and only one such number, and this is the total distance traveled.

■ 5.2 THE DEFINITE INTEGRAL OF A CONTINUOUS FUNCTION

The process we used to solve the two problems in Section 5.1 is called *integration*, and the end results of this process are called *definite integrals*. Our purpose here is to establish these notions in a more general way. First, some auxiliary notions.

(5.2.1) | By a *partition* of the closed interval $[a, b]$, we mean a finite subset of $[a, b]$ which contains the points a and b.

We index the elements of a partition according to their natural order. Thus, if we say that

$$P = \{x_0, x_1, x_2, \ldots, x_{n-1}, x_n\} \qquad \text{is a partition of } [a, b],$$

you can conclude that

$$a = x_0 < x_1 < \cdots < x_n = b.$$

Example 1 The sets

$$\{0, 1\}, \qquad \{0, \tfrac{1}{2}, 1\}, \qquad \{0, \tfrac{1}{4}, \tfrac{1}{2}, 1\}, \qquad \{0, \tfrac{1}{4}, \tfrac{1}{3}, \tfrac{1}{2}, \tfrac{5}{8}, 1\}$$

are all partitions of the interval $[0, 1]$. ❑

If $P = \{x_0, x_1, x_2, \ldots, x_{n-1}, x_n\}$ is a partition of $[a, b]$, then P breaks up $[a, b]$ into n subintervals

$$[x_0, x_1], [x_1, x_2], \ldots, [x_{n-1}, x_n] \qquad \text{of lengths} \qquad \Delta x_1, \Delta x_2, \ldots, \Delta x_n.$$

Suppose now that f is continuous on $[a, b]$. Then on each subinterval $[x_{i-1}, x_i]$ the function f takes on a maximum value, M_i, and a minimum value, m_i.

(5.2.2)

The number

$$U_f(P) = M_1 \Delta x_1 + M_2 \Delta x_2 + \cdots + M_n \Delta x_n$$

is called the *P upper sum for f,* and the number

$$L_f(P) = m_1 \Delta x_1 + m_2 \Delta x_2 + \cdots + m_n \Delta x_n$$

is called *the P lower sum for f.*

Example 2 The function $f(x) = 1 + x^2$ is continuous on $[0, 1]$. The partition $P = \{0, \frac{1}{2}, \frac{3}{4}, 1\}$ breaks up $[0, 1]$ into three subintervals

$$[x_0, x_1] = \left[0, \tfrac{1}{2}\right], \qquad [x_1, x_2] = \left[\tfrac{1}{2}, \tfrac{3}{4}\right], \qquad [x_2, x_3] = \left[\tfrac{3}{4}, 1\right]$$

of lengths

$$\Delta x_1 = \tfrac{1}{2} - 0 = \tfrac{1}{2}, \qquad \Delta x_2 = \tfrac{3}{4} - \tfrac{1}{2} = \tfrac{1}{4}, \qquad \Delta x_3 = 1 - \tfrac{3}{4} = \tfrac{1}{4}.$$

Since f increases on $[0, 1]$, it takes on its maximum value at the right endpoint of each subinterval:

$$M_1 = f\left(\tfrac{1}{2}\right) = \tfrac{5}{4}, \qquad M_2 = f\left(\tfrac{3}{4}\right) = \tfrac{25}{16}, \qquad M_3 = f(1) = 2.$$

The minimum values are taken on at the left endpoints:

$$m_1 = f(0) = 1, \qquad m_2 = f\left(\tfrac{1}{2}\right) = \tfrac{5}{4}, \qquad m_3 = f\left(\tfrac{3}{4}\right) = \tfrac{25}{16}.$$

Thus

$$U_f(P) = M_1 \Delta x_1 + M_2 \Delta x_2 + M_2 \Delta x_3 = \tfrac{5}{4}\left(\tfrac{1}{2}\right) + \tfrac{25}{16}\left(\tfrac{1}{4}\right) + 2\left(\tfrac{1}{4}\right) = \tfrac{97}{64} \cong 1.52$$

and

$$L_f(P) = m_1 \Delta x_1 + m_2 \Delta x_2 + m_3 \Delta x_3 = 1\left(\tfrac{1}{2}\right) + \tfrac{5}{4}\left(\tfrac{1}{4}\right) + \tfrac{25}{16}\left(\tfrac{1}{4}\right) = \tfrac{77}{64} \cong 1.20.$$

For a geometric interpretation of these sums, see Figure 5.2.1. ❑

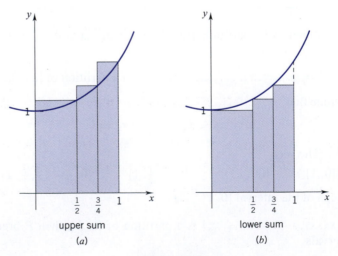

upper sum
(a)

lower sum
(b)

Figure 5.2.1

Example 3 The function $f(x) = \cos \pi x$ is continuous on $\left[0, \frac{3}{4}\right]$. The partition $P = \left\{0, \frac{1}{6}, \frac{1}{4}, \frac{1}{2}, \frac{2}{3}, \frac{3}{4}\right\}$ breaks up $\left[0, \frac{3}{4}\right]$ into five subintervals

$$[x_0, x_1] = \left[0, \tfrac{1}{6}\right], \qquad [x_1, x_2] = \left[\tfrac{1}{6}, \tfrac{1}{4}\right], \qquad [x_2, x_3] = \left[\tfrac{1}{4}, \tfrac{1}{2}\right],$$

$$[x_3, x_4] = \left[\tfrac{1}{2}, \tfrac{2}{3}\right], \qquad [x_4, x_5] = \left[\tfrac{2}{3}, \tfrac{3}{4}\right]$$

of lengths

$$\Delta x_1 = \tfrac{1}{6}, \qquad \Delta x_2 = \tfrac{1}{12}, \qquad \Delta x_3 = \tfrac{1}{4}, \qquad \Delta x_4 = \tfrac{1}{6}, \qquad \Delta x_5 = \tfrac{1}{12}.$$

See Figure 5.2.2.

The maximum values of f on these subintervals are as follows:

$$M_1 = f(0) = \cos 0 = 1, \qquad M_2 = f\left(\tfrac{1}{6}\right) = \cos \tfrac{1}{6}\pi = \tfrac{1}{2}\sqrt{3},$$

$$M_3 = f\left(\tfrac{1}{4}\right) = \cos \tfrac{1}{4}\pi = \tfrac{1}{2}\sqrt{2}, \qquad M_4 = f\left(\tfrac{1}{2}\right) = \cos \tfrac{1}{2}\pi = 0,$$

$$M_5 = f\left(\tfrac{2}{3}\right) = \cos \tfrac{2}{3}\pi = -\tfrac{1}{2}.$$

and the minimum values are as follows:

$$m_1 = f\left(\tfrac{1}{6}\right) = \cos \tfrac{1}{6}\pi = \tfrac{1}{2}\sqrt{3}, \qquad m_2 = f\left(\tfrac{1}{4}\right) = \cos \tfrac{1}{4}\pi = \tfrac{1}{2}\sqrt{2},$$

$$m_3 = f\left(\tfrac{1}{2}\right) = \cos \tfrac{1}{2}\pi = 0, \qquad m_4 = f\left(\tfrac{2}{3}\right) = \cos \tfrac{2}{3}\pi = -\tfrac{1}{2},$$

$$m_5 = f\left(\tfrac{3}{4}\right) = \cos \tfrac{3}{4}\pi = -\tfrac{1}{2}\sqrt{2}.$$

Therefore

$$U_f(P) = 1\left(\tfrac{1}{6}\right) + \tfrac{1}{2}\sqrt{3}\left(\tfrac{1}{12}\right) + \tfrac{1}{2}\sqrt{2}\left(\tfrac{1}{4}\right) + 0\left(\tfrac{1}{6}\right) + \left(-\tfrac{1}{2}\right)\left(\tfrac{1}{12}\right) \cong 0.37$$

and

$$L_f(P) = \tfrac{1}{2}\sqrt{3}\left(\tfrac{1}{6}\right) + \tfrac{1}{2}\sqrt{2}\left(\tfrac{1}{12}\right) + 0\left(\tfrac{1}{4}\right) + \left(-\tfrac{1}{2}\right)\left(\tfrac{1}{6}\right) + \left(-\tfrac{1}{2}\sqrt{2}\right)\left(\tfrac{1}{12}\right) \cong 0.06. \quad \square$$

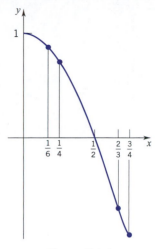

Figure 5.2.2

Both in Example 2 and in Example 3 the separation between $U_f(P)$ and $L_f(P)$ was quite large. Had we added more points to the partitions we chose, the upper sums would have been smaller, the lower sums would have been greater, and the separation between them would have been lessened.

By an argument that we omit here (it appears in Appendix B.4), it can be proved that, with f continuous on $[a, b]$, there is one and only one number I that satisfies the inequality

$$L_f(P) \leq I \leq U_f(P) \qquad \text{for } \textit{all } \text{partitions } P \text{ of } [a, b].$$

This is the number we want.

DEFINITION 5.2.3 THE DEFINITE INTEGRAL OF A CONTINUOUS FUNCTION

Let f be continuous on $[a, b]$. The unique number I that satisfies the inequality

$$L_f(P) \leq I \leq U_f(P) \qquad \text{for all partitions } P \text{ of } [a, b]$$

is called the *definite integral* (or more simply *the integral*) of f from a to b and is denoted by

$$\int_a^b f(x)\, dx.$$

The symbol $\int$ dates back to Leibniz and is called an *integral sign*. It is really an elongated *S*—as in *Sum*. The numbers a and b are called *the limits of integration* (a is the *lower limit* and b is the *upper limit*),[†] and we will speak of *integrating* a function f from a to b. The function f being integrated is called the *integrand*. This is not the only notation. Some mathematicians omit the dx and simply write $\int_a^b f$. We will keep the dx. As we go on, you will see that it does serve a useful purpose.

In the expression

$$\int_a^b f(x)\, dx$$

the letter x is a "dummy variable"; in other words, it can be replaced by any letter not already in use. Thus, for example,

$$\int_a^b f(x)\, d(x), \qquad \int_a^b f(t)\, dt, \qquad \int_a^b f(z)\, dz$$

all denote exactly the same quantity, the definite integral of f from a to b.

From the introduction to this chapter, you know that if f is nonnegative and continuous on $[a, b]$, then the integral of f from $x = a$ to $x = b$ gives the area below the graph of f from $x = a$ to $x = b$:

$$A = \int_a^b f(x)\, dx.$$

You also know that if an object moves with continuous speed $v(t) = |v(t)|$ from time $t = a$ to time $t = b$, then the integral of the speed function v gives the distance traveled by the object during that time period:

$$s = \int_a^b v(t)\, dt = \int_a^b |v(t)|\, dt.$$

We'll come back to these applications and introduce others as we go on. Right now we carry out some computations.

Example 4   (*The integral of a constant function*)

(5.2.4)
$$\boxed{\int_a^b k\, dx = k(b - a).}$$

In this case the integrand is the constant function $f(x) = k$. To verify the formula, we take $P = \{x_0, x_1, \ldots, x_n\}$ as an arbitrary partition of $[a, b]$. Since f is constantly k on $[a, b]$, f is constantly k on each subinterval $[x_{i-1}, x_i]$. Thus both m_i and M_i are k, and both $L_f(P)$ and $U_f(P)$ are

$$k\Delta x_1 + k\Delta x_2 + \cdots + k\Delta x_n = k(\Delta x_1 + \Delta x_2 + \cdots + \Delta x_n) = k(b - a).$$

explain $\longrightarrow$

Therefore it is certainly true that

$$L_f(P) \le k(b - a) \le U_f(P).$$

[†]There is no connection between the term "limit" as used here and the limits introduced in Chapter 2.

Since this inequality holds for all partitions P of $[a, b]$, we can conclude that

$$\int_a^b f(x)\,dx = k(b-a). \quad \square$$

For example,

$$\int_{-1}^1 3\,dx = 3[1-(-1)] = 3(2) = 6 \qquad \text{and}$$

$$\int_4^{10} -2\,dx = -2(10-4) = -2(6) = -12.$$

If $k > 0$, the region between the graph and the x-axis is a rectangle of height k erected on the interval $[a, b]$. (Figure 5.2.3.) The integral gives the area of this rectangle.

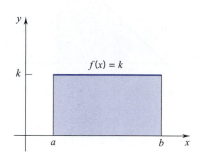

Figure 5.2.3

Example 5 *(The integral of the identity function)*

(5.2.5)

$$\int_a^b x\,dx = \tfrac{1}{2}(b^2 - a^2).$$

Here the integrand is the identity function $f(x) = x$. (Figure 5.2.4.) To verify the formula we take $P = \{x_0, x_1, \ldots, x_n\}$ as an arbitrary partition of $[a, b]$. On each subinterval $[x_{i-1}, x_i]$, the function $f(x) = x$ has a maximum value M_i and a minimum value m_i. Since f is an increasing function, the maximum value occurs at the right endpoint of the subinterval and the minimum value occurs at the left endpoint. Thus $M_i = x_i$ and $m_i = x_{i-1}$. It follows that

$$U_f(P) = x_1\Delta x_1 + x_2\Delta x_2 + \cdots + x_n\Delta x_n$$

and

$$L_f(P) = x_0\Delta x_1 + x_1\Delta x_2 + \cdots + x_{n-1}\Delta x_n.$$

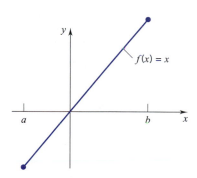

Figure 5.2.4

For each index i

$$(*) \qquad x_{i-1} \le \tfrac{1}{2}(x_i + x_{i-1}) \le x_i. \qquad \text{(explain)}$$

Multiplication by $\Delta x_i = x_i - x_{i-1}$ gives

$$x_{i-1}\Delta x_i \le \tfrac{1}{2}(x_i + x_{i-1})(x_i - x_{i-1}) \le x_i\Delta x_i,$$

which we write as

$$x_{i-1}\Delta x_i \le \tfrac{1}{2}\left(x_i^2 - x_{i-1}^2\right) \le x_i\Delta x_i.$$

Summing from $i = 1$ to $i = n$, we find that

$$(**) \qquad L_f(P) \le \tfrac{1}{2}\left(x_1^2 - x_0^2\right) + \tfrac{1}{2}\left(x_2^2 - x_1^2\right) + \cdots + \tfrac{1}{2}\left(x_n^2 - x_{n-1}^2\right) \le U_f(P).$$

The sum in the middle collapses to

$$\tfrac{1}{2}\left(x_n^2 - x_0^2\right) = \tfrac{1}{2}(b^2 - a^2).$$

Consequently

$$L_f(P) \le \tfrac{1}{2}(b^2 - a^2) \le U_f(P).$$

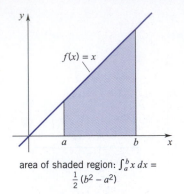

area of shaded region: $\int_a^b x \, dx = \frac{1}{2}(b^2 - a^2)$

Figure 5.2.5

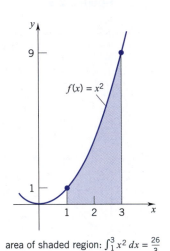

area of shaded region: $\int_1^3 x^2 \, dx = \frac{26}{3}$

Figure 5.2.6

Since P was chosen arbitrarily, we can conclude that this inequality holds for all partitions P of $[a, b]$. It follows that

$$\int_a^b x \, dx = \frac{1}{2}(b^2 - a^2). \quad \square$$

For example,

$$\int_{-1}^3 x \, dx = \frac{1}{2}[3^2 - (-1)^2] = \frac{1}{2}(8) = 4 \qquad \text{and} \qquad \int_{-2}^2 x \, dx = \frac{1}{2}[2^2 - (-2)^2] = 0.$$

If the interval $[a, b]$ lies to the right of the origin, then the region below the graph of

$$f(x) = x, \qquad x \in [a, b]$$

is the trapezoid shown in Figure 5.2.5. The integral

$$\int_a^b x \, dx$$

gives the area of this trapezoid: $A = (b - a)[\frac{1}{2}(a + b)] = \frac{1}{2}(b^2 - a^2)$.

Example 6

$$\int_1^3 x^2 \, dx = \frac{26}{3}. \qquad \text{(Figure 5.2.6)}$$

Let $P = \{x_0, x_1, \ldots, x_n\}$ be an arbitrary partition of $[1, 3]$. On each subinterval $[x_{i-1}, x_i]$ the function $f(x) = x^2$ has a maximum $M_i = x_i^2$ and a minimum $m_i = x_{i-1}^2$. It follows that

$$U_f(P) = x_1^2 \Delta x_1 + \cdots + x_n^2 \Delta x_n$$

and

$$L_f(P) = x_0^2 \Delta x_1 + \cdots + x_{n-1}^2 \Delta x_n.$$

For each index i, $1 \le i \le n$,

$$3x_{i-1}^2 \le x_{i-1}^2 + x_{i-1}x_i + x_i^2 \le 3x_i^2. \qquad \text{(Verify this)}$$

Division by 3 gives

$$x_{i-1}^2 \le \frac{1}{3}\left(x_{i-1}^2 + x_{i-1}x_i + x_i^2\right) \le x_i^2.$$

We now multiply this inequality by $\Delta x_i = x_i - x_{i-1}$. The middle term then becomes

$$\frac{1}{3}\left(x_{i-1}^2 + x_{i-1}x_i + x_i^2\right)(x_i - x_{i-1}) = \frac{1}{3}\left(x_i^3 - x_{i-1}^3\right),$$

and shows that

$$x_{i-1}^2 \Delta x_i \le \frac{1}{3}\left(x_i^3 - x_{i-1}^3\right) \le x_i^2 \Delta x_i.$$

The sum of the terms on the left is $L_f(P)$. The sum of all the middle terms collapses to $\frac{26}{3}$:

$$\frac{1}{3}\left(x_1^3 - x_0^3 + x_2^3 - x_1^3 + \cdots + x_n^3 - x_{n-1}^3\right) = \frac{1}{3}\left(x_n^3 - x_0^3\right) = \frac{1}{3}(3^3 - 1^3) = \frac{26}{3}.$$

The sum of the terms on the right is $U_f(P)$. Clearly, then,

$$L_f(P) \le \frac{26}{3} \le U_f(P).$$

Since P was chosen arbitrarily, we can conclude that this inequality holds for all partitions P of $[1, 3]$. It follows that

$$\int_1^3 x^2 \, dx = \tfrac{26}{3}. \quad ❑$$

The Integral as the Limit of Riemann Sums

For a function f continuous on $[a, b]$, we have defined the definite integral

$$\int_a^b f(x) \, dx$$

as the unique number that satisfies the inequality

$$L_f(P) \le \int_a^b f(x) \, dx \le U_f(P) \qquad \text{for all partitions } P \text{ of } [a, b].$$

This method of obtaining the definite integral (*squeezing* toward it with upper and lower sums) is called the *Darboux method*.[†]

There is another way to obtain the integral that, in some respects, has distinct advantages. Take a partition $P = \{x_0, x_1, \ldots, x_n\}$ of $[a, b]$. P breaks up $[a, b]$ into n subintervals

$$[x_0, x_1], [x_1, x_2], \ldots, [x_{n-1}, x_n]$$

of lengths

$$\Delta x_1, \Delta x_2, \ldots, \Delta x_n.$$

Now pick a point x_1^* from $[x_0, x_1]$ and form the product $f(x_1^*)\Delta x_1$; pick a point x_2^* from $[x_1, x_2]$ and form the product $f(x_2^*)\Delta x_2$; go on in this manner until you have formed the products

$$f(x_1^*)\Delta x_1, \quad f(x_2^*)\Delta x_2, \ldots, \quad f(x_n^*)\Delta x_n.$$

The sum of these products

$$S^*(P) = f(x_1^*)\Delta x_1 + f(x_2^*)\Delta x_2 + \cdots + f(x_n^*)\Delta x_n$$

is called a *Riemann sum*.[††] Since $m_i \le f(x_i^*) \le M_i$ for each index i, it's clear that

(5.2.6)

$$L_f(P) \le S^*(P) \le U_f(P).$$

This inequality holds for all partitions P of $[a, b]$.

Example 7 Let $f(x) = x^2$, $x \in [1, 3]$. Take $P = \left\{1, \tfrac{3}{2}, 2, 3\right\}$ and set

$$x_1^* = \tfrac{5}{4}, \qquad x_2^* = \tfrac{7}{4}, \qquad x_3^* = \tfrac{5}{2}. \qquad \text{(Figure 5.2.7)}$$

Here $\Delta x_1 = \tfrac{1}{2}$, $\Delta x_2 = \tfrac{1}{2}$, $\Delta x_3 = 1$. Therefore

$$S^*(P) = f\left(\tfrac{5}{4}\right) \cdot \tfrac{1}{2} + f\left(\tfrac{7}{4}\right) \cdot \tfrac{1}{2} + f\left(\tfrac{5}{2}\right) \cdot 1 = \tfrac{25}{16}\left(\tfrac{1}{2}\right) + \tfrac{49}{16}\left(\tfrac{1}{2}\right) + \tfrac{25}{4}(1) = \tfrac{137}{16} \cong 8.5625.$$

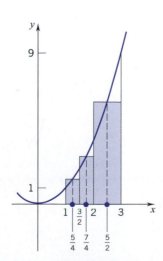

Figure 5.2.7

[†]After the French mathematician J. G. Darboux (1842–1917).

[††]After the German mathematician G. F. B. Riemann (1826–1866).

In Example 6 we showed that

$$\int_1^3 x^2 \, dx = \left[\tfrac{1}{3}x^3\right]_1^3 = \tfrac{27}{3} - \tfrac{1}{3} = \tfrac{26}{3} \cong 8.667.$$

Our Riemann approximation is pretty good. ❏

For each partition P of $[a, b]$, we define $\|P\|$, the *norm* of P, by setting

$$\|P\| = \max \Delta x_i, \qquad i = 1, 2, \dots, n.$$

The definite integral of f is the *limit* of Riemann sums in the following sense: given any $\epsilon > 0$, there exists a $\delta > 0$ such that

$$\text{if} \quad \|P\| < \delta, \quad \text{then} \quad \left| S^*(P) - \int_a^b f(x)\,dx \right| < \epsilon$$

no matter how the x_i^* are chosen within the $[x_{i-1}, x_i]$.

We can express this by writing

(5.2.7)

$$\int_a^b f(x)\,dx = \lim_{\|P\| \to 0} S^*(P),$$

which in expanded form reads

$$\int_a^b f(x)\,dx = \lim_{\|P\| \to 0} \left[f\!\left(x_1^*\right)\Delta x_1 + f\!\left(x_2^*\right)\Delta x_2 + \cdots + f\!\left(x_n^*\right)\Delta x_n \right].$$

A proof that the definite integral of a continuous function is the limit of Riemann sums in the sense just explained is given in Appendix B.5. Figure 5.2.8 illustrates the idea. Here the base interval is broken up into eight subintervals. The point x_1^* is chosen from $[x_0, x_1]$, x_2^* from $[x_1, x_2]$, and so on.

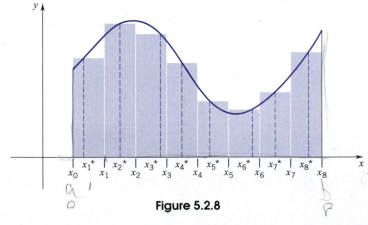

Figure 5.2.8

While the integral represents the area under the curve, the Riemann sum represents the sum of the areas of the shaded rectangles. The difference between the two can be made as small as we wish (less than ϵ) simply by making the maximum length of the base subintervals sufficiently small—that is, by making $\|P\|$ sufficiently small.

This approach to the definite integral was invented by Riemann some years before Darboux began his work. For this reason the integral we have been studying is called the *Riemann integral*.

Remark The process of integration can be extended to discontinuous functions so long as they are not "too" discontinuous.[†] Basically there are two ways to do this: One way is to extend the meaning of upper and lower sums. Another way, more accessible to us with the tools at hand, is to continue with Riemann sums. This is the course we'll follow when we return to this subject. (Project 5.5.) ❑

_____

[†]What we mean by this will be touched upon in Project 5.5.

EXERCISES 5.2

Exercises 1–10. Calculate $L_f(P)$ and $U_f(P)$.

1. $f(x) = 2x$, $x \in [0, 1]$; $P = \{0, \frac{1}{4}, \frac{1}{2}, 1\}$.

2. $f(x) = 1 - x$, $x \in [0, 2]$; $P = \{0, \frac{1}{3}, \frac{3}{4}, 1, 2\}$.

3. $f(x) = x^2$, $x \in [-1, 0]$; $P = \{-1, -\frac{1}{2}, -\frac{1}{4}, 0\}$.

4. $f(x) = 1 - x^2$, $x \in [0, 1]$; $P = \{0, \frac{1}{4}, \frac{1}{2}, 1\}$.

5. $f(x) = 1 + x^3$, $x \in [0, 1]$; $P = \{0, \frac{1}{2}, 1\}$.

6. $f(x) = \sqrt{x}$, $x \in [0, 1]$; $P = \{0, \frac{1}{25}, \frac{4}{25}, \frac{9}{25}, \frac{16}{25}, 1\}$.

7. $f(x) = x^2$, $x \in [-1, 1]$; $P = \{-1, -\frac{1}{4}, \frac{1}{4}, \frac{1}{2}, 1\}$.

8. $f(x) = x^2$, $x \in [-1, 1]$; $P = \{-1, -\frac{3}{4}, -\frac{1}{4}, \frac{1}{4}, \frac{1}{2}, 1\}$.

9. $f(x) = \sin x$, $x \in [0, \pi]$; $P = \{0, \frac{1}{6}\pi, \frac{1}{2}\pi, \pi\}$.

10. $f(x) = \cos x$, $x \in [0, \pi]$; $P = \{0, \frac{1}{3}\pi, \frac{1}{2}\pi, \pi\}$.

11. Let f be a function continuous on $[-1, 1]$ and take P as a partition of $[-1, 1]$. Show that each of the following three statements is false.

(a) $L_f(P) = 3$ and $U_f(P) = 2$.

(b) $L_f(P) = 3$, $U_f(P) = 6$, and $\int_{-1}^{1} f(x)\, dx = 2$.

(c) $L_f(P) = 3$, $U_f(P) = 6$, and $\int_{-1}^{1} f(x)\, dx = 10$.

12. (a) Given that $P = \{x_0, x_1, \ldots, x_n\}$ is an arbitrary partition of $[a, b]$, find $L_f(P)$ and $U_f(P)$ for $f(x) = x + 3$.

(b) Use your answers to part (a) to evaluate

$$\int_a^b f(x)\, dx.$$

13. Exercise 12 taking $f(x) = -3x$.

14. Exercise 12 taking $f(x) = 1 + 2x$.

Exercises 15–18. Express the limit as a definite integral over the indicated interval.

15. $\lim_{\|P\| \to 0} \left[(x_1^2 + 2x_1 - 3)\, \Delta x_1 + (x_2^2 + 2x_2 - 3)\, \Delta x_2 + \cdots + (x_n^2 + 2x_n - 3)\, \Delta x_n \right]$; $[-1, 2]$.

16. $\lim_{\|P\| \to 0} \left[(x_0^3 - 3x_0)\, \Delta x_1 + (x_1^3 - 3x_1)\, \Delta x_2 + \cdots + (x_{n-1}^3 - 3x_{n-1})\, \Delta x_n \right]$; $[0, 3]$.

17. $\lim_{\|P\| \to 0} \left[(t_1^*)^2 \sin(2t_1^* + 1)\, \Delta t_1 + (t_2^*)^2 \sin(2t_2^* + 1)\, \Delta t_2 + \cdots + (t_n^*)^2 \sin(2t_n^* + 1)\, \Delta t_n \right]$ where $t_i^* \in [t_{i-1}, t_i]$, $i = 1, 2, \ldots, n$; $[0, 2\pi]$.

18. $\lim_{\|P\| \to 0} \left[\frac{\sqrt{t_1^*}}{(t_1^*)^2 + 1} \Delta t_1 + \frac{\sqrt{t_2^*}}{(t_2^*)^2 + 1} \Delta t_2 + \cdots + \frac{\sqrt{t_n^*}}{(t_n^*)^2 + 1} \Delta t_n \right]$

where $t_i^* \in [t_{i-1}, t_i]$, $i = 1, 2, \ldots, n$; $[1, 4]$.

19. Let Ω be the region below the graph of $f(x) = x^2$, $x \in [0, 1]$. Draw a figure showing the Riemann sum $S^*(P)$ as an estimate for this area. Take $P = \{0, \frac{1}{4}, \frac{1}{2}, \frac{3}{4}, 1\}$ and set

$$x_1^* = \frac{1}{8}, \qquad x_2^* = \frac{3}{8}, \qquad x_3^* = \frac{5}{8}, \qquad x_4^* = \frac{7}{8}.$$

20. Let Ω be the region below the graph of $f(x) = \frac{3}{2}x + 1$, $x \in [0, 2]$. Draw a figure showing the Riemann sum $S^*(P)$ as an estimate for this area. Take $P = \{0, \frac{1}{4}, \frac{3}{4}, 1, \frac{3}{2}, 2\}$ and let the x_i^* be the midpoints of the subintervals.

21. Let $f(x) = 2x$, $x \in [0, 1]$. Take $P = \{0, \frac{1}{8}, \frac{1}{4}, \frac{1}{2}, \frac{3}{4}, 1\}$ and set

$$x_1^* = \frac{1}{16}, \quad x_2^* = \frac{3}{16}, \quad x_3^* = \frac{3}{8}, \quad x_4^* = \frac{5}{8}, \quad x_5^* = \frac{3}{4}.$$

Calculate the following:

(a) $L_f(P)$. (b) $S^*(P)$. (c) $U_f(P)$.

22. Taking f as in Exercise 21, determine

$$\int_0^1 f(x)\, dx.$$

23. Evaluate

$$\int_0^1 x^3\, dx$$

using upper and lower sums. HINT:

$$b^4 - a^4 = (b^3 + b^2 a + ba^2 + a^3)(b - a).$$

24. Evaluate

$$\int_0^1 x^4\, dx$$

using upper and lower sums.

Exercises 25–30. Assume that f and g are continuous, that $a < b$, and that $\int_a^b f(x)\, dx > \int_a^b g(x)\, dx$. Which of the statements necessarily holds for all partitions P of $[a, b]$? Justify your answer.

25. $L_g(P) < U_f(P)$. **26.** $L_g(P) < L_f(P)$.

27. $L_g(P) < \int_a^b f(x)\, dx$. **28.** $U_g(P) < U_f(P)$.

29. $U_f(P) > \int_a^b g(x)\, dx$. **30.** $U_g(P) < \int_a^b f(x)\, dx$.

31. A partition $P = \{x_0, x_1, x_2, \ldots, x_{n-1}, x_n\}$ of $[a, b]$ is said to be *regular* if the subintervals $[x_{i-1}, x_i]$ all have the same length $\Delta x = (b - a)/n$. Let $P = \{x_0, x_1, \ldots, x_{n-1}, x_n\}$ be a regular partition of $[a, b]$. Show that if f is continuous and increasing on $[a, b]$, then

$$U_f(P) - L_f(P) = [f(b) - f(a)] \Delta x.$$

32. Let $P = \{x_0, x_1, x_2, \ldots, x_{n-1}, x_n\}$ be a regular partition of the interval $[a, b]$. (See Exercise 31.) Show that if f is continuous and decreasing on $[a, b]$, then

$$U_f(P) - L_f(P) = [f(a) - f(b)] \Delta x.$$

▷ **33.** Set $f(x) = \sqrt{1 + x^2}$.

(a) Verify that f increases on $[0, 2]$.

(b) Let $P = \{x_0, x_1, \ldots, x_{n-1}, x_n\}$ be a regular partition of $[0, 2]$. Determine a value of n such that

$$0 \leq \int_0^2 f(x) \, dx - L_f(P) \leq 0.1.$$

(c) Use a programmable calculator or computer to calculate $\int_0^2 f(x) \, dx$ with an error of less than 0.1.

▷ **34.** Set $f(x) = 1/(1 + x^2)$.

(a) Verify that f decreases on $[0, 1]$.

(b) Let $P = \{x_0, x_1, \ldots, x_{n-1}, x_n\}$ be a regular partition of $[0, 2]$. Determine a value of n such that

$$0 \leq \int_0^2 f(x) \, dx - L_f(P) \leq 0.1.$$

(c) Use a programmable calculator or computer to calculate $\int_0^1 f(x) \, dx$ with an error of less than 0.05. NOTE: You will see in Chapter 7 that the exact value of this integral is $\pi/4$.

35. Show by induction that for each positive integer k,

$$1 + 2 + 3 + \cdots + k = \tfrac{1}{2} k(k + 1).$$

36. Show by induction that for each positive integer k,

$$1^2 + 2^2 + 3^2 + \cdots + k^2 = \tfrac{1}{6} k(k + 1)(2k + 1).$$

37. Let $P = \{x_0, x_1, x_2, \ldots, x_{n-1}, x_n\}$ be a regular partition of the interval $[0, b]$, and set $f(x) = x$.

(a) Show that

$$L_f(P) = \frac{b^2}{n^2} [0 + 1 + 2 + 3 + \cdots + (n - 1)].$$

(b) Show that

$$U_f(P) = \frac{b^2}{n^2} [1 + 2 + 3 + \cdots + n].$$

(c) Use Exercise 35 to show that

$$L_f(P) = \tfrac{1}{2} b^2 (1 - \|P\|) \quad \text{and} \quad U_f(P) = \tfrac{1}{2} b^2 (1 + \|P\|).$$

(d) Show that for all choices of x_i^*-points

$$\lim_{\|P\| \to 0} S^*(P) = \tfrac{1}{2} b^2 \quad \text{and therefore} \quad \int_0^b x \, dx = \tfrac{1}{2} b^2.$$

38. Let $P = \{x_0, x_1, x_2, \ldots, x_{n-1}, x_n\}$ be a regular partition of $[0, b]$, and let $f(x) = x^2$.

(a) Show that

$$L_f(P) = \frac{b^3}{n^3} [0^2 + 1^2 + 2^2 + \cdots + (n - 1)^2].$$

(b) Show that

$$U_f(P) = \frac{b^3}{n^3} [1^2 + 2^2 + 3^2 + \cdots + n^2].$$

(c) Use Exercise 36 to show that

$$L_f(P) = \tfrac{1}{6} b^3 (2 - 3\|P\| + \|P\|^2) \quad \text{and}$$

$$U_f(P) = \tfrac{1}{6} b^3 (2 + 3\|P\| + \|P\|^2).$$

(d) show that for all choices of x_i^*-points

$$\lim_{\|P\| \to 0} S(P) = \tfrac{1}{3} b^3 \quad \text{and therefore} \quad \int_0^b x^2 \, dx = \tfrac{1}{3} b^3.$$

39. Let f be a function continuous on $[a, b]$. Show that if P is a partition of $[a, b]$, then $L_f(P), U_f(P)$, and $\tfrac{1}{2}[L_f(P) + U_f(P)]$ are all Riemann sums.

Exercises 40–43. Using a regular partition P with 10 subintervals, estimate the integral

(a) by $L_f(P)$ and by $U_f(P)$, (b) by $\tfrac{1}{2}[L_f(P) + U_f(P)]$,

(c) by $S^*(P)$ using the midpoints of the subintervals. How does this result compare with your result in part (b)?

40. $\int_0^2 (x^3 + 2) \, dx.$ **41.** $\int_0^1 \sqrt{x} \, dx.$

42. $\int_0^2 \frac{1}{1 + x^2} \, dx.$ **43.** $\int_0^1 \sin \pi x \, dx.$

■ 5.3 THE FUNCTION $F(x) = \int_a^x f(t) \, dt$

The evaluation of the definite integral

$$\int_a^b f(x) \, dx$$

directly from upper and lower sums or from Riemann sums is usually a laborious and difficult process. Try, for example, to evaluate

$$\int_2^5 \left(x^3 + x^2 - \frac{2x}{1 - x^2} \right) dx \quad \text{or} \quad \int_{-1/2}^{1/4} \frac{x}{1 - x^2} \, dx$$

from such sums. Theorem 5.4.2, called the *fundamental theorem of integral calculus*, gives us another way to evaluate such integrals. This other way depends on a connection between integration and differentiation described in Theorem 5.3.5. Along the way we will pick up some information that is of interest in itself.

THEOREM 5.3.1

Suppose that f is continuous on $[a, b]$, and P and Q are partitions of $[a, b]$. If $Q \supseteq P$, then

$$L_f(P) \leq L_f(Q) \qquad \text{and} \qquad U_f(Q) \leq U_f(P).$$

This result can be justified as follows: By adding points to a partition, we make the subintervals $[x_{i-1}, x_i]$ smaller. This tends to make the minima, m_i, larger and the maxima, M_i, smaller. Thus the lower sums are made bigger, and the upper sums are made smaller. The idea is illustrated (for a positive function) in Figures 5.3.1 and 5.3.2.

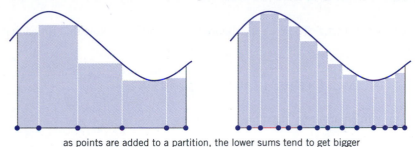

as points are added to a partition, the lower sums tend to get bigger

Figure 5.3.1

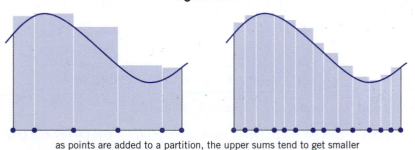

as points are added to a partition, the upper sums tend to get smaller

Figure 5.3.2

The next theorem says that the integral is *additive* on intervals.

THEOREM 5.3.2

If f is continuous on $[a, b]$ and $a < c < b$, then

$$\int_a^c f(t)\,dt + \int_c^b f(t)\,dt = \int_a^b f(t)\,dt.$$

For nonnegative functions f, this theorem is easily understood in terms of area. The area of part I in Figure 5.3.3 is given by

$$\int_a^c f(t)\,dt;$$

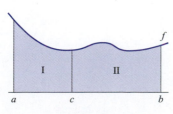

Figure 5.3.3

the area of part II by

$$\int_c^b f(t)\,dt;$$

and the area of the entire region by

$$\int_a^b f(t)\,dt.$$

The theorem says that

> the area of part I + the area of part II = the area of the entire region.   ❏

The fact that the additivity theorem is so easy to understand does not relieve us of the necessity to prove it. Here is a proof.

PROOF OF THEOREM 5.3.2   To prove the theorem, we need only show that for each partition P of $[a, b]$

$$L_f(P) \le \int_a^c f(t)\,dt + \int_c^b f(t)\,dt \le U_f(P). \qquad \text{(Why?)}$$

We begin with an arbitrary partition of $[a, b]$:

$$P = \{x_0, x_1, \dots, x_n\}.$$

Since the partition $Q = P \cup \{c\}$ contains P, we know from Theorem 5.3.1 that

(1) $\qquad\qquad L_f(P) \le L_f(Q) \qquad \text{and} \qquad U_f(Q) \le U_f(P).$

The sets

$$Q_1 = Q \cap [a, c] \qquad \text{and} \qquad Q_2 = Q \cap [c, b]$$

are partitions of $[a, c]$ and $[c, b]$, respectively. Moreover

(2) $\quad L_f(Q_1) + L_f(Q_2) = L_f(Q) \qquad \text{and} \qquad U_f(Q_1) + U_f(Q_2) = U_f(Q).$

Since

$$L_f(Q_1) \le \int_a^c f(t)\,dt \le U_f(Q_1) \qquad \text{and} \qquad L_f(Q_2) \le \int_c^b f(t)\,dt \le U_f(Q_2),$$

we have

$$L_f(Q_1) + L_f(Q_2) \le \int_a^c f(t)\,dt + \int_c^b f(t)\,dt \le U_f(Q_1) + U_f(Q_2),$$

and thus by (2),

$$L_f(Q) \le \int_a^c f(t)\,dt + \int_c^b f(t)\,dt \le U_f(Q).$$

Therefore, by (1),

$$L_f(P) \le \int_a^c f(t)\,dt + \int_c^b f(t)\,dt \le U_f(P). \quad ❏$$

Until now we have integrated only from left to right: from a number a to a number b greater than a. We integrate in the other direction by defining

(5.3.3)
$$\int_b^a f(t)\,dt = -\int_a^b f(t)\,dt.$$

The integral from any number to itself is defined to be zero:

(5.3.4)
$$\int_c^c f(t)\,dt = 0.$$

With these additional conventions, the additivity condition

$$\int_a^c f(t)\,dt + \int_c^b f(t)\,dt = \int_a^b f(t)\,dt$$

holds for all choices of a, b, c from an interval on which f is continuous, no matter what the order of a, b, c happens to be. We have left the proof of this to you as an exercise. (Exercise 16)

We are now ready to state the all-important connection that exists between integration and differentiation. Our first step is to point out that if f is continuous on $[a, b]$ and c is any number in $[a, b]$, then for each x in $[a, b]$, the integral

$$\int_c^x f(t)\,dt$$

is a number, and consequently we can define a function F on $[a, b]$ by setting

$$F(x) = \int_c^x f(t)\,dt.$$

THEOREM 5.3.5

Let f be continuous on $[a, b]$ and let c be any number in $[a, b]$. The function F defined on $[a, b]$ by setting

$$F(x) = \int_c^x f(t)\,dt$$

is continuous on $[a, b]$, differentiable on (a, b), and has derivative

$$F'(x) = f(x) \quad \text{for all } x \text{ in } (a, b).$$

PROOF We will prove the theorem for the special case where the integration that defines F is begun at the left endpoint a; namely, we will prove the theorem for the following function:

$$F(x) = \int_a^x f(t)\,dt. \qquad \text{(The more general case is left to you as Exercise 34.)}$$

We begin with x in the half-open interval $[a, b)$ and show that

$$\lim_{h \to 0^+} \frac{F(x + h) - F(x)}{h} = f(x).$$

A pictorial argument that applies to the case where $f > 0$ is roughed out in Figure 5.3.4.

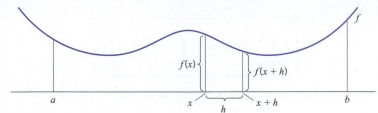

$F(x)$ = area from a to x and $F(x + h)$ = area from a to $x + h$. Therefore
$F(x + h) - F(x)$ = area from x to $x + h$. For small h this is approximately $f(x)\, h$. Thus

$$\frac{F(x + h) - F(x)}{h} \text{ is approximately } \frac{f(x)\, h}{h} = f(x).$$

Figure 5.3.4

Now to a proof. For $a \le x < x + h < b$,

$$\int_a^x f(t)\, dt + \int_x^{x+h} f(t)\, dt = \int_a^{x+h} f(t)\, dt.$$

Therefore

$$\int_a^{x+h} f(t)\, dt - \int_a^x f(t)\, dt = \int_x^{x+h} f(t)\, dt,$$

which, by the definition of F, gives

(1)
$$F(x + h) - F(x) = \int_x^{x+h} f(t)\, dt.$$

On the interval $[x, x + h]$, an interval of length h, f takes on a maximum value M_h and a minimum value m_h. On $[x, x + h]$, the product $M_h h$ is an upper sum for f and $m_h h$ is a lower sum for f. (Use the partition $\{x, x + h\}$.) Therefore

$$m_h \cdot h \le \int_x^{x+h} f(t)\, dt \le M_h \cdot h.$$

It follows from (1) and the fact that h is positive that

$$m_h \le \frac{F(x + h) - F(x)}{h} \le M_h.$$

Since f is continuous on $[x, x + h]$,

$$\lim_{h \to 0^+} m_h = f(x) = \lim_{h \to 0^+} M_h$$

and thus

(2)
$$\lim_{h \to 0^+} \frac{F(x + h) - F(x)}{h} = f(x).$$

This last statement follows from the "pinching theorem," Theorem 2.5.1, which, as we remarked in Section 2.5, applies also to one-sided limits. In a similar manner we can prove that, for x in the half-open interval $(a, b]$,

(3)
$$\lim_{h \to 0^-} \frac{F(x + h) - F(x)}{h} = f(x).$$

For x in the open interval (a, b), both (2) and (3) hold, and we have

$$F'(x) = \lim_{h \to 0} \frac{F(x + h) - F(x)}{h} = f(x).$$

This proves that F is differentiable on (a, b) and has derivative $F'(x) = f(x)$.

All that remains to be shown is that F is continuous from the right at a and continuous from the left at b. Limit (2) at $x = a$ gives

$$\lim_{h \to 0^+} \frac{F(a + h) - F(a)}{h} = f(a).$$

Now, for $h > 0$,

$$F(a + h) - F(a) = \frac{F(a + h) - F(a)}{h} \cdot h,$$

and so

$$\lim_{h \to 0^+} [F(a + h) - F(a)] = \lim_{h \to 0^+} \left(\frac{F(a + h) - F(a)}{h} \cdot h \right) = f(a) \cdot \lim_{h \to 0^+} h = 0.$$

Therefore

$$\lim_{h \to 0^+} F(a + h) = F(a).$$

This shows that F is continuous from the right at $x = a$. The continuity of F from the left at $x = b$ can be shown in a similar manner by applying limit (3) at $x = b$. ❑

Example 1 The function $F(x) = \int_{-1}^x (2t + t^2)\, dt$ for all $x \in [-1, 5]$ has derivative

$$F'(x) = 2x + x^2 \qquad \text{for all } x \in (-1, 5). \quad ❑$$

Example 2 For all real x, define

$$F(x) = \int_0^x \sin \pi t\, dt.$$

Find $F'(\tfrac{3}{4})$ and $F'(-\tfrac{1}{2})$.

SOLUTION By Theorem 5.3.5,

$$F'(x) = \sin \pi x \qquad \text{for all real } x.$$

Thus, $F'(\tfrac{3}{4}) = \sin\left(\tfrac{3}{4}\pi\right) = \tfrac{1}{2}\sqrt{2}$ and $F'\left(-\tfrac{1}{2}\right) = \sin\left(-\tfrac{1}{2}\pi\right) = -1.$ ❑

Example 3 Set

$$F(x) = \int_0^x \frac{1}{1 + t^2}\, dt \qquad \text{for all real numbers } x.$$

(a) Find the critical points of F and determine the intervals on which F increases and the intervals on which F decreases.

(b) Determine the concavity of the graph of F and find the points of inflection (if any).

(c) Sketch the graph of F.

SOLUTION

(a) To find the intervals on which F increases and the intervals on which F decreases, we examine the first derivative of F. By Theorem 5.3.5,

$$F'(x) = \frac{1}{1 + x^2} \qquad \text{for all real } x.$$

Since $F'(x) > 0$ for all real x, F increases on $(-\infty, \infty)$; there are no critical points.

(b) To determine the concavity of the graph and to find the points of inflection, we use the second derivative

$$F''(x) = \frac{-2x}{(1+x^2)^2}.$$

The sign of F'' and the behavior of the graph of F are as follows:

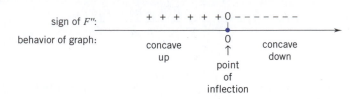

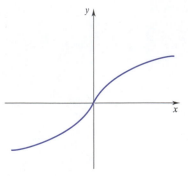

Figure 5.3.5

(c) Since $F(0) = 0$ and $F'(0) = 1$, the graph passes through the origin with slope 1. A sketch of the graph is shown in Figure 5.3.5. As you'll see in Chapter 7, the graph has two horizontal asymptotes: $y = \frac{1}{2}\pi$ and $y = -\frac{1}{2}\pi$. ❑

EXERCISES 5.3

1. Given that

$$\int_0^1 f(x)\,dx = 6, \qquad \int_0^2 f(x)\,dx = 4, \qquad \int_2^5 f(x)\,dx = 1,$$

find the following:

(a) $\displaystyle\int_0^5 f(x)\,dx$. (b) $\displaystyle\int_1^2 f(x)\,dx$. (c) $\displaystyle\int_1^5 f(x)\,dx$.

(d) $\displaystyle\int_0^0 f(x)\,dx$. (e) $\displaystyle\int_2^0 f(x)\,dx$. (f) $\displaystyle\int_5^1 f(x)\,dx$.

2. Given that

$$\int_1^4 f(x)\,dx = 5, \qquad \int_3^4 f(x)\,dx = 7, \qquad \int_1^8 f(x)\,dx = 11,$$

find the following:

(a) $\displaystyle\int_4^8 f(x)\,dx$. (b) $\displaystyle\int_4^3 f(x)\,dx$. (c) $\displaystyle\int_1^3 f(x)\,dx$.

(d) $\displaystyle\int_3^8 f(x)\,dx$. (e) $\displaystyle\int_8^4 f(x)\,dx$. (f) $\displaystyle\int_4^4 f(x)\,dx$.

3. Use upper and lower sums to show that

$$0.5 < \int_1^2 \frac{dx}{x} < 1.$$

4. Use upper and lower sums to show that

$$0.6 < \int_0^1 \frac{dx}{1+x^2} < 1.$$

5. For $x > -1$, set $F(x) = \int_0^x t\sqrt{t+1}\,dt$.

(a) Find $F(0)$. (b) Find $F'(x)$. (c) Find $F'(2)$.

(d) Express $F(2)$ as an integral of $t\sqrt{t+1}$.

(e) Express $-F(x)$ as an integral of $t\sqrt{t+1}$.

6. Let $F(x) = \int_\pi^x t\sin t\,dt$.

(a) Find $F(\pi)$. (b) Find $F'(x)$. (c) Find $F'(\frac{1}{2}\pi)$.

(d) Express $F(2\pi)$ as an integral of $t\sin t$.

(e) Express $-F(x)$ as an integral of $t\sin t$.

Exercises 7–12. Calculate the following for each F given below:

(a) $F'(-1)$. (b) $F'(0)$. (c) $F'(\frac{1}{2})$. (d) $F''(x)$.

7. $F(x) = \displaystyle\int_0^x \frac{dt}{t^2+9}$. 8. $F(x) = \displaystyle\int_x^0 \sqrt{t^2+1}\,dt$.

9. $F(x) = \displaystyle\int_x^1 t\sqrt{t^2+1}\,dt$. 10. $F(x) = \displaystyle\int_1^x \sin\pi t\,dt$.

11. $F(x) = \displaystyle\int_1^x \cos\pi t\,dt$. 12. $F(x) = \displaystyle\int_2^x (t+1)^3\,dt$.

13. Show that statements (a) and (b) are false.

(a) $U_f(P_1) = 4$ for the partition $P_1 = \{0, 1, \frac{3}{2}, 2\}$, and

$U_f(P_2) = 5$ for the partition $P_2 = \{0, \frac{1}{4}, 1, \frac{3}{2}, 2\}$.

(b) $L_f(P_1) = 5$ for the partition $P_1 = \{0, 1, \frac{3}{2}, 2\}$, and

$L_f(P_2) = 4$ for the partition $P_2 = \{0, \frac{1}{4}, 1, \frac{3}{2}, 2\}$.

14. (a) Which continuous functions f defined on $[a, b]$, have the property that $\mathcal{L}_f(P) = \mathcal{U}_f(P)$ for some partition P?

(b) Which continuous functions f defined on $[a, b]$ have the property that $\mathcal{L}_f(P) = \mathcal{U}_f(Q)$ for some partitions P and Q?

15. Which continuous functions f defined on $[a, b]$ have the property that all lower sums $\mathcal{L}_f(P)$ are equal?

16. Show that if f is continuous on an interval I, then

$$\int_a^c f(t)\,dt + \int_c^b f(t)\,dt = \int_a^b f(t)\,dt$$

for *every* choice of a, b, c from I. HINT: Assume $a < b$ and consider the four cases: $c = a, c = b, c < a, b < c$. Then consider what happens if $a > b$ or $a = b$.

Exercises 17 and 18. Find the critical points for F and, at each critical point, determine whether F has a local maximum, a local minimum, or neither.

17. $F(x) = \displaystyle\int_0^x \dfrac{t-1}{1+t^2}\, dt.$ 18. $F(x) = \displaystyle\int_0^x \dfrac{t-4}{1+t^2}\, dt.$

19. For $x > 0$, set $F(x) = \displaystyle\int_1^x (1/t)\, dt.$

   (a) Find the critical points for F, if any, and determine the intervals on which F increases and the intervals on which F decreases.
   (b) Determine the concavity of the graph of F and find the points of inflection, if any.
   (c) Sketch the graph of F.

20. Let $F(x) = \displaystyle\int_0^x t(t-3)^2\, dt.$

   (a) Find the critical points for F and determine the intervals on which F increases and the intervals on which F decreases.
   (b) Determine the concavity of the graph of F and find the points of inflection, if any.
   (c) Sketch the graph of F.

21. Suppose that f is differentiable with $f'(x) > 0$ for all x, and suppose that $f(1) = 0$. Set

$$F(x) = \int_0^x f(t)\, dt.$$

   Justify each statement.
   (a) F is continuous.
   (b) F is twice differentiable.
   (c) $x = 1$ is a critical point for F.
   (d) F takes on a local minimum at $x = 1$.
   (e) $F(1) < 0$.

   Make a rough sketch of the graph of F.

22. Suppose that g is differentiable with $g'(x) < 0$ for all $x < 1$, $g'(1) = 0$, and $g'(x) > 0$ for all $x > 1$, and suppose that $g(1) = 0$. Set

$$G(x) = \int_0^x g(t)\, dt.$$

   Justify each statement.
   (a) G is continuous.
   (b) G is twice differentiable.
   (c) $x = 1$ is a critical point for G.
   (d) The graph of G is concave down for $x < 1$ and concave up for $x > 1$.
   (e) G is an increasing function.

   Make a rough sketch of the graph of G.

23. (a) Sketch the graph of the function

$$f(x) = \begin{cases} 2 - x, & -1 \le x \le 0 \\ 2 + x, & 0 < x \le 3. \end{cases}$$

   (b) Calculate $F(x) = \displaystyle\int_{-1}^x f(t)\, dt$, $-1 \le x \le 3$, and sketch the graph of F.
   (c) What can you conclude about f and F at $x = 0$?

24. (a) Sketch the graph of the function

$$f(x) = \begin{cases} x^2 + x, & 0 \le x \le 1 \\ 2x, & 1 < x \le 3. \end{cases}$$

   (b) Calculate $F(x) = \displaystyle\int_0^x f(t)\, dt$, $0 \le x \le 3$, and sketch the graph of F.
   (c) What can you conclude about f and F at $x = 1$?

Exercises 25–28. Calculate $F'(x)$.

25. $F(x) \displaystyle\int_0^{x^3} t \cos t\, dt.$ HINT: Set $u = x^3$ and use the chain rule.

26. $F(x) = \displaystyle\int_1^{\cos x} \sqrt{1 - t^2}\, dt.$

27. $F(x) = \displaystyle\int_{x^2}^1 (t - \sin^2 t)\, dt.$

28. $F(x) = \displaystyle\int_0^{\sqrt{x}} \dfrac{t^2}{1 + t^4}\, dt.$

29. Set $F(x) = 2x + \displaystyle\int_0^x \dfrac{\sin 2t}{1 + t^2}\, dt.$ Determine

   (a) $F(0)$.   (b) $F'(0)$.   (c) $F''(0)$.

30. Set $F(x) = 2x + \displaystyle\int_0^{x^2} \dfrac{\sin 2t}{1 + t^2}\, dt.$ Determine

   (a) $F(0)$.   (b) $F'(x)$.

31. Assume that f is continuous and

$$\int_0^x f(t)\, dt = \dfrac{2x}{4 + x^2}.$$

   (a) Determine $f(0)$.
   (b) Find the zeros of f, if any.

32. Assume that f is continuous and

$$\int_0^x f(t)\, dt = \sin x - x \cos x.$$

   (a) Determine $f\left(\tfrac{1}{2}\pi\right)$.   (b) Find $f'(x)$.

33. (*A mean-value theorem for integrals*) Show that if f is continuous on $[a, b]$, then there is a least one number c in (a, b) for which

$$\int_a^b f(x)\, dx = f(c)(b - a).$$

34. We proved Theorem 5.3.5 only in the case that the integration which defines F is begun at the left endpoint a. Show that the result still holds if the integration is begun at an arbitrary point $c \in (a, b)$.

35. Let f be continuous on $[a, b]$. For each $x \in [a, b]$ set

$$F(x) = \int_c^x f(t)\, dt, \quad \text{and} \quad G(x) = \int_d^x f(t)\, dt$$

   taking c and d from $[a, b]$.
   (a) Show that F and G differ by a constant.
   (b) Show that $F(x) - G(x) = \int_c^d f(t)\, dt.$

36. Let f be everywhere continuous and set

$$F(x) = \int_0^x \left[t \int_1^t f(u)\, du \right] dt.$$

Find (a) $F'(x)$. (b) $F'(1)$. (c) $F''(x)$. (d) $F''(1)$.

▶ **Exercises 37–40.** Use a CAS to carry out the following steps:

(a) Solve the equation $F'(x) = 0$. Determine the intervals on which F increases and the intervals on which F decreases. Produce a figure that displays both the graph of F and the graph of F'.

(b) Solve the equation $F''(x) = 0$. Determine the intervals on which the graph of F is concave up and the intervals

on which the graph of F is concave down. Produce a figure that displays both the graph of F and the graph of F''.

37. $F(x) = \int_0^x (t^2 - 3t - 4)\, dt.$

38. $F(x) = \int_0^x (2 - 3\cos t)\, dt, \quad x \in [0, 2\pi]$

39. $F(x) = \int_x^0 \sin 2t\, dt, \quad x \in [0, 2\pi]$

40. $F(x) = \int_x^0 (2 - t)^2\, dt.$

■ 5.4 THE FUNDAMENTAL THEOREM OF INTEGRAL CALCULUS

The natural setting for differentiation is an open interval. For functions f defined on an open interval, the antiderivatives of f are simply the functions with derivative f. For continuous functions defined on a closed interval $[a, b]$, the term "antiderivative" takes into account the endpoints a and b.

DEFINITION 5.4.1   ANTIDERIVATIVE ON (a, b)

Let f be continuous on $[a, b]$. A function G is called an *antiderivative for f on $[a, b]$* if

  G is continuous on $[a, b]$ and $G'(x) = f(x)$ for all $x \in (a, b)$.

Theorem 5.3.5 tells us that if f is continuous on $[a, b]$, then

$$F(x) = \int_a^x f(t)\, dt$$

is an antiderivative for f on $[a, b]$. This gives us a prescription for constructing antiderivatives. It tells us that we can construct an antiderivative for f by integrating f.

The theorem below, called the "fundamental theorem," goes the other way. It gives us a prescription, not for finding antiderivatives, but for evaluating integrals. It tells us that we can evaluate the integral

$$\int_a^b f(t)\, dt$$

from any antiderivative of f by evaluating the antiderivative at b and at a.

THEOREM 5.4.2   THE FUNDAMENTAL THEOREM OF INTEGRAL CALCULUS

Let f be continuous on $[a, b]$. If G is any antiderivative for f on $[a, b]$, then

$$\int_a^b f(t)\, dt = G(b) - G(a).$$

PROOF From Theorem 5.3.5 we know that the function

$$F(x) = \int_a^x f(t)\,dt$$

is an antiderivative for f on $[a, b]$. If G is also an antiderivative for f on $[a, b]$, then both F and G are continuous on $[a, b]$ and satisfy $F'(x) = G'(x)$ for all x in (a, b). From Theorem 4.2.4 we know that there exists a constant C such that

$$F(x) = G(x) + C \qquad \text{for all } x \text{ in } [a, b].$$

Since $F(a) = 0$,

$$G(a) + C = 0 \qquad \text{and thus} \qquad C = -G(a).$$

It follows that

$$F(x) = G(x) - G(a) \qquad \text{for all } x \text{ in } [a, b].$$

In particular,

$$\int_a^b f(t)\,dt = F(b) = G(b) - G(a). \quad \square$$

We now evaluate some integrals by applying the fundamental theorem. In each case we use the simplest antiderivative we can think of.

Example 1 Evaluate $\int_1^4 x^2\,dx$.

SOLUTION As an antiderivative for $f(x) = x^2$, we can use the function

$$G(x) = \tfrac{1}{3}x^3. \qquad \text{(Verify this.)}$$

By the fundamental theorem,

$$\int_1^4 x^2\,dx = G(4) - G(1) = \tfrac{1}{3}(4)^3 - \tfrac{1}{3}(1)^3 = \tfrac{64}{3} - \tfrac{1}{3} = 21.$$

NOTE: Any other antiderivative of $f(x) = x^2$ has the form $H(x) = \tfrac{1}{3}x^3 + C$ for some constant C. Had we chosen such an H instead of G, then we would have had

$$\int_1^4 x^2\,dx = H(4) - H(1) = \left[\tfrac{1}{3}(4)^3 + C\right] - \left[\tfrac{1}{3}(1)^3 + C\right] = \tfrac{64}{3} + C - \tfrac{1}{3} - C = 21;$$

the C's would have canceled out. $\square$

Example 2 Evaluate $\int_0^{\pi/2} \sin x\,dx$.

SOLUTION Here we use the antiderivative $G(x) = -\cos x$:

$$\int_0^{\pi/2} \sin x\,dx = G(\pi/2) - G(0)$$
$$= -\cos(\pi/2) - [-\cos(0)] = 0 - (-1) = 1. \quad \square$$

Notation Expressions of the form $G(b) - G(a)$ are conveniently written

$$\Big[G(x)\Big]_a^b.$$

In this notation

$$\int_1^4 x^2\,dx = \left[\tfrac{1}{3}x^3\right]_1^4 = \tfrac{1}{3}(4)^3 - \tfrac{1}{3}(1)^3 = 21$$

and

$$\int_0^{\pi/2} \sin x\,dx = \Big[-\cos x\Big]_0^{\pi/2} = -\cos(\pi/2) - [-\cos(0)] = 1. \quad \square$$

To calculate

$$\int_a^b f(x)\,dx$$

by the fundamental theorem, we need to find an antiderivative for f. We do this by working back from the results of differentiation.

For rational r,

$$\frac{d}{dx}(x^{r+1}) = (r+1)x^r.$$

Thus, if $r \neq -1$,

$$\frac{d}{dx}\left(\frac{x^{r+1}}{r+1}\right) = x^r.$$

This tells us that

$$G(x) = \frac{x^{r+1}}{r+1} \qquad \text{is an antiderivative for } f(x) = x^r.$$

Some common trigonometric antiderivatives are listed in Table 5.4.1. Note that in each case the function on the left is the derivative of the function on the right.

■ **Table 5.4.1**

Function	Antiderivative	Function	Antiderivative
$\sin x$	$-\cos x$	$\cos x$	$\sin x$
$\sec^2 x$	$\tan x$	$\csc^2 x$	$-\cot x$
$\sec x \tan x$	$\sec x$	$\csc x \cot x$	$-\csc x$

We continue with computations.

$$\int_1^2 \frac{dx}{x^3} = \int_1^2 x^{-3}\,dx = \left[\frac{x^{-2}}{-2}\right]_1^2 = \left[-\frac{1}{2x^2}\right]_1^2 = -\tfrac{1}{8} - \left(-\tfrac{1}{2}\right) = \tfrac{3}{8},$$

$$\int_0^1 t^{5/3}\,dt = \left[\tfrac{3}{8}t^{8/3}\right]_0^1 = \tfrac{3}{8}(1)^{8/3} - \tfrac{3}{8}(0)^{8/3} = \tfrac{3}{8}.$$

$$\int_{-\pi/4}^{\pi/3} \sec^2 t\,dt = \left[\tan t\right]_{-\pi/4}^{\pi/3} = \tan\frac{\pi}{3} - \tan\frac{-\pi}{4} = \sqrt{3} - (-1) = \sqrt{3} + 1.$$

$$\int_{\pi/6}^{\pi/2} \csc x \cot x\,dx = \left[-\csc x\right]_{\pi/6}^{\pi/2} = -\csc\frac{\pi}{2} - \left[-\csc\frac{\pi}{6}\right] = -1 - (-2) = 1.$$

Example 3 Evaluate $\displaystyle\int_0^1 (2x - 6x^4 + 5)\,dx$.

SOLUTION As an antiderivative we use $G(x) = x^2 - \tfrac{6}{5}x^5 + 5x$:

$$\int_0^1 (2x - 6x^4 + 5)\,dx = \left[x^2 - \tfrac{6}{5}x^5 + 5x\right]_0^1 = 1 - \tfrac{6}{5} + 5 = \tfrac{24}{5}. \quad \square$$

Example 4 Evaluate $\displaystyle\int_{-1}^1 (x - 1)(x + 2)\,dx$.

SOLUTION First we carry out the indicated multiplication:

$$(x - 1)(x + 2) = x^2 + x - 2.$$

As an antiderivative we use $G(x) = \frac{1}{3}x^3 + \frac{1}{2}x^2 - 2x$:

$$\int_{-1}^{1}(x-1)(x+2)\,dx = \left[\frac{1}{3}x^3 + \frac{1}{2}x^2 - 2x\right]_{-1}^{1} = -\frac{10}{3}. \quad \square$$

We now give some slightly more complicated examples. The essential step in each case is the determination of an antiderivative. Check each computation in detail.

$$\int_{1}^{2}\frac{x^4+1}{x^2}\,dx = \int_{1}^{2}(x^2 + x^{-2})\,dx = \left[\frac{1}{3}x^3 - x^{-1}\right]_{1}^{2} = \frac{17}{6}.$$

$$\int_{1}^{5}\sqrt{x-1}\,dx = \int_{1}^{5}(x-1)^{1/2}\,dx = \left[\frac{2}{3}(x-1)^{3/2}\right]_{1}^{5} = \frac{16}{3}.$$

$$\int_{0}^{1}(4-\sqrt{x})^2\,dx = \int_{0}^{1}(16 - 8\sqrt{x} + x)\,dx = \left[16x - \frac{16}{3}x^{3/2} + \frac{1}{2}x^2\right]_{0}^{1} = \frac{67}{6}.$$

$$\int_{1}^{2}-\frac{dt}{(t+2)^2} = \int_{1}^{2}-(t+2)^{-2}\,dt = \left[(t+2)^{-1}\right]_{1}^{2} = -\frac{1}{12}.$$

The Linearity of the Integral

The preceding examples suggest some simple properties of the integral that are used regularly in computations. Throughout, take f and g as continuous functions and α and β as constants.

I. Constants may be factored through the integral sign:

(5.4.3)
$$\int_{a}^{b}\alpha f(x)\,dx = \alpha\int_{a}^{b}f(x)\,dx.$$

For example,

$$\int_{1}^{4}\frac{3}{7}\sqrt{x}\,dx = \frac{3}{7}\int_{1}^{4}x^{1/2}dx = \frac{3}{7}\left[\frac{x^{3/2}}{3/2}\right]_{1}^{4} = \frac{2}{7}\left[(4)^{3/2} - (1)^{3/2}\right] = \frac{2}{7}[8-1] = 2.$$

$$\int_{0}^{\pi/4}2\cos x\,dx = 2\int_{0}^{\pi/4}\cos x\,dx = 2\left[\sin x\right]_{0}^{\pi/4} = 2\left[\sin\frac{\pi}{4} - \sin 0\right]$$

$$= 2\frac{\sqrt{2}}{2} = \sqrt{2}.$$

II. The integral of a sum is the sum of the integrals:

(5.4.4)
$$\int_{a}^{b}[f(x) + g(x)]\,dx = \int_{a}^{b}f(x)\,dx + \int_{a}^{b}g(x)\,dx.$$

For example,

$$\int_{0}^{\pi/2}(\sin x + \cos x)\,dx = \int_{0}^{\pi/2}\sin x\,dx + \int_{0}^{\pi/2}\cos x\,dx$$

$$= \left[-\cos x\right]_{0}^{\pi/2} + \left[\sin x\right]_{0}^{\pi/2}$$

$$= (-\cos \pi/2) - (-\cos 0) + \sin \pi/2 - \sin 0$$

$$= 1 + 1 = 2.$$

III. The integral of a linear combination is the linear combination of the integrals:

(5.4.5)
$$\int_a^b [\alpha f(x) + \beta g(x)]\,dx = \alpha \int_a^b f(x)\,dx + \beta \int_a^b g(x)\,dx.$$

This applies to the linear combination of more than two functions. For example,

$$\int_0^1 (2x - 6x^4 + 5)\,dx = 2\int_0^1 x\,dx - 6\int_0^1 x^4 dx + \int_0^1 5\,dx$$

$$= 2\left[\frac{x^2}{2}\right]_0^1 - 6\left[\frac{x^5}{5}\right]_0^1 + \left[5x\right]_0^1 = 1 - \tfrac{6}{5} + 5 = \tfrac{24}{5}.$$

This is the result obtained in Example 3.

Properties I and II are particular instances of Property III. To prove III, let F be an antiderivative for f and let G be an antiderivative for g. Then, since

$$[\alpha F(x) + \beta G(x)]' = \alpha F'(x) + \beta G'(x) = \alpha f(x) + \beta g(x),$$

it follows that $\alpha F + \beta G$ is an antiderivative for $\alpha f + \beta g$. Therefore,

$$\int_a^b [\alpha f(x) + \beta g(x)]\,dx = \left[\alpha F(x) + \beta G(x)\right]_a^b$$

$$= [\alpha F(b) + \beta G(b)] - [\alpha F(a) + \beta G(a)]$$

$$= \alpha[F(b) - F(a)] + \beta[G(b) - G(a)]$$

$$= \alpha \int_a^b f(x)\,dx + \beta \int_a^b g(x)\,dx.$$

Example 5 Evaluate $\displaystyle\int_0^{\pi/4} \sec x[2\tan x - 5\sec x]\,dx.$

SOLUTION

$$\int_0^{\pi/4} \sec x[2\tan x - 5\sec x]\,dx = \int_0^{\pi/4} [2\sec x\tan x - 5\sec^2 x]\,dx$$

$$= 2\int_0^{\pi/4} \sec x\tan x\,dx - 5\int_0^{\pi/4} \sec^2 x\,dx$$

$$= 2\left[\sec x\right]_0^{\pi/4} - 5\left[\tan x\right]_0^{\pi/4}$$

$$= 2\left[\sec\frac{\pi}{4} - \sec 0\right] - 5\left[\tan\frac{\pi}{4} - \tan 0\right]$$

$$= 2[\sqrt{2} - 1] - 5[1 - 0] = 2\sqrt{2} - 7. \quad❏$$

EXERCISES 5.4

Exercises 1–34. Evaluate the integral.

1. $\displaystyle\int_0^1 (2x - 3)\,dx.$

2. $\displaystyle\int_0^1 (3x + 2)\,dx.$

5. $\displaystyle\int_1^4 2\sqrt{x}\,dx.$

6. $\displaystyle\int_0^4 \sqrt[3]{x}\,dx.$

3. $\displaystyle\int_{-1}^0 5x^4 dx.$

4. $\displaystyle\int_1^2 (2x + x^2)\,dx.$

7. $\displaystyle\int_1^5 2\sqrt{x - 1}\,dx.$

8. $\displaystyle\int_1^2 \left(\frac{3}{x^3} + 5x\right)\,dx.$

9. $\int_{-2}^{0} (x+1)(x-2)\,dx.$

10. $\int_{1}^{0} (t^3 + t^2)\,dt.$

11. $\int_{1}^{2} \left(3t + \frac{4}{t^2}\right) dt.$

12. $\int_{-1}^{-1} 7x^6 dx.$

13. $\int_{0}^{1} (x^{3/2} - x^{1/2})\,dx.$

14. $\int_{0}^{1} (x^{3/4} - 2x^{1/2})\,dx.$

15. $\int_{0}^{1} (x+1)^{17} dx.$

16. $\int_{0}^{a} (a^2 x - x^3)\,dx.$

17. $\int_{0}^{a} (\sqrt{a} - \sqrt{x})^2 dx.$

18. $\int_{-1}^{1} (x-2)^2 dx.$

19. $\int_{1}^{2} \frac{6-t}{t^3}\,dt.$

20. $\int_{1}^{3} \left(x^2 - \frac{1}{x^2}\right) dx.$

21. $\int_{1}^{2} 2x(x^2+1)\,dx.$

22. $\int_{0}^{1} 3x^2(x^3+1)\,dx.$

23. $\int_{0}^{\pi/2} \cos x\,dx.$

24. $\int_{0}^{\pi} 3\sin x\,dx.$

25. $\int_{0}^{\pi/4} 2\sec^2 x\,dx.$

26. $\int_{\pi/6}^{\pi/3} \sec x \tan x\,dx.$

27. $\int_{\pi/6}^{\pi/4} \csc u \cot u\,du.$

28. $\int_{\pi/4}^{\pi/3} -\csc^2 u\,du.$

29. $\int_{0}^{2\pi} \sin x\,dx.$

30. $\int_{0}^{\pi} \frac{1}{2}\cos x\,dx.$

31. $\int_{0}^{\pi/3} \left(\frac{2}{\pi}x - 2\sec^2 x\right) dx.$

32. $\int_{\pi/4}^{\pi/2} \csc x (\cot x - 3\csc x)\,dx.$

33. $\int_{0}^{3} \left[\frac{d}{dx}(\sqrt{4+x^2})\right] dx.$

34. $\int_{0}^{\pi/2} \left[\frac{d}{dx}(\sin^3 x)\right] dx.$

Exercises 35–38. Calculate the derivative with respect to x (a) without integrating; that is, using the results of Section 5.3; (b) by integrating and then differentiating the result.

35. $\int_{1}^{x} (t+2)^2\,dt.$

36. $\int_{0}^{x} (\cos t - \sin t)\,dt.$

37. $\int_{1}^{2x+1} \frac{1}{2}\sec u \tan u\,du.$

38. $\int_{x^2}^{2} t(t-1)\,dt.$

39. Define a function F on $[1, 8]$ such that $F'(x) = 1/x$ and (a) $F(2) = 0$; (b) $F(2) = -3$.

40. Define a function F on $[0, 4]$ such that $F'(x) = \sqrt{1+x^2}$ and (a) $F(3) = 0$; (b) $F(3) = 1$.

Exercises 41–44. Verify that the function is nonnegative on the given interval, and then calculate the area below the graph on that interval.

41. $f(x) = 4x - x^2;$ $[0, 4].$

42. $f(x) = x\sqrt{x} + 1;$ $[1, 9].$

43. $f(x) = 2\cos x;$ $[-\pi/2, \pi/4].$

44. $f(x) = \sec x \tan x;$ $[0, \pi/3].$

Exercises 45–48. Evaluate.

45. (a) $\int_{2}^{5} (x-3)\,dx.$ (b) $\int_{2}^{5} |x-3|\,dx.$

46. (a) $\int_{-4}^{2} (2x+3)\,dx.$ (b) $\int_{-4}^{2} |2x+3|\,dx.$

47. (a) $\int_{-2}^{2} (x^2-1)\,dx.$ (b) $\int_{-2}^{2} |x^2-1|\,dx.$

48. (a) $\int_{-\pi/2}^{\pi} \cos x\,dx.$ (b) $\int_{-\pi/2}^{\pi} |\cos x|\,dx.$

Exercises 49–52. Determine whether the calculation is valid. If it is not valid, explain why it is not valid.

49. $\int_{0}^{2\pi} x\cos x\,dx = \Big[x\sin x + \cos x\Big]_{0}^{2\pi} = 1 - 1 = 0.$

50. $\int_{0}^{2\pi} \sec^2 x\,dx = \Big[\tan x\Big]_{0}^{2\pi} = 0 - 0 = 0.$

51. $\int_{-2}^{2} \frac{1}{x^3}\,dx = \left[\frac{-1}{2x^2}\right]_{-2}^{2} = -\frac{1}{8} - \left(-\frac{1}{8}\right) = 0.$

52. $\int_{-2}^{2} |x|\,dx = \left[\frac{1}{2}x|x|\right]_{-2}^{2} = 2 - (-2) = 0.$

53. An object starts at the origin and moves along the x-axis with velocity

$$v(t) = 10t - t^2, \quad 0 \le t \le 10.$$

(a) What is the position of the object at any time t, $0 \le t \le 10$?

(b) When is the object's velocity a maximum, and what is its position at that time?

54. The velocity of a bob suspended on a spring is given:

$$v(t) = 3\sin t + 4\cos t, \quad t \ge 0.$$

At time $t = 0$, the bob is one unit below the equilibrium position. (See the figure.)

(a) Determine the position of the bob at each time $t \ge 0$.

(b) What is the bob's maximum displacement from the equilibrium position?

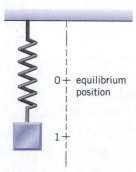

Exercises 55–58. Evaluate the integral.

55. $\int_{0}^{4} f(x)\,dx;$ $f(x) = \begin{cases} 2x+1, & 0 \le x \le 1 \\ 4-x, & 1 < x \le 4. \end{cases}$

56. $\int_{-2}^{4} f(x)\,dx;$ $f(x) = \begin{cases} 2+x^2, & -2 \le x < 0 \\ \frac{1}{2}x+2, & 0 \le x \le 4. \end{cases}$

57. $\int_{-\pi/2}^{\pi} f(x)dx; \quad f(x) = \begin{cases} 1 + 2\cos x, & -\pi/2 \le x \le \pi/3 \\ (3/\pi)x + 1, & \pi/3 < x \le \pi. \end{cases}$

58. $\int_{0}^{3\pi/2} f(x)dx; \quad f(x) = \begin{cases} 2\sin x, & 0 \le x \le \pi/2 \\ 2 + \cos x, & \pi/2 < x \le 3\pi/2. \end{cases}$

59. Let $f(x) = \begin{cases} x + 2, & -2 \le x \le 0 \\ 2, & 0 < x \le 1 \\ 4 - 2x, & 1 < x \le 2, \end{cases}$

    and set $g(x) = \int_{-2}^{x} f(t)\,dt.$

    (a) Carry out the integration.
    (b) Sketch the graphs of f and g.
    (c) Where is f continuous? Where is f differentiable? Where is g differentiable?

60. Let $f(x) = \begin{cases} 2 - x^2, & -1 \le x \le 1 \\ 1, & 1 < x < 3 \\ 2x - 5, & 3 \le x \le 5 \end{cases}$

    and let $g(x) = \int_{-1}^{x} f(t)\,dt.$

(a) Carry out the integration.
(b) Sketch the graphs of f and g.
(c) Where is f continuous? Where is f differentiable? Where is g differentiable?

61. (*Important*) If f is a function and its derivative f' is continuous on $[a, b]$, then

$$\int_{a}^{b} f'(t)\,dt = f(b) - f(a).$$

Explain the reasoning here.

62. Let f be a function such that f' is continuous on $[a, b]$. Show that

$$\int_{a}^{b} f(t)f'(t)\,dt = \frac{1}{2}[f^2(b) - f^2(a)].$$

63. Given that f has a continuous derivative, compare

$$\frac{d}{dx}\left[\int_{a}^{x} f(t)\,dt\right] \quad \text{to} \quad \int_{a}^{x} \frac{d}{dt}[f(t)]\,dt.$$

64. Given that f is a continuous function, set $F(x) = \int_{0}^{x} xf(t)\,dt$. Find $F'(x)$. HINT: The answer is not $xf(x)$.

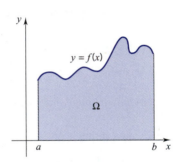

Figure 5.5.1

■ 5.5 SOME AREA PROBLEMS

The calculations of area that we carry out in this section are all based on what you already know: if f is continuous and nonnegative on $[a, b]$, then the area under the graph of f from $x = a$ to $x = b$ is given by the integral of f from $x = a$ to $x = b$; namely, with Ω as in Figure 5.5.1

(5.5.1)
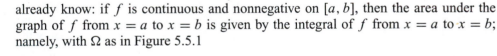
$$\text{area of } \Omega = \int_{a}^{b} f(x)\,dx.$$

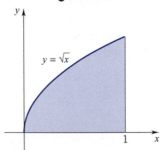

Figure 5.5.2

Example 1 Find the area below the graph of the square-root function from $x = 0$ to $x = 1$.

SOLUTION The graph is pictured in Figure 5.5.2. The area below the graph is $\frac{2}{3}$:

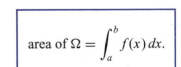
$$\int_{0}^{1} \sqrt{x}\,dx = \int_{0}^{1} x^{1/2}dx = \left[\frac{2}{3}x^{3/2}\right]_{0}^{1} = \frac{2}{3}. \quad \square$$

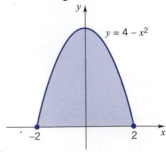

Figure 5.5.3

Example 2 Find the area of the region bounded above by the curve $y = 4 - x^2$ and below by the x-axis.

SOLUTION The curve intersects the x-axis at $x = -2$ and $x = 2$. See Figure 5.5.3. The area of the region is $\frac{32}{3}$:

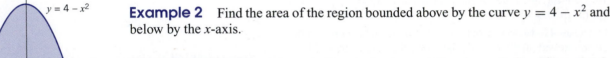
$$\int_{-2}^{2} (4 - x^2)dx = \left[4x - \frac{1}{3}x^3\right]_{-2}^{2} = \frac{32}{3}.$$

NOTE: The region is symmetric with respect to the y-axis. Therefore, the area of the region can be stated as $2 \int_0^2 (4 - x^2)\, dx$:

$$2 \int_0^2 (4 - x^2)dx = 2\left[4x - \tfrac{1}{3}x^3\right]_0^2 = 2\left(8 - \tfrac{8}{3}\right) = 2\left(\tfrac{16}{3}\right) = \tfrac{32}{3}.$$

We'll have more to say about the symmetry considerations in Section 5.8 ❏

Now we calculate the areas of somewhat more complicated regions. To avoid excessive repetitions, let's agree at the outset that throughout this section the symbols f, g, h represent continuous functions.

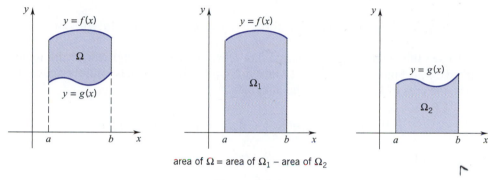

area of Ω = area of Ω_1 − area of Ω_2

Figure 5.5.4

Look at the region Ω shown in Figure 5.5.4. The upper boundary of Ω is the graph of a nonnegative function f and the lower boundary is the graph of a nonnegative function g. We can obtain the area of Ω by calculating the area of Ω_1 and subtracting off the area of Ω_2. Since

$$\text{area of } \Omega_1 = \int_a^b f(x)\, dx \quad \text{and} \quad \text{area of } \Omega_2 = \int_a^b g(x)\, dx,$$

we have

$$\text{area of } \Omega = \int_a^b f(x)\, dx - \int_a^b g(x)\, dx.$$

We can combine the two integrals and write

(5.5.2)
$$\boxed{\text{area of } \Omega = \int_a^b [f(x) - g(x)]\, dx.}$$

Example 3 Find the area of the region bounded above by the line $y = x + 2$ and bounded below by the parabola $y = x^2$.

SOLUTION The region is shown in Figure 5.5.5. The limits of integration were found by solving the two equations simultaneously:

$$x + 2 = x^2 \quad \text{iff} \quad x^2 - x - 2 = 0$$
$$\text{iff} \quad (x + 1)(x - 2) = 0$$
$$\text{iff} \quad x = -1 \quad \text{or} \quad x = 2.$$

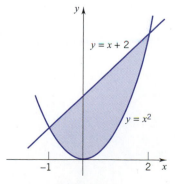

Figure 5.5.5

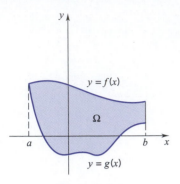

$y = f(x)$

Ω

$y = g(x)$

Figure 5.5.6

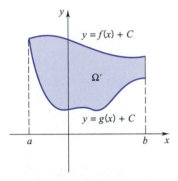

$y = f(x) + C$

Ω'

$y = g(x) + C$

Figure 5.5.7

The area of the region is given by the integral

$$\int_{-1}^{2} \left[(x+2) - x^2\right] dx = \left[\tfrac{1}{2}x^2 + 2x - \tfrac{1}{3}x^3\right]_{-1}^{2}$$

$$= \left(2 + 4 - \tfrac{8}{3}\right) - \left(\tfrac{1}{2} - 2 + \tfrac{1}{3}\right) = \tfrac{9}{2} \quad \square$$

We derived Formula 5.5.2 under the assumption that f and g were both nonnegative, but that assumption is unnecessary. The formula holds for any region Ω that has

an upper boundary of the form $\qquad y = f(x), \qquad x \in [a, b]$

and

a lower boundary of the form $\qquad y = g(x), \qquad x \in [a, b]$.

To see this, take Ω as in Figure 5.5.6. Obviously, Ω is congruent to the region marked Ω' in Figure 5.5.7; Ω' is Ω raised C units. Since Ω' lies entirely above the x-axis, the area of Ω' is given by the integral

$$\int_{a}^{b} \{[f(x) + C] - [g(x) + C]\} dx = \int_{a}^{b} [f(x) - g(x)] dx.$$

Since area of $\Omega =$ area of Ω',

$$\text{area of } \Omega = \int_{a}^{b} [f(x) - g(x)] dx$$

as asserted.

Example 4 Find the area of the region shown in Figure 5.5.8.

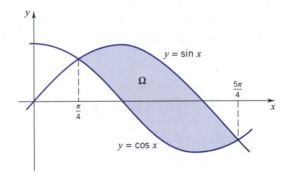

$y = \sin x$

Ω

$\dfrac{5\pi}{4}$

$\dfrac{\pi}{4}$

$y = \cos x$

Figure 5.5.8

SOLUTION From $x = \pi/4$ to $x = 5\pi/4$ the upper boundary is the curve $y = \sin x$ and the lower boundary is the curve $y = \cos x$. Therefore

$$\text{area of } \Omega = \int_{\pi/4}^{5\pi/4} [\sin x - \cos x] dx$$

$$= \left[-\cos x - \sin x\right]_{\pi/4}^{5\pi/4} = 2\sqrt{2}. \quad \square$$

Example 5 Find the area between

$$y = 4x \qquad \text{and} \qquad y = x^3$$

from $x = -2$ to $x = 2$.

SOLUTION A rough sketch of the region appears in Figure 5.5.9. The drawing is not to scale. What matters to us is that $y = x^3$ is the upper boundary from $x = -2$ to $x = 0$, but it is the lower boundary from $x = 0$ to $x = 2$. Therefore

$$\text{area} = \int_{-2}^{0} [x^3 - 4x]\,dx + \int_{0}^{2} [4x - x^3]\,dx$$

$$= \left[\tfrac{1}{4}x^4 - 2x^2\right]_{-2}^{0} + \left[2x^2 - \tfrac{1}{4}x^4\right]_{0}^{2}$$

$$= [0 - (-4)] + [4 - 0] = 8. \quad \square$$

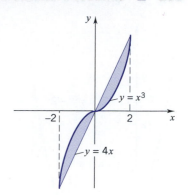

Figure 5.5.9

Example 6 Use integrals to represent the area of the region $\Omega = \Omega_1 \cup \Omega_2$ shaded in Figure 5.5.10.

SOLUTION From $x = a$ to $x = b$, the curve $y = f(x)$ is above the x-axis. Therefore

$$\text{area of } \Omega_1 = \int_{a}^{b} f(x)\,dx.$$

From $x = b$ to $x = c$, the curve $y = f(x)$ is below the x-axis. The upper boundary for Ω_2 is the curve $y = 0$ (the x-axis) and the lower boundary is the curve $y = f(x)$. Thus

$$\text{area of } \Omega_2 = \int_{b}^{c} [0 - f(x)]\,dx = -\int_{b}^{c} f(x)\,dx.$$

The area of Ω is the sum of these two areas:

$$\text{area of } \Omega = \int_{a}^{b} f(x)\,dx - \int_{b}^{c} f(x)\,dx. \quad \square$$

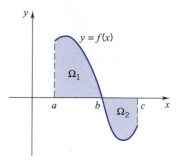

Figure 5.5.10

Figure 5.5.11 shows the graph of a function that crosses the x-axis repeatedly. The area between the graph of f and the x-axis from $x = a$ to $x = e$ is the sum

$$\text{area of } \Omega_1 + \text{area of } \Omega_2 + \text{area of } \Omega_3 + \text{area of } \Omega_4.$$

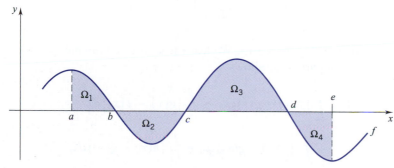

Figure 5.5.11

By the reasoning applied in Example 6, this area is

$$\int_{a}^{b} f(x)\,dx - \int_{b}^{c} f(x)\,dx + \int_{c}^{d} f(x)\,dx - \int_{d}^{e} f(x)\,dx.$$

What is the geometric significance of

$$\int_{a}^{e} f(x)\,dx?$$

Answer: Since

$$\int_{a}^{e} f(x)\,dx = \int_{a}^{b} f(x)\,dx + \int_{b}^{c} f(x)\,dx + \int_{c}^{d} f(x)\,dx + \int_{d}^{e} f(x)\,dx,$$

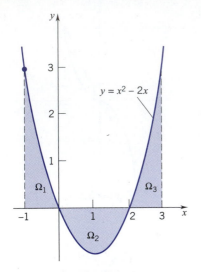

Figure 5.5.12

we have

$$\int_a^e f(x)\,dx = \text{area of } \Omega_1 - \text{area of } \Omega_2 + \text{area of } \Omega_3 - \text{area of } \Omega_4$$

$$= \text{area of } (\Omega_1 \cup \Omega_3) - \text{area of } (\Omega_2 \cup \Omega_4).$$

For a function that changes sign, the region between the graph and the x-axis has two parts: the part above the x-axis and the part below the x-axis. *The integral gives the area of the part above the x-axis minus the area of the part below the x-axis.*

Example 7 Evaluate $\int_{-1}^3 (x^2 - 2x)\,dx$ and interpret the result in terms of areas. Then find the area between the graph of $f(x) = x^2 - 2x$ and the x-axis from $x = -1$ to $x = 3$.

SOLUTION The graph of $f(x) = x^2 - 2x$ is shown in Figure 5.5.12. Routine calculation gives

$$\int_{-1}^3 (x^2 - 2x)\,dx = \left[\frac{1}{3}x^3 - x^2\right]_{-1}^3 = \frac{4}{3}.$$

This integral represents the area of $(\Omega_1 \cup \Omega_3)$ minus the area of Ω_2.

The area between the graph of f and the x-axis from $x = -1$ to $x = 3$ is the sum

$$A = \text{area of } \Omega_1 + \text{area of } \Omega_2 + \text{area of } \Omega_3$$

$$= \int_{-1}^0 (x^2 - 2x)\,dx + \left[-\int_0^2 (x^2 - 2x)\,dx\right] + \int_2^3 (x^2 - 2x)\,dx$$

$$= \int_{-1}^0 (x^2 - 2x)\,dx + \int_0^2 (2x - x^2)\,dx + \int_2^3 (x^2 - 2x)\,dx$$

$$= \left[\frac{1}{3}x^3 - x^2\right]_{-1}^0 + \left[x^2 - \frac{1}{3}x^3\right]_0^2 + \left[\frac{1}{3}x^3 - x^2\right]_2^3 = \frac{4}{3} + \frac{4}{3} + \frac{4}{3} = 4. \quad \square$$

We come now to Figure 5.5.13. We leave it to you to convince yourself that the area A of the shaded part is as follows:

$$A = \int_a^b [f(x) - g(x)]\,dx + \int_b^c [g(x) - f(x)]\,dx$$

$$+ \int_c^d [f(x) - g(x)]\,dx + \int_d^e [h(x) - g(x)]\,dx. \quad \square$$

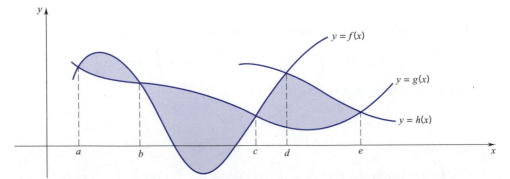

Figure 5.5.13

EXERCISES 5.5

Exercises 1–10. Find the area between the graph of f and the x-axis.

1. $f(x) = 2 + x^3$, $\quad x \in [0, 1]$.
2. $f(x) = (x + 2)^{-2}$, $\quad x \in [0, 2]$.
3. $f(x) = \sqrt{x + 1}$, $\quad x \in [3, 8]$.
4. $f(x) = x^2(3 + x)$, $\quad x \in [0, 8]$.
5. $f(x) = (2x^2 + 1)^2$, $\quad x \in [0, 1]$.
6. $f(x) = \frac{1}{2}(x + 1)^{-1/2}$, $\quad x \in [0, 8]$.
7. $f(x) = x^2 - 4$, $\quad x \in [1, 2]$.
8. $f(x) = \cos x$, $\quad x \in \left[\frac{1}{6}\pi, \frac{1}{3}\pi\right]$.
9. $f(x) = \sin x$, $\quad x \in \left[\frac{1}{3}\pi, \frac{1}{2}\pi\right]$.
10. $f(x) = x^3 + 1$, $\quad x \in [-2, -1]$.

Exercises 11–26. Sketch the region bounded by the curves and find its area.

11. $y = \sqrt{x}$, $\quad y = x^2$.
12. $y = 6x - x^2$, $\quad y = 2x$.
13. $y = 5 - x^2$, $\quad y = 3 - x$.
14. $y = 8$, $\quad y = x^2 + 2x$.
15. $y = 8 - x^2$, $\quad y = x^2$.
16. $y = \sqrt{x}$, $\quad y = \frac{1}{4}x$.
17. $x^3 - 10y^2 = 0$, $\quad x - y = 0$.
18. $y^2 - 27x = 0$, $\quad x + y = 0$.
19. $x - y^2 + 3 = 0$, $\quad x - 2y = 0$.
20. $y^2 = 2x$, $x - y = 4$.
21. $y = x$, $\quad y = 2x$, $\quad y = 4$.
22. $y = x^2$, $\quad y = -\sqrt{x}$, $\quad x = 4$.
23. $y = \cos x$, $\quad y = 4x^2 - \pi^2$.
24. $y = \sin x$, $\quad y = \pi x - x^2$.
25. $y = x$, $\quad y = \sin x$, $\quad x = \pi/2$.
26. $y = x + 1$, $\quad y = \cos x$, $\quad x = \pi$.

27. The graph of $f(x) = x^2 - x - 6$ is shown in the accompanying figure.

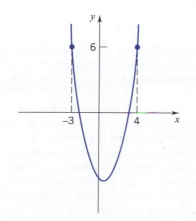

(a) Evaluate $\int_{-3}^{4} f(x)\, dx$ and interpret the result in terms of areas.
(b) Find the area between the graph of f and the x-axis from $x = -3$ to $x = 4$.
(c) Find the area between the graph of f and the x-axis from $x = -2$ to $x = 3$.

28. The graph of $f(x) = 2 \sin x$, $x \in [-\pi/2, 3\pi/4]$ is shown in the accompanying figure.

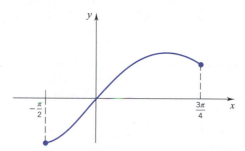

(a) Evaluate $\int_{-\pi/2}^{3\pi/4} f(x)\, dx$ and interpret the result in tems of areas.
(b) Find the area between the graph of f and the x-axis from $x = -\pi/2$ to $x = 3\pi/4$.
(c) Find the area between the graph of f and the x-axis from $x = -\pi/2$ to $x = 0$.

29. Set $f(x) = x^3 - x$.

(a) Evaluate $\int_{-2}^{2} f(x)\, dx$.
(b) Sketch the graph of f and find the area between the graph and the x-axis from $x = -2$ to $x = 2$.

30. Set $f(x) = \cos x + \sin x$.

(a) Evaluate $\int_{-\pi}^{\pi} f(x)\, dx$.
(b) Sketch the graph of f and find the area between the graph and the x-axis from $x = -\pi$ to $x = \pi$.

▶ 31. Set $f(x) = x^3 - 4x + 2$.

(a) Evaluate $\int_{-2}^{3} f(x)\, dx$.
(b) Use a graphing utility to graph f and estimate the area between the graph and the x-axis from $x = -2$ to $x = 3$. Use two decimal place accuracy in your approximations.
(c) Are your answers to parts (a) and (b) different? If so, explain why.

▶ 32. Set $f(x) = 3x^2 - 2 \cos x$.

(a) Evaluate $\int_{-\pi/2}^{\pi/2} f(x)\, dx$.
(b) Use a graphing utility to graph f and estimate the area between the graph and the x-axis from $x = -\pi/2$ to $x = \pi/2$. Use two decimal place accuracy in your approximations.
(c) Are your answers to parts (a) and (b) different? If so, explain why.

33. Set $f(x) = \begin{cases} x^2 + 1, & 0 \le x \le 1 \\ 3 - x, & 1 < x \le 3. \end{cases}$

Sketch the graph of f and find the area between the graph and the x-axis.

34. Set $f(x) = \begin{cases} 3\sqrt{x}, & 0 \le x \le 1 \\ 4 - x^2, & 1 < x \le 2. \end{cases}$

Sketch the graph of f and find the area between the graph and the x-axis.

35. Sketch the region bounded by the x-axis and the curves $y = \sin x$ and $y = \cos x$ with $x \in [0, \pi/2]$, and find its area.

36. Sketch the region bounded by $y = 1$ and $y = 1 + \cos x$ with $x \in [0, \pi]$, and find its area.

37. Use a graphing utility to sketch the region bounded by the curves $y = x^3 + 2x$ and $y = 3x + 1$ with $x \in [0, 2]$, and estimate its area. Use two decimal place accuracy in your approximations.

38. Use a graphing utility to sketch the region bounded by the curves $y = x^4 - 2x^2$ and $y = 4 - x^2$ with $x \in [-2, 2]$, and estimate its area. Use two decimal place accuracy in your approximations.

39. A sketch of the curves $y = x^4 - x^2 - 12$ and $y = h$ is shown in the figure.

   (a) Use a graphing utility to get an accurate drawing of $y = x^4 - x^2 - 12$.

   (b) Find the area of region II.

   (c) Estimate h so that region I and region II have equal areas.

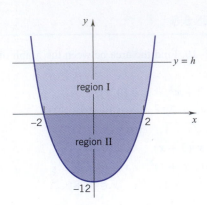

40. A sketch of the curves $y = x^3 - x^4$ and $y = h$ is shown in the figure. Estimate h so that region I and region II have equal areas.

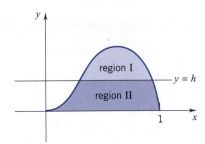

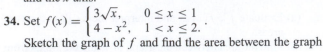

■ PROJECT 5.5 Integrability; Integrating Discontinuous Functions

Integrability

We begin with a function f defined on a closed interval $[a, b]$. Whether or not f is continuous on $[a, b]$, we can form arbitrary Riemann sums

$$S^*(P) = f(x_1^*)\Delta x_1 + f(x_1^*)\Delta x_1 + \cdots + f(x_n^*)\Delta x_n.$$

If these Riemann sums tend to a finite limit I in the sense already explained (5.2.7), then we say that f is (*Riemann*) *integrable* on $[a, b]$ and set

$$\int_a^b f(x)\, dx = I.$$

   A complete explanation of which functions are integrable and which functions are not integrable is beyond the scope of this text. Roughly speaking, a function is integrable iff it is not "too" discontinuous. Thus, for example, the Dirichlet function

$$f(x) = \begin{cases} 1, & x \text{ rational} \\ 0, & x \text{ irrational} \end{cases}$$

(which, as you know, is everywhere discontinuous) is not integrable on $[a, b]$: choosing the x_i^* to be rational, we have

$f(x_i^*) = 1$ for all i, and therefore

$$S^*(P) = (1)\Delta x_1 + (1)\Delta x_2 + \cdots + (1)\Delta x_n = b - a;$$

choosing the x_i^* to be irrational, we have $f(x_i^*) = 0$ for all i, and therefore

$$S^*(P) = (0)\Delta x_1 + (0)\Delta x_2 + \cdots + (0)\Delta x_n = 0.$$

Clearly the $S^*(P)$ do not tend to a limit as $\|P\| = \max \Delta x_i$ tends to 0. On the other hand, it can be shown that if f is bounded and has at most an enumerable set of discontinuities

$$x_1, x_2, \ldots, x_n, \ldots,$$

then f is integrable on $[a, b]$. In particular, bounded functions with only a finite number of discontinuities are integrable. These are the only functions we will be working with.

Remark   Were this a treatise in advanced mathematics, we would have to elaborate on the notion of integrability. But this is not a treatise in advanced mathematics; it is a text in calculus, and for calculus the integration of discontinuous functions is not very important. What is important to us in calculus is the link between integration and differentiation described in Theorem 5.3.5 and Theorem 5.4.2. This link is broken at the points where the integrand is discontinuous. ❑

Integrating Discontinuous Functions

Figure A shows three rectangles: the closed rectangle R_1, the rectangle R_2 obtained from R_1 by removing the rightmost side, and the rectangle R_3 obtained from R_1 by removing both sides.

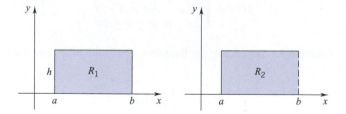

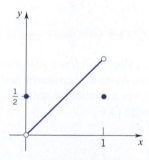

Figure A

The area of R_1 is $(b - a)h$. What is the area of R_2? We obtain R_2 by removing a line segment, which is a set of area 0. It follows that the area of R_2 is also $(b - a)h$. Thus R_3 also has area $(b - a)h$.

It is only a small step from these considerations to the following observation: *If a region Ω has area A, then every region which differs from Ω by only a finite number of line segments also has area A.*

In what follows we will begin by integrating over a closed interval $[a, b]$ functions g that differ from a continuous function f at only a finite number of points. By restricting ourselves to nonnegative functions, we can interpret the integral as the area under the graph and conclude that

$$\int_a^b g(x)\,dx = \int_a^b f(x)\,dx.$$

Figure B shows the graph of

$$g(x) = \begin{cases} \frac{1}{2}, & x = 0 \\ x, & 0 < x < 1 \\ \frac{1}{2}, & x = 1. \end{cases}$$

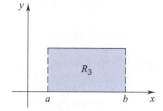

Figure B

On $[0, 1]$ g differs from the identity function $f(x) = x$ only at $x = 0$ and $x = 1$. Therefore

$$\int_0^1 g(x)\,dx = \int_0^1 x\,dx = \left[\tfrac{1}{2}x^2\right]_0^1 = \tfrac{1}{2}.$$

Figure C shows the graph of

$$g(x) = \begin{cases} 0, & x = -1 \\ x^2, & -1 < x < 0 \\ 1, & x = 0 \\ x^2, & 0 < x < 1 \\ 0, & x = 1. \end{cases}$$

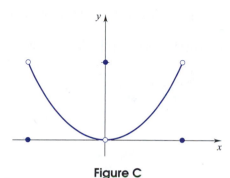

Figure C

On $[-1, 1]$ g differs from the squaring function $f(x) = x^2$ only at $x = -1, x = 0, x = 1$. Therefore

$$\int_{-1}^1 g(x)\,dx = \int_{-1}^1 x^2\,dx = \left[\tfrac{1}{3}x^3\right]_{-1}^1 = \tfrac{2}{3}.$$

We come now to a slightly different situation. Figure D shows the graph of

$$g(x) = \begin{cases} x, & x \in [0, 1) \\ x - 1, & x \in [1, 2) \\ x - 2, & x \in [2, 3) \\ x - 3, & x \in [3, 4) \\ 0, & x = 4. \end{cases}$$

This functions has jump discontinuities at $x = 1, x = 2, x = 3, x = 4$. We can integrate g on $[0, 4]$ by integrating from integer to integer and adding up the results:

$$\int_0^4 g(x)\,dx = \int_0^1 x\,dx + \int_1^2 (x - 1)\,dx + \int_2^3 (x - 2)\,dx$$

$$+ \int_3^4 (x - 3)\,dx.$$

Since the area under each line segment is $\tfrac{1}{2}$, the integral of g adds up to $4(\tfrac{1}{2}) = 2$. ❑

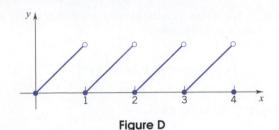

Figure D

Problem 1. (*The greatest integer functions*) The expression $[x]$ is used to denote the greatest integer less than x.

a. Sketch the graph of the function $g(x) = [x]$ and integrate g from $x = 0$ to $x = 5$.

The expression $[x]$ is used to denote the greatest integer less than or equal to x.

b. Sketch the graph of $g(x) = [x]$ and integrate g from $x = 0$ to $x = 5$.

c. Sketch the graph of $h(x) = [x] - [x]$ and integrate h from $x = 0$ to $x = 5$.

Problem 2. Graph the function g and evaluate the integral of g over the interval on which it is defined.

a.
$$g(x) = \begin{cases} 2 - x, & 0 \le x \le 1 \\ 2 + x, & 1 < x \le 2. \end{cases}$$

b.
$$g(x) = \begin{cases} x^2, & 0 \le x < 2 \\ x, & 2 \le x \le 5. \end{cases}$$

c.
$$g(x) = \begin{cases} \cos x, & 0 \le x < \frac{1}{2}\pi \\ \sin x, & \frac{1}{2}\pi \le x < \pi \\ \frac{1}{2}, & \pi \le x \le 2\pi. \end{cases}$$

Problem 3. For each of the functions g in Problem 2, form the integral

$$G(x) = \int_0^x g(t)\, dt.$$

a. Show that for the first function, G is not differentiable at $x = 1$.

b. Show that for the second function, G is not differentiable at $x = 2$.

c. Show that for the third function, G is not differentiable at $x = \frac{1}{2}\pi$ and not differentiable at $x = \pi$.

HINT: In each case show that at the selected value of x

$$\lim_{h \to 0^-} \frac{G(x+h) - G(x)}{h} \ne \lim_{h \to 0^+} \frac{G(x+h) - G(x)}{h}.$$

■ 5.6 INDEFINITE INTEGRALS

We begin with a continuous function f. If F is an antiderivative for f on $[a, b]$, then

(1)
$$\int_a^b f(x)\, dx = \Big[F(x)\Big]_a^b.$$

If C is a constant, then

$$\Big[F(x) + C\Big]_a^b = [F(b) + C] - [F(a) + C] = F(b) - F(a) = \Big[F(x)\Big]_a^b.$$

Thus we can replace (1) by writing

$$\int_a^b f(x)\, dx = \Big[F(x) + C\Big]_a^b.$$

If we have no particular interest in the interval $[a, b]$ but wish instead to emphasize that F is an antiderivative for f, which on open intervals simply means that $F' = f$, then we omit the a and the b and simply write

$$\int f(x)\, dx = F(x) + C.$$

Antiderivatives expressed in this manner are called *indefinite integrals*. The constant C is called the *constant of integration*; it is an *arbitrary* constant and we can assign to it any value we choose. Each value of C gives a particular antiderivative, and each antiderivative is obtained from a particular value of C.

For rational r different from -1 we have

$$\int x^r\, dx = \frac{x^{r+1}}{r+1} + C.$$

In particular,

$$\int x^2 dx = \tfrac{1}{3}x^3 + C \quad \text{and} \quad \int \sqrt{x}\, dx = \tfrac{2}{3}x^{3/2} + C.$$

Table 5.6.1 gives the antiderivatives of Table 5.4.1 expressed as indefinite integrals.

■ **Table 5.6.1**

$\int \sin x \, dx = -\cos x + C$	$\int \cos x \, dx = \sin x + C$
$\int \sec^2 x \, dx = \tan x + C$	$\int \csc^2 x \, dx = -\cot x + C$
$\int \sec x \tan x \, dx = \sec x + C$	$\int \csc x \cot x \, dx = -\csc x + C$

The calculation of indefinite integrals is a linear process. Unless α and β are both zero,

(5.6.1)
$$\int [\alpha f(x) + \beta g(x)] \, dx = \alpha \int f(x) \, dx + \beta \int g(x) \, dx.^{\dagger}$$

The equation holds in the following sense: if F and G are antiderivatives for f and g, then

$$\int [\alpha f(x) + \beta g(x)] \, dx = \alpha F(x) + \beta G(x) + C$$

and

$$\alpha \int f(x) \, dx + \beta \int g(x) \, dx = \alpha [F(x) + C_1] + \beta [G(x) + C_2]$$
$$= \alpha F(x) + \beta G(x) + \alpha C_1 + \beta C_2.$$

With α and β not both zero, $\alpha C_1 + \beta C_2$ is an arbitrary constant that we can denote by C thereby confirming (5.6.1). ❏

Example 1 Calculate $\int [5x^{3/2} - 2\csc^2 x] \, dx$.

SOLUTION

$$\int [5x^{3/2} - 2\csc^2 x] \, dx = 5 \int x^{3/2} dx - 2 \int \csc^2 x \, dx$$
$$= 5 \left(\tfrac{2}{5}\right) x^{5/2} + C_1 - 2(-\cot x) + C_2$$
$$= 2x^{5/2} + 2\cot x + C.$$

writing C for $C_1 + C_2$ ⬏

Example 2 Find f given that $f'(x) = x^3 + 2$ and $f(0) = 1$.

SOLUTION Since f' is the derivative of f, f is an antiderivative for f'. Thus

$$f(x) = \int (x^3 + 2) \, dx = \tfrac{1}{4}x^4 + 2x + C$$

---

† Explain how (5.6.1) fails if α and β are both zero.

for some value of the constant C. To evaluate C, we use the fact that $f(0) = 1$. Since

$$f(0) = 1 \quad \text{and} \quad f(0) = \tfrac{1}{4}(0)^4 + 2(0) + C = C,$$

we see that $C = 1$. Therefore

$$f(x) = \tfrac{1}{4}x^4 + 2x + 1. \quad \square$$

Example 3   Find f given that

$$f''(x) = 6x - 2, \qquad f'(1) = -5, \qquad \text{and } f(1) = 3.$$

SOLUTION   First we get f' by integrating f'':

$$f'(x) = \int (6x - 2)\, dx = 3x^2 - 2x + C.$$

Since

$$f'(1) = -5 \quad \text{and} \quad f'(1) = 3(1)^2 - 2(1) + C = 1 + C,$$

we have

$$-5 = 1 + C \qquad \text{and thus} \qquad C = -6.$$

Therefore

$$f'(x) = 3x^2 - 2x - 6.$$

Now we get f by integrating f':

$$f(x) = \int (3x^2 - 2x - 6)\, dx = x^3 - x^2 - 6x + K.$$

(We are writing the constant of integration as K because we used C before and it would be confusing to assign to C two different values in the same problem.) Since

$$f(1) = 3 \qquad \text{and} \qquad f(1) = (1)^3 - (1)^2 - 6(1) + K = -6 + K,$$

we have

$$3 = -6 + K \qquad \text{and thus} \qquad K = 9.$$

Therefore

$$f(x) = x^3 - x^2 - 6x + 9. \quad \square$$

Application to Motion

Example 4   An object moves along a coordinate line with velocity

$$v(t) = 2 - 3t + t^2 \quad \text{units per second.}$$

Its initial position (position at time $t = 0$) is 2 units to the right of the origin. Find the position of the object 4 seconds later.

SOLUTION   Let $x(t)$ be the position (coordinate) of the object at time t. We are given that $x(0) = 2$. Since $x'(t) = v(t)$,

$$x(t) = \int v(t)\, dt = \int (2 - 3t + t^2)\, dt = 2t - \tfrac{3}{2}t^2 + \tfrac{1}{3}t^3 + C.$$

Since $x(0) = 2$ and $x(0) = 2(0) - \frac{3}{2}(0)^2 + \frac{1}{3}(0)^3 + C = C$, we have $C = 2$ and

$$x(t) = 2t - \tfrac{3}{2}t^2 + \tfrac{1}{3}t^3 + 2.$$

The position of the object at time $t = 4$ is the value of this function at $t = 4$:

$$x(4) = 2(4) - \tfrac{3}{2}(4)^2 + \tfrac{1}{3}(4)^3 + 2 = 7\tfrac{1}{3}.$$

At the end of 4 seconds the object is $7\frac{1}{3}$ units to the right of the origin.
The motion of the object is represented schematically in Figure 5.6.1. ❏

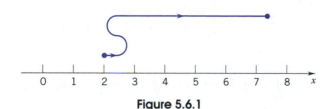

Figure 5.6.1

Recall that the speed v is the absolute value of velocity (Section 4.9):

$$\text{speed at time } t = v(t) = |v(t)|,$$

and the integral of the speed function gives the distance traveled (Section 5.1):

(5.6.2)
$$\int_a^b |v(t)| \, dt = \text{distance traveled from time } t = a \text{ to time } t = b.$$

Example 5   An object moves along the x-axis with acceleration $a(t) = 2t - 2$ units per second per second. Its initial position (position at time $t = 0$) is 5 units to the right of the origin. One second later the object is moving left at the rate of 4 units per second.

(a) Find the position of the object at time $t = 4$ seconds.
(b) How far does the object travel during these 4 seconds?

SOLUTION   (a) Let $x(t)$ and $v(t)$ denote the position and velocity of the object at time t. We are given that $x(0) = 5$ and $v(1) = -4$. Since $v'(t) = a(t)$,

$$v(t) = \int a(t) \, dt = \int (2t - 2) \, dt = t^2 - 2t + C.$$

Since

$$v(1) = -4 \quad \text{and} \quad v(1) = (1)^2 - 2(1) + C = -1 + C,$$

we have $C = -3$ and therefore

$$v(t) = t^2 - 2t - 3.$$

Since $x'(t) = v(t)$,

$$x(t) = \int v(t) \, dt = \int (t^2 - 2t - 3) \, dt = \tfrac{1}{3}t^3 - t^2 - 3t + K.$$

Since

$$x(0) = 5 \quad \text{and} \quad x(0) = \tfrac{1}{3}(0)^3 - (0)^2 - 3(0) + K = K,$$

we have $K = 5$. Therefore

$$x(t) = \tfrac{1}{3}t^3 - t^2 - 3t + 5.$$

As you can check, $x(4) = -\tfrac{5}{3}$. At time $t = 4$ the object is $\tfrac{5}{3}$ units to the left of the origin.

(b) The distance traveled from time $t = 0$ to $t = 4$ is given by the integral

$$s = \int_0^4 |v(t)|\, dt = \int_0^4 |t^2 - 2t - 3|\, dt.$$

To evaluate this integral, we first remove the absolute value sign. As you can verify,

$$|t^2 - 2t - 3| = \begin{cases} -(t^2 - 2t - 3), & 0 \le t < 3 \\ t^2 - 2t - 3, & 3 \le t \le 4 \end{cases}$$

Thus

$$s = \int_0^3 (3 + 2t - t^2)\, dt + \int_3^4 (t^2 - 2t - 3)\, dt$$

$$= \left[3t + t^2 - \frac{1}{3}t^3 \right]_0^3 + \left[\frac{1}{3}t^3 - t^2 - 3t \right]_3^4 = \frac{34}{3}.$$

During the 4 seconds the object travels a distance of $\tfrac{34}{3}$ units.

The motion of the object is represented schematically in Figure 5.6.2. ❏

Figure 5.6.2

QUESTION   The object in Example 5 leaves $x = 5$ at time $t = 0$ and arrives at $x = -\tfrac{5}{3}$ at time $t = 4$. The separation between $x = 5$ and $x = -\tfrac{5}{3}$ is only $|5 - (-\tfrac{5}{3})| = \tfrac{20}{3}$. How is it possible that the object travels a distance of $\tfrac{34}{3}$ units?

ANSWER   The object does not maintain a fixed direction. It changes direction at time $t = 3$. You can see this by noting that the velocity function

$$v(t) = t^2 - 2t - 3 = (t - 3)(t + 1)$$

changes signs at $t = 3$.

Example 6   Find the equation of motion for an object that moves along a straight line with constant acceleration a from an initial position x_0 with initial velocity v_0.

SOLUTION   Call the line of the motion the x-axis. Here $a(t) = a$ at all times t. To find the velocity we integrate the acceleration:

$$v(t) = \int a\, dt = at + C.$$

The constant C is the initial velocity v_0:

$$v_0 = v(0) = a \cdot 0 + C = C.$$

We see therefore that

$$v(t) = at + v_0.$$

To find the position function, we integrate the velocity:

$$x(t) = \int v(t)\,dt = \int (at + v_0)\,dt = \tfrac{1}{2}at^2 + v_0 t + K.$$

The constant K is the initial position x_0:

$$x_0 = x(0) = \tfrac{1}{2}a \cdot 0^2 + v_0 \cdot 0 + K = K.$$

The equation of motion can be written

(5.6.3) $x(t) = \tfrac{1}{2}at^2 + v_0 t + x_0.^\dagger$ ❑

---

†In the case of a free-falling body, $a = -g$ and we have Galileo's equation for free fall. See (4.9.5). There we denoted the initial position by y_0 (instead of by x_0) because there the motion was viewed as taking place along the y-axis.

EXERCISES 5.6

Exercises 1–18. Calculate.

1. $\displaystyle\int \frac{dx}{x^4}.$

2. $\displaystyle\int (x-1)^2\,dx.$

3. $\displaystyle\int (ax+b)\,dx.$

4. $\displaystyle\int (ax^2+b)\,dx.$

5. $\displaystyle\int \frac{dx}{\sqrt{1+x}}.$

6. $\displaystyle\int \left(\frac{x^3+1}{x^5}\right)dx.$

7. $\displaystyle\int \left(\frac{x^3-1}{x^2}\right)dx.$

8. $\displaystyle\int \left(\sqrt{x} - \frac{1}{\sqrt{x}}\right)dx.$

9. $\displaystyle\int (t-a)(t-b)\,dt.$

10. $\displaystyle\int (t^2-a)(t^2-b)\,dt.$

11. $\displaystyle\int \frac{(t^2-a)(t^2-b)}{\sqrt{t}}\,dt.$

12. $\displaystyle\int (2-\sqrt{x})(2+\sqrt{x})\,dx.$

13. $\displaystyle\int g(x)g'(x)\,dx.$

14. $\displaystyle\int \sin x \cos x\,dx.$

15. $\displaystyle\int \tan x \sec^2 x\,dx.$

16. $\displaystyle\int \frac{g'(x)}{[g(x)]^2}\,dx.$

17. $\displaystyle\int \frac{4}{(4x+1)^2}\,dx.$

18. $\displaystyle\int \frac{3x^2}{(x^3+1)^2}\,dx.$

Exercises 19–32. Find f from the information given.

19. $f'(x) = 2x - 1, \quad f(3) = 4.$

20. $f'(x) = 3 - 4x, \quad f(1) = 6.$

21. $f'(x) = ax + b, \quad f(2) = 0.$

22. $f'(x) = ax^2 + bx + c, \quad f(0) = 0.$

23. $f'(x) = \sin x, \quad f(0) = 2.$

24. $f'(x) = \cos x, \quad f(\pi) = 3.$

25. $f''(x) = 6x - 2, \quad f'(0) = 1, \quad f(0) = 2.$

26. $f''(x) = -12x^2, \quad f(0) = 1, \quad f(0) = 2.$

27. $f''(x) = x^2 - x, \quad f'(1) = 0, \quad f(1) = 2.$

28. $f''(x) = 1 - x, \quad f'(2) = 1, \quad f(2) = 0.$

29. $f''(x) = \cos x, \quad f'(0) = 1, \quad f(0) = 2.$

30. $f''(x) = \sin x, \quad f'(0) = -2, \quad f(0) = 1.$

31. $f''(x) = 2x - 3, \quad f(2) = -1, \quad f(0) = 3.$

32. $f''(x) = 5 - 4x, \quad f(1) = 1, \quad f(0) = -2.$

33. Compare $\dfrac{d}{dx}\left[\displaystyle\int f(x)\,dx\right]$ to $\displaystyle\int \dfrac{d}{dx}[f(x)]\,dx.$

34. Calculate

$$\int [f(x)g''(x) - g(x)f''(x)]\,dx.$$

35. An object moves along a coordinate line with velocity $v(t) = 6t^2 - 6$ units per second. Its initial position (position at time $t = 0$) is 2 units to the left of the origin. (a) Find the position of the object 3 seconds later. (b) Find the total distance traveled by the object during those 3 seconds.

36. An object moves along a coordinate line with acceleration $a(t) = (t+2)^3$ units per second per second. (a) Find the velocity function given that the initial velocity is 3 units per second. (b) Find the position function given that the initial velocity is 3 units per second and the initial position is the origin.

37. An object moves along a coordinate line with acceleration $a(t) = (t+1)^{-1/2}$ units per second per second. (a) Find the velocity function given that the initial velocity is 1 unit per second. (b) Find the position function given that the initial velocity is 1 unit per second and the initial position is the origin.

38. An object moves along a coordinate line with velocity $v(t) = t(1-t)$ units per second. Its initial position is 2 units to the left of the origin. (a) Find the position of the object 10 seconds later. (b) Find the total distance traveled by the object during those 10 seconds.

39. A car traveling at 60 mph decelerates at 20 feet per second per second. (a) How long does it take for the car to come to a complete stop? (b) What distance is required to bring the car to a complete stop?

40. An object moves along the x-axis with constant acceleration. Express the position $x(t)$ in terms of the initial position x_0,

the initial velocity v_0, the velocity $v(t)$, and the elapsed time t.

41. An object moves along the x-axis with constant acceleration a. Verify that

$$[v(t)]^2 = v_0^2 + 2a[x(t) - x_0].$$

42. A bobsled moving at 60 mph decelerates at a constant rate to 40 mph over a distance of 264 feet and continues to decelerate at that same rate until it comes to a full stop. (a) What is the acceleration of the sled in feet per second per second? (b) How long does it take to reduce the speed to 40 mph? (c) How long does it take to bring the sled to a complete stop from 60 mph? (d) Over what distance does the sled come to a complete stop from 60 mph?

43. In the AB-run, minicars start from a standstill at point A, race along a straight track, and come to a full stop at point B one-half mile away. Given that the cars can accelerate uniformly to a maximum speed of 60 mph in 20 seconds and can brake at a maximum rate of 22 feet per second per second, what is the best possible time for the completion of the AB-run?

Exercises 44–46. Find the general law of motion of an object that moves in a straight line with acceleration $a(t)$. Write x_0 for the initial position and v_0 for the initial velocity.

44. $a(t) = \sin t.$ **45.** $a(t) = 2A + 6Bt.$

46. $a(t) = \cos t.$

47. As a particle moves about the plane, its x-coordinate changes at the rate of $t^2 - 5$ units per second and its y-coordinate changes at the rate of $3t$ units per second. If the particle is at the point $(4, 2)$ when $t = 2$ seconds, where is the particle 4 seconds later?

48. As a particle moves about the plane, its x-coordinate changes at the rate of $t - 2$ units per second and its y-coordinate changes at the rate of $\sqrt{t}$ units per second. If the particle is at the point $(3, 1)$ when $t = 4$ seconds, where is the particle 5 second later?

49. A particle moves along the x-axis with velocity $v(t) = At + B$. Determine A and B given that the initial velocity of the particle is 2 units per second and the position of the

particle after 2 seconds of motion is 1 unit to the left of the initial position.

50. A particle moves along the x-axis with velocity $v(t) = At^2 + 1$. Determine A given that $x(1) = x(0)$. Compute the total distance traveled by the particle during the first second.

51. An object moves along a coordinate line with velocity $v(t) = \sin t$ units per second. The object passes through the origin at time $t = \pi/6$ seconds. When is the next time: (a) that the object passes through the origin? (b) that the object passes through the origin moving from left to right?

52. Exercise 51 with $v(t) = \cos t$.

53. An automobile with varying velocity $v(t)$ moves in a fixed direction for 5 minutes and covers a distance of 4 miles. What theorem would you invoke to argue that for at least one instant the speedometer must have read 48 miles per hour?

54. A speeding motorcyclist sees his way blocked by a haywagon some distance s ahead and slams on his brakes. Given that the brakes impart to the motorcycle a constant negative acceleration a and that the haywagon is moving with speed v_1 in the same direction as the motorcycle, show that the motorcyclist can avoid collision only if he is traveling at a speed less than $v_1 + \sqrt{2|a|s}$.

55. Find the velocity $v(t)$ given that $a(t) = 2[v(t)]^2$ and $v_0 \neq 0$.

Exercises 56 and 57. Find and compare

$$\frac{d}{dx}\left(\int f(x)\, dx \right) \quad \text{and} \quad \int \frac{d}{dx}[f(x)]\, dx.$$

56. $f(x) = \dfrac{x^2 - x^3 + x^4}{\sqrt{x}}.$ **57.** $f(x) = \cos x - 2\sin x.$

▶**Exercises 58–61.** Use a CAS to find f from the information given.

58. $f'(x) = \dfrac{\sqrt{x} + 1}{\sqrt{x}}; \quad f(4) = 2.$

59. $f'(x) = \cos x - 2\sin x; \quad f(\pi/2) = 2.$

60. $f''(x) = 3\sin x + 2\cos x; \quad f(0) = 0,\ f'(0) = 0.$

61. $f''(x) = 5 - 3x + x^2; \quad f(0) = -3,\ f'(0) = 4.$

■ 5.7 WORKING BACK FROM THE CHAIN RULE; THE *u*-SUBSTITUTION

To differentiate a composite function, we apply the chain rule. To integrate the outputs of the chain rule, we have to apply the chain rule in reverse. This process requires some ingenuity.

Example 1 Calculate

$$\int (x^2 - 1)^4 x\, dx.$$

SOLUTION   From the chain rule we know that

$$\frac{d}{dx}[(x^2 - 1)^5] = 5(x^2 - 1)^4 2x = 10(x^2 - 1)^4 x.$$

Working back from this, we have

$$\int (x^2 - 1)^4 x \, dx = \tfrac{1}{10} \int 10(x^2 - 1)^4 x \, dx = \tfrac{1}{10}(x^2 - 1)^5 + C.$$

You can check the result by differentiation. ❏

Example 2 Calculate

$$\int \sin^2 x \cos x \, dx.$$

SOLUTION Since

$$\frac{d}{dx}[\sin x] = \cos x,$$

we know from the chain rule that

$$\frac{d}{dx}[\sin^3 x] = 3 \sin^2 x \cos x.$$

Working back from this, we have

$$\int \sin^2 x \cos x \, dx = \tfrac{1}{3} \int 3 \sin^2 x \cos x \, dx = \tfrac{1}{3} \sin^3 x + C.$$

You can check the result by differentiation. ❏

Example 3 Calculate

$$\int 2x^2 \sin (x^3 + 1) \, dx.$$

SOLUTION Since

$$\frac{d}{dx}[\cos x] = -\sin x,$$

we know that

$$\frac{d}{dx}[\cos (x^3 + 1)] = -\sin (x^3 + 1) \, 3x^2.$$

Therefore

$$\int 2x^2 \sin (x^3 + 1) \, dx = -\tfrac{2}{3} \int -\sin (x^3 + 1) \, 3x^2 dx = -\tfrac{2}{3} \cos (x^3 + 1) + C.$$

You can check the result by differentiation. ❏

We carried out these integrations by making informed guesses based on our experience with the chain rule. The underlying principle can be stated as follows:

THEOREM 5.7.1

If f is a continuous function and $F' = f$, then

$$\int f(u(x))u'(x) \, dx = F(u(x)) + C$$

for all functions $u = u(x)$ which have values in the domain of f and continuous derivative u'

PROOF The key here is the chain rule. If f is continuous and $F' = f$, then

$$\int f(u(x))u'(x)\,dx = \int F'(u(x))u'(x)\,dx = \int \frac{d}{dx}[F(u(x))]\,dx = F(u(x)) + C.$$

by the chain rule ⤴

The *u-substitution*, described below, offers a somewhat mechanical way of carrying out such calculations. Set

$$u = u(x), \qquad du = u'(x)\,dx.$$

Then write

$$\int f(u(x))u'(x)\,dx = \int f(u)\,du = F(u) + C = F(u(x)) + C. \quad \square$$

⤴ where $F' = f$

Below we carry out some integrations by u-substitution. In each case the first step is to discern a function $u = u(x)$ which, up to a multiplicative constant, puts our integral in the form

$$\int f(u(x))u'(x)\,dx.$$

Example 4 Calculate

$$\int \frac{1}{(3+5x)^2}\,dx.$$

SOLUTION Set $u = 3 + 5x$, $du = 5\,dx$. Then

$$\frac{1}{(3+5x)^2}\,dx = \frac{1}{u^2}\left(\frac{1}{5}du\right) = \frac{1}{5}u^{-2}du$$

and

$$\int \frac{1}{(3+5x)^2}\,dx = \frac{1}{5}\int u^{-2}\,du = -\frac{1}{5}u^{-1} + C = -\frac{1}{5(3+5x)} + C. \quad \square$$

Example 5 Calculate $\displaystyle\int x^2\sqrt{4+x^3}\,dx$.

SOLUTION Set $u = 4 + x^3$, $du = 3x^2\,dx$. Then

$$x^2\sqrt{4+x^3}\,dx = \underbrace{(4+x^3)^{1/2}}_{u^{1/2}}\underbrace{x^2dx}_{\frac{1}{3}\,du} = \tfrac{1}{3}u^{1/2}du$$

and

$$\int x^2\sqrt{4+x^3}\,dx = \tfrac{1}{3}\int u^{1/2}\,du = \tfrac{2}{9}u^{3/2} + C = \tfrac{2}{9}(4+x^3)^{3/2} + C. \quad \square$$

Example 6 Calculate $\displaystyle\int 2x^3\sec^2(x^4+1)\,dx$.

SOLUTION Set $u = x^4 + 1$, $du = 4x^3\,dx$. Then

$$2x^3\sec^2(x^4+1)\,dx = 2\underbrace{\sec^2(x^4+1)}_{\sec^2 u}\underbrace{x^3dx}_{\frac{1}{4}\,du} = \tfrac{1}{2}\sec^2 u\,du$$

and

$$\int 2x^3\sec^2(x^4+1)\,dx = \tfrac{1}{2}\int \sec^2 u\,du = \tfrac{1}{2}\tan u + C = \tfrac{1}{2}\tan(x^4+1) + C. \quad \square$$

Example 7 Calculate $\int \sec^3 x \tan x \, dx$.

SOLUTION We can write $\sec^3 x \tan x \, dx$ as $\sec^2 x \sec x \tan x \, dx$. Setting

$$u = \sec x, \qquad du = \sec x \tan x \, dx,$$

we have

$$\sec^3 x \tan x \, dx = \underbrace{\sec^2 x}_{u^2} \underbrace{(\sec x \tan x) \, dx}_{du} = u^2 du.$$

Therefore

$$\int \sec^3 x \tan x \, dx = \int u^2 du = \tfrac{1}{3} u^3 + C = \tfrac{1}{3} \sec^3 x + C. \quad \square$$

Remark Every integral that we have calculated by a *u*-substitution can be calculated without it. All that's required is a firm grasp of the chain rule and some capacity for pattern recognition. Suggestion: redo these calculations without using a *u*-substitution. ❑

Example 8 Evaluate $\int_0^2 (x^2 - 1)(x^3 - 3x + 2)^3 \, dx$.

SOLUTION We need to find an antiderivative for the integrand. The indefinite integral

$$\int (x^2 - 1)(x^3 - 3x + 2)^3 dx$$

gives the set of all antiderivatives, and so we will calculate this first. Set

$$u = x^3 - 3x + 2, \qquad du = (3x^2 - 3) \, dx = 3(x^2 - 1) \, dx.$$

Then

$$(x^2 - 1)(x^3 - 3x + 2)^3 dx = \underbrace{(x^3 - 3x + 2)^3}_{u^3} \underbrace{(x^2 - 1) \, dx}_{\frac{1}{3} du} = \tfrac{1}{3} u^3 du.$$

It follows that

$$\int (x^2 - 1)(x^3 - 3x + 2)^3 dx = \tfrac{1}{3} \int u^3 du = \tfrac{1}{12} u^4 + C = \tfrac{1}{12}(x^3 - 3x + 2)^4 + C.$$

To evaluate the definite integral, we need only one antiderivative. We choose the one with $C = 0$. This gives

$$\int_0^2 (x^2 - 1)(x^3 - 3x + 2)^3 dx = \left[\tfrac{1}{12}(x^3 - 3x + 2)^4 \right]_0^2 = 20. \quad \square$$

The Definite Integral $\int_a^b f(u(x))u'(x) \, dx$

We can evaluate a definite integral of the form

$$\int_a^b f(u(x)) \, u'(x) \, dx$$

by first calculating the corresponding indefinite integral as we did in Example 8 or by employing the following formula:

(5.7.2)
$$\int_a^b f(u(x))u'(x)\,dx = \int_{u(a)}^{u(b)} f(u)\,du.$$

This formula is called the *change-of-variables formula*. The formula can be used to evaluate $\int_a^b f(u(x))\,u'(x)\,dx$ provided that u' is continuous on $[a, b]$ and f is continuous on the set of values taken on by u on $[a, b]$. Since u is continuous, this set is an interval that contains a and b.

PROOF Let F be an antiderivative for f. Then $F' = f$ and

$$\int_a^b f(u(x))u'(x)\,dx = \int_a^b F'(u(x))\,u'(x)\,dx$$

$$= \Big[F(u(x))\Big]_a^b = F(u(b)) - F(u(a)) = \int_{u(a)}^{u(b)} f(u)\,du.$$

We redo Example 8, this time using the change-of-variables formula.

Example 9 Evaluate $\int_0^2 (x^2 - 1)(x^3 - 3x + 2)^3\,dx$.

SOLUTION As before, set $u = x^3 - 3x + 2$, $du = 3(x^2 - 1)\,dx$. Then

$$(x^2 - 1)(x^3 - 3x + 2)^3 = \tfrac{1}{3}u^3\,du.$$

At $x = 0$, $u = 2$. At $x = 2$, $u = 4$. Therefore,

$$\int_0^2 (x^2 - 1)(x^3 - 3x + 2)^3\,dx = \tfrac{1}{3}\int_2^4 u^3\,du$$

$$= \Big[\tfrac{1}{12}u^4\Big]_2^4 = \tfrac{1}{12}(4)^4 - \tfrac{1}{12}(2)^4 = 20. \quad ❑$$

Example 10 Evaluate $\int_0^{1/2} \cos^3 \pi x \sin \pi x\,dx$.

SOLUTION Set $u = \cos \pi x$, $du = -\pi \sin \pi x\,dx$. Then

$$\cos^3 \pi x \sin \pi x\,dx = \underbrace{\cos^3 \pi x}_{u^3}\ \underbrace{\sin \pi x\,dx}_{-\frac{1}{\pi}\,du} = -\tfrac{1}{\pi}u^3\,du.$$

At $x = 0$, $u = 1$. At $x = 1/2$, $u = 0$. Therefore

$$\int_0^{1/2} \cos^3 \pi x \sin \pi x\,dx = -\frac{1}{\pi}\int_1^0 u^3\,du = \frac{1}{\pi}\int_0^1 u^3\,du = \frac{1}{\pi}\Big[\frac{1}{4}u^4\Big]_0^1 = \frac{1}{4\pi}. \quad ❑$$

The u-substitution can be applied to every integral with a continuous integrand:

$$\int f(x)\,dx = \int f(u(x))u'(x)\,dx.$$
$$\underset{\text{set } u(x)\,=\,x}{\overset{\uparrow}{}}$$

Of course there is no point to this. A u-substitution should be made only if it facilitates the integration. In the next two examples we have to use a little imagination to find a useful substitution.

Example 11 Calculate $\displaystyle\int x(x-3)^5 \, dx$.

SOLUTION Set $u = x - 3$. Then $du = dx$ and $x = u + 3$.
Now

$$x(x-3)^5 dx = (u+3)u^5 du = (u^6 + 3u^5) \, du$$

and

$$\int x(x-3)^5 dx = \int (u^6 + 3u^5) \, du$$

$$= \tfrac{1}{7}u^7 + \tfrac{1}{2}u^6 + C = \tfrac{1}{7}(x-3)^7 + \tfrac{1}{2}(x-3)^6 + C. \quad \square$$

Example 12 Evaluate $\displaystyle\int_0^{\sqrt{3}} x^5 \sqrt{x^2 + 1} \, dx$.

SOLUTION Set $u = x^2 + 1$. Then $du = 2x \, dx$ and $x^2 = u - 1$.
Now

$$x^5 \sqrt{x^2 + 1} \, dx = \underbrace{x^4}_{(u-1)^2} \underbrace{\sqrt{x^2 + 1}}_{\sqrt{u}} \underbrace{x \, dx}_{\frac{1}{2} du} = \tfrac{1}{2}(u-1)^2 \sqrt{u} \, du.$$

At $x = 0, u = 1$. At $x = \sqrt{3}, u = 4$. Thus

$$\int_0^{\sqrt{3}} x^5 \sqrt{x^2 + 1} \, dx = \tfrac{1}{2} \int_1^4 (u-1)^2 \sqrt{u} \, du$$

$$= \tfrac{1}{2} \int_1^4 (u^{5/2} - 2u^{3/2} + u^{1/2}) du$$

$$= \tfrac{1}{2} \left[\tfrac{2}{7}u^{7/2} - \tfrac{4}{5}u^{5/2} + \tfrac{2}{3}u^{3/2} \right]_1^4$$

$$= \left[u^{3/2}(\tfrac{1}{7}u^2 - \tfrac{2}{5}u + \tfrac{1}{3}) \right]_1^4 = \tfrac{848}{105}. \quad \square$$

EXERCISES 5.7

Exercises 1–20. Calculate.

1. $\displaystyle\int \frac{dx}{(2-3x)^2}.$

2. $\displaystyle\int \frac{dx}{\sqrt{2x+1}}.$

3. $\displaystyle\int \sqrt{2x+1} \, dx.$

4. $\displaystyle\int \sqrt{ax+b} \, dx.$

5. $\displaystyle\int (ax+b)^{3/4} dx.$

6. $\displaystyle\int 2ax(ax^2+b)^4 \, dx.$

7. $\displaystyle\int \frac{t}{(4t^2+9)^2} \, dt.$

8. $\displaystyle\int \frac{3t}{(t^2+1)^2} \, dt.$

9. $\displaystyle\int x^2(1+x^3)^{1/4} \, dx.$

10. $\displaystyle\int x^{n-1}\sqrt{a+bx^n} \, dx.$

11. $\displaystyle\int \frac{s}{(1+s^2)^3} \, ds.$

12. $\displaystyle\int \frac{2s}{\sqrt[3]{6-5s^2}} \, ds.$

13. $\displaystyle\int \frac{x}{\sqrt{x^2+1}} \, dx.$

14. $\displaystyle\int \frac{x^2}{(1-x^3)^{2/3}} \, dx.$

15. $\displaystyle\int 5x(x^2+1)^{-3} \, dx.$

16. $\displaystyle\int 2x^3(1-x^4)^{-1/4} \, dx.$

17. $\displaystyle\int x^{-3/4}(x^{1/4}+1)^{-2} dx.$

18. $\displaystyle\int \frac{4x+6}{\sqrt{x^2+3x+1}} \, dx.$

19. $\displaystyle\int \frac{b^3 x^3}{\sqrt{1-a^4 x^4}} \, dx.$

20. $\displaystyle\int \frac{x^{n-1}}{\sqrt{a+bx^n}} \, dx.$

Exercises 21–26. Evaluate.

21. $\displaystyle\int_0^1 x(x^2+1)^3 \, dx.$

22. $\displaystyle\int_{-1}^0 3x^2(4+2x^3)^2 \, dx.$

23. $\displaystyle\int_{-1}^1 \frac{r}{(1+r^2)^4} \, dr.$

24. $\displaystyle\int_0^3 \frac{r}{\sqrt{r^2+16}} \, dr.$

25. $\displaystyle\int_0^a y\sqrt{a^2-y^2} \, dy.$

26. $\displaystyle\int_{-a}^0 y^2 \left(1 - \frac{y^3}{a^2}\right)^{-2} dy.$

Exercises 27–30. Find the area below the graph of f.

27. $f(x) = x\sqrt{2x^2+1}, \quad x \in [0, 2].$

28. $f(x) = \dfrac{x}{(2x^2+1)^2}, \quad x \in [0, 2].$

29. $f(x) = x^{-3}(1+x^{-2})^{-3}, \quad x \in [1, 2].$

30. $f(x) = \dfrac{2x + 5}{(x + 2)^2(x + 3)^2}, \quad x \in [0, 1].$

Exercises 31–37. Calculate.

31. $\displaystyle\int x\sqrt{x + 1}\, dx. \quad [\text{set } u = x + 1]$

32. $\displaystyle\int 2x\sqrt{x - 1}\, dx.$

33. $\displaystyle\int x\sqrt{2x - 1}\, dx.$

34. $\displaystyle\int t(2t + 3)^8\, dt.$

35. $\displaystyle\int \frac{1}{\sqrt{x}\sqrt{\sqrt{x} + x}}\, dx.$

36. $\displaystyle\int_{-1}^{0} x^3(x^2 + 1)^6\, dx.$

37. $\displaystyle\int_{0}^{1} \frac{x + 3}{\sqrt{x + 1}}\, dx.$

38. $\displaystyle\int_{2}^{5} \frac{x^2}{\sqrt{x - 1}}\, dx.$

39. Find an equation $y = f(x)$ for the curve that passes through the point $(0, 1)$ and has slope

$$\frac{dy}{dx} = x\sqrt{x^2 + 1}.$$

40. Find an equation $y = f(x)$ for the curve that passes through the point $(4, \frac{1}{3})$ and has slope

$$\frac{dy}{dx} = -\frac{1}{2\sqrt{x}(1 + \sqrt{x})^2}.$$

Exercises 41–64. Calculate.

41. $\displaystyle\int \cos(3x + 1)\, dx.$

42. $\displaystyle\int \sin 2\pi x\, dx.$

43. $\displaystyle\int \csc^2 \pi x\, dx.$

44. $\displaystyle\int \sec 2x \tan 2x\, dx.$

45. $\displaystyle\int \sin(3 - 2x)\, dx.$

46. $\displaystyle\int \sin^2 x \cos x\, dx.$

47. $\displaystyle\int \cos^4 x \sin x\, dx.$

48. $\displaystyle\int x \sec^2 x^2\, dx.$

49. $\displaystyle\int \frac{\sin \sqrt{x}}{\sqrt{x}}\, dx.$

50. $\displaystyle\int \csc(1 - 2x) \cot(1 - 2x)\, dx.$

51. $\displaystyle\int \sqrt{1 + \sin x}\, \cos x\, dx.$

52. $\displaystyle\int \frac{\sin x}{\sqrt{1 + \cos x}}\, dx.$

53. $\displaystyle\int \sin \pi x \cos \pi x\, dx.$

54. $\displaystyle\int \sin^2 \pi x \cos \pi x\, dx.$

55. $\displaystyle\int \sin \pi x \cos^2 \pi x\, dx.$

56. $\displaystyle\int (1 + \tan^2 x) \sec^2 x\, dx.$

57. $\displaystyle\int x \sin^3 x^2 \cos x^2\, dx.$

58. $\displaystyle\int x \sin^4(x^2 - \pi) \cos(x^2 - \pi)\, dx.$

59. $\displaystyle\int \frac{\sec^2 x}{\sqrt{1 + \tan x}}\, dx.$

60. $\displaystyle\int \frac{\csc^2 2x}{\sqrt{2 + \cot 2x}}\, dx.$

61. $\displaystyle\int \frac{\cos(1/x)}{x^2}\, dx.$

62. $\displaystyle\int \frac{\sin(1/x)}{x^2}\, dx.$

63. $\displaystyle\int x^2 \tan(x^3 + \pi) \sec^2(x^3 + \pi)\, dx.$

64. $\displaystyle\int (x \sin^2 x + x^2 \sin x \cos x)\, dx.$

Exercises 65–70. Evaluate.

65. $\displaystyle\int_{-\pi}^{\pi} \sin^4 x \cos x\, dx.$

66. $\displaystyle\int_{-\pi/3}^{\pi/3} \sec x \tan x\, dx.$

67. $\displaystyle\int_{1/4}^{1/3} \sec^2 \pi x\, dx.$

68. $\displaystyle\int_{0}^{1} \cos^2 \frac{\pi}{2}x \sin \frac{\pi}{2}x\, dx.$

69. $\displaystyle\int_{0}^{\pi/2} \sin x \cos^3 x\, dx.$

70. $\displaystyle\int_{0}^{\pi} x \cos x^2\, dx.$

71. Derive the formula

$$\int \sin^2 x\, dx = \tfrac{1}{2}x - \tfrac{1}{4}\sin 2x + C.$$

HINT: Recall the half-angle formula
$\sin^2 \theta = \tfrac{1}{2}(1 - \cos 2\theta).$

72. Derive the formula

$$\int \cos^2 x\, dx = \tfrac{1}{2}x + \tfrac{1}{4}\sin 2x + C.$$

Calculate.

73. $\displaystyle\int \cos^2 5x\, dx.$

74. $\displaystyle\int \sin^2 3x\, dx.$

75. $\displaystyle\int_{0}^{\pi/2} \cos^2 2x\, dx.$

76. $\displaystyle\int_{0}^{2\pi} \sin^2 x\, dx.$

Exercises 77–81. Find the area between the curves.

77. $y = \cos x, \quad y = -\sin x, \quad x = 0, \quad x = \frac{\pi}{2}.$

78. $y = \cos \pi x, \quad y = \sin \pi x, \quad x = 0, \quad x = \frac{1}{4}.$

79. $y = \cos^2 \pi x, \quad y = \sin^2 \pi x, \quad x = 0, \quad x = \frac{1}{4}.$

80. $y = \cos^2 \pi x, \quad y = -\sin^2 \pi x, \quad x = 0, \quad x = \frac{1}{4}.$

81. $y = \csc^2 \pi x, \quad y = \sec^2 \pi x, \quad x = \frac{1}{6}, \quad x = \frac{1}{4}.$

82. Calculate

$$\int \sin x \cos x\, dx.$$

(a) Setting $u = \sin x$.
(b) Setting $u = \cos x$.
(c) Reconcile your answers to parts (a) and (b).

83. Calculate

$$\int \sec^2 x \tan x\, dx$$

(a) Setting $u = \sec x$.
(b) Setting $u = \tan x$.
(c) Reconcile your answers to parts (a) and (b).

84. Let f be a continuous function, c a real number. Show that

(a)
$$\int_{a+c}^{b+c} f(x - c)\, dx = \int_{a}^{b} f(x)\, dx,$$

and, if $c \neq 0$,

(b)
$$\frac{1}{c}\int_{ac}^{bc} f(x/c)\, dx = \int_{a}^{b} f(x)\, dx.$$

For Exercises 85 and 86 reverse the roles of x and u in (5.7.2) and write

$$\int_{x(a)}^{x(b)} f(x)\,dx = \int_{a}^{b} f(x(u))x'(u)\,du.$$

85. (*The area of a circular region*) The circle $x^2 + y^2 = r^2$ encloses a circular disc of radius r. Justify the familiar formula $A = \pi r^2$ by integration. HINT: The quarter-disk in the first quadrant is the region below the curve $y = \sqrt{r^2 - x^2}$, $x \in [0, r]$. Therefore

$$A = 4\int_{0}^{r} \sqrt{r^2 - x^2}\,dx.$$

Set $x = r \sin u$, $dx = r \cos u\,du$.

86. Find the area enclosed by the ellipse $b^2 x^2 + a^2 y^2 = a^2 b^2$.

■ 5.8 ADDITIONAL PROPERTIES OF THE DEFINITE INTEGRAL

We come now to some properties of the definite integral that we'll make use of time and time again. Some of the properties are pretty obvious; some are not. All are important.

I. The integral of a nonnegative continuous function is nonnegative:

(5.8.1)

$$\text{if} \quad f(x) \geq 0 \quad \text{for all } x \in [a, b], \quad \text{then} \quad \int_{a}^{b} f(x)\,dx \geq 0.$$

The integral of a positive continuous function is positive:

(5.8.2)

$$\text{if} \quad f(x) > 0 \quad \text{for all } x \in [a, b], \quad \text{then} \quad \int_{a}^{b} f(x)\,dx > 0.$$

Reasoning: (5.8.1) holds because in this case all of the lower sums $L_f(P)$ are nonnegative; (5.8.2) holds because in this case all the lower sums are positive. ❏

II. The integral is order-preserving: for continuous functions f and g,

(5.8.3)

$$\text{if} \quad f(x) \leq g(x) \text{ for all } x \in [a, b], \quad \text{then} \quad \int_{a}^{b} f(x)\,dx \leq \int_{a}^{b} g(x)\,dx$$

and

(5.8.4)

$$\text{if} \quad f(x) < g(x) \quad \text{for all } \ x \in [a, b], \quad \text{then} \quad \int_{a}^{b} f(x)\,dx < \int_{a}^{b} g(x)\,dx.$$

PROOF OF (5.8.3) If $f(x) \leq g(x)$ on $[a, b]$, then $f(x) - f(x) \geq 0$ on $[a, b]$. Thus by (5.8.1)

$$\int_{a}^{b} [g(x) - f(x)]\,dx \geq 0.$$

This gives

$$\int_{a}^{b} g(x)\,dx - \int_{a}^{b} f(x)\,dx \geq 0$$

and shows that

$$\int_{a}^{b} f(x)\,dx \leq \int_{a}^{b} g(x)\,dx.$$

The proof of (5.8.4) is similarly simple. ❏

III. Just as the absolute value of a sum of numbers is less than or equal to the sum of the absolute values of those numbers,

$$|x_1 + x_2 + \cdots + x_n| \le |x_1| + |x_2| + \cdots + |x_n|,$$

the absolute value of an integral of a continuous function is less than or equal to the integral of the absolute value of that function:

(5.8.5)

$$\left| \int_a^b f(x)\,dx \right| \le \int_a^b |f(x)|\,dx.$$

PROOF OF (5.8.5) Since $-|f(x)| \le f(x) \le |f(x)|$, it follows from (5.8.3) that

$$-\int_a^b |f(x)|\,dx \le \int_a^b f(x)\,dx \le \int_a^b |f(x)|\,dx.$$

This pair of inequalities is equivalent to (5.8.5). ❑

IV. If f is continuous on $[a, b]$, then

(5.8.6)

$$m(b - a) \le \int_a^b f(x)\,dx \le M(b - a)$$

where m is the minimum value of f on $[a, b]$ and M is the maximum.

Reasoning: $m(b - a)$ is a lower sum for f and $M(b - a)$ is an upper sum. ❑

You know from Theorem 5.3.5 that, if f is continuous on $[a, b]$, then for all $x \in (a, b)$

$$\frac{d}{dx}\left(\int_a^x f(t)\,dt \right) = f(x).$$

Below we give an extension of this result that plays a large role in Chapter 7.

V. If f is continuous on $[a, b]$ and u is a differentiable function of x with values in $[a, b]$, then for all $u(x) \in (a, b)$

(5.8.7)

$$\frac{d}{dx}\left(\int_a^{u(x)} f(t)\,dt \right) = f(u(x))u'(x).$$

PROOF OF (5.8.7) Since f is continuous on $[a, b]$, the function

$$F(u) = \int_a^u f(t)\,dt$$

is differentiable on (a, b) and

$$F'(u) = f(u).$$

This we know from Theorem 5.3.5. The result that we are trying to prove follows from noting that

$$\int_a^{u(x)} f(t)\,dt = F(u(x))$$

and applying the chain rule:

$$\frac{d}{dx}\left(\int_a^{u(x)} f(t)\,dt \right) = \frac{d}{dx}[F(u(x))] = F'(u(x))u'(x) = f(u(x))u'(x). \quad ❑$$

Example 1 Find $\dfrac{d}{dx}\left(\displaystyle\int_0^{x^3}\dfrac{1}{1+t}\,dt\right)$.

SOLUTION At this stage you probably cannot carry out the integration: it requires the natural logarithm function. (Not introduced in this text until Chapter 7.) But for our purposes, that doesn't matter. By (5.8.7),

$$\frac{d}{dx}\left(\int_0^{x^3}\frac{1}{1+t}\,dt\right) = \frac{1}{1+x^3}3x^2 = \frac{3x^2}{1+x^3}. \quad \square$$

Example 2 Find $\dfrac{d}{dx}\left(\displaystyle\int_x^{2x}\dfrac{1}{1+t^2}\,dt\right)$.

SOLUTION The idea is to express the integral in terms of integrals that have constant lower limits of integration. Once we have done that, we can apply (5.8.7). In this case, we choose 0 as a convenient lower limit. Then, by the additivity of the integral,

$$\int_0^x\frac{1}{1+t^2}\,dt + \int_x^{2x}\frac{1}{1+t^2}\,dt = \int_0^{2x}\frac{1}{1+t^2}\,dt.$$

Thus

$$\int_x^{2x}\frac{1}{1+t^2}\,dt = \int_0^{2x}\frac{1}{1+t^2}\,dt - \int_0^x\frac{1}{1+t^2}\,dt.$$

Differentiation gives

$$\frac{d}{dx}\left(\int_x^{2x}\frac{1}{1+t^2}\,dt\right) = \frac{d}{dx}\left(\int_0^{2x}\frac{1}{1+t^2}\,dt\right) - \frac{d}{dx}\left(\int_0^x\frac{1}{1+t^2}\,dt\right)$$

$$= \frac{1}{1+(2x)^2}(2) - \frac{1}{1+x^2}(1) = \frac{2}{1+4x^2} - \frac{1}{1+x^2}. \quad \square$$

by (5.8.7) ⟶

VI. Now a few words about the role of symmetry in integration. Suppose that f is continuous on an interval of the form $[-a, a]$, a closed interval symmetric about the origin.

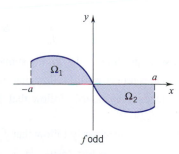

(5.8.8)

(a) if f is odd on $[-a, a]$, then $\displaystyle\int_{-a}^a f(x)\,dx = 0$.

(b) if f is even on $[-a, a]$, then $\displaystyle\int_{-a}^a f(x)\,dx = 2\int_0^a f(x)\,dx$.

f odd

Figure 5.8.1

These assertions can be verified by a simple change of variables. (Exercise 34.) Here we look at these assertions from the standpoint of area. For convenience we refer to Figures 5.8.1 and 5.8.2.

For the odd function,

$$\int_{-a}^a f(x)\,dx = \int_{-a}^0 f(x)\,dx + \int_0^a f(x)\,dx = \text{area of }\Omega_1 - \text{area of }\Omega_2 = 0.$$

For the even function,

$$\int_{-a}^a f(x)\,dx = \text{area of }\Omega_1 + \text{area of }\Omega_2 = 2(\text{area of }\Omega_2) = 2\int_0^a f(x)\,dx. \quad \square$$

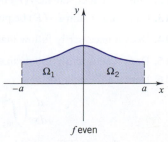

f even

Figure 5.8.2

Suppose we were asked to evaluate

$$\int_{-\pi}^{\pi} (\sin x - x \cos x)^3 \, dx.$$

A laborious calculation would show that this integral is zero. We don't have to carry out that calculation. The integrand is an odd function, and the interval of integration is symmetric about the origin. Thus we can tell immediately that the integral is zero:

$$\int_{-\pi}^{\pi} (\sin x - x \cos x)^3 \, dx = 0.$$

EXERCISES 5.8

Assume that f and g are continuous on $[a, b]$ and

$$\int_a^b f(x) \, dx > \int_a^b g(x) \, dx.$$

Answer questions 1–6, giving supporting reasons.

1. Does it necessarily follow that $\int_a^b [f(x) - g(x)] \, dx > 0$?
2. Does it necessarily follow that $f(x) > g(x)$ for all $x \in [a, b]$?
3. Does it necessarily follow that $f(x) > g(x)$ for at least some $x \in [a, b]$?
4. Does it necessarily follow that

$$\left| \int_a^b f(x) \, dx \right| > \left| \int_a^b g(x) \, dx \right|?$$

5. Does it necessarily follow that $\int_a^b |f(x)| \, dx > \int_a^b |g(x)| \, dx$?
6. Does it necessarily follow that $\int_a^b |f(x)| \, dx > \int_a^b g(x) \, dx$?

Assume that f is continuous on $[a, b]$ and

$$\int_a^b f(x) \, dx = 0.$$

Answer questions 7–15, giving supporting reasons.

7. Does it necessarily follow that $f(x) = 0$ for all $x \in [a, b]$?
8. Does it necessarily follow that $f(x) = 0$ for at least some $x \in [a, b]$?
9. Does it necessarily follow that $\int_a^b |f(x)| \, dx = 0$?
10. Does it necessarily follow that $|\int_a^b f(x) \, dx| = 0$?
11. Must all upper sums $U_f(P)$ be nonnegative?
12. Must all upper sums $U_f(P)$ be positive?
13. Can a lower sum $L_f(P)$ be positive?
14. Does it necessarily follow that $\int_a^b [f(x)]^2 \, dx = 0$?
15. Does it necessarily follow that $\int_a^b [f(x) + 1] \, dx = b - a$?
16. Derive a formula for

$$\frac{d}{dx} \left(\int_{u(x)}^b f(t) \, dt \right)$$

given that u is differentiable and f is continuous.

Exercises 17–23. Calculate.

17. $\dfrac{d}{dx} \left(\displaystyle\int_0^{1+x^2} \dfrac{dt}{\sqrt{2t+5}} \right).$ 18. $\dfrac{d}{dx} \left(\displaystyle\int_1^{x^2} \dfrac{dt}{t} \right).$

19. $\dfrac{d}{dx} \left(\displaystyle\int_x^a f(t) \, dt \right).$ 20. $\dfrac{d}{dx} \left(\displaystyle\int_0^{x^3} \dfrac{dt}{\sqrt{1+t^2}} \right).$

21. $\dfrac{d}{dx} \left(\displaystyle\int_{x^2}^3 \dfrac{\sin t}{t} \, dt \right).$ 22. $\dfrac{d}{dx} \left(\displaystyle\int_{\tan x}^4 \sin t^2 \, dt \right).$

23. $\dfrac{d}{dx} \left(\displaystyle\int_1^{\sqrt{x}} \dfrac{t^2}{1+t^2} \, dt \right).$

24. Show that

$$\frac{d}{dx} \left(\int_{u(x)}^{v(x)} f(t) \, dt \right) = f(v(x))v'(x) - f(u(x))u'(x),$$

given that u and v are differentiable and f is continuous.

Exercises 25–28. Calculate. HINT: Exercise 24.

25. $\dfrac{d}{dx} \left(\displaystyle\int_x^{x^2} \dfrac{dt}{t} \right).$ 26. $\dfrac{d}{dx} \left(\displaystyle\int_{\sqrt{x}}^{x^2+x} \dfrac{dt}{2+\sqrt{t}} \right).$

27. $\dfrac{d}{dx} \left(\displaystyle\int_{\tan x}^{2x} t\sqrt{1+t^2} \, dt \right).$ 28. $\dfrac{d}{dx} \left(\displaystyle\int_{3x}^{1/x} \cos 2t \, dt \right).$

29. Prove (5.8.4).

30. (*Important*) Prove that, if f is continuous on $[a, b]$ and

$$\int_a^b |f(x)| \, dx = 0,$$

then $f(x) = 0$ for all x in $[a, b]$. HINT: Exercise 50, Section 2.4.

31. Find $H'(2)$ given that

$$H(x) = \int_{2x}^{x^3-4} \frac{x}{1+\sqrt{t}} \, dt.$$

32. Find $H'(3)$ given that

$$H(x) = \frac{1}{x} \int_3^x [2t - 3H'(t)] \, dt.$$

33. (a) Let f be continuous on $[-a, 0]$. Use a change of variable to show that

$$\int_{-a}^{0} f(x)\, dx = \int_{0}^{a} f(-x)\, dx.$$

(b) Let f be continuous on $[-a, a]$. Show that

$$\int_{-a}^{a} f(x)\, dx = \int_{0}^{a} [f(x) + f(-x)]\, dx.$$

34. Let f be a function continuous on $[-a, a]$. Prove the statement basing your argument on Exercise 33.

(a) $\displaystyle\int_{-a}^{a} f(x)\, dx = 0$ if f is odd.

(b) $\displaystyle\int_{-a}^{a} f(x)\, dx = 2 \int_{0}^{a} f(x)\, dx$ if f is even.

Exercises 35–38. Evaluate using symmetry considerations.

35. $\displaystyle\int_{-\pi/4}^{\pi/4} (x + \sin 2x)\, dx.$ **36.** $\displaystyle\int_{-3}^{3} \frac{t^3}{1+t^2}\, dt.$

37. $\displaystyle\int_{-\pi/3}^{\pi/3} (1 + x^2 - \cos x)\, dx.$

38. $\displaystyle\int_{-\pi/4}^{\pi/4} (x^2 - 2x + \sin x + \cos 2x)\, dx.$

■ 5.9 MEAN-VALUE THEOREMS FOR INTEGRALS; AVERAGE VALUE OF A FUNCTION

We begin with a result that we asked you to prove earlier. (Exercise 33, Section 5.3.)

THEOREM 5.9.1   **THE FIRST MEAN-VALUE THEOREM FOR INTEGRALS**

If f is continuous on $[a, b]$, then there is at least one number c in (a, b) for which

$$\int_{a}^{b} f(x)\, dx = f(c)(b - a).$$

This number $f(c)$ is called *the average value* (or *mean value*) of f on $[a, b]$.

We now have the following identity:

(5.9.2)
$$\int_{a}^{b} f(x)\, dx = (\textit{the average value of } f \textit{ on } [a, b]) \cdot (b - a).$$

This identity provides a powerful, intuitive way of viewing the definite integral. Think for a moment about area. If f is constant and positive on $[a, b]$, then Ω, the region below the graph, is a rectangle. Its area is given by the formula

$$\text{area of } \Omega = (\text{the constant value of } f \text{ on } [a, b]) \cdot (b - a). \quad \text{(Figure 5.9.1)}$$

If f is now allowed to vary continuously on $[a, b]$, then we have

$$\text{area of } \Omega = \int_{a}^{b} f(x)\, dx,$$

and the area formula reads

$$\text{area of } \Omega = (\textit{the average value of } f \textit{ on } [a, b]) \cdot (b - a). \quad \text{(Figure 5.9.2)}$$

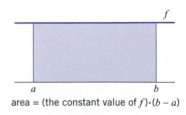

area = (the constant value of f)·$(b - a)$

Figure 5.9.1

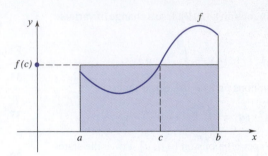

Figure 5.9.2

Think now about motion. If an object moves along a line with constant speed v during the time interval $[a, b]$, then

distance traveled = (the constant value of v on $[a, b]$) $\cdot$ $(b - a)$.

If the speed v varies, then we have

$$\text{distance traveled} = \int_a^b v(t)\, dt,$$

and the formula reads

> *distance traveled = (the average speed on $[a, b]$) $\cdot$ $(b - a)$.*

Let's calculate some simple averages. Writing f_{avg} for the average value of f on $[a, b]$, we have

$$f_{\text{avg}} = \frac{1}{b-a} \int_a^b f(x)\, dx.$$

The average value of a constant function $f(x) = k$ is, of course, k:

$$f_{\text{avg}} = \frac{1}{b-a} \int_a^b k\, dx = \frac{k}{b-a}\Big[x\Big]_a^b = \frac{k}{b-a}(b - a) = k.$$

The average value of a constant multiple of the identity function $f(x) = \alpha x$ is the arithmetical average of the values taken on by the function at the endpoints of the interval:

$$f_{\text{avg}} = \frac{1}{b-a} \int_a^b \alpha x\, dx = \frac{1}{b-a}\left[\frac{\alpha}{2}x^2\right]_a^b$$

$$= \frac{1}{b-a}\left[\frac{\alpha}{2}(b^2 - a^2)\right] = \frac{\alpha b + \alpha a}{2} = \frac{f(b) + f(a)}{2}.$$

What is the average value of the squaring function $f(x) = x^2$?

$$f_{\text{avg}} = \frac{1}{b-a} \int_a^b x^2 dx = \frac{1}{b-a}\left[\frac{x^3}{3}\right]_a^b = \frac{1}{b-a}\left(\frac{b^3 - a^3}{3}\right)$$

$$= \frac{1}{b-a}\left[\frac{(b^2 + ab + a^2)(b-a)}{3}\right] = \tfrac{1}{3}(b^2 + ab + a^2).$$

The average value of the squaring function on $[a, b]$ is not $\frac{1}{2}(b^2 + a^2)$; it is $\frac{1}{3}(b^2 + ab + a^2)$. On $[1, 3]$ the values of the squaring function range from 1 to 9. While the arithmetic average of these two values is 5, the average value of the squaring function on the entire interval $[1, 3]$ is not 5; it is $\frac{13}{3}$.

There is an extension of Theorem 5.9.1 which, as you'll see, is useful in applications.

THEOREM 5.9.3   THE SECOND MEAN-VALUE THEOREM FOR INTEGRALS

If f and g are continuous on $[a, b]$ and g is nonnegative, then there is a number c in (a, b) for which

$$\int_a^b f(x)g(x)\,dx = f(c) \int_a^b g(x)\,dx.$$

This number $f(c)$ is called the *g-weighted average of f on* $[a, b]$.

We will prove this theorem (and thereby obtain a proof of Theorem 5.9.1) at the end of this section. First, some physical considerations.

The Mass of a Rod   Imagine a thin rod (a straight material wire of negligible thickness) lying on the x-axis from $x = a$ to $x = b$. If the *mass density* of the rod (the mass per unit length) is constant, then the total mass M of the rod is simply the density λ times the length of the rod: $M = \lambda(b - a)$.[†] If the density λ varies continuously from point to point, say $\lambda = \lambda(x)$, then the mass of the rod is the average density of the rod times the length of the rod:

$$M = (\text{average density}) \cdot (\text{length}).$$

This is an integral:

(5.9.4)
$$M = \int_a^b \lambda(x)\,dx.$$

The Center of Mass of a Rod   Continue with that same rod. If the rod is homogeneous (constant density), then the center of mass of the rod (we denote this point by x_M) is simply the midpoint of the rod:

$$x_M = \tfrac{1}{2}(a + b). \qquad \text{(the average of } x \text{ from } a \text{ to } b)$$

If the rod is not homogeneous, the center of mass is still an average, but now a weighted average, *the density-weighted average of x from a to b*; namely, x_M is the point for which

$$x_M \int_a^b \lambda(x)\,dx = \int_a^b x\lambda(x)\,dx.$$

Since the integral on the left is M, we have

(5.9.5)
$$x_M M = \int_a^b x\lambda(x)\,dx.$$

Example 1   A rod of length L is placed on the x-axis from $x = 0$ to $x = L$. Find the mass of the rod and the center of mass given that the density of the rod varies directly as the distance from the $x = 0$ endpoint of the rod.

[†]The symbol λ is the Greek letter "lambda."

SOLUTION   Here $\lambda(x) = kx$ where k is some positive constant. Therefore

$$M = \int_0^L kx\,dx = \left[\tfrac{1}{2}kx^2\right]_0^L = \tfrac{1}{2}kL^2$$

and

$$x_M M = \int_0^L x(kx)\,dx = \int_0^L kx^2\,dx = \left[\tfrac{1}{3}kx^3\right]_0^L = \tfrac{1}{3}kL^3.$$

Division by M gives $x_M = \tfrac{2}{3}L$.

In this instance the center of mass is to the right of the midpoint. This makes sense. After all, the density increases from left to right. Thus mass accumulates near the right tip of the rod.   ❑

We know from physics that, close to the surface of the earth, where the force of gravity is given by the familiar formula $W = mg$, the center of mass is the *center of gravity*. This is the balance point. For the rod of Example 1, the balance point is at $x = \tfrac{2}{3}L$. Supported at that point, the rod will be in balance.

Later (in Project 10.6) you will see that a projectile fired at an angle follows a parabolic path. (Here we are disregarding air resistance.) Suppose that a rod is hurled into the air end over end. Certainly not every point of the rod can follow a parabolic path. What moves in a parabolic path is the center of mass of the rod.

We go back now to Theorem 5.9.3 and prove it. [There is no reason to construct a separate proof for Theorem 5.9.1. It is Theorem 5.9.3 with $g(x)$ identically 1.]

PROOF OF THEOREM 5.9.3   Since f is continuous on $[a, b]$, f takes on a minimum value m on $[a, b]$ and a maximum value M. Since g is nonnegative on $[a, b]$,

$$mg(x) \le f(x)g(x) \le Mg(x) \qquad \text{for all } x \text{ in } [a, b].$$

Therefore

$$\int_a^b mg(x)\,dx \le \int_a^b f(x)g(x)\,dx \le \int_a^b Mg(x)\,dx$$

and

$$m\int_a^b g(x)\,dx \le \int_a^b f(x)g(x)\,dx \le M\int_a^b g(x)\,dx.$$

We know that $\int_a^b g(x)\,dx \ge 0$. If $\int_a^b g(x)\,dx = 0$, then, by the inequality we just derived, $\int_a^b f(x)g(x)\,dx = 0$ and the theorem holds for all choices of c in (a, b). If $\int_a^b g(x)\,dx > 0$, then

$$m \le \frac{\displaystyle\int_a^b f(x)g(x)\,dx}{\displaystyle\int_a^b g(x)\,dx} \le M$$

and by the intermediate-value theorem (Theorem 2.6.1) there exists a number c in (a, b) for which

$$f(c) = \frac{\int_a^b f(x)g(x)\,dx}{\int_a^b g(x)\,dx}.$$

Obviously, then,

$$f(c)\int_a^b g(x)\,dx = \int_a^b f(x)g(x)\,dx. \quad ❑$$

EXERCISES 5.9

Exercises 1–12. Determine the average value of the function on the indicated interval and find an interior point of this interval at which the function takes on its average value.

1. $f(x) = mx + b, \quad x \in [0, c]$.

2. $f(x) = x^2, \quad x \in [-1, 1]$.

3. $f(x) = x^3, \quad x \in [-1, 1]$.

4. $f(x) = x^{-2}, \quad x \in [1, 4]$.

5. $f(x) = |x|, \quad x \in [-2, 2]$.

6. $f(x) = x^{1/3}, \quad x \in [-8, 8]$.

7. $f(x) = 2x - x^2, \quad x \in [0, 2]$.

8. $f(x) = 3 - 2x, \quad x \in [0, 3]$.

9. $f(x) = \sqrt{x}, \quad x \in [0, 9]$.

10. $f(x) = 4 - x^2, \quad x \in [-2, 2]$.

11. $f(x) = \sin x, \quad x \in [0, 2\pi]$.

12. $f(x) = \cos x, \quad x \in [0, \pi]$.

13. Let $f(x) = x^n$, n a positive integer. Determine the average value of f on the interval $[a, b]$.

14. Given that f is continuous on $[a, b]$, compare

$$f(b)(b - a) \quad \text{and} \quad \int_a^b f(x)\,dx.$$

(a) if f is constant on $[a, b]$; (b) if f increases on $[a, b]$; (c) if f decreases on $[a, b]$.

15. Suppose that f has a continuous derivative on $[a, b]$. What is the average value of f' on $[a, b]$?

16. Determine whether the assertion is true or false on an arbitrary interval $[a, b]$ on which f and g are continuous.

(a) $(f + g)_{avg} = f_{avg} + g_{avg}$.

(b) $(\alpha f)_{avg} = \alpha f_{avg}$.

(c) $(fg)_{avg} = (f_{avg})(g_{avg})$.

(d) $(fg)_{avg} = (f_{avg})/g_{avg}$.

17. Let $P(x, y)$ be an arbitrary point on the curve $y = x^2$. Express as a function of x the distance from P to the origin and calculate the average of this distance as x ranges from 0 to $\sqrt{3}$.

18. Let $P(x, y)$ be an arbitrary point on the line $y = mx$. Express as a function of x the distance from P to the origin

and calculate the average of this distance as x ranges from 0 to 1.

19. A stone falls from rest in a vacuum for t seconds. (Section 4.9). (a) Compare its terminal velocity to its average velocity; (b) compare its average velocity during the first $\frac{1}{2}t$ seconds to its average velocity during the next $\frac{1}{2}t$ seconds.

20. Let f be continuous. Show that, if f is an odd function, then its average value on every interval of the form $[-a, a]$ is zero.

21. Suppose that f is continuous on $[a, b]$ and $\int_a^b f(x)\,dx = 0$. Prove that there is at least one number c in (a, b) for which $f(c) = 0$.

22. Show that the average value of the functions $f(x) = \sin \pi x$ and $g(x) = \cos \pi x$ is zero on every interval of length $2n$, n a positive integer.

23. An object starts from rest at the point x_0 and moves along the x-axis with constant acceleration a.

(a) Derive formulas for the velocity and position of the object at each time $t \geq 0$.

(b) Show that the average velocity over any time interval $[t_1, t_2]$ is the arithmetic average of the initial and final velocities on that interval.

24. Find the point on the rod of Example 1 that breaks up that rod into two pieces of equal mass. (Observe that this point is not the center of mass.)

25. A rod 6 meters long is placed on the x-axis from $x = 0$ to $x = 6$. The mass density is $12/\sqrt{x + 1}$ kilograms per meter.

(a) Find the mass of the rod and the center of mass.

(b) What is the average mass density of the rod?

26. For a rod that extends from $x = a$ to $x = b$ and has mass density $\lambda = \lambda(x)$, the integral

$$\int_a^b (x - c)\lambda(x)\,dx$$

gives what is called the *mass moment* of the rod about the point $x = c$. Show that the mass moment about the center of mass is zero. (The center of mass can be defined as the point about which the mass moment is zero.)

27. A rod of length L is placed on the x-axis from $x = 0$ to $x = L$. Find the mass of the rod and the center of mass if the mass density of the rod varies directly: (a) as the square root of the distance from $x = 0$; (b) as the square of the distance from $x = L$.

28. A rod of varying mass density, mass M, and center of mass x_M, extends from $x = a$ to $x = b$. A partition $P = \{x_0, x_1, \ldots, x_n\}$ of $[a, b]$ decomposes the rod into n pieces in the obvious way. Show that, if the n pieces have masses $M_1, M_2, \ldots, M_n$ and centers of mass $x_{M_1}, x_{M_2}, \ldots, x_{M_n}$, then

$$x_M M = x_{M_1} M_1 + x_{m_2} M_2 + \cdots + x_{M_n} M_n.$$

29. A rod that has mass M and extends from $x = 0$ to $x = L$ consists of two pieces with masses M_1, M_2. Given that the center of mass of the entire rod is at $x = \frac{1}{4}L$ and the center of mass of the first piece is at $x = \frac{1}{8}L$, determine the center of mass of the second piece.

30. A rod that has mass M and extends from $x = 0$ to $x = L$ consists of two pieces. Find the mass of each piece given that the center of mass of the entire rod is at $x = \frac{2}{3}L$, the center of mass of the first piece is at $x = \frac{1}{4}L$, and the center of mass of the second piece is at $x = \frac{7}{8}L$.

31. A rod of mass M and length L is to be cut from a long piece that extends to the right from $x = 0$. Where should the cuts be made if the density of the long piece varies directly as the distance from $x = 0$? (Assume that $M \geq \frac{1}{2}kL^2$ where k is the constant of proportionality in the density function.)

32. Is the conclusion of Theorem 5.9.3 valid if g is negative throughout $[a, b]$? If so, prove it.

33. Prove Theorem 5.9.1 without invoking Theorem 5.9.3.

34. Let f be continuous on $[a, b]$. Let $a < c < b$. Prove that
$$f(c) = \lim_{h \to 0^+} (\text{average value of } f \text{ on } [c - h, c + h]).$$

35. Prove that two distinct continuous functions cannot have the same average on every interval.

36. The *arithmetic average* of n numbers is the sum of the numbers divided by n. Let f be a function continuous on $[a, b]$. Show that the average value of f on $[a, b]$ is the limit of arithmetic averages of values taken on by f on $[a, b]$ in the following sense: Partition $[a, b]$ into n subintervals of equal length $(b - a)/n$ and let $S^*(P)$ be a corresponding Riemann sum. Show that $S^*(P)/(b - a)$ is an arithmetic average of n values taken on by f and the limit of these arithmetic averages as $\|P\| \to 0$ is the average value of f on $[a, b]$.

37. A partition $P = \{x_0, x_1, x_2, \ldots, x_n\}$ of $[a, b]$ breaks up $[a, b]$ into n subintervals
$$[x_0, x_1], [x_1, x_2], \ldots, [x_{n-1}, x_n].$$
Show that if f is continuous on $[a, b]$, then there are n numbers $x_i^* \in [x_{i-1}, x_i]$ such that
$$\int_a^b f(x)\, dx = f(x_1^*)\Delta x_1 + f(x_2^*)\Delta x_2 + \cdots + f(x_n^*)\Delta x_n.$$
(Thus each partition P of $[a, b]$ gives rise to a Riemann sum which is exactly equal to the definite integral.)

▶ 38. Let $f(x) = x^3 - x + 1$ for $x \in [-1, 2]$.
(a) Find the average value of f on this interval.
(b) Estimate with three decimal place accuracy a number c in the interval at which f takes on its average value.
(c) Use a graphing utility to illustrate your results with a figure similar to Figure 5.9.2.

▶ 39. Exercise 38 taking $f(x) = \sin x$ with $x \in [0, \pi]$.

▶ 40. Exercise 38 taking $f(x) = 2\cos 2x$ with $x \in [-\pi/4, \pi/6]$.

▶ 41. Set $f(x) = -x^4 + 10x^2 + 25$.
(a) Estimate the numbers a and b with $a < b$ for which $f(a) = f(b) = 0$.
(b) Use a graphing utility to draw the graph of f on $[a, b]$.
(c) Estimate the numbers c in (a, b) for which
$$\int_a^b f(x)\, dx = f(c)(b - a).$$

▶ 42. Exercise 41 taking $f(x) = 8 + x^2 - x^4$.

■ CHAPTER 5. REVIEW EXERCISES

Exercises 1–22. Calculate.

1. $\displaystyle\int \frac{x^3 - 2x + 1}{\sqrt{x}}\, dx.$

2. $\displaystyle\int (x^{3/5} - 3x^{5/3})\, dx.$

3. $\displaystyle\int t^2(1 + t^3)^{10}\, dt.$

4. $\displaystyle\int (1 + 2\sqrt{x})^2\, dx.$

5. $\displaystyle\int \frac{(t^{2/3} - 1)^2}{t^{1/3}}\, dt.$

6. $\displaystyle\int x\sqrt{x^2 - 2}\, dx.$

7. $\displaystyle\int x\sqrt{2 - x}\, dx.$

8. $\displaystyle\int x^2(2 + 2x^3)^4\, dx.$

9. $\displaystyle\int \frac{(1 + \sqrt{x})^5}{\sqrt{x}}\, dx.$

10. $\displaystyle\int \frac{\sin(1/x)}{x^2}\, dx.$

11. $\displaystyle\int \frac{\cos x}{\sqrt{1 + \sin x}}\, dx.$

12. $\displaystyle\int (\sec\theta - \tan\theta)^2\, d\theta.$

13. $\displaystyle\int (\tan 3\theta - \cot 3\theta)^2\, d\theta.$

14. $\displaystyle\int x\sin^3 x^2 \cos x^2\, dx.$

15. $\displaystyle\int \frac{1}{1 + \cos 2x}\, dx.$

16. $\displaystyle\int \frac{1}{1 - \sin 2x}\, dx.$

17. $\displaystyle\int \sec^3 \pi x \tan \pi x\, dx.$

18. $\displaystyle\int ax\sqrt{1 + bx^2}\, dx.$

19. $\displaystyle\int ax\sqrt{1 + bx}\, dx.$

20. $\displaystyle\int ax^2\sqrt{1 + bx}\, dx.$

21. $\displaystyle\int \frac{g(x)g'(x)}{\sqrt{1 + g^2(x)}}\, dx.$

22. $\displaystyle\int \frac{g'(x)}{g^3(x)}\, dx.$

Exercises 23–28. Evaluate.

23. $\int_{-1}^{2} (x^2 - 2x + 3)\, dx.$

24. $\int_{0}^{1} \frac{x}{(x^2 + 1)^3}\, dx.$

25. $\int_{0}^{\pi/4} \sin^3 2x \cos 2x\, dx.$

26. $\int_{0}^{\pi/8} (\tan^2 2x + \sec^2 2x)\, dx.$

27. $\int_{0}^{2} (x^2 + 1)(x^3 + 3x - 6)^{1/3}\, dx.$

28. $\int_{1}^{8} \frac{(1 + x^{1/3})^2}{x^{2/3}}\, dx.$

29. Assume that f is a continuous function and that

$$\int_{0}^{2} f(x)\, dx = 3, \qquad \int_{0}^{3} f(x)\, dx = 1, \qquad \int_{3}^{5} f(x)\, dx = 8.$$

(a) Find $\int_{2}^{3} f(x)\, dx.$

(b) Find $\int_{2}^{5} f(x)\, dx.$

(c) Explain how we know that $f(x) \geq 4$ for at least one x in $[3, 5]$.

(d) Explain how we know that $f(x) < 0$ for at least one x in $[2, 3]$.

30. Let f be a function continuous on $[-2, 8]$ and let $g(x) = f(x) + 3$. If $\int_{-2}^{8} f(x)dx = 4$, what is $\int_{-2}^{8} g(x)dx$?

Exercises 31–36. Sketch the region bounded by the curves and find its area.

31. $y = 4 - x^2,\, y = x + 2.$

32. $y = 4 - x^2,\, x + y + 2 = 0.$

33. $y^2 = x,\quad x = 3y.$

34. $y = \sqrt{x}$, the x-axis, $y = 6 - x$.

35. $y = x^3$, the x-axis, $x + y = 2$.

36. $4y = x^2 - x^4,\, x + y + 1 = 0.$

Exercises 37–41. Carry out the differentiation.

37. $\dfrac{d}{dx}\left(\displaystyle\int_{0}^{x} \frac{dt}{1 + t^2} \right).$

38. $\dfrac{d}{dx}\left(\displaystyle\int_{0}^{x^2} \frac{dt}{1 + t^2} \right).$

39. $\dfrac{d}{dx}\left(\displaystyle\int_{x}^{x^2} \frac{dt}{1 + t^2} \right).$

40. $\dfrac{d}{dx}\left(\displaystyle\int_{0}^{\sin x} \frac{dt}{1 - t^2} \right).$

41. $\dfrac{d}{dx}\left(\displaystyle\int_{0}^{\cos x} \frac{dt}{1 - t^2} \right).$

42. At each point (x, y) of a curve γ the slope is $x\sqrt{x^2 + 1}$. Find an equation $y = f(x)$ for γ given that γ passes through the point $(0, 1)$.

43. Let $F(x) = \displaystyle\int_{0}^{x} \frac{1}{t^2 + 2t + 2}\, dt, \quad x$ real

(a) Does F take on the value 0? If so, where?

(b) Show that F increases $(-\infty, \infty)$.

(c) Determine the concavity of the graph of F.

(d) Sketch the graph of F.

44. Assume that f is a continuous function and that

$$\int_{0}^{x} t f(t)\, dt = x \sin x + \cos x - 1.$$

(a) Find $f(\pi)$. 　　(b) Calculate $f'(x)$.

Exercises 45–47. Find the average value of f on the indicated interval.

45. $f(x) = \dfrac{x}{\sqrt{x^2 + 9}}; \quad [0, 4].$

46. $f(x) = x + 2\sin x; \quad [0, \pi].$

47. Find the average value of $f(x) = \cos x$ on every closed interval of length 2π.

Exercises 48–53. Let f be a function continuous on $[\alpha, \beta]$ and let Ω be the region between the graph of f and the x-axis from $x = \alpha$ to $x = \beta$. Draw a figure. Do not assume that f keeps constant sign.

48. Write an integral over $[\alpha, \beta]$ that gives the area of the portion of Ω that lies above the x-axis minus the area of the portion of Ω that lies below the x-axis.

49. Write an integral over $[\alpha, \beta]$ that gives the area of Ω.

50. Write an integral over $[\alpha, \beta]$ that gives the area of the portion of Ω that lies above the x-axis.

51. Write an integral over $[\alpha, \beta]$ that gives the area of the portion of Ω that lies below the x-axis.

52. A rod extends from $x = 0$ to $x = a,\, a > 0$. Find the center of mass if the density of the rod varies directly as the distance from $x = 2a$.

53. A rod extends from $x = 0$ to $x = a,\, a > 0$. Find the center of mass if the density of the rod varies directly as the distance from $x = \frac{1}{4}a$.

CHAPTER 6

SOME

APPLICATIONS

OF THE INTEGRAL

■ **6.1 MORE ON AREA**

Representative Rectangles

You have seen that the definite integral can be viewed as the limit of Riemann sums:

(1) $$\int_a^b f(x)dx = \lim_{\|P\| \to 0} \left[f\left(x_1^*\right) \Delta x_1 + f\left(x_2^*\right) \Delta x_2 + \cdots + f\left(x_n^*\right) \Delta x_n \right].$$

With x_i^* chosen arbitrarily from $[x_{i-1}, x_i]$, you can think of $f(x_i^*)$ as a *representative* value of f for that interval. If f is positive, then the product

$$f\left(x_i^*\right) \Delta x_i$$

gives the area of the *representative rectangle* shown in Figure 6.1.1. Formula (1) tells us that we can approximate the area under the curve as closely as we wish by adding up the areas of representative rectangles. (Figure 6.1.2)

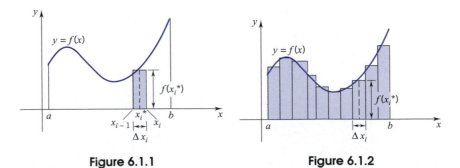

Figure 6.1.1   **Figure 6.1.2**

Figure 6.1.3 shows a region Ω bounded above by the graph of a function f and bounded below by the graph of a function g. As you know, we can obtain the area of Ω

by integrating the *vertical separation* $f(x) - g(x)$ from $x = a$ to $x = b$:

$$A = \int_a^b [f(x) - g(x)]\,dx.$$

In this case the approximating Riemann sums are of the form

$$[f(x_1^*) - g(x_1^*)]\Delta x_1 + [f(x_2^*) - g(x_2^*)]\Delta x_2 + \cdots + [f(x_n^*) - g(x_n^*)]\Delta x_n.$$

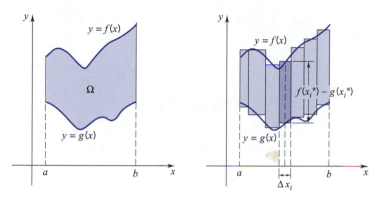

Figure 6.1.3

Here a representative rectangle has

$$\text{height} \quad f(x_i^*) - g(x_i^*), \quad \text{width} \quad \Delta x_i,$$

and area

$$[f(x_i^*) - g(x_i^*)]\Delta x_i.$$

Example 1 Find the area A of the set shaded in Figure 6.1.4.

SOLUTION From $x = -1$ to $x = 2$ the vertical separation is the difference $2x^2 - (x^4 - 2x^2)$. Therefore

$$A = \int_{-1}^2 [2x^2 - (x^4 - 2x^2)]\,dx = \int_{-1}^2 (4x^2 - x^4)\,dx$$

$$= \left[\tfrac{4}{3}x^3 - \tfrac{1}{5}x^5\right]_{-1}^2 = \left[\tfrac{32}{3} - \tfrac{32}{5}\right] - \left[-\tfrac{4}{3} + \tfrac{1}{5}\right] = \tfrac{27}{5} \quad ❏$$

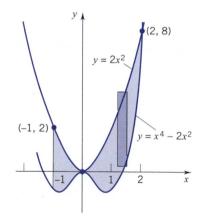

Figure 6.1.4

Areas Obtained by Integration with Respect to y

We can interchange the roles played by x and y. In Figure 6.1.5 you see a region Ω, the boundaries of which are given not in terms of x but in terms of y. Here we set the representative rectangles horizontally and calculate the area of the region as the limit of sums of the form

$$[F(y_1^*) - G(y_1^*)]\Delta y_1 + [F(y_2^*) - G(y_2^*)]\Delta y_2 + \cdots + [F(y_n^*) - G(y_n^*)]\Delta y_n.$$

These are Riemann sums for the integral of $F - G$. The area formula now reads

$$A = \int_c^d [F(y) - G(y)]\,dy.$$

In this case we are integrating with respect to y the *horizontal separation* $F(y) - G(y)$ from $y = c$ to $y = d$.

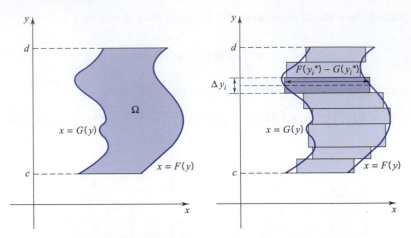

Figure 6.1.5

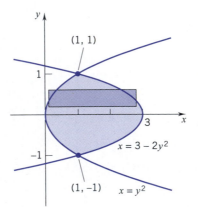

Figure 6.1.6

Example 2 Find the area of the region bounded on the left by the curve $x = y^2$ and bounded on the right by the curve $x = 3 - 2y^2$.

SOLUTION The region is sketched in Figure 6.1.6. The points of intersection can be found by solving the two equations simultaneously:

$$x = y^2 \quad \text{and} \quad x = 3 - 2y^2$$

together imply that

$$y = \pm 1.$$

The points of intersection are $(1, 1)$ and $(1, -1)$. The easiest way to calculate the area is to set our representative rectangles horizontally and integrate with respect to y. We then find the area of the region by integrating the horizontal separation

$$(3 - 2y^2) - y^2 = 3 - 3y^2$$

from $y = -1$ to $y = 1$:

$$A = \int_{-1}^{1} (3 - 3y^2)\, dy = \left[3y - y^3\right]_{-1}^{1} = 4.$$

NOTE: Our solution did not take advantage of the symmetry of the region. The region is symmetric about the x-axis (the integrand is an even function of y), and so

$$A = 2 \int_{0}^{1} (3 - 3y^2)\, dy = 2\left[3y - y^3\right]_{0}^{1} = 4. \quad ❑$$

Example 3 Calculate the area of the region bounded by the curves $x = y^2$ and $x - y = 2$ first **(a)** by integrating with respect to x and then **(b)** by integrating with respect to y.

SOLUTION Simple algebra shows that the two curves intersect at the points $(1, -1)$ and $(4, 2)$.

(a) To obtain the area of the region by integration with respect to x, we set the representative rectangles vertically and express the bounding curves as functions of x. Solving $x = y^2$ for y we get $y = \pm\sqrt{x}$; $y = \sqrt{x}$ is the upper half of the parabola and $y = -\sqrt{x}$ is the lower half. The equation of the line can be written $y = x - 2$. (See Figure 6.1.7.)

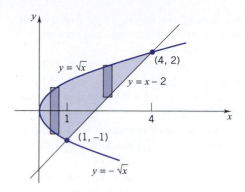

Figure 6.1.7

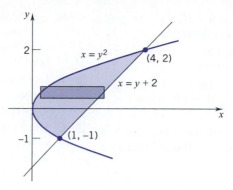

Figure 6.1.8

The upper boundary of the region is the curve $y = \sqrt{x}$. However, the lower boundary consists of two parts: $y = -\sqrt{x}$ from $x = 0$ to $x = 1$, and $y = x - 2$ from $x = 1$ to $x = 4$. Thus, we use two integrals:

$$A = \int_0^1 [\sqrt{x} - (-\sqrt{x})]\,dx + \int_1^4 [\sqrt{x} - (x - 2)]\,dx$$

$$= 2\int_0^1 \sqrt{x}\,dx + \int_1^4 (\sqrt{x} - x + 2)\,dx = \left[\tfrac{4}{3}x^{3/2}\right]_0^1 + \left[\tfrac{2}{3}x^{3/2} - \tfrac{1}{2}x^2 + 2x\right]_1^4 = \tfrac{9}{2}.$$

(b) To obtain the area by integration with respect to y, we set the representative rectangles horizontally. (See Figure 6.1.8.) The right boundary is the line $x = y + 2$ and the left boundary is the curve $x = y^2$. Since y ranges from -1 to 2,

$$A = \int_{-1}^2 [(y + 2) - y^2]\,dy = \left[\tfrac{1}{2}y^2 + 2y - \tfrac{1}{3}y^3\right]_{-1}^2 = \tfrac{9}{2}.$$

In this instance integration with respect to y was the more efficient route to take. ❑

EXERCISES 6.1

Exercises 1–14. Sketch the region bounded by the curves. Represent the area of the region by one or more integrals (a) in terms of x; (b) in terms of y. Evaluation not required.

1. $y = x^2$, $\quad y = x + 2$.
2. $y = x^2$, $\quad y = -4x$.
3. $y = x^3$, $\quad y = 2x^2$.
4. $y = \sqrt{x}$, $\quad y = x^3$.
5. $y = -\sqrt{x}$, $\quad y = x - 6$, $\quad y = 0$.
6. $x = y^3$, $\quad x = 3y + 2$.
7. $y = |x|$, $\quad 3y - x = 8$.
8. $y = x$, $\quad y = 2x$, $\quad y = 3$.
9. $x + 4 = y^2$, $\quad x = 5$.
10. $x = |y|$, $\quad x = 2$.
11. $y = 2x$, $\quad x + y = 9$, $\quad y = x - 1$.
12. $y = x^3$, $\quad y = x^2 + x - 1$.
13. $y = x^{1/3}$, $\quad y = x^2 + x - 1$.
14. $y = x + 1$, $\quad y + 3x = 13$, $\quad 3y + x + 1 = 0$.

Exercises 15–26. Sketch the region bounded by the curves and calculate the area of the region.

15. $4x = 4y - y^2$, $\quad 4x - y = 0$.
16. $x + y^2 - 4 = 0$, $\quad x + y = 2$.
17. $x = y^2$, $\quad x = 12 - 2y^2$.

18. $x + y = 2y^2$, $\quad y = x^3$.
19. $x + y - y^3 = 0$, $\quad x - y + y^2 = 0$.
20. $8x = y^3$, $\quad 8x = 2y^3 + y^2 - 2y$.
21. $y = \cos x$, $\quad y = \sec^2 x$, $\quad x \in [-\pi/4, \pi/4]$.
22. $y = \sin^2 x$, $\quad y = \tan^2 x$, $\quad x \in [-\pi/4, \pi/4]$.
    HINT: $\sin^2 x = \tfrac{1}{2}(1 - \cos 2x)$.
23. $y = 2\cos x$, $\quad y = \sin 2x$, $\quad x \in [-\pi, \pi]$.
24. $y = \sin x$, $\quad y = \sin 2x$, $\quad x \in [0, \pi/2]$.
25. $y = \sin^4 x \cos x$, $\quad x \in [0, \pi/2]$.
26. $y = \sin 2x$, $\quad y = \cos 2x$, $\quad x \in [0, \pi/4]$.

Exercises 27–28. Use integration to find the area of the triangle with the given vertices.

27. $(0, 0)$, $(1, 3)$, $(3, 1)$.
28. $(0, 1)$, $(2, 0)$, $(3, 4)$.
29. Use integration to find the area of the trapezoid with vertices $(-2, -2)$, $(1, 1)$, $(5, 1)$, $(7, -2)$.
30. Sketch the region bounded by $y = x^3$, $y = -x$, and $y = 1$. Find the area of the region.

31. Sketch the region bounded by $y = 6 - x^2$, $y = x (x \leq 0)$, and $y = -x (x \geq 0)$. Find the area of the region.

32. Find the area of the region bounded by the parabolas $x^2 = 4py$ and $y^2 = 4px$, p a positive constant.

33. Sketch the region bounded by $y = x^2$ and $y = 4$. This region is divided into two subregions of equal area by a line $y = c$. Find c.

34. The region between $y = \cos x$ and the x-axis for $x \in [0, \pi/2]$ is divided into two subregions of equal area by a line $x = c$. Find c.

Exercises 35–38. Represent the area of the given region by one or more integrals.

35. The region in the first quadrant bounded by the x-axis, the line $y = \sqrt{3}x$, and the circle $x^2 + y^2 = 4$.

36. The region in the first quadrant bounded by the y-axis, the line $y = \sqrt{3}x$, and the circle $x^2 + y^2 = 4$.

37. The region determined by the intersection of the circles $x^2 + y^2 = 4$ and $(x - 2)^2 + (y - 2)^2 = 4$.

38. The region in the first quadrant bounded by the x-axis, the parabola $y = x^2/3$, and the circle $x^2 + y^2 = 4$.

39. Take $a > 0, b > 0, n$ a positive integer. A rectangle with sides parallel to the coordinate axes has one vertex at the origin and opposite vertex on the curve $y = bx^n$ at a point where $x = a$. Calculate the area of the part of the rectangle that lies below the curve. Show that the ratio of this area to the area of the entire rectangle is independent of a and b, and depends solely on n.

40. (a) Calculate the area of the region in the first quadrant bounded by the coordinate axes and the parabola $y = 1 + a - ax^2, a > 0$.
(b) Determine the value of a that minimizes this area.

41. Use a graphing utility to draw the region bounded by the curves $y = x^4 - 2x^2$ and $y = x + 2$. Then find (approximately) the area of the region.

42. Use a graphing utility to sketch the region bounded by the curves $y = \sin x$ and $y = |x - 1|$. Then find (approximately) the area of the region.

43. A section of rain gutter is 8 feet long. Vertical cross sections of the gutter are in the shape of the parabolic region bounded by $y = \frac{4}{9}x^2$ and $y = 4$, with x and y measured in inches. What is the volume of the rain gutter?

HINT: $V = $ (cross-sectional area) $\times$ length.

44. (a) Calculate the area A of the region bounded by the graph of $f(x) = 1/x^2$ and the x-axis with $x \in [1, b]$.
(b) What happens to A as $b \to \infty$?

45. (a) Calculate the area A of the region bounded by the graph of $f(x) = 1/\sqrt{x}$ and the x-axis with $x \in [1, b]$.
(b) What happens to A as $b \to \infty$?

46. (a) Let $r > 1, r$ rational. Calculate the area A of the region bounded by the graph of $f(x) = 1/x^r$ and the x-axis with $x \in [1, b]$. What happens to A as $b \to \infty$?
(b) Let $0 < r < 1, r$ rational. Calculate the area A of the region bounded by the graph of $f(x) = 1/x^r$ and the x-axis with $x \in [1, b]$. What happens to A as $b \to \infty$?

■ 6.2 VOLUME BY PARALLEL CROSS SECTIONS; DISKS AND WASHERS

Figure 6.2.1 shows a plane region Ω and a solid formed by translating Ω along a line perpendicular to the plane of Ω. Such a solid is called a *right cylinder with cross section Ω.*

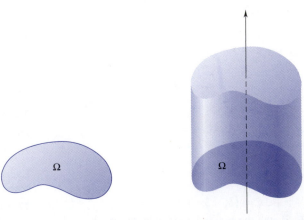

Figure 6.2.1

If Ω has area A and the solid has height h, then the volume of the solid is a simple product:

$$V = A \cdot h.$$ (cross-sectional area · height)

Two elementary examples are given in Figure 6.2.2.

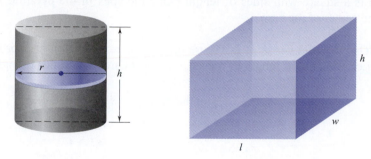

$V = (\pi r^2)h =$ (cross-sectional area) $\cdot$ height $V = (l \cdot w) \cdot h =$ (cross-sectional area) $\cdot$ height

Figure 6.2.2

To calculate the volume of a more general solid, we introduce a coordinate axis and then examine the cross sections of the solid that are perpendicular to that axis. In Figure 6.2.3 we depict a solid and a coordinate axis that we label the x-axis. As in the figure, we suppose that the solid lies entirely between $x = a$ and $x = b$. The figure shows an arbitrary cross section perpendicular to the x-axis. By $A(x)$ we mean the area of the cross section at coordinate x.

If the cross-sectional area $A(x)$ varies continuously with x, then we can find the volume V of the solid by integrating $A(x)$ from $x = a$ to $x = b$:

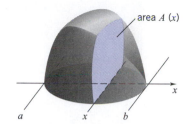

Figure 6.2.3

(6.2.1)

$$V = \int_a^b A(x)\,dx.$$

DERIVATION OF THE FORMULA   Let $P = \{x_0, x_1, x_2, \ldots, x_n\}$ be a partition of $[a, b]$. On each subinterval $[x_{i-1}, x_i]$ choose a point x_i^*. The solid from x_{i-1} to x_i can be approximated by a slab of cross-sectional area $A(x_i^*)$ and thickness Δx_i. The volume of this slab is the product

$$A\left(x_i^*\right)\Delta x_i. \qquad \text{(Figure 6.2.4)}$$

The sum of these products,

$$A\left(x_1^*\right)\Delta x_1 + A\left(x_2^*\right)\Delta x_2 + \cdots + A\left(x_n^*\right)\Delta x_n,$$

is a Riemann sum which approximates the volume of the entire solid. As $\|P\| \to 0$, such Riemann sums converge to

$$\int_a^b A(x)\,dx. \quad ❑$$

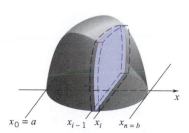

Figure 6.2.4

Remark   (*Average-value point of view*) Formula (6.2.1) can be written

(6.2.2)

$$V = \text{(average cross-sectional area)} \cdot (b - a).$$

Example 1   Find the volume of the pyramid of height h given that the base of the pyramid is a square with sides of length r and the apex of the pyramid lies directly above the center of the base

SOLUTION   Set the x-axis as in Figure 6.2.5. The cross section at coordinate x is a square. Let s denote the length of the side of that square. By similar triangles

$$\frac{\frac{1}{2}s}{h-x} = \frac{\frac{1}{2}r}{h} \quad \text{and therefore} \quad s = \frac{r}{h}(h-x).$$

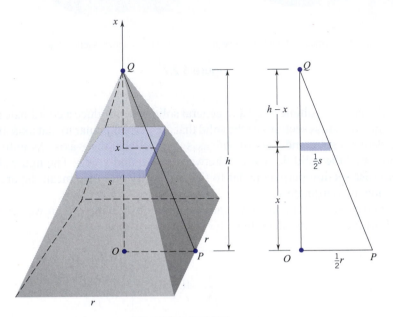

Figure 6.2.5

The area $A(x)$ of the square at coordinate x is $s^2 = (r^2/h^2)(h-x)^2$. Thus

$$V = \int_0^h A(x)\,dx = \frac{r^2}{h^2}\int_0^h (h-x)^2\,dx = \frac{r^2}{h^2}\left[-\frac{(h-x)^3}{3}\right]_0^h = \tfrac{1}{3}r^2 h. \quad \square$$

Example 2   The base of a solid is the region enclosed by the ellipse

$$\frac{x^2}{a^2} + \frac{y^2}{b^2} = 1.$$

Find the volume of the solid given that each cross section perpendicular to the x-axis is an isosceles triangle with base in the region and altitude equal to one-half the base.

SOLUTION   Set the x-axis as in Figure 6.2.6. The cross section at coordinate x is an isosceles triangle with base $\overline{PQ}$ and altitude $\frac{1}{2}\overline{PQ}$. The equation of the ellipse can be written

$$y^2 = \frac{b^2}{a^2}(a^2 - x^2).$$

Since

$$\text{length of } \overline{PQ} = 2y = \frac{2b}{a}\sqrt{a^2 - x^2},$$

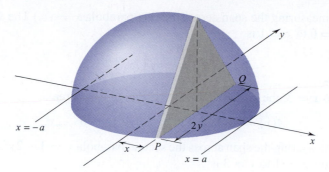

Figure 6.2.6

the isosceles triangle has area

$$A(x) = \tfrac{1}{2}bh = \tfrac{1}{2}\left(\frac{2b}{a}\sqrt{a^2 - x^2}\right)\left(\frac{b}{a}\sqrt{a^2 - x^2}\right) = \frac{b^2}{a^2}(a^2 - x^2).$$

We can find the volume of the solid by integrating $A(x)$ from $x = -a$ to $x = a$:

$$V = \int_{-a}^{a} A(x)\,dx = 2\int_{0}^{a} A(x)\,dx$$

⎿———— by symmetry

$$= \frac{2b^2}{a^2}\int_{0}^{a}(a^2 - x^2)\,dx = \frac{2b^2}{a^2}\left[a^2 x - \frac{x^3}{3}\right]_{0}^{a} = \tfrac{4}{3}ab^2. \quad \square$$

Example 3 The base of a solid is the region between the parabolas

$$x = y^2 \qquad \text{and} \qquad x = 3 - 2y^2.$$

Find the volume of the solid given that the cross sections perpendicular to the x-axis are squares.

SOLUTION The solid is pictured in Figure 6.2.7. The two parabolas intersect at $(1, 1)$ and $(1, -1)$. From $x = 0$ to $x = 1$ the cross section at coordinate x has area

$$A(x) = (2y)^2 = 4y^2 = 4x.$$

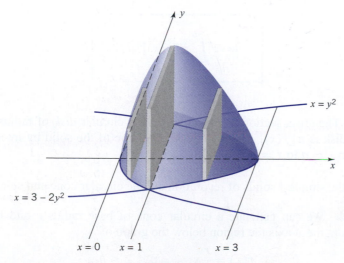

Figure 6.2.7

(Here we are measuring the span across the first parabola $x = y^2$.) The volume of the solid from $x = 0$ to $x = 1$ is

$$V_1 = \int_0^1 4x \, dx = \left[2x^2\right]_0^1 = 2.$$

From $x = 1$ to $x = 3$, the cross section at coordinate x has area

$$A(x) = (2y)^2 = 4y^2 = 2(3 - x) = 6 - 2x.$$

(Here we are measuring the span across the second parabola $x = 3 - 2y^2$.) The volume of the solid from $x = 1$ to $x = 3$ is

$$V_2 = \int_1^3 (6 - 2x) \, dx = \left[6x - x^2\right]_1^3 = 4.$$

The total volume is

$$V_1 + V_2 = 6. \quad ❏$$

Solids of Revolution: Disk Method

Suppose that f is nonnegative and continuous on $[a, b]$. (See Figure 6.2.8.) If we revolve about the x-axis the region bounded by the graph of f and the x-axis, we obtain a solid.

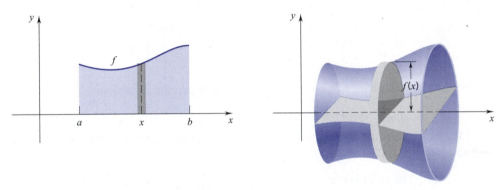

Figure 6.2.8

The volume of this solid is given by the formula

(6.2.3)
$$V = \int_a^b \pi [f(x)]^2 \, dx.$$

VERIFICATION   The cross section at coordinate x is a circular *disk* of radius $f(x)$. The area of this disk is $\pi [f(x)]^2$. We can get the volume of the solid by integrating this function from $x = a$ to $x = b$. ❏

Among the simplest solids of revolution are the circular cone and sphere.

Example 4   We can generate a circular cone of base radius r and height h by revolving about the x-axis the region below the graph of

$$f(x) = \frac{r}{h}x, \qquad 0 \le x \le h. \qquad \text{(Figure 6.2.9)}$$

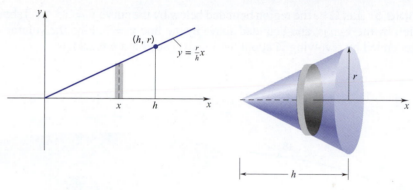

Figure 6.2.9

By (6.2.3),

$$\text{volume of cone} = \int_0^h \pi \left[\frac{r}{h}x\right]^2 dx = \frac{\pi r^2}{h^2} \int_0^h x^2 dx = \frac{\pi r^2}{h^2}\left[\frac{x^3}{3}\right]_0^h = \tfrac{1}{3}\pi r^2 h. \quad ❏$$

Example 5   A sphere of radius r can be obtained by revolving about the x-axis the region below the graph of

$$f(x) = \sqrt{r^2 - x^2}, \qquad -r \leq x \leq r. \qquad \text{(Draw a figure.)}$$

Therefore

$$\text{volume of sphere} = \int_{-r}^r \pi(r^2 - x^2)\, dx = \pi\left[r^2 x - \tfrac{1}{3}x^3\right]_{-r}^r = \tfrac{4}{3}\pi r^3.$$

NOTE: Archimedes derived this formula (by somewhat different methods) in the third century B.C.   ❏

We can interchange the roles played by x and y. By revolving about the y-axis the region of Figure 6.2.10, we obtain a solid of cross-sectional area $A(y) = \pi[g(y)]^2$ and volume

(6.2.4)

$$V = \int_c^d \pi[g(y)]^2\, dy.$$

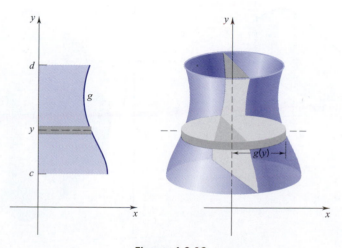

Figure 6.2.10

Example 6 Let Ω be the region bounded below by the curve $y = x^{2/3} + 1$, bounded to the left by the y-axis, and bounded above by the line $y = 5$. Find the volume of the solid generated by revolving Ω about the y-axis. (See Figure 6.2.11.)

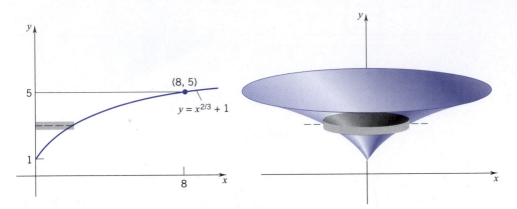

Figure 6.2.11

SOLUTION To apply (6.2.4) we need to express the right boundary of Ω as a function of y:

$$y = x^{2/3} + 1 \qquad \text{gives} \qquad x^{2/3} = y - 1 \qquad \text{and thus} \qquad x = (y - 1)^{3/2}.$$

The volume of the solid obtained by revolving Ω about the y-axis is given by the integral

$$V = \int_1^5 \pi [g(y)]^2 dy = \pi \int_1^5 [(y-1)^{3/2}]^2 dy$$

$$= \pi \int_1^5 (y-1)^3 dy = \pi \left[\frac{(y-1)^4}{4} \right]_1^5 = 64\pi \quad \square$$

Solids of Revolution: Washer Method

The washer method is a slight generalization of the disk method. Suppose that f and g are nonnegative continuous functions with $g(x) \leq f(x)$ for all x in $[a, b]$. (See Figure 6.2.12.) If we revolve the region Ω about the x-axis, we obtain a solid. The volume of

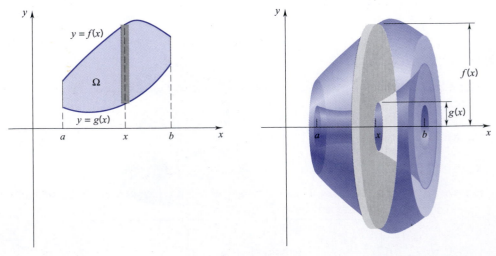

Figure 6.2.12

this solid is given by the formula

(6.2.5)
$$V = \int_a^b \pi([f(x)]^2 - [g(x)]^2)\,dx.$$
(washer method about the x-axis)

VERIFICATION The cross section at coordinate x is a circular ring (in this setting we call it a *washer*) of outer radius $f(x)$, inner radius $g(x)$, and area

$$A(x) = \pi[f(x)]^2 - \pi[g(x)]^2 = \pi([f(x)]^2 - [g(x)]^2).$$

We can get the volume of this solid by integrating $A(x)$ from $x = a$ to $x = b$.   ❏

As before, we can interchange the roles played by x and y. By revolving the region depicted in Figure 6.2.13 *about the y-axis*, we obtain a solid of volume

(6.2.6)
$$V = \int_c^d \pi([F(y)]^2 - [G(y)]^2)\,dy.$$
(washer method about the y-axis)

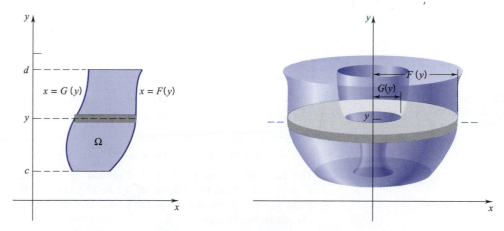

Figure 6.2.13

Example 7   Find the volume of the solid generated by revolving the region between
$$y = x^2 \qquad \text{and} \qquad y = 2x$$
(a) about the x-axis.   **(b)** about the y-axis.

SOLUTION   The curves intersect at the points $(0, 0)$ and $(2, 4)$.

(a) We refer to Figure 6.2.14. For each x from 0 to 2, the x cross section is a washer of outer radius $2x$ and inner radius x^2. By (6.2.5),

$$V = \int_0^2 \pi[(2x)^2 - (x^2)^2]\,dx = \pi \int_0^2 (4x^2 - x^4)\,dx = \pi\left[\tfrac{4}{3}x^3 - \tfrac{1}{5}x^5\right]_0^2 = \tfrac{64}{15}\pi.$$

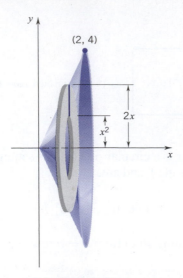

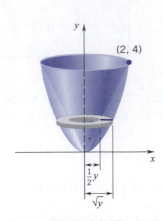

Figure 6.2.14 **Figure 6.2.15**

(b) The solid is depicted in Figure 6.2.15. For each y from 0 to 4, the y cross section is a washer of outer radius $\sqrt{y}$ and inner radius $\frac{1}{2}y$. By (6.2.6),

$$V = \int_0^4 \pi \left[(\sqrt{y})^2 - \left(\tfrac{1}{2}y\right)^2\right] dy = \pi \int_0^4 \left(y - \tfrac{1}{4}y^2\right) dy$$

$$= \pi \left[\tfrac{1}{2}y^2 - \tfrac{1}{12}y^3\right]_0^4 = \tfrac{8}{3}\pi. \quad \square$$

EXERCISES 6.2

Exercises 1–16. Sketch the region Ω bounded by the curves and find the volume of the solid generated by revolving this region about the x-axis.

1. $y = x$, $\quad y = 0$, $\quad x = 1$.

2. $x + y = 3$, $\quad y = 0$, $\quad x = 0$

3. $y = x^2$, $\quad y = 9$ **4.** $y = x^3$, $\quad y = 8$, $\quad x = 0$.

5. $y = \sqrt{x}$, $\quad y = x^3$. **6.** $y = x^2$, $\quad y = x^{1/3}$.

7. $y = x^3$, $\quad x + y = 10$, $\quad y = 1$.

8. $y = \sqrt{x}$, $\quad x + y = 6$, $\quad y = 1$.

9. $y = x^2$, $\quad y = x + 2$. **10.** $y = x^2$, $\quad y = 2 - x$.

11. $y = \sqrt{4 - x^2}$, $\quad y = 0$. **12.** $y = 1 - |x|$, $\quad y = 0$.

13. $y = \sec x$, $\quad x = 0$, $\quad x = \tfrac{1}{4}\pi$, $\quad y = 0$.

14. $y = \csc x$, $\quad x = \tfrac{1}{4}\pi$, $\quad x = \tfrac{3}{4}\pi$, $\quad y = 0$.

15. $y = \cos x$, $\quad y = x + 1$, $\quad x = \tfrac{1}{2}\pi$.

16. $y = \sin x$, $\quad x = \tfrac{1}{4}\pi$, $\quad x = \tfrac{1}{2}\pi$, $\quad y = 0$.

Exercises 17–26. Sketch the region Ω bounded by the curves and find the volume of the solid generated by revolving this region about the y-axis.

17. $y = 2x$, $\quad y = 4$, $\quad x = 0$.

18. $x + 3y = 6$, $\quad x = 0$, $\quad y = 0$.

19. $x = y^3$, $\quad x = 8$, $\quad y = 0$.

20. $x = y^2$, $\quad x = 4$. **21.** $y = \sqrt{x}$, $\quad y = x^3$.

22. $y = x^2$, $\quad y = x^{1/3}$. **23.** $y = x$, $\quad y = 2x$, $\quad x = 4$.

24. $x + y = 3$, $\quad 2x + y = 6$, $\quad x = 0$.

25. $x = y^2$, $\quad x = 2 - y^2$. **26.** $x = \sqrt{9 - y^2}$, $\quad x = 0$.

27. The base of a solid is the disk bounded by the circle $x^2 + y^2 = r^2$. Find the volume of the solid given that the cross sections perpendicular to the x-axis are: (a) squares; (b) equilateral triangles.

28. The base of a solid is the region bounded by the ellipse $4x^2 + 9y^2 = 36$. Find the volume of the solid given that cross sections perpendicular to the x-axis are: (a) equilateral triangles; (b) squares.

29. The base of a solid is the region bounded by $y = x^2$ and $y = 4$. Find the volume of the solid given that the cross sections perpendicular to the x-axis are: (a) squares; (b) semicircles; (c) equilateral triangles.

30. The base of a solid is the region between the parabolas $x = y^2$ and $x = 3 - 2y^2$. Find the volume of the solid given that the cross sections perpendicular to the x-axis are:

(a) rectangles of height h; (b) equilateral triangles;
(c) isosceles right triangles, hypotenuse on the xy-plane.

31. Carry out Exercise 29 with the cross sections perpendicular to the y-axis.

32. Carry out Exercise 30 with the cross sections perpendicular to the y-axis.

33. The base of a solid is the triangular region bounded by the y-axis and the lines $x + 2y = 4$, $x - 2y = 4$. Find the volume of the solid given that the cross sections perpendicular to the x-axis are: (a) squares; (b) isosceles right triangles with hypotenuse on the xy-plane.

34. The base of a solid is the region bounded by the ellipse $b^2x^2 + a^2y^2 = a^2b^2$. Find the volume of the solid given that the cross sections perpendicular to the x-axis are: (a) isosceles right triangles, hypotenuse on the xy-plane; (b) squares; (c) isosceles triangles of height 2.

35. The base of a solid is the region bounded by $y = 2\sqrt{\sin x}$ and the x-axis with $x \in [0, \pi/2]$. Find the volume of the solid given that cross sections perpendicular to the x-axis are: (a) equilateral triangles; (b) squares.

36. The base of a solid is the region bounded by $y = \sec x$ and $y = \tan x$ with $x \in [0, \pi/4]$. Find the volume of the solid given that cross sections perpendicular to the x-axis are: (a) semicircles; (b) squares.

37. Find the volume enclosed by the surface obtained by revolving the ellipse $b^2x^2 + a^2y^2 = a^2b^2$ about the x-axis.

38. Find the volume enclosed by the surface obtained by revolving the ellipse $b^2x^2 + a^2y^2 = a^2b^2$ about the y-axis.

39. Derive a formula for the volume of the frustum of a right circular cone in terms of the height h, the lower base radius R, and the upper base radius r. (See the figure.)

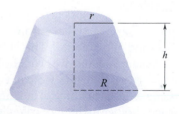

40. Find the volume enclosed by the surface obtained by revolving the equilateral triangle with vertices $(0, 0)$, $(a, 0)$, $(\frac{1}{2}a, \frac{1}{2}\sqrt{3}a)$ about the x-axis.

41. A hemispherical basin of radius r feet is being used to store water. To what percent of capacity is it filled when the water is:

(a) $\frac{1}{2}r$ feet deep?   (b) $\frac{1}{3}r$ feet deep?

42. A sphere of radius r is cut by two parallel planes: one, a units above the equator; the other, b units above the equator. Find the volume of the portion of the sphere that lies between the two planes. Assume that $a < b$.

43. A sphere of radius r is cut by a plane h units above the equator. Take $0 < h < r$. The top portion is called a *cap*. Derive the formula for the volume of a cap.

44. A hemispherical punch bowl 2 feet in diameter is filled to within 1 inch of the top. Thirty minutes after the party starts, there are only 2 inches of punch left at the bottom of the bowl.

(a) How much punch was there at the beginning?
(b) How much punch was consumed?

45. Let $f(x) = x^{-2/3}$ for $x > 0$.

(a) Sketch the graph of f.
(b) Calculate the area of the region bounded by the graph of f and the x-axis from $x = 1$ to $x = b$. Take $b > 1$.
(c) The region in part (b) is rotated about the x-axis. Find the volume of the resulting solid.
(d) What happens to the area of the region as $b \to \infty$? What happens to the volume of the solid?

46. This is a continuation of Exercise 45.

(a) Calculate the area of the region bounded by the graph of f and the x-axis from $x = c$ to $x = 1$. Take $0 < c < 1$.
(b) The region in part (a) is rotated about the x-axis. Find the volume of the resulting solid.
(c) What happens to the area of the region as $c \to 0^+$? What happens to the volume of the solid?

47. With x and y measured in feet, the configuration shown in the figure is revolved about the y-axis to form a parabolic container, no top. Given that a liquid is poured into the container at the rate of two cubic feet per minute, how fast is the level of the liquid rising when the depth of the liquid is 1 foot? 2 feet?

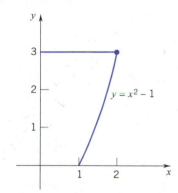

48. Let $x = f(y)$ be continuous and positive on the interval $[0, b]$. The configuration in the figure is revolved about the y-axis to form a container, no top. Suppose that the container is filled with water which then evaporates at a rate proportional to the area of the surface of the water. Show that the water level drops at a constant rate.

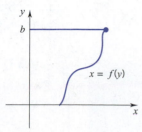

▶**49.** Set $f(x) = x^5$ and $g(x) = 2x$, $x \geq 0$.

(a) Use a graphing utility to display the graphs of f and g in one figure.
(b) Use a CAS to find the points of intersection of the two graphs.
(c) Use a CAS to find the area of the region bounded by the two graphs.

(d) The region in part (c) is revolved about the x-axis. Use a CAS to find the volume of the resulting solid.

50. Carry out Exercise 49 taking $f(x) = \sqrt{2x - 1}$ and $g(x) = x^2 - 4x + 4$.

51. The region between the graph of $f(x) = \sqrt{x}$ and the x-axis, $0 \le x \le 4$, is revolved about the line $y = 2$. Find the volume of the resulting solid.

52. The region bounded by the curves $y = (x - 1)^2$ and $y = x + 1$ is revolved about the line $y = -1$. Find the volume of the resulting solid.

53. The region between the graph of $y = \sin x$ and the x-axis, $0 \le x \le \pi$, is revolved about the line $y = 1$. Find the volume of the resulting solid.

54. The region bounded by $y = \sin x$ and $y = \cos x$, with $\pi/4 \le x \le \pi$, is revolved about the line $y = 1$. Find the volume of the resulting solid.

55. The region bounded by the curves $y = x^2 - 2x$ and $y = 3x$ is revolved about the line $y = -1$. Find the volume of the resulting solid.

56. Find the volume of the solid generated by revolving the region bounded by $y = x^2$ and $x = y^2$: (a) about the line $x = -2$; (b) about the line $x = 3$.

57. Find the volume of the solid generated by revolving the region bounded by $y^2 = 4x$ and $y = x$: (a) about the x-axis; (b) about the line $x = 4$.

58. Find the volume of the solid generated by revolving the region bounded by $y = x^2$ and $y = 4x$: (a) about the line $x = 5$; (b) about the line $x = -1$.

59. Find the volume of the solid generated by revolving the region OAB in the figure about: (a) the x-axis; (b) the line AB; (c) the line CA; (d) the y-axis.

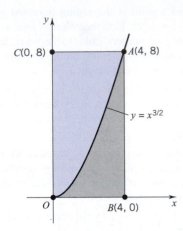

60. Find the volume of the solid generated by revolving the region OAC in the figure accompanying Exercise 59 about: (a) the y-axis; (b) the line CA; (c) the line AB; (d) the x-axis.

■ 6.3 VOLUME BY THE SHELL METHOD

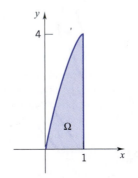

Figure 6.3.1

Figure 6.3.1 shows the region Ω below the curve $y = 5x - x^5$ from $x = 0$ to $x = 1$. By revolving Ω about the y-axis we obtain a solid of revolution. This solid has a certain volume. Call it V. To calculate V by the washer method we would have to express the curved boundary of Ω in the form $x = \phi(y)$, and this we can't do: given that $y = 5x - x^5$, we have no way of expressing x in terms of y. Thus, in this instance, the washer method fails. Below we introduce another method of calculating volume, a method by which we can avoid the difficulty just cited. It is called the *shell method.*

To describe the shell method of calculating volumes, we begin with a solid cylinder of radius R and height h, and from it we cut out a cylindrical core of radius r. (Figure 6.3.2)

Since the original cylinder has volume $\pi R^2 h$ and the piece removed has volume $\pi r^2 h$, the cylindrical shell that remains has volume

(6.3.1)
$$\pi R^2 h - \pi r^2 h = \pi h (R + r)(R - r).$$

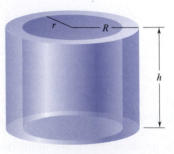

Figure 6.3.2

We will use this shortly.

Now let $[a, b]$ be an interval with $a \ge 0$ and let f be a nonnegative function continuous on $[a, b]$. If the region bounded by the graph of f and the x-axis is revolved about the y-axis, then a solid is generated. (Figure 6.3.3) The volume of this solid can be obtained from the formula

(6.3.2)
$$V = \int_a^b 2\pi x f(x)\, dx.$$

This is called the *shell-method formula*.

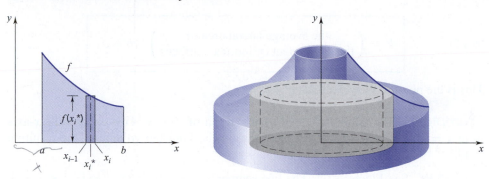

Figure 6.3.3

DERIVATION OF THE FORMULA We take a partition $P = \{x_0, x_1, \ldots, x_n\}$ of $[a, b]$ and concentrate on what happens on the *i*th subinterval $[x_{i-1}, x_i]$. Recall that when we form a Riemann sum we are free to choose x_i^* as any point from $[x_{i-1}, x_i]$. For convenience we take x_i^* as the midpoint $\frac{1}{2}(x_{i-1} + x_i)$. The representative rectangle of height $f(x_i^*)$ and base Δx_i (see Figure 6.3.2) generates a cylindrical shell of height $f(x_i^*)$, inner radius $r = x_{i-1}$, and outer radius $R = x_i$. We can calculate the volume of this shell by (6.3.1). Since

$$h = f\left(x_i^*\right) \qquad \text{and} \qquad R + r = x_i + x_{i-1} = 2x_i^* \qquad \text{and} \qquad R - r = \Delta x_i,$$

the volume of this shell is

$$\pi h(R + r)(R - r) = 2\pi x_i^* f\left(x_i^*\right) \Delta x_i.$$

The volume of the entire solid can be approximated by adding up the volumes of these shells:

$$V \cong 2\pi x_1^* f\left(x_1^*\right) \Delta x_1 + 2\pi x_2^* f\left(x_2^*\right) \Delta x_2 + \cdots + 2\pi x_n^* f\left(x_n^*\right) \Delta x_n.$$

The sum on the right is a Riemann sum. As $\|P\| \to 0$, such Riemann sums converge to

$$\int_a^b 2\pi x f(x)\, dx. \quad \square$$

Remark (*Average-value point of view*) To give some geometric insight into the shell-method formula, we refer to Figure 6.3.4. As the region below the graph of f is revolved about the y-axis, the vertical line segment at x generates a cylindrical surface of radius x, height $f(x)$, and lateral area $2\pi x f(x)$. As x ranges from $x = a$ to $x = b$, the cylindrical surfaces form a solid. The shell-method formula

$$V = \int_a^b 2\pi x f(x)\, dx$$

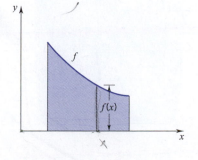

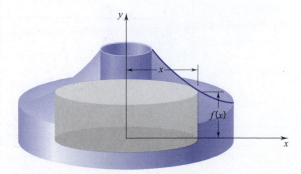

Figure 6.3.4

states that the volume of this solid can be expressed by writing

(6.3.3)
$$V = \left(\begin{array}{c} \text{the average lateral area of} \\ \text{the component cylindrical surfaces} \end{array} \right) \cdot (b - a)$$

This is the point of view we'll take. ❏

Example 1 The region bounded by the graph of $f(x) = 4x - x^2$ and the x-axis from $x = 1$ to $x = 4$ is revolved about the y-axis. Find the volume of the resulting solid.

SOLUTION See Figure 6.3.5. The line segment x units from the y-axis, $1 \le x \le 4$, generates a cylinder of radius x, height $f(x)$, and lateral area $2\pi x f(x)$. Thus

$$V = \int_1^4 2\pi x(4x - x^2)\,dx = 2\pi \int_1^4 (4x^2 - x^3)\,dx = 2\pi \left[\tfrac{4}{3}x^3 - \tfrac{1}{4}x^4 \right]_1^4 = \tfrac{81}{2}\pi. \quad ❏$$

The shell-method formula can be generalized. With Ω the region from $x = a$ to $x = b$ shown in Figure 6.3.6, the volume generated by revolving Ω about the y-axis is given by the formula

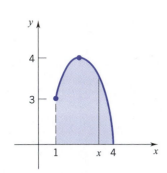

Figure 6.3.5

(6.3.4)
$$V = \int_a^b 2\pi x[f(x) - g(x)]\,dx.$$
(shell method about the y-axis)

The integrand $2\pi x[f(x) - g(x)]$ is the lateral area of the cylindrical surface, which is at a distance x from the axis of rotation.

As usual, we can interchange the roles played by x and y. With Ω the region from $y = c$ to $y = d$ shown in Figure 6.3.7, the volume generated by revolving Ω about the x-axis is given by the formula

(6.3.5)
$$V = \int_e^d 2\pi y[F(y) - G(y)]\,dy.$$
(shell method about the x-axis)

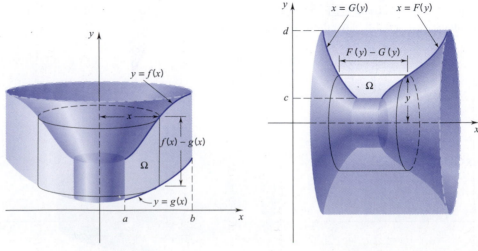

Figure 6.3.6

Figure 6.3.7

The integrand $2\pi y\,[F(y) - G(y)]$ is the lateral area of the cylindrical surface, which is at a distance y from the axis of rotation.

Example 2   Find the volume of the solid generated by revolving the region between

$$y = x^2 \qquad \text{and} \qquad y = 2x$$

(a) about the y-axis,   **(b)** about the x-axis.

SOLUTION   The curves intersect at the points $(0, 0)$ and $(2, 4)$.

(a) We refer to Figure 6.3.8. For each x from 0 to 2 the line segment at a distance x from the y-axis generates a cylindrical surface of radius x, height $(2x - x^2)$, and lateral area $2\pi x(2x - x^2)$. By (6.3.4),

$$V = \int_0^2 2\pi x(2x - x^2)\,dx = 2\pi \int_0^2 (2x^2 - x^3)\,dx = 2\pi \left[\tfrac{2}{3}x^3 - \tfrac{1}{4}x^4 \right]_0^2 = \tfrac{8}{3}\pi.$$

(b) We begin by expressing the bounding curves as functions of y. We write $x = \sqrt{y}$ for the right boundary and $x = \tfrac{1}{2}y$ for the left boundary. (See Figure 6.3.9.) For each y from 0 to 4 the line segment at a distance y from the x-axis generates a cylindrical surface of radius y, height $(\sqrt{y} - \tfrac{1}{2}y)$, and lateral area $2\pi y(\sqrt{y} - \tfrac{1}{2}y)$. By (6.3.5),

$$V = \int_0^4 2\pi y(\sqrt{y} - \tfrac{1}{2}y)\,dy = \pi \int_0^4 (2y^{3/2} - y^2)\,dy$$

$$= \pi \left[\tfrac{4}{5}y^{5/2} - \tfrac{1}{3}y^3 \right]_0^4 = \tfrac{64}{15}\pi. \quad \square$$

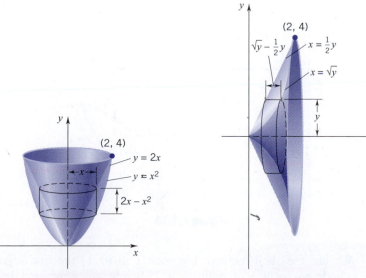

Figure 6.3.8   **Figure 6.3.9**

Example 3   A round hole of radius r is drilled through the center of a half-ball of radius a $(r < a)$. Find the volume of the remaining solid.

SOLUTION   A half-ball of radius a can be formed by revolving about the y-axis the first quadrant region bounded by $x^2 + y^2 = a^2$. What remains after the hole is drilled is the solid formed by revolving about the y-axis only that part of the region which is shaded in Figure 6.3.10.

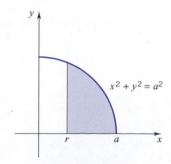

Figure 6.3.10

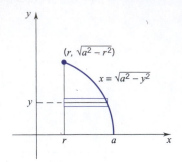

Figure 6.3.11

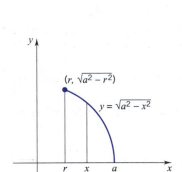

Figure 6.3.12

(a) By the washer method. We refer to Figure 6.3.11.

$$V = \int_0^{\sqrt{a^2-r^2}} \pi \left(\left[\sqrt{a^2-y^2} \right]^2 - r^2 \right) dy = \pi \int_0^{\sqrt{a^2-r^2}} (a^2 - r^2 - y^2) \, dy$$

$$= \pi \left[(a^2 - r^2)y - \tfrac{1}{3}y^3 \right]_0^{\sqrt{a^2-r^2}} = \tfrac{2}{3}\pi(a^2 - r^2)^{3/2}.$$

(b) By the shell method. We refer to Figure 6.3.12.

$$V = \int_r^a 2\pi x \sqrt{a^2 - x^2} \, dx.$$

Set $u = a^2 - x^2$, $du = -2x \, dx$. At $x = r$, $u = a^2 - r^2$; at $x = a$, $u = 0$. Therefore,

$$V = \int_r^a 2\pi x \sqrt{a^2 - x^2} \, dx = -\pi \int_{a^2-r^2}^0 u^{1/2} du = \pi \int_0^{a^2-r^2} u^{1/2} du$$

$$= \pi \left[\tfrac{2}{3}u^{3/2} \right]_0^{a^2-r^2} = \tfrac{2}{3}\pi(a^2 - r^2)^{3/2}.$$

If $r = 0$, no hole is drilled and $V = \tfrac{2}{3}\pi a^3$, the volume of the entire half-ball. ❑

In our last example we revolve a region about a line parallel to the y-axis.

Example 4 The region Ω between $y = \sqrt{x}$ and $y = x^2$, $0 \le x \le 1$, is revolved about the line $x = -2$. (See Figure 6.3.13.) Find the volume of the solid which is generated.

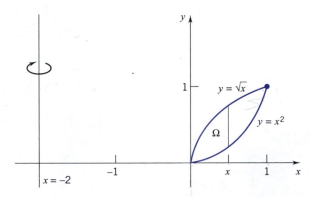

Figure 6.3.13

SOLUTION We use the shell method. For each x in $[0, 1]$ the line segment at x generates a cylindrical surface of radius $x + 2$, height $\sqrt{x} - x^2$, and lateral area $2\pi(x + 2)(\sqrt{x} - x^2)$. Therefore

$$V = \int_0^1 2\pi(x + 2)(\sqrt{x} - x^2) \, dx$$

$$= 2\pi \int_0^1 (x^{3/2} + 2x^{1/2} - x^3 - 2x^2) \, dx$$

$$= 2\pi \left[\tfrac{2}{5}x^{5/2} + \tfrac{4}{3}x^{3/2} - \tfrac{1}{4}x^4 - \tfrac{2}{3}x^3 \right]_0^1 = \tfrac{49}{30}\pi. \quad ❑$$

Remark We began this section by explaining that the washer method does not provide a way for us to calculate the volume generated by revolving about the y-axis the region shown in Figure 6.3.1. By the shell method we can easily calculate this volume:

$$V = \int_0^1 2\pi x(5x - x^5)\,dx = 2\pi \int_0^1 (5x^2 - x^6)\,dx = 2\pi\left[\tfrac{5}{3}x^3 - \tfrac{1}{7}x^7\right]_0^1 = \tfrac{64}{21}\pi. \quad \square$$

EXERCISES 6.3

Exercises 1–12. Sketch the region Ω bounded by the curves and use the shell method to find the volume of the solid generated by revolving Ω about the y-axis.

1. $y = x$, $\quad y = 0$, $\quad x = 1$.

2. $x + y = 3$, $\quad y = 0$, $\quad x = 0$.

3. $y = \sqrt{x}$, $\quad x = 4$, $\quad y = 0$.

4. $y = x^3$, $\quad x = 2$, $\quad y = 0$.

5. $y = \sqrt{x}$, $\quad y = x^3$. $\qquad$ **6.** $y = x^2$, $\quad y = x^{1/3}$.

7. $y = x$, $\quad y = 2x$, $\quad y = 4$.

8. $y = x$, $\quad y = 1$, $\quad x + y = 6$.

9. $x = y^2$, $\quad x = y + 2$. $\qquad$ **10.** $x = y^2$, $\quad x = 2 - y$.

11. $x = \sqrt{9 - y^2}$, $\quad x = 0$.

12. $x = |y|$, $\quad x = 2 - y^2$.

Exercises 13–24. Sketch the region Ω bounded by the curves and use the shell method to find the volume of the solid generated by revolving Ω about the x-axis.

13. $x + 3y = 6$, $\quad y = 0$, $\quad x = 0$.

14. $y = x$, $\quad y = 5$, $\quad x = 0$.

15. $y = x^2$, $\quad y = 9$. $\qquad$ **16.** $y = x^3$, $\quad y = 8$, $\quad x = 0$.

17. $y = \sqrt{x}$, $\quad y = x^3$. $\qquad$ **18.** $y = x^2$, $\quad y = x^{1/3}$.

19. $y = x^2$, $\quad y = x + 2$. $\qquad$ **20.** $y = x^2$, $\quad y = 2 - x$.

21. $y = x$, $\quad y = 2x$, $\quad x = 4$.

22. $y = x$, $\quad x + y = 8$, $\quad x = 1$.

23. $y = \sqrt{1 - x^2}$, $\quad x + y = 1$.

24. $y = x^2$, $\quad y = 2 - |x|$.

Exercises 25–30. The figure shows three regions within the unit square. Express the volume obtained by revolving the indicated region about the indicated line: (a) by an integral with respect to x; (b) by an integral with respect to y. Calculate each volume by evaluating one of these integrals

25. Ω_1, the y-axis. $\qquad$ **26.** Ω_1, the line $y = 2$.

27. Ω_2, the x-axis. $\qquad$ **28.** Ω_2, the line $x = -3$.

29. Ω_3, the y-axis. $\qquad$ **30.** Ω_3, the line $y = -1$.

31. Use the shell method to find the volume enclosed by the surface obtained by revolving the ellipse $b^2x^2 + a^2y^2 = a^2b^2$ about the y-axis.

32. Carry out Exercise 31 with the ellipse revolved about the x-axis.

33. Find the volume enclosed by the surface generated by revolving the equilateral triangle with vertices $(0, 0)$, $(a, 0)$, $(\tfrac{1}{2}a, \tfrac{1}{2}\sqrt{3}a)$ about the y-axis.

34. A ball of radius r is cut into two pieces by a horizontal plane a units above the center of the ball. Determine the volume of the upper piece by using the shell method.

35. Carry out Exercise 59 of Section 6.2, this time using the shell method.

36. Carry out Exercise 60 of Section 6.2, this time using the shell method.

37. (a) Verify that $F(x) = x \sin x + \cos x$ is an antiderivative of $f(x) = x \cos x$.
(b) Find the volume generated by revolving about the y-axis the region between $y = \cos x$ and the x-axis, $0 \leq x \leq \pi/2$.

38. (a) Sketch the region in the right half-plane that is outside the parabola $y = x^2$ and is between the lines $y = x + 2$ and $y = 2x - 2$.
(b) The region in part (a) is revolved about the y-axis. Use the method that you find most practical to calculate the volume of the solid generated.

For Exercises 39–42, set

$$f(x) = \begin{cases} \sqrt{3}x, & 0 \leq x < 1 \\ \sqrt{4 - x^2}, & 1 \leq x \leq 2, \end{cases}$$

and let Ω be the region between the graph of f and the x-axis. (See the figure.)

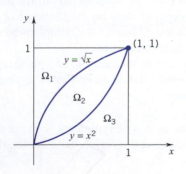

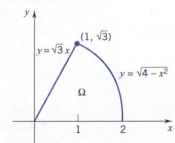

39. Revolve Ω about the y-axis.

(a) Express the volume of the resulting solid as an integral in x.

(b) Express the volume of the resulting solid as an integral in y.

(c) Calculate the volume of the solid by evaluating one of these integrals.

40. Carry out Exercise 39 for Ω revolved about the x-axis.

41. Carry out parts (a) and (b) of Exercise 39 for Ω revolved about the line $x = 2$.

42. Carry out parts (a) and (b) of Exercise 39 for Ω revolved about the line $y = -1$.

43. Let Ω be the circular disk $(x - b)^2 + y^2 \leq a^2, 0 < a < b$. The doughnut-shaped region generated by revolving Ω about the y-axis is called a *torus*. Express the volume of the torus as:

(a) a definite integral in x.

(b) a definite integral in y.

44. The circular disk $x^2 + y^2 \leq a^2, a > 0$, is revolved about the line $x = a$. Find the volume of the resulting solid.

45. Let r and h be positive numbers. The region in the first quadrant bounded by the line $x/r + y/h = 1$ and the coordinate axes is rotated about the y-axis. Use the shell method to derive the formula for the volume of a cone of radius r and height h.

46. A hole is drilled through the center of a ball of radius r, leaving a solid with a hollow cylindrical core of height h. Show that the volume of this solid is independent of the radius of the ball.

47. The region Ω in the first quadrant bounded by the parabola $y = r^2 - x^2$ and the coordinate axes is revolved about the y-axis. The resulting solid is called a *paraboloid*. A vertical hole of radius $a, a < r$, centered along the y-axis, is drilled through the paraboloid. Find the volume of the solid that remains: (a) by integrating with respect to x; (b) by integrating with respect to y.

48. (a) Draw the graph of $f(x) = \sin \pi x^2, x \in [-3, 3]$.

(b) Let Ω be the region bounded by the graph of f and the x-axis with $x \in [0, 1]$. If Ω is revolved about the x-axis and the disk method is used to calculate the volume, then the resulting integral *cannot be* readily evaluated by the fundamental theorem of calculus. Use a CAS to estimate this volume.

(c) If Ω is revolved about the y-axis and the shell method is used to calculate the volume, then the resulting integral *can be* evaluated by the fundamental theorem of calculus. Calculate this volume.

49. Set $f(x) = \sin x$ and $g(x) = \frac{1}{2}x$.

(a) Use a graphing utility to display the graphs of f and g in one figure.

(b) Use a CAS to find the points of intersection of the two graphs.

(c) Use a CAS to find the area of the region bounded by the two graphs.

(d) The region in part (c) is revolved about the y-axis. Use a CAS to find the volume of the resulting solid.

50. Set $f(x) = \dfrac{2}{(x + 1)^2}$ and $g(x) = \frac{3}{2} - \frac{1}{2}x$.

(a) Use a graphing utility to display the graphs of f and g in one figure.

(b) Use a CAS to find the points of intersection of the two graphs.

(c) Use a CAS to find the area of the region in the first quadrant bounded by the graphs.

(d) The region in part (c) is revolved about the y-axis. Use a CAS to find the volume of the resulting solid.

■ 6.4 THE CENTROID OF A REGION; PAPPUS'S THEOREM ON VOLUMES

The Centroid of a Region

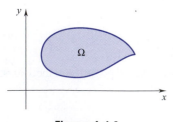

Figure 6.4.1

In Section 5.9 you saw how to locate the center of mass of a thin rod. Suppose now that we have a thin distribution of matter, a *plate*, laid out in the xy-plane in the shape of some region Ω. (Figure 6.4.1) If the mass density of the plate varies from point to point, then the determination of the center of mass of the plate requires the evaluation of a double integral. (Chapter 17) If, however, the mass density of the plate is constant throughout Ω, then the center of mass depends only on the shape of Ω and falls on a point $(\overline{x}, \overline{y})$ that we call the *centroid*. Unless Ω has a very complicated shape, we can locate the centroid by ordinary one-variable integration.

We will use two guiding principles to locate the centroid of a plane region. The first is obvious. The second we take from physics; the result conforms to physical intuition and is easily justified by double integration

Principle 1: Symmetry If the region has an axis of symmetry, then the centroid $(\overline{x}, \overline{y})$ lies somewhere along that axis. In particular, if the region has a center, then the center is the centroid.

Principle 2: Additivity If the region, having area A, consists of a finite number of pieces with areas $A_1, \ldots, A_n$ and centroids $(\overline{x}_1, \overline{y}_1), \ldots, (\overline{x}_n, \overline{y}_n)$, then

(6.4.1)
$$\overline{x}A = \overline{x}_1 A_1 + \cdots + \overline{x}_n A_n \quad \text{and} \quad \overline{y}A = \overline{y}_1 A_1 + \cdots + \overline{y}_n A_n.$$

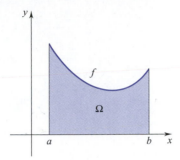

Figure 6.4.2

We are now ready to bring the techniques of calculus into play. Figure 6.4.2 shows the region Ω under the graph of a continuous function f. Denote the area of Ω by A. The centroid $(\overline{x}, \overline{y})$ of Ω can be obtained from the following formulas:

(6.4.2)
$$\overline{x}A = \int_a^b x f(x)\,dx, \quad \overline{y}A = \int_a^b \tfrac{1}{2}[f(x)]^2\,dx.$$

DERIVATION To derive these formulas we choose a partition $P = \{x_0, x_1, \ldots, x_n\}$ of $[a, b]$. This breaks up $[a, b]$ into n subintervals $[x_{i-1}, x_i]$. Choosing x_i^* as the midpoint of $[x_{i-1}, x_i]$, we form the midpoint rectangles R_i shown in Figure 6.4.3. The area of R_i is $f(x_i^*)\,\Delta x_i$, and the centroid of R_i is its center $(x_i^*, \tfrac{1}{2}f(x_i^*))$. By (6.4.1), the centroid $(\overline{x}_p, \overline{y}_p)$ of the union of all these rectangles satisfies the following equations:

$$\overline{x}_p A_p = x_1^* f(x_1^*)\,\Delta x_1 + \cdots + x_n^* f(x_n^*)\,\Delta x_n,$$

$$\overline{y}_p A_p = \tfrac{1}{2}[f(x_1^*)]^2 \Delta x_1 + \cdots + \tfrac{1}{2}[f(x_n^*)]^2 \Delta x_n.$$

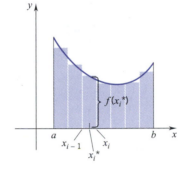

Figure 6.4.3

(Here A_P represents the area of the union of the n rectangles.) As $\|P\| \to 0$, the union of rectangles tends to the shape of Ω and the equations we just derived tend to the formulas given in (6.4.2). ❏

Before we start looking for centroids, we should explain what we are looking for. We learn from physics that, in our world of $W = mg$, the centroid of a plane region Ω is the balance point of the plate Ω, at least in the following sense: If Ω has centroid $(\overline{x}, \overline{y})$, then the plate Ω can be balanced on the line $x = \overline{x}$ and it can be balanced on the line $y = \overline{y}$. If $(\overline{x}, \overline{y})$ is actually in Ω, which is not necessarily the case, then the plate can be balanced at this point.

Example 1 Locate the centroid of the quarter-disk shown in Figure 6.4.4.

SOLUTION The quarter-disk is symmetric about the line $y = x$. Therefore we know that $\overline{x} = \overline{y}$. Here

$$\overline{y}A = \int_0^r \tfrac{1}{2}[f(x)]^2 dx = \int_0^r \tfrac{1}{2}(r^2 - x^2)\,dx = \tfrac{1}{2}\left[r^2 x - \tfrac{1}{3}x^3\right]_0^r = \tfrac{1}{3}r^3.$$

$$b(x) = \sqrt{r^2 - x^2} \longrightarrow$$

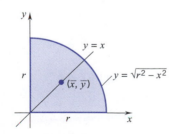

Figure 6.4.4

Since $A = \tfrac{1}{4}\pi r^2$,

$$\overline{y} = \frac{\tfrac{1}{3}r^3}{\tfrac{1}{4}\pi r^2} = \frac{4r}{3\pi}.$$

The centroid of the quarter-disk is the point

$$\left(\frac{4r}{3\pi}, \frac{4r}{3\pi}\right).$$

NOTE: It is almost as easy to calculate $\bar{x}A$:

$$\bar{x}A = \int_0^r xf(x)\,dx = \int_0^r x\sqrt{r^2 - x^2}\,dx$$

$$= -\frac{1}{2}\int_{r^2}^0 u^{1/2}\,du \qquad [u = (r^2 - x^2),\, du = -2x\,dx]$$

$$= -\frac{1}{2}\left[\frac{2}{3}u^{3/2}\right]_{r^2}^0 = \frac{1}{3}r^3,$$

and proceed from there. ❏

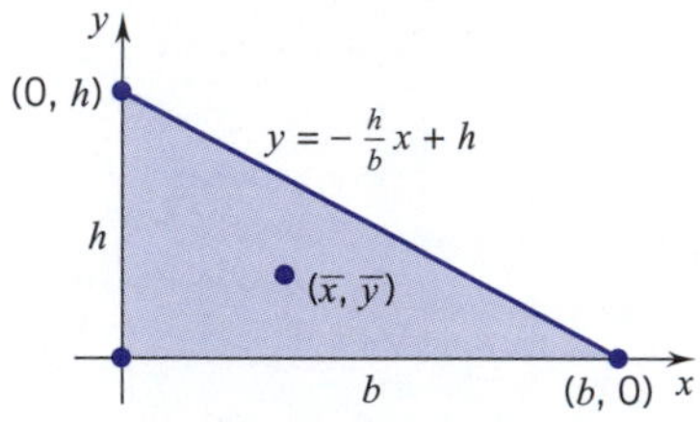

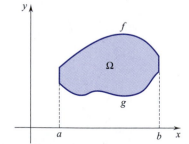

Figure 6.4.5

Example 2   Locate the centroid of the triangular region shown in Figure 6.4.5.

SOLUTION   The hypotenuse lies on the line

$$y = -\frac{h}{b}x + h\,.$$

Hence

$$\bar{x}A = \int_0^b xf(x)\,dx = \int_0^b \left(-\frac{h}{b}x^2 + hx\right)dx = \frac{1}{6}b^2h$$

and

$$\bar{y}A = \int_0^b \tfrac{1}{2}[f(x)]^2\,dx = \frac{1}{2}\int_0^b \left(\frac{h^2}{b^2}x^2 - \frac{2h^2}{b}x + h^2\right)dx = \tfrac{1}{6}bh^2.$$

Since $A = \frac{1}{2}bh$, we have

$$\bar{x} = \frac{\frac{1}{6}b^2h}{\frac{1}{2}bh} = \tfrac{1}{3}b \qquad \text{and} \qquad \bar{y} = \frac{\frac{1}{6}bh^2}{\frac{1}{2}bh} = \tfrac{1}{3}h\,.$$

The centroid is the point $(\tfrac{1}{3}b, \tfrac{1}{3}h)$. ❏

Figure 6.4.6 shows the region Ω between the graphs of two continuous functions f and g. In this case, if Ω has area A and centroid $(\bar{x}, \bar{y})$, then

(6.4.3) $$\bar{x}A = \int_a^b x[f(x) - g(x)]\,dx, \quad \bar{y}A = \int_a^b \tfrac{1}{2}([f(x)]^2 - [g(x)]^2)\,dx.$$

Figure 6.4.6

VERIFICATION   Let A_f be the area below the graph of f and let A_g be the area below the graph of g. Then, in obvious notation,

$$\bar{x}A + \bar{x}_g A_g = \bar{x}_f A_f \qquad \text{and} \qquad \bar{y}A + \bar{y}_g A_g = \bar{y}_f A_f$$

Therefore

$$\bar{x}A = \bar{x}_f A_f - \bar{x}_g A_g = \int_a^b xf(x)\,dx - \int_a^b xg(x)\,dx = \int_a^b x[f(x) - g(x)]\,dx$$

and

$$\bar{y}A = \bar{y}_f A_f - \bar{y}_g A_g = \int_a^b \tfrac{1}{2}[f(x)]^2\,dx - \int_a^b \tfrac{1}{2}[g(x)]^2\,dx$$

$$= \int_a^b \tfrac{1}{2}([f(x)]^2 - [g(x)]^2)\,dx. ❏$$

Example 3 Locate the centroid of the region shown in Figure 6.4.7.

SOLUTION Here there is no symmetry we can appeal to. We must carry out the calculations.

$$A = \int_0^2 [f(x) - g(x)]\, dx = \int_0^2 (2x - x^2)\, dx = \left[x^2 - \tfrac{1}{3}x^3\right]_0^2 = \tfrac{4}{3},$$

$$\overline{x} = \int_0^2 x[f(x) - g(x)]\, dx = \int_0^2 (2x^2 - x^3)\, dx = \left[\tfrac{2}{3}x^3 - \tfrac{1}{4}x^4\right]_0^2 = \tfrac{4}{3},$$

$$\overline{y}A = \int_0^2 \tfrac{1}{2}([f(x)]^2 - [g(x)]^2)\, dx = \tfrac{1}{2}\int_0^2 (4x^2 - x^4)\, dx = \tfrac{1}{2}\left[\tfrac{4}{3}x^3 - \tfrac{1}{5}x^5\right]_0^2 = \tfrac{32}{15}.$$

Therefore $\overline{x} = \tfrac{4}{3}/\tfrac{4}{3} = 1$ and $\overline{y} = \tfrac{32}{15}/\tfrac{4}{3} = \tfrac{8}{5}$. The centroid is the point $\left(1, \tfrac{8}{5}\right)$.

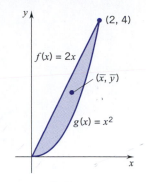

Figure 6.4.7

Pappus's Theorem on Volumes

All the formulas that we have derived for volumes of solids of revolution are simple corollaries to an observation made by a brilliant ancient Greek, Pappus of Alexandria (circa 300 A.D.).

THEOREM 6.4.4   PAPPUS'S THOREM ON VOLUMES[†]

A plane region is revolved about an axis that lies in its plane. If the region does not cross the axis, then the volume of the resulting solid of revolution is the area of the region multiplied by the circumference of the circle described by the centroid of the region:

$$V = 2\pi \overline{R} A$$

where A is the area of the region and $\overline{R}$ is the distance from the axis of revolution to the centroid of the region. (See Figure 6.4.8.)

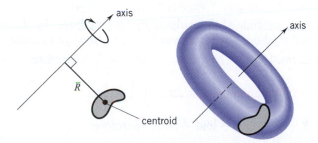

Figure 6.4.8

Basically we have derived only two formulas for the volumes of solids of revolution:

The Washer-Method Formula   If the region Ω of Figure 6.4.6 is revolved about the x-axis, the resulting solid has volume

$$V_x = \int_a^b \pi([f(x)]^2 - [g(x)]^2)\, dx.$$

[†]This theorem is found in Book VII of Pappus's *Mathematical Collection*, largely a survey of ancient geometry to which Pappus made many original contributions (among them this theorem). Much of what we know today of Greek geometry we owe to Pappus.

The Shell-Method Formula If the region Ω of Figure 6.4.6 is revolved about the y-axis, the resulting solid has volume

$$V_y = \int_a^b 2\pi x [f(x) - g(x)]\, dx.$$

Note that

$$V_x = \int_a^b \pi ([f(x)]^2 - [g(x)]^2)\, dx$$

$$= 2\pi \int_a^b \tfrac{1}{2} \left([f(x)]^2 - [g(x)]^2\right)\, dx = 2\pi \overline{y} A = 2\pi \overline{R} A$$

and

$$V_y = \int_a^b 2\pi x [f(x) - g(x)]\, dx = 2\pi \overline{x} A = 2\pi \overline{R} A,$$

as asserted by Pappus. ❏

Remark In stating Pappus's theorem, we assumed a complete revolution. If Ω is only partially revolved about a given axis, then the volume of the resulting solid is simply the area of Ω multiplied by the length of the circular arc described by the centroid. ❏

Applying Pappus's Theorem

Example 4 Earlier we saw that the region in Figure 6.4.7 has area $\frac{4}{3}$ and centroid $\left(1, \frac{8}{5}\right)$. Find the volumes of the solids formed by revolving this region **(a)** about the y-axis, **(b)** about the line $y = 5$.

SOLUTION

(a) We have already calculated this volume by two methods: by the washer method and by the shell method. The result was $V = \frac{8}{3}\pi$. Now we calculate the volume by Pappus's theorem. Here we have $\overline{R} = 1$ and $A = \frac{4}{3}$. Therefore

$$V = 2\pi (1) \left(\tfrac{4}{3}\right) = \tfrac{8}{3}\pi.$$

(b) In this case $\overline{R} = 5 - \frac{8}{5} = \frac{17}{5}$ and $A = \frac{4}{3}$. Therefore

$$V = 2\pi \left(\tfrac{17}{5}\right) \left(\tfrac{4}{3}\right) = \tfrac{136}{15}\pi.$$

Example 5 Find the volume of the torus generated by revolving the circular disk

$$(x - h)^2 + (y - k)^2 \leq r^2, \qquad h, k \geq r \qquad \text{(Figure 6.4.9)}$$

(a) about the x-axis, **(b)** about the y-axis.

SOLUTION The centroid of the disk is the center (h, k). This lies k units from the x-axis and h units from the y-axis. The area of the disk is πr^2. Therefore

(a) $V_x = 2\pi (k)(\pi r^2) = 2\pi^2 k r^2.$ **(b)** $V_y = 2\pi (h)(\pi r^2) = 2\pi^2 h r^2.$ ❏

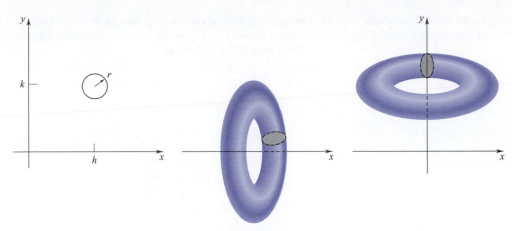

Figure 6.4.9

Example 6 Locate the centroid of the half-disk

$$x^2 + y^2 \leq r^2, \qquad y \geq 0$$

by appealing to Pappus's theorem.

SOLUTION Since the half-disk is symmetric about the y-axis, we know that $\overline{x} = 0$. All we need is $\overline{y}$.

If we revolve the half-disk about the x-axis, we obtain a solid ball of volume $\frac{4}{3}\pi r^3$. The area of the half-disk is $\frac{1}{2}\pi r^2$. By Pappus's theorem

$$\tfrac{4}{3}\pi r^3 = 2\pi \overline{y} \left(\tfrac{1}{2}\pi r^2 \right).$$

Simple division gives $\overline{y} = 4r/3\pi$. ❏

Remark Centroids of solids of revolution are introduced in Project 6.4. ❏

EXERCISES 6.4

Exercises 1–14. Sketch the region bounded by the curves. Locate the centroid of the region and find the volume generated by revolving the region about each of the coordinate axes.

1. $y = \sqrt{x}$, $y = 0$, $x = 4$. **2.** $y = x^3$, $y = 0$, $x = 2$.

3. $y = x^2$, $y = x^{1/3}$. **4.** $y = x^3$, $y = \sqrt{x}$.

5. $y = 2x$, $y = 2$, $x = 3$. **6.** $y = 3x$, $y = 6$, $x = 1$.

7. $y = x^2 + 2$, $y = 6$, $x = 0$.

8. $y = x^2 + 1$, $y = 1$, $x = 3$.

9. $\sqrt{x} + \sqrt{y} = 1$, $x + y = 1$.

10. $y = \sqrt{1 - x^2}$, $x + y = 1$.

11. $y = x^2$, $y = 0$, $x = 1$, $x = 2$.

12. $y = x^{1/3}$, $y = 1$, $x = 8$.

13. $y = x$, $x + y = 6$, $y = 1$.

14. $y = x$, $y = 2x$, $x = 3$.

Exercises 15–24 Locate the centroid of the bounded region determined by the curves.

15. $y = 6x - x^2$, $y = x$. **16.** $y = 4x - x^2$, $y = 2x - 3$.

17. $x^2 = 4y$, $x - 2y + 4 = 0$.

18. $y = x^2$, $2x - y + 3 = 0$.

19. $y^3 = x^2$, $2y = x$. **20.** $y^2 = 2x$, $y = x - x^2$.

21. $y = x^2 - 2x$, $y + 6x - x^2$.

22. $y = 6x - x^2$, $x + y = 6$.

23. $x + 1 = 0$, $x + y^2 = 0$.

24. $\sqrt{x} + \sqrt{y} = \sqrt{a}$, $x = 0$, $y = 0$.

25. Let Ω be the annular region (ring) formed by the circles

$$x^2 + y^2 = \tfrac{1}{4} \qquad x^2 + y^2 = 4.$$

(a) Locate the centroid of Ω. (b) Locate the centroid of the first-quadrant part of Ω. (c) Locate the centroid of the upper half of Ω.

26. The ellipse $b^2x^2 + a^2y^2 = a^2b^2$ encloses a region of area πab. Locate the centroid of the upper half of the region.

27. The rectangle in the accompanying figure is revolved about the line marked l. Find the volume of the resulting solid.

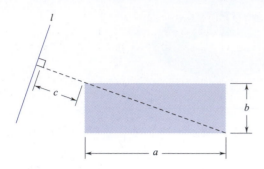

28. In Example 2 of this section you saw that the centroid of the triangle in Figure 6.4.5 is at the point $\left(\frac{1}{3}b, \frac{1}{3}h\right)$.

   (a) Verify that the line segments that join the centroid to the vertices divide the triangle into three triangles of equal area.

   (b) Find the distance d from the centroid of the triangle to the hypotenuse.

   (c) Find the volume generated by revolving the triangle about the hypotenuse.

29. The triangular region in the figure is the union of two right triangles Ω_1, Ω_2. Locate the centroid: (a) of Ω_1, (b) of Ω_2, (c) of the entire region.

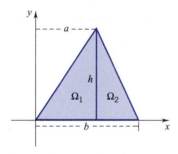

30. Find the volume of the solid generated by revolving the entire triangular region of Exercise 29. (a) about the x-axis; (b) about the y-axis.

31. (a) Find the volume of the ice-cream cone of Figure A. (A right circular cone topped by a solid hemisphere.)

   (b) Find $\bar{x}$ for the region Ω in Figure B.

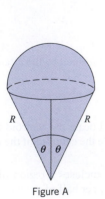

Figure A

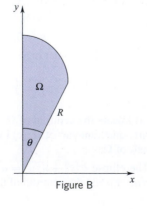

Figure B

32. The region Ω in the accompanying figure consists of a square S of side $2r$ and four semidisks of radius r. Locate the centroid of each of the following.

   (a) Ω.   (b) Ω_1.   (c) $S \cup \Omega_1$.   (d) $S \cup \Omega_3$.

   (e) $S \cup \Omega_1 \cup \Omega_3$.   (f) $S \cup \Omega_1 \cup \Omega_2$.

   (g) $S \cup \Omega_1 \cup \Omega_2 \cup \Omega_3$.

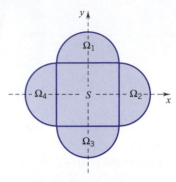

33. Give an example of a region that does not contain its centroid.

34. The centroid of a triangular region can be located without integration. Find the centroid of the region shown in the accompanying figure by applying Principles 1 and 2. Then verify that this point $\bar{x}, \bar{y}$ lies on each median of the triangle, two-thirds of the distance from the vertex to the opposite side.

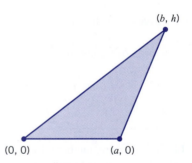

35. Use a graphing utility to draw the graphs of $y = \sqrt[3]{x}$ and $y = x^3$ for $x \geq 0$. Let Ω be the region bounded by the two curves. Use a CAS to find:

   (a) the area of Ω.

   (b) the centroid of Ω; plot the centroid.

   (c) the volume of the solid generated by revolving Ω about the x-axis.

   (d) the volume of the solid generated by revolving Ω about the y-axis.

36. Exercise 35 with $y = x^2 - 2x + 4$ and $y = 2x + 1$.

37. Use a graphing utility to draw the graphs of $y = 16 - 8x$ and $y = x^4 - 5x^2 + 4$. Let Ω be the region bounded by the two curves. Use a CAS to find:

   (a) the area of Ω.   (b) the centroid of Ω.

38. Exercise 37 with $y = 2 + \sqrt{x+2}$ and $y = \frac{1}{6}(5x^2 + 3x - 2)$.

■ PROJECT 6.4 Centroid of a Solid of Revolution

If a solid is *homogeneous* (constant mass density), then the center of mass depends only on the shape of the solid and is called the *centroid*. In general, determination of the centroid of a solid requires triple integration. (Chapter 17.) However, if the solid is a solid of revolution, then the centroid can be found by one-variable integration.

Let Ω be the region shown in the figure and let T be the solid generated by revolving Ω around the x-axis. By symmetry, the centroid of T is on the x-axis. Thus the centroid of T is determined solely by its x-coordinate $\overline{x}$.

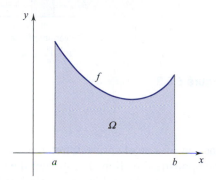

Problem 1. Show that $\overline{x}V = \int_a^b \pi x[f(x)]^2\,dx$ where V is the volume of T.

HINT: Use the following principle: if a solid of volume V consists of a finite number of pieces with volumes $V_1, V_2, \ldots, V_n$ and the pieces have centroids $\overline{x}_1, \overline{x}_2, \ldots, \overline{x}_n$, then $\overline{x}V = \overline{x}_1 V_1 + \overline{x}_2 V_2 + \cdots + \overline{x}_n V_n$.

Now revolve Ω around the y-axis and let S be the resulting solid. By symmetry, the centroid of S lies on the y-axis and is determined solely by its y-coordinate $\overline{y}$.

Problem 2. Show that $\overline{y}V = \int_a^b \pi x[f(x)]^2\,dx$ where V is the volume of S.

Problem 3. Use the results in Problems 1 and 2 to locate the centroid of each of the following solids:

(a) A solid cone of base radius r and height h.

(b) A ball of radius r.

(c) The solid generated by revolving about the x-axis the first-quadrant region bounded by the ellipse $b^2x^2 + a^2y^2 = a^2b^2$ and the coordinate axes.

(d) The solid generated by revolving the region below the graph of $f(x) = \sqrt{x}, x \in [0, 1]$, (i) about the x-axis; (ii) about the y-axis.

(e) The solid generated by revolving the region below the graph of $f(x) = 4 - x^2, x \in [0, 2]$, (i) about the x-axis; (ii) about the y-axis.

■ 6.5 THE NOTION OF WORK

We begin with a constant force F directed along some line that we call the x-axis. By convention we view F as positive if it acts in the direction of increasing x and negative if it acts in the direction of decreasing x. (Figure 6.5.1)

Figure 6.5.1

Suppose now that an object moves along the x-axis from $x = a$ to $x = b$ subject to this constant force F. The *work* done by F during the displacement is by definition the *force times the displacement*:

(6.5.1)
$$W = F \cdot (b - a).$$

It is not hard to see that, if F acts in the direction of the motion, then $W > 0$, but if F acts against the motion, then $W < 0$. Thus, for example, if an object slides off a table and falls to the floor, then the work done by gravity is positive (earth's gravity points down). But if an object is lifted from the floor and raised to tabletop level, then the work done by gravity is negative. However, the work done by the hand that lifts the object is positive.

To repeat, if an object moves from $x = a$ to $x = b$ subject to a constant force F, then the work done by F is the constant value of F times $b - a$. What is the work done by F if F does not remain constant but instead varies continuously as a function of x? As you would expect, we then define the work done by F as the *average value of F times b − a*:

(6.5.2)

$$W = \int_a^b f(x)\,dx.$$

(Figure 6.5.2)

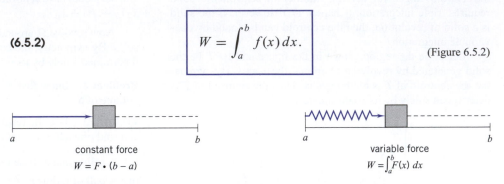

constant force
$W = F \cdot (b - a)$

variable force
$W = \int_a^b F(x)\,dx$

Figure 6.5.2

Hooke's Law

You can sense a variable force in the action of a steel spring. Stretch a spring within its elastic limit and you feel a pull in the opposite direction. The greater the stretching, the harder the pull of the spring. Compress a spring within its elastic limit and you feel a push against you. The greater the compression, the harder the push. According to Hooke's law (Robert Hooke, 1635–1703), the force exerted by the spring can be written

$$F(x) = -kx$$

where k is a positive number, called *the spring constant*, and x is the displacement from the equilibrium position. The minus sign indicates that the spring force always acts in the direction opposite to the direction in which the spring has been deformed (the force always acts so as to restore the spring to its equilibrium state).

Remark Hooke's law is only an approximation, but it is a good approximation for small displacements. In the problems that follow, we assume that the restoring force of the spring is given by Hooke's law. ❑

Example 1 A spring of natural length L, compressed to length $\frac{7}{8}L$, exerts a force F_0.

(a) Find the work done by the spring in restoring itself to natural length.

(b) What work must be done to stretch the spring to length $\frac{11}{10}L$?

SOLUTION Place the spring on the x-axis so that the equilibrium point falls at the origin. View compression as a move to the left. (See Figure 6.5.3.)

Our first step is to determine the spring constant. Compressed $\frac{1}{8}L$ units to the left, the spring exerts a force F_0. Thus, by Hooke's law

$$F_0 = F\left(-\tfrac{1}{8}L\right) = -k\left(-\tfrac{1}{8}L\right) = \tfrac{1}{8}kL.$$

Therefore $k = 8F_0/L$. The force law for this spring reads

$$F(x) = -\left(\frac{8F_0}{L}\right)x.$$

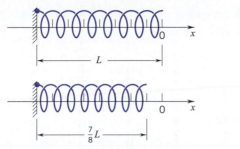

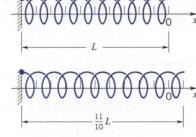

Figure 6.5.3   Figure 6.5.4

(a) To find the work done by this spring in restoring itself to equilibrium, we integrate $F(x)$ from $x = -\frac{1}{8}L$ to $x = 0$:

$$W = \int_{-L/8}^{0} F(x)\,dx = \int_{-L/8}^{0} -\left(\frac{8F_0}{L}\right) x\,dx = -\frac{8F_0}{L}\left[\frac{x^2}{2}\right]_{-L/8}^{0} = \frac{LF_0}{16}.$$

(b) We refer to Figure 6.5.4. To stretch the spring, we must counteract the force of the spring. The force exerted by the spring when stretched x units is

$$F(x) = -\left(\frac{8F_0}{L}\right) x.$$

To counter this force, we must apply the opposite force

$$-F(x) = \left(\frac{8F_0}{L}\right) x.$$

The work we must do to stretch the spring to length $\frac{11}{10}L$ can be found by integrating $-F(x)$ from $x = 0$ to $x = \frac{1}{10}L$:

$$W = \int_{0}^{L/10} -F(x)\,dx = \int_{0}^{L/10} \left(\frac{8F_0}{L}\right) x\,dx = \frac{8F_0}{L}\left[\frac{x^2}{2}\right]_{0}^{L/10} = \frac{LF_0}{25}. \quad ❑$$

Units   The unit of work is the work done by a unit force which displaces an object a unit distance in the direction of the force. If force is measured in *pounds* and distance is measured in *feet*, then the work is given in *foot-pounds*. In the SI system force is measured in *newtons*, distance is measured in *meters*, and work is given in *newton-meters*. These are called *joules*. There are other units used to quantify work, but for our purposes foot-pounds and joules are sufficient.[†]

Example 2   Stretched $\frac{1}{3}$ meter beyond its natural length, a certain spring exerts a restoring force with a magnitude of 10 newtons. What work must be done to stretch the spring an additional $\frac{1}{3}$ meter?

SOLUTION   Place the spring on the x-axis so that the equilibrium point falls at the origin. View stretching as a move to the right and assume Hooke's law: $F(x) = -kx$.

When the spring is stretched $\frac{1}{3}$ meter, it exerts a force of -10 newtons (10 newtons to the left). Therefore, $-10 = -k(\frac{1}{3})$ and $k = 30$.

___

[†]The term "*newton*" deserves definition. In general, force is measured by the acceleration that it imparts. The definition of a *newton* of force is made on that basis; namely, a force is said to measure 1 *newton* if it acts in the positive direction and imparts an acceleration of 1 meter per second per second to a mass of 1 kilogram.

To find the work necessary to stretch the spring an additional $\frac{1}{3}$ meter, we integrate the opposite force $-F(x) = 30x$ from $x = \frac{1}{3}$ to $x = \frac{2}{3}$:

$$W = \int_{1/3}^{2/3} 30x \, dx = 30 \left[\tfrac{1}{2}x^2 \right]_{1/3}^{2/3} = 5 \text{ joules.} \quad \square$$

Counteracting the Force of Gravity

To lift an object we must counteract the force of gravity. Therefore, the work required to lift an object is given by the equation

(6.5.3)

$$\boxed{\text{work} = (\text{weight of the object}) \times (\text{distance lifted}).^{\dagger}}$$

If an object is lifted from level $x = a$ to level $x = b$ and the weight of the object varies continuously with x—say the weight is $w(x)$—then the work done by the lifting force is given by the integral

(6.5.4)

$$\boxed{W = \int_a^b w(x) \, dx.}$$

This is just a special case of (6.5.2).

Example 3 A 150-pound bag of sand is hoisted from the ground to the top of a 50-foot building by a cable of negligible weight. Given that sand leaks out of the bag at the rate of 0.75 pounds for each foot that the bag is raised, find the work required to hoist the bag to the top of the building.

SOLUTION Once the bag has been raised x feet, the weight of the bag has been reduced to $150 - 0.75x$ pounds. Therefore

$$W = \int_0^{50} (150 - 0.75x) \, dx = \left[150x - \tfrac{1}{2}(0.75)x^2 \right]_0^{50}$$

$$= 150(50) - \tfrac{1}{2}(0.75)(50)^2 = 6562.5 \text{ foot-pounds} \quad \square$$

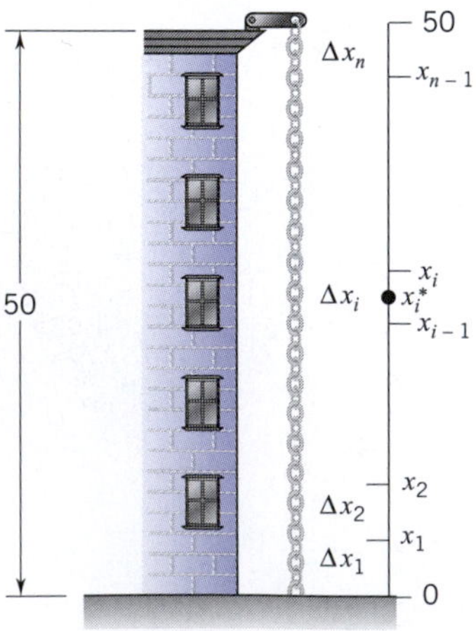

Example 4 What is the work required to hoist the sandbag of Example 3 given that the cable weighs 1.5 pounds per foot?

SOLUTION To the work required to hoist the sandbag of Example 3, which we found to be 6562.5 foot-pounds, we must add the work required to hoist the cable.

Instead of trying to apply (6.5.4), we go back to fundamentals. We partition the interval [0,50] as in Figure 6.5.5 and note that the ith piece of cable weighs $1.5\Delta x_i$ pounds and is approximately $50 - x_i^*$ feet from the top of the building. Thus the work required to lift this piece to the top is approximately

$$(1.5)\Delta x_i \left(50 - x_i^* \right) = 1.5\left(50 - x_i^* \right)\Delta x_i \text{ foot-pounds.}$$

Figure 6.5.5

---

†The weight of an object of mass m is the product mg where g is the magnitude of the acceleration due to gravity. The value of g is approximately 32 feet per second per second; in the metric system, approximately 9.8 meters per second per second.

It follows that the work required to hoist the entire cable is approximately

$$1.5(50 - x_1^*)\Delta x_1 + 1.5(50 - x_2^*)\Delta x_2 + \cdots + 1.5(50 - x_n^*)\Delta x_n \text{ foot-pounds.}$$

This sum is a Riemann sum which, as max $\Delta x_i \to 0$, converges to the definite integral

$$\int_0^{50} 1.5(50 - x)\,dx = 1.5\left[50x - \tfrac{1}{2}x^2\right]_0^{50} = 1875.$$

The work required to hoist the cable is 1875 foot-pounds.

The work required to hoist the sandbag by this cable is therefore

$$6562.5 \text{ foot-pounds} + 1875 \text{ foot-pounds} = 8437.5 \text{ foot-pounds.} \quad \square$$

Remark We just found that a hanging 50-foot cable that weighs 1.5 pounds per foot can be lifted to the point from which it hangs by doing 1875 foot-pounds of work. This result can be obtained by viewing the weight of the entire cable as concentrated at the center of mass of the cable: The cable weighs $1.5 \times 50 = 75$ pounds. Since the cable is homogeneous, the center of mass is at the midpoint of the cable, 25 feet below the suspension point. The work required to lift 75 pounds a distance of 25 feet is

$$75 \text{ pounds} \times 25 \text{ feet} = 1875 \text{ foot-pounds.}$$

We have found that this simplification works. But how come? To understand why this simplification works, we reason as follows: Initially the cable hangs from a suspension point 50 feet high. The work required to lift the bottom half of the cable to the 25-foot level can be offset exactly by the work done in lowering the top half of the cable to the 25-foot level. Thus, without doing any work (on a net basis), we can place the entire cable at the 25-foot level and proceed from there. $\quad \square$

(NOTE: In Exercise 31 you are asked to extend the center-of-mass argument to the nonhomogeneous case.)

Pumping Out a Tank Figure 6.5.6 depicts a storage tank filled to within a feet of the top with some liquid. Assume that the liquid is homogeneous and weighs $\sigma^\dagger$ pounds per cubic foot. Suppose now that this storage tank is pumped out from above so that the level of the liquid drops to b feet below the top of the tank. How much work has been done?

For each $x \in [a, b]$, we let

$$A(x) = \text{cross-sectional area } x \text{ feet below the top of the tank,}$$

$$s(x) = \text{distance that the } x\text{-level must be lifted.}$$

We let $P = \{x_0, x_1, \ldots, x_n\}$ be an arbitrary partition of $[a, b]$ and focus our attention on the ith subinterval $[x_{i-1}, x_i]$. (Figure 6.5.7) Taking x_i^* as an arbitrary point in the ith subinterval, we have

$$A(x_i^*)\Delta x_i = \text{approximate volume of the } i\text{th layer of liquid,}$$

$$\sigma A(x_i^*)\Delta x_i = \text{approximate weight of this volume,}$$

$$s(x_i^*) = \text{approximate distance this weight is to be lifted.}$$

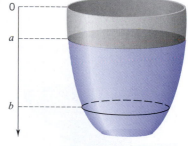

Figure 6.5.6

---

†The symbol σ is the lowercase Greek letter "sigma."

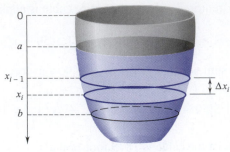

Figure 6.5.7

Therefore

$$\sigma s(x_i^*)A(x_i^*)\Delta x_i = \text{approximate work (weight} \times \text{distance) required to pump out}$$

this layer of liquid.

The work required to pump out all the liquid can be approximated by adding up all these terms:

$$W \cong \sigma s(x_1^*)A(x_1^*)\Delta x_1 + \sigma s(x_2^*)A(x_2^*)\Delta x_2 + \cdots + \sigma s(x_n^*)A(x_n^*)\Delta x_n.$$

The sum on the right is a Riemann sum. As $\|P\| \to 0$, such Riemann sums converge to give

(6.5.5)

$$W = \int_a^b \sigma s(x)\, A(x)\, dx.$$

We use this result in Example 5.

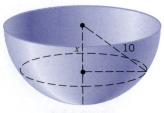

Figure 6.5.8

Example 5 A hemispherical water tank of radius 10 feet is being pumped out. (See Figure 6.5.8.) Find the work done in lowering the water level from 2 feet below the top of the tank to 4 feet below the top of the tank given that the pump is placed (**a**) at the top of the tank, (**b**) 3 feet above the top of the tank.

SOLUTION Take 62.5 pounds per cubic foot as the weight of water. From the figure you can see that the cross section x feet below the top of the tank is a disk of radius $\sqrt{100 - x^2}$. The area of this disk is

$$A(x) = \pi(100 - x^2).$$

In case (**a**) we have $s(x) = x$. Therefore

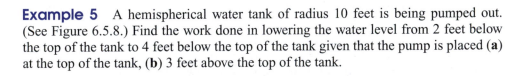

$$W = \int_2^4 62.5\pi x(100 - x^2)\, dx = 33{,}750\,\pi \cong 106{,}029 \text{ foot-pounds.}$$

In case (**b**) we have $s(x) = x + 3$. Therefore

$$W = \int_2^4 62.5\pi(x+3)(100-x^2)dx = 67{,}750\,\pi \cong 212{,}843 \text{ foot-pounds.} \quad \square$$

Suggestion: Work out Example 5 without invoking Formula (6.5.5). Simply construct the pertinent Riemann sums.

EXERCISES 6.5

Exercises 1–2. An object moves along the x-axis coordinatized in feet under the action of a force of $F(x)$ pounds. Find the work done by F in moving the object from $x = a$ to $x = b$.

1. $F(x) = x(x^2+1)^2$; $\quad a = 1, b = 4$.

2. $F(x) = 2x\sqrt{x+1}$; $\quad a = 3, b = 8$.

Exercises 3–6. An object moves along the x-axis coordinatized in meters under the action of a force of $F(x)$ newtons. Find the work done by F in moving the object from $x = a$ to $x = b$.

3. $F(x) = x\sqrt{x^2+7}$; $\quad a = 0, b = 3$.

4. $F(x) = x^2 + \cos 2x$; $\quad a = 0, b = \frac{1}{4}\pi$.

5. $F(x) = x + \sin 2x$; $\quad a = \frac{1}{6}\pi, b = \pi$.

6. $F(x) = \dfrac{\cos 2x}{\sqrt{2 + \sin 2x}}$; $\quad a = 0, b = \frac{1}{2}\pi$.

7. A 600-pound force compresses a 10-inch automobile coil exactly 1 inch. How much work must be done to compress that coil to 5 inches?

8. Five foot-pounds of work are needed to stretch a certain spring from 1 foot beyond natural length to 3 feet beyond natural length. How much stretching beyond natural length is achieved by a 6-pound force?

9. Stretched 4 feet beyond natural length, a certain spring exerts a restoring force of 200 pounds. How much work is required to stretch the spring: (a) 1 foot beyond natural length? (b) $1\frac{1}{2}$ feet beyond natural length?

10. A certain spring has natural length L. Given that W is the work required to stretch the spring from L feet to $L + a$ feet, find the work required to stretch the spring: (a) from L feet to $L + 2a$ feet; (b) from L feet to $L + na$ feet; (c) from $L + a$ feet to $L + 2a$ feet; (d) from $L + a$ feet to $L + na$ feet.

11. Find the natural length of a spring given that the work required to stretch it from 2 feet to 2.1 feet is one-half of the work required to stretch it from 2.1 feet to 2.2 feet.

12. A cylindrical tank of height 6 feet standing on a base of radius 2 feet is full of water. Find the work required to pump the water: (a) to an outlet at the top of the tank; (b) to a level of 5 feet above the top of the tank. (Take the weight of water as 62.5 pounds per cubic foot.)

13. A cylindrical tank of radius 3 feet and length 8 feet is laid out horizontally. The tank is half full of oil that weighs 60 pounds per cubic foot.

(a) Verify that the work done in pumping out the oil to the top of the tank is given by the integral

$$960 \int_0^3 (x+3)\sqrt{9-x^2}\,dx.$$

Evaluate this integral by evaluating the integrals

$$\int_0^3 x\sqrt{9-x^2}\,dx \quad \text{and} \quad \int_0^3 \sqrt{9-x^2}\,dx$$

separately.

(b) What is the work required to pump out the oil to a level 4 feet above the top of the tank?

▷14. Exercise 12 with the same tank laid out horizontally. Use a CAS for the integration.

15. Calculate the work required to hoist the cable of Example 4 by applying (6.5.4).

16. In the coordinate system used to derive (6.5.5) the liquid moves in the negative direction. How come W is positive?

17. A conical container (vertex down) of radius r feet and height h feet is full of liquid that weighs σ pounds per cubic foot. Find the work required to pump out the top $\frac{1}{2}h$ feet of liquid: (a) to the top of the tank; (b) to a level k feet above the top of the tank.

18. What is the work done by gravity if the tank of Exercise 17 is completely drained through an opening at the bottom?

19. A container of the form obtained by revolving the parabola $y = \frac{3}{4}x^2$, $0 \leq x \leq 4$, about the y-axis is full of water. Here x and y are given in meters. Find the work done in pumping the water: (a) to an outlet at the top of the tank; (b) to an outlet 1 meter above the top of the tank. Take $\sigma = 9800$.

20. The force of gravity exerted by the earth on a mass m at a distance r from the center of the earth is given by Newton's formula,

$$F = -G\frac{mM}{r^2},$$

where M is the mass of the earth and G is the universal gravitational constant. Find the work done by gravity in pulling a mass m from $r = r_1$ to $r = r_2$.

21. A chain that weighs 15 pounds per foot hangs to the ground from the top of an 80-foot building. How much work is required to pull the chain to the top of the building?

22. A box that weighs w pounds is dropped to the floor from a height of d feet. (a) What is the work done by gravity?

(b) Show that the work is the same if the box slides to the floor along a smooth inclined plane. (By saying "smooth," we are saying disregard friction.)

23. A 200-pound bag of sand is hoisted at a constant rate by a chain from ground level to the top of a building 100 feet high.
    (a) How much work is required to hoist the bag if the weight of the chain is negligible?
    (b) How much work is required to hoist the bag if the chain weighs 2 pounds per foot?

24. Suppose that the bag in Exercise 23 has a tear in the bottom and sand leaks out at a constant rate so that only 150 pounds of sand are left when the bag reaches the top.
    (a) How much work is required to hoist the bag if the weight of the chain is negligible?
    (b) How much work is required to hoist the bag if the chain weighs 2 pounds per foot?

25. A 100-pound bag of sand is lifted for 2 seconds at the rate of 4 feet per second. Find the work done in lifting the bag if the sand leaks out at the rate of half a pound per second.

26. A rope is used to pull up a bucket of water from the bottom of a 40-foot well. When the bucket is full of water, it weighs 40 pounds; however, there is a hole in the bottom, and the water leaks out at the constant rate of $\frac{1}{2}$ gallon for each 10 feet that the bucket is raised. Given that the weight of the rope is negligible, how much work is done in lifting the bucket to the top of the well? (Assume that water weighs 8.3 pounds per gallon.)

27. A rope of length l feet that weighs σ pounds per foot is lying on the ground. What is the work done in lifting the rope so that it hangs from a beam: (a) l feet high; (b) $2l$ feet high?

28. A load of weight w is lifted from the bottom of a shaft h feet deep. Find the work done given that the rope used to hoist the load weighs σ pounds per foot.

29. An 800-pound steel beam hangs from a 50-foot cable which weighs 6 pounds per foot. Find the work done in winding 20 feet of the cable about a steel drum.

30. A water container initially weighing w pounds is hoisted by a crane at the rate of n feet per second. What is the work done if the container is raised m feet and the water leaks out constantly at the rate of p gallons per second? (Assume that the water weighs 8.3 pounds per gallon.)

31. A chain of variable mass density hangs to the ground from the top of a building of height h. Show that the work required to pull the chain to the top of the building can be obtained by assuming that the weight of the entire chain is concentrated at the center of mass of the chain.

32. An object moves along the x-axis. At $x = a$ it has velocity v_a, and at $x = b$ it has velocity v_b. Use Newton's second law of motion, $F = ma = m(dv/dt)$, to show that

$$W = \int_a^b F(x)\,dx = \frac{1}{2}mv_b^2 - \frac{1}{2}mv_a^2.$$

The term $\frac{1}{2}mv^2$ is called the *kinetic energy* of the object. What you have been asked to show is that *the work done on*

an object equals the change in kinetic energy of that object. This is an important result.

33. An object of mass m is dropped from a height h. Express the impact velocity in terms of the gravitational constant g and the height h.

In Exercises 34–37 use the relation between work and kinetic energy given in Exercise 32.

34. The same amount of work on two objects results in the speed of one being three times that of the other. How are the masses of the two objects related?

35. A major league baseball weighs 5 oz. How much work is required to throw a baseball at a speed of 95 mph? (The ball's mass is its weight in pounds divided by 32 ft /sec², the acceleration due to gravity.)

36. How much work is required to increase the speed of a 2000-pound vehicle from 30 mph to 55 mph?

37. The speed of an earth satellite at an altitude of 100 miles is approximately 17,000 mph. How much work is required to launch a 1000-lb satellite into a 100-mile orbit?

(Power) Power is *work per unit time*. Suppose an object moves along the x-axis under the action of a force F. The work done by F in moving the object from $x = a$ to arbitrary x is given by the integral

$$W = \int_a^x F(u)\,du.$$

Viewing position as a function of time, setting $x = x(t)$, we have

$$P = \frac{dW}{dt} = F(x(t))\frac{dx}{dt} = F[x(t)]\,v(t).$$

This is called the *power* expended by the force F. If force is measured in pounds, distance in feet, and time in seconds, then power is given in foot-pounds per second. If force is measured in newtons, distance in meters, and time in seconds, then power is given in joules per second. These are called *watts*. Commonly used in engineering is the term *horsepower*:

1 horsepower = 550 foot-pounds per second

= 746 watts.

38. (a) Assume constant acceleration. What horsepower must an engine produce to accelerate a 3000-pound truck from 0 to 60 miles per hour (88 feet per second) in 15 seconds along a level road?
    (b) What horsepower must the engine produce if the road rises 4 feet for every 100 feet of road?
    HINT: Integration is not required to answer these questions.

39. A cylindrical tank set vertically with height 10 feet and radius 5 feet is half-filled with water. Given that a 1-horsepower pump can do 550 foot-pounds of work per second, how long will it take a $\frac{1}{2}$-horsepower pump:
    (a) to pump the water to an outlet at the top of the tank?

(b) to pump the water to a point 5 feet above the top of the tank?

40. A storage tank in the form of a hemisphere topped by a cylinder is filled with oil that weighs 60 pounds per cubic foot. The hemisphere has a 4-foot radius; the height of the cylinder is 8 feet.

(a) How much work is required to pump the oil to the top of the tank?

(b) How long would it take a $\frac{1}{2}$-horsepower motor to empty out the tank?

41. Show that *the rate of change of the kinetic energy of an object is the power of the force expended on it.*

■ *6.6 FLUID FORCE

If you pour oil into a container of water, you'll see that the oil soon rises to the top. Oil weighs less than water.

For any fluid, the weight per unit volume is called the *weight density* of the fluid. We'll denote this by the Greek letter σ.

An object submerged in a fluid experiences a *compressive force that acts at right angles to the surface of the body exposed to the fluid.* (It is to counter these compressive forces that submarines have to be built so structurally strong.)

Fluid in a container exerts a downward force on the base of the container. What is the magnitude of this force? It is the weight of the column of fluid directly above it. (Figure 6.6.1.) If a container with base area A is filled to a depth h by a fluid of weight density σ, the downward force on the base of the container is given by the product

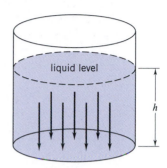

Figure 6.6.1

(6.6.1)
$$F = \sigma h A.$$

Fluid force acts not only on the base of the container but also on the walls of the container. In Figure 6.6.2, we have depicted a vertical wall standing against a body of liquid. (Think of it as the wall of a container or as a dam at the end of a lake.) We want to calculate the force exerted by the liquid on this wall.

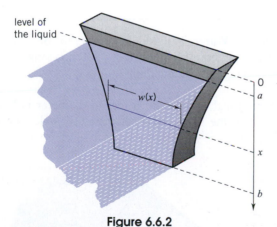

Figure 6.6.2

As in the figure, we assume that the liquid extends from depth a to depth b, and we let $w(x)$ denote the width of the wall at depth x. A partition $P = \{x_0, x_1, \ldots, x_n\}$ of $[a, b]$ of small norm subdivides the wall into n narrow horizontal strips. (See Figure 6.6.3.)

We can estimate the force on the ith strip by taking x_i^* as the midpoint of $[x_{i-1}, x_i]$. Then

$$w(x_i^*) = \text{the approximate width of the } i\text{th strip}$$

and

$$w(x_i^*)\Delta x_i = \text{the approximate area of the } i\text{th strip}.$$

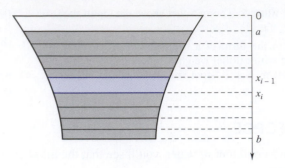

Figure 6.6.3

Since this strip is narrow, all the points of the strip are approximately at depth x_i^*. Thus, using (6.6.1), we can estimate the force on the ith strip by the product

$$\sigma x_i^* w(x_i^*) \Delta x_i.$$

Adding up all these estimates, we have an estimate for the force on the entire wall:

$$F \cong \sigma x_i^* w(x_i^*) \Delta x_1 + \sigma x_2^* w(x_2^*) \Delta x_2 + \cdots + \sigma x_n^* w(x_n^*) \Delta x_n.$$

The sum on the right is a Riemann sum for the integral

$$\int_a^b \sigma x w(x) \, dx.$$

As $\|P\| \to 0$, such Riemann sums converge to that integral. Thus we have

(6.6.2)

$$\boxed{\text{fluid force against the wall} = \int_a^b \sigma x w(x) \, dx.}$$

The Weight Density of Water The weight density σ of water is approximately 62.5 pounds per cubic foot; equivalently, about 9800 newtons per cubic meter. We'll use these values in carrying out computations.

Example 1 A cylindrical water main 6 feet in diameter is laid out horizontally. (Figure 6.6.4) Given that the main is capped half-full, calculate the fluid force on the cap.

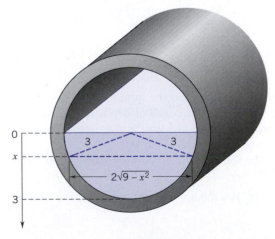

Figure 6.6.4

SOLUTION   Here $\sigma = 62.5$ pounds per cubic foot. From the figure we see that $w(x) = 2\sqrt{9 - x^2}$. The fluid force on the cap can be calculated as follows:

$$F = \int_0^3 (62.5)x(2\sqrt{9 - x^2})\,dx = 62.5 \int_0^3 2x\sqrt{9 - x^2}\,dx = 1125 \text{ pounds.} \quad \square$$

Example 2   A metal plate in the form of a trapezoid is affixed to a vertical dam as in Figure 6.6.5. The dimensions shown are in meters. Calculate the fluid force on the plate taking the weight density of water as 9800 newtons per cubic meter.

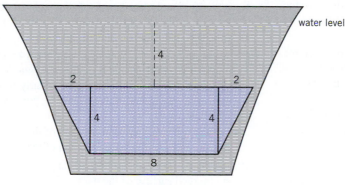

Figure 6.6.5

SOLUTION   First we find the width of the plate x meters below the water level. By similar triangles (see Figure 6.6.6),

$$t = \tfrac{1}{2}(8 - x) \qquad \text{so that} \qquad w(x) = 8 + 2t = 16 - x.$$

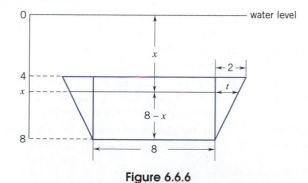

Figure 6.6.6

The fluid force against the plate is

$$\int_4^8 9800x(16 - x)\,dx = 9800 \int_4^8 (16x - x^2)\,dx$$

$$= 9800\left[8x^2 - \tfrac{1}{3}x^3\right]_4^8 \cong 2{,}300{,}000 \text{ newtons.} \quad \square$$

EXERCISES *6.6

1. A rectangular plate 8 feet by 6 feet is submerged vertically in a tank of water, an 8-foot edge at the surface of the water. Find the force of the water on each side of the plate.

2. A square plate 6 feet by 6 feet is submerged vertically in a tank of water, one edge parallel to the surface of the water. Calculate the fluid force on each side of the plate given that the center of the plate is 4 feet below the surface of the water.

3. A vertical dam at the end of a reservoir is in the form of an isosceles trapezoid: 100 meters across at the surface of the water, 60 feet across at the bottom. Given that the reservoir is 20 meters deep, calculate the force of the water on the dam.

4. A square metal plate 5 meters by 5 meters is affixed to the lowermost portion of the dam of Exercise 3. What is the force of the water on the plate?

5. A plate in the form of an isosceles trapezoid 4 meters at the top, 6 meters at the bottom, and 3 meters high has its upper edge 10 meters below the top of the dam of Exercise 3. Calculate the force of the water on this plate.

6. A vertical dam in the shape of a rectangle is 1000 feet wide and 100 feet high. Calculate the force on the dam given that
   (a) the water at the dam is 75 feet deep;
   (b) the water at the dam is 50 feet deep.

7. Each end of a horizontal oil tank is elliptical, with horizontal axis 12 feet long, vertical axis 6 feet long. Calculate the force on an end when the tank is half full of oil that weighs 60 pounds per cubic foot.

8. Each vertical end of a vat is a segment of a parabola (vertex down) 8 feet across the top and 16 feet deep. Calculate the force on an end when the vat is full of liquid that weighs 70 pounds per cubic foot.

9. The vertical ends of a water trough are isosceles right triangles with the 90° angle at the bottom. Calculate the force on an end of the trough when the trough is full of water given that the legs of the triangle are 8 feet long.

10. The vertical ends of a water trough are isosceles triangles 5 feet across the top and 5 feet deep. Calculate the force on an end when the trough is full of water.

11. The ends of a water trough are semicircular disks with radius 2 feet. Calculate the force of the water on an end given that the trough is full of water.

12. The ends of a water trough have the shape of the parabolic segment bounded by $y = x^2 - 4$ and $y = 0$; the measure-

ments are in feet. Assume that the trough is full of water and set up an integral that gives the force of the water on an end.

13. A horizontal cylindrical tank of diameter 8 feet is half full of oil that weighs 60 pounds per cubic foot. Calculate the force on an end.

14. Calculate the force on an end of the tank of Exercise 13 when the tank is full of oil.

15. A rectangular metal plate 10 feet by 6 feet is affixed to a vertical dam, the center of the plate 11 feet below water level. Calculate the force on the plate given that (a) the 10-foot sides are horizontal, (b) the 6-foot sides are horizontal.

16. A vertical cylindrical tank of diameter 30 feet and height 50 feet is full of oil that weighs 60 pounds per cubic foot. Calculate the force on the curved surface.

17. A swimming pool is 8 meters wide and 14 meters long. The pool is 1 meter deep at the shallow end and 3 meters deep at the deep end; the depth increases linearly from the shallow end to the deep end. Given that the pool is full of water, calculate
   (a) the force of the water on each of the sides,
   (b) the force of the water on each of the ends.

18. Relate the force on a vertical dam to the centroid of the submerged surface of the dam.

19. Two identical metal plates are affixed to a vertical dam. The centroid of the first plate is at depth h_1, and the centroid of the second plate is at depth h_2. Compare the forces on the two plates given that the two plates are completely submerged.

20. Show that if a plate submerged in a liquid makes an angle θ with the vertical, then the force on the plate is given by the formula

$$F = \int_a^b \sigma x \, w(x) \sec \theta \, dx$$

where σ is the weight density of the liquid and $w(x)$ is the width of the plate at depth x, $a \leq x \leq b$.

21. Find the force of the water on the bottom of the swimming pool of Exercise 17.

22. The face of a rectangular dam at the end of a reservoir is 1000 feet wide, 100 feet tall, and makes an angle of 30° with the vertical. Find the force of the water on the dam given that
   (a) the water level is at the top of the dam;
   (b) the water at the dam is 75 feet deep.

■ CHAPTER 6. REVIEW EXERCISES

Exercises 1–4. Sketch the region bounded by the curves. Represent the area of the region by one or more definite integrals (a) in terms of x; (b) in terms of y. Find the area of the region using the more convenient representation.

1. $y = 2 - x^2$, $y = -x$

2. $y = x^3$, $y = -x$, $y = 1$

3. $y^2 = 2(x - 1)$, $x - y = 5$

4. $y^3 = x^2$, $x - 3y + 4 = 0$

5. Find the area of the region bounded by $y = \sin x$ and $y = \cos x$ between consecutive intersections of the two graphs.

6. Find the area of the region bounded by $y = \tan^2 x$ and the x-axis from $x = 0$ to $x = \pi/4$.

7. The curve $y^2 = x(1 - x)^2$ is shown in the figure. Find the area of the loop.

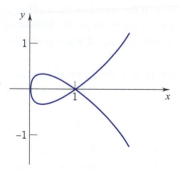

8. The curve $x^{1/2} + y^{1/2} = a^{1/2}$ is shown in the figure. Find the area of the region bounded by the curve and the coordinate axes.

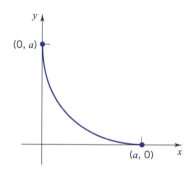

9. The base of a solid is the disk bounded by the circle $x^2 + y^2 = r^2$. Find the volume of the solid given that the cross sections perpendicular to the x-axis are: (a) semicircles; (b) isosceles right triangles with hypotenuse on the xy-plane.

10. The base of a solid is the region bounded by the equilateral triangle of side length a with one vertex at the origin and altitude along the positive x-axis. Find the volume of the solid given that cross-sections perpendicular to the x-axis are squares with one side on the base of the solid.

11. The base of a solid is the region in the first quadrant bounded by the coordinate axes and the line $2x + 3y = 6$. Find the volume of the solid given that the cross sections perpendicular to the x-axis are semicircles.

12. A solid in the shape of a right circular cylinder of radius 3 has its base on the xy-plane. A wedge is cut from the cylinder by a plane that passes through a diameter of the base and is inclined to the xy-plane at an angle of $30°$. Find the volume of the wedge.

Exercises 13–24. Sketch the region Ω bounded by the curves and find the volume of the solid generated by revolving Ω about the axis indicated.

13. $x^2 = 4y, \quad y = \frac{1}{2}x; \quad x$-axis.

14. $x^2 = 4y, \quad y = \frac{1}{2}x; \quad y$-axis.

15. $y = x^3, \quad y = 1, \quad x = 0; \quad x$-axis.

16. $y = x^3, \quad y = 1, \quad x = 0; \quad y$-axis.

17. $y = \sec x, \quad y = 0, \quad 0 \le x \le \pi/4; \quad x$-axis.

18. $y = \cos x, \quad -\pi/2 \le x \le \pi/2; \quad x$-axis.

19. $y = \sin x^2, \quad 0 \le x \le \sqrt{\pi}; \quad y$-axis.

20. $y = \cos x^2, \quad 0 \le x \le \sqrt{\pi/2}; \quad y$-axis.

21. $y = 3x - x^2, \quad y = x^2 - 3x; \quad y$-axis.

22. $y = 3x - x^2, \quad y = x^2 - 3x, \quad x = 4$.

23. $y = (x - 1)^2, \quad y = x + 1; \quad x$-axis.

24. $y = x^2 - 2x, \quad y = 3x; \quad y$-axis.

Exercises 25–30. The figure shows three regions within the rectangle bounded by the coordinate axes and the lines $x = 4$ and $y = 2$. Express the volume obtained by revolving the indicated region about the indicated line: (a) by an integral with respect to x; (b) by an integral with respect to y. Calculate each volume by evaluating one of these integrals

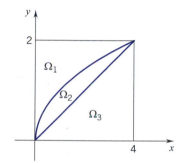

25. Ω_1; the x-axis.

26. Ω_1; the line $y = 2$.

27. Ω_2; the line $x = -1$.

28. Ω_2; the y-axis.

29. Ω_3; the y-axis.

30. Ω_3; the line $y = -2$.

Exercises 31–34. Find the centroid of the bounded region determined by the curves.

31. $y = 4 - x^2, \quad y = 0$.

32. $y = x^3, \quad y = 4x$.

33. $y = x^2 - 4, \quad y = 2x - x^2$.

34. $y = \cos x, \quad y = 0 \quad$ from $x = -\pi/2$ to $x = \pi/2$.

Exercises 35–36. Sketch the region bounded by the curves. Determine the centroid of the region and the volume of the solid generated by revolving the region about each of the coordinate axes.

35. $y = x, \quad y = 2 - x^2, \quad 0 \le x \le 1$

36. $y = x^3, \quad x = y^3, \quad 0 \le x \le 1$.

37. An object moves along the x-axis from $x = 0$ to $x = 3$ subject to a force $F(x) = x\sqrt{7 + x^2}$. Given that x is measured in feet and F in pounds, determine the work done by F.

38. One of the springs that supports a truck has a natural length of 12 inches. Given that a force of 8000 pounds compresses this spring $\frac{1}{2}$ inch, find the work required to compress the spring from 12 inches to 9 inches.

39. The work required to stretch a spring from 9 inches to 10 inches is 1.5 times the work needed to stretch the spring from 8 inches to 9 inches. What is the natural length of the spring?

40. A conical tank 10 feet deep and 8 feet across the top is filled with water to a depth of 5 feet. Find the work done in pumping the water (a) to an outlet at the top of the tank; (b) to an outlet 1 foot below the top of the tank. Take $\sigma = 62.5$ pounds per cubic foot as the weight density of water.

41. A 25-foot chain that weighs 4 pounds per foot hangs from the top of a tall building. How much work is required to pull the chain to the top of the building?

42. A bucket that weighs 5 pounds when empty rests on the ground filled with 60 pounds of sand. The bucket is lifted to the top of a 20 foot building at a constant rate. The sand leaks out of the bucket at a constant rate and only two-thirds of the sand remains when the bucket reaches the top. Find the work done in lifting the bucket of sand to the top of the building.

43. A spherical oil tank of radius 10 feet is half full of oil that weighs 60 pounds per cubic foot. Find the work required to pump the oil to an outlet at the top of the tank.

44. A rectangular fish tank has length 1 meter, width $\frac{1}{2}$ meter, depth $\frac{1}{2}$ meter. Given that the tank is full of water, find
    (a) the force of the water on each of the sides of the tank;
    (b) the force of the water on the bottom of the tank.
    Take the weight density of water as 9800 newtons per cubic meter.

45. A vertical dam is in the form of an isosceles trapezoid 300 meters across the top, 200 meters across the bottom, 50 meters high.
    (a) What is the force of the water on the face of the dam when the water level is even with the top of the dam?
    (b) What is the force of the water on the dam when the water level is 10 meters below the the top of the dam?
    Take the weight density of water as 9800 newtons per cubic meter.

CHAPTER 7

THE

TRANSCENDENTAL

FUNCTIONS

Some real numbers satisfy polynomial equations with integer coefficients:

$$\tfrac{3}{5} \text{ satisfies the equation } \quad 5x - 3 = 0;$$

$$\sqrt{2} \text{ satisfies the equation } \quad x^2 - 2 = 0.$$

Such numbers are called *algebraic*. There are, however, numbers that are not algebraic, among them π. Such numbers are called *transcendental*.

Some functions satisfy polynomial equations with polynomial coefficients:

$$f(x) = \frac{x}{\pi x + \sqrt{2}} \qquad \text{satisfies the equation} \qquad (\pi x + \sqrt{2})f(x) - x = 0;$$

$$f(x) = 2\sqrt{x} - 3x^2 \quad \text{satisfies the equation} \quad [f(x)]^2 + 6x^2 f(x) + (9x^4 - 4x) = 0.$$

Such functions are called *algebraic*. There are, however, functions that are not algebraic. Such functions are called *transcendental*. You are already familiar with some transcendental functions—the trigonometric functions. In this chapter we introduce other transcendental functions: the logarithm function, the exponential function, and the trigonometric inverses. But first, a little more on functions in general.

■ 7.1 ONE-TO-ONE FUNCTIONS; INVERSES

One-to-One Functions

A function can take on the same value at different points of its domain. Constant functions, for example, take on the same value at all points of their domains. The quadratic function $f(x) = x^2$ takes on the same value at $-c$ as it does at c; so does the absolute-value function $g(x) = |x|$. The function

$$f(x) = 1 + (x - 3)(x - 5)$$

takes on the same value at $x = 5$ as it does at $x = 3$:

$$f(3) = 1, \qquad f(5) = 1.$$

Functions for which this kind of repetition *does not* occur are called *one-to-one functions*.

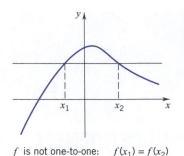

y

x_1 x_2 x

f is not one-to-one: $f(x_1) = f(x_2)$

Figure 7.1.1

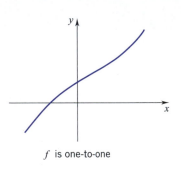

y

x

f is one-to-one

Figure 7.1.2

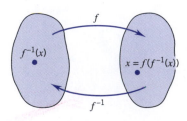

f

$f^{-1}(x)$

$x = f(f^{-1}(x))$

f^{-1}

Figure 7.1.3

> ### DEFINITION 7.1.1
>
> A function f is said to be *one-to-one* if there are no two distinct numbers in the domain of f at which f takes on the same value.
>
> $$f(x_1) = f(x_2) \quad \text{implies} \quad x_1 = x_2.$$

Thus, if f is one-to-one and x_1, x_2 are different points of the domain, then

$$f(x_1) \neq f(x_2).$$

The functions

$$f(x) = x^3 \quad \text{and} \quad f(x) = \sqrt{x}$$

are both one-to-one. The cubing function is one-to-one because no two distinct numbers have the same cube. The square-root function is one-to-one because no two distinct nonnegative numbers have the same square root.

There is a simple geometric test, called the *horizontal line test*, which can be used to determine whether a function is one-to-one. Look at the graph of the function. If some horizontal line intersects the graph more than once, then the function is not one-to-one. (Figure 7.1.1) If, on the other hand, no horizontal line intersects the graph more than once, then the function is one-to-one (Figure 7.1.2).

Inverses

We begin with a theorem about one-to-one functions.

> ### THEOREM 7.1.2
>
> If f is a one-to-one function, then there is one and only one function g defined on the range of f that satisfies the equation
>
> $$f(g(x)) = x \quad \text{for all } x \text{ in the range of } f.$$

PROOF The proof is straightforward. If x is in the range of f, then f must take on the value x at some number. Since f is one-to-one, there can be only one such number. We have called that number $g(x)$. ❑

The function that we have named g in the theorem is called the *inverse* of f and is usually denoted by the symbol f^{-1}.

> ### DEFINITION 7.1.3   INVERSE FUNCTION
>
> Let f be a one-to-one function. The *inverse* of f, denoted by f^{-1}, is the unique function defined on the range of f that satisfies the equation
>
> $$f(f^{-1}(x)) = x \quad \text{for all } x \text{ in the range of } f. \qquad \text{(Figure 7.1.3)}$$

Remark The notation f^{-1} for the inverse function is standard, at least in the United States. Unfortunately, there is the danger of confusing f^{-1} with the reciprocal of f, that is, with $1/f(x)$. The "-1" in the notation for the inverse of f is *not an exponent*; $f^{-1}(x)$ *does not mean* $1/f(x)$. On those occasions when we want to express $1/f(x)$ using the exponent -1, we will write $[f(x)]^{-1}$. ❑

Example 1 You have seen that the cubing function

$$f(x) = x^3$$

is one-to-one. Find the inverse.

SOLUTION We set $y = f^{-1}(x)$ and apply f to both sides:

$$f(y) = x$$
$$y^3 = x \qquad (f \text{ is the cubing function})$$
$$y = x^{1/3}.$$

Recalling that $y = f^{-1}(x)$, we have

$$f^{-1}(x) = x^{1/3}.$$

The inverse of the cubing function is the cube-root function. The graphs of $f(x) = x^3$ and $f^{-1}(x) = x^{1/3}$ are shown in Figure 7.1.4. ❑

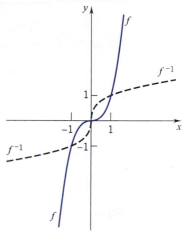

Figure 7.1.4

Remark We set $y = f^{-1}(x)$ to avoid clutter. It is easier to work with a single letter y than with the expression $f^{-1}(x)$. ❑

Example 2 Show that the linear function

$$y = 3x - 5$$

is one-to-one. Then find the inverse.

SOLUTION To show that f is one-to-one, let's suppose that

$$f(x_1) = f(x_2).$$

Then

$$3x_1 - 5 = 3x_2 - 5$$
$$3x_1 = 3x_2$$
$$x_1 = x_2.$$

The function is one-to-one since

$$f(x_1) = f(x_2) \qquad \text{implies} \qquad x_1 = x_2.$$

(Viewed geometrically, the result is obvious. The graph is a line with slope 3 and as such cannot be intersected by any horizontal line more than once.)

Now let's find the inverse. To do this, we set $y = f^{-1}(x)$ and apply f to both sides:

$$f(y) = x$$
$$3y - 5 = x$$
$$3y = x + 5$$
$$y = \tfrac{1}{3}x + \tfrac{5}{3}.$$

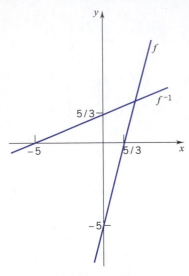

Figure 7.1.5

Recalling that $y = f^{-1}(x)$, we have

$$f^{-1}(x) = \tfrac{1}{3}x + \tfrac{5}{3}.$$

The graphs of f and f^{-1} are shown in Figure 7.1.5. ❏

Example 3   Find the inverse of the function

$$f(x) = (1 - x^3)^{1/5} + 2.$$

SOLUTION   We set $y = f^{-1}(x)$ and apply f to both sides:

$$f(y) = x$$
$$(1 - y^3)^{1/5} + 2 = x$$
$$(1 - y^3)^{1/5} = x - 2$$
$$1 - y^3 = (x - 2)^5$$
$$y^3 = 1 - (x - 2)^5$$
$$y = [1 - (x - 2)]^5]^{1/3}.$$

Recalling that $y = f^{-1}(x)$, we have

$$f^{-1}(x) = [1 - (x - 2)^5]^{1/3}.$$ ❏

Example 4   Show that the function

$$F(x) = x^5 + 2x^3 + 3x - 4$$

is one-to-one.

SOLUTION   Setting $F(x_1) = F(x_2)$, we have

$$x_1^5 + 2x_1^3 + 3x_1 - 4 = x_2^5 + 2x_2^3 + 3x_2 - 4$$
$$x_1^5 + 2x_1^3 + 3x_1 = x_2^5 + 2x_2^3 + 3x_2.$$

How to go on from here is far from clear. The algebra becomes complicated.
    Here is another approach. Differentiating F, we get

$$F'(x) = 5x^4 + 6x^2 + 3.$$

Note that $F'(x) > 0$ for all x and therefore F is an increasing function. Increasing functions are clearly one-to-one: $x_1 < x_2$ implies $F(x_1) < F(x_2)$, and so $F(x_1)$ cannot possibly equal $F(x_2)$. ❏

Remark   In Example 4 we used the sign of the derivative to test for one-to-oneness. For functions defined on an interval, the sign of the derivative and one-to-oneness can be summarized as follows: functions with positive derivative are increasing functions and therefore one-to-one; functions with negative derivative are decreasing functions and therefore one-to-one. ❏

Suppose that the function f has an inverse. Then, by definition, f^{-1} satisfies the equation

(7.1.4) | $f(f^{-1}(x)) = x$ for all x in the range of f.

It is also true that

(7.1.5)
$$f^{-1}(f(x)) = x \qquad \text{for all } x \text{ in the domain of } f.$$

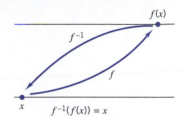

Figure 7.1.6

PROOF Take x in the domain of f and set $y = f(x)$. Since y is the range of f,

$$f(f^{-1}(y)) = y.$$

This means that

$$f(f^{-1}(f(x))) = f(x)$$

and tells us that f takes on the same value at $f^{-1}(f(x))$ as it does at x. With f one-to-one, this can only happen if

$$f^{-1}(f(x)) = x. \quad \Box$$

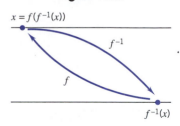

Figure 7.1.7

Equation (7.1.5) tells us that f^{-1} undoes what is done by f:

$$f \text{ takes } x \text{ to } f(x); \qquad f^{-1} \text{ takes } f(x) \text{ back to } x. \qquad \text{(Figure 7.1.6)}$$

Equation (7.1.4) tells us that f undoes what is done by f^{-1}:

$$f^{-1} \text{ takes } x \text{ to } f^{-1}(x); \qquad f \text{ takes } f^{-1}(x) \text{ back to } x. \qquad \text{(Figure 7.1.7)}$$

It is evident from this that

> domain of f^{-1} = range of f and range of f^{-1} = domain of f.

The Graphs of f and f^{-1}

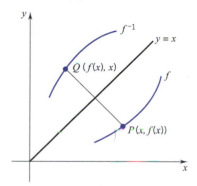

Figure 7.1.8

The graph of f consists of points $(x, f(x))$. Since f^{-1} takes on the value x at $f(x)$, the graph of f^{-1} consists of points $(f(x), x)$. If, as usual, we use the same scale on the y-axis as we do on the x-axis, then the points $(x, f(x))$ and $(f(x), x)$ are symmetric with respect to the line $y = x$. (Figure 7.1.8.) Thus we see that

> the graph of f^{-1} is the graph of f reflected in the line $y = x$.

This idea pervades all that follows.

Example 5 Sketch the graph of f^{-1} for the function f graphed in Figure 7.1.9.

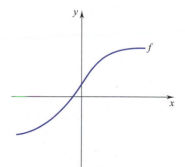

Figure 7.1.9

SOLUTION First we draw the line $y = x$. Then we reflect the graph of f in that line. The result is shown in Figure 7.1.10. ❏

Continuity and Differentiability of Inverses

Let f be a one-to-one function. Then f has an inverse, f^{-1}. Suppose, in addition, that f is continuous. Since the graph of f has no "holes" or "gaps," and since the graph of f^{-1} is simply the reflection of the graph of f in the line $y = x$, we can conclude that the graph of f^{-1} also has no holes or gaps; namely, we can conclude that f^{-1} is also continuous. We state this result formally; a proof is given in Appendix B.3.

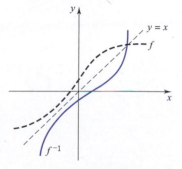

Figure 7.1.10

> **THEOREM 7.1.6**
>
> Let f be a one-to-one function defined on an open interval I. If f is continuous, then its inverse f^{-1} is also continuous.

Now suppose that f is differentiable. Is f^{-1} necessarily differentiable? Let's assume so for the moment.

From the definition of inverse, we know that

$$f(f^{-1}(x)) = x \qquad \text{for all } x \text{ in the range of } f.$$

Differentiation gives

$$\frac{d}{dx}[f(f^{-1}(x))] = 1.$$

However, by the chain rule,

$$\frac{d}{dx}[f(f^{-1}(x))] = f'(f^{-1}(x))(f^{-1})'(x).$$

Therefore

$$f'(f^{-1}(x))(f^{-1})'(x) = 1,$$

and, if $f'(f^{-1}(x)) \neq 0$,

(7.1.7)
$$(f^{-1})'(x) = \frac{1}{f'(f^{-1}(x))}.$$

For a geometric understanding of this relation, we refer you to Figure 7.1.11.

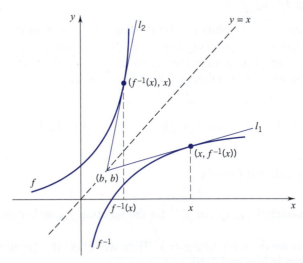

Figure 7.1.11

The graphs of f and f^{-1} are reflections of each other in the line $y = x$. The tangent lines l_1 and l_2 are also reflections of each other. From the figure,

$$(f^{-1})'(x) = \text{slope of } l_1 = \frac{f^{-1}(x) - b}{x - b}, \qquad f'(f^{-1}(x)) = \text{slope of } l_2 = \frac{x - b}{f^{-1}(x) - b},$$

so that $(f^{-1})'(x)$ and $f'(f^{-1}(x))$ are indeed reciprocals.

The figure shows two tangents intersecting the line $y = x$ at a common point. If the tangents have slope 1, they do not intersect that line at all. However, in that case, both graphs have slope 1, the derivatives are 1, and the relation holds. One more observation. We have assumed that $f'(f^{-1}(x)) \neq 0$. If $f'(f^{-1}(x)) = 0$, then the tangent to the graph of f at $(f^{-1}(x), x)$ is horizontal, and the tangent to the graph of f^{-1} at $(x, f^{-1}(x))$ is vertical. In this case f^{-1} is not differentiable at x.

Formula (7.1.7) has an unwieldy look about it; too many fussy little symbols. The following characterization of $(f^{-1})'$ may be easier to understand.

THEOREM 7.1.8

Let f be a one-to-one function differentiable on an open interval I. Let a be a point of I and let $f(a) = b$. If $f'(a) \neq 0$, then f^{-1} is differentiable at b and

$$(f^{-1})'(b) = \frac{1}{f'(a)}.$$

This theorem, proven in Appendix B.3, places our discussion on a firm footing.

Remark Note that $a = f^{-1}(b)$, and therefore

$$(f^{-1})'(b) = \frac{1}{f'(a)} = \frac{1}{f'(f^{-1}(b))}.$$

This is simply (7.1.7) at $x = b$. ❏

We rely on Theorem 7.1.8 when we cannot solve for f^{-1} explicitly and yet we want to evaluate $(f^{-1})'$ at a particular number.

Example 6 The function $f(x) = x^3 + \frac{1}{2}x$ is differentiable and has range $(-\infty, \infty)$.

(a) Show that f is one-to-one.

(b) Calculate $(f^{-1})'(9)$.

SOLUTION

(a) To show that f is one-to-one, we note that

$$f'(x) = 3x^2 + \frac{1}{2} > 0 \qquad \text{for all real } x.$$

Thus f is an increasing function and therefore one-to-one.

(b) To calculate $(f^{-1})'(9)$, we want to find a number a for which $f(a) = 9$. Then $(f^{-1})'(9)$ is simply $1/f'(a)$.

The assumption $f(a) = 9$ gives

$$a^3 + \frac{1}{2}a = 9$$

and tells us $a = 2$. (We must admit that this example was contrived so that the algebra would be easy to carry out.) Since $f'(2) = 3(2)^2 + \frac{1}{2} = \frac{25}{2}$, we conclude that

$$(f^{-1})'(9) = \frac{1}{f'(2)} = \frac{1}{\frac{25}{2}} = \frac{2}{25}. \qquad ❏$$

Finally, a few words about differentiating inverses in the Leibniz notation. Suppose that y is a one-to-one function of x:

$$y = y(x).$$

Then x is a one-to-one function of y:

$$x = x(y).$$

Moreover,

$$y(x(y)) = y \qquad \text{for all } y \text{ in the domain of } x.$$

Assuming that y is a differentiable function of x and x is a differentiable function of y, we have

$$y'(x(y))x'(y) = 1,$$

which, if $y'(x(y)) \neq 0$, gives

$$x'(y) = \frac{1}{y'(x(y))}.$$

In the Leibniz notation, we have

(7.1.9)

$$\frac{dx}{dy} = \frac{1}{dy/dx}.$$

The rate of change of x with respect to y is the reciprocal of the rate of change of y with respect to x.

Where are these rates of change to be evaluated? Given that $y(a) = b$, the right side is to be evaluated at $x = a$ and the left side at $y = b$.

EXERCISES 7.1

Exercises 1–26. Determine whether or not the function is one-to-one and, if so, find the inverse. If the function has an inverse, give the domain of the inverse.

1. $f(x) = 5x + 3.$

2. $f(x) = 3x + 5.$

3. $f(x) = 1 - x^2.$

4. $f(x) = x^5.$

5. $f(x) = x^5 + 1.$

6. $f(x) = x^2 - 3x + 2.$

7. $f(x) = 1 + 3x^3.$

8. $f(x) = x^3 - 1.$

9. $f(x) = (1 - x)^3.$

10. $f(x) = (1 - x)^4.$

11. $f(x) = (x + 1)^3 + 2.$

12. $f(x) = (4x - 1)^3.$

13. $f(x) = x^{3/5}.$

14. $f(x) = 1 - (x - 2)^{1/3}.$

15. $f(x) = (2 - 3x)^3.$

16. $f(x) = (2 - 3x^2)^3.$

17. $f(x) = \sin x, \ x \in \left[-\dfrac{\pi}{2}, \dfrac{\pi}{2}\right]$

18. $f(x) = \cos x, \ x \in \left[-\dfrac{\pi}{2}, \dfrac{\pi}{2}\right]$

19. $f(x) = \dfrac{1}{x}.$

20. $f(x) = \dfrac{1}{1 - x}.$

21. $f(x) = x + \dfrac{1}{x}.$

22. $f(x) = \dfrac{x}{|x|}.$

23. $f(x) = \dfrac{1}{x^3 + 1}.$

24. $f(x) = \dfrac{1}{1 - x} - 1.$

25. $f(x) = \dfrac{x + 2}{x + 1}.$

26. $f(x) = \dfrac{1}{(x + 1)^{2/3}}.$

27. What is the relation between a one-to-one function f and the function $(f^{-1})^{-1}$?

Exercises 28–31. Sketch the graph of the inverse of the function graphed below.

28.

29.

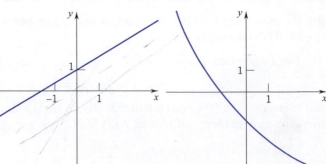

30.   **31.**

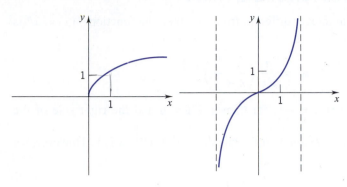

32. (a) Show that the composition of two one-to-one functions, f and g, is one-to-one.
  (b) Express $(f \circ g)^{-1}$ in terms of f^{-1} and g^{-1}.

33. (a) Let $f(x) = \frac{1}{3}x^3 + x^2 + kx$, k a constant. For what values of k is f one-to-one?
  (b) Let $g(x) = x^3 + kx^2 + x$, k a constant. For what values of k is g one-to-one?

34. (a) Suppose that f has an inverse, $f(2) = 5$, and $f'(2) = -\frac{3}{4}$. What is $(f^{-1})'(5)$?
  (b) Suppose that f has an inverse, $f(2) = -3$, and $f'(2) = \frac{2}{3}$. If $g = 1/f^{-1}$, what is $g'(-3)$?

Exercises 35–44. Verify that f has an inverse and find $(f^{-1})'(c)$.

35. $f(x) = x^3 + 1;\quad c = 9$.

36. $f(x) = 1 - 2x - x^3;\quad c = 4$.

37. $f(x) = x + 2\sqrt{x}, \quad x > 0;\quad c = 8$.

38. $f(x) = \sin x, -\frac{1}{2}\pi < x < \frac{1}{2}\pi;\quad c = -\frac{1}{2}$.

39. $f(x) = 2x + \cos x;\quad c = \pi$.

40. $f(x) = \dfrac{x+3}{x-1}, \quad x > 1;\quad c = 3$.

41. $f(x) = \tan x, -\frac{1}{2}\pi < x < \frac{1}{2}\pi;\quad c = \sqrt{3}$.

42. $f(x) = x^5 + 2x^3 + 2x;\quad c = -5$.

43. $f(x) = 3x - \dfrac{1}{x^3}, \quad x > 0;\quad c = 2$.

44. $f(x) = x - \pi + \cos x, \quad 0 < x < 2\pi;\quad c = -1$.

Exercises 45–47. Find a formula for $(f^{-1})'(x)$ given that f is one-to-one and its derivative satisfies the equation given.

45. $f'(x) = f(x)$.   **46.** $f'(x) = 1 + [f(x)]^2$.

47. $f'(x) = \sqrt{1 - [f(x)]^2}$.

48. Set

$$f(x) = \begin{cases} x^3 - 1, & x < 0 \\ x^2, & x \geq 0. \end{cases}$$

(a) Sketch the graph of f and verify that f is one-to-one.
(b) Find f^{-1}.

For Exercises 49 and 50, let $f(x) = \dfrac{ax + b}{cx + d}$.

49. (a) Show that f is one-to-one iff $ad - bc \neq 0$.
  (b) Suppose that $ad - bc \neq 0$. Find f^{-1}.

50. Determine the constants a, b, c, d for which $f = f^{-1}$.

51. Set

$$f(x) = \int_2^x \sqrt{1 + t^2}\, dt.$$

(a) Show that f has an inverse.
(b) Find $(f^{-1})'(0)$.

52. Set

$$f(x) = \int_1^{2x} \sqrt{16 + t^4}\, dt.$$

(a) Show that f has an inverse.
(b) Find $(f^{-1})'(0)$.

53. Let f be a twice differentiable one-to-one function and set $g = f^{-1}$.
  (a) Show that

$$g''(x) = -\frac{f''[g(x)]}{(f'[g(x)])^3}.$$

  (b) Suppose that the graph of f is concave up (down). What can you say then about the graph of f^{-1}?

54. Let P be a polynomial of degree n.
  (a) Can P have an inverse if n is even? Support your answer.
  (b) Can P have an inverse if n is odd? If so, give an example. Then give an example of a polynomial of odd degree that does not have an inverse.

55. The function $f(x) = \sin x, -\pi/2 < x < \pi/2$, is one-to-one, differentiable, and its derivative does not take on the value 0. Thus f has a differentiable inverse $y = f^{-1}(x)$. Find dy/dx by setting $f(y) = x$ and differentiating implicitly. Express the result as a function of x.

56. Exercise 55 for $f(x) = \tan x, -\pi/2 < x < \pi/2$.

Exercises 57–60. Find f^{-1}.

57. $f(x) = 4 + 3\sqrt{x - 1}, \quad x \geq 1$.

58. $f(x) = \dfrac{3x}{2x + 5}, \quad x \neq -5/2$.

59. $f(x) = \sqrt[3]{8 - x} + 2$.

60. $f(x) = \dfrac{1 - x}{1 + x}$.

▶ **Exercises 61–64.** Use a graphing utility to draw the graph of f. Show that f is one-to-one by consideration of f'. Draw a figure that displays both the graph of f and the graph of f^{-1}.

61. $f(x) = x^3 + 3x + 2$.   **62.** $f(x) = x^{3/5} - 1$.

63. $f(x) = 4\sin 2x, \quad -\pi/4 \leq x \leq \pi/4$.

64. $f(x) = 2 - \cos 3x, \quad 0 \leq x \leq \pi/3$.

■ 7.2 THE LOGARITHM FUNCTION, PART I

You have seen that if n is an integer different from -1, then the function $f(x) = x^n$ is a derivative:

$$x^n = \frac{d}{dx}\left(\frac{x^{n+1}}{n+1}\right).$$

This formula breaks down if $n = -1$, for then $n + 1 = 0$ and the right side of the formula is meaningless.

No function that we have studied so far has derivative $x^{-1} = 1/x$. However, we can easily construct one: set

$$L(x) = \int_1^x \frac{1}{t}dt.$$

From Theorem 5.3.5 we know that L is differentiable and

$$L'(x) = \frac{1}{x} \qquad \text{for all } x > 0.$$

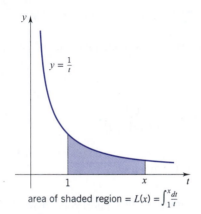

area of shaded region = $L(x) = \int_1^x \frac{dt}{t}$

Figure 7.2.1

This function has a remarkable property that we'll get to in a moment. First some preliminary observations: Make sure you understand them.

(1) L is defined for all $x > 0$.

(2) Since

$$L'(x) = \frac{1}{x} \qquad \text{for all } x > 0,$$

L increases on $(0, \infty)$.

(3) $L(x)$ is negative if $0 < x < 1$, $\quad L(1) = 0$, $\quad L(x)$ is positive for $x > 1$.

(4) For $x > 1$, $L(x)$ gives the area of the region shaded in Figure 7.2.1.

Now to the remarkable property.

THEOREM 7.2.1

For all positive numbers a and b,

$$L(ab) = L(a) + L(b).$$

PROOF Set $b > 0$. For all $x > 0$, $L(xb)$ and $L(x)$ have the same derivative:

$$\frac{d}{dx}[L(xb)] = \underset{\substack{\big\uparrow \\ \text{chain rule}}}{\frac{1}{xb}} \cdot b = \frac{1}{x} = \frac{d}{dx}[L(x)].$$

Therefore the two functions differ by some constant C:

$$L(xb) = L(x) + C. \qquad \text{(Theorem 4.2.4)}$$

We can evaluate C by setting $x = 1$:

$$L(b) = L(1 \cdot b) = \underset{\substack{\big\uparrow \\ L(1) = 0}}{L(1)} + C = C.$$

It follows that, for all $x > 0$,

$$L(x \cdot b) = L(x) + L(b).$$

We get the statement made in the theorem by setting $x = a$. ❑

From Theorem 7.2.1 and the fact that $L(1) = 0$, it readily follows that

(7.2.2)

> (1) for all positive numbers b, $\qquad L(1/b) = -L(b)$
>
> and
>
> (2) for all positive numbers a and b, $\qquad L(a/b) = L(a) - L(b)$.

PROOF

(1) $0 = L(1) = L(b \cdot 1/b) = L(b) + L(1/b)$ $\quad$ and therefore $\quad L(1/b) = -L(b)$;

(2) $L(a/b) = L(a \cdot 1/b) = L(a) + L(1/b) = L(a) - L(b)$. ❑

We now prove that

(7.2.3)

> for all positive numbers a and all rational numbers p/q,
>
> $$L(a^{p/q}) = \frac{p}{q} L(a).$$

PROOF You have seen that $d[L(x)]/dx = 1/x$. By the chain rule,

$$\frac{d}{dx}[L(x^{p/q})] = \frac{1}{x^{p/q}} \frac{d}{dx}(x^{p/q}) = \frac{1}{x^{p/q}} \left(\frac{p}{q}\right) x^{(p/q)-1} = \frac{p}{q}\left(\frac{1}{x}\right) = \frac{d}{dx}\left[\frac{p}{q}L(x)\right].$$

$$\underset{(3.7.1)}{\uparrow}$$

Since $L(x^{p/q})$ and $\dfrac{p}{q}L(x)$ have the same derivative, they differ by a constant:

$$L(x^{p/q}) = \frac{p}{q}L(x) + C.$$

Since both functions are zero at $x = 1, C = 0$. Therefore $L(x^{p/q}) = \dfrac{p}{q}L(x)$ for all $x > 0$. We get the theorem as stated by setting $x = a$. ❑

The domain of L is $(0, \infty)$. What is the range of L?

(7.2.4)

> The range of L is $(-\infty, \infty)$.

PROOF Since L is continuous on $(0, \infty)$, we know from the intermediate-value theorem that it "skips" no values. Thus, the range of L is an interval. To show that the interval is $(-\infty, \infty)$, we need only show that the interval is unbounded above and unbounded below. We can do this by taking M as an arbitrary positive number and showing that L takes on values greater than M and values less than $-M$.

Let M be an arbitrary positive number. Since

$$L(2) = \int_1^2 \frac{1}{t}\, dt$$

is positive (explain), we know that some positive multiple of $L(2)$ must be greater than M; namely, we know that there exists a positive integer n such that

$$nL(2) > M.$$

Multiplying this equation by -1, we have

$$-nL(2) < -M.$$

Since

$$nL(2) = L(2^n) \qquad \text{and} \qquad -nL(2) = L(2^{-n}), \qquad (7.2.3)$$

we have

$$L(2^n) > M \qquad \text{and} \qquad L(2^{-n}) < -M.$$

This proves that the range of L is unbounded in both directions. Since the range of L is an interval, it must be $(-\infty, \infty)$, the set of all real numbers. ❑

The Number e

Since the range of L is $(-\infty, \infty)$ and L is an increasing function (and therefore one-to-one), we know that L takes on as a value every real number and it does so only once. In particular, there is one and only one real number at which L takes on the value 1. *This unique number is denoted throughout the world by the letter e[†].*

Figure 7.2.2 locates e on the number line: the area under the curve $y = 1/t$ from $t = 1$ to $t = e$ is exactly 1.

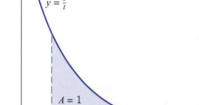

Figure 7.2.2

The Logarithm Function

Since

$$L(e) = \int_1^e \frac{1}{t}\, dt = 1,$$

we see from (7.2.3) that

(7.2.5)

for all rational numbers p/q

$$L(e^{p/q}) = \frac{p}{q}.$$

The function that we have labeled L is known as the *natural logarithm function*, or more simply as the *logarithm function*, and from now on $L(x)$ will be written $\ln x$. Here are the arithmetic properties of the logarithm function that we have already established. Both a and b represent arbitrary positive real numbers.

(7.2.6)

$\ln(1) = 0,$	$\ln(e) = 1,$
$\ln(ab) = \ln a + \ln b,$	$\ln(1/b) = -\ln b,$
$\ln(a/b) = \ln a - \ln b,$	$\ln a^{p/q} = \dfrac{p}{q}\ln a.$

[†]After the celebrated Swiss mathematician Leonhard Euler (1707–1783), considered by many the greatest mathematician of the eighteenth century.

The Graph of the Logarithm Function

You know that the logarithm function

$$\ln x = \int_1^x \frac{1}{t}\, dt$$

has domain $(0, \infty)$, range $(-\infty, \infty)$, and derivative

$$\frac{d}{dx}(\ln x) = \frac{1}{x}.$$

For small x the derivative is large (near 0, the curve is steep); for large x the derivative is small (far out, the curve flattens out). At $x = 1$ the logarithm is 0 and its derivative $1/x$ is 1. [The graph crosses the x-axis at the point $(1, 0)$, and the tangent line at that point is parallel to the line $y = x$.] The second derivative,

$$\frac{d^2}{dx^2}(\ln x) = -\frac{1}{x^2},$$

is negative on $(0, \infty)$. (The graph is concave down throughout.) We have sketched the graph in Figure 7.2.3. The y-axis is a vertical asymptote:

$$\text{as } x \to 0^+, \qquad \ln x \to -\infty.$$

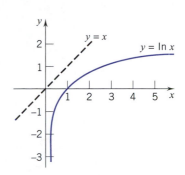

Figure 7.2.3

Example 1 We use upper and lower sums to estimate

$$\ln 2 = \int_1^2 \frac{dt}{t} \qquad \text{(Figure 7.2.4)}$$

from the partition

$$P = \{1 = \tfrac{10}{10}, \tfrac{11}{10}, \tfrac{12}{10}, \tfrac{13}{10}, \tfrac{14}{10}, \tfrac{15}{10}, \tfrac{16}{10}, \tfrac{17}{10}, \tfrac{18}{10}, \tfrac{19}{10}, \tfrac{20}{10} = 2\}.$$

Using a calculator, we find that

$$L_f(P) = \tfrac{1}{10}\left(\tfrac{10}{11} + \tfrac{10}{12} + \tfrac{10}{13} + \tfrac{10}{14} + \tfrac{10}{15} + \tfrac{10}{16} + \tfrac{10}{17} + \tfrac{10}{18} + \tfrac{10}{19} + \tfrac{10}{20}\right)$$

$$= \tfrac{1}{11} + \tfrac{1}{12} + \tfrac{1}{13} + \tfrac{1}{14} + \tfrac{1}{15} + \tfrac{1}{16} + \tfrac{1}{17} + \tfrac{1}{18} + \tfrac{1}{19} + \tfrac{1}{20} > 0.668$$

and

$$U_f(P) = \tfrac{1}{10}\left(\tfrac{10}{10} + \tfrac{10}{11} + \tfrac{10}{12} + \tfrac{10}{13} + \tfrac{10}{14} + \tfrac{10}{15} + \tfrac{10}{16} + \tfrac{10}{17} + \tfrac{10}{18} + \tfrac{10}{19}\right)$$

$$= \tfrac{1}{10} + \tfrac{1}{11} + \tfrac{1}{12} + \tfrac{1}{13} + \tfrac{1}{14} + \tfrac{1}{15} + \tfrac{1}{16} + \tfrac{1}{17} + \tfrac{1}{18} + \tfrac{1}{19} < 0.719.$$

We know then that

$$0.668 < L_f(P) \le \ln 2 \le U_f(P) < 0.719.$$

The average of these two estimates,

$$\tfrac{1}{2}(0.668 + 0.719) = 0.6935,$$

is not far off. Rounded off to four decimal places, our calculator gives $\ln 2 \cong 0.6931$. ❏

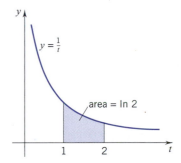

Figure 7.2.4

Table 7.2.1 gives the natural logarithms of the integers 1 through 10 rounded off to the nearest hundredth.

Example 2 Use the properties of logarithms and Table 7.2.1 to estimate the following:

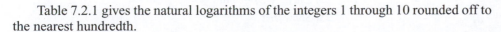

(a) $\ln 0.2$.   **(b)** $\ln 0.25$.   **(c)** $\ln 2.4$.   **(d)** $\ln 90$.

■ **Table 7.2.1**

n	$\ln n$	n	$\ln n$
1	0.00	6	1.79
2	0.69	7	1.95
3	1.10	8	2.08
4	1.39	9	2.20
5	1.61	10	2.30

SOLUTION

(a) $\ln 0.2 = \ln \frac{1}{5} = -\ln 5 \cong -1.61.$ **(b)** $\ln 0.25 = \ln \frac{1}{4} = -\ln 4 \cong -1.39.$

(c) $\ln 2.4 = \ln \frac{12}{5} = \ln \frac{(3)(4)}{5} = \ln 3 + \ln 4 - \ln 5 \cong 0.88.$

(d) $\ln 90 = \ln [(9)(10)] = \ln 9 + \ln 10 \cong 4.50.$ ❏

Example 3   Estimate e on the basis of Table 7.2.1.

SOLUTION   We know that $\ln e = 1$. From the table you can see that

$$3 \ln 3 - \ln 10 \cong 1.$$

The expression on the left can be written

$$\ln 3^3 - \ln 10 = \ln 27 - \ln 10 = \ln \frac{27}{10} = \ln 2.7.$$

This tells us that $\ln 2.7 \cong 1$ and therefore $e \cong 2.7.$ ❏

Remark   It can be shown that e is an irrational number, in fact a transcendental number. The decimal expansion of e to twelve decimal places reads

$$e \cong 2.718281828459.^\dagger \quad ❏$$

[†] Exercise 66 in Section 12.6 guides you through a proof of the irrationality of e. A proof that e is transcendental is beyond the reach of this text.

EXERCISES 7.2

Exercises 1–10. Estimate the logarithm on the basis of Table 7.2.1; check your results on a calculator.

1. $\ln 20.$ **2.** $\ln 16.$

3. $\ln 1.6.$ **4.** $\ln 3^4.$

5. $\ln 0.1.$ **6.** $\ln 2.5.$

7. $\ln 7.2.$ **8.** $\ln \sqrt{630}.$

9. $\ln \sqrt{2}$ **10.** $\ln 0.4.$

11. Verify that the area under the curve $y = 1/x$ from $x = 1$ to $x = 2$ equals the area from $x = 2$ to $x = 4$, the area from $x = 3$ to $x = 6$, the area from $x = 4$ to $x = 8$, and, more generally, the area from $x = k$ to $x = 2k$. Draw some figures.

12. Verify that the area under the curve $y = 1/x$ from $x = 1$ to $x = m$ equals the area from $x = 2$ to $x = 2m$, the area from $x = 3$ to $x = 3m$, and, more generally, the area from $x = k$ to $x = km$.

13. Estimate

$$\ln 1.5 = \int_1^{1.5} \frac{dt}{t}$$

by using the approximation $\frac{1}{2}[L_f(P) + U_f(P)]$ with

$$P = \{1 = \tfrac{8}{8}, \tfrac{9}{8}, \tfrac{10}{8}, \tfrac{11}{8}, \tfrac{12}{8} = 1.5\}.$$

14. Estimate

$$\ln 2.5 = \int_1^{2.5} \frac{dt}{t}$$

by using the approximation $\frac{1}{2}[L_f(P) + U_f(P)]$ with

$$P = \{1 = \tfrac{4}{4}, \tfrac{5}{4}, \tfrac{6}{4}, \tfrac{7}{4}, \tfrac{8}{4}, \tfrac{9}{4}, \tfrac{10}{4} = 2.5\}.$$

15. Taking $\ln 5 \cong 1.61$, use differentials to estimate
(a) $\ln 5.2$,   (b) $\ln 4.8$,   (c) $\ln 5.5$.

16. Taking $\ln 10 \cong 2.30$, use differentials to estimate
(a) $\ln 10.3$,   (b) $\ln 9.6$,   (c) $\ln 11$.

Exercises 17–22. Solve the equation for x.

17. $\ln x = 2.$ **18.** $\ln x = -1.$

19. $(2 - \ln x) \ln x = 0.$ **20.** $\frac{1}{2} \ln x = \ln (2x - 1).$

21. $\ln [(2x + 1)(x + 2)] = 2 \ln (x + 2).$

22. $2 \ln (x + 2) - \frac{1}{2} \ln x^4 = 1.$

23. Show that

$$\lim_{x \to 1} \frac{\ln x}{x - 1} = 1.$$

HINT: Note that $\dfrac{\ln x}{x - 1} = \dfrac{\ln x - \ln 1}{x - 1}$ and interpret the limit as a derivative.

Exercises 24–25. Let n be a positive integer greater than 2. Draw relevant figures.

24. Find the greatest integer k for which

$$\frac{1}{2} + \frac{1}{3} + \cdots + \frac{1}{k} < \ln n.$$

25. Find the least integer k for which

$$\ln n < 1 + \frac{1}{2} + \frac{1}{3} + \cdots + \frac{1}{k}.$$

▶ **Exercises 26–28.** A function g is given. (i) Use the intermediate-value theorem to conclude that there is a number r in the indicated interval at which $g(r) = \ln r$. (ii) Use a graphing utility to draw a figure that displays both the graph of the logarithm and the graph

of g on the indicated interval. Find r accurate to four decimal places.

26. $g(x) = 2x - 3$; [1, 2].

27. $g(x) = \sin x$; [2, 3].

28. $g(x) = \dfrac{1}{x^2}$; [1, 2].

▶ **Exercises 29–30.** Estimate the limit numerically by evaluating the function at the indicated values of x. Then use a graphing utility to zoom in on the graph and justify your estimate.

29. $\lim\limits_{x \to 1} \dfrac{\ln x}{x - 1}$; $x = 1 \pm 0.5, 1 \pm 0.1, 1 \pm 0.01, 1 \pm 0.001,$ $1 \pm 0.0001.$

30. $\lim\limits_{x \to 0^+} \sqrt{x} \ln x$; $x = 0.5, \ 0.1, \ 0.01, \ 0.001, \ 0.0001.$

■ 7.3 THE LOGARITHM FUNCTION, PART II

Differentiating and Graphing

We know that for $x > 0$

$$\frac{d}{dx}(\ln x) = \frac{1}{x}.$$

As usual, we differentiate composite functions by the chain rule. Thus

$$\frac{d}{dx}[\ln(1 + x^2)] = \frac{1}{1 + x^2}\frac{d}{dx}(1 + x^2) = \frac{2x}{1 + x^2} \qquad \text{for all real } x$$

and

$$\frac{d}{dx}[\ln(1 + 3x)] = \frac{1}{1 + 3x}\frac{d}{dx}(1 + 3x) = \frac{3}{1 + 3x} \qquad \text{for all } x > -\tfrac{1}{3}.$$

Example 1 Determine the domain and find $f'(x)$ for

$$f(x) = \ln(x\sqrt{4 + x^2}).$$

SOLUTION For x to be in the domain of f, we must have $x\sqrt{4 + x^2} > 0$, and thus we must have $x > 0$. The domain of f is the set of positive numbers.

Before differentiating f, we make use of the special properties of the logarithm:

$$f(x) = \ln(x\sqrt{4 + x^2}) = \ln x + \ln[(4 + x^2)^{1/2}] = \ln x + \tfrac{1}{2}\ln(4 + x^2).$$

From this we see that

$$f'(x) = \frac{1}{x} + \frac{1}{2} \cdot \frac{1}{4 + x^2} \cdot 2x = \frac{1}{x} + \frac{x}{4 + x^2} = \frac{4 + 2x^2}{x(4 + x^2)}. \quad \square$$

Example 2 Sketch the graph of

$$f(x) = \ln|x|.$$

SOLUTION The function, defined at all $x \neq 0$, is an even function: $f(-x) = f(x)$ for all $x \neq 0$. The graph has two branches:

$$y = \ln(-x), \qquad x < 0 \qquad \text{and} \qquad y = \ln x, \quad x > 0.$$

Each branch is the mirror image of the other. (Figure 7.3.1.) ❑

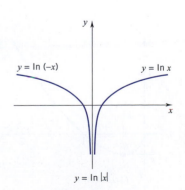

Figure 7.3.1

Example 3 (*Important*) Show that

(7.3.1)

$$\frac{d}{dx}(\ln |x|) = \frac{1}{x} \qquad \text{for all} \quad x \neq 0.$$

SOLUTION   For $x > 0$,

$$\frac{d}{dx}(\ln |x|) = \frac{d}{dx}(\ln x) = \frac{1}{x}.$$

For $x < 0$, we have $|x| = -x > 0$, and therefore

$$\frac{d}{dx}(\ln |x|) = \frac{d}{dx}[\ln(-x)] = \frac{1}{-x}\frac{d}{dx}(-x) = \left(\frac{1}{-x}\right)(-1) = \frac{1}{x}. \quad ❑$$

Applying the chain rule, we have

$$\frac{d}{dx}(\ln |1 - x^3|) = \frac{1}{1 - x^3}\frac{d}{dx}(1 - x^3) = \frac{-3x^2}{1 - x^3} = \frac{3x^2}{x^3 - 1}$$

and

$$\frac{d}{dx}\left(\ln \left|\frac{x - 1}{x - 2}\right|\right) = \frac{d}{dx}(\ln |x - 1|) - \frac{d}{dx}(\ln |x - 2|) = \frac{1}{x - 1} - \frac{1}{x - 2}.$$

Example 4   Set $f(x) = x \ln x$.
(a) Give the domain of f and indicate where f takes on the value 0. **(b)** On what intervals does f increase? decrease? **(c)** Find the extreme values of f. **(d)** Determine the concavity of the graph and give the points of inflection. **(e)** Sketch the graph of f.

SOLUTION   Since the logarithm function is defined only for positive numbers, the domain of f is $(0, \infty)$. The function takes on the value 0 at $x = 1$: $f(1) = 1 \ln 1 = 0$.
    Differentiating f, we have

$$f'(x) = x \cdot \frac{1}{x} + \ln x = 1 + \ln x.$$

To find the critical points of f, we set $f'(x) = 0$:

$$1 + \ln x = 0, \qquad \ln x = -1, \qquad x = \frac{1}{e}. \qquad \text{(verify this)}$$

Since the logarithm is an increasing function, the sign chart for f' looks like this:

sign of f': $- - - - - - - -0+ +$

behavior of f: decreases $\frac{1}{e}$ increases x

f decreases on $(0, 1/e]$ and increases on $[1/e, \infty)$. Therefore

$$f(1/e) = \frac{1}{e} \ln \left(\frac{1}{e}\right) = \frac{1}{e}(\ln 1 - \ln e) = -\frac{1}{e} \cong -\frac{1}{2.72} \cong -0.368$$

is a local minimum for f and the absolute minimum.
    Since $f''(x) = 1/x > 0$ for $x > 0$, the graph of f is concave up throughout. There are no points of inflection.
    You can verify numerically that $\lim_{x \to 0^+} x \ln x = 0$. Finally note that as $x \to \infty$, $x \ln x \to \infty$.
    A sketch of the graph of f is shown in Figure 7.3.2.   ❑

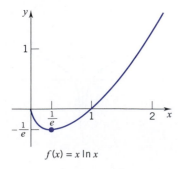

$f(x) = x \ln x$

Figure 7.3.2

Example 5   Set $f(x) = \ln\left(\dfrac{x^4}{x-1}\right)$.

(a) Specify the domain of f. **(b)** On what intervals does f increase? decrease? **(c)** Find the extreme values of f. **(d)** Determine the concavity of the graph and find the points of inflection. **(e)** Sketch the graph, specifying the asymptotes if any.

SOLUTION   Since the logarithm function is defined only for positive numbers, the domain of f is the open interval $(1, \infty)$.

Making use of the special properties of the logarithm, we write

$$f(x) = \ln x^4 - \ln(x - 1) = 4\ln x - \ln(x - 1).$$

Differentiation gives

$$f'(x) = \frac{4}{x} - \frac{1}{x-1} = \frac{3x - 4}{x(x-1)}$$

$$f''(x) = -\frac{4}{x^2} + \frac{1}{(x-1)^2} = -\frac{(x-2)(3x-2)}{x^2(x-1)^2}.$$

Since f is defined only for $x > 1$, we disregard all $x \le 1$. Note that $f'(x) = 0$ at $x = 4/3$ (critical point) and we have:

sign of f': $------0+++++++++++++++$

behavior of f: 1 $\frac{4}{3}$ x

decreases   increases

Thus f decreases on $(1, \frac{4}{3}]$ and increases on $[\frac{4}{3}, \infty)$. The number

$$f\left(\tfrac{4}{3}\right) = 4\ln 4 - 3\ln 3 \cong 2.25$$

is a local minimum and the absolute minimum. There are no other extreme values.

Testing for concavity: observe that $f''(x) = 0$ at $x = 2$. (We ignore $x = 2/3$ since $2/3$ is not part of the domain of f.) The sign chart for f'' looks like this:

sign of f'': $+++++++++++++++++++++0-----------------$

concavity: 1   concave up   2   concave down x

The graph is concave up on $(1, 2)$ and concave down on $(2, \infty)$. The point

$$(2, f(2)) = (2, 4\ln 2) \cong (2, 2.77)$$

is a point of inflection, the only point of inflection.

Before sketching the graph, we note that the derivative

$$f'(x) = \frac{4}{x} - \frac{1}{x-1}$$

is very large negative for x close to 1 and very close to 0 for x large. This tells us that the graph is very steep for x close to 1 and very flat for x large. See Figure 7.3.3. The line $x = 1$ is a vertical asymptote: as $x \to 1^+$, $f(x) \to \infty$.   ❑

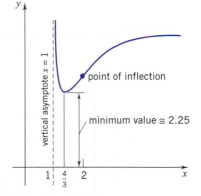

point of inflection

minimum value ≅ 2.25

vertical asymptote $x = 1$

Figure 7.3.3

Integration

The integral counterpart of (7.3.1) takes the form

(7.3.2)
$$\int \frac{1}{x}\,dx = \ln|x| + C.$$

The relation is valid on every interval that does not include 0.

Integrals of the form

$$\int \frac{u'(x)}{u(x)}dx \qquad \text{with } u(x) \neq 0 \qquad \text{can be written} \qquad \int \frac{1}{u}\,du$$

by setting

$$u = u(x), \qquad du = u'(x)\,dx.$$

Example 6 Calculate $\displaystyle\int \frac{x^2}{1 - 4x^3}\,dx$.

SOLUTION Up to a constant factor, x^2 is the derivative of $1 - 4x^3$. Therefore, we set

$$u = 1 - 4x^3, \qquad du = -12x^2 dx.$$

$$\int \frac{x^2}{1 - 4x^3}dx = -\tfrac{1}{12}\int \frac{du}{u} = -\tfrac{1}{12}\ln|u| + C = -\tfrac{1}{12}\ln|1 - 4x^3| + C. \quad \square$$

Example 7 Evaluate $\displaystyle\int_1^2 \frac{6x^2 + 2}{x^3 + x + 1}\,dx$.

SOLUTION Set $u = x^3 + x + 1, \quad du = (3x^2 + 1)\,dx.$
At $x = 1$, $u = 3$; at $x = 2$, $u = 11$.

$$\int_1^2 \frac{6x^2 + 2}{x^3 + x + 1}dx = 2\int_3^{11} \frac{du}{u} = 2\Big[\ln|u|\Big]_3^{11}$$

$$= 2(\ln 11 - \ln 3) = 2\ln\left(\tfrac{11}{3}\right). \quad \square$$

Here is an example of a different sort.

Example 8 Calculate $\displaystyle\int \frac{\ln x}{x}\,dx$.

SOLUTION Since $1/x$ is the derivative of $\ln x$, we set

$$u = \ln x, \qquad du = \frac{1}{x}\,dx.$$

This gives

$$\int \frac{\ln x}{x}dx = \int u\,du = \tfrac{1}{2}u^2 + C = \tfrac{1}{2}(\ln x)^2 + C. \quad \square$$

Integration of the Trigonometric Functions

We repeat Table 5.6.1:

$$\int \sin x\,dx = -\cos x + C \qquad \int \cos x\,dx = \sin x + C$$

$$\int \sec^2 x\,dx = \tan x + C \qquad \int \csc^2 x\,dx = -\cot x + C$$

$$\int \sec x \tan x\,dx = \sec x + C \qquad \int \csc x \cot x\,dx = -\csc x + C$$

Now that you are familiar with the logarithm function, we can add four more basic formulas to the table:

(7.3.3)

$$\int \tan x\, dx = -\ln|\cos x| + C = \ln|\sec x| + C$$

$$\int \cot x\, dx = \ln|\sin x| + C$$

$$\int \sec x\, dx = \ln|\sec x + \tan x| + C$$

$$\int \csc x\, dx = \ln|\csc x - \cot x| + C$$

The derivation of these formulas runs as follows:

$$\int \tan x\, dx = \int \frac{\sin x}{\cos x}\, dx \qquad (\text{set } u = \cos x, \quad du = -\sin x\, dx)$$

$$= -\int \frac{du}{u} = -\ln|u| + C$$

$$= -\ln|\cos x| + C = \ln\left|\frac{1}{\cos x}\right| + C$$

$$= \ln|\sec x| + C.$$

$$\int \cot x\, dx = \int \frac{\cos x}{\sin x}\, dx \qquad (\text{set } u = \sin x, \quad du = \cos x\, dx)$$

$$= \int \frac{du}{u} = \ln|u| + C = \ln|\sin x| + C.$$

$$\int \sec x\, dx \overset{\dagger}{=} \int \sec x \frac{\sec x + \tan x}{\sec x + \tan x}\, dx$$

$$= \int \frac{\sec x \tan x + \sec^2 x}{\sec x + \tan x}\, dx$$

$$[\text{set } u = \sec x + \tan x, \quad du = (\sec x \tan x + \sec^2 x)\, dx]$$

$$= \int \frac{du}{u} = \ln|u| + C = \ln|\sec x + \tan x| + C.$$

The derivation of the formula for $\int \csc x\, dx$ is left to you.

Example 9 Calculate $\int \cot \pi x\, dx$.

SOLUTION Set $u = \pi x$, $du = \pi\, dx$.

$$\int \cot \pi x\, dx = \frac{1}{\pi} \int \cot u\, du = \frac{1}{\pi} \ln|\sin u| + C = \frac{1}{\pi} \ln|\sin \pi x| + C. \quad \square$$

---

†Only experience prompts us to multiply numerator and denominator by $\sec x + \tan x$.

Remark   The u-substitution simplifies many calculations, but you will find with experience that you can carry out many of these integrations without it.   ❏

Example 10   Evaluate $\displaystyle\int_0^{\pi/8} \sec 2x \; dx$.

SOLUTION   As you can check, $\frac{1}{2}\ln|\sec 2x + \tan 2x|$ is an antiderivative for $\sec 2x$. Therefore

$$\int_0^{\pi/8} \sec 2x \; dx = \frac{1}{2}\Big[\ln|\sec 2x + \tan 2x|\Big]_0^{\pi/8}$$

$$= \tfrac{1}{2}[\ln(\sqrt{2}+1) - \ln 1] = \tfrac{1}{2}\ln(\sqrt{2}+1) \cong 0.44 \quad ❏$$

Example 11   Calculate $\displaystyle\int \frac{\sec^2 3x}{1+\tan 3x} dx$.

SOLUTION   Set $u = 1 + \tan 3x, \quad du = 3\sec^2 3x \; dx$.

$$\int \frac{\sec^2 3x}{1+\tan 3x} dx = \tfrac{1}{3}\int \frac{du}{u} = \tfrac{1}{3}\ln|u| + C = \tfrac{1}{3}\ln|1+\tan 3x| + C. \quad ❏$$

Logarithmic Differentiation

We can differentiate a lengthy product

$$g(x) = g_1(x)g_2(x)\cdots g_n(x)$$

by first writing

$$\ln|g(x)| = \ln(|g_1(x)||g_2(x)|\cdots|g_n(x)|)$$

$$= \ln|g_1(x)| + \ln|g_2(x)| + \cdots + \ln|g_n(x)|$$

and then differentiating:

$$\frac{g'(x)}{g(x)} = \frac{g_1'(x)}{g_1(x)} + \frac{g_2'(x)}{g_2(x)} + \cdots + \frac{g_n'(x)}{g_n(x)}.$$

Multiplication by $g(x)$ then gives

(7.3.4)

$$g'(x) = g(x)\left(\frac{g_1'(x)}{g_1(x)} + \frac{g_2'(x)}{g_2(x)} + \cdots + \frac{g_n'(x)}{g_n(x)}\right).$$

The process by which $g'(x)$ was obtained is called *logarithmic differentiation*. Logarithmic differentiation is valid at all points x where $g(x) \neq 0$. At points x where $g(x) = 0$, the process fails.

A product of n factors,

$$g(x) = g_1(x)g_2(x)\cdots g_n(x)$$

can, of course, also be differentiated by repeated applications of the product rule, Theorem 3.2.6. The great advantage of logarithmic differentiation is that it readily gives us an explicit formula for the derivative, a formula that's easy to remember and easy to work with.

Example 12   Calculate the derivative of

$$g(x) = x(x-1)(x-2)(x-3)$$

by logarithmic differentiation.

SOLUTION We can write down $g'(x)$ directly from Formula (7.3.4):

$$g'(x) = x(x - 1)(x - 2)(x - 3) \left(\frac{1}{x} + \frac{1}{x - 1} + \frac{1}{x - 2} + \frac{1}{x - 3} \right);$$

or we can go through the process by which we derived Formula (7.3.4):

$$\ln |g(x)| = \ln |x| + \ln |x - 1| + \ln |x - 2| + \ln |x - 3|,$$

$$\frac{g'(x)}{g(x)} = \frac{1}{x} + \frac{1}{x - 1} + \frac{1}{x - 2} + \frac{1}{x - 3}$$

$$g'(x) = x(x - 1)(x - 2)(x - 3) \left(\frac{1}{x} + \frac{1}{x - 1} + \frac{1}{x - 2} + \frac{1}{x - 3} \right). \quad ❏$$

The result is valid at all numbers x other than 0, 1, 2, 3. These are the numbers where $g(x) = 0$. ❏

Logarithmic differentiation can be applied to quotients.

Example 13 Calculate the derivative of

$$g(x) = \frac{(x^2 + 1)^3 (2x - 5)^2}{(x^2 + 5)^2}$$

by logarithmic differentiation.

SOLUTION Our first step is to write

$$g(x) = (x^2 + 1)^3 (2x - 5)^2 (x^2 + 5)^{-2}.$$

Then, according to (7.3.4),

$$g'(x) = \frac{(x^2 + 1)^3 (2x - 5)^2}{(x^2 + 5)^2} \left[\frac{3(x^2 + 1)^2 (2x)}{(x^2 + 1)^3} + \frac{2(2x - 5)(2)}{(2x - 5)^2} + \frac{(-2)(x^2 + 5)^{-3}(2x)}{(x^2 + 5)^{-2}} \right]$$

$$= \frac{(x^2 + 1)^3 (2x - 5)^2}{(x^2 + 5)^2} \left(\frac{6x}{x^2 + 1} + \frac{4}{2x - 5} - \frac{4x}{x^2 + 5} \right).$$

We don't have to rely on (7.3.4). We can simply write

$$\ln |g(x)| = \ln |(x^2 + 3)^3| + \ln |(2x - 5)^2| - \ln |(x^2 + 5)^2|$$

$$= 3 \ln |x^2 + 1| + 2 \ln |2x - 5| - 2 \ln |x^2 + 5|$$

and go on from there:

$$\frac{g'(x)}{g(x)} = \frac{3(2x)}{x^2 + 1} + \frac{2(2)}{2x - 5} - \frac{2(2x)}{x^2 + 5}$$

$$g'(x) = g(x) \left(\frac{6x}{x^2 + 1} + \frac{4}{2x - 5} - \frac{4x}{x^2 + 5} \right).$$

The result is valid at all numbers x other than $\frac{5}{2}$. At this number $g(x) = 0$. ❏

That logarithmic differentiation fails at the points where a product $g(x)$ is 0 is not a serious deficiency because at these points we can easily apply the product rule. For example, suppose that $g(a) = 0$. Then one of the factors of $g(x)$ is 0 at $x = a$. We write that factor in front and call it $g_1(x)$. We then have

$$g(x) = g_1(x)[g_2(x) \cdots g_n(x)] \qquad \text{with} \qquad g_1(a) = 0.$$

By the product rule,

$$g'(x) = g_1(x)\frac{d}{dx}[g_2(x) \cdots g_n(x)] + g_1'(x)[g_2(x) \cdots g_n(x)].$$

Since $g_1(a) = 0$,

$$g'(a) = g_1'(a)[g_2(a) \cdots g_n(a)].$$

We go back to the function of Example 12 and calculate the derivative of

$$g(x) = x(x - 1)(x - 2)(x - 3)$$

at $x = 3$ by the method just described. Since it is the factor $x - 3$ that is 0 at $x = 3$, we write

$$g(x) = (x - 3)[x(x - 1)(x - 2)].$$

By the product rule,

$$g'(x) = (x - 3)\frac{d}{dx}[x(x - 1)(x - 2)] + 1[x(x - 1)(x - 2)].$$

Therefore

$$g'(3) = 3(3 - 1)(3 - 2) = 6.$$

EXERCISES 7.3

Exercises 1–14. Determine the domain and find the derivative.

1. $f(x) = \ln 4x$.

2. $f(x) = \ln(2x + 1)$.

3. $f(x) = \ln(x^3 + 1)$.

4. $f(x) = \ln[(x + 1)^3]$.

5. $f(x) = \ln\sqrt{1 + x^2}$.

6. $f(x) = (\ln x)^3$.

7. $f(x) = \ln|x^4 - 1|$.

8. $f(x) = \ln(\ln x)$.

9. $f(x) = (2x + 1)^2 \ln(2x + 1)$.

10. $f(x) = \ln\left|\dfrac{x + 2}{x^3 - 1}\right|$.

11. $f(x) = \dfrac{1}{\ln x}$.

12. $f(x) = \ln\sqrt[4]{x^2 + 1}$.

13. $f(x) = \sin(\ln x)$.

14. $f(x) = \cos(\ln x)$.

Exercises 15–36. Calculate.

15. $\displaystyle\int \frac{dx}{x + 1}$.

16. $\displaystyle\int \frac{dx}{3 - x}$.

17. $\displaystyle\int \frac{x}{3 - x^2}\,dx$.

18. $\displaystyle\int \frac{x + 1}{x^2}\,dx$.

19. $\displaystyle\int \tan 3x\,dx$.

20. $\displaystyle\int \sec\tfrac{1}{2}\pi x\,dx$.

21. $\displaystyle\int x \sec x^2\,dx$.

22. $\displaystyle\int \frac{\csc^2 x}{2 + \cot x}\,dx$.

23. $\displaystyle\int \frac{x}{(3 - x^2)^2}\,dx$.

24. $\displaystyle\int \frac{\ln(x + a)}{x + a}\,dx$.

25. $\displaystyle\int \frac{\sin x}{2 + \cos x}\,dx$.

26. $\displaystyle\int \frac{\sec^2 2x}{4 - \tan 2x}\,dx$.

27. $\displaystyle\int \frac{1}{x \ln x}\,dx$.

28. $\displaystyle\int \frac{x^2}{2x^3 - 1}\,dx$.

29. $\displaystyle\int \frac{dx}{x(\ln x)^2}$.

30. $\displaystyle\int \frac{\sec 2x \tan 2x}{1 + \sec 2x}\,dx$.

31. $\displaystyle\int \frac{\sin x - \cos x}{\sin x + \cos x}\,dx$.

32. $\displaystyle\int \frac{1}{\sqrt{x}(1 + \sqrt{x})}\,dx$. HINT: Set $u = 1 + \sqrt{x}$.

33. $\displaystyle\int \frac{\sqrt{x}}{1 + x\sqrt{x}}\,dx$.

34. $\displaystyle\int \frac{\tan(\ln x)}{x}\,dx$.

35. $\displaystyle\int (1 + \sec x)^2\,dx$.

36. $\displaystyle\int (3 - \csc x)^2\,dx$.

Exercises 37–46. Evaluate.

37. $\displaystyle\int_1^e \frac{dx}{x}$.

38. $\displaystyle\int_1^{e^2} \frac{dx}{x}$.

39. $\displaystyle\int_e^{e^2} \frac{dx}{x}$.

40. $\displaystyle\int_0^1 \left(\frac{1}{x + 1} - \frac{1}{x + 2}\right)dx$.

41. $\displaystyle\int_4^5 \frac{x}{x^2 - 1}\,dx$.

42. $\displaystyle\int_{1/4}^{1/3} \tan \pi x\,dx$.

43. $\displaystyle\int_{\pi/6}^{\pi/2} \frac{\cos x}{1 + \sin x}\, dx.$

44. $\displaystyle\int_{\pi/4}^{\pi/2} (1 + \csc x)^2\, dx.$

45. $\displaystyle\int_{\pi/4}^{\pi/2} \cot x\, dx.$

46. $\displaystyle\int_{1}^{e} \frac{\ln x}{x}\, dx.$

47. Pinpoint the error in the following:

$$\int_{1}^{5} \frac{1}{x-2}\, dx = \left[\ln|x-2|\right]_{1}^{5} = \ln 3.$$

48. Show that $\displaystyle\lim_{x\to 0} \frac{\ln(1+x)}{x} = 1$ from the definition of derivative.

Exercises 49–52. Calculate the derivative by logarithmic differentiation and then evaluate g' at the indicated value of x.

49. $g(x) = (x^2 + 1)^2(x-1)^5 x^3;\quad x = 1.$

50. $g(x) = x(x+a)(x+b)(x+c);\quad x = -b.$

51. $g(x) = \dfrac{x^4(x-1)}{(x+2)(x^2+1)};\quad x = 0.$

52. $g(x) = \left[\dfrac{(x-1)(x-2)}{(x-3)(x-4)}\right]^2;\quad x = 2.$

Exercises 53–56. Sketch the region bounded by the curves and find its area.

53. $y = \sec x,\quad y = 2,\quad x = 0,\quad x = \pi/6.$

54. $y = \csc \frac{1}{2}\pi x,\quad y = x,\quad x = \frac{1}{2}.$

55. $y = \tan x,\quad y = 1,\quad x = 0.$

56. $y = \sec x,\quad y = \cos x,\quad x = 0,\quad x = \frac{\pi}{4}.$

Exercises 57–58. Find the area of the part of the first quadrant that lies between the curves.

57. $x + 4y - 5 = 0$ and $xy = 1.$

58. $x + y - 3 = 0$ and $xy = 2.$

59. The region bounded by the graph of $f(x) = 1/\sqrt{1+x}$ and the x-axis for $0 \le x \le 8$ is revolved about the x-axis. Find the volume of the resulting solid.

60. The region bounded by the graph of $f(x) = 3/(1+x^2)$ and the x-axis for $0 \le x \le 3$ is revolved about the y-axis. Find the volume of the resulting solid.

61. The region bounded by the graph of $f(x) = \sqrt{\sec x}$ and the x-axis for $-\pi/3 \le x \le \pi/3$ is revolved about the x-axis. Find the volume of the resulting solid.

62. The region bounded by the graph of $f(x) = \tan x$ and the x-axis for $0 \le x \le \pi/4$ is revolved about the x-axis. Find the volume of the resulting solid.

63. A particle moves along a coordinate line with acceleration $a(t) = -(t+1)^{-2}$ feet per second per second. Find the distance traveled by the particle during the time interval $[0, 4]$ given that the initial velocity $v(0)$ is 1 foot per second.

64. Exercise 63 taking $v(0)$ as 2 feet per second.

Exercises 65–66. Find a formula for the nth derivative.

65. $\dfrac{d^n}{dx^n}(\ln x).$

66. $\dfrac{d^n}{dx^n}[\ln(1-x)].$

67. Show that $\int \csc x\, dx = \ln|\csc x - \cot x| + C$ using the methods of this section.

68. (a) Show that for $n = 2$, (7.3.4) reduces to the product rule (3.2.6) except at those points where $g(x) = 0$.
(b) Show that (7.3.4) applied to

$$g(x) = \frac{g_1(x)}{g_2(x)}$$

reduces to the quotient rule (3.2.10) except at those points where $g(x) = 0$.

Exercises 69–74. (i) Find the domain of f, (ii) find the intervals on which the function increases and the intervals on which it decreases, (iii) find the extreme values, (iv) determine the concavity of the graph and find the points of inflection, and, finally, (v) sketch the graph, indicating asymptotes.

69. $f(x) = \ln(4-x).$

70. $f(x) = x - \ln x.$

71. $f(x) = x^2 \ln x.$

72. $f(x) = \ln(4 - x^2).$

73. $f(x) = \ln\left[\dfrac{x}{1+x^2}\right].$

74. $f(x) = \ln\left[\dfrac{x^3}{x-1}\right].$

75. Show that the average slope of the logarithm curve from $x = a$ to $x = b$ is

$$\frac{1}{b-a} \ln\left(\frac{b}{a}\right).$$

76. (a) Show that $f(x) = \ln 2x$ and $g(x) = \ln 3x$ have the same derivative.
(b) Calculate the derivative of $F(x) = \ln kx$, where k is any positive number.
(c) Explain these results in terms of the properties of logarithms.

▶**Exercises 77–80.** Use a graphing utility to graph f on the indicated interval. Estimate the x-intercepts of the graph of f and the values of x where f has either a local or absolute extreme value. Use four decimal place accuracy in your answers.

77. $f(x) = \sqrt{x}\, \ln x;\qquad (0, 10].$

78. $f(x) = x^3 \ln x;\qquad (0, 2].$

79. $f(x) = \sin(\ln x);\qquad (1, 100].$

80. $f(x) = x^2 \ln(\sin x);\qquad (0, 2].$

▶**81.** A particle moves along a coordinate line with acceleration $a(t) = 4 - 2(t+1) + 3/(t+1)$ feet per second per second from $t = 0$ to $t = 3$.
(a) Find the velocity v of the particle at each time t during the motion given that $v(0) = 2$.
(b) Use a graphing utility to graph v and a together.
(c) Estimate the time t at which the particle has maximum velocity and the time at which it has minimum velocity. Use four decimal place accuracy.

▶**82.** Exercise 81 with $a(t) = 2\cos 2(t+1) + 2/(t+1)$ feet per second per second from $t = 0$ to $t = 7$.

▶**83.** Set $f(x) = 1/x$ and $g(x) = -x^2 + 4x - 2.$
(a) Use a graphing utility to graph f and g together.

(b) Use a CAS to find the points where the two graphs intersect.

(c) Use a CAS to find the area of the region bounded by the two graphs.

▷ **84.** Exercise 83 taking $f(x) = \dfrac{x-1}{x}$ and $g(x) = |x - 2|$.

▷ **Exercises 85–86.** Use a CAS to find (i) $f'(x)$ and $f''(x)$; (ii) the points where f, f' and f'' are zero; (iii) the intervals on which f, f' and f'' are positive, negative; (iv) the extreme values of f.

85. $f(x) = \dfrac{\ln x}{x^2}$.

86. $f(x) = \dfrac{1 + 2\ln x}{2\sqrt{\ln x}}$.

■ 7.4 THE EXPONENTIAL FUNCTION

Rational powers of e already have an established meaning: by $e^{p/q}$ we mean the qth root of e raised to the pth power. But what is meant by $e^{\sqrt{2}}$ or e^{π}?

Earlier we proved that each rational power $e^{p/q}$ has logarithm p/q:

(7.4.1)
$$\ln e^{p/q} = \frac{p}{q}.$$

The definition of e^z for z irrational is patterned after this relation.

DEFINITION 7.4.2

If z is irrational, then by e^z we mean the unique number that has logarithm z:
$$\ln e^z = z.$$

What is $e^{\sqrt{2}}$? It is the unique number that has logarithm $\sqrt{2}$. What is e^{π}? It is the unique number that has logarithm π. Note that e^x now has meaning for every real value of x: it is the unique number that has logarithm x.

DEFINITION 7.4.3

The function
$$E(x) = e^x \qquad \text{for all real } x$$
is called the *exponential function*.

Some properties of the exponential function are listed below.

(1) In the first place,

(7.4.4)
$$\ln e^x = x \qquad \text{for all real } x$$

Writing $L(x) = \ln x$ and $E(x) = e^x$, we have
$$L(E(x)) = x \qquad \text{for all real } x.$$

This says that the *exponential function is the inverse of the logarithm function*.

(2) The graph of the exponential function appears in Figure 7.4.1. It can be obtained from the graph of the logarithm by reflection in the line $y = x$.

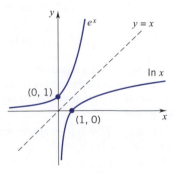

Figure 7.4.1

(3) Since the graph of the logarithm lies to the right of the y-axis, the graph of the exponential function lies above the x-axis:

(7.4.5)
$$e^x > 0 \qquad \text{for all real } x.$$

(4) Since the graph of the logarithm crosses the x-axis at $(1, 0)$, the graph of the exponential function crosses the y-axis at $(0, 1)$:

$$\ln 1 = 0 \qquad \text{gives} \qquad e^0 = 1.$$

(5) Since the y-axis is a vertical asymptote for the graph of the logarithm function, the x-axis is a horizontal asymptote for the graph of the exponential function:

$$\text{as } x \to -\infty, \quad e^x \to 0.$$

(6) Since the exponential function is the inverse of the logarithm function, the logarithm function is the inverse of the exponential function; thus

(7.4.6)
$$e^{\ln x} = x \qquad \text{for all } x > 0.$$

You can verify this equation directly by observing that both sides have the same logarithm:

$$\ln \left(e^{\ln x} \right) = \ln x$$

since, for all real t, $\ln e^t = t$.

You know that for rational exponents

$$e^{(p/q + r/s)} = e^{p/q} \cdot e^{r/s}.$$

This property holds for all exponents, including irrational exponents

---

THEOREM 7.4.7

$$e^{a+b} = e^a \cdot e^b \qquad \text{for all real } a \text{ and } b.$$

PROOF

$$\ln e^{a+b} = a + b = \ln e^a + \ln e^b = \ln \left(e^a \cdot e^b \right).$$

The one-to-oneness of the logarithm function gives

$$e^{a+b} = e^a \cdot e^b. \quad \square$$

We leave it to you to verify that

(7.4.8)
$$e^{-b} = \frac{1}{e^b} \qquad \text{and} \qquad e^{a-b} = \frac{e^a}{e^b}.$$

We come now to one of the most important results in calculus. It is marvelously simple.

THEOREM 7.4.9

The exponential function is its own derivative: for all real x,

$$\frac{d}{dx}(e^x) = e^x.$$

PROOF The logarithm function is differentiable, and its derivative is never 0. It follows (Section 7.1) that its inverse, the exponential function, is also differentiable. Knowing this, we can show that

$$\frac{d}{dx}(e^x) = e^x$$

by differentiating both sides of the identity

$$\ln e^x = x.$$

On the left-hand side, the chain rule gives

$$\frac{d}{dx}(\ln e^x) = \frac{1}{e^x}\frac{d}{dx}(e^x).$$

On the right-hand side, the derivative is 1:

$$\frac{d}{dx}(x) = 1.$$

Equating these derivatives, we have

$$\frac{1}{e^x}\frac{d}{dx}(e^x) = 1 \qquad \text{and thus} \qquad \frac{d}{dx}(e^x) = e^x. \quad ❏$$

Compositions are differentiated by the chain rule.

Example 1

(a) $\dfrac{d}{dx}(e^{kx}) = e^{kx}\dfrac{d}{dx}(kx) = e^{kx}k = ke^{kx}.$

(b) $\dfrac{d}{dx}(e^{\sqrt{x}}) = e^{\sqrt{x}}\dfrac{d}{dx}(\sqrt{x}) = e^{\sqrt{x}}\left(\dfrac{1}{2\sqrt{x}}\right) = \dfrac{1}{2\sqrt{x}}e^{\sqrt{x}}.$

(c) $\dfrac{d}{dx}(e^{-x^2}) = e^{-x^2}\dfrac{d}{dx}(-x^2) = e^{-x^2}(-2x) = -2x\,e^{-x^2}. \quad ❏$

The relation

$$\frac{d}{dx}(e^x) = e^x \qquad \text{and its corollary} \qquad \frac{d}{dx}(e^{kx}) = k\,e^{kx}$$

have important applications to engineering, physics, chemistry, biology, and economics. We take up some of these applications in Section 7.6.

Example 2 Let $f(x) = xe^{-x}$ for all real x.

(a) On what intervals does f increase? decrease?

(b) Find the extreme values of f.

(c) Determine the concavity of the graph and find the points of inflection.

(d) Sketch the graph indicating the asymptotes if any.

SOLUTION

$$f(x) = xe^{-x},$$

$$f'(x) = xe^{-x}(-1) + e^{-x} = (1-x)e^{-x},$$

$$f''(x) = (1-x)e^{-x}(-1) - e^{-x} = (x-2)e^{-x}.$$

Since $e^{-x} > 0$ for all x, we have $f'(x) = 0$ only at $x = 1$. (critical point) The sign of f' and the behavior of f are as follows:

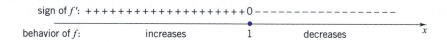

The function f increases on $(-\infty, 1]$ and decreases on $[1, \infty)$. The number

$$f(1) = \frac{1}{e} \cong \frac{1}{2.72} \cong 0.368$$

is a local maximum and the absolute maximum. The function has no other extreme values.

The sign of f'' and the concavity of the graph of f are as follows:

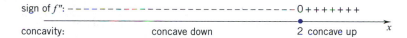

The graph is concave down on $(-\infty, 2)$ and concave up on $(2, \infty)$. The point

$$(2, f(2)) = (2, 2e^{-2}) \cong \left(2, \frac{2}{(2.72)^2}\right) \cong (2, 0.27)$$

is a point of inflection, the only point of inflection. In Section 11.6 we show that as $x \to \infty$, $f(x) = x/e^x \to 0$. Accepting this result for now, we conclude that the x-axis is a horizontal asymptote. The graph is given in Figure 7.4.2. ❏

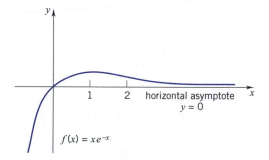

Figure 7.4.2

Example 3   Let $f(x) = e^{-x^2/2}$ for all real x.

(a) Determine the symmetry of the graph and find the asymptotes.
(b) On what intervals does f increase? decrease?
(c) Find the extreme values.
(d) Determine the concavity of the graph and find the points of inflection.
(e) Sketch the graph.

SOLUTION Since $f(-x) = e^{-(-x)^2/2} = e^{-x^2/2} = f(x)$, f is an even function. Thus the graph is symmetric about the y-axis. As $x \to \pm\infty$, $e^{-x^2/2} \to 0$. Therefore, the x-axis is a horizontal asymptote. There are no vertical asymptotes.

Differentiating f, we have

$$f'(x) = e^{-x^2/2}(-x) = -xe^{-x^2/2}$$

$$f''(x) = -x(-xe^{-x^2/2}) - e^{-x^2/2} = (x^2 - 1)e^{-x^2/2}.$$

Since $e^{-x^2/2} > 0$ for all x, we have $f'(x) = 0$ only at $x = 0$ (critical point). The sign of f' and the behavior of f are as follows:

The function increases on $(-\infty, 0]$ and decreases $[0, \infty)$. The number

$$f(0) = e^0 = 1$$

is a local maximum and the absolute maximum. The function has no other extreme values.

Now consider $f''(x) = (x^2 - 1)e^{-x^2/2}$. The sign of f'' and the concavity of the graph of f are as follows:

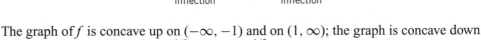

The graph of f is concave up on $(-\infty, -1)$ and on $(1, \infty)$; the graph is concave down on $(-1, 1)$. The points $(-1, e^{-1/2})$ and $(1, e^{-1/2})$ are points of inflection.

The graph of f is the bell-shaped curve sketched in Figure 7.4.3.[†] ❑

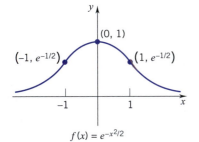

$f(x) = e^{-x^2/2}$

Figure 7.4.3

The integral counterpart of Theorem 7.4.9 takes the form

(7.4.10)

$$\boxed{\int e^x \, dx = e^x + C.}$$

In practice

$$\int e^{u(x)} u'(x) \, dx \qquad \text{is reduced to} \qquad \int e^u \, du$$

by setting

$$u = u(x), \qquad du = u'(x) \, dx.$$

Example 4 Find $\displaystyle\int 9 \, e^{3x} \, dx$.

SOLUTION Set $u = 3x$, $du = 3 \, dx$.

$$\int 9 \, e^{3x} \, dx = 3 \int e^u \, du = 3e^u + C = 3 \, e^{3x} + C.$$

_____

[†]Bell-shaped curves play a big role in probability and statistics.

If you recognize at the very beginning that

$$3e^{3x} = \frac{d}{dx}(e^{3x}),$$

then you can dispense with the u-substitution and simply write

$$\int 9e^{3x}dx = 3\int 3e^{3x}dx = 3e^{3x} + C. \quad \square$$

Example 5 Find $\int \dfrac{e^{\sqrt{x}}}{\sqrt{x}}dx.$

SOLUTION Set $u = \sqrt{x}, \quad du = \dfrac{1}{2\sqrt{x}}dx.$

$$\int \frac{e^{\sqrt{x}}}{\sqrt{x}}dx = 2\int e^u\,du = 2e^u + C = 2e^{\sqrt{x}} + C.$$

If you recognize from the start that

$$\frac{1}{2}\left(\frac{e^{\sqrt{x}}}{\sqrt{x}}\right) = \frac{d}{dx}\left(e^{\sqrt{x}}\right),$$

then you can dispense with the u-substitution and integrate directly:

$$\int \frac{e^{\sqrt{x}}}{\sqrt{x}}dx = 2\int \frac{1}{2}\left(\frac{e^{\sqrt{x}}}{\sqrt{x}}\right)dx = 2e^{\sqrt{x}} + C. \quad \square$$

Example 6 Find $\int \dfrac{e^{3x}}{e^{3x}+1}dx.$

SOLUTION We can put this integral in the form

$$\int \frac{1}{u}\,du$$

by setting

$$u = e^{3x} + 1, \qquad du = 3e^{3x}dx.$$

Then

$$\int \frac{e^{3x}}{e^{3x}+1}dx = \tfrac{1}{3}\int \frac{1}{u}\,du = \tfrac{1}{3}\ln|u| + C = \tfrac{1}{3}\ln(e^{3x}+1) + C. \quad \square$$

Example 7 Evaluate

$$\int_0^{\sqrt{2\ln 3}} xe^{-x^2/2}dx.$$

SOLUTION Set $u = -\tfrac{1}{2}x^2, \quad du = -x\,dx.$

At $x = 0, \; u = 0;$ at $x = \sqrt{2\ln 3}, \quad u = -\ln 3.$ Thus

$$\int_0^{\sqrt{2\ln 3}} xe^{-x^2/2}dx = -\int_0^{-\ln 3} e^u\,du = -\left[e^u\right]_0^{-\ln 3} = 1 - e^{-\ln 3} = 1 - \tfrac{1}{3} = \tfrac{2}{3}. \quad \square$$

Example 8 Evaluate $\int_0^1 e^x(e^x + 1)^{1/5}dx$.

SOLUTION Set $u = e^x + 1$, $du = e^x dx$.

At $x = 0$, $u = 2$; at $x = 1$, $u = e + 1$. Thus

$$\int_0^1 e^x(e^x + 1)^{1/5}dx = \int_2^{e+1} u^{1/5}du = \left[\tfrac{5}{6}u^{6/5}\right]_2^{e+1} = \tfrac{5}{6}[(e + 1)^{6/5} - 2^{6/5}]. \quad \square$$

EXERCISES 7.4

Exercises 1–24. Differentiate.

1. $y = e^{-2x}$.
2. $y = 3e^{2x+1}$.
3. $y = e^{x^2-1}$.
4. $y = 2e^{-4x}$.
5. $y = e^x \ln x$.
6. $y = x^2 e^x$.
7. $y = x^{-1}e^{-x}$.
8. $y = e^{\sqrt{x}+1}$.
9. $y = \frac{1}{2}(e^x + e^{-x})$.
10. $y = \frac{1}{2}(e^x - e^{-x})$.
11. $y = e^{\sqrt{x}} \ln \sqrt{x}$.
12. $y = (3 - 2e^{-x})^3$.
13. $y = (e^{x^2} + 1)^2$.
14. $y = (e^{2x} - e^{-2x})^2$.
15. $y = (x^2 - 2x + 2)e^x$.
16. $y = x^2 e^x - xe^{x^2}$.
17. $y = \dfrac{e^x - 1}{e^x + 1}$.
18. $y = \dfrac{e^{2x} - 1}{e^{2x} + 1}$.
19. $y = e^{4\ln x}$.
20. $y = \ln e^{3x}$.
21. $f(x) = \sin(e^{2x})$.
22. $f(x) = e^{\sin 2x}$.
23. $f(x) = e^{-2x} \cos x$.
24. $f(x) = \ln(\cos e^{2x})$.

Exercises 25–42. Calculate.

25. $\int e^{2x} dx$.
26. $\int e^{-2x} dx$.
27. $\int e^{kx} dx$.
28. $\int e^{ax+b} dx$.
29. $\int xe^{x^2} dx$.
30. $\int xe^{-x^2} dx$.
31. $\int \dfrac{e^{1/x}}{x^2} dx$.
32. $\int \dfrac{e^{2\sqrt{x}}}{\sqrt{x}} dx$.
33. $\int \ln e^x dx$.
34. $\int e^{\ln x} dx$.
35. $\int \dfrac{4}{\sqrt{e^x}} dx$.
36. $\int \dfrac{e^x}{e^x + 1} dx$.
37. $\int \dfrac{e^x}{\sqrt{e^x + 1}} dx$.
38. $\int \dfrac{xe^{ax^2}}{e^{ax^2} + 1} dx$.
39. $\int \dfrac{e^{2x}}{2e^{2x} + 3} dx$.
40. $\int \dfrac{\sin(e^{-2x})}{e^{2x}} dx$.
41. $\int \cos xe^{\sin x} dx$.
42. $\int e^{-x}[1 + \cos(e^{-x})] dx$.

Exercises 43–52. Evaluate.

43. $\int_0^1 e^x dx$.
44. $\int_0^1 e^{-kx} dx$.

45. $\int_0^{\ln \pi} e^{-6x} dx$.
46. $\int_0^1 xe^{-x^2} dx$.
47. $\int_0^1 \dfrac{e^x + 1}{e^x} dx$.
48. $\int_0^1 \dfrac{4 - e^x}{e^x} dx$.
49. $\int_0^{\ln 2} \dfrac{e^x}{e^x + 1} dx$.
50. $\int_0^1 \dfrac{e^x}{4 - e^x} dx$.
51. $\int_0^1 x(e^{x^2} + 2) dx$.
52. $\int_0^{\ln \pi/4} e^x \sec e^x dx$.

53. Let a be a positive constant.
(a) Find a formula for the nth derivative of $f(x) = e^{ax}$.
(b) Find a formula for the nth derivative of $f(x) = e^{-ax}$.

54. A particle moves along a coordinate line, its position at time t given by the function
$$x(t) = Ae^{kt} + Be^{-kt}. \quad (A > 0, B > 0, k > 0)$$
(a) Find the times t at which the particle is closest to the origin.
(b) Show that the acceleration of the particle is proportional to the position coordinate. What is the constant of proportionality?

55. A rectangle has one side on the x-axis and the upper two vertices on the graph of $y = e^{-x^2}$. Where should the vertices be placed so as to maximize the area of the rectangle?

56. A rectangle has two sides on the positive x- and y-axes and one vertex at a point P that moves along the curve $y = e^x$ in such a way that y increases at the rate of $\frac{1}{2}$ unit per minute. How is the area of the rectangle changing when $y = 3$?

57. Set $f(x) = e^{-x^2}$.
(a) What is the symmetry of the graph?
(b) On what intervals does the function increase? decrease?
(c) What are the extreme values of the function?
(d) Determine the concavity of the graph and find the points of inflection.
(e) The graph has a horizontal asymptote. What is it?
(f) Sketch the graph.

58. Let Ω be the region below the graph of $y = e^x$ from $x = 0$ to $x = 1$.
(a) Find the volume of the solid generated by revolving Ω about the x-axis.

(b) Set up the definite integral that gives the volume of the solid generated by revolving Ω about the y-axis using the shell method. (You will see how to evaluate this integral in Section 8.2.)

59. Let Ω be the region below the graph of $y = e^{-x^2}$ from $x = 0$ to $x = 1$.

(a) Find the volume of the solid generated by revolving Ω about the y-axis.

(b) Form the definite integral that gives the volume of the solid generated by revolving Ω about the x-axis using the disk method. (At this point we cannot carry out the integration.)

Exercises 60–63. Sketch the region bounded by the curves and find its area.

60. $x = e^{2y}$, $x = e^{-y}$, $x = 4$.

61. $y = e^x$, $y = e^{2x}$, $y = e^4$.

62. $y = e^x$, $y = e$, $y = x$, $x = 0$.

63. $x = e^y$, $y = 1$, $y = 2$, $x = 2$.

Exercises 64–68. Determine the following: (i) the domain; (ii) the intervals on which f increases, decreases; (iii) the extreme values; (iv) the concavity of the graph and the points of inflection. Then sketch the graph, indicating all asymptotes.

64. $f(x) = (1-x)e^x$.

65. $f(x) = e^{(1/x^2)}$.

66. $f(x) = x^2 e^{-x}$.

67. $f(x) = x^2 \ln x$.

68. $f(x) = (x - x^2)e^{-x}$

69. For each positive integer n find the number x_n for which $\int_0^{x_n} e^x \, dx = n$.

70. Find the critical points and the extreme values. Take k as a positive integer.

(a) $f(x) = x^k \ln x$, $x > 0$.

(b) $f(x) = x^k e^{-x}$, x real.

71. Take $a > 0$ and refer to the figure.

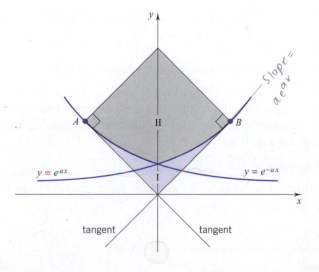

(a) Find the points of tangency, marked A and B.

(b) Find the area of region I.

(c) Find the area of region II.

72. Prove that for all $x > 0$ and all positive integers n

$$e^x > 1 + x + \frac{x^2}{2!} + \frac{x^3}{3!} + \cdots + \frac{x^n}{n!}.$$

Recall that $n! = n(n-1)(n-2) \cdots 3 \cdot 2 \cdot 1$.

HINT: $e^x = 1 + \int_0^x e^t \, dt > 1 + \int_0^x dt = 1 + x$

$$e^x = 1 + \int_0^x e^t \, dt > 1 + \int_0^x (1+t) dt$$

$$= 1 + x + \frac{x^2}{2}, \quad \text{and so on.}$$

73. Prove that, if n is a positive integer, then

$$e^x > x^n \quad \text{for all } x \text{ sufficiently large.}$$

HINT: Exercise 72.

74. Set $f(x) = e^{-x^2}$ and $g(x) = x^2$.

(a) Use a graphing utility to draw a figure that displays the graphs of f and g.

(b) Estimate the x-coordinates a and b ($a < b$) of the two points where the curves intersect. Use four decimal place accuracy.

(c) Estimate the area between the two curves from $x = a$ to $x = b$.

75. Exercise 74 with $f(x) = e^x$ and $g(x) = 4 - x^2$.

Exercises 76–78. Use a graphing utility to draw a figure that displays the graphs of f and g. The figure should suggest that f and g are inverses. Show that this is true by verifying that $f(g(x)) = x$ for each x in the domain of g.

76. $f(x) = e^{2x}$, $g(x) = \ln \sqrt{x}$; $x > 0$.

77. $f(x) = e^{x^2}$, $g(x) = \sqrt{\ln x}$; $x \geq 1$.

78. $f(x) = e^{x-2}$, $g(x) = 2 + \ln x$; $x > 0$.

79. Set $f(x) = \sin e^x$. (a) Find the zeros of f. (b) Use a graphing utility to graph f.

80. Exercise 79 with $f(x) = e^{\sin x} - 1$.

81. Set $f(x) = e^{-x}$ and $g(x) = \ln x$.

(a) Use a graphing utility to draw a figure that displays the graphs of f and g.

(b) Estimate the x-coordinate of the point where the two graphs intersect.

(c) Estimate the slopes at the point of intersection.

(d) Are the curves perpendicular to each other?

82. (a) Use a graphing utility to draw a figure that displays the graphs of $f(x) = 10e^{-x}$ and $g(x) = 7 - e^x$.

(b) Find the x-coordinates a and b ($a < b$) of the two points where the curves intersect.

(c) Use a CAS to find the area between the two curves from $x = a$ to $x = b$.

83. Use a CAS to calculate the integral.

(a) $\displaystyle\int \frac{1}{1 - e^x} \, dx$.

(b) $\displaystyle\int e^{-x} \left(\frac{1 - e^x}{e^x} \right)^4 dx$.

(c) $\displaystyle\int \frac{e^{\tan x}}{\cos^2 x} \, dx$.

■ **PROJECT 7.4   Some Rational Bounds for the Number _e_**

The purpose of this project is to lead you through a proof that, for each positive integer n

(7.4.11)
$$\left(1 + \frac{1}{n}\right)^n \le e \le \left(1 + \frac{1}{n}\right)^{n+1}.$$

It will follow that

$$(1 + \tfrac{1}{2})^2 \le e \le \left(1 + \tfrac{1}{2}\right)^3,$$

$$(1 + \tfrac{1}{3})^3 \le e \le \left(1 + \tfrac{1}{3}\right)^4,$$

$$(1 + \tfrac{1}{4})^4 \le e \le \left(1 + \tfrac{1}{4}\right)^5, \qquad \text{and so on.}$$

The proof outlined below is based directly on the definition of the logarithm function

$$\ln x = \int_1^x \frac{1}{t}\, dt, \qquad x > 0$$

and on the characterization of e as the unique number for which

$$\int_1^e \frac{1}{t}\, dt = 1.$$

The proof has two steps.

Step 1. Show that for each positive integer n,

$$\frac{1}{n+1} \le \ln\left(1 + \frac{1}{n}\right) \le \frac{1}{n}.$$

HINT: For all numbers t in $\left[1, 1 + \frac{1}{n}\right]$,

$$\frac{1}{1 + \frac{1}{n}} \le \frac{1}{t} \le 1.$$

Step 2. Show that

$$\left(1 + \frac{1}{n}\right)^n \le e \le \left(1 + \frac{1}{n}\right)^{n+1}$$

by applying the exponential function to each entry in the inequality derived in Step 1.

The bounds that we have derived for e are simple, elegant, and easy to remember, but they do not provide a very efficient method for calculating e. For example, rounded off to seven decimal places,

$$\left(1 + \frac{1}{100}\right)^{100} \cong 2.7048138 \quad \text{and} \quad \left(1 + \frac{1}{100}\right)^{101}$$
$$\cong 2.7318620.$$

Apparently a lot of accuracy here, but it doesn't help us much in finding a decimal expansion for e. It tells us only that, rounded off to one decimal place, $e \cong 2.7$. For a more accurate decimal expansion of e, we need to resort to very large values of n. A much more efficient way of calculating e is given in Section 12.6.

■ 7.5 ARBITRARY POWERS; OTHER BASES

Arbitrary Powers: The Function $f(x) = x^r$

The elementary notion of exponent applies only to rational numbers. Expressions such as

$$10^5, \qquad 2^{1/3}, \qquad 7^{-4/5}, \qquad \pi^{-1/2}$$

make sense, but so far we have attached no meaning to expressions such as

$$10^{\sqrt{2}}, \qquad 2^\pi, \qquad 7^{-\sqrt{3}}, \qquad \pi^e.$$

The extension of our sense of exponent to allow for irrational exponents is conveniently done by making use of the logarithm function and the exponential function. The heart of the matter is to observe that for $x > 0$ and p/q rational,

$$x^{p/q} = e^{(p/q)\ln x}.$$

(To verify this, take the logarithm of both sides.) We *define* x^z for irrational z by setting

$$x^z = e^{z \ln x}.$$

We can now state that

(7.5.1)
$$\begin{array}{l} \text{if} \qquad x > 0, \qquad \text{then} \\ x^r = e^{r \ln x} \qquad \text{for all real numbers } r. \end{array}$$

In particular,

$$10^{\sqrt{2}} = e^{\sqrt{2}\,\ln 10}, \qquad 2^{\pi} = e^{\pi\,\ln 2}, \qquad 7^{-\sqrt{3}} = e^{-\sqrt{3}\,\ln 7}, \qquad \pi^{e} = e^{e\,\ln \pi}.$$

With this extended sense of exponent, the usual laws of exponents still hold:

(7.5.2)

$$x^{r+s} = x^r x^s, \qquad x^{r-s} = \frac{x^r}{x^s}, \qquad (x^r)^s = x^{rs}$$

PROOF

$$x^{r+s} = e^{(r+s)\ln x} = e^{r\ln x} \cdot e^{s\ln x} = x^r x^s,$$

$$x^{r-s} = e^{(r-s)\ln x} = e^{r\ln x} \cdot e^{-s\ln x} = \frac{e^{r\ln x}}{e^{s\ln x}} = \frac{x^r}{x^s},$$

$$(x^r)^s = e^{s\ln x^r} = e^{rs\ln x} = x^{rs}. \quad \square$$

The differentiation of arbitrary powers follows the pattern established for rational powers; namely, for each real number r and each $x > 0$

(7.5.3)

$$\frac{d}{dx}(x^r) = rx^{r-1}.$$

PROOF

$$\frac{d}{dx}(x^r) = \frac{d}{dx}(e^{r\ln x}) = e^{r\ln x}\frac{d}{dx}(r\ln x) = x^r \cdot \frac{r}{x} = rx^{r-1}.$$

Another way to see this is to write $f(x) = x^r$ and use logarithmic differentiation:

$$\ln f(x) = r\ln x$$

$$\frac{f'(x)}{f(x)} = \frac{r}{x}$$

$$f'(x) = \frac{rf(x)}{x} = \frac{rx^r}{x} = rx^{r-1}. \quad \square$$

Thus

$$\frac{d}{dx}\left(x^{\sqrt{2}}\right) = \sqrt{2}\,x^{\sqrt{2}-1} \qquad \text{and} \qquad \frac{d}{dx}(x^{\pi}) = \pi x^{\pi-1}.$$

As usual, we differentiate compositions by the chain rule. Thus

$$\frac{d}{dx}\left[(x^2+5)^{\sqrt{3}}\right] = \sqrt{3}(x^2+5)^{\sqrt{3}-1}\frac{d}{dx}(x^2+5) = 2\sqrt{3}\,x(x^2+5)^{\sqrt{3}-1}.$$

Example 1 Find $\dfrac{d}{dx}\left[(x^2+1)^{3x}\right]$.

SOLUTION One way to find this derivative is to observe that $(x^2+1)^{3x} = e^{3x\ln(x^2+1)}$ and then differentiate:

$$\frac{d}{dx}\left[(x^2+1)^{3x}\right] = \frac{d}{dx}\left[e^{3x\ln(x^2+1)}\right] = e^{3x\ln(x^2+1)}\left[3x \cdot \frac{2x}{x^2+1} + 3\ln(x^2+1)\right]$$

$$= (x^2+1)^{3x}\left[\frac{6x^2}{x^2+1} + 3\ln(x^2+1)\right].$$

Another way to find this derivative is to set $f(x) = (x^2 + 1)^{3x}$, take the logarithm of both sides, and proceed from there

$$\ln f(x) = 3x \cdot \ln(x^2 + 1)$$

$$\frac{f'(x)}{f(x)} = 3x \cdot \frac{2x}{x^2 + 1} + [\ln(x^2 + 1)](3) = \frac{6x^2}{x^2 + 1} + 3\ln(x^2 + 1)$$

$$f'(x) = f(x)\left[\frac{6x^2}{x^2 + 1} + 3\ln(x^2 + 1)\right]$$

$$= (x^2 + 1)^{3x}\left[\frac{6x^2}{x^2 + 1} + 3\ln(x^2 + 1)\right]. \quad \square$$

Each derivative formula gives rise to a companion integral formula. The integral version of (7.5.3) takes the form

(7.5.4)
$$\int x^r \, dx = \frac{x^{r+1}}{r + 1} + C, \quad \text{for} \quad r \neq -1.$$

Note the exclusion of $r = -1$. What is the integral if $r = -1$?

Example 2 Find $\displaystyle\int \frac{x^3}{(2x^4 + 1)^\pi} \, dx$.

SOLUTION Set $u = 2x^4 + 1$, $\quad du = 8x^3 \, dx$.

$$\int \frac{x^3}{(2x^4 + 1)^\pi} \, dx = \frac{1}{8}\int u^{-\pi} \, du = \frac{1}{8}\left(\frac{u^{1-\pi}}{1 - \pi}\right) + C = \frac{(2x^4 + 1)^{1-\pi}}{8(1 - \pi)} + C. \quad \square$$

Base p: The Function $f(x) = p^x$

To form the function $f(x) = x^r$, we take a positive variable x and raise it to a constant power r. To form the function $f(x) = p^x$, we take a positive constant p and raise it to a variable power x. Since $1^x = 1$ for all x, the function is of interest only if $p \neq 1$.

Functions of the form $f(x) = p^x$ are called *exponential functions with base p*. The high status enjoyed by Euler's number e comes from the fact that

$$\frac{d}{dx}(e^x) = e^x.$$

For other bases the derivative has an extra factor:

(7.5.5)
$$\frac{d}{dx}(p^x) = p^x \ln p.$$

PROOF

$$\frac{d}{dx}(p^x) = \frac{d}{dx}(e^{x \ln p}) = e^{x \ln p} \ln p = p^x \ln p. \quad \square$$

For example,

$$\frac{d}{dx}(2^x) = 2^x \ln 2 \quad \text{and} \quad \frac{d}{dx}(10^x) = 10^x \ln 10.$$

The next differentiation requires the chain rule:

$$\frac{d}{dx}\left(2^{3x^2}\right) = 2^{3x^2}(\ln 2)\frac{d}{dx}(3x^2) = 6x2^{3x^2}\ln 2.$$

The integral version of (7.5.5) reads

(7.5.6)
$$\int p^x\,dx = \frac{1}{\ln p}p^x + C.$$

The formula holds for all positive numbers p different from 1. For example,

$$\int 2^x\,dx = \frac{1}{\ln 2}2^x + C.$$

Example 3 Find $\int x5^{-x^2}\,dx$.

SOLUTION Set $u = -x^2$, $du = -2x\,dx$.

$$\int x5^{-x^2}\,dx = -\frac{1}{2}\int 5^u\,du = -\frac{1}{2}\left(\frac{1}{\ln 5}\right)5^u + C$$

$$= \frac{-1}{2\ln 5}5^{-x^2} + C. \quad \square$$

Example 4 Evaluate $\int_1^2 3^{2x-1}\,dx$.

SOLUTION Set $u = 2x - 1$, $du = 2\,dx$.
At $x = 1$, $u = 1$; at $x = 2$, $u = 3$. Thus

$$\int_1^2 3^{2x-1}\,dx = \frac{1}{2}\int_1^3 3^u\,du = \frac{1}{2}\left[\frac{1}{\ln 3}\cdot 3^u\right]_1^3 = \frac{12}{\ln 3} \cong 10.923. \quad \square$$

Base p: The Function $f(x) = \log_p x$

If $p > 0$, then

$$\ln p^t = t\ln p \qquad \text{for all } t.$$

If p is also different from 1, then $\ln p \neq 0$, and we have

$$\frac{\ln p^t}{\ln p} = t.$$

This indicates that the function

$$f(x) = \frac{\ln x}{\ln p}$$

satisfies the relation

$$f(p^t) = t \qquad \text{for all real } t.$$

In view of this, we call

$$\frac{\ln x}{\ln p}$$

the logarithm of x to the base p and write

(7.5.7)
$$\log_p x = \frac{\ln x}{\ln p}.$$

The relation holds for all $x > 0$ and assumes that p is a positive number different from 1. For example,

$$\log_2 32 = \frac{\ln 32}{\ln 2} = \frac{\ln 2^5}{\ln 2} = \frac{5 \ln 2}{\ln 2} = 5$$

and

$$\log_{100}\left(\tfrac{1}{10}\right) = \frac{\ln\left(\tfrac{1}{10}\right)}{\ln 100} = \frac{\ln 10^{-1}}{\ln 10^2} = \frac{-\ln 10}{2 \ln 10} = -\frac{1}{2}. \quad ❑$$

We can obtain these same results more directly from the relation

(7.5.8)
$$\log_p p^t = t.$$

Accordingly

$$\log_2 32 = \log_2 2^5 = 5 \quad \text{and} \quad \log_{100}\left(\tfrac{1}{10}\right) = \log_{100}(100^{-1/2}) = -\tfrac{1}{2}.$$

Since $\log_p x$ and $\ln x$ differ only by a constant factor, there is no reason to introduce new differentiation and integration formulas. For the record, we simply point out that

$$\frac{d}{dx}(\log_p x) = \frac{d}{dx}\left(\frac{\ln x}{\ln p}\right) = \frac{1}{x \ln p}.$$

If p is e, the factor $\ln p$ is 1 and we have

$$\frac{d}{dx}(\log_e x) = \frac{1}{x}.$$

The logarithm to the base e, $\ln = \log_e$, is called *the natural logarithm* (or simply *the logarithm*) because it is the logarithm with the simplest derivative.

Example 5 Calculate

(a) $\dfrac{d}{dx}(\log_5 |x|),$ **(b)** $\dfrac{d}{dx}[\log_2(3x^2 + 1)],$ **(c)** $\displaystyle\int \frac{1}{x \ln 10} dx.$

SOLUTION

(a) $\dfrac{d}{dx}(\log_5 |x|) = \dfrac{d}{dx}\left[\dfrac{\ln |x|}{\ln 5}\right] = \dfrac{1}{x \ln 5}.$

(b) $\quad \dfrac{d}{dx}[\log_2(3x^2 + 1)] = \dfrac{d}{dx}\left[\dfrac{\ln(3x^2 + 1)}{\ln 2}\right]$

$$= \frac{1}{(3x^2 + 1) \ln 2} \frac{d}{dx}(3x^2 + 1) = \frac{6x}{(3x^2 + 1) \ln 2}.$$

by the chain rule ⟶

(c) $\displaystyle\int \frac{1}{x \ln 10} dx = \frac{1}{\ln 10} \int \frac{1}{x} dx = \frac{\ln |x|}{\ln 10} + C = \log_{10} |x| + C. \quad ❑$

EXERCISES 7.5

Exercises 1–8. Evaluate.

1. $\log_2 64$.

2. $\log_2 \frac{1}{64}$.

3. $\log_{64} \frac{1}{2}$.

4. $\log_{10} 0.01$.

5. $\log_5 1$.

6. $\log_5 0.2$.

7. $\log_5 125$.

8. $\log_2 4^3$.

Exercises 9–12. Show that the identity holds.

9. $\log_p xy = \log_p x + \log_p y$.

10. $\log_p \frac{1}{x} = -\log_p x$.

11. $\log_p x^y = y\log_p x$.

12. $\log_p \frac{x}{y} = \log_p x - \log_p y$.

Exercises 13–16. Find the numbers x which satisfy the equation.

13. $10^x = e^x$.

14. $\log_5 x = 0.04$.

15. $\log_x 10 = \log_4 100$.

16. $\log_x 2 = \log_3 x$.

17. Estimate $\ln a$ given that $e^{t_1} < a < e^{t_2}$.

18. Estimate e^b given that $\ln x_1 < b < \ln x_2$.

Exercises 19–28. Differentiate.

19. $f(x) = 3^{2x}$.

20. $g(x) = 4^{3x^2}$.

21. $f(x) = 2^{5x}3^{\ln x}$.

22. $F(x) = 5^{-2x^2+x}$.

23. $g(x) = \sqrt{\log_3 x}$.

24. $h(x) = 7^{\sin x^2}$.

25. $f(x) = \tan(\log_5 x)$.

26. $g(x) = \frac{\log_{10} x}{x^2}$.

27. $F(x) = \cos(2^x + 2^{-x})$.

28. $h(x) = a^{-x}\cos bx$.

Exercises 29–35. Calculate.

29. $\int 3^x\,dx$.

30. $\int 2^{-x}\,dx$.

31. $\int (x^3 + 3^{-x})\,dx$.

32. $\int x10^{-x^2}\,dx$.

33. $\int \frac{dx}{x\ln 5}$.

34. $\int \frac{\log_5 x}{x}\,dx$.

35. $\int \frac{\log_2 x^3}{x}\,dx$.

36. Show that, if a, b, c are positive, then

$$\log_a c = \log_a b\,\log_b c$$

provided that a and b are both different from 1.

Exercises 37–40. Find $f'(e)$.

37. $f(x) = \log_3 x$.

38. $f(x) = x\log_3 x$.

39. $f(x) = \ln(\ln x)$.

40. $f(x) = \log_3(\log_2 x)$.

Exercises 41–42. Calculate $f'(x)$ by first taking the logarithm of both sides.

41. $f(x) = p^x$.

42. $f(x) = p^{g(x)}$.

Exercises 43–52. Calculate.

43. $\frac{d}{dx}[(x+1)^x]$.

44. $\frac{d}{dx}[(\ln x)^x]$.

45. $\frac{d}{dx}[(\ln x)^{\ln x}]$.

46. $\frac{d}{dx}\left[\left(\frac{1}{x}\right)^x\right]$.

47. $\frac{d}{dx}[x^{\sin x}]$.

48. $\frac{d}{dx}[(\cos x)^{(x^2+1)}]$.

49. $\frac{d}{dx}[(\sin x)^{\cos x}]$.

50. $\frac{d}{dx}[x^{(x^2)}]$.

51. $\frac{d}{dx}[x^{(2^x)}]$.

52. $\frac{d}{dx}[(\tan x)^{\sec x}]$.

53. Show that

$$\text{as } x \to \infty, \qquad \left(1 + \frac{1}{x}\right)^x \to e.$$

HINT: Since the logarithm function has derivative 1 at $x = 1$,

$$\text{as } h \to 0, \qquad \frac{\ln(1+h) - \ln 1}{h} = \frac{\ln(1+h)}{h} \to 1.$$

Exercises 54–58. Draw a figure that displays the graphs of both functions.

54. $f(x) = e^x$ and $g(x) = 3^x$.

55. $f(x) = e^x$ and $g(x) = 2^x$.

56. $f(x) = \ln x$ and $g(x) = \log_3 x$.

57. $f(x) = 2^x$ and $g(x) = \log_2 x$.

58. $f(x) = \ln x$ and $g(x) = \log_2 x$.

Exercises 59–65. Evaluate.

59. $\int_1^2 2^{-x}\,dx$.

60. $\int_0^1 4^x\,dx$.

61. $\int_1^4 \frac{dx}{x\ln 2}$.

62. $\int_0^2 p^{x/2}\,dx$.

63. $\int_0^1 x10^{1+x^2}\,dx$.

64. $\int_0^1 \frac{5p^{\sqrt{x+1}}}{\sqrt{x+1}}\,dx$.

65. $\int_0^1 (2^x + x^2)\,dx$.

Exercises 66–68. Give the exact value.

66. $7^{1/\ln 7}$.

67. $5^{(\ln 17)/(\ln 5)}$.

68. $(16)^{1/\ln 2}$.

69. (a) Use a graphing utility to draw a figure that displays the graphs of both $f(x) = 2^x$ and $g(x) = x^2 - 1$.
(b) Use a CAS to find the x-coordinates of the three points where the curves intersect.
(c) Use a CAS to find the area of the bounded region that lies between the two curves.

70. Exercise 69 for $f(x) = 2^{-x}$ and $g(x) = 1/x^2$.

■ 7.6 EXPONENTIAL GROWTH AND DECAY

We begin by comparing exponential change to linear change. Let $y = y(t)$ be a function of time t.

If y is a linear function, a function of the form

$$y(t) = kt + C,$$ (k, C constants)

then y changes by the *same additive amount during all periods of the same duration*:

$$y(t + \Delta t) = k(t + \Delta t) + C = (kt + C) + k\Delta t = y(t) + k\Delta t.$$

During every period of length Δt, y changes by the same amount $k\Delta t$.

If y is a function of the form

$$y(t) = Ce^{kt},$$ (k, C constants)

then y changes by the same *multiplicative factor during all periods of the same duration*:

$$y(t + \Delta t) = Ce^{k(t+\Delta t)} = Ce^{kt}e^{k\Delta t} = e^{k\Delta t}y(t).$$

During every period of length Δt, y changes by the factor $e^{k\Delta t}$.

Functions of the form

$$f(t) = Ce^{kt}$$

have the property that the derivative $f'(t)$ is proportional to $f(t)$:

$$f'(t) = Cke^{kt} = kCe^{kt} = kf(t).$$

Moreover, they are the only such functions:

THEOREM 7.6.1

If

$$f'(t) = kf(t) \quad \text{for all } t \text{ in some interval,}$$

then there is a constant C such that

$$f(t) = Ce^{kt} \quad \text{for all } t \text{ in that interval.}$$

PROOF We assume that

$$f'(t) = kf(t)$$

and write

$$f'(t) - kf(t) = 0.$$

Multiplying this equation by e^{-kt}, we have

(∗) $$e^{-kt}f'(t) - ke^{-kt}f(t) = 0.$$

Observe now that the left side of this equation is the derivative

$$\frac{d}{dt}[e^{-kt}f(t)].$$ (Verify this.)

Equation (∗) can therefore be written

$$\frac{d}{dt}[e^{-kt}f(t)] = 0.$$

It follows that

$$e^{-kt}f(t) = C \quad \text{for some constant } C.$$

Multiplication by e^{kt} gives

$$f(t) = Ce^{kt}. \quad \square$$

Remark In the study of exponential growth or decay, time is usually measured from time $t = 0$. The constant C is the value of f at time $t = 0$:

$$f(0) = Ce^0 = C.$$

This is called the *initial value of f*. Thus the exponential $f(t) = Ce^{kt}$ can be written

$$f(t) = f(0)e^{kt}. \quad \square$$

Example 1 Find $f(t)$ given that $f'(t) = 2f(t)$ for all t and $f(0) = 5$.

SOLUTION The fact that $f'(t) = 2f(t)$ tells us that $f(t) = Ce^{2t}$ where C is some constant. Since $f(0) = C = 5$, we have $f(t) = 5e^{2t}$. $\quad \square$

Population Growth

Under ideal conditions (unlimited space, adequate food supply, immunity to disease, and so on), the rate of increase of a population P at time t is proportional to the size of the population at time t. That is,

$$P'(t) = kP(t)$$

where $k > 0$ is a constant, called the *growth constant*. Thus, by our theorem, the size of the population at any time t is given by

$$P(t) = P(0)e^{kt},$$

and the population is said to grow *exponentially*. This is a model of uninhibited growth. In reality, the rate of increase of a population does not continue to be proportional to the size of the population. After some time has passed, factors such as limitations on space or food supply, diseases, and so forth set in and affect the growth rate of the population.

Example 2 In 1980 the world population was approximately 4.5 billion and in the year 2000 it was approximately 6 billion. Assume that the world population at each time t increases at a rate proportional to the world population at time t. Measure t in years after 1980.

(a) Determine the growth constant and derive a formula for the population at time t.

(b) Estimate how long it will take for the world population to reach 9 billion (double the 1980 population).

(c) The world population for 2002 was reported to be about 6.2 billion. What population did the formula in part (a) predict for the year 2002?

SOLUTION Let $P(t)$ be the world population in billions t years after 1980. Since $P(0) = 4.5 = \frac{9}{2}$, the basic equation $P'(t) = kP(t)$ gives

$$P(t) = \tfrac{9}{2}e^{kt}.$$

(a) Since $P(20) = 6$, we have

$$\tfrac{9}{2}e^{20k} = 6, \qquad 20k = \ln\tfrac{12}{9} = \ln\tfrac{4}{3}, \qquad k = \tfrac{1}{20}\ln\tfrac{4}{3} \cong 0.0143.$$

The growth constant k is approximately 0.0143. The population t years after 1980 is

$$P(t) \cong \tfrac{9}{2}e^{0.0143t}.$$

(b) To find the value of t for which $P(t) = 9$, we set $\frac{9}{2}e^{0.0143t} = 9$:

$$e^{0.0143t} = 2, \qquad 0.0143t = \ln 2, \qquad \text{and} \qquad t = \frac{\ln 2}{0.0143} \cong 48.47.$$

Based on the data given, the world population should reach 9 billion approximately $48\frac{1}{2}$ years after 1980—around midyear 2028. (As of January 1, 2002, demographers were predicting that the world population would peak at 9 billion in the year 2070 and then start to decline.)

(c) The population predicted for the year 2002 is

$$P(22) \cong \frac{9}{2}e^{0.0143(22)} = \frac{9}{2}e^{0.3146} \cong 6.164$$

billion, not far off the reported figure of 6.2 billion. ❏

Bacterial Colonies

Example 3 The size of a bacterial colony increases at a rate proportional to the size of the colony. Suppose that when the first measurement is taken, time $t = 0$, the colony occupies an area of 0.25 square centimeters and 8 hours later the colony occupies 0.35 square centimeters.

(a) Estimate the size of the colony t hours after the initial measurement is taken. What is the expected size of the colony at the end of 12 hours?

(b) Find the *doubling time*, the time it takes for the colony to double in size.

SOLUTION Let $S(t)$ be the size of the colony at time t, size measured in square centimeters, t measured in hours. The basic equation $S'(t) = kS(t)$ gives

$$S(t) = S(0)\,e^{kt}.$$

Since $S(0) = 0.25$, we have

$$S(t) = (0.25)\,e^{kt}.$$

We can evaluate the growth constant k from the fact that $S(8) = 0.35$:

$$0.35 = (0.25)\,e^{8k}, \qquad e^{8k} = 1.4, \qquad 8k = \ln(1.4)$$

and therefore

$$k = \tfrac{1}{8}\ln(1.4) \cong 0.042.$$

(a) The size of the colony at time t is

$$S(t) \cong (0.25)\,e^{0.042t} \quad \text{square centimeters.}$$

The expected size of the colony at the end of 12 hours is

$$S(12) \cong (0.25)\,e^{0.042(12)} \cong (0.25)\,e^{0.504} \cong 0.41 \text{ square centimeters.}$$

(b) To find the doubling time, we seek the value of t for which $S(t) = 2(0.25) = 0.50$. Thus we set

$$(0.25)\,e^{0.042t} = 0.50$$

and solve for t:

$$e^{0.042t} = 2, \qquad 0.042t = \ln 2, \qquad t = \frac{\ln 2}{0.042} \cong 16.50.$$

The doubling time is approximately $16\frac{1}{2}$ hours. ❏

Remark   There is a way of expressing $S(t)$ that uses the exact value of k. We have seen that $k = \frac{1}{8} \ln(1.4)$. Therefore

$$S(t) = (0.25)e^{(t/8)\ln(1.4)} = (0.25)\,e^{\ln(1.4)^{t/8}} = (0.25)(1.4)^{t/8}.$$

We leave it to you as an exercise to verify that the population function derived in Example 2 can be written $P(t) = \frac{9}{2}\left(\frac{4}{3}\right)^{t/20}$. ❏

Radioactive Decay

Although different radioactive substances decay at different rates, each radioactive substance decays at a rate proportional to the amount of the substance present: if $A(t)$ is the amount present at time t, then

$$A'(t) = kA(t) \qquad \text{for some constant } k.$$

Since A decreases, the constant k, called the *decay constant*, is a negative number. From general considerations already explained, we know that

$$A(t) = A(0)\,e^{kt}$$

where $A(0)$ is the amount present at time $t = 0$.

The *half-life* of a radioactive substance is the time T it takes for half of the substance to decay. The decay constant k and the half-life T are related by the equation

(7.6.2)
$$\boxed{kT = -\ln 2.}$$

PROOF   The relation $A(T) = \frac{1}{2}A(0)$ gives

$$\tfrac{1}{2}A(0) = A(0)e^{kT}, \qquad e^{kT} = \tfrac{1}{2}, \qquad kT = -\ln 2. \quad ❏$$

Example 4   Today we have A_0 grams of a radioactive substance with a half-life of 8 years.

(a) How much of this substance will remain in 16 years?

(b) How much of the substance will remain in 4 years?

(c) What is the decay constant?

(d) How much of the substance will remain in t years?

SOLUTION   We know that exponentials change by the same factor during all time periods of the same length.

(a) During the first 8 years A_0 will decrease to $\frac{1}{2}A_0$, and during the following 8 years it will decrease to $\frac{1}{2}(\frac{1}{2}A_0) = \frac{1}{4}A_0$. Answer: $\frac{1}{4}A_0$ grams.

(b) In 4 years A_0 will decrease to some fractional multiple αA_0 and in the following 4 years to $\alpha^2 A_0$. Since $\alpha^2 = \frac{1}{2}$, $\alpha = \sqrt{2}/2$. Answer: $(\sqrt{2}/2)A_0$ grams.

(c) In general, $kT = -\ln 2$. Here $T = 8$ years. Answer: $k = -\frac{1}{8}\ln 2$.

(d) In general, $A(t) = A(0)e^{kt}$. Here $A(0) = A_0$ and $k = -\frac{1}{8}\ln 2$. Answer: $A(t) = A_0 e^{-\frac{1}{8}(\ln 2)t}$. ❏

Example 5   Cobalt-60 is a radioactive substance used extensively in radiology. It has a half-life of 5.3 years. Today we have a sample of 100 grams.

(a) Determine the decay constant of cobalt-60.

(b) How much of the 100 grams will remain in t years?

(c) How long will it take for 90% of the sample to decay?

SOLUTION

(a) Equation (7.6.2) gives

$$k = \frac{-\ln 2}{T} = \frac{-\ln 2}{5.3} \cong -0.131.$$

(b) Given that $A(0) = 100$, the amount that will remain in t years is

$$A(t) = 100e^{-0.131t}.$$

(c) If 90% of the sample decays, then 10%, which is 10 grams, remains. We seek the time t at which

$$100e^{-0.131t} = 10.$$

We solve this equation for t:

$$e^{-0.131t} = 0.1, \qquad -0.131t = \ln(0.1), \qquad t = \frac{\ln(0.1)}{-0.131} \cong 17.6.$$

It will take approximately 17.6 years for 90% of the sample to decay. ❏

Compound Interest

Consider money invested at annual interest rate r. If the accumulated interest is credited once a year, then the interest is said to be compounded annually; if twice a year, then semiannually; if four times a year, then quarterly. The idea can be pursued further. Interest can be credited every day, every hour, every second, every half-second, and so on. In the limiting case, interest is credited instantaneously. Economists call this *continuous compounding*.

The economists' formula for continuous compounding is a simple exponential:

(7.6.3)
$$A(t) = A_0 e^{rt}.$$

Here t is measured in years,

$$A_0 = A(0) = \text{the initial investment,}$$

$$r = \text{the annual interest rate expressed as a decimal,}$$

$$A(t) = \text{the principal at time } t.$$

A DERIVATION OF THE COMPOUND INTEREST FORMULA Fix t and take h as a small time increment. Then

$$A(t+h) - A(t) = \text{interest earned from time } t \text{ to time } t + h.$$

Had the principal remained $A(t)$ from time t to time $t + h$, the interest earned during this time period would have been

$$rh\,A(t).$$

Had the principal been $A(t + h)$ throughout the time interval, the interest earned would have been

$$rh\,A(t+h).$$

The actual interest earned must be somewhere in between:

$$rh\, A(t) \le A(t + h) - A(t) \le rh\, A(t + h).$$

Dividing by h, we get

$$rA(t) \le \frac{A(t + h) - A(t)}{h} \le rA(t + h).$$

If A varies continuously, then, as h tends to zero, $rA(t + h)$ tends to $rA(t)$ and (by the pinching theorem) the difference quotient in the middle must also tend to $rA(t)$:

$$\lim_{h \to 0} \frac{A(t + h) - A(t)}{h} = rA(t).$$

This says that

$$A'(t) = rA(t).$$

Thus, with continuous compounding, the principal increases at a rate proportional to the amount present and the growth constant is the interest rate r. Now, it follows that

$$A(t) = Ce^{rt}.$$

If A_0 is the initial investment, we have $C = A_0$ and therefore $A(t) = A_0 e^{rt}$. ❏

Remark   Frequency of compounding affects the return on principal, but (on modest sums) not very much. Listed below are the year-end values of $1000 invested at 6% under various forms of compounding:

(a) Annual compounding: $1000(1 + 0.06) = \$1060$.
(b) Quarterly compounding: $1000[1 + (.06/4)]^4 = \$1061.36$.
(c) Monthly compounding: $1000[1 + (.06/12)]^{12} = \$1061.67$.
(d) Continuous compounding: $1000\, e^{0.06} \cong \$1061.84$. ❏

Example 6 $1000 is deposited in a bank account that yields 5% compounded continuously. Estimate the value of the account 6 years later. How much interest will have been earned during that 6-year period?

SOLUTION   Here $A_0 = 1000$ and $r = 0.05$. The value of the account t years after the deposit is made is given by the function

$$A(t) = 1000\, e^{0.05t}.$$

At the end of the sixth year, the value of the account will be

$$A(6) = 1000\, e^{0.05(6)} = 1000\, e^{0.3} \cong 1349.86.$$

Interest earned: $349.86. ❏

Example 7   How long does it take to double your money at interest rate r compounded continuously?

SOLUTION   During t years an initial investment A_0 grows in value to

$$A(t) = A_0\, e^{rt}.$$

You double your money once you have reached the time period t for which

$$A_0\, e^{rt} = 2A_0.$$

Solving this equation for t, we have

$$e^{rt} = 2, \qquad rt = \ln 2, \qquad t = \frac{\ln 2}{r} \cong \frac{0.69}{r}. \quad ❏$$

For example, at 8% an investment doubles in value in $\frac{0.69}{0.08} = 8.625$ years.

Remark A popular estimate for the doubling time at an interest rate α% is the *rule of 72:*

$$\text{doubling time} \cong \frac{72}{\alpha}.$$

According to this rule, the doubling time at 8% is approximately $\frac{72}{8} = 9$ years. Here is how the rule originated:

$$\frac{0.69}{\alpha/100} = \frac{69}{\alpha} \cong \frac{72}{\alpha}.$$

For rough calculations 72 is preferred to 69 because 72 has more divisors.[†] ❑

_____

[†]This way of calculating doubling time is too inaccurate for our purposes. We will not use it.

EXERCISES 7.6

▶NOTE: Some of these exercises require a calculator or graphing utility.

1. Find the amount of interest earned by $500 compounded continuously for 10 years:
   (a) at 6%,   (b) at 8%,   (c) at 10%.

2. How long does it take for a sum of money to double if compounded continuously:
   (a) at 6%?   (b) at 8%?   (c) at 10%?

3. At what rate r of continuous compounding does a sum of money triple in 20 years?

4. At what rate r of continuous compounding does a sum of money double in 10 years?

5. Show that the population function derived in Example 2 can be written $P(t) = \frac{9}{2} \left(\frac{4}{3}\right)^{t/20}$.

6. A biologist observes that a certain bacterial colony triples every 4 hours and after 12 hours occupies 1 square centimeter.
   (a) How much area was occupied by the colony when first observed?
   (b) What is the doubling time for the colony?

7. A population P of insects increases at a rate proportional to the current population. Suppose there are 10,000 insects at time $t = 0$ and 20,000 insects a week later.
   (a) Find an expression for the number $P(t)$ of insects at each time $t > 0$.
   (b) How many insects will there be in $\frac{1}{2}$ year? In 1 year?

8. Determine the time period in which $y = Ce^{kt}$ changes by a factor of q.

9. The population of a certain country increases at the rate of 3.5% per year. By what factor does it increase every 10 years? What percentage increase per year will double the population every 15 years?

10. According to the Bureau of the Census, the population of the United States in 1990 was approximately 249 million

and in 2000, 281 million. Use this information to estimate the population in 1980. (The actual figure was about 227 million.)

11. Use the data of Exercise 10 to predict the population for 2010. Compare the prediction for 2001 with the actual reported figure of 284.8 million.

12. Use the data of Exercise 10 to estimate how long it will take for the U.S population to double.

13. It is estimated that the arable land on earth can support a maximum of 30 billion people. Extrapolate from the data given in Example 2 to estimate the year when the food supply will become insufficient to support the world population. (Rest assured that there are strong reasons to believe that such extrapolations are invalid. Conditions change.)

14. Water is pumped into a tank to dilute a saline solution. The volume of the solution, call it V, is kept constant by continuous outflow. The amount of salt in the tank, call it s, depends on the amount of water that has been pumped in; call this x. Given that

$$\frac{ds}{dx} = -\frac{s}{V},$$

find the amount of water that must be pumped into the tank to eliminate 50% of the salt. Take V as 10,000 gallons.

15. A 200-liter tank initially full of water develops a leak at the bottom. Given that 20% of the water leaks out in the first 5 minutes, find the amount of water left in the tank t minutes after the leak develops if the water drains off at a rate proportional to the amount of water present.

16. What is the half-life of a radioactive substance if it takes 5 years for one-third of the substance to decay?

17. Two years ago there were 5 grams of a radioactive substance. Now there are 4 grams. How much will remain 3 years from now?

18. A year ago there were 4 grams of a radioactive substance. Now there are 3 grams. How much was there 10 years ago?

19. Suppose the half-life of a radioactive substance is n years. What percentage of the substance present at the start of a year will decay during the ensuing year?

20. A radioactive substance weighed n grams at time $t = 0$. Today, 5 years later, the substance weighs m grams. How much will it weigh 5 years from now?

21. The half-life of radium-226 is 1620 years. What percentage of a given amount of the radium will remain after 500 years? How long will it take for the original amount to be reduced by 75%?

22. Cobalt-60 has a half-life of 5.3 years. What percentage of a given amount of cobalt will remain after 8 years? If you have 100 grams of cobalt now, how much was there 3 years ago?

23. (*The power of exponential growth*) Imagine two racers competing on the x-axis (which has been calibrated in meters), a linear racer LIN [position function of the form $x_1(t) = kt + C$] and an exponential racer EXP [position function of the form $x_2(t) = e^{kt} + C$]. Suppose that both racers start out simultaneously from the origin, LIN at 1 million meters per second, EXP at only 1 meter per second. In the early stages of the race, fast-starting LIN will move far ahead of EXP, but in time EXP will catch up to LIN, pass her, and leave her hopelessly behind. Show that this is true as follows:
    (a) Express the position of each racer as a function of time, measuring t in seconds.
    (b) Show that LIN's lead over EXP starts to decline about 13.8 seconds into the race.
    (c) Show that LIN is still ahead of EXP some 15 seconds into the race but far behind 3 seconds later.
    (d) Show that, once EXP passes LIN, LIN can never catch up.

24. (*The weakness of logarithmic growth*) Having been soundly beaten in the race of Exercise 23, LIN finds an opponent she can beat, LOG, the logarithmic racer [position function $x_3(t) = k \ln(t + 1) + C$]. Once again the racetrack is the x-axis calibrated in meters. Both racers start out at the origin, LOG at 1 million meters per second, LIN at only 1 meter per second. (LIN is tired from the previous race.) In this race LOG will shoot ahead and remain ahead for a long time, but eventually LIN will catch up to LOG, pass her, and leave her permanently behind. Show that this is true as follows:
    (a) Express the position of each racer as a function of time t, measuring t in seconds.
    (b) Show that LOG's lead over LIN starts to decline $10^6 - 1$ seconds into the race.
    (c) Show that LOG is still ahead of LIN $10^7 - 1$ seconds into the race but behind LIN $10^8 - 1$ seconds into the race.
    (d) Show that, once LIN passes LOG, LOG can never catch up.

25. Atmospheric pressure p varies with altitude h according to the equation

$$\frac{dp}{dh} = kp \qquad \text{where } k \text{ is a constant.}$$

Given that p is 15 pounds per square inch at sea level and 10 pounds per square inch at 10,000 feet, find p at: (a) 5000 feet; (b) 15,000 feet.

26. The compound interest formula

$$Q = Pe^{rt}$$

can be written

$$P = Qe^{-rt}.$$

In this formulation we have P as the investment required today to obtain Q in t years. In this sense P dollars is *present value* of Q dollars to be received t years from now. Find the present value of $20,000 to be received 4 years from now. Assume continuous compounding at 4%.

27. Find the interest rate r needed for $6000 to be the present value of $10,000 8 years from now.

28. You are 45 years old and are looking forward to an annual pension of $50,000 per year at age 65. What is the present-day purchasing power (present value) of your pension if money can be invested over this period at a continuously compounded interest rate of: (a) 4%? (b) 6%? (c) 8%?

29. The cost of the tuition, fees, room, and board at XYZ College is currently $25,000 per year. What would you expect to pay 3 years from now if the costs at XYZ are rising at the continuously compounded rate of: (a) 5%? (b) 8%? (c) 12%?

30. A boat moving in still water is subject to a retardation proportional to its velocity. Show that the velocity t seconds after the power is shut off is given by the formula $v = \alpha e^{-kt}$ where α is the velocity at the instant the power is shut off.

31. A boat is drifting in still water at 4 miles per hour; 1 minute later, at 2 miles per hour. How far has the boat drifted in that 1 minute? (See Exercise 30.)

32. During the process of inversion, the amount A of raw sugar present decreases at a rate proportional to A. During the first 10 hours, 1000 pounds of raw sugar have been reduced to 800 pounds. How many pounds will remain after 10 more hours of inversion?

33. The method of *carbon dating* makes use of the fact that all living organisms contain two isotopes of carbon, carbon-12, denoted ^{12}C (a stable isotope), and carbon-14, denoted ^{14}C (a radioactive isotope). The ratio of the amount of ^{14}C to the amount of ^{12}C is essentially constant (approximately 1/10,000). When an organism dies, the amount of ^{12}C present remains unchanged, but the ^{14}C decays at a rate proportional to the amount present with a half-life of approximately 5700 years. This change in the amount of ^{14}C relative to the amount of ^{12}C makes it possible to estimate the time at which the organism lived. A fossil found in an archaeological dig was found to contain 25% of the original amount of ^{14}C. What is the approximate age of the fossil?

34. The Dead Sea Scrolls are approximately 2000 years old. How much of the original ^{14}C remains in them?

Exercises 35–37. Find all the functions f that satisfy the equation for all real t.

35. $f'(t) = tf(t)$. HINT: Write $f'(t) - tf(t) = 0$ and multiply the equation by $e^{-t^2/2}$.

36. $f'(t) = \sin t f(t)$.   **37.** $f'(t) = \cos t f(t)$.

38. Let g be a function everywhere continuous and not identically zero. Show that if $f'(t) = g(t)f(t)$ for all real t, then either f is identically zero or f does not take on the value zero.

■ 7.7 THE INVERSE TRIGONOMETRIC FUNCTIONS

Arc Sine

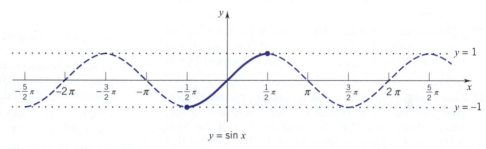

$y = \sin x$

Figure 7.7.1

Figure 7.7.1 shows the sine wave. Clearly the sine function is not one-to-one: it takes on every value from -1 to 1 an infinite number of times. However, on the interval $[-\frac{1}{2}\pi, \frac{1}{2}\pi]$ it takes on every value from -1 to 1, but only once. (See the solid part of the wave.) Thus the function

$$y = \sin x, \qquad x \in \left[-\tfrac{1}{2}\pi, \tfrac{1}{2}\pi\right]$$

maps the interval $[-\frac{1}{2}\pi, \frac{1}{2}\pi]$ onto $[-1, 1]$ in a one-to-one manner and has an inverse that maps $[-1, 1]$ back to $[-\frac{1}{2}\pi, \frac{1}{2}\pi]$, also in a one-to-one manner. The inverse is called the *arc sine function*:

$$y = \arcsin x, \qquad x \in [-1, 1]$$

is the inverse of the function

$$y = \sin x, \qquad x \in \left[-\tfrac{1}{2}\pi, \tfrac{1}{2}\pi\right].$$

These functions are graphed in Figure 7.7.2. Each graph is the reflection of the other in the line $y = x$.

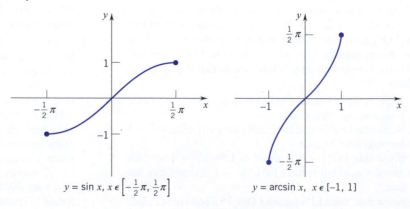

$y = \sin x, \; x \in \left[-\frac{1}{2}\pi, \frac{1}{2}\pi\right]$ $y = \arcsin x, \; x \in [-1, 1]$

Figure 7.7.2

Since these functions are inverses,

(7.7.1)

$$\text{for all } x \in [-1, 1], \qquad \sin(\arcsin x) = x$$

and

(7.7.2)

$$\text{for all } x \in \left[-\tfrac{1}{2}\pi, \tfrac{1}{2}\pi\right], \qquad \arcsin(\sin x) = x.$$

Table 7.7.1 gives some representative values of the sine function from $x = -\tfrac{1}{2}\pi$ to $x = \tfrac{1}{2}\pi$. Reversing the order of the columns, we have a table for the arc sine. (Table 7.7.2.)

On the basis of Table 7.7.2 one could guess that for all $x \in [-1, 1]$

$$\arcsin(-x) = -\arcsin(x).$$

This is indeed the case. Being the inverse of an odd function ($\sin(-x) = -\sin x$ for all $x \in [-\tfrac{1}{2}\pi, \tfrac{1}{2}\pi]$), the arc sine is itself an odd function. (We leave it to you to verify this.)

Example 1 Calculate if defined:

(a) $\arcsin(\sin \tfrac{1}{16}\pi)$ **(b)** $\arcsin(\sin \tfrac{7}{3}\pi)$

(c) $\sin(\arcsin \tfrac{1}{3})$ **(d)** $\arcsin(\sin \tfrac{9}{5}\pi)$

(e) $\sin(\arcsin 2)$.

SOLUTION

(a) Since $\tfrac{1}{16}\pi$ is in the interval $[-\tfrac{1}{2}\pi, \tfrac{1}{2}\pi]$, we know from (7.7.2) that

$$\arcsin\left(\sin \tfrac{1}{16}\pi\right) = \tfrac{1}{16}\pi.$$

(b) Since $\tfrac{7}{3}\pi$ is not in the interval $[-\tfrac{1}{2}\pi, \tfrac{1}{2}\pi]$, we cannot apply (7.7.2) directly. However, $\tfrac{7}{3}\pi = \tfrac{1}{3}\pi + 2\pi$ and $\sin(\tfrac{1}{3}\pi + 2\pi) = \sin(\tfrac{1}{3}\pi)$. Therefore

$$\arcsin\left(\sin \tfrac{7}{3}\pi\right) = \underset{\underset{\text{from (7.7.2)}}{\uparrow}}{\arcsin\left(\sin \tfrac{1}{3}\pi\right)} = \tfrac{1}{3}\pi.$$

(c) From (7.7.1),

$$\sin\left(\arcsin \tfrac{1}{3}\right) = \tfrac{1}{3}.$$

(d) Since $\tfrac{9}{5}\pi$ is not within the interval $[-\tfrac{1}{2}\pi, \tfrac{1}{2}\pi]$, we cannot apply (7.7.2) directly. However, $\tfrac{9}{5}\pi = 2\pi - \tfrac{1}{5}\pi$. Therefore

$$\arcsin\left(\sin \tfrac{9}{5}\pi\right) = \underset{\underset{\text{from (7.7.2)}}{\uparrow}}{\arcsin\left[\sin\left(-\tfrac{1}{5}\pi\right)\right]} = -\tfrac{1}{5}\pi.$$

(e) The expression $\sin(\arcsin 2)$ makes no sense since 2 is not in the domain of the arc sine. (There is *no* angle with sine 2.) The arc sine is defined only on $[-1, 1]$. ❏

■ **Table 7.7.1**

x	$\sin x$
$-\tfrac{1}{2}\pi$	-1
$-\tfrac{1}{3}\pi$	$-\tfrac{1}{2}\sqrt{3}$
$-\tfrac{1}{4}\pi$	$-\tfrac{1}{2}\sqrt{2}$
$-\tfrac{1}{6}\pi$	$-\tfrac{1}{2}$
0	0
$\tfrac{1}{6}\pi$	$\tfrac{1}{2}$
$\tfrac{1}{4}\pi$	$\tfrac{1}{2}\sqrt{2}$
$\tfrac{1}{3}\pi$	$\tfrac{1}{2}\sqrt{3}$
$\tfrac{1}{2}\pi$	1

■ **Table 7.7.2**

x	$\arcsin$
-1	$-\tfrac{1}{2}\pi$
$-\tfrac{1}{2}\sqrt{3}$	$-\tfrac{1}{3}\pi$
$-\tfrac{1}{2}\sqrt{2}$	$-\tfrac{1}{4}\pi$
$-\tfrac{1}{2}$	$-\tfrac{1}{6}\pi$
0	0
$\tfrac{1}{2}$	$\tfrac{1}{6}\pi$
$\tfrac{1}{2}\sqrt{2}$	$\tfrac{1}{4}\pi$
$\tfrac{1}{2}\sqrt{3}$	$\tfrac{1}{3}\pi$
1	$\tfrac{1}{2}\pi$

Since the derivative of the sine function,

$$\frac{d}{dx}(\sin x) = \cos x,$$

is nonzero on $(-\frac{1}{2}\pi, \frac{1}{2}\pi)$, the arc sine function is differentiable on the open interval $(-1, 1)^{\dagger}$. We can find the derivative as follows: reading from the accompanying figure

$$y = \arcsin x$$

$$\sin y = x$$

$$\cos y \frac{dy}{dx} = 1$$

$$\frac{dy}{dx} = \frac{1}{\cos y} = \frac{1}{\sqrt{1 - x^2}}.$$

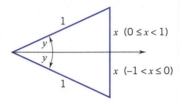

Thus

(7.7.3)
$$\boxed{\frac{d}{dx}(\arcsin x) = \frac{1}{\sqrt{1 - x^2}}.}$$

Example 2

$$\frac{d}{dx}(\arcsin 3x^2) = \underbrace{\frac{1}{\sqrt{1 - (3x^2)^2}}}_{\text{the chain rule}} \cdot \frac{d}{dx}(3x^2) = \frac{6x}{\sqrt{1 - 9x^4}}.$$

NOTE: We continue with the convention that if the domain of a function f is not specified explicitly, then it is understood to be the maximal set of real numbers x for which $f(x)$ is a real number. In this case, the domain is the set of real numbers x for which $-1 \leq 3x^2 \leq 1$. This is the interval $[-1/\sqrt{3}, \ 1/\sqrt{3}]$. ❏

The integral counterpart of (7.7.3) reads

(7.7.4)
$$\boxed{\int \frac{dx}{\sqrt{1 - x^2}} = \arcsin x + C.}$$

Example 3   Show that for $a > 0$

(7.7.5)
$$\boxed{\int \frac{dx}{\sqrt{a^2 - x^2}} = \arcsin \frac{x}{a} + C.}$$

SOLUTION   We change variables so that a^2 is replaced by 1 and we can use (7.7.4). To this end we set

$$au = x, \quad a\,du = dx.$$

---
†Section 7.1.

Then

$$\int \frac{dx}{\sqrt{a^2 - x^2}} = \int \frac{a\, du}{\sqrt{a^2 - a^2 u^2}} = \frac{a\, du}{a\sqrt{1 - u^2}}$$

since $a > 0$

$$= \int \frac{du}{\sqrt{1 - u^2}} = \arcsin u + C = \arcsin \frac{x}{a} + C. \quad \square$$

Example 4   Evaluate $\displaystyle\int_0^{\sqrt{3}} \frac{dx}{\sqrt{4 - x^2}}\, dx.$

SOLUTION   By (7.7.5),

$$\int \frac{dx}{\sqrt{4 - x^2}} = \arcsin \frac{x}{2} + C.$$

It follows that

$$\int_0^{\sqrt{3}} \frac{dx}{\sqrt{4 - x^2}}\, dx = \left[\arcsin \frac{x}{2}\right]_0^{\sqrt{3}} = \arcsin \frac{\sqrt{3}}{2} - \arcsin 0 = \frac{\pi}{3} - 0 = \frac{\pi}{3}. \quad \square$$

Arc Tangent

Although not one-to-one on its full domain, the tangent function is one-to-one on the open interval $\left(-\frac{1}{2}\pi, \frac{1}{2}\pi\right)$ and on that interval the function takes on as a value every real number. (See Figure 7.7.3.) Thus the function

$$y = \tan x, \qquad x \in \left(-\tfrac{1}{2}\pi, \tfrac{1}{2}\pi\right)$$

maps the interval $\left(-\frac{1}{2}\pi, \frac{1}{2}\pi\right)$ onto $(-\infty, \infty)$ in a one-to-one manner and has an inverse that maps $(-\infty, \infty)$ back to $\left(-\frac{1}{2}\pi, \frac{1}{2}\pi\right)$, also in a one-to-one manner. This inverse is called the *arc tangent*: the *arc tangent function*

$$y = \arctan x, \qquad x \in (-\infty, \infty)$$

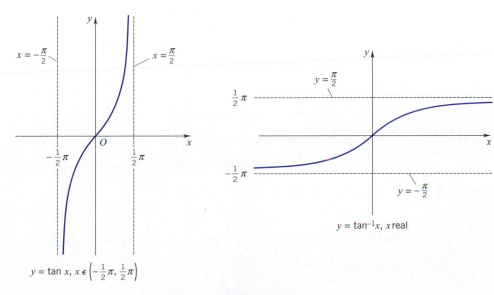

Figure 7.7.3

is the inverse of the function

$$y = \tan x, \quad x \in \left(-\tfrac{1}{2}\pi, \tfrac{1}{2}\pi\right).$$

These functions are graphed in Figure 7.7.3.

Each graph is the reflection of the other in the line $y = x$. While the tangent has vertical asymptotes, the inverse tangent has horizontal asymptotes. Both functions are odd functions.

Since these functions are inverses,

(7.7.6)

$$\boxed{\text{for all real numbers } x \qquad \tan(\arctan x) = x}$$

and

(7.7.7)

$$\boxed{\text{for all } x \in \left(-\tfrac{1}{2}\pi, \tfrac{1}{2}\pi\right), \qquad \arctan(\tan x) = x.}$$

It is hard to make a mistake with (7.7.6) since that relation holds for all real numbers, but the application of (7.7.7) requires some care since it applies only to x in $\left(-\tfrac{1}{2}\pi, \tfrac{1}{2}\pi\right)$. Thus, while $\arctan(\tan\tfrac{1}{4}\pi) = \tfrac{1}{4}\pi$, $\arctan(\tan\tfrac{7}{5}\pi) \neq \tfrac{7}{5}\pi$. To calculate $\arctan(\tan\tfrac{7}{5}\pi)$, we use the fact that the tangent function has period π. Therefore

$$\arctan\left(\tan\tfrac{7}{5}\pi\right) = \arctan\left(\tan\tfrac{2}{5}\pi\right) = \tfrac{2}{5}\pi.$$

The final equality holds since $\tfrac{2}{5}\pi \in \left(-\tfrac{1}{2}\pi, \tfrac{1}{2}\pi\right)$.

Since the derivative of the tangent function,

$$\frac{d}{dx}(\tan x) = \sec^2 x = \frac{1}{\cos^2 x},$$

is never 0 on $\left(-\tfrac{1}{2}\pi, \tfrac{1}{2}\pi\right)$, the arc tangent function is everywhere differentiable. (Section 7.1) We can find the derivative as we did for the arc sine: reading from the figure

$$y = \arctan x$$

$$\tan y = x$$

$$\sec^2 y \, \frac{dy}{dx} = 1$$

$$\frac{dy}{dx} = \frac{1}{\sec^2 y} = \cos^2 y = \frac{1}{1 + x^2}.$$

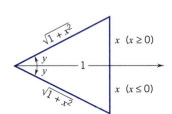

We have found that

(7.7.8)

$$\boxed{\frac{d}{dx}(\arctan x) = \frac{1}{1 + x^2}.}$$

Example 5

$$\frac{d}{dx}[\arctan(ax^2 + bx + c)] = \frac{1}{1 + (ax^2 + bx + c)^2} \cdot \frac{d}{dx}(ax^2 + bx + c)$$

by the chain rule ⟶

$$= \frac{2ax + b}{1 + (ax^2 + bx + c)^2}. \quad ❑$$

The integral counterpart of (7.7.8) reads

(7.7.9)
$$\int \frac{dx}{1+x^2} = \arctan x + C.$$

Example 6 Show that, for $a \neq 0$,

(7.7.10)
$$\int \frac{dx}{a^2 + x^2} = \frac{1}{a} \arctan \frac{x}{a} + C.$$

SOLUTION We change variables so that a^2 is replaced by 1 and we can use (7.7.9). We set

$$au = x, \qquad a\,du = dx.$$

Then

$$\int \frac{dx}{a^2 + x^2} = \int \frac{a\,du}{a^2 + a^2 u^2} = \frac{1}{a} \int \frac{du}{1 + u^2}$$

$$= \frac{1}{a} \arctan u + C = \frac{1}{a} \arctan \frac{x}{a} + C. \quad ❑$$

(7.7.9) ⟶

Example 7 Evaluate $\displaystyle\int_0^2 \frac{dx}{4 + x^2}$.

SOLUTION By (7.7.10),

$$\int \frac{dx}{4 + x^2} = \int \frac{dx}{2^2 + x^2} = \frac{1}{2} \arctan \frac{x}{2} + C,$$

and therefore

$$\int_0^2 \frac{dx}{4 + x^2} = \left[\frac{1}{2} \arctan \frac{x}{2} \right]_0^2 = \frac{1}{2} \arctan 1 - \frac{1}{2} \arctan 0 = \frac{\pi}{8}. \quad ❑$$

Arc Cosine, Arc Cotangent, Arc Secant, Arc Cosecant

These functions are not as important to us as the arc sine and arc tangent, but they do deserve some attention.

Arc Cosine While the cosine function is not one-to-one, it is one-to-one on $[0, \pi]$ and maps that interval onto $[-1, 1]$. (Figure 1.6.13) The *arc cosine* function

$$y = \arccos x, \qquad x \in [-1, 1]$$

is the inverse of the function

$$y = \cos x, \qquad x \in [0, \pi].$$

Arc Cotangent The cotangent function is one-to-one on $(0, \pi)$ and maps that interval onto $(-\infty, \infty)$. The *arc cotangent* function

$$y = \operatorname{arccot} x, \qquad x \in (-\infty, \infty)$$

is the inverse of the function

$$y = \cot x, \qquad x \in (0, \pi).$$

Arc Secant, Arc Cosecant These functions can be defined explicitly in terms of the arc cosine and the arc sine. For $|x| \geq 1$, we set

$$\operatorname{arcsec} x = \arccos(1/x), \qquad \operatorname{arccsc} x = \arcsin(1/x).$$

In the Exercises you are asked to show that for all $|x| \geq 1$

$$\sec(\operatorname{arcsec} x) = x \qquad \text{and} \qquad \csc(\operatorname{arccsc} x) = x.$$

Relations to $\frac{1}{2}\pi$

Where defined

(7.7.11)

$$
\begin{aligned}
\arcsin x + \arccos x &= \tfrac{1}{2}\pi, \\
\arctan x + \operatorname{arccot} x &= \tfrac{1}{2}\pi, \\
\operatorname{arcsec} x + \operatorname{arccsc} x &= \tfrac{1}{2}\pi.
\end{aligned}
$$

We derive the first relation; the other two we leave to you. (Exercises 73, 74.)

Our derivation is based on the identity

$$\cos\theta = \sin\left(\tfrac{1}{2}\pi - \theta\right). \qquad \text{(Section 1.6)}$$

Suppose that $y = \arccos x$. Then

$$\cos y = x \qquad \text{with} \qquad y \in [0, \pi]$$

and therefore

$$\sin\left(\tfrac{1}{2}\pi - y\right) = x \qquad \text{with} \qquad \left(\tfrac{1}{2}\pi - y\right) \in \left[-\tfrac{1}{2}\pi, \tfrac{1}{2}\pi\right].$$

It follows that

$$\arcsin x = \tfrac{1}{2}\pi - y, \qquad \arcsin x + y = \tfrac{1}{2}\pi, \qquad \arcsin x + \arccos x = \tfrac{1}{2}\pi$$

as asserted. ❏

Derivatives

(7.7.12)

$$
\begin{aligned}
\frac{d}{dx}(\arcsin x) &= \frac{1}{\sqrt{1-x^2}}, & \frac{d}{dx}(\arccos x) &= -\frac{1}{\sqrt{1-x^2}} \\[2mm]
\frac{d}{dx}(\arctan x) &= \frac{1}{1+x^2}, & \frac{d}{dx}(\operatorname{arccot} x) &= -\frac{1}{1+x^2} \\[2mm]
\frac{d}{dx}(\operatorname{arcsec} x) &= \frac{1}{|x|\sqrt{x^2-1}}, & \frac{d}{dx}(\operatorname{arccsc} x) &= -\frac{1}{|x|\sqrt{x^2-1}}.
\end{aligned}
$$

VERIFICATION The derivatives of the arc sine and the arc tangent were calculated earlier. That the derivatives of the arc cosine and the arc cotangent are as stated follows immediately from (7.7.11). Once we show that

$$\frac{d}{dx}(\operatorname{arcsec} x) = \frac{1}{|x|\sqrt{x^2-1}},$$

the last formula will follow from (7.7.11). Hence we focus on the arc secant. Since

$$\operatorname{arcsec} x = \arccos (1/x),$$

the chain rule gives

$$\frac{d}{dx}(\operatorname{arcsec} x) = -\frac{1}{\sqrt{1 - (1/x)^2}} \cdot \frac{d}{dx}\left(\frac{1}{x}\right)$$

$$= -\frac{\sqrt{x^2}}{\sqrt{x^2 - 1}}\left(-\frac{1}{x^2}\right) = \frac{\sqrt{x^2}}{x^2\sqrt{x^2 - 1}}.$$

This tells us that

$$\frac{d}{dx}(\operatorname{arcsec} x) = \begin{cases} \dfrac{1}{x\sqrt{x^2 - 1}}, & \text{for } x > 1 \\ \dfrac{1}{-x\sqrt{x^2 - 1}}, & \text{for } x < -1. \end{cases}$$

The statement

$$\frac{d}{dx}(\operatorname{arcsec} x) = \frac{1}{|x|\sqrt{x^2 - 1}}$$

is just a summary of this result.   ❏

Remark on Notation   The expressions $\arcsin x$, $\arctan x$, $\arccos x$, and so on are sometimes written $\sin^{-1} x$, $\tan^{-1} x$, $\cos^{-1} x$, and so on.   ❏

EXERCISES 7.7

Exercises 1–9. Determine the exact value.

1. (a) $\arctan 0$;   (b) $\arcsin (-\sqrt{3}/2)$.
2. (a) $\operatorname{arcsec} 2$;   (b) $\arctan (\sqrt{3})$.
3. (a) $\arccos (-\frac{1}{2})$;   (b) $\operatorname{arcsec} (-\sqrt{2})$.
4. (a) $\sec (\operatorname{arcsec} [-2/\sqrt{3}])$;   (b) $\sec (\arccos [-\frac{1}{2}])$.
5. (a) $\cos (\operatorname{arcsec} 2)$;   (b) $\arctan (\sec 0)$.
6. (a) $\arcsin (\sin [11\pi/6])$;   (b) $\arctan (\tan [11\pi/4])$.
7. (a) $\arccos (\sec [7\pi]/6)$;   (b) $\operatorname{arcsec} (\sin [13\pi/6])$.
8. (a) $\cos (\arcsin [\frac{3}{5}])$;   (b) $\sec (\arctan [\frac{4}{3}])$.
9. (a) $\sin (2 \arccos [\frac{1}{2}])$;   (b) $\cos (2 \arcsin [\frac{4}{5}])$.

10. (a) What are the domain and range of the arc cosine?
    (b) What are the domain and range of the arc cotangent?

Exercises 11–32. Differentiate.

11. $y = \arctan (x + 1)$.   12. $y = \arctan \sqrt{x}$.
13. $f(x) = \operatorname{arcsec} (2x^2)$.   14. $f(x) = e^x \arcsin x$.
15. $f(x) = x \arcsin 2x$.   16. $f(x) = e^{\arctan x}$.
17. $u = (\arcsin x)^2$.   18. $v = \arctan e^x$.
19. $y = \dfrac{\arctan x}{x}$.   20. $y = \operatorname{arcsec} \sqrt{x^2 + 2}$.
21. $f(x) = \sqrt{\arctan 2x}$.   22. $f(x) = \ln (\arctan x)$.
23. $y = \arctan (\ln x)$.   24. $g(x) = \operatorname{arcsec} (\cos x + 2)$.
25. $\theta = \arcsin(\sqrt{1 - r^2})$ 26. $\theta = \arcsin \left(\dfrac{r}{r + 1}\right)$.

27. $g(x) = x^2 \operatorname{arcsec}^{-1}\left(\dfrac{1}{x}\right)$.   28. $\theta = \arctan \left(\dfrac{1}{1 + r^2}\right)$.
29. $y = \sin [\operatorname{arcsec} (\ln x)]$.   30. $f(x) = e^{\sec^{-1} x}$.
31. $f(x) = \sqrt{c^2 - x^2} + c \arcsin \left(\dfrac{x}{c}\right)$. Take $c > 0$.
32. $y = \dfrac{x}{\sqrt{c^2 - x^2}} - \arcsin \left(\dfrac{x}{c}\right)$. Take $c > 0$.

33. If $0 < x < 1$, then $\arcsin x$ is the radian measure of the acute angle that has sine x. We can construct an angle of radian measure $\arcsin x$ by drawing a right triangle with a side of length x and hypotenuse of length 1. Use the accompanying figure to determine the following:

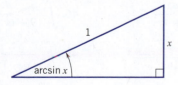

(a) $\sin (\arcsin x)$.
(b) $\cos (\arcsin x)$.
(c) $\tan (\arcsin x)$.
(d) $\cot (\arcsin x)$.
(e) $\sec (\arcsin x)$.
(f) $\csc (\arcsin x)$.

34. If $0 < x < 1$, then $\arctan x$ is the radian measure of the acute angle with tangent x. We can construct an angle of radian

measure arctan x by drawing a right triangle with legs of length x and 1. Use the accompanying figure to determine the following:

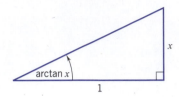

(a) $\tan(\arctan x)$ (b) $\cot(\arctan x)$
(c) $\sin(\arctan x)$ (d) $\cos(\arctan x)$
(e) $\sec(\arctan x)$ (f) $\csc(\arctan x)$.

35. Calculate $\displaystyle\int \frac{1}{\sqrt{a^2 - (x+b)^2}}\,dx$ taking $a > 0$.

HINT: Set $au = x + b$.

36. Calculate $\displaystyle\int \frac{1}{a^2 + (x+b)^2}\,dx$ taking $a > 0$.

37. Show that $\displaystyle\int \frac{1}{|x|\sqrt{x^2 - a^2}}\,dx = \left|\frac{1}{a}\right| \text{arcsec}\left|\frac{x}{a}\right| + C$,
taking $a > 0$.

38. (a) Verify, without reference to right triangles, that for all $|x| \geq 1$

$$\sec(\text{arcsec}\, x) = x \quad \text{and} \quad \csc(\text{arcsec}\, x) = x.$$

(b) What is the range of the arc secant? (The arc secant is the inverse of the secant restricted to this set.)
(c) What is the range of the arc cosecant? (The arc cosecant is the inverse of the cosecant restricted to this set.)

Exercises 39–52. Evaluate.

39. $\displaystyle\int_0^1 \frac{dx}{1 + x^2}$.

40. $\displaystyle\int_{-1}^1 \frac{dx}{1 + x^2}$.

41. $\displaystyle\int_0^{1/\sqrt{2}} \frac{dx}{\sqrt{1 - x^2}}$.

42. $\displaystyle\int_0^1 \frac{dx}{\sqrt{4 - x^2}}$.

43. $\displaystyle\int_0^5 \frac{dx}{25 + x^2}$.

44. $\displaystyle\int_5^8 \frac{dx}{x\sqrt{x^2 - 16}}$.

45. $\displaystyle\int_0^{3/2} \frac{dx}{9 + 4x^2}$.

46. $\displaystyle\int_2^5 \frac{dx}{9 + (x-2)^2}$.

47. $\displaystyle\int_{3/2}^3 \frac{dx}{x\sqrt{16x^2 - 9}}$.

48. $\displaystyle\int_4^6 \frac{dx}{(x-3)\sqrt{x^2 - 6x + 8}}$.

49. $\displaystyle\int_{-3}^{-2} \frac{dx}{\sqrt{4 - (x+3)^2}}$.

50. $\displaystyle\int_{\ln 2}^{\ln 3} \frac{e^{-x}}{\sqrt{1 - e^{-2x}}}\,dx$.

51. $\displaystyle\int_0^{\ln 2} \frac{e^x}{1 + e^{2x}}\,dx$.

52. $\displaystyle\int_0^{1/2} \frac{1}{\sqrt{3 - 4x^2}}\,dx$.

Exercise 53–62. Calculate.

53. $\displaystyle\int \frac{x}{\sqrt{1 - x^4}}\,dx$.

54. $\displaystyle\int \frac{\sec^2 x}{\sqrt{9 - \tan^2 x}}\,dx$.

55. $\displaystyle\int \frac{x}{1 + x^4}\,dx$.

56. $\displaystyle\int \frac{dx}{\sqrt{4x - x^2}}$.

57. $\displaystyle\int \frac{\sec^2 x}{9 + \tan^2 x}\,dx$.

58. $\displaystyle\int \frac{\cos x}{3 + \sin^2 x}\,dx$.

59. $\displaystyle\int \frac{\arcsin x}{\sqrt{1 - x^2}}\,dx$.

60. $\displaystyle\int \frac{\arctan x}{1 + x^2}\,dx$.

61. $\displaystyle\int \frac{dx}{x\sqrt{1 - (\ln x)^2}}$.

62. $\displaystyle\int \frac{dx}{x[1 + (\ln x)^2]}$.

63. Find the area below the curve $y = 1/\sqrt{4 - x^2}$ from $x = -1$ to $x = 1$.

64. Find the area below the curve $y = 3/(9 + x^2)$ from $x = -3$ to $x = 3$.

65. Sketch the region bounded above by $y = 8/(x^2 + 4)$ and bounded below by $4y = x^2$. What is the area of this region?

66. The region below the curve $y = 1/\sqrt{4 + x^2}$ from $x = 0$ to $x = 2$ is revolved about the x-axis. Find the volume of the resulting solid.

67. The region in Exercise 66 is revolved about the y-axis. Find the volume of the resulting solid.

68. The region below the curve $y = 1/x^2\sqrt{x^2 - 9}$ from $x = 2\sqrt{3}$ to $x = 6$ is revolved about the y-axis. Find the volume of the resulting solid.

69. A billboard k feet wide is perpendicular to a straight road and is s feet from the road. From what point on the road would a motorist have the best view of the billboard; that is, at what point on the road (see the figure) is the angle θ subtended by the billboard a maximum?

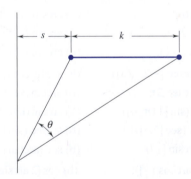

70. A person walking along a straight path at the rate of 6 feet per second is followed by a spotlight that is located 30 feet from the path. How fast is the spotlight turning at the instant the person is 50 feet past the point on the path that is closest to the spotlight?

71. (a) Show that

$$F(x) = \frac{x}{2}\sqrt{a^2 - x^2} + \frac{a^2}{2}\arcsin\left(\frac{x}{a}\right), a > 0$$

is an antiderivative for $f(x) = \sqrt{a^2 - x^2}$.

(b) Use the result in part (a) to calculate $\displaystyle\int_{-a}^a \sqrt{a^2 - x^2}\,dx$ and interpret your result as an area.

72. Set

$$f(x) = \arctan\left(\frac{a+x}{1-ax}\right), \quad x \neq 1/a.$$

(a) Show that $f'(x) = \dfrac{1}{1+x^2}, \quad x \neq 1/a.$

(b) Show that there is no constant C such that $f(x) = \arctan x + C$ for all $x \neq 1/a$.

(c) Find constants C_1 and C_2 such that

$$\begin{aligned} f(x) &= \arctan x + C_1 \quad \text{for } x < 1/a \\ f(x) &= \arctan x + C_2 \quad \text{for } x > 1/a. \end{aligned}$$

73. Show, without reference to right triangles, that

$$\arctan x + \operatorname{arccot} x = \tfrac{1}{2}\pi \quad \text{for all real } x.$$

HINT: Use the identity $\cot\theta = \tan\left(\tfrac{1}{2}\pi - \theta\right)$.

74. Show, without reference to right triangles, that

$$\operatorname{arcsec} x + \operatorname{arccsc} x = \tfrac{1}{2}\pi \quad \text{for } |x| \geq 1.$$

75. The statement

$$\int_0^3 \frac{1}{\sqrt{1-x^2}}\,dx = \arcsin 3 - \arcsin 0 = \arcsin 3$$

is nonsensical since the sine function does not take on the value 3. Where did we go wrong here?

▶**76.** Evaluate

$$\lim_{x\to 0} \frac{\arcsin x}{x}$$

numerically. Justify your answer by other means.

▶**77.** Estimate the integral

$$\int_0^{0.5} \frac{1}{\sqrt{1-x^2}}\,dx$$

by using the partition $\{0, 0.1, 0.2, 0.3, 0.4, 0.5\}$ and the intermediate points

$$x_1^* = 0.05, \quad x_2^* = 0.15, \quad x_3^* = 0.25,$$

$$x_4^* = 0.35, \quad x_5^* = 0.45.$$

Note that the sine of your estimate is close to 0.5. Explain the reason for this.

▶**78.** Use a graphing utility to draw the graph of $f(x) = \dfrac{1}{1+x^2}$ on $[0, 10]$.

(a) Calculate $\int_0^n f(x)\,dx$ for $n = 1000, 2500, 5000, 10,000$.

(b) What number are these integrals approaching?

(c) Determine the value of

$$\lim_{t\to\infty} \int_0^t \frac{1}{1+x^2}\,dx.$$

■ PROJECT 7.7 Refraction

Dip a straight stick in a pool of water and it appears to bend. Only in a vacuum does light travel at speed c (the famous c of $E = mc^2$). Light does not travel as fast through a material medium. The *index of refraction n* of a medium relates the speed of light in that medium to c:

$$n = \frac{c}{\text{speed of light in the medium}}$$

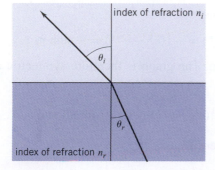

When light travels from one medium to another, it changes direction; we say that light is *refracted*. Experiment shows that the *angle of refraction* θ_r is related to the *angle of incidence* θ_i by Snell's law:

$$n_i \sin\theta_i = n_r \sin\theta_r.$$

Like the law of reflection (see Example 5, Section 4.5), Snell's law of refraction can be derived from Fermat's principle of least time.

Problem 1. A light beam passes from a medium with index of refraction n_1 through a plane sheet of material with top and bottom faces parallel and then out into some other medium with index of refraction n_2. Show that Snell's law implies the $n_1 \sin\theta_1 = n_2 \sin\theta_2$ regardless of the thickness of the sheet or its index of refraction.

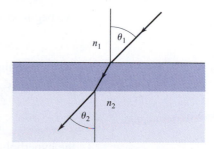

A star is not where it is supposed to be. The index of refraction of the atmosphere varies with height above the earth's surface, $n = n(y)$, and light that passes through the atmosphere follows some curved path, $y = y(x)$. Think of the atmosphere as a succession of thin parallel slabs. When a light ray strikes a slab at height y, it is traveling at some angle θ to the vertical; when it emerges at height $y + \Delta y$, it is traveling at a slightly different angle, $\theta + \Delta\theta$.

Problem 2.

(a) Use the result in Problem 1 to show that

$$\frac{1}{n}\frac{dn}{dy} = -\cot\theta\,\frac{d\theta}{dy} = \frac{d^2y/dx^2}{1+(dy/dx)^2}.$$

(b) Verify that the slope of the light path must vary in such a way that

$$1 + (dy/dx)^2 = (\text{constant})\,[n(y)]^2.$$

(c) How must n vary with height y for light to travel along a circular arc?

■ 7.8 THE HYPERBOLIC SINE AND COSINE

Certain combinations of the exponentials e^x and e^{-x} occur so frequently in mathematical applications that they are given special names. The *hyperbolic sine* (sinh) and *hyperbolic cosine* (cosh) are the functions defined as follows:

(7.8.1)
$$\sinh x = \tfrac{1}{2}(e^x - e^{-x}), \qquad \cosh x = \tfrac{1}{2}(e^x + e^{-x}).$$

The reason for these names will become apparent as we go on.

Since

$$\frac{d}{dx}(\sinh x) = \frac{d}{dx}\left[\tfrac{1}{2}(e^x - e^{-x})\right] = \tfrac{1}{2}(e^x + e^{-x})$$

and

$$\frac{d}{dx}(\cosh x) = \frac{d}{dx}\left[\tfrac{1}{2}(e^x + e^{-x})\right] = \tfrac{1}{2}(e^x - e^{-x}),$$

we have

(7.8.2)
$$\frac{d}{dx}(\sinh x) = \cosh x, \qquad \frac{d}{dx}(\cosh x) = \sinh\ x.$$

In short, each of these functions is the derivative of the other.

The Graphs

We begin with the hyperbolic sine. Since

$$\sinh(-x) = \tfrac{1}{2}(e^{-x} - e^x) = -\tfrac{1}{2}(e^x - e^{-x}) = -\sinh x,$$

the hyperbolic sine is an odd function. The graph is therefore symmetric about the origin. Since

$$\frac{d}{dx}(\sinh x) = \cosh x = \tfrac{1}{2}(e^x + e^{-x}) > 0 \qquad \text{for all real } x,$$

the hyperbolic sine increases everywhere. Since

$$\frac{d^2}{dx^2}(\sinh x) = \frac{d}{dx}(\cosh x) = \sinh x = \tfrac{1}{2}(e^x - e^{-x}),$$

you can see that

$$\frac{d^2}{dx^2}(\sinh x) \quad \text{is} \quad \begin{cases} \text{negative,} & \text{for} \quad x < 0 \\ 0, & \text{at} \quad x = 0 \\ \text{positive,} & \text{for} \quad x > 0. \end{cases}$$

The graph is therefore concave down on $(-\infty, 0)$ and concave up on $(0, \infty)$. The point $(0, \sinh 0) = (0, 0)$ is a point of inflection, the only point of inflection. The slope at the origin is $\cosh 0 = 1$. A sketch of the graph appears in Figure 7.8.1.

We turn now to the hyperbolic cosine. Since

$$\cosh(-x) = \tfrac{1}{2}(e^{-x} + e^x) = \tfrac{1}{2}(e^x + e^{-x}) = \cosh x,$$

the hyperbolic cosine is an even function. The graph is therefore symmetric about the y-axis. Since

$$\frac{d}{dx}(\cosh\ x) = \sinh\ x,$$

you can see that

$$\frac{d}{dx}(\cosh\ x) \text{ is } \begin{cases} \text{negative,} & \text{for } x < 0 \\ 0, & \text{at } x = 0 \\ \text{positive,} & \text{for } x > 0. \end{cases}$$

The function therefore decreases on $(-\infty, 0]$ and increases on $[0, \infty)$. The number

$$\cosh 0 = \tfrac{1}{2}(e^0 + e^{-0}) = \tfrac{1}{2}(1 + 1) = 1$$

is a local and absolute minimum. There are no other extreme values. Since

$$\frac{d^2}{dx^2}(\cosh\ x) = \frac{d}{dx}(\sinh\ x) = \cosh x > 0 \qquad \text{for all real } x.$$

the graph is everywhere concave up. (See Figure 7.8.2.)

Figure 7.8.3 shows the graphs of three functions:

$$y = \sinh x = \tfrac{1}{2}(e^x - e^{-x}), \qquad y = \tfrac{1}{2}e^x, \qquad y = \cosh x = \tfrac{1}{2}(e^x + e^{-x}).$$

Since $e^{-x} > 0$, it follows that, for all real x,

$$\sinh x < \tfrac{1}{2}e^x < \cosh x. \qquad \text{for all real } x.$$

Although markedly different for negative x, these functions are almost indistinguishable for large positive x. This follows from the fact that, as $x \to \infty$, $e^{-x} \to 0$.

The Catenary

A preliminary point: in what follows we use the fact that for all real numbers t

$$\cosh^2 t = 1 + \sinh^2 t.$$

The verification of this identity is left to you as an exercise.

Figure 7.8.4 depicts a flexible cable of uniform density supported from two points of equal height. The cable sags under its own weight and so forms a curve. Such a curve is called a *catenary*. (After the Latin word for chain.)

To obtain a mathematical characterization of the catenary, we introduce an x,y-coordinate system so that the lowest point of the chain falls on the positive y-axis (Figure 7.8.5). An engineering analysis of the forces that act on the cable shows that the shape of the catenary, call it $y = f(x)$, is such that

$$(*) \qquad \frac{d^2y}{dx^2} = \frac{1}{a}\sqrt{1 + \left(\frac{dy}{dx}\right)^2}$$

where a is a positive constant that depends on the length of the cable and on its mass density. As we show below, curves of the form

$$(**) \qquad y = a \cosh \frac{x}{a} + C \qquad (C \text{ constant})$$

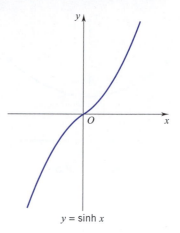

$y = \sinh x$

Figure 7.8.1

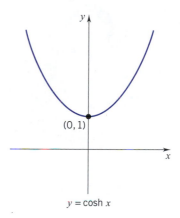

$(0, 1)$

$y = \cosh x$

Figure 7.8.2

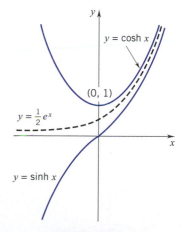

$y = \cosh x$

$(0, 1)$

$y = \tfrac{1}{2}e^x$

$y = \sinh x$

Figure 7.8.3

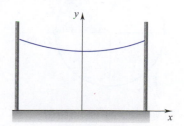

Figure 7.8.4

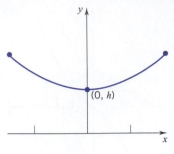

Figure 7.8.5

meet this condition exactly:

$$\frac{dy}{dx} = a\left(\sinh \frac{x}{a}\right)\frac{1}{a} = \sinh \frac{x}{a}$$

$$\frac{d^2y}{dx^2} = \left(\cosh \frac{x}{a}\right)\frac{1}{a} = \frac{1}{a}\cosh \frac{x}{a}$$

$$\frac{d^2y}{dx^2} = \frac{1}{a}\cosh \frac{x}{a} = \frac{1}{a}\sqrt{1+\sinh^2 \frac{x}{a}} = \frac{1}{a}\sqrt{1+\left(\frac{dy}{dx}\right)^2}.$$

└── follows from $\cosh^2 t = 1 + \sinh^2 t$

The cable of Figure 7.8.5 is of the form

$$y = a\cosh \frac{x}{a} + (h-a).$$

[This assertion is based on the fact that only curves of the form (∗∗) satisfy (∗) and the conditions imposed by Figure 7.8.5. This can be proven.]

The Gateway Arch in St. Louis, Missouri, is in the shape of an inverted catenary (see Figure 7.8.6). This arch is 630 feet high at its center, and it measures 630 feet across the base. The value of the constant a for this arch is approximately 127.7, and its equation takes the form

$$y = -127.7\cosh(x/127.7) + 757.7.$$

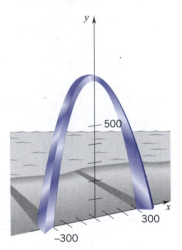

Figure 7.8.6

Identities

The hyperbolic sine and cosine functions satisfy identities similar to those satisfied by the "circular" sine and cosine.

(7.8.3)

$$\cosh^2 t - \sinh^2 t = 1,$$

$$\sinh(t+s) = \sinh t \,\cosh s + \cosh t \,\sinh s,$$

$$\cosh(t+s) = \cosh t \,\cosh s + \sinh t \,\sinh s,$$

$$\sinh 2t = 2\sinh t \cosh t,$$

$$\cosh 2t = \cosh^2 t + \sinh^2 t.$$

The verification of these identities is left to you as a collection of exercises.

Relation to the Hyperbola $x^2 - y^2 = 1$

The hyperbolic sine and cosine are related to the hyperbola $x^2 - y^2 = 1$ much as the "circular" sine and cosine are related to the circle $x^2 + y^2 = 1$:

1. For each real t,

$$\cos^2 t + \sin^2 t = 1,$$

and thus the point $(\cos t, \sin t)$ lies on the circle $x^2 + y^2 = 1$. For each real t,

$$\cosh^2 t - \sinh^2 t = 1,$$

and thus the point $(\cosh t, \sinh t)$ lies on the hyperbola $x^2 - y^2 = 1$.

2. For each t in $[0, 2\pi]$ (see Figure 7.8.7), the number $\frac{1}{2}t$ gives the area of the circular sector generated by the circular arc that begins at $(1, 0)$ and ends at $(\cos t, \sin t)$. As

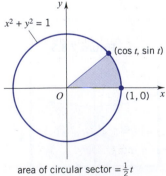

area of circular sector $= \frac{1}{2}t$

Figure 7.8.7

we prove below, for each $t > 0$ (see Figure 7.8.8), the number $\frac{1}{2}t$ gives the area of the hyperbolic sector generated by the hyperbolic arc that begins at $(1, 0)$ and ends at $(\cosh t, \sinh t)$.

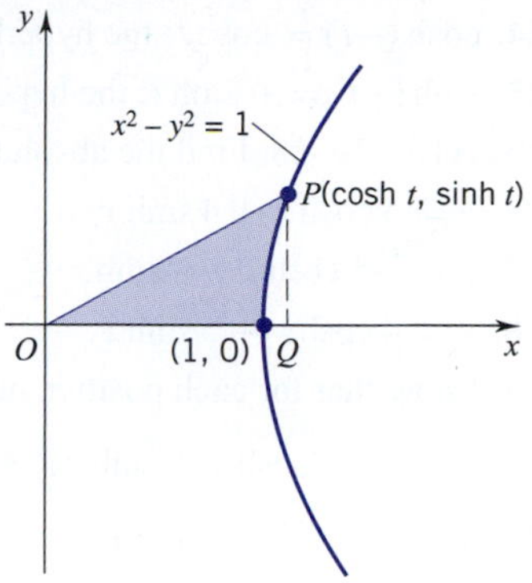

PROOF Let $A(t)$ be the area of the hyperbolic sector. Then,

$$A(t) = \tfrac{1}{2} \cosh t \, \sinh t - \int_1^{\cosh t} \sqrt{x^2 - 1} \, dx.$$

The first term, $\frac{1}{2} \cosh t \, \sinh t$, gives the area of the triangle OPQ, and the integral

$$\int_1^{\cosh t} \sqrt{x^2 - 1} \, dx$$

gives the area of the unshaded portion of the triangle. We wish to show that

$$A(t) = \tfrac{1}{2}t \qquad \text{for all } t \geq 0.$$

We will do so by showing that

$$A'(t) = \tfrac{1}{2} \qquad \text{for all } t > 0 \qquad \text{and} \qquad A(0) = 0.$$

Differentiating $A(t)$, we have

$$A'(t) = \tfrac{1}{2}\left[\cosh t \frac{d}{dt}(\sinh t) + \sinh t \frac{d}{dt}(\cosh t)\right] - \frac{d}{dt}\left(\int_1^{\cosh t} \sqrt{x^2 - 1}\, dx\right),$$

and therefore

(1) $$A'(t) = \tfrac{1}{2}(\cosh^2 t + \sinh^2 t) - \frac{d}{dt}\left(\int_1^{\cosh t} \sqrt{x^2 - 1}\, dx\right),$$

Now we differentiate the integral:

$$\frac{d}{dt}\left(\int_1^{\cosh t} \sqrt{x^2 - 1}\, dx\right) = \sqrt{\cosh^2 t - 1}\frac{d}{dt}(\cosh t) = \sinh t \cdot \sinh t = \sinh^2 t.$$

$$\underset{(5.8.7)}{\uparrow}$$

Substituting this last expression into (1), we have

$$A'(t) = \tfrac{1}{2}(\cosh^2 t + \sinh^2 t) - \sinh^2 t = \tfrac{1}{2}(\cosh^2 t - \sinh^2 t) = \tfrac{1}{2}.$$

It is not hard to see that $A(0) = 0$:

$$A(0) = \tfrac{1}{2} \cosh 0 \, \sinh 0 - \int_1^{\cosh 0} \sqrt{x^2 - 1}\, dx = \tfrac{1}{2}(1)(0) - \int_1^1 \sqrt{x^2 - 1}\, dx = 0. \quad \square$$

$x^2 - y^2 = 1$

$P(\cosh t, \sinh t)$

$(1, 0)$ Q

area of hyperbolic sector $= \frac{1}{2}t$

Figure 7.8.8

EXERCISES 7.8

Exercises 1–18. Differentiate.

1. $y = \sinh x^2$.

2. $y = \cosh(x + a)$.

3. $y = \sqrt{\cosh ax}$.

4. $y = (\sinh ax)(\cosh ax)$.

5. $y = \dfrac{\sinh x}{\cosh x - 1}$.

6. $y = \dfrac{\sinh x}{x}$.

7. $y = a \sinh bx - b \cosh ax$.

8. $y = e^x(\cosh x + \sinh x)$.

9. $y = \ln |\sinh ax|$.

10. $y = \ln |1 - \cosh ax|$.

11. $y = \sinh(e^{2x})$.

12. $y = \cosh(\ln x^3)$.

13. $y = e^{-x} \cosh 2x$.

14. $y = \arctan(\sinh x)$.

15. $y = \ln(\cosh x)$.

16. $y = \ln(\sinh x)$.

17. $y = (\sinh x)^x$.

18. $y = x^{\cosh x}$.

Exercises 19–25. Verify the identity.

19. $\cosh^2 t - \sinh^2 t = 1$.

20. $\sinh(t + s) = \sinh t \, \cosh s + \cosh t \, \sinh s$.

21. $\cosh(t + s) = \cosh t \, \cosh s + \sinh t \, \sinh s$.

22. $\sinh 2t = 2 \sinh t \, \cosh t$.

23. $\cosh 2t = \cosh^2 t + \sinh^2 t = 2\cosh^2 t - 1 = 2 \sinh^2 t + 1$.

24. $\cosh(-t) = \cosh t$; the hyperbolic cosine function is even.

25. $\sinh(-t) = -\sinh t$; the hyperbolic sine function is odd.

Exercises 26–28. Find the absolute extreme values.

26. $y = 5\cosh x + 4\sinh x$.

27. $y = -5\cosh x + 4\sinh x$.

28. $y = 4\cosh x + 5\sinh x$.

29. Show that for each positive integer n

$$(\cosh x + \sinh x)^n = \cosh nx + \sinh nx.$$

30. Verify that $y = A\cosh cx + B\sinh cx$ satisfies the equation $y'' - c^2 y = 0$.

31. Determine A, B, and c so that $y = A\cosh cx + B\sinh cx$ satisfies the conditions $y'' - 9y = 0$, $y(0) = 2$, $y'(0) = 1$. Take $c > 0$.

32. Determine A, B, and c so that $y = A\cosh cx + B\sinh cx$ satisfies the conditions $4y'' - y = 0$, $y(0) = 1$, $y'(0) = 2$. Take $c > 0$.

Exercises 33–44. Calculate.

33. $\displaystyle\int \cosh ax\, dx$.

34. $\displaystyle\int \sinh ax\, dx$.

35. $\displaystyle\int \sinh^2 ax \cosh ax\, dx$.

36. $\displaystyle\int \sinh ax \cosh^2 ax\, dx$.

37. $\displaystyle\int \frac{\sinh ax}{\cosh ax}\, dx$.

38. $\displaystyle\int \frac{\cosh ax}{\sinh ax}\, dx$.

39. $\displaystyle\int \frac{\sinh ax}{\cosh^2 ax}\, dx$.

40. $\displaystyle\int \sinh^2 x\, dx$.

41. $\displaystyle\int \cosh^2 x\, dx$.

42. $\displaystyle\int \sinh 2x\, e^{\cosh 2x}\, dx$.

43. $\displaystyle\int \frac{\sinh \sqrt{x}}{\sqrt{x}}\, dx$.

44. $\displaystyle\int \frac{\sinh x}{1 + \cosh x}\, dx$.

Exercises 45 and 46. Find the average value of the function on the interval indicated.

45. $f(x) = \cosh x$, $x \in [-1, 1]$.

46. $f(x) = \sinh 2x$, $x \in [0, 4]$.

47. Find the area below the curve $y = \sinh x$ from $x = 0$ to $x = \ln 10$.

48. Find the area below the curve $y = \cosh 2x$ from $x = -\ln 5$ to $x = \ln 5$.

49. Find the volume of the solid generated by revolving about the x-axis the region between $y = \cosh x$ and $y = \sinh x$ from $x = 0$ to $x = 1$.

50. The region below the curve $y = \sinh x$ from $x = 0$ to $x = \ln 5$ is revolved about the x-axis. Find the volume of the resulting solid.

51. The region below the curve $y = \cosh 2x$ from $x = -\ln 5$ to $x = \ln 5$ is revolved about the x-axis. Find the volume of the resulting solid.

52. (a) Evaluate

$$\lim_{x \to \infty} \frac{\sinh x}{e^x}.$$

(b) Evaluate

$$\lim_{x \to \infty} \frac{\cosh x}{e^{ax}}$$

if $0 < a < 1$ and if $a > 1$.

C▶ 53. Use a graphing utility to sketch in one figure the graphs of $f(x) = 2 - \sinh x$ and $g(x) = \cosh x$.

(a) Use a CAS to find the point in the first quadrant where the two graphs intersect.

(b) Use a CAS to find the area of the region in the first quadrant bounded by the graphs of f and g and the y-axis.

C▶ 54. Use a graphing utility to sketch in one figure the graphs of $f(x) = \cosh x - 1$ and $g(x) = 1/\cosh x$.

(a) Use a CAS to find the points where the two graphs intersect.

(b) Use a CAS to find the area of the region bounded by the graphs of f and g.

■ *7.9 THE OTHER HYPERBOLIC FUNCTIONS

The hyperbolic tangent is defined by setting

$$\tanh x = \frac{\sinh x}{\cosh x} = \frac{e^x - e^{-x}}{e^x + e^{-x}}.$$

There is also a *hyperbolic cotangent*, a *hyperbolic secant*, and a *hyperbolic cosecant*:

$$\coth x = \frac{\cosh x}{\sinh x}, \qquad \operatorname{sech} x = \frac{1}{\cosh x}, \qquad \operatorname{csch} x = \frac{1}{\sinh x}.$$

The derivatives are as follows:

(7.9.1)

$$\frac{d}{dx}(\tanh x) = \operatorname{sech}^2 x, \qquad\qquad \frac{d}{dx}(\coth x) = -\operatorname{csch}^2 x,$$

$$\frac{d}{dx}(\operatorname{sech} x) = -\operatorname{sech} x \tanh x, \qquad \frac{d}{dx}(\operatorname{csch} x) = -\operatorname{csch} x \coth x.$$

These formulas are easy to verify. For instance,

$$\frac{d}{dx}(\tanh x) = \frac{d}{dx}\left(\frac{\sinh x}{\cosh x}\right) = \frac{\cosh x \frac{d}{dx}(\sinh x) - \sinh x \frac{d}{dx}(\cosh x)}{\cosh^2 x}$$

$$= \frac{\cosh^2 x - \sinh^2 x}{\cosh^2 x} = \frac{1}{\cosh^2 x} = \text{sech}^2 x.$$

We leave it to you to verify the other formulas.

Let's examine the hyperbolic tangent a little further. Since

$$\tanh(-x) = \frac{\sinh(-x)}{\cosh(-x)} = \frac{-\sinh x}{\cosh x} = -\tanh x,$$

the hyperbolic tangent is an odd function and thus the graph is symmetric about the origin. Since

$$\frac{d}{dx}(\tanh x) = \text{sech}^2 x > 0 \qquad \text{for all real } x,$$

the function is everywhere increasing. From the relation

$$\tanh x = \frac{e^x - e^{-x}}{e^x + e^{-x}} = \frac{e^x - e^{-x}}{e^x + e^{-x}} \cdot \frac{e^x}{e^x} = \frac{e^{2x} - 1}{e^{2x} + 1} = \frac{e^{2x} + 1 - 2}{e^{2x} + 1} = 1 - \frac{2}{e^{2x} + 1},$$

you can see that $\tanh x$ always remains between -1 and 1. Moreover,

$$\text{as } x \to \infty, \qquad \tanh x \to 1 \qquad \text{and} \qquad \text{as } x \to -\infty, \qquad \tanh x \to -1.$$

The lines $y = 1$ and $y = -1$ are horizontal asymptotes. To check on the concavity of the graph, we take the second derivative:

$$\frac{d^2}{dx^2}(\tanh x) = \frac{d}{dx}(\text{sech}^2 x) = 2 \text{ sech } x \frac{d}{dx}(\text{sech } x)$$

$$= 2 \text{ sech } x (-\text{sech } x \ \tanh x)$$

$$= -2 \text{ sech}^2 x \ \tanh x.$$

Since

$$\tanh x = \frac{e^x - e^{-x}}{e^x + e^{-x}} \quad \text{is} \quad \begin{cases} \text{negative,} & \text{for} \quad x < 0 \\ 0, & \text{at} \quad x = 0 \\ \text{positive,} & \text{for} \quad x > 0, \end{cases}$$

you can see that

$$\frac{d^2}{dx^2}(\tanh x) \quad \text{is} \quad \begin{cases} \text{positive,} & \text{for} \quad x < 0 \\ 0, & \text{at} \quad x = 0 \\ \text{negative,} & \text{for} \quad x > 0. \end{cases}$$

The graph is therefore concave up on $(-\infty, 0)$ and concave down on $(0, \infty)$. The point $(0, \tanh 0) = (0, 0)$ is a point of inflection. At the origin the slope is

$$\text{sech}^2 0 = \frac{1}{\cosh^2 0} = 1.$$

The graph is shown in Figure 7.9.1.

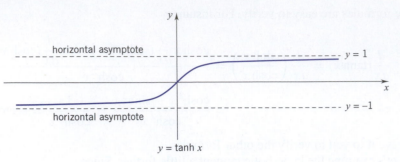

Figure 7.9.1

The Hyperbolic Inverses

Of the six hyperbolic functions, only the hyperbolic cosine and its reciprocal, hyperbolic secant, fail to be one-to-one (refer to the graphs of $y = \sinh x$, $y = \cosh x$, and $y = \tanh x$). Thus, the hyperbolic sine, hyperbolic tangent, hyperbolic cosecant, and hyperbolic cotangent functions all have inverses. If we restrict the domains of the hyperbolic cosine and hyperbolic secant functions to $x \geq 0$, then these functions will also have inverses. The hyperbolic inverses that are important to us are the *inverse hyperbolic sine*, the *inverse hyperbolic cosine*, and the *inverse hyperbolic tangent*. These functions,

$$y = \sinh^{-1} x, \qquad y = \cosh^{-1} x, \qquad y = \tanh^{-1} x,$$

are the inverses of

$$y = \sinh x, \qquad y = \cosh x \quad (x \geq 0), \qquad y = \tanh x$$

respectively.[†]

THEOREM 7.9.2

(i) $\sinh^{-1} x = \ln(x + \sqrt{x^2 + 1}), \qquad x$ real

(ii) $\cosh^{-1} x = \ln(x + \sqrt{x^2 - 1}), \qquad x \geq 1$

(iii) $\tanh^{-1} x = \frac{1}{2} \ln\left(\frac{1 + x}{1 - x}\right), \qquad -1 < x < 1.$

PROOF To prove (i), we set $y = \sinh^{-1} x$ and note that

$$\sinh y = x.$$

This gives in sequence:

$$\tfrac{1}{2}(e^y - e^{-y}) = x, \qquad e^y - e^{-y} = 2x \qquad e^y - 2x - e^{-y} = 0, \qquad e^{2y} - 2x\, e^y - 1 = 0.$$

This last equation is a quadratic equation in e^y. From the general quadratic formula, we find that

$$e^y = \tfrac{1}{2}(2x \pm \sqrt{4x^2 + 4}) = x \pm \sqrt{x^2 + 1}.$$

Since $e^y > 0$, the minus sign on the right is impossible. Consequently, we have

$$e^y = x + \sqrt{x^2 + 1},$$

[†]The expressions $\sinh^{-1} x$, $\cosh^{-1} x$, $\tan^{-1} x$ can be written $\operatorname{arcsinh} x$, $\operatorname{arccosh} x$, $\operatorname{arctanh} x$. However the "-1" notation is more common.

and, taking the natural log of both sides,

$$y = \ln(x + \sqrt{x^2 + 1}).$$

To prove (ii), we set

$$y = \cosh^{-1} x, \qquad x \geq 1$$

and note that

$$\cosh y = x \qquad \text{and} \qquad y \geq 0.$$

This gives in sequence:

$$\tfrac{1}{2}(e^y + e^{-y}) = x, \qquad e^y + e^{-y} = 2x, \qquad e^{2y} - 2xe^y + 1 = 0.$$

Again we have a quadratic in e^y. Here the general quadratic formula gives

$$e^y = \tfrac{1}{2}(2x \pm \sqrt{4x^2 - 4}) = x \pm \sqrt{x^2 - 1}.$$

Since y is nonnegative,

$$e^y = x \pm \sqrt{x^2 - 1}$$

cannot be less than 1. This renders the negative sign impossible (check this out) and leaves

$$e^y = x + \sqrt{x^2 - 1}$$

as the only possibility. Taking the natural log of both sides, we get

$$y = \ln(x + \sqrt{x^2 - 1}).$$

The proof of (iii) is left as an exercise. ❑

EXERCISES 7.9

Exercises 1–10. Differentiate.

1. $y = \tanh^2 x$.

2. $y = \tanh^2 3x$.

3. $y = \ln(\tanh x)$.

4. $y = \tanh(\ln x)$.

5. $y = \sinh(\arctan e^{2x})$.

6. $y = \text{sech}(3x^2 + 1)$.

7. $y = \coth(\sqrt{x^2 + 1})$.

8. $y = \ln(\text{sech } x)$.

9. $y = \dfrac{\text{sech } x}{1 + \cosh x}$.

10. $y = \dfrac{\cosh x}{1 + \text{sech } x}$.

Exercises 11–13. Verify the formula.

11. $\dfrac{d}{dx}(\coth x) = -\text{csch}^2 x$.

12. $\dfrac{d}{dx}(\text{sech } x) = -\text{sech } x \tanh x$.

13. $\dfrac{d}{dx}(\text{csch } x) = -\text{csch } x \coth x$.

14. Show that

$$\tanh(t + s) = \frac{\tanh t + \tanh s}{1 + \tanh t \tanh s}.$$

15. Given that $\tanh x_0 = \tfrac{4}{5}$, find (a) $\text{sech } x_0$.
HINT: $1 - \tanh^2 x = \text{sech}^2 x$. Then find (b) $\cosh x_0$, (c) $\sinh x_0$, (d) $\coth x_0$, (e) $\text{csch } x_0$.

16. Given that $\tanh t_0 = -\tfrac{5}{12}$, evaluate the remaining hyperbolic functions at t_0.

17. Show that, if $x^2 \geq 1$, then $x - \sqrt{x^2 - 1} \leq 1$.

18. Show that

$$\tanh^{-1} x = \frac{1}{2} \ln\left(\frac{1+x}{1-x}\right), \qquad -1 < x < 1.$$

19. Show that

(7.9.3) $\quad \dfrac{d}{dx}(\sinh^{-1} x) = \dfrac{1}{\sqrt{x^2 + 1}}, \qquad x$ real.

20. Show that

(7.9.4) $\quad \dfrac{d}{dx}(\cosh^{-1} x) = \dfrac{1}{\sqrt{x^2 - 1}}, \qquad x > 1.$

21. Show that

(7.9.5) $\quad \dfrac{d}{dx}(\tanh^{-1} x) = \dfrac{1}{1 - x^2}, \qquad -1 < x < 1.$

22. Show that

$$\frac{d}{dx}(\operatorname{sech}^{-1}x) = \frac{-1}{x\sqrt{1-x^2}}, \qquad 0 < x < 1.$$

23. Show that

$$\frac{d}{dx}(\operatorname{csch}^{-1}x) = \frac{-1}{|x|\sqrt{1+x^2}}, \qquad x \neq 0.$$

24. Show that

$$\frac{d}{dx}(\coth^{-1}x) \doteq \frac{1}{1-x^2}, \qquad |x| > 1.$$

25. Sketch the graph of $y = \operatorname{sech} x$, giving: (a) the extreme values; (b) the points of inflection; and (c) the concavity.

26. Sketch the graphs of (a) $y = \coth x$, (b) $y = \operatorname{csch} x$.

27. Graph $y = \sinh x$ and $y = \sinh^{-1} x$ in the same coordinate system. Find all points of inflection.

28. Sketch the graphs of (a) $y = \cosh^{-1} x$, (b) $y = \tanh^{-1} x$.

29. Given that $\tan\phi = \sinh x$, show that

(a) $\dfrac{d\phi}{dx} = \operatorname{sech} x.$

(b) $x = \ln(\sec\phi + \tan\phi).$

(c) $\dfrac{dx}{d\phi} = \sec\phi.$

30. The region bounded by the graph of $y = \operatorname{sech} x$ between $x = -1$ and $x = 1$ is revolved about the x-axis. Find the volume of the solid generated.

Exercises 31–40. Calculate.

31. $\displaystyle\int \tanh x\, dx.$

32. $\displaystyle\int \coth x\, dx.$

33. $\displaystyle\int \operatorname{sech} x\, dx.$

34. $\displaystyle\int \operatorname{csch} x\, dx.$

35. $\displaystyle\int \operatorname{sech}^3 x \tanh x\, dx.$

36. $\displaystyle\int x\operatorname{sech}^2 x^2\, dx.$

37. $\displaystyle\int \tanh x \ln(\cosh x)\, dx.$

38. $\displaystyle\int \frac{1 + \tanh x}{\cosh^2 x}\, dx.$

39. $\displaystyle\int \frac{\operatorname{sech}^2 x}{1 + \tanh x}\, dx.$

40. $\displaystyle\int \tanh^5 x\operatorname{sech}^2 x\, dx.$

Exercises 41–43. Verify the formula. In each case, take $a > 0$.

41. $\displaystyle\int \frac{1}{\sqrt{a^2 + x^2}}\, dx = \sinh^{-1}\left(\frac{x}{a}\right) + C.$

42. $\displaystyle\int \frac{1}{\sqrt{x^2 - a^2}}\, dx = \cosh^{-1}\left(\frac{x}{a}\right) + C.$

43. $\displaystyle\int \frac{1}{a^2 - x^2}\, dx = \begin{cases} \dfrac{1}{a}\tanh^{-1}\left(\dfrac{x}{a}\right) + C & \text{if } |x| < a. \\[2mm] \dfrac{1}{a}\coth^{-1}\left(\dfrac{x}{a}\right) + C & \text{if } |x| > a. \end{cases}$

44. If an object of mass m falling from rest under the action of gravity encounters air resistance that is proportional to the square of its velocity, then the velocity $v(t)$ of the object at time t satisfies the equation

$$m\frac{dv}{dt} = mg - kv^2$$

where $k > 0$ is the constant of proportionality and g is the gravitational constant.

(a) Show that

$$v(t) = \sqrt{\frac{mg}{k}}\tanh\left(\sqrt{\frac{gk}{m}}\,t\right)$$

is a solution of the equation which satisfies $v(0) = 0$.

(b) Find

$$\lim_{t\to\infty} v(t).$$

This limit is called the *terminal velocity* of the body.

■ CHAPTER 7. REVIEW EXERCISES

Exercises 1–8. Determine whether the function f is one-to-one and, if so, find the inverse.

1. $f(x) = x^{1/3} + 2.$

2. $f(x) = x^2 - x - 6.$

3. $f(x) = \dfrac{x+1}{x-1}.$

4. $f(x) = (2x+1)^3.$

5. $f(x) = e^{1/x}.$

6. $f(x) = \sin 2x + \cos x$

7. $f(x) = x\ln x.$

8. $f(x) = \dfrac{2x+1}{3-2x}.$

Exercises 9–12. Show that f has an inverse and find $(f^{-1})'(c).$

9. $f(x) = \dfrac{1}{1+e^x}; \quad c = \frac{1}{2}.$

10. $f(x) = 3x - \dfrac{1}{x^3}, x > 0; \quad c = 2.$

11. $f(x) = \displaystyle\int_0^x \sqrt{4 + t^2}\, dt; \quad c = 0.$

12. $f(x) = x - \pi + \cos x; \quad c = -1.$

Exercises 13–22. Calculate the derivative.

13. $f(x) = (\ln x^2)^3.$

14. $y = 2\sin(e^{3x}).$

15. $g(x) = \dfrac{e^x}{1 + e^{2x}}.$

16. $f(x) = (x^2 + 1)^{\sinh x}.$

17. $y = \ln(x^3 + 3^x).$

18. $g(x) = \arctan(\cosh x).$

19. $f(x) = (\cosh x)^{1/x}.$

20. $f(x) = 2x^3\arcsin(x^2).$

21. $f(x) = \log_3\left(\dfrac{1+x}{1-x}\right).$ **22.** $f(x) = \text{arcsec}\,\sqrt{x^2+4}.$

Exercises 23–38. Calculate.

23. $\displaystyle\int \frac{e^x}{\sqrt{1-e^{2x}}}\,dx.$

24. $\displaystyle\int_1^e \frac{\sqrt{\ln x}}{x}\,dx.$

25. $\displaystyle\int \frac{\cos x}{4+\sin^2 x}\,dx.$

26. $\displaystyle\int \tan x \ln(\cos x)\,dx.$

27. $\displaystyle\int \frac{\sec\sqrt{x}}{\sqrt{x}}\,dx.$

28. $\displaystyle\int \frac{1}{x\sqrt{x^4-9}}\,dx.$

29. $\displaystyle\int \frac{5^{\ln x}}{x}\,dx.$

30. $\displaystyle\int_0^2 x^2 e^{x^3}\,dx.$

31. $\displaystyle\int_1^8 \frac{x^{1/3}}{x^{4/3}+1}\,dx.$

32. $\displaystyle\int \frac{\sec x\tan x}{1+\sec^2 x}\,dx.$

33. $\displaystyle\int 2^x \sinh 2^x\,dx.$

34. $\displaystyle\int \frac{e^x}{e^x+e^{-x}}\,dx.$

35. $\displaystyle\int_2^5 \frac{1}{x^2-4x+13}\,dx.$

36. $\displaystyle\int \frac{1}{\sqrt{15+2x-x^2}}\,dx.$

37. $\displaystyle\int_0^2 \text{sech}^2\left(\frac{x}{2}\right)\,dx.$

38. $\displaystyle\int \tanh^2 2x\,dx.$

Exercises 39–42. Find the area below the graph.

39. $y = \dfrac{x}{x^2+1},\quad x \in [0,1].$ **40.** $y = \dfrac{1}{x^2+1},\quad x \in [0,1].$

41. $y = \dfrac{1}{\sqrt{1-x^2}},\quad x \in \left[0,\tfrac{1}{2}\right].$

42. $y = \dfrac{x}{\sqrt{1-x^2}},\quad x \in \left[0,\tfrac{1}{2}\right].$

43. (a) Apply the mean-value theorem to the function $f(x) = \ln(1+x)$ to show that for all $x > -1$

$$\frac{x}{1+x} < \ln(1+x) < x.$$

(b) Use the result in part (a) to show that $\displaystyle\lim_{x\to 0}\frac{\ln(1+x)}{x} = 1.$

44. Show that for all positive integers m and n with $1 < m < n$

$$\ln\frac{n+1}{m} < \frac{1}{m} + \frac{1}{m+1} + \cdots + \frac{1}{n} < \ln\frac{n}{m-1}.$$

HINT: $\dfrac{1}{k+1} < \displaystyle\int_k^{k+1}\frac{dx}{x} < \dfrac{1}{k}.$

45. Find the area of the region between the curve $xy = a^2$, the x-axis, and the vertical lines $x = a, x = 2a$.

46. Find the area of the region between the curve $y = \sec\frac{1}{2}\pi x$ and the x-axis from $x = -\frac{1}{2}$ and $x = \frac{1}{2}$.

47. Let Ω be the region between the graph of $y = (1+x^2)^{-1/2}$ and the x-axis, from $x = 0$ to $x = \sqrt{3}$. Find the volume of the solid generated by revolving Ω (a) about the x-axis; (b) about the y-axis.

48. Let Ω be the region between the graph of $y = (1+x^2)^{-1/4}$ and the x-axis, from $x = 0$ to $x = \frac{1}{2}$. Find the volume of the solid generated by revolving Ω (a) about the x-axis; (b) about the y-axis.

49. Let $f(x) = \dfrac{\ln x}{x}$ on $(0,\infty)$. (a) Find the intervals where f increases and the intervals where it decreases; (b) find the extreme values; (c) determine the concavity of the graph and find the points of inflection; (d) sketch the graph, including all asymptotes.

50. Exercise 49 for the function $f(x) = x^2 e^{-x^2}$.

51. Given that $|a| < 1$, find the value of b for which

$$\int_0^1 \frac{b}{\sqrt{1-b^2x^2}}\,dx = \int_0^a \frac{1}{\sqrt{1-x^2}}\,dx.$$

52. Show that

$$\int_0^1 \frac{a}{1+a^2x^2}\,dx = \int_0^a \frac{1}{1+x^2}\,dx \quad \text{for all real numbers } a.$$

53. A certain bacterial culture, growing exponentially, increases from 20 grams to 40 grams in the period from 6 a.m. to 8 a.m.

(a) How many grams will be present at noon?
(b) How long will it take for the culture to reach 200 grams?

54. A certain radioactive substance loses 20% of its mass per year. What is the half-life of the substance?

55. Polonium-210 decays exponentially with a half-life of 140 days.

(a) At time $t = 0$ a sample of polonium-210 has a mass of 100 grams. Find an expression that gives the mass at an arbitrary time t.
(b) How long will it take for the 100-gram mass to decay to 75 grams?

56. From 1980 to 1990 the population of the United States grew from 227 million to 249 million. During that same period the population of Mexico grew from 62 million to 79 million. If the populations of the United States and Mexico continue to grow at these rates, when will the two populations be equal?

57. The population of a suburb of a large city is increasing at a rate proportional to the number of people currently living in the suburb. If, after two years, the population has doubled and after four years the population is 25,000, find:

(a) the number of people living in the suburb initially;
(b) the length of time for the population to quadruple.

58. Let p and q be numbers greater than 1 which satisfy the condition $1/p + 1/q = 1$. Show that for all positive a and b

$$ab \le \frac{a^p}{p} + \frac{b^q}{q}.$$

HINT: Let C be the curve $y = x^{p-1}, x \ge 0$. Let Ω_1 be the region between C and the x-axis from $x = 0$ to $x = a$. Let Ω_2 be the region between C and the y-axis from $y = 0$ to $y = b$. Argue that

$$ab \le \text{area of } \Omega_1 + \text{area of } \Omega_2.$$

CHAPTER

8

TECHNIQUES OF

INTEGRATION

■ 8.1 INTEGRAL TABLES AND REVIEW

We begin by listing the more important integrals with which you are familiar.

1. $\displaystyle\int k\,du = ku + C,$ k constant.

2. $\displaystyle\int u^r\,du = \frac{u^{r+1}}{r+1} + C,$ r constant, $r \neq -1$.

3. $\displaystyle\int \frac{1}{u}\,du = \ln|u| + C.$ 4. $\displaystyle\int e^u\,du = e^u + C.$

5. $\displaystyle\int p^u\,du = \frac{p^u}{\ln p} + C,$ $p > 0$ constant, $p \neq 1$.

6. $\displaystyle\int \sin u\,du = -\cos u + C.$ 7. $\displaystyle\int \cos u\,du = \sin u + C.$

8. $\displaystyle\int \tan u\,du = \ln|\sec u| + C.$ 9. $\displaystyle\int \cot u\,du = \ln|\sin u| + C.$

10. $\displaystyle\int \sec u\,du = \ln|\sec u + \tan u| + C.$

11. $\displaystyle\int \csc u\,du = \ln|\csc u - \cot u| + C.$

12. $\displaystyle\int \sec u \tan u\,du = \sec u + C.$ 13. $\displaystyle\int \csc u \cot u\,du = -\csc u + C.$

14. $\displaystyle\int \sec^2 u\,du = \tan u + C.$ 15. $\displaystyle\int \csc^2 u\,du = -\cot u + C.$

16. $\displaystyle\int \frac{du}{\sqrt{a^2 - u^2}}\,du = \arcsin\frac{u}{a} + C,$ $a > 0$ constant.

17. $\displaystyle\int \frac{du}{a^2 + u^2} = \frac{1}{a}\arctan\frac{u}{a} + C,$ $a > 0$ constant.

18. $\int \dfrac{du}{|u|\sqrt{u^2-a^2}}\,du = \dfrac{1}{a}\operatorname{arcsec}\dfrac{u}{a} + C,\qquad a>0$ constant.

19. $\int \sinh u\,du = \cosh u + C.$ **20.** $\int \cosh u\,du = \sinh u + C.$

For review we work out a few integrals by *u*-substitution.

Example 1 Calculate $\int x\tan x^2\,dx.$

SOLUTION Set $u = x^2,\, du = 2x\,dx.$ Then

$$\int x\tan x^2\,dx = \tfrac{1}{2}\int \tan u\,du = \tfrac{1}{2}\ln|\sec u| + C = \tfrac{1}{2}\ln|\sec x^2| + C. \quad \square$$

$\underset{\text{Formula 8}}{\uparrow}$

Example 2 Calculate $\displaystyle\int_0^1 \dfrac{e^x}{e^x+2}\,dx.$

SOLUTION Set

$$u = e^x + 2,\quad du = e^x\,dx.\quad \text{At } x=0, u=3;\ at\ x=1, u=e+2.$$

Thus

$$\int_0^1 \dfrac{e^x}{e^x+2}\,dx = \int_3^{e+2}\dfrac{du}{u} = \Big[\ln|u|\Big]_3^{e+2}$$

Formula 3

$$= \ln(e+2) - \ln 3 = \ln\big[\tfrac{1}{3}(e+2)\big] \cong 0.45. \quad \square$$

Example 3 Calculate $\displaystyle\int \dfrac{\cos 2x}{(2+\sin 2x)^{1/3}}\,dx.$

SOLUTION Set $u = 2+\sin 2x,\quad du = 2\cos 2x\,dx.$ Then

$$\int \dfrac{\cos 2x}{(2+\sin 2x)^{1/3}}\,dx = \tfrac{1}{2}\int \dfrac{1}{u^{1/3}}\,du = \tfrac{1}{2}\int u^{-1/3}\,du = \tfrac{1}{2}\left(\tfrac{3}{2}\right)u^{2/3} + C$$

Formula 2

$$= \tfrac{3}{4}(2+\sin 2x)^{2/3} + C. \quad \square$$

The final example requires a little algebra.

Example 4 Calculate $\displaystyle\int \dfrac{dx}{x^2+2x+5}.$

SOLUTION First we complete the square in the denominator:

$$\int \dfrac{dx}{x^2+2x+5} = \int \dfrac{dx}{(x^2+2x+1)+4} = \int \dfrac{dx}{(x+1)^2+4}.$$

We know that

$$\int \dfrac{du}{u^2+4} = \dfrac{1}{2}\arctan\dfrac{u}{2} + C.$$

Setting

$$u = x+1,\quad du = dx,$$

we have

$$\int \frac{dx}{x^2 + 2x + 5} = \int \frac{du}{u^2 + 2^2} = \frac{1}{2} \arctan \frac{u}{2} + C = \frac{1}{2} \arctan \left(\frac{x + 1}{2} \right) + C. \quad \square$$

Using a Table of Integrals A table of over 100 intergrals, including those listed at the beginning of this section, appears on the inside covers of this text. This is a relatively short list. Mathematical handbooks such as *Burington's Handbook of Mathematical Tables and Formulas* and *CRC Standard Mathematical Tables* contain extensive tables; the table in the CRC reference lists 600 integrals.

The entries in a table of integrals are grouped by the form of the integrand: "forms containing $a + bu$," "forms containing $\sqrt{a^2 - u^2}$," "trigonometric forms," and so forth. The table on the inside covers is grouped in this manner. This is the only table of integrals we'll refer to in this text.

Example 5 We use the table to calculate

$$\int \frac{dx}{\sqrt{4 + x^2}}.$$

SOLUTION Of the integrals containing $\sqrt{a^2 + u^2}$, the one that fits our needs is Formula 77:

$$\int \frac{du}{\sqrt{a^2 + u^2}} = \ln \left| u + \sqrt{a^2 + u^2} \right| + C.$$

In our case, $a = 2$ and $u = x$. Therefore

$$\int \frac{dx}{\sqrt{4 + x^2}} = \ln \left| x + \sqrt{4 + x^2} \right| + C. \quad \square$$

Example 6 We use the table to calculate

$$\int \frac{dx}{3x^2(2x - 1)}.$$

SOLUTION The presence of the linear expression $2x - 1$ prompts us to look in the $a + bu$ grouping. The formula that applies is Formula 109:

$$\int \frac{du}{u^2(a + bu)} = -\frac{1}{au} + \frac{b}{a^2} \ln \left| \frac{a + bu}{u} \right| + C.$$

In our case $a = -1, b = 2, u = x$. Therefore

$$\int \frac{dx}{3x^2(2x - 1)} = \frac{1}{3} \int \frac{dx}{x^2(2x - 1)} = \frac{1}{3} \left[\frac{1}{x} + 2 \ln \left| \frac{2x - 1}{x} \right| \right] + C. \quad \square$$

Example 7 We use the table to calculate

$$\int \frac{\sqrt{9 - 4x^2}}{x^2} dx.$$

SOLUTION Closest to what we need is Formula 90:

$$\int \frac{\sqrt{a^2 - u^2}}{u^2} du = -\frac{1}{u} \sqrt{a^2 - u^2} - \arcsin \frac{u}{a} + C.$$

We can write our integral to fit the formula by setting

$$u = 2x, \quad du = 2\,dx.$$

Doing this, we have

$$\int \frac{\sqrt{9 - 4x^2}}{x^2}\,dx = 2\int \frac{\sqrt{9 - u^2}}{u^2}\,du = 2\left[-\frac{1}{u}\sqrt{9 - u^2} - \arcsin\frac{u}{3}\right] + C$$

Check this out. ⟶

$$= 2\left[-\frac{1}{2x}\sqrt{9 - 4x^2} - \arcsin\frac{2x}{3}\right] + C. \quad ❑$$

EXERCISES 8.1

Exercises 1–38. Calculate.

1. $\int e^{2-x}\,dx.$

2. $\int \cos\frac{2}{3}x\,dx.$

3. $\int_0^1 \sin \pi x\,dx.$

4. $\int_0^1 \sec \pi x \tan \pi x\,dx.$

5. $\int \sec^2(1 - x)\,dx.$

6. $\int \frac{dx}{5^x}.$

7. $\int_{\pi/6}^{\pi/3} \cot x\,dx.$

8. $\int_0^1 \frac{x^3}{1 + x^4}\,dx.$

9. $\int \frac{x}{\sqrt{1 - x^2}}\,dx.$

10. $\int_{-\pi/4}^{\pi/4} \frac{dx}{\cos^2 x}.$

11. $\int_{-\pi/4}^{\pi/4} \frac{\sin x}{\cos^2 x}\,dx.$

12. $\int \frac{e^{\sqrt{x}}}{\sqrt{x}}\,dx.$

13. $\int_1^2 \frac{e^{1/x}}{x^2}\,dx.$

14. $\int \frac{x^3}{\sqrt{1 - x^4}}\,dx.$

15. $\int_0^c \frac{dx}{x^2 + c^2}.$

16. $\int a^x e^x\,dx.$

17. $\int \frac{\sec^2\theta}{\sqrt{3\tan\theta + 1}}\,d\theta.$

18. $\int \frac{\sin\phi}{3 - 2\cos\phi}\,d\phi.$

19. $\int \frac{e^x}{ae^x - b}\,dx.$

20. $\int \frac{dx}{x^2 - 4x + 13}.$

21. $\int \frac{x}{(x + 1)^2 + 4}\,dx.$

22. $\int \frac{\ln x}{x}\,dx.$

23. $\int \frac{x}{\sqrt{1 - x^4}}\,dx.$

24. $\int \frac{e^x}{1 + e^{2x}}\,dx.$

25. $\int \frac{dx}{x^2 + 6x + 10}.$

26. $\int e^x \tan e^x\,dx.$

27. $\int x \sin x^2\,dx.$

28. $\int \frac{x}{9 + x^4}\,dx.$

29. $\int \tan^2 x\,dx.$

30. $\int \cosh 2x \sinh^3 2x\,dx.$

31. $\int_1^c \frac{\ln x^3}{x}\,dx.$

32. $\int_0^{\pi/4} \frac{\arctan x}{1 + x^2}\,dx.$

33. $\int \frac{\arcsin x}{\sqrt{1 - x^2}}\,dx.$

34. $\int e^x \cosh(2 - e^x)\,dx.$

35. $\int \frac{1}{x\ln x}\,dx.$

36. $\int_{-1}^1 \frac{x^2}{x^2 + 1}\,dx.$

37. $\int_0^{\pi/4} \frac{1 + \sin x}{\cos^2 x}\,dx.$

38. $\int_0^{1/2} \frac{1 + x}{\sqrt{1 - x^2}}\,dx.$

Exercises 39–48. Calculate using our table of integrals.

39. $\int \sqrt{x^2 - 4}\,dx.$

40. $\int \sqrt{4 - x^2}\,dx.$

41. $\int \cos^3 2t\,dt.$

42. $\int \sec^4 t\,dt.$

43. $\int \frac{dx}{x(2x + 3)}.$

44. $\int \frac{x\,dx}{2 + 3x}.$

45. $\int \frac{\sqrt{x^2 + 9}}{x^2}\,dx.$

46. $\int \frac{dx}{x^2\sqrt{x^2 - 2}}.$

47. $\int x^3 \ln x\,dx.$

48. $\int x^3 \sin x\,dx.$

49. Evaluate $\int_0^\pi \sqrt{1 + \cos x}\,dx.$

HINT: $\cos x = 2\cos^2 \frac{1}{2}x - 1.$

50. Calculate $\int \sec^2 x \tan x\,dx$ in two ways.

(a) Set $u = \tan x$ and verify that

$$\int \sec^2 x \tan x\,dx = \tfrac{1}{2}\tan^2 x + C_1.$$

(b) Set $u = \sec x$ and verify that

$$\int \sec^2 x \tan x\,dx = \tfrac{1}{2}\sec^2 x + C_2.$$

(c) Reconcile the results in parts (a) and (b).

51. Verify that, for each positive integer n:

(a) $\int_0^\pi \sin^2 nx\,dx = \tfrac{1}{2}\pi.$

HINT: $\sin^2\theta = \tfrac{1}{2}(1 - \cos 2\theta)$

(b) $\int_0^\pi \sin nx \cos nx\,dx = 0.$

(c) $\int_0^{\pi/n} \sin nx \cos nx\,dx = 0.$

52. (a) Calculate $\int \sin^3 x \, dx$. HINT: $\sin^2 x = 1 - \cos^2 x$.
 (b) Calculate $\int \sin^5 x \, dx$.
 (c) Explain how to calculate $\int \sin^{2k-1} x \, dx$ for an arbitrary positive integer k.

53. (a) Calculate $\int \tan^3 x \, dx$. HINT: $\tan^2 x = \sec^2 x - 1$.
 (b) Calculate $\int \tan^5 x \, dx$.
 (c) Calculate $\int \tan^7 x \, dx$.
 (d) Explain how to calculate $\int \tan^{2k+1} x \, dx$ for an arbitrary positive integer k.

54. (a) Sketch the region between the curves $y = \csc x$ and $y = \sin x$ over the interval $[\frac{1}{6}\pi, \frac{1}{2}\pi]$.
 (b) Calculate the area of that region.
 (c) The region is revolved about the x-axis. Find the volume of the resulting solid.

55. (a) Use a graphing utility to sketch the graph of

$$f(x) = \frac{1}{\sin x + \cos x} \qquad \text{for} \qquad 0 \le x \le \frac{\pi}{2}.$$

 (b) Find A and B such that $\sin x + \cos x = A \sin(x + B)$.
 (c) Find the area of the region between the graph of f and the x-axis.

56. (a) Use a graphing utility to sketch the graph of $f(x) = e^{-x^2}$.
 (b) Let $a > 0$. The region between the graph of f and the y-axis from $x = 0$ to $x = a$ is revolved about the y-axis. Find the volume of the resulting solid.
 (c) Find the value of a for which the solid in part (b) has a volume of 2 cubic units.

57. (a) Use a graphing utility to draw the curves

$$y = \frac{x^2 + 1}{x + 1} \qquad \text{for} \quad x > -1 \quad \text{and} \quad x + 2y = 16$$

 in the same coordinate system.
 (b) These curves intersect at two points and determine a bounded region Ω. Estimate the x-coordinates of the two points of intersection accurate to two decimal places.
 (c) Determine the approximate area of the region Ω.

58. (a) Use a graphing utility to draw the curve

$$y^2 = x^2(1 - x).$$

 (b) Your drawing in part (a) should show that the curve forms a loop for $0 \le x \le 1$. Calculate the area of the loop. HINT: Use the symmetry of the curve.

■ 8.2 INTEGRATION BY PARTS

We begin with the formula for the derivative of a product:

$$u(x)v'(x) + v(x)u'(x) = (u \cdot v)'(x).$$

Integrating both sides, we get

$$\int u(x)v'(x) \, dx + \int v(x)u'(x) \, dx = \int (u \cdot v)'(x) \, dx.$$

Since

$$\int (u \cdot v)'(x) \, dx = u(x)v(x) + C,$$

we have

$$\int u(x)v'(x) \, dx + \int v(x)u'(x) \, dx = u(x)v(x) + C$$

and therefore

$$\int u(x)v'(x) \, dx = u(x)v(x) - \int v(x)u'(x) \, dx + C.$$

Since the calculation of

$$\int v(x)u'(x) \, dx$$

will yield its own arbitrary constant, there is no reason to keep the constant C. We therefore drop it and write

(8.2.1)
$$\int u(x)v'(x) \, dx = u(x)v(x) - \int v(x)u'(x) \, dx.$$

The process of finding

$$\int u(x)v'(x)\,dx$$

by calculating

$$\int v(x)u'(x)\,dx$$

and then using (8.2.1) is called *integration by parts*.

Usually we write

$$u = u(x), \qquad dv = v'(x)\,dx$$
$$du = u'(x)\,dx, \qquad v = v(x).$$

Then the formula for integration by parts reads

(8.2.2)

$$\int u\,dv = uv - \int v\,du.$$

Integration by parts is a very versatile tool. According to (8.2.2) we can calculate $\int u\,dv$ by calculating $\int v\,du$ instead. The payoff is immediate in those cases where we can choose u and v so that

$$\int v\,du \qquad \text{is easier to calculate than} \qquad \int u\,dv.$$

Example 1 Calculate $\displaystyle\int x\,e^x\,dx$.

SOLUTION We want to separate x from e^x. Setting

$$u = x, \qquad dv = e^x\,dx$$

we have

$$du = dx, \qquad v = e^x.$$

Accordingly,

$$\int x\,e^x\,dx = \int u\,dv = uv - \int v\,du = xe^x - \int e^x\,dx = xe^x - e^x + C.$$

Our choice of u and dv worked out well. Does the choice of u and dv make a difference? Suppose we had set

$$u = e^x, \qquad dv = x\,dx.$$

Then we would have had

$$du = e^x\,dx, \qquad v = \tfrac{1}{2}x^2.$$

In this case integration by parts would have led to

$$\int x\,e^x\,dx = \int u\,dv = uv - \int v\,du = \tfrac{1}{2}x^2 e^x - \tfrac{1}{2}\int x^2 e^x\,dx,$$

giving us an integral which at this stage is difficult for us to deal with. This choice of u and dv would not have been helpful. ❑

Example 2 Calculate $\int x \sin 2x \, dx$.

SOLUTION Setting

$$u = x, \qquad dv = \sin 2x \, dx,$$

we have

$$du = dx, \qquad v = -\tfrac{1}{2} \cos 2x.$$

Therefore,

$$\int x \sin 2x \, dx = -\tfrac{1}{2} x \cos 2x - \int -\tfrac{1}{2} \cos 2x \, dx = -\tfrac{1}{2} x \cos 2x + \tfrac{1}{4} \sin 2x + C.$$

As you can verify, had we set

$$u = \sin 2x, \qquad dv = x \, dx,$$

then we would have run into an integral more difficult to evaluate than the integral with which we started. ❏

In Examples 1 and 2 there was only one effective way of choosing u and dv. With some integrals we have more latitude.

Example 3 Calculate $\int x \ln x \, dx$.

SOLUTION Setting

$$u = \ln x, \qquad dv = x \, dx,$$

we have

$$du = \frac{1}{x} \, dx, \qquad v = \frac{x^2}{2}.$$

The substitution gives

$$\int x \ln x \, dx = \int u \, dv = uv - \int v \, du$$

$$= \frac{x^2}{2} \ln x - \int \frac{1}{x} \frac{x^2}{2} \, dx = \tfrac{1}{2} x^2 \ln x - \tfrac{1}{2} \int x \, dx = \tfrac{1}{2} x^2 \ln x - \tfrac{1}{4} x^2 + C.$$

ANOTHER APPROACH This time we set

$$u = x \ln x, \qquad\qquad dv = dx$$

so that

$$du = (1 + \ln x) \, dx, \qquad v = x.$$

In this case the relation

$$\int u \, dv = uv - \int v \, du$$

gives

$$\int x \ln x \, dx = x^2 \ln x - \int x(1 + \ln x) \, dx.$$

The new integral is more complicated than the one with which we started. It may therefore look like we are worse off than when we began, but that is not the case. Going on, we have

$$\int x \ln x \, dx = x^2 \ln x - \int x \, dx - \int x \ln x \, dx$$

$$2 \int x \ln x \, dx = x^2 \ln x - \int x \, dx$$

$$= x^2 \ln x - \tfrac{1}{2}x^2 + C$$

$$\int x \ln x \, dx = \tfrac{1}{2}x^2 \ln x - \tfrac{1}{4}x^2 + C.$$

This is the result we obtained before. In the last step we wrote $C/2$ as C. We can do this because C represents an arbitrary constant. ❑

Remark   As you just saw, integration by parts can be useful even if $\int v \, du$ is not easier to calculate than $\int u \, dv$. What matters to us is the interplay between the two integrals. ❑

To calculate some integrals, we have to integrate by parts more than once.

Example 4   Calculate $\int x^2 e^{-x} dx$.

SOLUTION   Setting

$$u = x^2, \qquad dv = e^{-x} \, dx,$$

we have

$$du = 2x \, dx \qquad v = -e^{-x}.$$

This gives

$$\int x^2 e^{-x} dx = \int u \, dv = uv - \int v \, du = -x^2 e^{-x} - \int -2xe^{-x} dx$$

$$= -x^2 e^{-x} + \int 2xe^{-x} \, dx.$$

We now calculate the integral on the right, again by parts. This time we set

$$u = 2x, \qquad dv = e^{-x} \, dx.$$

This gives

$$du = 2 \, dx, \qquad v = -e^{-x}$$

and thus

$$\int 2xe^{-x} dx = \int u \, dv = uv - \int v \, du = -2xe^{-x} - \int -2e^{-x} \, dx$$

$$= -2xe^{-x} + \int 2e^{-x} \, dx = -2xe^{-x} - 2e^{-x} + C.$$

Combining this with our earlier calculations, we have

$$\int x^2 e^{-x} \, dx = -x^2 e^{-x} - 2xe^{-x} - 2e^{-x} + C = -(x^2 + 2x + 2)e^{-x} + C. \quad ❑$$

Example 5 Calculate $\int e^x \cos x \, dx$.

SOLUTION Once again we'll need to integrate by parts twice. First we write

$$u = e^x, \qquad dv = \cos x \, dx$$
$$du = e^x \, dx, \qquad v = \sin x.$$

This gives

(1) $$\int e^x \cos x \, dx = \int u \, dv = uv - \int v \, du = e^x \sin x - \int e^x \sin x \, dx.$$

Now we work with the integral on the right. Setting

$$u = e^x, \qquad dv = \sin x \, dx$$
$$du = e^x \, dx, \qquad v = -\cos x,$$

we have

(2) $$\int e^x \sin x \, dx = \int u \, dv = uv - \int v \, du = -e^x \cos x + \int e^x \cos x \, dx.$$

Substituting (2) into (1), we get

$$\int e^x \cos x \, dx = e^x \sin x + e^x \cos x - \int e^x \cos x \, dx$$

$$2 \int e^x \cos x \, dx = e^x (\sin x + \cos x)$$

$$\int e^x \cos x \, dx = \tfrac{1}{2} e^x (\sin x + \cos x).$$

Since this is an indefinite integral, we add an arbitrary constant C:

$$\int e^x \cos x \, dx = \tfrac{1}{2} e^x (\sin x + \cos x) + C.$$

(We began this example by setting $u = e^x$, $dv = \cos x \, dx$. As you can check, the substitution $u = \cos x$, $dv = e^x \, dx$ would have worked out just as well.) ❏

Integration by parts is often used to calculate integrals where the integrand is a mixture of function types; for example, polynomials mixed with exponentials, polynomials mixed with trigonometric functions, and so forth. Some integrands, however, are better left as mixtures; for example,

$$\int 2x e^{x^2} dx = e^{x^2} + C \qquad \text{and} \qquad \int 3x^2 \cos x^3 dx = \sin x^3 + C.$$

Any attempt to separate these integrands for integration by parts is counterproductive. The mixtures in these integrands arise from the chain rule, and we need these mixtures to calculate the integrals.

Example 6 Calculate $\int x^5 \cos x^3 \, dx$.

SOLUTION To integrate $\cos x^3$, we need an x^2 factor. So we'll keep x^2 together with $\cos x^3$ and set

$$u = x^3, \qquad dv = x^2 \cos x^3 \, dx.$$

Then

$$du = 3x^2\, dx, \qquad v = \tfrac{1}{3}\sin x^3$$

and

$$\int x^5 \cos x^3\, dx = \tfrac{1}{3}x^3 \sin x^3 - \int x^2 \sin x^3\, dx$$

$$= \tfrac{1}{3}x^3 \sin x^3 + \tfrac{1}{3}\cos x^3 + C. \quad \square$$

The counterpart to (8.2.1) for definite integrals reads

(8.2.3)
$$\int_a^b u(x)v'(x)\, dx = \Big[u(x)v(x)\Big]_a^b - \int_a^b v(x)u'(x)\, dx.$$

This follows directly from writing the product rule as

$$u(x)v'(x) = (u \cdot v)'(x) - v(x)u'(x).$$

Just integrate from $x = a$ to $x = b$.

We can circumvent this formula by working with indefinite integrals and bringing in the limits of integration only at the end. This is the course we will follow in the next example.

Example 7 Evaluate $\int_1^2 x^3 \ln x\, dx$.

SOLUTION First we calculate the indefinite integral, proceeding by parts. We set

$$u = \ln x, \qquad dv = x^3\, dx$$

$$du = \frac{1}{x}\, dx, \qquad v = \frac{1}{4}x^4.$$

This gives

$$\int x^3 \ln x\, dx = \tfrac{1}{4}x^4 \ln x - \tfrac{1}{4}\int x^3 dx = \tfrac{1}{4}x^4 \ln x - \tfrac{1}{16}x^4 + C.$$

To evaluate the definite integral, we need only one antiderivative. We choose the one with $C = 0$. This gives

$$\int_1^2 x^3 \ln x\, dx = \Big[\tfrac{1}{4}x^4 \ln x - \tfrac{1}{16}x^4\Big]_1^2 = 4\ln 2 - \tfrac{15}{16}. \quad \square$$

Through integration by parts, we construct an antiderivative for the logarithm, for the arc sine, and for the arc tangent.

(8.2.4)
$$\int \ln x\, dx = x \ln x - x + C.$$

(8.2.5)
$$\int \arcsin x\, dx = x \arcsin x + \sqrt{1 - x^2} + C.$$

(8.2.6)

$$\int \arctan x \, dx = x \arctan x - \tfrac{1}{2} \ln(1 + x^2) + C.$$

We will work with the arc sine. The logarithm and the arc tangent formulas are left to the Exercises.

To find the integral of the arc sine, we set

$$u = \arcsin x, \qquad dv = dx$$

$$du = \frac{1}{\sqrt{1 - x^2}} \, dx, \qquad v = x.$$

This gives

$$\int \arcsin x \, dx = x \arcsin x - \int \frac{x}{\sqrt{1 - x^2}} \, dx = x \arcsin x + \sqrt{1 - x^2} + C. \quad \square$$

EXERCISES 8.2

Exercises 1–40. Calculate.

1. $\displaystyle\int x e^{-x} \, dx.$

2. $\displaystyle\int_0^2 x 2^x \, dx.$

3. $\displaystyle\int x^2 e^{-x^3} \, dx.$

4. $\displaystyle\int x \ln x^2 \, dx.$

5. $\displaystyle\int_0^1 x^2 e^{-x} \, dx.$

6. $\displaystyle\int x^3 e^{-x^2} \, dx.$

7. $\displaystyle\int \frac{x^2}{\sqrt{1 - x}} \, dx.$

8. $\displaystyle\int \frac{dx}{x(\ln x)^3}.$

9. $\displaystyle\int_1^{c^2} x \ln \sqrt{x} \, dx.$

10. $\displaystyle\int_0^3 x\sqrt{x + 1} \, dx.$

11. $\displaystyle\int \frac{\ln(x + 1)}{\sqrt{x + 1}} \, dx.$

12. $\displaystyle\int x^2(e^x - 1) \, dx.$

13. $\displaystyle\int (\ln x)^2 \, dx.$

14. $\displaystyle\int x(x + 5)^{-14} \, dx.$

15. $\displaystyle\int x^3 3^x \, dx.$

16. $\displaystyle\int \sqrt{x} \ln x \, dx.$

17. $\displaystyle\int x(x + 5)^{14} \, dx.$

18. $\displaystyle\int (2^x + x^2)^2 \, dx.$

19. $\displaystyle\int_0^{1/2} x \cos \pi x \, dx.$

20. $\displaystyle\int_0^{\pi/2} x^2 \sin x \, dx.$

21. $\displaystyle\int x^2(x + 1)^9 \, dx.$

22. $\displaystyle\int x^2(2x - 1)^{-7} \, dx.$

23. $\displaystyle\int e^x \sin x \, dx.$

24. $\displaystyle\int (e^x + 2x)^2 \, dx.$

25. $\displaystyle\int_0^1 \ln(1 + x^2) \, dx.$

26. $\displaystyle\int x \ln(x + 1) \, dx.$

27. $\displaystyle\int x^n \ln x \, dx, \quad n \neq -1.$

28. $\displaystyle\int e^{3x} \cos 2x \, dx.$

29. $\displaystyle\int x^3 \sin x^2 \, dx.$

30. $\displaystyle\int x^3 \sin x \, dx.$

31. $\displaystyle\int_0^{1/4} \arcsin 2x \, dx.$

32. $\displaystyle\int \frac{\arcsin 2x}{\sqrt{1 - 4x^2}} \, dx.$

33. $\displaystyle\int_0^1 x \arctan x^2 \, dx.$

34. $\displaystyle\int \cos \sqrt{x} \, dx.$ HINT: Set $u = \sqrt{x}, dv = \dfrac{\cos \sqrt{x}}{\sqrt{x}} \, dx.$

35. $\displaystyle\int x^2 \cosh 2x \, dx.$

36. $\displaystyle\int_{-1}^1 x \sinh 2x^2 \, dx.$

37. $\displaystyle\int \frac{1}{x} \arcsin(\ln x) \, dx.$

38. $\displaystyle\int \cos(\ln x) \, dx.$ HINT: Integrate by parts twice.

39. $\displaystyle\int \sin(\ln x) \, dx.$

40. $\displaystyle\int_1^{2e} x^2(\ln x)^2 \, dx.$

41. Derive (8.2.4): $\displaystyle\int \ln x \, dx = x \ln x - x + C.$

42. Derive (8.2.6):

$$\int \arctan x \, dx = x \arctan x - \tfrac{1}{2} \ln(1 + x^2) + C.$$

Derive the following three formulas.

43. $\displaystyle\int x^k \ln x \, dx = \frac{x^{k+1}}{k + 1} \ln x - \frac{x^{k+1}}{(k + 1)^2} + C, k \neq -1.$

44. $\displaystyle\int e^{ax} \cos bx \, dx = \frac{e^{ax}(a \cos bx + b \sin bx)}{a^2 + b^2} + C.$

45. $\displaystyle\int e^{ax} \sin bx \, dx = \frac{e^{ax}(a \sin bx - b \cos bx)}{a^2 + b^2} + C.$

46. What happens if you try integration by parts to calculate $\int e^{ax} \cosh ax \, dx$? Calculate this integral by some other method.

47. Set $f(x) = x \sin x$. Find the area between the graph of f and the x-axis from $x = 0$ to $x = \pi$.

48. Set $g(x) = x \cos \frac{1}{2}x$. Find the area between the graph of g and the x-axis from $x = 0$ to $x = \pi$.

Exercises 49–50. Find the area between the graph of f and the x-axis.

49. $f(x) = \arcsin x$, $x \in \left[0, \frac{1}{2}\right]$.

50. $f(x) = xe^{-2x}$, $x \in [0, 2]$.

51. Let Ω be the region between the graph of the logarithm function and the x-axis from $x = 1$ to $x = e$. (a) Find the area of Ω. (b) Find the centroid of Ω. (c) Find the volume of the solids generated by revolving Ω about each of the coordinate axes.

52. Let $f(x) = \dfrac{\ln x}{x}$, $x \in [1, 2e]$.

(a) Find the area of the region Ω bounded by the graph of f and the x-axis.

(b) Find the volume of the solid generated by revolving Ω about the x-axis.

Exercises 53–56. Find the centroid of the region under the graph.

53. $f(x) = e^x$, $x \in [0, 1]$.

54. $f(x) = e^{-x}$, $x \in [0, 1]$.

55. $f(x) = \sin x$, $x \in [0, \pi]$.

56. $f(x) = \cos x$, $x \in \left[0, \frac{1}{2}\pi\right]$.

57. The mass density of a rod that extends from $x = 0$ to $x = 1$ is given by the function $\lambda(x) = e^{kx}$ where k is a constant. (a) Calculate the mass of the rod. (b) Find the center of mass of the rod.

58. The mass density of a rod that extends from $x = 2$ to $x = 3$ is given by the logarithm function $f(x) = \ln x$. (a) Calculate the mass of the rod. (b) Find the center of mass of the rod.

Exercises 59–62. Find the volume generated by revolving the region under the graph about the y-axis.

59. $f(x) = \cos \frac{1}{2}\pi x$, $x \in [0, 1]$.

60. $f(x) = x \sin x$, $x \in [0, \pi]$.

61. $f(x) = x e^x$, $x \in [0, 1]$.

62. $f(x) = x \cos x$, $x \in \left[0, \frac{1}{2}\pi\right]$.

63. Let Ω be the region under the curve $y = e^x$, $x \in [0, 1]$. Find the centroid of the solid generated by revolving Ω about the x-axis. (For the appropriate formula, see Project 6.4.)

64. Let Ω be the region under the graph of $y = \sin x$, $x \in \left[0, \frac{1}{2}\pi\right]$. Find the centroid of the solid generated by revolving Ω about the x-axis. (For the appropriate formula, see Project 6.4.)

65. Let Ω be the region between the curve $y = \cosh x$ and the x-axis from $x = 0$ to $x = 1$. Find the area of Ω and determine the centroid.

66. Let Ω be the region given in Exercise 65. Find the centroid of the solid generated by revolving Ω:

(a) about the x-axis;   (b) about the y-axis

67. Let n be a positive integer. Use integration by parts to show that

$$\int x^n e^{ax} \, dx = \frac{x^n e^{ax}}{a} - \frac{n}{a} \int x^{n-1} e^{ax} dx, \quad a \neq 0.$$

68. Let n be a positive integer. Show that

$$\int (\ln x)^n dx = x(\ln x)^n - n \int (\ln x)^{n-1} \, dx.$$

The formula given in Exercise 67 reduces the calculation of $\int x^n e^{ax} \, dx$ to the calculation of $\int x^{n-1} e^{ax} \, dx$. The formula given in Exercise 68 reduces the calculation of $\int (\ln x)^n \, dx$ to the calculation of $\int (\ln x)^{n-1} \, dx$. Formulas (such as these) which reduce the calculation of an expression in n to the calculation of the corresponding expression in $n - 1$ are called *reduction formulas*.

Exercises 69–72. Calculate the following integrals by using the appropriate reduction formulas.

69. $\displaystyle\int x^3 e^{2x} \, dx.$ **70.** $\displaystyle\int x^2 e^{-x} \, dx.$

71. $\displaystyle\int (\ln x)^3 \, dx.$ **72.** $\displaystyle\int (\ln x)^4 \, dx.$

73. (a) As you can probably see, were you to integrate $\int x^3 e^x \, dx$ by parts, the result would be of the form

$$\int x^3 e^x \, dx = Ax^3 e^x + Bx^2 e^x + Cxe^x + De^x + E.$$

Differentiate both sides of this equation and solve for the coefficients A, B, C, D. In this manner you can calculate the integral without actually carrying out the integration.

(b) Calculate $\int x^3 e^x \, dx$ by using the appropriate reduction formula.

74. If P is a polynomial of degree k, then

$$\int P(x) e^x dx = [P(x) - P'(x) + \cdots \pm P^{(k)}(x)]e^x + C.$$

Verify this statement. For simplicity, take $k = 4$.

75. Use the statement in Exercise 74 to calculate:

(a) $\displaystyle\int (x^2 - 3x + 1)e^x \, dx.$ (b) $\displaystyle\int (x^3 - 2x)e^x \, dx.$

76. Use integration by parts to show that if f has an inverse with continuous first derivative, then

$$\int f^{-1}(x) \, dx = xf^{-1}(x) - \int x(f^{-1})'(x) \, dx.$$

77. Show that if f and g have continuous second derivatives and $f(a) = g(a) = f(b) = g(b) = 0$, then

$$\int_a^b f(x)g''(x) \, dx = \int_a^b g(x)f''(x) \, dx.$$

78. You are familiar with the identity

$$f(b) - f(a) = \int_a^b f'(x) \, dx.$$

(a) Assume that f has a continuous second derivative. Use integration by parts to derive the identity

$$f(b) - f(a) = f'(a)(b - a) - \int_a^b f''(x)(x - b) \, dx.$$

(b) Assume that f has a continuous third derivative. Use the result in part (a) and integration by parts to derive the identity

$$f(b) - f(a) = f'(a)(b - a) + \frac{f''(a)}{2}(b - a)^2$$
$$- \int_a^b \frac{f'''(x)}{2}(x - b)^2 dx.$$

Going on in this manner, we are led to what are called Taylor series (Chapter 12).

79. Use a graphing utility to draw the curve $y = x \sin x$ for $x \geq 0$. Then use a CAS to calculate the area between the curve and the x-axis

(a) from $x = 0$ to $x = \pi$.
(b) from $x = \pi$ to $x = 2\pi$.
(c) from $x = 2\pi$ to $x = 3\pi$.
(d) What is the area between the curve and the x-axis from $x = n\pi$ to $x = (n + 1)\pi$? Take n an arbitrary nonnegative integer.

80. Use a graphing utility to draw the curve $y = x \cos x$ for $x \geq 0$. Then use a CAS to calculate the area between the curve and the x-axis

(a) from $x = \frac{1}{2}\pi$ to $x = \frac{3}{2}\pi$.
(b) from $x = \frac{3}{2}\pi$ to $x = \frac{5}{2}\pi$.
(c) from $x = \frac{5}{2}\pi$ to $x = \frac{7}{2}\pi$.
(d) What is the area between the curve and the x-axis from $x = \frac{1}{2}(2n - 1)\pi$ to $x = \frac{1}{2}(2n + 1)\pi$? Take n an arbitrary positive integer.

81. Use a graphing utility to draw the curve $y = 1 - \sin x$ from $x = 0$ to $x = \pi$. Then use a CAS

(a) to find the area of the region Ω between the curve and the x-axis.
(b) to find the volume of the solid generated by revolving Ω about the y-axis.
(c) to find the centroid of Ω.

82. Use a graphing utility to draw the curve $y = xe^x$ from $x = 0$ to $x = 10$. Then use a CAS

(a) to find the centroid of the region Ω between the curve and the x-axis.
(b) to find the volume of the solid generated by revolving Ω about the x-axis.
(c) to find the volume of the solid generated by revolving Ω about the y-axis.

■ PROJECT 8.2 Sine Waves $y = \sin nx$ and Cosine Waves $y = \cos nx$

Problem 1. Show that for each positive integer n,

$$\int_0^{2\pi} \sin^2 nx \, dx = \pi \quad \text{and} \quad \int_0^{2\pi} \cos^2 nx \, dx = \pi.$$

HINT: Use the identities

$$\sin^2 \theta = \tfrac{1}{2} - \tfrac{1}{2}\cos 2\theta, \qquad \cos^2 \theta = \tfrac{1}{2} + \tfrac{1}{2}\cos 2\theta.$$

Problem 2. Show that for $m \neq n$,

$$\int_0^{2\pi} \sin mx \sin nx \, dx = 0$$

and

$$\int_0^{2\pi} \cos mx \cos nx \, dx = 0.$$

HINT: Verify that

$$\int_0^{2\pi} \cos [(m + n)x] \, dx = 0.$$

Then use the addition formula

$$\cos (\alpha + \beta) = \cos \alpha \cos \beta - \sin \alpha \sin \beta$$

to show that

$$\int_0^{2\pi} \cos mx \cos nx \, dx = \int_0^{2\pi} \sin mx \sin nx \, dx.$$

Evaluate the cosine integral by parts.

Problem 3. Show that for $m \neq n$,

$$\int_0^{2\pi} \sin mx \cos nx \, dx = 0.$$

HINT: Verify that

> $\cos 2\alpha = \cos 0 = 0$

$$\int_0^{2\pi} \sin [(m + n)x] \, dx = 0.$$

Then use the addition formula

$$\sin (\alpha + \beta) = \sin \alpha \cos \beta + \cos \alpha \sin \beta$$

to show that

$$\int_0^{2\pi} \sin mx \cos nx \, dx + \int_0^{2\pi} \cos mx \sin nx \, dx = 0.$$

Evaluate the second integral by parts.

Summary

For each positive integer n

$$\int_0^{2\pi} \sin^2 nx \, dx = \pi \quad \text{and}$$

$$\int_0^{2\pi} \cos^2 nx \, dx = \pi,$$

(8.2.6)   and for positive integers $m \neq n$

$$\int_0^{2\pi} \sin mx \sin nx \, dx = 0,$$

$$\int_0^{2\pi} \cos mx \cos nx \, dx = 0,$$

$$\int_0^{2\pi} \sin mx \cos nx \, dx = 0.$$

These relations lie at the heart of wave theory.

Problem 4. (*The superposition of waves*) A function of the form

$$f(x) = a_1 \sin x + a_2 \sin 2x + \cdots + a_n \sin nx + b_1 \cos x$$

$$+ b_2 \cos 2x + \cdots + b_n \cos nx$$

is called a *trigonometric polynomial*, and the coefficients a_k, b_k are called the *Fourier coefficients*.[†] Determine the a_k and b_k from $k = 1$ to $k = n$.

HINT: Evaluate

$$\int_0^{2\pi} f(x) \sin kx \, dx \quad \text{and} \quad \int_0^{2\pi} f(x) \cos kx \, dx$$

using the relations just summarized.

[†]After the French mathematician J. B. J. Fourier (1768–1830), who was the first to use such polynomials to closely approximate functions of great generality.

■ 8.3 POWERS AND PRODUCTS OF TRIGONOMETRIC FUNCTIONS

Integrals of trigonometric powers and products can usually be reduced to elementary integrals by the imaginative use of the basic trigonometric identities and, here and there, some integration by parts.

These are the identities that we'll rely on:

Unit circle relations

$$\sin^2 \theta + \cos^2 \theta = 1, \qquad \tan^2 \theta + 1 = \sec^2 \theta, \qquad \cot^2 \theta + 1 = \csc^2 \theta$$

Addition formulas

$$\sin(\alpha + \beta) = \sin \alpha \cos \beta + \cos \alpha \sin \beta$$

$$\sin(\alpha - \beta) = \sin \alpha \cos \beta - \cos \alpha \sin \beta$$

$$\cos(\alpha + \beta) = \cos \alpha \cos \beta - \sin \alpha \sin \beta$$

$$\cos(\alpha - \beta) = \cos \alpha \cos \beta + \sin \alpha \sin \beta$$

Double-angle formulas

$$\sin 2\theta = 2 \sin \theta \cos \theta, \qquad \cos 2\theta = 1 - 2 \sin^2 \theta$$

Half-angle formulas

$$\sin^2 \theta = \tfrac{1}{2} - \tfrac{1}{2} \cos 2\theta, \qquad \cos^2 \theta = \tfrac{1}{2} + \tfrac{1}{2} \cos 2\theta.$$

Sines and Cosines

Example 1   Calculate

$$\int \sin^2 x \cos^5 x \, dx.$$

SOLUTION   The relation $\cos^2 x = 1 - \sin^2 x$ enables us to express $\cos^4 x$ in terms of $\sin x$. The integrand then becomes

$$\text{(a polynomial in } \sin x) \cos x,$$

an expression that we can integrate by the chain rule.

$$\int \sin^2 x \cos^5 x \, dx = \int \sin^2 x \cos^4 x \cos x \, dx$$

$$= \int \sin^2 x (1 - \sin^2 x)^2 \cos x \, dx$$

$$= \int (\sin^2 x - 2\sin^4 x + \sin^6 x) \cos x \, dx$$

$$= \tfrac{1}{3} \sin^3 x - \tfrac{2}{5} \sin^5 x + \tfrac{1}{7} \sin^7 x + C. \quad \square$$

Example 2   Calculate

$$\int \sin^5 x \, dx$$

SOLUTION   The relation $\sin^2 x = 1 - \cos^2 x$ enables us to express $\sin^4 x$ in terms of $\cos x$. The integrand then becomes

$$\text{(a polynomial in } \cos x) \sin x,$$

an expression that we can integrate by the chain rule:

$$\int \sin^5 x \, dx = \int \sin^4 x \sin x \, dx$$

$$= \int (1 - \cos^2 x)^2 \sin x \, dx$$

$$= \int (1 - 2\cos^2 x + \cos^4 x) \sin x \, dx$$

$$= \int \sin x \, dx - 2 \int \cos^2 x \sin x \, dx + \int \cos^4 x \sin x \, dx$$

$$= -\cos x + \tfrac{2}{3} \cos^3 x - \tfrac{1}{5} \cos^5 x + C. \quad \square$$

Example 3   Calculate

$$\int \sin^2 x \, dx \qquad \text{and} \qquad \int \cos^2 x \, dx.$$

SOLUTION   Since $\sin^2 x = \tfrac{1}{2} - \tfrac{1}{2} \cos 2x$ and $\cos^2 x = \tfrac{1}{2} + \tfrac{1}{2} \cos 2x$,

$$\int \sin^2 x \, dx = \int \left(\tfrac{1}{2} - \tfrac{1}{2} \cos 2x \right) dx = \tfrac{1}{2} x - \tfrac{1}{4} \sin 2x + C$$

and

$$\int \cos^2 x \, dx = \int \left(\tfrac{1}{2} + \tfrac{1}{2} \cos 2x \right) dx = \tfrac{1}{2} x + \tfrac{1}{4} \sin 2x + C. \quad \square$$

Example 4   Calculate

$$\int \sin^2 x \cos^2 x \, dx.$$

SOLUTION   The relation $2 \sin x \cos x = \sin 2x$ gives $\sin^2 x \cos^2 x = \frac{1}{4} \sin^2 2x$ and that we can integrate:

$$\int \sin^2 x \cos^2 x \, dx = \frac{1}{4} \int \sin^2 2x \, dx$$

$$= \frac{1}{4} \int \left(\frac{1}{2} - \frac{1}{2} \cos 4x\right) dx$$

$$= \frac{1}{8} \int dx - \frac{1}{8} \int \cos 4x \, dx = \frac{1}{8}x - \frac{1}{32} \sin 4x + C. \quad \square$$

Example 5   Calculate

$$\int \sin^4 x \, dx.$$

SOLUTION

$$\sin^4 x = (\sin^2 x)^2 = \left(\frac{1}{2} - \frac{1}{2} \cos 2x\right)^2 = \frac{1}{4} - \frac{1}{2} \cos 2x + \frac{1}{4} \cos^2 2x$$

$$= \frac{1}{4} - \frac{1}{2} \cos 2x + \frac{1}{8} + \frac{1}{8} \cos 4x$$

$$= \frac{3}{8} - \frac{1}{2} \cos 2x + \frac{1}{8} \cos 4x.$$

Therefore

$$\int \sin^4 x \, dx = \int \left(\frac{3}{8} - \frac{1}{2} \cos 2x + \frac{1}{8} \cos 4x\right) dx$$

$$= \frac{3}{8}x - \frac{1}{4} \sin 2x + \frac{1}{32} \sin 4x + C. \quad \square$$

Example 6   Calculate

$$\int \sin 5x \sin 3x \, dx.$$

SOLUTION   The only identities that feature the product of sines with different arguments are the addition formulas for the cosine:

$$\cos(\alpha - \beta) = \cos \alpha \cos \beta + \sin \alpha \sin \beta, \qquad \cos(\alpha + \beta) = \cos \alpha \cos \beta - \sin \alpha \sin \beta.$$

We can express $\sin \alpha \sin \beta$ in terms of something we can integrate by subtracting the second equation from the first one.

   In our case we have

$$\cos 2x = \cos(5x - 3x) = \cos 5x \cos 3x + \sin 5x \sin 3x$$

$$\cos 8x = \cos(5x + 3x) = \cos 5x \cos 3x - \sin 5x \sin 3x$$

and therefore

$$\sin 5x \sin 3x = \frac{1}{2}(\cos 2x - \cos 8x).$$

Using this relation, we have

$$\int \sin 5x \sin 3x \, dx = \frac{1}{2} \int \cos 2x \, dx - \frac{1}{2} \int \cos 8x \, dx$$

$$= \frac{1}{4} \sin 2x - \frac{1}{16} \sin 8x + C. \quad \square$$

Tangents and Secants

Example 7 Calculate

$$\int \tan^4 x \, dx.$$

SOLUTION The relation $\tan^2 x = \sec^2 x - 1$ gives

$$\tan^4 x = \tan^2 x \sec^2 x - \tan^2 x = \tan^2 x \sec^2 x - \sec^2 x + 1.$$

Therefore

$$\int \tan^4 x \, dx = \int (\tan^2 x \sec^2 x - \sec^2 x + 1) \, dx$$

$$= \frac{1}{3} \tan^3 x - \tan x + x + C. \quad \square$$

Example 8 Calculate

$$\int \sec^3 x \, dx.$$

SOLUTION The relation $\sec^2 x = \tan^2 x + 1$ gives

$$\int \sec^3 x \, dx = \int \sec x \, (\tan^2 x + 1) \, dx = \int \sec x \tan^2 x \, dx + \int \sec x \, dx.$$

We know the second integral, but the first integral gives us problems. (Here the relation $\tan^2 x = \sec^2 x - 1$ doesn't help, for, as you can check, that takes us right back to where we started.)

Not seeing any other way to proceed, we try integration by parts on the original integral. Setting

$$u = \sec x, \qquad\qquad dv = \sec^2 x \, dx$$
$$du = \sec x \tan x \, dx, \qquad v = \tan x,$$

we have

$$\int \sec^3 x \, dx = \sec x \tan x - \int \sec x \tan^2 x \, dx$$

$$\underset{\tan^2 x = \sec^2 x - 1 \longrightarrow}{} = \sec x \tan x - \int \sec^3 x \, dx + \int \sec x \, dx$$

$$2 \int \sec^3 x \, dx = \sec x \tan x + \int \sec x \, dx$$

$$\int \sec^3 x \, dx = \frac{1}{2} \sec x \tan x + \frac{1}{2} \ln | \sec x + \tan x | + C. \quad \square$$

Example 9 Calculate

$$\int \sec^6 x \, dx.$$

SOLUTION Applying the relation $\sec^2 x = \tan^2 x + 1$ to $\sec^4 x$, we can express the integrand as

$$(\text{a polynomial in } \tan x) \sec^2 x.$$

We can integrate this by the chain rule:

$$\sec^6 x = \sec^4 x \sec^2 x = (\tan^2 x + 1)^2 \sec^2 x = (\tan^4 x + 2\tan^2 x + 1) \sec^2 x.$$

Therefore

$$\int \sec^6 x \, dx = \int (\tan^4 x \sec^2 x + 2\tan^2 x \sec^2 x + \sec^2 x) \, dx$$

$$= \tfrac{1}{5} \tan^5 x + \tfrac{2}{3} \tan^3 x + \tan x + C. \quad \square$$

Example 10 Calculate

$$\int \tan^5 x \sec^3 x \, dx.$$

SOLUTION Applying the relation $\tan^2 x = \sec^2 x - 1$ to $\tan^4 x$, we can express the integrand as

$$(\text{a polynomial in } \sec x) \sec x \tan x.$$

We can integrate this by the chain rule:

$$\int \tan^5 x \sec^3 x \, dx = \int \tan^4 x \sec^2 x \sec x \tan x \, dx$$

$$= \int (\sec^2 x - 1)^2 \sec^2 x \sec x \tan x \, dx$$

$$= \int (\sec^6 x - 2\sec^4 x + \sec^2 x) \sec x \tan x \, dx$$

$$= \tfrac{1}{7} \sec^7 x - \tfrac{2}{5} \sec^5 x + \tfrac{1}{3} \sec^3 x + C. \quad \square$$

Cotangents and Cosecants

The integrals in Examples 7–10 featured tangents and secants. In carrying out the integrations, we relied on the identity $\tan^2 x + 1 = \sec^2 x$ and in one instance resorted to integration by parts. To calculate integrals that feature cotangents and cosecants, use the identity $\cot^2 x + 1 = \csc^2 x$ and, if necessary, integration by parts.

EXERCISES 8.3

Exercises 1–44. Calculate. (If you run out of ideas, use the examples as models.)

1. $\displaystyle\int \sin^3 x \, dx.$

2. $\displaystyle\int_0^{\pi/8} \cos^2 4x \, dx.$

3. $\displaystyle\int_0^{\pi/6} \sin^2 3x \, dx.$

4. $\displaystyle\int \cos^3 x \, dx.$

5. $\displaystyle\int \cos^4 x \sin^3 x \, dx.$

6. $\displaystyle\int \sin^3 x \cos^2 x \, dx.$

7. $\displaystyle\int \sin^3 x \cos^3 x \, dx.$

8. $\displaystyle\int \sin^2 x \cos^4 x \, dx.$

9. $\displaystyle\int \sec^2 \pi x \, dx.$

10. $\displaystyle\int \csc^2 2x \, dx.$

11. $\int \tan^3 x \, dx.$

12. $\int \cot^3 x \, dx.$

13. $\int_0^\pi \sin^4 x \, dx.$

14. $\int \cos^3 x \cos 2x \, dx.$

15. $\int \sin 2x \cos 3x \, dx.$

16. $\int_0^{\pi/2} \cos 2x \sin 3x \, dx.$

17. $\int \tan^2 x \sec^2 x \, dx.$

18. $\int \cot^2 x \csc^2 x \, dx.$

19. $\int \sin^2 x \sin 2x \, dx.$

20. $\int_0^{\pi/2} \cos^4 x \, dx.$

21. $\int \sin^6 x \, dx.$

22. $\int \cos^5 x \sin^5 x \, dx.$

23. $\int_{\pi/6}^{\pi/2} \cot^2 x \, dx.$

24. $\int \tan^6 x \, dx.$

25. $\int \cot^3 x \csc^3 x \, dx.$

26. $\int \tan^3 x \sec^3 x \, dx.$

27. $\int \sin 5x \sin 2x \, dx.$

28. $\int \sec^4 3x \, dx.$

29. $\int \sin^{5/2} x \cos^3 x \, dx.$

30. $\int \frac{\sin^3 x}{\cos x} \, dx.$

31. $\int \tan^5 3x \, dx.$

32. $\int \cot^5 2x \, dx.$

33. $\int_{-1/6}^{1/3} \sin^4 3\pi x \cos^3 3\pi x \, dx.$

34. $\int_0^{1/2} \cos \pi x \cos \frac{1}{2}\pi x \, dx.$

35. $\int_0^{\pi/4} \cos 4x \sin 2x \, dx.$

36. $\int (\sin 3x - \sin x)^2 \, dx.$

37. $\int \tan^4 x \sec^4 x \, dx.$

38. $\int \cot^4 x \csc^4 x \, dx.$

39. $\int \sin \frac{1}{2}x \cos 2x \, dx.$

40. $\int_0^{2\pi} \sin^2 ax \, dx, \ a \neq 0.$

41. $\int_0^{\pi/4} \tan^3 x \sec^2 x \, dx.$

42. $\int_{\pi/4}^{\pi/2} \csc^3 x \cot x \, dx.$

43. $\int_0^{\pi/6} \tan^2 2x \, dx.$

44. $\int_0^{\pi/3} \tan x \sec^{3/2} x \, dx.$

45. Find the area between the curve $y = \sin^2 x$ and the x-axis from $x = 0$ to $x = \pi$.

46. The region between the curve $y = \cos x$ and the x-axis from $x = -\pi/2$ to $x = \pi/2$ is revolved about the x-axis. Find the volume of the resulting solid.

47. The region of Exercise 45 is revolved about the x-axis. Find the volume of the resulting solid.

48. The region bounded by the y-axis and the curves $y = \sin x$ and $y = \cos x$, $0 \le x \le \pi/4$, is revolved about the x-axis. Find the volume of the resulting solid.

49. The region bounded by the y-axis, the line $y = 1$, and the curve $y = \tan x$, $x \in [0, \pi/4]$, is revolved about the x-axis. Find the volume of the resulting solid.

50. The region between the curve $y = \tan^2 x$ and the x-axis from $x = 0$ to $x = \pi/4$ is revolved about the x-axis. Find the volume of the resulting solid.

51. The region between the curve $y = \tan x$ and the x-axis from $x = 0$ to $x = \pi/4$ is revolved about the line $y = -1$. Find the volume of the resulting solid.

52. The region between the curve $y = \sec^2 x$ and the x-axis from $x = 0$ to $x = \pi/4$ is revolved about the x axis. Find the volume of the resulting solid.

53. (a) Use integration by parts to show that for $n > 2$

$$\int \sin^n x \, dx = -\frac{1}{n} \sin^{n-1} x \cos x + \frac{n-1}{n} \int \sin^{n-2} x \, dx.$$

(b) Then show that

$$\int_0^{\pi/2} \sin^n x \, dx = \frac{n-1}{n} \int_0^{\pi/2} \sin^{n-2} x \, dx.$$

(c) Verify the *Wallis sine formulas*:
for even $n \ge 2$,

$$\int_0^{\pi/2} \sin^n x \, dx = \frac{(n-1)\cdots 5 \cdot 3 \cdot 1}{n \cdots 6 \cdot 4 \cdot 2} \cdot \frac{\pi}{2};$$

for odd $n \ge 3$,

$$\int_0^{\pi/2} \sin^n x \, dx = \frac{(n-1)\cdots 4 \cdot 2}{n \cdots 5 \cdot 3}.$$

54. Use Exercise 53 to show that

$$\int_0^{\pi/2} \cos^n x \, dx = \int_0^{\pi/2} \sin^n x \, dx.$$

55. Evaluate by the Wallis formulas:

(a) $\int_0^{\pi/2} \sin^7 x \, dx.$ (b) $\int_0^{\pi/2} \cos^6 x \, dx.$

56. Use a graphing utility to draw the graph of the function $f(x) = x + \sin 2x$, $x \in [0, \pi]$. The region between the graph of f and the x-axis is revolved about the x-axis.

(a) Use a CAS to find the volume of the resulting solid.

(b) Calculate the volume exactly by carrying out the integration.

57. Use a graphing utility to draw the graph of the function $g(x) = \sin^2 x^2$, $x \in [0, \sqrt{\pi}]$. The region between the graph of g and the x-axis is revolved about the y-axis.

(a) Use a CAS to find the volume of the resulting solid.

(b) Calculate the volume exactly by carrying out the integration.

58. Use a graphing utility to draw in one figure the graphs of both $f(x) = 1 + \cos x$ and $g(x) = \sin \frac{1}{2}x$ from $x = 0$ to $x = 2\pi$.

(a) Use a CAS to find the points where the two curves intersect; then find the area between the two curves.

(b) The region between the two curves is revolved about the x-axis. Use a CAS to find the volume of the resulting solid.

■ 8.4 INTEGRALS FEATURING $\sqrt{a^2-x^2}, \sqrt{a^2+x^2}, \sqrt{x^2-a^2}$

Preliminary Remark By reversing the roles played by x and u in the statement of Theorem 5.7.1, we can conclude that

(8.4.1)

$$\text{if } F' = f, \qquad \text{then} \qquad \int f(x(u))x'(u)\,du = F(x(u)) + C$$

and

$$\int_{x(a)}^{x(b)} f(x)\,dx = \int_a^b f(x(u))x'(u)\,du.$$

These are the substitution rules that we will follow in this section. ❑

Integrals that feature $\sqrt{a^2-x^2}$, $\sqrt{a^2+x^2}$, or $\sqrt{x^2-a^2}$ can often be calculated by a *trigonometric substitution*. Taking $a > 0$, we proceed as follows:

$$\text{for } \sqrt{a^2-x^2} \qquad \text{we set} \qquad x = a\sin u;$$
$$\text{for } \sqrt{a^2+x^2} \qquad \text{we set} \qquad x = a\tan u;$$
$$\text{for } \sqrt{x^2-a^2} \qquad \text{we set} \qquad x = a\sec u.$$

In making such substitutions, we must make clear exactly what values of u we are using. Failure to do this can lead to nonsensical results.

We begin with a familiar integral.

Example 1 You know that

$$\int_{-a}^{a} \sqrt{a^2 - x^2}\,dx$$

represents the area of the half-disk of radius a and is therefore $\frac{1}{2}\pi a^2$. (Figure 8.4.1) We confirm this by a trigonometric substitution.

For x from $-a$ to a, we set

$$x = a\sin u, \qquad dx = a\cos u\,du,$$

taking u from $-\frac{1}{2}\pi$ to $\frac{1}{2}\pi$. For such u, $\cos u \geq 0$ and

$$\sqrt{a^2 - x^2} = \sqrt{a^2 - a^2\sin^2 u} = a\cos u.$$

At $u = -\frac{1}{2}\pi, x = -a;$ at $u = \frac{1}{2}\pi, x = a$. Therefore

$$\int_{-a}^{a} \sqrt{a^2 - x^2}\,dx = \int_{-\pi/2}^{\pi/2} a^2\cos^2 u\,du = a^2 \int_{-\pi/2}^{\pi/2} \left(\tfrac{1}{2} + \tfrac{1}{2}\cos 2u\right)\,du = \tfrac{1}{2}\pi a^2. \quad ❑$$

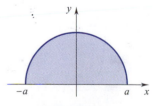

Figure 8.4.1

To a limited extent we can give a geometric view of a trigonometric substitution by drawing a suitable right triangle. Since the right triangle interpretation applies only to u between 0 and $\frac{1}{2}\pi$, we will not use it as the basis of our calculations.

Example 2 To calculate

$$\int \frac{dx}{(a^2 - x^2)^{3/2}} \, dx$$

we note that the integral can be written

$$\int \left(\frac{1}{\sqrt{a^2 - x^2}} \right)^3 dx.$$

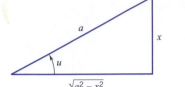

This integral features $\sqrt{a^2 - x^2}$. For each x between $-a$ and a, we set

$$x = a \sin u, \quad dx = a \cos u \, du,$$

taking u between $-\frac{1}{2}\pi$ and $\frac{1}{2}\pi$. For such u, $\cos u > 0$ and

$$\sqrt{a^2 - x^2} = a \cos u.$$

Therefore

$$\int \frac{dx}{(a^2 - x^2)^{3/2}} = \int \frac{a \cos u}{(a \cos u)^3} \, du$$

$$= \frac{1}{a^2} \int \frac{1}{\cos^2 u} \, du$$

$$= \frac{1}{a^2} \int \sec^2 u \, du$$

$$= \frac{1}{a^2} \tan u + C = \frac{x}{a^2 \sqrt{a^2 - x^2}} + C.$$

$$\tan u = \frac{\sin u}{\cos u}$$

Check the result by differentiation. ❏

Before giving the next example, we point out that, at those numbers u where $\cos u \neq 0$, $\sec u$ and $\cos u$ have the same sign ($\sec u = 1/\cos u$).

Example 3 We calculate

$$\int \sqrt{a^2 + x^2} \, dx.$$

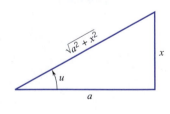

For each real number x, we set

$$x = a \tan u, \qquad dx = a \sec^2 u \, du,$$

taking u between $-\frac{1}{2}\pi$ and $\frac{1}{2}\pi$. For such u, $\sec u > 0$ and

$$\sqrt{a^2 + x^2} = a \sec u.$$

└── Check this out.

Therefore

$$\int \sqrt{a^2 + x^2} \, dx = \int (a \sec u) \, a \sec^2 u \, du$$

$$= a^2 \int \sec^3 u \, du$$

$$= \frac{a^2}{2}(\sec u \tan u + \ln|\sec u + \tan u|) + C$$

Example 7, Section 8.3 ⟶

$$= \frac{a^2}{2}\left[\frac{\sqrt{a^2 + x^2}}{a}\left(\frac{x}{a}\right) + \ln\left|\frac{\sqrt{a^2 + x^2}}{a} + \frac{x}{a}\right|\right] + C$$

$$= \tfrac{1}{2}x\sqrt{a^2 + x^2} + \tfrac{1}{2}a^2 \ln|x + \sqrt{a^2 + x^2}| - \tfrac{1}{2}a^2 \ln a + C.$$

Noting that $x + \sqrt{a^2 + x^2} > 0$ and, absorbing the constant $-\tfrac{1}{2}a^2 \ln a$ in C, we can write

(8.4.2)
$$\int \sqrt{a^2 + x^2}\, dx = \tfrac{1}{2}x\sqrt{a^2 + x^2} + \tfrac{1}{2}a^2 \ln\left(x + \sqrt{a^2 + x^2}\right) + C.$$

This is a standard integration formula. (Formula 78) ❏

Example 4 We calculate

$$\int \frac{dx}{\sqrt{x^2 - 1}}.$$

The domain of the integrand consists of two separated sets: all $x > 1$ and all $x < -1$.
    Both for $x > 1$ and $x < -1$, we set

$$x = \sec u, \qquad dx = \sec u \tan u\, du.$$

For $x > 1$ we take u between 0 and $\tfrac{1}{2}\pi$; for $x < -1$ we take u between π and $\tfrac{3}{2}\pi$. For such u, $\tan u > 0$ and

$$\sqrt{x^2 - 1} = \tan u.$$

Therefore,

$$\int \frac{dx}{\sqrt{x^2 - 1}} = \int \frac{\sec u \tan u}{\tan u}\, du = \int \sec u\, du$$

$$= \ln|\sec u + \tan u| + C$$

$$= \ln|x + \sqrt{x^2 - 1}| + C.$$

Check the result by differentiation. ❏

Example 5 We calculate

$$\int \frac{dx}{x^2\sqrt{x^2 - 4}}.$$

For $x > 2$ and $x < -2$, we set

$$x = 2\sec u, \qquad dx = 2\sec u \tan u\, du.$$

For $x > 2$ we take u between 0 and $\tfrac{1}{2}\pi$; for $x < -2$ we take u between π and $\tfrac{3}{2}\pi$. For such u, $\tan u > 0$ and

$$\sqrt{x^2 - 4} = 2\tan u.$$

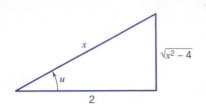

With this substitution

$$\int \frac{dx}{x^2\sqrt{x^2-4}} = \int \frac{2\sec u \tan u}{4\sec^2 u \cdot 2\tan u}\,du$$

$$= \frac{1}{4}\int \frac{1}{\sec u}\,du$$

$$= \frac{1}{4}\int \cos u\,du$$

$$= \frac{1}{4}\sin u + C = \frac{\sqrt{x^2-4}}{4x} + C.$$

Check the result by differentiation. ❏

Remark   Before rushing into a trigonometric substitution, look at the integral carefully to see whether there is a simpler way to proceed. For instance, you can calculate

$$\int \frac{x}{\sqrt{a^2-x^2}}\,dx$$

by setting $x = a\sin u$, and so on, but you can also carry out the integration by setting $u = a^2 - x^2$. Try both substitutions and decide which is the more effective. ❏

Example 6   Calculate

$$\int \frac{dx}{\sqrt{x^2+2x+5}}.$$

By completing the square under the radical, we can write the integral as

$$\int \frac{dx}{\sqrt{(x+1)^2+4}}.$$

For each real number x, we set

$$x + 1 = 2\tan u, \qquad dx = 2\sec^2 u\,du,$$

taking u between $-\frac{1}{2}\pi$ and $\frac{1}{2}\pi$. For such u, $\sec u > 0$ and

$$\sqrt{(x+1)^2+4} = \sqrt{4\tan^2 u + 4} = 2\sqrt{\tan^2 u + 1} = 2\sec u.$$

Therefore

$$\int \frac{dx}{\sqrt{x^2+2x+5}} = \int \frac{2\sec^2 u}{2\sec u}\,du$$

$$= \int \sec u\,du = \ln|\sec u + \tan u| + C$$

$$= \ln\left|\tfrac{1}{2}\sqrt{x^2+2x+5} + \tfrac{1}{2}(x+1)\right| + C$$

$$= \ln\tfrac{1}{2} + \ln\left|\sqrt{x^2+2x+5} + x + 1\right| + C.$$

Absorbing the constant $\ln\frac{1}{2}$ in C and noting that the expression within the absolute value signs is positive, we have

$$\int \frac{dx}{\sqrt{x^2+2x+5}} = \ln\left(\sqrt{x^2+2x+5} + x + 1\right) + C.$$

Check the result by differentiation. ❏

Trigonometric substitutions can be effective in cases where the quadratic is not under a radical sign. In particular, the reduction formula

(8.4.3)

$$\int \frac{dx}{(x^2 + a^2)^n} \, dx = \frac{1}{a^{2n-1}} \int \cos^{2(n-1)} u \, du$$

(a very useful little formula) can be obtained by setting $x = a \tan u$, taking u between $-\frac{1}{2}\pi$ and $\frac{1}{2}\pi$. The derivation of this formula is left to you as an exercise.

EXERCISES 8.4

Exercises 1–34. Calculate.

1. $\int \frac{dx}{\sqrt{a^2 - x^2}}$.

2. $\int_{5/2}^{4} \frac{x}{\sqrt{x^2 - 4}} \, dx$.

3. $\int \sqrt{x^2 - 1} \, dx$.

4. $\int \frac{x}{\sqrt{4 - x^2}} \, dx$.

5. $\int \frac{x^2}{\sqrt{4 - x^2}} \, dx$.

6. $\int \frac{x^2}{\sqrt{x^2 - 4}} \, dx$.

7. $\int \frac{x}{(1 - x^2)^{3/2}} \, dx$.

8. $\int \frac{x^2}{\sqrt{4 + x^2}} \, dx$.

9. $\int_{0}^{1/2} \frac{x^2}{(1 - x^2)^{3/2}} \, dx$.

10. $\int \frac{x}{a^2 + x^2} \, dx$. ← x^3 in answer key

11. $\int x\sqrt{4 - x^2} \, dx$.

12. $\int_{0}^{2} \frac{x^2}{\sqrt{16 - x^2}} \, dx$.

13. $\int_{0}^{5} x^2\sqrt{25 - x^2} \, dx$.

14. $\int \frac{\sqrt{1 - x^2}}{x^4} \, dx$.

15. $\int \frac{x^2}{(x^2 + 8)^{3/2}} \, dx$.

16. $\int_{0}^{a} \sqrt{a^2 - x^2} \, dx$.

17. $\int \frac{dx}{x\sqrt{a^2 - x^2}}$.

18. $\int \frac{\sqrt{x^2 - 1}}{x} \, dx$.

19. $\int_{0}^{3} \frac{x^3}{\sqrt{9 + x^2}} \, dx$.

20. $\int \frac{dx}{x^2\sqrt{a^2 - x^2}}$.

21. $\int \frac{dx}{x^2\sqrt{a^2 + x^2}}$.

22. $\int \frac{dx}{(x^2 + 2)^{3/2}}$.

23. $\int_{0}^{1} \frac{dx}{(5 - x^2)^{3/2}}$.

24. $\int \frac{dx}{e^x\sqrt{4 + e^{2x}}}$.

25. $\int \frac{dx}{x^2\sqrt{x^2 - a^2}}$.

26. $\int \frac{e^x}{\sqrt{9 - e^{2x}}} \, dx$.

27. $\int \frac{dx}{e^x\sqrt{e^{2x} - 9}}$.

28. $\int \frac{dx}{\sqrt{x^2 - 2x - 3}}$.

29. $\int \frac{dx}{(x^2 - 4x + 4)^{3/2}}$.

30. $\int \frac{x}{\sqrt{6x - x^2}} \, dx$.

31. $\int x\sqrt{6x - x^2 - 8} \, dx$.

32. $\int \frac{x + 2}{\sqrt{x^2 + 4x + 13}} \, dx$.

33. $\int \frac{x}{(x^2 + 2x + 5)^2} \, dx$.

34. $\int \frac{x}{\sqrt{x^2 - 2x + 3}} \, dx$.

35. Use integration by parts to derive the formula

$$\int \text{arcsec } x \, dx = x \, \text{arcsec } x - \ln\left|x + \sqrt{x^2 - 1}\right| + C.$$

36. Calculate $\int \frac{1}{x}\sqrt{a^2 - x^2} \, dx$
    (a) by setting $u = \sqrt{a^2 - x^2}$.
    (b) by a trigonometric substitution.
    (c) Then reconcile the results.

37. Verify (8.4.3).

Exercises 38–39. Use (8.4.3) to calculate the integral.

38. $\int \frac{1}{(x^2 + 1)^2} \, dx$.

39. $\int \frac{1}{(x^2 + 1)^3} \, dx$.

Exercises 40–41. Calculate the integral: (a) by integrating by parts, (b) by applying a trigonometric substitution.

40. $\int x \arctan x \, dx$.

41. $\int x \arcsin x \, dx$.

42. Find the area under the curve $y = (\sqrt{x^2 - 9})/x$ from $x = 3$ to $x = 5$.

43. The region under the curve $y = 1/(1 + x^2)$ from $x = 0$ to $x = 1$ is revolved about the x-axis. Find the volume of the resulting solid.

44. The shaded part in the figure is called a *circular segment*. Calculate the area of the segment:

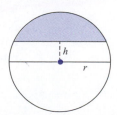

45. Show that in a disk of radius r a sector with central angle of radian measure θ has area $A = \frac{1}{2}r^2\theta$. HINT: Assume first

that $0 < \theta < \frac{1}{2}\pi$ and subdivide the region as indicated in the figure. Then verify that the formula holds for any sector.

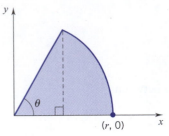

46. Find the area of the region bounded on the left and right by the two branches of the hyperbola $(x^2/a^2) - (y^2/b^2) = 1$, and above and below by the lines $y = \pm b$.

47. Find the area between the right branch of the hyperbola $(x^2/9) - (y^2/16) = 1$ and the line $x = 5$.

48. If the circle $(x - b)^2 + y^2 = a^2, b > a > 0$, is revolved about the y-axis, the resulting "doughnut-shaped" solid is called a *torus*. Use the shell method to find the formula for the volume of the torus.

49. Calculate the mass and the center of mass of a rod that extends from $x = 0$ to $x = a > 0$ and has mass density $\lambda(x) = (x^2 + a^2)^{-1/2}$.

50. Calculate the mass and center of mass of the rod of Exercise 49 given that the rod has mass density $\lambda(x) = (x^2 + a^2)^{-3/2}$.

For Exercises 51–53, let Ω be the region under the curve $y = \sqrt{x^2 - a^2}$ from $x = a$ to $x = \sqrt{2}\,a$.

51. Sketch Ω, find the area of Ω, and locate the centroid.

52. Find the volume of the solid generated by revolving Ω about the x-axis and determine the centroid of that solid.

53. Find the volume of the solid generated by revolving Ω about the y-axis and determine the centroid of that solid.

54. Use a trigonometric substitution to derive the formula

$$\int \frac{1}{\sqrt{a^2 + x^2}}\, dx = \ln\left(x + \sqrt{a^2 + x^2}\right) + C.$$

55. Use a trigonometric substitution to derive the formula

$$\int \frac{1}{\sqrt{x^2 - a^2}}\, dx = \ln\left|x + \sqrt{x^2 - a^2}\right| + C.$$

⊳56. Set $f(x) = \dfrac{x^2}{\sqrt{1 - x^2}}$. Use a CAS

(a) to draw the graph of f;
(b) to find the area between the graph of f and the x-axis from $x = 0$ to $x = \frac{1}{2}$;
(c) to find the volume of the solid generated by revolving about the y-axis the region described in part (b).

⊳57. Set $f(x) = \dfrac{\sqrt{x^2 - 9}}{x^2}, x \geq 3$. Use a CAS

(a) to draw the graph of f;
(b) to find the area between the graph of f and the x-axis from $x = 3$ to $x = 6$;
(c) to locate the centroid of the region described in part (b).

■ 8.5 RATIONAL FUNCTIONS; PARTIAL FRACTIONS

In this section we present a method for integrating rational functions. Recall that a rational function is, by definition, the quotient of two polynomials. For example,

$$f(x) = \frac{1}{x^2 - 4}, \qquad g(x) = \frac{2x^2 + 3}{x(x - 1)^2}, \qquad h(x) = \frac{3x^4 - 20x^2 + 17}{x^3 + 2x^2 - 7}$$

are rational functions, but

$$f(x) = \frac{1}{\sqrt{x}}, \qquad g(x) = \frac{x^2 + 1}{\ln x}, \qquad h(x) = \frac{\sin x}{x}$$

are not rational functions.

A rational function $R(x) = P(x)/Q(x)$ is said to be *proper* if the degree of the numerator is less than the degree of the denominator. If the degree of the numerator is greater than or equal to the degree of the denominator, then the rational function is called *improper*.[†] We will focus our attention on *proper rational functions* because any improper rational function can be written as a sum of a polynomial and a proper rational function:

$$\frac{P(x)}{Q(x)} = p(x) + \frac{r(x)}{Q(x)}. \quad^{††}$$

[†]These terms are taken from the familiar terms used to describe rational numbers p/q.

[††]This is analogous to writing an improper fraction as a so-called *mixed number*.

This is accomplished simply by dividing the denominator into the numerator.

As is shown in algebra, every proper rational function can be written as the sum of *partial fractions*, fractions of the form

(8.5.1)
$$\frac{A}{(x - \alpha)^k} \quad \text{and} \quad \frac{Bx + C}{(x^2 + \beta x + \gamma)^k}$$

with $x^2 + \beta x + \gamma$ irreducible.[†] Such a sum is called a *partial fraction decomposition*.

We begin by calculating some partial fraction decompositions. Later we will use these decompostions to calculate integrals.

Example 1 (*The denominator splits into distinct linear factors.*) For

$$\frac{2x}{x^2 - x - 2} = \frac{2x}{(x - 2)(x + 1)}$$

we write

$$\frac{2x}{x^2 - x - 2} = \frac{A}{x - 2} + \frac{B}{x + 1},$$

with the constants A and B to be determined.

Multiplication by $(x - 2)(x + 1)$ yields the equation

(1) $$2x = A(x + 1) + B(x - 2)$$

We illustrate two methods for finding A and B.

METHOD 1   We substitute numbers for x in (1):

Setting $x = 2$, we get $4 = 3A$, which gives $A = \frac{4}{3}$,

Setting $x = -1$, we get $-2 = -3B$, which gives $B = \frac{2}{3}$.

The desired decomposition reads

$$\frac{2x}{x^2 - x - 2} = \frac{4}{3(x - 2)} + \frac{2}{3(x + 1)}.$$

You can verify this by carrying out the addition on the right.

METHOD 2   (This method is based on the observation that two polynomials are equal iff their coefficients are equal.)

We rewrite (1) as

$$2x = (A + B)x + (A - 2B).$$

Equating coefficients, we have

$$A + B = 2$$
$$A - 2B = 0.$$

We can find A and B by solving these equations simultaneously. The solutions are again: $A = \frac{4}{3}, B = \frac{2}{3}$.   ❏

[†] Not factorable into linear expressions with real coefficients. This is the case if $\beta^2 - 4\gamma < 0$.

In general, each distinct linear factor $x - \alpha$ in the denominator gives rise to a term of the form

$$\frac{A}{x - \alpha}.$$

Example 2 (*The denominator has a repeated linear factor.*) For

$$\frac{2x^2 + 3}{x(x - 1)^2},$$

we write

$$\frac{2x^2 + 3}{x(x - 1)^2} = \frac{A}{x} + \frac{B}{x - 1} + \frac{C}{(x - 1)^2}.$$

This leads to

$$2x^2 + 3 = A(x - 1)^2 + Bx(x - 1) + Cx.$$

To determine A, B, C, we substitute three values for x. (Method 1.) We select 0 and 1 because for these values of x several terms on the right will drop out. As a third value of x, any other number will do; we select 2 just to keep the arithmetic simple.

Setting $x = 0$, we get $3 = A$.

Setting $x = 1$, we get $5 = C$.

Setting $x = 2$, we get $11 = A + 2B + 2C$,

which, with $A = 3$ and $C = 5$, gives $B = -1$.

This gives us

$$\frac{2x^2 + 3}{x(x - 1)^2} = \frac{3}{x} - \frac{1}{x - 1} + \frac{5}{(x - 1)^2}. \quad \square$$

In general, each factor of the form $(x - a)^k$ in the denominator gives rise to an expression of the form

$$\frac{A_1}{x - \alpha} + \frac{A_2}{(x - \alpha)^2} + \cdots + \frac{A_k}{(x - \alpha)^k}.$$

Example 3 (*The denominator has an irreducible quadratic factor.*) For

$$\frac{x^2 + 5x + 2}{(x + 1)(x^2 + 1)},$$

we write

$$\frac{x^2 + 5x + 2}{(x + 1)(x^2 + 1)} = \frac{A}{x + 1} + \frac{Bx + C}{x^2 + 1}$$

and obtain

$$x^2 + 5x + 2 = A(x^2 + 1) + (Bx + C)(x + 1) = (A + B)x^2 + (B + C)x + A + C.$$

Equating coefficients (Method 2), we have

$$A + B = 1$$
$$B + C = 5$$
$$A + C = 2.$$

This system of equations is satisfied by $A = -1$, $B = 2$, $C = 3$. (Check this out.) The decomposition reads

$$\frac{x^2 + 5x + 2}{(x + 1)(x^2 + 1)} = \frac{-1}{x + 1} + \frac{2x + 3}{x^2 + 1}.$$

(We could have obtained this result by using Method 1; for example, by setting $x = -1$, $x = 0$, $x = 1$.) ❑

In the examples that follow, we'll use Method 1.

Example 4 (*The denominator has an irreducible quadratic factor.*) For

$$\frac{1}{x(x^2 + x + 1)},$$

we write

$$\frac{1}{x(x^2 + x + 1)} = \frac{A}{x} + \frac{Bx + C}{x^2 + x + 1}$$

and obtain

$$1 = A(x^2 + x + 1) + (Bx + C)x.$$

Again, we select values of x that produce simple arithmetic.

$$\begin{array}{ll} 1 = A & (x = 0) \\ 1 = 3A + B + C & (x = 1) \\ 1 = A + B - C & (x = -1). \end{array}$$

From this we find that

$$A = 1, \qquad B = -1, \qquad C = -1,$$

and therefore

$$\frac{1}{x(x^2 + x + 1)} = \frac{1}{x} - \frac{x + 1}{x^2 + x + 1}. \quad ❑$$

> In general, each irreducible quadratic factor $x^2 + \beta x + \gamma$ in the denominator gives rise to a term of the form
>
> $$\frac{Ax + B}{x^2 + \beta x + \gamma}.$$

Example 5 (*The denominator has a repeated irreducible quadratic factor.*) For

$$\frac{3x^4 + x^3 + 20x^2 + 3x + 31}{(x + 1)(x^2 + 4)^2},$$

we write

$$\frac{3x^4 + x^3 + 20x^2 + 3x + 31}{(x+1)(x^2+4)^2} = \frac{A}{x+1} + \frac{Bx+C}{x^2+4} + \frac{Dx+E}{(x^2+4)^2}.$$

This gives

$$3x^4 + x^3 + 20x^2 + 3x + 31 = A(x^2+4)^2 + (Bx+C)(x+1)(x^2+4)$$
$$+ (Dx+E)(x+1).$$

This time we use $x = -1, 0, 1, 2, -2$:

$$50 = 25A \qquad\qquad\qquad\qquad (x = -1)$$
$$31 = 16A \qquad + 4C \qquad + E \quad (x = 0)$$
$$58 = 25A + 10B + 10C + 2D + 2E \quad (x = 1)$$
$$173 = 64A + 48B + 24C + 6D + 3E \quad (x = 2)$$
$$145 = 64A + 16B - 8C + 2D - E \quad (x = -2).$$

With a little patience, you can determine that

$$A = 2, \qquad B = 1, \qquad C = 0, \qquad D = 0, \qquad E = -1.$$

This gives the decomposition

$$\frac{3x^4 + x^3 + 20x^2 + 3x + 31}{(x+1)(x^2+4)^2} = \frac{2}{x+1} + \frac{x}{x^2+4} - \frac{1}{(x^2+4)^2}. \quad ❏$$

In general, each multiple irreducible quadratic factor $(x^2 + \beta x + \gamma)^k$ in the denominator gives rise to an expression of the form

$$\frac{A_1 x + B_1}{x^2 + \beta x + \gamma} + \frac{A_2 x + B_2}{(x^2 + \beta x + \gamma)^2} + \cdots + \frac{A_k x + B_k}{(x^2 + \beta x + \gamma)^k}.$$

As indicated at the beginning of this section, if the rational function is improper, then a polynomial will appear in the decomposition.

Example 6 *(An improper rational function.)* The quotient

$$\frac{x^5 + 2}{x^2 - 1}$$

is improper. Dividing the denominator into the numerator, we find that

$$\frac{x^5 + 2}{x^2 - 1} = x^3 + x + \frac{x+2}{x^2-1}. \qquad\qquad \text{(Verify this.)}$$

The decomposition of the remaining fraction reads

$$\frac{x+2}{x^2-1} = \frac{A}{x+1} + \frac{B}{x-1}.$$

As you can verify, $A = -\frac{1}{2}$, $B = \frac{3}{2}$. Therefore

$$\frac{x^5 + 2}{x^2 - 1} = x^3 + x - \frac{1}{2(x+1)} + \frac{3}{2(x-1)}. \quad ❏$$

We have been decomposing quotients into partial fractions in order to integrate them. Here we carry out the integrations, leaving some of the details to you.

Example 1′

$$\int \frac{2x}{x^2 - x - 2}\, dx = \int \left[\frac{4}{3(x - 2)} + \frac{2}{3(x + 1)} \right] dx$$

$$= \tfrac{4}{3} \ln |x - 2| + \tfrac{2}{3} \ln |x + 1| + C$$

$$= \tfrac{1}{3} \ln \left[(x - 2)^4 (x + 1)^2 \right] + C. \quad \square$$

Example 2′

$$\int \frac{2x^2 + 3}{x(x - 1)^2}\, dx = \int \left[\frac{3}{x} - \frac{1}{x - 1} + \frac{5}{(x - 1)^2} \right] dx$$

$$= 3 \ln |x| - \ln |x - 1| - \frac{5}{x - 1} + C$$

$$= \ln \left| \frac{x^3}{x - 1} \right| - \frac{5}{x - 1} + C \quad \square$$

Example 3′

$$\int \frac{x^2 + 5x + 2}{(x + 1)(x^2 + 1)}\, dx = \int \left(\frac{-1}{x + 1} + \frac{2x + 3}{x^2 + 1} \right) dx$$

$$= -\int \frac{1}{x + 1}\, dx + \int \frac{2x + 3}{x^2 + 1}\, dx.$$

Since

$$-\int \frac{1}{x + 1}\, dx = -\ln |x + 1| + C_1$$

and

$$\int \frac{2x + 3}{x^2 + 1}\, dx = \int \frac{2x}{x^2 + 1}\, dx + 3\int \frac{1}{x^2 + 1}\, dx = \ln (x^2 + 1) + 3 \arctan x + C_2,$$

we have

$$\int \frac{x^2 + 5x + 2}{(x + 1)(x^2 + 1)}\, dx = -\ln |x + 1| + \ln (x^2 + 1) + 3 \arctan x + C$$

$$= \ln \left| \frac{x^2 + 1}{x - 1} \right| + 3 \arctan x + C. \quad \square$$

Example 4′

$$I = \int \frac{dx}{x(x^2 + x + 1)} = \int \left(\frac{1}{x} - \frac{x + 1}{x^2 + x + 1} \right) dx = \ln |x| - \int \frac{x + 1}{x^2 + x + 1}\, dx.$$

To calculate the remaining integral, note that $(d/dx)(x^2 + x + 1) = 2x + 1$. We can manipulate the integrand to get a term of the form du/u with $u = x^2 + x + 1$:

$$\frac{x + 1}{x^2 + x + 1} = \frac{\tfrac{1}{2}[2x + 1] + \tfrac{1}{2}}{x^2 + x + 1} = \frac{1}{2} \left(\frac{2x + 1}{x^2 + x + 1} + \frac{1}{x^2 + x + 1} \right).$$

This gives us

$$\int \frac{x+1}{x^2+x+1}\,dx = \frac{1}{2}\int \frac{2x+1}{x^2+x+1}\,dx + \frac{1}{2}\int \frac{1}{x^2+x+1}\,dx.$$

The first integral is a logarithm:

$$\frac{1}{2}\int \frac{2x+1}{x^2+x+1}\,dx = \frac{1}{2}\ln(x^2+x+1) + C_1. \qquad (x^2+x+1 > 0 \text{ for all } x)$$

The second integral is an arc tangent:

$$\frac{1}{2}\int \frac{1}{x^2+x+1}\,dx = \frac{1}{2}\int \frac{dx}{(x+\frac{1}{2})^2+(\sqrt{3}/2)^2} = \frac{1}{\sqrt{3}}\arctan\left[\frac{2}{\sqrt{3}}\left(x+\frac{1}{2}\right)\right] + C_2.$$

Combining the results, we have the integral we want:

$$I = \ln|x| - \frac{1}{2}\ln(x^2+x+1) - \frac{1}{\sqrt{3}}\arctan\left[\frac{2}{\sqrt{3}}\left(x+\frac{1}{2}\right)\right] + C. \quad \square$$

Example 5′

$$\int \frac{3x^4+x^3+20x^2+3x+31}{(x+1)(x^2+4)^2}\,dx = \int \left[\frac{2}{x+1} + \frac{x}{x^2+4} - \frac{1}{(x^2+4)^2}\right]dx.$$

The first two fractions are easy to integrate:

$$\int \frac{2}{x+1}\,dx = 2\ln|x+1| + C_1,$$

$$\int \frac{x}{x^2+4}\,dx = \frac{1}{2}\int \frac{2x}{x^2+4}\,dx = \frac{1}{2}\ln(x^2+4) + C_2.$$

The integral of the last fraction is of the form

$$\int \frac{dx}{(x^2+a^2)^n}.$$

As you saw in the preceding section, such integrals can be calculated by setting $x = a\tan u$, $u \in (-\frac{1}{2}\pi, \frac{1}{2}\pi)$. [(8.4.3).] In this case

$$\int \frac{dx}{(x^2+4)^2} = \frac{1}{8}\int \cos^2 u\,du$$
$$\hookleftarrow x = 2\tan u$$

$$= \frac{1}{16}\int (1+\cos 2u)\,du$$
half-angle formula $\longrightarrow$

$$= \frac{1}{16}u + \frac{1}{32}\sin 2u + C_3$$

$$= \frac{1}{16}u + \frac{1}{16}\sin u\cos u + C_3$$
$\sin 2u = 2\sin u\cos u \longrightarrow$

$$= \frac{1}{16}\arctan\frac{x}{2} + \frac{1}{16}\left(\frac{x}{\sqrt{x^2+4}}\right)\left(\frac{2}{\sqrt{x^2+4}}\right) + C_3$$

$$= \frac{1}{16}\arctan\frac{x}{2} + \frac{1}{8}\left(\frac{x}{x^2+4}\right) + C_3.$$

The integral we want is equal to

$$2 \ln |x + 1| + \frac{1}{2} \ln (x^2 + 4) - \frac{1}{8} \left(\frac{x}{x^2 + 4} \right) - \frac{1}{16} \arctan \frac{x}{2} + C. \quad \square$$

Example 6′

$$\int \frac{x^5 + 2}{x^2 - 1} dx = \int \left[x^3 + x - \frac{1}{2(x + 1)} + \frac{3}{2(x - 1)} \right] dx$$

$$= \frac{1}{4} x^4 + \frac{1}{2} x^2 - \frac{1}{2} \ln |x + 1| + \frac{3}{2} \ln |x - 1| + C. \quad \square$$

EXERCISES 8.5

Exercises 1–8. Decompose into partial fractions.

1. $\dfrac{1}{x^2 + 7x + 6}.$

2. $\dfrac{x^2}{(x - 1)(x^2 + 4x + 5)}.$

3. $\dfrac{x}{x^4 - 1}.$

4. $\dfrac{x^4}{(x - 1)^3}.$

5. $\dfrac{x^2 - 3x - 1}{x^3 + x^2 - 2x}.$

6. $\dfrac{x^3 + x^2 + x + 2}{x^4 + 3x^2 + 2}.$

7. $\dfrac{2x^2 + 1}{x^3 - 6x^2 + 11x - 6}.$

8. $\dfrac{1}{x(x^2 + 1)^2}.$

Exercises 9–30. Calculate.

9. $\displaystyle\int \frac{7}{(x - 2)(x + 5)} \, dx.$

10. $\displaystyle\int \frac{x}{(x + 1)(x + 2)(x + 3)} \, dx.$

11. $\displaystyle\int \frac{2x^4 - 4x^3 + 4x^2 + 3}{x^3 - x^2} \, dx.$

12. $\displaystyle\int \frac{x^2 + 1}{x(x^2 - 1)} \, dx.$

13. $\displaystyle\int \frac{x^5}{(x - 2)^2} \, dx.$

14. $\displaystyle\int \frac{x^5}{x - 2} \, dx.$

15. $\displaystyle\int \frac{x + 3}{x^2 - 3x + 2} \, dx.$

16. $\displaystyle\int \frac{x^2 + 3}{x^2 - 3x + 2} \, dx.$

17. $\displaystyle\int \frac{dx}{(x - 1)^3}.$

18. $\displaystyle\int \frac{dx}{x^2 + 2x + 2}.$

19. $\displaystyle\int \frac{x^2}{(x - 1)^2(x + 1)} \, dx.$

20. $\displaystyle\int \frac{2x - 1}{(x + 1)^2(x - 2)^2} \, dx.$

21. $\displaystyle\int \frac{dx}{x^4 - 16}.$

22. $\displaystyle\int \frac{3x^5 - 3x^2 + x}{x^3 - 1} \, dx.$

23. $\displaystyle\int \frac{x^3 + 4x^2 - 4x - 1}{(x^2 + 1)^2} \, dx.$

24. $\displaystyle\int \frac{dx}{(x^2 + 16)^2}.$

25. $\displaystyle\int \frac{dx}{x^4 + 4}.^{\dagger}$

26. $\displaystyle\int \frac{dx}{x^4 + 16}.^{\dagger}$

27. $\displaystyle\int \frac{x - 3}{x^3 + x^2} \, dx.$

28. $\displaystyle\int \frac{1}{(x - 1)(x^2 + 1)^2} \, dx.$

29. $\displaystyle\int \frac{x + 1}{x^3 + x^2 - 6x} \, dx.$

30. $\displaystyle\int \frac{x^3 + x^2 + x + 3}{(x^2 + 1)(x^2 + 3)} \, dx.$

Exercises 31–34. Evaluate.

31. $\displaystyle\int_0^2 \frac{x}{x^2 + 5x + 6} \, dx.$

32. $\displaystyle\int_1^3 \frac{1}{x^3 + x} \, dx.$

33. $\displaystyle\int_1^3 \frac{x^2 - 4x + 3}{x^3 + 2x^2 + x} \, dx.$

34. $\displaystyle\int_0^2 \frac{x^3}{(x^2 + 2)^2} \, dx.$

Exercises 35–38. Calculate.

35. $\displaystyle\int \frac{\cos \theta}{\sin^2 \theta - 2 \sin \theta - 8} \, d\theta.$

36. $\displaystyle\int \frac{e^t}{e^{2t} + 5e^t + 6} \, dt.$

37. $\displaystyle\int \frac{1}{t([\ln t]^2 - 4)} \, dt.$

38. $\displaystyle\int \frac{\sec^2 \theta}{\tan^3 \theta - \tan^2 \theta} \, d\theta.$

Exercises 39–45. Derive the formula.

39. $\displaystyle\int \frac{u}{a + bu} \, du = \frac{1}{b^2} (a + bu - a \ln |a + bu|) + C.$

40. $\displaystyle\int \frac{du}{u(a + bu)} = \frac{1}{a} \ln \left| \frac{u}{a + bu} \right| + C.$

41. $\displaystyle\int \frac{du}{u^2(a + bu)} = -\frac{1}{au} + \frac{b}{a^2} \ln \left| \frac{a + bu}{u} \right| + C.$

42. $\displaystyle\int \frac{du}{u(a + bu)^2} = \frac{1}{a(a + bu)} + \frac{1}{a^2} \ln \left| \frac{a + bu}{u} \right| + C.$

43. $\displaystyle\int \frac{du}{a^2 - u^2} = \frac{1}{2a} \ln \left| \frac{a + u}{a - u} \right| + C.$

44. $\displaystyle\int \frac{u \, du}{a^2 - u^2} = -\frac{1}{2} \ln |a^2 - u^2| + C.$

45. $\displaystyle\int \frac{u^2 \, du}{a^2 - u^2} = -u + \frac{a}{2} \ln \left| \frac{a + u}{a - u} \right| + C.$

Exercises 46–47. Calculate

$$\int \frac{du}{(a + bu)(c + du)}$$

with the coefficients as stipulated.

46. a, b, c, d all different from 0, $\quad ad = bc.$

47. a, b, c, d all different from 0, $\quad ad \neq bc.$

†HINT: With $a > 0$, $x^4 + a^2 = (x^2 + \sqrt{2a}x + a)(x^2 - \sqrt{2a}x + a)$.

48. Show that for $y = \dfrac{1}{x^2 - 1}$,

$$\frac{d^n y}{dx^n} = \frac{(-1)^n n!}{2} \left[\frac{1}{(x-1)^{n+1}} - \frac{1}{(x+1)^{n+1}} \right].$$

49. Find the volume of the solid generated by revolving the region between the curve $y = 1/\sqrt{4 - x^2}$ and the x-axis from $x = 0$ to $x = 3/2$: (a) about the x-axis; (b) about the y-axis.

50. Calculate

$$\int x^3 \arctan x \, dx.$$

51. Find the centroid of the region under the curve $y = (x^2 + 1)^{-1}$ from $x = 0$ to $x = 1$.

52. Find the centroid of the solid generated by revolving the region of Exercise 51 about: (a) the x-axis; (b) the y-axis.

▶ 53. Use a CAS to decompose into partial fractions.

(a) $\dfrac{6x^4 + 11x^3 - 2x^2 - 5x - 2}{x^2(x+1)^3}$.

(b) $-\dfrac{x^3 + 20x^2 + 4x + 93}{(x^2 + 4)(x^2 - 9)}$.

(c) $\dfrac{x^2 + 7x + 12}{x(x^2 + 2x + 4)}$.

▶ Exercises 54–55. Use a CAS to decompose the integrand into partial fractions. Use the decomposition to evaluate the integral.

54. $\displaystyle\int \frac{2x^6 - 13x^5 + 23x^4 - 15x^3 + 40x^2 - 24x + 9}{x^5 - 6x^4 + 9x^2} \, dx.$

55. $\displaystyle\int \frac{x^8 + 2x^7 + 7x^6 + 23x^5 + 10x^4 + 95x^3 - 19x^2 + 133x - 52}{x^6 + 2x^5 + 5x^4 - 16x^3 + 8x^2 + 32x - 48} \, dx.$

▶ 56. Use a CAS to calculate the integrals

$$\int \frac{1}{x^2 + 2x + n} \, dx, \quad n = 0, 1, 2.$$

Verify your results by differentiation.

▶ 57. Set

$$f(x) = \frac{x}{x^2 + 5x + 6}.$$

(a) Use a graphing utility to draw the graph of f.
(b) Calculate the area of the region that lies between the graph of f and the x-axis from $x = 0$ to $x = 4$.

58. (a) The region of Exercise 57 is revolved about the y-axis. Find the volume of the solid generated.
(b) Find the centroid of the solid described in part (a).

▶ 59. Set

$$f(x) = \frac{9 - x}{(x + 3)^2}.$$

(a) Use a graphing utility to draw the graph of f.
(b) Find the area of the region that lies between the graph of f and the x-axis from $x = -2$ to $x = 9$.

60. (a) The region of Exercise 59 is revolved about the x-axis. Find the volume of the solid generated.
(b) Find the centroid of the solid described in part (a).

■ *8.6 SOME RATIONALIZING SUBSTITUTIONS

There are integrands which are not rational functions but can be transformed into rational functions by a suitable substitution. Such substitutions are known as *rationalizing substitutions*.

First we consider integrals in which the integrand contains an expression of the form $\sqrt[n]{f(x)}$. In such cases, setting $u = \sqrt[n]{f(x)}$, which is equivalent to setting $u^n = f(x)$, is sometimes effective. The idea is to replace fractional exponents by integer exponents; integer exponents are usually easier to work with.

Example 1 Find $\displaystyle\int \frac{dx}{1 + \sqrt{x}}$.

SOLUTION To rationalize the integrand, we set

$$u^2 = x, \qquad 2u \, du = dx,$$

taking $u \geq 0$. Then $u = \sqrt{x}$ and

$$\int \frac{dx}{1 + \sqrt{x}} = \int \frac{2u}{1 + u} \, du = \int \left(2 - \frac{2}{1 + u} \right) du$$

$$\underset{\text{divide} \longrightarrow}{}$$

$$= 2u - 2 \ln(1 + u) + C$$

$$\underset{1 + u > 0 \longrightarrow}{}$$

$$= 2\sqrt{x} - 2 \ln(1 + \sqrt{x}) + C. \quad \square$$

Example 2 Find $\displaystyle\int \frac{dx}{\sqrt[3]{x} + \sqrt{x}}$.

SOLUTION Here the integrand contains two distinct roots, $\sqrt[3]{x}$ and $\sqrt{x}$. We can eliminate both radicals by setting

$$u^6 = x, \qquad 6u^5 du = dx,$$

taking $u > 0$. This substitution gives $\sqrt[3]{x} = u^2$ and $\sqrt{x} = u^3$. Therefore

$$\int \frac{dx}{\sqrt[3]{x} + \sqrt{x}} = \int \frac{6u^5}{u^2 + u^3} \, du = 6 \int \frac{u^3}{1 + u} \, du$$

$$\underset{\text{divide} \longrightarrow}{} = 6 \int \left(u^2 - u + 1 - \frac{1}{1+u} \right) du$$

$$\underset{1 + u > 0 \longrightarrow}{} = 6[\tfrac{1}{3}u^3 - \tfrac{1}{2}u^2 + u - \ln(1+u) + C]$$

$$= 2\sqrt{x} - 3\sqrt[3]{x} + 6\sqrt[6]{x} - 6\ln(1 + \sqrt[6]{x}) + C. \quad ❏$$

Example 3 Find $\displaystyle\int \sqrt{1 - e^x} \, dx$.

SOLUTION To rationalize the integrand, we set

$$u = \sqrt{1 - e^x}.$$

Then $0 \le u < 1$. To express dx in terms of u and du, we solve the equation for x:

$$u^2 = 1 - e^x, \qquad 1 - u^2 = e^x, \qquad \ln(1 - u^2) = x, \qquad -\frac{2u}{1 - u^2} \, du = dx.$$

The rest is straightforward:

$$\int \sqrt{1 - e^x} \, dx = \int u \left(-\frac{2u}{1 - u^2} \right) du$$

$$= \int \frac{2u^2}{u^2 - 1} \, du = \int \left(2 + \frac{1}{u - 1} - \frac{1}{u + 1} \right) du$$

$$\underset{\substack{\text{divide; then use} \\ \text{partial fractions}}}{}$$

$$= 2u + \ln|u - 1| - \ln|u + 1| + C$$

$$= 2u + \ln\left| \frac{u - 1}{u + 1} \right| + C$$

$$= 2\sqrt{1 - e^x} + \ln\left| \frac{\sqrt{1 - e^x} - 1}{\sqrt{1 - e^x} + 1} \right| + C. \quad ❏$$

Now we consider rational expressions in $\sin x$ and $\cos x$. Suppose, for example, that we want to calculate

$$\int \frac{1}{3\sin x - 4\cos x} \, dx.$$

We can convert the integrand into a rational function of u by setting

$$u \tan \tfrac{1}{2}x \qquad \text{taking } x \text{ between } -\pi \text{ and } \pi.$$

This gives

$$\cos \tfrac{1}{2}x = \frac{1}{\sec \tfrac{1}{2}x} = \frac{1}{\sqrt{1 + \tan^2 \tfrac{1}{2}x}} = \frac{1}{\sqrt{1 + u^2}}$$

and

$$\sin \tfrac{1}{2}x = \cos \tfrac{1}{2}x \, \tan \tfrac{1}{2}x = \frac{u}{\sqrt{1 + u^2}}.$$

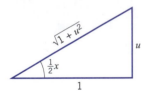

The right triangle illustrates these relations for $x \in (0, \pi)$.

Note that

$$\sin x = 2 \sin \tfrac{1}{2}x \cos \tfrac{1}{2}x = \frac{2u}{1 + u^2}$$

and

$$\cos x = \cos^2 \tfrac{1}{2}x - \sin^2 \tfrac{1}{2}x = \frac{1 - u^2}{1 + u^2}.$$

Since $u = \tan \tfrac{1}{2}x$ with $\tfrac{1}{2}x \in (-\tfrac{1}{2}\pi, \tfrac{1}{2}\pi)$, $\tfrac{1}{2}x = \arctan u$ and $x = 2 \arctan u$. Therefore

$$dx = \frac{2}{1 + u^2} \, du.$$

In summary, if the integrand is a rational expression in $\sin x$ and $\cos x$, then the substitution $u = \tan \tfrac{1}{2}x$, $x \in (-\pi, \pi)$, gives

$$\sin x = \frac{2u}{1 + u^2}, \qquad \cos x = \frac{1 - u^2}{1 + u^2}, \qquad dx = \frac{2}{1 + u^2} \, du$$

and converts the integrand into a rational function of u. The resulting integral can then be calculated by the methods of Section 8.5.

Example 4 Find $\displaystyle \int \frac{1}{3 \sin x - 4 \cos x} \, dx$.

SOLUTION Set $u = \tan \tfrac{1}{2}x$ with $x \in (-\pi, \pi)$. Then

$$\frac{1}{3 \sin x - 4 \cos x} = \frac{1}{[6u/(1 + u^2)] - [4(1 - u^2)/(1 + u^2)]} = \frac{1 + u^2}{4u^2 + 6u - 4}$$

and

$$\int \frac{1}{3 \sin x - 4 \cos x} \, dx = \int \frac{1 + u^2}{4u^2 + 6u - 4} \cdot \frac{2}{1 + u^2} \, du = \int \frac{1}{2u^2 + 3u - 2} \, du.$$

Since

$$\frac{1}{2u^2 + 3u - 2} = \frac{1}{(u + 2)(2u - 1)} = \frac{-1/5}{u + 2} + \frac{2/5}{2u - 1}, \qquad \text{(partial fractions)}$$

we have

$$\int \frac{1}{2u^2 + 3u - 2} \, du = -\frac{1}{5} \int \frac{1}{u + 2} \, du + \frac{2}{5} \int \frac{1}{2u - 1} \, du$$

$$= -\tfrac{1}{5} \ln |u + 2| + \tfrac{1}{5} \ln |2u - 1| + C$$

$$= -\tfrac{1}{5} \ln |\tan \tfrac{1}{2}x + 2| + \tfrac{1}{5} \ln |2 \tan \tfrac{1}{2}x - 1| + C. \quad \square$$

EXERCISES *8.6

Exercises 1–30. Calculate.

1. $\displaystyle\int \frac{dx}{1-\sqrt{x}}.$

2. $\displaystyle\int \frac{\sqrt{x}}{1+x}\,dx.$

3. $\displaystyle\int \sqrt{1+e^x}\,dx.$

4. $\displaystyle\int \frac{dx}{x(x^{1/3}-1)}.$

5. $\displaystyle\int x\sqrt{1+x}\,dx.$

6. $\displaystyle\int x^2\sqrt{1+x}\,dx.$

7. $\displaystyle\int (x+2)\sqrt{x-1}\,dx.$

8. $\displaystyle\int (x-1)\sqrt{x+2}\,dx.$

9. $\displaystyle\int \frac{x^3}{(1+x^2)^3}\,dx.$

10. $\displaystyle\int x(1+x)^{1/3}\,dx.$

11. $\displaystyle\int \frac{\sqrt{x}}{\sqrt{x}-1}\,dx.$

12. $\displaystyle\int \frac{x}{\sqrt{x}+1}\,dx.$

13. $\displaystyle\int \frac{\sqrt{x-1}+1}{\sqrt{x-1}-1}\,dx.$

14. $\displaystyle\int \frac{1-e^x}{1+e^x}\,dx.$

15. $\displaystyle\int \frac{dx}{\sqrt{1+e^x}}.$

16. $\displaystyle\int \frac{dx}{1+e^{-x}}.$

17. $\displaystyle\int \frac{x}{\sqrt{x+4}}\,dx.$

18. $\displaystyle\int \frac{x+1}{x\sqrt{x-2}}\,dx.$

19. $\displaystyle\int 2x^2(4x+1)^{-5/2}\,dx.$

20. $\displaystyle\int x^2\sqrt{x-1}\,dx.$

21. $\displaystyle\int \frac{x}{(ax+b)^{3/2}}\,dx.$

22. $\displaystyle\int \frac{x}{\sqrt{ax+b}}\,dx.$

23. $\displaystyle\int \frac{1}{1+\cos x-\sin x}\,dx.$

24. $\displaystyle\int \frac{1}{2+\cos x}\,dx.$

25. $\displaystyle\int \frac{1}{2+\sin x}\,dx.$

26. $\displaystyle\int \frac{\sin x}{1+\sin^2 x}\,dx.$

27. $\displaystyle\int \frac{1}{\sin x+\tan x}\,dx.$

28. $\displaystyle\int \frac{1}{1+\sin x+\cos x}\,dx.$

29. $\displaystyle\int \frac{1-\cos x}{1+\sin x}\,dx.$

30. $\displaystyle\int \frac{1}{5+3\sin x}\,dx.$

Exercises 31–36. Evaluate.

31. $\displaystyle\int_0^4 \frac{x^{3/2}}{x+1}\,dx.$

32. $\displaystyle\int_0^8 \frac{1}{1+\sqrt[3]{x}}\,dx.$

33. $\displaystyle\int_0^{\pi/2} \frac{\sin 2x}{2+\cos x}\,dx.$

34. $\displaystyle\int_0^{\pi/2} \frac{1}{1+\sin x}\,dx.$

35. $\displaystyle\int_0^{\pi/3} \frac{1}{\sin x-\cos x-1}\,dx.$ 36. $\displaystyle\int_0^1 \frac{\sqrt{x}}{1+\sqrt{x}}\,dx.$

37. Use the method of this section to show that
$$\int \sec x\,dx = \int \frac{1}{\cos x}\,dx = \ln\left|\frac{1+\tan\frac{1}{2}x}{1-\tan\frac{1}{2}x}\right|+C.$$

38. (a) Another way to calculate $\int \sec x\,dx$ is to write
$$\int \sec x\,dx = \int \frac{\cos x}{\cos^2 x}\,dx = \int \frac{\cos x}{1-\sin^2 x}\,dx.$$
Use the method of this section to show that
$$\int \sec x\,dx = \ln\sqrt{\frac{1+\sin x}{1-\sin x}}+C.$$

(b) Show that the result in part (a) is equivalent to the familiar formula
$$\int \sec x\,dx = \ln|\sec x+\tan x|+C.$$

39. (a) Use the approach given in Exercise 38 (a) to show that
$$\int \csc x\,dx = \ln\sqrt{\frac{1-\cos x}{1+\cos x}}+C.$$

(b) Show that the result in part (a) is equivalent to the formula
$$\int \csc x\,dx = \ln|\csc x-\cot x|+C.$$

40. The integral of a rational function of $\sinh x$ and $\cosh x$ can be transformed into a rational function of u by means of the substitution $u=\tanh\frac{1}{2}x$. Show that this substitution gives
$$\sinh x = \frac{2u}{1-u^2}, \quad \cosh x = \frac{1+u^2}{1-u^2}, \quad dx = \frac{2}{1-u^2}\,du.$$

Exercises 41–44. Integrate by setting $u=\tanh\frac{1}{2}x$.

41. $\displaystyle\int \operatorname{sech} x\,dx.$

42. $\displaystyle\int \frac{1}{1+\cosh x}\,dx.$

43. $\displaystyle\int \frac{1}{\sinh x+\cosh x}\,dx.$

44. $\displaystyle\int \frac{1-e^x}{1+e^x}\,dx.$

■ 8.7 NUMERICAL INTEGRATION

To evaluate a definite integral of a continuous function by the formula
$$\int_a^b f(x)\,dx = F(b)-F(a),$$

we must be able to find an antiderivative F and we must to able to evaluate this antiderivative both at a and at b. When this is not feasible, the method fails.

The method fails even for such simple-looking integrals as

$$\int_0^1 \sqrt{x}\,\sin x\,dx \qquad \text{or} \qquad \int_0^1 e^{-x^2}\,dx.$$

There are no *elementary functions* with derivative $\sqrt{x}\,\sin x$ or e^{-x^2}.

Here we take up some simple numerical methods for estimating definite integrals—methods that you can use whether or not you can find an antiderivative. All the methods we describe involve only simple arithmetic and are ideally suited to the computer.

We focus now on

$$\int_a^b f(x)\,dx.$$

We suppose that f is continuous on $[a, b]$ and, for pictorial convenience, assume that f is positive. Take a regular partition $P = \{x_0, x_1, x_2, \ldots, x_{n-1}, x_n\}$ of $[a, b]$, subdividing the interval into n subintervals each of length $(b - a)/n$:

$$[a, b] = [x_0, x_1] \cup \cdots \cup [x_{i-1}, x_i] \cup \cdots \cup [x_{n-1}, x_n]$$

with

$$\Delta x_i = \frac{b - a}{n}.$$

The region Ω_i pictured in Figure 8.7.1, can be approximated in many ways.

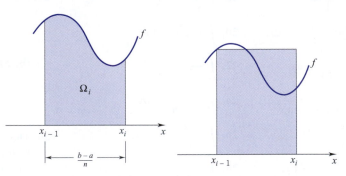

Figure 8.7.1   **Figure 8.7.2**

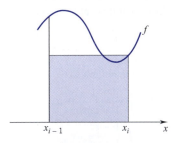

Figure 8.7.3

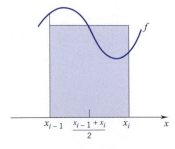

Figure 8.7.4

(1) By the left-endpoint rectangle (Figure 8.7.2):

$$\text{area} \cong f(x_{i-1})\Delta x_i = f(x_{i-1})\left(\frac{b - a}{n}\right).$$

(2) By the right-endpoint rectangle (Figure 8.7.3):

$$\text{area} \cong f(x_i)\Delta x_i = f(x_i)\left(\frac{b - a}{n}\right).$$

(3) By the midpoint rectangle (Figure 8.7.4):

$$\text{area} \cong f\left(\frac{x_{i-1} + x_i}{2}\right)\Delta x_i = f\left(\frac{x_{i-1} + x_i}{2}\right)\left(\frac{b - a}{n}\right).$$

(4) By a trapezoid (Figure 8.7.5):

$$\text{area} \cong \frac{1}{2}[f(x_{i-1}) + f(x_i)]\Delta x_i = \frac{1}{2}[f(x_{i-1}) + f(x_i)]\left(\frac{b - a}{n}\right).$$

(5) By a parabolic region (Figure 8.7.6): take the parabola $y = Ax^2 + Bx + C$ that passes through the three points indicated.

$$\text{area} \cong \frac{1}{6}\left[f(x_{i-1}) + 4f\left(\frac{x_{i-1} + x_i}{2}\right) + f(x_i)\right]\Delta x_i$$

$$= \left[f(x_{i-1}) + 4f\left(\frac{x_{i-1} + x_i}{2}\right) + f(x_i)\right]\left(\frac{b-a}{6n}\right).$$

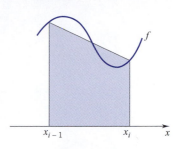

Figure 8.7.5

You can verify this formula for the area under the parabola by doing Exercises 11 and 12. (If the three points are collinear, the parabola degenerates to a straight line and the parabolic region becomes a trapezoid. The formula then gives the area of that trapezoid.)

The approximations to Ω_i just considered yield the following estimates for

$$\int_a^b f(x)\,dx.$$

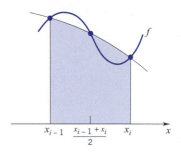

Figure 8.7.6

(1) The left-endpoint estimate:

$$L_n = \frac{b-a}{n}[f(x_0) + f(x_1) + \cdots + f(x_{n-1})].$$

(2) The right-endpoint estimate:

$$R_n = \frac{b-a}{n}[f(x_1) + f(x_2) + \cdots + f(x_n)].$$

(3) The midpoint estimate:

$$M_n = \frac{b-a}{n}\left[f\left(\frac{x_0 + x_1}{2}\right) + \cdots + f\left(\frac{x_{n-1} + x_n}{2}\right)\right].$$

(4) The trapezoidal estimate (*trapezoidal rule*):

$$T_n = \frac{b-a}{n}\left[\frac{f(x_0) + f(x_1)}{2} + \frac{f(x_1) + f(x_2)}{2} + \cdots + \frac{f(x_{n-1}) + f(x_n)}{2}\right]$$

$$= \frac{b-a}{2n}[f(x_0) + 2f(x_1) + \cdots + 2f(x_{n-1}) + f(x_n)].$$

(5) The parabolic estimate (*Simpson's rule*):

$$S_n = \frac{b-a}{6n}\left\{f(x_0) + f(x_n) + 2[f(x_1) + \cdots + f(x_{n-1})]\right.$$

$$\left. + 4\left[f\left(\frac{x_0 + x_1}{2}\right) + \cdots + f\left(\frac{x_{n-1} + x_n}{2}\right)\right]\right\}.$$

The first three estimates, L_n, R_n, M_n, are Riemann sums (Section 5.2); T_n and S_n, although not explicitly written as Riemann sums, can be written as Riemann sums. (Exercise 26.) It follows from (5.2.6) that any one of these estimates can be used to approximate the integral as closely as we may wish. All we have to do is take n sufficiently large.

As an example, we will find the approximate value of

$$\ln 2 = \int_1^2 \frac{dx}{x}$$

by applying each of the five estimates. Here

$$f(x) = \frac{1}{x}, \qquad [a, b] = [1, 2].$$

Taking $n = 5$, we have

$$\frac{b-a}{n} = \frac{2-1}{5} = \frac{1}{5}.$$

The partition points are

$$x_0 = \tfrac{5}{5}, \quad x_1 = \tfrac{6}{5}, \quad x_2 = \tfrac{7}{5}, \quad x_3 = \tfrac{8}{5}, \quad x_4 = \tfrac{9}{5}, \quad x_5 = \tfrac{10}{5}. \qquad \text{(Figure 8.7.7)}$$

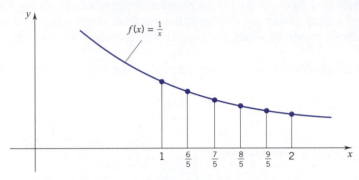

Figure 8.7.7

Using a calculator and rounding off to four decimal places, we have the following estimates:

$$L_5 = \tfrac{1}{5}\left(\tfrac{5}{5} + \tfrac{5}{6} + \tfrac{5}{7} + \tfrac{5}{8} + \tfrac{5}{9}\right) = \left(\tfrac{1}{5} + \tfrac{1}{6} + \tfrac{1}{7} + \tfrac{1}{8} + \tfrac{1}{9}\right) \cong 0.7456.$$

$$R_5 = \tfrac{1}{5}\left(\tfrac{5}{6} + \tfrac{5}{7} + \tfrac{5}{8} + \tfrac{5}{9} + \tfrac{5}{10}\right) = \left(\tfrac{1}{6} + \tfrac{1}{7} + \tfrac{1}{8} + \tfrac{1}{9} + \tfrac{1}{10}\right) \cong 0.6456.$$

$$M_5 = \tfrac{1}{5}\left(\tfrac{10}{11} + \tfrac{10}{13} + \tfrac{10}{15} + \tfrac{10}{17} + \tfrac{10}{19}\right) = 2\left(\tfrac{1}{11} + \tfrac{1}{13} + \tfrac{1}{15} + \tfrac{1}{17} + \tfrac{1}{19}\right) \cong 0.6919.$$

$$T_5 = \tfrac{1}{10}\left(\tfrac{5}{5} + \tfrac{10}{6} + \tfrac{10}{7} + \tfrac{10}{8} + \tfrac{10}{9} + \tfrac{5}{10}\right) = \left(\tfrac{1}{10} + \tfrac{1}{6} + \tfrac{1}{7} + \tfrac{1}{8} + \tfrac{1}{9} + \tfrac{1}{20}\right) \cong 0.6956.$$

$$S_5 = \tfrac{1}{30}\left[\tfrac{5}{5} + \tfrac{5}{10} + 2\left(\tfrac{5}{6} + \tfrac{5}{7} + \tfrac{5}{8} + \tfrac{5}{9}\right) + 4\left(\tfrac{10}{11} + \tfrac{10}{13} + \tfrac{10}{15} + \tfrac{10}{17} + \tfrac{10}{19}\right)\right] \cong 0.6932.$$

Since the integrand $1/x$ decreases throughout the interval $[1, 2]$, you can expect the left-endpoint estimate, 0.7456, to be too large, and you can expect the right-endpoint estimate, 0.6456, to be too small. The other estimates should be better.

The value of $\ln 2$ given on a calculator is $\ln 2 \cong 0.69314718$, which is 0.6931 rounded off to four decimal places. Thus S_5 is correct to the nearest thousandth.

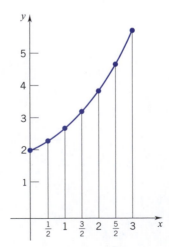

Figure 8.7.8

Example 1 Find the approximate value of $\displaystyle\int_0^3 \sqrt{4 + x^3}\,dx$ by the trapezoidal rule. Take $n = 6$.

SOLUTION Each subinterval has length $\dfrac{b-a}{n} = \dfrac{3-0}{6} = \dfrac{1}{2}$. The partition points are

$$x_0 = 0, \quad x_1 = \tfrac{1}{2}, \quad x_2 = 1, \quad x_3 = \tfrac{3}{2}, \quad x_4 = 2, \quad x_5 = \tfrac{5}{2}, \quad x_6 = 3. \quad \text{(Figure 8.7.8)}$$

Then

$$T_6 \cong \tfrac{1}{4}\left[f(0) + 2f\left(\tfrac{1}{2}\right) + 2f(1) + 2f\left(\tfrac{3}{2}\right) + 2f(2) + 2f\left(\tfrac{5}{2}\right) + f(3)\right],$$

with $f(x) = \sqrt{4 + x^3}$. Using a calculator and rounding off to three decimal places, we have

$$f(0) = 2.000, \qquad f\left(\tfrac{1}{2}\right) \cong 2.031, \qquad f(1) \cong 2.236, \qquad f\left(\tfrac{3}{2}\right) \cong 2.716,$$

$$f(2) \cong 3.464, \qquad f\left(\tfrac{5}{2}\right) \cong 4.430, \qquad f(3) \cong 5.568.$$

Thus

$$T_6 \cong \tfrac{1}{4}(2.000 + 4.062 + 4.472 + 5.432 + 6.928 + 8.860 + 5.568) \cong 9.331. \quad \square$$

Example 2 Find the approximate value of

$$\int_0^3 \sqrt{4 + x^3}\, dx$$

by Simpson's rule. Take $n = 3$.

SOLUTION There are three subintervals, each of length

$$\frac{b - a}{n} = \frac{3 - 0}{3} = 1.$$

Here

$$x_0 = 0, \qquad x_1 = 1, \qquad x_2 = 2, \qquad x_3 = 3,$$

$$\frac{x_0 + x_1}{2} = \frac{1}{2}, \qquad \frac{x_1 + x_2}{2} = \frac{3}{2}, \qquad \frac{x_2 + x_3}{2} = \frac{5}{2}.$$

Simpson's rule yields

$$S_3 = \tfrac{1}{6}\left[f(0) + f(3) + 2f(1) + 2f(2) + 4f\left(\tfrac{1}{2}\right) + 4f\left(\tfrac{3}{2}\right) + 4f\left(\tfrac{5}{2}\right) \right],$$

with $f(x) = \sqrt{4 + x^3}$. Taking the values of f as calculated in Example 1, we have

$$S_3 = \tfrac{1}{6}(2.000 + 5.568 + 4.472 + 6.928 + 8.124 + 10.864 + 17.72) \cong 9.279.$$

For comparison, the value of this integral accurate to five decimal places is 9.27972. $\square$

Error Estimates

A numerical estimate is useful only to the extent that we can gauge its accuracy. When we use any kind of approximation method, we face two forms of error: the error inherent in the method we use (we call this the *theoretical error*) and the error that accumulates from rounding off the decimals that arise during the course of computation (we call this the *round-off error*). The effect of round-off error is obvious: if at each step we round off too crudely, then we can hardly expect an accurate final result. We will examine theoretical error.

We begin with a function f continuous and increasing on $[a, b]$. We subdivide $[a, b]$ into n nonoverlapping intervals, each of length $(b - a)/n$. We want to estimate

$$\int_a^b f(x)\, dx$$

by the left-endpoint method. What is the theoretical error? It should be clear from Figure 8.7.9 that the theoretical error does not exceed

$$[f(b) - f(a)]\left(\frac{b - a}{n}\right).$$

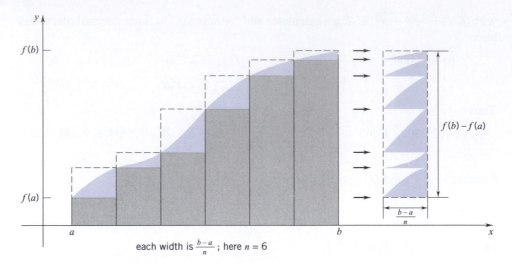

each width is $\frac{b-a}{n}$; here $n = 6$

Figure 8.7.9

The error is represented by the sum of the areas of the shaded regions. These regions, when shifted to the right, all fit together within a rectangle of height $f(b) - f(a)$ and base $(b-a)/n$.

Similar reasoning shows that, under the same circumstances, the theoretical error associated with the trapezoidal method does not exceed

(∗)
$$\tfrac{1}{2}[f(b) - f(a)]\left(\frac{b-a}{n}\right).$$

In this setting, at least, the trapezoidal estimate does a better job than the left-endpoint estimate.

The trapezoidal rule is more accurate than (∗) suggests. As is shown in texts on numerical analysis, if f is continuous on $[a, b]$ and twice differentiable on (a, b), then the theoretical error of the trapezoidal rule,

$$E_n^T = \int_a^b f(x)\,dx - T_n,$$

can be written

(8.7.1)
$$E_n^T = -\frac{(b-a)^3}{12\,n^2}f''(c),$$

where c is some number between a and b. Usually we cannot pinpoint c any further. However, if f'' is bounded on $[a, b]$, say $|f''(x)| \leq M$ for $a \leq x \leq b$, then

(8.7.2)
$$|E_n^T| \leq \frac{(b-a)^3}{12\,n^2}M.$$

Recall the trapezoidal-rule estimate of ln 2 derived at the beginning of this section:

$$\ln 2 = \int_1^2 \frac{1}{x}\,dx \cong 0.696.$$

To find the theoretical error, we apply (8.72). Here

$$f(x) = \frac{1}{x}, \qquad f'(x) = -\frac{1}{x^2}, \qquad f''(x) = \frac{2}{x^3}.$$

Since f'' is a decreasing function, it takes on its maximum value at the left endpoint of the interval. Thus $|f''(x)| \le f''(1) = 2$ on $[1, 2]$. Therefore, with $a = 1$, $b = 2$, and $n = 5$, we have

$$|E_5^T| \le \frac{(2-1)^3}{12 \cdot 5^2} \cdot 2 = \frac{1}{150} < 0.007.$$

The estimate 0.696 is in theoretical error by less than 0.007.

To get an estimate for

$$\ln 2 = \int_1^2 \frac{1}{x}\, dx$$

that is accurate to four decimal places, we need

(1)
$$\frac{(b-a)^3}{12n^2} M < 0.00005.$$

Since

$$\frac{(b-a)^3}{12\,n^2} M \le \frac{(2-1)^3}{12\,n^2} \cdot 2 = \frac{1}{6\,n^2},$$

we can satisfy (1) by having

$$\frac{1}{6n^2} < 0.00005,$$

which is to say, by having

$$n^2 > 3333.$$

As you can check, $n = 58$ is the smallest integer that satisfies this inequality. In this case, the trapezoidal rule requires a regular partition with at least 58 subintervals to guarantee four decimal place accuracy.

Simpson's rule is more effective than the trapezoidal rule. If f is continuous on $[a, b]$ and $f^{(4)}$ exists on (a, b), then the theoretical error for Simpson's rule,

$$E_n^S = \int_a^b f(x)\, dx - S_n,$$

can be written

(8.7.3)
$$E_n^S = -\frac{(b-a)^5}{2880n^4} f^{(4)}(c),$$

where, as before, c is some number between a and b. Whereas (8.7.1) varies as $1/n^2$, this quantity varies as $1/n^4$. Thus, for comparable n, we can expect greater accuracy from Simpson's rule. In addition, if we assume that $f^{(4)}(x)$ is bounded on $[a, b]$, say $|f^{(4)}(x)| \le M$ for $a \le x \le b$, then

(8.7.4)
$$|E_n^S| \le \frac{(b-a)^5}{2880n^4} M.$$

To estimate

$$\ln 2 = \int_1^2 \frac{1}{x}\, dx$$

by the trapezoidal rule with theoretical error less than 0.00005, we needed to subdivide the interval [1, 2] into at least fifty-eight subintervals of equal length. To achieve the same degree of accuracy with Simpson's rule, we need only four subintervals:

$$\text{for } f(x) = 1/x, \qquad f^{(4)}(x) = 24/x^4.$$

Therefore $|f^{(4)}(x)| \le 24$ for all $x \in [1, 2]$ and

$$|E_n^S| \le \frac{(2-1)^5}{2880\, n^4}\, 24 = \frac{1}{120\, n^4}.$$

This quantity is less than 0.00005 provided only that $n^4 > 167$. This condition is met by $n = 4$.

Remark If your calculus course is computer related, you will no doubt see these and other numerical methods more thoroughly applied. ❏

EXERCISES 8.7

▶ In Exercises 1–10 round off your calculations to four decimal places.

1. Estimate

$$\int_0^{12} x^2\, dx$$

by: (a) the left-cndpoint estimate, $n = 12$; (b) the right-endpoint estimate, $n = 12$; (c) the midpoint estimate, $n = 6$; (d) the trapezoidal rule, $n = 12$; (e) Simpson's rule, $n = 6$. Check your results by performing the integration.

2. Estimate

$$\int_0^1 \sin^2 \pi x\, dx$$

by: (a) the midpoint estimate, $n = 3$; (b) the trapezoidal rule, $n = 6$; (c) Simpson's rule, $n = 3$. Check your results by performing the integration.

3. Estimate

$$\int_0^3 \frac{dx}{1+x^3}$$

by: (a) the left-endpoint estimate, $n = 6$; (b) the right-endpoint estimate, $n = 6$; (c) the midpoint estimate, $n = 3$; (d) the trapezoidal rule, $n = 6$; (e) Simpson's rule, $n = 3$.

4. Estimate

$$\int_0^\pi \frac{\sin x}{\pi + x}\, dx$$

by: (a) the trapezoidal rule, $n = 6$; (b) Simpson's rule, $n = 3$. (Note the superiority of Simpson's rule.)

5. Estimate the value of π by estimating the integral

$$\int_0^1 \frac{dx}{1+x^2} = \arctan 1 = \frac{\pi}{4}$$

by (a) the trapezoidal rule, $n = 6$; (b) Simpson's rule, $n = 3$.

6. Estimate

$$\int_0^2 \frac{dx}{\sqrt{4+x^3}}$$

by: (a) the trapezoidal rule, $n = 4$; (b) Simpson's rule, $n = 2$.

7. Estimate

$$\int_{-1}^1 \cos x^2\, dx$$

by: (a) the midpoint estimate, $n = 4$; (b) the trapezoidal rule, $n = 8$; (c) Simpson's rule, $n = 4$.

8. Estimate

$$\int_1^2 \frac{e^x}{x}\, dx$$

by: (a) the midpoint estimate, $n = 4$; (b) the trapezoidal rule, $n = 8$; (c) Simpson's rule, $n = 4$.

9. Estimate

$$\int_0^2 e^{-x^2}\, dx$$

by: (a) the trapezoidal rule, $n = 10$; (b) Simpson's rule, $n = 5$.

10. Estimate

$$\int_2^4 \frac{1}{\ln x}\, dx$$

by: (a) the midpoint estimate, $n = 4$; (b) the trapezoidal rule, $n = 8$; (c) Simpson's rule, $n = 4$.

11. Show that there is a unique parabola of the form $y = Ax^2 + Bx + C$ through three distinct noncollinear points with different x-coordinates.

12. Show that the function $g(x) = Ax^2 + Bx + C$ satisfies the condition

$$\int_a^b g(x)\, dx = \frac{b-a}{6}\left[g(a) + 4g\left(\frac{a+b}{2}\right) + g(b)\right]$$

for every interval $[a, b]$.

▷ **Exercises 13–22.** Determine the values of n which guarantee a theoretical error less than ϵ if the integral is estimated by: (a) the trapezoidal rule; (b) Simpson's rule.

13. $\int_1^4 \sqrt{x}\, dx$; $\epsilon = 0.01$. **14.** $\int_1^3 x^5\, dx$; $\epsilon = 0.01$.

15. $\int_1^4 \sqrt{x}\, dx$; $\epsilon = 0.00001$. **16.** $\int_1^3 x^5\, dx$; $\epsilon = 0.00001$.

17. $\int_0^\pi \sin x\, dx$; $\epsilon = 0.001$. **18.** $\int_0^\pi \cos x\, dx$; $\epsilon = 0.001$.

19. $\int_1^3 e^x\, dx$; $\epsilon = 0.01$. **20.** $\int_1^e \ln x\, dx$; $\epsilon = 0.01$.

21. $\int_0^2 e^{-x^2}\, dx$; $\epsilon = 0.0001$. **22.** $\int_0^2 e^x\, dx$; $\epsilon = 0.00001$.

23. Show that Simpson's rule is exact (theoretical error zero) for every polynomial of degree 3 or less.

24. Show that the trapezoidal rule is exact (theoretical error zero) if f is linear.

25. (a) Set $f(x) = x^2$. Let $[a, b] = [0, 1]$ and take $n = 2$. Show that in this case the theoretical error inequality

$$|E_n^T| \le \frac{(b-a)^3}{12n^2} M$$

is replaced by equality if M is taken as the maximum value of f'' on $[a, b]$.

(b) Set $f(x) = x^4$. Let $[a, b] = [0, 1]$ and take $n = 1$. Show that in this case the theoretical error inequality

$$|E_n^S| \le \frac{(b-a)^5}{2880n^4} M$$

is replaced by equality if M is taken as the maximum value of $f^{(4)}$ on $[a, b]$.

26. Show that, if f is continuous, then T_n and S_n can both be written as Riemann sums.

27. Let f be a function positive on $[a, b]$. Compare

$$M_n, \quad T_n, \quad \int_a^b f(x)\, dx$$

given that the graph of f is: (a) concave up; (b) concave down.

28. Show that $\frac{1}{3}T_n + \frac{2}{3}M_n = S_n$.

▷ **29.** Use a CAS and the trapezoidal rule to estimate:

(a) $\int_0^{10} (x + \cos x)\, dx$, $n = 50$.

(b) $\int_{-4}^7 (x^5 - 5x^4 + x^3 - 3x^2 - x + 4)\, dx$, $n = 30$.

▷ **30.** Use a CAS and Simpson's rule to estimate:

(a) $\int_{-4}^3 \frac{x^2}{x^2 + 4}\, dx$, $n = 50$.

(b) $\int_0^{\pi/6} (x + \tan x)\, dx$, $n = 25$.

31. Estimate the theoretical error if Simpson's rule with $n = 20$ is used to approximate

$$\int_1^5 \frac{x^2 - 4}{x^2 + 9}\, dx.$$

32. Estimate the theoretical error if the trapezoidal rule with $n = 30$ is used to approximate

$$\int_2^7 \frac{x^2}{x^2 + 1}\, dx.$$

■ **CHAPTER 8. REVIEW EXERCISES**

Exercises 1–40. Calculate.

1. $\int \frac{\cos x}{4 + \sin^2 x}\, dx$. **2.** $\int_0^{\pi/4} \frac{x^2}{1 + x^2}\, dx$.

3. $\int 2x \sinh x\, dx$. **4.** $\int (\tan x + \cot x)^2\, dx$.

5. $\int \frac{x-3}{x^2(x+1)}\, dx$. **6.** $\int x \arctan x\, dx$.

7. $\int \sin 2x \cos x\, dx$. **8.** $\int 3x\, e^{-3x}\, dx$.

9. $\int_0^3 \ln\sqrt{x+1}\, dx$. **10.** $\int \frac{2}{x(1+x^2)}\, dx$.

11. $\int \frac{\sin^3 x}{\cos x}\, dx$. **12.** $\int \frac{\cos x}{\sin^3 x}\, dx$.

13. $\int_0^1 e^{-x} \cosh x\, dx$. **14.** $\int \frac{x^2 + 3}{\sqrt{x^2 + 9}}\, dx$.

15. $\int \frac{dx}{e^x - 4e^{-x}}$. **16.** $\int \frac{1}{x^3 - 1}\, dx$,

17. $\displaystyle\int x2^x \, dx.$

18. $\displaystyle\int \ln(x\sqrt{x}) \, dx.$

19. $\displaystyle\int \frac{\sqrt{a^2 - x^2}}{x^2} \, dx.$

20. $\displaystyle\int_0^2 x^2 e^{x^3} \, dx.$

21. $\displaystyle\int x^3 e^{x^2} \, dx.$

22. $\displaystyle\int \sin 2x \sin 3x \, dx.$

23. $\displaystyle\int \frac{\sin^5 x}{\cos^7 x} \, dx.$

24. $\displaystyle\int \left(\frac{\sqrt{4 + x^2}}{x} - \frac{x}{\sqrt{4 + x^2}} \right) dx.$

25. $\displaystyle\int_{\pi/6}^{\pi/3} \frac{\sin x}{\sin 2x} \, dx.$

26. $\displaystyle\int \frac{x + 3}{\sqrt{x^2 + 2x - 8}} \, dx.$

27. $\displaystyle\int \frac{x^2 + x}{\sqrt{1 - x^2}} \, dx.$

28. $\displaystyle\int x \tan^2 2x \, dx.$

29. $\displaystyle\int \frac{\cos^4 x}{\sin^2 x} \, dx.$

30. $\displaystyle\int_0^3 x \ln\sqrt{x^2 + 1} \, dx.$

31. $\displaystyle\int (\sin 2x + \cos 2x)^2 \, dx.$

32. $\displaystyle\int \sqrt{\cos x} \sin^3 x \, dx.$

33. $\displaystyle\int \frac{5x + 1}{(x + 2)(x^2 - 2x + 1)} \, dx.$

34. $\displaystyle\int \tan^{3/2} x \sec^4 x \, dx.$

35. $\displaystyle\int \frac{1}{\sqrt{x + 1} - \sqrt{x}} \, dx.$

36. $\displaystyle\int_0^{1/2} \cos \pi x \cos \tfrac{1}{2} \pi x \, dx.$

37. $\displaystyle\int x^2 \cos 2x \, dx.$

38. $\displaystyle\int e^{2x} \sin 4x \, dx.$

39. $\displaystyle\int \frac{1 - \sin 2x}{1 + \sin 2x} \, dx.$

40. $\displaystyle\int \frac{5x + 3}{(x - 1)(x^2 + 2x + 5)} \, dx.$

41. Show that, for $a \neq 0$,

(a) $\displaystyle\int x^n \cos ax \, dx = \frac{x^n \sin ax}{a} - \frac{n}{a} \int x^{n-1} \sin ax \, dx,$

(b) $\displaystyle\int x^n \sin ax \, dx = -\frac{x^n \cos ax}{a} + \frac{n}{a} \int x^{n-1} \cos ax \, dx.$

42. Use the formulas in Exercise 41 to calculate

(a) $\displaystyle\int x^2 \cos 3x \, dx.$

(b) $\displaystyle\int x^3 \sin 4x \, dx.$

43. (a) Show that

$$\int x^m (\ln x)^n \, dx = \frac{x^{m+1}(\ln x)^n}{m + 1} - \frac{n}{m + 1}$$

$$\times \int x^m (\ln x)^{n-1} \, dx.$$

(b) Calculate $\displaystyle\int x^4 (\ln x)^3 \, dx.$

44. Calculate the area of the region between the curve $y = x^2 \arctan x$ and the x-axis from $x = 0$ to $x = 1$.

45. Find the centroid of the region bounded by the graph of $y = (1 - x^2)^{-1/2}$ and the x-axis, $x \in \left[0, \tfrac{1}{2}\right]$.

46. Let $f(x) = x + \sin x$ and $g(x) = x$ both for $x \in [0, \pi]$.

(a) Sketch the graphs of f and g in the same coordinate system.

(b) Calculate the area of the region Ω between the graphs of f and g.

(c) Calculate the centroid of Ω.

47. (a) Find the volume of the solid generated by revolving about the x-axis the region Ω of Exercise 45.

(b) Find the volume of the solid generated by revolving about the y-axis the region Ω of Exercise 45.

48. The region between the curve $y = \ln 2x$ and the x-axis from $x = 1$ to $x = e$ is revolved about the y-axis. Find the volume of the solid generated.

49. Estimate $\displaystyle\int_0^2 \sqrt{x^3 + x} \, dx$ by: (a) the midpoint estimate, $n = 4$; (b) the trapezoidal rule, $n = 8$; (c) Simpson's rule, $n = 4$. Round off your calculations to four decimal places.

50. Estimate $\displaystyle\int_0^2 \sqrt{1 + 3x} \, dx$ by: (a) the trapezoidal rule, $n = 8$; (b) Simpson's rule, $n = 4$. Round off your calculations to four decimal places. (c) Find the exact value of the integral and compare it to your results in (a) and (b).

51. For the integral of Exercise 50, determine the values of n which guarantee a theoretical error of less than 0.00001 if the integral is estimated by: (a) the trapezoidal rule; (b) Simpson's rule.

52. Estimate $\displaystyle\int_1^3 \frac{e^{-x}}{x} \, dx$ by: (a) the trapezoidal rule, $n = 8$; (c) Simpson's rule, $n = 4$. Round off your calculations to four decimal places.

CHAPTER 9

SOME DIFFERENTIAL EQUATIONS

Introduction

An equation that relates an unknown function to one or more of its derivatives is called a *differential equation*. We have already introduced some differential equations. In Chapter 7 we used the differential equation

(1)
$$\frac{dy}{dt} = ky \qquad \text{[there written } f'(t) = kf(t)\text{]}$$

to model exponential growth and decay. In various exercises (Section 3.6 and 4.9) we used the differential equation

(2)
$$\frac{d^2y}{dt^2} + \omega^2 y = 0,$$

the equation of simple harmonic motion, to model the motion of a simple pendulum and the oscillation of a weight suspended at the end of a spring.

The *order* of a differential equation is the order of the highest derivative that appears in the equation. Thus (1) is a *first-order* equation and (2) is a *second-order* equation.

A function that satisfies a differential equation is called a *solution* of the equation. Finding the solutions of a differential equation is called *solving* the equation.

All functions $y = Ce^{kt}$ where C is a constant are solutions of equation (1):

$$\frac{dy}{dt} = kCe^{kt} = ky.$$

All functions of the form $y = C_1 \cos \omega t + C_2 \sin \omega t$, where C_1 and C_2 constants are solutions of equation (2):

$$y = C_1 \cos \omega t + C_2 \sin \omega t$$

$$\frac{dy}{dt} = -\omega C_1 \sin \omega t + \omega C_2 \cos \omega t$$

$$\frac{d^2y}{dt^2} = -\omega^2 C_1 \cos \omega t - \omega^2 C_2 \sin \omega t = -\omega^2 y$$

and therefore

$$\frac{d^2 y}{dt^2} + \omega^2 y = 0.$$

Remark Differential equations reach far beyond the boundaries of pure mathematics. Countless processes in the physical sciences, in the life sciences, in engineering, and in the social sciences are modeled by differential equations.

The study of differential equations is a huge subject, certainly beyond the scope of this text or any text on calculus. In this little chapter we examine some simple, but useful, differential equations. We continue the study of differential equations in Chapter 19.

One more point. In (1) and (2) we used the letter t to indicate the independent variable because we're looking at changes with respect to time. In much of what follows, we'll use the letter x. Whether we use x or t doesn't matter. What matters is the structure of the equation ❑

■ 9.1 FIRST-ORDER LINEAR EQUATIONS

A differential equation of the form

(9.1.1)

$$y' + p(x)y = q(x)$$

is called a *first-order linear differential equation*. Here p and q are given functions defined and continuous on some interval I.

[In the simplest case, $p(x) = 0$ for all x, the equation reduces to

$$y' = q(x).$$

The solutions of this equation are the antiderivatives of q.]

Solving Equations $y' + p(x)y = q(x)$ First we calculate

$$H(x) = \int p(x)\, dx,$$

omitting the constant of integration. (We want one antiderivative for p, not a whole collection of them.) We form $e^{H(x)}$, multiply the equation by $e^{H(x)}$, and obtain

$$e^{H(x)}y' + e^{H(x)}p(x)y = e^{H(x)}q(x).$$

The left side of this equation is the derivative of $e^{H(x)}y$. (Verify this.) Thus, we have

$$\frac{d}{dx}\left[e^{H(x)}y\right] = e^{H(x)}q(x).$$

Integration gives

$$e^{H(x)}y = \int e^{H(x)}q(x)\, dx + C$$

and yields

(9.1.2)

$$y(x) = e^{-H(x)}\left[\int e^{H(x)}q(x)\, dx + C\right]. \qquad ❑$$

Remark   There is no reason to commit this formula to memory. What is important here is the method of solution. The key step is multiplication by $e^{H(x)}$, where $H(x) = \int p(x)\,dx$. It is multiplication by this factor, called an *integrating factor*, that enables us to write the left side in a form that we can integrate directly.   ❏

Note that (9.1.2) contains an arbitrary constant C. A close look at the steps taken to obtain this solution makes it clear that that this solution includes *all* the functions that satisfy the differential equation. For this reason we call it the *general solution*. By assigning a particular value to the constant C, we obtain what is called a *particular solution*.

Example 1   Find the general solution of the equation

$$y' + ay = b, \qquad a, b \text{ constants,} \qquad a \neq 0.$$

SOLUTION   First we calculate an integrating factor:

$$H(x) = \int a\,dx = ax \qquad \text{and therefore} \qquad e^{H(x)} = e^{ax}.$$

Multiplying the differential equation by e^{ax}, we get

$$e^{ax}y' + a\,e^{ax}y = b\,e^{ax}.$$

The left-hand side is the derivative of $e^{ax}y$. (Verify this.) Thus, we have

$$\frac{d}{dx}[e^{ax}y] = b\,e^{ax}.$$

We integrate this equation and find that

$$e^{ax}y = \frac{b}{a}e^{ax} + C.$$

It follows that

$$y = \frac{b}{a} + Ce^{-ax}.$$

This is the general solution.   ❏

Example 2   Find the general solution of the equation

$$y' + 2xy = x$$

and then find the particular solution y for which $y(0) = 2$.

SOLUTION   This equation has the form (9.1.1). To solve the equation, we calculate the integrating factor $e^{H(x)}$:

$$H(x) = \int 2x\,dx = x^2 \qquad \text{and so} \qquad e^{H(x)} = e^{x^2}.$$

Multiplication by e^{x^2} gives

$$e^{x^2}y' + 2xe^{x^2}y = xe^{x^2}$$

$$\frac{d}{dx}\left[e^{x^2}y\right] = xe^{x^2}.$$

Integrating this equation, we get

$$e^{x^2}y = \tfrac{1}{2}e^{x^2} + C,$$

which we write as

$$y = \tfrac{1}{2} + Ce^{-x^2}.$$

This is the general solution. To find the solution y for which $y(0) = 2$, we set $x = 0$, $y = 2$ and solve for C:

$$2 = \tfrac{1}{2} + Ce^0 = \tfrac{1}{2} + C \qquad \text{and so} \qquad C = \tfrac{3}{2}.$$

The function

$$y = \tfrac{1}{2} + \tfrac{3}{2}e^{-x^2}$$

is the particular solution that satisfies the given condition. ❏

Remark When a differential equation is used as a mathematical model in some application, there is usually an *initial condition* $y(x_0) = y_0$ that makes it possible to evaluate the arbitrary constant that appears in the general solution. The problem of finding a particular solution that satisfies a given condition is called an *initial-value problem.* ❏

Example 3 Solve the initial-value problem:

$$xy' - 2y = 3x^4, \qquad y(-1) = 2.$$

SOLUTION This differential equation does not have the form of (9.1.1), but we can put it in that form by dividing the equation by x:

$$y' - \frac{2}{x}y = 3x^3.$$

Now we set

$$H(x) = \int -\frac{2}{x}\,dx = -2\ln x = \ln x^{-2}$$

and get the integrating factor

$$e^{H(x)} = e^{\ln x^{-2}} = x^{-2}.$$

Multiplication by x^{-2} gives

$$x^{-2}y' - 2x^{-3}y = 3x$$

$$\frac{d}{dx}[x^{-2}y] = 3x.$$

Integrating this equation, we get

$$x^{-2}y = \tfrac{3}{2}x^2 + C,$$

which we write as

$$y = \tfrac{3}{2}x^4 + Cx^2.$$

This is the general solution. Applying the initial condition, we have

$$y(-1) = 2 = \tfrac{3}{2}(-1)^4 + C(-1)^2 = \tfrac{3}{2} + C.$$

This gives $C = \tfrac{1}{2}$. The function $y = \tfrac{3}{2}x^4 + \tfrac{1}{2}x^2$ is the solution of the initial-value problem. (Check this out.) ❏

Applications

Newton's Law of Cooling Newton's law of cooling states that the rate of change of the temperature T of an object is proportional to the difference between T and the (assumed constant) temperature τ of the surrounding medium, called the *ambient temperature*. The mathematical formulation of Newton's law takes the form

$$\frac{dT}{dt} = m(T - \tau) \qquad \text{where } m \text{ is a constant.}$$

Remark The constant m in this model must be negative; for if the object is warmer than the ambient temperature $(T - \tau > 0)$, then its temperature will decrease $(dT/dt < 0)$, which implies $m < 0$; if the object is colder than the ambient temperature $(T - \tau < 0)$, its temperature will increase $(dT/dt > 0)$, which again implies $m < 0$. ❏

To emphasize that the constant of proportionality is negative, we write Newton's law of cooling as

(9.1.3)
$$\frac{dT}{dt} = -k(T - \tau), \qquad k > 0 \text{ constant.}$$

This equation can be rewritten as

$$\frac{dT}{dt} + kT = k\tau,$$

a first-order linear equation with $p(t) = k$ and $q(t) = k\tau$ constant. From the result in Example 1, we see that

$$T = \frac{k\tau}{k} + Ce^{-kt} = \tau + Ce^{-kt}.$$

The constant C is determined by the initial temperature $T(0)$:

$$T(0) = \tau + Ce^0 = \tau + C \qquad \text{so that} \qquad C = T(0) - \tau.$$

The temperature of the object at any time t is given by the function

(9.1.4)
$$T(t) = \tau + [T(0) - \tau]e^{-kt}.$$

Example 4 A cup of coffee is served to you at 185°F in a room where the temperature is 65°F. Two minutes later, the temperature of the coffee has dropped to 155°F. How many more minutes would you expect to wait for the coffee to cool to 105°F?

SOLUTION In this case $\tau = 65$ and $T(0) = 185$. Therefore

$$T(t) = 65 + [185 - 65]e^{-kt} = 65 + 120e^{-kt}.$$

To determine the constant of proportionality k, we use the fact that $T(2) = 155$:

$$T(2) = 65 + 120e^{-2k} = 155, \qquad e^{-2k} = \tfrac{90}{120} = \tfrac{3}{4}, \qquad k = -\tfrac{1}{2}\ln(3/4) \cong 0.144.$$

Taking k as 0.144, we write $T(t) = 65 + 120e^{-0.144t}$ for the temperature of the coffee at time t.

Now we want to find the value of t for which $T(t) = 105°F$. To do this, we solve the equation

$$65 + 120e^{-0.144t} = 105$$

for t:

$$120e^{-0.144t} = 40, \qquad -0.144t = \ln(1/3), \qquad t \cong 7.63 \text{ min}.$$

Therefore you would expect to wait another 5.63 minutes. ❏

Remark In arriving at the function $T(t) = 65 + 120e^{-0.144t}$, we used 0.144 for k. But this is only an approximation to k. We can avoid this slight inaccuracy by stopping at $e^{-2k} = \frac{3}{4}$ and writing

$$e^{-k} = \left(\tfrac{3}{4}\right)^{1/2}, \qquad e^{-kt} = \left(\tfrac{3}{4}\right)^{t/2}, \qquad T(t) = 65 + 120\left(\tfrac{3}{4}\right)^{t/2}.$$

This version of T is exact and therefore preferable on a theoretical basis. However, it has a disadvantage: it's harder to use in computations. For our purposes, both versions of T are acceptable. ❏

A Mixing Problem

Example 5 A chemical manufacturing company has a 1000-gallon holding tank which it uses to control the release of pollutants into a sewage system. Initially the tank has 360 gallons of fluid containing 2 pounds of pollutant per gallon. Fluid containing 3 pounds of pollutant per gallon enters the tank at the rate of 80 gallons per hour and is uniformly mixed with the fluid already in the tank. Simultaneously, fluid is released from the tank at the rate of 40 gallons per hour. Determine the rate (lbs/gal) at which the pollutant is being released after 10 hours of this process.

SOLUTION Let $P(t)$ be the amount of pollutant (in pounds) in the tank at time t. The rate of change of pollutant in the tank, dP/dt, satisfies the condition

$$\frac{dP}{dt} = (\text{rate in}) - (\text{rate out}).$$

The pollutant is entering the tank at the rate of $3 \times 80 = 240$ pounds per hour (rate in).

Fluid is entering the tank at the rate of 80 gallons per hour and is leaving at the rate of 40 gallons per hour. The amount of fluid in the tank is increasing at the rate of 40 gallons per hour, and so there are $360 + 40t$ gallons of fluid in the tank at time t. We can now conclude that the amount of pollutant per gallon in the tank at time t is given by the function

$$\frac{P(t)}{360 + 40t},$$

and the rate at which pollutant is leaving the tank is

$$40\frac{P(t)}{360 + 40t} = \frac{P(t)}{9 + t} \quad (\text{rate out}).$$

Therefore, our differential equation reads

$$\frac{dP}{dt} = (\text{rate in}) - (\text{rate out}) = 240 - \frac{P}{9 + t},$$

which we can write as

$$\frac{dP}{dt} + \frac{1}{9+t}P = 240.$$

This is a first-order linear differential equation. Here we have

$$p(t) = \frac{1}{9+t} \quad \text{and} \quad H(t) = \int \frac{1}{9+t}\,dt = \ln|9+t| = \ln(9+t). \quad (9+t>0)$$

As an integrating factor, we use

$$e^{H(t)} = e^{\ln(9+t)} = 9+t.$$

Multiplying the differential equation by $9+t$, we have

$$(9+t)\frac{dP}{dt} + P = 240(9+t)$$

$$\frac{d}{dt}[(9+t)P] = 240(9+t)$$

$$(9+t)P = 120(9+t)^2 + C$$

$$P(t) = 120(9+t) + \frac{C}{9+t}.$$

Since the amount of pollutant in the tank is initially $2 \times 360 = 720$ (pounds), we see that

$$P(0) = 120(9) + \frac{C}{9} = 720, \quad \text{which implies that} \quad C = -3240.$$

Thus, the function

$$P(t) = 120(9+t) - \frac{3240}{9+t} \quad \text{(pounds)}$$

gives the amount of pollutant in the tank at any time t. After 10 hours there are $360 + 40(10) = 760$ gallons of fluid in the tank, and there are

$$P(10) = 120(19) - \tfrac{3240}{19} \cong 2109$$

pounds of pollutant. Therefore, the rate at which pollutant is being released into the sewage system after 10 hours is $\frac{2109}{760} \cong 2.78$ pounds per gallon. ❑

EXERCISES 9.1

Exercises 1–6. Determine whether the functions satisfy the differential equation.

1. $2y' - y = 0$; $\quad y_1(x) = e^{x/2}, \quad y_2(x) = x^2 + 2e^{x/2}$.

2. $y' + xy = x$; $\quad y_1(x) = e^{-x^2/2}, \quad y_2(x) = 1 + Ce^{-x^2/2}$.

3. $y' + y = y^2$; $\quad y_1(x) = \dfrac{1}{e^x + 1}, \quad y_2(x) = \dfrac{1}{Ce^x + 1}$.

4. $y'' + 4y = 0$; $\quad y_1(x) = 2\sin 2x, \quad y_2(x) = 2\cos x$.

5. $y'' - 4y = 0$; $\quad y_1(x) = e^{2x}, \quad y_2(x) = C\sinh 2x$.

6. $y'' - 2y' - 3y = 7e^{3x}$; $\quad y_1(x) = e^{-x} + 2e^{3x}$,
$$y_2(x) = \tfrac{7}{4}xe^{3x}.$$

Exercises 7–22. Find the general solution.

7. $y' - 2y = 1$.

8. $xy' - 2y = -x$.

9. $2y' + 5y = 2$.

10. $y' - y = -2e^{-x}$.

11. $y' - 2y = 1 - 2x$.

12. $xy' + 2y = \dfrac{\cos x}{x}$.

13. $xy' - 4y = -2nx$.

14. $y' + y = 2 + 2x$.

15. $y' - e^x y = 0$.

16. $y' - y = e^x$.

17. $(1 + e^x)y' + y = 1$.

18. $xy' + y = (1+x)e^x$.

19. $y' + 2xy = xe^{-x^2}$.

20. $xy' - y = 2x\ln x$.

21. $y' + \dfrac{2}{x+1}y = 0$.

22. $y' + \dfrac{2}{x+1}y = (x+1)^{5/2}$.

Exercises 23–28. Find the particular solution determined by the initial condition.

23. $y' + y = x$, $\quad y(0) = 1$.

24. $y' - y = e^{2x}$, $\quad y(1) = 1$.

25. $y' + y = \dfrac{1}{1 + e^x}$, $\quad y(0) = e$.

26. $y' + y = \dfrac{1}{1 + 2e^x}$, $\quad y(0) = e$

27. $xy' - 2y = x^3 e^x$, $\quad y(1) = 0$.

28. $xy' + 2y = xe^{-x}$, $\quad y(1) = -1$.

29. Find all functions that satisfy the differential equation $y' - y = y'' - y'$ HINT: Set $z = y' - y$.

30. Find the general solution of $y' + ry = 0$, r constant.
   (a) Show that if y is a solution and $y(a) = 0$ at some number $a \geq 0$, then $y(x) = 0$ for all x. (Thus a solution y is either identically zero or never zero.)
   (b) Show that if $r < 0$, then all nonzero solutions are unbounded.
   (c) Show that if $r > 0$, then all solutions tend to 0 as $x \to \infty$.
   (d) What are the solutions if $r = 0$?

Exercises 31 and 32 are given in reference to the differential equation

$$y' + p(x)y = 0$$

with p continuous on an interval I.

31. (a) Show that if y_1 and y_2 are solutions, then $u = y_1 + y_2$ is also a solution.
   (b) Show that if y is a solution and C is a constant, then $u = Cy$ is also a solution.

32. (a) Let $a \in I$. Show that the general solution can be written

$$y(x) = Ce^{-\int_a^x p(t)\,dt}$$

   (b) Show that if y is a solution and $y(b) = 0$ for some $b \in I$, then $y(x) = 0$ for all $x \in I$.
   (c) Show that if y_1 and y_2 are solutions and $y_1(b) = y_2(b)$ for some $b \in I$, then $y_1(x) = y_2(x)$ for all $x \in I$.

Exercises 33 and 34 are given in reference to the differential equation (9.1.1)

$$y' + p(x)y = q(x),$$

with p and q continuous on some interval I.

33. Let $a \in I$ and let $H(x) = \int_a^x p(t)\,dt$. Show that

$$y(x) = e^{-H(x)} \int_a^x q(t)e^{H(t)}\,dt$$

is the solution of the differential equation that satisfies the initial condition $y(a) = 0$.

34. Show that if y_1 and y_2 are solutions of (9.1.1), then $y = y_1 - y_2$ is a solution of $y' + p(x)y = 0$.

35. A thermometer is taken from a room where the temperature is 72°F to the outside, where the temperature is 32°F. Outside for $\frac{1}{2}$ minute, the thermometer reads 50°F. What will the thermometer read after it has been outside for 1 minute? How many minutes does the thermometer have to be outside for it to read 35°F?

36. A small metal ball at room temperature 20°C is dropped into a large container of boiling water (100°C). Given that the temperature of the ball increases 2° in 2 seconds, what will be the temperature of the ball 6 seconds after immer-

sion? How long does the ball have to remain in the boiling water for the temperature of the ball to reach 90°C?

37. An object falling from rest in air is subject not only to gravitational force but also to air resistance. Assume that the air resistance is proportional to the velocity and acts in a direction opposite to the motion. Then the velocity of the object at time t satisfies an equation of the form

$$v' = 32 - kv,$$

where k is a positive constant and $v(0) = 0$. Here we are measuring distance in feet and the positive direction is down.
   (a) Find $v(t)$.
   (b) Show that $v(t)$ cannot exceed $32/k$ and that $v(t) \to 32/k$ as $t \to \infty$.
   (c) Sketch the graph of v.

38. Suppose that a certain population P has a birth rate dB/dt and a death rate dD/dt. Then the rate of change of P is the difference

$$\frac{dP}{dt} = \frac{dB}{dt} - \frac{dD}{dt}$$

   (a) Assume that $dB/dt = aP$ and $dD/dt = bP$, where a and b are constants. Find $P(t)$ if $P(0) = P_0 > 0$.
   (b) What happens to $P(t)$ as $t \to \infty$ if

   (i) $a > b$, (ii) $a = b$, (iii) $a < b$?

39. The current i in an electrical circuit consisting of a resistance R, inductance L, and voltage E varies with time according to the formula

$$L\frac{di}{dt} + Ri = E.$$

Take R, L, E as positive constants.
   (a) Find $i(t)$ if $i(0) = 0$.
   (b) What limit does the current approach as $t \to \infty$?
   (c) In how many seconds does the current reach 90% of this limit?

40. The current i in an electrical circuit consisting of a resistance R, inductance L, and a voltage $E\sin\omega t$ varies with time according to the formula

$$L\frac{di}{dt} + Ri = E\sin\omega t.$$

Take R, L, E as positive constants.
   (a) Find $i(t)$ if $i(0) = i_0$.
   (b) What happens to the current in this case as $t \to \infty$?
   (c) Sketch the graph of i.

41. A 200-liter tank, initially full of water, develops a leak at the bottom. Given that 20% of the water leaks out in the first 5 minutes, find the amount of water left in the tank t minutes after the leak develops:
   (a) if the water drains off at a rate proportional to the amount of water present.
   (b) if the water drains off at a rate proportional to the product of the time elapsed and the amount of water present.

42. At a certain moment a 100-gallon mixing tank is full of brine containing 0.25 pound of salt per gallon. Find the amount of salt present t minutes later if the brine is being continuously drawn off at the rate of 3 gallons per minute and replaced by brine containing 0.2 pound of salt per gallon.

43. An advertising company introduces a new product to a metropolitan area of population M. Let $P = P(t)$ denote the number of people who become aware of the product by time t. Suppose that P increases at a rate which is proportional to the number of people still unaware of the product. The company determines that no one was aware of the product at the beginning of the campaign $[P(0) = 0]$ and that 30% of the people were aware of the product after 10 days of advertising.
    (a) Give the differential equation that describes the number of people who become aware of the product by time t.
    (b) Determine the solution of the differential equation from part (a) that satisfies the initial condition $P(0) = 0$.
    (c) How long does it take for 90% of the population to become aware of the product?

44. A drug is fed intravenously into a patient's bloodstream at a constant rate r. Simultaneously, the drug diffuses into the patient's body at a rate proportional to the amount of drug present.
    (a) Determine the differential equation that describes the amount $Q(t)$ of the drug in the patient's bloodstream at time t.
    (b) Determine the solution $Q = Q(t)$ of the differential equation found in part (a) that satisfies the initial condition $Q(0) = 0$.
    (c) What happens to $Q(t)$ as $t \to \infty$?

▷45. (a) The differential equation

$$\frac{dP}{dt} = (2 \cos 2\pi t)P$$

models a population that undergoes periodic fluctuations. Assume that $P(0) = 1000$ and find $P(t)$. Use a graphing utility to draw the graph of P.
    (b) The differential equation

$$\frac{dP}{dt} = (2 \cos 2\pi t)P + 2000 \cos 2\pi t$$

models a population that undergoes periodic fluctuations as well as periodic migration. Continue to assume that $P(0) = 1000$ and find $P(t)$ in this case. Use a graphing utility to draw the graph of P and estimate the maximum value of P.

▷46. The *Gompertz equation*

$$\frac{dP}{dt} = P(a - b \ln P),$$

where a and b are positive constants, is another model of population growth.
    (a) Find the solution of this differential equation that satisfies the initial condition $P(0) = P_0$. HINT: Define a new dependent variable Q by setting $Q = \ln P$.
    (b) What happens to $P(t)$ as $t \to \infty$?
    (c) Determine the concavity of the graph of P.
    (d) Use a graphing utility to draw the graph of P in the case where $a = 4$, $b = 2$, and $P_0 = \frac{1}{2}e^2$. Does the graph confirm your result in part (c)?

■ 9.2 INTEGRAL CURVES; SEPARABLE EQUATIONS

Introduction

We begin with a pair of functions P and Q which have continuous derivatives $P' = p$, $Q' = q$ and form the equation

(1) $$P(x) + Q(y) = C, \qquad C \text{ constant.}$$

Note that in this equation the x's and y's are not intermingled; they are separated.

Equation (1) represents a one-parameter family of curves. Different values of the parameter C give different curves.

If $y = y(x)$ is a differentiable function which on its domain satisfies (1), then by implicit differentiation we find that

$$P'(x) + Q'(y)y' = 0.$$

Since $P' = p$ and $Q' = q$, we have

(2) $$p(x) + q(y)y' = 0.$$

In this sense, curves (1) satisfy differential equation (2).

What does this mean? It means that if $y = y(x)$ is a differentiable function and its graph lies on one of the curves

$$P(x) + Q(y) = C,$$

then along the graph of the function the numbers $x, y(x), y'(x)$ are related by the equation

$$p(x) + q(y(x))y'(x) = 0.$$

Separable Equations

In our introduction we started with a family of curves in the xy-plane and obtained the differential equation satisfied by these curves in the sense explained. Now we reverse the process. We start with a differential equation and obtain the family of curves which satisfy it. These curves are called the *integral curves* (or *solution curves*) of the differential equation

Our starting point is a differential equation of the form

(9.2.1)
$$p(x) + q(y)y' = 0,$$

with p and q continuous. A differential equation which can be written in this form is called *separable*.

To find the integral curves of (9.2.1), we expand the equation to read

$$p(x) + q(y(x))y'(x) = 0.$$

Integrating this equation with respect to x, we find that

$$\int p(x)\,dx + \int q(y(x))y'(x)\,dx = C,$$

where C is an arbitrary constant. From $y = y(x)$, we have $dy = y'(x)\,dx$. Therefore

$$\int p(x)\,dx + \int q(y)\,dy = C.$$

The variables have been separated. Now, if P is an antiderivative of p and Q is an antiderivative of q, then this last equation can be written

(9.2.2)
$$P(x) + Q(y) = C.$$

This equation represents a one-parameter family of curves, and these curves are the integral curves (solution curves) of the differential equation.

Example 1 The differential equation

$$x + yy' = 0$$

is separable. (The variables are already separated.) We can find the integral curves by writing

$$\int x\,dx + \int y\,dy = C.$$

Carrying out the integration, we have

$$\tfrac{1}{2}x^2 + \tfrac{1}{2}y^2 + C,$$

which, since C is arbitrary, we can write as

$$x^2 + y^2 = C.$$

For this equation to give us a curve, we must take $C > 0$. (If $C < 0$, there are no real numbers x, y which satisfy the equation. If $C = 0$, there is no curve—just a point, the origin.) For $C > 0$, the equation gives the circle of radius $\sqrt{C}$ centered at the origin. The family of integral curves consists of all circles centered at the origin. ❏

Example 2 Show that the differential equation

$$y + yy' = xy - y'$$

is separable and find an integral curve which satisfies the initial condition $y(2) = 1$.

SOLUTION First we show that the equation is separable. To do this, we write

$$(y + 1)y' = y(x - 1)$$
$$y(1 - x) + (y + 1)y' = 0.$$

Next we divide the equation by y. As you can verify, the horizontal line $y = 0$ is an integral curve. However, we can ignore it because it doesn't satisfy the initial condition $y(2) = 1$. With $y \neq 0$, we can write

$$1 - x + \frac{y + 1}{y}y' = 0,$$

$$1 - x + \left(1 + \frac{1}{y}\right)y' = 0.$$

The equation is separable. Writing

$$\int (1 - x)\,dx + \int \left(1 + \frac{1}{y}\right)\,dy = C,$$

we find that the integral curves take the form

$$x - \tfrac{1}{2}x^2 + y + \ln|y| = C.$$

The condition $y(2) = 1$ forces

$$2 - \tfrac{1}{2}(2)^2 + 1 + \ln(1) = C, \qquad \text{which implies} \qquad C = 1.$$

The integral curve that satisfies the initial condition is the curve

$$x - \tfrac{1}{2}x^2 + y + \ln|y| = 1.$$

Figure 9.2.1 shows the curve for $x > 0$. ❏

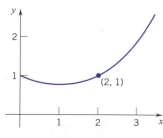

Figure 9.2.1

Functions as Solutions

If the equations of the integral curves can be solved for y in terms of x, then the integral curves are graphs of functions $y = f(x)$. Functions so obtained satisfy the differential equation and are therefore solutions of the differential equation in the ordinary sense.

Example 3 Show that the differential equation

$$y' = xe^{y-x}$$

is separable and find the integral curves. Show that these curves are the graphs of functions. Verify by differentiation that these functions are solutions of the differential equation

SOLUTION The equation is separable since it can be written as

$$xe^{-x} - e^{-y}y' = 0.$$

(Verify this.)

Setting

$$\int xe^{-x}\,dx - \int e^{-y}\,dy = C,$$

we have $-xe^{-x} - e^{-x} + e^{-y} = C$ and therefore

$$e^{-y} = xe^{-x} + e^{-x} + C.$$

These equations give the integral curves.

To show that these curves are the graphs of functions, we take the logarithm of both sides:

$$\ln(e^{-y}) = \ln(xe^{-x} + e^{-x} + C)$$

$$-y = \ln(xe^{-x} + e^{-x} + C)$$

$$y = -\ln(xe^{-x} + e^{-x} + C).$$

The integral curves are the graphs of the functions

$$y = -\ln(xe^{-x} + e^{-x} + C).$$

Since

$$y' = -\frac{-xe^{-x} + e^{-x} - e^{-x}}{xe^{-x} + e^{-x} + C} = \frac{xe^{-x}}{xe^{-x} + e^{-x} + C} = \frac{xe^{-x}}{e^{-y}},$$

we have

$$y' = xe^{y-x}.$$

This shows that the functions satisfy the differential equation. ❑

Applications

In the mid-nineteenth century the Belgian mathematician P. F. Verhulst used the differential equation

(9.2.3)

$$\frac{dy}{dt} = ky(M - y),$$

where k and M are positive constants, to model population growth. This equation is now known as the *logistic equation*, and its solutions are functions called *logistic functions*. Life scientists have used this equation to model the spread of disease, and social scientists have used it to study the dissemination of information. In the case of disease, if M denotes the total number of people in the population under consideration and $y(t)$ is the number of infected people at time t, then the differential equation states that the rate of growth of the disease is proportional to the product of the number of people who are infected and the number who are not infected.

The differential equation is separable, since it can be written in the form

$$k - \frac{1}{y(M - y)}y' = 0.$$

Integrating this equation, we have

$$\int k\, dt - \int \frac{1}{y(M - y)}\, dy = C$$

$$kt - \int \left(\frac{1/M}{y} + \frac{1/M}{M - y}\right) dy = C \qquad \text{(partial fraction decomposition)}$$

and therefore

$$\frac{1}{M}\ln|y| - \frac{1}{M}\ln|M - y| = kt + C.$$

These are the integral curves.

It is a good exercise in manipulating logarithms, exponentials, and arbitrary constants to show that such equations can be solved for y in terms of t, and thus the solutions can be expressed as functions of t. The result can be written as

$$y = \frac{CM}{C + e^{-Mkt}}. \qquad \text{(not the same C as above)}$$

If R is the number of people initially infected, then the solution function $y = y(t)$ satisfies the initial condition $y(0) = R$. From this we see that

$$R = \frac{CM}{C + 1} \qquad \text{and therefore} \qquad C = \frac{R}{M - R}.$$

As you can check, this value of C gives

(9.2.4)

$$\boxed{\; y(t) = \frac{MR}{R + (M - R)e^{-Mkt}}. \;}$$

This particular solution is shown graphically in Figure 9.2.2. Note that y is an increasing function. In the Exercises you are asked to show that the graph is concave up on $[0, t_1]$ and concave down on $[t_1, \infty)$. In the case of a disease, this means that the disease spreads at an increasing rate up to time $t = t_1$; and after t_1 the disease is still spreading, but at a decreasing rate. As $t \to \infty$, $e^{-Mkt} \to 0$, and therefore $y^{(t)} \to M$. Over time, according to this model, the disease will tend to infect the entire population.

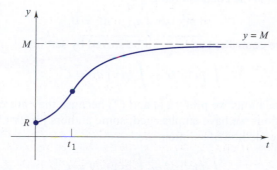

Figure 9.2.2

Example 4 Assume that a rumor spreads through a population of 5000 people at a rate proportional to the product of the number of people who have heard it and the number who have not heard it. Suppose that by a certain day, call it time $t = 0$, 100 people have heard the rumor and 2 days later the rumor is known to 500 people. How long will it take for the rumor to spread to half of the population?

SOLUTION Let $y(t)$ denote the number of people aware of the rumor by time t. Then y satisfies the logistic equation with $M = 5000$ and the initial condition $R = y(0) = 100$. Thus, by (9.2.4),

$$y(t) = \frac{100(5000)}{100 + 4900e^{-5000kt}} = \frac{5000}{1 + 49e^{-5000kt}}.$$

The constant of proportionality k can be determined from the condition $y(2) = 500$:

$$500 = \frac{5000}{1 + 49e^{-10,000k}}, \qquad 1 + 49e^{-10,000k} = 10, \qquad e^{-10,000k} = \frac{9}{49}.$$

Therefore

$$-10,000k = \ln(9/49) \qquad \text{and} \qquad k \cong 0.00017.$$

Taking k as 0.00017, we have

$$y(t) = \frac{5000}{1 + 49e^{-0.85t}}.$$

To determine how long it will take for the rumor to spread to half of the population, we seek the value of t for which

$$2500 = \frac{5000}{1 + 49e^{-0.85t}}.$$

Solving this equation for t, we get

$$1 + 49e^{-0.85t} = 2, \qquad e^{-0.85t} = \frac{1}{49}, \qquad t = \frac{\ln(1/49)}{-0.85} \cong 4.58.$$

It will take slightly more than $4\frac{1}{2}$ days for the rumor to spread to half of the population. ❏

Differential Notation

The equation

(1)
$$p(x) + q(y)y' = 0$$

can obviously be written

(2)
$$p(x) + q(y)\frac{dy}{dx} = 0.$$

In differential notation the equation reads

(3)
$$p(x)\,dx + q(y)\,dy = 0.$$

These equations are equivalent and can all be solved by setting

$$\int p(x)\,dx + \int q(y)\,dy = C.$$

Of these formulations we prefer (1) and (2) because they are easy to understand in terms of the notions we have emphasized; some authors prefer (3) because it leads more directly to the method of solution:

$$\int p(x)\,dx + \int q(y)\,dy = C.$$

EXERCISES 9.2

Exercises 1–12. Find the integral curves. If the curves are the graphs of functions $y = f(x)$, determine all the functions that satisfy the equation.

1. $y' = y \sin(2x + 3)$.

2. $y' = (x^2 + 1)(y^2 + y)$.

3. $y' = (xy)^3$.

4. $y' = 3x^2(1 + y^2)$.

5. $\dfrac{dy}{dx} = \dfrac{\sin 1/x}{x^2 y \cos y}$.

6. $\dfrac{dy}{dx} = \dfrac{y^2 + 1}{y + xy}$.

7. $y' = xe^{x+y}$.

8. $y' = xy^2 - x - y^2 + 1$.

9. $(y \ln x) y' = \dfrac{(y + 1)^2}{x}$.

10. $e^y \sin 2x\, dx + \cos x\, (e^{2y} - y)\, dy = 0$.

11. $(y \ln x) y' = \dfrac{y^2 + 1}{x}$.

12. $y' = \dfrac{1 + 2y^2}{y \sin x}$.

Exercises 13–20. Solve the initial-value problem.

13. $\dfrac{dy}{dx} = x\sqrt{\dfrac{1 - y^2}{1 - x^2}}$, $y(0) = 0$.

14. $\dfrac{dy}{dx} = \dfrac{e^{x-y}}{1 + e^x}$, $y(1) = 0$.

15. $\dfrac{dy}{dx} = \dfrac{x^2 y - y}{y + 1}$, $y(3) = 1$.

16. $x^2 y' = y - xy$, $y(-1) = -1$.

17. $(xy^2 + y^2 + x + 1)\, dx + (y - 1)\, dy = 0$, $y(2) = 0$.

18. $\cos y\, dx + (1 + e^{-x}) \sin y\, dy = 0$, $y(0) = \pi/4$.

19. $y' = 6e^{2x-y}$, $y(0) = 0$.

20. $xy' - y = 2x^2 y$, $y(1) = 1$.

21. Suppose that a chemical A combines with a chemical B to form a compound C. In addition, suppose that the rate at which C is produced at time t varies directly with the amounts of A and B present at time t. With this model, if A_0 grams of A are mixed with B_0 grams of B, then

$$\dfrac{dC}{dt} = k(A_0 - C)(B_0 - C).$$

(a) Find the amount of compound C present at time t if $A_0 = B_0$.
(b) Find the amount of compound C present at time t if $A_0 \neq B_0$.

22. Assume that the growth of a certain biological culture is modeled by the differential equation

$$\dfrac{dS}{dt} = 0.0020 S\,(800 - S),$$

where $S = S(t)$ denotes the size of the culture, measured in square millimeters, at time t. When first observed, at time $t = 0$, the culture occupied an area of 100 square millimeters.

(a) Determine the size of the culture at each later time t.
(b) Use a graphing utility to draw the graphs of S and dS/dt.
(c) At what time t does the culture grow most rapidly? Use three decimal place accuracy.

23. When an object of mass m moves through air or a viscous medium, it is acted on by a frictional force that acts in the direction opposite to its motion. This frictional force depends on the velocity of the object and (within close approximation) is given by

$$F(v) = -\alpha v - \beta v^2,$$

where α and β are positive constants.

(a) From Newton's second law, $F = ma$, we have

$$m\dfrac{dv}{dt} = -\alpha v - \beta v^2.$$

Solve this differential equation to find $v = v(t)$.
(b) Find v if the object has initial velocity $v(0) = v_0$.
(c) What happens to $v(t)$ as $t \to \infty$?

24. A descending parachutist is acted on by two forces: a constant downward force mg and the upward force of air resistance, which (within close approximation) is of the form $-\beta v^2$ where β is a positive constant. (In this problem we are taking the downward direction as positive.)

(a) Express t in terms of the velocity v, the initial velocity v_0, and the constant $v_c = \sqrt{mg/\beta}$.
(b) Express v as a function of t.
(c) Express the acceleration a as a function of t. Verify that the acceleration never changes sign and in time tends to zero.
(d) Show that in time v tends to v_c. (This number v_c is called the *terminal velocity*.)

25. A flu virus is spreading rapidly through a small town with a population of 25,000. The virus is spreading at a rate proportional to the product of the number of people who have been infected and the number who haven't been infected. When first reported, at time $t = 0$, 100 people had been infected and 10 days later, 400 people.

(a) How many people will have been infected by time t? (Measure t in days.)
(b) How long will it take for the infection to reach half of the population?
(c) Use a graphing utility to graph the function you found in part (a).

26. Let y be the logistic function (9.2.4). Show that dy/dt increases for $y < M/2$ and decreases for $y > M/2$. What can you conclude about dy/dt when $y = M/2$? Explain.

27. A rescue package of mass 100 kilograms is dropped from a height of 4000 meters. As the object falls, the air resistance is equal to twice its velocity. After 10 seconds, the package's parachute opens and the air resistance is now four times the square of its velocity.

(a) What is the velocity of the package the instant the parachute opens?
(b) What is the velocity of the package t seconds after the parachute opens?
(c) What is the terminal velocity of the package?

HINT: There are two differential equations that govern the package's velocity and position: one for the free-fall period and one for the period after the parachute opens.

28. It is known that m parts of chemical A combine with n parts of chemical B to produce a compound C. Suppose that the rate at which C is produced varies directly with the product of the amounts of A and B present at that instant. Find the amount of C produced in t minutes from an initial mixing of A_0 pounds of A with B_0 pounds of B, given that:
    (a) $n = m$, $A_0 = B_0$, and A_0 pounds of C are produced in the first minute.
    (b) $n = m$, $A_0 = \frac{1}{2}B_0$, and A_0 pounds of C are produced in the first minute.

(c) $n \neq m$, $A_0 = B_0$, and A_0 pounds of C are produced in the first minute.

HINT: Denote by $A(t)$, $B(t)$, $C(t)$ the amounts of A, B, C present at time t. Observe that $C'(t) = kA(t)B(t)$. Then note that

$$A_0 - A(t) = \frac{m}{m+n}C(t) \quad \text{and} \quad B_0 - B(t) = \frac{n}{m+n}C(t)$$

and thus

$$C'(t) = k\left[A_0 - \frac{m}{m+n}C(t)\right]\left[B_0 - \frac{n}{m+n}C(t)\right].$$

■ PROJECT 9.2 ORTHOGONAL TRAJECTORIES

If two curves intersect at right angles, one with slope m_1 and the other with slope m_2, then $m_1m_2 = -1$. A curve that intersects every member of a family of curves at right angles is called an *orthogonal trajectory* for that family of curves.

A differential equation of the form

$$y' = f(x, y),$$

where $f(x, y)$ is an expression in x and y, generates a family of curves: all the integral curves of that particular equation. The orthogonal trajectories of this family are the integral curves of the differential equation

$$y' = -\frac{1}{f(x, y)}.$$

The figure indicates a family of parabolas $y = Cx^2$. The orthogonal trajectories of this family of parabolas constitute a family of ellipses.

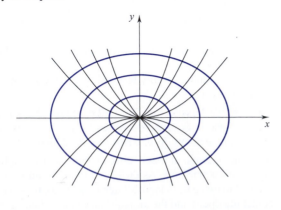

We can establish this by starting with the equation

$$y = Cx^2$$

and differentiating. Differentiation gives

$$y' = 2Cx.$$

Since $y = Cx^2$, we have $C = y/x^2$, and thus

$$y' = 2\left(\frac{y}{x^2}\right)x = \frac{2y}{x}.$$

The orthogonal trajectories are the integral curves of the differential equation

$$y' = -\frac{x}{2y}.$$

As you can check, the solutions of this equation can be written in the form

$$x^2 + 2y^2 = K^2.$$

Problem 1. Find the orthogonal trajectories of the following families of curves. In each case draw several of the curves and several of the orthogonal trajectories.

  a. $2x + 3y = C$.   **b.** $y = Cx$.

  c. $xy = C$.   **d.** $y = Cx^3$.

  e. $y = Ce^x$.   **f.** $x = Cy^4$.

Problem 2. Find the orthogonal trajectories of the following families of curves. In each case use a CAS to draw several of the curves and several of the orthogonal trajectories. If you can, graph the curves in one color and the orthogonal trajectories in another.

  a. $y^2 - x^2 = C$.   **b.** $y^2 = Cx^3$.

  c. $y = \dfrac{Ce^x}{x}$.   **d.** $e^x \sin y = C$.

Problem 3. Show that the given family of curves is *self-orthogonal*. Use a graphing utility to graph at least four members of the family.

  a. $y^2 = 4C(x + C)$.   **b.** $\dfrac{x^2}{C^2} + \dfrac{y^2}{C^2 - 4} = 1$.

■ 9.3 THE EQUATION $y'' + ay' + by = 0$

A differential equation of the form

(9.3.1)
$$y'' + ay' + by = 0,$$

where a and b are real numbers, is called a *homogeneous second-order linear differential equation with constant coefficients*.

By a *solution* of this equation, we mean a function $y = y(x)$ that satisfies the equation for all real x.

The Characteristic Equation

As you can readily verify, the function $y = e^{-ax}$ satisfies the differential equation

$$y' + ay = 0. \qquad \text{(Verify this.)}$$

This suggests that the differential equation

$$y'' + ay' + by = 0$$

may have a solution of the form $y = e^{rx}$.

If $y = e^{rx}$, then

$$y' = re^{rx} \qquad \text{and} \qquad y'' = r^2 e^{rx}.$$

Substitution into the differential equation gives

$$r^2 e^{rx} + are^{rx} + b\,e^{rx} = e^{rx}(r^2 + ar + b) = 0,$$

and since $e^{rx} \neq 0$,

$$r^2 + ar + b = 0.$$

This shows that the function $y = e^{rx}$ satisfies the differential equation iff

$$r^2 + ar + b = 0.$$

This quadratic equation in r is called the *characteristic equation*.

The nature of the solutions of the differential equation

$$y'' + ay' + by = 0$$

depends on the nature of the characteristic equation. There are three cases to be considered: the characteristic equation has two distinct real roots; it has only one real root; it has no real roots. We'll consider these cases one at a time.

Case 1: The characteristic equation has two distinct real roots r_1 and r_2. In this case both

$$y_1 = e^{r_1 x} \qquad \text{and} \qquad y_2 = e^{r_2 x}$$

are solutions of the differential equation.

Case 2: The characteristic equation has only one real root $r = \alpha$. In this case the characteristic equation takes the form

$$(r - \alpha)^2 = 0.$$

This can be written

$$r^2 - 2\alpha r + \alpha^2 = 0.$$

The differential equation then reads

$$y'' - 2\alpha y' + \alpha^2 y = 0.$$

As you are asked to show in Exercise 33, the substitution $y = ue^{ax}$ gives

$$u'' = 0.$$

This equation is satisfied by

the constant function $u_1 = 1$ and   the identity function $u_2 = x$.

Thus the original differential equation is satisfied by the products

$$y_1 = e^{\alpha x} \qquad \text{and} \qquad y_2 = xe^{\alpha x}. \qquad\qquad \text{(Verify this.)}$$

Case 3: The characteristic equation has two complex roots

$$r_1 = \alpha + i\beta, \qquad r_2 = \alpha - i\beta \qquad \text{with } \beta \neq 0.$$

In this case the characteristic equation takes the form

$$(r - \alpha - i\beta)(r - \alpha + i\beta) = 0,$$

which, multiplied out, becomes

$$r^2 - 2\alpha r + (\alpha^2 + \beta^2) = 0.$$

The differential equation thus reads

$$y'' - 2\alpha y' + (\alpha^2 + \beta^2)y = 0.$$

As you are asked to show in Exercise 33, the substitution $y = ue^{\alpha x}$ eliminates α and gives

$$u'' + \beta^2 u = 0.$$

This equation, the equation of simple harmonic motion, is satisfied by the functions

$$u_1 = \cos \beta x \qquad \text{and} \qquad u_2 = \sin \beta x.^\dagger$$

Thus, the original differential equation is satisfied by the products

$$y_1 = e^{\alpha x} \cos \beta x \qquad \text{and} \qquad y_2 = e^{\alpha x} \sin \beta x.$$

Linear Combinations of Solutions; Existence and Uniqueness of Solutions; Wronskians

Observe that *if y_1 and y_2 are both solutions of the homogeneous equation, then every linear combination*

$$u(x) = C_1 y_1(x) + C_2 y_2(x)$$

is also a solution.

PROOF Set

$$u = C_1 y_1 + C_2 y_2$$

and observe that

$$u' = C_1 y_1' + C_2 y_2' \qquad \text{and} \qquad u'' = C_1 y_1'' + C_2 y_2''.$$

---

†You have seen this before. In any case, you can easily verify it by differentiation.

Since y_1 and y_2 are solutions of (9.3.1),

$$y_1'' + ay_1' + by_1 = 0 \quad \text{and} \quad y_2'' + ay_2' + by_2 = 0.$$

Therefore,

$$\begin{aligned} u'' + au' + bu &= (C_1 y_1'' + C_2 y_2'') + a(C_1 y_1' + C_2 y_2') + b(C_1 y_1 + C_2 y_2) \\ &= C_1(y_1'' + ay_1' + by_1) + C_2(y_2'' + ay_2' + by_2) \\ &= C_1(0) + C_2(0) = 0. \quad \square \end{aligned}$$

You have seen how to obtain solutions of the differential equation

$$y'' + ay' + by = 0$$

from the characteristic equation

$$r^2 + ar + b = 0.$$

We can form more solutions by taking linear combinations of these solutions. Question: Are there still other solutions or do all solutions arise in this manner? Answer: All solutions of the homogeneous equation are linear combinations of the solutions that we have already found.

To show this, we have to go a little deeper into the theory. Our point of departure is a result that we prove in a supplement to this section.

THEOREM 9.3.2

EXISTENCE AND UNIQUENESS THEOREM

Let x_0, α_0, α_1 be arbitrary real numbers. The homogeneous equation

$$y'' + ay' + by = 0$$

has a unique solution $y = y(x)$ that satisfies the initial conditions

$$y(x_0) = \alpha_0, \qquad y'(x_0) = \alpha_1.$$

Geometrically, the theorem says that there is one and only one solution the graph of which passes through a prescribed point (x_0, α_0) with prescribed slope α_1. We assume the result and go on from there.

DEFINITION 9.3.3

Let y_1 and y_2 be two solutions of

$$y'' + ay' + by = 0.$$

The *Wronskian*[†] of y_1 and y_2 is the function W defined for all real x by

$$W(x) = y_1(x)y_2'(x) - y_2(x)y_1'(x).$$

[†]Named after Count Hoene Wronski, a Polish mathematician (1776–1853).

Note that the Wronskian can be written as the 2×2 determinant

$$W(x) = \begin{vmatrix} y_1(x) & y_2(x) \\ y_1'(x) & y_2'(x) \end{vmatrix}. \qquad \text{(Appendix A.2)}$$

Wronskians have a very special property.

THEOREM 9.3.4

If both y_1 and y_2 are solutions of

$$y'' + ay' + by = 0,$$

then their Wronskian W is either identically zero or never zero.

PROOF Assume that both y_1 and y_2 are solutions of the equation and set

$$W = y_1 y_2' - y_2 y_1'.$$

Differentiation gives

$$W' = y_1 y_2'' + y_1' y_2' - y_1' y_2' - y_1'' y_2 = y_1 y_2'' - y_1'' y_2.$$

Since y_1 and y_2 are solutions, we know that

$$y_1'' + ay_1' + by_1 = 0$$

and

$$y_2'' + ay_2' + by_2 = 0.$$

Multiplying the first equation by $-y_2$ and the second equation by y_1, we have

$$-y_1'' y_2 - ay_1' y_2 - by_1 y_2 = 0$$
$$y_2'' y_1 + ay_2' y_1 + by_2 y_1 = 0.$$

We now add these two equations and obtain

$$(y_1 y_2'' - y_2 y_1'') + a(y_1 y_2' - y_2 y_1') = 0.$$

In terms of the Wronskian, we have

$$W' + aW = 0.$$

This is a first-order linear differential equation with general solution

$$W(x) = Ce^{-ax}.$$

If $C = 0$, then W is identically 0; if $C \neq 0$, then W is never zero. ❏

THEOREM 9.3.5

Every solution of the homogeneous equation

$$y'' + ay' + by = 0$$

can be expressed in a unique manner as the linear combination of any two solutions with a nonzero Wronskian.

PROOF Let u be any solution of the equation and let y_1, y_2 be any two solutions with nonzero Wronskian. Choose a number x_0 and form the equations

$$C_1 y_1(x_0) + C_2 y_2(x_0) = u(x_0)$$

$$C_1 y_1'(x_0) + C_2 y_2'(x_0) = u'(x_0).$$

The Wronskian of y_1 and y_2 at x_0,

$$W(x_0) = y_1(x_0)y_2'(x_0) - y_2(x_0)y_1'(x_0),$$

is different from zero. This guarantees that the system of equations (1) has a unique solution given by

$$C_1 = \frac{u(x_0)y_2'(x_0) - y_2(x_0)u'(x_0)}{y_1(x_0)y_2'(x_0) - y_2(x_0)y_1'(x_0)}, \qquad C_2 = \frac{y_1(x_0)u'(x_0) - u(x_0)y_1'(x_0)}{y_1(x_0)y_2'(x_0) - y_2(x_0)y_1'(x_0)}.$$

Our work is finished. The function $C_1 y_1 + C_2 y_2$ is a solution of the equation which by (1) has the same value as u at x_0 and the same derivative. Thus, by Theorem 9.3.2, $C_1 y_1 + C_2 y_2$ and u cannot be different functions; that is,

$$u = C_1 y_1 + C_2 y_2.$$

This proves the theorem. ❏

The General Solution

The arbitrary linear combination $y = C_1 y_1 + C_2 y_2$ of any two solutions with nonzero Wronskian is called the *general solution*. By the theorem we just proved, we can obtain any *particular* solution by adjusting C_1 and C_2.

We now return to the solutions obtained earlier and prove the final result, for practical purposes the summarizing result.

THEOREM 9.3.6

Given the equation

$$y'' + ay' + by = 0,$$

we form the characteristic equation

$$r^2 + ar + b = 0.$$

I. If the characteristic equation has two distinct real roots r_1 and r_2, then the general solution takes the form

$$y = C_1 e^{r_1 x} + C_2 e^{r_2 x}.$$

II. If the characteristic equation has only one real root $r = \alpha$, then the general solution takes the form

$$y = C_1 e^{\alpha x} + C_2 x e^{\alpha x} = (C_1 + C_2 x)e^{\alpha x}.$$

III. If the characteristic equation has two complex roots,

$$r_1 = \alpha + i\beta \qquad \text{and} \qquad r_2 = \alpha - i\beta,$$

then the general solution takes the form

$$y = C_1 e^{\alpha x} \cos \beta x + C_2 e^{\alpha x} \sin \beta x = e^{\alpha x}(C_1 \cos \beta x + C_2 \sin \beta x).$$

PROOF To prove this theorem, it is enough to show that the three solution pairs

$$e^{r_1 x}, e^{r_2 x} \qquad e^{\alpha x}, x\, e^{\alpha x} \qquad e^{\alpha x} \cos \beta x, e^{\alpha x} \sin \beta x$$

all have nonzero Wronskians. The Wronskian of the first pair is the function

$$W(x) = e^{r_1 x} \frac{d}{dx}(e^{r_2 x}) - \frac{d}{dx}(e^{r_1 x}) e^{r_2 x}$$

$$= e^{r_1 x} r_2\, e^{r_2 x} - r_1\, e^{r_1 x}\, e^{r_2 x} = (r_2 - r_1)\, e^{(r_1 + r_2) x}.$$

$W(x)$ is different from zero since, by assumption, $r_2 \neq r_1$.

We leave it to you to verify that the other pairs also have nonzero Wronskians. ❏

It is time to use what we have learned.

Example 1 Find the general solution of the equation $y'' + 2y' - 15y = 0$. Then find the particular solution that satisfies the initial conditions

$$y(0) = 0, \qquad y'(0) = -1.$$

SOLUTION The characteristic equation is the quadratic $r^2 + 2r - 15 = 0$. Factoring the left side, we have

$$(r + 5)(r - 3) = 0.$$

There are two real roots: -5 and 3. The general solution takes the form

$$y = C_1\, e^{-5x} + C_2\, e^{3x}.$$

Differentiating the general solution, we have

$$y' = -5C_1\, e^{-5x} + 3C_2\, e^{3x}.$$

The conditions

$$y(0) = 0, \qquad y'(0) = -1$$

are satisfied iff

$$C_1 + C_2 = 0 \qquad \text{and} \qquad -5C_1 + 3C_2 = -1.$$

Solving these two equations simultaneously, we find that

$$C_1 = \tfrac{1}{8}, \qquad C_2 = -\tfrac{1}{8}.$$

The solution that satisfies the prescribed initial conditions is the function

$$y = \tfrac{1}{8} e^{-5x} - \tfrac{1}{8} e^{3x}. ❏$$

Example 2 Find the general solution of the equation $y'' + 4y' + 4y = 0$.

SOLUTION The characteristic equation takes the form $r^2 + 4r + 4 = 0$, which can be written

$$(r + 2)^2 = 0.$$

The number -2 is the only root and

$$y = C_1 e^{-2x} + C_2 x e^{-2x}$$

is the general solution. ❏

Example 3 Find the general solution of the equation $y'' + y + 3y = 0$.

SOLUTION The characteristic equation is $r^2 + r + 3 = 0$. The quadratic formula shows that there are two complex roots:

$$r_1 = -\tfrac{1}{2} + i\tfrac{1}{2}\sqrt{11}, \qquad r_2 = -\tfrac{1}{2} - i\tfrac{1}{2}\sqrt{11}.$$

The general solution takes the form

$$y = e^{-x/2}[C_1 \cos(\tfrac{1}{2}\sqrt{11}x) + C_2 \sin(\tfrac{1}{2}\sqrt{11}x)]. \quad ❑$$

In our final example we revisit the equation of simple harmonic motion.

Example 4 Find the general solution of the equation

$$y'' + \omega^2 y = 0. \qquad (\omega \neq 0).$$

SOLUTION The characteristic equation is $r^2 + \omega^2 = 0$ and the roots are

$$r_1 = \omega i, \qquad r_2 = -\omega i.$$

Thus the general solution is

$$y = C_1 \cos \omega x + C_2 \sin \omega x. \quad ❑$$

Remark As you probably recall, the equation in Example 4 describes the oscillatory motion of an object suspended by a spring under the assumption that there are no forces acting on the spring-mass system other than the restoring force of the spring. This spring-mass problem and some generalizations of it are studied in Section 19.5. In the Exercises you are asked to show that the general solution that we gave above can be written

$$y = A \sin(\omega x + \phi_0),$$

where A and ϕ_0 are constants with $A > 0$ and $\phi_0 \in [0, 2\pi)$. ❑

EXERCISES 9.3

Exercises 1–18. Find the general solution.

1. $y'' + 2y' - 8y = 0$.

2. $y'' - 13y' + 42y = 0$.

3. $y'' + 8y' + 16y = 0$.

4. $y'' + 7y' + 3y = 0$.

5. $y'' + 2y' + 5y = 0$.

6. $y'' - 3y' + 8y = 0$.

7. $2y'' + 5y' - 3y = 0$.

8. $y'' - 12y = 0$.

9. $y'' + 12y = 0$.

10. $y'' - 3y' + \tfrac{9}{4}y = 0$.

11. $5y'' + \tfrac{11}{4}y' - \tfrac{3}{4}y = 0$.

12. $2y'' + 3y' = 0$.

13. $y'' + 9y = 0$.

14. $y'' - y' - 30y = 0$.

15. $2y'' + 2y' + y' = 0$.

16. $y'' - 4y' + 4y = 0$.

17. $8y'' + 2y' - y = 0$.

18. $5y'' - 2y' + y = 0$.

Exercises 19–24. Solve the initial-value problem.

19. $y'' - 5y' + 6y = 0, \quad y(0) = 1, \quad y'(0) = 1$.

20. $y'' + 2y' + y = 0, \quad y(2) = 1, \quad y'(2) = 2$.

21. $y'' + \tfrac{1}{4}y = 0, \quad y(\pi) = 1, \quad y'(\pi) = -1$.

22. $y'' - 2y' + 2y = 0, \quad y(0) = -1, \quad y'(0) = -1$.

23. $y'' + 4y' + 4y = 0, \quad y(-1) = 2, \quad y'(-1) = 1$.

24. $y'' - 2y' + 5y = 0, \quad y(\pi/2) = 0, \quad y'(\pi/2) = 2$.

25. Find all solutions of the equation $y'' - y' - 2y = 0$ that satisfy the given conditions:

(a) $y(0) = 1$.

(b) $y'(0) = 1$.

(c) $y(0) = 1, \quad y'(0) = 1$.

26. Show that the general solution of the differential equation

$$y'' - \omega^2 y = 0 \qquad (\omega > 0)$$

can be written

$$y = C_1 \cosh \omega x + C_2 \sinh \omega x.$$

27. Suppose that the roots r_1 and r_2 of the characteristic equation are real and distinct. Then they can be written as $r_1 = \alpha + \beta$ and $r_2 = \alpha - \beta$, where α and β are real. Show that the general solution of the homogeneous equation can be expressed in the form

$$y = e^{\alpha x}(C_1 \cosh \beta x + C_2 \sinh \beta x).$$

28. Show that the general solution of the differential equation

$$y'' + \omega^2 y = 0$$

can be written

$$y = A \sin(\omega x + \phi_0)$$

where A and ϕ_0 are constants with $A \geq 0$ and $\phi_0 \in [0, 2\pi)$.

29. Complete the proof of Theorem 9.3.6 by showing that the following solutions have nonzero Wronskians.

(a) $y_1 = e^{\alpha x}$,　$y_2 = x e^{\alpha x}$.　　　　(one root case)
(b) $y_1 = e^{\alpha x} \cos \beta x$,　$y_2 = e^{\alpha x} \sin \beta x$.　(complex root case)

30. In the absence of any external electromotive force, the current i in a simple electrical circuit varies with time t according to the formula

$$L\frac{d^2 i}{dt^2} + R\frac{di}{dt} + \frac{1}{C}i = 0.　　(L, R, C \text{ constants})^\dagger$$

Find the general solution of this equation given that $L = 1$, $R = 10^3$, and

(a) $C = 5 \times 10^{-6}$.
(b) $C = 4 \times 10^{-6}$.
(c) $C = 2 \times 10^{-6}$.

31. Find a differential equation $y'' + ay' + by = 0$ that is satisfied by both functions.

(a) $y_1 = e^{2x}$,　$y_2 = e^{-4x}$.
(b) $y_1 = 3 e^{-x}$,　$y_2 = 4 e^{5x}$.
(c) $y_1 = 2 e^{3x}$,　$y_2 = xe^{3x}$.

32. Find a differential equation $y'' + ay' + by = 0$ that is satisfied by both functions.

(a) $y_1 = 2 \cos 2x$,　$y_2 = -\sin 2x$.
(b) $y_1 = e^{-2x} \cos 3x$,　$y_2 = 2e^{-2x} \sin 3x$.

33. (a) Show that the substitution $y = e^{\alpha x} u$ transforms

$$y'' - 2\alpha y' + \alpha^2 y = 0　　\text{into}　　u'' = 0.$$

(b) Show that the substitution $y = e^{\alpha x} u$ transforms

$$y'' - 2\alpha y' + (\alpha^2 + \beta^2)y = 0　\text{into}　u'' + \beta^2 u = 0.$$

Exercises 34 and 35 relate to the differential equation $y'' + ay' + by = 0$, where a and b are nonnegative constants.

$^\dagger L$ is inductance, R is resistance, and C is capacitance. If L is given in henrys, R in ohms, C in farads, and t in seconds, then the current is given in amperes.

34. Prove that if a and b are both positive, then $y(x) \to 0$ as $x \to \infty$ for all solutions y of the equation.

35. (a) Prove that if $a = 0$ and $b > 0$, then all solutions of the equation are bounded.
(b) Suppose that $a > 0$, $b = 0$, and $y = y(x)$ is a solution of the equation. Prove that

$$\lim_{x \to \infty} y(x) = k$$

for some constant k. Determine k for the solution that satisfies the initial conditions: $y(0) = y_0$, $y'(0) = y_1$.

36. (*Important*) Let y_1, y_2 be solutions of the homogeneous equation. Show that the Wronskian of y_1, y_2 is zero iff one of these functions is a scalar multiple of the other.

37. Let y_1, y_2 be solutions of the homogeneous equation. Show that if $y_1(x_0) = y_2(x_0) = 0$ for some number x_0, then one of these functions is a scalar multiple of the other.

(Euler equation) An equation of the form

$$(*)　　　　x^2 y'' + \alpha x y' + \beta y = 0,$$

where α and β are real numbers, is called an *Euler equation*.

38. Show that the Euler equation $(*)$ can be transformed into an equation of the form

$$\frac{d^2 y}{dz^2} + a\frac{dy}{dz} + by = 0$$

where a and b are real numbers, by means of the change of variable $z = \ln x$.　　HINT:　If $z = \ln x$, then by the chain rule,

$$\frac{dy}{dx} = \frac{dy}{dz}\frac{dz}{dx} = \frac{dy}{dz}\frac{1}{x}.$$

Now calculate $d^2 y/dx^2$ and substitute the result into the differential equation.

Exercises 39–42. Use the change of variable indicated in Exercise 38 to transform the given equation into an equation with constant coefficients. Find the general solution of that equation, and then express it in terms of x.

39. $x^2 y'' - xy' - 8y = 0$.　　　**40.** $x^2 y'' - 2xy' + 2y = 0$.
41. $x^2 y'' - 3xy' + 4y = 0$.　　**42.** $x^2 y'' - xy' + 5y = 0$.

---

*SUPPLEMENT TO SECTION 9.3

PROOF OF THEOREM 9.3.2

Existence: Take two solutions y_1, y_2 with nonzero Wronskian

$$W(x) = y_1(x)y_2'(x) - y_2(x)y_1'(x).$$

For any numbers x_0, α_0, α_1 the equations

$$C_1 y_1(x_0) + C_2 y_2(x_0) = \alpha_0$$

$$C_1 y_1'(x_0) + C_2 y_2'(x_0) = \alpha_1$$

can be solved for C_1 and C_2. For those values of C_1 and C_2, the function

$$y = C_1 y_1 + C_2 y_2$$

is a solution of the homogeneous equation that satisfies the prescribed initial conditions.

Uniqueness: Let us assume that there are two distinct solutions y_1, y_2 that satisfy the same prescribed initial conditions

$$y_1(x_0) = \alpha_0 = y_2(x_0) \quad \text{and} \quad y_1'(x_0) = \alpha_1 = y_2'(x_0).$$

Then the solution $y = y_1 - y_2$ satisfies the initial conditions

$$y(x_0) = 0, y'(x_0) = 0.$$

Since y_1 and y_2 are, by assumption, distinct functions, there is at least one number x_1 at which y is not zero. Therefore, by the continuity of y there exists an interval I on which y does not take on the value zero.

Now let u be *any* solution of the homogeneous equation. The Wronskian of y and u is zero at x_0:

$$W(x_0) = y(x_0)u'(x_0) - u(x_0)y'(x_0) = (0)u'(x_0) - u(x_0)(0) = 0.$$

Therefore the Wronskian of y and u is everywhere zero. Since $y(x) \neq 0$ for all $x \in I$, the quotient u/y is defined on I, and on that interval

$$\frac{d}{dx}\left(\frac{u}{y}\right) = \frac{yu' - uy'}{y^2} = \frac{W}{y^2} = 0, \qquad \frac{u}{y} = C, \qquad \text{and} \qquad u = Cy.$$

We have shown that on the interval I every solution is some scalar multiple of y.

Now let u_1 and u_2 be any two solutions with a nonzero Wronskian W. From what we have just shown, there are constants C_1 and C_2 such that on I

$$u_1 = C_1 y \qquad \text{and} \qquad u_2 = C_2 y.$$

Then on I

$$W = u_1 u_2' - u_2 u_1' = (C_1 y)(C_2 y') - (C_2 y)(C_1 y') = C_1 C_2 (yy' - yy') = 0.$$

This contradicts the statement that $W \neq 0$.

The assumption that there are two distinct solutions that satisfy the same prescribed initial conditions has led to a contradiction. This proves uniqueness.

■ CHAPTER 9. REVIEW EXERCISES

Exercises 1–10. Find the general solution.

1. $y' + y = 2e^{-2x}$.

2. $\dfrac{dy}{dx} = \dfrac{2y \cos 2x}{2y^2 + 1}$.

3. $y^2 + 1 = yy' \sec^2 x$.

4. $\dfrac{y}{x}y' = \dfrac{e^x}{\ln y}$.

5. $x^2 y' + 3xy = \sin 2x$.

6. $xy' + 2y = 2e^{x^2}$.

7. $\dfrac{dy}{dx} = 1 + x^2 + y^2 + x^2 y^2$.

8. $y' = \dfrac{x^2 y - y}{y + 1}$.

9. $\dfrac{dy}{dx} = \dfrac{x^3 - 2y}{x}$.

10. $\dfrac{dy}{dx} = xy^2 \sqrt{1 + x^2}$.

Exercises 11–14. Solve the initial-value problem.

11. $x^2 y' + xy = 2 + x^2$; $\quad y(1) = 2$.

12. $yy' = 4x\sqrt{y^2 + 1}$; $\quad y(0) = 1$.

13. $e^{-x}y' = e^x - 2e^x y$; $\quad y(0) = \dfrac{1}{2} + \dfrac{1}{e}$.

14. $\dfrac{dy}{dx} = \sec y \tan x$; $\quad y(0) = \dfrac{\pi}{2}$.

Exercises 15–22. Find the general solution.

15. $y'' - 2y' + 2y = 0$.

16. $y'' + y' + \frac{1}{4}y = 0$.

17. $y'' - y' - 2y = 0$.

18. $y'' - 4y' = 0$.

19. $y'' - 6y' + 9y = 0$.

20. $y'' + 4y = 0$.

21. $y'' + 4y' + 13y = 0$.

22. $3y'' - 5y' - 2y = 0$.

Exercises 23–26. Solve the initial-value problem.

23. $y'' - y' = 0$; $\quad y(0) = 1$, $y'(0) = 0$.

24. $y'' + 7y' + 12y = 0$; $\quad y(0) = 2$, $y'(0) = 8$.

25. $y'' - 6y' + 13y = 0$; $\quad y(0) = 2$, $y'(0) = 2$.

26. $y'' + 4y' + 4y = 0$; $\quad y(-1) = 2$, $y'(-1) = 1$.

Exercises 27–28. Find the orthogonal trajectories of the family of curves.

27. $y = Ce^{2x}$.

28. $y = \dfrac{C}{1 + x^2}$.

Exercises 29–30. Find the values of r, if any, for which $y = x^r$ is a solution of the equation.

29. $x^2 y'' + 4xy + 2y = 0$.   **30.** $x^2 y'' - xy' - 8y = 0$.

31. An investor has found a business that is increasing in value at a rate proportional to the square of its present value. If the business was worth 1 million dollars one year ago and is worth 1.5 million dollars today, how much will it be worth 1 year from now? One and one-half years from now? Two years from now?

32. An investor has found a business that is increasing in value at a rate proportional to the square root of its present value. If the business was worth 1 million dollars two years ago and is worth 1.44 million dollars today, how much will it be worth 5 years from now? When will the business be worth 4 million dollars?

33. The rate at which a certain drug is absorbed into the bloodstream is described by the differential equation

$$\frac{dy}{dt} = a - by$$

where a and b are positive constants and $y = y(t)$ is the amount of the drug in the bloodstream at time t (hours).

(a) Find the solution of the differential equation that satisfies the initial condition $y(0) = 0$.

(b) Show that $\lim_{t \to \infty} y(t)$ exists and give the value of this limit.

(c) How long will it take for the concentration of the drug to reach 90% of its limiting value?

34. A metal bar at a temperature of $100°C$ is placed in a freezer kept at a constant temperature of $0°C$. After 20 minutes the temperature of the bar is $50°C$.

(a) What is the temperature of the bar after 30 minutes?

(b) How long will it take for the temperature of the bar to reach $25°C$?

35. An object at an unknown temperature is placed in a room which is held at a constant temperature of $70°F$. After 10 minutes the temperature of the object is $20°F$ and after 20 minutes its temperature is $35°F$.

(a) Find an expression for the temperature of the object at time t.

(b) What was the temperature of the object when it was placed in the room?

36. A 1200-gallon tank initially contains 40 pounds of salt dissolved in 600 gallons of water. Water containing 1/2 pound of salt per gallon is poured into the tank at the rate of 6 gallons per minute. The mixture is continually stirred and is drained from the tank at the rate of 4 gallon per minute.

(a) Find T, the length of time needed to fill the tank.

(b) Find the amount of salt in the tank at any time t, $0 \leq t \leq T$.

(c) Find the amount of salt in the tank at the instant it overflows.

37. A tank initially holds 80 gallons of a brine solution containing 1/8 pounds of salt per gallon. Another brine solution containing 1 pound of salt per gallon is poured into the tank at the rate of 4 gallons per minute. The mixture is continuously stirred and is drained from the tank at the rate of 8 gallons per minute.

(a) Find T, the length of time needed to empty the tank.

(b) Find the amount of salt in the tank at any time t, $0 \leq t \leq T$.

(c) Find the amount of salt in the tank at the instant it contains exactly 40 gallons of the solution.

▶ 38. The differential equation

$$\frac{dP}{dt} = P(10^{-1} - 10^{-5}P)$$

models the population of a certain community. Assume that $P(0) = 2000$ and that time t is measured in months.

(a) Find $P(t)$ and show that $\lim_{t \to \infty} P(t)$ exists.

(b) Use a graphing utility to draw the graph of P and estimate how long it will take for the population to reach 90% of its limiting value.

39. A rumor is spreading through a town with a population of 20,000. The rumor is spreading at a rate proportional to the product of the number of people who have heard it and the number of people who have not heard it. Ten days ago 500 people had heard the rumor; today 1200 have heard it.

(a) How many people will have heard the rumor 10 days from now?

(b) At what time will the rumor be spreading the fastest?

40. If a flexible cable of uniform density is suspended between two fixed points at equal height, then the shape $y = y(x)$ of the cable must satisfy the initial-value problem

(1) $\dfrac{d^2 y}{dx^2} = \dfrac{1}{a}\sqrt{1 + \left[\dfrac{dy}{dx}\right]^2}$, $y(0) = a$, $y'(0) = 0$.

Setting $u = dy/dx$, we have the equation

(2) $\dfrac{du}{dx} = \dfrac{1}{a}\sqrt{1 + u^2}$.

(a) Solve equation (2).

(b) Integrate to find the shape of the cable.

CHAPTER 10

THE CONIC SECTIONS;

POLAR COORDINATES;

PARAMETRIC EQUATIONS

■ 10.1 GEOMETRY OF PARABOLA, ELLIPSE, HYPERBOLA

You are familiar with parabola, ellipse, hyperbola in the sense that you recognize the equations of these curves and the general shape. Here we define these curves geometrically (Figure 10.1.1), derive the equations from the geometric definition, and explain the role played by these curves in the reflection of light and sound. (See Figure 10.1.1.)

Geometric Definition

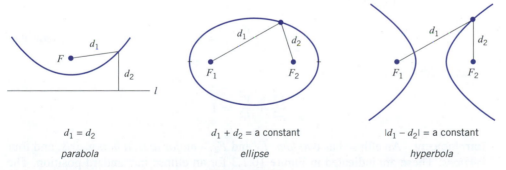

$d_1 = d_2$	$d_1 + d_2 = $ a constant	$|d_1 - d_2| = $ a constant
parabola	*ellipse*	*hyperbola*

Figure 10.1.1

Parabola

Standard Position F on the positive y-axis, l horizontal. Then F has coordinates of the form $(c, 0)$ with $c > 0$ and l has equation $x = -c$.

Equation

$$x^2 = 4cy, \qquad c > 0.$$

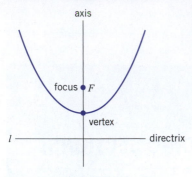

Figure 10.1.2

Derivation of the Equation A point $P(x, y)$ lies on the parabola iff $d_1 = d_2$, which here means

$$\sqrt{x^2 + (y - c)^2} = y + c.$$

This equation simplifies to

$$x^2 = 4cy. \qquad \text{(verify this)}$$

Terminology A parabola has a *focus*, a *directrix*, a *vertex*, and an *axis*. These are indicated in Figure 10.1.2.

Ellipse

Standard Position F_1 and F_2 on the x-axis at equal distances c from the origin. Then F_1 is at $(-c, 0)$ and F_2 at $(c, 0)$. With d_1 and d_2 as in the defining figure, set $d_1 + d_2 = 2a$.

Equation

$$\frac{x^2}{a^2} + \frac{y^2}{a^2 - c^2} = 1.$$

Setting $b = \sqrt{a^2 - c^2}$, we have

$$\boxed{\frac{x^2}{a^2} + \frac{y^2}{b^2} = 1.} \qquad \text{(the familiar equation)}$$

Derivation of the Equation A point $P(x, y)$ lies on the ellipse iff $d_1 + d_2 = 2a$, which here means iff

$$\sqrt{(x + c)^2 + y^2} + \sqrt{(x - c)^2 + y^2} = 2a.$$

This equation simplifies to

$$\frac{x^2}{a^2} + \frac{y^2}{a^2 - c^2} = 1, \qquad \text{(verify this)}$$

which, with $b = \sqrt{a^2 - c^2}$, is

$$\frac{x^2}{a^2} + \frac{y^2}{b^2} = 1.$$

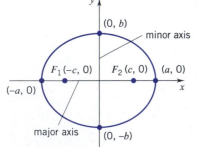

Figure 10.1.3

Terminology An ellipse has two *foci*, F_1 and F_2, a *major axis, a minor axis*, and four *vertices*. These are indicated in Figure 10.1.3 for an ellipse in standard position. The point at which the axes of the ellipse intersect is called the *center* of the ellipse.

Hyperbola

Standard Position F_1 and F_2 on the x-axis at equal distances c from the origin. Then F_1 is at $(-c, 0)$ and F_2 at $(c, 0)$. With d_1 and d_2 as in the defining figure, set $|d_1 - d_2| = 2a$.

Equation

$$\frac{x^2}{a^2} - \frac{y^2}{c^2 - a^2} = 1.$$

Setting $b = \sqrt{c^2 - a^2}$, we have

$$\boxed{\frac{x^2}{a^2} - \frac{y^2}{b^2} = 1.}$$ (the familiar equation)

Derivation of the Equation A point $P(x, y)$ lies on the hyperbola iff $|d_1 - d_2| = 2a$, which here means iff

$$\sqrt{(x + c)^2 + y^2} - \sqrt{(x - c)^2 + y^2} = \pm 2a.$$

This equation simplifies to

$$\frac{x^2}{a^2} - \frac{y^2}{c^2 - a^2} = 1,$$ (verify this)

which, with $b = \sqrt{c^2 - a^2}$, is

$$\frac{x^2}{a^2} - \frac{y^2}{b^2} = 1.$$

Terminology A hyperbola has two *foci*, F_1 and F_2, two *vertices*, a *transverse axis* that joins the two vertices, and two *asymptotes*. These are indicated in Figure 10.1.4 for a hyperbola in standard position. The midpoint of the transverse axis is called the center of the hyperbola.

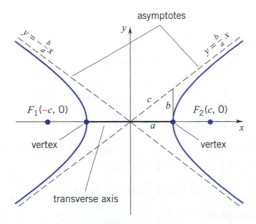

Figure 10.1.4

Translations

Suppose that x_0 and y_0 are real numbers and S is a set in the xy-plane. By replacing each point (x, y) of S by $(x + x_0, y + y_0)$, we obtain a set S' which is congruent to S and obtained from S without any rotation. (Figure 10.1.5.) Such a displacement is called a *translation*.

The translation

$$(x, y) \longrightarrow (x + x_0, y + y_0)$$

applied to a curve C with equation $E(x, y) = 0^\dagger$ results in a curve C' with equation $E(x - x_0, y - y_0) = 0$.

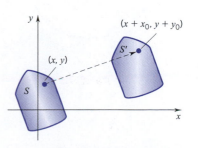

Figure 10.1.5

---

†Every equation in x and y can be written in this manner: simply transfer the right side of the equation to the left side.

PROOF The coordinates of (x, y) satisfy the equation $E(x, y) = 0$

iff

the coordinates of $(x + x_0, y + y_0)$ satisfy the equation $E(x - x_0, y - y_0) = 0$. ❏

Examples

(1) The translation $(x, y) \rightarrow (x - 1, y + 3)$ moves points one unit left and three units up. Applying this translation to the parabola with equation $y = \frac{1}{2}x^2$, we obtain the parabola with equation $y - 3 = \frac{1}{2}(x + 1)^2$.

(2) The translation $(x, y) \rightarrow (x + 5, y - 4)$ moves points five units right and four units down. Applying this translation to the ellipse with equation

$$\frac{x^2}{9} + \frac{y^2}{4} = 1,$$

we obtain the ellipse with equation

$$\frac{(x - 5)^2}{9} + \frac{(y + 4)^2}{4} = 1. \quad ❏$$

Earlier (Exercise 57, Section 1.4) you were asked to show that the distance between the origin and any line $l : Ax + By + C = 0$ is given by the formula

$(*)$ $$d(0, l) = \frac{|C|}{\sqrt{A^2 + B^2}}.$$

By means of a translation we can show that the distance between any point $P(x_0, y_0)$ and the line $l : Ax + By + C = 0$ is given by the formula

(10.1.1)
$$d(P, l) = \frac{|Ax_0 + By_0 + C|}{\sqrt{A^2 + B^2}}.$$

PROOF The translation $(x, y) \rightarrow (x - x_0, y - y_0)$ takes the point $P(x_0, y_0)$ to the origin O and the line $l : Ax + By + C = 0$ to the line

$$l' : A(x + x_0) + B(y + y_0) + C = 0.$$

We can write this equation as

$$Ax + By + K = 0 \quad \text{with} \quad K = \overbrace{Ax_0 + By_0 + C}^{\text{(a constant)}}.$$

Applying $(*)$ to l', we have

$$d(O, l') = \frac{|K|}{\sqrt{A^2 + B^2}} = \frac{|Ax_0 + By_0 + C|}{\sqrt{A^2 + B^2}}.$$

Since P and l have been moved the same distance in the same direction, P to O and l to l', we can conclude that $d(P, l) = d(O, l')$. This confirms (10.1.1). ❏

We will rely on this result later in the section.

Parabolic Mirrors

The discussion below is based on the geometric principle of reflected light (introduced in Example 5, Section 4.5): *the angle of incidence equals the angle of reflection.*

Take a parabola and revolve it about its axis. This gives you a parabolic surface. A curved mirror of this form is called a *parabolic mirror*. Such mirrors are used in searchlights (automotive headlights, flashlights, etc.) and in reflecting telescopes. Our purpose here is to explain the reason for this.

We begin with a parabola and choose the coordinate system so that the equation takes the form $x^2 = 4cy$ with $c > 0$. We can express y in terms of x by writing

$$y = \frac{x^2}{4c}.$$

Since

$$\frac{dy}{dx} = \frac{2x}{4c} = \frac{x}{2c},$$

the tangent line at the point $P(x_0, y_0)$ has slope $m = x_0/2c$ and has equation

(1)
$$(y - y_0) = \frac{x_0}{2c}(x - x_0).$$

For the rest, we refer to Figure 10.1.6.

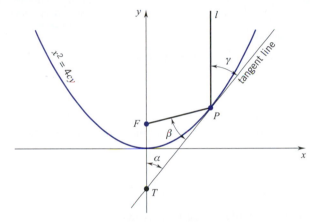

Figure 10.1.6

In the figure we have drawn a ray (a half-line) l parallel to the axis of the parabola, which in this setup is the y-axis. We want to show that the angles marked β and γ are equal.

Since the tangent line at $P(x_0, y_0)$ is not vertical, it intersects the y-axis at some point T. To find the coordinates of T, we set $x = 0$ in (1) and solve for y. This gives

$$y = y_0 - \frac{x_0^2}{2c}.$$

Since the point (x_0, y_0) lies on the parabola, we know that $x_0^2 = 4cy_0$ and therefore

$$y = y_0 - \frac{x_0^2}{2c} = y_0 - \frac{4cy_0}{2c} = -y_0.$$

The y-coordinate of T is $-y_0$. Since the focus F is at $(c, 0)$,

$$d(F, T) = y_0 + c.$$

The distance between F and P is also $y_0 + c$:

$$d(F, P) = \sqrt{x_0^2 + (y_0 - c)^2} = \sqrt{4cy_0 + (y_0 - c)^2} = \sqrt{(y_0 + c)^2} = y_0 + c.$$

$$(x_0^2 = 4cy_0) \longrightarrow \qquad\qquad \longrightarrow (y_0 + c > 0)$$

Since $d(F, T) = d(F, P)$, the triangle TFP is isosceles and the angles marked α and β are equal. Since l is parallel to the y-axis, $\alpha = \gamma$ and thus (and this is what we wanted to show) $\beta = \gamma$.

The fact that $\beta = \gamma$ has important optical consequences. It means (Figure 10.1.7) that light emitted from a source at the focus of a parabolic mirror is reflected in a beam parallel to the axis of that mirror; this is the principle of the searchlight. It also means that light coming to a parabolic mirror in a beam parallel to the axis of the mirror is reflected entirely to the focus; this is the principle of the reflecting telescope.

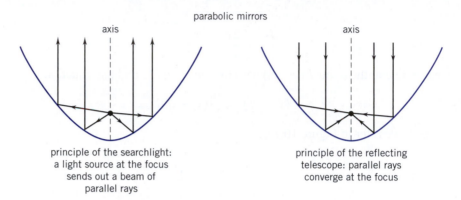

principle of the searchlight: a light source at the focus sends out a beam of parallel rays

principle of the reflecting telescope: parallel rays converge at the focus

Figure 10.1.7

Elliptical Reflectors

Like the parabola, the ellipse has an interesting reflecting property. To derive it, we work with the ellipse

$$\frac{x^2}{a^2} + \frac{y^2}{b^2} = 1.$$

Differentiating implicitly with respect to x, we get

$$\frac{2x}{a^2} + \frac{2y}{b^2}\frac{dy}{dx} = 0 \qquad \text{and thus} \qquad \frac{dy}{dx} = -\frac{b^2 x}{a^2 y}.$$

The slope of the ellipse at a point $P(x_0, y_0)$ not on the x-axis is therefore

$$-\frac{b^2 x_0}{a^2 y_0},$$

and the tangent line at that point has equation

$$y - y_0 = -\frac{b^2 x_0}{a^2 y_0}(x - x_0).$$

We can rewrite this last equation as

$$(b^2 x_0)x + (a^2 y_0)y - a^2 b^2 = 0.$$

We can now show the following:

(10.1.2)

At each point P of the ellipse, the focal radii $\overline{F_1 P}$ and $\overline{F_2 P}$ make equal angles with the tangent.

PROOF If P lies on the x-axis, the focal radii both lie on the x-axis and the result is clear. To visualize the argument for a point $P = P(x_0, y_0)$ not on the x-axis, see Figure 10.1.8.

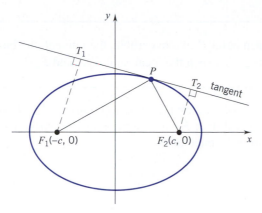

Figure 10.1.8

To show that $\overline{F_1 P}$ and $\overline{F_2 P}$ make equal angles with the tangent, we need only show that the triangles $P T_1 F_1$ and $P T_2 F_2$ are similar. We can do this by showing that

$$\frac{d(T_1, F_1)}{d(F_1, P)} = \frac{d(T_2, F_2)}{d(F_2, P)}$$

which, in view of (10.1.1), can be done by showing that

$$\frac{|-b^2 x_0 c - a^2 b^2|}{\sqrt{(x_0 + c)^2 + y_0^2}} = \frac{|b^2 x_0 c - a^2 b^2|}{\sqrt{(x_0 - c)^2 + y_0^2}}. \qquad \text{(verify this)}$$

This equation simplifies to

$$\frac{(x_0 c + a^2)^2}{(x_0 + c)^2 + y_0^2} = \frac{(x_0 c - a^2)^2}{(x_0 - c)^2 + y_0^2}$$

$$(a^2 - c^2) x_0^2 + a^2 y_0^2 = a^2 (a^2 - c^2)$$

$$\frac{x_0^2}{a^2} + \frac{y_0^2}{b^2} = 1. \qquad (b = \sqrt{a^2 - c^2})$$

This last equation holds since the point $P(x_0, y_0)$ is on the ellipse. ❑

The result we just proved has the following physical consequence:

(10.1.3) An elliptical reflector takes light or sound originating at one focus and reflects it to the other focus.

In elliptical rooms called "whispering chambers," a whisper at one focus, inaudible nearby, is easily heard at the other focus. You will experience this phenomenon if you visit the Statuary Room in the Capitol in Washington, D.C. In many hospitals there are elliptical water tubs designed to break up kidney stones. The patient is positioned so that the stone is at one focus. Small vibrations set off at the other focus are so efficiently concentrated that the stone is shattered.

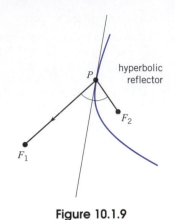

Figure 10.1.9

Hyperbolic Reflectors

A straightforward calculation that you are asked to carry out in the Exercises shows that:

(10.1.4)

> At each point P of a hyperbola, the tangent line bisects the angle between the focal radii $\overline{F_1 P}$ and $\overline{F_2 P}$.

For some consequences of this, we refer you to Figure 10.1.9. There you see the right branch of a hyperbola with foci F_1, F_2. Light or sound aimed at F_2 from any point to the left of the reflector is beamed to F_1.

EXERCISES 10.1

Exercises 1–8. Find the vertex, focus, axis, and directrix of the given parabola. Then sketch the parabola.

1. $y = \frac{1}{2}x^2$.

2. $y = -\frac{1}{2}x^2$.

3. $y = \frac{1}{2}(x-1)^2$.

4. $y = -\frac{1}{2}(x-1)^2$.

5. $y + 2 = \frac{1}{4}(x-2)^2$.

6. $y - 2 = \frac{1}{4}(x+2)^2$.

7. $y = x^2 - 4x$.

8. $y = x^2 + x + 1$.

Exercises 9–16. An ellipse is given. Find the center, the foci, the length of the major axis, and the length of the minor axis. Then sketch the ellipse.

9. $x^2/9 + y^2/4 = 1$.

10. $x^2/4 + y^2/9 = 1$.

11. $3x^2 + 2y^2 = 12$.

12. $3x^2 + 4y^2 - 12 = 0$.

13. $4x^2 + 9y^2 - 18y = 27$.

14. $4x^2 + y^2 - 6y + 5 = 0$.

15. $4(x-1)^2 + y^2 = 64$.

16. $16(x-2)^2 + 25(y-3)^2 = 400$.

Exercises 17–26. A hyperbola is given. Find the center, the vertices, the foci, the asymptotes, and the length of the transverse axis. Then sketch the hyperbola.

17. $x^2 - y^2 = 1$.

18. $y^2 - x^2 = 1$.

19. $x^2/9 - y^2/16 = 1$.

20. $x^2/16 - y^2/9 = 1$.

21. $y^2/16 - x^2/9 = 1$.

22. $y^2/9 - x^2/16 = 1$.

23. $(x-1)^2/9 - (y-3)^2/16 = 1$.

24. $(x-1)^2/16 - (y-3)^2/9 = 1$.

25. $4x^2 - 8x - y^2 + 6y - 1 = 0$.

26. $-3x^2 + y^2 - 6x = 0$.

27. A parabola intersects a rectangle of area A at two opposite vertices. Show that, if one side of the rectangle falls on the axis of the parabola, then the parabola subdivides the rectangle into two pieces, one of area $\frac{1}{3}A$, the other of area $\frac{2}{3}A$.

28. A line through the focus of a parabola intersects the parabola at two points P and Q. Show that the tangent line through P is perpendicular to the tangent line through Q.

29. Show that the graph of every quadratic function $y = ax^2 + bx + c$ is a parabola. Find the vertex, the focus, the axis, and the directrix.

30. Find the centroid of the first-quadrant portion of the elliptical region $b^2x^2 + a^2y^2 \leq a^2b^2$.

31. Find the center, the vertices, the foci, the asymptotes, and the length of the transverse axis of the hyperbola with equation $xy = 1$. HINT: Define new XY-coordinates by setting $x = X + Y$ and $y = X - Y$.

32. As t ranges from 0 to 2π, the points $(a\cos t, b\sin t)$ generate a curve in the xy-plane. Identify the curve.

33. An ellipse has area A and major axis of length $2a$. What is the distance between the foci?

34. A searchlight reflector is in the shape of a parabolic mirror. If it is 5 feet in diameter and 2 feet deep at the center, how far is the focus from the vertex of the mirror?

The line that passes through the focus of a parabola and is parallel to the directrix intersects the parabola at two points A and B. The line segment $\overline{AB}$ is called the *latus rectum* of the parabola.

In Exercises 35–38 we work with the parabola $x^2 = 4cy$, $c > 0$. By Ω we mean the region bounded below by the parabola and above by the latus rectum.

35. Find the length of the latus rectum.

36. What is the slope of the parabola at the endpoints of the latus rectum?

37. Determine the area of Ω and locate the centroid.

38. Find the volume of the solid generated by revolving Ω about the y-axis and locate the centroid of the solid. (For the centroid formulas, see Project 6.4.)

39. Suppose that a flexible inelastic cable (see the figure) fixed at the ends supports a horizontal load. (Imagine a suspension bridge and think of the load on the cable as the roadway.) Show that, if the load has constant weight per unit length, then the cable hangs in the form of a parabola.

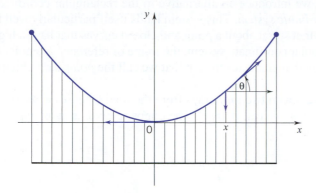

HINT: The part of the cable that supports the load from 0 to x is subject to the following forces:

(1) the weight of the load, which in this case is proportional to x.
(2) the horizontal pull at $0 : p(0)$.
(3) the tangential pull at $x : p(x)$.

Balancing the vertical forces, we have

$$kx = p(x)\sin\theta. \qquad \text{(weight = vertical pull at } x)$$

Balancing the horizontal forces, we have

$$p(0) = p(x)\cos\theta. \qquad \text{(pull at 0 = horizontal pull at } x)$$

40. A lighting panel is perpendicular to the axis of a parabolic mirror. Show that all light rays beamed parallel to this axis are reflected to the focus of the mirror *in paths of the same length.*

41. All equilateral triangles are similar; they differ only in scale. Show that the same is true of all parabolas.

For Exercises 42–44 we refer to a hyperbola in standard position.

42. Find functions $x = x(t)$, $y = y(t)$ such that, as t ranges over the set of real numbers, the points $(x(t), y(t))$ traverse

   (a) the right branch of the hyperbola.
   (b) the left branch of the hyperbola.

43. Find the area of the region between the right branch of the hyperbola and the vertical line $x = 2a$.

44. Show that at each point P of the hyperbola the tangent line bisects the angle between the focal radii $\overline{F_1 P}$ and $\overline{F_2 P}$.

Although all parabolas have exactly the same shape (Exercise 41), ellipses come in different shapes. The shape of an ellipse depends on its *eccentricity e.* This is half the distance between the foci divided by half the length of the major axis:

(10.1.5)

$$e = c/a.$$

For every ellipse, $0 < e < 1$.

Exercises 45–48. Determine the eccentricity of the ellipse.

45. $x^2/25 + y^2/16 = 1$.   **46.** $x^2/16 + y^2/25 = 1$.

47. $(x - 1)^2/25 + (y + 2)^2/9 = 1$.

48. $(x + 1)^2/169 + (y - 1)^2/144 = 1$.

49. Suppose that E_1 and E_2 are both ellipses with the same major axis. Compare the shape of E_1 to the shape of E_2 if $e_1 < e_2$.

50. What happens to an ellipse with major axis $2a$ if e tends to 0?

51. What happens to an ellipse with major axis $2a$ if e tends to 1?

Exercises 52–53. Write an equation for the ellipse.

52. Major axis from $(-3, 0)$ to $(3, 0)$, eccentricity $\frac{1}{3}$.

53. Major axis from $(-3, 0)$ to $(3, 0)$, eccentricity $\frac{2}{3}\sqrt{2}$.

54. Let l be a line and let F be a point not on l. You have seen that the set of points P for which

$$d(F, P) = d(l, P)$$

is a parabola. Show that, if $0 < e < 1$, then the set of all points P for which

$$d(F, P) = e\, d(l, P)$$

is an ellipse of eccentricity e. HINT: Begin by choosing a coordinate system whereby F falls on the origin and l is a vertical line $x = k$.

The shape of a hyperbola is determined by its *eccentricity e.* This is half the distance between the foci divided by half the length of the transverse axis:

(10.1.6)

$$e = c/a.$$

For all hyperbolas, $e > 1$.

Exercises 55–58. Determine the eccentricity of the hyperbola.

55. $x^2/9 - y^2/16 = 1$.   **56.** $x^2/16 - y^2/9 = 1$.

57. $x^2 - y^2 = 1$.   **58.** $x^2/25 - y^2/144 = 1$.

59. Suppose H_1 and H_2 are both hyperbolas with the same transverse axis. Compare the shape of H_1 to the shape of H_2 if $e_1 < e_2$.

60. What happens to a hyperbola if e tends to 1?

61. What happens to a hyperbola if e increases without bound?

62. (Compare to Exercise 54.) Let l be a line and let F be a point not on l. Show that, if $e > 1$, then the set of all points P for which

$$d(F, P) = e\, d(l, P)$$

is a hyperbola of eccentricity, e. HINT: Begin by choosing a coordinate system whereby F falls on the origin and l is a vertical line $x = k$.

63. Show that every parabola has an equation of the form

$$(\alpha x + \beta y)^2 = \gamma x + \delta y + \epsilon \qquad \text{with} \qquad \alpha^2 + \beta^2 \neq 0.$$

HINT: Take $l : Ax + By + C = 0$ as the directrix, $F(a, b)$ as the focus.

■ 10.2 POLAR COORDINATES

We use coordinates to indicate position with respect to a frame of reference. When we use rectangular coordinates, our frame of reference is a pair of lines that intersect at right angles. In this section we introduce an alternative to the rectangular coordinate system called the *polar coordinate system*. This system lends itself particularly well to the representation of curves that spiral about a point and closed curves that have a high degree of symmetry. In the polar coordinate system, the frame of reference is a point O that we call the *pole* and a ray that emanates from it that we call the *polar axis*. (Figure 10.2.1)

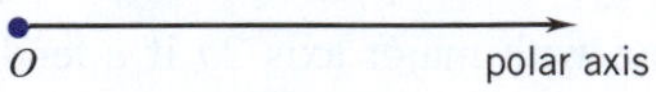

In Figure 10.2.2 we have drawn two more rays from the pole One lies at an angle of θ radians from the polar axis; we call it ray θ. The opposite ray lies at an angle of $\theta + \pi$ radians; we call it ray $\theta + \pi$.

Figure 10.2.1

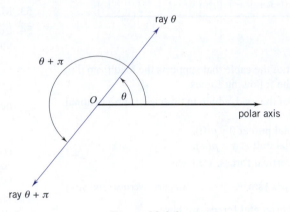

Figure 10.2.2

Figure 10.2.3 shows some points along these rays, labeled with *polar coordinates*.

(10.2.1)

In general, a point is assigned *polar coordinates* $[r, \theta]$ if it lies at a distance $|r|$ from the pole

on the ray θ, if $r \geq 0$ and   on the ray $\theta + \pi$ if $r < 0$.

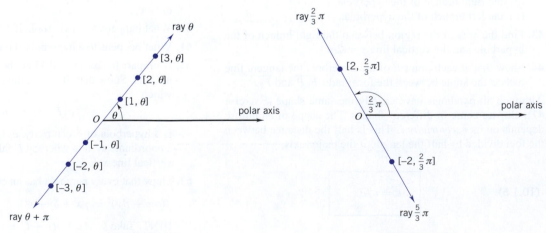

Figure 10.2.3   **Figure 10.2.4**

Figure 10.2.4 shows the point $[2, \frac{2}{3}\pi]$. The point lies two units from the pole on the ray $\frac{2}{3}\pi$. The point $[-2, \frac{2}{3}\pi]$ also lies two units from the pole, not on the ray $\frac{2}{3}\pi$, but on the opposite ray, the ray $\frac{5}{3}\pi$.

Polar coordinates are not unique. Many pairs $[r, \theta]$ can represent the same point.

(1) If $r = 0$, it does not matter how we choose θ. The resulting point is still the pole:

(10.2.2)
$$O = [0, \theta] \qquad \text{for all } \theta.$$

(2) Geometrically there is no distinction between angles that differ by an integer multiple of 2π. Consequently, as suggested in Figure 10.2.5,

(10.2.3)
$$[r, \theta] = [r, \theta + 2n\pi], \qquad \text{for all integers } n.$$

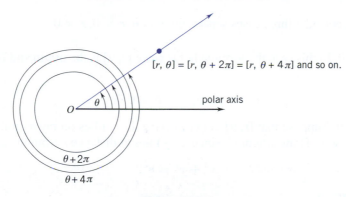

Figure 10.2.5

(3) Adding π to the second coordinate is equivalent to changing the sign of the first coordinate:

(10.2.4)
$$[r, \theta + \pi] = [-r, \theta]. \qquad \text{(Figure 10.2.6)}$$

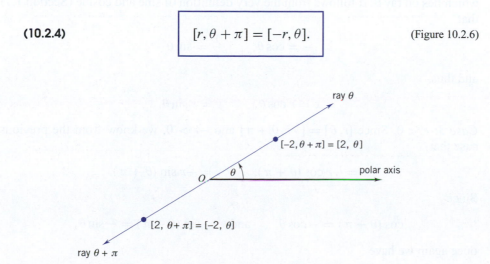

Figure 10.2.6

Remark Some authors do not allow r to take on negative values. There are some advantages to this approach. For example, the polar coordinates of a point are unique if θ is restricted to the interval $[0, 2\pi)$ or to $(-\pi, \pi]$. On the other hand, there are advantages in graphing and finding points of intersection that follow from letting r take on any real value, and this is the approach that we have adopted. Since there is no convention on this issue, you should be aware of the two approaches. ❏

Relation to Rectangular Coordinates

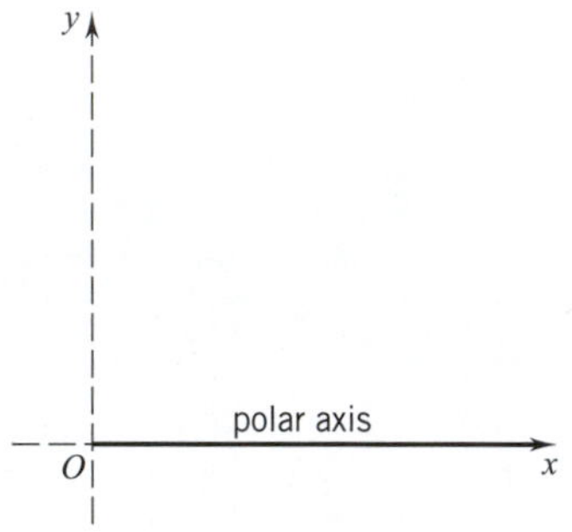

In Figure 10.2.7 we have superimposed a polar coordinate system on a rectangular coordinate system. We have placed the pole at the origin and directed the polar axis along the positive x-axis.

The relation between polar coordinates $[r, \theta]$ and rectangular coordinates (x, y) is given by the following equations:

Figure 10.2.7

(10.2.5)
$$x = r \cos \theta, \qquad y = r \sin \theta.$$

PROOF We'll consider three cases separately: $r = 0, r > 0, r < 0$.

Case 1: $r = 0$. In this case the formula holds since $[r, \theta]$ is the origin and both x and y are 0:

$$0 = 0 \cos \theta, \qquad 0 = 0 \sin \theta.$$

Case 2: $r > 0$. Suppose that $[r, \theta] = (x, y)$. Then (x, y) lies on ray θ, and $(x/r, y/r)$ also lies on ray θ. (Draw a figure.) Since (x, y) lies at distance r from the origin,

$$x^2 + y^2 = r^2.$$

It follows that

$$\left(\frac{x^2}{r^2}\right) + \left(\frac{y^2}{r^2}\right) = 1.$$

This places $(x/r, y/r)$ on the unit circle. Thus $(x/r, y/r)$ is the point on the unit circle which lies on ray θ. It follows from the very definition of sine and cosine (Section 1.7) that

$$\frac{x}{r} = \cos \theta, \qquad \frac{y}{r} = \sin \theta$$

and thus

$$x = r \cos \theta, \qquad y = r \sin \theta.$$

Case 3: $r < 0$. Since $[r, \theta] = [-r, \theta + \pi]$ and $-r > 0$, we know from the previous case that

$$x = -r \cos (\theta + \pi), \qquad y = -r \sin (\theta + \pi).$$

Since

$$\cos (\theta + \pi) = -\cos \theta \qquad \text{and} \qquad \sin (\theta + \pi) = -\sin \theta,$$

once again we have

$$x = r \cos \theta, \qquad y = r \sin \theta. \quad ❏$$

From the relations we just proved, it should be clear that, unless $x = 0$,

(10.2.6)
$$\tan\theta = \frac{y}{x},$$

and, under all circumstances,

(10.2.7)
$$x^2 + y^2 = r^2.$$

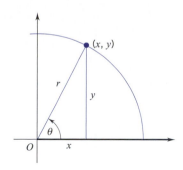

Figure 10.2.8

Figure 10.2.8 illustrates all this for points in the first quadrant.

Example 1 Find the rectangular coordinates of the point P with polar coordinates $[-2, \frac{1}{3}\pi]$.

SOLUTION The relations $x = r\cos\theta$, $y = r\sin\theta$ give

$$x = -2\cos\tfrac{1}{3}\pi = -2(\tfrac{1}{2}) = -1, \qquad y = -2\sin\tfrac{1}{3}\pi = -2(\tfrac{1}{2}\sqrt{3}) = -\sqrt{3}.$$

The point P has rectangular coordinates $(-1, -\sqrt{3})$. ❏

Example 2 Find all possible polar coordinates for the point P that has rectangular coordinates $(-2, 2\sqrt{3})$.

SOLUTION We know that $r\cos\theta = -2$, $r\sin\theta = 2\sqrt{3}$. It follows that
$$r^2 = r^2\cos^2\theta + r^2\sin^2\theta = (-2)^2 + (2\sqrt{3})^2 = 16,$$

so that $r = \pm 4$.

Taking $r = 4$, we have

$$4\cos\theta = -2 \qquad 4\sin\theta = 2\sqrt{3}$$
$$\cos\theta = -\tfrac{1}{2} \qquad \sin\theta = \tfrac{1}{2}\sqrt{3}.$$

These equations are satisfied by setting $\theta = \frac{2}{3}\pi$, or more generally by setting

$$\theta = \tfrac{2}{3}\pi + 2n\pi.$$

The polar coordinates of P with first coordinate $r = 4$ are all pairs of the form

$$[4, \tfrac{2}{3}\pi + 2n\pi],$$

where n ranges over the set of integers.

We could go through the same process again, this time taking $r = -4$, but there is no need to do so. Since $[r, \theta] = [-r, \theta + \pi]$, we know that

$$[4, \tfrac{2}{3}\pi + 2n\pi] = [-4, (\tfrac{2}{3}\pi + \pi) + 2n\pi].$$

The polar coordinates of P with first coordinate $r = -4$ are thus all pairs of the form

$$[-4, \tfrac{5}{3}\pi + 2n\pi]$$

where again n ranges over the set of integers. ❏

Here are some simple sets specified in polar coordinates. We leave it to you to draw appropriate figures.

(1) The circle of radius a centered at the origin is given by the equation

$$r = a.$$

The interior of the circle is given by $r < a$ and the exterior by $r > a$.

(2) The line that passes through the origin with an inclination of α radians has polar equation

$$\theta = \alpha.$$

(3) For $a \neq 0$, the vertical line $x = a$ has polar equation

$$r \cos \theta = a \qquad \text{or, equivalently,} \qquad r = a \sec \theta.$$

(What's the equation if $a = 0$?)

(4) For $b \neq 0$, the horizontal line $y = b$ has polar equation

$$r \sin \theta = b \qquad \text{or, equivalently,} \qquad r = b \csc \theta.$$

(What's the equation if $b = 0$?)

Example 3 Find an equation in polar coordinates for the hyperbola $x^2 - y^2 = a^2$.

SOLUTION Setting $x = r \cos \theta$ and $y = r \sin \theta$, we have

$$r^2 \cos^2 \theta - r^2 \sin^2 \theta = a^2$$
$$r^2(\cos^2 \theta - \sin^2 \theta) = a^2$$
$$r^2 \cos 2\theta = a^2. \quad \square$$

Example 4 Show that the equation $r = 2a \cos \theta$ represents a circle. (Take $a > 0$ and see Figure 10.2.9.)

SOLUTION Multiplication by r gives

$$r^2 = 2ar \cos \theta$$
$$x^2 + y^2 = 2ax$$
$$x^2 - 2ax + y^2 = 0$$
$$x^2 - 2ax + a^2 + y^2 = a^2$$
$$(x - a)^2 + y^2 = a^2.$$

This is the circle of radius a centered at the point with rectangular coordinates $(a, 0)$. $\quad \square$

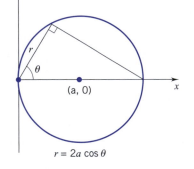

$r = 2a \cos \theta$

Figure 10.2.9

Symmetry

Symmetry about the x- and y-axes and symmetry about the origin are illustrated in Figure 10.2.10. The coordinates marked are, of course, not the only ones possible. (The difficulties that can arise from this are explored in Section 10.3.)

Example 5 Test the curve $r^2 = \cos 2\theta$ for symmetry.

SOLUTION Since $\cos [2(-\theta)] = \cos (-2\theta) = \cos 2\theta$, you can see that if $[r, \theta]$ is on the curve, then so is $[r, -\theta]$. This tells us that the curve is symmetric about the x-axis. Since

$$\cos [2(\pi - \theta)] = \cos (2\pi - 2\theta) = \cos (-2\theta) = \cos 2\theta,$$

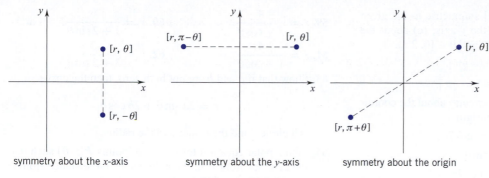

Figure 10.2.10

you can see that if $[r, \theta]$ is on the curve, then so is $[r, \pi - \theta]$. The curve is therefore symmetric about the y-axis.

Being symmetric about both axes, the curve must be symmetric about the origin. You can verify this directly by noting that

$$\cos [2(\pi + \theta)] = \cos (2\pi + 2\theta) = \cos 2\theta,$$

so that if $[r, \theta]$ lies on the curve, then so does $[r, \pi + \theta]$. A sketch of the curve appears in Figure 10.2.11. Such a curve is called a *lemniscate*. ❑

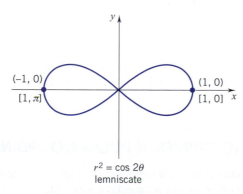

$r^2 = \cos 2\theta$
lemniscate

Figure 10.2.11

EXERCISES 10.2

Exercises 1–8. Plot the point with these polar coordinates.

1. $[1, \frac{1}{3}\pi]$.   **2.** $[1, \frac{1}{2}\pi]$.

3. $[-1, \frac{1}{3}\pi]$.   **4.** $[-1, -\frac{1}{3}\pi]$.

5. $[4, \frac{5}{4}\pi]$.   **6.** $[-2, 0]$.

7. $[-\frac{1}{2}, \pi]$.   **8.** $[\frac{1}{3}, \frac{2}{3}\pi]$.

Exercises 9–16. Give the rectangular coordinates of the point.

9. $[3, \frac{1}{2}\pi]$.   **10.** $[4, \frac{1}{6}\pi]$.

11. $[-1, -\pi]$.   **12.** $[-1, \frac{1}{4}\pi]$.

13. $[-3, -\frac{1}{3}\pi]$ **14.** $[2, 0]$.

15. $[3, -\frac{1}{2}\pi]$.   **16.** $[2, 3\pi]$.

Exercises 17–24. Below some points are specified in rectangular coordinates. Give all possible polar coordinates for each point.

17. $(0, 1)$.   **18.** $(1, 0)$.

19. $(-3, 0)$.   **20.** $(4, 4)$.

21. $(2, -2)$.   **22.** $(3, -3\sqrt{3})$.

23. $(4\sqrt{3}, 4)$.   **24.** $(\sqrt{3}, -1)$.

25. Find a formula for the distance between $[r_1, \theta_1]$ and $[r_2, \theta_2]$.

26. Show that for $r_1 > 0, r_2 > 0, |\theta_1 - \theta_2| < \pi$ the distance formula you found in Exercise 25 reduces to the law of cosines.

Exercises 27–30. Find the point $[r, \theta]$ symmetric to the given point (a) about the x-axis; (b) about the y-axis; (c) about the origin. Express your answer with $r > 0$ and $\theta \in [0, 2\pi)$.

27. $[\frac{1}{2}, \frac{1}{6}\pi]$.   **28.** $[3, -\frac{5}{4}\pi]$.

29. $[-2, \frac{1}{3}\pi]$.   **30.** $[-3, -\frac{7}{4}\pi]$.

Exercises 31–36. Test the curve for symmetry about the coordinate axes and for symmetry about the origin.

31. $r = 2 + \cos\theta$.   **32.** $r = \cos 2\theta$.

33. $r(\sin\theta + \cos\theta) = 1$.   **34.** $r\sin\theta = 1$.

35. $r^2\sin 2\theta = 1$.   **36.** $r^2\cos 2\theta = 1$.

Exercises 37–48. Write the equation in polar coordinates.

37. $x = 2$.   **38.** $y = 3$.

39. $2xy = 1$.   **40.** $x^2 + y^2 = 9$.

41. $x^2 + (y - 2)^2 = 4$.   **42.** $(x - a)^2 + y^2 = a^2$.

43. $y = x$.   **44.** $x^2 - y^2 = 4$.

45. $x^2 + y^2 + x = \sqrt{x^2 + y^2}$.

46. $y = mx$.   **47.** $(x^2 + y^2)^2 = 2xy$.

48. $(x^2 + y^2)^2 = x^2 - y^2$.

Exercises 49–58. Identify the curve and write the equation in rectangular coordinates.

49. $r\sin\theta = 4$.   **50.** $r\cos\theta = 4$.

51. $\theta = \frac{1}{3}\pi$.   **52.** $\theta^2 = \frac{1}{9}\pi^2$.

53. $r = 2(1 - \cos\theta)^{-1}$.   **54.** $r = 2\sin\theta$.

55. $r = 6\cos\theta$.   **56.** $\theta = -\frac{1}{2}\pi$.

57. $\tan\theta = 2$.   **58.** $r = 4\sin(\theta + \pi)$.

Exercises 59–62. Write the equation in rectangular coordinates and identify the curve.

59. $r = \dfrac{4}{2 - \cos\theta}$.   **60.** $r = \dfrac{6}{1 + 2\sin\theta}$.

61. $r = \dfrac{4}{1 - \cos\theta}$.   **62.** $r = \dfrac{2}{3 + 2\sin\theta}$.

63. Show that if a and b are not both zero, then the curve

$$r = 2a\sin\theta + 2b\cos\theta$$

is a circle. Find the center and the radius.

64. Find a polar equation for the set of points $P[r, \theta]$ such that the distance from P to the pole equals the distance from P to the line $x = -d$. Take $d \geq 0$. See the figure.

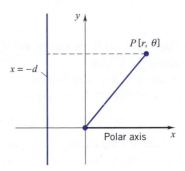

65. Find a polar equation for the set of points $P[r, \theta]$ such that the distance from P to the pole is half the distance from P to the line $x = -d$. Take $d > 0$.

66. Find a polar equation for the set of points $P[r, \theta]$ such that the distance from P to the pole is twice the distance from P to the line $x = -d$. Take $d > 0$.

■ 10.3 SKETCHING CURVES IN POLAR COORDINATES

Here we sketch some curves that are (relatively) simple in polar coordinates $[r, \theta]$ but devilishly difficult to work with in rectangular coordinates (x, y).

Example 1   Sketch the curve $r = \theta, \theta \geq 0$ in polar coordinates.

SOLUTION   At $\theta = 0, r = 0$; at $\theta = \frac{1}{4}\pi, r = \frac{1}{4}\pi$; at $\theta = \frac{1}{2}\pi, r = \frac{1}{2}\pi$; and so on. The curve is shown in detail from $\theta = 0$ to $\theta = 2\pi$ in Figure 10.3.1. It is an unending spiral, *the spiral of Archimedes*. More of the spiral is shown on a smaller scale in the right part of the figure.   ❑

Now to some closed curves.

Example 2   Sketch the curve $r = 1 - 2\cos\theta$ in polar coordinates.

SOLUTION   Since the cosine function is periodic with period 2π, the curve $r = 1 - 2\cos\theta$ is a closed curve which repeats itself on every θ-interval of length 2π. We will draw the curve from $\theta = 0$ to $\theta = 2\pi$. That will account for r in every direction.

We begin by representing the function $r = 1 - 2\cos\theta$ in rectangular coordinates (θ, r). This puts us in familiar territory and enables us to see at a glance how r varies with θ.

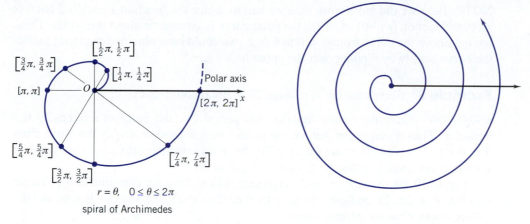

$r = \theta, \quad 0 \le \theta \le 2\pi$

spiral of Archimedes

Figure 10.3.1

In Figure 10.3.2 we have marked the values of θ where r is zero and the values of θ where r takes on an extreme value.

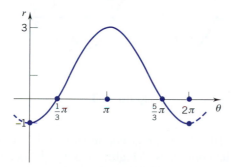

Figure 10.3.2

Reading from the figure we have the following: as θ increases from 0 to $\frac{1}{3}\pi$, r increases from -1 to 0; as θ increases from $\frac{1}{3}\pi$ to π, r increases from 0 to 3; as θ increases from π to $\frac{5}{3}\pi$, r decreases from 3 to 0; finally, as θ increases from $\frac{5}{3}\pi$ to 2π, r decreases from 0 to -1.

By applying this information step by step, we develop a sketch of the curve $r = 1 - 2\cos\theta$ in polar coordinates. (Figure 10.3.3.)

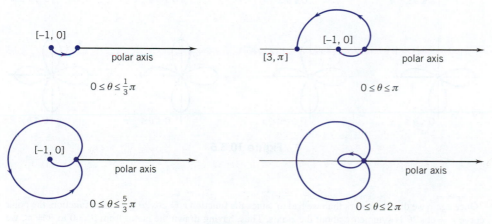

Figure 10.3.3

NOTE: Here we did more work then we had to. Since the function $r = 1 - 2\cos\theta$ is an even function $[r(-\theta) = r(\theta)]$, the polar curve is symmetric about the x-axis. Then, having drawn the curve from $\theta = 0$ to $\theta = \pi$, we could have obtained the lower half of the curve simply by flipping over the upper half. ❏

Example 3 Sketch the curve $r = \cos 2\theta$ in polar coordinates.

SOLUTION Since the cosine function has period 2π, the function $r = \cos 2\theta$ has period π. Thus it may seem that we can restrict ourselves to sketching the curve from $\theta = 0$ to $\theta = \pi$. But this is not the case. To obtain the complete curve, we must account for r in every direction; that is, from $\theta = 0$ to $\theta = 2\pi$.

Figure 10.3.4 shows $r = \cos 2\theta$ represented in rectangular coordinates (θ, r) from $\theta = 0$ to $\theta = 2\pi$. In the figure we have marked the values of θ where r is zero and the values where r has an extreme value.

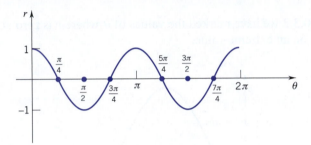

Figure 10.3.4

Translating Figure 10.3.4 into polar coordinates $[r, \theta]$, we obtain a sketch of the curve $r = \cos 2\theta$ in polar coordinates. (Figure 10.3.5.) The sketch is developed in eight stages. These stages are determined by the values of θ marked in Figure 10.3.4.[†] ❏

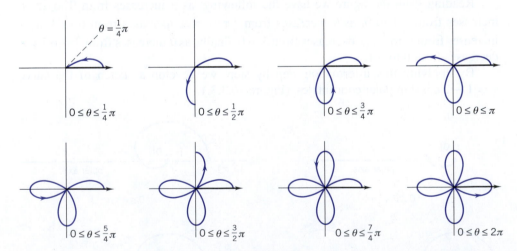

Figure 10.3.5

[†]Once again we did more work than we had to. Since the function $r = \cos 2\theta$ is an even function, the polar curve $r = \cos 2\theta$ is symmetric about the x-axis. Thus, having drawn the curve from $\theta = 0$ to $\theta = \pi$, we could have obtained the rest of the curve by reflection in the horizontal axis.

Example 4 Figure 10.3.6 shows four *cardioids*, heart-shaped curves. Rotation of $r = 1 + \cos\theta$ by $\frac{1}{2}\pi$ radians (measured in the counterclockwise direction) gives

$$r = 1 + \cos\left(\theta - \tfrac{1}{2}\pi\right) = 1 + \sin\theta.$$

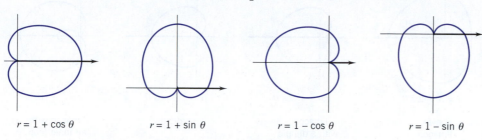

$r = 1 + \cos\theta$ $r = 1 + \sin\theta$ $r = 1 - \cos\theta$ $r = 1 - \sin\theta$

Figure 10.3.6

Rotation by another $\frac{1}{2}\pi$ radians gives

$$r = 1 + \cos(\theta - \pi) = 1 - \cos\theta.$$

Rotation by yet another $\frac{1}{2}\pi$ radians gives

$$r = 1 + \cos\left(\theta - \tfrac{3}{2}\pi\right) = 1 - \sin\theta.$$

Note how easy it is to rotate axes in polar coordinates: each change

$$\cos\theta \rightarrow \sin\theta \rightarrow -\cos\theta \rightarrow -\sin\theta$$

represents a counterclockwise rotation by $\frac{1}{2}\pi$ radians. ❏

At this point we will try to give you a brief survey of some of the basic polar curves. (The numbers a and b that appear below are to be interpreted as nonzero constants.)

Lines : $\theta = a$, $r = a\sec\theta$, $r = a\csc\theta$.   (Figure 10.3.7)

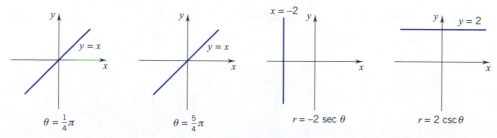

$\theta = \frac{1}{4}\pi$ $\theta = \frac{5}{4}\pi$ $r = -2\sec\theta$ $r = 2\csc\theta$

Figure 10.3.7

Circles : $r = a$, $r = a\sin\theta$, $r = a\cos\theta$.   (Figure 10.3.8)

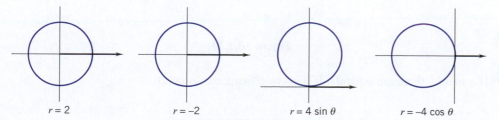

$r = 2$ $r = -2$ $r = 4\sin\theta$ $r = -4\cos\theta$

Figure 10.3.8

Limaçons[†]: $r = a + b \sin \theta$, $\qquad r = a + b \cos \theta$. $\hspace{2cm}$ (Figure 10.3.9)

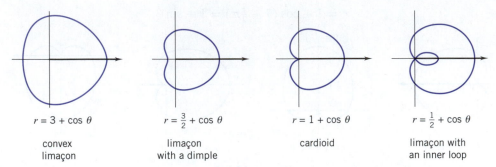

$r = 3 + \cos \theta$	$r = \frac{3}{2} + \cos \theta$	$r = 1 + \cos \theta$	$r = \frac{1}{2} + \cos \theta$
convex limaçon	limaçon with a dimple	cardioid	limaçon with an inner loop

Figure 10.3.9

The general shape of the curve depends on the relative magnitudes of a and b.

Lemniscates[††]: $r^2 = a \sin 2\theta$, $\qquad r^2 = a \cos 2\theta$ $\hspace{2cm}$ (Figure 10.3.10)

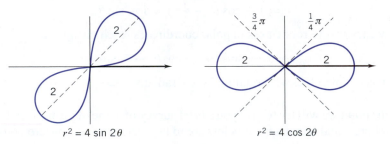

$r^2 = 4 \sin 2\theta$ $\hspace{3cm}$ $r^2 = 4 \cos 2\theta$

Figure 10.3.10

Petal Curves: $r = a \sin n\theta$, $\qquad r = a \cos n\theta$, $\qquad$ integer n. $\hspace{1.5cm}$ (Figure 10.3.11)

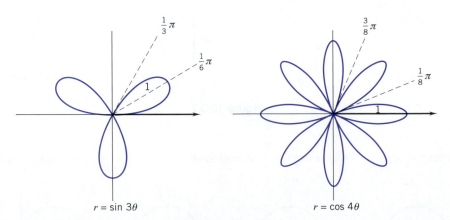

$r = \sin 3\theta$ $\hspace{4cm}$ $r = \cos 4\theta$

Figure 10.3.11

If n is *odd*, there are n petals. If n is *even*, there are $2n$ petals.

[†]From the French term for "snail". The word is pronounced with a soft c.
[††]From the Latin *lemniscatus*, meaning "adorned with pendant ribbons."

The Intersection of Polar Curves

The fact that a single point has many pairs of polar coordinates can cause complications. In particular, it means that a point $[r_1, \theta_1]$ can lie on a curve given by a polar equation although the coordinates r_1 and θ_1 do not satisfy the equation. For example, the coordinates of $[2, \pi]$ do not satisfy the equation $r^2 = 4\cos\theta$:

$$r^2 = 2^2 = 4 \qquad \text{but} \qquad 4\cos\theta = 4\cos\pi = -4.$$

Nevertheless the point $[2, \pi]$ does lie on the curve $r^2 = 4\cos\theta$. We know this because $[2, \pi] = [-2, 0]$ and the coordinates of $[-2, 0]$ do satisfy the equation:

$$r^2 = (-2)^2 = 4, \qquad 4\cos\theta = 4\cos 0 = 4$$

In general, a point $P[r_1, \theta_1]$ lies on a curve given by a polar equation if it has at least one polar coordinate representation $[r, \theta]$ with coordinates that satisfy the equation. The difficulties are compounded when we deal with two or more curves. Here is an example.

Example 5 Find the points where the cardioids

$$r = a(1 - \cos\theta) \qquad \text{and} \qquad r = a(1 + \cos\theta) \qquad\qquad (a > 0)$$

intersect.

SOLUTION We begin by solving the two equations simultaneously. Adding these. equations, we get $2r = 2a$ and thus $r = a$, Given that $r = a$, we can conclude that $\cos\theta = 0$ and therefore $\theta = \frac{1}{2}\pi + n\pi$. The points $[a, \frac{1}{2}\pi + n\pi]$ all lie on both curves. Not all of these points are distinct:

for n even, $\left[a, \frac{1}{2}\pi + n\pi\right] = \left[a, \frac{1}{2}\pi\right]$:   for n odd, $\left[a, \frac{1}{2}\pi + n\pi\right] = \left[a, \frac{3}{2}\pi\right].$

In short, by solving the two equations simultaneously we have arrived at two common points:

$$\left[a, \tfrac{1}{2}\pi\right] = (0, a) \qquad \text{and} \qquad \left[a, \tfrac{3}{2}\pi\right] = (0, -a).$$

However, by sketching the two curves (see Figure 10.3.12), we see that there is a third point at which the curves intersect; the two curves intersect at the origin, which clearly lies on both curves:

for $r = a(1 - \cos\theta)$ take $\theta = 0, 2\pi, \ldots,$
for $r = a(1 + \cos\theta)$ take $\theta = \pi, 3\pi, \ldots,$

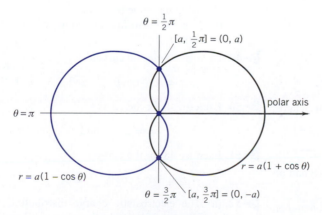

Figure 10.3.12

The reason that the origin does not appear when we solve the two equations simultaneously is that the curves do not pass through the origin "simultaneously"; that is, they do not pass through the origin for the same values of θ. Think of each of the equations

$$r = a(1 - \cos\theta) \qquad \text{and} \qquad r = a(1 + \cos\theta)$$

as giving the position of an object at time θ. At the points we found by solving the two equations simultaneously, the objects collide. (They both arrive there at the same time.) At the origin the situation is different. Both objects pass through the origin, but no collision takes place because the objects pass through the origin at *different* times. ❏

Remark Problems of incidence [does the point P lie on the curve $r = \rho(\theta)$?] and problems of intersection [where do the polar curves $r = \rho_1(\theta)$ and $r = \rho_2(\theta)$ intersect?] can usually be analyzed by sketching the curves. However, there are situations where such problems can be handled more readily by first changing to rectangular coordinates (x, y). ❏

EXERCISES 10.3

Exercises 1–32. Sketch the polar curve.

1. $\theta = -\frac{1}{4}\pi$.

2. $r = -3$.

3. $r = 4$.

4. $r = 3\cos\theta$.

5. $r = -2\sin\theta$.

6. $\theta = \frac{2}{3}\pi$.

7. $r\csc\theta = 3$.

8. $r = 1 - \cos\theta$.

9. $r = \theta, \quad -\frac{1}{2}\pi \leq \theta \leq \pi$.

10. $r\sec\theta = -2$.

11. $r = \sin 3\theta$.

12. $r^2 = \cos 2\theta$.

13. $r^2 = \sin 2\theta$.

14. $r = \cos 2\theta$.

15. $r^2 = 4, \quad 0 \leq \theta \leq \frac{3}{4}\pi$.

16. $r = \sin\theta$.

17. $r^3 = 9r$.

18. $\theta = -\frac{1}{4}, \quad 1 \leq r < 2$.

19. $r = -1 + \sin\theta$.

20. $r^2 = 4r$.

21. $r = \sin 2\theta$.

22. $r = \cos 3\theta, \quad 0 \leq \theta \leq \frac{1}{2}\pi$.

23. $r = \cos 5\theta, \quad 0 \leq \theta \leq \frac{1}{2}\pi$.

24. $r = e^\theta, \quad -\pi \leq \theta \leq \pi$.

25. $r = 2 + \sin\theta$.

26. $r = \cot\theta$.

27. $r = \tan\theta$.

28. $r = 2 - \cos\theta$.

29. $r = 2 + \sec\theta$.

30. $r = 3 - \csc\theta$.

31. $r = -1 + 2\cos\theta$.

32. $r = 1 + 2\sin\theta$.

Exercises 33–36. Determine whether the point lies on the curve.

33. $r^2\cos\theta = 1; \quad [1, \pi]$.

34. $r^2 = \cos 2\theta; \quad [1, \frac{1}{4}\pi]$.

35. $r = \sin\frac{1}{3}\theta; \quad [\frac{1}{2}, \frac{1}{2}\pi]$

36. $r^2 = \sin 3\theta; \quad [1, -\frac{5}{6}\pi]$.

37. Show that the point $[2, \pi]$ lies both on $r^2 = 4\cos\theta$ and on $r = 3 + \cos\theta$.

38. Show that the point $[2, \frac{1}{2}\pi]$ lies both on $r^2\sin\theta = 4$ and on $r = 2\cos 2\theta$.

Exercises 39–46. Sketch the curves and find the points at which they intersect. Express your answers in rectangular coordinates.

39. $r = \sin\theta, \quad r = -\cos\theta$.

40. $r^2 = \sin\theta, \quad r = 2 - \sin\theta$.

41. $r = \cos^2\theta, \quad r = -1$.

42. $r = 2\sin\theta, \quad r = 2\cos\theta$.

43. $r = 1 - \cos\theta, \quad r = \cos\theta$.

44. $r = 1 - \cos\theta, \quad r = \sin\theta$.

45. $r = \sin 2\theta, \quad r = \sin\theta$.

46. $r = 1 - \cos\theta, \quad r = 1 + \sin\theta$.

47. (a) Use a graphing utility to draw the curves

$$r = 1 + \cos\left(\theta - \frac{1}{3}\pi\right) \quad \text{and} \quad r = 1 + \cos\left(\theta + \frac{1}{6}\pi\right).$$

Compare these curves to the curve $r = 1 + \cos\theta$.

(b) More generally, compare the curve $r = f(\theta - \alpha)$ to the curve $r = f(\theta)$. (Take $\alpha > 0$.)

48. (a) Use a graphing utility to draw the curves

$$r = 1 + \sin\theta \qquad \text{and} \qquad r^2 = 4\sin 2\theta$$

using the same polar axis.

(b) Use a CAS to find the points where the two curves intersect.

49. Exercise 48 for the curves

$$r = 1 + \cos\theta \qquad \text{and} \qquad r = 1 + \cos\frac{1}{2}\theta.$$

50. Exercise 48 for the curves

$$r = 2 \qquad \text{and} \qquad r = 2\sin 3\theta.$$

51. Exercise 48 for the curves

$$r = 1 - 3\cos\theta \qquad \text{and} \qquad r = 2 - 5\sin\theta.$$

52. (a) The electrostatic charge distribution consisting of a charge q ($q > 0$) at the point $[r, 0]$ and a charge $-q$ at $[r, \pi]$ is called a *dipole*. The *lines of force* for the dipole are given by the equations

$$r = k\sin^2\theta.$$

Use a graphing utility to draw the lines of force for $k = 1, 2, 3$.

(b) The *equipotential lines* (the set of points with equal electric potential) for the dipole are given by the equations

$$r^2 = m \cos \theta$$

Use a graphing utility to draw the equipotential lines for $m = 1, 2, 3$.

(c) Draw the curves $r = 2 \sin^2 \theta$ and $r^2 = 2 \cos \theta$ using the same polar axis. Estimate the xy-coordinates of the points where the two curves intersect.

▶ 53. Use a graphing utility to draw the curves

$$r = 1 + \sin k\theta + \cos^2 2k\theta$$

for $k = 1, 2, 3, 4, 5$. Suggest a name for such curves.

▶ 54. Use a graphing utility to draw the curve $r = e^{\cos\theta} - 2 \cos 4\theta$. Suggest a name for this curve.

▶ 55. The curves $r = A \cos k\theta$ and $r = A \sin k\theta$ are known as *petal curves*. (See Figure 10.3.11.) Use a graphing utility to draw the curves

$$r = 2 \cos k\theta \qquad \text{and} \qquad r = 2 \sin k\theta$$

for $k = \frac{3}{2}$ and $k = \frac{5}{2}$. Form a conjecture about the shape of these curves for numbers k of the form $(2m + 1)/2$.

▶ 56. Use a graphing utility to draw the curves

$$r = 2 \cos k\theta \qquad \text{and} \qquad r = 2 \sin k\theta$$

for $k = \frac{4}{3}$ and $k = \frac{5}{3}$. Make a conjecture about the shape of these curves for $k = m/3$ (a) m even, not a multiple of 3; (b) m odd, not a multiple of 3.

■ PROJECT 10.3 Parabola, Ellipse, Hyperbola in Polar Coordinates

In Section 10.1 we defined a parabola in terms of a focus and a directrix, but our definitions of the ellipse and hyperbola were given in terms of two foci; there was no mention of a directrix. In this project we give a unified approach to the conic sections that involves a focus and a directrix in all three cases.

Let F be a point of the plane and l a line which does not pass through F. We call F the *focus* and l the *directrix*. Let e be a positive number (the *eccentricity*) and consider the set of points P that satisfy the condition

(1) $$\frac{\text{distance from } P \text{ to } F}{\text{distance from } P \text{ to } l} = e.$$

In the figure, we have superimposed a rectangular coordinate system on a polar coordinate system. We have placed F at the origin and taken l as the vertical line $x = d, d > 0$.

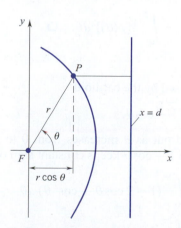

Problem 1. Show that the set of points P that satisfy condition (1) is given by the polar equation

(2) $$r = \frac{ed}{1 + e \cos \theta}.$$

Problem 2. Verify the following statements.

a. If $0 < e < 1$, equation (2) gives an ellipse of eccentricity e with right focus at the origin, major axis horizontal:

$$\frac{(x + c)^2}{a^2} + \frac{y^2}{a^2 - c^2} = 1 \quad \text{with} \quad a = \frac{ed}{1 - e^2}, \quad c = ea.$$

b. If $e = 1$, equation (2) gives a parabola with focus at the origin and directrix $x = d$:

$$y^2 = -4\frac{d}{2}\left(x - \frac{d}{2}\right).$$

c. If $e > 1$, equation (2) gives a hyperbola of eccentricity e with left focus at the origin, transverse axis horizontal:

$$\frac{(x - c)^2}{a^2} - \frac{y^2}{c^2 - a^2} = 1 \quad \text{with} \quad a = \frac{ed}{e^2 - 1}, \quad c = ea.$$

Problem 3. Identify the curve and write the equation in rectangular coordinates.

a. $r = \dfrac{8}{4 + 3 \cos \theta}$.

b. $r = \dfrac{6}{1 + 2 \cos \theta}$.

c. $r = \dfrac{6}{2 + 2 \cos \theta}$.

Problem 4. Taking α and β as positive constants, relate the curves

$$r = \frac{\alpha}{1 + \beta \sin \theta}, \qquad r = \frac{\alpha}{1 - \beta \cos \theta}, \qquad r = \frac{\alpha}{1 - \beta \sin \theta}$$

to the curve

$$r = \frac{\alpha}{1 + \beta \cos \theta}.$$

HINT: Example 4.

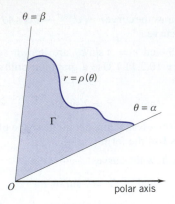

Figure 10.4.1

■ 10.4 AREA IN POLAR COORDINATES

Here we develop a way of calculating the area of a region with boundary given in polar coordinates.

As a start, we suppose that α and β are two real numbers with $\alpha < \beta \leq \alpha + 2\pi$. We take $\rho^\dagger$ as a function that is continuous on $[\alpha, \beta]$ and keeps constant sign on that interval. We want the area of the polar region Γ generated by the curve

$$r = \rho(\theta), \qquad \alpha \leq \theta \leq \beta.$$

Such a region is portrayed in Figure 10.4.1.

In the figure $\rho(\theta)$ remains nonnegative. If $\rho(\theta)$ were negative, the region Γ would appear on the opposite side of the pole. In either case, the area of Γ is given by the formula

(10.4.1)

$$A = \int_\alpha^\beta \tfrac{1}{2}[\rho(\theta)]^2 d\theta.$$

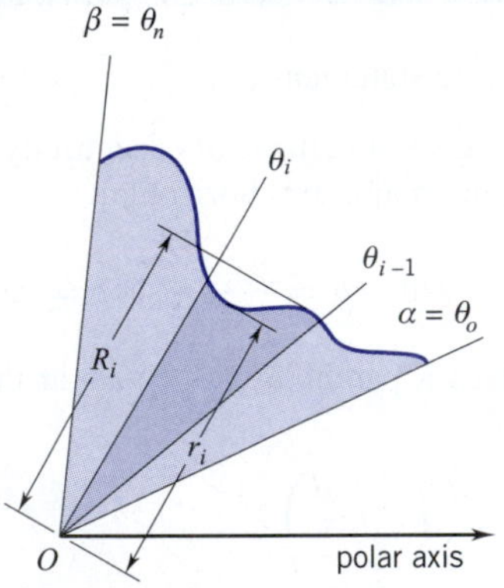

PROOF We consider the case where $\rho(\theta) \geq 0$. We take $P = \{\theta_1, \theta_2, \ldots, \theta_n\}$ as a partition of $[\alpha, \beta]$ and direct our attention to the region from θ_{i-1} to θ_i. We set

$$r_i = \text{min value of } \rho \text{ on } [\theta_{i-1}, \theta_i] \qquad \text{and} \qquad R_i = \text{max value of } \rho \text{ on } [\theta_{i-1}, \theta_i].$$

The part of Γ that lies from θ_{i-1} to θ_i contains a circular sector of radius r_i and central angle $\Delta\theta_i - \theta_{i-1}$ and is contained in a circular sector of radius R_i with the same central angle $\Delta\theta_i$. (See Figure 10.4.2.) Its area A_i must therefore satisfy the inequality

$$\tfrac{1}{2}r_i^2\Delta\theta_i \leq A \leq \tfrac{1}{2}R_i^2\Delta\theta_i.^{\dagger\dagger}$$

By adding up these inequalities from $i = 1$ to $1 = n$, we can see that the total area A of Γ must satisfy the inequality

(1)

$$L_f(P) \leq A \leq U_f(P)$$

where $f(\theta) = \tfrac{1}{2}[\rho(\theta)]^2$. Since f is continuous and (1) holds for every partition P of $[\alpha, \beta]$, we can conclude that

$$A = \int_\alpha^\beta f(\theta)\, d\theta = \int_\alpha^\beta \tfrac{1}{2}[\rho(\theta)]^2 d\theta. \quad \square$$

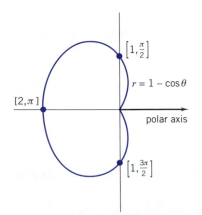

Figure 10.4.2

Figure 10.4.3

Example 1 Calculate the area enclosed by the cardioid

$$r = 1 - \cos\theta, \qquad \text{(Figure 10.4.3)}$$

SOLUTION The entire curve is traced out as θ increases from 0 to 2π. Note that $1 - \cos\theta \geq 0$ for all θ in $[0, 2\pi]$. Since $1 - \cos\theta$ keeps constant sign on $[0, 2\pi]$,

$$A = \int_0^{2\pi} \tfrac{1}{2}(1 - \cos\theta)^2 d\theta = \tfrac{1}{2} \int_0^{2\pi} (1 - 2\cos\theta + \cos^2\theta)\, d\theta$$

$$= \tfrac{1}{2} \int_0^{2\pi} (\tfrac{3}{2} - 2\cos\theta + \tfrac{1}{2}\cos 2\theta)\, d\theta.$$

↑———— half-angle formula: $\cos^2\theta = \tfrac{1}{2} + \tfrac{1}{2}\cos 2\theta$

---

†The symbol ρ is the lower case Greek letter "rho."

†† The area of a circular sector of radius r and central angle α is $\tfrac{1}{2}r^2\alpha$.

Since

$$\int_0^{2\pi} \cos\theta\, d\theta = \sin\theta \Big|_0^{2\pi} = 0 \qquad \text{and} \qquad \int_0^{2\pi} \cos 2\theta\, d\theta = \tfrac{1}{2}\sin 2\theta \Big|_0^{2\pi} = 0,$$

we have

$$A = \tfrac{1}{2}\int_0^{2\pi} \tfrac{3}{2}\, d\theta = \tfrac{3}{4}\int_0^{2\pi} d\theta = \tfrac{3}{2}\pi. \quad \square$$

A slightly more complicated type of region is pictured in Figure 10.4.4. We approach the problem of calculating the area of the region Ω in the same way that we calculated the area between two curves in Section 5.5; that is, we calculate the area out to $r = \rho_2(\theta)$ and subtract from it the area out to $r = \rho_1(\theta)$. This gives

$$\text{area of } \Omega = \int_\alpha^\beta \tfrac{1}{2}[\rho_2(\theta)]^2 d\theta - \int_\alpha^\beta \tfrac{1}{2}[\rho_1(\theta)]^2 d\theta,$$

which can be written

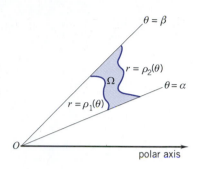

Figure 10.4.4

(10.4.2)

$$\text{area of } \Omega = \int_\alpha^\beta \tfrac{1}{2}([\rho_2(\theta)]^2 - [\rho_1(\theta)]^2)\, d\theta.$$

To find the area between two polar curves, we first determine the curves that serve as outer and inner boundaries of the region and the intervals of θ values over which these boundaries are traced out. Since the polar coordinates of a point are not unique, extra care must be used to determine these intervals of θ values.

Example 2 Find the area of the region that lies within the circle $r = 2\cos\theta$ but is outside the circle $r = 1$.

SOLUTION The region is shown in Figure 10.4.5. Our first step is to find the values of θ for the two points where the circles intersect:

$$2\cos\theta = 1, \qquad \cos\theta = \tfrac{1}{2}, \qquad \theta = \tfrac{1}{3}\pi, \tfrac{5}{3}\pi.$$

Since the region is symmetric about the polar axis, the area below the polar axis equals the area above the polar axis. Thus

$$A = 2\int_0^{\pi/3} \tfrac{1}{2}([2\cos\theta]^2 - [1]^2)\, d\theta.$$

Carry out the integration and you will see that $A = \tfrac{1}{3}\pi + \tfrac{1}{2}\sqrt{3} \cong 1.91$. $\square$

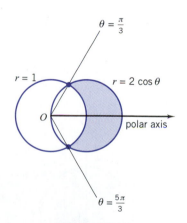

Figure 10.4.5

Example 3 Find the area A of the region between the inner and outer loops of the limaçon

$$r = 1 - 2\cos\theta \qquad \text{(Figure 10.4.6)}$$

SOLUTION We find that $r = 0$ at $\theta = \pi/3$ and at $\theta = 5\pi/3$. The outer loop is traced out as θ increases from $\pi/3$ to $\theta = 5\pi/3$. Thus

$$\text{area within outer loop} = A_1 = \int_{\pi/3}^{5\pi/3} \tfrac{1}{2}[1 - 2\cos\theta]^2\, d\theta.$$

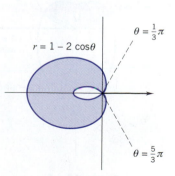

Figure 10.4.6

The lower half of the inner loop is traced out as θ increases from 0 to $\pi/3$ and the upper half as θ increases from $5\pi/3$ to 2π. (Verify this.) Therefore

$$\text{area within inner loop} = A_2 = \int_0^{\pi/3} \tfrac{1}{2}[1 - 2\cos\theta]^2 d\theta + \int_{5\pi/3}^{2\pi} \tfrac{1}{2}[1 - 2\cos\theta]^2\, d\theta.$$

Note that

$$\int \tfrac{1}{2}[1 - 2\cos\theta]^2 d\theta = \int \tfrac{1}{2}[1 - 4\cos\theta + 4\cos^2\theta]\, d\theta$$

$$= \tfrac{1}{2}\int [1 - 4\cos\theta + 2(1 + \cos 2\theta)]\, d\theta$$

$$= \tfrac{1}{2}\int [3 - 4\cos\theta + 2\cos 2\theta]\, d\theta$$

$$= \tfrac{1}{2}[3\theta - 4\sin\theta + \sin 2\theta] + C.$$

Therefore

$$A_1 = \tfrac{1}{2}\left[3\theta - 4\sin\theta + \sin 2\theta\right]_{\pi/3}^{5\pi/3} = 2\pi + \tfrac{3}{2}\sqrt{3},$$

$$A_2 = \tfrac{1}{2}\left[3\theta - 4\sin\theta + \sin 2\theta\right]_0^{\pi/3} + \tfrac{1}{2}\left[3\theta - 4\sin\theta + \sin 2\theta\right]_{5\pi/3}^{2\pi}$$

$$= \tfrac{1}{2}\pi - \tfrac{3}{4}\sqrt{3} + \tfrac{1}{2}\pi - \tfrac{3}{4}\sqrt{3} = \pi - \tfrac{3}{2}\sqrt{3},$$

and

$$A = A_1 - A_2 = 2\pi + \tfrac{3}{2}\sqrt{3} - \left(\pi - \tfrac{3}{2}\sqrt{3}\right) = \pi + 3\sqrt{3} \cong 8.34. \quad ❏$$

Remark   We could have done Example 3 more efficiently by exploiting the symmetry of the region. The region is symmetric about the x-axis. Therefore

$$A = 2\int_{\pi/3}^{\pi} \tfrac{1}{2}[1 - 2\cos\theta]^2\, d\theta - 2\int_0^{\pi/3} \tfrac{1}{2}[1 - 2\cos\theta]^2\, d\theta. \quad ❏$$

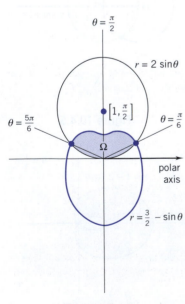

$\theta = \frac{\pi}{2}$

$r = 2\sin\theta$

$\left[1, \frac{\pi}{2}\right]$

$\theta = \frac{5\pi}{6}$

$\theta = \frac{\pi}{6}$

Ω

polar axis

$r = \frac{3}{2} - \sin\theta$

Figure 10.4.7

Example 4   The region Ω common to the circle $r = 2\sin\theta$ and the limaçon $r = \frac{3}{2} - \sin\theta$ is indicated in Figure 10.4.7. The θ-coordinates from 0 to 2π of the points of intersection can be found by solving the two equations simultaneously:

$$2\sin\theta = \tfrac{3}{2} - \sin\theta, \qquad \sin\theta = \tfrac{1}{2}, \qquad \theta = \tfrac{1}{6}\pi,\ \tfrac{5}{6}\pi.$$

Therefore the area of Ω can be represented as follows:

$$\text{area of } \Omega = \int_0^{\pi/6} \tfrac{1}{2}[2\sin\theta]^2 d\theta + \int_{\pi/6}^{5\pi/6} \tfrac{1}{2}[\tfrac{3}{2} - \sin\theta]^2 d\theta + \int_{5\pi/6}^{\pi} \tfrac{1}{2}[2\sin\theta]^2 d\theta;$$

or, by the symmetry of the region,

$$\text{area of } \Omega = 2\int_0^{\pi/6} \tfrac{1}{2}[2\sin\theta]^2 d\theta + 2\int_{\pi/6}^{\pi/2} \tfrac{1}{2}[\tfrac{3}{2} - \sin\theta]^2 d\theta.$$

As you can verify, the area of Ω is $\tfrac{5}{4} - \tfrac{15}{8}\sqrt{3} \cong 0.68. \quad ❏$

EXERCISES 10.4

Exercises 1–6. Calculate the area enclosed by the curve. Take $a > 0$.

1. $r = a \cos \theta$ from $\theta = -\frac{1}{2}\pi$ to $\theta = \frac{1}{2}\pi$.

2. $r = a \cos 3\theta$ from $\theta = -\frac{1}{6}\pi$ to $\theta = \frac{1}{6}\pi$.

3. $r = a\sqrt{\cos 2\theta}$ from $\theta = -\frac{1}{4}\pi$ to $\theta = \frac{1}{4}\pi$.

4. $r = a(1 + \cos 3\theta)$ from $\theta = -\frac{1}{3}\pi$ to $\theta = \frac{1}{3}\pi$.

5. $r^2 = a^2 \sin^2 \theta$.   **6.** $r^2 = a^2 \sin^2 2\theta$.

Exercises 7–16. Find the area between the curves.

7. $r = \tan 2\theta$ and the rays $\theta = 0$, $\theta = \frac{1}{8}\pi$.

8. $r = \cos \theta$, $r = \sin \theta$, and the rays $\theta = 0$, $\theta = \frac{1}{4}\pi$.

9. $r = 2\cos \theta$, $r = \cos \theta$, and the rays $\theta = 0$, $\theta = \frac{1}{4}\pi$.

10. $r = 1 + \cos \theta$, $r = \cos \theta$, and the rays $\theta = 0$, $\theta = \frac{1}{2}\pi$.

11. $r = a(4 \cos \theta - \sec \theta)$ and the rays $\theta = 0$, $\theta = \frac{1}{4}\pi$.

12. $r = \frac{1}{2}\sec^2 \frac{1}{2}\theta$ and the vertical line through the origin.

13. $r = e^\theta$, $0 \le \theta \le \pi$; $r = \theta$, $0 \le \theta \le \pi$; the rays $\theta = 0$, $\theta = \pi$.

14. $r = e^\theta$, $2\pi \le \theta \le 3\pi$; $r = \theta$, $0 \le \theta \le \pi$; the rays $\theta = 0$, $\theta = \pi$.

15. $r = e^\theta$, $0 \le \theta \le \pi$; $r = e^{\theta/2}$, $0 \le \theta \le \pi$; the rays $\theta = 2\pi$, $\theta = 3\pi$.

16. $r = e^\theta$, $0 \le \theta \le \pi$; $r = e^\theta$, $2\pi \le \theta \le 3\pi$; the rays $\theta = 0$, $\theta = \pi$.

Exercises 17–28. Represent the area by one or more integrals.

17. Outside $r = 2$, but inside $r = 4 \sin \theta$.

18. Outside $r = 1 - \cos \theta$, but inside $r = 1 + \cos \theta$.

19. Inside $r = 4$, but to the right of $r = 2 \sec \theta$.

20. Inside $r = 2$, but outside $r = 4 \cos \theta$.

21. Inside $r = 4$, but between the lines $\theta = \frac{1}{2}\pi$ and $r = 2 \sec \theta$.

22. Inside the inner loop of $r = 1 - 2 \sin \theta$.

23. Inside one petal of $r = 2 \sin 3\theta$.

24. Outside $r = 1 + \cos \theta$, but inside $r = 2 - \cos \theta$.

25. Interior to both $r = 1 - \sin \theta$ and $r = \sin \theta$.

26. Inside one petal of $r = 5 \cos 6\theta$.

27. Outside $r = \cos 2\theta$, but inside $r = 1$.

28. Interior to both $r = 2a \cos \theta$ and $r = 2a \sin \theta$, $a > 0$.

29. Find the area within the three circles: $r = 1, r = 2 \cos \theta$, $r = 2 \sin \theta$.

30. Find the area outside the circle $r = a$ but inside the lemniscate $r^2 = 2a^2 \cos 2\theta$.

31. Fix $a > 0$ and let n range over the set of positive integers. Show that the petal curves $r = a \cos 2n\theta$ and $r = a \sin 2n\theta$ all enclose exactly the same area. What is this area?

32. Fix $a > 0$ and let n range over the set of positive integers. Show that the petal curves $r = a \cos([2n + 1]\theta)$ and $r = a \sin([2n + 1]\theta)$ all enclose exactly the same area. What is this area?

Centroids in Polar Coordinates

The accompanying figure shows a region Ω bounded by a polar curve $r = \rho(\theta)$ from $\theta = \alpha$ to $\theta = \beta$ with $\alpha < \beta \le \alpha + 2\pi$. We take $\rho(\theta)$ as nonnegative. A partition of $[\alpha, \beta]$, $\{\alpha = \theta_0, \theta_1, \theta_2, \dots, \theta_n = \beta\}$, breaks up Ω into n subregions with areas A_i and centroids $(\overline{x}_i, \overline{y}_i)$. As explained in Section 6-4, if Ω has centroid $(\overline{x}, \overline{y})$ and area A, then

$$\overline{x}A = \overline{x_1}A_1 + \overline{x_2}A_2 + \cdots + \overline{x_n}A_n,$$
$$\overline{y}A = \overline{y_1}A_1 + \overline{y_2}A_2 + \cdots + \overline{y_n}A_n.$$

Approximate $(\overline{x}_i, \overline{y}_i)$ by the centroid (x_i^*, y_i^*) of the triangle shown in the figure and approximate A_i by the area of the circular sector of radius $\rho(x_i^*)$ from $\theta = \theta_{i-1}$ to $\theta = \theta_i$. (Recall that the centroid of a triangle lies on each median, two-thirds of the distance from the vertex to the opposite side. Exercise 34, Section 6.4.)

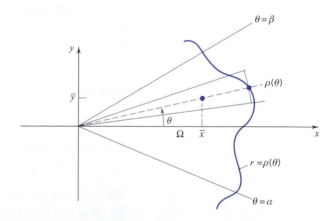

33. Go on to show that the rectangular coordinates $\overline{x}, \overline{y}$ of the centroid of Ω are given by the equations

(10.4.3)

$$\overline{x}A = \frac{1}{3}\int_\alpha^\beta \rho^3(\theta) \cos \theta \, d\theta,$$

$$\overline{y}A = \frac{1}{3}\int_\alpha^\beta \rho^3(\theta) \sin \theta \, d\theta$$

where A is the area of Ω.

34. In Section 6.4 we showed that the quarter-disk of Figure 6.4.4 has centroid $(4r/3\pi, 4r/3\pi)$. Obtain this result from (10.4.3).

35. Find the rectangular coordinates of the centroid of the region enclosed by the cardioid $r = 1 + \cos \theta$.

36. Find the rectangular coordinates of the centroid of the region enclosed by the cardioid $r = 2 + \sin\theta$.

▷ **Exercises 37–38.** Use a graphing utility to draw the polar curve. Then use a CAS to find the area of the region it encloses.

37. $r = 2 + \cos\theta$.　　　　**38.** $r = 2\cos 3\theta$.

▷ **Exercises 39–40.** Use a graphing utility to draw the polar curve. Then use a CAS to find the area inside the first curve but outside the second curve.

39. $r = 4\cos 3\theta$,　　$r = 2$.

40. $r = 2\cos\theta$,　　$r = 1 - \cos\theta$.

▷ **41.** The curve

$$y^2 = x^2\left(\frac{a-x}{a+x}\right), \quad a > 0$$

　is called a *strophoid*.

(a) Show that in polar coordinates the equation can be written

$$r = a\cos 2\theta\sec\theta.$$

(b) Draw the curve for $a = 1, 2, 4$. Use a graphing utility.
(c) Setting $a = 2$, find the area inside the loop.

▷ **42.** The curve

$$(x^2 + y^2)^2 = ax^2 y, \quad a > 0$$

　is called a *bifolium*.

(a) Show that in polar coordinates the equation can be written

$$r = a\sin\theta\cos^2\theta.$$

(b) Draw the curve for $a = 1, 2, 4$. Use a graphing utility.
(c) Setting $a = 2$, find the area inside one of the loops.

■ 10.5 CURVES GIVEN PARAMETRICALLY

So far we have specified curves by equations in rectangular coordinates or by equations in polar coordinates. Here we introduce a more general method. We begin with a pair of functions $x = x(t), y = y(t)$ differentiable on the interior of an interval I. At the endpoints of I (if any) we require only one-sided continuity.

For each number t in I we can interpret $(x(t), y(t))$ as the point with x-coordinate $x(t)$ and y-coordinate $y(t)$. Then, as t ranges over I, the point $(x(t), y(t))$ traces out a path in the xy-plane. (Figure 10.5.1.) We call such a path a *parametrized curve* and refer to t as the *parameter*.

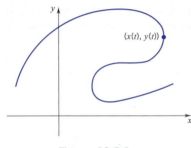

Figure 10.5.1

Example 1　Identify the curve parametrized by the functions

$$x(t) = t + 1, \qquad y(t) = 2t - 5 \qquad t \in (-\infty, \infty).$$

SOLUTION　We can express $y(t)$ in terms of $x(t)$:

$$y(t) = 2[x(t) - 1] - 5 = 2x(t) - 7.$$

The functions parametrize the line $y = 2x - 7$: as t ranges over the set of real numbers, the point $(x(t), y(t))$ traces out the line $y = 2x - 7$.　❑

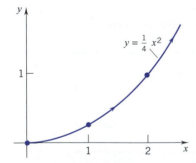

Figure 10.5.2

Example 2　Identify the curve parametrized by the functions

$$x(t) = 2t, \qquad y(t) = t^2 \qquad t \in [0, \infty).$$

SOLUTION　For each $t \in [0, \infty)$, both $x(t)$ and $y(t)$ are nonnegative. Therefore the curve lies in the first quadrant.

From the first equation we have $t = \frac{1}{2}x(t)$ and therefore

$$y(t) = \frac{1}{4}[x(t)]^2.$$

The functions parametrize the right half of the parabola $y = \frac{1}{4}x^2$: as t ranges over the interval $[0, \infty)$, the point $(x(t), y(t))$ traces out the parabolic arc $y = \frac{1}{4}x^2, x \geq 0$. (Figure 10.5.2.)　❑

Example 3　Identify the curve parametrized by the functions

$$x(t) = \sin^2 t, \qquad y(t) = \cos t \qquad t \in [0, \pi].$$

SOLUTION   Since $x(t) \geq 0$, the curve lies in the right half-plane. Since

$$x(t) = \sin^2 t = 1 - \cos^2 t = 1 - [y(t)]^2,$$

the points $(x(t), y(t))$ all lie on the parabola

$$x = 1 - y^2. \qquad \text{(Figure 10.5.3)}$$

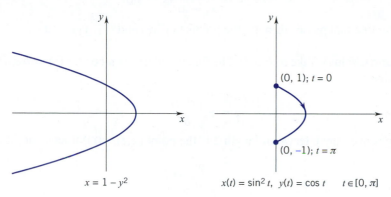

$$x = 1 - y^2 \qquad\qquad x(t) = \sin^2 t, \quad y(t) = \cos t \qquad t \in [0, \pi]$$

Figure 10.5.3

At $t = 0$, $x = 0$ and $y = 1$; at $t = \pi$, $x = 0$ and $y = -1$. As t ranges from 0 to π, the point $(x(t), y(t))$ traverses the parabolic arc

$$x = 1 - y^2, \qquad -1 \leq y \leq 1$$

from the point $(0, 1)$ to the point $(0, -1)$. ❏

Remark   Changing the domain in Example 3 to all real t does *not* give us any more of the parabola. For any given t we still have

$$0 \leq x(t) \leq 1 \qquad \text{and} \qquad -1 \leq y(t) \leq 1.$$

As t ranges over the set of real numbers, the point $(x(t), y(t))$ traces out that same parabolic arc back and forth an infinite number of times. ❏

Straight Lines   Given that $(x_0, y_0) \neq (x_1, y_1)$, the functions

(10.5.1) $\quad x(t) = x_0 + t(x_1 - x_0), \qquad y(t) = y_0 + t(y_1 - y_0) \qquad t \in (-\infty, \infty)$

parametrize the line that passes through the points (x_0, y_0) and (x_1, y_1).

PROOF   If $x_1 = x_0$, then we have

$$x(t) = x_0, \qquad y(t) = y_0 + t(y_1 - y_0).$$

As t ranges over the set of real numbers, $x(t)$ remains constantly x_0 and $y(t)$ ranges over the set of real numbers. The functions parametrize the vertical line $x = x_0$.

If $x_1 \neq x_0$, then we can solve the first equation for t:

$$t = \frac{x(t) - x_0}{x_1 - x_0}.$$

Substituting this into the second equation, we find that

$$y(t) - y_0 = \frac{y_1 - y_0}{x_1 - x_0}[x(t) - x_0].$$

The functions parametrize the line with equation

$$y - y_0 = \frac{y_1 - y_0}{x_1 - x_0}(x - x_0).$$

This is the line that passes through the points (x_0, y_0) and (x_1, y_1). ❏

Ellipses and Circles Take $a, b > 0$. The functions $x(t) = a \cos t$, $y(t) = b \sin t$ satisfy the identity

$$\frac{[x(t)]^2}{a^2} + \frac{[y(t)]^2}{b^2} = 1.$$

As t ranges over any interval of length 2π, the point $(x(t), y(t))$ traces out the ellipse

$$\frac{x^2}{a^2} + \frac{y^2}{b^2} = 1.$$

Usually we let t range from 0 to 2π and parametrize the ellipse by setting

(10.5.2)
$$x(t) = a \cos t, \qquad y(t) = b \sin t \qquad t \in [0, 2\pi].$$

If $b = a$, we have a circle. We can parametrize the circle

$$x^2 + y^2 = a^2$$

by setting

(10.5.3)
$$x(t) = a \cos t, \qquad y(t) = a \sin t \qquad t \in [0, 2\pi].$$

Hyperbolas Take $a, b > 0$. The functions $x(t) = a \cosh t$, $y(t) = b \sinh t$ satisfy the identity

$$\frac{[x(t)]^2}{a^2} - \frac{[y(t)]^2}{b^2} = 1.$$

Since $x(t) = a \cosh t > 0$ for all t, as t ranges over the set of real numbers, the point $(x(t), y(t))$ traces out the right branch of the hyperbola

$$\frac{x^2}{a^2} - \frac{y^2}{b^2} = 1. \qquad\qquad \text{(Figure 10.5.4)}$$

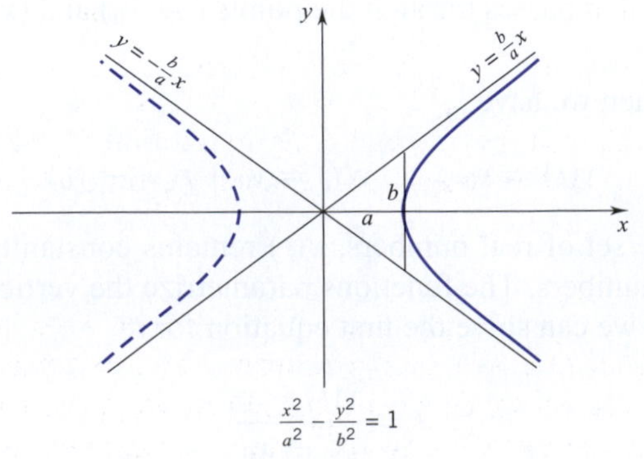

Figure 10.5.4

Two observations.

(1) Usually we think of t as representing time. Then $(x(t), y(t))$ gives position at time t. As time progresses, $(x(t), y(t))$ traces out a path in the xy-plane. Different parametrizations of the same path give different ways of traversing that path. (Examples 4 and 5.)

(2) On occasion we'll find it useful to parametrize the graph of a function $y = f(x)$ defined on an interval I. We can do this by setting

$$x(t) = t, \qquad y(t) = f(t) \qquad t \in I,$$

As t ranges over I, the point $(t, f(t))$ traces out the graph of f.

Example 4 The line that passes through the points $(1, 2)$ and $(3, 6)$ has equation $y = 2x$. The line segment that joins these points is the graph of the function

$$y = 2x, \qquad 1 \le x \le 3.$$

We will parametrize this line segment in different ways using the parameter t to indicate time measured in seconds.

We begin by setting

$$x(t) = t, \qquad y(t) = 2t \qquad t \in [1, 3].$$

At time $t = 1$, the particle is at the point $(1, 2)$. It traverses the line segment and arrives at the point $(3, 6)$ at time $t = 3$.

Now we set

$$x(t) = t + 1, \qquad y(t) = 2t + 2 \qquad t \in [0, 2].$$

At time $t = 0$, the particle is at the point $(1, 2)$. It traverses the line segment and arrives at the point $(3, 6)$ at time $t = 2$.

The equations

$$x(t) = 3 - t, \qquad y(t) = 6 - 2t \qquad t \in [0, 2]$$

represent a traversal of that same line segment but in the opposite direction. At time $t = 0$, the particle is at the point $(3, 6)$. It arrives at $(1, 2)$ at time $t = 2$.

Set

$$x(t) = 3 - 4t, \qquad y(t) = 6 - 8t \qquad t \in \left[0, \tfrac{1}{2}\right].$$

Now the particle traverses the same line segment in only half a second. At time $t = 0$, the particle is at the point $(3, 6)$. It arrives at $(1, 2)$ at time $t = \tfrac{1}{2}$.

Finally we set

$$x(t) = 2 - \cos t, \qquad y(t) = 4 - 2 \cos t \qquad t \in [0, 4\pi].$$

In this instance the particle begins and ends its motion at the point $(1, 2)$, having traced and retraced the line segment twice during a span of 4π seconds. (See Figure 10.5.5.) ❏

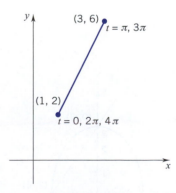

Figure 10.5.5

Remark If the path of an object is given in terms of a time parameter t and we eliminate the parameter to obtain an equation in x and y, it may be that we obtain a clearer view of the path, but we do so at considerable expense. The equation in x and y does not tell us where the particle is at any time t. The parametric equations do. ❏

Example 5 We return to the ellipse $\dfrac{x^2}{a^2} + \dfrac{y^2}{b^2} = 1$ and again use the parameter t to measure time measured in seconds.

A particle with position given by the equations

$$x(t) = a \cos t, \qquad y(t) = b \sin t \qquad t \in [0, 2\pi]$$

traverses the ellipse in a counterclockwise manner. It begins at the point $(a, 0)$ and makes a full circuit in 2π seconds. If the equations of motion are

$$x(t) = a \cos 2\pi t, \qquad y(t) = -b \sin 2\pi t \qquad t \in [0, 1],$$

the particle still travels the same ellipse, but in a different manner. Once again it starts at $(a, 0)$, but this time it moves clockwise and makes a full circuit in only 1 second. If the equations of motion are

$$x(t) = a \sin 4\pi t, \qquad y(t) = -b \cos 4\pi t \qquad t \in [0, \infty],$$

the motion begins at $(0, b)$ and goes on in perpetuity. The motion is clockwise, a complete circuit taking place every half second. ❑

Intersections and Collisions

Example 6 Two particles start at the same instant, the first along the linear path

$$x_1(t) = \tfrac{16}{3} - \tfrac{8}{3}t, \qquad y_1(t) = 4t - 5 \qquad t \geq 0$$

and the second along the elliptical path

$$x_2(t) = 2 \sin \tfrac{1}{2}\pi t, \qquad y_2(t) = -3 \cos \tfrac{1}{2}\pi t \qquad t \geq 0.$$

(a) At what points, if any, do the paths intersect?

(b) At what points, if any, do the particles collide?

SOLUTION To see where the paths intersect, we find equations for them in x and y. The linear path can be written

$$3x + 2y - 6 = 0, \qquad x \leq \tfrac{16}{3}$$

and the elliptical path

$$\frac{x^2}{4} + \frac{y^2}{9} = 1.$$

Solving the two equations simultaneously, we get

$$x = 2, \quad y = 0 \qquad \text{and} \qquad x = 0, \quad y = 3.$$

This means that the paths intersect at the points $(2, 0)$ and $(0, 3)$. This answers part (a). Now for part (b). The first particle passes through the point $(2, 0)$ only when

$$x_1(t) = \frac{16}{3} - \frac{8}{3}t = 2 \qquad \text{and} \qquad y_1(t) = 4t - 5 = 0.$$

As you can check, this happens only when $t = \tfrac{5}{4}$. When $t = \tfrac{5}{4}$, the second particle is elsewhere. Hence no collision takes place at $(2, 0)$ There is, however, a collision at $(0,3)$ because both particles get there at exactly the same time, $t = 2$:

$$x_1(2) = 0 = x_2(2), \qquad y_1(2) = 3 = y_2(2). \qquad \text{(See Figure 10.5.6.)} \quad ❑$$

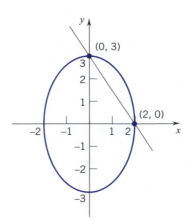

Figure 10.5.6

EXERCISES 10.5

Exercises 1–12. Express the curve by an equation in x and y.

1. $x(t) = t^2$, $\quad y(t) = 2t + 1$.

2. $x(t) = 3t - 1$, $\quad y(t) = 5 - 2t$.

3. $x(t) = t^2$, $\quad y(t) = 4t^4 + 1$.

4. $x(t) = 2t - 1$, $\quad y(t) = 8t^3 - 5$.

5. $x(t) = 2 \cos t$, $\quad y(t) = 3 \sin t$.

6. $x(t) = \sec^2 t$, $\quad y(t) = 2 + \tan t$.

7. $x(t) = \tan t$, $\quad y(t) = \sec t$.

8. $x(t) = 2 - \sin t$, $\quad y(t) = \cos t$.

9. $x(t) = \sin t$, $\quad y(t) = 1 + \cos^2 t$.

10. $x(t) = e^t$, $\quad y(t) = 4 - e^{2t}$.

11. $x(t) = 4 \sin t$, $\quad y(t) = 3 + 2 \sin t$.

12. $x(t) = \csc t$, $\quad y(t) = \cot t$.

Exercises 13–21. Express the curve by an equation in x and y; then sketch the curve.

13. $x(t) = e^{2t}$, $\quad y(t) = e^{2t} - 1 \quad t \le 0$.

14. $x(t) = 3 \cos t$, $\quad y(t) = 2 - \cos t \quad 0 \le t \le \pi$.

15. $x(t) = \sin t$, $\quad y(t) = \csc t \quad 0 < t \le \frac{1}{4}\pi$.

16. $x(t) = 1/t$, $\quad y(t) = 1/t^2 \quad 0 < t < 3$.

17. $x(t) = 3 + 2t$, $\quad y(t) = 5 - 4t \quad -1 \le t \le 2$.

18. $x(t) = \sec t$, $\quad y(t) = \tan t \quad 0 \le t \le \frac{1}{4}\pi$.

19. $x(t) = \sin \pi t$, $\quad y(t) = 2t \quad 0 \le t \le 4$.

20. $x(t) = 2 \sin t$, $\quad y(t) = \cos t \quad 0 \le t \le \frac{1}{2}\pi$.

21. $x(t) = \cot t$, $\quad y(t) = \csc t \quad \frac{1}{4}\pi \le t < \frac{1}{2}\pi$.

22. (*Important*) Parametrize the polar curve $r = \rho(\theta)$, $\theta \in [\alpha, \beta]$.

23. A particle with position given by the equations

$$x(t) = \sin 2\pi t, \quad y(t) = \cos 2\pi t \quad t \in [0, 1].$$

starts at the point $(0, 1)$ and traverses the unit circle $x^2 + y^2 = 1$ once in a clockwise manner. Write equations in the form

$$x(t) = f(t), \quad y(t) = g(t) \quad t \in [0, 1].$$

so that the particle
(a) begins at $(0, 1)$ and traverses the circle once in a counterclockwise manner;
(b) begins at $(0, 1)$ and traverses the circle twice in a clockwise manner;
(c) traverses the quarter circle from $(1, 0)$ to $(0, 1)$;
(d) traverses the three-quarter circle from $(1, 0)$ to $(0, 1)$.

24. A particle with position given by the equations

$$x(t) = 3 \cos 2\pi t, \quad y(t) = 4 \sin 2\pi t \quad t \in [0, 1].$$

starts at the point $(3, 0)$ and traverses the ellipse $16x^2 + 9y^2 = 144$ once in a counterclockwise manner. Write equations of the form

$$x(t) = f(t), \quad y(t) = g(t) \quad t \in [0, 1],$$

so that the particle
(a) begins at $(3, 0)$ and traverses the ellipse once in a clockwise manner;
(b) begins at $(0, 4)$ and traverses the ellipse once in a clockwise manner;
(c) begins at $(-3, 0)$ and traverses the ellipse twice in a counterclockwise manner;
(d) traverses the upper half of the ellipse from $(3, 0)$ to $(0, 3)$.

25. Find a parametrization

$$x = x(t), \quad y = y(t) \quad t \in (-1, 1),$$

for the horizontal line $y = 2$.

26. Find a parametrization

$$x(t) = \sin f(t), \quad y(t) = \cos f(t) \quad t \in (0, 1),$$

which traces out the unit circle infinitely often.

Exercises 27–32. Find a parametrization

$$x = x(t), \quad y = y(t) \quad t \in [0, 1]$$

for the given curve.

27. The line segment from $(3, 7)$ to $(8, 5)$.

28. The line segment from $(2, 6)$ to $(6, 3)$.

29. The parabolic arc $x = 1 - y^2$ from $(0, -1)$ to $(0,1)$.

30. The parabolic arc $x = y^2$ from $(4, 2)$ to $(0, 0)$.

31. The curve $y^2 = x^3$ from $(4, 8)$ to $(1, 1)$.

32. The curve $y^3 = x^2$ from $(1, 1)$ to $(8, 4)$.

(*Important*) For Exercises 33–36 assume that the curve

$$C : x = x(t), \quad y = y(t) \quad t \in [c, d],$$

is the graph of a nonnegative function $y = f(x)$ over an interval $[a, b]$. Assume that $x'(t)$ and $y(t)$ are continuous, $x(c) = a$ and $x(d) = b$.

33. (*The area under a parametrized curve*) Show that

(10.5.4)

$$\text{the area below } C = \int_c^d y(t) \, x'(t) \, dt.$$

HINT: Since C is the graph of f, $y(t) = f(x(t))$.

34. (*The centroid of a region under a parametrized curve*). Show that, if the region under C has area A and centroid $(\bar{x}, \bar{y})$, then

(10.5.5)
$$\bar{x}A = \int_c^d x(t)\, y(t)\, x'(t)\, dt,$$
$$\bar{y}A = \int_c^d \tfrac{1}{2}[y(t)]^2 x'(t)\, dt.$$

35. (*The volume of the solid generated by revolving about a coordinate axis the region under a parametrized curve*) Show that

(10.5.6)
$$V_x = \int_c^d \pi[y(t)]^2 x'(t)\, dt,$$
$$V_y = \int_c^d 2\pi x(t) y(t)\, x'(t)\, dt.$$
provided $x(c) \geq 0$

36. (*The centroid of the solid generated by revolving about a coordinate axis the region under a parametrized curve*) Show that

(10.5.7)
$$\bar{x}V_x = \int_c^d \pi x(t)[y(t)]^2 x'(t)\, dt,$$
$$\bar{y}V_y = \int_c^d \pi x(t)[y(t)]^2 x'(t)\, dt.$$
provided $x(c) \geq 0$

37. Sketch the curve
$$x(t) = at, \quad y(t) = a(1 - \cos t) \quad t \in [0, 2\pi]$$
and find the area below it. Take $a > 0$.

38. Determine the centroid of the region under the curve of Exercise 37.

39. Find the volume generated by revolving the region of Exercise 38 about: (a) the x-axis; (b) the y-axis.

40. Find the centroid of the solid generated by revolving the region of Exercise 38 about: (a) the x-axis; (b) the y-axis.

41. Give a parametrization for the upper half of the ellipse $b^2x^2 + a^2y^2 = a^2b^2$ that satisfies the assumptions made for Exercises 33–36.

42. Use the parametrization you chose for Exercise 41 to find (a) the area of the region enclosed by the ellipse; (b) the centroid of the upper half of that region.

43. Two particles start at the same instant, the first along the ray
$$x(t) = 2t + 6, \quad y(t) = 5 - 4t \quad t \geq 0$$
and the second along the circular path
$$x(t) = 3 - 5\cos \pi t, \quad y(t) = 1 + 5\sin \pi t \quad t \geq 0.$$
(a) At what points, if any, do these paths intersect?
(b) At what points, if any, do the particles collide?

44. Two particles start at the same instant, the first along the elliptical path
$$x_1(t) = 2 - 3\cos \pi t, \quad y_1(t) = 3 + 7\sin \pi t \quad t \geq 0.$$
and the second along the parabolic path
$$x_2(t) = 3t + 2, \quad y_2(t) = -\tfrac{7}{15}(3t + 1)^2 + \tfrac{157}{15} \quad t \geq 0.$$
(a) At what points, if any, do these paths intersect?
(b) At what points, if any, do the particles collide?

We can determine the points where a parametrized curve
$$C: \quad x = x(t), \quad y = y(t) \quad t \in I$$
intersects itself by finding the numbers r and s in $I (r \neq s)$ for which
$$x(r) = x(s) \quad \text{and} \quad y(r) = y(s).$$
Use this method to find the point(s) of self-intersection of each of the following curves.

45. $x(t) = t^2 - 2t, \quad y(t) = t^3 - 3t^2 + 2t \quad t$ real.

46. $x(t) = \cos t(1 - 2\sin t), \quad y(t) = \sin t\,(1 - 2\sin t)$
$t \in [0, \pi]$.

47. $x(t) = \sin 2\pi t, \quad y(t) = 2t - t^2 \quad t \in [0, 4]$.

48. $x(t) = t^3 - 4t, \quad y(t) = t^3 - 3t^2 + 2t \quad t$ real.

Exercises 49–52. A particle moves along the curve described by the parametric equations $x = f(t), y = g(t)$. Use a graphing utility to draw the path of the particle and describe the motion of the particle as it moves along the curve.

49. $x = 2t, \quad y = 4t - t^2 \quad 0 \leq t \leq 6$.

50. $x = 3(t^2 - 3), \quad y = t^3 - 3t \quad -3 \leq t \leq 3$.

51. $x = \cos(t^2 + t), \quad y = \sin(t^2 + t) \quad 0 \leq t \leq 2.1$.

52. $x = \cos(\ln t), \quad y = \sin(\ln t) \quad 1 \leq t \leq e^{2\pi}$.

53. Use a graphing utility to draw the curve
$$x(\theta) = \cos \theta(a - b \sin \theta), \quad y(\theta) = \sin \theta\,(a - b \sin \theta)$$
from $\theta = 0$ to $\theta = 2\pi$ given that
(a) $a = 1, b = 2$.
(b) $a = 2, b = 2$.
(c) $a = 2, b = 1$.
(d) In general, what can you say about the curve if $a < b$? $a > b$?
(e) Express the curve in the form $r = \rho(\theta)$.

■ PROJECT 10.5 Parabolic Trajectories

In the early part of the seventeenth century Galileo Galilei observed the motion of stones projected from the tower of Pisa and concluded that their trajectory was parabolic. Using calculus, together with some simplifying assumptions, we obtain results that agree with Galileo's observations.

Consider a projectile fired at an angle θ, $0 < \theta < \pi/2$, from a point (x_0, y_0) with initial velocity v_0. (Figure A.) The horizontal component of v_0 is $v_0 \cos\theta$, and the vertical component is $v_0 \sin\theta$. (Figure B.) Let $x = x(t)$, $y = y(t)$ be parametric equations for the path of the projectile.

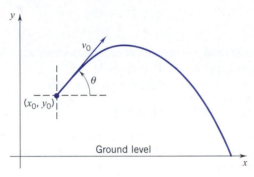

Figure A

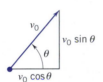

Figure B

We neglect air resistance and the curvature of the earth. Under these circumstances there is no horizontal acceleration and

$$x''(t) = 0.$$

The only vertical acceleration is due to gravity; therefore

$$y''(t) = -g.$$

Problem 1. Show that the path of the projectile (the trajectory) is given parametrically by the functions

$$x(t) = (v_0 \cos\theta)t + x_0, \quad y(t) = -\tfrac{1}{2}gt^2 + (v_0 \sin\theta)t + y_0.$$

Problem 2. Show that in rectangular coordinates the equation of the trajectory can be written

$$y = -\frac{g}{2v_0^2}\sec^2\theta \,(x - x_0)^2 + \tan\theta(x - x_0) + y_0.$$

Problem 3. Measure distances in feet, time in seconds, and set $g = 32\text{ft/sec}^2$. Take (x_0, y_0) as the origin and the x-axis as ground level. Consider a projectile fired at an angle θ with initial velocity v_0.

a. Give parametric equations for the trajectory; give an equation in x and y for the trajectory.

b. Find the range of the projectile, which in this case is the x-coordinate of the point of impact.

c. How many seconds after firing does the impact take place?

d. Choose θ so as to maximize the range.

e. Choose θ so that the projectile lands at $x = b$.

Problem 4.

a. Use a graphing utility to draw the path of the projectile fired at an angle of $30°$ with initial velocity $v_0 = 1500$ ft/sec. Determine the range of the projectile and the height reached.

b. Keeping $v_0 = 1500$ ft/sec, experiment with several values of θ. Confirm that $\theta = \pi/4$ maximizes the range. What angle maximizes the height reached?

■ 10.6 TANGENTS TO CURVES GIVEN PARAMETRICALLY

Let C be a curve parametrized by the functions

$$x = x(t), \quad y = y(t)$$

defined on some interval I. We will assume that I is an open interval and the parametrizing functions are differentiable.

Since a parametrized curve can intersect itself, at a point of C there can be

(i) one tangent,   (ii) two or more tangents,   or (iii) no tangent at all. (Figure 10.6.1)

To make sure that there is at least one tangent line at each point of C, we will make the additional assumption that

(10.6.1)
$$[x'(t)]^2 + [y'(t)]^2 \neq 0.$$

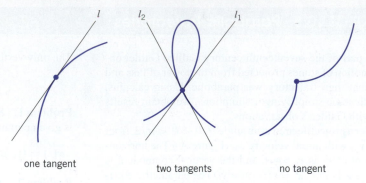

one tangent   two tangents   no tangent

Figure 10.6.1

This is equivalent to assuming that $x'(t)$ and $y'(t)$ are never simultaneously zero. Without this assumption almost anything can happen. See Exercises 31–35.

Now choose a point (x_0, y_0) on the curve C and a time t_0 at which

$$x(t_0) = x_0 \qquad \text{and} \qquad y(t_0) = y_0.$$

We want the slope m of the curve as it passes through the point (x_0, y_0) at time t_0[†]. To find this slope, we assume that $x'(t_0) \neq 0$. With $x'(t_0) \neq 0$, we can be sure that, for h sufficiently small but different from zero,

$$x(t_0 + h) - x(t_0) \neq 0. \qquad \text{(explain)}$$

For such h we can form the quotient

$$\frac{y(t_0 + h) - y(t_0)}{x(t_0 + h) - x(t_0)}.$$

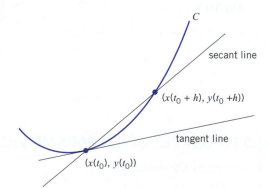

Figure 10.6.2

This quotient is the slope of the secant line pictured in Figure 10.6.2. The limit of this quotient as h tends to zero is the slope of the tangent line and thus the slope of the curve. Since

$$\frac{y(t_0 + h) - y(t_0)}{x(t_0 + h) - x(t_0)} = \frac{(1/h)[y(t_0 + h) - y(t_0)]}{(1/h)[x(t_0 + h) - x(t_0)]} \to \frac{y'(t_0)}{x'(t_0)} \quad \text{as } h \to 0,$$

[†] It could pass through the point (x_0, y_0) at other times also.

you can see that

(10.6.2)
$$m = \frac{y'(t_0)}{x'(t_0)}.$$

As an equation of the tangent line, we can write

$$y - y(t_0) = \frac{y'(t_0)}{x'(t_0)}[x - x(t_0)]. \qquad \text{(point-slope form)}$$

Multiplication by $x'(t_0)$ gives

$$y'(t_0)[x - x(t_0)] - x'(t_0)[y - y(t_0)] = 0$$

and thus

(10.6.3)
$$y'(t_0)[x - x_0] - x'(t_0)[y - y_0] = 0.$$

We derived this equation under the assumption that $x'(t_0) \neq 0$. If $x'(t_0) = 0$, (10.6.3) still makes sense. It is simply $y'(t_0)[x - x_0] = 0$, which, since $y'(t_0) \neq 0,$[†] can be simplified to read

(10.6.4)
$$x = x_0.$$

In this case the tangent line is vertical.

Example 1 Find an equation for each tangent to the curve

$$x(t) = t^3, \qquad y(t) = 1 - t \qquad t \in (-\infty, \infty)$$

at the point $(8, -1)$.

SOLUTION Since the curve passes through the point $(8, -1)$ only when $t = 2$, there can be only one tangent line at that point. Differentiating $x(t)$ and $y(t)$, we have

$$x'(t) = 3t^2, \qquad y'(t) = -1$$

and therefore

$$x'(2) = 12, \qquad y'(2) = -1.$$

The tangent line has equation

$$(-1)[x - 8] - 12[y - (-1)] = 0. \qquad \text{[by (10.6.3)]}$$

This reduces to

$$x + 12y + 4 = 0. \quad ❏$$

Example 2 Find the points on the curve

$$x(t) = 3 - 4\sin t, \qquad y(t) = 4 + 3\cos t \qquad t \in (-\infty, \infty)$$

at which there is (i) a horizontal tangent, (ii) a vertical tangent.

[†] We are assuming that $x'(t)$ and $y'(t)$ are not simultaneously zero.

SOLUTION Since the derivatives

$$x'(t) = -4\cos t \qquad \text{and} \qquad y'(t) = -3\sin t$$

are never simultaneously zero, the curve does have at least one tangent line at each of its points.

To find the points at which there is a horizontal tangent, we set $y'(t) = 0$. This gives $t = n\pi$, $n = 0, \pm 1, \pm 2, \ldots$ Horizontal tangents occur at all points of the form $(x(n\pi), y(n\pi))$. Since

$$x(n\pi) = 3 - 4\sin n\pi = 3 \quad \text{and} \quad y(n\pi) = 4 + 3\cos n\pi = \begin{cases} 7, & n \text{ even} \\ 1, & n \text{ odd,} \end{cases}$$

horizontal tangents occur only at $(3,7)$ and $(3,1)$.

To find the points at which there is a vertical tangent, we set $x'(t) = 0$. This gives $t = \frac{1}{2}\pi + n\pi$, $n = 0, \pm 1, \pm 2, \cdots$. Vertical tangents occur at all points of the form $(x(\frac{1}{2}\pi + n\pi), y(\frac{1}{2}\pi + n\pi))$. Since

$$x\left(\tfrac{1}{2}\pi + n\pi\right) = 3 - 4\sin\left(\tfrac{1}{2}\pi + n\pi\right) = \begin{cases} -1, & n \text{ even} \\ 7, & n \text{ odd} \end{cases}$$

and

$$y\left(\tfrac{1}{2}\pi + n\pi\right) = 4 + 3\cos\left(\tfrac{1}{2}\pi + n\pi\right) = 4,$$

vertical tangents occur only at $(-1, 4)$ and $(7, 4)$. ❑

Remark The results obtained in Example 2 become obvious once you recognize the parametrized curve as the ellipse

$$\frac{(x-3)^2}{16} + \frac{(y-4)^2}{9} = 1.$$

The ellipse is sketched in Figure 10.6.3 ❑

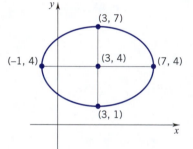

Figure 10.6.3

Example 3 The curve parametrized by the functions

$$x(t) = \frac{1-t^2}{1+t^2}, \qquad y(t) = \frac{t(1-t^2)}{1+t^2} \qquad t \in (-\infty, \infty)$$

is called a *strophoid*. The curve is shown in Figure 10.6.4. Find equations for the lines tangent to the curve at the origin. Then find the points at which there is a horizontal tangent.

SOLUTION The curve passes through the origin when $t = -1$ and when $t = 1$. (Verify this.) Differentiating $x(t)$ and $y(t)$, we have

$$x'(t) = \frac{(1+t^2)(-2t) - (1-t^2)(2t)}{(1+t^2)^2} = \frac{-4t}{(1+t^2)^2}$$

$$y'(t) = \frac{(1+t^2)(1-3t^2) - t(1-t^2)(2t)}{(1+t^2)^2} = \frac{1 - 4t^2 - t^4}{(1+t^2)^2}.$$

At time $t = -1$, the curve passes through the origin with slope

$$\frac{y'(-1)}{x'(-1)} = \frac{-1}{1} = -1.$$

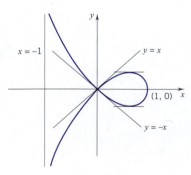

Figure 10.6.4

Therefore, the tangent line has equation $y = -x$.

At time $t = 1$, the curve passes through the origin with slope

$$\frac{y'(1)}{x'(1)} = \frac{-1}{-1} = 1.$$

Therefore, the tangent line has equation $y = x$.

To find the points at which there is a horizontal tangent, we set $y'(t) = 0$. this gives

$$1 - 4t^2 - t^4 = 0.$$

This equation is a quadratic in t^2. By the quadratic formula,

$$t^2 = \frac{-4 + \sqrt{16 + 4}}{2} = -2 \pm \sqrt{5}.$$

Since $t^2 \geq 0$, we exclude the possibility that $t^2 = -2 - \sqrt{5}$. We are left with $t^2 = \sqrt{5} - 2$. This gives $t = \pm\sqrt{\sqrt{5} - 2}$. Note that

$$x\left(\pm\sqrt{\sqrt{5} - 2}\right) = \frac{1 - (\sqrt{5} - 2)}{1 + (\sqrt{5} - 2)} = \frac{\sqrt{5} - 1}{2} \cong 0.62$$

and

$$y\left(\pm\sqrt{\sqrt{5} - 2}\right) = \left(\pm\sqrt{\sqrt{5} - 2}\right)x = \left(\pm\sqrt{\sqrt{5} - 2}\right)\frac{\sqrt{5} - 1}{2} \cong \pm 0.30.$$

There is a horizontal tangent line at the points $(0.62, \pm 0.30)$. (The coordinates are approximations.) ❏

We can apply these ideas to a curve given in polar coordinates by an equation of the form $r = \rho(\theta)$. The coordinate transformations

$$x = r\cos\theta, \qquad y = r\sin\theta$$

enable us to parametrize such a curve by setting

$$x(\theta) = \rho(\theta)\cos\theta, \qquad y(\theta) = \rho(\theta)\sin\theta.$$

Example 4 Take $a > 0$. Find the slope of the spiral $r = a\theta$ at $\theta = \frac{1}{2}\pi$. (The curve is shown for $\theta \geq 0$ in Figure 10.6.5.)

SOLUTION We write

$$x(\theta) = r\cos\theta = a\theta\cos\theta, \qquad y(\theta) = r\sin\theta = a\theta\sin\theta.$$

Now we differentiate:

$$x'(\theta) = -a\theta\sin\theta + a\cos\theta, \qquad y'(\theta) = a\theta\cos\theta + a\sin\theta.$$

Since

$$x'\left(\tfrac{1}{2}\pi\right) = -\tfrac{1}{2}\pi a \qquad \text{and} \qquad y'\left(\tfrac{1}{2}\pi\right) = a,$$

the slope of the curve at $\theta = \frac{1}{2}\pi$ is

$$\frac{y'\left(\tfrac{1}{2}\pi\right)}{x'\left(\tfrac{1}{2}\pi\right)} = -\frac{2}{\pi} \cong -0.64. \quad ❏$$

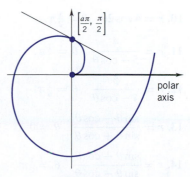

Figure 10.6.5

Example 5 Find the points of the cardioid $r = 1 - \cos\theta$ at which the tangent line is vertical.

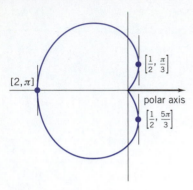

Figure 10.6.6

SOLUTION Since the cosine function has period 2π, we need only concern ourselves with θ in $[0, 2\pi)$. The curve can be parametrized by setting

$$x(\theta) = (1 - \cos\theta)\cos\theta, \qquad y(\theta) = (1 - \cos\theta)\sin\theta.$$

Differentiating and simplifying, we find that

$$x'(\theta) = (2\cos\theta - 1)\sin\theta, \qquad y'(\theta) = (1 - \cos\theta)(1 + 2\cos\theta).$$

The only numbers in the interval $[0, 2\pi)$ at which x' is zero and y' is not zero are $\frac{1}{3}\pi, \pi, \frac{5}{3}\pi$. The tangent line is vertical at

$$\left[\tfrac{1}{2}, \tfrac{1}{3}\pi\right], \qquad [2, \pi], \qquad \left[\tfrac{1}{2}, \tfrac{5}{3}\pi\right]. \qquad \text{(Figure 10.6.6)}$$

These points have rectangular coordinates $(\frac{1}{4}, \frac{1}{4}\sqrt{3}), (-2, 0), (\frac{1}{4}, -\frac{1}{4}\sqrt{3})$. ❏

EXERCISES 10.6

Exercises 1–8. Find an equation in x and y for the line tangent to the curve.

1. $x(t) = t$, $\quad y(t) = t^3 - 1$ $\quad$ at $t = 1$.

2. $x(t) = t^2$, $\quad y(t) = t + 5$ $\quad$ at $t = 2$.

3. $x(t) = 2t$, $\quad y(t) = \cos\pi t$ $\quad$ at $t = 0$.

4. $x(t) = 2t - 1$, $\quad y(t) = t^4$ $\quad$ at $t = 1$.

5. $x(t) = t^2$, $\quad y(t) = (2 - t)^2$ $\quad$ at $t = \frac{1}{2}$.

6. $x(t) = 1/t$, $\quad y(t) = t^2 + 1$ $\quad$ at $t = 1$.

7. $x(t) = \cos^3 t$, $\quad y(t) = \sin^3 t$ $\quad$ at $t = \frac{1}{4}\pi$.

8. $x(t) = e^t$, $\quad y(t) = 3e^{-t}$ $\quad$ at $t = 0$.

Exercises 9–14. Find an equation in x and y for the line tangent to the polar curve at the indicated value of θ.

9. $r = 4 - 2\sin\theta$ $\quad \theta = 0$.

10. $r = 4\cos 2\theta$ $\quad \theta = \frac{1}{2}\pi$.

11. $r = \dfrac{4}{5 - \cos\theta}$ $\quad \theta = \frac{1}{2}\pi$.

12. $r = \dfrac{5}{4 - \cos\theta}$ $\quad \theta = \frac{1}{6}\pi$.

13. $r = \dfrac{\sin\theta - \cos\theta}{\sin\theta + \cos\theta}$ $\quad \theta = 0$.

14. $r = \dfrac{\sin\theta + \cos\theta}{\sin\theta - \cos\theta}$ $\quad \theta = \frac{1}{2}\pi$.

Exercises 15–18. Parametrize the curve by a pair of differentiable functions

$$x = x(t), \qquad y = y(t) \quad \text{with} \quad [x'(t)]^2 + [y'(t)]^2 \neq 0.$$

Sketch the curve and determine the tangent line at the origin from the parametrization that you selected.

15. $y = x^3$.

16. $x = y^3$.

17. $y^5 = x^3$.

18. $y^3 = x^5$.

Exercises 19–26. Find the points (x, y) at which the curve has: (a) a horizontal tangent; (b) a vertical tangent. Then sketch the curve.

19. $x(t) = 3t - t^3$, $\quad y(t) = t + 1$.

20. $x(t) = t^2 - 2t$, $\quad y(t) = t^3 + 12t$.

21. $x(t) = 3 - 4\sin t$, $\quad y(t) = 4 + 3\cos t$.

22. $x(t) = \sin 2t$, $\quad y(t) = \sin t$.

23. $x(t) = t^2 - 2t$, $\quad y(t) = t^3 - 3t^2 + 2t$.

24. $x(t) = 2 - 5\cos t$, $\quad y(t) = 3 + \sin t$.

25. $x(t) = \cos t$, $\quad y(t) = \sin 2t$.

26. $x(t) = 3 + 2\sin t$, $\quad y(t) = 2 + 5\sin t$.

27. Find the tangent(s) to the curve

$$x(t) = -t + 2\cos\tfrac{1}{4}\pi t, \quad y(t) = t^4 - 4t^2$$

at the point $(2, 0)$.

28. Find the tangent(s) to the curve

$$x(t) = t^3 - t, \quad y(t) = t\sin\tfrac{1}{2}\pi t$$

at the point $(0, 1)$.

29. Let $P[r_1, \theta_1]$ be a point on a polar curve $r = f(\theta)$ as in the figure. Show that if $f'(\theta_1) = 0$ but $f(\theta_1) \neq 0$, then the tangent line at P is perpendicular to the line segment $\overline{OP}$.

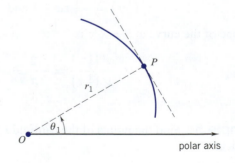

30. If $0 < a < 1$, the polar curve $r = a - \cos\theta$ is a limaçon with an inner loop. Choose a so that the curve will intersect itself at the pole in a right angle.

Exercises 31–35. Verify that $x'(0) = y'(0) = 0$ and that the given description holds at the point where $t = 0$. Sketch the curve.

31. $x(t) = t^3$, $y(t) = t^2$; cusp.

32. $x(t) = t^3$, $y(t) = t^5$; horizontal tangent.

33. $x(t) = t^5$, $y(t) = t^3$; vertical tangent.

34. $x(t) = t^3 - 1$, $y(t) = 2t^3$; tangent with slope 2.

35. $x(t) = t^2$, $y(t) = t^2 + 1$; no tangent line.

36. Suppose that $x = x(t), y = y(t)$ are twice differentiable functions that parametrize a curve. Take a point on the curve at which $x'(t) \neq 0$ and d^2y/dx^2 exists. Show that

(10.6.5)
$$\frac{d^2y}{dx^2} = \frac{x'(t)y''(t) - y'(t)x''(t)}{[x'(t)]^3}.$$

Exercises 37–40. Calculate d^2y/dx^2 at the indicated point without eliminating the parameter t.

37. $x(t) = \cos t$, $y(t) = \sin t$ at $t = \frac{1}{6}\pi$.

38. $x(t) = t^3$, $y(t) = t - 2$ at $t = 1$.

39. $x(t) = e^t$, $y(t) = e^{-t}$ at $t = 0$.

40. $x(t) = \sin^2 t$, $y(t) = \cos t$ at $t = \frac{1}{4}\pi$.

41. Let $x = 2 + \sec t$, $y = 2 - \tan t$. Use a CAS to find d^2y/dx^2.

42. Use a CAS to find an equation in x and y for the line tangent to the curve

$$x = \sin^2 t \quad y = \cos^2 t \qquad \text{at } t = \frac{1}{4}\pi.$$

Then use a graphing utility to sketch a figure that shows the curve and the tangent line.

43. Exercise 42 for $x = e^{-3t}$, $y = e^t$ at $t = \ln 2$.

44. Use a CAS to find an equation in x and y for the line tangent to the polar curve

$$r = \frac{4}{2 + \sin\theta} \qquad \text{at } \theta = \frac{1}{3}\pi.$$

Then use a graphing utility to sketch a figure that shows the curve and the tangent line.

■ 10.7 ARC LENGTH AND SPEED

Figure 10.7.1 represents a curve C parametrized by a pair of functions

$$x = x(t), \qquad y = y(t) \qquad t \in [a, b].$$

We will assume that the functions are *continuously differentiable* on $[a, b]$ (have first derivatives which are continuous on $[a, b]$). We want to determine the length of C.

Here our experience in Chapter 5 can be used as a model. To decide what should be meant by the area of a region Ω, we approximated Ω by the union of a finite number of rectangles. To decide what should be meant by the length of C, we approximate C by the union of a finite number of line segments.

Each number t in $[a, b]$ gives rise to a point $P = P(x(t), y(t))$ that lies on C. By choosing a finite number of points in $[a, b]$,

$$a = t_0 < t_1 < \cdots < t_{i-1} < t_i < \cdots < t_{n-1} < t_n = b,$$

we obtain a finite number of points on C,

$$P_0, P_1, \ldots, P_{i-1}, P_i, \ldots, P_{n-1}, P_n.$$

We join these points consecutively by line segments and call the resulting path,

$$\gamma = \overline{P_0 P_1} \cup \cdots \cup \overline{P_{i-1}, P_i} \cup \cdots \cup \overline{P_{n-1} P_n},$$

a polygonal path inscribed in C. (See Figure 10.7.2.)

The length of such a polygonal path is the sum of the distances between consecutive vertices:

$$\text{length of } \gamma = L(\gamma) = d(P_0, P_1) + \cdots + d(P_{i-1}, P_i) + \cdots + d(p_{n-1}, P_n).$$

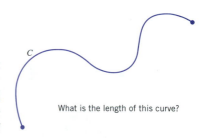

What is the length of this curve?

Figure 10.7.1

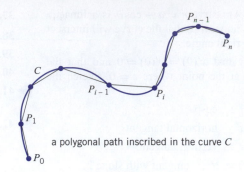

a polygonal path inscribed in the curve C

Figure 10.7.2

The ith line segment $\overline{P_{i-1}P_i}$ has length

$$d(P_{i-1}, P_i) = \sqrt{[x(t_i) - x(t_{i-1})]^2 + [y(t_i) - y(t_{i-1})]^2}$$

$$= \sqrt{\left[\frac{x(t_i) - x(t_{i-1})}{t_i - t_{t-1}}\right]^2 + \left[\frac{y(t_i) - y(t_{i-1})}{t_i - t_{i-1}}\right]^2}(t_i - t_{i-1}).$$

By the mean-value theorem, there exist points t_i^* and t_i^{**}, both in the interval (t_{i-1}, t_i), for which

$$\frac{x(t_i) - x(t_{i-1})}{t_i - t_{i-1}} = x'(t_i^*) \qquad \text{and} \qquad \frac{y(t_i) - y(t_{t-1})}{t_i - t_{i-1}} = y'(t_i^{**}).$$

Letting $\Delta t_i = t_i - t_{i-t}$, we have

$$d(P_{i-1}, P_i) = \sqrt{[x'(t_i^*)]^2 + [y'(t_i^{**})]^2}\,\Delta t_i.$$

Adding up these terms, we obtain

$$L(\gamma) = \sqrt{[x'(t_1^*)]^2 + [y'(t_1^{**})]^2}\,\Delta t_1 + \cdots + \sqrt{[x'(t_n^*)]^2 + [y'(t_n^{**})]^2}\,\Delta t_n.$$

As written, $L(\gamma)$ is not a Riemann sum: in general, $t_i^* \neq t_i^{**}$. It is nevertheless true (and at the moment we ask you to take on faith) that, as max $\Delta t_i \to 0$, $L(\gamma)$ approaches the integral

$$\int_a^b \sqrt{[x'(t)]^2 + [y'(t)]^2}\,dt.$$

Arc Length Formulas

By the argument just given, admittedly incomplete, we have obtained a way to calculate arc length. The length of the path C traced out by a pair of continuously differentiable functions

$$x = x(t), \quad y = y(t) \qquad t \in [a, b]$$

is given by the formula

(10.7.1)

$$L(C) = \int_a^b \sqrt{[x'(t)]^2 + [y'(t)]^2}\,dt.$$

More insight into this formula is provided in Chapter 14.

Let's use (10.7.1) to obtain the circumference of the unit circle. Parametrizing the unit circle by setting

$$x(t) = \cos t, \qquad y(t) = \sin t \qquad t \in [0, 2\pi],$$

we have

$$x'(t) = -\sin t, \qquad y'(t) = \cos t$$

and thus

$$\text{circumference} = \int_0^{2\pi} \sqrt{\sin^2 t + \cos^2 t} \; dt = \int_0^{2\pi} 1 \, dt = 2\pi.$$

Nothing surprising here. But suppose we parametrize the unit circle by setting

$$x(t) = \cos 2t, \qquad y(t) = \sin 2t \qquad t \in [0, 2\pi].$$

Then we have

$$x'(t) = -2\sin 2t, \qquad y'(t) = 2\cos 2t$$

and the arc length formula gives

$$L(C) = \int_0^{2\pi} \sqrt{4\sin^2 2t + 4\cos^2 2t} \; dt = \int_0^{2\pi} 2 \, dt = 4\pi.$$

This is not the circumference of the unit circle. What's wrong here? There is nothing wrong here. Formula (10.7.1) gives the length of the path traced out by the parametrizing functions. The functions $x(t) = \cos 2t$, $y(t) = \sin 2t$ with $t \in [0, 2\pi]$ trace out the unit circle not once, but twice. Hence the discrepancy.

When applying (10.7.1) to calculate the arc length of a curve given to us geometrically, we must make sure that the functions that we use to parametrize the curve trace out each arc of the curve only once.

Suppose now that C is the graph of a continuously differentiable function

$$y = f(x), \qquad x \in [a, b].$$

We can parametrize C by setting

$$x(t) = t, \qquad y(t) = f(t) \qquad t \in [a, b].$$

Since

$$x'(t) = 1 \qquad \text{and} \qquad y'(t) = f'(t),$$

(10.7.1) gives

$$L(C) = \int_a^b \sqrt{1 + [f'(t)]^2} \; dt.$$

Replacing t by x, we can write:

(10.7.2) The length of the graph of $f = \displaystyle\int_a^b \sqrt{1 + [f'(x)]^2} \; dx.$

A direct derivation of this formula is outlined in Exercise 52.

Example 1 The function $f(x) = \frac{1}{6}x^3 + \frac{1}{2}x^{-1}$ has derivative

$$f'(x) = \frac{1}{2}x^2 - \frac{1}{2}x^{-2}.$$

In this case

$$1 + [f'(x)]^2 = 1 + \left(\tfrac{1}{4}x^4 - \tfrac{1}{2} + \tfrac{1}{4}x^{-4}\right) = \tfrac{1}{4}x^4 + \tfrac{1}{2} + \tfrac{1}{4}x^{-4} = \left(\tfrac{1}{2}x^2 + \tfrac{1}{2}x^{-2}\right)^2.$$

The length of the graph from $x = 1$ to $x = 3$ is

$$\int_1^3 \sqrt{1 + [f'(x)]^2}\, dx = \int_1^3 \sqrt{\left(\tfrac{1}{2}x^2 + \tfrac{1}{2}x^{-2}\right)^2}\, dx$$

$$= \int_1^3 \left(\tfrac{1}{2}x^2 + \tfrac{1}{2}x^{-2}\right) dx = \left[\tfrac{1}{6}x^3 - \tfrac{1}{2}x^{-1}\right]_1^3 = \tfrac{14}{3}. \quad \Box$$

Example 2 The graph of the function $f(x) = x^2$ from $x = 0$ to $x = 1$ is a parabolic arc. The length of this arc is given by

$$\int_0^1 \sqrt{1 + [f'(x)]^2}\, dx = \int_0^1 \sqrt{1 + 4x^2}\, dx = 2\int_0^1 \sqrt{\left(\tfrac{1}{2}\right)^2 + x^2}\, dx$$

$$= \left[x\sqrt{\left(\tfrac{1}{2}\right)^2 + x^2} + \left(\tfrac{1}{2}\right)^2 \ln\left(x + \sqrt{\left(\tfrac{1}{2}\right)^2 + x^2}\right)\right]_0^1$$

by (8.4.1) $\longrightarrow$

$$= \tfrac{1}{2}\sqrt{5} + \tfrac{1}{4}\ln\left(2 + \sqrt{5}\right) \cong 1.48. \quad \Box$$

Suppose now that C is the graph of a continuously differentiable polar function

$$r = \rho(\theta), \qquad \alpha \le \theta \le \beta.$$

We can parametrize C by setting

$$x(\theta) = \rho(\theta)\cos\theta, \qquad y(\theta) = \rho(\theta)\sin\theta \qquad \theta \in [\alpha, \beta].$$

A straightforward calculation that we leave to you shows that

$$[x'(\theta)]^2 + [y'(\theta)]^2 = [\rho(\theta)]^2 + [\rho'(\theta)]^2.$$

The arc length formula then reads

(10.7.3)

$$\boxed{\; L(C) = \int_\alpha^\beta \sqrt{[\rho(\theta)]^2 + [\rho'(\theta)]^2}\, d\theta. \;}$$

Example 3 For fixed $a > 0$, the equation $r = a$ represents a circle of radius a. Here

$$\rho(\theta) = a \qquad \text{and} \qquad \rho'(\theta) = 0.$$

The circle is traced out once as θ ranges from 0 to 2π. Therefore the length of the curve (the circumference of the circle) is given by

$$\int_0^{2\pi} \sqrt{[\rho(\theta)]^2 + [\rho'(\theta)^2]}\, d\theta = \int_0^{2\pi} \sqrt{a^2 + 0^2}\, d\theta = \int_0^{2\pi} a\, d\theta = 2\pi a. \quad \Box$$

Example 4 We calculate the arc length of the cardioid $r = a(1 - \cos\theta)$. We take $a > 0$. To make sure that no arc of the curve is traced out more than once, we restrict

θ to the interval $[0, 2\pi]$. Here

$$\rho(\theta) = a(1 - \cos\theta) \qquad \text{and} \qquad \rho'(\theta) = a\sin\theta,$$

so that

$$[\rho(\theta)]^2 + [\rho'(\theta)]^2 = a^2(1 - 2\cos\theta + \cos^2\theta) + a^2\sin^2\theta = 2a^2(1 - \cos\theta).$$

The identity $\frac{1}{2}(1 - \cos\theta) = \sin^2\frac{1}{2}\theta$ gives

$$[\rho(\theta)]^2 + [\rho'(\theta)]^2 = 4a^2\sin^2\frac{1}{2}\theta.$$

The length of the cardioid is $8a$:

$$\int_0^{2\pi} \sqrt{[\rho(\theta)]^2 + [\rho'(\theta)]^2}\, d\theta = \int_0^{2\pi} 2a\sin\frac{1}{2}\theta\, d\theta = 4a\left[-\cos\frac{1}{2}\theta\right]_0^{2\pi} = 8a. \quad \square$$

$\underset{\uparrow}{}$ $\sin\frac{1}{2}\theta \geq 0$ for $\theta \in [0, 2\pi]$

The Geometric Significance of dx/ds and dy/ds

Figure 10.7.3 shows the graph of a function $y = f(x)$ which we assume to be continuously differentiable. At the point (x, y) the tangent line has an inclination marked α_x. Note that $\alpha_x \in (-\frac{1}{2}\pi, \frac{1}{2}\pi)$.

The length of the arc from a to x can be written

$$s(x) = \int_a^x \sqrt{1 + [f'(t)]^2}\, dt.$$

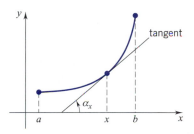

Figure 10.7.3

Differentiation with respect to x gives $s'(x) = \sqrt{1 + [f'(x)]^2}$. (Theorem 5.3.5.) Using the Leibniz notation, we have

$$\frac{ds}{dx} = \sqrt{1 + \left(\frac{dy}{dx}\right)^2} = \sqrt{1 + \tan^2\alpha_x} = \sec\alpha_x.$$

$\underset{\uparrow}{}$ $\sec\alpha_x > 0$ for $\alpha_x \in (-\frac{1}{2}\pi, \frac{1}{2}\pi)$

Note that s is a one-to-one function of x with nonzero derivative (the derivative is at least 1) and therefore has a differentiable inverse. By (7.1.9)

$$\frac{dx}{ds} = \frac{1}{\sec\alpha_x} = \cos\alpha_x.$$

To find dy/ds, we note that

$$\tan\alpha_x = \frac{dy}{dx} = \frac{dy}{ds}\frac{ds}{dx} = \frac{dy}{ds}\sec\alpha_x.$$

$\underset{\uparrow}{}$ chain rule

Multiplication by $\cos\alpha_x$ gives

$$\frac{dy}{ds} = \sin\alpha_x.$$

For the record,

$$\textbf{(10.7.4)} \qquad \frac{dx}{ds} = \cos\alpha_x \quad \text{and} \quad \frac{dy}{ds} = \sin\alpha_x \qquad \text{where } \alpha_x \text{ is the inclination of the tangent line at the point } (x, y).$$

Speed Along a Plane Curve

So far we have talked about speed only in connection with straight-line motion. How can we calculate the speed of an object that moves along a curve? Imagine an object moving along some curved path. Suppose that $(x(t), y(t))$ gives the position of the object at time t. The distance traveled by the object from time zero to any later time t is simply the length of the path up to time t:

$$s(t) = \int_0^t \sqrt{[x'(u)]^2 + [y'(u)]^2} \, du.$$

The time rate of change of this distance is what we call the *speed* of the object. Denoting the speed of the object at time t by $v(t)$, we have

(10.7.5)
$$v(t) = s'(t) = \sqrt{[x'(t)]^2 + [y'(t)]^2}.$$

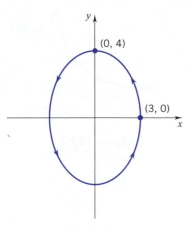

Figure 10.7.4

Example 5 The position of a particle at time t is given by the parametric equations

$$x(t) = 3\cos 2t, \qquad y(t) = 4\sin 2t \qquad t \in [0, 2\pi].$$

Find the speed of the particle at time t and determine the times when the speed is a maximum and when it is a minimum.

SOLUTION The path of the particle is the ellipse

$$\frac{x^2}{9} + \frac{y^2}{16} = 1. \qquad \text{(Figure 10.7.4)}$$

The particle moves around the curve in the counterclockwise direction, the direction indicated by the arrows. The speed of the particle at time t is

$$v(t) = \sqrt{[x'(t)]^2 + [y'(t)]^2} = \sqrt{(-6\sin 2t)^2 + (8\cos 2t)^2} = \sqrt{36\sin^2 2t + 64\cos^2 2t}$$
$$= \sqrt{36 + 28\cos^2 2t}. \qquad (\sin^2 2t = 1 - \cos^2 2t)$$

The maximum speed is 8, and this occurs when $\cos^2 2t = 1$; that is, when $t = 0, \pi/2, \pi, 3\pi/2, 2\pi$. At these times, $\sin^2 2t = 0$ and the particle is at an end of the minor axis. The minimum speed is 6, which occurs when $\cos^2 2t = 0$ (and $\sin^2 2t = 1$); that is, when $t = \pi/4, 3\pi/4, 5\pi/4, 7\pi/4$. At these times, the particle is at an end of the major axis. ❏

In the Leibniz notation the formula for speed reads

(10.7.6)
$$v = \frac{dx}{dt} = \sqrt{\left(\frac{dx}{dt}\right)^2 + \left(\frac{dy}{dt}\right)^2}.$$

If we know the speed of an object and we know its mass, then we can calculate its kinetic energy.

Example 6 A particle of mass m slides down a frictionless curve (see Figure 10.7.5) from a point (x_0, y_0) to a point (x_1, y_1) under the force of gravity. As discussed in Project 4.9B, the particle has two forms of energy during the motion: gravitational potential energy mgy and kinetic energy $\frac{1}{2}mv^2$. Show that the sum of these two quantities remains constant:

$$GPE + KE = C.$$

In terms of the variables of motion this reads

$$mgy + \frac{1}{2}mv^2 = C.$$

SOLUTION The particle is subjected to a vertical force $-mg$ (a downward force of magnitude mg). Since the particle is constrained to remain on the curve, the effective force on the particle is tangential. The tangential component of the vertical force is $-mg \sin \alpha$ (see Figure 10.7.5.) The speed of the particle is ds/dt and the tangential acceleration is d^2s/dt^2. (It is as if the particle were moving along the tangent line.)

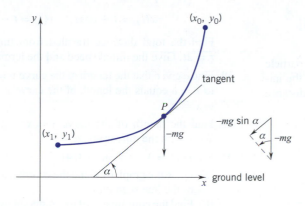

Figure 10.7.5

Therefore, by Newton's law $F = ma$, we have

$$m\frac{d^2s}{dt^2} = -mg \sin \alpha = -mg\frac{dy}{ds}.$$

by (10.7.4) ⟶

Thus we can write

$$mg\frac{dy}{ds} + m\frac{d^2s}{dt^2} = 0$$

$$mg\frac{dy}{ds}\frac{ds}{dt} + m\frac{ds}{dt}\frac{d^2s}{dt^2} = 0 \qquad \text{(we multiplied by } \tfrac{ds}{dt})$$

$$mg\frac{dy}{dt} + mv\frac{dv}{dt} = 0. \qquad \text{(chain rule)}$$

Integrating with respect to t, we have

$$mgy + \tfrac{1}{2}mv^2 = C,$$

as asserted. ❑

EXERCISES 10.7

Exercises 1–18. Find the length of the graph and compare it to the straight-line distance between the endpoints of the graph.

1. $f(x) = 2x + 3, \quad x \in [0, 1]$.

2. $f(x) = 3x + 2, \quad x \in [0, 1]$.

3. $f(x) = (x - \frac{4}{9})^{3/2}, \quad x \in [1, 4]$.

4. $f(x) = x^{3/2}, \quad x \in [0, 44]$.

5. $f(x) = \frac{1}{3}\sqrt{x}(x - 3), \quad x \in [0, 3]$.

6. $f(x) = \frac{2}{3}(x - 1)^{3/2}, \quad x \in [1, 2]$.

7. $f(x) = \frac{1}{3}(x^2 + 2)^{3/2}, \quad x \in [0, 1]$.

8. $f(x) = \frac{1}{3}(x^2 - 2)^{3/2}, \quad x \in [2, 4]$.

9. $f(x) = \frac{1}{4}x^2 - \frac{1}{2}\ln x$, $x \in [1, 5]$.

10. $f(x) = \frac{1}{8}x^2 - \ln x$, $x \in [1, 4]$.

11. $f(x) = \frac{3}{8}x^{4/3} - \frac{3}{4}x^{2/3}$, $x \in [1, 8]$.

12. $f(x) = \frac{1}{10}x^5 + \frac{1}{6}x^{-3}$, $x \in [1, 2]$.

13. $f(x) = \ln(\sec x)$, $x \in [0, \frac{1}{4}\pi]$.

14. $f(x) = \frac{1}{2}x^2$, $x \in [0, 1]$.

15. $f(x) = \frac{1}{2}x\sqrt{x^2 - 1} - \frac{1}{2}\ln(x + \sqrt{x^2 - 1})$, $x \in [1, 2]$.

16. $f(x) = \cosh x$, $x \in [0, \ln 2]$.

17. $f(x) = \frac{1}{2}x\sqrt{3 - x^2} + \frac{3}{2}\arcsin(\frac{1}{3}\sqrt{3}x)$, $x \in [0, 1]$.

18. $f(x) = \ln(\sin x)$, $x \in [\frac{1}{6}\pi, \frac{1}{2}\pi]$.

Exercises 19–24. The equations give the position of a particle at each time t during the time interval specified. Find the initial speed of the particle, the terminal speed, and the distance traveled.

19. $x(t) = t^2$, $y(t) = 2t$ from $t = 0$ to $t = \sqrt{3}$.

20. $x(t) = t - 1$, $y(t) = \frac{1}{2}t^2$ from $t = 0$ to $t = 1$.

21. $x(t) = t^2$, $y(t) = t^3$ from $t = 0$ to $t = 1$.

22. $x(t) = a\cos^3 t$, $y(t) = a\sin^3 t$ from $t = 0$ to $t = \frac{1}{2}\pi$.

23. $x(t) = e^t \sin t$, $y(t) = e^t \cos t$ from $t = 0$ to $t = \pi$.

24. $x(t) = \cos t + t\sin t$, $y(t) = \sin t - t\cos t$ from $t = 0$ to $t = \pi$.

25. Let $a > 0$. Find the length of the path traced out by

$$x(\theta) = a(\theta - \sin\theta), \quad y(\theta) = a(1 - \cos\theta)$$

as θ ranges from 0 to 2π.

26. Let $a > 0$. Find the length of the path traced out by

$$x(\theta) = 2a\cos\theta - a\cos 2\theta,$$

$$y(\theta) = 2a\sin\theta - a\sin 2\theta$$

as θ ranges from 0 to 2π.

27. (a) Let $a > 0$. Find the length of the path traced out by

$$x(\theta) = 3a\cos\theta + a\cos 3\theta,$$

$$y(\theta) = 3a\sin\theta - a\sin 3\theta$$

as θ ranges from 0 to 2π.

(b) Show that this path can also be parametrized by

$$x(\theta) = 4a\cos^3\theta, \quad y(\theta) = 4a\sin^3\theta \quad 0 \le \theta \le 2\pi.$$

28. The curve defined parametrically by

$$x(\theta) = \theta\cos\theta, \quad y(\theta) = \theta\sin\theta.$$

is called an *Archimedean spiral*. Find the length of the arc traced out as θ ranges from 0 to 2π.

Exercises 29–36. Find the length of the polar curve.

29. $r = 1$ from $\theta = 0$ to $\theta = 2\pi$.

30. $r = 3$ from $\theta = 0$ to $\theta = \pi$.

31. $r = e^\theta$ from $\theta = 0$ to $\theta = 4\pi$. (logarithmic spiral)

32. $r = ae^\theta, a > 0$, from $\theta = -2\pi$ to $\theta = 2\pi$.

33. $r = e^{2\theta}$ from $\theta = 0$ to $\theta = 2\pi$.

34. $r = 1 + \cos\theta$ from $\theta = 0$ to $\theta = 2\pi$.

35. $r = 1 - \cos\theta$ from $\theta = 0$ to $\theta = \frac{1}{2}\pi$.

36. $r = 2a\sec\theta, a > 0$, from $\theta = 0$ to $\theta = \frac{1}{4}\pi$.

37. At time t a particle has position

$$x(t) = 1 + \arctan t, \quad y(t) = 1 - \ln\sqrt{1 + t^2}.$$

Find the total distance traveled from time $t = 0$ to time $t = 1$. Give the initial speed and the terminal speed.

38. At time t a particle has position

$$x(t) = 1 + \cos t, \quad y(t) = t - \sin t.$$

Find the total distance traveled from time $t = 0$ to time $t = 2t$. Give the initial speed and the terminal speed.

39. Find c given that the length of the curve $y = \ln x$ from $x = 1$ to $x = e$ equals the length of the curve $y = e^x$ from $x = 0$ to $x = c$.

40. Find the length of the curve $y = x^{2/3}, x \in [1, 8]$. HINT: Work with the mirror image $y = x^{3/2}, x \in [1, 4]$.

41. Set $f(x) = 3x - 5$ on $[-3, 4]$.

(a) Draw the graph of f and the coordinates of the midpoint of the line segment.

(b) Find the coordinates of the midpoint by finding the number c for which

$$\int_{-3}^{c}\sqrt{1 + [f'(x)]^2}\,dx = \int_{c}^{4}\sqrt{1 + [f'(x)]^2}\,dx.$$

▷ 42. Set $f(x) = x^{3/2}$ on $[1, 5]$. Use a graphing utility to draw the graph of f and a CAS to find the coordinates of the midpoint. HINT: Find the number c for which

$$\int_{1}^{c}\sqrt{1 + [f'(x)]^2}\,dx = \frac{1}{2}\int_{1}^{5}\sqrt{1 + [f'(x)]^2}\,dx.$$

43. Show that the curve $y = \cosh x$ has the property that for every interval $[a, b]$ the length of the curve from $x = a$ to $x = b$ equals the area under the curve from $x = a$ to $x = b$.

▷ 44. Let $f(x) = 2\ln x$ on $[1, e]$. Draw the graph of f and use a CAS to estimate the length of the graph.

▷ 45. Let $f(x) = \sin x - x\cos x$ on $[0, \pi]$. Use a graphing utility to draw the graph of f and use a CAS to estimate the length of the graph.

▷ 46. (a) Use a graphing utility to draw the curve

$$x(t) = e^{2t}\cos 2t, \quad y(t) = e^{2t}\sin 2t \quad 0 \le t \le \pi/3.$$

(b) Use a CAS to estimate the length of the curve. Round off your answer to four decimal places.

▷ 47. (a) Use a graphing utility to draw the curve

$$x(t) = t^2, \quad y(t) = t^3 - t \quad t \text{ real}.$$

(b) Your drawing should show that the curve has a loop. Use a CAS to estimate the length of the loop. Round off your answer to four decimal places.

▷ 48. The curve

$$x(t) = \frac{3t}{t^3 + 1}, \quad y(t) = \frac{3t^2}{t^3 + 1} \quad t \ne -1.$$

is called the *folium of Descartes*.

(a) Use a graphing utility to draw this curve.

(b) Your drawing in part (a) should show that the curve has a loop in the first quadrant. Use a CAS to estimate the length of the loop. Round off your answer to four decimal places. HINT: Use symmetry.

49. Sketch the polar curve

$$r = 1 - \cos\theta, \qquad 0 \le \theta \le \pi$$

and calculate the length of the curve.

▷**50.** Use a graphing utility to draw the polar curve

$$r = \sin 5\theta, \qquad 0 \le \theta \le 2\pi$$

and use a CAS to calculate the length of the curve to four decimal place accuracy.

▷**51.** (a) Let $a > b > 0$. Show that the arc length of the ellipse

$$x(t) = a\cos t, \quad y(t) = b\sin t \qquad 0 \le t \le 2\pi$$

is given by the formula

$$L = 4a \int_0^{\pi/2} \sqrt{1 - e^2 \cos^2 t}\, dt.$$

where $e = \sqrt{a^2 - b^2}/a$ is the eccentricity. The integrand does not have an elementary antiderivative.

(b) Set $a = 5$ and $b = 4$. Approximate the arc length of the ellipse using a CAS. Round off your answer to two decimal places.

52. The figure shows the graph of a function f continuously differentiable from $x = a$ to $x = b$ together with a polynomial approximation. Show that the length of this polygonal approximation can be written as the following Riemann sum:

$$\sqrt{1 + [f'(x_1^*)]^2}\,\Delta x_1 + \cdots + \sqrt{1 + [f'(x_n^*)]^2}\,\Delta x_n.$$

As $\|P\| = \max \Delta x_i$ tends to 0, such Riemann sums tend to

$$\int_a^b \sqrt{1 + [f'(x)]^2}\, dx.$$

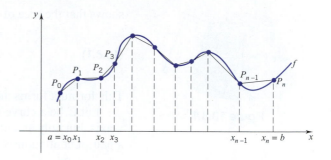

53. Suppose that f is continuously differentiable from $x = a$ to $x = b$. Show that the

(10.7.7)

length of the graph of $f = \displaystyle\int_a^b |\sec[\alpha(x)]|\, dx$

where $\alpha(x)$ is the inclination of the tangent line at $(x, f(x))$.

54. Show that a homogeneous, flexible, inelastic rope hanging from two fixed points assumes the shape of a *catenary*:

$$f(x) = a \cosh\left(\frac{x}{a}\right) = \frac{a}{2}(e^{x/a} + e^{-x/a}). \qquad (a > 0)$$

HINT: Refer to the figure. The part of the rope that corresponds to the interval $[0, x]$ is subject to the following forces:

(1) its weight, which is proportional to its length;
(2) a horizontal pull at 0, $p(0)$;
(3) a tangential pull at x, $p(x)$.

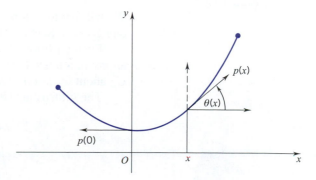

■ 10.8 THE AREA OF A SURFACE OF REVOLUTION; THE CENTROID OF A CURVE; PAPPUS'S THEOREM ON SURFACE AREA

The Area of a Surface of Revolution

In Figure 10.8.1 you can see the frustum of a cone; one radius marked r, the other R. The slant height is marked s. An interesting elementary calculation that we leave to you

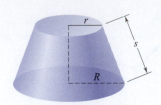

Figure 10.8.1

shows that the area of this slanted surface is given by the formula

(10.8.1)

$$A = \pi(r + R)s.$$

(Exercise 21)

This formula forms the basis for all that follows.

Let C be a curve in the upper half-plane (Figure 10.8.2). The curve can meet the x-axis, but only at a finite number of points. We will assume that C is parametrized by a pair of continuously differentiable functions

$$x = x(t), \qquad y = y(t) \qquad t \in [c, d].$$

Furthermore, we will assume that C is *simple*: no two values of t between c and d give rise to the same point of C; that is, the curve does not intersect itself.

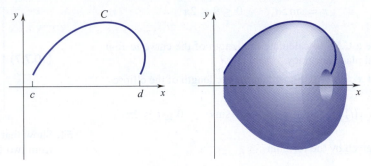

Figure 10.8.2

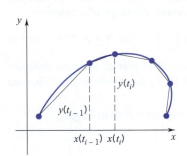

Figure 10.8.3

If we revolve C about the x-axis, we obtain a surface of revolution. The area of that surface is given by the formula

(10.8.2)

$$A = \int_c^d 2\pi y(t)\sqrt{[x'(t)]^2 + [y'(t)]^2}\, dt.$$

We will try to outline how this formula comes about. The argument is similar to the one given in Section 10.7 for the length of a curve.

Each partition $P = \{c = t_0 < t_1 < \cdots < t_n = d\}$ of $[c, d]$ generates a polygonal approximation to C. (Figure 10.8.3) Call this polygonal approximation C_p. By revolving C_p about the x-axis, we get a surface made up of n conical frustums.

The ith frustum (Figure 10.8.4) has slant height

$$s_i = \sqrt{[x(t_i) - x(t_{i-1})]^2 + [y(t_i) - y(t_{i-1})]^2}$$

$$= \sqrt{\left[\frac{x(t_i) - x(t_{i-1})}{t_i - t_{i-1}}\right]^2 \left[\frac{y(t_i) - y(t_{i-1})}{t_i - t_{i-1}}\right]^2}\,(t_i - t_{i-1}).$$

The lateral area $\pi[y(t_{i-1}) + y(t_i)]s_i$ [see 10.8.1] can be written

$$\pi[y(t_{i-1}) + y(t_i)]\sqrt{\left[\frac{x(t_i) - x(t_{i-1})}{t_i - t_{i-1}}\right]^2 + \left[\frac{y(t_i) - y(t_{i-1})}{t_i - t_{i-1}}\right]^2}\,(t_i - t_{i-1}).$$

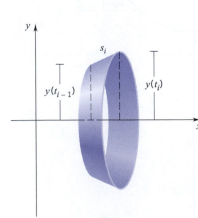

Figure 10.8.4

There exist points $t_i^*, t_i^{**}, t_i^{***}$, all in $[t_{i-1}, t_i]$, such that

$$y(t_i) + y(t_{i-1}) = 2y(t_i^*), \qquad \frac{x(t_i) - x(t_{i-1})}{t_i - t_{i-1}} = x'(t_i^{**}), \qquad \frac{y(t_i) - y(t_{i-1})}{t_i - t_{i-1}} = y'(t_i^{***}).$$

↑
intermediate-value theorem   └──── mean-value theorem ────┘

Let $\Delta t_i = t_i - t_{i-1}$. We can now write the lateral area of the ith frustum as

$$2\pi y(t_i^*)\sqrt{[\lambda'(t_i^{**})]^2 + [y'(t^{***})]^2}\,\Delta t_i.$$

The area generated by revolving all of C_p is the sum of these terms:

$$2\pi y(t_1^*)\sqrt{[x'(t_1^{**})]^2 + [y'(t_1^{***})]^2}\,\Delta t_i + \cdots + 2\pi y(t_n^*)\sqrt{[x'(t_n^{**})]^2 + [y'(t_n^{***})]^2}\,\Delta t_n.$$

This is not a Riemann sum: we don't know that $t_i^* = t^{**} = t_i^{***}$. But it is "close" to a Riemann sum. Close enough that, as $\|P\| \to 0$, this "almost" Riemann sum tends to the integral:

$$\int_c^d 2\pi y(t)\sqrt{[x'(t)]^2 + [y'(t)]^2}\,dt.$$

That this is so follows from a theorem of advanced calculus known as Duhamel's principle. We will not attempt to fill in the details. ❏

Example 1 Derive a formula for the surface area of a sphere from (10.8.2.)

SOLUTION We can generate a sphere of radius r by revolving the arc

$$x(t) = r\cos t, \qquad y(t) = r\sin t \qquad t \in [0, \pi]$$

about the x-axis. Differentiation gives

$$x'(t) = -r\sin t, \quad y'(t) = r\cos t.$$

By (10.8.2),

$$A = 2\pi \int_0^\pi r\sin t\sqrt{r^2(\sin^2 t + \cos^2 t)}\,dt$$

$$= 2\pi r^2 \int_0^\pi \sin t\,dt = 2\pi r^2\Big[-\cos t\Big]_0^\pi = 4\pi r^2. \quad ❏$$

Example 2 Find the area of the surface generated by revolving about the x-axis the curve $y^2 - 2\ln y = 4x$ from $y = 1$ to $y = 2$.

SOLUTION We can represent the curve parametrically by setting

$$x(t) = \tfrac{1}{4}(t^2 - 2\ln t), \qquad y(t) = t \qquad t \in [1, 2].$$

Here

$$x'(t) = \tfrac{1}{2}(t - t^{-1}), \qquad y'(t) = 1$$

and

$$[x'(t)]^2 + [y'(t)]^2 = \big[\tfrac{1}{2}(t + t^{-1})\big]^2. \qquad \text{(check this)}$$

It follows that

$$A = \int_1^2 2\pi t\Big[\tfrac{1}{2}(t + t^{-1})\Big]dt = \int_1^2 \pi(t^2 + 1)\,dt = \pi\Big[\tfrac{1}{3}t^3 + t\Big]_1^2 = \tfrac{10}{3}\pi. \quad ❏$$

Suppose now that C is the graph of a continuously differentiable nonnegative function $y = f(x)$, $x \in [a, b]$. The area of the surface generated by revolving C about the x-axis is given by the formula

(10.8.3)

$$A = \int_a^b 2\pi f(x)\sqrt{1 + [f'(x)]^2}\, dx.$$

This follows readily from (10.8.2). Set

$$x = t, \quad y(t) = f(t) \qquad t \in [a, b].$$

Apply (10.3.2) and then replace the dummy variable t by x.

Example 3 Find the area of the surface generated by revolving about the x-axis the graph of the sine function from $x = 0$ to $x = \frac{1}{2}\pi$.

SOLUTION Setting $f(x) = \sin x$, we have $f'(x) = \cos x$ and therefore

$$A = \int_0^{\pi/2} 2\pi \sin x \sqrt{1 + \cos^2 x}\; dx.$$

To calculate this integral, we set

$$u = \cos x, \quad du = -\sin x\, dx.$$

At $\quad x = 0, \quad u = 1;\quad$ at $\quad x = \frac{1}{2}\pi, u = 0.$ Therefore

$$A = -2\pi \int_1^0 \sqrt{1 + u^2}\, du = 2\pi \int_0^1 \sqrt{1 + u^2}\, du$$

$$= 2\pi \left[\tfrac{1}{2} u\sqrt{1 + u^2} + \tfrac{1}{2} \ln\left(u + \sqrt{1 + u^2}\right) \right]_0^1$$

by (8.4.1) ⟶

$$= \pi[\sqrt{2} + \ln(1 + \sqrt{2})] \cong 2.3\pi \cong 7.23. \quad ❏$$

Centroid of a Curve

The centroid of a plane region Ω is the center of mass of a homogeneous plate in the shape of Ω. Likewise, the centroid of a solid of revolution T is the center of mass of a homogeneous solid in the shape of T. All this was covered in Section 6.4.

What do we mean by the centroid of a plane curve C? Exactly what you would expect. By the *centroid* of a plane curve C, we mean the center of mass of a homogeneous wire in the shape of C. (There is no suggestion here that the centroid of a curve lies on the curve itself. In general, it does not.)

We can locate the centroid of a curve from the following principles, which we take from physics.

Principle 1: Symmetry. If a curve has an axis of symmetry, then the centroid $(\bar{x}, \bar{y})$ lies somewhere along that axis.

Principle 2: Additivity. If a curve with length L is broken up into a finite number of pieces with arc lengths $\Delta s_1, \ldots, \Delta s_n$ and centroids $(\bar{x}_1, \bar{y}_1), \cdots, (\bar{x}_n \bar{y}_n)$, then

$$\bar{x}L = \bar{x}_1 \Delta s_1 + \cdots + \bar{x}_n \Delta s_n \text{ and } \bar{y}L = \bar{y}_1 \Delta s_1 + \cdots + \bar{y}_n \Delta s_n.$$

Figure 10.8.5 shows a curve C that begins at A and ends at B. Let's suppose that the curve is continuously differentiable (can be parametrized by continuously differentiable functions) and that the length of the curve is L. We want a formula for the centroid $(\bar{x}, \bar{y})$.

Let $(X(s), Y(s))$ be the point on C that is at an arc distance s from the initial point A. (What we are doing here is called *parametrizing C by arc length*.) A partition

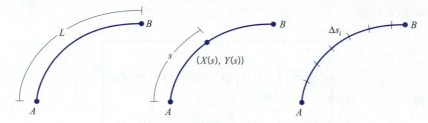

Figure 10.8.5

$P = \{0 = s_0 < s_1 < \cdots < s_n = L\}$ of $[0, L]$ breaks up C into n little pieces of lengths $\Delta s_1, \ldots, \Delta s_n$ and centroids $(\overline{x}_1, \overline{y}_1), \ldots, (\overline{x}_n, \overline{y}_n)$. From Principle 2 we know that

$$\overline{x}L = \overline{x}_1 \Delta s_1 + \cdots + \overline{x}_n \Delta s_n \qquad \text{and} \qquad \overline{y}L = \overline{y}_1 \Delta s_i + \cdots + \overline{y}_n \Delta s_n.$$

Now for each i let $s_i^* = \frac{1}{2}(s_{i-1} + s_i)$. Then $\overline{x}_i \cong X(s_i^*)$ and $\overline{y}_i \cong Y(s_i^*)$. [We are approximating $(\overline{x}_i, \overline{y}_i)$ by the center of the ith little piece.] We can therefore write

$$\overline{x}L \cong \overline{X}_1(s_1^*)\Delta s_1 + \cdots + \overline{X}_n(s_n^*)\Delta s_n, \qquad \overline{y}L \cong \overline{Y}_1(s_1^*)\Delta s_1 + \cdots + \overline{Y}_n(s_n^*)\Delta s_n.$$

The sums on the right are Riemann sums tending to easily recognizable limits: letting $\|P\| \to 0$, we have

(10.8.4)
$$\overline{x}L \int_0^L X(s)\, ds \qquad \text{and} \qquad \overline{y}L = \int_0^L Y(s)\, ds.$$

These formulas give the centroid of a curve in terms of the arc length parameter. It is but a short step from here to formulas stated in terms of a more general parameter.

Suppose that the curve C is given parametrically by the functions

$$x = x(t), \qquad y = y(t) \qquad t \in [c, d]$$

where t is now an arbitrary parameter. Then

$$s(t) = \int_c^t \sqrt{[x'(u)]^2 + [y'(u)]^2}\, du, \qquad ds = s'(t)\, dt = \sqrt{[x'(t)]^2 + [y'(t)]^2}\, dt.$$

At $s = 0, t = c$; at $s = L, t = d$. Changing variables in (10.8.4) from s to t, we have

$$\overline{x}L = \int_c^d X(s(t))\, s'(t)\, dt = \int_c^d X(s(t))\sqrt{[x'(t)]^2 + [y'(t)]^2}\, dt$$

$$\overline{y}L = \int_c^d Y(s(t))\, s'(t)\, dt = \int_c^d Y(s(t))\sqrt{[x'(t)]^2 + [y'(t)]^2}\, dt.$$

A moment's reflection shows that

$$X(s(t)) = x(t) \qquad \text{and} \qquad Y(s(t)) = y(t).$$

We can then write

(10.8.5)

$$\bar{x}L = \int_c^d x(t)\sqrt{[x'(t)]^2 + [y'(t)]^2}\, dt,$$

$$\bar{y}L = \int_c^d y(t)\sqrt{[x'(t)]^2 + [y'(t)]^2}\, dt.$$

These are the centroid formulas stated in terms of an arbitrary parameter t.

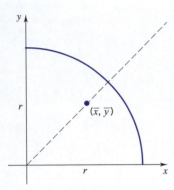

Figure 10.8.6

Example 4 Locate the centroid of the quarter-circle shown in Figure 10.8.6.

SOLUTION We can parametrize that quarter-circle by setting

$$x(t) = r\cos t, \qquad y(t) = r\sin t \qquad t \in [0, \pi/2].$$

Since the curve is symmetric about the line $x = y$, we know that $\bar{x} = \bar{y}$. Here $x'(t) = -r\sin t$ and $y'(t) = r\cos t$. Therefore

$$\sqrt{[x'(t)]^2 + [y'(t)]^2} = \sqrt{r^2\sin^2 t + r^2\cos^2 t} = r.$$

By (10.8.5)

$$\bar{y}L = \int_0^{\pi/2} (r\,\sin t)r\,dt = r^2\int_0^{\pi/2}\sin t\,dt = r^2\Big[-\cos t\Big]_0^{\pi/2} = r^2.$$

Note that $L = \pi r/2$. Therefore $\bar{y} = r^2/L = 2r/\pi$. The centroid of the quarter-circle is at the point $(2r/\pi, 2r/\pi)$. ❑

Remark In Section 6.4 we found that the centroid of the quarter-disk is at the point $(4r/3\pi, 4r/3\pi)$. This point is closer to the origin than the centroid of the quarter-circle. To be expected, since on average the points of the quarter-disk are closer to the origin than the points of the quarter-circle. ❑

Example 5 Take $a > 0$. Locate the centroid of the cardioid $r = a(1 - \cos\theta)$.

SOLUTION The curve (see Figure 10.3.6) is symmetric about the x-axis. Thus $\bar{y} = 0$.
To find $\bar{x}$ we parametrize the curve as follows: taking $\theta \in [0, 2\pi]$, we set

$$x(\theta) = r\cos\theta = a(1 - \cos\theta)\cos\theta,$$

$$y(\theta) = r\sin\theta = a(1 - \cos\theta)\sin\theta.$$

A straightforward calculation shows that

$$[x'(\theta)]^2 + [y'(\theta)]^2 = 4a^2\sin^2\tfrac{1}{2}\theta.$$

Applying (10.8.5), we have

$$\bar{x}L = \int_0^{2\pi} [a(1 - \cos\theta)\cos\theta]\left[2a\sin\tfrac{1}{2}\theta\right] d\theta = -\tfrac{32}{5}a^2.$$

(check this out) $\longrightarrow$

By Example 4 of Section 10.7, $L = 8a$. Thus $\bar{x} = \left(-\tfrac{32}{5}a^2\right)/8a = -\tfrac{4}{5}a$. The centroid of the curve is at the point $\left(-\tfrac{4}{5}a, 0\right)$. ❑

If C is a continuously differentiable curve of the form

$$y = f(x), \qquad x \in [a, b],$$

then (10.8.5) reduces to

(10.8.6)
$$\overline{x}L = \int_a^b x\sqrt{1 + [f'(x)]^2}\, dx, \qquad \overline{y}L = \int_a^b f(x)\sqrt{1 + [f'(x)]^2}\, dx.$$

You can easily verify this.

Pappus's Theorem on Surface Area

That same Pappus who gave us that wonderful theorem on volumes of solids of revolution (Theorem 6.4.4) gave us the following equally marvelous result on surface area:

THEOREM 10.8.7 PAPPUS'S THEOREM ON SURFACE AREA

A plane curve is revolved about an axis that lies in its plane. The curve may meet the axis but, if so, only at a finite number of points: If the curve does not cross the axis, then the area of the resulting surface of revolution is the length of the curve multiplied by the circumference of the circle described by the centroid of the curve:

$$A = 2\pi \overline{R}L$$

where L is the length of the curve and $\overline{R}$ is the distance from the axis of revolution to the centroid of the curve.

Pappus did not have calculus to help him when he made his inspired guesses; he did his work thirteen centuries before Newton or Leibniz was born. With the formulas that we have developed through calculus (through Newton and Leibniz, that is), Pappus's theorem is easily verified. Call the plane of the curve the xy-plane and call the axis of rotation the x-axis. Then $\overline{R} = \overline{y}$ and

$$A = \int_c^d 2\pi y(t)\sqrt{[x'(t)]^2 + [y'(t)]^2}\, dt$$

$$= 2\pi \int_c^d y(t)\sqrt{[x'(t)]^2 + [y'(t)]^2}\, dt = 2\pi \overline{y}L = 2\pi \overline{R}L. \quad ❏$$

EXERCISES 10.8

Exercises 1–10. Find the length of the curve, locate the centroid, and determine the area of the surface generated by revolving the curve about the x-axis.

1. $f(x) = 4, \quad x \in [0, 1]$.

2. $f(x) = 2x, \quad x \in [0, 1]$.

3. $y = \frac{4}{3}x, \quad x \in [0, 3]$.

4. $y = -\frac{12}{5}x + 12, \quad x \in [0, 5]$.

5. $x(t) = 3t, \quad y(t) = 4t \quad t \in [0, 2]$.

6. $r = 5, \quad \theta \in [0, \frac{1}{4}\pi]$.

7. $x(t) = 2\cos t, \quad y(t) = 2\sin t \quad t \in [0, \frac{1}{6}\pi]$.

8. $x(t) = \cos^3 t, \quad y(t) = \sin^3 t \quad t \in [0, \frac{1}{2}\pi]$.

9. $x^2 + y^2 = a^2$ with $x \in [-\frac{1}{2}a, \frac{1}{2}a]$ and $y > 0$.

10. $r = 1 + \cos\theta, \quad \theta \in [0, \pi]$.

Exercises 11–18. Find the area of the surface generated by revolving the curve about the x-axis.

11. $f(x) = \frac{1}{3}x^3, \quad x \in [0, 2].$

12. $f(x) = \sqrt{x}, \quad x \in [1, 2].$

13. $4y = x^3, \quad x \in [0, 1].$

14. $y^2 = 9x, \quad x \in [0, 4].$

15. $y = \cos x, x \in \left[0, \frac{1}{2}\pi\right].$

16. $f(x) = 2\sqrt{1 - x}, \quad x \in [-1, 0].$

17. $r = e^\theta, \quad \theta \in \left[0, \frac{1}{2}\pi\right].$

18. $y = \cosh x, \quad x \in [0, \ln 2].$

19. Take $a > 0$. The curve

$$x(\theta) = a(\theta - \sin\theta), \quad y(\theta) = a(1 - \cos\theta) \quad \theta \text{ real}$$

is called a *cycloid*.

(a) Find the area under the curve from $\theta = 0$ to $\theta = 2\pi$.

(b) Find the area of the surface generated by revolving this part of the curve about the x-axis.

20. Take $a > 0$. The curve

$$x(\theta) = 3a\cos\theta + a\cos 3\theta,$$

$$y(\theta) = 3a\sin\theta - a\sin 3\theta$$

is called a *hypocycloid*.

(a) Use a graphing utility to draw the curves with $a = 1$, $2, \frac{1}{2}$.

(b) Take $a = 1$. Find the area enclosed by the curve.

(c) Take $a = 1$. Set up a definite integral that gives the area of the surface generated by revolving the curve about the x-axis.

21. By cutting a cone of slant height s and base radius r along a lateral edge and laying the surface flat, we can form a sector of a circle of radius s. (See the figure.) Use this idea to verify (10.8.1).

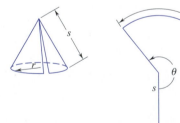

area $= \frac{1}{2}\theta s^2$, θ in radians

22. The figure shows a ring formed by two quarter-circles. Call the corresponding quarter-discs Ω_a and Ω_r. From Section 6.4 we know that Ω_a has centroid at $(4a/3\pi, 4a/3\pi)$ and Ω_r has centroid at $(4r/3\pi, 4r/3\pi)$.

(a) Locate the centroid of the ring without integration.

(b) Locate the centroid of the outer arc from your answer to part (a) by letting a tend to r.

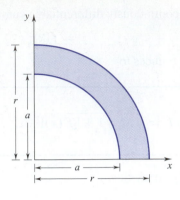

23. (a) Locate the centroid of each side of the triangle in the figure.

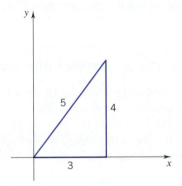

(b) Use your answers in part (a) to calculate the centroid of the triangle.

(c) Where is the centroid of the triangular region?

(d) Where is the centroid of the curve consisting of sides 4 and 5?

(e) Use Pappus's theorem to find the lateral surface area of a cone of base radius 4 and height 3.

24. Find the area of the surface generated by revolving the curve about the x-axis.

(a) $2x = y\sqrt{y^2 - 1} + \ln|y - \sqrt{y^2 - 1}|, y \in [2, 5].$

(b) $6a^2xy = y^4 + 3a^4, \quad y \in [a, 3a].$

25. Use Pappus's theorem to find the surface area of the *torus* generated by revolving about the x-axis the circle $x^2 + (y - b^2) = a^2$. $(0 < a \le b)$

26. (a) We calculated the surface area of a sphere from (10.8.2), not from (10.8.3). Could we just as well have used (10.8.3)? Explain.

(b) Verify that (10.8.2) applied to

$$C: \quad x(t) = \cos t, \quad y(t) = r \quad t \in [0, 2\pi]$$

gives $A = 8\pi r$. However, the surface obtained by revolving C about the x-axis is a cylinder of base radius r and height 2, and therefore A should be $4\pi r$. What's wrong here?

27. (*An interesting property of the sphere*) Slice a sphere along two parallel planes a fixed distance apart. Show that the surface area of the band so obtained depends only on the distance between the planes, not on their location.

28. Locate the centroid of a first-quadrant circular arc

$$C : x(t) = r \cos t, \quad y(t) = r \sin t, \quad t \in [\theta_1, \theta_2].$$

29. Find the surface area of the ellipsoid obtained by revolving the ellipse

$$\frac{x^2}{a^2} + \frac{y^2}{b^2} = 1 \qquad (0 < b < a)$$

(a) about its major axis; (b) about its minor axis.

The Centroid of a Surface of Revolution

If a material surface is homogeneous (constant mass density), then the center of mass of that material surface is called the *centroid*. In general, the determination of the centroid of a surface requires the use of *surface integrals*. (Chapter 18) However, if the surface is a surface of revolution, then the centroid can be found by ordinary one-variable integration.

30. Let C be a simple curve in the upper half-plane parametrized by a pair of continuously differentiable functions.

$$x = x(t), \quad y = y(t) \qquad t \in [c, d].$$

By revolving C about the x-axis, we obtain a surface of revolution, the area of which we denote by A. By symmetry, the centroid of the surface lies on the x-axis. Thus the centroid is completely determined by its x-coordinate $\bar{x}$. Show that

$$(10.8.8) \qquad \bar{x} A = \int_c^d 2\pi x(t) y(t) \sqrt{[x'(t)]^2 + [y'(t)]^2} \, dt$$

by assuming the following additivity principle: if the surface is broken up into n surfaces of revolution with areas $A_1, \ldots, A_n$ and the centroids of the surfaces have x-coordinates $\bar{x}_1, \ldots, \bar{x}_n$, then

$$\bar{x} A = \bar{x}_1 A_1 + \cdots + \bar{x}_n A_n.$$

31. Locate the centroid of a hemisphere of radius r.

32. Locate the centroid of a conical surface of base radius r and height h.

33. Locate the centroid of the lateral surface of the frustum of a cone of height h, base radius R, upper radius r.

■ **PROJECT 10.8 The Cycloid**

Take a wheel (a roll of tape will do) and mark a point on the rim. Call that point P. Now roll the wheel slowly, keeping your eyes on P. The jumping-kangaroo path described by P is called a *cycloid*. To obtain a mathematical characterization of the cycloid, let the radius of the wheel be R and set the wheel on the x-axis so that the point P starts out at the origin. The figure shows P after a turn of θ radians.

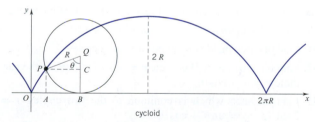
cycloid

Problem 1. Show that the cycloid can be parametrized by the functions

$$x(\theta) = R(\theta - \sin \theta), \quad y(\theta) = R(1 - \cos \theta).$$

HINT: Length of OB = length of $\overset{\frown}{PB} = R\theta$.

Problem 2.

a. At the end of each arch, the cycloid comes to a cusp. Show that x' and y' are both 0 at the end of each arch.

b. Show that the area under an arch of the cycloid is three times the area of the rolling circle.

c. Find the length of an arch of the cycloid.

Problem 3.

a. Locate the centroid of the region under the first arch of the cycloid.

b. Find the volume of the solid generated by revolving the region under an arch of the cycloid about the x-axis.

c. Find the volume of the solid generated by revolving the region under an arch of the cycloid about the y-axis.

The curve given parametrically by

$$x(\phi) = R(\phi + \sin \phi), \quad y(\phi) = R(1 - \cos \phi), \quad \phi \in [-\pi, \pi]$$

is called the *inverted cycloid*. It is obtained by reflecting one arch of the cycloid in the x-axis and then translating the resulting curve so that the low point is at the origin.

Problem 4. Use a graphing utility to draw the inverted cycloid.

Problem 5.

a. Find the inclination α of the line tangent to the inverted cycloid at the point $(x\phi, y(\phi))$.

b. Let s be the arc distance from the low point of the inverted cycloid to the point $(x(\phi), y(\phi))$. Show that $s = 4R \sin \frac{1}{2}\phi = 4R \sin \alpha$, where α is the inclination of the tangent line at $(x(\phi), y(\phi))$.

Visualize two particles sliding without friction down an arch of the inverted cycloid. If the two particles are released at the same time from different positions, which one will reach the bottom first? Neither—they will both get there at exactly the same time. Being the only curve that has this property, the inverted arch of a cycloid is known as the *tautochrone*, the *same-time* curve.

Problem 6. Verify that the inverted arch of a cycloid has the tautochrone property by showing that:

a. The effective gravitational force on a particle of mass m is $-mg \sin \alpha$, where α is the inclination of the tangent line at the position of the particle. From this, conclude that

$$(*) \qquad \frac{d^2 s}{dt^2} = -g \sin \alpha.$$

b. Combine (*) with Problem 5(b) to show that each descending particle is the downward phase of simple harmonic motion with period

$$T = 4\pi \sqrt{R/g}.$$

Thus, while the amplitude of the motion depends on the point of release, the frequency does not. Two particles released simultaneously from different points of the curve will reach the low point of the curve in exactly the same amount of time: $T/4 = \pi \sqrt{R/g}$.

Suppose now that a particle descends without friction along a curve from point A to a point B not directly below it. What should be the shape of the curve so that the particle descends from A to B in the least possible time? This question was first formulated by Johann Bernoulli and posed by him as a challenge to the scientific community in 1696. The challenge was readily accepted and within months the answer was found—by Johann Bernoulli himself, by his brother Jacob, by Newton, by Leibniz, and by L'Hôpital. The answer? Part of an inverted cycloid. Because of this, the inverted cycloid is heralded as the *brachystochrone*, the *least-time* curve.

A proof that the inverted cycloid is the least-time curve, the curve of quickest descent, is beyond our reach. The argument requires a sophisticated variant of calculus known as *the calculus of variations*. We can, however, compare the time of descent along a cycloid to the time of descent along a straight-line path.

Problem 7. You have seen that a particle descends along the inverted arch of a cycloid from $(\pi R, 2R)$ to $(0, 0)$ in time $t = T/4 = \pi \sqrt{R/g}$. What is the time of descent along a straight-line path?

■ CHAPTER 10. REVIEW EXERCISES

Exercises 1–8. Identify the conic section. If the curve is a parabola, find the vertex, focus, axis, and directrix; if the curve is an ellipse, find the center, the foci, and the lengths of the major and minor axes; if the curve is a hyperbola, find the center, vertices, the foci, the asymptotes, and the length of the transverse axis. Then sketch the curve.

1. $x^2 - 4y - 4 = 0$.

2. $x^2 - 4y^2 - 10x + 41 = 0$.

3. $x^2 + 4y^2 + 6x + 8 = 0$.

4. $9x^2 + 4y^2 - 18x - 8y = 23$.

5. $4x^2 - 9y^2 - 8x - 18y + 31 = 0$.

6. $x^2 - 10x - 8y + 41 = 0$.

7. $9x^2 + 25y^2 - 18x + 100y - 116 = 0$.

8. $2x^2 - 3y^2 + 4\sqrt{3}x - 6\sqrt{3}y = 9$.

Exercises 9–10. Give the rectangular coordinates of the point.

9. $[-4, \frac{1}{3}\pi]$.

10. $[\sqrt{2}, -\frac{5}{4}\pi]$.

Exercises 11–14. Points are specified in rectangular coordinates. Give all possible polar coordinates for each point.

11. $(0, -4)$.

12. $(-\frac{1}{2}, -\frac{1}{2})$.

13. $(2\sqrt{3}, -2)$.

14. $(-1, \sqrt{3})$.

Exercises 15–18. Write the equation in polar coordinates.

15. $y = x^2$.

16. $x^2 + y^2 - 2x = 0$.

17. $x^2 + y^2 - 4x + 2y = 0$.

18. $y^2 = \dfrac{x^4}{1 - x^2}$.

Exercises 19–22. Write the equation in rectangular coordinates.

19. $r = 5 \sec \theta$.

20. $r = -4 \cos \theta$.

21. $r = 3 \cos \theta + 4 \sin \theta$.

22. $r^2 = \sec 2\theta$.

Exercises 23–26. Sketch the polar curve.

23. $r = 2(1 - \sin \theta)$.

24. $r = 2 + \cos 2\theta$.

25. $r = \sin^2(\theta/2)$.

26. $r = 1 - 2 \sin \theta$.

Exercises 27–28. Sketch the curves and find the points at which they intersect.

27. $r = \sqrt{2} \sin \theta$, $r^2 = \cos 2\theta$.

28. $r = 2 \cos \theta$, $r = 2\sqrt{3} \sin \theta$.

Exercises 29–30. Sketch the curve and find the area it encloses.

29. $r = 2(1 - \cos \theta)$.

30. $r^2 = 4 \sin 2\theta$.

31. Find the area inside one petal of $r = 2 \cos 4\theta$.

32. Find the area inside the circle $r = \sin \theta$ but outside the cardioid $r = 1 - \cos \theta$.

33. Find the area which is common to the two circles: $r = 2 \sin \theta$, $r = \sin \theta + \cos \theta$.

Exercises 34–38. Express the curve by an equation in x and y; then sketch the curve.

34. $x = 1/t$, $y = t^2 + 1$, $t \neq 0$.

35. $x = \sin t$, $y = \cos 2t$, $0 \leq t \leq \pi$.

36. $x = \cosh t$, $y = \sinh t$, $-\infty < t < \infty$.

37. $x = e^t$, $y = e^{2t} - 2e^t + 1$, $t \geq 0$.

38. $x = t + 2$, $y = t^2 + 4t + 8$, $t \leq 0$.

39. Find a parametrization $x = x(t)$, $y = y(t)$, $t \in [0, 1]$, for the line segment from $(1, 4)$ to $(5, 6)$.

40. Find a parametrization $x = x(t)$, $y = y(t)$, $t \in [0, 1]$, for the line segment from $(2, -1)$ to $(-2, 3)$.

41. A particle traverses the ellipse $4x^2 + 9y^2 = 36$ in a clockwise manner beginning at the point $(0, 2)$. Find a parametrization $x = x(t)$, $y = y(t)$, $t \in [0, 2\pi]$, for the path of the particle.

Exercises 42–43. Find an equation in x and y for the line tangent to the curve.

42. $x(t) = 3t - 1$, $y(t) = 9t^2 - 3t$, $t = 1$.

43. $x(t) = 3e^t$, $y(t) = 5e^{-t}$, $t = 0$.

44. Find an equation in x and y for the line tangent to the polar curve $C : r = 2\sin 2\theta$, at $\theta = \pi/4$.

Exercises 45–46. Find the points (x, y) at which the given curve has (a) a horizontal tangent, (b) a vertical tangent.

45. $x(t) = t^2 + 2t$, $y(t) = t^3 + \frac{3}{2}t^2 - 6t$.

46. $x(t) = 2\cos t - \cos 2t$, $y(t) = 2\sin t - \sin 2t$.

47. Find dy/dx and d^2y/dx^2 for the curve $C : x = 3t^2$, $y = 4t^3$.

48. An object moves in a plane so that dx/dt and d^2y/dt^2 are nonzero constants. Identify the path of the object.

Exercises 49–54. Find the length of the curve.

49. $y = x^{3/2} + 2$ from $x = 0$ to $x = \frac{5}{9}$.

50. $y = \ln(1 - x^2)$ from $x = 0$ to $x = \frac{1}{2}$.

51. $x(t) = \cos t$, $y(t) = \sin^2 t$ from $t = 0$ to $t = \frac{1}{2}\pi$.

52. $x(t) = \frac{1}{2}t^2$, $y(t) = \frac{2}{3}(6t + 9)^{3/2}$ from $t = 0$ to $t = 4$.

53. $r = 1 - \sin\theta$ from $\theta = 0$ to $\theta = 2\pi$.

54. $r = \theta^2$ from $\theta = 0$ to $\theta = \sqrt{5}$.

Exercises 55–58. Find the area of the surface generated by revolving the curve about the x-axis.

55. $y^2 = 4x$ from $x = 0$ to $x = 24$.

56. $x(t) = \frac{2}{3}t^{3/2}$, $y(t) = t$ from $t = 3$ to $t = 8$.

57. $6xy = x^4 + 3$ from $x = 1$ to $x = 3$.

58. $3x^2 + 4y^2 = 3a^2$, $y \geq 0$.

59. Show that, if f is continuously differentiable on $[a, b]$ and f' is never zero, then the length of the graph of $f^{-1} =$ the length of the graph of f.

60. A particle starting at time $t = 0$ at the point $(4, 2)$ moves until time $t = 1$ with $x'(t) = x(t)$ and $y'(t) = 2y(t)$. Find the initial speed v_0, the terminal speed v_1, and the distance traveled s.

Exercises 61–63 concern the *astroid*: the curve

$$x^{2/3} + y^{2/3} = a^{2/3}.$$

61. Sketch the astroid and show that the curve can be parametrized by setting

$$x(\theta) = a\cos^3\theta, \quad y(\theta) = a\sin^3\theta \quad \theta \in [0, 2\pi].$$

62. Find the length of the astroid.

63. Find the centroid of the first-quadrant part of the astroid.

64. A particle moves from time $t = 0$ to $t = 1$ so that $x(t) = 4t - \sin \pi t$, $y(t) = 4t + \cos \pi t$.

(a) When does the particle have minimum speed? When does it have maximum speed?

(b) What is the slope of the tangent line at the point where $t = \frac{1}{4}$?

SEQUENCES;

INDETERMINATE FORMS;

IMPROPER INTEGRALS

■ 11.1 THE LEAST UPPER BOUND AXIOM

So far our approach to the real number system has been somewhat primitive. We have simply taken the point of view that there is a one-to-one correspondence between the set of points on a line and the set of real numbers, and that this enables us to measure all distances, take all roots of nonnegative numbers, and, in short, fill in all the gaps left by the set of rational numbers. This point of view is basically correct and has served us well, but it is not sufficiently sharp to put our theorems on a sound basis, nor is it sufficiently sharp for the work that lies ahead.

We begin with a nonempty set S of real numbers. As indicated in Section 1.2, a number M is an *upper bound* for S if

$$x \leq M \qquad \text{for all} \qquad x \in S.$$

It follows that if M is an upper bound for S, then every number in $[M, \infty)$ is also an upper bound for S. Of course, not all sets of real numbers have upper bounds. Those that do are said to be *bounded above*.

It is clear that every set that has a largest element has an upper bound: if b is the largest element of S, then $x \leq b$ for all $x \in S$. This makes b an upper bound for S. The converse is false: the sets

$$S_1 = (-\infty, 0) \qquad \text{and} \qquad S_2 = \{1/2, 2/3, 3/4, \cdots, n/(n+1), \cdots\}$$

both have upper bounds (for instance, 2 is an upper bound for each set), but neither has a largest element.

Let's return to the first set, S_1. While $(-\infty, 0)$ does not have a largest element, the set of its upper bounds, namely $[0, \infty)$, does have a smallest element, namely 0. We call 0 the *least upper bound* of $(-\infty, 0)$.

Now let's reexamine S_2. While the set of quotients

$$\frac{n}{n+1} = 1 - \frac{1}{n+1}, n = 1, 2, 3, \cdots,$$

does not have a greatest element, the set of its upper bounds, $[1, \infty)$, does have a least element, 1. The number 1 is the *least upper bound* of that set of quotients.

In general, if S is a nonempty set of numbers which is bounded above, then the *least upper bound* of S is the least number which is an upper bound for S.

We now state explicitly one of the key *assumptions* that we make about the real number system. This assumption, called the *least upper bound axiom*, provides the sharpness and clarity that we require.

AXIOM 11.1.1 THE LEAST UPPER BOUND AXIOM

Every nonempty set of real numbers that has an upper bound has a *least upper bound*.

Some find this axiom obvious; some find it unintelligible. For those of you who find it obvious, note that the axiom is not satisfied by the rational number system; namely, it is not true that every nonempty set of rational numbers that has a rational upper bound has a least rational upper bound. (For a detailed illustration of this, we refer you to Exercise 33.) Those who find the axiom unintelligible will come to understand it by working with it.

We indicate the least upper bound of a set S by writing lub S. As you will see from the examples below, the least upper bound idea has wide applicability.

(1) lub $(-\infty, 0) = 0$,   lub$(-\infty, 0] = 0$.

(2) lub $(-4, -1) = -1$,   lub$(-4, -1] = -1$.

(3) lub $\{1/2, 2/3, 3/4, \cdots, n/(n + 1), \cdots\} = 1$.

(4) lub $\{-1/2, -1/8, -1/27, \cdots, -1/n^3, \cdots\} = 0$.

(5) lub $\{x : x^2 < 3\} = $ lub$\{x : -\sqrt{3} < x < \sqrt{3}\} = \sqrt{3}$.

(6) For each decimal fraction

$$b = 0.b_1 b_2 b_3, \cdots,$$

we have

$$b = \text{lub } \{0.b_1, 0.b_1 b_2, 0.b_1 b_2 b_3, \cdots\}.$$

(7) If S consists of the lengths of all polygonal paths inscribed in a semicircle of radius 1, then lub $S = \pi$ (half the circumference of the unit circle).

The least upper bound of a set has a special property that deserves particular attention. The idea is this: the fact that M is the least upper bound of set S does not guarantee that M is in S (indeed, it need not be, as illustrated in the preceding examples), but it guarantees that we can approximate M as closely as we wish by elements of S.

THEOREM 11.1.2

If M is the least upper bound of the set S and ϵ is a positive number, then there is at least one number s in S such that

$$M - \epsilon < s \leq M.$$

PROOF Let $\epsilon > 0$. Since M is an upper bound for S, the condition $s \leq M$ is satisfied by all numbers s in S. All we have to show therefore is that there is some number s in S such that

$$M - \epsilon < s.$$

Suppose on the contrary that there is no such number in S. We then have

$$x \leq M - \epsilon \qquad \text{for all} \qquad x \in S.$$

This makes $M - \epsilon$ an upper bound for S. But this cannot be, for then $M - \epsilon$ is an upper bound for S that is *less* than M, which contradicts the assumption that M is the *least* upper bound. ❏

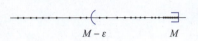

Figure 11.1.1

The theorem we just proved is illustrated in Figure 11.1.1. Take S as the set of points marked in the figure. If $M = \text{lub } S$, then S has at least one element in every half-open interval of the form $(M - \epsilon, M]$.

Example 1

(a) Let $S = \{1/2, 2/3, 3/4, \cdots, n/(n+1), \cdots\}$ and take $\epsilon = 0.0001$. Since 1 is the least upper bound of S, there must be a number $s \in S$ such that

$$1 - 0.0001 < s < 1.$$

There is: take, for example, $s = \dfrac{99,999}{100,000}$.

(b) Let $S = \{0, 1, 2, 3\}$ and take $\epsilon = 0.00001$. It is clear that 3 is the least upper bound of S. Therefore, there must be a number $s \in S$ such that

$$3 - 0.00001 < s \leq 3.$$

There is: $s = 3$. ❏

We come now to lower bounds. Recall that a number m is a *lower bound* for a nonempty set S if

$$m \leq x \qquad \text{for all} \qquad x \in S.$$

Sets that have lower bounds are said to be *bounded below*. Not all sets have lower bounds, but those that do have *greatest lower bounds*. We don't have to assume this. We can prove it by using the least upper bound axiom.

> **THEOREM 11.1.3**
>
> Every nonempty set of real numbers that has a lower bound has a *greatest lower bound*.

PROOF Suppose that S is nonempty and that it has a lower bound k. Then

$$k \leq s \qquad \text{for all} \qquad s \in S.$$

It follows that $-s \leq -k$ for all $s \in S$; that is,

$$\{-s : s \in S\} \qquad \text{has an upper bound} - k.$$

From the least upper bound axiom we conclude that $\{-s : s \in S\}$ has a least upper bound; call it m. Since $-s \leq m$ for all $s \in S$, we can see that

$$-m \leq s \qquad \text{for all} \qquad s \in S,$$

and thus $-m$ is a lower bound for S. We now assert that $-m$ is the greatest lower bound of the set S. To see this, note that, if there existed a number m_1 satisfying

$$-m < m_1 \leq s \qquad \text{for all} \qquad s \in S,$$

then we would have

$$-s \leq -m_1 < m \qquad \text{for all} \qquad s \in S,$$

and thus m would not be the *least* upper bound of $\{-s : s \in S\}$.[†] ❑

The greatest lower bound, although not necessarily in the set, can be approximated as closely as we wish by members of the set. In short, we have the following theorem, the proof of which is left as an exercise.

> ### THEOREM 11.1.4
>
> If m is the greatest lower bound of the set S and ϵ is a positive number, then there is at least one number s in S such that
>
> $$m \leq s < m + \epsilon.$$

The theorem is illustrated in Figure 11.1.2. If $m = \text{glb } S$ (that is, if m is the greatest lower bound of the set S), then S has at least one element in every half-open interval of the form $[m, m + \epsilon)$.

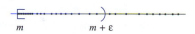

Figure 11.1.2

Remark Given that a function f is defined and continuous on $[a, b]$, what do we know about f? Certainly the following:

(1) We know that if f takes on two values, then it takes on every value in between (the intermediate-value theorem). Thus f maps intervals onto intervals.
(2) We know that f takes on both a maximum value and minimum value (the extreme-value theorem).

We "know" this, but actually we have proven none of it. With the least upper bound axiom in hand, we can prove both theorems. (Appendix B.) ❑

[†]We proved Theorem 11.1.3 by assuming the least upper bound axiom. We could have proceeded the other way. We could have set Theorem 11.1.3 as an axiom and then proved the least upper bound axiom as a theorem.

EXERCISES 11.1

Exercises 1–20. Find the least upper bound (if it exists) and the greatest lower bound (if it exists).

1. $(0, 2)$.

2. $[0, 2]$.

3. $(0, \infty)$.

4. $(-\infty, 1)$.

5. $\{x : x^2 < 4\}$.

6. $\{x : |x - 1| < 2\}$.

7. $\{x : x^3 \geq 8\}$.

8. $\{x : x^4 \leq 16\}$.

9. $\{2\frac{1}{2}, 2\frac{1}{3}, 2\frac{1}{4}, \cdots\}$.

10. $\{-1, -\frac{1}{2}, -\frac{1}{3}, -\frac{1}{4}, \cdots\}$.

11. $\{0.9, 0.99, 0.999, \cdots\}$

12. $\{-2, 2, -2.1, 2.1, -2.11, 2.11, \cdots\}$.

13. $\{x : \ln x < 1\}$.

14. $\{x : \ln x > 0\}$.

15. $\{x : x^2 + x - 1 < 0\}$.

16. $\{x : x^2 + x + 2 \geq 0\}$.

17. $\{x : x^2 > 4\}$.

18. $\{x : |x - 1| > 2\}$.

19. $\{x : \sin x \geq -1\}$.

20. $\{x : e^x < 1\}$.

Exercises 21–24. Find a number s that satisfies the assertion made in Theorem 11.1.4 for S and ϵ as given below.

21. $S = \left\{\frac{1}{11}, \left(\frac{1}{11}\right)^2, \left(\frac{1}{11}\right)^3, \cdots, \left(\frac{1}{11}\right)^n, \cdots\right\}$, $\epsilon = 0.001$.

22. $S = \{1, 2, 3, 4\}$, $\epsilon = 0.0001$.

23. $S = \left\{\frac{1}{10}, \frac{1}{1000}, \frac{1}{100,000}, \cdots, \left(\frac{1}{10}\right)^{2n-1}, \cdots\right\}$, $\epsilon = \left(\frac{1}{10}\right)^k (k \geq 1)$.

24. $S = \left\{\frac{1}{2}, \frac{1}{4}, \frac{1}{8}, \cdots, \left(\frac{1}{2}\right)^n, \cdots\right\}$, $\epsilon = \left(\frac{1}{4}\right)^k (k \geq 1)$.

25. Prove Theorem 11.1.4 by imitating the proof of Theorem 11.1.2.

26. Let $S = \{a_1, a_2, a_3, \cdots, a_n\}$ be a finite nonempty set of real numbers.

    (a) Show that S is bounded.
    (b) Show that lub S and glb S are elements of S.

27. Suppose that b is an upper bound for a set S of real numbers. Prove that if $b \in S$, then $b =$ lub S.

28. Let S be a bounded set of real numbers and suppose that lub $S =$ glb S. What can you conclude about S?

29. Suppose that S is a nonempty bounded set of real numbers and T is a nonempty subset of S.

    (a) Show that T is bounded.
    (b) Show that glb $S \leq$ glb $T \leq$ lub $T \leq$ lub S.

30. Show by example

    (a) that the least upper bound of a set of rational numbers need not be rational.
    (b) that the least upper bound of a set of irrational numbers need not be irrational.

31. Let c be a positive number. Prove that the set $S = \{c, 2c, 3c, \cdots, nc, \cdots\}$ is not bounded above.

32. (a) Show that the least upper bound of a set of negative numbers cannot be positive.
    (b) Show that the greatest lower bound of a set of positive numbers cannot be negative.

33. The set S of rational numbers x with $x^2 < 2$ has rational upper bounds but no least rational upper bound. The argument goes like this. Suppose that S has a least rational upper bound and call it x_0. Then either

$$x_0^2 = 2, \quad \text{or} \quad x_0^2 > 2, \quad \text{or} \quad x_0^2 < 2.$$

    (a) Show that $x_0^2 = 2$ is impossible by showing that if $x_0^2 = 2$, then x_0 is not rational.
    (b) Show that $x_0^2 > 2$ is impossible by showing that if $x_0^2 > 2$, then there is a positive integer n for which $(x_0 - \frac{1}{n})^2 > 2$, which makes $x_0 - \frac{1}{n}$ a rational upper bound for S that is less than the least rational upper bound x_0.
    (c) Show that $x_0^2 < 2$ is impossible by showing that if $x_0^2 < 2$, then there is a positive integer n for which $(x_0 + \frac{1}{n})^2 < 2$. This places $x_0 + \frac{1}{n}$ in S and shows that x_0 cannot be an upper bound for S.

34. Recall that a *prime number* is an integer $p > 1$ that has no positive integer divisors other than 1 and p. A famous theorem of Euclid states that there are an infinite number of primes (and therefore the set of primes is unbounded above). Prove that there are an infinite number of primes. HINT: Following the way of Euclid, assume that there are only a finite number of primes $p_1, p_2, \cdots, p_n$ and examine the number $x = (p_1 p_2 \cdots p_n) + 1$.

▷ 35. Let $S = \{2, (\frac{3}{2})^2, (\frac{4}{3})^3, (\frac{5}{4})^4, \cdots, (\frac{n+1}{n})^n, \cdots\}$.

    (a) Use a graphing utility or CAS to calculate $(\frac{n+1}{n})^n$ for $n = 5, 10, 100, 1000, 10,000$.
    (b) Does S have a least upper bound? If so, what is it? Does S have a greatest lower bound? If so, what is it?

▷ 36. Let $S = \{a_1, a_2, a_3, \cdots, a_n, \cdots\}$ with $a_1 = 4$ and for further subscripts $a_{n+1} = 3 - 3/a_n$.

    (a) Calculate the numbers $a_2, a_3, a_4, \cdots, a_{10}$.
    (b) Use a graphing utility or CAS to calculate $a_{20}, a_{30}, \cdots, a_{50}$.
    (c) Does S have a least upper bound? If so, what is it? Does S have a greatest lower bound? If so, what is it?

▷ 37. Let $S = \{\sqrt{2}, \sqrt{2\sqrt{2}}, \sqrt{2\sqrt{2\sqrt{2}}}, \cdots\}$. Thus $S = \{a_1, a_2, a_3, \cdots, a_n, \cdots\}$ with $a_1 = \sqrt{2}$ and for further subscripts $a_{n+1} = \sqrt{2a_n}$.

    (a) Use a graphing utility or CAS to calculate the numbers $a_1, a_2, a_3, \cdots, a_{10}$.
    (b) Show by induction that $a_n < 2$ for all n.
    (c) What is the least upper bound of S?
    (d) In the definition of S, replace 2 by an arbitrary positive number c. What is the least upper bound in this case?

▷ 38. Let $S = \{\sqrt{2}, \sqrt{2 + \sqrt{2}}, \sqrt{2 + \sqrt{2 + \sqrt{2}}}, \cdots\}$. Thus $a_1 = \sqrt{2}$ and for further subscripts $a_{n+1} = \sqrt{2 + a_n}$.

    (a) Use a graphing utility or CAS to calculate the numbers $a_1, a_2, a_3, \cdots, a_{10}$.
    (b) Show by induction that $a_n < 2$ for all n.
    (c) What is the least upper bound of S?
    (d) In the definition of S, replace 2 by an arbitrary positive number c. What is the least upper bound in this case?

■ 11.2 SEQUENCES OF REAL NUMBERS

To this point we have considered sequences only in a peripheral manner. Here we focus on them.

What is a sequence of real numbers?

DEFINITION 11.2.1   SEQUENCE OF REAL NUMBERS

A *sequence of real numbers* is a real-valued function defined on the set of positive integers.

You may find this definition somewhat surprising, but in a moment you will see that it makes sense.

Suppose we have a sequence of real numbers

$$a_1, a_2, a_3, \cdots, a_n, \cdots$$

What is a_1? It is the image of 1. What is a_2? It is the image of 2. What is a_3? It is the image of 3. In general, a_n is the image of n, the value that the function takes on at the integer n.

By convention, a_1 is called the *first term* of the sequence, a_2 the *second term*, and so on. More generally, a_n, the term with *index n*, is called the *nth term*.

Sequences can be defined by giving the law of formation. For example:

$$a_n = \frac{1}{n} \qquad \text{is the sequence} \qquad 1, \tfrac{1}{2}, \tfrac{1}{3}, \tfrac{1}{4}, \cdots;$$

$$b_n = \frac{n}{n+1} \qquad \text{is the sequence} \qquad \tfrac{1}{2}, \tfrac{2}{3}, \tfrac{3}{4}, \tfrac{4}{5}, \cdots;$$

$$c_n = n^2 \qquad \text{is the sequence} \qquad 1, 4, 9, 16, \cdots.$$

It's like defining f by giving $f(x)$.

Sequences can be multiplied by constants; they can be added, subtracted, and multiplied. From

$$a_1, a_2, a_3, \cdots, a_n, \cdots \qquad \text{and} \qquad b_1, b_2, b_3, \cdots, b_n, \cdots$$

we can form

the scalar product sequence : $\alpha a_1, \alpha a_2, \alpha a_3, \cdots, \alpha a_n, \cdots,$

the sum sequence : $a_1 + b_1, a_2 + b_2, a_3 + b_3, \cdots, a_n + b_n, \cdots,$

the difference sequence : $a_1 - b_1, a_2 - b_2, a_3 - b_3, \cdots, a_n - b_n, \cdots,$

the product sequence : $a_1 b_1, a_2 b_2, a_3 b_3, \cdots, a_n b_n, \cdots.$

If the b_i's are all different from zero, we can form

the reciprocal sequence : $\dfrac{1}{b_1}, \dfrac{1}{b_2}, \dfrac{1}{b_3}, \cdots, \dfrac{1}{b_n}, \cdots,$

the quotient sequence : $\dfrac{a_1}{b_1}, \dfrac{a_2}{b_2}, \dfrac{a_3}{b_3}, \cdots, \dfrac{a_n}{b_n}, \cdots.$

The *range* of a sequence is the set of values taken on by the sequence. While there can be repetition in a sequence, there can be no repetition in the statement of its range. The range is a set, and in a set there is no repetition. A number is either in a particular set or it's not. It can't be there more than once. The sequences

$$0, 1, 0, 1, 0, 1, 0, 1, \cdots \qquad \text{and} \qquad 0, 0, 1, 1, 0, 0, 1, 1, \cdots$$

both have the same range: the set $\{0, 1\}$. The range of the sequence

$$0, 1, -1, 2, 2, -2, 3, 3, 3, -3, 4, 4, 4, 4, -4, \cdots$$

is the set of integers.

Boundedness and unboundedness for sequences are what they are for other real-valued functions. Thus the sequence $a_n = 2^n$ is bounded below (with greatest lower bound 2) but unbounded above. The sequence $b_n = 2^{-n}$ is bounded. It is bounded below with greatest lower bound 0, and it is bounded above with least upper bound $\tfrac{1}{2}$.

Many of the sequences we work with have some regularity. They either have an upward tendency or they have a downward tendency. The following terminology is standard. The sequence with terms a_n is said to be

$$
\begin{array}{llll}
\textit{increasing} & \text{if} & a_n < a_{n+1} & \text{for all } n, \\
\textit{nondecreasing} & \text{if} & a_n \le a_{n+1} & \text{for all } n, \\
\textit{decreasing} & \text{if} & a_n > a_{n+1} & \text{for all } n, \\
\textit{nonincreasing} & \text{if} & a_n \ge a_{n+1} & \text{for all } n.
\end{array}
$$

A sequence that satisfies any of these conditions is called *monotonic*.
The sequences

$$1, \tfrac{1}{2}, \tfrac{1}{3}, \tfrac{1}{4}, \cdots, \tfrac{1}{n}, \cdots$$

$$2, 4, 8, 16, \cdots, 2^n, \cdots$$

$$2, 2, 4, 4, 6, 6, \cdots, 2n, 2n, \cdots$$

are monotonic. The sequence

$$1, \tfrac{1}{2}, 1, \tfrac{1}{3}, 1, \tfrac{1}{4}, 1, \cdots$$

is not monotonic.

Now to some examples that are less trivial.

Example 1 The sequence $a_n = \dfrac{n}{n+1}$ is increasing. It is bounded below by $\tfrac{1}{2}$ (the greatest lower bound) and above by 1 (the least upper bound).

PROOF Since

$$
\frac{a_{n+1}}{a_n} = \frac{(n+1)/(n+2)}{n/(n+1)} = \frac{n+1}{n+2} \cdot \frac{n+1}{n} = \frac{n^2 + 2n + 1}{n^2 + 2n} > 1,
$$

we have $a_n < a_{n+1}$. This confirms that the sequence is increasing. The sequence can be displayed as

$$\tfrac{1}{2}, \tfrac{2}{3}, \tfrac{3}{4}, \tfrac{4}{5}, \cdots, \tfrac{98}{99}, \tfrac{99}{100}, \cdots$$

It is clear that $\tfrac{1}{2}$ is the greatest lower bound and 1 is the least upper bound. (See the two representations in Figure 11.2.1.) ❏

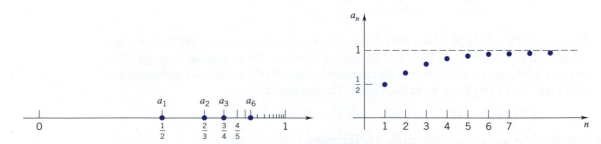

Figure 11.2.1

Example 2 The sequence $a_n = \dfrac{2^n}{n!}$ is nonincreasing and starts decreasing at $n = 2$.[†]

[†] Recall that $n! = n(n-1)(n-2) \cdots 3 \cdot 2 \cdot 1$.

PROOF The first two terms are equal:

$$a_1 = \frac{2^1}{1!} = 2 = \frac{2^2}{2!} = a_2.$$

For $n \geq 2$ the sequence decreases:

$$\frac{a_{n+1}}{a_n} = \frac{2^{n+1}}{(n+1)!} \cdot \frac{n!}{2^n} = \frac{2}{n+1} < 1. \qquad \text{(see Figure 11.2.2)} \quad ❏$$

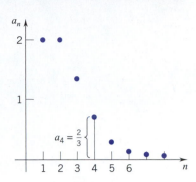

Figure 11.2.2

Example 3 For $c > 1$, the sequence $a_n = c^n$ increases without bound.

PROOF Choose a number $c > 1$. Then

$$\frac{a_{n+1}}{a_n} = \frac{c^{n+1}}{c^n} = c > 1.$$

This shows that the sequence increases. To show the unboundedness, we take an arbitrary positive number M and show that there exists a positive integer k for which

$$c^k \geq M.$$

A suitable k is one for which

$$k \geq \frac{\ln M}{\ln c},$$

for then

$$k \ln c \geq \ln M, \qquad \ln c^k \geq \ln M, \qquad c^k \geq M. \quad ❏$$

Since sequences are defined on the set of positive integers and not on an interval, they are not directly susceptible to the methods of calculus. Fortunately, we can sometimes circumvent this difficulty by dealing initially, not with the sequence itself, but with a differentiable function of a real variable x that agrees with the given sequence at the positive integers n.

Example 4 The sequence $a_n = \dfrac{n}{e^n}$ is decreasing. It is bounded above by $1/e$ and below by 0.

PROOF We will work with the function

$$f(x) = \frac{x}{e^x}.$$

Note that $f(1) = 1/e = a_1$, $f(2) = 2/e^2 = a_2$, $f(3) = 3/e^3 = a_3$, and so on.

Differentiating f, we get

$$f'(x) = \frac{e^x - xe^x}{e^{2x}} = \frac{1-x}{e^x}.$$

Since $f'(x) < 0$ for $x > 1$, f decreases on $[1, \infty)$. Thus $f(1) > f(2) > f(3) > \cdots$. Thus $a_1 > a_2 > a_3 > \cdots$. The sequence is decreasing.

The first term $a_1 = 1/e$ is the least upper bound of the sequence. Since all the terms of the sequence are positive, 0 is a lower bound for the sequence.

In Figure 11.2.3 we have sketched the graph of $f(x) = x/e^x$ and marked some of the points (n, a_n). As the figure suggests, the a_n decrease toward 0. The number 0 is the greatest lower bound of the sequence.[†] ❏

[†]Proof of this is readily obtained by a method outlined in Section 11.6.

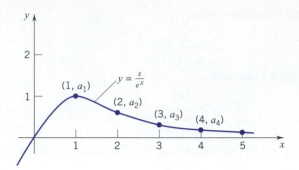

Figure 11.2.3

Example 5 The sequence $a_n = n^{1/n}$ decreases for $n \geq 3$.

PROOF We could compare a_n with a_{n+1} directly, but it is easier to consider the function

$$f(x) = x^{1/x}$$

instead. Since $f(x) = e^{(1/x)\ln x}$, we have

$$f'(x) = e^{(1/x)\ln x} \frac{d}{dx}\left(\frac{1}{x}\ln x\right) = x^{1/x}\left(\frac{1 - \ln x}{x^2}\right).$$

For $x > e$, $f'(x) < 0$. This shows that f decreases on $[e, \infty)$. Since $3 > e$, the function f decreases on $[3, \infty)$, and the sequence decreases for $n \geq 3$. ❏

EXERCISES 11.2

Exercises 1–8. The first several terms of a sequence $a_1, a_2 \cdots$ are given. Assume that the pattern continues as indicated and find an explicit formula for the a_n.

1. $2, 5, 8, 11, 14, \cdots$.

2. $2, 0, 2, 0, 2, \cdots$.

3. $1, -\frac{1}{3}, \frac{1}{5}, -\frac{1}{7}, \frac{1}{9}, \cdots$

4. $\frac{1}{2}, \frac{3}{4}, \frac{7}{8}, \frac{15}{16}, \frac{31}{32}, \cdots$.

5. $2, \frac{5}{2}, \frac{10}{3}, \frac{17}{4}, \frac{26}{5}, \cdots$.

6. $-\frac{1}{4}, \frac{2}{9}, -\frac{3}{16}, \frac{4}{25}, -\frac{5}{36}, \cdots$.

7. $1, \frac{1}{2}, 3, \frac{1}{4}, 5, \frac{1}{6}, \cdots$.

8. $1, 2, \frac{1}{9}, 4, \frac{1}{25}, 6, \frac{1}{49}, \cdots$.

Exercises 9–40. Determine the boundedness and monotonicity of the sequence with a_n, as indicated.

9. $\dfrac{2}{n}$.

10. $\dfrac{(-1)^n}{n}$.

11. $\dfrac{n + (-1)^n}{n}$.

12. $(1.001)^n$.

13. $(0.9)^n$.

14. $\dfrac{n - 1}{n}$.

15. $\dfrac{n^2}{n + 1}$.

16. $\sqrt{n^2 + 1}$.

17. $\dfrac{4n}{\sqrt{4n^2 + 1}}$.

18. $\dfrac{2^n}{4^n + 1}$.

19. $\dfrac{4^n}{2^n + 100}$.

20. $\dfrac{n^2}{\sqrt{n^3 + 1}}$.

21. $\ln\left(\dfrac{2n}{n + 1}\right)$.

22. $\dfrac{n + 2}{3^{10}\sqrt{n}}$.

23. $\dfrac{(n + 1)^2}{n^2}$.

24. $(-1)^n\sqrt{n}$.

25. $\sqrt{4 - \dfrac{1}{n}}$.

26. $\ln\left(\dfrac{n + 1}{n}\right)$.

27. $(-1)^{2n+1}\sqrt{n}$.

28. $\dfrac{\sqrt{n + 1}}{\sqrt{n}}$.

29. $\dfrac{2^n - 1}{2^n}$.

30. $\dfrac{1}{2n} - \dfrac{1}{2n + 3}$.

31. $\sin\dfrac{\pi}{n + 1}$.

32. $(-\frac{1}{2})^n$.

33. $(1.2)^{-n}$.

34. $\dfrac{n + 3}{\ln(n + 3)}$.

35. $\dfrac{1}{n} - \dfrac{1}{n + 1}$.

36. $\cos n\pi$.

37. $\dfrac{\ln(n + 2)}{n + 2}$.

38. $\dfrac{(-2)^n}{n^{10}}$.

39. $\dfrac{3^n}{(n + 1)^2}$.

40. $\dfrac{1 - (\frac{1}{2})^n}{(\frac{1}{2})^n}$.

41. Show that the sequence $a_n = 5^n/n!$ decreases for $n \geq 5$. Is the sequence nonincreasing?

42. Let M be a positive integer. Show that $a_n = M^n/n!$ decreases for $n \geq M$.

43. Show that, if $0 < c < d$, then the sequence

$$a_n = (c^n + d^n)^{1/n}$$

is bounded and monotonic.

44. Show that linear combinations and products of bounded sequences are bounded.

Sequences can be defined *recursively*: one or more terms are given explicitly; the remaining ones are then defined in terms of their predecessors.

Exercises 45–56. Give the first six terms of the sequence and then give the nth term.

45. $a_1 = 1$; $\quad a_{n+1} = \dfrac{1}{n+1} a_n$.

46. $a_1 = 1$; $\quad a_{n+1} = a_n + 3n(n+1) + 1$.

47. $a_1 = 1$; $\quad a_{n+1} = \frac{1}{2}(a_n + 1)$.

48. $a_1 = 1$; $\quad a_{n+1} = \frac{1}{2} a_n + 1$.

49. $a_1 = 1$; $\quad a_{n+1} = a_n + 2$.

50. $a_1 = 1$; $\quad a_{n+1} = \dfrac{n}{n+1} a_n$.

51. $a_1 = 1$; $\quad a_{n+1} = a_n + 2n + 1$.

52. $a_1 = 1$; $\quad a_{n+1} = 2a_n + 1$.

53. $a_1 = 1$; $\quad a_{n+1} = a_n + \cdots + a_1$.

54. $a_1 = 3$; $\quad a_{n+1} = 4 - a_n$.

55. $a_1 = 1$, $\quad a_2 = 3$; $\quad a_{n+1} = 2a_n - a_{n-1}$, $\quad n \geq 2$.

56. $a_1 = 1$, $\quad a_2 = 3$; $\quad a_{n+1} = 3a_n - 2n - 1$, $\quad n \geq 2$.

Exercises 57–60. Use mathematical induction to prove the following assertions.

57. If $a_1 = 1$ and $a_{n+1} = 2a_n + 1$, then $a_n = 2^n - 1$.

58. If $a_1 = 3$ and $a_{n+1} = a_n + 5$, then $a_n = 5n - 2$.

59. If $a_1 = 1$ and $a_{n+1} = \dfrac{n+1}{2n} a_n$, then $a_n = \dfrac{n}{2^{n-1}}$.

60. If $a_1 = 1$ and $a_{n+1} = a_n - \dfrac{1}{n(n+1)}$, then $a_n = \dfrac{1}{n}$.

61. Let r be a real number, $r \neq 0$. Define

$$\begin{aligned} S_1 &= 1 \\ S_2 &= 1 + r \\ S_3 &= 1 + r + r^2 \\ &\cdot \\ &\cdot \\ &\cdot \\ S_n &= 1 + r + r^2 + \cdots + r^{n-1} \\ &\cdot \\ &\cdot \\ &\cdot \end{aligned}$$

(a) Suppose that $r = 1$. What is S_n for $n = 1, 2, 3, \cdots$?

(b) Suppose that $r \neq 1$. Find a formula for S_n that does not involve adding up the powers of r. HINT: Calculate $S_n - r S_n$.

62. Set $a_n = \dfrac{1}{n(n+1)}$, $n = 1, 2, 3, \cdots$, and form the sequence

$$\begin{aligned} S_1 &= a_1 \\ S_2 &= a_1 + a_2 \\ S_3 &= a_1 + a_2 + a_3 \\ &\cdot \\ &\cdot \\ &\cdot \\ S_n &= a_1 + a_2 + a_3 + \cdots + a_n \\ &\cdot \\ &\cdot \\ &\cdot \end{aligned}$$

Find a formula for S_n that does not involve adding up the terms $a_1, a_2, a_3, \cdots$. HINT: Use partial fractions to write $1/[k(k+1)]$ as the sum of two fractions.

63. A ball is dropped from a height of 100 feet. Each time it hits the ground, it rebounds to 75% of its previous height.

(a) Let S_n be the distance that the ball travels between the nth and the $(n+1)$st bounce. Find a formula for S_n.

(b) Let T_n be the time that the ball is in the air between the nth and the $(n+1)$st bounce. Find a formula for T_n.

64. Suppose that a bacterial culture is growing exponentially (Section 7.6) and that the culture doubles every 12 hours. Obtain a formula for the number P_n of square millimeters of culture expected in n hours given that the culture measured 5 square millimeters when first observed.

65. Define a sequence by setting $a_n = \sqrt{n^2 + n} - n$, $n = 1, 2, 3, \cdots$. Use a graphing utility or CAS to plot the first 15 terms of the sequence. Determine whether the sequence is monotonic. If so, in what sense?

66. Exercise 65 for the sequence defined recursively by setting

$$a_1 = 100, \qquad a_{n+1} = \sqrt{2 + a_n}, \qquad n = 1, 2, 3, \cdots.$$

67. Define a sequence recursively by setting

$$a_1 = 1; \qquad a_{n+1} = 1 + \sqrt{a_n}, \qquad n = 1, 2, 3, \cdots.$$

(a) Show by induction that this is an increasing sequence.

(b) Show by induction that the sequence is bounded above.

(c) Use a graphing utility or CAS to calculate $a_2, a_3, a_4, \cdots, a_{15}$.

(d) Use a graphing utility or CAS to plot the first 15 terms of the sequence.

(e) Estimate the least upper bound of the sequence.

68. Define a sequence recursively by setting

$$a_1 = 1; \qquad a_{n+1} = \sqrt{3a_n}, \qquad n = 1, 2, 3, \cdots.$$

(a) Show by induction that this is an increasing sequence.

(b) Show by induction that the sequence is bounded above.

(c) Use a graphing utility or CAS to calculate $a_2, a_3, a_4, \cdots, a_{15}$.

(d) Use a graphing utility or CAS to plot the first 15 terms of the sequence.

(e) Estimate the least upper bound of the sequence.

■ 11.3 LIMIT OF A SEQUENCE

You have seen the limit process applied in various settings. The limit process applied to sequences is particularly simple and exactly what you might expect.

DEFINITION 11.3.1 LIMIT OF A SEQUENCE

$$\lim_{n \to \infty} a_n = L$$

if for each $\epsilon > 0$, there exists a positive integer K such that

$$\text{if} \quad n \geq K, \quad \text{then} \quad |a_n - L| < \epsilon.$$

Example 1 Since

$$\frac{4n - 1}{n} = 4 - \frac{1}{n},$$

it is intuitively clear that

$$\lim_{n \to \infty} \frac{4n - 1}{n} = 4.$$

To verify that this statement conforms to the definition of limit of a sequence, we must show that for each $\epsilon > 0$ there exists a positive integer K such that

$$\text{if} \quad n \geq K, \quad \text{then} \quad \left| \frac{4n - 1}{n} - 4 \right| < \epsilon.$$

To do this, we fix $\epsilon > 0$ and note that

$$\left| \frac{4n - 1}{n} - 4 \right| = \left| \left(4 - \frac{1}{n} \right) - 4 \right| = \frac{1}{n}.$$

We now choose K sufficiently large that $1/K < \epsilon$. If $n \geq K$, then $1/n \leq 1/K < \epsilon$ and consequently

$$\left| \frac{4n - 1}{n} - 4 \right| = \frac{1}{n} < \epsilon. \quad \square$$

Example 2 Since

$$\frac{2\sqrt{n}}{\sqrt{n} + 1} = \frac{2}{1 + 1/\sqrt{n}},$$

it is intuitively clear that

$$\lim_{n \to \infty} \frac{2\sqrt{n}}{\sqrt{n} + 1} = 2.$$

To verify that this statement conforms to the definition of limit of a sequence, we must show that for each $\epsilon > 0$ there exists a positive integer K such that

$$\text{if} \quad n \geq K, \quad \text{then} \quad \left| \frac{2\sqrt{n}}{\sqrt{n} + 1} - 2 \right| < \epsilon.$$

To do this, we fix $\epsilon > 0$ and note that

$$\left| \frac{2\sqrt{n}}{\sqrt{n}+1} - 2 \right| = \left| \frac{2\sqrt{n} - 2(\sqrt{n}+1)}{\sqrt{n}+1} \right| = \left| \frac{-2}{\sqrt{n}+1} \right| = \frac{2}{\sqrt{n}+1} < \frac{2}{\sqrt{n}}.$$

We now choose K sufficiently large that $2/\sqrt{K} < \epsilon$. If $n \geq K$, then $2/\sqrt{n} \leq 2/\sqrt{K} < \epsilon$ and consequently

$$\left| \frac{2\sqrt{n}}{\sqrt{n}+1} - 2 \right| < \frac{2}{\sqrt{n}} < \epsilon. \quad \square$$

In Section 11.1 we gave meaning to decimal expansion

$$b = 0.b_1 b_2 b_3 \cdots$$

through the least upper bound axiom. We stated that

$$b = \text{lub}\{0.b_1, \, 0.b_1 b_2, \, 0.b_1 b_2 b_3, \, \cdots\}.$$

We can view b as the limit of a sequence: set $a_n = 0.b_1 b_2 \cdots b_n$ and we have

$$\lim_{n \to \infty} a_n = b.$$

Example 3 You are familiar with the assertion

$$\frac{1}{3} = 0.333 \cdots.$$

Here we justify this assertion by showing that the sequence with terms $a_n = 0.\overbrace{333 \cdots 3}^{n}$ satisfies the limit condition

$$\lim_{n \to \infty} a_n = \frac{1}{3}.$$

To this end, we fix $\epsilon > 0$ and observe that

$$\left| a_n - \frac{1}{3} \right| = \left| 0.\overbrace{333 \cdots 3}^{n} - \frac{1}{3} \right| = \left| \frac{\overbrace{0.999 \cdots 9}^{n} - 1}{3} \right| = \frac{1}{3} \cdot \frac{1}{10^n} < \frac{1}{10^n}.$$

We now choose K sufficiently large that $1/10^K < \epsilon$. If $n \geq K$, then $1/10^n \leq 1/10^K < \epsilon$ and therefore

$$\left| a_n - \frac{1}{3} \right| < \frac{1}{10^n} < \epsilon. \quad \square$$

Limit Theorems

The limit process for sequences is so similar to the limit processes you have already studied that you may find you can prove many of the limit theorems yourself. In any case, try to come up with your own proofs and refer to these only if necessary.

THEOREM 11.3.2 UNIQUENESS OF LIMIT

If $\quad \lim_{n \to \infty} a_n = L \quad$ and $\quad \lim_{n \to \infty} a_n = M, \quad$ then $\quad L = M.$

A proof along the lines of our proof of Theorem 2.3.1 is given in the supplement at the end of this section. Try to construct your own proof.

DEFINITION 11.3.3

A sequence that has a limit is said to be *convergent*. A sequence that has no limit is said to be *divergent*.

Instead of writing

$$\lim_{n \to \infty} a_n = L,$$

we will often write

$$a_n \to L \qquad \text{(read "} a_n \text{ converges to } L\text{")}$$

or more fully,

$$a_n \to L \qquad \text{as} \qquad n \to \infty.$$

THEOREM 11.3.4

Every convergent sequence is bounded.

PROOF Assume that $a_n \to L$ and choose any positive number: 1, for instance. Using 1 as ϵ, you can see that there must be a positive integer K such that

$$|a_n - L| < 1 \qquad \text{for all} \qquad n \geq K.$$

Since $|a_n| - |L| \leq \big||a_n| - |L|\big| \leq |a_n - L|$, we have

$$|a_n| < 1 + |L| \qquad \text{for all} \qquad n \geq K.$$

It follows that

$$|a_n| \leq \max\{|a_1|, |a_2|, \ldots, |a_{K-1}|, 1 + |L|\} \qquad \text{for all } n.$$

This proves that the sequence is bounded. ❏

Since every convergent sequence is bounded, a sequence that is not bounded cannot be convergent; namely:

(11.3.5) | Every unbounded sequence is divergent.

The sequences

$$a_n = \tfrac{1}{2}n, \qquad b_n = \frac{n^2}{n+1}, \qquad c_n = n \ln n$$

are all unbounded. Each of these sequences is therefore divergent.

Boundedness does not imply convergence. As a counterexample, consider the "oscillating" sequence

$$\{1, 0, 1, 0, 1, 0, \cdots\}.$$

This sequence is certainly bounded (above by 1 and below by 0), but it does not converge: the limit would have to be arbitrarily close both to 0 and to 1.

Boundedness together with monotonicity does imply convergence.

THEOREM 11.3.6

A nondecreasing sequence which is bounded above converges to the least upper bound of its range.

A nonincreasing sequence which is bounded below converges to the greatest lower bound of its range.

PROOF Suppose the sequence with terms a_n is bounded above and nondecreasing. If L is the least upper bound of this sequence, then

$$a_n \leq L \qquad \text{for all } n.$$

Now let ϵ be an arbitrary positive number. By Theorem 11.1.2 there exists a_K such that

$$L - \epsilon < a_K.$$

Since the sequence is nondecreasing,

$$a_K \leq a_n \qquad \text{for all} \qquad n \geq K.$$

It follows that

$$L - \epsilon < a_n \leq L \qquad \text{for all} \qquad n \geq K.$$

This shows that

$$|a_n - L| < \epsilon \qquad \text{for all} \qquad n \geq K$$

and proves that

$$a_n \to L.$$

The nonincreasing case can be handled in a similar manner. ❏

Since $b = 0.b_1 b_2 b_3 \cdots$ is the least upper bound of the sequence with terms $a_n = 0.b_1 b_2 \cdots b_n$, and this sequence is bounded and nondecreasing, the theorem confirms that

$$\lim_{n \to \infty} a_n = b.$$

Example 4 We shall show that the sequence $a_n = (3^n + 4^n)^{1/n}$ is convergent. Since

$$3 = (3^n)^{1/n} < (3^n + 4^n)^{1/n} < (4^n + 4^n)^{1/n} = (2 \cdot 4^n)^{1/n} = 2^{1/n} \cdot 4 \leq 8,$$

the sequence is bounded. Note that

$$(3^n + 4^n)^{(n+1)/n} = (3^n + 4^n)^{1/n}(3^n + 4^n)$$
$$= (3^n + 4^n)^{1/n} 3^n + (3^n + 4^n)^{1/n} 4^n.$$

Since

$$(3^n + 4^n)^{1/n} > (3^n)^{1/n} = 3 \qquad \text{and} \qquad (3^n + 4^n)^{1/n} > (4^n)^{1/n} = 4,$$

we have

$$(3^n + 4^n)^{(n+1)/n} > 3 \cdot (3^n) + 4 \cdot (4^n) = 3^{n+1} + 4^{n+1}.$$

Taking the $(n + 1)$st root of the left and right sides of this inequality, we obtain

$$(3^n + 4^n)^{1/n} > (3^{n+1} + 4^{n+1})^{1/(n+1)}.$$

The sequence is decreasing. Being also bounded, it must be convergent. (For the limit, see Exercise 43.) ❏

The theorem which follows will help us work with limits of sequences without having to resort so frequently to the ϵ, K details set forth in Definition 11.3.1.

THEOREM 11.3.7

Let α be a real number. If $a_n \to L$ and $b_n \to M$, then

(i) $a_n + b_n \to L + M$,   (ii) $\alpha a_n \to \alpha L$,   (iii) $a_n b_n \to LM$.

If, in addition, $M \neq 0$ and each $b_n \neq 0$, then

(iv) $\dfrac{1}{b_n} \to \dfrac{1}{M}$ and   (v) $\dfrac{a_n}{b_n} \to \dfrac{L}{M}$.

Proofs of parts (i) and (ii) are left as exercises. For proofs of parts (iii)–(v) see the supplement at the end of this section.

We are now in a position to handle any rational sequence

$$a_n = \frac{\alpha_k n^k + \alpha_{k-1} n^{k-1} + \cdots + \alpha_0}{\beta_j n^j + \beta_{j-1} n^{j-1} + \cdots + \beta_0}.$$

To determine the behavior of such a sequence, we need only divide both numerator and denominator by the highest power of n that occurs.

Example 5

$$\frac{3n^4 - 2n^2 + 1}{n^5 - 3n^3} = \frac{3/n - 2/n^3 + 1/n^5}{1 - 3/n^2} \to \frac{0}{1} = 0. \quad ❏$$
$$\text{divide by } n^5$$

Example 6

$$\frac{1 - 4n^7}{n^7 + 12n} = \frac{1/n^7 - 4}{1 + 12/n^6} \to \frac{-4}{1} = -4. \quad ❏$$
$$\text{divide by } n^7$$

Example 7

$$\frac{n^4 - 3n^2 + n + 2}{n^3 + 7n} = \frac{1 - 3/n^2 + 1/n^3 + 2/n^4}{1/n + 7/n^3}.$$
$$\text{divide by } n^4$$

Since the numerator tends to 1 and the denominator tends to 0, the sequence is unbounded. Therefore, it cannot converge. ❏

THEOREM 11.3.8

$$a_n \to L \quad \text{iff} \quad a_n - L \to 0 \quad \text{iff} \quad |a_n - L| \to 0.$$

We leave the proof to you.

THEOREM 11.3.9   THE PINCHING THEOREM FOR SEQUENCES

Suppose that for all n sufficiently large

$$a_n \le b_n \le c_n.$$

If $a_n \to L$ and $c_n \to L$, then $b_n \to L$.

Once again the proof is left to you.

The following is an obvious corollary to the pinching theorem.

(11.3.10)

Suppose that for all n sufficiently large

$$|b_n| \le c_n.$$

If $c_n \to 0$, then $b_n \to 0$.

Example 8

$$\frac{\cos n}{n} \to 0 \quad \text{since} \quad \left|\frac{\cos n}{n}\right| \le \frac{1}{n} \quad \text{and} \quad \frac{1}{n} \to 0. \quad \square$$

Example 9

$$\sqrt{4 + (1/n)^2} \to 2$$

since

$$2 \le \sqrt{4 + (1/n)^2} \le \sqrt{4 + 4(1/n) + (1/n)^2} = 2 + 1/n \quad \text{and} \quad 2 + 1/n \to 2. \quad \square$$

Example 10   (*A limit to remember*)

(11.3.11)

$$\lim_{n \to \infty} \left(1 + \frac{1}{n}\right)^n = e.$$

PROOF In Project 7.4 you were asked to show that

$$\left(1 + \frac{1}{n}\right)^n \le e \le \left(1 + \frac{1}{n}\right)^{n+1} \qquad \text{for all positive integers } n.$$

Dividing the right inequality by $1 + 1/n$, we have

$$\frac{e}{1 + 1/n} \leq \left(1 + \frac{1}{n}\right)^n .$$

Combining this with the left inequality, we can write

$$\frac{e}{1 + 1/n} \leq \left(1 + \frac{1}{n}\right)^n \leq e.$$

Since

$$\frac{e}{1 + 1/n} \to \frac{e}{1} = e,$$

the pinching theorem guarantees that

$$\left(1 + \frac{1}{n}\right)^n \to e. \quad \square$$

Continuous Functions Applied to Convergent Sequences

The continuous image of a convergent sequence is a convergent sequence. More precisely, we have the following theorem:

THEOREM 11.3.12

Suppose that

$$c_n \to c.$$

If a function f, defined at all the c_n, is continuous at c, then

$$f(c_n) \to f(c).$$

PROOF We assume that f is continuous at c and take $\epsilon > 0$. From the continuity of f at c we know that there exists a number $\delta > 0$ such that

$$\text{if} \quad |x - c| < \delta, \quad \text{then} \quad |f(x) - f(c)| < \epsilon.$$

Since $c_n \to c$, we know that there exists a positive integer K such that

$$\text{if} \quad n \geq K, \quad \text{then} \quad |c_n - c| < \delta.$$

It follows that

$$\text{if} \quad n \geq K, \quad \text{then} \quad |f(c_n) - f(c)| < \epsilon. \quad \square$$

Example 11 It is clear that

$$\frac{\pi}{n} \to 0.$$

Since the functions

$$f(x) = \sin x, \qquad f(x) = e^x, \qquad f(x) = \arctan x$$

are defined at all the π/n and are continuous at 0, we can conclude that

$$\sin \pi/n \to \sin 0 = 0, \qquad e^{\pi/n} \to e^0 = 1, \qquad \arctan \pi/n \to \arctan 0 = 0. \quad \square$$

Example 12

$$\frac{2n-1}{n} = 2 - \frac{1}{n} \to 2.$$

Since the functions

$$f(x) = \sqrt{x}, \qquad f(x) = \ln x, \qquad f(x) = \frac{1}{x}$$

are defined at all the terms of the sequence and are continuous at 2, we know that

$$\sqrt{\frac{2n-1}{n}} \to \sqrt{2}, \qquad \ln\left(\frac{2n-1}{n}\right) \to \ln 2, \qquad \frac{n}{2n-1} \to \frac{1}{2}. \quad \square$$

Example 13

$$\frac{\pi^2 n^2 - 8}{16n^2} = \frac{\pi^2 - 8/n^2}{16} \to \frac{\pi^2}{16}.$$

Since $f(x) = \tan\sqrt{x}$ is defined at all the terms of the sequence and is continuous at $\pi^2/16$,

$$\tan\sqrt{\frac{\pi^2 n^2 - 8}{16n^2}} \to \tan\sqrt{\frac{\pi^2}{16}} = \tan\frac{\pi}{4} = 1. \quad \square$$

Example 14 Since the absolute-value function is everywhere defined and everywhere continuous,

$$a_n \to L \qquad \text{implies} \qquad |a_n| \to |L|. \quad \square$$

Stability of Limit

Start with a sequence

$$a_1, a_2, \cdots, a_n, \cdots.$$

Now change a finite number of terms of this sequence, say the first million terms. We now have a sequence

$$b_1, b_2, \cdots, b_n, \cdots.$$

We know nothing about the b_n from $n = 1$ to $n = 1,000,000$. But we know that

$$b_n = a_n \qquad \text{for} \qquad n > 1,000,000.$$

This is enough to guarantee that

$$\text{if} \qquad \lim_{n\to\infty} a_n = L, \qquad \text{then} \qquad \lim_{n\to\infty} b_n = L.$$

PROOF If the positive integer K is such that

$$|a_n - L| < \epsilon \qquad \text{for all} \qquad n \geq K,$$

then the positive integer $K_1 = \max\{K, 1,000,000\}$ is such that

$$|b_n - L| < \epsilon \qquad \text{for all } n \geq K_1. \quad \square$$

Similar reasoning shows that

$$\text{if} \qquad \lim_{n\to\infty} b_n = L, \qquad \text{then} \qquad \lim_{n\to\infty} a_n = L.$$

The following should now be clear:

We cannot deflect a convergent sequence from its limit by changing a finite number of terms; nor can we force a divergent sequence to converge by changing a finite number of terms. (For if we could, the convergence of the second sequence would force the convergence of the first sequence.)

Remark For some time now we have asked you to take on faith two fundamentals of integration: that continuous functions do have definite integrals and that these integrals can be expressed as limits of Riemann sums. We could not give you proofs of these assertions because we did not have the necessary tools. Now we do. Proofs are given in Appendix B. ❑

EXERCISES 11.3

Exercises 1–40. State whether the sequence converges and, if it does, find the limit.

1. 2^n.

2. $\dfrac{2}{n}$.

3. $\dfrac{(-1)^n}{n}$.

4. $\sqrt{n}$.

5. $\dfrac{n-1}{n}$.

6. $\dfrac{n+(-1)^n}{n}$.

7. $\dfrac{n+1}{n^2}$.

8. $\sin\dfrac{\pi}{2n}$.

9. $\dfrac{2^n}{4^n+1}$.

10. $\dfrac{n^2}{n+1}$.

11. $(-1)^n\sqrt{n}$.

12. $\dfrac{4^n}{\sqrt{n^2+1}}$.

13. $(-\frac{1}{2})^n$.

14. $\dfrac{4n}{2^n+10^6}$.

15. $\tan\dfrac{n\pi}{4n+1}$.

16. $\dfrac{10^{10}\sqrt{n}}{n+1}$.

17. $\dfrac{(2n+1)^2}{(3n-1)^2}$.

18. $\ln\left(\dfrac{2n}{n+1}\right)$.

19. $\dfrac{n^2}{\sqrt{2n^4+1}}$.

20. $\dfrac{n^4-1}{n^4+n-6}$.

21. $\cos n\pi$.

22. $\dfrac{n^5}{17n^4+12}$.

23. $e^{1/\sqrt{n}}$.

24. $\sqrt{4-1/n}$.

25. $\ln n-\ln(n+1)$.

26. $\dfrac{2^n-1}{2^n}$.

27. $\dfrac{\sqrt{n+1}}{2\sqrt{n}}$.

28. $\dfrac{1}{n}-\dfrac{1}{n+1}$.

29. $\left(1+\dfrac{1}{n}\right)^{2n}$.

30. $\left(1+\dfrac{1}{n}\right)^{n/2}$.

31. $\dfrac{2^n}{n^2}$.

32. $2\ln 3n-\ln(n^2+1)$.

33. $\dfrac{\sin n}{\sqrt{n}}$.

34. $\arctan\left(\dfrac{n}{n+1}\right)$.

35. $\sqrt{n^2+n}-n$.

36. $\dfrac{\sqrt{4n^2+n}}{n}$.

37. $\arctan n$.

38. $3.\overbrace{444\cdots4}^{n}$.

39. $\arcsin\left(\dfrac{1-n}{n}\right)$.

40. $\dfrac{(n+1)(n+4)}{(n+2)(n+3)}$.

▶**Exercises 41–42.** Use a graphing utility or CAS to plot the first 15 terms of the sequence. Determine whether the sequence converges, and if it does, give the limit.

41. (a) $\sqrt[n]{n}$.   (b) $\dfrac{3^n}{n!}$.

42. (a) $\dfrac{n}{\sqrt[n]{n!}}$.   (b) $n^2\sin(\pi/n)$.

43. Show that if $0<a<b$, then $(a^n+b^n)^{1/n}\to b$.

44. (a) Determine the values of r for which r^n converges.
(b) Determine the values of r for which nr^n converges.

45. Prove that
if $a_n\to L$ and $b_n\to M$, then $a_n+b_n\to L+M$.

46. Let α be a real number. Prove that
if $a_n\to L$, then $\alpha a_n\to\alpha L$.

47. Given that
$\left(1+\dfrac{1}{n}\right)^n\to e$ show that $\left(1+\dfrac{1}{n}\right)^{n+1}\to e$.

48. Determine the convergence or divergence of a rational sequence
$$a_n=\dfrac{\alpha_k n^k+\alpha_{k-1}n^{k-1}+\cdots+\alpha_0}{\beta_j n^j+\beta_{j-1}n^{j-1}+\cdots+\beta_0}$$
given that: (a) $k=j$; (b) $k<j$; (c) $k>j$. Justify your answers. Assume that $\alpha_k\neq0$ and $\beta_j\neq0$.

49. Prove that a bounded nonincreasing sequence converges to its greatest lower bound.

50. From a sequence with terms a_n, collect the even-numbered terms $e_n=a_{2n}$ and the odd-numbered terms $o_n=a_{2n-1}$. Show that
$a_n\to L$ iff $e_n\to L$ and $o_n\to L$.

51. Prove the pinching theorem for sequences.

52. Show that if $a_n \to 0$ and the sequence with terms b_n is bounded, then $a_n b_n \to 0$.

53. Suppose that $a_n \to L$. Show that if $a_n \leq M$ for all n, then $L \leq M$.

54. Earlier you saw that if $a_n \to L$, then $|a_n| \to |L|$. Is the converse true? Namely, if $|a_n| \to |L|$, does it follow that $a_n \to L$? Prove this, or give a counterexample.

55. Prove that $a_n \to 0$ iff $|a_n| \to 0$.

56. Suppose that $a_n \to L$ and $b_n \to L$. Show that the sequence

$$a_1, b_1, a_2, b_2, a_3, b_3, \cdots$$

converges to L.

57. Let f be a function continuous everywhere and let r be a real number. Define a sequence as follows:

$$a_1 = r, \quad a_2 = f(r), \quad a_3 = f[f(r)], \quad a_4 = f\{f[f(r)]\}, \cdots.$$

Prove that if $a_n \to L$, then L is a fixed point of $f : f(L) = L$.

58. Show that

$$\frac{2^n}{n!} \to 0$$

by showing that

$$\frac{2^n}{n!} \leq \frac{4}{n}.$$

59. Prove that $(1/n)^{1/p} \to 0$ for all positive integers p.

60. Prove Theorem 11.3.8.

Exercises 61–66. Below are some sequences defined recursively.[†] Determine in each case whether the sequence converges and, if so, find the limit. Start each sequence with $a_1 = 1$.

61. $a_{n+1} = \frac{1}{e} a_n.$

62. $a_{n+1} = 2^{n+1} a_n.$

63. $a_{n+1} = \frac{1}{n+1} a_n.$

64. $a_{n+1} = \frac{n}{n+1} a_n.$

65. $a_{n+1} = 1 - a_n.$

66. $a_{n+1} = \frac{1}{2} a_n + 1.$

[†]The notion was introduced in Exercises 11.2.

Exercises 67–74. Evaluate numerically the limit of each sequence as $n \to \infty$. Some of these sequences converge more rapidly than others. Determine for each sequence the least value of n for which the nth term differs from the limit by less than 0.001.

67. $\frac{1}{n^2}.$

68. $\frac{1}{\sqrt{n}}.$

69. $\frac{n}{10^n}.$

70. $\frac{n^{10}}{10^n}.$

71. $\frac{1}{n!}.$

72. $\frac{2^n}{n!}.$

73. $\frac{\ln n}{n^2}.$

74. $\frac{\ln n}{n}.$

75. (a) Find the limit of the sequence defined in Exercise 67, Section 11.2.
   (b) Find the limit of the sequence defined in Exercise 68, Section 11.2.

76. Define a sequence recursively by setting

$$a_1 = 1, \quad a_n = \sqrt{6 + a_{n-1}}, \quad n = 2, 3, 4, \cdots.$$

   (a) Estimate a_2, a_3, a_4, a_5, a_6, rounding off your answers to four decimal places.
   (b) Show by induction that $a_n \leq 3$ for all n.
   (c) Show that the a_n constitute an increasing sequence. HINT: $a_{n+1}^2 - a_n^2 = (3 - a_n)(2 + a_n)$.
   (d) What is the limit of this sequence?

77. Define a sequence recursively by setting

$$a_1 = 1, \quad a_n = \cos a_{n-1}, \quad n = 2, 3, 4, \cdots.$$

   (a) Estimate $a_2, a_3, a_4, \cdots, a_{10}$, rounding off your answers to four decimal places.
   (b) Assume that the sequence converges and estimate the limit to four decimal places.

78. Define a sequence recursively by setting

$$a_1 = 1, \quad a_n = a_{n-1} + \cos a_{n-1}, \quad n = 2, 3, 4, \cdots.$$

   (a) Estimate $a_2, a_3, a_4, \cdots, a_{10}$, rounding off your answers to four decimal places.
   (b) Assume that the sequence converges and estimate the limit to four decimal places.

■ PROJECT 11.3 Sequences and the Newton-Raphson Method

Let R be a positive number. The sequence defined recursively by setting

$$(1) \quad a_1 = 1, \quad a_n = \frac{1}{2}\left(a_{n-1} + \frac{R}{a_{n-1}}\right), \quad n = 2, 3, 4, \cdots.$$

can be used to approximate $\sqrt{R}$.

Problem 1. Let $R = 3$.

a. Calculate $a_2, a_3, \cdots, a_8$. Round off your answers to six decimal places.

b. Show that if $a_n \to L$, then $L = \sqrt{3}$.

The Newton-Raphson method (Section 4.12) applied to a differentiable function f generates a sequence which, under certain conditions, converges to a zero of f. The recurrence relation is of the form

$$x_{n+1} = x_n - \frac{f(x_n)}{f'(x_n)}, \quad n = 1, 2, 3, \cdots.$$

Problem 2. Show that recurrence relation (1) is the recurrence relation generated by the Newton-Raphson method applied to the function $f(x) = x^2 - R$.

Problem 3. Each of the following recurrence relations is based on the Newton-Raphson method. Determine whether the sequence converges and, if so, give the limit.

a. $x_1 = 1, \quad x_{n+1} = x_n - \dfrac{x_n^3 - 8}{3x_n^2}$.

b. $x_1 = 0, \quad x_{n+1} = x_n - \dfrac{\sin x_n - 0.5}{\cos x_n}$.

c. $x_1 = 1, \quad x_{n+1} = x_n - \dfrac{\ln x_n - 1}{1/x_n}$.

Problem 4. For each sequence in Problem 3 find a function f that generates the sequence. Then check each of your answers by evaluating f.

---

*SUPPLEMENT TO SECTION 11.3

PROOF OF THEOREM 11.3.2

If $L \neq M$, then

$$\tfrac{1}{2}|L - M| > 0.$$

The assumption that $\lim\limits_{n \to \infty} a_n = L$ and $\lim\limits_{n \to \infty} a_n = M$ gives the existence of K_1 such that

$$\text{if} \quad n \geq K_1, \quad\quad \text{then} \quad\quad |a_n - L| < \tfrac{1}{2}|L - M|$$

and the existence of K_2 such that

$$\text{if} \quad n \geq K_2, \quad \text{then} \quad |a_n - M| < \tfrac{1}{2}|L - M|.^\dagger$$

For $n \geq \max\{K_1, K_2\}$, we have

$$|a_n - L| + |a_n - M| < |L - M|.$$

By the triangle inequality we have

$$|L - M| = |(L - a_n) + (a_n - M)| \leq |L - a_n| + |a_n - M| = |a_n - L| + |a_n - M|.$$

Combining the last two statements, we have

$$|L - M| < |L - M|.$$

The hypothesis $L \neq M$ has led to an absurdity. We conclude that $L = M$. ❏

PROOF OF THEOREM 11.3.7 (III)–(V)

To prove (iii), we set $\epsilon > 0$. For each n,

$$|a_n b_n - LM| = |(a_n b_n - a_n M) + (a_n M - LM)|$$

$$\leq |a_n||b_n - M| + |M||a_n - L|.$$

Since convergent sequences are bounded, there exists $Q > 0$ such that

$$|a_n| \leq Q \quad\quad \text{for all } n.$$

Since $|M| < |M| + 1$, we have

$$(1) \quad\quad\quad |a_n b_n - LM| \leq Q|b_n - M| + (|M| + 1)|a_n - L|.^{\dagger\dagger}$$

---

†We can reach these conclusions from Definition 11.3.1 by taking $\tfrac{1}{2}|L - M|$ as ϵ.

††Soon we will want to divide by the coefficient of $|a_n - L|$. We have replaced $|M|$ by $|M| + 1$ because $|M|$ can be zero.

Since $b_n \to M$, we know that there exists K_1 such that

$$\text{if} \quad n \geq K_1, \quad \text{then} \quad |b_n - M| < \frac{\epsilon}{2Q}.$$

Since $a_n \to L$, we know that there exists K_2 such that

$$\text{if} \quad n \geq K_2, \quad \text{then} \quad |a_n - L| < \frac{\epsilon}{2(|M| + 1)}.$$

For $n \geq \max\{K_1, K_2\}$, both conditions hold, and consequently

$$Q|b_n - M| + (|M| + 1)|a_n - L| < \frac{\epsilon}{2} + \frac{\epsilon}{2} = \epsilon.$$

In view of (1), we can conclude that

$$\text{if} \quad n \geq \max\{K_1, K_2\}, \quad \text{then} \quad |a_n b_n - LM| < \epsilon.$$

This proves that

$$a_n b_n \to LM. \quad \square$$

To prove (iv), once again we set $\epsilon > 0$. In the first place

$$\left| \frac{1}{b_n} - \frac{1}{M} \right| = \left| \frac{M - b_n}{b_n M} \right| = \frac{|b_n - M|}{|b_n||M|}.$$

Since $b_n \to M$ and $|M|/2 > 0$, there exists K_1 such that

$$\text{if} \quad n \geq K_1, \quad \text{then} \quad |b_n - M| < \frac{|M|}{2}.$$

This tells us that for $n \geq K_1$ we have

$$|b_n| > \frac{|M|}{2} \quad \text{and thus} \quad \frac{1}{|b_n|} < \frac{2}{|M|}.$$

Thus for $n \geq K_1$ we have

(2)
$$\left| \frac{1}{b_n} - \frac{1}{M} \right| \leq \frac{2}{|M|^2} |b_n - M|.$$

Since $b_n \to M$, there exists K_2 such that

$$\text{if} \quad n \geq K_2, \quad \text{then} \quad |b_n - M| < \frac{\epsilon |M|^2}{2}.$$

Thus for $n \geq K_2$ we have

$$\frac{2}{|M|^2} |b_n - M| < \epsilon.$$

In view of (2), we can be sure that

$$\text{if} \quad n \geq \max\{K_1, K_2\}, \quad \text{then} \quad \left| \frac{1}{b_n} - \frac{1}{M} \right| < \epsilon.$$

This proves that

$$\frac{1}{b_n} \to \frac{1}{M}. \quad \square$$

The proof of (v) is now easy:

$$\frac{a_n}{b_n} = a_n \cdot \frac{1}{b_n} \to L \cdot \frac{1}{M} = \frac{L}{M}. \quad \square$$

■ 11.4 SOME IMPORTANT LIMITS

Our purpose here is to direct your attention to some limits that are particularly important in calculus and to give you more experience with limit arguments. Before looking at the proofs given, try to construct you own proofs.

(11.4.1)

> Each limit is taken as $n \to \infty$.
>
> **(1)** $x^{1/n} \to 1$ for all $x > 0$.
>
> **(2)** $x^n \to 0$ if $|x| < 1$.
>
> **(3)** $\dfrac{1}{n^\alpha} \to 0$ for all $\alpha > 0$.
>
> **(4)** $\dfrac{x^n}{n!} \to 0$ for all real x.
>
> **(5)** $\dfrac{\ln n}{n} \to 0$.
>
> **(6)** $n^{1/n} \to 1$.
>
> **(7)** $\left(1 + \dfrac{x}{n}\right)^n \to e^x$ for all real x.

(1)

> If $x > 0$, then
>
> $x^{1/n} \to 1$ as $n \to \infty$.

PROOF Fix any $x > 0$. Since

$$\ln(x^{1/n}) = \frac{1}{n} \ln x,$$

we see that

$$\ln(x^{1/n}) \to 0.$$

We reduce $\ln(x^{1/n})$ to $x^{1/n}$ by applying the exponential function. Since the exponential function is defined at the terms $\ln(x^{1/n})$ and is continuous at 0, it follows from Theorem 11.3.12 that

$$x^{1/n} = e^{\ln(x^{1/n})} \to e^0 = 1. \quad \Box$$

(2)

> If $|x| < 1$, then
>
> $x^n \to 0$ as $n \to \infty$.

PROOF The result clearly holds for $x = 0$. Now fix any $x \neq 0$ with $|x| < 1$ and observe that the sequence $a_n = |x|^n$ is a decreasing sequence:

$$|x|^{n+1} = |x||x|^n < |x|^n.$$

Let $\epsilon > 0$. By (11.4.1), $\epsilon^{1/n} \to 1$. Thus there exists an integer $K > 0$ for which

$$|x| < \epsilon^{1/K}. \qquad \text{(explain)}$$

This implies that $|x|^K < \epsilon$. Since the $|x|^n$ form a decreasing sequence, we have

$$|x^n| = |x|^n < \epsilon \qquad \text{for all} \qquad n \geq K. \quad \square$$

(3)

> For each $\alpha > 0$
>
> $$\frac{1}{n^\alpha} \to 0 \qquad \text{as} \qquad n \to \infty.$$

PROOF Take $\alpha > 0$. There exists an odd positive integer p for which $1/p < \alpha$. Then

$$0 < \frac{1}{n^\alpha} = \left(\frac{1}{n}\right)^\alpha \leq \left(\frac{1}{n}\right)^{1/p}.$$

Since $1/n \to 0$ and the function $f(x) = x^{1/p}$ is defined at the terms $1/n$ and is continuous at 0, we have

$$\left(\frac{1}{n}\right)^{1/p} \to 0, \qquad \text{and thus by the pinching theorem,} \qquad \frac{1}{n^\alpha} \to 0. \quad \square$$

(4)

> For each real x
>
> $$\frac{x^n}{n!} \to 0 \qquad \text{as} \qquad n \to \infty.$$

PROOF Fix any real number x and choose an integer $k > |x|$. For $n > k + 1$,

$$\frac{k^n}{n!} = \left(\frac{k^k}{k!}\right) \left[\frac{k}{k+1}\frac{k}{k+2} \cdots \frac{k}{n-1}\right] \left(\frac{k}{n}\right) < \left(\frac{k^{k+1}}{k!}\right)\left(\frac{1}{n}\right).$$

$$\text{The middle term is less than 1.} \longrightarrow$$

Since $k > |x|$, we have

$$0 < \frac{|x|^n}{n!} < \frac{k^n}{n!} < \left(\frac{k^{k+1}}{k!}\right)\left(\frac{1}{n}\right).$$

Since k is fixed and $1/n \to 0$, it follows from the pinching theorem that

$$\frac{|x|^n}{n!} \to 0 \qquad \text{and thus} \qquad \frac{x^n}{n!} \to 0. \quad \square$$

(5)

> $$\frac{\ln n}{n} \to 0 \qquad \text{as} \qquad n \to \infty.$$

PROOF A routine proof can be based on L'Hôpital's rule (Theorem 11.6.1), but that is not available to us yet. We will appeal to the pinching theorem and base our argument on the integral representation of the logarithm:

$$0 \leq \frac{\ln n}{n} = \frac{1}{n}\int_1^n \frac{1}{t}\,dt \leq \frac{1}{n}\int_1^n \frac{1}{\sqrt{t}}\,dt = \frac{2}{n}(\sqrt{n} - 1) = 2\left(\frac{1}{\sqrt{n}} - \frac{1}{n}\right) \to 0. \quad \square$$

(6)

> $$n^{1/n} \to 1 \qquad \text{as} \qquad n \to \infty.$$

PROOF We know that

$$\ln n^{1/n} = \frac{\ln n}{n} \to 0.$$

Applying the exponential function, we have

$$n^{1/n} \to e^0 = 1. \quad \square$$

---

(7)

For each real x

$$\left(1 + \frac{x}{n}\right)^n \to e^x \qquad \text{as} \qquad n \to \infty.$$

PROOF For $x = 0$, the result is obvious. For $x \neq 0$,

$$\ln\left(1 + \frac{x}{n}\right)^n = n \ln\left(1 + \frac{x}{n}\right) = x \left[\frac{\ln(1 + x/n) - \ln 1}{x/n}\right].$$

The crux here is to recognize that the bracketed expression is a difference quotient for the logarithm function. Once we see this, we let $h = x/n$ and write

$$\lim_{n \to \infty}\left[\frac{\ln(1 + x/n) - \ln 1}{x/n}\right] = \lim_{h \to 0}\left[\frac{\ln(1 + h) - \ln 1}{h}\right] = 1.^{\dagger}$$

It follows that

$$\ln\left(1 + \frac{x}{n}\right)^n \to x.$$

Applying the exponential function, we have

$$\left(1 + \frac{x}{n}\right)^n \to e^x. \quad \square$$

---

†For each $t > 0$

$$\lim_{h \to 0}\left[\frac{\ln(t + h) - \ln t}{h}\right] = \frac{d}{dt}(\ln t) = \frac{1}{t}.$$

At $t = 1, 1/t = 1$.

EXERCISES 11.4

Exercises 1–38. State whether the sequence converges as $n \to \infty$; if it does, find the limit.

1. $2^{2/n}$.

2. $e^{-\alpha/n}$.

3. $\left(\dfrac{2}{n}\right)^n$.

4. $\dfrac{\log_{10} n}{n}$.

5. $\dfrac{\ln(n + 1)}{n}$.

6. $\dfrac{3^n}{4^n}$.

7. $\dfrac{x^{100n}}{n!}$.

8. $n^{1/(n+2)}$.

9. $n^{\alpha/n}, \ \alpha > 0$.

10. $\ln\left(\dfrac{n + 1}{n}\right)$.

11. $\dfrac{3^{n+1}}{4^{n-1}}$.

12. $\displaystyle\int_{-n}^{0} e^{2x}\,dx$.

13. $(n + 2)^{1/n}$.

14. $\left(1 - \dfrac{1}{n}\right)^n$.

15. $\displaystyle\int_{0}^{n} e^{-x}\,dx$.

16. $\dfrac{2^{3n-1}}{7^{n+2}}$.

17. $\displaystyle\int_{-n}^{n} \dfrac{dx}{1 + x^2}$.

18. $\displaystyle\int_{0}^{n} e^{-nx}\,dx$.

19. $(n + 2)^{1/(n+2)}$.

20. $n^2 \sin n\pi$.

21. $\dfrac{\ln n^2}{n}$.

22. $\displaystyle\int_{-1+1/n}^{1+1/n} \dfrac{dx}{\sqrt{1 - x^2}}$.

23. $n^2 \sin \dfrac{\pi}{n}$.

24. $\dfrac{n!}{2n}$.

25. $\dfrac{5^{n+1}}{4^{2n-1}}$.

26. $\left(1 + \dfrac{x}{n}\right)^{3n}$.

27. $\left(\dfrac{n + 1}{n + 2}\right)^n$.

28. $\displaystyle\int_{1/n}^{1} \dfrac{dx}{\sqrt{x}}$.

29. $\displaystyle\int_n^{n+1} e^{-x^2}\,dx.$

30. $\displaystyle\left(1 + \frac{1}{n^2}\right)^n.$

31. $\displaystyle\frac{n^n}{2n^2}.$

32. $\displaystyle\int_0^{1/n} \cos e^x\,dx.$

33. $\displaystyle\left(1 + \frac{x}{2n}\right)^{2n}.$

34. $\displaystyle\left(1 + \frac{1}{n}\right)^{n^2}.$

35. $\displaystyle\int_{-1/n}^{1/n} \sin x^2\,dx.$

36. $\displaystyle\left(|t| + \frac{x}{n}\right)^n, t > 0, x \ge 0.$

37. $\displaystyle\frac{\sin(6/n)}{\sin(3/n)}.$

38. $\displaystyle\frac{\arctan n}{n}.$

39. Show that $\displaystyle\lim_{n\to\infty}\left[(n+1)^{1/2} - n^{1/2}\right] = 0.$

40. Show that $\displaystyle\lim_{n\to\infty}\left[(n^2+n)^{1/2} - n\right] = \frac{1}{2}.$

41. (a) Show that a regular polygon of n sides inscribed in a circle of radius r has perimeter $p_n = 2rn\sin(\pi/n)$.
(b) Find

$$\lim_{n\to\infty} p_n$$

and give a geometric interpretation to your result.

42. Show that

$$\text{if} \quad 0 < c < d, \quad \text{then} \quad (c^n + d^n)^{1/n} \to d.$$

Exercises 43–45. Find the indicated limit.

43. $\displaystyle\lim_{n\to\infty}\frac{1 + 2 + \cdots + n}{n^2}.$

HINT: $1 + 2 + \cdots + n = \dfrac{n(n+1)}{2}.$

44. $\displaystyle\lim_{n\to\infty}\frac{1^2 + 2^2 + \cdots + n^2}{(1+n)(2+n)}.$

HINT: $1^2 + 2^2 + \cdots + n^2 = \dfrac{n(n+1)(2n+1)}{6}.$

45. $\displaystyle\lim_{n\to\infty}\frac{1^3 + 2^3 + \cdots + n^3}{2n^4 + n - 1}.$

HINT: $1^3 + 2^3 + \cdots + n^3 = \dfrac{n^2(n+1)^2}{4}.$

46. A sequence $a_1, a_2, \cdots$ is called a *Cauchy sequence*[†] if

$$(11.4.2)$$

for each $\epsilon > 0$ there exists an index k such that $\|a_n - a_m\| < \epsilon \quad$ for all $m, n \ge k$.

Show that

$$(11.4.3)$$

every convergent sequence is a Cauchy sequence.

[†]After the French baron Augustin Louis Cauchy (1789–1857), one of the most prolific mathematicians of all time.

It is also true that every Cauchy sequence is convergent, but that is more difficult to prove.

47. (*Arithmetic means*) For a sequence $a_1, a_2, \cdots$, set

$$m_n = \frac{1}{n}(a_1 + a_2 + \cdots + a_n).$$

(a) Prove that if the a_n form an increasing sequence, then the m_n form an increasing sequence.
(b) Prove that if $a_n \to 0$, then $m_n \to 0$.

48. (a) Let $a_1, a_2, \cdots$ be a convergent sequence. Prove that

$$\lim_{n\to\infty}(a_n - a_{n-1}) = 0.$$

(b) What can you say about the converse? That is, suppose that $a_1, a_2, \cdots$ is a sequence for which

$$\lim_{n\to\infty}(a_n - a_{n-1}) = 0.$$

Does $a_1, a_2, \cdots$ necessarily converge? If so, prove it; if not, give a counterexample.

49. Starting with $0 < a < b$, form the arithmetic mean $a_1 = \frac{1}{2}(a+b)$ and the geometric mean $b_1 = \sqrt{ab}$. For $n = 2, 3, 4, \cdots$ set

$$a_n = \tfrac{1}{2}(a_{n-1} + b_{n-1}) \quad \text{and} \quad b_n = \sqrt{a_{n-1}b_{n-1}}.$$

(a) Show by induction on n that

$$a_{n-1} > a_n > b_n > b_{n-1} \quad \text{for } n = 2, 3, 4, \cdots.$$

(b) Show that the two sequences converge and $\displaystyle\lim_{n\to\infty} a_n = \lim_{n\to\infty} b_n$. The common value of this limit is called the *arithmetic-geometric mean of a and b*.

◐ 50. You have seen that for all real x

$$\lim_{n\to\infty}\left(1 + \frac{x}{n}\right)^n = e^x.$$

However, the rate of convergence is different at different x. Verify that with $n = 100$, $(1 + 1/n)^n$ is within 1% of its limit, while $(1 + 5/n)^n$ is still about 12% from its limit. Give comparable accuracy estimates at $x = 1$ and at $x = 5$ for $n = 1000$.

◐ 51. Evaluate

$$\lim_{n\to\infty}\left(\sin\tfrac{1}{n}\right)^{1/n}$$

numerically and by graphing. Justify your answer by other means.

◐ 52. We have stated that

$$\lim_{n\to\infty}\left[(n^2+n)^{1/2} - n\right] = \tfrac{1}{2}. \quad \text{(Exercise 40)}$$

Evaluate

$$\lim_{n\to\infty}\left[(n^3 + n^2)^{1/3} - n\right]$$

numerically. Then formulate a conjecture about

$$\lim_{n\to\infty}\left[(n^k + n^{k-1})^{1/k} - n\right] \quad \text{for } k = 1, 2, 3, \cdots$$

and prove that your conjecture is valid.

53. The sequence defined recursively by setting

$$a_{n+2} = a_{n+1} + a_n \qquad \text{starting with} \qquad a_1 = a_2 = 1$$

is called the *Fibonacci sequence*.

(a) Calculate $a_3, a_4, \cdots, a_{10}$.

(b) Define

$$r_n = \frac{a_{n+1}}{a_n}.$$

Calculate $r_1, r_2, \cdots, r_6$.

(c) Assume that $r_n \to L$, and find L. HINT: Relate r_n to r_{n-1}.

54. Set

$$a_n = \frac{1}{n^2} + \frac{2}{n^2} + \frac{3}{n^2} + \cdots + \frac{n}{n^2}.$$

Show that a_n is a Riemann sum for $\int_0^1 x \, dx$. Does the sequence $a_1, a_2, \cdots$ converge? If so, to what?

■ 11.5 THE INDETERMINATE FORM (0/0)

For each x-limit process that we have considered,

$$\text{as } x \to c, \qquad \text{as } x \to c^+, \qquad \text{as } x \to c^-, \qquad \text{as } x \to \infty, \qquad \text{as } x \to -\infty,$$

if $f(x) \to L$ and $g(x) \to M \neq 0$, then

$$\frac{f(x)}{g(x)} \to \frac{L}{M}.$$

The attempt to extend this algebra of limits to the case where $f(x) \to 0$ and $g(x) \to 0$ leads to the nonsensical result

$$\frac{f(x)}{g(x)} \to \frac{0}{0}.$$

Knowing only that $f(x) \to 0$ and $g(x) \to 0$, we can conclude nothing about the limit behavior of the quotient $f(x)/g(x)$. The quotient can tend to a finite limit:

$$\text{as} \qquad x \to 0, \qquad \frac{\sin x}{x} \to 1;$$

it can tend to $\pm\infty$:

$$\text{as} \qquad x \to 0, \qquad \frac{|\sin x|}{x^2} = \frac{1}{|x|} \left| \frac{\sin x}{x} \right| = \infty;$$

it can tend to no limit at all:

$$\text{as} \qquad x \to 0, \qquad \frac{|x|}{x} = \begin{cases} 1, & x > 0 \\ -1, & x < 0 \end{cases} \qquad \text{tends to no limit.}$$

For whatever limit process is being used, if $f(x) \to 0$ and $g(x) \to 0$, the quotient $f(x)/g(x)$ is called an *indeterminate of the form* $0/0$.

Such indeterminates can usually be handled by elementary methods. (We have done this right along.) Where elementary methods are difficult to apply, L'Hôpital's rule[†] (explained below) can be decisive. The rule applies equally well to all forms of x-approach:

$$\text{as } x \to c, \qquad \text{as } x \to c^+, \qquad \text{as } x \to c^-, \qquad \text{as } x \to \infty, \qquad \text{as } x \to -\infty.$$

As used in the statement of the rule, the symbol γ (the Greek letter gamma) can represent a real number, it can represent ∞, it can represent $-\infty$. By allowing this flexibility to γ, we avoid tiresome repetitions.

[†]Attributed to the Frenchman G. F. A. L'Hôpital (1661–1704). The result was actually discovered by his teacher John Bernoulli (1667–1748).

(11.5.1)

We defer consideration of the proof of this rule to the end of the section. First we demonstrate the usefulness of the rule.

Example 1   Find $\displaystyle\lim_{x \to \pi/2} \frac{\cos x}{\pi - 2x}$.

SOLUTION   As $x \to \pi/2$, both numerator and denominator tend to zero and it is not at all obvious what happens to the quotient

$$\frac{f(x)}{g(x)} = \frac{\cos x}{\pi - 2x}.$$

Therefore we test the quotient of derivatives:

$$\text{as} \quad x \to \frac{\pi}{2}, \quad \frac{f'(x)}{g'(x)} = \frac{-\sin x}{-2} = \frac{\sin x}{2} \to \frac{1}{2}.$$

It follows from L'Hôpital's rule that

$$\text{as} \quad x \to \frac{\pi}{2}, \quad \frac{\cos x}{\pi - 2x} \to \frac{1}{2}.$$

We can express all this on just one line using $*$ to indicate the differentiation of numerator and denominator:

$$\lim_{x \to \pi/2} \frac{\cos x}{\pi - 2x} \overset{*}{=} \lim_{x \to \pi/2} \frac{-\sin x}{-2} = \lim_{x \to \pi/2} \frac{\sin x}{2} = \frac{1}{2}. \quad \square$$

Example 2   Find $\displaystyle\lim_{x \to 0^+} \frac{x}{\sin \sqrt{x}}$.

SOLUTION   As $x \to 0^+$, both numerator and denominator tend to 0 and

$$\frac{f'(x)}{g'(x)} = \frac{1}{(\cos \sqrt{x})(1/2[\sqrt{x}])} = \frac{2\sqrt{x}}{\cos \sqrt{x}} \to \frac{0}{1} = 0.$$

It follows from L'Hôpital's rule that

$$\lim_{x \to 0^+} \frac{x}{\sin \sqrt{x}} \to 0.$$

For short, we can write

$$\lim_{x \to 0^+} \frac{x}{\sin \sqrt{x}} \overset{*}{=} \lim_{x \to 0^+} \frac{2\sqrt{x}}{\cos \sqrt{x}} = 0. \quad \square$$

Remark   There is a tendency to abuse this limit-finding technique. L'Hôpital's rule *does not apply* to cases where numerator or denominator has a finite nonzero limit. For example,

$$\lim_{x \to 0} \frac{x}{x + \cos x} = \frac{0}{1} = 0.$$

A blind application of L'Hôpital's rule leads to

$$\lim_{x \to 0} \frac{x}{x + \cos x} \overset{*}{=} \lim_{x \to 0} \frac{1}{1 - \sin x} = 1.$$

This is nonsense. ❏

Sometimes it is necessary to differentiate numerator and denominator more than once. For two differentiations we require that $g(x)$, $g'(x)$, $g''(x)$ never be zero in the approach. For three differentiations we require that $g(x)$, $g'(x)$, $g''(x)$, $g'''(x)$ never be zero in the approach. And so on.

Example 3 Find $\lim\limits_{x \to 0} \dfrac{e^x - x - 1}{x^2}$.

SOLUTION As $x \to 0$, both numerator and denominator tend to 0. Here

$$\frac{f'(x)}{g'(x)} = \frac{e^x - 1}{2x}.$$

Since both numerator and denominator still tend to 0, we differentiate again:

$$\frac{f''(x)}{g''(x)} = \frac{e^x}{2}.$$

Since this last quotient tends to $\frac{1}{2}$, we can conclude that

$$\frac{e^x - 1}{2x} \to \frac{1}{2} \qquad \text{and therefore} \qquad \frac{e^x - x - 1}{x^2} \to \frac{1}{2}.$$

We abbreviate this argument by writing

$$\lim_{x \to 0} \frac{e^x - x - 1}{x^2} \overset{*}{=} \lim_{x \to 0} \frac{e^x - 1}{2x} \overset{*}{=} \lim_{x \to 0} \frac{e^x}{2} = \frac{1}{2}. ❏$$

L'Hôpital's rule can be used to find the limit of a sequence.

Example 4 Find $\lim\limits_{n \to \infty} \dfrac{e^{2/n} - 1}{1/n}$.

SOLUTION The quotient is an indeterminate of the form 0/0. To apply the methods of this section, we replace the integer variable n by the real variable x and examine the behavior of

$$\frac{e^{2/x} - 1}{1/x} \qquad \text{as} \qquad x \to \infty.$$

Applying L'Hôpital's rule, we have

$$\lim_{x \to \infty} \frac{e^{2/x} - 1}{1/x} \overset{*}{=} \lim_{x \to \infty} \frac{e^{2/x}(-2/x^2)}{(-1/x^2)} = 2 \lim_{x \to \infty} e^{2/x} = 2.$$

It follows that

$$\text{as } n \to \infty, \qquad \frac{e^{2/n} - 1}{1/n} \to 2. ❏$$

To derive L'Hôpital's rule, we need a generalization of the mean-value theorem.

THEOREM 11.5.2   THE CAUCHY MEAN-VALUE THEOREM[†]

Suppose that f and g are differentiable on (a, b) and continuous on $[a, b]$. If g' is never 0 in (a, b), then there is a number r in (a, b) for which

$$\frac{f'(r)}{g'(r)} = \frac{f(b) - f(a)}{g(b) - g(a)}.$$

PROOF We can prove this by applying Rolle's theorem (4.1.3) to the function

$$G(x) = [g(b) - g(a)][f(x) - f(a)] - [g(x) - g(a)][f(b) - f(a)].$$

Since

$$G(a) = 0 \quad \text{and} \quad G(b) = 0,$$

there exists (by Rolle's theorem) a number r in (a, b) for which $G'(r) = 0$.
Differentiation gives

$$G'(x) = [g(b) - g(a)]f'(x) - g'(x)[f(b) - f(a)].$$

Setting $x = r$, we have

$$[g(b) - g(a)]f'(r) - g'(r)[f(b) - f(a)] = 0,$$

and thus

$$[g(b) - g(a)]f'(r) = g'(r)[f(b) - f(a)].$$

Since g' is never 0 in (a, b),

$$g'(r) \neq 0 \quad \text{and} \quad g(b) - g(a) \neq 0.$$
$$\uparrow\!\rule{1em}{0.4pt}\text{ explain}$$

We can therefore divide by these numbers and obtain

$$\frac{f'(r)}{g'(r)} = \frac{f(b) - f(a)}{g(b) - g(a)}. \quad \square$$

Now we prove L'Hôpital's rule for the case $x \to c^+$. This requires that in the approach $g(x) \neq 0$ and $g'(x) \neq 0$. We assume that, as $x \to c^+$,

$$f(x) \to 0, \qquad g(x) \to 0, \qquad \text{and} \qquad \frac{f'(x)}{g'(x)} \to \gamma.$$

We want to show that

$$\frac{f(x)}{g(x)} \to \gamma.$$

PROOF By defining $f(c) = 0$ and $g(c) = 0$, we make f and g continuous on an interval $[c, c + h]$. For each $x \in (c, c + h)$

$$\frac{f(x)}{g(x)} = \frac{f(x) - f(c)}{g(x) - g(c)} = \frac{f'(r)}{g'(r)}$$

[†]The same Cauchy who gave us Cauchy sequences.

with r between c and x. (The Cauchy mean-value theorem applied on the interval $[c, x]$.) If, as $x \to c^+$,

$$\frac{f'(x)}{g'(x)} \to \gamma, \qquad \text{then} \qquad \frac{f(x)}{g(x)} = \frac{f'(r)}{g'(r)} \to \gamma. \quad \Box$$

The case $x \to c^-$ can be handled in a similar manner. The two cases together prove the rule for the case $x \to c$.

Here is an outline of the proof of L'Hôpital's rule for the case $x \to \infty$.

PROOF The key here is to set $x = 1/t$:

$$\lim_{x \to \infty} \frac{f'(x)}{g'(x)} = \lim_{t \to 0^+} \frac{[f(1/t)]'}{[g(1/t)]'} = \lim_{t \to 0^+} \frac{-t^{-2} f'(1/t)}{-t^{-2} g'(1/t)}$$

$$= \lim_{t \to 0^+} \frac{f'(1/t)}{g'(1/t)} = \lim_{t \to 0^+} \frac{f(1/t)}{g(1/t)} = \lim_{x \to \infty} \frac{f(x)}{g(x)}. \quad \Box$$

↑———— by L'Hôpital's rule for the case $t \to 0^+$

EXERCISES 11.5

Exercises 1–32. Calculate.

1. $\displaystyle \lim_{x \to 0^+} \frac{\sin x}{\sqrt{x}}$.

2. $\displaystyle \lim_{x \to 1} \frac{\ln x}{1 - x}$.

3. $\displaystyle \lim_{x \to 0} \frac{e^x - 1}{\ln(1 + x)}$.

4. $\displaystyle \lim_{x \to 4} \frac{\sqrt{x} - 2}{x - 4}$.

5. $\displaystyle \lim_{x \to \pi/2} \frac{\cos x}{\sin 2x}$.

6. $\displaystyle \lim_{x \to a} \frac{x - a}{x^n - a^n}$.

7. $\displaystyle \lim_{x \to 0} \frac{2^x - 1}{x}$.

8. $\displaystyle \lim_{x \to 0} \frac{\arctan x}{x}$.

9. $\displaystyle \lim_{x \to 1} \frac{x^{1/2} - x^{1/4}}{x - 1}$.

10. $\displaystyle \lim_{x \to 0} \frac{e^x - 1}{x(1 + x)}$.

11. $\displaystyle \lim_{x \to 0} \frac{e^x - e^{-x}}{\sin x}$.

12. $\displaystyle \lim_{x \to 0} \frac{1 - \cos x}{3x}$.

13. $\displaystyle \lim_{x \to 0} \frac{x + \sin \pi x}{x - \sin \pi x}$.

14. $\displaystyle \lim_{x \to 0} \frac{a^x - (a + 1)^x}{x}$.

15. $\displaystyle \lim_{x \to 0} \frac{e^x + e^{-x} - 2}{1 - \cos 2x}$.

16. $\displaystyle \lim_{x \to 0} \frac{x - \ln(x + 1)}{1 - \cos 2x}$.

17. $\displaystyle \lim_{x \to 0} \frac{\tan \pi x}{e^x - 1}$.

18. $\displaystyle \lim_{x \to 0} \frac{\cos x - 1 + x^2/2}{x^4}$.

19. $\displaystyle \lim_{x \to 0} \frac{1 + x - e^x}{x(e^x - 1)}$.

20. $\displaystyle \lim_{x \to 0} \frac{\ln(\sec x)}{x^2}$.

21. $\displaystyle \lim_{x \to 0} \frac{x - \tan x}{x - \sin x}$.

22. $\displaystyle \lim_{x \to 0} \frac{xe^{nx} - x}{1 - \cos nx}$.

23. $\displaystyle \lim_{x \to 1^-} \frac{\sqrt{1 - x^2}}{\sqrt{1 - x^3}}$.

24. $\displaystyle \lim_{x \to 0} \frac{2x - \sin \pi x}{4x^2 - 1}$.

25. $\displaystyle \lim_{x \to \pi/2} \frac{\ln(\sin x)}{(\pi - 2x)^2}$.

26. $\displaystyle \lim_{x \to 0^+} \frac{\sqrt{x}}{\sqrt{x} + \sin \sqrt{x}}$.

27. $\displaystyle \lim_{x \to 0} \frac{\cos x - \cos 3x}{\sin(x^2)}$.

28. $\displaystyle \lim_{x \to 0} \frac{\sqrt{a + x} - \sqrt{a - x}}{x}$.

29. $\displaystyle \lim_{x \to \pi/4} \frac{\sec^2 x - 2 \tan x}{1 + \cos 4x}$.

30. $\displaystyle \lim_{x \to 0} \frac{x - \arcsin x}{\sin^3 x}$.

31. $\displaystyle \lim_{x \to 0} \frac{\arctan x}{\arctan 2x}$.

32. $\displaystyle \lim_{x \to 0} \frac{\arcsin x}{x}$.

Exercises 33–36. Find the limit of the sequence.

33. $\displaystyle \lim_{n \to \infty} \frac{(\pi/2 - \arctan n)}{1/n}$.

34. $\displaystyle \lim_{n \to \infty} \frac{\ln(1 - 1/n)}{\sin(1/n)}$.

35. $\displaystyle \lim_{n \to \infty} \frac{1}{n[\ln(n + 1) - \ln n]}$.

36. $\displaystyle \lim_{n \to \infty} \frac{\sinh \pi/n - \sin \pi/n}{\sin^3 \pi/n}$.

▶ **Exercises 37–42.** Use technology (graphing utility or CAS) to calculate the limit.

37. $\displaystyle \lim_{x \to 0} \frac{x^3}{4^x - 1}$.

38. $\displaystyle \lim_{x \to 0} \frac{4x}{\sin^2 x}$.

39. $\displaystyle \lim_{x \to 0} \frac{x}{\frac{\pi}{2} - \arccos x}$.

40. $\displaystyle \lim_{x \to 2^+} \frac{\sqrt{2x} - 2}{\sqrt{x} - 2}$.

41. $\displaystyle \lim_{x \to 0} \frac{\tanh x}{x}$.

42. $\displaystyle \lim_{x \to \frac{\pi}{2}} \frac{1 + \cos 2x}{1 - \sin x}$.

43. Find the fallacy:

$$\lim_{x \to 0} \frac{2 + x + \sin x}{x^3 + x - \cos x} \overset{*}{=} \lim_{x \to 0} \frac{1 + \cos x}{3x^2 + 1 + \sin x}$$

$$\overset{*}{=} \lim_{x \to 0} \frac{-\sin x}{6x + \cos x} = \frac{0}{1} = 0.$$

44. Show that, if $a > 0$, then

$$\lim_{n \to \infty} n(a^{1/n} - 1) = \ln a.$$

45. Find values of a and b for which

$$\lim_{x \to 0} \frac{\cos ax - b}{2x^2} = -4.$$

46. Find values of a and b for which

$$\lim_{x \to 0} \frac{\sin 2x + ax + bx^3}{x^3} = 0.$$

47. Calculate $\displaystyle\lim_{x \to 0} \frac{(1+x)^{1/x} - e}{x}$.

48. Let f be a twice differentiable function and fix a value of x.

(a) Show that

$$\lim_{h \to 0} \frac{f(x+h) - f(x-h)}{2h} = f'(x).$$

(b) Show that

$$\lim_{h \to 0} \frac{f(x+h) - 2f(x) + f(x-h)}{h^2} = f''(x).$$

49. Given that f is continuous, use L'Hôpital's rule to determine

$$\lim_{x \to 0} \left(\frac{1}{x} \int_0^x f(t)\, dt \right).$$

50. The integral $Si(x) = \displaystyle\int_0^x \frac{\sin t}{t}\, dt$ plays a role in applied mathematics. Calculate the following limits:

(a) $\displaystyle\lim_{x \to 0} \frac{Si(x)}{x}$.

(b) $\displaystyle\lim_{x \to 0} \frac{Si(x) - x}{x^3}$.

51. The *Fresnel function* $C(x) = \displaystyle\int_0^x \cos^2 t\, dt$ arises in the study of the diffraction of light. Calculate the following limits:

(a) $\displaystyle\lim_{x \to 0} \frac{C(x)}{x}$.

(b) $\displaystyle\lim_{x \to 0} \frac{C(x) - x}{x^3}$.

52. (a) Given that the function f is differentiable, $f(a) = 0$ and $f'(a) \neq 0$, determine

$$\lim_{x \to a} \frac{\int_a^x f(t)\, dt}{f(x)}.$$

(b) Suppose f is k-times differentiable, $f(a) = f'(a) = \cdots = f^{k-1}(a) = 0$, and $f^k(a) \neq 0$. Calculate

$$\lim_{x \to a} \frac{\int_a^x f(t)\, dt}{f(x)}.$$

53. Let $A(b)$ be the area of the region bounded by the parabola $y = x^2$ and the horizontal line $y = b\,(b > 0)$, and let $T(b)$ be the area of triangle AOB. (See the figure.) Find $\displaystyle\lim_{b \to 0^+} T(b)/A(b)$.

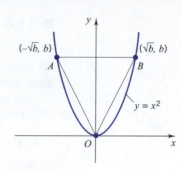

54. The figure shows an angle θ between 0 and $\pi/2$. Let $T(\theta)$ be the area of triangle ABC, and let $S(\theta)$ be the area of the segment of the circle cut by the chord AB. Find $\displaystyle\lim_{\theta \to 0^+} T(\theta)/S(\theta)$.

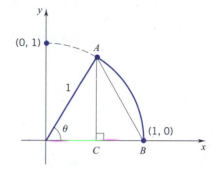

▷55. Set

$$f(x) = \frac{x^2 - 16}{\sqrt{x^2 + 9} - 5}.$$

(a) Use a graphing utility to graph f. What is the behavior of $f(x)$ as $x \to \infty$? as $x \to -\infty$?

(b) What is the behavior of f as $x \to 4$? Confirm your answer by applying L'Hôpital's rule.

▷56. Set $f(x) = \dfrac{x - \sin x}{x^3}$.

(a) Use a graphing utility to graph f. What is the behavior of $f(x)$ as $x \to \infty$? as $x \to -\infty$?

(b) What is the behavior of f as $x \to 0$? Confirm your answer by applying L'Hôpital's rule.

▷57. Set $f(x) = \dfrac{2^{\sin x} - 1}{x}$.

(a) Use a graphing utility to graph f. Estimate

$$\lim_{x \to 0} f(x).$$

(b) Use L'Hôpital's rule to confirm your estimate.

▷58. Set $g(x) = \dfrac{3^{\cos x} - 3}{x^2}$.

(a) Use a graphing utility to graph g. Estimate

$$\lim_{x \to 0} g(x).$$

(b) Use L'Hôpital's rule to confirm your estimate.

■ 11.6 THE INDETERMINATE FORM (∞/∞); OTHER INDETERMINATE FORMS

We come now to limits of quotients where numerator and denominator both tend to ∞. Such quotients are called *indeterminates of the form* ∞/∞ for the limit process used.

As in the case of L'Hôpital's rule (0/0), the result we state applies equally well to all forms of x-approach:

$$\text{as } x \to c, \qquad \text{as } x \to c^+, \qquad \text{as } x \to c^-, \qquad \text{as } x \to \infty, \qquad \text{as } x \to -\infty.$$

As before, the symbol γ can represent a real number, it can represent ∞, it can represent $-\infty$.

(11.6.1)

> **L'HÔPITAL'S RULE (∞/∞)**
>
> Suppose that
> $$f(x) \to \pm\infty \qquad \text{and} \qquad g(x) \to \pm\infty$$
> and in the approach $g'(x) \neq 0$.
>
> If $\qquad \dfrac{f'(x)}{g'(x)} \to \gamma, \qquad$ then $\qquad \dfrac{f(x)}{g(x)} \to \gamma.$

While the proof of L'Hôpital's rule in this setting is a little more complicated than it was in the (0/0) case,[†] the application of the rule is much the same.

Example 1 Let α be any positive number. Show that

(11.6.2)

> $$\text{as } x \to \infty, \qquad \frac{\ln x}{x^\alpha} \to 0.$$

SOLUTION As $x \to \infty$, both numerator and denominator tend to ∞. L'Hôpital's rule gives

$$\lim_{x\to\infty} \frac{\ln x}{x^\alpha} \overset{*}{=} \lim_{x\to\infty} \frac{1/x}{\alpha x^{\alpha-1}} = \lim_{x\to\infty} \frac{1}{\alpha x^\alpha} = 0. \quad \square$$

For example, as $x \to \infty$,

$$\frac{\ln x}{x^2}, \qquad \frac{\ln x}{x}, \qquad \frac{\ln x}{x^{0.01}}, \qquad \frac{\ln x}{x^{0.001}}$$

all tend to zero.

Example 2 Let α be any positive number. Show that

(11.6.3)

> $$\text{as } x \to \infty, \qquad \frac{e^x}{x^\alpha} \to \infty.$$

[†]We omit the proof.

SOLUTION Choose a positive integer $k > \alpha$. For large $x > 0$

$$\frac{e^x}{x^k} < \frac{e^x}{x^\alpha}.$$

We will prove (11.6.3) by proving that

$$\text{as } x \to \infty, \qquad \frac{e^x}{x^k} \to \infty.$$

We do this by L'Hôpital's rule. We differentiate numerator and denominator k-times:

$$\lim_{x\to\infty} \frac{e^x}{x^k} \overset{*}{=} \lim_{x\to\infty} \frac{e^x}{kx^{k-1}} \overset{*}{=} \lim_{x\to\infty} \frac{e^x}{k(k-1)x^{k-2}} \overset{*}{=} \cdots \overset{*}{=} \lim_{x\to\infty} \frac{e^x}{k!} = \infty. \quad \square$$

For example, as $x \to \infty$,

$$\frac{e^x}{x^\pi}, \qquad \frac{e^x}{x^{100}}, \qquad \frac{e^x}{x^{10,000}} \qquad \text{all tend to } \infty.$$

Remark Limit (11.6.2) tells us that $\ln x$ tends to infinity more slowly than any positive power of x. Limit (11.6.3) tells us that e^x tends to infinity more quickly than any positive power of x. $\square$

Example 3 Determine the behavior of $a_n = \dfrac{2^n}{n^2}$ as $n \to \infty$.

SOLUTION To use the methods of calculus, we investigate $\lim_{x\to\infty} \dfrac{2^x}{x^2}$. Since both numerator and denominator tend to ∞ with x, we try L'Hôpital's rule:

$$\lim_{x\to\infty} \frac{2^x}{x^2} \overset{*}{=} \lim_{x\to\infty} \frac{2^x \ln 2}{2x} \overset{*}{=} \lim_{x\to\infty} \frac{2^x (\ln 2)^2}{2} = \infty.$$

Therefore the sequence must also diverge to ∞. $\square$

The Indeterminates $0 \cdot \infty, \infty - \infty$

The usual way to deal with such indeterminates is to try to write them as quotients to which we can apply one of L'Hôpital's rules.

Example 4 $(0 \cdot \infty)$ Find $\lim_{x\to 0^+} \sqrt{x} \ln x$.

SOLUTION As $x \to 0^+$, $\sqrt{x} \to 0$ and $\ln x \to -\infty$. Thus we are dealing with an indeterminate which (up to sign) is of the form $0 \cdot \infty$.

Writing $\sqrt{x} \ln x$ as

$$\frac{\ln x}{1/\sqrt{x}},$$

we can apply L'Hôpital's rule ∞/∞:

$$\lim_{x\to 0^+} \sqrt{x} \ln x = \lim_{x\to 0^+} \frac{\ln x}{1/\sqrt{x}} \overset{*}{=} \lim_{x\to 0^+} \frac{1/x}{-\frac{1}{2}x^{-3/2}} = \lim_{x\to 0^+} -2\sqrt{x} = 0.$$

We could have chosen to write $\sqrt{x} \ln x$ as

$$\frac{\sqrt{x}}{1/\ln x},$$

but, as you can check, the calculations would not have worked out very well. $\square$

Example 5 $(\infty - \infty)$ Find $\lim\limits_{x \to (\pi/2)^-} (\tan x - \sec x)$.

SOLUTION As $x \to (\pi/2)^-$, $\tan x \to \infty$ and $\sec x \to \infty$. Thus we are dealing with an indeterminate of the form $\infty - \infty$.

We proceed to write $\tan x - \sec x$ as a quotient:

$$\tan x - \sec x = \frac{\sin x}{\cos x} - \frac{1}{\cos x} = \frac{\sin x - 1}{\cos x}.$$

Since the function on the right is an indeterminate of the form $0/0$, we can write

$$\lim_{x \to (\pi/2)^-} \frac{\sin x - 1}{\cos x} \overset{*}{=} \lim_{x \to (\pi/2)^-} \frac{\cos x}{-\sin x} = \frac{0}{-1} = 0.$$

This shows that $\lim\limits_{x \to (\pi/2)^-} (\tan x - \sec x) = 0$. ❏

The Indeterminates $0^0, 1^\infty, \infty^0$

These indeterminates all arise in an obvious manner from consideration of expressions of the form $[f(x)]^{g(x)}$. Here we require that $f(x)$ remain positive. (Arbitrary powers are defined only for positive numbers.) Such indeterminates are usually handled by first applying the logarithm function:

$$y = [f(x)]^{g(x)} \qquad \text{gives} \qquad \ln y = g(x) \ln f(x).$$

Example 6 (0^0) Show that

(11.6.4)
$$\lim_{x \to 0^+} x^x = 1.$$

SOLUTION Here we are dealing with an indeterminate of the form 0^0. Our first step is to take the logarithm of x^x. Then we apply L'Hôpital's rule:

$$\lim_{x \to 0^+} \ln(x^x) = \lim_{x \to 0^+} (x \ln x) = \lim_{x \to 0^+} \frac{\ln x}{1/x} \overset{*}{=} \lim_{x \to 0^+} \frac{1/x}{-1/x^2} = \lim_{x \to 0^+} (-x) = 0.$$

Thus, as $x \to 0$, $\ln(x^x) \to 0$ and $x^x = e^{\ln(x^x)} \to e^0 = 1$. ❏

Example 7 (1^∞) Find $\lim\limits_{x \to 0^+} (1 + x)^{1/x}$.

SOLUTION Here we are dealing with an indeterminate of the form 1^∞: as $x \to 0^+$, $1 + x \to 1$ and $1/x \to \infty$. Taking the logarithm and then applying L'Hôpital's rule, we have

$$\lim_{x \to 0^+} \ln(1 + x)^{1/x} = \lim_{x \to 0^+} \frac{\ln(1 + x)}{x} \overset{*}{=} \lim_{x \to 0^+} \frac{1}{1 + x} = 1.$$

As $x \to 0^+$, $\ln(1 + x)^{1/x} \to 1$ and $(1 + x)^{1/x} = e^{\ln(1+x)^{1/x}} \to e^1 = e$. Set $x = 1/n$ and we have the familiar result: as $n \to \infty$, $[1 + (1/n)]^n \to e$. ❏

Example 8 (∞^0) Show that

$$\text{if} \qquad 1 < a < b \qquad \text{then} \qquad \lim_{x \to \infty} (a^x + b^x)^{1/x} = b.$$

SOLUTION Here we are dealing with an indeterminate of the form ∞^0: as $x \to \infty$, $a^x + b^x \to \infty$ and $1/x \to 0$. Taking the logarithm and then applying L'Hôpital's rule, we find that

$$\lim_{x \to \infty} \ln (a^x + b^x)^{1/x} = \lim_{x \to \infty} \frac{\ln (a^x + b^x)}{x}$$

$$\overset{*}{=} \lim_{x \to \infty} \frac{a^x \ln a + b^x \ln b}{a^x + b^x} = \lim_{x \to \infty} \frac{(a/b)^x \ln a + \ln b}{(a/b)^x + 1} = \ln b.$$

This gives $\lim_{x \to \infty} (a^x + b^x)^{1/x} = b.$ ❑

The Misuse of L'Hôpital's Rules

The calculation of limits by differentiation of numerator and denominator is so compellingly easy that there is a tendency to abuse the method. But note: not all quotients are amenable to L'Hôpital's rules — only those which are indeterminates 0/0 or ∞/∞.

Take, for example, the following (mindless) attempt to find

$$\lim_{x \to 0^+} x^{1/x}.$$

Taking the logarithm of $x^{1/x}$ and (mis)applying L'Hôpital's rules, we have

$$\lim_{x \to 0^+} \ln x^{1/x} = \lim_{x \to 0^+} \frac{\ln x}{x} \overset{*}{=} \lim_{x \to 0^+} \frac{1}{x} = \infty.$$

This seems to indicate that as $x \to 0^+$, $\ln x^{1/x} \to \infty$ and therefore $x^{1/x} = e^{\ln x^{1/x}} \to \infty$.
   This may look fine, but it's wrong:

$$\left(\tfrac{1}{2}\right)^2 = \tfrac{1}{4}, \qquad \left(\tfrac{1}{3}\right)^3 = \tfrac{1}{27}, \qquad \left(\tfrac{1}{4}\right)^4 = \tfrac{1}{256}, \cdots.$$

The limit of $x^{1/x}$ as $x \to 0^+$ is clearly 0, not ∞.
   Where did we go wrong? We went wrong in applying L'Hôpital's method to calculate

$$\lim_{x \to 0^+} \frac{\ln x}{x}.$$

As $x \to 0^+$, $\ln x \to -\infty$ and $x \to 0$. L'Hôpital's rules do not apply.

EXERCISES 11.6

Exercises 1–34. Calculate.

1. $\lim_{x \to -\infty} \dfrac{x^2 + 1}{1 - x}.$

2. $\lim_{x \to \infty} \dfrac{20x}{x^2 + 1}.$

3. $\lim_{x \to \infty} \dfrac{x^3}{1 - x^3}.$

4. $\lim_{x \to \infty} \dfrac{x^3 - 1}{2 - x}.$

5. $\lim_{x \to \infty} \left(x^2 \sin \dfrac{1}{x}\right).$

6. $\lim_{x \to \infty} \dfrac{\ln x^k}{x}.$

7. $\lim_{x \to \pi/2^-} \dfrac{\tan 5x}{\tan x}.$

8. $\lim_{x \to 0} (x \ln |\sin x|).$

9. $\lim_{x \to 0^+} x^{2x}.$

10. $\lim_{x \to \infty} \left(x \sin \dfrac{\pi}{x}\right).$

11. $\lim_{x \to 0} [x (\ln |x|)^2].$

12. $\lim_{x \to 0^+} \dfrac{\ln x}{\cot x}.$

13. $\lim_{x \to \infty} \left(\dfrac{1}{x} \displaystyle\int_0^x e^{t^2} dt\right).$

14. $\lim_{x \to \infty} \dfrac{\sqrt{1 + x^2}}{x^2}.$

15. $\lim_{x \to 0} \left[\dfrac{1}{\sin^2 x} - \dfrac{1}{x^2}\right].$

16. $\lim_{x \to 0} |\sin x|^x.$

17. $\lim_{x \to 1} x^{1/(x-1)}.$

18. $\lim_{x \to 0^+} x^{\sin x}.$

19. $\lim_{x \to \infty} \left(\cos \dfrac{1}{x}\right)^x.$

20. $\lim_{x \to \pi/2} |\sec x|^{\cos x}.$

21. $\lim_{x \to 0} \left[\dfrac{1}{\ln (1 + x)} - \dfrac{1}{x}\right].$

22. $\lim_{x \to \infty} (x^2 + a^2)^{(1/x)^2}.$

23. $\lim_{x \to 0} \left(\dfrac{1}{x} - \cot x\right).$

24. $\lim_{x \to \infty} \ln \left(\dfrac{x^2 - 1}{x^2 + 1}\right)^3.$

25. $\lim_{x \to \infty} (\sqrt{x^2 + 2x} - x).$

26. $\lim_{x \to \infty} \left(1 + \dfrac{a}{x}\right)^{bx}.$

27. $\lim_{x \to \infty} (x^3 + 1)^{1/\ln x}.$

28. $\lim_{x \to \infty} (e^x + 1)^{1/x}.$

29. $\lim_{x\to\infty} (\cosh x)^{1/x}$.

30. $\lim_{x\to\infty} \left(1 + \dfrac{1}{x}\right)^{3x}$.

31. $\lim_{x\to 0} \left(\dfrac{1}{\sin x} - \dfrac{1}{x}\right)$.

32. $\lim_{x\to 0}(e^x + 3x)^{1/x}$.

33. $\lim_{x\to 1} \left(\dfrac{1}{\ln x} - \dfrac{x}{x-1}\right)$.

34. $\lim_{x\to 0} \left(\dfrac{1 + 2^x}{2}\right)^{1/x}$.

Exercises 35–42. Find the limit of the sequence.

35. $\lim_{n\to\infty} \left(\dfrac{1}{n} \ln \dfrac{1}{n}\right)$.

36. $\lim_{n\to\infty} \dfrac{n^k}{2^n}$.

37. $\lim_{n\to\infty} (\ln n)^{1/n}$.

38. $\lim_{n\to\infty} \dfrac{\ln n}{n^p}\, (p > 0)$.

39. $\lim_{n\to\infty} (n^2 + n)^{1/n}$.

40. $\lim_{n\to\infty} n^{\sin(\pi/n)}$.

41. $\lim_{n\to\infty} \dfrac{n^2 \ln n}{e^n}$.

42. $\lim_{n\to\infty} (\sqrt{n} - 1)^{1/\sqrt{n}}$.

▶**Exercises 43–46.** Use technology (graphing utility or CAS) to calculate the limit.

43. $\lim_{x\to 0}(\sin x)^x$.

44. $\lim_{x\to(\frac{\pi}{4})^-} (\tan x)^{\tan 2x}$.

45. $\lim_{x\to 0} \left(\dfrac{1}{\sin x} - \dfrac{1}{\tan x}\right)$.

46. $\lim_{x\to 0^+} (\sinh x)^{-x}$.

Exercises 47–52. Sketch the curve, specifying all vertical and horizontal asymptotes.

47. $y = x^2 - \dfrac{1}{x^3}$.

48. $y = \sqrt{\dfrac{x}{x-1}}$.

49. $y = xe^x$.

50. $y = xe^{-x}$.

51. $y = x^2 e^{-x}$.

52. $y = \dfrac{\ln x}{x}$.

The graphs of two functions $y = f(x)$ and $y = g(x)$ are said to be *asymptotic as $x \to \infty$* if

$$\lim_{x\to\infty} [f(x) - g(x)] = 0;$$

they are said to be *asymptotic as x to $-\infty$* if

$$\lim_{x\to-\infty} [f(x) - g(x)] = 0.$$

These ideas are implemented in Exercises 53–56.

53. Show that the hyperbolic arc $y = (b/a)\sqrt{x^2 - a^2}$ is asymptotic to the line $y = (b/a)x$ as $x \to \infty$.

54. Show that the graphs of $y = \cosh x$ and $y = \sinh x$ are asymptotic as $x \to \infty$.

55. Give an example of a function the graph of which is asymptotic to the parabola $y = x^2$ as $x \to \infty$ and crosses the graph of the parabola exactly twice.

56. Give an example of a function the graph of which is asymptotic to the line $y = x$ as $x \to \infty$ and crosses the graph of the line infinitely often.

57. Find the fallacy:

$$\lim_{x\to 0^+} \dfrac{x^2}{\sin x} \overset{*}{=} \lim_{x\to 0^+} \dfrac{2x}{\cos x} \overset{*}{=} \lim_{x\to 0^+} \dfrac{2}{-\sin x} = -\infty.$$

58. (a) Show by induction that, for each positive integer k,

$$\lim_{x\to\infty} \dfrac{(\ln x)^k}{x} = 0.$$

(b) Show that, for each positive number α,

$$\lim_{x\to\infty} \dfrac{(\ln x)^\alpha}{x} = 0.$$

59. The *geometric mean* of two positive numbers a and b is $\sqrt{ab}$. Show that

$$\sqrt{ab} = \lim_{x\to\infty} \left[\tfrac{1}{2}(a^{1/x} + b^{1/x})\right]^x.$$

60. The differential equation satisfied by the velocity of an object of mass m dropped from rest under the influence of gravity with air resistance directly proportional to the velocity can be written

(*) $$m\dfrac{dv}{dt} + kv = mg,$$

where $k > 0$ is the constant of proportionality, g is the gravitational constant and $v(0) = 0$. The velocity of the object at time t is given by

$$v(t) = (mg/k)(1 - e^{-(k/m)t}).$$

(a) Fix t and find $\lim_{k\to 0^+} v(t)$.

(b) Set $k = 0$ in (*) and solve the initial-value problem

$$m\dfrac{dv}{dt} = mg, \qquad v(0) = 0.$$

Does this result fit in with what you found in part (a)?

▶**Exercises 61–62.** Set $f(x) = xe^{-x}$. Use a graphing utility to draw the graph of f on $[0, 20]$. For what follows, take f on the interval $[0, b]$.

61. (a) Find the area A_b of the region Ω_b that lies between the graph of f and the x-axis.

(b) Find the centroid $(\bar{x}_b, \bar{y}_b)$ of Ω_b.

(c) Find the limit of $A_b, \bar{x}_b, \bar{y}_b$ as $b \to \infty$. Interpret your results geometrically.

62. (a) Find the volume of the solid generated by revolving Ω_b about the x-axis.

(b) Find the volume of the solid generated by revolving Ω_b about the y-axis.

(c) Find the limit of each of these volumes as $b \to \infty$. Interpret your results geometrically.

▶63. Let $f(x) = (1 + x)^{1/x}$ and $g(x) = (1 + x^2)^{1/x}$ on $(0, \infty)$.

(a) Use a graphing utility to graph f and g in the same coordinate system. Estimate

$$\lim_{x\to 0^+} g(x).$$

(b) Use L'Hôpital's rule to obtain the exact value of this limit.

▶64. Set $f(x) = \sqrt{x^2 + 3x + 1} - x$.

(a) Use a graphing utility to graph f. Then use your graph to estimate

$$\lim_{x\to\infty} f(x).$$

(b) Use L'Hôpital's rule to obtain the exact value of this limit.

65. Set $g(x) = \sqrt[3]{x^3 - 5x^2 + 2x + 1} - x$.

(a) Use a graphing utility to graph g. Then use your graph to estimate

$$\lim_{x \to \infty} g(x).$$

(b) Use L'Hôpital's rule to obtain the exact value of this limit.

66. The results obtained in Exercises 64 and 65 can be generalized: Let n be a positive integer and let P be the polynomial

$$P(x) = x^n + b_1 x^{n-1} + b_2 x^{n-2} + \cdots + b_{n-1} x + b_n.$$

Show that $\lim_{x \to \infty} ([P(x)]^{1/n} - x) = \dfrac{b_1}{n}$.

■ 11.7 IMPROPER INTEGRALS

In the definition of the definite integral

$$\int_a^b f(x)\,dx$$

it is assumed that $[a,b]$ is a bounded interval and f is a bounded function. In this section we use the limit process to investigate integrals in which either the interval is unbounded or the integrand is an unbounded function. Such integrals are called *improper integrals*.

Integrals over Unbounded Intervals

We begin with a function f which is continuous on an unbounded interval $[a, \infty)$. For each number $b > a$ we can form the definite integral

$$\int_a^b f(x)\,dx.$$

If, as b tends to ∞, this integral tends to a finite limit L,

$$\lim_{b \to \infty} \int_a^b f(x)\,dx = L.$$

then we write

$$\int_a^\infty f(x)\,dx = L$$

and say that

the improper integral $\qquad \displaystyle\int_a^\infty f(x)\,dx \qquad$ *converges to L.*

Otherwise, we say that

the improper integral $\qquad \displaystyle\int_a^\infty f(x)\,dx \qquad$ *diverges.*

In a similar manner,

improper integrals $\quad \displaystyle\int_{-\infty}^b f(x)\,dx \quad$ arise as limits of the form $\quad \displaystyle\lim_{a \to -\infty} \int_a^b f(x)\,dx.$

Example 1

(a) $\displaystyle\int_0^\infty e^{-2x}\,dx = \tfrac{1}{2}.$

(b) $\displaystyle\int_1^\infty \frac{dx}{x}$ diverges.

(c) $\displaystyle\int_1^\infty \frac{dx}{x^2} = 1.$

(d) $\displaystyle\int_{-\infty}^1 \cos \pi x\,dx$ diverges.

VERIFICATION

(a) $\displaystyle\int_0^\infty e^{-2x}\,dx = \lim_{b\to\infty}\int_0^b e^{-2x}\,dx = \lim_{b\to\infty}\left[-\frac{e^{-2x}}{2}\right]_0^b = \lim_{b\to\infty}\left(\frac{1}{2}-\frac{1}{2e^{2b}}\right)=\frac{1}{2}.$

(b) $\displaystyle\int_1^\infty \frac{dx}{x} = \lim_{b\to\infty}\int_1^b \frac{dx}{x} = \lim_{b\to\infty}\ln b = \infty.$

(c) $\displaystyle\int_1^\infty \frac{dx}{x^2} = \lim_{b\to\infty}\int_1^b \frac{dx}{x^2} = \lim_{b\to\infty}\left[-\frac{1}{x}\right]_1^b = \lim_{b\to\infty}\left(1-\frac{1}{b}\right)=1.$

(d) Note first that

$$\int_a^1 \cos\pi x\,dx = \left[\frac{1}{\pi}\sin\pi x\right]_a^1 = -\frac{1}{\pi}\sin\pi a.$$

As a tends to $-\infty$, $\sin\pi a$ oscillates between -1 and 1. Therefore the integral oscillates between $1/\pi$ and $-1/\pi$ and does not converge. ❏

The usual formulas for area and volume are extended to the unbounded case by the use of improper integrals.

Example 2 Fix $p > 0$ and let Ω be the region below the graph of

$$f(x) = \frac{1}{x^p}, \qquad x \geq 1. \qquad\text{(Figure 11.7.1)}$$

As we show below,

$$\text{area of }\Omega = \begin{cases} \dfrac{1}{p-1}, & \text{if } p > 1 \\ \infty, & \text{if } p \leq 1. \end{cases}$$

This comes about from setting

$$\text{area of }\Omega = \lim_{b\to\infty}\int_1^b \frac{dx}{x^p} = \int_1^\infty \frac{dx}{x^p}.$$

For $p \neq 1$,

$$\int_1^\infty \frac{dx}{x^p} = \lim_{b\to\infty}\int_1^b \frac{dx}{x^p} = \lim_{b\to\infty}\frac{1}{1-p}(b^{1-p}-1) = \begin{cases} \dfrac{1}{p-1}, & \text{if } p > 1 \\ \infty, & \text{if } p < 1. \end{cases}$$

For $p = 1$,

$$\int_1^\infty \frac{dx}{x^p} = \int_1^\infty \frac{dx}{x} = \infty,$$

as you have seen already. ❏

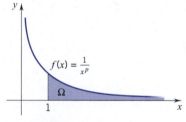

Figure 11.7.1

For future reference we record the following:

(11.7.1) $\displaystyle\int_1^\infty \frac{dx}{x^p}$ converges if $p > 1$ and diverges if $0 < p \leq 1$.

Example 3 *A configuration of finite volume with infinite surface area.* You have seen that the region below the graph of $f(x) = 1/x$ with $x \geq 1$ has infinite area. Suppose that this region of infinite area is revolved about the x-axis. See Figure 11.7.2. What is the volume of the resulting configuration? It may surprise you somewhat, but the volume is not infinite. It is in fact π: using the disk method to calculate volume (Section 6.2), we have

$$V = \int_1^\infty \pi[f(x)]^2 \, dx = \pi \int_1^\infty \frac{dx}{x^2} = \pi \lim_{b\to\infty} \int_1^b \frac{dx}{x^2}$$

$$= \pi \lim_{b\to\infty} \left[\frac{-1}{x} \right]_1^b = \pi \cdot 1 = \pi.$$

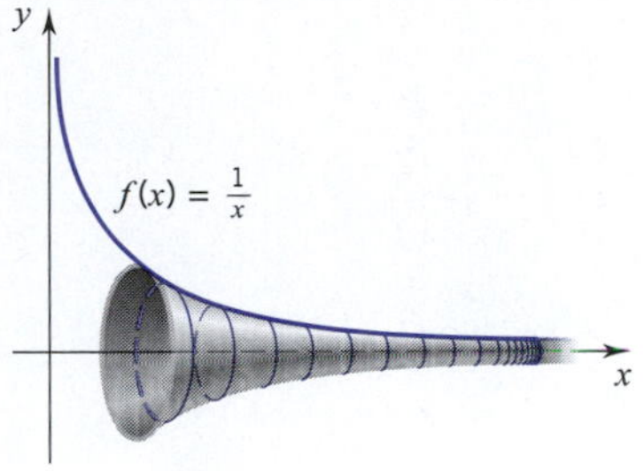

$f(x) = \frac{1}{x}$

Figure 11.7.2

However, the surface that bounds this configuration does have infinite area (Exercise 41). This surface is known as *Gabriel's horn.* ❏

When it proves difficult to determine the convergence or divergence of an improper integral, we try comparison with improper integrals of known behavior.

(11.7.2)

(*A comparison test*) Suppose that f and g are continuous and
$$0 \le f(x) \le g(x) \quad \text{for all } x \in [a, \infty). \qquad \text{(Figure 11.7.3)}$$

(i) If $\displaystyle\int_a^\infty g(x)\,dx$ converges, then $\displaystyle\int_a^\infty f(x)\,dx$ converges.

(ii) If $\displaystyle\int_a^\infty f(x)\,dx$ diverges, then $\displaystyle\int_a^\infty g(x)\,dx$ diverges.

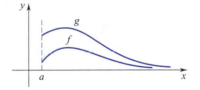

Figure 11.7.3

A similar result holds for integrals from $-\infty$ to b. The proof of (11.7.2) is left to you as an exercise.

Example 4 The improper integral $\displaystyle\int_1^\infty \frac{dx}{\sqrt{1+x^3}}$ converges since

$$\frac{1}{\sqrt{1+x^3}} < \frac{1}{x^{3/2}} \quad \text{for } x \in [1, \infty) \quad \text{and} \quad \int_1^\infty \frac{dx}{x^{3/2}} \text{ converges.}$$

To evaluate

$$\lim_{b\to\infty} \int_1^b \frac{dx}{\sqrt{1+x^3}}$$

directly, we would first have to evaluate

$$\int_1^b \frac{dx}{\sqrt{1+x^3}} \quad \text{for each } b > 1,$$

and this we can't do because we have no way of calculating $\displaystyle\int \frac{dx}{\sqrt{1+x^3}}$. ❏

Example 5 The improper integral $\displaystyle\int_1^\infty \frac{dx}{\sqrt{1+x^2}}$ diverges since

$$\frac{1}{1+x} \le \frac{1}{\sqrt{1+x^2}} \quad \text{for } x \in [1, \infty) \quad \text{and} \quad \int_1^\infty \frac{dx}{1+x} \text{ diverges.}$$

We can obtain this result by evaluating

$$\int_1^b \frac{dx}{\sqrt{1+x^2}}$$

and letting b tend to ∞. Try to carry out the calculation in this manner. ❏

Suppose now that f is continuous on $(-\infty, \infty)$. The *improper integral*

$$\int_{-\infty}^{\infty} f(x)\, dx$$

is said to *converge* if

$$\int_{-\infty}^{0} f(x)\, dx \qquad \text{and} \qquad \int_{0}^{\infty} f(x)\, dx$$

both converge. In this case we set

$$\int_{-\infty}^{\infty} f(x) = L + M$$

where

$$\int_{-\infty}^{0} f(x)\, dx = L \qquad \text{and} \qquad \int_{0}^{\infty} f(x)\, dx = M.$$

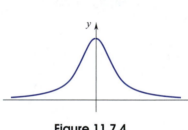

Figure 11.7.4

Example 6 Let $r > 0$. Determine whether the improper integral

$$\int_{-\infty}^{\infty} \frac{r}{r^2 + x^2}\, dx \qquad \text{(Figure 11.7.4)}$$

converges or diverges. If it converges, give the value of the integral.

SOLUTION According to the definition, we need to determine the convergence or divergence of both

$$\int_{-\infty}^{0} \frac{r}{r^2 + x^2}\, dx \qquad \text{and} \qquad \int_{0}^{\infty} \frac{r}{r^2 + x^2}\, dx.$$

For the first integral:

$$\int_{-\infty}^{0} \frac{r}{r^2 + x^2}\, dx = \lim_{a \to -\infty} \int_{a}^{0} \frac{r}{r^2 + x^2}\, dx = \lim_{a \to -\infty} \left[\arctan\left(\frac{x}{r}\right) \right]_a^0$$

$$= -\lim_{a \to -\infty} \arctan\left(\frac{a}{r}\right) = -\left(-\frac{\pi}{2}\right) = \frac{\pi}{2}.$$

For the second integral:

$$\int_{0}^{\infty} \frac{r}{r^2 + x^2}\, dx = \lim_{b \to \infty} \int_{0}^{b} \frac{r}{r^2 + x^2}\, dx = \lim_{b \to \infty} \left[\arctan\left(\frac{x}{r}\right) \right]_0^b$$

$$= \lim_{b \to \infty} \arctan\left(\frac{b}{r}\right) = \frac{\pi}{2}.$$

Since both of these integrals converge, the improper integral

$$\int_{-\infty}^{\infty} \frac{r}{r^2 + x^2}\, dx$$

converges. The value of the integral is $\frac{1}{2}\pi + \frac{1}{2}\pi = \pi$. ❏

Remark Note that we did not define

(1)
$$\int_{-\infty}^{\infty} f(x)\,dx$$

as

(2)
$$\lim_{b\to\infty} \int_{-b}^{b} f(x)\,dx.$$

It is easy to show that if (1) exists, then (2) exists and (1) = (2). However, the existence of (2) does not imply the existence of (1). For example, (2) exists and is 0 for every odd function f, but this is certainly not the case for (1). See Exercises 57 and 58. ❏

Integrals of Unbounded Functions

Improper integrals can arise on bounded intervals. Suppose that f is continuous on the half-open interval $[a, b)$ but is unbounded there. See Figure 11.7.5. For each number $c < b$, we can form the definite integral

$$\int_{a}^{c} f(x)\,dx.$$

If, as $c \to b^-$, the integral tends to a finite limit L, namely, if

$$\lim_{c\to b^-} \int_{a}^{c} f(x)\,dx = L,$$

then we write

$$\int_{a}^{b} f(x)\,dx = L$$

and say that

the improper integral $\quad \int_{a}^{b} f(x)\,dx \quad$ converges to L.

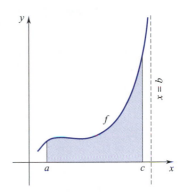

Figure 11.7.5

Otherwise, we say that *the improper integral diverges*.

Similarly, functions which are continuous but unbounded on half-open intervals of the form $(a, b]$ lead to consideration of

$$\lim_{c\to a^+} \int_{c}^{b} f(x)\,dx.$$

If this limit exists and is L, then we write

$$\int_{a}^{b} f(x)\,dx = L$$

and say that

the improper integral $\quad \int_{a}^{b} f(x)\,dx \quad$ converges to L.

Otherwise, we say that *the improper integral diverges*.

Example 7

(a) $\int_{0}^{1} (1-x)^{-2/3}\,dx = 3.$ \qquad (b) $\int_{0}^{2} \frac{dx}{x}$ diverges.

VERIFICATION

(a)
$$\int_0^1 (1-x)^{-2/3}dx = \lim_{c \to 1^-} \int_0^c (1-x)^{-2/3}dx$$

$$= \lim_{c \to 1^-} \left[-3(1-x)^{1/3} \right]_0^c = \lim_{c \to 1^-} [-3(1-c)^{1/3} + 3] = 3.$$

(b)
$$\int_0^2 \frac{dx}{x} = \lim_{c \to 0^+} \int_c^2 \frac{dx}{x} = \lim_{c \to 0^+} \left[\ln x \right]_c^2 = \lim_{c \to 0^+} [\ln 2 - \ln c] = \infty. \quad \square$$

Finally, suppose that f is continuous at each point of $[a, b]$ except at some interior point c where f has an infinite discontinuity. We say that the *improper integral*

$$\int_a^b f(x)\,dx$$

converges if *both*

$$\int_a^c f(x)\,dx \qquad \text{and} \qquad \int_c^b f(x)\,dx$$

converge. If

$$\int_a^c f(x)\,dx = L \qquad \text{and} \qquad \int_c^b f(x)\,dx = M,$$

we set

$$\int_a^b f(x)\,dx = L + M.$$

Example 8 Test

(∗)
$$\int_1^4 \frac{dx}{(x-2)^2}$$

for convergence.

SOLUTION The integrand has an infinite discontinuity at $x = 2$. See Figure 11.7.6.

For integral (∗) to converge both

$$\int_1^2 \frac{dx}{(x-2)^2} \qquad \text{and} \qquad \int_2^4 \frac{dx}{(x-2)^2}$$

must converge. Neither does. For instance, as $c \to 2^-$,

$$\int_1^c \frac{dx}{(x-2)^2} = \left[-\frac{1}{x-2} \right]_1^c = -\frac{1}{c-2} - 1 \to \infty.$$

This tells us that

$$\int_1^2 \frac{dx}{(x-2)^2}$$

diverges and shows that (∗) diverges.

If we overlook the infinite discontinuity at $x = 2$, we can be led to the *incorrect conclusion* that

$$\int_1^4 \frac{dx}{(x-2)^2} = \left[-\frac{1}{x-2} \right]_1^4 = -\frac{3}{2}. \quad \square$$

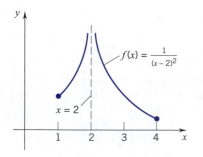

$f(x) = \dfrac{1}{(x-2)^2}$

$x = 2$

Figure 11.7.6

Example 9 Evaluate $\displaystyle\int_{-2}^{1} \frac{dx}{x^{4/5}}$.

SOLUTION Since the integrand has an infinite discontinuity at $x = 0$, the integral is improper. Thus we need to evaluate both

$$\int_{-2}^{0} \frac{dx}{x^{4/5}} \quad \text{and} \quad \int_{0}^{1} \frac{dx}{x^{4/5}}.$$

Note that

$$\int_{-2}^{0} \frac{dx}{x^{4/5}} = \lim_{c \to 0^-} \int_{-2}^{c} \frac{dx}{x^{4/5}} = \lim_{c \to 0^-} \left[5x^{1/5} \right]_{-2}^{c} = \lim_{c \to 0^-} \left[5c^{1/5} - 5(-2)^{1/5} \right] = 5\left(2^{1/5}\right)$$

and

$$\int_{0}^{1} \frac{dx}{x^{4/5}} = \lim_{c \to 0^+} \int_{c}^{1} \frac{dx}{x^{4/5}} = \lim_{c \to 0^+} \left[5x^{1/5} \right]_{c}^{1} = \lim_{c \to 0^+} \left[5 - 5c^{1/5} \right] = 5.$$

It follows that the integral from $x = -2$ to $x = 1$ converges and we have

$$\int_{-2}^{1} \frac{dx}{x^{4/5}} = 5 + 5\left(2^{1/5}\right) \cong 10.74 \quad \square$$

EXERCISES 11.7

Exercises 1–34. Below we list some improper integrals. Determine whether the integral converges and, if so, evaluate the integral.

1. $\displaystyle\int_{1}^{\infty} \frac{dx}{x^2}$.

2. $\displaystyle\int_{0}^{\infty} \frac{dx}{1 + x^2}$.

3. $\displaystyle\int_{0}^{\infty} \frac{dx}{4 + x^2}$.

4. $\displaystyle\int_{0}^{\infty} e^{-px} dx, \quad p > 0$.

5. $\displaystyle\int_{0}^{\infty} e^{px} dx, \quad p > 0$.

6. $\displaystyle\int_{0}^{1} \frac{dx}{\sqrt{x}}$.

7. $\displaystyle\int_{0}^{8} \frac{dx}{x^{2/3}}$.

8. $\displaystyle\int_{0}^{1} \frac{dx}{x^2}$.

9. $\displaystyle\int_{0}^{1} \frac{dx}{\sqrt{1 - x^2}}$.

10. $\displaystyle\int_{0}^{1} \frac{dx}{\sqrt{1 - x}}$.

11. $\displaystyle\int_{0}^{2} \frac{x}{\sqrt{4 - x^2}} dx$.

12. $\displaystyle\int_{0}^{a} \frac{dx}{\sqrt{a^2 - x^2}}$.

13. $\displaystyle\int_{e}^{\infty} \frac{\ln x}{x} dx$.

14. $\displaystyle\int_{e}^{\infty} \frac{dx}{x \ln x}$.

15. $\displaystyle\int_{0}^{1} x \ln x \, dx$.

16. $\displaystyle\int_{e}^{\infty} \frac{dx}{x(\ln x)^2}$.

17. $\displaystyle\int_{-\infty}^{\infty} \frac{dx}{1 + x^2}$.

18. $\displaystyle\int_{2}^{\infty} \frac{dx}{x^2 - 1}$.

19. $\displaystyle\int_{-\infty}^{\infty} \frac{dx}{x^2}$.

20. $\displaystyle\int_{1/3}^{3} \frac{dx}{\sqrt[3]{3x - 1}}$.

21. $\displaystyle\int_{1}^{\infty} \frac{dx}{x(x + 1)}$.

22. $\displaystyle\int_{-\infty}^{0} x e^x dx$.

23. $\displaystyle\int_{3}^{5} \frac{x}{\sqrt{x^2 - 9}} dx$.

24. $\displaystyle\int_{1}^{4} \frac{dx}{x^2 - 4}$.

25. $\displaystyle\int_{-3}^{3} \frac{dx}{x(x + 1)}$.

26. $\displaystyle\int_{1}^{\infty} \frac{x}{(1 + x^2)^2} dx$.

27. $\displaystyle\int_{-3}^{1} \frac{dx}{x^2 - 4}$.

28. $\displaystyle\int_{-\infty}^{\infty} \frac{1}{e^x + e^{-x}} dx$.

29. $\displaystyle\int_{0}^{\infty} \cosh x \, dx$.

30. $\displaystyle\int_{1}^{4} \frac{dx}{x^2 - 5x + 6}$.

31. $\displaystyle\int_{0}^{\infty} e^{-x} \sin x \, dx$.

32. $\displaystyle\int_{0}^{\infty} \cos^2 x \, dx$.

33. $\displaystyle\int_{0}^{1} \frac{e^{\sqrt{x}}}{\sqrt{x}} dx$.

34. $\displaystyle\int_{0}^{\pi/2} \frac{\cos x}{\sqrt{\sin x}} dx$.

▶**Exercise 35–36.** Use a graphing utility to draw the graph of the integrand. Then use a CAS to determine whether the integral converges or diverges.

35. (a) $\displaystyle\int_{0}^{\infty} \frac{x}{(16 + x^2)^2} dx$.

(b) $\displaystyle\int_{0}^{\infty} \frac{x^2}{(16 + x^2)^2} dx$.

(c) $\displaystyle\int_{0}^{\infty} \frac{x}{16 + x^4} dx$.

(d) $\displaystyle\int_{0}^{\infty} \frac{x}{16 + x^2} dx$.

36. (a) $\displaystyle\int_{0}^{2} \frac{x^3}{\sqrt[3]{2 - x}} dx$.

(b) $\displaystyle\int_{0}^{2} \frac{1}{\sqrt{2 - x}} dx$.

(c) $\displaystyle\int_{0}^{2} \frac{x}{\sqrt{2 - x}} dx$.

(d) $\displaystyle\int_{0}^{2} \frac{1}{\sqrt{2x - x^2}} dx$.

37. Evaluate

$$\int_{0}^{1} \arcsin x \, dx$$

using integration by parts even though the technique leads to an improper integral.

38. (a) For what values of r is

$$\int_0^\infty x^r e^{-x} dx$$

convergent?

(b) Show by induction that

$$\int_0^\infty x^n e^{-x} dx = n!, \quad n = 1, 2, 3, \cdots.$$

39. The integral

$$\int_0^\infty \frac{1}{\sqrt{x}(1+x)} dx$$

is improper in two distinct ways: the interval of integration is unbounded and the integrand is unbounded. If we rewrite the integral as

$$\int_0^1 \frac{1}{\sqrt{x}(1+x)} dx + \int_1^\infty \frac{1}{\sqrt{x}(1+x)} dx,$$

then we have two improper integrals, the first having an unbounded integrand and the second defined on an unbounded interval. If each of these integrals converges with values L_1 and L_2, then the original integral converges and has value $L_1 + L_2$. Evaluate the original integral.

40. Evaluate

$$\int_1^\infty \frac{1}{x\sqrt{x^2-1}} dx$$

by the method outlined in Exercise 39.

41. The graph of $f(x) = 1/x$ with $x \geq 1$ is revolved about the x-axis. Show that the resulting surface has infinite area. (We stated this in Example 3.)

42. Sketch the curves $y = \sec x$ and $y = \tan x$ for $0 \leq x < \pi/2$. Calculate the area between the two curves.

43. Let Ω be the region bounded by the coordinate axes, the curve $y = 1/\sqrt{x}$, and the line $x = 1$. (a) Sketch Ω. (b) Show that Ω has finite area and find it. (c) Show that if Ω is revolved about the x-axis, the configuration obtained does not have finite volume.

44. Let Ω be the region between the curve $y = 1/(1+x^2)$ and the x-axis, $x \geq 0$. (a) Sketch Ω. (b) Find the area of Ω. (c) Find the volume obtained by revolving Ω about the x-axis. (d) Find the volume obtained by revolving Ω about the y-axis.

45. Let Ω be the region bounded by the curve $y = e^{-x}$ and the x-axis, $x \geq 0$. (a) Sketch Ω. (b) Find the area of Ω. (c) Find the volume obtained by revolving Ω about the x-axis. (d) Find the volume obtained by revolving Ω about the y-axis. (e) Find the surface area of the configuration in part (c).

46. What point would you call the centroid of the region in Exercise 45? Does Pappus's theorem work in this instance?

47. Let Ω be the region bounded by the curve $y = e^{-x^2}$ and the x-axis, $x \geq 0$. (a) Show that Ω has finite area. (The area

is $\frac{1}{2}\sqrt{\pi}$, as you will see in Chapter 17.) (b) Calculate the volume generated by revolving Ω about the y-axis.

48. Let Ω be the region bounded below by $y(x^2 + 1) = x$, above by $xy = 1$, and to the left by $x = 1$. (a) Find the area of Ω. (b) Show that the configuration obtained by revolving Ω about the x-axis has finite volume. (c) Calculate the volume generated by revolving Ω about the y-axis.

49. Let Ω be the region bounded by the curve $y = x^{-1/4}$ and the x-axis, $0 < x \leq 1$. (a) Sketch Ω. (b) Find the area of Ω. (c) Find the volume obtained by revolving Ω about the x-axis. (d) Find the volume obtained by revolving Ω about the y-axis.

50. Prove the validity of comparison test (11.7.2).

Exercises 51–56. Use comparison test (11.7.2) to determine whether the integral converges.

51. $\displaystyle\int_1^\infty \frac{x}{\sqrt{1+x^5}} dx.$ **52.** $\displaystyle\int_1^\infty 2^{-x^2} dx.$

53. $\displaystyle\int_0^\infty (1+x^5)^{-1/6} dx.$ **54.** $\displaystyle\int_\pi^\infty \frac{\sin^2 2x}{x^2} dx.$

55. $\displaystyle\int_1^\infty \frac{\ln x}{x^2} dx.$ **56.** $\displaystyle\int_e^\infty \frac{dx}{\sqrt{x+1}\ln x}.$

57. (a) Show that

$$\int_{-\infty}^\infty \frac{2x}{1+x^2} dx$$

diverges by showing that

$$\int_0^\infty \frac{2x}{1+x^2} dx$$

diverges.

(b) Then show that $\displaystyle\lim_{b\to\infty} \int_{-b}^b \frac{2x}{1+x^2} dx = 0.$

58. Show that

(a) $\displaystyle\int_{-\infty}^\infty \sin x \, dx$ diverges

although

(b) $\displaystyle\lim_{b\to\infty} \int_{-b}^b \sin x \, dx = 0.$

59. Calculate the arc distance from the origin to the point $(x(\theta_1), y(\theta_1))$ along the exponential spiral $r = ae^{c\theta}$. (Take $a > 0, c > 0.$)

60. The function

$$f(x) = \frac{1}{\sqrt{2\pi}} \int_{-\infty}^x e^{-t^2/2} dt$$

is important in statistics. Prove that the integral on the right converges for all real x.

Exercises 61–64: *Laplace transforms.* Let f be continuous on $[0, \infty)$. The *Laplace transform* of f is the function F defined by setting

$$F(s) = \int_0^\infty e^{-sx} f(x) \, dx.$$

The domain of F is the set of numbers s for which the improper integral converges. Find the Laplace transform F of each of the following functions specifying the domain of F.

61. $f(x) = 1$.

62. $f(x) = x$.

63. $f(x) = \cos 2x$.

64. $f(x) = e^{ax}$.

Exercises 65–68: *Probability density functions.* A nonnegative function f defined on $(-\infty, \infty)$ is called a *probability density function* if

$$\int_{-\infty}^{\infty} f(x)\,dx = 1.$$

65. Show that the function f defined by

$$f(x) = \begin{cases} 6x/(1+3x^2)^2, & x \geq 0 \\ 0, & x < 0 \end{cases}$$

is a probability density function.

66. Let $k > 0$. Show that the function

$$f(x) = \begin{cases} ke^{-kx}, & x \geq 0 \\ 0, & x < 0 \end{cases}$$

is a probability density function. It is called the *exponential density function*.

67. The *mean* of a probability density function f is defined as the number

$$\mu = \int_{-\infty}^{\infty} x f(x)\,dx.$$

Calculate the mean for the exponential density function.

68. The *standard deviation* of a probability density function f is defined as the number

$$\sigma = \left[\int_{-\infty}^{\infty} (x - \mu)^2 f(x)\,dx \right]^{1/2}$$

where μ is the mean. Calculate the standard deviation for the exponential density function.

69. (*Useful later*) Let f be a continuous, positive, decreasing function defined on $[1, \infty)$. Show that

$$\int_{1}^{\infty} f(x)\,dx$$

converges iff the sequence

$$a_n = \int_{1}^{n} f(x)\,dx$$

converges.

■ CHAPTER 11. REVIEW EXERCISES

Exercises 1–6. Find the least upper bound (if it exists) and the greatest lower bound (if it exists).

1. $\{x : |x - 2| \leq 3\}$.

2. $\{x : x^2 > 3\}$.

3. $\{x : x^2 - x - 2 \leq 0\}$.

4. $\{x : \cos x \leq 1\}$.

5. $\{x : e^{-x^2} \leq 2\}$.

6. $\{x : \ln x < e\}$.

Exercises 7–12. Determine the boundedness and monotonicity of the sequence with a_n as indicated.

7. $\dfrac{2n}{3n+1}$.

8. $\dfrac{n^2 - 1}{n}$.

9. $1 + \dfrac{(-1)^n}{n}$.

10. $\dfrac{4^n}{1 + 4^n}$.

11. $\dfrac{2^n}{n^2}$.

12. $\dfrac{\sin(n\pi/2)}{n^2}$.

Exercises 13–26. State whether the sequence converges and if it does, find the limit.

13. $n\, 2^{1/n}$.

14. $\dfrac{(n+1)(n+2)}{(n+3)(n+4)}$.

15. $\left(\dfrac{n}{1+n}\right)^{1/n}$.

16. $\dfrac{4n^2 + 5n + 1}{n^3 + 1}$.

17. $\cos(\pi/n)\sin(\pi/n)$.

18. $\left(2 + \dfrac{1}{n}\right)^n$.

19. $\left[\ln\left(1 + \dfrac{1}{n}\right)\right]^n$.

20. $3\ln 2n - \ln(n^3 + 1)$.

21. $\dfrac{3n^2 - 1}{\sqrt{4n^4 + 2n^2 + 3}}$.

22. $\dfrac{\sqrt[3]{n^2 + 4}}{2n + 1}$.

23. $(\pi/n)\cos(\pi/n)$.

24. $(n/\pi)\sin(n\pi)$.

25. $\displaystyle\int_{n}^{n+1} e^{-x}\,dx$.

26. $\displaystyle\int_{1}^{n} \dfrac{1}{\sqrt{x}}\,dx$.

27. Show that, if $a_n \to L$, then $a_{n+1} \to L$.

28. Suppose that the sequence a_n converges to L. Define the sequence m_n by

$$m_n = \dfrac{a_1 + a_2 + \cdots + a_n}{n}.$$

Prove that $m_n \to L$.

29. Choose any real number a and form the sequence

$$\cos a,\ \cos(\cos a),\ \cos(\cos(\cos a)),\ \cdots.$$

Convince yourself numerically that this sequence converges to some number L. Determine L and verify that $\cos L = L$. (This is an effective numerical method for solving the equation $\cos x = x$.)

30. Find a numerical solution to the equation $\sin(\cos x) = x$. HINT: Use the method of Exercise 31.

Exercises 31–40. Calculate.

31. $\displaystyle\lim_{x\to\infty} \dfrac{5x + 2\ln x}{x + 3\ln x}$.

32. $\displaystyle\lim_{x\to 0} \dfrac{e^x - 1}{\tan 2x}$.

33. $\displaystyle\lim_{x\to 0} \dfrac{\ln(\cos x)}{x^2}$.

34. $\displaystyle\lim_{x\to 1} x^{1/(x-1)}$.

35. $\displaystyle\lim_{x\to 0} \left(1 + \dfrac{4}{x}\right)^{2x}$.

36. $\displaystyle\lim_{x\to 0} \dfrac{e^{2x} - e^{-2x}}{\sin x}$.

37. $\lim\limits_{x\to 0^+} x^2 \ln x$.

38. $\lim\limits_{x\to\infty} \dfrac{(10)^x}{x^{10}}$.

39. $\lim\limits_{x\to 0} \dfrac{e^x + e^{-x} - x^2 - 2}{\sin^2 x - x^2}$.

40. $\lim\limits_{x\to 1} \csc(\pi x) \ln x$.

41. Calculate $\lim\limits_{x\to\infty} xe^{-x^2} \int_0^x e^{t^2} dx$.

42. Let n be a positive integer. Calculate $\lim\limits_{x\to 0} \dfrac{e^{-1/x^2}}{x^n}$.

Exercises 43–50. Determine whether the integral converges and, if so, evaluate the integral.

43. $\displaystyle\int_1^\infty \dfrac{e^{-\sqrt{x}}}{\sqrt{x}} dx$.

44. $\displaystyle\int_0^1 \dfrac{x}{\sqrt{1 - x^2}} dx$.

45. $\displaystyle\int_0^1 \dfrac{1}{1 - x^2} dx$.

46. $\displaystyle\int_0^{\pi/2} \sec x \, dx$.

47. $\displaystyle\int_1^\infty \dfrac{\sin(\pi/x)}{x^2} dx$.

48. $\displaystyle\int_0^9 \dfrac{1}{(x - 1)^{2/3}} dx$.

49. $\displaystyle\int_0^\infty \dfrac{1}{e^x + e^{-x}} dx$.

50. $\displaystyle\int_2^\infty \dfrac{1}{x(\ln x)^k} dx$.

51. Evaluate $\displaystyle\int_0^a \ln(1/x) \, dx$ for $a > 0$.

52. Find the length of the curve $y = (a^{2/3} - x^{2/3})^{3/2}$ from $x = 0$ to $x = a, a > 0$.

53. Let S and T be nonempty sets of real numbers which are bounded above. Let $S + T$ be the set defined by

$$S + T = \{x + y : x \in S \text{ and } y \in T\}.$$

Prove that lub $(S + T) = $ lub $S + $ lub T.

54. Let S be a nonempty set which is bounded below. Let $B = \{b : b$ is a lower bound of $S\}$. Show that (a) B is nonempty; (b) B is bounded above; (c) lub $B = $ glb S.

55. Let f be a function continuous on $(-\infty, \infty)$ and L a real number.

(a) Show that

$$\text{if} \quad \int_{-\infty}^\infty f(x)\,dx = L \quad \text{then} \quad \lim_{c\to\infty} \int_{-c}^c f(x)\,dx = L.$$

(b) Find an example which shows that the converse of (a) is false.

56. Show that

$$\int_{-\infty}^\infty f(x)\,dx = L \quad \text{iff} \quad \lim_{c\to\infty} \int_{-c}^c f(x)\,dx = L$$

in the event that f is (a) nonnegative or (b) even.

57. In general the least upper bound of a set of numbers need not be in the set. Show that the least upper bound of a set of integers must be in the set.

58. Let f be a function continuous on $[a,b]$. As usual, denote by $L_f(P)$ and $U_f(P)$ the upper and lower sums that correspond to the partition P. What is the least upper bound of all $L_f(P)$? What is the greatest lower bound of all $U_f(P)$?

Mathematics was not invented overnight. Some of the most fruitful mathematical ideas have their origin in ancient times.

Some 500 years before the birth of Christ, a self-taught country boy, Zeno of Elea, invented four paradoxes that rocked the intellectual establishment of his day. Here is one of Zeno's paradoxes clothed in modern terminology.

Suppose that a particle moves along a coordinate line at constant speed. Suppose that it starts at $x = 1$ and heads toward the point $x = 0$. If the particle reaches the halfway mark $x = \frac{1}{2}$ in t seconds, then it will reach the point $x = \frac{1}{4}$ in $t + \frac{1}{2}t$ seconds, the point $x = \frac{1}{8}$ in $t + \frac{1}{2}t + \frac{1}{4}t$ seconds, and so on. More generally, it will reach the point $x = \frac{1}{2^{n+1}}$ in

$$t + \tfrac{1}{2}t + \tfrac{1}{4}t + \cdots + \tfrac{1}{2^n}t \qquad \text{seconds.}$$

This line of reasoning suggests that the particle cannot reach $x = 0$ until it has traveled through the sum of an infinite number of little time segments:

$$t + \tfrac{1}{2}t + \tfrac{1}{4}t + \cdots + \tfrac{1}{2^n}t + \cdots \qquad \text{seconds.}$$

But what sense does this make? Even if we were very swift of mind and could add at the rate of one term per nanosecond, we would still never finish. An infinite number of nanoseconds is still an infinity of time, and that is more time than we have.

Mathematicians sidestep paradoxes of this sort by using infinite series. Before we begin the study of infinite series, we introduce some notation.

■ 12.1 SIGMA NOTATION

We can indicate the sequence

$$1, \tfrac{1}{2}, \tfrac{1}{4}, \tfrac{1}{8}, \tfrac{1}{16}, \cdots$$

by setting $a_n = \left(\frac{1}{2}\right)^{n-1}$ and writing

$$a_1, a_2, a_3, a_4, a_5, \cdots.$$

We can indicate that same sequence beginning with index 0: set $b_n = \left(\frac{1}{2}\right)^n$ and write

$$b_0, b_1, b_2, b_3, b_4, \cdots.$$

More generally, we can set $c_n = \left(\frac{1}{2}\right)^{n-p}$ and write

$$c_p, c_{p+1}, c_{p+2}, c_{p+3}, c_{p+4}, \cdots.$$

We have the same sequence but have begun with index p. We will often begin with an index other than 1.

The symbol $\sum$ is the capital Greek letter "sigma." We write

(1)
$$\sum_{k=0}^{n} a_k$$

("the sum of the a_k from k equals 0 to k equals n") to indicate the sum

$$a_0 + a_1 + \cdots + a_n.$$

More generally, for $n \geq m$, we write

(2)
$$\sum_{k=m}^{n} a_k$$

to indicate the sum

$$a_m + a_{m+1} + \cdots + a_n.$$

In (1) and (2) the letter "k" is being used as a "dummy" variable; namely, it can be replaced by any other letter not already engaged. For instance,

$$\sum_{i=3}^{7} a_i, \qquad \sum_{j=3}^{7} a_j, \qquad \sum_{k=3}^{7} a_k$$

all mean the same thing:

$$a_3 + a_4 + a_5 + a_6 + a_7.$$

You know that

$$(a_0 + \cdots + a_n) + (b_0 + \cdots + b_n) = (a_0 + b_0) + \cdots + (a_n + b_n),$$

$$\alpha(a_0 + \cdots + a_n) = \alpha a_0 + \cdots + \alpha a_n,$$

$$(a_0 + \cdots + a_m) + (a_{m+1} + \cdots + a_n) = a_0 + \cdots + a_n.$$

We can make these statements in $\sum$ notation by writing

$$\sum_{k=0}^{n} a_k + \sum_{k=0}^{n} b_k = \sum_{k=0}^{n} (a_k + b_k),$$

$$\alpha \sum_{k=0}^{n} a_k = \sum_{k=0}^{n} \alpha a_k,$$

$$\sum_{k=0}^{m} a_k + \sum_{k=m+1}^{n} a_k = \sum_{k=0}^{n} a_k.$$

At times we'll find it convenient to change indices. Observe that

$$\sum_{k=j}^{n} a_k = \sum_{i=0}^{n-j} a_{i+j}. \qquad\qquad \text{(Set } i = k - j.\text{)}$$

You can learn to master this notation by doing the Exercises at the end of the section. First, one more remark. If all the a_k are equal to some fixed number r, then

$$\sum_{k=0}^{n} a_k \qquad \text{can be written} \qquad \sum_{k=0}^{n} r.$$

Obviously,

$$\sum_{k=0}^{n} r = \overbrace{r + r + \cdots + r}^{n+1} = (n+1)r.$$

In particular,

$$\sum_{k=0}^{n} 1 = n + 1.$$

EXERCISES 12.1

Exercises 1–10. Evaluate.

1. $\displaystyle\sum_{k=0}^{2} (3k + 1)$.

2. $\displaystyle\sum_{k=1}^{4} (3k - 1)$.

3. $\displaystyle\sum_{k=0}^{3} 2^k$.

4. $\displaystyle\sum_{k=1}^{4} \frac{1}{2^k}$.

5. $\displaystyle\sum_{k=0}^{3} (-1)^k 2^k$.

6. $\displaystyle\sum_{k=0}^{3} (-1)^k 2^{k+1}$.

7. $\displaystyle\sum_{k=2}^{4} \frac{1}{3^{k-1}}$.

8. $\displaystyle\sum_{k=3}^{5} \frac{(-1)^k}{k!}$.

9. $\displaystyle\sum_{k=0}^{3} \left(\frac{1}{2}\right)^{2k}$.

10. $\displaystyle\sum_{k=0}^{3} (-1)^k \left(\frac{1}{2}\right)^{2k}$.

Exercises 11–16. Express in sigma notation.

11. $1 + 3 + 5 + 7 + \cdots + 21$.

12. $1 - 3 + 5 - 7 + \cdots - 19$.

13. $1 \cdot 2 + 2 \cdot 3 + 3 \cdot 4 + \cdots + 35 \cdot 36$.

14. The lower sum $m_1 \Delta x_1 + m_2 \Delta x_2 + \cdots + m_n \Delta x_n$.

15. The upper sum $M_1 \Delta x_1 + M_2 \Delta x_2 + \cdots + M_n \Delta x_n$.

16. The Riemann sum $f(x_1^*)\Delta x_1 + f(x_2^*)\Delta x_2 + \cdots + f(x_n^*)\Delta x_n$.

Exercises 17–20. Write the given sums as $\displaystyle\sum_{k=3}^{10} a_k$ and as $\displaystyle\sum_{i=0}^{7} a_{i+3}$.

17. $\dfrac{1}{2^3} + \dfrac{1}{2^4} + \cdots + \dfrac{1}{2^{10}}$.

18. $\dfrac{3^3}{3!} + \dfrac{4^4}{4!} + \cdots + \dfrac{10^{10}}{10!}$.

19. $\dfrac{3}{4} - \dfrac{4}{5} + \cdots - \dfrac{10}{11}$.

20. $\dfrac{1}{3} + \dfrac{1}{5} + \dfrac{1}{7} + \cdots + \dfrac{1}{17}$.

Exercises 21–24. Transform the first expression into the second by a change of indices.

21. $\displaystyle\sum_{k=2}^{10} \frac{k}{k^2 + 1}$; $\displaystyle\sum_{n=-1}^{7} \frac{n + 3}{n^2 + 6n + 10}$.

22. $\displaystyle\sum_{n=2}^{12} \frac{(-1)^n}{n - 1}$; $\displaystyle\sum_{k=1}^{11} \frac{(-1)^{k+1}}{k}$.

23. $\displaystyle\sum_{k=4}^{25} \frac{1}{k^2 - 9}$; $\displaystyle\sum_{n=7}^{28} \frac{1}{n^2 - 6n}$.

24. $\displaystyle\sum_{k=0}^{15} \frac{3^{2k}}{k!}$; $81 \displaystyle\sum_{n=-2}^{13} \frac{3^{2n}}{(n + 2)!}$.

25. Express the decimal fraction $0. a_1 a_2 \cdots a_n$ in sigma notation using powers of $1/10$.

26. Show that $\displaystyle\sum_{k=1}^{n} \frac{1}{\sqrt{k}} \geq \sqrt{n}$.

Exercises 27–30. Use a graphing utility or CAS to evaluate the sum.

27. $\displaystyle\sum_{k=0}^{50} \frac{1}{4^k}$.

28. $\displaystyle\sum_{k=1}^{50} \frac{1}{k^2}$.

29. $\displaystyle\sum_{k=0}^{50} \frac{1}{k!}$.

30. $\displaystyle\sum_{k=0}^{50} \left(\frac{2}{3}\right)^k$.

■ 12.2 INFINITE SERIES

While it is possible to add two numbers, three numbers, a hundred numbers, or even a million numbers, it is impossible to add an infinite number of numbers. The theory of *infinite series* arose from attempts to circumvent this impossibility.

Introduction; Definition

To form an infinite series, we begin with an infinite sequence of real numbers: $a_0, a_1, a_2 \cdots$. We can't form the sum of all the a_k (there are an infinite number of them), but we can form the *partial sums*:

$$s_0 = a_0 = \sum_{k=0}^{0} a_k,$$

$$s_1 = a_0 + a_1 = \sum_{k=0}^{1} a_k,$$

$$s_2 = a_0 + a_1 + a_2 = \sum_{k=0}^{2} a_k,$$

$$s_3 = a_0 + a_1 + a_2 + a_3 = \sum_{k=0}^{3} a_k,$$

$$s_n = a_0 + a_1 + a_2 + a_3 + \cdots + a_n = \sum_{k=0}^{n} a_k,$$

and so on.

DEFINITION 12.2.1

If, as $n \to \infty$, the sequence of partial sums

$$s_n = \sum_{k=0}^{n} a_k$$

tends to a finite limit L, we write

$$\sum_{k=0}^{\infty} a_k = L$$

and say that

the series $\qquad \displaystyle\sum_{k=0}^{\infty} a_k \qquad$ *converges to L.*

We call L the *sum* of the series. If the sequence of partial sums diverges, we say that

the series $\qquad \displaystyle\sum_{k=0}^{\infty} a_k \qquad$ *diverges.*

Remark   NOTE: *The sum of a series is not a sum in the ordinary sense. It is a limit.* ❑

Here are some examples.

Example 1   We begin with the series

$$\sum_{k=0}^{\infty} \frac{1}{(k+1)(k+2)}.$$

To determine whether this series converges, we examine the partial sums.

Since

$$\frac{1}{(k+1)(k+2)} = \frac{1}{k+1} - \frac{1}{k+2},$$

you can see that

$$s_n = \frac{1}{1\cdot 2} + \frac{1}{2\cdot 3} + \cdots + \frac{1}{n(n+1)} + \frac{1}{(n+1)(n+2)}$$

$$= \left(\frac{1}{1} - \frac{1}{2}\right) + \left(\frac{1}{2} - \frac{1}{3}\right) + \cdots + \left(\frac{1}{n} - \frac{1}{n+1}\right) + \left(\frac{1}{n+1} - \frac{1}{n+2}\right)$$

$$= 1 - \frac{1}{2} + \frac{1}{2} - \frac{1}{3} + \cdots + \frac{1}{n} - \frac{1}{n+1} + \frac{1}{n+1} - \frac{1}{n+2}.$$

Since all but the first and last terms occur in pairs with opposite signs, the sum "telescopes" to give

$$s_n = 1 - \frac{1}{n+2}.$$

As $n \to \infty$, $s_n \to 1$. This indicates that the series converges to 1:

$$\sum_{k=0}^{\infty} \frac{1}{(k+1)(k+2)} = 1. \quad \square$$

Example 2 The series

$$\sum_{k=0}^{\infty} (-1)^k \quad \text{and} \quad \sum_{k=0}^{\infty} 2^k$$

illustrate two forms of divergence: *bounded divergence, unbounded divergence.*

For the first series,

$$s_n = 1 - 1 + 1 - 1 + \cdots + (-1)^n.$$

Here

$$s_n = \begin{cases} 1, & \text{if } n \text{ is even} \\ 0, & \text{if } n \text{ is odd.} \end{cases}$$

The sequence of partial sums reduces to $1, 0, 1, 0, \cdots$. Since the sequence diverges, the series diverges. This is an example of bounded divergence.

For the second series,

$$s_n = \sum_{k=0}^{n} 2^k = 1 + 2 + 2^2 + \cdots + 2^n.$$

Since $s_n > 2^n$, the sum tends to ∞, and the series diverges. This is an example of unbounded divergence. $\square$

The Geometric Series

The *geometric progression*

$$1, x, x^2, x^3, \cdots$$

gives rise to the numbers

$$1, \quad 1+x, \quad 1+x+x^2, \quad 1+x+x^2+x^3, \cdots.$$

These numbers are the partial sums of what is called the *geometric series*:

$$\sum_{k=0}^{\infty} x^k.$$

This series is so important that we will give it special attention.

The following result is fundamental:

(12.2.2)

> (i) If $|x| < 1$, then $\displaystyle\sum_{k=0}^{\infty} x^k = \frac{1}{1-x}$.
>
> (ii) If $|x| \geq 1$, then $\displaystyle\sum_{k=0}^{\infty} x^k$ diverges.

PROOF The nth partial sum of the geometric series

$$\sum_{k=0}^{n} x^k$$

takes the form

(1) $$s_n = 1 + x + x^2 + \cdots + x^n.$$

Multiplication by x gives

$$xs_n = x + x^2 + x^3 + \cdots + x^{n+1}.$$

Subtracting the second equation from the first, we have

$$(1-x)s_n = 1 - x^{n+1}.$$

For $x \neq 1$, this gives

(2) $$s_n = \frac{1 - x^{n+1}}{1 - x}.$$

If $|x| < 1$, then $x^{n+1} \to 0$ as $n \to \infty$ and thus [by (2)]

$$s_n \to \frac{1}{1-x}.$$

This proves (i).

Now we prove (ii). If $x = 1$, then [by (1)] $s_n = n + 1 \to \infty$ and the series diverges. If $x = -1$, then [by (1)] s_n alternates between 1 and 0 and the series diverges. If $|x| > 1$, then x^{n+1} diverges and [by (2)] the sequence of partial sums also diverges. Thus the series diverges. ❏

Setting $x = \frac{1}{2}$ in (12.2.2), we have

$$\sum_{k=0}^{n} \frac{1}{2^k} = \frac{1}{1 - \frac{1}{2}} = 2.$$

By beginning the summation at $k = 1$ instead of at $k = 0$, we drop the term $1/2^0 = 1$ and obtain

(12.2.3)

$$\sum_{k=1}^{\infty} \frac{1}{2^k} = 1.$$

(This result is so useful that it should be committed to memory.) The partial sums of this series

$$s_1 = \tfrac{1}{2}$$

$$s_2 = \tfrac{1}{2} + \tfrac{1}{4} = \tfrac{3}{4}$$

$$s_3 = \tfrac{1}{2} + \tfrac{1}{4} + \tfrac{1}{8} = \tfrac{7}{8}$$

$$s_4 = \tfrac{1}{2} + \tfrac{1}{4} + \tfrac{1}{8} + \tfrac{1}{16} = \tfrac{15}{16}$$

$$s_5 = \tfrac{1}{2} + \tfrac{1}{4} + \tfrac{1}{8} + \tfrac{1}{16} + \tfrac{1}{32} = \tfrac{31}{32}$$

and so on

are illustrated in Figure 12.2.1. After s_1, each new partial sum lies halfway between the previous partial sum and the number 1.

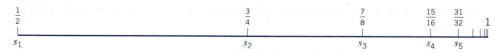

Figure 12.2.1

The convergence of the geometric series at $x = \tfrac{1}{10}$ gives us one more way to assign precise meaning to infinite decimals. (We have briefly dealt with this matter before.) Begin with the fact that

$$\sum_{k=0}^{\infty} \frac{1}{10^k} = \sum_{k=0}^{\infty} \left(\frac{1}{10}\right)^k = \frac{1}{1 - \frac{1}{10}} = \frac{10}{9}.$$

This gives

$$\sum_{k=1}^{\infty} \frac{1}{10^k} = \left(\sum_{k=0}^{\infty} \frac{1}{10^k}\right) - 1 = \frac{1}{9}$$

and shows that the partial sums

$$s_n = \frac{1}{10} + \frac{1}{10^2} + \cdots + \frac{1}{10^n}$$

are all less than $\tfrac{1}{9}$. Now take a series of the form

$$\sum_{k=1}^{\infty} \frac{a_k}{10^k} \qquad \text{with} \qquad a_k = 0, \text{ or } 1, \cdots, \text{ or } 9.$$

Its partial sums

$$t_n = \frac{a_1}{10} + \frac{a_2}{10^2} + \cdots + \frac{a_n}{10^n}$$

are all less than 1:

$$t_n = \frac{a_1}{10} + \frac{a_2}{10^2} + \cdots + \frac{a_n}{10^n} \leq 9\left(\frac{1}{10} + \frac{1}{10^2} + \cdots + \frac{1}{10^n}\right) = 9s_n < 9\left(\frac{1}{9}\right) = 1.$$

Since the sequence of partial sums is nondecreasing and is bounded above, it is convergent. (Theorem 11.2.6.) This tells us that the series

$$\sum_{k=1}^{\infty} \frac{a_k}{10^k}$$

is convergent. The sum of this series is what we mean by the infinite decimal

$$0.a_1 a_2 a_3 \cdots a_n \cdots.$$

Example 3 A ball dropped from height h hits the floor and rebounds to a height proportional to h, that is, to a height σh with $\sigma < 1$. It then falls from height σh, hits the floor and rebounds to the height $\sigma(\sigma h) = \sigma^2 h$, and so on. Find the length of the path of the ball.

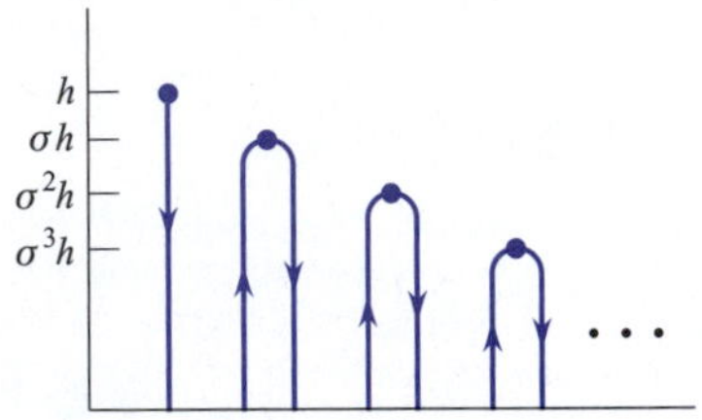

SOLUTION The motion of the ball is illustrated in Figure 12.2.2. The length of the path, call it s, is the sum of the series

$$s = h + 2\sigma h + 2\sigma^2 h + 2\sigma^3 h + \cdots$$

$$= h + 2\sigma h[1 + \sigma + \sigma^2 + \cdots] = h + 2\sigma h \sum_{k=0}^{\infty} \sigma^k.$$

The series in σ is a geometric series which, since $|\sigma| < 1$, converges to $1/(1 - \sigma)$. Therefore

$$s = h + 2\sigma h \frac{1}{1 - \sigma}.$$

If h is 6 feet and $\sigma = \frac{2}{3}$, then

$$s = 6 + 2\left(\tfrac{2}{3}\right)6\frac{1}{1 - \tfrac{2}{3}} = 6 + 24 = 30 \text{ feet.}^\dagger \quad \square$$

Figure 12.2.2

We will return to the geometric series later. Right now we turn our attention to series in general.

Some Basic Results

> **THEOREM 12.2.4**
>
> 1. If $\displaystyle\sum_{k=0}^{\infty} a_k$ converges and $\displaystyle\sum_{k=0}^{\infty} b_k$ converges, then $\displaystyle\sum_{k=0}^{\infty}(a_k + b_k)$ converges.
>
>   Moreover, if $\displaystyle\sum_{k=0}^{\infty} a_k = L$ and $\displaystyle\sum_{k=0}^{\infty} b_k = M$, then $\displaystyle\sum_{k=0}^{\infty}(a_k + b_k) = L + M$.
>
> 2. If $\displaystyle\sum_{k=0}^{\infty} a_k$ converges, then $\displaystyle\sum_{k=0}^{\infty} \alpha a_k$ converges for each real number α.
>
>   Moreover, if $\displaystyle\sum_{k=0}^{\infty} a_k = L$, then $\displaystyle\sum_{k=0}^{\infty} \alpha a_k = \alpha L$.

PROOF Let

$$s_n = \sum_{k=0}^{n} a_k, \qquad t_n = \sum_{k=0}^{n} b_k, \qquad u_n = \sum_{k=0}^{n}(a_k + b_k), \qquad v_n = \sum_{k=0}^{n} \alpha a_k.$$

Note that

$$u_n = s_n + t_n \qquad \text{and} \qquad v_n = \alpha s_n.$$

---

†Of course, this is an idealization of what happens in practice. In the world we live in, balls do not keep bouncing forever. Friction dampens the action.

If $s_n \to L$ and $t_n \to M$, then

$$u_n \to L + M \quad \text{and} \quad v_n \to \alpha L. \quad \square$$

THEOREM 12.2.5

The *kth term* of a convergent series tends to 0; namely,

$$\text{if} \quad \sum_{k=0}^{\infty} a_k \quad \text{converges,} \quad \text{then} \quad a_k \to 0 \quad \text{as} \quad k \to \infty.$$

PROOF Suppose that the series converges. Then the partial sums tend to some number L:

$$s_n = \sum_{k=0}^{n} a_k \to L.$$

Only one step behind, the s_{n-1} also tend to L: $s_{n-1} \to L$. Since $a_n = s_n - s_{n-1}$, we have $a_n \to L - L = 0$. A change in notation gives $a_k \to 0$. $\quad \square$

(12.2.6)

$$\text{if} \quad a_k \not\to 0 \quad \text{then} \quad \sum_{k=0}^{\infty} a_k \quad \text{diverges.}$$

This is a very useful observation.

Example 4

(a) As $k \to \infty$, $\dfrac{k}{k+1} \to 1$. Since $\dfrac{k}{k+1} \not\to 0$, the series

$$\sum_{k=0}^{\infty} \frac{k}{k+1} = 0 + \frac{1}{2} + \frac{2}{3} + \frac{3}{4} + \frac{4}{5} + \cdots \qquad \text{diverges.}$$

(b) Since $\sin k \not\to 0$ as $k \to \infty$, the series

$$\sum_{k=0}^{\infty} \sin k = \sin 0 + \sin 1 + \sin 2 + \sin 3 + \cdots \qquad \text{diverges.} \quad \square$$

CAUTION Theorem 12.2.5 does *not* say that, if $a_k \to 0$, then $\sum_{k=0}^{\infty} a_k$ converges. There are divergent series for which $a_k \to 0$. (One such series appears in Example 5.) $\quad \square$

Example 5 The *kth* term of the series

$$\sum_{k=1}^{\infty} \frac{1}{\sqrt{k}} = \frac{1}{\sqrt{1}} + \frac{1}{\sqrt{2}} + \frac{1}{\sqrt{3}} + \frac{1}{\sqrt{4}} + \cdots$$

tends to zero:

$$a_k = \frac{1}{\sqrt{k}} \to 0 \quad \text{as} \quad k \to \infty.$$

However, the series diverges:

$$s_n = \frac{1}{\sqrt{1}} + \cdots + \frac{1}{\sqrt{n}} \geq \underbrace{\frac{1}{\sqrt{n}} + \cdots + \frac{1}{\sqrt{n}}}_{n \text{ terms}} = \frac{n}{\sqrt{n}} = \sqrt{n} \to \infty. \quad \square$$

Back for a moment to Zeno's paradox. Since the particle moves at constant speed and reaches the halfway point in t seconds, it is expected to reach its destination in $2t$ seconds. The series with partial sums

$$s_n = t + \tfrac{1}{2}t + \tfrac{1}{4}t + \cdots + \tfrac{1}{2^n}t = \left(1 + \tfrac{1}{2} + \tfrac{1}{4} + \cdots + \tfrac{1}{2^n}\right)t$$

does converge to $2t$.

EXERCISES 12.2

Exercises 1–10. Find the sum of the series.

1. $\displaystyle\sum_{k=1}^{\infty} \frac{1}{2k(k+1)}.$

2. $\displaystyle\sum_{k=3}^{\infty} \frac{1}{k^2 - k}.$

3. $\displaystyle\sum_{k=1}^{\infty} \frac{1}{k(k+3)}.$

4. $\displaystyle\sum_{k=0}^{\infty} \frac{1}{(k+1)(k+3)}.$

5. $\displaystyle\sum_{k=0}^{\infty} \frac{3}{10^k}.$

6. $\displaystyle\sum_{k=0}^{\infty} \frac{(-1)^k}{5^k}.$

7. $\displaystyle\sum_{k=0}^{\infty} \frac{1 - 2^k}{3^k}.$

8. $\displaystyle\sum_{k=0}^{\infty} \frac{1}{2^{k+3}}.$

9. $\displaystyle\sum_{k=0}^{\infty} \frac{2^{k+3}}{3^k}.$

10. $\displaystyle\sum_{k=0}^{\infty} \frac{3^{k-1}}{4^{3k+1}}.$

11. Use series to show that every repeating decimal fraction represents a rational number (the quotient of two integers).

12. (a) Let j be a positive integer. Show that

$$\sum_{k=0}^{\infty} a_k \quad \text{converges} \quad \text{iff} \quad \sum_{k=j}^{\infty} a_k \quad \text{converges}.$$

(b) Show that if $\displaystyle\sum_{k=0}^{\infty} a_k = L$, then $\displaystyle\sum_{k=j}^{\infty} a_k = L - \sum_{k=0}^{j-1} a_k.$

(c) Show that if $\displaystyle\sum_{k=j}^{\infty} a_k = M$, then $\displaystyle\sum_{k=0}^{\infty} a_k = M + \sum_{k=0}^{j-1} a_k.$

Exercises 13–14. Derive the indicated result by appealing to the geometric series.

13. $\displaystyle\sum_{k=0}^{\infty} (-1)^k x^k = \frac{1}{1+x}, \quad |x| < 1.$

14. $\displaystyle\sum_{k=0}^{\infty} (-1)^k x^{2k} = \frac{1}{1+x^2}, \quad |x| < 1.$

Exercises 15–18. Find a series expansion for the expression.

15. $\dfrac{x}{1-x} \quad$ for $|x| < 1.$

16. $\dfrac{x}{1+x} \quad$ for $|x| < 1.$

17. $\dfrac{x}{x+x^2} \quad$ for $|x| < 1.$

18. $\dfrac{x}{1+4x^2} \quad$ for $|x| < \dfrac{1}{2}.$

Exercises 19–22. Show that the series diverges.

19. $1 + \dfrac{3}{2} + \dfrac{9}{4} + \dfrac{27}{8} + \dfrac{81}{16} + \cdots.$

20. $\displaystyle\sum_{k=0}^{\infty} \frac{(-5)^k}{4^{k+1}}.$

21. $\displaystyle\sum_{k=1}^{\infty} \left(\frac{k+1}{k}\right)^k.$

22. $\displaystyle\sum_{k=2}^{\infty} \frac{k^{k-2}}{3k}.$

23. Assume that a ball dropped to the floor rebounds to a height proportional to the height from which it was dropped. Find the total length of the path of a ball dropped from a height of 6 feet, given it rebounds initially to a height of 3 feet.

24. In the setting of Exercise 23, to what height does the ball rebound initially if the total length of the path of the ball is 21 feet?

25. How much money must you deposit at $r\%$ interest compounded annually to enable your descendants to withdraw n_1 dollars at the end of the first year, n_2 dollars at the end of the second year, n_3 dollars at the end of the third year, and so on in perpetuity? Assume that the set of n_k is bounded above, $n_k \leq N$ for all k, and express your answer as an infinite series.

26. Sum the series you obtained in Exercise 25, setting

(a) $r = 5, n_k = 5000 \left(\frac{1}{2}\right)^{k-1}.$

(b) $r = 6, n_k = 1000(0.8)^{k+1}.$

(c) $r = 5, n_k = N.$

27. Suppose that only 90% of the outstanding currency is recirculated into the economy: then 90% of that is spent, and so on. Under this hypothesis, what is the long-term economic value of a dollar?

28. Consider the following sequence of steps. First, take the unit interval $[0, 1]$ and delete the open interval $(\frac{1}{3}, \frac{2}{3})$. Next, delete the two open intervals $(\frac{1}{9}, \frac{2}{9})$ and $(\frac{7}{9}, \frac{8}{9})$ from the two intervals that remain after the first step. For the third step, delete the middle thirds from the four intervals that remain after the second step. Continue on in this manner. What is the sum of the lengths of the intervals that have been deleted? The set that remains after all of the "middle thirds" have been deleted is called the *Cantor middle third set*. Give some points that are in the Cantor set.

29. Start with a square that has sides four units long. Join the midpoints of the sides of the square to form a second square inside the first. Then join the midpoints of the sides of the second square to form a third square, and so on. See the figure. Find the sum of the areas of the squares.

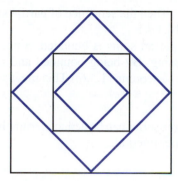

30. (a) Show that if the series $\sum a_k$ converges and the series $\sum b_k$ diverges, then the series $\sum (a_k + b_k)$ diverges.

(b) Give examples to show that if $\sum a_k$ and $\sum b_k$ both diverge, then each of the series

$$\sum (a_k + b_k) \qquad \text{and} \qquad \sum (a_k - b_k)$$

may converge or may diverge.

31. Let $\sum_{k=0}^{\infty} a_k$ be a convergent series and let $R_n = \sum_{k=n+1}^{\infty} a_k$. Prove that $R_n \to 0$ as $n \to \infty$. Note that if s_n is the nth partial sum of the series, then $\sum_{k=0}^{\infty} a_k = s_n + R_n$; R_n is called the *remainder*.

32. (a) Prove that if $\sum_{k=0}^{\infty} a_k$ is a convergent series with all terms nonzero, then $\sum_{k=0}^{\infty} (1/a_k)$ diverges.

(b) Suppose that $a_k > 0$ for all k and $\sum_{k=0}^{\infty} a_k$ diverges. Show by example that $\sum_{k=0}^{\infty} (1/a_k)$ may converge and it may diverge.

33. Show that

$$\sum_{k=1}^{\infty} \ln \left(\frac{k+1}{k} \right) \qquad \text{diverges}$$

although

$$\ln \left(\frac{k+1}{k} \right) \to 0.$$

34. Show that

$$\sum_{k=1}^{\infty} \left(\frac{k+1}{k} \right)^k \qquad \text{diverges.}$$

35. (a) Assume that $d_k \to 0$ and show that

$$\sum_{k=1}^{\infty} (d_k - d_{k+1}) = d_1.$$

(b) Sum the following series:

(i) $\sum_{k=1}^{\infty} \frac{\sqrt{k+1} - \sqrt{k}}{\sqrt{k(k+1)}}.$ (ii) $\sum_{k=1}^{\infty} \frac{2k+1}{2k^2(k+1)^2}.$

36. Show that

$$\sum_{k=1}^{\infty} kx^{k-1} = \frac{1}{(1-x)^2} \qquad \text{for} \qquad |x| < 1.$$

HINT: Verify that s_n, the nth partial sum of the series, satisfies the identity

$$(1-x)^2 s_n = 1 - (n+1)x^n + nx^{n+1}.$$

▷ **(Exercises 37–40.** *Speed of convergence*) Find the least integer N for which the nth partial sum of the series differs from the sum of the series by less than 0.0001.

37. $\sum_{k=0}^{\infty} \frac{1}{4^k}.$ **38.** $\sum_{k=0}^{\infty} (0.9)^k.$

39. $\sum_{k=1}^{\infty} \frac{1}{k(k+2)}.$ **40.** $\sum_{k=0}^{\infty} \left(\frac{2}{3} \right)^k.$

41. Start with the geometric series $\sum_{k=0}^{\infty} x^k$ with $|x| < 1$ and a positive number ϵ. Determine the least positive integer N for which $|L - s_N| < \epsilon$ given that the sum of the series is L and s_N is the Nth partial sum.

42. Prove that the series $\sum_{k=1}^{\infty} (a_{k+1} - a_k)$ converges iff the a_n tend to a finite limit.

■ 12.3 THE INTEGRAL TEST; BASIC COMPARISON, LIMIT COMPARISON

Here we begin our study of *series with nonnegative terms*: $a_k \geq 0$ for all k. For such series the sequence of partial sums is nondecreasing:

$$s_{n+1} = \sum_{k=0}^{n+1} a_k = a_{n+1} + \sum_{k=0}^{n} a_k \geq \sum_{k=0}^{n} a_k = s_n.$$

The following simple theorem is fundamental.

> **THEOREM 12.3.1**
>
> A series with nonnegative terms converges iff the sequence of partial sums is bounded.

PROOF Assume that the series converges. Then the sequence of partial sums is convergent and therefore bounded. (Theorem 11.3.4.)

Suppose now that the sequence of partial sums is bounded. Since the terms are nonnegative, the sequence is nondecreasing. By being bounded and nondecreasing, the sequence of partial sums converges. (Theorem 11.3.6.) This means that the series converges. ❏

The convergence or divergence of some series can be deduced from the convergence or divergence of a closely related improper integral.

> **THEOREM 12.3.2 THE INTEGRAL TEST**
>
> If f is continuous, positive, and decreasing on $[1, \infty)$, then
>
> $$\sum_{k=1}^{\infty} f(k) \quad \text{converges} \quad \text{iff} \quad \int_1^{\infty} f(x)\,dx \quad \text{converges.}$$

PROOF In Exercise 69, Section 11.7, you were asked to show that if f is continuous, positive, and decreasing on $[1, \infty)$, then

$$\int_1^{\infty} f(x)\,dx \quad \text{converges} \quad \text{iff} \quad \text{the sequence} \quad a_n = \int_1^{n} f(x)\,dx \text{ converges.}$$

We assume this result and base our proof on the behavior of the sequence of integrals. To visualize our argument, see Figure 12.3.1.

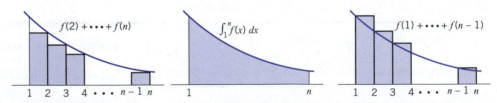

Figure 12.3.1

Let's suppose that f is continuous, positive, and decreasing on $[1, \infty)$. Since f decreases on the interval $[1, n]$,

$$f(2) + \cdots + f(n) \qquad \text{is a lower sum for } f \text{ on } [1, n],$$

and

$$f(1) + \cdots + f(n - 1) \qquad \text{is an upper sum for } f \text{ on } [1, n].$$

Consequently,

$$f(2) + \cdots + f(n) \leq \int_1^{n} f(x)\,dx \leq f(1) + \cdots + f(n - 1).$$

If the sequence of integrals converges, it is bounded. Then, by the left inequality, the sequence of partial sums is bounded and the series converges.

Suppose now that the sequence of integrals diverges. Since f is positive, the sequence of integrals increases:

$$\int_1^n f(x)\,dx < \int_1^{n+1} f(x)\,dx.$$

Since this sequence diverges, it must be unbounded. Then, by the right inequality, the sequence of partial sums is unbounded and the series diverges. ❏

Applying the Integral Test

Example 1 (*The harmonic series*)

(12.3.3)

$$\sum_{k=1}^{\infty} \frac{1}{k} = 1 + \frac{1}{2} + \frac{1}{3} + \frac{1}{4} + \cdots \qquad \text{diverges.}$$

PROOF The function $f(x) = 1/x$ is continuous, positive, and decreasing on $[1, \infty)$. We know that

$$\int_1^{\infty} \frac{dx}{x} \qquad \text{diverges.} \qquad (11.7.1)$$

By the integral test,

$$\sum_{k=1}^{\infty} \frac{1}{k} \qquad \text{diverges.} \quad ❏$$

The next example generalizes on this.

Example 2 (*The p-series*)

(12.3.4)

$$\sum_{k=1}^{\infty} \frac{1}{k^p} = 1 + \frac{1}{2^p} + \frac{1}{3^p} + \frac{1}{4^p} + \cdots \qquad \text{converges} \qquad \text{iff} \quad p > 1.$$

PROOF If $p \le 0$, then the terms of the series are all greater than or equal to 1. Therefore, the terms do not tend to zero and the series cannot converge. We assume therefore that $p > 0$. The function $f(x) = 1/x^p$ is then continuous, positive, and decreasing on $[1, \infty)$. Thus, by the integral test,

$$\sum_{k=1}^{\infty} \frac{1}{k^p} \qquad \text{converges} \qquad \text{iff} \qquad \int_1^{\infty} \frac{dx}{x^p} \qquad \text{converges.}$$

Earlier you saw that

$$\int_1^{\infty} \frac{dx}{x^p} \qquad \text{converges} \qquad \text{iff} \qquad p > 1. \qquad (11.7.1)$$

It follows that

$$\sum_{k=1}^{\infty} \frac{1}{k^p} \quad \text{converges} \quad \text{iff} \quad p > 1. \quad \square$$

Example 3 Show that the series

$$\sum_{k=1}^{\infty} \frac{1}{k \ln(k+1)} = \frac{1}{\ln 2} + \frac{1}{2 \ln 3} + \frac{1}{3 \ln 4} + \cdots \quad \text{diverges.}$$

SOLUTION We begin by setting $f(x) = \dfrac{1}{x \ln(x+1)}$. Since f is continuous, positive, and decreasing on $[1, \infty)$, we can use the integral test. Note first that for all $x \in [1, \infty)$

$$\frac{1}{x \ln(x+1)} > \frac{1}{(x+1) \ln(x+1)}.$$

Therefore

$$\int_1^b \frac{1}{x \ln(x+1)} \, dx > \int_1^b \frac{1}{(x+1)\ln(x+1)} \, dx = \Big[\ln[\ln(x+1)] \Big]_1^b$$

$$= \ln[\ln(b+1)] - \ln[\ln 2].$$

As $b \to \infty$, $\ln[\ln(b+1)] \to \infty$. This shows that

$$\int_1^{\infty} \frac{1}{x \ln(x+1)} \, dx$$

diverges. Therefore the series diverges. $\quad \square$

Remark on Notation You have seen that for each $j \geq 0$

$$\sum_{k=0}^{\infty} a_k \quad \text{converges} \quad \text{iff} \quad \sum_{k=j}^{\infty} a_k \quad \text{converges.}$$

(Exercise 12, Section 12.2.) This tells us that in deciding whether a series converges, it does not matter where we begin the summation. Where detailed indexing would contribute nothing, we will omit it and write $\sum a_k$ without specifying where the summation begins.

For instance, it makes sense to say that

$$\sum \frac{1}{k^2} \quad \text{converges} \quad \text{and} \quad \sum \frac{1}{k} \quad \text{diverges}$$

without specifying where the summation begins. $\quad \square$

In the absence of detailed indexing, we cannot know definite limits, but we can be sure of the following:

$$\boxed{\begin{array}{l} \text{(1) if } \sum a_k \text{ and } \sum b_k \text{ converge,} \qquad \text{then} \quad \sum a_k + b_k \qquad \text{converges;} \\ \text{(2) for each } \alpha \neq 0 \\ \qquad\qquad \sum \alpha a_k \quad \text{converges} \quad \text{iff} \quad \sum a_k \quad \text{converges} \\ \text{and, equivalently,} \\ \qquad\qquad \sum \alpha a_k \quad \text{diverges} \quad \text{iff} \quad \sum a_k \quad \text{diverges.} \end{array}}$$

(12.3.5)

The convergence or divergence of a series with nonnegative terms can often be determined by comparison with a series of known behavior.

THEOREM 12.3.6 THE BASIC COMPARISON THEOREM

Suppose that $\sum a_k$ and $\sum b_k$ are series with nonnegative terms and

$$\sum a_k \leq \sum b_k \qquad \text{for all } k \text{ sufficiently large.}$$

 (i) If $\sum b_k$ converges, then $\sum a_k$ converges

 (ii) If $\sum a_k$ diverges, then $\sum b_k$ diverges.

PROOF The proof is just a matter of noting that, in the first case, the partial sums of $\sum a_k$ form a bounded increasing sequence and, in the second case, they form an unbounded increasing sequence. ❑

Applying the Basic Comparison Theorem

Example 4

(a) $\sum \dfrac{1}{2k^3 + 1}$ converges by comparison with $\sum \dfrac{1}{k^3}$:

$$\frac{1}{2k^3 + 1} < \frac{1}{k^3} \qquad \text{and} \qquad \sum \frac{1}{k^3} \qquad \text{converges.}$$

(b) $\sum \dfrac{k^3}{k^5 + 5k^4 + 7}$ converges by comparison with $\sum \dfrac{1}{k^2}$:

$$\frac{k^3}{k^5 + 5k^4 + 7} < \frac{k^3}{k^5} = \frac{1}{k^2} \qquad \text{and} \qquad \sum \frac{1}{k^2} \qquad \text{converges.} ❑$$

Example 5 Show that $\sum \dfrac{1}{3k + 1}$ diverges.

SOLUTION Since

$$\frac{1}{4k} = \frac{1}{3k + k} < \frac{1}{3k + 1},$$

all we have to do is show that

$$\sum \frac{1}{4k} \qquad \text{diverges.}$$

This follows by (12.3.5) from the divergence of

$$\sum \frac{1}{k}. ❑$$

Example 6 Show that $\sum \dfrac{1}{\ln(k + 6)}$ diverges.

SOLUTION Since the graph of $y = \ln x$ stays below the line $y = x$, we know that, for all real x, $\ln x < x$. It follows that, for all k sufficiently large,

$$\ln(k + 6) < \ln(2k) < 2k$$

and

$$\frac{1}{2k} < \frac{1}{\ln(k+6)}.$$

Since $\sum 1/k$ diverges, $\sum 1/2k$ diverges. This tells us that the original series diverges. ❏

When direct comparison is difficult to apply, the following test can be useful.

THEOREM 12.3.7 THE LIMIT COMPARISON THEOREM

Let $\sum a_k$ and $\sum b_k$ be series with *positive terms*. If $a_k/b_k \to L$, and L is *positive*, then

$$\sum a_k \qquad \text{converges} \qquad \text{iff} \qquad \sum b_k \qquad \text{converges.}$$

PROOF Choose ϵ between 0 and L. Since $a_k/b_k \to L$, we know that for all k sufficiently large (for all k greater than some k_0)

$$\left| \frac{a_k}{b_k} - L \right| < \epsilon.$$

For such k we have

$$L - \epsilon < \frac{a_k}{b_k} < L + \epsilon,$$

and thus

$$(L - \epsilon)b_k < a_k < (L + \epsilon)b_k.$$

This last inequality is what we need:

if $\sum a_k$ converges, then $\sum (L - \epsilon)b_k$ converges, and thus $\sum b_k$ converges;

if $\sum b_k$ converges, then $\sum (L + \epsilon)b_k$ converges, and thus $\sum a_k$ converges. ❏

Applying the Limit Comparison Theorem

To apply the limit comparison theorem to a series $\sum a_k$, we must find a series $\sum b_k$ of known behavior for which a_k/b_k converges to a positive number.

Example 7 Determine whether the series

$$\sum_{k=1}^{\infty} \frac{1}{5^k - 3}$$

converges or diverges.

SOLUTION For large k

$$\frac{1}{5^k - 3} \qquad \text{differs little from} \qquad \frac{1}{5^k}.$$

As $k \to \infty$,

$$\frac{1}{5^k - 3} \div \frac{1}{5^k} = \frac{5^k}{5^k - 3} = \frac{1}{1 - 3/5^k} \to 1.$$

Since

$$\sum \frac{1}{5^k} \qquad \text{converges,}$$

(it is a convergent geometric series), the original series converges. ❑

Example 8 Determine whether the series

$$\sum \frac{3k^2 + 2k + 1}{k^3 + 1}$$

converges or diverges.

SOLUTION For large k the terms with the highest powers of k dominate. Here $3k^2$ dominates the numerator and k^3 dominates the denominator. Thus, for large k,

$$\frac{3k^2 + 2k + 1}{k^3 + 1} \qquad \text{differs little from} \qquad \frac{3k^2}{k^3} = \frac{3}{k}.$$

As $k \to \infty$,

$$\frac{3k^2 + 2k + 1}{k^3 + 1} \div \frac{3}{k} = \frac{3k^3 + 2k^2 + k}{3k^3 + 3} = \frac{1 + 2/(3k) + 1/(3k^2)}{1 + 1/k^3} \to 1.$$

Since

$$\sum \frac{1}{k} \qquad \text{diverges,} \qquad \sum \frac{3}{k} \qquad \text{diverges.}$$

It follows that the original series diverges. ❑

Example 9 Determine whether the series

$$\sum \frac{2k + 5}{\sqrt{k^6 + 3k^3}}$$

converges or diverges.

SOLUTION For large k, $2k$ dominates the numerator and $\sqrt{k^6} = k^3$ dominates the denominator. Thus, for such k,

$$\frac{2k + 5}{\sqrt{k^6 + 3k^3}} \qquad \text{differs little from} \qquad \frac{2k}{\sqrt{k^6}} = \frac{2k}{k^3} = \frac{2}{k^2}.$$

As $k \to \infty$,

$$\frac{2k + 5}{\sqrt{k^6 + 3k^3}} \div \frac{2}{k^2} = \frac{2k^3 + 5k^2}{2\sqrt{k^6 + 3k^3}} = \frac{2k^3 + 5k^2}{2k^3\sqrt{1 + 3/k^3}} = \frac{1 + 5/2k}{\sqrt{1 + 3/k^3}} \to 1.$$

Since

$$\sum \frac{1}{k^2} \qquad \text{converges,} \qquad \sum \frac{2}{k^2} \qquad \text{converges.}$$

It follows that the original series converges. ❑

Remark The question of what we can and cannot conclude by limit comparison if $a_k/b_k \to 0$ or if $a_k/b_k \to \infty$ is taken up in Exercises 47 and 48. ❑

EXERCISES 12.3

Exercises 1–36. Determine whether the series converges or diverges.

1. $\sum \dfrac{k}{k^3 + 1}$.

2. $\sum \dfrac{1}{3k + 2}$.

3. $\sum \dfrac{1}{(2k + 1)^2}$.

4. $\sum \dfrac{\ln k}{k}$.

5. $\sum \dfrac{1}{\sqrt{k + 1}}$.

6. $\sum \dfrac{1}{k^2 + 1}$.

7. $\sum \dfrac{1}{\sqrt{2k^2 - k}}$.

8. $\sum \left(\dfrac{2}{5}\right)^k$.

9. $\sum \dfrac{\arctan k}{1 + k^2}$.

10. $\sum \dfrac{\ln k}{k^3}$.

11. $\sum \dfrac{1}{k^{2/3}}$.

12. $\sum \dfrac{1}{k(k + 1)(k + 2)}$.

13. $\sum \left(\dfrac{4}{3}\right)^k$.

14. $\sum \dfrac{1}{1 + 2\ln k}$.

15. $\sum \dfrac{\ln \sqrt{k}}{k}$.

16. $\sum \dfrac{2}{k(\ln k)^2}$.

17. $\sum \dfrac{1}{2 + 3^{-k}}$.

18. $\sum \dfrac{7k + 2}{2k^5 + 7}$.

19. $\sum \dfrac{2k + 5}{5k^3 + 3k^2}$.

20. $\sum \dfrac{k^4 - 1}{3k^2 + 5}$.

21. $\sum \dfrac{1}{k \ln k}$.

22. $\sum \dfrac{1}{2^{k+1} - 1}$.

23. $\sum \dfrac{1 + 2^k}{1 + 5^k}$.

24. $\sum \dfrac{k^{3/2}}{k^{5/2} + 2k - 1}$.

25. $\sum \dfrac{2k + 1}{\sqrt{k^4 + 1}}$.

26. $\sum \dfrac{2k + 1}{\sqrt{k^3 + 1}}$.

27. $\sum \dfrac{2k + 1}{\sqrt{k^5 + 1}}$.

28. $\sum \dfrac{1}{\sqrt{2k(k + 1)}}$.

29. $\sum ke^{-k^2}$.

30. $\sum k^2 2^{-k^3}$.

31. $\sum \dfrac{2 + \sin k}{k^2}$.

32. $\sum \dfrac{2 + \cos k}{\sqrt{k + 1}}$.

33. $\sum \dfrac{1}{1 + 2 + \cdots + k}$.

34. $\sum \dfrac{k}{1 + 2^2 + \cdots + k^2}$.

35. $\sum \dfrac{2k}{(2k)!}$.

36. $\sum \dfrac{2k!}{(2k)!}$.

37. Find the values of p for which $\displaystyle\sum_{k=2}^{\infty} \dfrac{1}{k(\ln k)^p}$ converges.

38. Find the values of p for which $\displaystyle\sum_{k=2}^{\infty} \dfrac{\ln k}{k^p}$ converges.

39. (a) Show that $\displaystyle\sum_{k=0}^{\infty} e^{-\alpha k}$ converges for each $\alpha > 0$.

(b) Show that $\displaystyle\sum_{k=0}^{\infty} ke^{-\alpha k}$ converges for each $\alpha > 0$.

(c) Show that, more generally, $\displaystyle\sum_{k=0}^{\infty} k^n e^{-\alpha k}$ converges for each nonnegative integer n and each $\alpha > 0$.

40. Let $p > 1$. Use the integral test to show that

$$\dfrac{1}{(p - 1)(n + 1)^{p-1}} < \sum_{k=1}^{\infty} \dfrac{1}{k^p} - \sum_{k=1}^{n} \dfrac{1}{k^p} < \dfrac{1}{(p - 1)n^{p-1}}.$$

This result gives bounds on the *error* (the remainder) R_n that results from using s_n to approximate the sum of the p-series.

▷**Exercises 41–42.** (a) Use a CAS or graphing utility. Calculate the sum of the first 100 terms of the series. (b) Use the inequalities given in Exercise 40 to obtain upper and lower bounds for R_{100}. (c) Use parts (a) and (b) to estimate the sum of the series.

41. $\displaystyle\sum_{k=1}^{\infty} \dfrac{1}{k^3}$.

42. $\displaystyle\sum_{k=1}^{\infty} \dfrac{1}{k^4}$.

For Exercises 43–46, use the error bounds given in Exercise 40.

▷**43.** (a) If you were to use s_{100} to approximate $\displaystyle\sum_{k=1}^{\infty} \dfrac{1}{k^2}$, what would be the bounds on your error?

(b) How large would you have to choose n to ensure that R_n is less than 0.0001?

▷**44.** (a) If you were to use s_{100} to approximate $\displaystyle\sum_{k=1}^{\infty} \dfrac{1}{k^3}$, what would be the bounds on your error?

(b) How large would you have to choose n to ensure that R_n is less than 0.0001?

(c) Use the result of part (b) to estimate $\displaystyle\sum_{k=1}^{\infty} \dfrac{1}{k^3}$.

▷**45.** (a) How many terms of the series $\displaystyle\sum_{k=1}^{\infty} \dfrac{1}{k^4}$ must you use to ensure that R_n is less than 0.0001?

(b) How large do you have to choose n to ensure that R_n is less than 0.001?

(c) Use the result of part (b) to estimate $\displaystyle\sum_{k=1}^{\infty} \dfrac{1}{k^4}$.

▷**46.** Exercise 45 for the series $\displaystyle\sum_{k=1}^{\infty} \dfrac{1}{k^5}$.

Exercises 47 and 48 complete the limit comparison test.

47. Let $\sum a_k$ and $\sum b_k$ be series with positive terms. Suppose that $a_k/b_k \to 0$.

(a) Show that if $\sum b_k$ converges, then $\sum a_k$ converges.

(b) Show that if $\sum a_k$ diverges, then $\sum b_k$ diverges.

(c) Show by example that if $\sum a_k$ converges, then $\sum b_k$ may converge or diverge.

(d) Show by example that if $\sum b_k$ diverges, then $\sum a_k$ may converge or diverge.

[Parts (c) and (d) explain why we stipulated $L > 0$ in Theorem 12.3.7.]

48. Let $\sum a_k$ and $\sum b_k$ be series with positive terms. Suppose that $a_k/b_k \to \infty$.

(a) Show that if $\sum b_k$ diverges, then $\sum a_k$ diverges.

(b) Show that if $\sum a_k$ converges, then $\sum b_k$ converges.

(c) Show by example that if $\sum a_k$ diverges, then $\sum b_k$ may converge or diverge.

(d) Show by example that if $\sum b_k$ converges, then $\sum a_k$ may converge or may diverge.

49. Let $\sum a_k$ be a series with nonnegative terms.

(a) Show that if $\sum a_k$ converges, then $\sum a_k^2$ converges.

(b) Give an example where $\sum a_k^2$ converges and $\sum a_k$ converges; give an example where $\sum a_k^2$ converges but $\sum a_k$ diverges.

50. Let $\sum a_k$ be a series with nonnegative terms. Show that if $\sum a_k^2$ converges, then $\sum (a_k/k)$ converges.

51. Let f be a continuous, positive, decreasing function on $[1, \infty)$ for which $\int_1^\infty f(x)\,dx$ converges. Then we know that the series $\sum_{k=1}^\infty f(k)$ also converges. Show that

$$0 < L - s_n < \int_n^\infty f(x)\,dx$$

where L is the sum of the series and s_n is the nth partial sum.

▶**Exercises 52–53.** Use a CAS and the result obtained in Exercise 51 to find the least integer n for which the difference between the sum of the series and the nth partial sum is less than 0.001. Find s_n.

52. $\displaystyle\sum_{k=1}^\infty \frac{1}{k^2 + 1}.$

53. $\displaystyle\sum_{k=1}^\infty k\,e^{-k^2}.$

54. Let s_n be the nth partial sum of the harmonic series, a series which you know diverges.

(a) Show that

$$\ln(n + 1) < s_n < 1 + \ln n.$$

(b) Find the least integer n for which $s_n > 100$.

55. Show that

$$\sum \frac{\ln k}{k\sqrt{k}} \quad \text{converges by comparison with} \quad \sum \frac{1}{k^{5/4}}.$$

56. Let p and q be polynomials with nonnegative coefficients. Give necessary and sufficient conditions on p and q for the convergence of

$$\sum \frac{p(k)}{q(k)}.$$

■ 12.4 THE ROOT TEST; THE RATIO TEST

We continue our study of series with nonnegative terms. Comparison with the geometric series

$$\sum x^k$$

leads to two important tests for convergence: the root test and the ratio test.

THEOREM 12.4.1 THE ROOT TEST

Let $\sum a_k$ be a series with nonnegative terms, and suppose that

$$(a_k)^{1/k} \to \rho.$$

(i) If $\rho < 1$, then $\sum a_k$ converges.

(ii) If $\rho > 1$, then $\sum a_k$ diverges.

(iii) If $\rho = 1$, then the test is inconclusive. The series may converge; it may diverge.

PROOF We suppose first that $\rho < 1$ and choose μ, so that

$$\rho < \mu < 1.$$

Since $(a_k)^{1/k} \to \rho$,

$$(a_k)^{1/k} < \mu \qquad \text{for all} \quad k \quad \text{sufficiently large.}$$

Thus

$$a_k < \mu^k \qquad \text{for all} \quad k \quad \text{sufficiently large.}$$

Since $\sum \mu^k$ converges (a geometric series with $0 < \mu < 1$), we know by the basic comparison theorem that $\sum a_k$ converges.

We suppose now that $\rho > 1$ and choose μ so that

$$\rho > \mu > 1.$$

Since $(a_k)^{1/k} \to \rho$,

$$(a_k)^{1/k} > \mu \qquad \text{for all} \quad k \quad \text{sufficiently large.}$$

Thus

$$a_k > \mu^k \qquad \text{for all} \quad k \quad \text{sufficiently large.}$$

Since $\sum \mu^k$ diverges (a geometric series with $\mu > 1$), we know by the basic comparison theorem that $\sum a_k$ diverges.

To see the inconclusiveness of the root test when $\rho = 1$, consider the series $\sum (1/k)$ and $\sum (1/k^2)$. The first series diverges while the second series converges, but in each case $(a_k)^{1/k} \to 1$:

$$(a_k)^{1/k} = \left(\frac{1}{k}\right)^{1/k} = \frac{1}{k^{1/k}} \to 1,$$

$$(a_k)^{1/k} = \left(\frac{1}{k^2}\right)^{1/k} = \left(\frac{1}{k^{1/k}}\right)^2 \to 1^2 = 1.$$

[Earlier we showed that, as $k \to \infty$, $k^{1/k} \to 1$. (11.4.1)] ❑

Applying the Root Test

Example 1 For the series $\sum \dfrac{1}{(\ln k)^k}$

$$(a_k)^{1/k} = \frac{1}{\ln k} \to 0.$$

The series converges. ❑

Example 2 For the series $\sum \dfrac{2^k}{k^3}$

$$(a_k)^{1/k} = 2 \left(\frac{1}{k}\right)^{3/k} = 2 \left[\left(\frac{1}{k}\right)^{1/k}\right]^3 = 2 \left[\frac{1}{k^{1/k}}\right]^3 \to 2 \cdot 1^3 = 2.$$

The series diverges. ❑

Example 3 In the case of $\sum \left(1 - \dfrac{1}{k}\right)^k$

$$(a_k)^{1/k} = 1 - \frac{1}{k} \to 1.$$

Here the root test is inconclusive. It is also unnecessary: since $a_k = (1 - 1/k)^k$ tends to $1/e$ and not to 0, the series diverges. ❑

> **THEOREM 12.4.2   THE RATIO TEST**
>
> Let $\sum a_k$ be a series with positive terms and suppose that
> $$\frac{a_{k+1}}{a_k} \to \lambda.$$
>
> **(i)** If $\lambda < 1$, then $\sum a_k$ converges.
>
> **(ii)** If $\lambda > 1$, then $\sum a_k$ diverges.
>
> **(iii)** If $\lambda = 1$, then the test is inconclusive. The series may converge; it may diverge.

PROOF   We suppose first that $\lambda < 1$ and choose μ so that $\lambda < \mu < 1$. Since
$$\frac{a_{k+1}}{a_k} \to \lambda,$$
we know that there exists $k_0 > 0$ such that

$$\text{if} \quad k \geq k_0, \quad \text{then} \quad \frac{a_{k+1}}{a_k} < \mu. \qquad \text{(explain)}$$

This gives
$$a_{k_0+1} < \mu a_{k_0}, \qquad a_{k_0+2} < \mu a_{k_0+1} < \mu^2 a_{k_0},$$
and more generally,
$$a_{k_0+j} < \mu^j a_{k_0}, \quad j = 1, 2, \cdots.$$

For $k > k_0$ we have

(1)
$$a_k < \mu^{k-k_0} a_{k_0} = \frac{a_{k_0}}{\mu^{k_0}} \mu^k.$$
$$\uparrow \underline{\qquad} \text{Set } j = k - k_0.$$

Since $\mu < 1$,
$$\sum \frac{a_{k_0}}{\mu^{k_0}} \mu^k = \frac{a_{k_0}}{\mu^{k_0}} \sum \mu^k \qquad \text{converges.}$$

It follows from (1) and the basic comparison theorem that $\sum a_k$ converges. The proof of the rest of the theorem is left to the Exercises. ❏

Remark   Contrary to some people's intuition, the root and ratio tests are *not* equivalent. See Exercise 48. ❏

Applying the Ratio Test

Example 4   The ratio test shows that the series $\sum \dfrac{1}{k!}$ converges:
$$\frac{a_{k+1}}{a_k} = \frac{1}{(k+1)!} \cdot \frac{k!}{1} = \frac{1}{k+1} \to 0. \quad ❏$$

Example 5   For the series $\sum \dfrac{k}{10^k}$
$$\frac{a_{k+1}}{a_k} = \frac{k+1}{10^{k+1}} \cdot \frac{10^k}{k} = \frac{1}{10} \frac{k+1}{k} \to \frac{1}{10}.$$
The series converges.[†] ❏

[†]You are asked to find the sum of this series in Exercise 41.

Example 6 For the series $\sum \dfrac{k^k}{k!}$

$$\frac{a_{k+1}}{a_k} = \frac{(k+1)^{k+1}}{(k+1)!} \cdot \frac{k!}{k^k} = \left(\frac{k+1}{k}\right)^k = \left(1 + \frac{1}{k}\right)^k \to e.$$

Since $e > 1$, the series diverges. ❏

Example 7 For the series $\sum \dfrac{1}{2k+1}$, the ratio test is inconclusive:

$$\frac{a_{k+1}}{a_k} = \frac{1}{2(k+1)+1} \cdot \frac{2k+1}{1} = \frac{2k+1}{2k+3} = \frac{2+1/k}{2+3/k} \to 1.$$

Therefore, we have to look further. Comparison with the harmonic series shows that the series diverges:

$$\frac{1}{2k+1} > \frac{1}{3k} \quad \text{and} \quad \sum \frac{1}{3k} = \frac{1}{3}\sum \frac{1}{k} \quad \text{diverges.} \quad ❏$$

Summary on Convergence Tests

In general, the root test is used only if powers are involved. The ratio test is particularly effective with factorials and with combinations of powers and factorials. If the terms are rational functions of k, the ratio test is inconclusive and the root test is difficult to apply. Series with rational terms are most easily handled by limit comparison with a p-series, a series of the form $\sum 1/k^p$. If the terms have the configuration of a derivative, you may be able to apply the integral test. Finally, keep in mind that, if $a_k \not\to 0$, then there is no reason to try any convergence test; the series diverges. [(12.2.6).]

EXERCISES 12.4

Exercises 1–40. Determine whether the series converges or diverges.

1. $\sum \dfrac{10^k}{k!}$.

2. $\sum \dfrac{1}{k2^k}$.

3. $\sum \dfrac{1}{k^k}$.

4. $\sum \left(\dfrac{k}{2k+1}\right)^k$.

5. $\sum \dfrac{k!}{100^k}$.

6. $\sum \dfrac{(\ln k)^2}{k}$.

7. $\sum \dfrac{k^2+2}{k^3+6k}$.

8. $\sum \dfrac{1}{(\ln k)^k}$.

9. $\sum k\left(\dfrac{2}{3}\right)^k$.

10. $\sum \dfrac{1}{(\ln k)^{10}}$.

11. $\sum \dfrac{1}{1+\sqrt{k}}$.

12. $\sum \dfrac{2k+\sqrt{k}}{k^3+\sqrt{k}}$.

13. $\sum \dfrac{k!}{10^{4k}}$.

14. $\sum \dfrac{k^2}{e^k}$.

15. $\sum \dfrac{\sqrt{k}}{k^2+1}$.

16. $\sum \dfrac{2^k k!}{k^k}$.

17. $\sum \dfrac{k!}{(k+2)!}$.

18. $\sum \dfrac{1}{k}\left(\dfrac{1}{\ln k}\right)^{3/2}$.

19. $\sum \dfrac{1}{k}\left(\dfrac{1}{\ln k}\right)^{1/2}$.

20. $\sum \dfrac{1}{\sqrt{k^3-1}}$.

21. $\sum \left(\dfrac{k}{k+100}\right)^k$.

22. $\sum \dfrac{(k!)^2}{(2k)!}$.

23. $\sum k^{-(1+1/k)}$.

24. $\sum \dfrac{11}{1+100^{-k}}$.

25. $\sum \dfrac{\ln k}{e^k}$.

26. $\sum \dfrac{k!}{k^k}$.

27. $\sum \dfrac{\ln k}{k^2}$.

28. $\sum \dfrac{k!}{1\cdot 3\cdots(2k-1)}$.

29. $\sum \dfrac{2\cdot 4\cdots 2k}{(2k)!}$.

30. $\sum \dfrac{(2k+1)^{2k}}{(5k^2+1)^k}$.

31. $\sum \dfrac{k!(2k)!}{(3k)!}$.

32. $\sum \dfrac{\ln k}{k^{5/4}}$.

33. $\sum \dfrac{k^{k/2}}{k!}$.

34. $\sum \dfrac{k^k}{(3^k)^2}$.

35. $\sum \dfrac{k^k}{3^{k^2}}$.

36. $\sum \left(\sqrt{k}-\sqrt{k-1}\right)^k$.

37. $\dfrac{1}{2} + \dfrac{2}{3^2} + \dfrac{4}{4^3} + \dfrac{8}{5^4} + \cdots$.

38. $1 + \dfrac{1 \cdot 2}{1 \cdot 3} + \dfrac{1 \cdot 2 \cdot 3}{1 \cdot 3 \cdot 5} + \dfrac{1 \cdot 2 \cdot 3 \cdot 4}{1 \cdot 3 \cdot 5 \cdot 7} + \cdots.$

39. $\dfrac{1}{4} + \dfrac{1 \cdot 3}{4 \cdot 7} + \dfrac{1 \cdot 3 \cdot 5}{4 \cdot 7 \cdot 10} + \dfrac{1 \cdot 3 \cdot 5 \cdot 7}{4 \cdot 7 \cdot 10 \cdot 13} + \cdots.$

40. $\dfrac{2}{3} + \dfrac{2 \cdot 4}{3 \cdot 7} + \dfrac{2 \cdot 4 \cdot 6}{3 \cdot 7 \cdot 11} + \dfrac{2 \cdot 4 \cdot 6 \cdot 8}{3 \cdot 7 \cdot 11 \cdot 15} + \cdots.$

41. Find the sum of the series

$$\sum_{k=1}^{\infty} \frac{k}{10^k}.$$

HINT: Exercise 36 of Section 12.2.

42. Complete the proof of the ratio test by proving that
(a) if $\lambda > 1$, then $\sum a_k$ diverges, and
(b) if $\lambda = 1$, the ratio test is inconclusive.

43. Let r be a positive number. Show that $a_k = r^k/k! \to 0$ by considering the series $\sum a_k$.

44. Show that $a_k = k!/k^k \to 0$ by appealing to the series $\sum a_k$ of Exercise 26.

45. Find the integers $p \geq 2$ for which $\sum \dfrac{(k!)^2}{(pk)!}$ converges.

46. Find the positive numbers r for which $\sum \dfrac{r^k}{k^r}$ converges.

47. Take $r > 0$ and let the a_k be positive. Use the root test to show that, if $(a_k)^{1/k} \to \rho$ and $\rho < 1/r$, then $\sum a_k r^k$ converges.

48. Set

$$a_k = \begin{cases} \dfrac{1}{2^k}, & \text{for odd } k \\[2mm] \dfrac{1}{2^{k-2}}, & \text{for even } k. \end{cases}$$

The resulting series

$$\sum_{k=1}^{\infty} a^k = \frac{1}{2} + 1 + \frac{1}{8} + \frac{1}{4} + \cdots$$

is a rearrangement of the geometric series

$$\sum_{k=0}^{\infty} \frac{1}{2^k} = 1 + \frac{1}{2} + \frac{1}{4} + \frac{1}{8} + \cdots.$$

(a) Use the root test to show that $\sum a_k$ converges.
(b) Show that the ratio test does not apply.

■ 12.5 ABSOLUTE CONVERGENCE AND CONDITIONAL CONVERGENCE; ALTERNATING SERIES

In this section we consider series that have both positive and negative terms.

Absolute Convergence and Conditional Convergence

Let $\sum a_k$ be a series with positive and negative terms. One way to show that $\sum a_k$ converges is to show that the series of absolute values $\sum |a_k|$ converges.

THEOREM 12.5.1

If $\quad \sum |a_k| \quad$ converges, $\quad$ then $\quad \sum a_k \quad$ converges.

PROOF For each k

$$-|a_k| \leq a_k \leq |a_k| \qquad \text{and therefore} \qquad 0 \leq a_k + |a_k| \leq 2|a_k|.$$

If $\sum |a_k|$ converges, then $\sum 2|a_k| = 2 \sum |a_k|$ converges, and therefore, by basic comparison, $\sum(a_k + |a_k|)$ converges. Since

$$a_k = (a_k + |a_k|) - |a_k|,$$

we know that $\sum a_k$ converges. ❏

Series $\sum a_k$ for which $\sum |a_k|$ converges are called *absolutely convergent*. The theorem we have just proved says that

(12.5.2) absolutely convergent series are convergent.

As we will show presently, the converse is false. There are convergent series that are not absolutely convergent. Such series are called *conditionally convergent*.

Example 1   Consider the series

$$\sum_{k=1}^{\infty} \frac{(-1)^{k+1}}{k^2} = 1 - \frac{1}{2^2} + \frac{1}{3^2} - \frac{1}{4^2} + \frac{1}{5^2} - \cdots .$$

If we replace each term by its absolute value, we obtain the series

$$\sum_{k=1}^{\infty} \frac{1}{k^2} = 1 + \frac{1}{2^2} + \frac{1}{3^2} + \frac{1}{4^2} + \frac{1}{5^2} + \cdots .$$

This is a *p*-series with $p = 2$. It is therefore convergent. This means that the initial series is absolutely convergent.   ❏

Example 2   Consider the series

$$1 - \frac{1}{2} - \frac{1}{2^2} + \frac{1}{2^3} - \frac{1}{2^4} + \frac{1}{2^5} + \frac{1}{2^6} - \frac{1}{2^7} - \frac{1}{2^8} + \cdots .$$

If we replace each term by its absolute value, we obtain the series

$$1 + \frac{1}{2} + \frac{1}{2^2} + \frac{1}{2^3} + \frac{1}{2^4} + \frac{1}{2^5} + \frac{1}{2^6} + \frac{1}{2^7} + \frac{1}{2^8} + \cdots .$$

This is a convergent geometric series. The initial series is therefore absolutely convergent.   ❏

Example 3   The series

$$1 - \frac{1}{2} + \frac{1}{3} - \frac{1}{4} + \frac{1}{5} - \frac{1}{6} + \cdots$$

is only conditionally convergent.

It is convergent (see the next theorem), but it is not absolutely convergent: if we replace each term by its absolute value, we get the harmonic series

$$1 + \frac{1}{2} + \frac{1}{3} + \frac{1}{4} + \frac{1}{5} + \frac{1}{6} + \cdots .$$ ❏

Alternating Series

A series such as

$$1 - \frac{1}{2} + \frac{1}{3} - \frac{1}{4} + \frac{1}{5} - \frac{1}{6} + \cdots$$

in which consecutive terms have opposite signs is a called an *alternating series*. As in our example, we shall follow custom and begin all alternating series with a positive term. In general, then, an alternating series will look like this:

$$a_0 - a_1 + a_2 - a_3 + \cdots = \sum_{k=0}^{\infty} (-1)^k a_k$$

with all the a_k positive. In this setup the partial sums of even index end with a positive term and the partial sums of odd index end with a negative term.

12.5.3   THE BASIC THEOREM ON ALTERNATING SERIES

Let $a_0, a_1, a_2,$ be a decreasing sequence of positive numbers. The series

$$a_0 - a_1 + a_2 - a_3 + \cdots = \sum_{k=0}^{\infty} (-1)^k a_k \qquad \text{converges} \qquad \text{iff} \qquad a_k \to 0$$

PROOF If the series converges, then its terms must tend to zero. This implies that $a_k \to 0$.

Now we assume that $a_k \to 0$ and show that the series converges. First we look at the partial sums of even index, s_{2m}. Since

$$s_{2m} = (a_0 - a_1) + (a_2 - a_3) + \cdots + (a_{2m-2} - a_{2m-1}) + a_{2m}$$

is a sum of positive numbers, $s_{2m} > 0$. Since

$$s_{2m+2} = s_{2m} - (a_{2m+1} - a_{2m+2}) \qquad \text{and} \qquad a_{2m+1} - a_{2m+2} > 0,$$

it's clear that

$$s_{2m+2} < s_{2m}.$$

This shows that the sequence of partial sums of even index is a decreasing sequence of positive numbers. Being bounded below by 0, the sequence converges. (Theorem 11.3.6.) We call the limit L and write

$$s_{2m} \to L.$$

Now we look at the partial sums of odd index, s_{2m+1}, and note that

$$s_{2m+1} = s_{2m} - a_{2m+1}.$$

Since $s_{2m} \to L$ and $a_{2m+1} \to 0$,

$$s_{2m+1} \to L.$$

Since the partial sums of even index and the partial sums of odd index both tend to L, the sequence of all partial sums tends to L. (Exercise 50, Section 11.3.) ❑

It is clear from the theorem that the following series all converge:

$$1 - \frac{1}{2} + \frac{1}{3} - \frac{1}{4} + \frac{1}{5} - \frac{1}{6} + \cdots, \qquad 1 - \frac{1}{\sqrt{2}} + \frac{1}{\sqrt{3}} - \frac{1}{\sqrt{4}} + \frac{1}{\sqrt{5}} - \frac{1}{\sqrt{6}} + \cdots,$$

$$1 - \frac{1}{2!} + \frac{1}{3!} - \frac{1}{4!} + \frac{1}{5!} - \frac{1}{6!} + \cdots.$$

The first two series converge only conditionally; the third converges absolutely.

Estimating the Sum of an Alternating Series You have seen that if $a_0, a_1, a_2, \cdots$ is a decreasing sequence of positive numbers that tends to 0, then

$$\sum_{k=0}^{\infty} (-1)^k a_k \qquad \text{converges to some sum} \qquad L.$$

(12.5.4)

> The number L lies between all consecutive partial sums, s_n, s_{n+1}. From this it follows that s_n approximates L to within a_{n+1}:
>
> $$|s_n - L| < a_{n+1}.$$

PROOF We have shown that the partial sums of even index decrease toward L. As you can show, the partial sums of odd index increase toward L. It follows that L lies between all consecutive partial sums s_n and s_{n+1}: one of these is greater than L; the other is less than L. Therefore, for each index n,

$$|s_n - L| < |s_n - s_{n+1}| = a_{n+1}. \qquad ❑$$

Example 4   Both

$$1 - \frac{1}{2} + \frac{1}{3} - \frac{1}{4} + \frac{1}{5} - \frac{1}{6} + \cdots \quad \text{and} \quad 1 - \frac{1}{2^2} + \frac{1}{3^2} - \frac{1}{4^2} + \frac{1}{5^2} - \frac{1}{6^2} + \cdots$$

are convergent alternating series. The nth partial sum of the first series approximates the sum of that series within $1/(n+1)$; the nth partial sum of the second series approximates the sum of that series within $1/(n+1)^2$. The second series converges more rapidly than the first series.   ❑

Example 5   Give a numerical estimate for the sum of the series

$$\sum_{k=0}^{\infty} \frac{(-1)^k}{(2k+1)!} = 1 - \frac{1}{3!} + \frac{1}{5!} - \frac{1}{7!} + \cdots$$

correct to three decimal places.

SOLUTION   By Theorem 12.5.3, the series converges to a sum L. For s_n to approximate L to three decimal places, we must have $|s_n - L| < 0.0005$. Writing out the first few terms of the series, we have

$$1 - \frac{1}{3!} + \frac{1}{5!} - \frac{1}{7!} + \cdots = 1 - \frac{1}{6} + \frac{1}{120} - \frac{1}{5040} + \cdots.$$

Since $a_3 = \dfrac{1}{7!} = \dfrac{1}{5040} < \dfrac{1}{5000} = 0.0002 < 0.0005$,

$$s_2 = 1 - \frac{1}{3!} + \frac{1}{5!} = 1 - \frac{1}{6} + \frac{1}{120} = \frac{101}{120}$$

approximates L to within 0.0005.

We can obtain a decimal estimate for L by noting that $0.8416 < \frac{101}{120} < 0.8417$. To three decimal places, L is 0.842.   ❑

Rearrangements

A *rearrangement* of a series $\sum a_k$ is a series that has exactly the same terms but in a different order. Thus, for example,

$$1 + \frac{1}{3^3} - \frac{1}{2^2} + \frac{1}{5^5} - \frac{1}{4^4} + \frac{1}{7^7} - \frac{1}{6^6} + \cdots$$

and

$$1 + \frac{1}{3^3} + \frac{1}{5^5} - \frac{1}{2^2} - \frac{1}{4^4} + \frac{1}{7^7} + \frac{1}{9^9} - \cdots$$

are both rearrangements of

$$1 - \frac{1}{2^2} + \frac{1}{3^3} - \frac{1}{4^4} + \frac{1}{5^5} - \frac{1}{6^6} + \frac{1}{7^7} - \cdots.$$

In 1867 Riemann published a theorem on rearrangements of series that underscores the importance of distinguishing between absolute convergence and conditional convergence. According to this theorem, all rearrangements of an absolutely convergent series converge absolutely to the same sum. In sharp contrast, a series that is only conditionally convergent can be rearranged to converge to any number we please. It can be arranged to diverge to $+\infty$, to diverge to $-\infty$, even to oscillate between any two bounds we choose.

EXERCISES 12.5

Exercises 1–31. Test these series for (a) absolute convergence, (b) conditional convergence.

1. $1 + (-1) + 1 + \cdots + (-1)^k + \cdots$.

2. $\dfrac{1}{4} - \dfrac{1}{6} + \dfrac{1}{8} - \dfrac{1}{10} + \cdots + \dfrac{(-1)^k}{2k} + \cdots$.

3. $\dfrac{1}{2} - \dfrac{2}{3} + \dfrac{3}{4} - \dfrac{4}{5} + \cdots + (-1)^{k+1}\dfrac{k}{(k+1)} + \cdots$.

4. $\dfrac{1}{2\ln 2} - \dfrac{1}{3\ln 3} + \dfrac{1}{4\ln 4} - \dfrac{1}{5\ln 5} + \cdots + (-1)^k\dfrac{1}{k\ln k} + \cdots$.

5. $\sum (-1)^k \dfrac{\ln k}{k}$.

6. $\sum (-1)^k \dfrac{k}{\ln k}$.

7. $\sum \left(\dfrac{1}{k} - \dfrac{1}{k!} \right)$.

8. $\sum \dfrac{k^3}{2^k}$.

9. $\sum (-1)^k \dfrac{1}{2k+1}$.

10. $\sum (-1)^k \dfrac{(k!)^2}{(2k)!}$.

11. $\sum \dfrac{k!}{(-2)^k}$.

12. $\sum \sin \left(\dfrac{k\pi}{4} \right)$.

13. $\sum (-1)^k \left(\sqrt{k+1} - \sqrt{k} \right)$.

14. $\sum (-1)^k \dfrac{k}{k^2+1}$.

15. $\sum \sin \left(\dfrac{\pi}{4k^2} \right)$.

16. $\sum \dfrac{(-1)^k}{\sqrt{k(k+1)}}$.

17. $\sum (-1)^k \dfrac{k}{2^k}$.

18. $\sum \left(\dfrac{1}{\sqrt{k}} - \dfrac{1}{\sqrt{k+1}} \right)$.

19. $\sum \dfrac{(-1)^k}{k - 2\sqrt{k}}$.

20. $\sum (-1)^k \dfrac{k+2}{k^2+k}$.

21. $\sum (-1)^k \dfrac{4^{k-2}}{e^k}$.

22. $\sum (-1)^k \dfrac{k^2}{2^k}$.

23. $\sum (-1)^k k \sin (1/k)$.

24. $\sum (-1)^{k+1} \dfrac{k^k}{k!}$.

25. $\sum (-1)^k k e^{-k}$.

26. $\sum \dfrac{\cos \pi k}{k}$.

27. $\sum (-1)^k \dfrac{\cos \pi k}{k}$.

28. $\sum \dfrac{\sin (\pi k/2)}{k\sqrt{k}}$.

29. $\sum \dfrac{\sin (\pi k/4)}{k^2}$.

30. $\dfrac{1}{2} - \dfrac{1}{3} + \dfrac{1}{4} + \cdots + \dfrac{1}{3k+2} - \dfrac{1}{3k+3} - \dfrac{1}{3k+4} + \cdots$.

31. $\dfrac{2 \cdot 3}{4 \cdot 5} - \dfrac{5 \cdot 6}{7 \cdot 8} + \cdots + (-1)^k \dfrac{(3k+2)(3k+3)}{(3k+4)(3k+5)} + \cdots$.

Exercises 32–35. The partial sum indicated is used to estimate the sum of the series. Estimate the error.

32. $\displaystyle\sum_{k=1}^{\infty} (-1)^{k+1} \dfrac{1}{k}$; s_{20}.

33. $\displaystyle\sum_{k=0}^{\infty} (-1)^k \dfrac{1}{\sqrt{k+1}}$; s_{80}.

34. $\displaystyle\sum_{k=0}^{\infty} (-1)^k \dfrac{1}{(10)^k}$; s_4.

35. $\displaystyle\sum_{k=1}^{\infty} (-1)^{k+1} \dfrac{1}{k^3}$; s_9.

36. Let s_n be the nth partial sum of the series

$$\sum_{k=0}^{\infty} (-1)^k \dfrac{1}{10^k}.$$

Find the least value of n for which s_n approximates the sum of the series within (a) 0.001, (b) 0.0001.

37. Find the sum of the series of Exercise 36.

▶**Exercises 38–39.** Use a CAS to find the least integer n for which s_n approximates the sum of the series to the indicated accuracy. Find s_n.

38. $\displaystyle\sum_{k=1}^{\infty} (-1)^k \dfrac{(0.9)^k}{k}$; 0.001.

39. $\displaystyle\sum_{k=0}^{\infty} (-1)^k \dfrac{1}{\sqrt{k+1}}$; 0.005.

40. Verify that the series

$$1 - \dfrac{1}{2} + \dfrac{1}{2} - \dfrac{1}{3} + \dfrac{1}{2} - \dfrac{1}{3} + \dfrac{1}{4} + \dfrac{1}{3} - \dfrac{1}{4} + \dfrac{1}{3} - \dfrac{1}{4} + \cdots$$

diverges and explain how this does not violate the basic theorem on alternating series.

41. Let L be the sum of the series

$$\sum_{k=0}^{\infty} (-1)^k \dfrac{1}{k!}$$

and let s_n be the nth partial sum. Find the least value of n for which s_n approximates L to within (a) 0.01, (b) 0.001.

42. Let $a_0, a_1, a_2, \cdots$ be a nonincreasing sequence of positive numbers that converges to 0. Does the alternating series $\sum (-1)^k a_k$ necessarily converge?

43. Can the hypothesis of Theorem 12.5.3 be relaxed to require only that the a_{2k} and the a_{2k+1} form decreasing sequences of positive numbers with limit 0?

44. Show that if $\sum a_k$ is absolutely convergent and $|b_k| \le |a_k|$ for all k, then $\sum b_k$ is absolutely convergent.

45. (a) Show that if $\sum a_k$ is absolutely convergent, then $\sum a_k^2$ is convergent.
    (b) Show by means of an example that the converse of the result in part (a) is false.

46. Let $\displaystyle\sum_{k=0}^{\infty} (-1)^k a_k$ be an alternating series with the a_k forming a decreasing sequence of positive numbers. Show that the sequence of partial sums of odd index increases and is bounded above.

47. In Section 12.8 we prove that, if $\sum a_k c^k$ converges, then $\sum a_k x^k$ converges absolutely for all x such that $|x| < |c|$. Try to prove this now.

48. Form the series

$$a - \tfrac{1}{2}b + \tfrac{1}{3}a - \tfrac{1}{4}b + \tfrac{1}{5}a - \tfrac{1}{6}b + \cdots.$$

(a) Express this series in $\sum$ notation.
(b) For what positive values of a and b is this series absolutely convergent? conditionally convergent?

■ 12.6 TAYLOR POLYNOMIALS IN x; TAYLOR SERIES IN x

We begin with a function f continuous at 0 and form the constant polynomial $P_0(x) = f(0)$. If f is differentiable at 0, the linear function that best approximates f at points close to 0 is the linear polynomial

$$P_1(x) = f(0) + f'(0)x;$$

P_1 has the same value as f at 0 and also has the same first derivative (the same rate of change):

$$P_1(0) = f(0), \qquad P_1'(0) = f'(0).$$

If f is twice differentiable at 0, then we can get a better approximation to f by using the quadratic polynomial

$$P_2(x) = f(0) + f'(0)x + \frac{f''(0)}{2!}x^2;$$

P_2 has the same value as f at 0 and the same first two derivatives:

$$P_2(0) = f(0), \qquad P_2'(0) = f'(0), \qquad P_2''(0) = f''(0).$$

If f has three derivatives at 0, we can form the cubic polynomial

$$P_3(x) = f(0) + f'(0)x + \frac{f''(0)}{2!}x^2 + \frac{f'''(0)}{3!}x^3;$$

P_3 has the same value as f at 0 and the same first three derivatives:

$$P_3(0) = f(0), \qquad P_3'(0) = f'(0), \qquad P_3''(0) = f''(0), \qquad P_3'''(0) = f'''(0).$$

In general, if f has n derivatives at 0, we can form the polynomial

$$P_n(x) = f(0) + f'(0)x + \frac{f''(0)}{2!}x^2 + \cdots + \frac{f^{(n)}(0)}{n!}x^n;$$

P_n has the same value as f at 0 and the same first n derivatives:

$$P_n(0) = f(0), \qquad P_n'(0) = f'(0), \qquad P_n''(0) = f''(0), \ldots, \qquad P_n^{(n)}(0) = f^{(n)}(0).$$

These approximating polynomials $P_0(x), P_1(x), P_2(x), \cdots, P_n(x)$ are called *Taylor polynomials* after the English mathematician Brook Taylor (1685–1731). Taylor introduced these polynomials in 1712.

Example 1   The exponential function $f(x) = e^x$ has derivatives

$$f'(x) = e^x, \qquad f''(x) = e^x, \qquad f'''(x) = e^x, \qquad \text{and so on.}$$

Thus

$$f(0) = 1, \qquad f'(0) = 1, \qquad f''(0) = 1, \qquad f'''(0) = 1, \cdots, \qquad f^{(n)}(0) = 1.$$

The nth Taylor polynomial takes the form

$$P_n(x) = 1 + x + \frac{x^2}{2!} + \frac{x^3}{3!} + \cdots + \frac{x^n}{n!}.$$

Figure 12.6.1 shows the graph of the exponential function and the graphs of the first four approximating polynomials. ❑

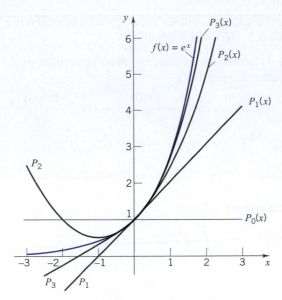

Figure 12.6.1

Example 2 To find the Taylor polynomials that approximate the sine function, we write

$$f(x) = \sin x, \qquad f'(x) = \cos x, \qquad f''(x) = -\sin x, \qquad f'''(x) = -\cos x.$$

The pattern now repeats itself:

$$f^{(4)}(x) = \sin x, \qquad f^{(5)}(x) = \cos x, \qquad f^{(6)}(x) = -\sin x, \qquad f^{(7)}(x) = -\cos x.$$

At $x = 0$, the sine function and all its even derivatives are 0. The odd derivatives are alternately 1 and -1:

$$f'(0) = 1, \qquad f'''(0) = -1, \qquad f^{(5)}(0) = 1, \qquad f^{(7)}(0) = -1, \qquad \text{and so on.}$$

Therefore the Taylor polynomials are as follows:

$$P_0(x) = 0$$

$$P_1(x) = P_2(x) = x$$

$$P_3(x) = P_4(x) = x - \frac{x^3}{3!}$$

$$P_5(x) = P_6(x) = x - \frac{x^3}{3!} + \frac{x^5}{5!}$$

$$P_7(x) = P_8(x) = x - \frac{x^3}{3!} + \frac{x^5}{5!} - \frac{x^7}{7!} \qquad \text{and so on.}$$

Only odd powers appear. This is not surprising since the sine function is an odd function. ❏

It is not enough to say that the Taylor polynomials

$$P_n(x) = f(0) + f'(0)x + \frac{f''(0)}{2!}x^2 + \cdots + \frac{f^{(n)}(0)}{n!}x^n$$

approximate $f(x)$. We must describe the accuracy of the approximation.

Our first step is to prove a result known as Taylor's theorem.

THEOREM 12.6.1 TAYLOR'S THEOREM

If f has $n + 1$ continuous derivatives on an open interval I that contains 0, then for each $x \in I$

$$f(x) = f(0) + f'(0)x + \frac{f''(0)}{2!}x^2 + \cdots + \frac{f^{(n)}(0)}{n!}x^n + R_n(x)$$

with $R_n(x) = \dfrac{1}{n!} \displaystyle\int_0^x f^{(n+1)}(t)(x - t)^n \, dt$. We call $R_n(x)$ the *remainder*.

PROOF Fix x in the interval I. Since

$$\int_0^x f'(t)\,dt = f(x) - f(0),$$

we have

$$f(x) = f(0) + \int_0^x f'(t)\,dt.$$

We now integrate by parts. We set

$$u = f'(t) \qquad \text{and} \qquad dv = dt.$$

This forces

$$du = f''(t)\,dt.$$

For v we may choose any function of t with derivative identically 1. To suit our purpose, we choose

$$v = -(x - t).$$

We carry out the integration by parts and get

$$\int_0^x f'(t) = \Big[-f'(t)(x - t)\Big]_0^x + \int_0^x f''(t)(x - t)\,dt$$

$$= f'(0)x + \int_0^x f''(t)(x - t)\,dt,$$

which with (1) gives

$$f(x) = f(0) + f'(0)x + \int_0^x f''(t)(x - t)\,dt.$$

This completes the first step toward the proof of the theorem.

Integrating by parts again [set $u = f''(t), dv = (x - t)\,dt$, leading to $du = f'''(t)\,dt, v = -\frac{1}{2}(x - t)^2$], we get

$$f(x) = f(0) + f'(0)x + \frac{f''(0)}{2!}x^2 + \frac{1}{2!}\int_0^x f'''(t)(x - t)^2 \, dt.$$

Continuing to integrate by parts (see Exercise 61), we get after n steps

$$f(x) = f(0) + f'(0)x + \frac{f''(0)}{2!}x^2 + \cdots + \frac{f^{(n)}(0)}{n!}x^n + \frac{1}{n!}\int_0^x f^{(n+1)}(t)(x - t)^n \, dt,$$

which is what we have been trying to prove. ❏

To see how closely

$$P_n(x) = f(0) + f'(0)x + \frac{f''(0)}{2!}x^2 + \cdots + \frac{f^{(n)}(0)}{n!}x^n$$

approximates $f(x)$, we need an estimate for the remainder term $R_n(x)$. The following corollary to Taylor's theorem gives a more convenient form of the remainder. It was established by Joseph Lagrange in 1797, and it is known as the Lagrange formula for the remainder. The proof is left to you as an exercise.

COROLLARY 12.6.2   LAGRANGE FORMULA FOR THE REMAINDER

The remainder in Taylor's theorem can be written

$$R_n(x) = \frac{f^{(n+1)}(c)}{(n+1)!}x^{n+1}$$

with c some number between 0 and x.

Remark   If we write Taylor's theorem using the Lagrange formula for the remainder, we have

$$f(x) = f(0) + f'(0)x + \frac{f''(0)}{2!}x^2 + \cdots + \frac{f^{(n)}(0)}{n!}x^n + \frac{f^{(n+1)}(c)}{(n+1)!}x^{n+1}$$

with c some number between 0 and x. This result is an extension of the mean-value theorem (Theorem 4.1.1), for if we let $n = 0$ and take $x > 0$, we get

$$f(x) = f(0) + f'(c)x, \qquad \text{which can be written} \qquad f(x) - f(0) = f'(c)(x - 0)$$

with $c \in (0, x)$. This is the mean-value theorem for f on the interval $[0, x]$. ❏

The following estimate for $R_n(x)$ is an immediate consequence of Corollary 12.6.2:

(12.6.3)
$$\left| R_n(x) \right| \leq \left(\max_{t \in J} |f^{(n+1)}(t)| \right) \frac{|x|^{n+1}}{(n+1)!}$$

where J is the closed interval that joins 0 to x.

Example 3   The Taylor polynomials for the exponential function $f(x) = e^x$ take the form

$$P_n(x) = 1 + x + \frac{x^2}{2!} + \cdots + \frac{x^n}{n!}. \qquad \text{(Example 1)}$$

We will show with our remainder estimate that for each real x

$$R_n(x) \to 0 \qquad \text{as} \qquad n \to \infty,$$

and therefore we can approximate e^x as closely as we wish by Taylor polynomials.

We begin by fixing x and letting M be the maximum value of the exponential function on the closed interval J that joins 0 to x. (If $x > 0$, then $M = e^x$; if $x < 0$, $M = e^0 = 1$.) Since

$$f^{(n+1)}(t) = e^t \qquad \text{for all } n,$$

we have

$$\max_{t \in J} |f^{(n+1)}(t)| = M \qquad \text{for all } n.$$

Thus, by (12.6.3)

$$|R_n(x)| \le M \frac{|x|^{n+1}}{(n+1)!}.$$

From (11.4.4), we know that, as $n \to \infty$,

$$\frac{|x|^{n+1}}{(n+1)!} \to 0.$$

It follows then that $R_n(x) \to 0$ as asserted. ❏

Example 4 We return to the sine function $f(x) = \sin x$ and its Taylor polynomials

$$P_1(x) = P_2(x) = x,$$

$$P_3(x) = P_4(x) = x - \frac{x^3}{3!},$$

$$P_5(x) = P_6(x) = x - \frac{x^3}{3!} + \frac{x^5}{5!}, \qquad \text{and so on.}$$

The pattern of derivatives was established in Example 2; namely, for all k,

$$f^{(4k)}(x) = \sin x, \qquad f^{(4k+1)}(x) = \cos x,$$

$$f^{(4k+2)}(x) = -\sin x, \qquad f^{(4k-3)}(x) = -\cos x.$$

Thus, for all n and all real t,

$$|f^{(n+1)}(t)| \le 1.$$

It follows from our remainder estimate (12.6.3) that

$$|R_n(x)| \le \frac{|x|^{n+1}}{(n+1)!}$$

Since

$$\frac{|x|^{n+1}}{(n+1)!} \to 0 \qquad \text{for all real } x,$$

we see that $R_n(x) \to 0$ for all real x. Thus the sequence of Taylor polynomials converges to the sine function and therefore can be used to approximate $\sin x$ for any real number x with as much accuracy as we wish. ❏

Taylor Series in x

By definition $0! = 1$. Adopting the convention that $f^{(0)} = f$, we can write Taylor polynomials

$$P_n(x) = f(0) + f'(0)x + \frac{f''(0)}{2!}x^2 + \cdots + \frac{f^{(n)}(0)}{n!}x^n$$

in $\sum$ notation:

$$P_n(x) = \sum_{k=0}^{n} \frac{f^{(k)}(0)}{k!}x^k.$$

If f is infinitely differentiable on an open interval I that contains 0, then we have

$$f(x) = \sum_{k=0}^{n} \frac{f^{(k)}(0)}{k!} x^k + R_n(x), \qquad x \in I,$$

for all positive integers n. If, as in the case of the exponential function and the sine function, $R_n(x) \to 0$ for each $x \in I$, then for all such x

$$\sum_{k=0}^{n} \frac{f^{(k)}(0)}{k!} x^k \to f(x).$$

In this case, we say that $f(x)$ can be expanded as a *Taylor series in x* and write

(12.6.4)

$$f(x) = \sum_{k=0}^{\infty} \frac{f^{(k)}(0)}{k!} x^k.$$

Taylor series in x are sometimes called Maclaurin series after Colin Maclaurin, a Scottish mathematician (1698–1746). In some circles the name Maclaurin remains attached to these series, although Taylor considered them some twenty years before Maclaurin.

It follows from Example 3 that

(12.6.5)

$$e^x = \sum_{k=0}^{\infty} \frac{x^k}{k!} = 1 + x + \frac{x^2}{2!} + \frac{x^3}{3!} + \cdots \qquad \text{for all real } x.$$

From Example 4 we have

(12.6.6)

$$\sin x = \sum_{k=0}^{\infty} \frac{(-1)^k}{(2k+1)!} x^{2k+1} = x - \frac{x^3}{3!} + \frac{x^5}{5!} - \frac{x^7}{7!} + \cdots \qquad \text{for all real } x.$$

We leave it to you as an exercise to show that

(12.6.7)

$$\cos x = \sum_{k=0}^{\infty} \frac{(-1)^k}{(2k)!} x^{2k} = 1 - \frac{x^2}{2!} + \frac{x^4}{4!} - \frac{x^6}{6!} + \cdots \qquad \text{for all real } x.$$

We come now to the logarithm function. Since $\ln x$ is not defined at $x = 0$, we cannot expand $\ln x$ in powers of x. We work instead with $\ln(1 + x)$.

(12.6.8)

$$\ln(1 + x) = \sum_{k=1}^{\infty} \frac{(-1)^{k+1}}{k} x^k = x - \frac{x^2}{2} + \frac{x^3}{3} - \cdots \qquad \text{for } -1 < x \le 1.$$

PROOF[†] The function

$$f(x) = \ln(1 + x)$$

[†]The proof we give here illustrates the methods of this section. A much simpler way of obtaining this series expansion is given in Section 12.9.

is defined on $(-1, \infty)$ and on that interval has derivatives of all orders:

$$f'(x) = \frac{1}{1+x}, \qquad f''(x) = -\frac{1}{(1+x)^2}, \qquad f'''(x) = \frac{2}{(1+x)^3},$$

$$f^{(4)}(x) = -\frac{3!}{(1+x)^4}, \qquad f^{(5)}(x) = \frac{4!}{(1+x)^5}, \qquad \text{and so on.}$$

The pattern is now clear: for $k \geq 1$

$$f^{(k)}(x) = (-1)^{k+1}\frac{(k-1)!}{(1+x)^k}.$$

Evaluation at $x = 0$ gives $f^{(k)}(0) = (-1)^{k+1}(k-1)!$ and

$$\frac{f^{(k)}(0)}{k!} = \frac{(-1)^{k+1}}{k}.$$

Since $f(0) = 0$, the nth Taylor polynomial takes the form

$$P_n(x) = \sum_{k=1}^{n}(-1)^{k+1}\frac{x^k}{k} = x - \frac{x^2}{2} + \cdots + (-1)^{n+1}\frac{x^n}{n}.$$

Therefore, all we have to show is that

$$R_n(x) \to 0 \qquad \text{for} \quad -1 < x \leq 1.$$

Instead of trying to apply our usual remainder estimate [in this case, that estimate is not delicate enough to show that $R_n(x) \to 0$ for x under consideration], we keep the remainder in its integral form. From Taylor's theorem,

$$R_n(x) = \frac{1}{n!}\int_0^x f^{(n+1)}(t)(x-t)^n\, dt.$$

Therefore in this case

$$R_n(x) = \frac{1}{n!}\int_0^x (-1)^{n+2}\frac{n!}{(1+t)^{n+1}}(x-t)^n dt = (-1)^n\int_0^x \frac{(x-t)^n}{(1+t)^{n+1}}\, dt.$$

For $0 \leq x \leq 1$ we have

$$|R_n(x)| = \int_0^x \frac{(x-t)^n}{(1+t)^{n+1}}\, dt \leq \int_0^x (x-t)^n\, dt = \frac{x^{n+1}}{n+1} \to 0.$$

$$\underset{\text{explain}}{\uparrow}$$

For $-1 < x < 0$ we have

$$|R_n(x)| = \left|\int_0^x \frac{(x-t)^n}{(1+t)^{n+1}}\, dt\right| = \int_x^0 \left(\frac{t-x}{1+t}\right)^n \frac{1}{1+t}\, dt.$$

By the first mean-value theorem for integrals (Theorem 5.9.1), there exists a number x_n between x and 0 such that

$$\int_x^0 \left(\frac{t-x}{1+t}\right)^n \frac{1}{1+t}\, dt = \left(\frac{x_n-x}{1+x_n}\right)^n \left(\frac{1}{1+x_n}\right)(-x).$$

Since $-x = |x|$ and $0 < 1+x < 1+x_n$, we can conclude that

$$|R_n(x)| < \left(\frac{x_n+|x|}{1+x_n}\right)^n \left(\frac{|x|}{1+x}\right).$$

Since $|x| < 1$ and $x_n < 0$, we have

$$x_n < |x|x_n, \qquad x_n + |x| < |x|x_n + |x| = |x|(1+x_n),$$

and thus

$$\frac{x_n + |x|}{1 + x_n} < |x|.$$

It follows that

$$|R_n(x)| < |x|^n \left(\frac{|x|}{1 + x} \right).$$

Since $|x| < 1$, $R_n(x) \to 0$. ❑

Remark The series expansion for $\ln(1 + x)$ that we have just verified for $-1 < x \leq 1$ cannot be extended to other values of x. For $x \leq -1$ neither side makes sense: $\ln(1 + x)$ is not defined, and the series on the right diverges. For $x > 1$, $\ln(1 + x)$ is defined, but the series on the right diverges and hence does not represent the function. At $x = 1$, the series gives the intriguing result

$$\ln 2 = 1 - \tfrac{1}{2} + \tfrac{1}{3} - \tfrac{1}{4} + \cdots. \quad ❑$$

We want to emphasize again the role played by the remainder term $R_n(x)$. We can form a Taylor series

$$\sum_{k=0}^{\infty} \frac{f^{(k)}(0)}{k!} x^k$$

for any function f with derivatives of all orders at $x = 0$, but such a series need not converge at any number $x \neq 0$. Even if it does converge, the sum need not be $f(x)$. (See Exercise 63.) *The Taylor series converges to $f(x)$ iff the remainder term $R_n(x)$ tends to 0.*

Some Numerical Calculations

If the Taylor series converges to $f(x)$, we can use the partial sums (the Taylor polynomials) to calculate $f(x)$ as accurately as we wish. In what follows we show some sample calculations. For ready reference we list some values of $k!$ in Table 12.6.1.

■ **Table 12.6.1**

		$k!$
2!	=	2
3!	=	6
4!	=	24
5!	=	120
6!	=	720
7!	=	5,040
8!	=	40,320

Example 5 Determine the maximum possible error we incur by using $P_6(x)$ to estimate $f(x) = e^x$ for x in the interval $[0, 1]$.

SOLUTION For all x,

$$e^x = 1 + x + \frac{x^2}{2!} + \frac{x^3}{3!} + \cdots + \frac{x^n}{n!} + \cdots.$$

Now fix any x in the interval $[0, 1]$. From Example 3 we have

$$P_6(x) = 1 + x + \frac{x^2}{2!} + \frac{x^3}{3!} + \frac{x^4}{4!} + \frac{x^5}{5!} + \frac{x^6}{6!}$$

and

$$|R_6(x)| \leq \max_{0 \leq t \leq x} |f^{(7)}(t)| \frac{|x|^7}{7!} = \max_{0 \leq t \leq x} e^t \frac{1}{7!} \leq e^x \frac{1}{7!} \leq \frac{e}{7!} < \frac{3}{7!} = \frac{1}{1680} < 0.0006.$$

$$e^x \leq e \quad \longrightarrow \qquad \longleftarrow \quad e < 3$$

The maximum possible error we incur by using $P_6(x)$ to estimate e^x on $[0, 1]$ is less than 0.0006. In particular, we can be sure that

$$P_6(1) = 1 + 1 + \frac{1}{2!} + \frac{1}{3!} + \frac{1}{4!} + \frac{1}{5!} + \frac{1}{6!} = \frac{1957}{720}$$

differs from e by less than 0.0006. Our calculator gives $\frac{1957}{720} \cong 2.7180556$ and $e \cong 2.7182818$. This difference is actually less than 0.0003. ❑

Example 6 Give an estimate for $e^{0.2}$ correct to three decimal places (that is, remainder less than 0.0005).

SOLUTION From Example 3 we know that the nth Taylor polynomial for e^x evaluated at $x = 0.2 = \frac{1}{5}$,

$$P_n(0.2) = 1 + 0.2 + \frac{(0.2)^2}{2!} + \frac{(0.2)^3}{3!} + \cdots + \frac{(0.2)^n}{n!},$$

approximates $e^{0.2}$ to within

$$|R_n(0.2)| \le e^{0.2}\frac{|0.2|^{n+1}}{(n+1)!} < e\frac{(0.2)^{n+1}}{(n+1)!} < 3\frac{1}{5^{n+1}(n+1)!}.$$

$$\uparrow \quad e < 3$$

We require

$$\frac{3}{5^{n+1}(n+1)!} < 0.0005,$$

which is equivalent to requiring

$$5^{n+1}(n+1)! > 6000.$$

The least integer that satisfies this inequality is $n = 3$. $[5^4(4!) = 15,000.]$ Thus

$$P_3(0.2) = 1 + 0.2 + \frac{(0.2)^2}{2!} + \frac{(0.2)^3}{3!} = \frac{7.328}{6} \cong 1.22133$$

differs from $e^{0.2}$ by less than 0.0005. Our calculator gives $e^{0.2} \cong 1.2214028$, so in fact our estimate differs by less than 0.00008. ❑

Example 7 Use the sine series to estimate $\sin 0.5$ within 0.001.

SOLUTION At $x = 0.5 = \frac{1}{2}$, the sine series gives

$$\sin 0.5 = 0.5 - \frac{(0.5)^3}{3!} + \frac{(0.5)^5}{5!} - \frac{(0.5)^7}{7!} + \cdots.$$

From Example 4, we know that $P_n(0.5)$ approximates $\sin 0.5$ to within

$$|R_n(0.5)| \le \frac{(0.5)^{n+1}}{(n+1)!} = \frac{1}{2^{n+1}(n+1)!}.$$

Note that

$$\frac{1}{2^{n+1}(n+1)!} < 0.001 \qquad \text{iff} \qquad 2^{(n+1)}(n+1)! > 1000.$$

The least integer that satisfies this inequality is $n = 4$. $[2^5(5!) = 3840.]$ Thus

$$\overset{\text{the coefficient of } x^4 \text{ is 0}}{\underset{\downarrow}{}}$$

$$P_4(0.5) = P_3(0.5) = 0.5 - \frac{(0.5)^3}{3!} = \frac{23}{48}$$

approximates $\sin 0.5$ within 0.001.

Our calculator gives

$$\frac{23}{48} \cong 0.4791667 \qquad \text{and} \qquad \sin 0.5 \cong 0.4794255. \quad \square$$

Remark   We could have solved the last problem without reference to the remainder estimate derived in Example 4. The series for $\sin 0.5$ is a convergent alternating series of the form stipulated in Theorem 12.5.3. By (12.5.4) we can conclude immediately that $\sin 0.5$ lies between every two consecutive partial sums. In particular

$$0.4791667 \cong 0.5 - \frac{(0.5)^3}{3!} < \sin 0.5 < 0.5 - \frac{(0.5)^3}{3!} + \frac{(0.5)^5}{5!} \cong 0.4794271. \quad \square$$

Example 8   Use the series for $\ln(1 + x)$ to estimate $\ln 1.4$ within 0.01.

SOLUTION   By (12.6.8),

$$\ln 1.4 = \ln(1 + 0.4) = 0.4 - \tfrac{1}{2}(0.4)^2 + \tfrac{1}{3}(0.4)^3 - \tfrac{1}{4}(0.4)^4 + \cdots.$$

This is a convergent alternating series of the form stipulated in Theorem 12.5.3. Therefore $\ln 1.4$ lies between every two consecutive partial sums.
   The first term less than 0.01 is

$$\tfrac{1}{4}(0.4)^4 = \tfrac{1}{4}(0.0256) = 0.0064.$$

The relation

$$0.4 - \tfrac{1}{2}(0.4)^2 + \tfrac{1}{3}(0.4)^3 - \tfrac{1}{4}(0.4)^4 < \ln 1.4 < 0.4 - \tfrac{1}{2}(0.4)^2 + \tfrac{1}{3}(0.4)^3$$

gives

$$0.335 < \ln 1.4 < 0.341.$$

Within the prescribed limits of accuracy, we can take $\ln 1.4 \cong 0.34.^{\dagger} \quad \square$

---

†A much more effective way of estimating logarithms is given in the Exercises.

EXERCISES 12.6

Exercises 1–4. Find the Taylor polynomial P_4 for the function f.

1. $f(x) = x - \cos x$.

2. $f(x) = \sqrt{1 + x}$.

3. $f(x) = \ln \cos x$.

4. $f(x) = \sec x$.

Exercises 5–8. Find the Taylor polynomial P_5 for the given function f.

5. $f(x) = (1 + x)^{-1}$.

6. $f(x) = e^x \sin x$.

7. $f(x) = \tan x$.

8. $f(x) = x \cos x^2$.

9. Determine $P_0(x)$, $P_1(x)$, $P_2(x)$, $P_3(x)$ for

$$f(x) = 1 - x + 3x^2 + 5x^3.$$

10. Determine $P_0(x)$, $P_1(x)$, $P_2(x)$, $P_3(x)$ for $f(x) = (x + 1)^3$.

Exercises 11–16. Determine the nth Taylor polynomial P_n for the function f.

11. $f(x) = e^{-x}$.

12. $f(x) = \sinh x$.

13. $f(x) = \cosh x$.

14. $f(x) = \ln(1 - x)$.

15. $f(x) = e^{rx}$, $\quad r$ a real number.

16. $f(x) = \cos bx$, $\quad b$ a real number.

Exercises 17–20. Assume that f is a function with $|f^{(n)}(x)| \leq 1$ for all n and all real x. (The sine and cosine functions have this property.)

17. Estimate the maximum possible error if $P_5(1/2)$ is used to estimate $f(1/2)$.

18. Estimate the maximum possible error if $P_7(-2)$ is used to estimate $f(-2)$.

19. Find the least integer n for which you can be sure that $P_n(2)$ approximates $f(2)$ within 0.001.

20. Find the least integer n for which you can be sure that $P_n(-4)$ approximates $f(-4)$ within 0.001.

Exercises 21–24. Assume that f is a function with $|f^{(n)}(x)| \leq 3$ for all n and all real x.

21. Find the least integer n for which you can be sure that $P_n(1/2)$ approximates $f(1/2)$ with four decimal place accuracy.

22. Find the least integer n for which you can be sure that $P_n(2)$ approximates $f(2)$ with three decimal place accuracy.

23. Find the values of x for which you can be sure that $P_5(x)$ approximates $f(x)$ within 0.05.

24. Find the values of x for which you can be sure that $P_9(x)$ approximates $f(x)$ within 0.05.

Exercises 25–32. Use Taylor polynomials to estimate the following within 0.01.

25. $\sqrt{e}$.   26. $\sin 0.3$.

27. $\sin 1$.   28. $\ln 1.2$.

29. $\cos 1$.   30. $e^{0.8}$.

31. $\sin 10°$.   32. $\cos 6°$.

Exercises 33–40. Find the Lagrange form of the remainder $R_n(x)$.

33. $f(x) = e^{2x}$; $n = 4$.

34. $f(x) = \ln(1+x)$; $n = 5$.

35. $f(x) = \cos 2x$; $n = 4$.   36. $f(x) = \sqrt{x+1}$; $n = 3$.

37. $f(x) = \tan x$; $n = 2$.   38. $f(x) = \sin x$; $n = 5$.

39. $f(x) = \arctan x$; $n = 2$.   40. $f(x) = \dfrac{1}{1+x}$; $n = 4$.

Exercises 41–44. Write the remainder $R_n(x)$ in Lagrange form.

41. $f(x) = e^{-x}$.   42. $f(x) = \sin 2x$.

43. $f(x) = \dfrac{1}{1-x}$.   44. $f(x) = \ln(1+x)$.

45. Let P_n be the nth Taylor polynomial for the function

$$f(x) = \ln(1+x).$$

Find the least integer n for which: (a) $P_n(0.5)$ approximates $\ln 1.5$ within 0.01; (b) $P_n(0.3)$ approximates $\ln 1.3$ within 0.01; (c) $P_n(1)$ approximates $\ln 2$ within 0.001.

46. Let P_n be the nth Taylor polynomial for the function

$$f(x) = \sin x.$$

Find the least integer n for which: (a) $P_n(1)$ approximates $\sin 1$ within 0.001; (b) $P_n(2)$ approximates $\sin 2$ within 0.001. (c) $P_n(3)$ approximate $\sin 3$ within 0.001.

▶ 47. Set $f(x) = e^x$.

   (a) Find the Taylor polynomial P_n for f of least degree that approximates e^x at $x = \frac{1}{2}$ with four decimal place accuracy. Then use that polynomial to obtain an estimate of $\sqrt{e}$.

   (b) Find the Taylor polynomial P_n for f of least degree that approximates e^x at $x = -1$ with four decimal place accuracy. Then use that polynomial to obtain an estimate of $1/e$.

▶ 48. Set $f(x) = \cos x$.

   (a) Find the Taylor polynomial P_n for f of least degree that approximates $\cos x$ at $x = \pi/30$ with three decimal place accuracy. Then use that polynomial to obtain an estimate of $\cos(\pi/30)$.

   (b) Find the Taylor polynomial P_n for f of least degree that approximates $\cos x°$ at $x = 9$ with four decimal place accuracy. Then use that polynomial to obtain an estimate of $\cos 9°$.

49. Show that a polynomial $P(x) = a_0 + a_1 x + \cdots + a_n x^n$ is its own Taylor series.

50. Show that

$$\cos x = \sum_{k=0}^{\infty} \frac{(-1)^k}{(2k)!} x^{2k} \quad \text{for all real } x.$$

51. Show that

$$\sinh x = \sum_{k=0}^{\infty} \frac{1}{(2k+1)!} x^{2k+1} \quad \text{for all real } x.$$

52. Show that

$$\cosh x = \sum_{k=0}^{\infty} \frac{1}{(2k)!} x^{2k} \quad \text{for all real } x.$$

Exercises 53–57. Derive a series expansion in x for the function and specify the numbers x for which the expansion is valid. Take $a > 0$.

53. $f(x) = e^{ax}$.   54. $f(x) = \sin ax$.

55. $f(x) = \cos ax$.   56. $f(x) = \ln(1 - ax)$.

57. $f(x) = \ln(a + x)$.

58. The series we derived for $\ln(1 + x)$ converges too slowly to be of much practical use. The following series expansion is much more effective:

(12.6.9)

$$\ln\left(\frac{1+x}{1-x}\right) = 2\left(x + \frac{x^3}{3} + \frac{x^5}{5} + \cdots\right)$$

$$\text{for} \quad -1 < x < 1.$$

Derive this series expansion.

59. Set $x = \frac{1}{3}$ and use the first three nonzero terms of (12.6.9) to estimate $\ln 2$.

60. Use the first two nonzero terms of (12.6.9) to estimate $\ln 1.4$.

61. Verify the identity

$$\frac{f^{(k)}(0)}{k!} x^k = \frac{1}{(k-1)!} \int_0^x f^{(k)}(t)(x-t)^{k-1}\, dt$$

$$-\frac{1}{k!} \int_0^x f^{(k+1)}(t)(x-t)^k\, dt$$

by using integration by parts on the second integral.

62. Prove Corollary 12.6.2 and then derive remainder estimate (12.6.3).

▶ 63. (a) Use a graphing utility to draw the graph of the function

$$f(x) = \begin{cases} e^{-1/x^2}, & x \neq 0. \\ 0, & x = 0. \end{cases}$$

   (b) Use L' Hôpital's rule to show that for every positive integer n,

$$\lim_{x \to 0} \frac{e^{-1/x^2}}{x^n} = 0.$$

   (c) Prove by induction that $f^{(n)}(0) = 0$ for all $n \geq 1$.

   (d) What is the Taylor series of f?

(e) For what values of x does the Taylor series of f actually converge to $f(x)$?

64. Set $f(x) = \cos x$. Using a graphing utility or a CAS, draw a figure that gives the graph of f and the graphs of the Taylor polynomials P_2, P_4, P_6, P_8.

65. Set $f(x) = \ln(1+x)$. Using a graphing utility or a CAS, draw a figure that gives the graph of f and the graphs of the Taylor polynomials P_2, P_3, P_4, P_5.

66. Show that e is irrational by taking the following steps.

(1) Begin with the expansion

$$e = \sum_{k=0}^{\infty} \frac{1}{k!}$$

and show that the qth partial sum

$$s_q = \sum_{k=0}^{\infty} \frac{1}{k!}$$

satisfies the inequality

$$0 < q!(e - s_q) < \frac{1}{q}.$$

(2) Show that $q!s_q$ is an integer and argue that, if e were of the form p/q, then $q!(e - s_q)$ would be a positive integer less than 1.

■ 12.7 TAYLOR POLYNOMIALS AND TAYLOR SERIES IN $x - a$

So far we have carried out series expansions only in powers of x. Here we generalize to series expansions in powers of $x - a$ where a is an arbitrary real number. We begin with a more general version of Taylor's theorem.

THEOREM 12.7.1 TAYLOR'S THEOREM

If g has $n + 1$ continuous derivatives on an open interval I that contains the point a, then for each $x \in I$

$$g(x) = g(a) + g'(a)(x - a) + \frac{g''(a)}{2!}(x - a)^2 + \cdots + \frac{g^{(n)}(a)}{n!}(x - a)^n + R_n(x)$$

with

$$R_n(x) = \frac{1}{n!} \int_a^x g^{(n+1)}(t)(x - t)^n \, dt.$$

The polynomial

$$P_n(x) = g(a) + g'(a)(x - a) + \frac{g''(a)}{2!}(x - a)^2 + \cdots + \frac{g^{(n)}(a)}{n!}(x - a)^n$$

is called the *nth Taylor polynomial for g in powers of $x - a$*. In this more general setting, the *Lagrange formula for the remainder*, $R_n(x)$, takes the form

(12.7.2)
$$R_n(x) = \frac{g^{(n+1)}(c)}{(n + 1)!}(x - a)^{n+1}$$

where c is some number between a and x.

Now let $x \in I$, $x \neq a$, and let J be the closed interval that joins a to x. Then

(12.7.3)
$$|R_n(x)| \leq \left(\max_{t \in J} |g^{(n+1)}(t)| \right) \frac{|x - a|^{n+1}}{(n + 1)!}.$$

If $R_n(x) \to 0$, then we have the series representation

$$g(x) = g(a) + g'(a)(x - a) + \frac{g''(a)}{2!}(x - a)^2 + \cdots + \frac{g^{(n)}(a)}{n!}(x - a)^n + \cdots,$$

which, in sigma notation, takes the form

(12.7.4)

$$g(x) = \sum_{k=0}^{\infty} \frac{g^{(k)}(a)}{k!}(x - a)^k.$$

This is known as the Taylor expansion of $g(x)$ in powers of $x - a$. The series on the right is called a *Taylor series in $x - a$*.

All this differs from what you saw before only by a translation. Define

$$f(x) = g(x + a).$$

Then

$$f^{(k)}(x) = g^{(k)}(x + a) \quad \text{and} \quad f^{(k)}(0) = g^{(k)}(a).$$

The results of this section as stated for g can all be derived by applying the results of Section 12.6 to the function f.

Example 1 Expand $g(x) = 4x^3 - 3x^2 + 5x - 1$ in powers of $x - 2$.

SOLUTION We need to evaluate g and its derivatives at $x = 2$:

$$g(x) = 4x^3 - 3x^2 + 5x - 1$$
$$g'(x) = 12x^2 - 6x + 5$$
$$g''(x) = 24x - 6$$
$$g'''(x) = 24.$$

All higher derivatives are identically 0.

Substitution gives $g(2) = 29$, $g'(2) = 41$, $g''(2) = 42$, $g'''(2) = 24$, and $g^{(k)}(2) = 0$ for all $k \geq 4$. Thus, from (12.7.4),

$$g(x) = 29 + 41(x - 2) + \frac{42}{2!}(x - 2)^2 + \frac{24}{3!}(x - 2)^3$$

$$= 29 + 41(x - 2) + 21(x - 2)^2 + 4(x - 2)^3. \quad \square$$

Example 2 Expand $g(x) = x^2 \ln x$ in powers of $x - 1$.

SOLUTION We need to evaluate g and its derivatives at $x = 1$:

$$g(x) = x^2 \ln x$$
$$g'(x) = x + 2x \ln x$$
$$g''(x) = 3 + 2 \ln x$$
$$g'''(x) = 2x^{-1}$$
$$g^{(4)}(x) = -2x^{-2}$$
$$g^{(5)}(x) = (2)(2)x^{-3}$$
$$g^{(6)}(x) = -(2)(2)(3)x^{-4} = -2(3!)x^{-4}$$
$$g^{(7)}(x) = (2)(2)(3)(4)x^{-5} = (2)(4!)x^{-5} \quad \text{and so on.}$$

The pattern is now clear: for $k \geq 3$

$$g^{(k)}(x) = (-1)^{k+1}2(k-3)!\, x^{-k+2}.$$

Evaluation at $x = 1$ gives $g(1) = 0,\; g'(1) = 1,\; g''(1) = 3$ and, for $k \geq 3$,

$$g^{(k)}(1) = (-1)^{k+1}\, 2(k-3)!.$$

The expansion in powers of $x - 1$ can be written

$$g(x) = (x-1) + \frac{3}{2!}(x-1)^2 + \sum_{k=3}^{\infty} \frac{(-1)^{k+1}(2)(k-3)!}{k!}(x-1)^k$$

$$= (x-1) + \frac{3}{2}(x-1)^2 + 2\sum_{k=3}^{\infty} \frac{(-1)^{k+1}}{k(k-1)(k-2)}(x-1)^k. \quad \square$$

Another way to expand $g(x)$ in powers of $x - a$ is to expand $g(t + a)$ in powers of t and then set $t = x - a$. This is the approach we take when the expansion in t is either known to us or is readily available.

Example 3 We can expand $g(x) = e^{x/2}$ in powers of $x - 3$ by expanding

$$g(t + 3) = e^{(t+3)/2}$$

in powers of t and then setting $t = x - 3$.

Note that

$$g(t+3) = e^{3/2}e^{t/2} = e^{3/2}\sum_{k=0}^{\infty} \frac{(t/2)^k}{k!} = e^{3/2}\sum_{k=0}^{\infty} \frac{1}{2^k k!}t^k.$$

exponential series ——↑

Setting $t = x - 3$, we have

$$g(x) = e^{3/2}\sum_{k=0}^{\infty} \frac{1}{2^k k!}(x-3)^k.$$

Since the expansion of $g(t + 3)$ is valid for all real t, the expansion of $g(x)$ is valid for all real x. ❑

Now we prove that

(12.7.5)

the series expansion

$$\ln x = \ln a + \frac{1}{a}(x-a) - \frac{1}{2a^2}(x-a)^2 + \frac{1}{3a^3}(x-a)^3 - \cdots$$

is valid for $0 < x \leq 2a$.

PROOF We expand $\ln(a + t)$ in powers of t and then set $t = x - a$. Note first that

$$\ln(a+t) = \ln\left[a\left(1 + \frac{t}{a}\right)\right] = \ln a + \ln\left(1 + \frac{t}{a}\right).$$

From (12.6.8) we know that the expansion

$$\ln\left(1 + \frac{t}{a}\right) = \frac{t}{a} - \frac{1}{2}\left(\frac{t}{a}\right)^2 + \frac{1}{3}\left(\frac{t}{a}\right)^3 - \cdots$$

holds for $-1 < t/a \le 1$; that is, for $-a \le t \le a$. Adding $\ln a$ to both sides, we have

$$\ln(a+t) = \ln a + \frac{t}{a} - \frac{1}{2}\left(\frac{t}{a}\right)^2 + \frac{1}{3}\left(\frac{t}{a}\right)^3 - \cdots \qquad \text{for} \quad -a < t \le a.$$

Setting $t = x - a$, we find that

$$\ln x = \ln a + \frac{1}{a}(x-a) - \frac{1}{2a^2}(x-a)^2 + \frac{1}{3a^3}(x-a)^3 - \cdots$$

for $-a < x - a \le a$ and thus for $0 < x \le 2a$. ❏

EXERCISES 12.7

Exercises 1–6. Find the Taylor polynomial of the function f for the given values of a and n and give the Lagrange form of the remainder.

1. $f(x) = \sqrt{x}$; $a = 4$, $n = 3$.
2. $f(x) = \cos x$; $a = \pi/3$, $n = 4$.
3. $f(x) = \sin x$; $a = \pi/4$, $n = 4$.
4. $f(x) = \ln x$; $a = 1$, $n = 5$.
5. $f(x) = \arctan x$; $a = 1$, $n = 3$.
6. $f(x) = \cos \pi x$; $a = \frac{1}{2}$, $n = 4$.

Exercises 7–22. Expand $g(x)$ as indicated and specify the values of x for which the expansion is valid.

7. $g(x) = 3x^3 - 2x^2 + 4x + 1$ in powers of $x - 1$.
8. $g(x) = x^4 - x^3 + x^2 - x + 1$ in powers of $x - 2$.
9. $g(x) = 2x^5 + x^2 - 3x - 5$ in powers of $x + 1$.
10. $g(x) = x^{-1}$ in powers of $x - 1$.
11. $g(x) = (1+x)^{-1}$ in powers of $x - 1$.
12. $g(x) = (b+x)^{-1}$ in powers of $x - a$, $a \ne -b$.
13. $g(x) = (1-2x)^{-1}$ in powers of $x + 2$.
14. $g(x) = e^{-4x}$ in powers of $x + 1$.
15. $g(x) = \sin x$ in powers of $x - \pi$.
16. $g(x) = \sin x$ in powers of $x - \frac{1}{2}\pi$.
17. $g(x) = \cos x$ in powers of $x - \pi$.
18. $g(x) = \cos x$ in powers of $x - \frac{1}{2}\pi$.
19. $g(x) = \sin \frac{1}{2}\pi x$ in powers of $x - 1$.
20. $g(x) = \sin \pi x$ in powers of $x - 1$.
21. $g(x) = \ln(1 + 2x)$ in powers of $x - 1$.
22. $g(x) = \ln(2 + 3x)$ in powers of $x - 4$.

Exercises 23–32. Expand $g(x)$ as indicated.

23. $g(x) = x \ln x$ in powers of $x - 2$.
24. $g(x) = x^2 + e^{3x}$ in powers of $x - 2$.
25. $g(x) = x \sin x$ in powers of x.
26. $g(x) = \ln(x^2)$ in powers of $x - 1$.
27. $g(x) = (1 - 2x)^{-3}$ in powers of $x + 2$.
28. $g(x) = \sin^2 x$ in powers of $x - \frac{1}{2}\pi$.
29. $g(x) = \cos^2 x$ in powers of $x - \pi$.
30. $g(x) = (1 + 2x)^{-4}$ in powers of $x - 2$.
31. $g(x) = x^n$ in powers of $x - 1$.
32. $g(x) = (x - 1)^n$ in powers of x.
33. (a) Expand e^x in powers of $x - a$.
    (b) Use the expansion to show that $e^{x_1 + x_2} = e^{x_1} e^{x_2}$.
    (c) Expand e^{-x} in powers of $x - a$.
34. (a) Expand $\sin x$ and $\cos x$ in powers of $x - a$.
    (b) Show that both series are absolutely convergent for all real x.
    (c) As noted earlier (Section 12.5), Riemann proved that the order of the terms of an absolutely convergent series may be changed without altering the sum of the series. Use Riemann's discovery and the Taylor expansions of part (a) to derive the addition formulas

$$\sin(x_1 + x_2) = \sin x_1 \cos x_2 + \cos x_1 \sin x_2,$$

$$\cos(x_1 + x_2) = \cos x_1 \cos x_2 - \sin x_1 \sin x_2.$$

▷35. Use a CAS to determine the Taylor polynomial P_6 in powers of $(x - 1)$ for $f(x) = \arctan x$.
▷36. Use a CAS to determine the Taylor polynomial P_8 in powers of $(x - 2)$ for $f(x) = \cosh 2x$.

■ 12.8 POWER SERIES

You have become familiar with Taylor series

$$\sum_{k=0}^{\infty} \frac{f^{(k)}(0)}{k!} x^k \qquad \text{and} \qquad \sum_{k=0}^{\infty} \frac{f^{(k)}(a)}{k!} (x - a)^k.$$

Here we study series of the form

$$\sum_{k=0}^{\infty} a_k x^k \quad \text{and} \quad \sum_{k=0}^{\infty} a_k(x-a)^k$$

without regard to how the coefficients a_k have been generated. Such series are called *power series*: the first is a *power series in x*; the second is a *power series in x − a*.

Since a simple translation converts

$$\sum_{k=0}^{\infty} a_k(x-a)^k \quad \text{into} \quad \sum_{k=0}^{\infty} a_k x^k,$$

we can focus our attention on power series of the form

$$\sum_{k=0}^{\infty} a_k x^k.$$

When detailed indexing is unnecessary, we will omit it and write

$$\sum a_k x^k.$$

DEFINITION 12.8.1

A power series $\sum a_k x^k$ is said to converge

(i) at c if $\sum a_k c^k$ converges;

(ii) on the set S if $\sum a_k x^k$ converges at each $x \in S$.

The following result is fundamental.

THEOREM 12.8.2

If $\sum a_k x^k$ converges at $c \neq 0$, it converges absolutely at all x with $|x| < |c|$.

If $\sum a_k x^k$ diverges at d, then it diverges at all x with $|x| > |d|$.

PROOF Suppose that $\sum a_k c^k$ converges. Then $a_k c^k \to 0$, and, for k sufficiently large,

$$|a_k c^k| \leq 1.$$

This gives

$$|a_k x^k| = |a_k c^k| \left|\frac{x}{c}\right|^k \leq \left|\frac{x}{c}\right|^k.$$

For $|x| < |c|$, we have

$$\left|\frac{x}{c}\right| < 1.$$

The convergence of $\sum |a_k x^k|$ follows by comparison with the geometric series. This proves the first statement.

Suppose now that $\sum a_k d^k$ diverges. There cannot exist x with $|x| > d$ at which $\sum a_k x^k$ converges, for the existence of such an x would imply the absolute convergence of $\sum a_k d^k$. This proves the second statement. ❏

It follows from the theorem we just proved that there are exactly three possibilities for a power series:

Case 1. The series converges only at $x = 0$. This is what happens with

$$\sum k^k x^k.$$

For $x \neq 0$, $k^k x^k \nrightarrow 0$, and so the series cannot converge.

Case 2. The series converges absolutely at all real numbers x. This is what happens with the exponential series

$$\sum \frac{x^k}{k!}.$$

Case 3. There exists a positive number r such that the series converges absolutely for $|x| < r$ and diverges for $|x| > r$. This is what happens with the geometric series

$$\sum x^k.$$

Here there is absolute convergence for $|x| < 1$ and divergence for $|x| > 1$.

Associated with each case is a *radius of convergence*:

In Case 1, we say that the radius of convergence is 0.

In Case 2, we say that the radius of convergence is ∞.

In Case 3, we say that the radius of convergence is r.

The three cases are pictured in Figure 12.8.1.

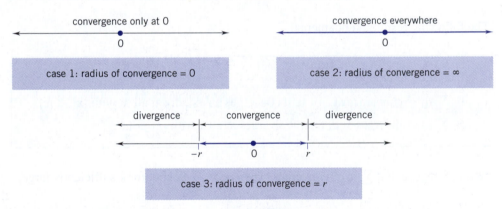

Figure 12.8.1

In general, the behavior of a power series at $-r$ and at r is not predictable. For example, the series

$$\sum x^k, \qquad \sum \frac{(-1)^k}{k} x^k, \qquad \sum \frac{1}{k} x^k, \qquad \sum \frac{1}{k^2} x^k$$

all have radius of convergence 1, but the first series converges only on $(-1, 1)$, the second series converges on $(-1, 1]$, the third on $[-1, 1)$, and the fourth on $[-1, 1]$.

The maximal interval on which a power series converges is called the *interval of convergence*. For a series with infinite radius of convergence, the interval of convergence is $(-\infty, \infty)$. For a series with radius of convergence r, the interval of convergence can be $[-r, r]$, $(-r, r]$, $[-r, r)$, or $(-r, r)$. For a series with radius of convergence 0, the interval of convergence collapses to $\{0\}$.

Example 1 Verify that the series

(1)
$$\sum \frac{(-1)^k}{k} x^k$$

has interval of convergence $(-1, 1]$.

SOLUTION First we show that the radius of convergence is 1 (namely, that the series converges absolutely for $|x| < 1$ and diverges for $|x| > 1$). We do this by forming the series

(2)
$$\sum \left| \frac{(-1)^k}{k} x^k \right| = \sum \frac{1}{k} |x|^k$$

and applying the ratio test.

We set

$$b_k = \frac{1}{k} |x|^k$$

and note that

$$\frac{b_{k+1}}{b_k} = \frac{|x|^{k+1}/(k+1)}{|x|^k/k} = \frac{k}{k+1} \frac{|x|^{k+1}}{|x|^k} = \frac{k}{k+1} |x| \to |x|.$$

By the ratio test, series (2) converges for $|x| < 1$ and diverges for $|x| > 1$.[†] It follows that series (1) converges absolutely for $|x| < 1$ and diverges for $|x| > 1$. The radius of convergence is therefore 1.

Now we test the endpoints $x = -1$ and $x = 1$. At $x = -1$,

$$\sum \frac{(-1)^k}{k} x^k \qquad \text{becomes} \qquad \sum \frac{(-1)^k}{k} (-1)^k = \sum \frac{1}{k}.$$

This is the harmonic series, which, as you know, diverges. At $x = 1$,

$$\sum \frac{(-1)^k}{k} x^k \qquad \text{becomes} \qquad \sum \frac{(-1)^k}{k}.$$

This is a convergent alternating series.

We have shown that series (1) converges absolutely for $|x| < 1$, diverges at -1, and converges at 1. The interval of convergence is $(-1, 1]$. ❏

Example 2 Verify that the series

(1)
$$\sum \frac{1}{k^2} x^k$$

has interval of convergence $[-1, 1]$.

SOLUTION First we examine the series

(2)
$$\sum \left| \frac{1}{k^2} x^k \right| = \sum \frac{1}{k^2} |x|^k.$$

Again we use the ratio test. (The root test also works.) We set

$$b_k = \frac{1}{k^2} |x|^k$$

[†]We could also have used the root test:

$$(b_k)^{1/k} = \left| \frac{1}{k} \right|^{1/k} |x| = \frac{1}{k^{1/k}} |x| \to |x|.$$

and note that

$$\frac{b_{k+1}}{b_k} = \frac{k^2}{(k+1)^2}\frac{|x|^{k+1}}{|x|^k} = \left(\frac{k}{k+1}\right)^2 |x| \to |x|.$$

By the ratio test, (2) converges for $|x| < 1$ and diverges for $|x| > 1$. This shows that (1) converges absolutely for $|x| < 1$ and diverges for $|x| > 1$. The radius of convergence is therefore 1.

Now for the endpoints. At $x = -1$,

$$\sum \frac{1}{k^2}x^k \qquad \text{takes the form} \qquad \sum \frac{(-1)^k}{k^2} = -1 + \tfrac{1}{4} - \tfrac{1}{9} + \tfrac{1}{16} - \cdots.$$

This is a convergent alternating series. At $x = 1$,

$$\sum \frac{1}{k^2}x^k \qquad \text{becomes} \qquad \sum \frac{1}{k^2}.$$

This is a convergent p-series. The interval of convergence is therefore the closed interval $[-1, 1]$. ❏

Example 3 Find the interval of convergence of the series

(1)
$$\sum \frac{k}{6^k}x^k.$$

SOLUTION We begin by examining the series

(2)
$$\sum \left|\frac{k}{6^k}x^k\right| = \sum \frac{k}{6^k}|x|^k.$$

We set

$$b_k = \frac{k}{6^k}|x|^k$$

and apply the root test. (The ratio test also works.) Since

$$(b_k)^{1/k} = \tfrac{1}{6}k^{1/k}|x| \to \tfrac{1}{6}|x|, \qquad\qquad \text{(Recall that } k^{1/k} \to 1.)$$

you can see that (2) converges

$$\text{for} \quad \tfrac{1}{6}|x| < 1, \qquad \text{that is, for} \quad |x| < 6,$$

and diverges

$$\text{for} \quad \tfrac{1}{6}|x| > 1, \qquad \text{that is, for} \quad |x| > 6.$$

Testing the endpoints:

$$\text{at} \quad x = 6, \quad \sum \frac{k}{6^k}6^k = \sum k, \qquad \text{which is divergent;}$$

$$\text{at} \quad x = -6 \quad \sum \frac{k}{6^k}(-6)^k = \sum(-1)^k k, \qquad \text{which is also divergent.}$$

The interval of convergence is the open interval $(-6, 6)$. ❏

Example 4 Find the interval of convergence of the series

$$(1) \qquad \sum \frac{k!}{(3k)!} x^k.$$

SOLUTION Proceeding as before, we begin by examining the series

$$(2) \qquad \sum \left| \frac{k!}{(3k)!} x^k \right| = \sum \frac{k!}{(3k)!} |x|^k.$$

We set

$$b_k = \frac{k!}{(3k)!} |x|^k.$$

Since factorials are involved, we use the ratio test. Note that

$$\frac{b_{k+1}}{b_k} = \frac{(k+1)!}{[3(k+1)]!} \cdot \frac{(3k)!}{k!} \cdot \frac{|x|^{k+1}}{|x|^k} = \frac{k+1}{(3k+3)(3k+2)(3k+1)} |x|$$

$$= \frac{1}{3(3k+2)(3k+1)} |x|.$$

Since

$$\frac{1}{3(3k+2)(3k+1)} \to 0,$$

the ratio b_{k+1}/b_k tends to 0 no matter what x is. By the ratio test, (2) converges at all x and therefore (1) converges absolutely at all x. The radius of convergence is ∞ and the interval of convergence is $(-\infty, \infty)$. ❏

Example 5 Find the interval of convergence of the series $\sum \frac{k^k}{2^k} x^k$.

SOLUTION We set $b_k = \frac{k^k}{2^k} |x|^k$ and apply the root test. Since $(b_k)^{1/k} = \frac{1}{2}k|x| \to \infty$ for every $x \neq 0$, the series diverges for all $x \neq 0$; the series converges only at $x = 0$. ❏

Example 6 Find the interval of convergence of the series

$$\sum \frac{(-1)^k}{k^2 3^k} (x+2)^k.$$

SOLUTION We consider the series

$$\sum \left| \frac{(-1)^k}{k^2 3^k} (x+2)^k \right| = \sum \frac{1}{k^2 3^k} |x+2|^k.$$

We set

$$b_k = \frac{1}{k^2 3^k} |x+2|^k$$

and apply the ratio test (the root test will work equally well):

$$\frac{b_{k+1}}{b_k} = \frac{k^2 3^k}{(k+1)^2 3^{k+1}} \cdot \frac{|x+2|^{k+1}}{|x+2|^k} = \frac{k^2}{3(k+1)^2} |x+2| \to \frac{1}{3} |x+2|.$$

The series is absolutely convergent for $-5 < x < 1$:

$$\tfrac{1}{3}|x+2| < 1 \qquad \text{iff} \qquad |x+2| < 3 \qquad \text{iff} \qquad -5 < x < 1.$$

We now check the endpoints. At $x = -5$,

$$\sum \frac{(-1)^k}{k^2 3^k}(-3)^k = \sum \frac{1}{k^2}.$$

This is a convergent p-series. At $x = 1$,

$$\sum \frac{(-1)^k}{k^2 3^k}(3)^k = \sum \frac{(-1)^k}{k^2}.$$

This is a convergent alternating series. The interval of convergence is the closed interval $[-5, 1]$. ❑

EXERCISES 12.8

1. Suppose that the series $\sum_{k=0}^{\infty} a_k x^k$ converges at $x = 3$. What can you conclude about the convergence or divergence of the following series?

(a) $\sum_{k=0}^{\infty} a_k 2^k$.

(b) $\sum_{k=0}^{\infty} a_k (-2)^k$.

(c) $\sum_{k=0}^{\infty} a_k (-3)^k$.

(d) $\sum_{k=0}^{\infty} a_k 4^k$.

2. Suppose that the series $\sum_{k=0}^{\infty} a_k x^k$ converges at $x = -3$ and diverges at $x = 5$. What can you conclude about the convergence or divergence of the following series?

(a) $\sum_{k=0}^{\infty} a_k 2^k$.

(b) $\sum_{k=0}^{\infty} a_k (-6)^k$.

(c) $\sum_{k=0}^{\infty} a_k 4^k$.

(d) $\sum_{k=0}^{\infty} (-1)^k a_k 3^k$.

Exercises 3–40. Find the interval of convergence.

3. $\sum k x^k$.

4. $\sum \frac{1}{k} x^k$.

5. $\sum \frac{1}{(2k)!} x^4$.

6. $\sum \frac{2^k}{k^2} x^k$.

7. $\sum (-k)^{2k} x^{2k}$.

8. $\sum \frac{(-1)^k}{\sqrt{k}} x^k$.

9. $\sum \frac{1}{k 2^k} x^k$.

10. $\sum \frac{1}{k^2 2^k} x^k$.

11. $\sum \left(\frac{k}{100} \right)^k x^k$.

12. $\sum \frac{k^2}{1 + k^2} x^k$.

13. $\sum \frac{2^k}{\sqrt{k}} x^k$.

14. $\sum \frac{1}{\ln k} x^k$.

15. $\sum \frac{k-1}{k} x^k$.

16. $\sum k a^k x^k$.

17. $\sum \frac{k}{10^k} x^k$.

18. $\sum \frac{3k^2}{e^k} x^k$.

19. $\sum \frac{x^k}{k^k}$.

20. $\sum \frac{7^k}{k!} x^k$.

21. $\sum \frac{(-)^k}{k^k}(x - 2)^k$.

22. $\sum k! x^k$.

23. $\sum (-1)^k \frac{2^k}{3^{k+1}} x^k$.

24. $\sum \frac{2^k}{(2k)!} x^k$.

25. $\sum (-1)^k \frac{k!}{k^3}(x - 1)^k$.

26. $\sum \frac{(-e)^k}{k^2} x^k$.

27. $\sum \left(\frac{k}{k-1} \right) \frac{(x+2)^k}{2^k}$.

28. $\sum \frac{\ln k}{k}(x + 1)^k$.

29. $\sum (-1)^k \frac{k^2}{(k+1)!}(x + 3)^k$.

30. $\sum \frac{k^3}{e^k}(x - 4)^k$.

31. $\sum \left(1 + \frac{1}{k} \right)^k x^k$.

32. $\sum \frac{(-1)^k a^k}{k^2}(x - a)^k$.

33. $\sum \frac{\ln k}{2^k}(x - 2)^k$.

34. $\sum \frac{1}{(\ln k)^k}(x - 1)^k$.

35. $\sum (-1)^k (\frac{2}{3})^k (x + 1)^k$.

36. $\sum \frac{2^{1/k} \pi^k}{k(k+1)(k+2)}(x - 2)^k$.

37. $1 - \frac{x}{2} + \frac{2x^2}{4} - \frac{3x^2}{8} + \frac{4x^4}{16} - \cdots$.

38. $\frac{(x-1)}{5^2} + \frac{4}{5^4}(x - 1)^2 + \frac{9}{5^6}(x - 1)^3 + \frac{16}{5^8}(x - 1)^4 + \cdots$.

39. $\frac{3x^2}{4} + \frac{9x^4}{9} + \frac{27x^6}{16} + \frac{81x^8}{25} + \cdots$.

40. $\frac{1}{16}(x + 1) - \frac{2}{25}(x + 1)^2 + \frac{3}{36}(x + 1)^3 - \frac{4}{49}(x + 1)^4 + \cdots$.

41. Suppose that the series $\sum_{k=0}^{\infty} a_k (x - 1)^k$ converges at $x = 3$. What can you conclude about the convergence or divergence of the following series?

(a) $\sum_{k=0}^{\infty} a_k$.

(b) $\sum_{k=0}^{\infty} (-1)^k a_k$.

(c) $\sum_{k=0}^{\infty} (-1)^k a_k 2^k$.

42. Suppose that the series $\sum_{k=0}^{\infty} a_k (x + 2)^k$ converges at $x = 4$. At what other values of x must the series converge ? Does the series necessarily converge at $x = -8$?

43. Let $\sum a_k x^k$ be a series with radius of convergence $r > 0$.

(a) Show that if the series is absolutely convergent at one endpoint of its interval of convergence, then it is absolutely convergent at the other endpoint.

(b) Show that if the interval of convergence is $(-r, r]$ then the series is only conditionally convergent at r.

44. Let $r > 0$ be arbitrary. Give an example of a power series $\sum a_k x^k$ with radius of convergence r.

45. The power series $\sum_{k=0}^{\infty} a_k x^k$ has the property that $a_{k+3} = a_k$ for all $k \geq 0$.

(a) What is the radius of convergence of the series?

(b) What is the sum of the series?

46. Find the interval of convergence of the series $\sum s_k x^k$ where s_k is the kth partial sum of the series

$$\sum_{n=1}^{\infty} \frac{1}{n}.$$

47. Let $\sum_{k=0}^{\infty} a_k x^k$ be a power series with radius of convergence r, r possibly infinite.

(a) Given that $|a_k|^{1/k} \to \rho$ show that, if $\rho \neq 0$, then $r = 1/\rho$ and, if $\rho = 0$, then $r = \infty$.

(b) Given that $|a_{k+1}/a_k| \to \lambda$ show that, if $\lambda \neq 0$, then $r = 1/\lambda$ and, if $\lambda = 0$, then $r = \infty$.

48. Let $\sum a_k x^k$ be a power series with finite radius of convergence r. Show that the power series $\sum a_k x^{2k}$ has radius of convergence $\sqrt{r}$.

■ 12.9 DIFFERENTIATION AND INTEGRATION OF POWER SERIES

Differentiation of Power Series

We begin with a simple but important result.

> **THEOREM 12.9.1**
>
> If
>
> $$\sum_{k=0}^{\infty} a_k x^k = a_0 + a_1 x + a_2 x^2 + \cdots + a_n x^n + \cdots$$
>
> converges on $(-c, c)$, then
>
> $$\sum_{k=0}^{\infty} \frac{d}{dx}(a_k x^k) = a_1 + 2a_2 x + \cdots + na_n x^{n-1} + \cdots$$
>
> also converges on $(-c, c)$.

PROOF Assume that

$$\sum_{k=0}^{\infty} a_k x^k \qquad \text{converges on } (-c, c).$$

Then it converges absolutely on this interval. (You can reason this out from Theorem 12.8.2.)

Now let x be some number in $(-c, c)$ and choose $\epsilon > 0$ such that

$$|x| < |x| + \epsilon < c.$$

Since $|x| + \epsilon$ lies within the interval of convergence,

$$\sum_{k=0}^{\infty} |a_k(|x| + \epsilon)^k| \qquad \text{converges.}$$

As you are asked to show in Exercise 48, for all k sufficiently large,

$$|k\, x^{k-1}| \leq (|x| + \epsilon)^k.$$

It follows that for all such k,

$$|ka_k x^{k-1}| \le |a_k(|x| + \epsilon)^k|.$$

Since

$$\sum_{k=0}^{\infty} |a_k(|x| + \epsilon)^k| \qquad \text{converges,}$$

we can conclude that

$$\sum_{k=0}^{\infty} \left| \frac{d}{dx}(a_k x^k) \right| = \sum_{k=1}^{\infty} |ka_k x^{k-1}| \qquad \text{converges}$$

and thus that

$$\sum_{k=0}^{\infty} \frac{d}{dx}(a_k x^k) = \sum_{k=1}^{\infty} ka_k x^{k-1} \qquad \text{converges.} \quad \square$$

Repeated application of the theorem shows that

$$\sum_{k=0}^{\infty} \frac{d^2}{dx^2}(a_k x^k), \qquad \sum_{k=0}^{\infty} \frac{d^3}{dx^3}(a_k x^k), \qquad \sum_{k=0}^{\infty} \frac{d^4}{dx^4}(a_k x^k), \qquad \text{and so on}$$

all converge on $(-c, c)$.

Example 1 Since the geometric series

$$\sum_{k=0}^{\infty} x^k = 1 + x + x^2 + x^3 + x^4 + x^5 + x^6 + \cdots.$$

converges on $(-1, 1)$, the series

$$\sum_{k=0}^{\infty} \frac{d}{dx}(x^k) = \sum_{k=1}^{\infty} kx^{k-1} = 1 + 2x + 3x^2 + 4x^3 + 5x^4 + 6x^5 + \cdots$$

$$\sum_{k=0}^{\infty} \frac{d^2}{dx^2}(x^k) = \sum_{k=2}^{\infty} k(k-1)x^{k-2} = 2 + 6x + 12x^2 + 20x^3 + 30x^4 + \cdots$$

$$\sum_{k=0}^{\infty} \frac{d^3}{dx^3}(x^k) = \sum_{k=3}^{\infty} k(k-1)(k-2)x^{k-3} = 6 + 24x + 60x^2 + 120x^3 + \cdots$$

and so on

all converge on $(-1, 1)$. $\quad \square$

Remark That $\sum a_k x^k$ and $\sum ka_k x^k$ have the same radius of convergence does not indicate that they converge at exactly the same points. If the series have a finite radius of convergence r, it can happen that the first series converges at r and/or $-r$ and the second series does not. For example

$$\sum_{k=1}^{\infty} \frac{1}{k^2} x^k$$

converges on $[-1, 1]$, but the derived series

$$\sum_{k=1}^{\infty} \frac{1}{k} x^k$$

converges only on $[-1, 1)$. At $x = 1$ this series diverges. ❏

Suppose now that

$$\sum_{k=0}^{\infty} a_k x^k \qquad \text{converges on } (-c, c).$$

Then, as we have seen,

$$\sum_{k=0}^{\infty} \frac{d}{dx}(a_k x^k) \qquad \text{also converges on } (-c, c).$$

Using the first series, we can define a function f on $(-c, c)$ by setting

$$f(x) = \sum_{k=0}^{\infty} a_k x^k.$$

Using the second series, we can define a function g on $(-c, c)$ by setting

$$g(x) = \sum_{k=0}^{\infty} \frac{d}{dx}(a_k x^k).$$

The crucial point is that

$$f'(x) = g(x).$$

THEOREM 12.9.2 THE DIFFERENTIABILITY THEOREM

If

$$f(x) = \sum_{k=0}^{\infty} a_k x^k \qquad \text{for all } x \text{ in } (-c, c),$$

then f is differentiable on $(-c, c)$ and

$$f'(x) = \sum_{k=0}^{\infty} \frac{d}{dx}(a_k x^k) \qquad \text{for all } x \text{ in } (-c, c).$$

By applying this theorem to f', you can see that f' is itself differentiable. This in turn implies that f'' is differentiable, and so on. In short, f has derivatives of all orders. The discussion up to this point can be summarized as follows:

In the interior of its interval of convergence a power series defines an infinitely differentiable function the derivatives of which can be obtained by differentiating term by term:

$$\frac{d^n}{dx^n}\left(\sum_{k=0}^{\infty} a_k x^k\right) = \sum_{k=0}^{\infty} \frac{d^n}{dx^n}(a_k x^k) \qquad \text{for all } n.$$

For a detailed proof of the differentiability theorem, see the supplement at the end of this section. We go on to examples.

Example 2 You know that $\dfrac{d}{dx}(e^x) = e^x$. You can see this directly by differentiating the exponential series:

$$\frac{d}{dx}(e^x) = \frac{d}{dx}\left(\sum_{k=0}^{\infty}\frac{x^k}{k!}\right) = \sum_{k=0}^{\infty}\frac{d}{dx}\left(\frac{x^k}{k!}\right) = \sum_{k=1}^{\infty}\frac{x^{k-1}}{(k-1)!} = \sum_{n=0}^{\infty}\frac{x^n}{n!} = e^x. \quad \square$$

$$\underset{\text{set } n = k-1}{}$$

Example 3 You have seen that

$$\sin x = x - \frac{x^3}{3!} + \frac{x^5}{5!} - \frac{x^7}{7!} + \frac{x^9}{9!} - \cdots, \qquad \cos x = 1 - \frac{x^2}{2!} + \frac{x^4}{4!} - \frac{x^6}{6!} + \frac{x^8}{8!} - \cdots.$$

The relations

$$\frac{d}{dx}(\sin x) = \cos x, \qquad \frac{d}{dx}(\cos x) = -\sin x$$

can be confirmed by differentiating these series term by term:

$$\frac{d}{dx}(\sin x) = 1 - \frac{3x^2}{3!} + \frac{5x^4}{5!} - \frac{7x^6}{7!} + \frac{9x^8}{9!} - \cdots$$

$$= 1 - \frac{x^2}{2!} + \frac{x^4}{4!} - \frac{x^6}{6!} + \frac{x^8}{8!} - \cdots = \cos x$$

$$\frac{d}{dx}(\cos x) = -\frac{2x}{2!} + \frac{4x^3}{4!} - \frac{6x^5}{6!} + \frac{8x^7}{8!} - \cdots$$

$$= -x + \frac{x^3}{3!} - \frac{x^5}{5!} + \frac{x^7}{7!} - \cdots$$

$$= -\left(x - \frac{x^3}{3!} + \frac{x^5}{5!} - \frac{x^7}{7!} + \cdots\right) = -\sin x. \quad \square$$

Example 4 We can sum the series $\displaystyle\sum_{k=1}^{\infty}\frac{x^k}{k}$ for all x in $(-1, 1)$ by setting

$$g(x) = \sum_{k=1}^{\infty}\frac{x^k}{k} \qquad \text{for all } x \text{ in } (-1, 1)$$

and noting that

$$g'(x) = \sum_{k=1}^{\infty}\frac{kx^{k-1}}{k} = \sum_{k=1}^{\infty}x^{k-1} = \sum_{n=0}^{\infty}x^n = \frac{1}{1-x}.$$

$$\underset{\text{the geometric series}}{}$$

Since

$$g'(x) = \frac{1}{1-x} \qquad \text{and} \qquad g(0) = 0,$$

we know that

$$g(x) = -\ln(1-x) = \ln\left(\frac{1}{1-x}\right).$$

It follows that

$$\sum_{k=1}^{\infty}\frac{x^k}{k} = \ln\left(\frac{1}{1-x}\right) \qquad \text{for all } x \text{ in } (-1, 1). \quad \square$$

Integration of Power Series

Power series can be integrated term by term.

THEOREM 12.9.3 TERM-BY-TERM INTEGRATION

If $f(x) = \sum_{k=0}^{\infty} a_k x^k$ converges on $(-c, c)$,   then

$F(x) = \sum_{k=0}^{\infty} \frac{a_k}{k+1} x^{k+1}$ converges on $(-c, c)$ and $\int f(x)\,dx = F(x) + C.$

PROOF Suppose that $\sum_{k=0}^{\infty} a_k x^k$ converges on $(-c, c)$. Then $\sum_{k=0}^{\infty} |a_k x^k|$ also converges on $(-c, c)$. (Theorem 12.8.2) Since

$$\left| \frac{a_k}{k+1} x^k \right| \leq |a_k x^k| \qquad \text{for all } k,$$

we know by basic comparison that

$$\sum_{k=0}^{\infty} \left| \frac{a_k}{k+1} x^k \right| \qquad \text{converges on } (-c, c).$$

It follows that

$$x \sum \frac{a_k}{k+1} x^k = \sum_{k=0}^{\infty} \frac{a_k}{k+1} x^{k+1} \qquad \text{converges on } (-c, c).$$

Since

$$f(x) = \sum_{k=0}^{\infty} a_k x^k \qquad \text{and} \qquad F(x) = \sum_{k=0}^{\infty} \frac{a_k}{k+1} x^{k+1},$$

we know from the differentiability theorem that

$$F'(x) = f(x) \qquad \text{and thus} \qquad \int f(x) = F(x) + C. \quad \square$$

Term-by-term integration can be expressed by writing

(12.9.4)
$$\int \left(\sum_{k=0}^{\infty} a_k x^k \right) dx = \left(\sum_{k=0}^{\infty} \frac{a_k}{k+1} x^{k+1} \right) + C.$$

Example 5 You are familiar with the series expansion

$$\frac{1}{1+x} = \frac{1}{1-(-x)} = \sum_{k=0}^{\infty} (-1)^k x^k.$$

It is valid for all x in $(-1, 1)$ and for no other x. Integrating term by term, we have

$$\ln(1 + x) = \int \left(\sum_{k=0}^{\infty} (-1)^k x^k \right) dx = \left(\sum_{k=0}^{\infty} \frac{(-1)^k}{k+1} x^{k+1} \right) + C$$

for all x in $(-1, 1)$. At $x = 0$ both $\ln(1 + x)$ and the series on the right are 0. It follows that $C = 0$ and thus

$$\ln(1 + x) = \sum_{k=1}^{\infty} \frac{(-1)^k}{k+1} x^{k+1} = x - \frac{x^2}{2} + \frac{x^3}{3} - \frac{x^4}{4} + \cdots$$

for all x in $(-1, 1)$. ❏

In Section 12.6 we asserted that this expansion for $\ln(1 + x)$ was valid on the half-closed interval $(-1, 1]$. This gave us an expansion for $\ln 2$. Term-by-term integration gives us only the open interval $(-1, 1)$. Well, you may say, it's easy to see that the logarithm series also converges at $x = 1$.[†] True enough, but how do we know that it converges to $\ln 2$? The answer lies in a theorem proved by the Norwegian mathematician Niels Henrik Abel (1802–1829).

THEOREM 12.9.5   ABEL'S THEOREM

Suppose that $\sum_{k=0}^{\infty} a_k x^k$ converges on $(-c, c)$ and $f(x) = \sum_{k=0}^{\infty} a_k x^k$ on this interval.

If f is left continuous at c and $\sum_{k=0}^{\infty} a_k c^k$ converges, then

$$f(c) = \sum_{k=0}^{\infty} a_k c^k.$$

If f is right continuous at $-c$ and $\sum_{k=0}^{\infty} a_k(-c)^k$ converges, then

$$f(-c) = \sum_{k=0}^{\infty} a_k(-c)^k.$$

From Abel's theorem, it is evident that the series for $\ln(1 + x)$ does represent the function at $x = 1$ and therefore

$$1 - \frac{1}{2} + \frac{1}{3} - \frac{1}{4} + \cdots = \ln 2.$$

We come now to the arc tangent:

(12.9.6) $\boxed{\arctan x = x - \frac{x^3}{3} + \frac{x^5}{5} - \frac{x^7}{7} + \cdots \qquad \text{for } -1 \leq x \leq 1.}$

[†] An alternating series with $a_k \to 0$.

PROOF For all x in $(-1, 1)$

$$\frac{1}{1 + x^2} = \frac{1}{1 - (-x^2)} = \sum_{k=0}^{\infty} (-1)^k x^{2k}.$$

Integration gives

$$\arctan x = \int \left(\sum_{k=0}^{\infty} (-1)^k x^{2k} \right) dx = \left(\sum_{k=0}^{\infty} \frac{(-1)^k}{2k + 1} x^{2k+1} \right) + C.$$

The constant C is 0, as we can see by noting that both the series on the right and the arc tangent are 0 at $x = 0$. Thus, for all x in $(-1, 1)$, we have

$$\arctan x = \sum_{k=0}^{\infty} \frac{(-1)^k}{2k + 1} x^{2k+1} = x - \frac{x^3}{3} + \frac{x^5}{5} - \frac{x^7}{7} + \cdots.$$

That the series also represents the function at $x = -1$, and $x = 1$ follows directly from Abel's theorem: at both points the arc tangent is continuous in the sense required; at both points the series converges. ❑

Since $\arctan 1 = \frac{1}{4}\pi$, we have

$$\frac{1}{4}\pi = 1 - \frac{1}{3} + \frac{1}{5} - \frac{1}{7} + \frac{1}{9} - \cdots.$$

This series, known to the Scottish mathematician James Gregory by 1671, is an elegant formula for π, but it converges too slowly for computational purposes. A much more effective way of calculating π is outlined in Project 12.9.B.

Since term-by-term integration can be used to obtain an antiderivative, it can be used to obtain a definite integral.

Example 6 For all real x

$$e^x = 1 + x + \frac{x^2}{2!} + \frac{x^3}{3!} + \frac{x^4}{4!} + \frac{x^5}{5!} + \frac{x^6}{6!} + \cdots.$$

It follows that for all real x

$$e^{-x^2} = 1 - x^2 + \frac{x^4}{2!} - \frac{x^6}{3!} + \frac{x^8}{4!} - \frac{x^{10}}{5!} + \frac{x^{12}}{6!} - \cdots$$

and

$$\int_0^1 e^{-x^2} dx = \left[x - \frac{x^3}{3} + \frac{x^5}{5(2!)} - \frac{x^7}{7(3!)} + \frac{x^9}{7(4!)} - \frac{x^{11}}{11(5!)} + \frac{x^{13}}{13(6!)} - \cdots \right]_0^1$$

$$= 1 - \frac{1}{3} + \frac{1}{5(2!)} - \frac{1}{7(3!)} + \frac{1}{9(4!)} - \frac{1}{11(5!)} + \frac{1}{13(6!)} - \cdots.$$

The integral on the left is the sum of the series on the right.

To obtain a decimal estimate for the integral, we work with the series on the right. The series is an alternating series with terms that tend to 0 and have decreasing magnitude. Therefore the integral lies between consecutive partial sums. In particular it lies between

$$1 - \frac{1}{3} + \frac{1}{5(2!)} - \frac{1}{7(3!)} + \frac{1}{9(4!)} - \frac{1}{11(5!)}$$

and

$$\left[1 - \frac{1}{3} + \frac{1}{5(2!)} - \frac{1}{7(3!)} + \frac{1}{9(4!)} - \frac{1}{11(5!)} \right] + \frac{1}{13(6!)}.$$

Simple arithmetic shows that the first sum is greater than 0.74672 and the second sum is less than 0.74684. It follows that

$$0.74672 < \int_0^1 e^{-x^2} dx < 0.74684.$$

Within 0.0001 the integral is 0.7468. ❏

The integral of Example 6 was easy to estimate numerically because it could be expressed as an alternating series to which we could apply the basic theorem on alternating series. The next example requires more subtlety and illustrates a method more general than that used in Example 6.

Example 7 We want a numerical estimate for $\int_0^1 e^{x^2} dx$. Proceeding as we did in Example 6, we find that

$$\int_0^1 e^{x^2} dx = 1 + \frac{1}{3} + \frac{1}{5(2!)} + \frac{1}{7(3!)} + \frac{1}{9(4!)} + \frac{1}{11(5!)} + \frac{1}{13(6!)} + \cdots.$$

We now have a series expansion for the integral, but that expansion does not take us directly to a numerical estimate for the integral. We know that s_n, the nth partial sum of the series, approximates the integral, but we don't know the accuracy of the approximation. We have no handle on the remainder left by s_n.

We start again, this time keeping track of the remainder. For $x \in [0, 1]$,

$$0 \le e^x - \left(1 + x + \frac{x^2}{2!} + \cdots + \frac{x^n}{n!}\right) = R_n(x) \le e\left[\frac{x^{n+1}}{(n+1)!}\right] < \frac{3}{(n+1)!}. \qquad (12.6.3)$$

If $x \in [0, 1]$, then $x^2 \in [0, 1]$ and therefore

$$0 \le e^{x^2} - \left(1 + x^2 + \frac{x^4}{2!} + \cdots + \frac{x^{2n}}{n!}\right) < \frac{3}{(n+1)!}$$

Integrating this inequality from $x = 0$ to $x = 1$, we have

$$0 \le \int_0^1 \left[e^{x^2} - \left(1 + x^2 + \frac{x^4}{2!} + \cdots + \frac{x^{2n}}{n!}\right)\right] dx < \int_0^1 \frac{3}{(n+1)!} dx.$$

Carrying out the integration where possible, we see that

$$0 \le \int_0^1 e^{x^2} dx - \left[1 + \frac{1}{3} + \frac{1}{5(2!)} + \cdots + \frac{1}{(2n+1)(n!)}\right] < \frac{3}{(n+1)!}.$$

We can use this inequality to estimate the integral as closely as we wish. Since

$$\frac{3}{7!} = \frac{1}{1680} < 0.0006, \qquad \text{(we took } n = 6\text{)}$$

we see that

$$\alpha = 1 + \frac{1}{3} + \frac{1}{5(2!)} + \frac{1}{7(3!)} + \frac{1}{9(4!)} + \frac{1}{11(5!)} + \frac{1}{13(6!)}$$

approximates the integral within 0.0006. Arithmetical computation shows that

$$1.4626 \le \alpha \le 1.4627.$$

It follows that

$$1.4626 \leq \int_0^1 e^{x^2}\, dx \leq 1.4627 + 0.0006 = 1.4633.$$

The estimate 1.463 approximates the integral within 0.0004. ❑

Power Series; Taylor Series

It is time to relate Taylor series

$$\sum_{k=0}^{\infty} \frac{f^{(k)}(0)}{k!} x^k$$

to power series in general. The relationship is very simple.

> On its interval of convergence, a power series is the Taylor series of its sum.

To see this, all you have to do is differentiate

$$f(x) = a_0 + a_1 x + a_2 x^2 + \cdots + a_k x^k + \cdots$$

term by term. Do this and you will find that $f^{(k)}(0) = k! a_k$, and therefore

$$a_k = \frac{f^{(k)}(0)}{k!}.$$

The a_k are the Taylor coefficients of f.

We end this section by carrying out a few simple expansions.

Example 8 Expand $\cosh x$ and $\sinh x$ in powers of x.

SOLUTION There is no need to go through the labor of computing the Taylor coefficients

$$\frac{f^{(k)}(0)}{k!}.$$

By definition,

$$\cosh x = \tfrac{1}{2}(e^x + e^{-x}) \quad \text{and} \quad \sinh x = \tfrac{1}{2}(e^x - e^{-x}). \qquad [(7.8.1)]$$

Since

$$e^x = 1 + x + \frac{x^2}{2!} + \frac{x^3}{3!} + \frac{x^4}{4!} + \frac{x^5}{5!} + \cdots,$$

we have

$$e^{-x} = 1 - x + \frac{x^2}{2!} - \frac{x^3}{3!} + \frac{x^4}{4!} - \frac{x^5}{5!} + \cdots.$$

Thus

$$\cosh x = \frac{1}{2}\left(2 + 2\frac{x^2}{2!} + 2\frac{x^4}{4!} + \cdots \right) = 1 + \frac{x^2}{2!} + \frac{x^4}{4!} + \cdots = \sum_{k=0}^{\infty} \frac{x^{2k}}{(2k)!}$$

and

$$\sinh x = \frac{1}{2}\left(2x + 2\frac{x^3}{3!} + 2\frac{x^5}{5!} + \cdots\right) = x + \frac{x^3}{3!} + \frac{x^5}{5!} + \cdots = \sum_{k=0}^{\infty} \frac{x^{2k+1}}{(2k+1)!}.$$

Both expansions are valid for all real x, since the exponential expansions are valid for all real x. ❑

Example 9 Expand $x^2 \cos x^3$ in powers of x.

SOLUTION

$$\cos x = 1 - \frac{x^2}{2!} + \frac{x^4}{4!} - \frac{x^6}{6!} + \cdots.$$

Thus

$$\cos x^3 = 1 - \frac{(x^3)^2}{2!} + \frac{(x^3)^4}{4!} - \frac{(x^3)^6}{6!} + \cdots = 1 - \frac{x^6}{2!} + \frac{x^{12}}{4!} - \frac{x^{18}}{6!} + \cdots$$

and

$$x^2 \cos x^3 = x^2 - \frac{x^8}{2!} + \frac{x^{14}}{4!} - \frac{x^{20}}{6!} + \cdots.$$

This expansion is valid for all real x, since the expansion for $\cos x$ is valid for all real x.

ALTERNATIVE SOLUTION Since

$$x^2 \cos x^3 = \frac{d}{dx}\left(\tfrac{1}{3}\sin x^3\right),$$

we can obtain the expansion for $x^2 \cos x^3$ by expanding $\frac{1}{3}\sin x^3$ and then differentiating term by term. ❑

EXERCISES 12.9

Exercises 1–6. Expand $f(x)$ in powers of x, basing your calculations on the geometric series

$$\frac{1}{1-x} = 1 + x + x^2 + \cdots + x^n + \cdots.$$

1. $f(x) = \dfrac{1}{(1-x)^2}.$

2. $f(x) = \dfrac{1}{(1-x)^3}.$

3. $f(x) = \dfrac{1}{(1-x)^k}.$

4. $f(x) = \ln(1-x).$

5. $f(x) = \ln(1-x^2).$

6. $f(x) = \ln(2-3x).$

Exercises 7–8. Expand $f(x)$ in powers of x, basing your calculations on the tangent series

$$\tan x = x + \tfrac{1}{3}x^3 + \tfrac{2}{15}x^5 + \tfrac{17}{315}x^7 + \cdots.$$

7. $f(x) = \sec^2 x.$

8. $f(x) = \ln\cos x.$

Exercises 9–10. Find $f^{(9)}(0)$.

9. $f(x) = x^2 \sin x.$

10. $f(x) = x\cos x^2.$

Exercises 11–22. Expand $f(x)$ in powers of x.

11. $f(x) = \sin x^2.$

12. $f(x) = x^2 \arctan x.$

13. $f(x) = e^{3x^3}.$

14. $f(x) = \dfrac{1-x}{1+x}.$

15. $f(x) = \dfrac{2x}{1-x^2}.$

16. $f(x) = x\sinh x^2.$

17. $f(x) = \dfrac{1}{1-x} + e^x.$

18. $f(x) = \cosh x \sinh x.$

19. $f(x) = x\ln(1+x^3).$

20. $f(x) = (x^2+x)\ln(1+x).$

21. $f(x) = x^3 e^{-x^3}.$

22. $f(x) = x^5(\sin x + \cos 2x).$

Exercises 23–26. Evaluate the limit (i) by using L'Hôpital's rule, (ii) by using power series.

23. $\displaystyle\lim_{x\to 0} \frac{1-\cos x}{x^2}.$

24. $\displaystyle\lim_{x\to 0} \frac{\sin x - x}{x^2}.$

25. $\displaystyle\lim_{x\to 0} \frac{\cos x - 1}{x\sin x}.$

26. $\displaystyle\lim_{x\to 0} \frac{e^x - 1 - x}{x\arctan x}.$

Exercises 27–30. Find a power series representation for the improper integral.

27. $\displaystyle\int_0^x \frac{\ln(1+t)}{t}\,dt.$

28. $\displaystyle\int_0^x \frac{1-\cos t}{t^2}\,dt.$

29. $\displaystyle\int_0^x \frac{\arctan t}{t}\,dt.$

30. $\displaystyle\int_0^x \frac{\sinh t}{t}\,dt.$

Exercises 31–36. Estimate to within 0.01 by using series.

31. $\int_0^1 e^{-x^3}\,dx.$

32. $\int_0^1 \sin x^2\,dx.$

33. $\int_0^1 \sin\sqrt{x}\,dx.$

34. $\int_0^1 x^4 e^{-x^2}\,dx.$

35. $\int_0^1 \arctan x^2\,dx.$

36. $\int_1^2 \dfrac{1-\cos x}{x}\,dx.$

Exercises 37–40. Use a power series to estimate the integral within 0.0001.

37. $\int_0^1 \dfrac{\sin x}{x}\,dx.$

38. $\int_0^{0.5} \dfrac{1-\cos x}{x^2}\,dx.$

39. $\int_0^{0.5} \dfrac{\ln(1+x)}{x}\,dx.$

40. $\int_0^{0.2} x\sin x\,dx.$

Exercises 41–43. Sum the series.

41. $\displaystyle\sum_{k=0}^\infty \dfrac{1}{k!}x^{3k}.$

42. $\displaystyle\sum_{k=0}^\infty \dfrac{1}{k!}x^{3k+1}.$

43. $\displaystyle\sum_{k=0}^\infty \dfrac{3k}{k!}x^{3k-1}.$

44. Set $f(x) = \dfrac{e^x - 1}{x}.$

    (a) Expand $f(x)$ in a power series.

    (b) Differentiate the series and show that

$$\sum_{n=1}^\infty \dfrac{n}{(n+1)!} = 1.$$

45. Set $f(x) = xe^x.$

    (a) Expand $f(x)$ in a power series.

    (b) Integrate the series and show that

$$\sum_{n=1}^\infty \dfrac{1}{n!(n+2)} = \dfrac{1}{2}.$$

46. Deduce the differentiation formulas

$$\dfrac{d}{dx}(\sinh x) = \cosh x, \qquad \dfrac{d}{dx}(\cosh x) = \sinh x$$

from the expansions of $\sinh x$ and $\cosh x$ in powers of x.

47. Show that, if $\sum a_k x^k$ and $\sum b_k x^k$ both converge to the same sum on some interval, then $a_k = b_k$ for each k.

48. Show that, if $\epsilon > 0$, then

$$|kx^{k-1}| < (|x| + \epsilon)^k \text{ for all } k \text{ sufficiently large.}$$

49. Suppose that the function f has the power series representation $f(x) = \sum_{k=0}^\infty a_k x^k$.

    (a) Show that if f is an even function, then $a_{2k+1} = 0$ for all k.

    (b) Show that if f is an odd function, then $a_{2k} = 0$ for all k.

50. Suppose that the function f is infinitely differentiable on an open interval that contains 0, and suppose that $f'(x) = -2f(x)$ and $f(0) = 1$. Express $f(x)$ as a power series in x. What is the sum of this series?

51. Suppose that the function f is infinitely differentiable on an open interval that contains 0, and suppose that $f''(x) = -2f(x)$ for all x and $f(0) = 0$, $f'(0) = 1$. Express $f(x)$ as a power series in x. What is the sum of this series?

52. Expand $f(x)$, $f'(x)$, and $\int f(x)\,dx$ in power series

    (a) $f(x) = x2^{-x^2}.$

    (b) $f(x) = x\arctan x.$

Exercises 53–55. Estimate within 0.001 by series expansion and check your result by carrying out the integration directly.

53. $\int_0^{1/2} x\ln(1+x)\,dx.$

54. $\int_0^1 x\sin x\,dx.$

55. $\int_0^1 xe^{-x}\,dx.$

56. Show that

$$0 \le \int_0^2 e^{x^2}\,dx - \left[2 + \dfrac{2^3}{3} + \dfrac{2^5}{5(2!)} + \cdots + \dfrac{2^{2n+1}}{(2n+1)n!}\right]$$

$$< \dfrac{e^4 2^{2n+3}}{(n+1)!}.$$

■ **PROJECT 12.9A The Binomial Series**

Starting with the binomial $1 + x$ and raising it to the power α, we obtain the function

$$f(x) = (1+x)^\alpha.$$

Here α is an arbitrary real number different from 0. It can be positive or negative. It can be rational or irrational.

Problem 1. Show that the Taylor series of this function can be written

$$1 + \alpha x + \dfrac{\alpha(\alpha-1)}{2!}x^2 + \dfrac{\alpha(\alpha-1)(\alpha-2)}{3!}x^3 + \cdots.$$

This series is called the *binomial series*.

Problem 2. Show that the binomial series converges absolutely on $(-1, 1)$. HINT: Use the ratio test.

    Denote the sum of the binomial series by $\varphi(x)$.

Problem 3. Use term-by-term differentiation to show that

$$(1+x)\varphi'(x) = \alpha\varphi(x) \qquad \text{for all } x \in (-1, 1).$$

HINT: Compare coefficients.

Problem 4. Form the function

$$g(x) = \dfrac{\varphi(x)}{(1+x)^\alpha}.$$

and show that $g'(x) = 0$ for all $x \in (-1, 1)$. HINT: Differentiate by the quotient rule and apply the identity derived in Problem 3.

From Problem 4 you know that g is constant on $(-1, 1)$. Since $g(0) = 1$, $g(x) = 1$ for all $x \in (-1, 1)$. This tells us

$$\varphi(x) = (1 + x)^\alpha \qquad \text{for all } x \in (-1, 1)$$

and therefore

(12.9.7)
$$(1 + x)^\alpha = 1 + \alpha x + \frac{\alpha(\alpha - 1)}{2!}x^2 + \frac{\alpha(\alpha - 1)(\alpha - 2)}{3!}x^3 + \cdots.$$

This is one of the most important series expansions in mathematics. Replacing x by $-x$ and setting $\alpha = -1$, we obtain the geometric series. The binomial series associated with positive integer values of α have only a finite number of nonzero terms (check this out), and these give the binomial expansions of elementary algebra:

$$(1 + x)^1 = 1 + x$$

$$(1 + x)^2 = 1 + 2x + x^2$$

$$(1 + x)^3 = 1 + 3x + 3x^2 + x^3$$

$$(1 + x)^4 = 1 + 4x + 6x^2 + 4x^3 + x^4$$

and so on.

Problem 5. Use the binomial series to obtain a power series in x for the function given.

a. $f(x) = \sqrt{1 + x}$.   b. $f(x) = \sqrt{1 - x}$.

c. $f(x) = \sqrt{1 + x^2}$.   d. $f(x) = \sqrt{1 - x^2}$.

e. $f(x) = \dfrac{1}{\sqrt{1 + x}}$.   f. $f(x) = \dfrac{1}{\sqrt[4]{1 + x}}$.

Problem 6.

a. Use the binomial series to obtain a power series in x for the function $f(x) = 1/\sqrt{1 - x^2}$.

b. Use the series you found for f to construct a power series for the function $F(x) = \arcsin x$ and give the radius of convergence.

Problem 7.

a. Use the binomial series to obtain a power series in x for the function $f(x) = 1/\sqrt{1 + x^2}$.

b. Use the series you found for f to construct a power series for the function $F(x) = \sinh^{-1} x$ and give the radius of convergence. [Use (7.9.3).]

■ PROJECT 12.9B   Estimating π

The value of π to twenty decimal places is

$$\pi \approx 3.14159\ 26535\ 89793\ 23846.$$

In Exercises 8.7 you were asked to estimate π by estimating the integral

$$\int_0^1 \frac{4}{1 + x^2}\, dx = \frac{\pi}{4}$$

using the trapezoidal rule and Simpson's rule.

In this project we estimate π (much more effectively) by using the arc tangent series

$$\arctan x = x - \frac{x^3}{3} + \frac{x^5}{5} - \frac{x^7}{7} + \cdots \qquad \text{for } -1 \leq x \leq 1$$

and the relation

(1)
$$\frac{\pi}{4} = 4 \arctan \tfrac{1}{5} - \arctan \tfrac{1}{239}.$$

(This relation was discovered in 1706 by John Machin, a Scotsman.)

Problem 1. Verify (1). HINT: Using the addition formula for the tangent function, calculate $\tan(2 \arctan \tfrac{1}{5})$, then $\tan(4 \arctan \tfrac{1}{5})$, and finally $\tan(4 \arctan \tfrac{1}{5} - \arctan \tfrac{1}{239})$.

The arc tangent series gives

$$\arctan \tfrac{1}{5} = \tfrac{1}{5} - \tfrac{1}{3}\left(\tfrac{1}{5}\right)^3 + \tfrac{1}{5}\left(\tfrac{1}{5}\right)^5 - \tfrac{1}{7}\left(\tfrac{1}{5}\right)^7 + \cdots$$

and

$$\arctan \tfrac{1}{239} = \tfrac{1}{239} - \tfrac{1}{3}\left(\tfrac{1}{239}\right)^3 + \tfrac{1}{5}\left(\tfrac{1}{239}\right)^5 - \tfrac{1}{7}\left(\tfrac{1}{239}\right)^7 + \cdots.$$

These are convergent alternating series. (The terms alternate in sign, have decreasing magnitude, and tend to 0.) Thus we know, for example, that

$$\tfrac{1}{5} - \tfrac{1}{3}\left(\tfrac{1}{5}\right)^3 \leq \arctan \tfrac{1}{5} \leq \tfrac{1}{5} - \tfrac{1}{3}\left(\tfrac{1}{5}\right)^3 + \tfrac{1}{5}\left(\tfrac{1}{5}\right)^5$$

and

$$\tfrac{1}{239} - \tfrac{1}{3}\left(\tfrac{1}{239}\right)^3 \leq \arctan \tfrac{1}{239} \leq \tfrac{1}{239}.$$

Problem 2. Show that $3.1459262 < \pi < 3.14159267$ by using six terms of the series for $\arctan \tfrac{1}{5}$ and two terms of the series for $\arctan \tfrac{1}{239}$.

Greater accuracy can be obtained by using more terms. For example, fifteen terms of the series for $\arctan \tfrac{1}{5}$ and just four terms of the series for $\arctan \tfrac{1}{239}$ determine π accurately to twenty decimal places.

Problem 3.

a. Use a CAS to obtain the sum of the first fifteen terms of the series for arctan $\frac{1}{5}$ and the first four terms of the series for arctan $\frac{1}{239}$.

b. Use the result in part (a) to estimate π. Compare your estimate to the twenty-place estimate given at the beginning of this project.

---

*SUPPLEMENT TO SECTION 12.9

PROOF OF THEOREM 12.9.2

Set $\quad f(x) = \sum_{k=0}^{\infty} a_k x^k \quad$ and $\quad g(x) = \sum_{k=0}^{\infty} \frac{d}{dx}(a_k x^k) = \sum_{k=1}^{\infty} k a_k x^{k-1}.$

Select x from $(-c, c)$. We want to show that

$$f'(x) = \lim_{h \to 0} \frac{f(x+h) - f(x)}{h} = g(x).$$

For $x + h$ in $(-c, c)$, $h \neq 0$, we have

$$\left| g(x) - \frac{f(x+h) - f(x)}{h} \right| = \left| \sum_{k=1}^{\infty} k a_k x^{k-1} - \sum_{k=0}^{\infty} \frac{a_k(x+h)^k - a_k x^k}{h} \right|$$

$$= \left| \sum_{k=1}^{\infty} k a_k x^{k-1} - \sum_{k=1}^{\infty} a_k \left[\frac{(x+h)^k - x^k}{h} \right] \right|.$$

By the mean-value theorem,

$$\frac{(x+h)^k - x^k}{h} = k(t_k)^{k-1}$$

for some number t_k between x and $x + h$. Thus we can write

$$\left| g(x) - \frac{f(x+h) - f(x)}{h} \right| = \left| \sum_{k=1}^{\infty} k a_k x^{k-1} - \sum_{k=0}^{\infty} k a_k (t_k)^{k-1} \right|$$

$$= \left| \sum_{k=1}^{\infty} k a_k [x^{k-1} - (t_k)^{k-1}] \right|$$

$$= \left| \sum_{k=2}^{\infty} k a_k [x^{k-1} - (t_k)^{k-1}] \right|.$$

By the mean-value theorem,

$$\frac{x^{k-1} - (t_k)^{k-1}}{x - t_k} = (k-1)(p_{k-1})^{k-2}$$

for some number p_{k-1} between x and t_k. It follows that

$$|x^{k-1} - (t_k)^{k-1}| = |x - t_k||(k-1)(p_{k-1})^{k-2}|.$$

Since $|x - t_k| < |h|$ and $|p_{k-1}| \leq |\alpha|$ where $|\alpha| = \max\{|x|, |x + h|\}$,

$$|x^{k-1} - (t_k)^{k-1}| \leq |h||(k-1)\alpha^{k-2}|.$$

Thus

$$\left| g(x) - \frac{f(x+h) - f(x)}{h} \right| \leq |h| \sum_{k=2}^{\infty} |k(k-1)a_k \alpha^{k-2}|.$$

Since this series converges,

$$\lim_{h \to 0} \left(|h| \sum_{k=2}^{\infty} |k(k-1)a_k \alpha^{k-2}| \right) = 0.$$

This gives

$$\lim_{h \to 0} \left| g(x) - \frac{f(x+h) - f(x)}{h} \right| = 0 \quad \text{and thus} \quad f'(x) = \lim_{h \to 0} \frac{f(x+h) - f(x)}{h} = g(x). \quad \square$$

■ CHAPTER 12. REVIEW EXERCISES

Exercises 1–4. Find the sum of the series.

1. $\sum_{k=0}^{\infty} \left(\frac{3}{4} \right)^k.$

2. $\sum_{k=0}^{\infty} (-1)^k \left(\frac{1}{2} \right)^k.$

3. $\sum_{k=0}^{\infty} \frac{(\ln 2)^k}{k!}.$

4. $\sum_{k=1}^{\infty} \frac{1}{k(k+1)}.$

Exercises 5–20. Determine convergence or divergence if the series has only nonnegative terms; determine whether the series is absolutely convergent, conditionally convergent, or divergent if it contains both positive and negative terms.

5. $\sum_{k=0}^{\infty} \frac{1}{2k+1}.$

6. $\sum_{k=0}^{\infty} \frac{1}{(2k+1)(2k+3)}.$

7. $\sum_{k=0}^{\infty} \frac{(-1)^k}{(k+1)(k+2)}.$

8. $\sum_{k=0}^{\infty} \frac{(-1)^k}{(2k+1)}.$

9. $\sum_{k=0}^{\infty} \frac{(-1)^k (100)^k}{k!}.$

10. $\sum_{k=0}^{\infty} \frac{k+1}{3^k}$

11. $\sum_{k=1}^{\infty} \frac{k!}{k^{k/2}}.$

12. $\sum_{k=0}^{\infty} \frac{k + \cos k}{k^3 + 1}.$

13. $\sum_{k=0}^{\infty} \frac{(-1)^k}{\sqrt{(k+1)(k+2)}}.$

14. $\sum_{k=1}^{\infty} k \left(\frac{3}{4} \right)^k.$

15. $\sum_{k=0}^{\infty} \frac{k^e}{e^k}.$

16. $\sum_{k=1}^{\infty} (-1)^{k-1} \frac{\ln k}{\sqrt{k}}.$

17. $\sum_{k=0}^{\infty} \frac{(2k)!}{2^k k!}.$

18. $\sum_{k=0}^{\infty} \frac{(-1)^k}{\sqrt{k^3 + 1}}.$

19. $\sum_{k=0}^{\infty} \frac{(\arctan k)^2}{1 + k^2}.$

20. $\sum_{k=0}^{\infty} \frac{2^k + k^4}{3^k}.$

Exercises 21–28. Find the Taylor series expansion in powers of x.

21. $f(x) = xe^{2x^2}.$

22. $f(x) = \ln(1 + x^2).$

23. $f(x) = \sqrt{x} \arctan \sqrt{x}.$

24. $f(x) = a^x, a > 0.$

25. $f(x) = x \ln \left(\frac{1 + x^2}{1 - x^2} \right).$

26. $f(x) = (x + x^2) \sin x^2.$

27. $f(x) = (1 - x)^{1/3}$ up to x^3.

28. $f(x) = \arcsin x$ up to x^4.

Exercises 29–36. Find the interval of convergence.

29. $\sum \frac{5^k}{k} x^k.$

30. $\sum \frac{(-1)^k}{3^k} x^{k+1}.$

31. $\sum \frac{2^k}{(2k)!} (x-1)^{2k}.$

32. $\sum \frac{1}{2^k} (x-2)^k.$

33. $\sum \frac{(-1)^k k}{3^{2k}} x^k.$

34. $\sum \frac{k}{2k+1} x^{2k+1}.$

35. $\sum \frac{(-1)^k}{\sqrt{k}} (x+3)^k.$

36. $\sum \frac{k!}{2} (x+1)^k.$

Exercises 37–40. Find the Taylor series expansion of f and give the radius of convergence.

37. $f(x) = e^{-2x}$ in powers of $(x + 1)$.

38. $f(x) = \sin 2x$ in powers of $(x - \pi/4)$.

39. $f(x) = \ln x$ in powers of $(x - 1)$.

40. $f(x) = \sqrt{x + 1}$ in powers of x.

Exercises 41–46. Estimate within the accuracy indicated from a series expansion.

41. $\int_0^{1/2} \frac{dx}{1 + x^4}, \quad 0.01.$

42. $e^{2/3}, \quad 0.01.$

43. $\sqrt[3]{68}, \quad 0.01.$

44. $\int_0^1 x \sin x^4 \, dx, \quad 0.01.$

45. $\sin 48°, \quad 0.0001.$

46. $\int_0^1 x^2 e^{-x^2} \, dx, \quad 0.001.$

47. Use the Lagrange form of the remainder to show that the approximation

$$\sin x \cong x - \tfrac{1}{6} x^3 + \tfrac{1}{120} x^5$$

is accurate to four decimal places for $0 \le x \le \pi/4$.

48. Use the Lagrange form of the remainder to show that the approximation

$$\cos x \cong 1 - \tfrac{1}{2} x^2 + \tfrac{1}{24} x^4 - \tfrac{1}{720} x^6$$

is accurate to five decimal places for $0 \le x \le \pi/4$.

49. Find the sum of the series

$$\sum_{k=1}^{\infty} a_k \quad \text{given that } a_k = \int_{k}^{k+1} xe^{-x}\, dx.$$

50. Show that every sequence of real numbers can be covered by a sequence of open intervals of arbitrarily small total length; namely, show that if $x_1, x_2, x_3, \ldots$ is a sequence of real numbers and ϵ is positive, then there exists a sequence of open intervals (a_n, b_n) with $a_n < x_n < b_n$ such that

$$\sum_{n=1}^{\infty}(b_n - a_n) < \epsilon.$$

51. Prove that the series $\sum_{k=1}^{\infty}(a_{k+1} - a_k)$ converges iff the sequence a_k converges.

52. Determine whether or not the series $\sum_{k=2}^{\infty} a_k$ converges or diverges. If it converges, find the sum.

(a) $a_k = \sum_{n=0}^{\infty}\left(\dfrac{1}{k}\right)^n.$

(b) $a_k = \sum_{n=1}^{\infty}\left(\dfrac{1}{k}\right)^n.$

(c) $a_k = \sum_{n=2}^{\infty}\left(\dfrac{1}{k}\right)^n.$

APPENDIX A

SOME ADDITIONAL TOPICS

■ A.1 ROTATION OF AXES; ELIMINATING THE xy-TERM

Rotation of Axes

We begin with a rectangular coordinate system O-xy. By rotating this system about the origin counterclockwise through an angle of α radians, we obtain a new coordinate system O-XY. See Figure A.1.1.

A point P now has two pairs of rectangular coordinates:

$$(x, y) \text{ in the } O\text{-}xy \text{ system} \qquad \text{and} \qquad (X, Y) \text{ in the } O\text{-}XY \text{ system.}$$

Here we investigate the relation between (x, y) and (X, Y). With P as in Figure A.1.2,

$$x = r \cos(\alpha + \beta), \qquad y = r \sin(\alpha + \beta) \qquad \text{and} \qquad X = r \cos\beta, \quad Y = r \sin\beta.$$

(This follows from 10.2.5.) The addition formulas

$$\cos(\alpha + \beta) = \cos\alpha \cos\beta - \sin\alpha \sin\beta, \qquad \sin(\alpha + \beta) = \sin\alpha \cos\beta + \cos\alpha \sin\beta$$

give

$$x = r\cos(\alpha + \beta) = (\cos\alpha)r\cos\beta - (\sin\alpha)r\sin\beta,$$
$$y = r\sin(\alpha + \beta) = (\sin\alpha)r\cos\beta + (\cos\alpha)r\sin\beta,$$

and therefore

(A.1.1) $\quad x = (\cos\alpha)X - (\sin\alpha)Y, \qquad y = (\sin\alpha)X + (\cos\alpha)Y.$

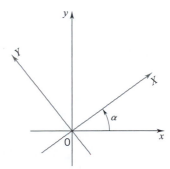

Figure A.1.1

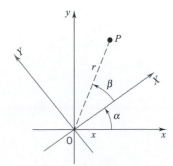

Figure A.1.2

These formulas give the algebraic consequences of a counterclockwise rotation of α radians.

Equations of Second Degree

As you know, the graph of an equation of the form

$$ax^2 + cy^2 + dx + ey + f = 0 \qquad \text{with } a, c \text{ not both } 0$$

is a conic section: except in degenerate cases, a circle, an ellipse, a parabola, or a hyperbola.

The *general equation of second degree in x and y* is an equation of the form

(A.1.2) $\quad ax^2 + bxy + cy^2 + dx + ey + f = 0 \qquad$ with a, b, c not all 0.

The graph is still a conic section, but, in the presence of an xy-term, more difficult to visualize.

Eliminating the *xy*-Term

A rotation of the coordinate system enables us to eliminate the xy-term. As we'll show, if in an O-xy coordinate system a curve γ has an equation of the form

$$(1) \qquad ax^2 + bxy + cy^2 + dx + ey + f = 0 \qquad \text{with } b \neq 0,$$

then there exists a coordinate system O-XY obtainable from the O-xy system by a counterclockwise rotation α of less than $\pi/2$ radians such that in the O-XY system γ has an equation of the form

$$(2) \qquad AX^2 + CY^2 + DX + EY + F = 0 \qquad \text{with } A, C \text{ not both 0.}$$

To see this, substitute

$$x = (\cos \alpha) X - (\sin \alpha)Y, \qquad y = (\sin \alpha) X + (\cos \alpha)Y$$

into equation (1). This will give you a second-degree equation in X and Y in which the coefficient of XY is

$$-2a \cos \alpha \sin \alpha + b(\cos^2 \alpha - \sin^2 \alpha) + 2c \cos \alpha \sin \alpha.$$

This can be simplified to read

$$(c - a) \sin 2\alpha + b \cos 2\alpha.$$

To eliminate the XY-term, we must have this coefficient equal to zero; that is, we must have

$$b \cos 2\alpha = (a - c) \sin 2\alpha.$$

Since we are assuming that $b \neq 0$, we can divide by b and obtain

$$\cos 2\alpha = \left(\frac{a - c}{b} \right) \sin 2\alpha.$$

We know that $\sin 2\alpha \neq 0$ because, if it were 0, then $\cos 2\alpha$ would be 0 and we know that can't be because sine and cosine are never simultaneously 0. Thus we can divide by $\sin 2\alpha$ and obtain

$$\cot 2\alpha = \frac{a - c}{b}.$$

We can find a rotation α that eliminates the xy-term by applying the arc cotangent function:

$$2\alpha = \text{arccot}\left(\frac{a - c}{b} \right) \qquad \text{which gives} \qquad \alpha = \frac{1}{2}\text{arccot}\left(\frac{a - c}{b} \right).$$

In summary, we have shown that every equation of form (1) can be transformed into an equation of form (2) by rotating axes through the angle

(A.1.3)
$$\alpha = \frac{1}{2}\operatorname{arccot}\left(\frac{a-c}{b}\right).$$

Note that, as promised, $\alpha \in (0, \pi/2)$. We leave it to you to show that A and C in (2) are not both zero.

Example 1 In the case of $xy - 2 = 0$, $a = c = 0$, $b = 1$. Setting $\alpha = \frac{1}{2}\operatorname{arccot} 0 = \frac{1}{2}\left(\frac{1}{2}\pi\right) = \frac{1}{4}\pi$, we have

$$x = (\cos \tfrac{1}{4}\pi) X - (\sin \tfrac{1}{4}\pi) Y = \tfrac{1}{2}\sqrt{2}(X - Y),$$
$$y = (\sin \tfrac{1}{4}\pi) X + (\cos \tfrac{1}{4}\pi) Y = \tfrac{1}{2}\sqrt{2}(X + Y).$$

Equation $xy - 2 = 0$ becomes

$$\tfrac{1}{2}(X^2 - Y^2) - 2 = 0,$$

which can be written

$$\frac{X^2}{2^2} - \frac{Y^2}{2^2} = 1.$$

This is the equation of a hyperbola in standard position in the $O\text{-}XY$ system. The hyperbola is sketched in Figure A.1.3. ❑

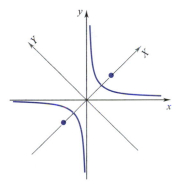

Figure A.1.3

Example 2 In the case of $11x^2 + 4\sqrt{3}xy + 7y^2 - 1 = 0$, we have $a = 11$, $b = 4\sqrt{3}$, $c = 7$. Thus we set

$$\alpha = \tfrac{1}{2}\operatorname{arccot}\left(\frac{11-4}{4\sqrt{3}}\right) = \tfrac{1}{2}\operatorname{arccot}\left(\frac{1}{\sqrt{3}}\right) = \tfrac{1}{6}\pi.$$

As you can verify, equations

$$x = (\cos \tfrac{1}{6}\pi) X - (\sin \tfrac{1}{6}\pi) Y = \tfrac{1}{2}(\sqrt{3}X - Y),$$
$$y = (\sin \tfrac{1}{6}\pi) X + (\cos \tfrac{1}{6}\pi) Y = \tfrac{1}{2}(X + \sqrt{3}Y)$$

transform the initial equation into $13X^2 + 5Y^2 - 1 = 0$. This we write as

$$\frac{X^2}{(1/\sqrt{13})^2} + \frac{Y^2}{(1/\sqrt{5})^2} = 1.$$

This is the equation of an ellipse in standard position in the $O\text{-}XY$ system. The ellipse is sketched in Figure A.1.4. ❑

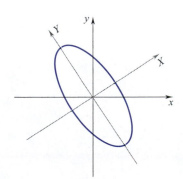

Figure A.1.4

■ A.2 DETERMINANTS

By a *matrix* we mean a rectangular arrangement of numbers enclosed in parentheses. For example,

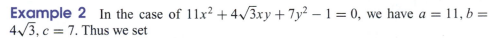

$$\begin{pmatrix} 2 & 4 \\ 3 & 1 \end{pmatrix} \quad \begin{pmatrix} 1 & 6 & 3 \\ 5 & 2 & 2 \end{pmatrix} \quad \begin{pmatrix} 2 & 4 & 0 \\ 4 & 7 & 1 \\ 0 & 1 & 1 \end{pmatrix}$$

are all matrices. The numbers that appear in a matrix are called the *entries*.

Each matrix has a certain number of rows and a certain number of columns. A matrix with m rows and n columns is called an $m \times n$ *matrix*. Thus the first matrix above is a 2×2 matrix, the second a 2×3 matrix, the third a 3×3 matrix. The first and third matrices are called *square*; they have the same number of rows as columns. Here we will be working with square matrices as these are the only ones that have determinants.

We could give a definition of determinant that is applicable to all square matrices, but the definition is complicated and would serve little purpose at this point. Our interest here is in the 2×2 case and in the 3×3 case. We begin with the 2×2 case.

(A.2.1)

> The *determinant* of the matrix
> $$\begin{pmatrix} a_1 & a_2 \\ b_1 & b_2 \end{pmatrix}$$
> is the number $a_1 b_2 - a_2 b_1$.

We have a special notation for the determinant. We change the parentheses of the matrix to vertical bars:

$$\text{determinant of } \begin{pmatrix} a_1 & a_2 \\ b_1 & b_2 \end{pmatrix} = \begin{vmatrix} a_1 & a_2 \\ b_1 & b_2 \end{vmatrix} = a_1 b_2 - a_2 b_1.$$

Thus, for example,

$$\begin{vmatrix} 5 & 8 \\ 4 & 2 \end{vmatrix} = (5 \cdot 2) - (8 \cdot 4) = 10 - 32 = -22,$$

$$\begin{vmatrix} 4 & 0 \\ 0 & \frac{1}{4} \end{vmatrix} = \left(4 \cdot \frac{1}{4}\right) - (0 \cdot 0) = 1.$$

We remark on three properties of 2×2 determinants:

1. If the rows or columns of a 2×2 determinant are interchanged, the determinant changes sign:

$$\begin{vmatrix} b_1 & b_2 \\ a_1 & a_2 \end{vmatrix} = - \begin{vmatrix} a_1 & a_2 \\ b_1 & b_2 \end{vmatrix}, \qquad \begin{vmatrix} a_2 & a_1 \\ b_2 & b_1 \end{vmatrix} = - \begin{vmatrix} a_1 & a_2 \\ b_1 & b_2 \end{vmatrix}.$$

PROOF Just note that

$$b_1 a_2 - b_2 a_1 = -(a_1 b_2 - a_2 b_1) \qquad \text{and} \qquad a_2 b_1 - a_1 b_2 = -(a_1 b_2 - a_2 b_1). \quad ❑$$

2. A common factor can be removed from any row or column and placed as a factor in front of the determinant:

$$\begin{vmatrix} \lambda a_1 & \lambda a_2 \\ b_1 & b_2 \end{vmatrix} = \lambda \begin{vmatrix} a_1 & a_2 \\ b_1 & b_2 \end{vmatrix}, \qquad \begin{vmatrix} \lambda a_1 & a_2 \\ \lambda b_1 & b_2 \end{vmatrix} = \lambda \begin{vmatrix} a_1 & a_2 \\ b_1 & b_2 \end{vmatrix}.$$

PROOF Just note that

$$(\lambda a_1)b_2 - (\lambda a_2)b_1 = \lambda(a_1 b_2 - a_2 b_1)$$

and

$$(\lambda a_1)b_2 - a_2(\lambda b_1) = \lambda(a_1 b_2 - a_2 b_1). \quad \square$$

3. If the rows or columns of a 2×2 determinant are the same, the determinant is 0.

PROOF

$$\begin{vmatrix} a_1 & a_2 \\ a_1 & a_2 \end{vmatrix} = a_1 a_2 - a_2 a_1 = 0, \qquad \begin{vmatrix} a_1 & a_1 \\ b_1 & b_1 \end{vmatrix} = a_1 b_1 - a_1 b_1 = 0. \quad \square$$

The determinant of a 3×3 matrix is harder to define. One definition is this:

$$\begin{vmatrix} a_1 & a_2 & a_3 \\ b_1 & b_2 & b_3 \\ c_1 & c_2 & c_3 \end{vmatrix} = a_1 b_2 c_3 - a_1 b_3 c_2 + a_2 b_3 c_1 - a_2 b_1 c_3 + a_3 b_1 c_2 - a_3 b_2 c_1.$$

The problem with this definition is that it is hard to remember. What saves us is that the expansion on the right can be conveniently written in terms of 2×2 determinants; namely, the expression on the right can be written

$$a_1(b_2 c_3 - b_3 c_2) - a_2(b_1 c_3 - b_3 c_1) + a_3(b_1 c_2 - b_2 c_1),$$

which turns into

$$a_1 \begin{vmatrix} b_2 & b_3 \\ c_2 & c_3 \end{vmatrix} - a_2 \begin{vmatrix} b_1 & b_3 \\ c_1 & c_3 \end{vmatrix} + a_3 \begin{vmatrix} b_1 & b_2 \\ c_1 & c_2 \end{vmatrix}.$$

Thus we have

(A.2.2) $$\begin{vmatrix} a_1 & a_2 & a_3 \\ b_1 & b_2 & b_3 \\ c_1 & c_2 & c_3 \end{vmatrix} = a_1 \begin{vmatrix} b_2 & b_3 \\ c_2 & c_3 \end{vmatrix} - a_2 \begin{vmatrix} b_1 & b_3 \\ c_1 & c_3 \end{vmatrix} + a_3 \begin{vmatrix} b_1 & b_2 \\ c_1 & c_2 \end{vmatrix}.$$

We will take this as our definition. It is called the *expansion of the determinant by elements of the first row*. Note that the coefficients are the entries a_1, a_2, a_3 of the first row, that they occur alternately with $+$ and $-$ signs, and that each is multiplied by a determinant. You can remember which determinant goes with which entry a_i as follows: in the original matrix, mentally cross out the row and column in which the entry a_i is found, and take the determinant of the remaining 2×2 matrix. For example, the determinant that goes with a_3 is

$$\begin{vmatrix} a_1 & a_2 & a_3 \\ b_1 & b_2 & b_3 \\ c_1 & c_2 & c_3 \end{vmatrix} = \begin{vmatrix} b_1 & b_2 \\ c_1 & c_2 \end{vmatrix}.$$

When first starting to work with specific 3×3 determinants, it is a good idea to set up the formula with blank 2×2 determinants:

$$\begin{vmatrix} a_1 & a_2 & a_3 \\ b_1 & b_2 & b_3 \\ c_1 & c_2 & c_3 \end{vmatrix} = a_1 \begin{vmatrix} & \\ & \end{vmatrix} - a_2 \begin{vmatrix} & \\ & \end{vmatrix} + a_3 \begin{vmatrix} & \\ & \end{vmatrix}$$

and then fill in the 2×2 determinants by using the "crossing out" rule explained above.

Example 1

$$\begin{vmatrix} 1 & 2 & 1 \\ 0 & 3 & 4 \\ 6 & 2 & 5 \end{vmatrix} = 1 \begin{vmatrix} 3 & 4 \\ 2 & 5 \end{vmatrix} - 2 \begin{vmatrix} 0 & 4 \\ 6 & 5 \end{vmatrix} + 1 \begin{vmatrix} 0 & 3 \\ 6 & 2 \end{vmatrix}$$

$$= 1(15 - 8) - 2(0 - 24) + 1(0 - 18)$$

$$= 7 + 48 - 18 = 37. \quad \square$$

A straightforward (but somewhat laborious) calculation shows that 3×3 determinants have the three properties we proved earlier for 2×2 determinants.

1. If two rows or columns are interchanged, the determinant changes sign.
2. A common factor can be removed from any row or column and placed as a factor in front of the determinant.
3. If two rows or columns are the same, the determinant is 0.

EXERCISES A.2

Exercises 1–8. Evaluate the following determinants.

1. $\begin{vmatrix} 1 & 2 \\ 3 & 4 \end{vmatrix}$.

2. $\begin{vmatrix} 1 & -1 \\ -1 & 1 \end{vmatrix}$.

3. $\begin{vmatrix} 1 & 1 \\ a & a \end{vmatrix}$.

4. $\begin{vmatrix} a & b \\ b & d \end{vmatrix}$.

5. $\begin{vmatrix} 1 & 0 & 3 \\ 2 & 4 & 1 \\ 0 & 1 & 0 \end{vmatrix}$.

6. $\begin{vmatrix} 1 & 0 & 0 \\ 0 & 2 & 0 \\ 0 & 0 & 3 \end{vmatrix}$.

7. $\begin{vmatrix} 0 & 0 & 1 \\ 0 & 2 & 0 \\ 3 & 0 & 0 \end{vmatrix}$.

8. $\begin{vmatrix} a & 0 & 0 \\ b & c & 0 \\ d & e & f \end{vmatrix}$.

9. If A is a matrix, its *transpose* A^T is obtained by interchanging the rows and columns. Thus

$$\begin{pmatrix} a_1 & a_2 \\ b_1 & b_2 \end{pmatrix}^T = \begin{pmatrix} a_1 & b_1 \\ a_2 & b_2 \end{pmatrix}$$

and

$$\begin{pmatrix} a_1 & a_2 & a_3 \\ b_1 & b_2 & b_3 \\ c_1 & c_2 & c_3 \end{pmatrix}^T = \begin{pmatrix} a_1 & b_1 & c_1 \\ a_2 & b_2 & c_2 \\ a_3 & b_3 & c_3 \end{pmatrix}.$$

Show that the determinant of a matrix equals the determinant of its transpose: (a) for the 2×2 case; (b) for the 3×3 case.

Exercises 10–14. Justify the assertions by invoking the relevant properties of determinants.

10. $\begin{vmatrix} 1 & 2 & 3 \\ 4 & 5 & 6 \\ 7 & 8 & 9 \end{vmatrix} + \begin{vmatrix} 4 & 5 & 6 \\ 1 & 2 & 3 \\ 7 & 8 & 9 \end{vmatrix} = 0.$

11. $\begin{vmatrix} 1 & 2 & 3 \\ 4 & 5 & 6 \\ 7 & 8 & 9 \end{vmatrix} = \begin{vmatrix} 4 & 5 & 6 \\ 7 & 8 & 9 \\ 1 & 2 & 3 \end{vmatrix}.$

12. $\begin{vmatrix} 1 & 2 & 3 \\ 4 & 5 & 6 \\ 7 & 8 & 9 \end{vmatrix} + \begin{vmatrix} 1 & 2 & 3 \\ 1 & 2 & 3 \\ 7 & 8 & 9 \end{vmatrix} = \begin{vmatrix} 1 & 2 & 3 \\ 4 & 5 & 6 \\ 7 & 8 & 9 \end{vmatrix}.$

13. $\frac{1}{2} \begin{vmatrix} 1 & 0 & 7 \\ 3 & 4 & 5 \\ 2 & 4 & 6 \end{vmatrix} = \begin{vmatrix} 1 & 0 & 7 \\ 3 & 4 & 5 \\ 1 & 2 & 3 \end{vmatrix}.$

14. $\begin{vmatrix} 1 & 2 & 3 \\ x & 2x & 3x \\ 4 & 5 & 6 \end{vmatrix} = 0.$

15. (a) Verify that the equations

$$3x + 4y = 6$$
$$2x - 3y = 7$$

can be solved by the prescription

$$x = \frac{\begin{vmatrix} 6 & 4 \\ 7 & -3 \end{vmatrix}}{\begin{vmatrix} 3 & 4 \\ 2 & -3 \end{vmatrix}}, \qquad y = \frac{\begin{vmatrix} 3 & 6 \\ 2 & 7 \end{vmatrix}}{\begin{vmatrix} 3 & 4 \\ 2 & -3 \end{vmatrix}}.$$

(b) More generally, verify that the equations

$$a_1 x + a_2 y = d$$
$$b_1 x + b_2 y = e$$

can be solved by the prescription

$$x = \frac{\begin{vmatrix} d & a_2 \\ e & b_2 \end{vmatrix}}{\begin{vmatrix} a_1 & a_2 \\ b_1 & b_2 \end{vmatrix}}, \qquad y = \frac{\begin{vmatrix} a_1 & d \\ b_1 & e \end{vmatrix}}{\begin{vmatrix} a_1 & a_2 \\ b_1 & b_2 \end{vmatrix}}$$

provided that the determinant in the denominator is different from 0.

(c) Devise an analogous rule for solving three linear equations in three unknowns.

16. Show that a 3×3 determinant can be "expanded by the elements of the bottom row" as follows:

$$\begin{vmatrix} a_1 & a_2 & a_3 \\ b_1 & b_2 & b_3 \\ c_1 & c_2 & c_3 \end{vmatrix} = c_1 \begin{vmatrix} a_2 & a_3 \\ b_2 & b_3 \end{vmatrix} - c_2 \begin{vmatrix} a_1 & a_3 \\ b_1 & b_3 \end{vmatrix} + c_3 \begin{vmatrix} a_1 & a_2 \\ b_1 & b_2 \end{vmatrix}.$$

HINT: You can check this directly by writing out the values of the determinants on the right, or you can interchange rows twice to bring the bottom row to the top and then expand by the elements of the top row.

APPENDIX B

SOME ADDITIONAL PROOFS

In this appendix we give some proofs that many would consider too advanced for the main body of the text. Some details are omitted. These are left to you.

The arguments presented in Sections B.1, B.2, and B.4 require some familiarity with the *least upper bound axiom*. This is discussed in Section 11.1. In addition, Section B.4 requires some understanding of *sequences*, for which we refer you to Sections 11.2 and 11.3.

■ B.1 THE INTERMEDIATE-VALUE THEOREM

> **LEMMA B.1.1**
>
> Let f be continuous on [a, b]. If $f(a) < 0 < f(b)$ or $f(b) < 0 < f(a)$, then there is a number c between a and b for which $f(c) = 0$.

PROOF Suppose that $f(a) < 0 < f(b)$. (The other case can be treated in a similar manner.) Since $f(a) < 0$, we know from the continuity of f that there exists a number ξ such that f is negative on $[a, \xi)$. Let

$$c = \text{lub} \{\xi : f \text{ is negative on } [a, \xi)\}.$$

Clearly, $c \le b$. We cannot have $f(c) > 0$, for then f would be positive on some interval extending to the left of c, and we know that, to the left of c, f is negative. Incidentally, this argument excludes the possibility that $c = b$ and means that $c < b$. We cannot have $f(c) < 0$, for then there would be an interval $[a, t)$, with $t > c$, on which f is negative, and this would contradict the definition of c. It follows that $f(c) = 0$. ❑

> **THEOREM B.1.2 THE INTERMEDIATE-VALUE THEOREM**
>
> If f is continuous on $[a, b]$ and K is a number between $f(a)$ and $f(b)$, then there is at least one number c between a and b for which $f(c) = K$.

PROOF Suppose, for example, that

$$f(a) < K < f(b).$$

(The other possibility can be handled in a similar manner.) The function

$$g(x) = f(x) - K$$

is continuous on $[a, b]$. Since

$$g(a) = f(a) - K < 0 \quad \text{and} \quad g(b) = f(b) - K > 0,$$

we know from the lemma that there is a number c between a and b for which $g(c) = 0$. Obviously, then, $f(c) = K$. ❑

■ B.2 BOUNDEDNESS; EXTREME-VALUE THEOREM

LEMMA B.2.1

If f is continuous on $[a, b]$, then f is bounded on $[a, b]$.

PROOF Consider

$$\{x : x \in [a, b] \text{ and } f \text{ is bounded on } [a, x]\}.$$

It is easy to see that this set is nonempty and bounded above by b. Thus we can set

$$c = \text{lub } \{x : f \text{ is bounded on } [a, x]\}.$$

Now we argue that $c = b$. To do so, we suppose that $c < b$. From the continuity of f at c, it is easy to see that f is bounded on $[c - \epsilon, c + \epsilon]$ for some $\epsilon > 0$. Being bounded on $[a, c - \epsilon]$ and on $[c - \epsilon, c + \epsilon]$, it is obviously bounded on $[a, c + \epsilon]$. This contradicts our choice of c. We can therefore conclude that $c = b$. This tells us that f is bounded on $[a, x]$ for all $x < b$. We are now almost through. From the continuity of f, we know that f is bounded on some interval of the form $[b - \epsilon, b]$. Since $b - \epsilon < b$, we know from what we have just proved that f is bounded on $[a, b - \epsilon]$. Being bounded on $[a, b - \epsilon]$ and bounded on $[b - \epsilon, b]$, it is bounded on $[a, b]$. ❑

THEOREM B.2.2 THE EXTREME-VALUE THEOREM

If f is continuous on a bounded closed interval $[a, b]$, then on that interval f takes on both a maximum value M and a minimum value m.

PROOF By the lemma, f is bounded on $[a, b]$. Set

$$M = \text{lub } \{f(x) : x \in [a, b]\}.$$

We must show that there exists c in $[a, b]$ such that $f(c) = M$. To do this, we set

$$g(x) = \frac{1}{M - f(x)}.$$

If f does not take on the value M, then g is continuous on $[a, b]$ and thus, by the lemma, bounded on $[a, b]$. A look at the definition of g makes it clear that g cannot be bounded on $[a, b]$. The assumption that f does not take on the value M has led to a contradiction. (That f takes a minimum value m can be proved in a similar manner.) ❑

■ B.3 INVERSES

> **THEOREM B.3.1 CONTINUITY OF THE INVERSE**
>
> Let f be a one-to-one function defined on an interval (a, b). If f is continuous, then its inverse f^{-1} is also continuous.

PROOF If f is continuous, then, being one-to-one, f either increases throughout (a, b) or it decreases throughout (a, b). The proof of this assertion we leave to you.

Suppose now that f increases throughout (a, b). Let's take c in the domain of f^{-1} and show that f^{-1} is continuous at c.

We first observe that $f^{-1}(c)$ lies in (a, b) and choose $\epsilon > 0$ sufficiently small so that $f^{-1}(c) - \epsilon$ and $f^{-1}(c) + \epsilon$ also lie in (a, b). We seek a $\delta > 0$ such that

$$\text{if } c - \delta < x < c + \delta, \qquad \text{then} \qquad f^{-1}(c) - \epsilon < f^{-1}(x) < f^{-1}(c) + \epsilon.$$

This condition can be met by choosing δ to satisfy

$$f(f^{-1}(c) - \epsilon) < c - \delta \qquad \text{and} \qquad c + \delta < f(f^{-1}(c) + \epsilon),$$

for then, if $c - \delta < x < c + \delta$,

$$f(f^{-1}(c) - \epsilon) < x < f(f^{-1}(c) + \epsilon),$$

and, since f^{-1} also increases,

$$f^{-1}(c) - \epsilon < f^{-1}(x) < f^{-1}(c) + \epsilon.$$

The case where f decreases throughout (a, b) can be handled in a similar manner. ❑

> **THEOREM B.3.2 DIFFERENTIABILITY OF THE INVERSE**
>
> Let f be a one-to-one function differentiable on an open interval I. Let a be a point of I and let $f(a) = b$. If $f'(a) \neq 0$, then f^{-1} is differentiable at b and
>
> $$(f^{-1})'(b) = \frac{1}{f'(a)}.$$

PROOF (Here we use the characterization of derivative spelled out in Theorem 3.5.7.) We take $\epsilon > 0$ and show that there exists a $\delta > 0$ such that

$$\text{if } \quad 0 < |t - b| < \delta, \qquad \text{then} \qquad \left| \frac{f^{-1}(t) - f^{-1}(b)}{t - b} - \frac{1}{f'(a)} \right| < \epsilon.$$

Since f is differentiable at a, there exists a $\delta_1 > 0$ such that

$$\text{if } \quad 0 < |x - a| < \delta_1, \qquad \text{then} \qquad \left| \frac{1}{\dfrac{f(x) - f(a)}{x - a}} - \frac{1}{f'(a)} \right| < \epsilon,$$

and therefore

$$\left| \frac{x - a}{f(x) - f(a)} - \frac{1}{f'(a)} \right| < \epsilon.$$

By the previous theorem, f^{-1} is continuous at b. Hence there exists a $\delta > 0$ such that

$$\text{if} \quad 0 < |t - b| < \delta, \quad \text{then} \quad 0 < |f^{-1}(t) - f^{-1}(b)| < \delta_1,$$

and therefore

$$\left| \frac{f^{-1}(t) - f^{-1}(b)}{t - b} - \frac{1}{f'(a)} \right| < \epsilon. \quad \square$$

■ B.4 THE INTEGRABILITY OF CONTINUOUS FUNCTIONS

The aim here is to prove that, if f is continuous on $[a, b]$, then there is one and only one number I that satisfies the inequality

$$L_f(P) \le I \le U_f(P) \quad \text{for all partitions } P \text{ of } [a, b].$$

DEFINITION B.4.1

A function f is said to be *uniformly continuous* on $[a, b]$ if, for each $\epsilon > 0$, there exists $\delta > 0$ such that

$$\text{if} \quad x, y \in [a, b] \quad \text{and} \quad |x - y| < \delta, \quad \text{then} \quad |f(x) - f(y)| < \epsilon.$$

For convenience, let's agree to say that *the interval $[a, b]$ has the property P_ϵ* if there exist sequences $x_1, x_2, x_3, \cdots$ and $y_1, y_2, y_3, \cdots$ with

$$x_n, y_n \in [a, b], \quad |x_n - y_n| < 1/n, \quad |f(x_n) - f(y_n)| > \epsilon$$

for all indices n.

LEMMA B.4.2

If f is not uniformly continuous on $[a, b]$, then $[a, b]$ has the property P_ϵ for some $\epsilon > 0$.

PROOF If f is not uniformly continuous on $[a, b]$, then there is at least one $\epsilon > 0$ for which there is no $\delta > 0$ such that

$$\text{if} \quad x, y \in [a, b] \quad \text{and} \quad |x - y| < \delta, \quad \text{then} \quad |f(x) - f(y)| < \epsilon.$$

The interval $[a, b]$ has the property P_ϵ for that choice of ϵ. The details of the argument are left to you. $\square$

LEMMA B.4.3

Let f be continuous on $[a, b]$. If $[a, b]$ has the property P_ϵ, then at least one of the subintervals $[a, \frac{1}{2}(a + b)]$, $[\frac{1}{2}(a + b), b]$ has the property P_ϵ.

PROOF Let's suppose that the lemma is false. For convenience, we let $c = \frac{1}{2}(a + b)$, so that the halves become $[a, c]$ and $[c, b]$. Since $[a, c]$ fails to have the property P_ϵ, there exists an integer p such that

if $x, y \in [a, c]$ and $|x - y| < 1/p$,   then $|f(x) - f(y)| < \epsilon$.

Since $[c, b]$ fails to have the property P_ϵ, there exists an integer q such that

if $x, y \in [c, b]$ and $|x - y| < 1/q$,   then $|f(x) - f(y)| < \epsilon$.

Since f is continuous at c, there exists an integer r such that, if $|x - c| < 1/r$, then $|f(x) - f(c)| < \frac{1}{2}\epsilon$. Set $s = \max[p, q, r]$ and suppose that

$$x, y \in [a, b], \qquad |x - y| < 1/s.$$

If x, y are both in $[a, c]$ or both in $[c, b]$, then

$$|f(x) - f(y)| < \epsilon.$$

The only other possibility is that $x \in [a, c]$ and $y \in [c, b]$. In this case we have

$$|x - c| < 1/r, \qquad |y - c| < 1/r,$$

and thus

$$|f(x) - f(c)| < \tfrac{1}{2}\epsilon, \qquad |f(y) - f(c)| < \tfrac{1}{2}\epsilon.$$

By the triangle inequality, we again have

$$|f(x) - f(y)| < \epsilon.$$

In summary, we have obtained the existence of an integer s with the property that

$$x, y \in [a, b], \qquad |x - y| < 1/s \quad \text{implies} \quad |f(x) - f(y)| < \epsilon.$$

Hence $[a, b]$ does not have property P_ϵ. This is a contradiction and proves the lemma. ❑

THEOREM B.4.4

If f is continuous on $[a, b]$, then f is uniformly continuous on $[a, b]$.

PROOF We suppose that f is not uniformly continuous on $[a, b]$ and base our argument on a mathematical version of the classical maxim "Divide and conquer".

By the first lemma of this section, we know that $[a, b]$ has property P_ϵ for some $\epsilon > 0$. We bisect $[a, b]$ and note by the second lemma that one of the halves, say $[a_1, b_1]$, has property P_ϵ. We then bisect $[a_1, b_1]$ and note that one of the halves, say $[a_2, b_2]$, has property P_ϵ. Continuing in this manner, we obtain a sequence of intervals $[a_n, b_n]$, each with property P_ϵ. Then for each n we can choose $x_n, y_n \in [a_n, b_n]$ such that

$$|x_n - y_n| < 1/n \qquad \text{and} \qquad |f(x_n) - f(y_n)| \geq \epsilon.$$

Since

$$a \leq a_n \leq a_{n+1} < b_{n+1} \leq b_n \leq b,$$

we see that sequences $a_1, a_2, \cdots$ and $b_1, b_2, \cdots$ are both bounded and monotonic. Thus they are convergent. Since $b_n - a_n \to 0$, we see that both sequences converge to the same limit, say L. From the inequality

$$a_n \le x_n \le y_n \le b_n,$$

we conclude that

$$x_n \to L \qquad \text{and} \qquad y_n \to L.$$

This tells us that

$$|f(x_n) - f(y_n)| \to |f(L) - f(L)| = 0.$$

and contradicts the statement that $|f(x_n) - f(y_n)| \ge \epsilon$ for all n. ❏

LEMMA B.4.5

If P and Q are partitions of $[a, b]$, then $L_f(P) \le U_f(Q)$.

PROOF $P \cup Q$ is a partition of $[a, b]$ that contains both P and Q. It is obvious then that

$$L_f(P) \le L_f(P \cup Q) \le U_f(P \cup Q) \le U_f(Q). ❏$$

From this lemma it follows that the set of all lower sums is bounded above and has a least upper bound L. The number L satisfies the inequality

$$L_f(P) \le L \le U_f(P) \qquad \text{for all partitions } P$$

and is clearly the least of such numbers. Similarly, we find that the set of all upper sums is bounded below and has a greatest lower bound U. The number U satisfies the inequality

$$L_f(P) \le U \le U_f(P) \qquad \text{for all partitions } P$$

and is clearly the largest of such numbers.

We are now ready to prove the basic theorem.

THEOREM B.4.6   THE INTEGRABILITY THEOREM

If f is continuous on $[a, b]$, then there exists one and only one number I that satisfies the inequality.

$$L_f(P) \le I \le U_f(P), \qquad \text{for all partitions } P \text{ of } [a, b].$$

PROOF We know that

$$L_f(P) \le L \le U \le U_f(P) \qquad \text{for all } P,$$

so that existence is no problem. We will have uniqueness if we can prove that

$$L = U.$$

To do this, we take $\epsilon > 0$ and note that f, being continuous on $[a, b]$, is uniformly continuous on $[a, b]$. Thus there exists a $\delta > 0$ such that, if

$$x, y \in [a, b] \qquad \text{and} \qquad |x - y| < \delta, \qquad \text{then} \qquad |f(x) - f(y)| < \frac{\epsilon}{b - a}.$$

We now choose a partition $P = \{x_0, x_1, \ldots, x_n\}$ for which max $\Delta x_i < \delta$. For this partition P, we have

$$U_f(P) - L_f(P) = \sum_{i=1}^{n} M_i \Delta x_i - \sum_{i=1}^{n} m_i \Delta x_i$$

$$= \sum_{i=1}^{n} (M_i - m_i) \Delta x_i$$

$$< \sum_{i=1}^{n} \frac{\epsilon}{b-a} \Delta x_i = \frac{\epsilon}{b-a} \sum_{i=1}^{n} \Delta x_i = \frac{\epsilon}{b-a}(b-a) = \epsilon.$$

Since

$$U_f(P) - L_f(P) < \epsilon \qquad \text{and} \qquad 0 \leq U - L \leq U_f(P) - L_f(P),$$

you can see that

$$0 \leq U - L < \epsilon.$$

Since ϵ was chosen arbitrarily, we must have $U - L = 0$ and $L = U$. ❏

■ B.5 THE INTEGRAL AS THE LIMIT OF RIEMANN SUMS

For the notation we refer to Section 5.2.

THEOREM B.5.1

If f is continuous on $[a, b]$, then

$$\int_a^b f(x)dx = \lim_{\|P\| \to 0} S^*(P).$$

PROOF Let $\epsilon > 0$. We must show that there exists a $\delta > 0$ such that

$$\text{if} \qquad \|P\| < \delta, \qquad \text{then} \qquad \left| S^*(P) - \int_a^b f(x)dx \right| < \epsilon.$$

From the proof of Theorem B.4.6 we know that there exists a $\delta > 0$ such that

$$\text{if} \qquad \|P\| < \delta, \qquad \text{then} \qquad U_f(P) - L_f(P) < \epsilon.$$

For such P we have

$$U_f(P) - \epsilon < L_f(P) \leq S^*(P) \leq U_f(P) < L_f(P) + \epsilon.$$

This gives

$$\int_a^b f(x)dx - \epsilon < S^*(P) < \int_a^b f(x)\,dx + \epsilon,$$

and therefore

$$\left| S^*(P) - \int_a^b f(x)dx \right| < \epsilon. \qquad ❏$$

CHAPTER 1

SECTION 1.2

1. rational   **3.** rational   **5.** rational   **7.** rational   **9.** rational   **11.** $=$

13. $>$ **15.** $<$ **17.** 6   **19.** 10   **21.** 13   **23.** $5 - \sqrt{5}$

25.   **27.**   **29.**

31.   **33.**   **35.**

37.   **39.**

41. bounded; lower bound 0, upper bound 4   **43.** not bounded   **45.** not bounded   **47.** bounded above; $\sqrt{2}$ is an upper bound

49. $x_0 = 2$, $x_1 \cong 2.75$, $x_2 \cong 2.58264$, $x_3 \cong 2.57133$, $x_4 \cong 2.57128$, $x_5 \cong 2.57128$; bounded; lower bound 2, upper bound 3 (the smallest upper bound $\cong 2.57128\ldots$); $x_n \cong 2.5712815907$ (10 decimal places).

51. $(x - 5)^2$ **53.** $8(x^2 + 2)(x^4 - 2x^2 + 4)$ **55.** $(2x + 3)^2$

57. $2, -1$ **59.** 3   **61.** no real roots   **63.** no real zeros   **65.** 120   **67.** 56   **69.** 1

71. If r and $r + s$ are rational, then $s = (r + s) - r$ is rational.

73. The product could be either rational or irrational; $0 \cdot \sqrt{2} = 0$ is rational, $1 \cdot \sqrt{2} = \sqrt{2}$ is irrational.   **79.** $b - a = 0$; division by 0

SECTION 1.3

1. $(-\infty, 1)$ **3.** $(-\infty, -3)$ **5.** $\left(-\infty, -\frac{1}{5}\right)$ **7.** $(-1, 1)$ **9.** $(-\infty, -2] \cup [3, \infty)$ **11.** $\left[-1, \frac{1}{2}\right]$

13. $(0, 1) \cup (2, \infty)$ **15.** $[0, \infty)$ **17.** $(0, 2)$ **19.** $(-\infty, -6) \cup (2, \infty)$ **21.** $(-2, 2)$ **23.** $(-\infty, -3) \cup (3, \infty)$

25. $\left(\frac{3}{2}, \frac{5}{2}\right)$ **27.** $(-1, 0) \cup (0, 1)$ **29.** $\left(\frac{3}{2}, 2\right) \cup \left(2, \frac{5}{2}\right)$ **31.** $(-5, 3) \cup (3, 11)$ **33.** $\left(-\frac{5}{8}, -\frac{3}{8}\right)$

35. $(-\infty, -4) \cup (-1, \infty)$ **37.** $|x| < 3$ **39.** $|x - 2| < 5$ **41.** $|x + 2| < 5$

43. $A \geq 2$ **45.** $A \leq \frac{4}{3}$ **47.** (a) $\dfrac{1}{x} < \dfrac{1}{\sqrt{x}} < 1 < \sqrt{x} < x$ (b) $x < \sqrt{x} < 1 < \dfrac{1}{\sqrt{x}} < \dfrac{1}{x}$

SECTION 1.4

1. 10   **3.** $4\sqrt{5}$ **5.** $(4, 6)$ **7.** $\left(\frac{9}{2}, -3\right)$ **9.** $-\frac{2}{3}$ **11.** -1 **13.** $-y_0/x_0$

15. slope 2, y-intercept -4 **17.** slope $\frac{1}{3}$, y-intercept 2   **19.** slope $\frac{7}{3}$, y-intercept $\frac{4}{3}$ **21.** $y = 5x + 2$ **23.** $y = -5x + 2$ **25.** $y = 3$

27. $x = -3$ **29.** $y = 7$ **31.** $3y - 2x - 17 = 0$ **33.** $2y + 3x - 20 = 0$ **35.** $\left(\frac{1}{2}\sqrt{2}, \frac{1}{2}\sqrt{2}\right), \left(-\frac{1}{2}\sqrt{2}, -\frac{1}{2}\sqrt{2}\right)$ **37.** $(3, 4), \left(\frac{117}{25}, \frac{44}{25}\right)$

39. $(1, 1)$ **41.** $\left(-\frac{2}{23}, \frac{38}{23}\right)$ **43.** $\frac{17}{2}$ **45.** $-\frac{5}{12}$ **47.** $x - 2y - 3 = 0$ **49.** $(2.36, -0.21)$ **51.** $(0.61, 2.94), (2.64, 1.42)$

53. $3x + 13y - 40 = 0$ **55.** isosceles right triangle   **61.** $\left(1, \frac{10}{3}\right)$ **65.** $F = \frac{9}{5}C + 32; -40°$

SECTION 1.5

1. $f(0) = 2$, $f(1) = 1$, $f(-2) = 16$, $f\left(\frac{3}{2}\right) = 2$ **3.** $f(0) = 0$, $f(1) = \sqrt{3}$, $f(-2) = 0$, $f\left(\frac{3}{2}\right) = \dfrac{\sqrt{21}}{2}$

5. $f(0) = 0$, $f(1) = \frac{1}{2}$, $f(-2) = -1$, $f(\frac{3}{2}) = \frac{12}{23}$ **7.** $f(-x) = x^2 + 2x$, $f\left(\dfrac{1}{x}\right) = \dfrac{1 - 2x}{x^2}$, $f(a + b) = a^2 + 2ab + b^2 - 2a - 2b$

9. $f(-x) = \sqrt{1 + x^2}$, $f\left(\dfrac{1}{x}\right) = \dfrac{\sqrt{x^2 + 1}}{|x|}$, $f(a + b) = \sqrt{a^2 + 2ab + b^2 + 1}$ **11.** $2a^2 + 4ah + 2h^2 - 3a - 3h; 4a - 3 + 2h$

13. $1, 3$ **15.** -2 **17.** $3, -3$ **19.** dom $(f) = (-\infty, \infty)$; range $(f) = (0, \infty)$ **21.** dom $(f) = (-\infty, \infty)$; range $(f) = (-\infty, \infty)$

23. dom $(f) = (-\infty, 0) \cup (0, \infty)$; range $(f) = (0, \infty)$ **25.** dom $(f) = (-\infty, 1]$; range $(f) = [0, \infty)$

27. dom $(f) = (-\infty, 7]$; range $(f) = [-1, \infty)$ **29.** dom $(f) = (-\infty, 2)$; range $(f) = (0, \infty)$

31. horizontal line one unit above x-axis   **33.** line through the origin with slope 2   **35.** line through $(0, 2)$ with slope $\frac{1}{2}$

37. upper semicircle of radius 2 centered at the origin   **39.**

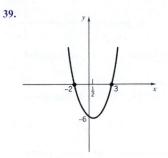

41. dom $(f) = (-\infty, 0) \cup (0, \infty)$; range $(f) = [-1, 1]$ **43.** dom $(f) = [0, \infty)$; range $(f) = [1, \infty)$

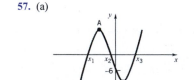

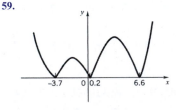

45. yes, dom $(f) = [-2, 2]$; range $(f) = [-2, 2]$ **47.** no   **49.** odd   **51.** neither   **53.** even   **55.** odd

57. (a)

(b) $x_1 = -6.566$, $x_2 = -0.493$, $x_3 = 5.559$

(c) $A(-4, 28.667)$, $B(3, -28.500)$

59.

$-5 \le x \le 8,\, 0 \le y \le 100$

61. range: $[-9, \infty)$ **63.** $A = \dfrac{C^2}{4\pi}$, where C is the circumference; dom $(A) = [0, \infty)$

65. $V = s^{3/2}$, where s is the area of a face; dom $V = [0, \infty)$ **67.** $S = 3d^2$, where d is the diagonal of a face; dom $(S) = [0, \infty)$

69. $A = \dfrac{\sqrt{3}}{4}x^2$, where x is the length of a side; dom $(A) = [0, \infty)$ **71.** $A = \dfrac{15x}{2} - \dfrac{x^2}{2} - \dfrac{\pi x^4}{8}$, $0 \le x \le \dfrac{30}{\pi + 2}$

73. $A = bx - \dfrac{b}{a}x^2$, $0 \le x \le a$ **75.** $A = \dfrac{P^2}{16} + \dfrac{(28 - P)^2}{4x}$, $0 \le P \le 28$ **77.** $V = \pi r^2(108 - 2\pi r)$

SECTION 1.6

1. polynomial, degree 0   **3.** rational function   **5.** neither   **7.** neither   **9.** neither

11. dom $(f) = (-\infty, \infty)$ **13.** dom $(f) = (-\infty, \infty)$ **15.** dom $(f) = (-\infty, -2) \cup (-2, 2) \cup (2, \infty)$

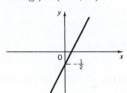   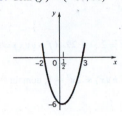   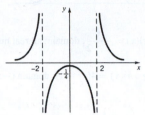

17. $\dfrac{5\pi}{4}$ **19.** $-\dfrac{5\pi}{3}$ **21.** $\dfrac{\pi}{12}$ **23.** $-270°$ **25.** $300°$ **27.** $114.59°$ **31.** $\dfrac{\pi}{6}, \dfrac{5\pi}{6}$ **33.** $\dfrac{\pi}{2}$ **35.** $\dfrac{\pi}{4}, \dfrac{7\pi}{4}$

37. $\dfrac{\pi}{4}, \dfrac{3\pi}{4}, \dfrac{5\pi}{4}, \dfrac{7\pi}{4}$ **39.** 0.7771 **41.** 0.7101 **43.** 3.1524 **45.** $0.5505, \pi - 0.5505$

47. $1.4231, \pi + 1.4231$ **49.** $1.7997, 2\pi - 1.7997$ **51.** $x \cong 1.31, 1.83, 3.40, 3.93, 5.50, 6.02$

53. dom $(f) = (-\infty, \infty)$; range $(f) = [0, 1]$ **55.** dom $(f) = (-\infty, \infty)$; range $(f) = [-2, 2]$

57. dom $(f) = \left(k\pi - \dfrac{\pi}{2}, k\pi + \dfrac{\pi}{2}\right), k = 0 \pm 1, \pm 2, \ldots$; range $(f) = [1, \infty)$ **59.** 2   **61.** 6π

63.   **65.**   **67.**

   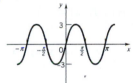   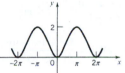

69. odd   **71.** even   **73.** odd   **77.** $\left(\dfrac{23}{37}, \dfrac{116}{37}\right)$; approx $73°$ **79.** $\left(\dfrac{-17}{73}, \dfrac{-2}{73}\right)$; approx $82°$

91.   **93.** (c) A changes the amplitude; B stretches or compresses the graph horizontally.

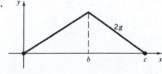

SECTION 1.7

1. $\frac{15}{2}$ **3.** $\frac{105}{2}$ **5.** $\frac{-27}{4}$ **7.** 3

9. $(f + g)(x) = x - 1$; domain $(-\infty, \infty)$

$(f - g)(x) = 3x - 5$; domain $(-\infty, \infty)$

$(f \cdot g)(x) = -2x^2 + 7x - 6$; domain $(-\infty, \infty)$

$\left(\dfrac{f}{g}\right)(x) = \dfrac{2x - 3}{2 - x}$; domain: all real numbers except $x = 2$

11. $(f + g)(x) = x + \sqrt{x - 1} - \sqrt{x + 1}$; domain $[1, \infty)$

$(f - g)(x) = \sqrt{x - 1} + \sqrt{x + 1} - x$; domain $[1, \infty)$

$(f \cdot g)(x) = \sqrt{x - 1}(x - \sqrt{x + 1}) = x\sqrt{x - 1} - \sqrt{x^2 - 1}$;

domain $(1, \infty]$

$\left(\dfrac{f}{g}\right)(x) = \dfrac{\sqrt{x - 1}}{x - \sqrt{x + 1}}$; domain $\left[1, \dfrac{1 + \sqrt{5}}{2}\right) \cup \left(\dfrac{1 + \sqrt{5}}{2}, \infty\right)$

13. (a) $(6f + 3g)(x) = 6x + 3\sqrt{x}, x > 0$ (b) $(f - g)(x) = x + \dfrac{3}{\sqrt{x}} - \sqrt{x}, x > 0$ (c) $(f/g)(x) = \dfrac{x\sqrt{x} + 1}{x - 2}, x > 0, x \neq 2$

15.   **17.**   **19.**

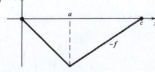   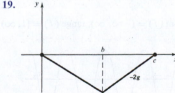

21.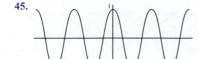

23. $(f \circ g)(x) = 2x^2 + 5$; domain $(-\infty, \infty)$ **25.** $(f \circ g)(x) = \sqrt{x^2 + 5}$; domain $(-\infty, \infty)$

27. $(f \circ g)(x) = \dfrac{x}{x-2}$; domain: all real numbers except $x = 0, x = 2$ **29.** $(f \circ g)(x) = |\sin 2x|$; domain $(-\infty, \infty)$

31. $(f \circ g \circ h)(x) = 4(x^2 - 1)$; domain $(-\infty, \infty)$ **33.** $(f \circ g \circ h)(x) = 2x^2 + 1$; domain $(-\infty, \infty)$ **35.** $f(x) = \dfrac{1}{x}$ **37.** $f(x) = 2\sin x$

39. $g(x) = \left(1 - \dfrac{1}{x^4}\right)^{2/3}$ **41.** $g(x) = 2x^3 - 1$ **43.** $(f \circ g)(x) = |x|; (g \circ f)(x) = x$ **45.** $(f \circ g)(x) = \cos^2 x; (g \circ f)(x) = \sin(1 - x^2)$

47. $g(x) = c$ **49.** $g(x) = c$

51. (a) dom $g = [3, a + 3]$; range $g = [0, b]$
   (b) dom $g = [-4, a - 4]$; range $g = [0, 3b]$
   (c) dom $g = [0, \frac{1}{2}a]$; range $g = [0, b]$
   (d) dom $g = [0, 2a]$; range $g = [0, b]$

53. fg is an even function since $(fg)(-x) = f(-x)g(-x) = f(x)g(x) = (fg)(x)$

55. (a) $f(x) = \begin{cases} -x, & -1 \le x < 0 \\ 1, & x < -1 \end{cases}$ (b) $f(x) = \begin{cases} x, & -1 \le x < 0 \\ -1, & x < -1 \end{cases}$ **57.** $g(-x) = f(-x) + f(x) = f(x) + f(-x) = g(x)$

59. $f(x) = \dfrac{f(x) + f(-x)}{2} + \dfrac{f(x) - f(-x)}{2}$ **61.** (a) $(f \circ g)(x) = \dfrac{5x^2 + 16x - 16}{(2 - x)^2}$ (b) $(g \circ k)(x) = x$ (c) $(f \circ k \circ g)(x) = x^2 - 4$

63. (a) For fixed a, varying b varies the y-coordinate of the vertex of the parabola.
   (b) For fixed b, varying a varies the x-coordinate of the vertex of the parabola.
   (c) The graph of $-F$ is the reflection of the graph of F in the x-axis.

65. (a) For $c > 0$, the graph of cf is the graph of f scaled vertically by the factor c.
   For $c < 0$, the graph of cf is the graph of f scaled vertically by the factor $|c|$ and then reflected in the x-axis.
   (b) For $c > 1$, the graph of g is the graph of f compressed horizontally;
   For $0 < c < 1$, the graph of g is the graph of f stretched horizontally;
   For $-1 < c < 0$, the graph of g is the graph of f stretched horizontally and reflected in the y-axis;
   For $c < -1$, the graph of g is the graph of f compressed horizontally and reflected in the y-axis

SECTION 1.8

11. all positive integers n **13.** $\left(1 - \frac{1}{2}\right)\left(1 - \frac{1}{3}\right) \cdots \left(1 - \frac{1}{n}\right) = \frac{1}{n}$ **19.** $n = 41$

Chapter 1 Review Exercises

1. rational   **3.** irrational   **5.** bounded below; 1 is a lower bound   **7.** bounded; 1 is an upper bound, -5 is a lower bound

9. $-1, \frac{1}{2}$ **11.** 5   **13.** $(-\infty, 2/5)$ **15.** $(-\infty, -2] \cup [3, \infty)$ **17.** $(-2, -1) \cup (2, \infty)$ **19.** $(1, 3)$

21. $(-5, -4) \cup (-4, -3)$ **23.** (a) $5\sqrt{2}$ (b) $\left(\frac{3}{2}, \frac{1}{2}\right)$ **25.** $x = 2$ **27.** $y = -\frac{3}{2}x$ **29.** $(-1, 3/2)$ **31.** $(1, 2), (3, 18)$

33. dom $(f) = (-\infty, \infty)$; range $(f) = (-\infty, 4]$ **35.** dom $(f) = [4, \infty)$; range $(f) = [0, \infty)$

37. dom $(f) = (-\infty, \infty)$; range $(f) = [1, \infty)$ **39.** dom $(f) = (-\infty, \infty)$; range $(f) = [0, \infty)$ **41.** $\frac{7}{6}\pi, \frac{11}{6}\pi$ **43.** $\frac{3}{2}\pi$

45.   **47.**

49. $(f + g)(x) = x^2 + 3x + 1$; domain $(-\infty, \infty)$

$\quad (f - g)(x) = 3 + 3x - x^2$; domain $(-\infty, \infty)$

$\quad (fg)(x) = 3x^3 + 2x^2 - 3x - 2$; domain $(-\infty, \infty)$

$\quad (f/g)(x) = \dfrac{3x + 2}{x^2 - 1}$; domain $x \neq -1, 1$

51. $(f + g)(x) = \cos^2 x + \sin 2x$; domain $[0, 2\pi]$

$\quad (f - g)(x) = \cos^2 x - \sin 2x$; domain $[0, 2\pi]$

$\quad (fg)(x) = \sin 2x \cos^2 x$; domain $(-\infty, \infty)$

$\quad (f/g)(x) = \dfrac{\cos^2 x}{\sin 2x}$; domain $x \in (0, 2\pi), x \neq \pi/2, \pi, 3\pi/2$

53. $(f \circ g)(x) = \sqrt{x^2 - 4}$; domain $(-\infty, -2] \cup [2, \infty)$

$\quad (g \circ f)(x) = x - 4$; domain $[-1, \infty)$

55. (a) $y = kx$ (c) if $\alpha > 0$, (a, b) and $(\alpha a, \alpha b)$ are on the same side of the origin; if $\alpha < 0$, (a, b) and $(\alpha a, \alpha b)$ are on opposite sides of the origin.

CHAPTER 2

SECTION 2.1

1. (a) 2 (b) −1 (c) does not exist (d) −3 **3.** (a) does not exist (b) −3 (c) does not exist (d) −3

5. (a) does not exist (b) does not exist (c) does not exist (d) 1 **7.** (a) 2 (b) 2 (c) 2 (d) −1 **9.** (a) 0 (b) 0 (c) 0 (d) 0

11. $c = 0, 6$ **13.** −1 **15.** 12 **17.** 1 **19.** $\frac{3}{2}$ **21.** does not exist **23.** 2 **25.** does not exist **27.** 1

29. does not exist **31.** 2 **33.** 2 **35.** 0 **37.** 1 **39.** 16 **41.** does not exist **43.** 4 **45.** does not exist

47. $1/\sqrt{2}$ **49.** 4 **51.** 2 **53.** $\frac{3}{2}$ **55.** (a) (i) 5 (ii) does not exist **57.** (a) (i) $-5/4$ (ii) 0 **59.** $c = -1$ **61.** $c = -2$

63. $f(x) \to 1, g(x) \to 0$ as $x \to 0$

SECTION 2.2

1. $\frac{1}{2}$ **3.** does not exist **5.** 4 **7.** does not exist **9.** −1 **11.** does not exist **13.** 0 **15.** 2 **17.** 1 **19.** 1

21. δ_1 **23.** 0.05 **25.** 0.02 **27.** $\delta = 1.75$ **29.** for $\epsilon = 0.5$, take $\delta = 0.24$; for $\epsilon = 0.25$, take $\delta = 0.1$

31. for $\epsilon = 0.5$, take $\delta = 0.75$; for $\epsilon = 0.25$, take $\delta = 0.43$ **33.** for $\epsilon = 0.25$, take $\delta = 0.23$; for $\epsilon = 0.1$, take $\delta = 0.14$

35. Take $\delta = \frac{1}{2}\epsilon$. If $0 < |x - 4| < \frac{1}{2}\epsilon$, then $|(2x - 5) - 3| = 2|x - 4| < \epsilon$.

37. Take $\delta = \frac{1}{6}\epsilon$. If $0 < |x - 3| < \frac{1}{6}\epsilon$, then $|(6x - 7) - 11| = 6|x - 3| < \epsilon$.

39. Take $\delta = \frac{1}{3}\epsilon$. If $0 < |x - 2| < \frac{1}{3}\epsilon$, then $||1 - 3x| - 5| \leq 3|x - 2| < \epsilon$. **41.** Statements (b), (e), (g), and (i) are necessarily true.

43. (i) $\lim\limits_{x \to 3} \dfrac{1}{x - 1} = \dfrac{1}{2}$ (ii) $\lim\limits_{h \to 0} \dfrac{1}{(3 + h) - 1} = \dfrac{1}{2}$ (iii) $\lim\limits_{x \to 3} \left(\dfrac{1}{x - 1} - \dfrac{1}{2} \right) = 0$ (iv) $\lim\limits_{x \to 3} \left| \dfrac{1}{x - 1} - \dfrac{1}{2} \right| = 0$

45. (i) and (iv) of (2.2.6) with $L = 0$

61. (b) No. Consider $f(x) = 1 - x^2$ and $g(x) = 1 + x^2$ on $(-1, 1)$ and let $c = 0$

SECTION 2.3

1. (a) 3 (b) 4 (c) −2 (d) 0 (e) does not exist (f) $\frac{1}{3}$

3. $\lim\limits_{x \to 4} \left[\left(\dfrac{1}{x} - \dfrac{1}{4} \right) \left(\dfrac{1}{x - 4} \right) \right] = \lim\limits_{x \to 4} \left[\left(\dfrac{4 - x}{4x} \right) \left(\dfrac{1}{x - 4} \right) \right] = \lim\limits_{x \to 4} \dfrac{-1}{4x} = -\dfrac{1}{16}$; Theorem 2.3.2 does not apply since $\lim\limits_{x \to 4} \dfrac{1}{x - 4}$ does not exist.

5. 3 **7.** −3 **9.** 5 **11.** does not exist **13.** −1 **15.** does not exist **17.** 1 **19.** 4 **21.** $\frac{1}{4}$ **23.** $-\frac{2}{3}$

25. does not exist **27.** −1 **29.** 4 **31.** a/b **33.** 5/4 **35.** does not exist **37.** 2

39. (a) 0 (b) $-\frac{1}{16}$ (c) 0 (d) does not exist **41.** (a) 4 (b) −2 (c) 2 (d) does not exist **43.** $f(x) = 1/x, g(x) = -1/x, c = 0$

45. True. **47.** True. **49.** False. **51.** False. **55.** (b) $\lim\limits_{x \to 0} g(x) = \dfrac{1}{L}$ **57.** 5 **59.** $\frac{1}{4}$

61. (a) 1 (b) $2x$ (c) $3x^2$ (d) $4x^3$ (e) nx^{n-1}

SECTION 2.4

1. (a) $x = -3, x = 0, x = 2, x = 6$ (b) At -3, neither; at 0, continuous from the right; at 2, neither; at 6, neither
(c) removable discontinuity at $x = 2$; jump discontinuity at $x = 0$

3. continuous   **5.** continuous   **7.** continuous   **9.** removable discontinuity   **11.** jump discontinuity

13. continuous   **15.** infinite discontinuity   **17.** no discontinuities   **19.** no discontinuities

21. infinite discontinuity at $x = 3$ **23.** no discontinuities   **25.** jump discontinuities at 0 and 2

27. removable discontinuity at -2; jump discontinuity at 3   **29.**

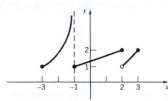

31. $f(1) = 2$ **33.** impossible   **35.** 4   **37.** $A - B = 3$ with $B \neq 3$ **39.** $c = -3$ **41.** $f(5) = \frac{1}{6}$ **43.** $f(5) = \frac{1}{3}$

45. nowhere   **47.** $x = 0, x = 2$, and all nonintegral values of x

57. $k = 5/2$ **59.** $\lim\limits_{x \to 0^-} f(x) = -1$, $\lim\limits_{x \to 0^+} f(x) = 1$; f is discontinuous at 0 for all k.

SECTION 2.5

1. 3   **3.** $\frac{3}{5}$ **5.** 2   **7.** 0   **9.** does not exist   **11.** $\frac{9}{5}$ **13.** $\frac{2}{3}$ **15.** 1   **17.** $\frac{1}{2}$ **19.** -4 **21.** 1

23. $\frac{3}{5}$ **25.** 0   **27.** $\dfrac{2}{\pi}\sqrt{2}$ **29.** -1 **31.** 0   **35.** 0   **37.** 1   **39.** $\dfrac{\sqrt{2}}{2}$;   **41.** $-\sqrt{3}$;   **51.** 10   **53.** 1/3

SECTION 2.6

1. $f(1) = -1 < 0$, $f(2) = 6 > 0$ **3.** $f(0) = 2 > 0$, $f(\pi/2) = 1 - \pi^2/4 < 0$ **5.** $f\left(\frac{1}{4}\right) = \frac{1}{16} > 0$, $f(1) = -\frac{1}{2} < 0$

7. Let $f(x) = x^3 - \sqrt{x+2}$; $f(1) = 1 - \sqrt{3} < 0$, $f(2) = 6 > 0$

9. Let $F(x) = x^5 - 2x^2 + 5x - 1$; $F(0) = -1 < 0$, $F(1) = 3 > 0$. Therefore there is a number $c \in (0, 1)$ such that $F(c) = 0$ which implies $f(c) = 1$.

11. $f(-3) = -13 < 0$, $f(-2) = 2 > 0$; $f(0) = 2 > 0$, $f(1) = -1 < 0$, $f(2) = 2 > 0$; f has a root in $(-3, -2)$, $(0, 1)$, and $(1, 2)$

13.   **15.**   **17.**   **19.**

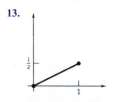 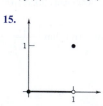 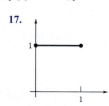 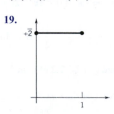

21. impossible by the intermediate-value theorem   **23.**

37. f has a zero on $(-3, -2), (0, 1), (1, 2)$; the zeros of f are $r_1 = -2.4909, r_2 = 0.6566, r_3 = 1.8343$

39. f has a zero on $(-2, -1), (0, 1), (1, 2)$; the zeros of f are $r_1 = -1.3482, r_2 = 0.2620, r_3 = 1.0816$

41. f is not continuous at $x = 1$. Therefore f does not satisfy the hypothesis of the intermediate-value theorem.

43. f satisfies the hypothesis of the intermediate-value theorem.

$$\frac{f(\pi/2) + f(2\pi)}{2} = \frac{1}{2} = f(c) \text{ for } c \cong 2.38, 4.16, 5.25.$$

45. f is bounded; the maximum value of f is 1, the minimum value is -1

47. f is bounded; no maximum value, the minimum value is approximately 0.3540

Chapter 2 Review Exercises

1. 1   **3.** 0   **5.** 1   **7.** 1   **9.** 0   **11.** 1/2   **13.** does not exist   **15.** 3/5   **17.** 4/3   **19.** -3

21. $-3/4$ **23.** -1 **25.** $-1/4$ **27.** 0   **29.** 2   **31.** (c), (e)

33.

(b) (i) 1   (ii) 0   (iii) does not exist   (iv) -6 (v) 4   (vi) does not exist

(c) (i) yes; no   (ii) no; yes

35. $A = 7, B = 1/4$ **37.** $f(-3) = -8$ **39.** $f(0) = \pi$ **41.** (b)

43. $f(x) = 2\cos x - x + 1$; $f(1) = 2\cos 1 > 0$, $f(2) = 2\cos 2 - 1 < 0$ **49.** (a) yes; $f(x) = \begin{cases} \dfrac{\sin \pi x}{x}, & x > 0 \\ \pi, & x = 0 \end{cases}$ (b) no

CHAPTER 3

SECTION 3.1

1. -3 **3.** $5 - 2x$ **5.** $4x^3$ **7.** $\dfrac{1}{2\sqrt{x-1}}$ **9.** $-2x^{-3}$ **11.** 2   **13.** 6   **15.** -2

17. $y + 3x - 16 = 0$ **19.** $x - 4y + 3 = 0$

21. (a) Removable discontinuity at $c = -1$; jump discontinuity at $c = 1$ (b) f is continuous but not differentiable at $c = 0$ and $c = 3$

23. $x = -1$ **25.** $x = 0$ **27.** $x = 1$ **29.** 4   **31.** does not exist

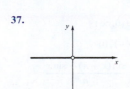

33.

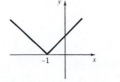

35.

37.

39. f is not continuous at 1   **41.** $A = 3, B = -2$

In 43–48, there are many possible answers. Here are some.

43. $f(x) = c, c$ any constant   **45.** $f(x) = |x + 1|$; $f(x) = \begin{cases} 0, & x \ne -1 \\ 1, & x = -1 \end{cases}$ **47.** $f(x) = 2x + 5$

49. (b) f is not differentiable at 2   **51.** (a) no   (b)

53. (a) $x = c$;   (b) $y - f(c) = \dfrac{-1}{f'(c)}(x - c)$;   (c) $y = f(c)$ **55.** $y - 2 = -4(x - 4)$ **57.** $\frac{111}{2}$

59. (c) $g'(0) = 0$ **61.** $f'(1) = -1$ **63.** $f'(-1) = \frac{1}{3}$ **65.** $f'(2) \cong 7.071$

67. (a) $f'(x) = \dfrac{5}{2\sqrt{5x - 4}}$; $f'(3) = \dfrac{5}{2\sqrt{11}}$ **69.** (c) $f'(x) = 10x - 21x^2$

(b) $f'(x) = -2x + 16x^3 - 6x^5$; $f'(-2) = 68$ (d) $f'(x) = 0$ at $x = 0, 10/21$

(c) $f'(x) = -\dfrac{3(3 - 2x)}{(2 + 3x)^2} - \dfrac{2}{(2 + 3x)}$; $f'(-1) = -13$

71. (a) tangent: $y - \frac{21}{8} = -\frac{11}{4}(x - \frac{3}{2})$;   normal: $y - \frac{21}{8} = \frac{4}{11}(x - \frac{3}{2})$

(c) $(1.453, 1.547)$

SECTION 3.2

1. -1 **3.** $55x^4 - 18x^2$ **5.** $2ax + b$ **7.** $\dfrac{2}{x^3}$ **9.** $3x^2 - 6x - 1$ **11.** $\dfrac{3x^2 - 2x^3}{(1 - x)^2}$ **13.** $\dfrac{2(x^2 + 3x + 1)}{(2x + 3)^2}$

15. $8x^3 + 15x^2 - 8x - 10$ **17.** $-\dfrac{2(3x^2 - x + 1)}{x^2(x - 2)^2}$ **19.** $-80x^9 + 81x^8 - 64x^7 + 63x^6$ **21.** $f'(0) = -\frac{1}{4}$; $f'(1) = -1$

23. $f'(0) = 0$; $f'(1) = -1$ **25.** $f'(0) = \dfrac{ad - bc}{d^2}$; $f'(1) = \dfrac{ad - bc}{(c + d)^2}$ **27.** $f'(0) = 3$ **29.** $f'(0) = \frac{20}{9}$ **31.** $2y - x - 8 = 0$

33. $y - 4x + 12 = 0$ **35.** $(-1, 27), (3, -5)$ **37.** $(-1, -\frac{5}{2}), (1, \frac{5}{2})$ **39.** (a) $-2, 0, 2$ **41.** (a) 2

(b) $(-2, 0) \cup (2, \infty)$ (b) $(-\infty, 0) \cup (2, \infty)$

(c) $(-\infty, -2) \cup (0, 2)$ (c) $(0, 2)$

43. $(-2, -10)$ **45.** $f(x) = x^3 + x^2 + x + C, C$ any constant   **47.** $f(x) = \frac{2}{3}x^3 - \frac{3}{2}x^2 + 1/x + C, C$ any constant   **49.** $A = -2, B = -8$

51. $\frac{425}{8}$ **53.** $A = -1, B = 0, C = 4$ **55.** $x = -\dfrac{b}{2a}$ **57.** $c = -1, 1$ **61.** $y = -x, y + 24 = 26(x + 3)$

63. $f(x) = |x|$ and $g(x) = -|x|$ are not differentiable at 0; their sum $h(x) = 0$ is differentiable everywhere.

67. $F'(x) = 2x \left(1 + \dfrac{1}{x}\right)(2x^3 - x + 1) + (x^2 + 1)\left(-\dfrac{1}{x^2}\right)(2x^3 - x + 1) + (x^2 + 1)\left(1 + \dfrac{1}{x}\right)(6x^2 - 1)$

71. (a) $0, -2$ **73.** (a) $f'(x) \neq 0$ for all $x \neq 0$

(b) $(-\infty, -2) \cup (0, \infty)$ (b) $(0, \infty)$

(c) $(-2, -1) \cup (-1, 0)$ (c) $(-\infty, 0)$

75. (a) Let $D(h) = \dfrac{\sin(x + h) - \sin x}{h}$.

At $x = 0$, $D(0.001) \cong 0.99999$, $D(-0.001) \cong 0.99999$; at $x = \pi/6$, $D(0.001) \cong 0.86578$, $D(-0.001) \cong 0.86628$; at $x = \pi/4$, $D(0.001) \cong 0.70675$, $D(-0.001) \cong 0.70746$; at $x = \pi/3$, $D(-0.001) \cong 0.49957$, $D(-0.001) \cong 0.50043$; at $x = \pi/2$, $D(0.001) \cong -0.0005$, $D(-0.001) \cong 0.0005$

(b) $\cos(0) = 1, \cos(\pi/6) \cong 0.866025, \cos(\pi/4) \cong 0.707107, \cos(\pi/3) = 0.5, \cos(\pi/2) = 0$ (c) $f'(x) = \cos x$.

SECTION 3.3

1. $\dfrac{dy}{dx} = 12x^3 - 2x$ 3. $\dfrac{dy}{dx} = 1 + \dfrac{1}{x^2}$ 5. $\dfrac{dy}{dx} = \dfrac{1 - x^2}{(1 + x^2)^2}$ 7. $\dfrac{dy}{dx} = \dfrac{2x - x^2}{(1 - x)^2}$ 9. $\dfrac{dy}{dx} = \dfrac{-6x^2}{(x^3 - 1)^2}$ 11. 2

13. $18x^2 + 30x + 5x^{-2}$ 15. $\dfrac{2t^3(t^3 - 2)}{(2t^3 - 1)^2}$ 17. $\dfrac{2}{(1 - 2u)^2}$ 19. $-\left[\dfrac{1}{(u-1)^2} + \dfrac{1}{(u+1)^2}\right] = -2\left[\dfrac{u^2 + 1}{(u^2 - 1)^2}\right]$

21. $\dfrac{-2}{(x-1)^2}$ 23. 47 25. $\frac{1}{4}$ 27. $42x - 120x^3$ 29. $-6x^{-3}$ 31. $4 - 12x^{-4}$ 33. 2 35. 0 37. $6 + 60x^{-6}$

39. $1 - 4x$ 41. -24 43. -24 45. $y = x^4 - \frac{1}{3}x^3 + 2x^2 + C$, C any constant   47. $y = x^5 - x^{-4} + C$, C any constant

49. $p(x) = 2x^2 - 6x + 7$ 51. (a) $n!$ (b) 0 (c) $f^{(k)}(x) = n(n-1)\cdots(n-k+1)x^{n-k}$

53. (a) $f'(0) = 0$.

(b) $f'(x) = \begin{cases} 2x, & x \geq 0 \\ 0, & x < 0 \end{cases}$

(c) $f''(x) = \begin{cases} 2, & x > 0 \\ 0, & x < 0 \end{cases}$

(d)

57. (a) $x = 0$ (b) $x > 0$ (c) $x < 0$ 59. (a) $x = -2, x = 1$ (b) $x < -2, x > 1$ (c) $-2 < x < 1$

63. $\dfrac{d}{dx}(uvw) = vw\dfrac{du}{dx} + uw\dfrac{dv}{dx} + uv\dfrac{dw}{dx}$ 65. $\dfrac{d^n}{dx^n}[f(x)] = n!(1-x)^{-n-1} = \dfrac{n!}{(1-x)^{n+1}}$ 67. (b) $\left(-\dfrac{1}{\sqrt{2}}, \dfrac{1}{2\sqrt{2}}\right), \left(\dfrac{1}{\sqrt{2}}, -\dfrac{1}{2\sqrt{2}}\right)$

69. (a) $f'(x) = 3x^2 + 2x - 4$ (c) The graph is "falling" when $f'(x) < 0$; the graph is rising when $f'(x) > 0$.

71. (a) $f'(x) = \frac{3}{2}x^2 - 6x + 3$ (c) the tangent line is horizontal.   (d) $x_1 \cong 0.586$, $x_2 \cong 3.414$

SECTION 3.4

1. $\dfrac{dA}{dr} = 2\pi r$, 4π 3. $\dfrac{dA}{dz} = z$, 4 5. $-\dfrac{5}{36}$ 7. $\dfrac{dV}{dr} = 4\pi r^2$ 9. $x_0 = \frac{3}{4}$ 11. (a) $\dfrac{3\sqrt{2}}{4}w^2$ (b) $\dfrac{\sqrt{3}}{3}z^2$.

13. (a) $\frac{1}{2}r^2$ (b) $r\theta$ (c) $-4Ar^{-3} = -2\theta/r$ 15. $x = \frac{1}{2}$

SECTION 3.5

1. $y = x^4 + 2x^2 + 1$; $y' = 4x^3 + 4x = 4x(x^2 + 1)$ 3. $y = 8x^3 + 12x^2 + 6x + 1$; $y' = 24x^2 + 24x + 6 = 6(2x + 1)^2$

   $y = (x^2 + 1)^2$; $y' = 2(x^2 + 1)(2x) = 4x(x^2 + 1)$ $y = (2x + 1)^3$; $y' = 3(2x + 1)^2(2) = 6(2x + 1)^2$

5. $f(x) = x^2 + 2 + x^{-2}$; $f'(x) = 2x - 2x^{-3} = 2x(1 - x^{-4})$

   $f(x) = (x + x^{-1})^2$; $f'(x) = 2(x + x^{-1})(1 - x^{-2}) = 2x(1 + x^{-2})(1 - x^{-2}) = 2x(1 - x^{-4})$

7. $2(1 - 2x)^{-2}$ 9. $20(x^5 - x^{10})^{19}(5x^4 - 10x^9)$ 11. $4\left(x - \dfrac{1}{x}\right)^3\left(1 + \dfrac{1}{x^2}\right)$ 13. $4(x - x^3 + x^5)^3(1 - 3x^2 + 5x^4)$

15. $-4(t^{-1} + t^{-2})^3(t^{-2} + 2t^{-3})$ 17. $324x^3\left[\dfrac{1 - x^2}{(x^2 + 1)^5}\right]$ 19. $-\left(\dfrac{x^3}{3} + \dfrac{x^2}{2} + \dfrac{x}{1}\right)^{-2}(x^2 + x + 1)$

21. -1 23. 0 25. $\dfrac{dy}{dt} = \dfrac{dy}{du}\dfrac{du}{dx}\dfrac{dx}{dt} = \dfrac{7(2t - 5)^4 + 12(2t + 5)^2 - 2}{[(2t - 5)^4 + 2(2t - 5)^2 + 2]^2}[4(2t - 5)]$ 27. 16 29. 1 31. 1 33. 1 35. 2

37. 0 39. $12(x^3 + x)^2[(3x^2 + 1)^2 + 2x(x^3 + x)]$ 41. $\dfrac{6x(1 + x)}{(1 - x)^5}$ 43. $2xf'(x^2 + 1)$ 45. $2f(x)f'(x)$

47. (a) $x = 0$ (b) $x < 0$ (c) $x > 0$ 49. (a) $x = -1, x = 1$ (b) $-1 < x < 1$ (c) $x < -1, x > 1$

51. $y^{(n)} = \dfrac{n!}{(1-x)^{n+1}}$ **53.** $y^{(n)} = n!\,b^n$ **55.** $y = (x^2+1)^3 + C, C$ any constant   **57.** $y = (x^3-2)^2 + C, C$ any constant

59. $L'(x^2+1) = \dfrac{2x}{x^2+1}$ **61.** $T'(x) = 0$

65. Let $P(x)$ be a polynomial function of degree n. The number a is a zero of multiplicity k for P, $k < n$, iff $P(a) = p'(a) = \cdots = P^{(k-1)}(a) = 0$ and $P^{(k)}(a) \neq 0$.

67. $800\,\pi$ cm^3/sec   **69.** (a) $\dfrac{dF}{dt} = \dfrac{2k}{r^3}(49 - 9.8t)$

71. (a) $f'(1) = -\frac{1}{2}$; $x + 2y - 2 = 0$ **73.** (a) $-\dfrac{f'(1/x)}{x^2}$ **75.** $[g'(x)]^2 f''[g(x)] + f'[g(x)]g''(x)$

   (c) $(0.797, 1.202)$ (b) $\dfrac{4xf'[(x^2-1)/(x^2+1)]}{(1+x^2)^2}$

   (c) $\dfrac{f'(x)}{[1+f(x)]^2}$

SECTION 3.6

1. $\dfrac{dy}{dx} = -3\sin x - 4\sec x \tan x$ **3.** $\dfrac{dy}{dx} = 3x^2 \csc x - x^3 \csc x \cot x$ **5.** $\dfrac{dy}{dt} = -2\cos t \sin t$ **7.** $\dfrac{dy}{du} = 2u^{-1/2}\sin^3\sqrt{u}\cos\sqrt{u}$

9. $\dfrac{dy}{dx} = 2x\sec^2 x^2$ **11.** $\dfrac{dy}{dx} = 4(1 - \pi\csc^2\pi x)(x + \cot\pi x)^3$ **13.** $\dfrac{d^2y}{dx^2} = -\sin x$ **15.** $\dfrac{d^2y}{dx^2} = \cos x(1 + \sin x)^{-2}$

17. $\dfrac{d^2y}{du^2} = 12\cos 2u(2\sin^2 2u - \cos^2 2u)$ **19.** $\dfrac{d^2y}{dt^2} = 8\sec^2 2t \tan 2t$ **21.** $\dfrac{d^2y}{dx^2} = (2-9x^2)\sin 3x + 12x\cos 3x$

23. $\dfrac{d^2y}{dx^2} = 0$ **25.** $\sin x$ **27.** $(27t^3 - 12t)\sin 3t - 45t^2 \cos 3t$ **29.** $3\cos 3x f'(\sin 3x)$ **31.** $y = x$ **33.** $y - \sqrt{3} = -4(x - \frac{1}{6}\pi)$

35. $y - \sqrt{2} = \sqrt{2}(x - \frac{1}{4}\pi)$ **37.** at π **39.** at $\frac{1}{6}\pi, \frac{7}{6}\pi$ **41.** at $\frac{1}{2}\pi, \pi, \frac{3}{2}\pi$ **43.** at $\frac{1}{4}\pi, \frac{3}{4}\pi, \frac{5}{4}\pi, \frac{7}{4}\pi$ **45.** at $\frac{7}{6}\pi, \frac{11}{6}\pi$

47. (a) $f'(x) = 0$ at $x = \frac{1}{6}\pi, \frac{5}{6}\pi$;   (b) $f'(x) > 0$ on $\left(0, \frac{1}{6}\pi\right) \cup \left(\frac{5}{6}\pi, 2\pi\right)$ (c) $f'(x) < 0$ on $\left(\frac{1}{6}\pi, \frac{5}{6}\pi\right)$

49. (a) $f'(x) = 0$ at $x = \frac{1}{4}\pi, \frac{5}{4}\pi$;   (b) $f'(x) > 0$ on $\left(0, \frac{1}{4}\pi\right) \cup \left(\frac{5}{4}\pi, 2\pi\right)$ (c) $f'(x) < 0$ on $\left(\frac{1}{4}\pi, \frac{5}{4}\pi\right)$

51. (a) $\dfrac{dy}{dt} = \dfrac{dy}{du}\dfrac{du}{dx}\dfrac{dx}{dt} = (2u)(\sec x \tan x)(\pi) = 2\pi\sec^2\pi t \tan\pi t$ (b) $y = \sec^2\pi t - 1$; $\dfrac{dy}{dt} = 2\sec\pi t(\sec\pi t \tan\pi t)\pi = 2\pi\sec^2\pi t \tan\pi t$

53. (a) $\dfrac{dy}{dt} = \dfrac{dy}{du}\dfrac{du}{dx}\dfrac{dx}{dt} = 4[\frac{1}{2}(1-u)]^3(-\frac{1}{2}) \cdot (-\sin x) \cdot 2 = 4[\frac{1}{2}(1-\cos 2t)]^3 \sin 2t = (4\sin^6 t)(2\sin t \cos t) = 8\sin^7 t \cos t$

   (b) $y = [\frac{1}{2}(1-\cos 2t)]^4 = \sin^8 t$; $\dfrac{dy}{dt} = 8\sin^7 t \cos t$

55. $\dfrac{d^n}{dx^n}(\cos x) = \begin{cases} (-1)^{(n+1)/2}\sin x, & n \text{ odd} \\ (-1)^{n/2}\cos x, & n \text{ even} \end{cases}$ **61.** $f(x) = 2\sin x + 3\cos x + C, C$ any constant

63. $f(x) = \sin 2x + \sec x + C, C$ any constant   **65.** $f(x) = \sin(x^2) + \cos 2x + C, C$ any constant

67. (a) $f'(x) = \sin\dfrac{1}{x} - \dfrac{1}{x}\cos\dfrac{1}{x}$; $g'(x) = 2x\sin\dfrac{1}{x} - \cos\dfrac{1}{x}$ (b) $\lim_{x\to 0} g'(x) = \lim_{x\to 0}\left(2x\sin\dfrac{1}{x} - \cos\dfrac{1}{x}\right) = \lim_{x\to 0}\cos\dfrac{1}{x}$ does not exist

69. (a) $a = -\frac{1}{2}$, $b = \frac{\sqrt{3}}{2} + \frac{\pi}{3}$ (b)

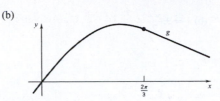

73. $A(x) = \frac{1}{2}c^2 \sin x;\ A'(x) = \frac{1}{2}c^2 \cos x$

75. (a) $f^{(4p)}(x) = k^{4p} \cos kx,\quad f^{(4p+1)}(x) = -k^{4p+1} \sin kx,$ (b) $m = k^2$
$f^{(4p+2)}(x) = -k^{4p+2} \cos kx,\quad f^{(4p+3)}(x) = k^{4p+3} \sin kx$

77. f has a horizontal tangent at $x = \frac{1}{2}\pi, x = \frac{3}{2}\pi, x \cong 3.39, x \cong 6.03$ **79.** $y = x;\quad |\sin x - x| < 0.01$ on $(-0.3924, 0.3924)$

SECTION 3.7

1. $-\dfrac{x}{y}$ **3.** $-\dfrac{4x}{9y}$ **5.** $-\dfrac{x^2(x+3y)}{x^3+y^3}$ **7.** $\dfrac{2(x-y)}{2(x-y)+1}$ **9.** $\dfrac{y - \cos(x+y)}{\cos(x+y) - x}$ **11.** $\dfrac{16}{(x+y)^3}$ **13.** $\dfrac{90}{(2y+x)^3}$

15. $\dfrac{d^2y}{dx^2} = \frac{3}{2}x\cos^2 y - \frac{9}{8}x^4 \sin y \cos^3 y$ **17.** $\dfrac{dy}{dx} = \frac{5}{8},\quad \dfrac{d^2y}{dx^2} = -\dfrac{9}{128}$ **19.** $\dfrac{dy}{dx} = -\frac{1}{2},\quad \dfrac{d^2y}{dx^2} = 0$

21. tangent $2x + 3y - 5 = 0$; normal $3x - 2y + 12 = 0$ **23.** tangent $x + 2y + 8 = 0$; normal $2x - y + 1 = 0$

25. tangent: $y = \dfrac{-2}{\sqrt{3}}x + \left(\dfrac{1}{\sqrt{3}} + \dfrac{\pi}{3}\right)$; normal: $y = \dfrac{\sqrt{3}}{2}x + \left(\dfrac{\pi}{3} - \dfrac{\sqrt{3}}{4}\right)$ **27.** $\frac{3}{2}x^2(x^3+1)^{-1/2}$ **29.** $\dfrac{x}{(\sqrt[4]{2x^2+1})^3}$

31. $\dfrac{x(2x^2-5)}{\sqrt{2-x^2}\sqrt{3-x^2}}$ **33.** $\dfrac{1}{2}\left(\dfrac{1}{\sqrt{x}} - \dfrac{1}{x\sqrt{x}}\right)$ **35.** $\dfrac{1}{(\sqrt{x^2+1})^3}$

37. (a)   (b)   (c)   **39.** $-\dfrac{2b^2}{9(\sqrt[3]{a+bx})^5}$

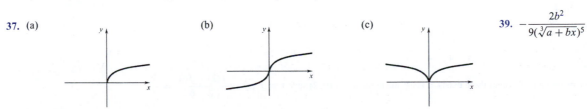

41. $\dfrac{\sqrt{x}\sec^2\sqrt{x} - \tan\sqrt{x} + 2x\sec^2\sqrt{x}\tan\sqrt{x}}{4x\sqrt{x}}$ **45.** at right angles **47.** at $(1,1), \alpha = \pi/4;$ at $(0,0), \alpha = \pi/2$

53. $y - 2x + 12 = 0,\ y - 2x - 12 = 0$ **55.** $\left(-\dfrac{\sqrt{6}}{4}, \pm\dfrac{\sqrt{2}}{4}\right), \left(\dfrac{\sqrt{6}}{4}, \pm\dfrac{\sqrt{2}}{4}\right)$

59. (b) $d(h) \to \infty$ as $h \to 0^-$ and as $h \to 0^+$ **61.** $f'(x) > 0$ on $(-\infty, \infty)$

(c) The graph of f has a vertical tangent at $(0,0)$.

63.   **65.** $y'|_{(3,4)} = 3$ **67.** $\dfrac{dy}{dx}\Big|_{(1,3\sqrt{3})} = -\sqrt{3}$

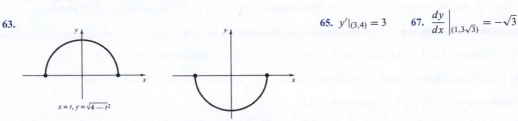

$x = t, y = \sqrt{4 - t^2}$

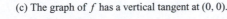

$x = t, y = -\sqrt{4 - t^2}$

69. (a)

(b) $x = \pm\dfrac{\sqrt{2}}{2}$

Chapter 3 Review Exercises

1. $f'(x) = 3x^2 - 4$ **3.** $g'(x) = -\dfrac{-1}{(x-2)^2}$ **5.** $y' = \dfrac{2}{3x^{1/3}}$ **7.** $y' = -\dfrac{x^4 + 4x^3 + 3x^2 + 2x + 2}{(x^3 - 1)^2}$ **9.** $f'(x) = \dfrac{x}{(a^2 - x^2)^{3/2}}$

11. $y' = -\dfrac{6b}{x^3}\left(a + \dfrac{b}{x^2}\right)^2$ **13.** $y' = \dfrac{\sec^2\sqrt{2x+1}}{\sqrt{2x+1}}$ **15.** $F'(x) = \dfrac{3x^3 + 8x^2 + 8x + 8}{\sqrt{x^2+2}}$ **17.** $h'(t) = \sec t^2 + 2t^2\sec t^2\tan t^2 + 6t^2$

19. $\dfrac{ds}{dt} = \dfrac{-4}{(2+3t)^{4/3}(2-3t)^{2/3}}$ **21.** $f'(\theta) = -3\csc^2(3\theta + \pi)$ **23.** $1/12$ **25.** $\dfrac{6 + \pi\sqrt{3}}{72}$ **27.** tangent: $y = 4x$;   normal: $x + 4y = 17$

29. tangent: $y = 2x$;   normal: $y = -\frac{1}{2}x$ **31.** $f''(x) = -\cos(2-x)$ **33.** $y'' = 2\cos x - x\sin x$ **35.** $y^{(n)} = (-1)^n n! b^n$

37. $\dfrac{dy}{dx} = -\dfrac{3x^2 y + y^3}{x^3 + 3xy^2}$ **39.** $\dfrac{dy}{dx} = \dfrac{6x^2 + 3\cos y - 2y}{2x + 3x\sin y}$ **41.** tangent: $5x - 3y = 9$;   normal: $3x + 5y = 19$

43. (a) $x = 2, 4$;   (b) $(-\infty, 2) \cup (4, \infty)$;   (c) $(2, 4)$

45. (a) $x = \frac{1}{3}\pi, \frac{2}{3}\pi, \frac{4}{3}\pi, \frac{5}{3}\pi$;   (b) $(0, \frac{1}{3}\pi) \cup (\frac{2}{3}\pi, \frac{4}{3}\pi) \cup (\frac{5}{3}\pi, 2\pi)$;   (c) $(\frac{1}{3}\pi, \frac{2}{3}\pi) \cup (\frac{4}{3}\pi, \frac{5}{3}\pi)$

47. (a) $x = 1$;   (b) $x = 3$;   (c) $x = 1/3$ **49.** $y = -x, y = 26x + 54$ **51.** $A = 1, B = -1, C = 4, D = -3$

53. 0 **55.** $\frac{1}{2}\sqrt{3}$ **57.** -1

CHAPTER 4

SECTION 4.1

1. $c = \dfrac{\sqrt{3}}{3} \cong 0.577$ **3.** $c = \dfrac{\pi}{4}, \dfrac{3\pi}{4}, \dfrac{5\pi}{4}, \dfrac{7\pi}{4}$ **5.** $c = \frac{3}{2}$ **7.** $c = \frac{1}{3}\sqrt{39}$ **9.** $c = \frac{1}{2}\sqrt{2}$ **11.** $c = 0$

13. No. By mean-value theorem there exists at least one number $c \in (0, 2)$ such that $f'(c) = \dfrac{f(2) - f(0)}{2 - 0} = \dfrac{3}{2}$.

17. $f'(x) = \begin{cases} 2, & x \leq -1 \\ 3x^2 - 1, & x > -1; \end{cases}$ $-3 < c \leq -1$ and $c = 1$ **21.** $\dfrac{f(1) - f(-1)}{(1) - (-1)} = 0$ and $f'(x)$ is never zero; f is not differentiable at 0.

31. (a) $f'(x) = 3x^2 - 3 = 3(x^2 - 1) \neq 0$ on $(-1, 1)$ **37.** Set, for instance, $f(x) = \begin{cases} 1, & a < x < b \\ 0, & x = a, b \end{cases}$

    (b) $-2 < b < 2$

47. $c = 0.676$ **49.** $c = 0.3045$ **51.** $c = 0$ **53.** $c = 2.205$ **55.** $c = {}^8/3$

SECTION 4.2

1. increases on $(-\infty, -1]$ and $[1, \infty]$, decreases on $[-1, 1]$ **3.** increases on $(-\infty, -1]$ and $[1, \infty)$, decreases on $[-1, 0)$ and $(0, 1]$

5. increases on $\left[-\frac{3}{4}, \infty\right)$, decreases on $\left(-\infty, -\frac{3}{4}\right]$ **7.** increases on $[-1, \infty)$, decreases on $(-\infty, -1]$

9. increases on $(-\infty, 2)$, decreases on $(2, \infty)$ **11.** increases on $(-\infty, -1)$ and $(-1, 0]$, decreases on $[0, 1)$ and $(1, \infty)$

13. increases on $[-\sqrt{5}, 0]$ and $[\sqrt{5}, \infty)$, decreases on $(-\infty, -\sqrt{5}]$ and $[0, \sqrt{5}]$ **15.** increases on $(-\infty, -1)$ and $(-1, \infty)$

17. increases on $[0, \infty)$, decreases on $(-\infty, 0]$ **19.** increases on $[0, 2\pi]$

21. increases on $\left[\frac{2}{3}\pi, \pi\right]$, decreases on $\left[0, \frac{2}{3}\pi\right]$ **23.** increases on $\left[0, \frac{2}{3}\pi\right]$ and $\left[\frac{5}{6}\pi, \pi\right]$, decreases on $\left[\frac{2}{3}\pi, \frac{5}{6}\pi\right]$

25. $f(x) = \frac{1}{3}x^3 - x + \frac{8}{3}$ **27.** $f(x) = x^5 + x^4 + x^3 + x^2 + x + 5$ **29.** $f(x) = \frac{3}{4}x^{4/3} - \frac{2}{3}x^{3/2} + 1, x \geq 0$ **31.** $f(x) = 2x - \cos x + 4$

33. increases on $(-\infty, -3)$ and $[-1, 1]$, decreases on $[-3, -1]$ and $[1, \infty)$ **35.** increases on $(-\infty, 0]$ and $[3, \infty)$, decreases on $[0, 1)$ and $[1, 3]$

37.

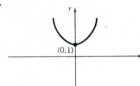

39.

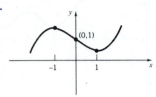

41.

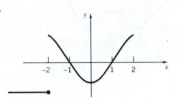

43.

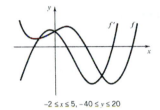

45. Not possible; f is increasing, so $f(2)$ must be greater than $f(-1)$.

47. (a) true   (b) false; see Figure 4.2.10

49. (a) true   (b) false; $f(x) = x^3$ on $(-1, 1)$

55. (b) $C = 1$ (c) $f(x) = \sin x$, $g(x) = \cos x$ **61.** $0.06975 < \sin 4° < 0.06981$

63. $f'(x) = 0$ at $x = -0.633, 0.5, 2.633$

f is decreasing on $[-2, -0.633]$ and $[0.5, 2.633]$

f is increasing on $[-0.633, 0.5]$ and $[2.633, 5]$

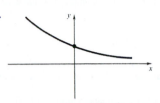

$-2 \leq x \leq 5, -40 \leq y \leq 20$

65. $f'(x) = 0$ at $x = 0.770, 2.155, 3.798, 5.812$

f is decreasing on $[0, 0.770], [2.155, 3.798],$ and $[5.812, 6]$

f is increasing on $[0.770, 2.155]$ and $[3.798, 5.812]$

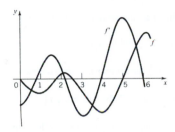

67. (a) $0, \frac{\pi}{2}, \pi, \frac{3\pi}{2}, 2\pi$

(b) $(\pi, \frac{3\pi}{2}) \cup (\frac{3\pi}{2}, 2\pi)$

(c) $(0, \frac{\pi}{2}) \cup (\frac{\pi}{2}, \pi)$

69. (a) 0

(b) $(0, \infty)$

(c) $(-\infty, 0)$

71. $f = C$ constant; $f'(x) = 0$

SECTION 4.3

1. no critical pts; no local extreme values   **3.** critical pts. ± 1; local max $f(-1) = -2$, local min $f(1) = 2$

5. critical pts. $0, \frac{2}{3}$; $f(0) = 0$ local min, $f(\frac{2}{3}) = \frac{4}{27}$ local max   **7.** no critical pts; no local extreme values

9. critical pt. $-\frac{1}{2}$; local max $f(-\frac{1}{2}) = -8$ **11.** critical pts. $0, \frac{3}{5}, 1$; local max $f(\frac{3}{5}) = 2^2 3^3/5^5$, local min $f(1) = 0$

13. critical pts. $\frac{5}{8}, 1$; local max $f(\frac{5}{8}) = \frac{27}{2048}$ **15.** critical pts. $-2, 0$; local max $f(-2) = -4$, local min $f(0) = 0$

17. critical pts. $-2, -\frac{12}{7}, 0$; local max $f(-\frac{12}{7}) = \frac{144}{49}(\frac{2}{7})^{1/3}$, local min $f(0) = 0$ **19.** critical pts. $-\frac{1}{2}, 3$; local min $f(-\frac{1}{2}) = \frac{7}{2}$

21. critical pt. 1; local min $f(1) = 3$ **23.** critical pts. $\frac{1}{4}\pi, \frac{5}{4}\pi$; local max $f(\frac{1}{4}\pi) = \sqrt{2}$, local min $f(\frac{5}{4}\pi) = -\sqrt{2}$

25. critical pts. $\frac{1}{3}\pi, \frac{1}{2}\pi, \frac{2}{3}\pi$; local max $f(\frac{1}{2}\pi) = 1 - \sqrt{3}$, local min $f(\frac{1}{3}\pi) = f(\frac{2}{3}\pi) = -\frac{3}{4}$

27. critical pts. $\frac{1}{3}\pi, \frac{5}{3}\pi$; local max $f(\frac{5}{3}\pi) = \frac{5}{4}\sqrt{3} + \frac{10}{3}\pi$, local min $f(\frac{1}{3}\pi) = -\frac{5}{4}\sqrt{3} + \frac{2}{3}\pi$

29. (a) f increases on $[-2, 0] \cup [3, \infty)$; f decreases on $(-\infty, -2] \cup [0, 3]$
(b) $f(0)$ is a local max; $f(-2)$ and $f(3)$ are local minima.

35. critical points $1, 2, 3$; local max $P(2) = -4$, local min $P(1) = P(3) = -5$

37.

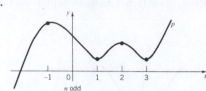
n odd

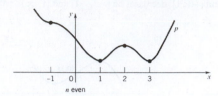
n even

39. $a = 4, b = \pm 2$.

43. (a) $f'(x) = 4x^3 - 14x - 8$; $f'(2) = -4$, $f'(3) = 58$. Therefore f' has a zero in $(2, 3)$. Since $f''(x) = 12x^2 - 14 > 0$ on $[2, 3]$, f' has exactly one zero in $(2, 3)$.

47. (a) critical pts. $-0.692, 2.248$; local max $f(-0.692) \cong 29.342$, local min $f(2.248) \cong -8.766$
(b) f is increasing on $[-3, -0.692]$ and on $[2.248, 4]$; f is decreasing on $[-0.692, 2.248]$

49. (a) critical pts. $-2.201, -0.654, 0.654, 2.201$; $f(-2.201) \cong 2.226$ and $f(0.654) \cong 6.634$ are local maxima, $f(-0.654) \cong -6.634$ and $f(2.201) \cong -2.226$ are local minima
(b) f is increasing on $[-3, -2.201]$, $[-0.654, 0.654]$, and $[2.201, 3]$; f is decreasing on $[-2.201, -0.654]$ and $[0.654, 2.201]$

51. $f'(x) > 0$ on $(\frac{2}{3}, \infty)$; no local extrema.   **53.** critical pts. $-1.326, 0, 1.816$; local maxima at -1.326 and 1.816, local minimum at 0.

SECTION 4.4

1. $f(-2) = 0$ endpt min and absolute min

3. critical pt. 2; $f(0) = 1$ endpt max and absolute max, $f(2) = -3$ local min and absolute min, $f(3) = -2$ endpt max.

5. critical pt. $2^{-1/3}$; $f(2^{-2/3}) = 3 \cdot 2^{-2/3}$ local min

7. critical pt. $2^{-1/3}$; $f(\frac{1}{10}) = 10\frac{1}{100}$ endpt max and absolute max; $f(2^{-1/3}) = 3 \cdot 2^{-2/3}$ local min and absolute min, $f(2) = 4\frac{1}{2}$ endpt max

9. critical pt. $\frac{3}{2}$; $f(0) = 2$ endpt max and absolute max, $f(\frac{3}{2}) = -\frac{1}{4}$ local min and absolute min, $f(2) = 0$ endpt max

11. critical pt. -2; $f(-3) = -\frac{3}{13}$ endpt max, $f(-2) = -\frac{1}{4}$ local min and absolute min, $f(1) = \frac{1}{5}$ endpt max and absolute max

13. critical pts. $\frac{1}{4}, 1$; $f(0) = 0$ endpt min and absolute min, $f(\frac{1}{4}) = \frac{1}{16}$ local max, $f(1) = 0$ local min and absolute min

15. critical pt. 2; $f(2) = 2$ local max and absolute max, $f(3) = 0$ endpt min   **17.** critical pt. 1; no extreme values

19. critical pt. $\frac{5}{6}\pi$; $f(0) = -\sqrt{3}$ endpt min and absolute min, $f(\frac{5}{6}\pi) = \frac{7}{4}$ local max and absolute max, $f(\pi) = \sqrt{3}$ endpt min

21. $f(0) = 5$ endpt max and absolute max, $f(\pi) = -5$ endpt min and absolute min

23. critical pt. 0; $f(-\frac{1}{3}\pi) = \frac{1}{3}\pi - \sqrt{3}$ endpt min and absolute min, no absolute max

25. critical pts. $1, 4$; $f(0) = 0$ endpt max, $f(1) = -2$ local min and absolute min, $f(4) = 1$ local max and absolute max, $f(7) = -2$ endpt min and absolute min

27. critical pts. $-1, 1, 3$; $f(-2) = 5$ endpt max, $f(-1) = 2$ local min and absolute min, $f(1) = 6$ local max and absolute max, $f(3) = 2$ local min and absolute min

29. critical pts. $-1, 0, 2$; $f(-3) = 2$ endpt max and absolute max, $f(-1) = 0$ local min, $f(0) = 2$ local max and absolute max, $f(2) = -2$ local min and absolute min

31.

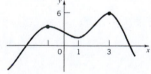

33. Not possible: $f(1) = 0$, f increasing implies $f(3) > 0$.

41. $f(x) = \begin{cases} 1 & \text{if } x \text{ is a rational number} \\ 0 & \text{if } x \text{ is an irrational number.} \end{cases}$ **45.** $\left(\dfrac{p}{p+q}\right)^p \left(\dfrac{q}{p+q}\right)^q$

47. $x = \pi/4$ **49.** critical pts.: -1.452, local max; 0.727, local min; $f(-2.5)$ absolute min; $f(3)$ absolute max

51. critical pts.: -1.683, local max; -0.284, local min; 0.645, local max; 1.760, local min; $f(-\pi)$ absolute min; $f(\pi)$ absolute max.

53. Yes; $M = f(2) = 1$; $m = f(1) = f(3) = 0$ **55.** Yes; $M = f(6) = 2 + \sqrt{3}$; $m = f(1) = \frac{3}{2}$

SECTION 4.5

1. 400 **3.** 20 by 10 ft **5.** 32 **7.** 600 ft **9.** radius of semi-circle $\dfrac{3p}{12 + 5\pi}$ **11.** $x = 2$, $y = \frac{3}{2}$ **13.** $-\frac{5}{2}$

15. $\frac{10}{3}\sqrt{3}$ by $\frac{5}{3}\sqrt{3}$ in. **17.** equilateral triangle with side 4 **19.** $(1, 1)$ **21.** 24 sq. units

23. height of rectangle $\frac{15}{11}(5 - \sqrt{3}) \cong 4.46$ in; side of triangle $\frac{10}{11}(6 + \sqrt{3}) \cong 7.03$ in.

25. $\frac{5}{3} \times \frac{5}{3}$ **27.** $(0, \sqrt{3})$ **29.** $5\sqrt{5}$ ft **31.** 54 by 72 in. **33.** (a) use it all for the circle

(b) use $28\pi/(4 + \pi) \cong 12.32$ in. for the circle

35. base radius $\frac{10}{3}$ and height $\frac{8}{3}$ **37.** 10 by 10 by 12.5 ft **39.** equilateral triangle with side $2r\sqrt{3}$

41. base radius $\frac{1}{3}R\sqrt{6}$ and height $\frac{2}{3}R\sqrt{3}$ **43.** base radius $\frac{2}{3}R\sqrt{2}$ and height $\frac{4}{3}R$ **45.** \$160,000 **47.** $\tan\theta = m$

49. $x = \dfrac{a^{1/3}s}{a^{1/3} + b^{1/3}}$ **53.** $n = 1$ **55.** 125 customers **57.** $m = 1$ **59.** 59 mph **61.** no minimum exists

63. walk along the shore

65. (b) $(1 + \sqrt{2}, 2 + \sqrt{2})$ **67.** $\left(\frac{21}{10}, \frac{7}{10}\right)$

(c) $y - (2 + \sqrt{2}) = \dfrac{1 - \sqrt{2}}{3 - \sqrt{2}}(x - [1 + \sqrt{2}])$

(e) $l_{PQ} = l_N$

SECTION 4.6

1. (a) increasing on $[a, b]$, $[d, n]$, decreasing on $[b, d]$, $[n, p]$

(b) concave up on (c, k), (l, m), concave down on (a, c), (k, l), (m, p); points of inflection at $x = c, k, l,$ and m.

3. (i) f', (ii) f, (iii) f'' **5.** concave down on $(-\infty, 0)$, concave up on $(0, \infty)$

7. concave down on $(-\infty, 0)$, concave up on $(0, \infty)$; pt of inflection $(0, 2)$

9. concave up on $(-\infty, -\frac{1}{3}\sqrt{3})$, concave down on $(-\frac{1}{3}\sqrt{3}, \frac{1}{3}\sqrt{3})$, concave up on $(\frac{1}{3}\sqrt{3}, \infty)$; pts of inflection $(-\frac{1}{3}\sqrt{3}, -\frac{5}{36})$, $(\frac{1}{3}\sqrt{3}, -\frac{5}{36})$

11. concave down on $(-\infty, -1)$ and on $(0, 1)$, concave up on $(-1, 0)$ and on $(1, \infty)$; pt of inflection $(0, 0)$

13. concave up on $(-\infty, -\frac{1}{3}\sqrt{3})$, concave down on $(-\frac{1}{3}\sqrt{3}, \frac{1}{3}\sqrt{3})$, concave up on $(\frac{1}{3}\sqrt{3}, \infty)$; pts of inflection $(-\frac{1}{3}\sqrt{3}, \frac{4}{9})$, $(\frac{1}{3}\sqrt{3}, \frac{4}{9})$

15. concave up on $(0, \infty)$ **17.** concave down on $(-\infty, -2)$, concave up on $(-2, \infty)$; pt of inflection $(-2, 0)$

19. concave up on $(0, \frac{1}{4}\pi)$, concave down on $(\frac{1}{4}\pi, \frac{3}{4}\pi)$, concave up on $(\frac{3}{4}\pi, \pi)$; pts of inflection $(\frac{1}{4}\pi, \frac{1}{2})$ and $(\frac{3}{4}\pi, \frac{1}{2})$

21. concave up on $(0, \frac{1}{12}\pi)$, concave down on $(\frac{1}{12}\pi, \frac{5}{12}\pi)$, concave up on $(\frac{5}{12}\pi, \pi)$; pts of inflection $(\frac{1}{12}\pi, \frac{1}{2} + \frac{1}{144}\pi^2)$ and $(\frac{5}{12}\pi, \frac{1}{2} + \frac{25}{144}\pi^2)$

23. $(\pm 3.94822, 10.39228)$ **25.** $(-3, 0), (-2.11652, 2.39953), (-0.28349, -18.43523)$

27. (a) f increases on $(-\infty, -\sqrt{3}]$ and $[\sqrt{3}, \infty)$, decreases on $[-\sqrt{3}, \sqrt{3}]$

(b) $f(-\sqrt{3}) \cong 10.39$ local max; $f(\sqrt{3}) \cong -10.39$ local min.

(c) concave down on $(-\infty, 0)$, concave up on $(0, \infty)$

(d) $(0, 0)$ is a point of inflection

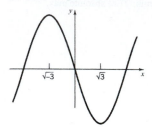

29. (a) f decreases on $(-\infty, -1]$ and $[1, \infty)$, increases on $[-1, 1]$

(b) $f(-1) = -1$ local min; $f(1) = 1$ local max

(c) concave down on $(-\infty, -\sqrt{3})$ and $(0, \sqrt{3})$, concave up on $(-\sqrt{3}, 0)$ and $(\sqrt{3}, \infty)$

(d) points of inflection $\left(-\sqrt{3}, -\dfrac{\sqrt{3}}{2}\right)$, $(0, 0)$, $\left(\sqrt{3}, \dfrac{\sqrt{3}}{2}\right)$

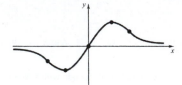

31. (a) increasing on $[-\pi, \pi]$

(b) no local max or min

(c) concave up on $(-\pi, 0)$, concave down on $(0, \pi)$

(d) point of inflection $(0, 0)$

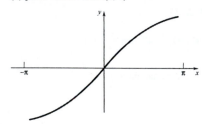

33. (a) increasing on $(-\infty, \infty)$

(b) no local max or min

(c) concave down on $(-\infty, 0)$, concave up on $(0, 1)$, no concavity on $(1, \infty)$

(d) point of inflection $(0, 0)$

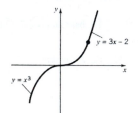

35.

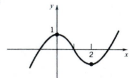

37.

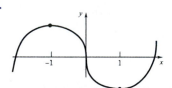

39. $d = \frac{1}{3}(a + b + c)$ **41.** $a = -\frac{1}{2}, \; b = \frac{1}{2}$ **43.** $A = 18, \, B = -4$ **45.** $f(x) = x^3 - 3x^2 + 3x - 3$; one

47. (a) $p''(x) = 6x + 2a$ has exactly one zero; $x = -\frac{1}{3}a$.

49. (a)

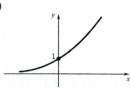

(b) No. If $f''(x) < 0$ and $f'(x) < 0$ for all x, then $f(x) < f'(0)x + f(0)$ on $(0, \infty)$ which implies $f(x) \to -\infty$ as $x \to \infty$.

51. (a) concave up on $(-4, -0.913)$ and $(0.913, 4)$; concave down on $(-0.913, 0.913)$

(b) points of inflection at $x \cong -0.913, 0.913$

53. (a) concave up on $(-\pi, -1.996)$ and $(-0.345, 2.550)$; concave down on $(-1.996, -0.345)$ and $(2.550, \pi)$

(b) points of inflection at $x \cong -1.996, -0.345, 2.550$

55. (a) $0.68824, 2.27492, 4.00827, 5.59494$

(b) $(0.68824, 2.27492) \cup (4.00827, 5.59494)$

(c) $(0, 0.68824) \cup (2.27492, 4.00827) \cup (5.59494, \pi)$

57. (a) ± 1, ± 0.71523, ± 0.32654

(b) $(-0.71523, -0.32654) \cup (0, 0.32654) \cup (0.71523, 1), (1, \infty)$

(c) $(-\infty, -1) \cup (-1, -0.71523) \cup (-0.32654, 0) \cup (0.32654, 0.71523)$

SECTION 4.7

1. (a) ∞ (b) $-\infty$ (c) ∞ (d) 1 (e) 0 (f) $x = -1, x = 1$ (g) $y = 0, y = 1$ 3. vertical; $x = \frac{1}{3}$; horizontal; $y = \frac{1}{3}$

5. vertical: $x = 2$; horizontal: none   7. vertical: $x = \pm 3$; horizontal: $y = 0$ 9. vertical: $x = -\frac{4}{3}$; horizontal: $y = \frac{4}{9}$

11. vertical: $x = \frac{5}{2}$; horizontal; $y = 0$ 13. vertical: none; horizontal: $y = \pm \frac{3}{2}$ 15. vertical: $x = 1$; horizontal: $y = 0$

17. vertical: none; horizontal: $y = 0$ 19. vertical: $x = (2n + \frac{1}{2})\pi$; horizontal: none   21. neither   23. cusp

25. tangent   27. neither   29. cusp   31. cusp   33. neither; f not continuous at $x = 0$

35.

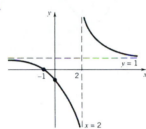

37.

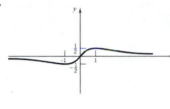

39. (a) increasing on $(-\infty, -1], [1, \infty)$; decreasing on $[-1, 1]$

(b) concave down on $(-\infty, 0)$; concave up on $(0, \infty)$

vertical tangent at $x = 0$

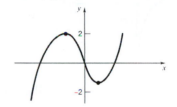

41. (a) decreasing on $[0, 2]$; increasing on $(-\infty, 0], [2, \infty)$

(b) concave up on $(-1, 0), (0, \infty)$; concave down on $(-\infty, -1)$ vertical cusp at $x = 0$

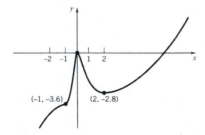

43. vertical cusp at $x = 0$

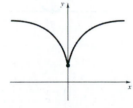

45. vertical tangent line at $x = 0$

$y = 1$ and $y = -1$ horizontal asymptotes

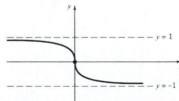

47. (a) p odd

(b) p even

49.

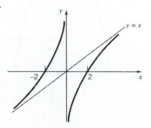

51.

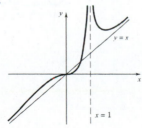

53. $y = 3x - 4$ 55. $y = 1$

SECTION 4.8

1.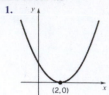

3.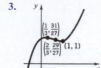

5.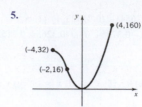

7.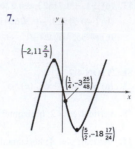

9.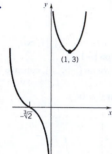

11.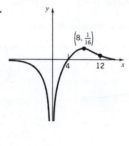

13.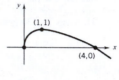

15.

17.

19.

21.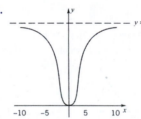

23.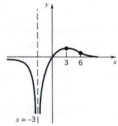

25.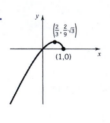

27.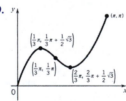

29.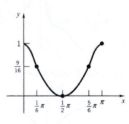

31.

33.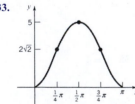

35.

37.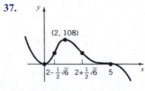

39.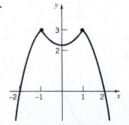

41.

43.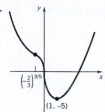

45.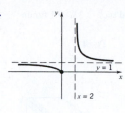

47.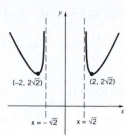

49.

51.

53.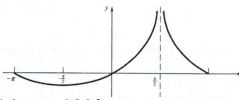

55. (a) increasing on $(-\infty, -1], (0, 1], [3, \infty)$; decreasing on $[-1, 0), [1, 3]$ critical pts. $x = -1, 0, 1, 3$

(b) concave up on $(-\infty, -3), (2, \infty)$ concave down on $(-3, 0), (0, 2)$ (c)

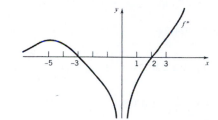

SECTION 4.9

1. $x(5) = -6, v(5) = -7, a(5) = -2$, speed $= 7$ **3.** $x(1) = 6, v(1) = -2, a(1) = \frac{4}{3}$, speed $= 2$

5. $x(1) = 0, v(1) = 18, a(1) = 54$, speed $= 18$ **7.** (a) none   (b) $a(t) = 0$ for all $t \geq 0$ **9.** (a) $t = 1, 3$ (b) $t = 2$

11. A **13.** A **15.** A and B **17.** A **19.** A and C **21.** $(0, 2), (7, \infty)$ **23.** $(0, 3), (4, \infty)$ **25.** $(2, 5)$

27. $(0, 2 - \frac{2}{3}\sqrt{3}), (4, \infty)$ **29.** $t = 3$ **31.** $t = 2, 2\sqrt{3}$ **33.** $\frac{1}{2}\pi \leq t \leq \frac{2}{3}\pi, \frac{7}{6}\pi \leq t \leq \frac{4}{3}\pi, \frac{11}{6}\pi \leq t \leq 2\pi$

35. $0 \leq t \leq \frac{1}{4}\pi, \frac{7}{4}\pi \leq t \leq 2\pi$ **37.** $\frac{5}{6}\pi \leq t \leq \frac{3}{2}\pi$ **39.** 576 ft   **41.** $v_0^2/2g$ **45.** 9 ft/sec

47. (a) 2 sec   (b) 16 ft   (c) 48 ft/sec   **49.** (a) $\frac{1625}{16}$ ft   (b) $\frac{6475}{64}$ ft   (c) 100 ft   **51.** 984 ft

55. Yes. by the mean-value theorem there is at least one time $t = c$ such that

$$v(c) = s'(c) = \frac{120 - 0}{1.67 - 0} \cong 71.86.$$

57. If the speed of the car is less than 60 mi/hr $= 1$ mi/min, then the distance traveled in 20 minutes is less than 20 miles. Therefore, the car must have gone at least 1 mi/min at some time $t < 20$. Let t_1 be the first instant the car's speed is 1 mi/min. Then the car traveled $r < t_1$ miles during the time interval $[0, t_1)$. Now apply the mean-value theorem on the interval $[t_1, 20]$.

59. The bob attains maximum speed of A_ω at the equilibrium point.

SECTION 4.10

1. (a) -2 units/sec   (b) 4 units/sec   **3.** 9 units/sec   **5.** $\dfrac{-12 \sin t \cos t}{\sqrt{16 \cos^2 t + 4 \sin^2 t}}; -\frac{3}{5}\sqrt{10}$ **7.** $-\frac{2}{27}$ m/min, $-\frac{8}{3}$ m²/min

9. $\frac{1}{2}\alpha k\sqrt{3}$ cm²/min   **11.** decreasing 7 in.²/sec   **13.** boat A **15.** $-\frac{119}{5}$ ft²/sec   **17.** 10 ft³/hr

19. $\frac{1600}{3}$ ft/min, $\frac{2800}{3}$ ft/min   **21.** 0.5634 lb/sec   **23.** dropping $1/2\pi$ in./min   **25.** $10/\pi$ cm³/min   **27.** decreasing 0.04 rad/min

29. 5π mi/min   **31.** $\frac{4}{130} \cong 0.031$ rad/sec   **33.** 0.1 ft/min   **35.** decreasing 0.12 rad/sec   **37.** increasing $\frac{4}{101}$ rad/min

39. $-\dfrac{15}{\sqrt{82}}$ ft/sec   **41.** 4.961 m/sec   **43.** 4

SECTION 4.11

1. $dV = 3x^2 h$
$\Delta V - dV = 3xh^2 + h^3$ (see figure)

3. $10 + \frac{1}{150}$ taking $x = 1000$; 10.0067   **5.** $2 - \frac{1}{64}$ taking $x = 16$; 1.9844   **7.** 8.15 taking $x = 32$; 8.1491   **9.** 0.719; 0.7193

11. 0.531; 0.5317   **13.** 1.6   **15.** $2\pi rht$ **17.** error ≤ 0.01 ft   **19.** 98 gallons   **23.** 0.00307 sec   **25.** within $\frac{1}{2}\%$

27. (a) and (b) are true   **31.** $m = f'(x)$

SECTION 4.12

1. (a) $x_{n+1} = \dfrac{x_n}{2} + \dfrac{12}{x_n}$ (b) $x_4 \cong 4.89898$ **3.** (a) $x_{n+1} = \dfrac{2x_n}{3} + \dfrac{25}{3}\left(\dfrac{1}{x_n}\right)^2$ (b) $x_4 \cong 2.92402$

5. (a) $x_{n+1} = \dfrac{x_n \sin x_n + \cos x_n}{\sin x_n + 1}$ (b) $x_4 \cong 0.73909$ **7.** (a) $x_{n+1} = \dfrac{6 + x_n}{2\sqrt{x_n + 3} - 1}$ (b) $x_4 \cong 2.30278$ **9.** $x_n = (-1)^{n-1} 2^n x_1$

11. (c) $x_4 \cong 1.6777$, $f(x_4) \cong 0.00020$ **13.** (b) $x_4 \cong 2.23607$, $f(x_4) \cong 0.00001$ **15.** (b) $\dfrac{1}{2.7153} \cong 0.36828$

17. (b) $x_3 \cong 2.98827$; local min   **19.** $r_1 \cong 2.029$, $r_2 \cong 4.913$

Chapter 4. Review Exercises

1. $c = \pm\frac{1}{3}\sqrt{3}$ **3.** $c = \pm\sqrt{7/3}$ **5.** $c = 1 + \sqrt{3}$ **7.** f is not differentiable on $(-1, 1)$.   **9.** violates the mean-value theorem

11. increasing on $(-\infty, -1] \cup [0, \infty)$; decreasing on $[-1, 0]$
critical pts. $-1, 0$; $f(-1) = 2$ local max, $f(0) = 1$ local min

13. increasing on $(-\infty, -2] \cup [-\frac{4}{5}, \infty)$; decreasing on $[-2, -\frac{4}{5}]$
critical pts. $-2, -\frac{4}{5}, 1$; $f(-2) = 0$ local max, $f(-\frac{4}{5}) \cong -8.381$ local min

15. increasing on $[-1, 1]$; decreasing on $(-\infty, -1] \cup [1, \infty)$
critical pts. $-1, 1$; $f(-1) = -\frac{1}{2}$ local min, $f(1) = \frac{1}{2}$ local max

17. critical pts. $-1, -\frac{1}{3}$; $f(-1) = 1$ local max, $f(-\frac{1}{3}) = \frac{23}{27}$ local min;
$f(-2) = -1$ endpoint and absolute min, $f(1) = 5$ endpoint and absolute max

19. critical pt. $\sqrt{2}$; $f(\sqrt{2}) = 4$ absolute min, $f(4) = 16.25$ endpoint and absolute max

21. critical pt. $\frac{2}{3}$; $f(\frac{2}{3}) = \frac{2}{9}\sqrt{3}$ absolute max, $f(1) = 0$ endpoint min; no local or absolute min

23. vertical: $x = 4$, $x = -3$; horizontal: $y = 3$ **25.** vertical: $x = 1$; oblique: $y = x$ **27.** vertical cusp

29.   **31.**   **33.**

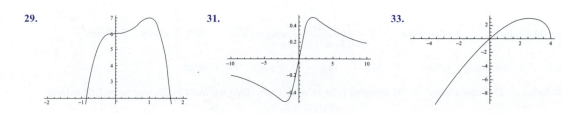

35.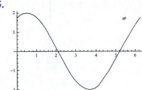

37.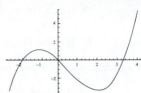

39. (a) $3 \times 3 \times 3$ (b) $3\sqrt{2} \times 3\sqrt{2} \times 3/2$ **41.** $P/3 \times P/6$ **43.** $[0, \frac{1}{6}\pi] \cup [\frac{1}{2}\pi, \frac{5}{6}\pi] \cup [\frac{3}{2}\pi, 2\pi]$

45. (b) $x(17) \cong 4.25$, $v(17) \cong 0.1172$, $a(17) \cong 0.0032$ **47.** 1520 ft.   **49.** 120 ft/min   **51.** 15 mi/hr   **53.** $\frac{1}{6}$ ft/min

55. $f(4.2) \cong 2.5375$ **57.** $\tan 43° \cong 0.9302$ **59.** (a) $x_{n+1} = \dfrac{2x_n^3 + 10}{3x_n^2}$ (b) $x_4 \cong 2.15443$, $f(2.15443) \cong 0.00007$

CHAPTER 5

SECTION 5.2

1. $L_f(P) = \frac{5}{8}, U_f(P) = \frac{11}{8}$ **3.** $L_f(P) = \frac{9}{64}, U_f(P) = \frac{37}{64}$ **5.** $L_f(P) = \frac{17}{16}, U_f(P) = \frac{25}{16}$ **7.** $L_f(P) = \frac{3}{16}, U_f(P) = \frac{43}{32}$

9. $L_f(P) = \frac{1}{6}\pi, U_f(P) = \frac{11}{12}\pi$ **11.** (a) $L_f(P) \leq U_f(P)$ but $3 \nleq 2$

$\quad$ (b) $L_f(P) \leq \displaystyle\int_{-1}^{1} f(x)\,dx \leq U_f(P)$ but $3 \nleq 2 \leq 6$

$\quad$ (c) $L_f(P) \leq \displaystyle\int_{-1}^{1} f(x)\,dx \leq U_f(P)$ but $3 \leq 10 \nleq 6$

13. (a) $L_f(P) = -3x_1(x_1 - x_0) - 3x_2(x_2 - x_1) - \cdots - 3x_n(x_n - x_{n-1})$ (b) $-\frac{3}{2}(b^2 - a^2)$ **15.** $\displaystyle\int_{-1}^{2} (x^2 + 2x - 3)\,dx$

$\quad U_f(P) = -3x_0(x_1 - x_0) - 3x_1(x_2 - x_1) - \cdots - 3x_{n-1}(x_n - x_{n-1})$

17. $\displaystyle\int_{0}^{2\pi} t^2 \sin(2t + 1)\,dt$ **19.**

21. (a) $L_f(P) = \frac{25}{32}$ **23.** $\frac{1}{4}$

$\quad$ (b) $S^*(P) = \frac{15}{16}$

$\quad$ (c) $U_f(P) = \frac{39}{32}$

25. necessarily holds: $L_g(P) \leq \displaystyle\int_{a}^{b} g(x)\,dx < \int_{a}^{b} f(x)\,dx \leq U_f(P)$ **27.** necessarily holds: $L_g(P) \leq \displaystyle\int_{a}^{b} g(x)\,dx < \int_{a}^{b} f(x)\,dx$

29. necessarily holds: $U_f(P) \geq \displaystyle\int_{a}^{b} f(x)\,dx > \int_{a}^{b} g(x)\,dx$ **33.** (b) $n = 25$ (c) 3.0

41. (a) $L_f(P) \cong 0.6105$, $U_f(P) \cong 0.7105$ (b) $1/2[L_f(P) + U_f(P)] = 0.6605$ (c) $S^*(P) \cong 0.6684$

43. (a) $L_f(P) \cong 0.53138$, $U_f(P) \cong 0.73138$ (b) $\frac{1}{2}[L_f(P) + U_f(P)] = 0.63138$ (c) $S^*(P) \cong 0.63926$

SECTION 5.3

1. (a) 5   (b) -2 (c) -1 (d) 0   (e) -4 (f) 1

5. (a) $F(0) = 0$ (b) $F'(x) = x\sqrt{x + 1}$ (c) $F'(2) = 2\sqrt{3}$ (d) $F(2) = \displaystyle\int_{0}^{2} t\sqrt{t + 1}\,dt$ (e) $-F(x) = \displaystyle\int_{x}^{0} t\sqrt{t + 1}\,dt$

7. (a) $\frac{1}{10}$ (b) $\frac{1}{9}$ (c) $\frac{4}{37}$ (d) $\dfrac{-2x}{(x^2+9)^2}$ **9.** (a) $\sqrt{2}$ (b) 0 (c) $-\frac{1}{4}\sqrt{5}$ (d) $-(\sqrt{x^2+1}+x^2/\sqrt{x^2+1})$

11. (a) -1 (b) 1 (c) 0 (d) $-\pi \sin \pi x$ **13.** (a) Since $P_1 \subseteq P_2$, $U_f(P_2) \le U_f(P_1)$. (b) Since $P_1 \subseteq P_2$, $L_f(P_1) \le L_f(P_2)$.

15. constant functions   **17.** $x=1$ is a critical number; F has a local minimum at $x=1$.

19. (a) F is increasing on $(0,\infty)$.   (b) The graph of F is concave down on $(0,\infty)$.   (c)

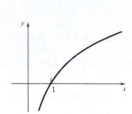

21. (a) f is continuous; Theorem 5.3.5   (e) $F(0)=0$, $F'(x)<0$ on $(0,1)$, $F''(x)>0$ on $(0,\infty)$

(b) $F'(x)=f(x)$ and f is differentiable; $F''(x)=f'(x)$

(c) $F'(1)=f(1)=0$

(d) $F''(1)=f'(1)>0$

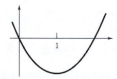

23. (a)   (b) $F(x)=\begin{cases} 2x-\frac{1}{2}x^2+\frac{5}{2}, & -1\le x\le 0 \\ 2x+\frac{1}{2}x^2+\frac{5}{2}, & 0<x\le 3 \end{cases}$

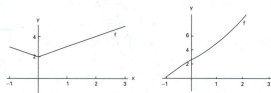

(c) f is continuous at $x=0$, but not differentiable; F is continuous and differentiable at $x=0$.

25. $F'(x)=3x^5\cos(x^3)$.   **27.** $F'(x)=2x[\sin^2(x^2)-x^2]$.   **29.** (a) 0   (b) 2   (c) 2   **31.** (a) $f(0)=\frac{1}{2}$ (b) 2, -2

37. (a) $F'(x)=0$ at $x=-1,4$;   **39.** (a) $F'(x)=0$ at $x=0,\frac{\pi}{2},\pi,\frac{3\pi}{2},2\pi$;

F increasing on $(-\infty,-1]$, $[4,\infty)$; F increasing on $[\frac{\pi}{2},\pi]$, $[\frac{3\pi}{2},2\pi]$;

F decreasing on $[-1,4]$ F decreasing on $[0,\frac{\pi}{2}]$, $[\pi,\frac{3\pi}{2}]$

(b) $F''(x)=0$ at $x=\frac{3}{2}$;   (b) $F''(x)=0$ at $x=\frac{\pi}{4},\frac{3\pi}{4},\frac{5\pi}{4},\frac{7\pi}{4}$;

the graph of F is concave up on $(\frac{3}{2},\infty)$; concave down on $(-\infty,\frac{3}{2})$ the graph of F is concave up on $(\frac{\pi}{4},\frac{3\pi}{4})$, $(\frac{5\pi}{4},\frac{7\pi}{4})$;

the graph of F is concave down on $(0,\frac{\pi}{4})$, $(\frac{3\pi}{4},\frac{5\pi}{4})$, $(\frac{7\pi}{4},2\pi)$

SECTION 5.4

1. -2 **3.** 1   **5.** $\frac{28}{3}$ **7.** $\frac{32}{3}$ **9.** $\frac{2}{3}$ **11.** $\frac{13}{2}$ **13.** $-\frac{4}{15}$ **15.** $\frac{1}{18}(2^{18}-1)$ **17.** $\frac{1}{6}a^2$ **19.** $\frac{7}{4}$

21. $\frac{21}{2}$ **23.** 1   **25.** 2   **27.** $2-\sqrt{2}$ **29.** 0   **31.** $\frac{\pi}{9}-2\sqrt{3}$ **33.** $\sqrt{13}-2$ **35.** $F'(x)=(x+2)^2$

37. $F'(x)=\sec(2x+1)\tan(2x+1)$ **39.** (a) $\displaystyle\int_2^x \frac{dt}{t}$ (b) $-3+\displaystyle\int_2^x \frac{dt}{t}$ **41.** $\dfrac{32}{2}$ **43.** $2+\sqrt{2}$ **45.** (a) 3/2 (b) 5/2

47. (a) 4/3   (b) 4   **49.** valid   **51.** not valid; $1/x^3$ is not defined at $x=0$.

53. (a) $x(t) = 5t^2 - \frac{1}{3}t^3$, $0 \leq t \leq 10$ (b) At $t = 5$; $x(5) = \dfrac{250}{3}$ **55.** $\dfrac{13}{2}$ **57.** $2 + \sqrt{3} + \frac{11\pi}{6}$

59. (a)
$$g(x) = \begin{cases} \frac{1}{2}x^2 + 2x + 2, & -2 \leq x \leq 0 \\ 2x + 2, & 0 < x \leq 1 \\ -x^2 + 4x + 1, & 1 < x \leq 2 \end{cases}$$

(b)

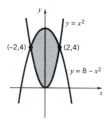

(c) f is continuous on $[-2, 2]$; f is differentiable on $(-2, 0)$, $(0, 1)$, $(1, 2)$; g is differentiable on $(-2, 2)$.

63. $f(x)$ and $f(x) - f(a)$, respectively

SECTION 5.5

1. $\dfrac{9}{4}$ **3.** $\dfrac{38}{3}$ **5.** $\dfrac{47}{15}$ **7.** $\dfrac{5}{3}$ **9.** $\dfrac{1}{2}$

11. area $= \dfrac{1}{3}$

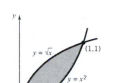

13. area $= \dfrac{9}{2}$

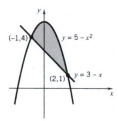

15. area $= \dfrac{64}{3}$

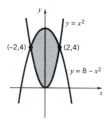

17. area $= 10$

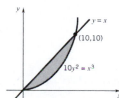

19. area $= \dfrac{32}{3}$

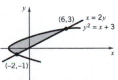

21. area $= 4$

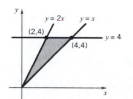

23. area $= 2 + \dfrac{2}{3}\pi^3$

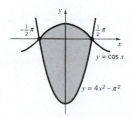

25. area $= \dfrac{1}{8}\pi^2 - 1$

27. (a) $-\dfrac{91}{6}$, the area of the region bounded by f and the x-axis for $x \in [-3, -2] \cup [3, 4]$ minus the area of the region bounded by f and the x-axis for $x \in [-2, 3]$.

(b) $\dfrac{53}{2}$ (c) $\dfrac{125}{6}$

29. (a) 0   (b) 5   **31.** (a) $\dfrac{65}{4}$ (b) 17.87   **33.** $\dfrac{10}{3}$ **35.** area $= 2 - \sqrt{2}$ **37.** 2.86

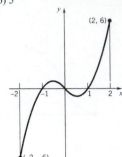

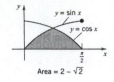

Area $= 2 - \sqrt{2}$

39. (b) $\dfrac{608}{15}$ (c) $h \cong 9.4489$

SECTION 5.6

1. $-\dfrac{1}{3x^3} + C$ **3.** $\frac{1}{2}ax^2 + bx + C$ **5.** $2\sqrt{1+x} + C$ **7.** $\dfrac{1}{2}x^2 + \dfrac{1}{x} + C$ **9.** $\frac{1}{3}t^3 - \frac{1}{2}(a+b)t^2 + abt + C$

11. $\frac{2}{9}t^{9/2} - \frac{2}{5}(a+b)t^{5/2} + 2abt^{1/2} + C$ **13.** $\frac{1}{2}[g(x)]^2 + C$ **15.** $\frac{1}{2}\sec^2 x + C$ or $\frac{1}{2}\tan^2 x + C$ **17.** $-\dfrac{1}{4x+1} + C$ **19.** $x^2 - x - 2$

21. $\frac{1}{2}ax^2 + bx - 2a - 2b$ **23.** $3 - \cos x$ **25.** $x^3 - x^2 + x + 2$ **27.** $\frac{1}{12}(x^4 - 2x^3 + 2x + 23)$ **29.** $x - \cos x + 3$

31. $\frac{1}{3}x^3 - \frac{3}{2}x^2 - \frac{1}{3}x + 3$ **33.** $\dfrac{d}{dx}\left(\displaystyle\int f(x)\,dx\right) = f(x)$; $\displaystyle\int \dfrac{d}{dx}[f(x)]\,dx = f(x) + C$

35. (a) 34 units to the right of the origin   (b) 44 units   **37.** (a) $v(t) = 2(t+1)^{1/2} - 1$ (b) $x(t) = \frac{4}{3}(t+1)^{3/2} - t - \frac{4}{3}$

39. (a) 4.4 sec   (b) 193.6 ft   **43.** 42 sec   **45.** $x(t) = x_0 + v_0 t + At^2 + Bt^3$ **47.** at $\left(\frac{160}{3}, 50\right)$ **49.** $A = -\frac{5}{2}, B = 2$

51. (a) at $t = \frac{11}{6}\pi$ sec (b) at $t = \frac{13}{6}\pi$ sec   **53.** mean-value theorem   **55.** $v(t) = v_0(1 - 2tv_0)^{-1}$

57. $\dfrac{d}{dx}\left[\displaystyle\int (\cos x - 2\sin x)\,dx\right] = \cos x - 2\sin x$; $\displaystyle\int \dfrac{d}{dx}[f(x)]\,dx = \cos x - 2\sin x + C$ **59.** $f(x) = \sin x + 2\cos x + 1$

61. $f(x) = \frac{1}{12}x^4 - \frac{1}{2}x^3 + \frac{5}{2}x^2 + 4x - 3$

SECTION 5.7

1. $\dfrac{1}{3(2 - 3x)} + C$ **3.** $\frac{1}{3}(2x + 1)^{3/2} + C$ **5.** $\dfrac{4}{7a}(ax + b)^{7/4} + C$ **7.** $-\dfrac{1}{8(4t^2 + 9)} + C$ **9.** $\frac{4}{15}(1 + x^3)^{5/4} + C$

11. $-\dfrac{1}{4(2 + s^2)^2} + C$ **13.** $\sqrt{x^2 + 1} + C$ **15.** $-\frac{5}{4}(x^2 + 1)^{-2} + C$ **17.** $-4(x^{1/4} + 1)^{-1} + C$ **19.** $-\dfrac{b^3}{2a^4}\sqrt{1 - a^4 x^4} + C$

21. $\frac{15}{8}$ **23.** 0   **25.** $\frac{1}{3}[a]^3$ **27.** $\frac{13}{3}$ **29.** $\frac{39}{400}$ **31.** $\frac{2}{5}(x + 1)^{5/2} - \frac{2}{3}(x + 1)^{3/2} + C$ **33.** $\frac{1}{10}(2x - 1)^{5/2} + \frac{1}{6}(2x - 1)^{3/2} + C$

35. $4\sqrt{1 + \sqrt{x}} + C$ **37.** $\frac{16}{3}\sqrt{2} - \frac{14}{3}$ **39.** $y = \frac{1}{3}(x^2 + 1)^{3/2} + \frac{2}{3}$ **41.** $\frac{1}{3}\sin(3x + 1) + C$ **43.** $-(\cot \pi x)/\pi + C$

45. $\frac{1}{2}\cos(3 - 2x) + C$ **47.** $-\frac{1}{5}\cos^5 x + C$ **49.** $-2\cos x^{1/2} + C$ **51.** $\frac{2}{3}(1 + \sin x)^{3/2} + C$ **53.** $\frac{1}{2}\pi \sin^2 \pi x + C$

55. $-\frac{1}{3\pi}\cos^3 \pi x + C$ **57.** $\frac{1}{8}\sin^4 x^2 + C$ **59.** $2(1 + \tan x)^{1/2} + C$ **61.** $-\sin(1/x) + C$ **63.** $\frac{1}{6}\tan^2(x^3 + \pi) + C$

65. 0   **67.** $(\sqrt{3} - 1)/\pi$ **69.** $\frac{1}{4}$ **73.** $\frac{1}{2}x + \frac{1}{20}\sin 10x + C$ **75.** $\frac{\pi}{4}$ **77.** 2   **79.** $1/2\pi$ **81.** $(4\sqrt{3} - 6)/3\pi$

83. (a) $\frac{1}{2}\sec^2 x + C$ (b) $\frac{1}{2}\tan^2 x + C'$ (c) $\frac{1}{2}\sec^2 x + C = \frac{1}{2}(1 + \tan^2 x) + C = \frac{1}{2}\tan^2 x + C'$

SECTION 5.8

1. yes; $\int_a^b [f(x) - g(x)]\,dx = \int_a^b f(x)\,dx - \int_a^b g(x)\,dx > 0$ 3. yes; if $f(x) \le g(x)$ for all $x \in [a, b]$ then
$$\int_a^b f(x)\,dx \le \int_a^b g(x)\,dx$$

5. no; take $f(x) = 0, g(x) = -1$ on $[0, 1]$ 7. no; take any odd function on an interval of the form $[-c, c]$

9. no; $\int_{-1}^1 x\,dx = 0$, but $\int_{-1}^1 |x|\,dx = 1$ 11. yes; $U_f(P) \ge \int_a^b f(x)\,dx = 0$ 13. no; $L_f(P) \le \int_a^b f(x)\,dx = 0$ 15. yes

17. $\dfrac{2x}{\sqrt{2x^2 + 7}}$ 19. $-f(x)$ 21. $-\dfrac{2\sin(x^2)}{x}$ 23. $\dfrac{\sqrt{x}}{2(1 + x)}$ 25. $\dfrac{1}{x}$ 27. $4x\sqrt{1 + 4x^2} - \tan x \sec^2 x |\sec x|$ 31. $\frac{20}{3}$

35. 0   37. $\frac{2}{3}\pi + \frac{2}{81}\pi^3 - \sqrt{3}$

SECTION 5.9

1. $A.V. = \frac{1}{2}mc + b, \quad x = \frac{1}{2}c$ 3. $A.V. = 0, \quad x = 0$ 5. $A.V = 1, \quad x = \pm 1$ 7. $A.V. = \frac{2}{3}, \quad x = 1 \pm \frac{1}{3}\sqrt{3}$

9. $A.V. = 2, \quad x = 4$ 11. $A.V. = 0, \quad x = \pi$ 13. $A.V. = \dfrac{b^{n+1} - a^{n+1}}{(n + 1)(b - a)}$ 15. $\dfrac{f(b) - f(a)}{b - a}$

17. $D = \sqrt{x^2 + x^4}; A.V. = \frac{7}{9}\sqrt{3}$ 19. (a) The terminal velocity is twice the average velocity.

(b) The average velocity during the first $\frac{1}{2}t$ seconds is one-third of the average velocity during the next $\frac{1}{2}t$ seconds.

23. (a) $v(t) = at, x(t) = \frac{1}{2}at^2 + x_0$ (b) $V_{avg} = \dfrac{1}{t_2 - t_1}\int_{t_1}^{t_2} at\,dt = \dfrac{at_1 + at_2}{2} = \dfrac{v(t_1) + v(t_2)}{2}$

25. (a) $M = 24\left(\sqrt{7} - 1\right); x_M = \dfrac{4\sqrt{7} + 2}{3\sqrt{7} - 3}$ (b) $A.V. = 4\left(\sqrt{7} - 1\right)$ 27. (a) $M = \frac{2}{3}kL^{3/2}, x_M = \frac{3}{5}L$ (b) $M = \frac{1}{3}kL^3, x_M = \frac{1}{4}L$

29. $x_{M_2} = (2M - M_1)L/8M_2$ 31. $x = \dfrac{2M \pm kL^2}{2kL}$ 39. (a) $A.V. = 2/\pi$ (b) $c \cong 0.691$

41. (a) $a \cong -3.4743, b \cong 3.4743$
(c) $f(c) = A.V. \cong 36.0948 \quad c \cong \pm 2.9545$ or $c \cong \pm 1.1274$,

Chapter 5. Review Exercises

1. $\frac{2}{7}x^{7/2} - \frac{4}{3}x^{3/2} + 2x^{1/2} + C$ 3. $\frac{1}{33}(1 + t^3)^{11} + C$ 5. $\frac{1}{2}(t^{3/2} - 1)^3 + C$ 7. $-\frac{2}{15}(2 - x)^{3/2}(4 + 3x) + C$

9. $\frac{1}{3}(1 + \sqrt{x})^6 + C$ 11. $2\sqrt{1 + \sin x} + C$ 13. $\frac{1}{3}\tan 3\theta - \frac{1}{3}\cot 3\theta - 4\theta + C$ 15. $\frac{1}{2}\tan x + C$ 17. $\frac{1}{3\pi}\sec^3 \pi x + C$

19. $\dfrac{2a}{15b^2}(1 + bx)^{3/2}(3bx - 2)$ 21. $\sqrt{1 + g^2(x)} + C$ 23. 9   25. $\frac{1}{8}$ 27. $4 - \frac{1}{4}(6)^{4/3}$

29. (a) -2 (b) 6   (c) $f_{avg} = 4$ (d) $\int_2^3 f = -2$.   31. $\frac{9}{2}$ 33. $\frac{9}{2}$ 35. $\frac{3}{4}$ 37. $\dfrac{1}{1 + x^2}$ 39. $\dfrac{2x}{1 + x^4} - \dfrac{1}{1 + x^2}$

41. $-\csc x$ 43. (a) $x = 0$ (b) $F'(x) = \dfrac{1}{x^2 + 2x + 2} > 0$ (c) concave up on $(-\infty, -1)$, concave down on $(-1, \infty)$

(d)

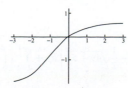

45. $\frac{1}{2}$ 47. 0   49. $\int_\alpha^\beta |f(x)|\,dx$ 51. $\frac{1}{2}\left[\int_\alpha^\beta |f(x)|\,dx - \int_\alpha^\beta f(x)\,dx\right]$ 53. $x_H = \frac{41a}{60}$

CHAPTER 6

SECTION 6.1

1.

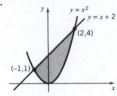

(a) $\displaystyle\int_{-1}^{2}[(x+2)-x^2]\,dx$

(b) $\displaystyle\int_{0}^{1}[\sqrt{y}-(-\sqrt{y})]\,dy + \int_{1}^{4}[\sqrt{y}-(y-2)]\,dy$

3.

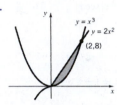

(a) $\displaystyle\int_{0}^{2}[2x^2-x^3]\,dx$

(b) $\displaystyle\int_{0}^{8}\left[y^{1/3}-\left(\frac{1}{2}y\right)^{1/2}\right]\,dy$

5.

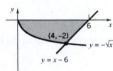

(a) $\displaystyle\int_{0}^{4}[0-(-\sqrt{x})]\,dx + \int_{4}^{6}[0-(x-6)]\,dx$

(b) $\displaystyle\int_{-2}^{0}[(y+6)-y^2]\,dy$

7.

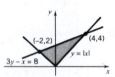

(a) $\displaystyle\int_{-2}^{0}\left[\frac{8+x}{3}-(-x)\right]\,dx + \int_{0}^{4}\left[\frac{8+x}{3}-x\right]\,dx$

(b) $\displaystyle\int_{0}^{2}[y-(-y)]\,dy + \int_{2}^{4}[y-(3y-8)]\,dy$

9.

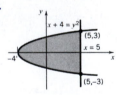

(a) $\displaystyle\int_{-4}^{5}[\sqrt{4+x}-(-\sqrt{4+x})]\,dx$

(b) $\displaystyle\int_{-3}^{3}[5-(y^2-4)]\,dy$

11.

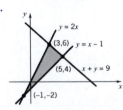

(a) $\displaystyle\int_{-1}^{3}[2x-(x-1)]\,dx + \int_{3}^{5}[(9-x)-(x-1)]\,dx$

(b) $\displaystyle\int_{-2}^{4}\left[(y+1)-\frac{1}{2}y\right]\,dy + \int_{4}^{6}\left[(9-y)-\frac{1}{2}y\right]\,dy$

13.

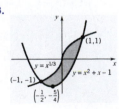

(a) $\displaystyle\int_{-1}^{1}[x^{1/3}-(x^2+x-1)]\,dx$

(b) $\displaystyle\int_{-5/4}^{-1}\left[\left(-\frac{1}{2}+\frac{1}{2}\sqrt{4y+5}\right)-\left(-\frac{1}{2}-\frac{1}{2}\sqrt{4y+5}\right)\right]\,dy + \int_{-1}^{1}\left[\left(-\frac{1}{2}+\frac{1}{2}\sqrt{4y+5}\right)-y^3\right]\,dy$

15. area $= \frac{9}{8}$

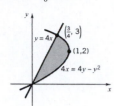

17. area $= 32$

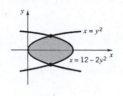

19. area $= \frac{37}{12}$

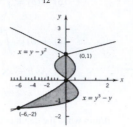

21. area $= 2 - \sqrt{2}$

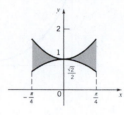

23. area $= 8$

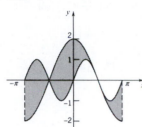

25. area $= \frac{1}{5}$

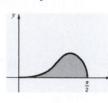

27. 4

29. $\frac{39}{2}$

31. area $= 27$

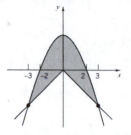

33. $c = 4^{2/3}$

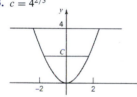

35. $A = \int_0^{\sqrt{3}} \left[\sqrt{4 - y^2} - \frac{1}{\sqrt{3}} y \right] dy$

37. $A = \int_0^2 [\sqrt{4 - x^2} - (2 - \sqrt{4x - x^2})]\, dx$

39. The ratio is $\dfrac{1}{n+1}$.

41. area $= 7.93$

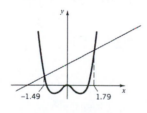

43. 1536 cu in. $\cong 0.89$ cu ft

45. (a) $A(b) = 2\sqrt{b} - 2$

(b) $A(b) \to \infty$ as $x \to \infty$

SECTION 6.2

1. $\frac{1}{3}\pi$ **3.** $\frac{1944}{5}\pi$ **5.** $\frac{5}{14}\pi$ **7.** $\frac{3790}{21}\pi$ **9.** $\frac{72}{5}\pi$ **11.** $\frac{32}{3}\pi$ **13.** π **15.** $\dfrac{\pi^2}{24}(\pi^2 + 6\pi + 6)$

17. $\frac{16}{3}\pi$ **19.** $\frac{768}{7}\pi$ **21.** $\frac{2}{5}\pi$ **23.** $\frac{128}{3}\pi$ **25.** $\frac{16}{3}\pi$ **27.** (a) $\frac{16}{3}r^3$ (b) $\frac{4}{3}\sqrt{3}\, r^3$ **29.** (a) $\frac{512}{15}$ (b) $\frac{64}{15}\pi$ (c) $\frac{128}{15}\sqrt{3}$

31. (a) 32   (b) 4π (c) $8\sqrt{3}$ **33.** (a) $\frac{64}{3}$ (b) $\frac{16}{3}$ **35.** (a) $\sqrt{3}$ (b) 4   **37.** $\frac{4}{3}\pi a b^2$ **39.** $\frac{1}{3}\pi h(R^2 + rR + r^2)$

41. (a) $31\frac{1}{4}\%$ (b) $14\frac{22}{27}\%$ **43.** $V = \frac{1}{3}\pi(2r^3 - 3r^2 h + h^3)$

45. (a)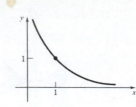

(b) $A(b) = \int_1^b x^{-2/3} dx = 3(b^{1/3} - 1)$

(c) $V(b) = \int_1^b \pi (x^{-2/3})^2 dx = 3\pi(1 - b^{-1/3})$

(d) As $b \to \infty$, $A(b) \to \infty$ and $V(b) \to 3\pi$

47. $\dfrac{1}{\pi}$ ft/min when the depth is 1 foot; $\dfrac{2}{3\pi}$ ft/min when the depth is 2 feet.   **49.** (b) $x = 0$, $x = 2^{1/4} \cong 1.1892$;   (c) 0.9428;   (d) 5.1234

51. $\frac{40}{3}\pi$ **53.** $4\pi - \frac{1}{2}\pi^2$ **55.** 250π **57.** (a) $\frac{32}{3}\pi$ (b) $\frac{64}{5}\pi$ **59.** (a) 64π (b) $\frac{1024}{35}\pi$ (c) $\frac{704}{5}\pi$ (d) $\frac{512}{7}\pi$

SECTION 6.3

1. $\frac{2}{3}\pi$ **3.** $\frac{128}{5}\pi$ **5.** $\frac{2}{5}\pi$ **7.** 16π **9.** $\frac{72}{5}\pi$ **11.** 36π **13.** 8π **15.** $\frac{1944}{5}\pi$ **17.** $\frac{5}{14}\pi$

19. $\frac{72}{5}\pi$ **21.** 64π **23.** $\frac{1}{3}\pi$ **25.** (a) $V = \int_0^1 2\pi x(1 - \sqrt{x})\,dx$ (b) $V = \int_0^1 \pi y^4 dy$; $V = \frac{1}{5}\pi$

27. (a) $V = \int_0^1 \pi(x - x^4)\,dx$ (b) $V = \int_0^1 2\pi y(\sqrt{y} - y^2)\,dy$; $V = \frac{3}{10}\pi$

29. (a) $V = \int_0^1 2\pi x^3\,dx$ (b) $V = \int_0^1 \pi(1 - y)\,dy$; $V = \frac{\pi}{2}$ **31.** $\frac{4}{3}\pi ba^2$ **33.** $\frac{1}{4}\pi a^3 \sqrt{3}$

35. (a) 64π (b) $\frac{1024}{35}\pi$ (c) $\frac{704}{5}\pi$ (d) $\frac{512}{7}\pi$ **37.** (a) $F'(x) = x\cos x$ (b) $V = \pi^2 - 2\pi$

39. (a) $V = \int_0^1 2\sqrt{3}\,\pi x^2 dx + \int_1^2 2\pi x\sqrt{4 - x^2}\,dx$ (b) $V = \int_0^{\sqrt{3}} \pi\left(4 - \frac{4}{3}y^2\right)dy$ (c) $V = \frac{8\pi\sqrt{3}}{3}$

41. (a) $V = \int_0^1 2\sqrt{3}\pi x(2 - x)\,dx + \int_1^2 2\pi(2 - x)\sqrt{4 - x^2}\,dx$ (b) $V = \int_0^{\sqrt{3}} \pi\left[\left(2 - \frac{y}{\sqrt{3}}\right)^2 - \left(2 - \sqrt{4 - y^2}\right)^2\right]dy$

43. (a) $V = 2\int_{b-a}^{b+a} 2\pi x\sqrt{a^2 - (x - b)^2}\,dx$ (b) $V = \int_{-a}^{a} \pi\left[\left(b + \sqrt{a^2 - y^2}\right)^2 - \left(b - \sqrt{a^2 - y^2}\right)^2\right]dy$

45. $V = \int_0^r 2\pi x\left(-\left(\frac{h}{r}\right)x + h\right)dx = \frac{1}{3}\pi r^2 h$ **47.** $\frac{\pi r^4}{2} - \pi a^2 r^2 + \frac{\pi a^4}{2} = \frac{\pi}{2}(r^2 - a^2)^2$

49. (b) $x = 0, x = 1.8955$ (c) 0.4208   (d) 2.6226

SECTION 6.4

1. $\left(\frac{12}{5}, \frac{3}{4}\right)$, $V_x = 8\pi$, $V_y = \frac{128}{5}\pi$ **3.** $\left(\frac{3}{7}, \frac{12}{25}\right)$, $V_x = \frac{2}{3}\pi$, $V_y = \frac{5}{14}\pi$ **5.** $\left(\frac{7}{3}, \frac{10}{3}\right)$, $V_x = \frac{80}{3}\pi$, $V_y = \frac{56}{3}\pi$ **7.** $\left(\frac{3}{4}, \frac{22}{5}\right)$, $V_x = \frac{704}{15}\pi$, $V_y = 8\pi$

9. $\left(\frac{2}{5}, \frac{2}{5}\right)$, $V_x = \frac{4}{15}\pi$, $V_y = \frac{4}{15}\pi$ **11.** $\left(\frac{45}{28}, \frac{93}{70}\right)$, $V_x = \frac{31}{5}\pi$, $V_y = \frac{15}{2}\pi$ **13.** $\left(3, \frac{5}{3}\right)$, $V_x = \frac{40}{3}\pi$, $V_y = 24\pi$ **15.** $\left(\frac{5}{2}, 5\right)$ **17.** $\left(1, \frac{8}{5}\right)$

19. $\left(\frac{10}{3}, \frac{40}{21}\right)$ **21.** $(2, 4)$ **23.** $\left(-\frac{3}{5}, 0\right)$ **25.** (a) $(0, 0)$ (b) $\left(\frac{14}{5\pi}, \frac{14}{5\pi}\right)$ (c) $\left(0, \frac{14}{5\pi}\right)$ **27.** $V = \pi ab(2c + \sqrt{a^2 + b^2})$

29. (a) $\left(\frac{2}{3}a, \frac{1}{3}h\right)$ (b) $\left(\frac{2}{3}a + \frac{1}{3}b, \frac{1}{3}h\right)$ (c) $\left(\frac{1}{3}a + \frac{1}{3}b, \frac{1}{3}h\right)$ **31.** (a) $\frac{1}{3}\pi R^3 \sin^2\theta(2\sin\theta + \cos\theta)$ (b) $\dfrac{2R\sin\theta(2\sin\theta + \cos\theta)}{3(\pi\sin\theta + 2\cos\theta)}$

33. An annular region; see Exercise 25(a).   **35.** (a) $A = \frac{1}{2}$ (b) $\left(\frac{16}{35}, \frac{16}{35}\right)$ (c) $V = \frac{16}{35}\pi$ (d) $V = \frac{16}{35}\pi$

37. (a) $A = \frac{250}{3}$ (b) $\left(-\frac{9}{8}, \frac{290}{21}\right) \cong (-1.125, 13.8095)$

SECTION 6.5

1. 817.5 ft-lb   **3.** $\frac{1}{3}(64 - 7^{3/2})$ ft-lb   **5.** $\frac{35\pi^2}{72} - \frac{1}{4}$ newton-meters   **7.** 625 ft-lb   **9.** (a) 25-ft-lb   (b) $\frac{225}{4}$ ft-lb

11. 1.95 ft   **13.** (a) $(6480\pi + 8640)$ ft-lb   (b) $(15, 120\pi + 8640)$ ft-lb   **15.** 8437.5 ft-lb

17. (a) $\frac{11}{192}\pi r^2 h^2 \sigma$ ft-lb   (b) $\left(\frac{11}{192}\pi r^2 h^2 \sigma + \frac{7}{24}\pi r^2 hk\sigma\right)$ ft-lb   **19.** (a) $384\pi\sigma$ newton-meters   (b) $480\pi\sigma$ newton-meters

21. 48,000 ft-lb   **23.** (a) 20,000 ft-lb   (b) 30,000 ft-lb   **25.** 796 ft-lb   **27.** (a) $\frac{1}{2}\sigma l^2$ ft-lb   (b) $\frac{3}{2}l^2\sigma$ ft-lb   **29.** 20,800 ft-lb

33. $v = -\sqrt{2gh}$ **35.** 94.8 ft-lb   **37.** 9.714×10^9 ft-lb   **39.** (a) 670 sec or 11 min, 10 sec   (b) 1116 sec or 18 min, 36 sec

SECTION 6.6

1. 9000 lb   **3.** 1.437×10^8 newtons   **5.** 1.7052×10^6 newtons   **7.** 2160 lb   **9.** $\frac{8000}{3}\sqrt{2}$ lb   **11.** 333.33 lb   **13.** 2560 lb

15. (a) 41,250 lb   (b) 41,250 lb   **17.** (a) 297,267 newtons   (b) 39,200 newtons at the shallow end; 352,800 newtons at the deep end

19. $F_2 = \dfrac{h_2}{h_1} F_1$ **21.** 2.217×10^6 newtons

Chapter 6. Review Exercises

1. (a) $\displaystyle\int_{-1}^{2} (2 - x^2 + x)\, dx$ (b) $\displaystyle\int_{-2}^{1} (\sqrt{2 - y} + y)\, dy$; $\frac{9}{2}$

3. (a) $\displaystyle\int_{1}^{3} 2\sqrt{2x - 2}\, dx + \int_{3}^{9}\left(\sqrt{2x - 2} - x + 5\right) dx$ (b) $\displaystyle\int_{-2}^{4}\left(y - \frac{1}{2}y^2 + 4\right) dy$;   18

5. $2\sqrt{2}$ **7.** $\frac{8}{15}$ **9.** (a) $\frac{2}{3}\pi r^3$ (b) $\frac{4}{3}r^3$ **11.** $\frac{1}{2}\pi$ **13.** $\frac{4}{15}\pi$ **15.** $\frac{6}{7}\pi$ **17.** π **19.** 2π **21.** 27π

23. $\frac{72}{5}\pi$ **25.** 8π **27.** $\frac{104}{15}\pi$ **29.** $\frac{64}{3}\pi$ **31.** $\left(0, \frac{8}{5}\right)$ **33.** $\left(\frac{1}{2}, -\frac{3}{2}\right)$ **35.** $\left(\frac{5}{14}, \frac{38}{35}\right)$; around x-axis: $\frac{38}{15}\pi$, around y-axis: $\frac{5}{6}\pi$

37. $\frac{1}{3}(64 - 7\sqrt{7})$ **39.** 6.5 inches   **41.** 1250 ft-lbs   **43.** $550,000\pi$ ft-lbs

45. (a) $\frac{1225}{3} \times 10^6$ newtons   (b) $\frac{10976}{3} \times 10^5$ newtons

CHAPTER 7

SECTION 7.1

1. $f^{-1}(x) = \frac{1}{5}(x - 3)$ **3.** not one-to-one   **5.** $f^{-1}(x) = (x - 1)^{1/5}$ **7.** $f^{-1}(x) = [\frac{1}{3}(x - 1)]^{1/3}$ **9.** $f^{-1}(x) = 1 - x^{1/3}$

11. $f^{-1}(x) = (x - 2)^{1/3} - 1$ **13.** $f^{-1}(x) = x^{5/3}$ **15.** $f^{-1}(x) = \frac{1}{3}(2 - x^{1/3})$ **17.** $f^{-1}(x) = \arcsin x$ (to be studied in Section 7.7)

19. $f^{-1}(x) = 1/x$ **21.** not one-to-one   **23.** $f^{-1}(x) = \left(\dfrac{1 - x}{x}\right)^{1/3}$ **25.** $f^{-1}(x) = (2 - x)/(x - 1)$ **27.** they are equal

29.   **31.**   **33.** (a) $k \geq 1$

                                       (b) $-\sqrt{3} \leq k \leq \sqrt{3}$

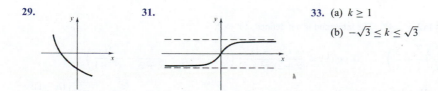

35. $f'(x) = 3x^2 \geq 0$ on $(-\infty, \infty)$, $f'(x) = 0$ only at $x = 0$; $(f^{-1})'(9) = \frac{1}{12}$ **37.** $f'(x) = 1 + \dfrac{1}{\sqrt{x}} > 0$ on $(0, \infty)$; $(f^{-1})'(8) = \frac{2}{3}$

39. $f'(x) = 2 - \sin x > 0$ on $(-\infty, \infty)$; $(f^{-1})'(\pi) = 1$ **41.** $f'(x) = \sec^2 x > 0$ on $(-\pi/2, \pi/2)$; $(f^{-1})'(\sqrt{3}) = \frac{1}{4}$

43. $f'(x) = 3 + \dfrac{3}{x^4} > 0$ on $(0, \infty)$; $(f^{-1})'(2) = \frac{1}{6}$ **45.** $(f^{-1})'(x) = \dfrac{1}{x}$ **47.** $(f^{-1})'(x) = \dfrac{1}{\sqrt{1 - x^2}}$

49. (a) $f'(x) = \dfrac{ad - bc}{(cx + d)^2} \neq 0$ iff $ad - bc \neq 0$ (b) $f^{-1}(x) = \dfrac{dx - b}{a - cx}$ **51.** (a) $f'(x) = \sqrt{1 + x^2} > 0$ (b) $(f^{-1})'(0) = \dfrac{1}{\sqrt{5}}$

53. (b) If f is increasing, then the graphs of f and g have opposite concavity; if f is decreasing, then the graphs of f and g have the same concavity.

55. $(f^{-1})'(x) = \dfrac{1}{\sqrt{1 - x^2}}$ **57.** $f^{-1}(x) = \dfrac{x^2 - 8x + 25}{9}$, $x \geq 4$ **59.** $f^{-1}(x) = 16 - 12x + 6x^2 - x^3$

61.

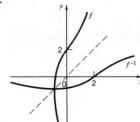

63.

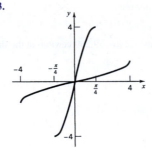

SECTION 7.2

1. $\ln 2 + \ln 10 \cong 2.99$ **3.** $2\ln 4 - \ln 10 \cong 0.48$ **5.** $-\ln 10 \cong -2.30$ **7.** $\ln 8 + \ln 9 - \ln 10 \cong 1.98$ **9.** $\frac{1}{2}\ln 2 \cong 0.35$

11. $\displaystyle\int_k^{2k} \frac{1}{x}\,dx = \ln 2$ **13.** 0.406 **15.** (a) 1.65 (b) 1.57 (c) 1.71 **17.** $x = e^2$ **19.** $x = 1, e^2$ **21.** $x = 1$

23. $\displaystyle\lim_{x \to 1} \frac{\ln x}{x - 1} = \dfrac{d(\ln x)}{dx}\bigg|_{x=1} = 1$ **25.** $k = n - 1$ **27.** (a) $\ln 3 - \sin 3 \cong 0.96 > 0$; $\ln 2 - \sin 2 \cong -0.22 < 0$ **29.** 1
                                                     (b) $r \cong 2.2191$

SECTION 7.3

1. domain $(0, \infty)$, $f'(x) = \dfrac{1}{x}$ **3.** domain $(-1, \infty)$, $f'(x) = \dfrac{3x^2}{x^3 + 1}$ **5.** domain $(-\infty, \infty)$, $f'(x) = \dfrac{x}{1 + x^2}$

7. domain all $x \neq \pm 1$, $f'(x) = \dfrac{4x^3}{x^4 - 1}$ **9.** domain $(-\frac{1}{2}, \infty)$, $f'(x) = 2(2x + 1)[1 + 2\ln(2x + 1)]$

11. domain $(0, 1) \cup (1, \infty)$, $f'(x) = -\dfrac{1}{x(\ln x)^2}$ **13.** domain $(0, \infty)$; $f'(x) = \dfrac{1}{x}\cos(\ln x)$ **15.** $\ln|x + 1| + C$ **17.** $-\frac{1}{2}\ln|3 - x^2| + C$

19. $\frac{1}{3}\ln|\sec 3x| + C$ **21.** $\frac{1}{2}\ln|\sec x^2 + \tan x^2| + C$ **23.** $\dfrac{1}{2(3 - x^2)} + C$ **25.** $-\ln|2 + \cos x| + C$ **27.** $\ln|\ln x| + C$

29. $\dfrac{-1}{\ln x} + C$ **31.** $-\ln|\sin x + \cos x| + C$ **33.** $\frac{2}{3}\ln\left[1 + x\sqrt{x}\right] + C$ **35.** $x + 2\ln|\sec x + \tan x| + \tan x + C$ **37.** 1

39. 1 **41.** $\frac{1}{2}\ln\frac{8}{5}$ **43.** $\ln\frac{4}{3}$ **45.** $\frac{1}{2}\ln 2$ **47.** The integrand is not defined at $x = 2$.

49. $g'(x) = (x^2 + 1)^2(x - 1)^5 x^3 \left(\dfrac{4x}{x^2 + 1} + \dfrac{5}{x - 1} + \dfrac{3}{x}\right)$ **51.** $g'(x) = \dfrac{x^4(x - 1)}{(x + 2)(x^2 + 1)}\left(\dfrac{4}{x} + \dfrac{1}{x - 1} - \dfrac{1}{x + 2} - \dfrac{2x}{x^2 + 1}\right)$

53. $\frac{1}{3}\pi - \frac{1}{2}\ln 3$ **55.** $\frac{1}{4}\pi - \frac{1}{2}\ln 2$ **57.** $\frac{15}{8} - \ln 4$ **59.** $\pi \ln 9$ **61.** $2\pi \ln(2 + \sqrt{3})$ **63.** $\ln 5$ ft   **65.** $(-1)^{n-1}\dfrac{(n - 1)!}{x^n}$

69. (i) domain $(-\infty, 4)$

   (ii) decreases throughout

   (iii) no extreme values

   (iv) concave down throughout; no pts of inflection

(v)

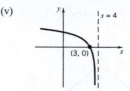

71. (i) domain $(0, \infty)$

   (ii) decreases on $(0, e^{-1/2})$, increases on $[e^{-1/2}, \infty]$

   (iii) $f(e^{-1/2}) = -\frac{1}{2}e^{-1}$ local and absolute min.

   (iv) concave down on $(0, e^{-3/2})$; concave up on $(e^{3/2}, \infty)$ pt of inflection at $(e^{-3/2}, -\frac{3}{2}e^{-3})$

(v)

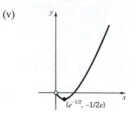

73. (i) domain $(0, \infty)$

   (ii) increases on $(0, 1]$; decreases on $[1, \infty)$

   (iii) $f(1) = \ln \frac{1}{2}$ local and absolute max

   (iv) concave down on $(0, 2.0582)$; concave up on $(2.0582, \infty)$; point of inflection $(2.0582, -0.9338)$ (approx.)

(v)

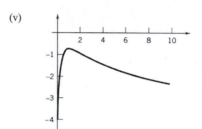

75. average slope $= \dfrac{1}{b-a} \displaystyle\int_a^b \dfrac{1}{x}\,dx = \dfrac{1}{b-a}\ln(b/a)$ **77.** x-intercept: 1; absolute min at $x = e^{-2}$; absolute max at $x = 10$

79. x-intercepts: $1, 23.1407$; absolute max at $x \cong 4.8105$, absolute min at $x = 100$

81. (a) $v(t) = 2 + 2t - t^2 + 3\ln(t+1)$ (c) max velocity at $t \cong 1.5811$; min velocity at $t = 0$

83. (b) x-coordinates of points of intersection: $x = 1, 3.30278$

   (c) $A \cong 2.34042$

85. (a) $f'(x) = \dfrac{1 - 2\ln x}{x^3}$; $f''(x) = \dfrac{-5 + 6\ln x}{x^4}$

   (b) $f(1) = 0$; $f'(e^{1/2}) = 0$; $f''(e^{5/6}) = 0$

   (c) $f(x) > 0$ on $(1, \infty)$; $f'(x) > 0$ on $(0, e^{1/2})$; $f''(x) > 0$ on $(e^{5/6}, \infty)$;

     $f(x) < 0$ on $(0, 1)$; $f'(x) < 0$ on $(e^{1/2}, \infty)$; $f''(x) < 0$ on $(0, e^{5/6})$ (d) $f(e^{1/2})$ local and absolute maximum

SECTION 7.4

1. $\dfrac{dy}{dx} = -2e^{-2x}$ **3.** $\dfrac{dy}{dx} = 2x\,e^{x^2-1}$ **5.** $\dfrac{dy}{dx} = e^x\left(\dfrac{1}{x} + \ln x\right)$ **7.** $\dfrac{dy}{dx} = -(x^{-1} + x^{-2})e^{-x}$ **9.** $\dfrac{dy}{dx} = \dfrac{1}{2}(e^x - e^{-x})$

11. $\dfrac{dy}{dx} = \dfrac{1}{2}e^{\sqrt{x}}\left(\dfrac{1}{x} + \dfrac{\ln\sqrt{x}}{\sqrt{x}}\right)$ **13.** $\dfrac{dy}{dx} = 4x\,e^{x^2}(e^{x^2} + 1)$ **15.** $\dfrac{dy}{dx} = x^2\,e^x$ **17.** $\dfrac{dy}{dx} = \dfrac{2e^x}{(e^x + 1)^2}$ **19.** $\dfrac{dy}{dx} = 4x^3$

21. $2e^{2x}\cos(e^{2x})$ **23.** $f'(x) = -e^{-2x}(2\cos x + \sin x)$ **25.** $\frac{1}{2}e^{2x} + C$ **27.** $\frac{1}{k}e^{kx} + C$ **29.** $\frac{1}{2}e^{x^2} + C$ **31.** $-e^{1/x} + C$

33. $\frac{1}{2}x^2 + C$ **35.** $-8e^{-x/2} + C$ **37.** $2\sqrt{e^x + 1} + C$ **39.** $\frac{1}{4}\ln(2e^{2x} + 3) + C$ **41.** $e^{\sin x} + C$ **43.** $e - 1$ **45.** $\frac{1}{6}(1 - \pi^{-6})$

47. $2 - \dfrac{1}{e}$ **49.** $\ln\frac{3}{2}$ **51.** $\frac{1}{2}e + \frac{1}{2}$ **53.** (a) $f^{(n)}(x) = a^n e^{ax}$ (b) $f^{(n)}(x) = (-1)^n a^n e^{-ax}$ **55.** at $\left(\pm\frac{1}{\sqrt{2}}, \frac{1}{\sqrt{e}}\right)$

57. (a) f is an even function; symmetric with respect to the y-axis.

   (b) f increases on $(-\infty, 0]$; f decreases on $[0, \infty)$.

   (c) $f(0) = 1$ is a local and absolute maximum.

(d) the graph is concave up on $(-\infty, -1/\sqrt{2}) \cup (1/\sqrt{2}, \infty)$; the graph is concave down on $(-1/\sqrt{2}, 1/\sqrt{2})$; points of inflection at $(-1/\sqrt{2}, e^{-1/2})$ and $(1/\sqrt{2}, e^{-1/2})$

(e) the x-axis   (f)

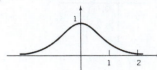

59. (a) $\pi(1 - e^{-1})$ (b) $\displaystyle\int_0^1 \pi e^{-2x^2} dx$ **61.** $\frac{1}{2}(3e^4 + 1)$ **63.** $e^2 - e - 2$

65. (a) domain $(-\infty, 0) \cup (0, \infty)$

    (b) increases on $(-\infty, 0)$, decreases on $(0, \infty)$

    (c) no extreme values

    (d) concave up on $(-\infty, 0)$ and on $(0, \infty)$

(e)

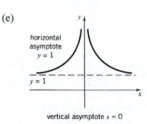

67. (a) domain $(0, \infty)$

    (b) f increases on $(e^{-1/2}, \infty)$; f decreases on $(0, e^{-1/2})$.

    (c) $f(e^{-1/2}) = -1/2e$ is a local and absolute minimum.

    (d) the graph is concave down on $(0, e^{-3/2})$; the graph is concave up on $(e^{-3/2}, \infty)$; point of inflection at $(e^{-3/2}, -3/2e^3)$

(e)

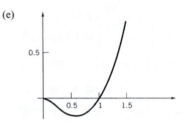

69. $x_n = \ln(n + 1)$ **71.** (a) $\left(\pm\dfrac{1}{a}, e\right)$ (b) $\dfrac{1}{a}(e - 2)$ (c) $\dfrac{1 + 2a^2 e}{a^3 e}$

75. (a)

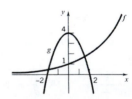

    (b) $x = -1.9646; x = 1.0580$

    (c) 6.4240

79. (a) $x = \ln(|n\pi|), n = \pm 1, \pm 2, \ldots$

81. (b) $x \cong 1.3098$

    (c) $f'(1.3098) \cong -0.26987; g'(1.3098) \cong 0.76348$

    (d) no

77.

83. (a) $x - \ln|e^x - 1|, + C$

    (b) $-\frac{1}{5}e^{-5x} + e^{-4x} - 2e^{-3x} + 2e^{-2x} - e^{-x} + C$

    (c) $e^{\tan x} + C$

SECTION 7.5

1. 6   **3.** $-\frac{1}{6}$ **5.** 0   **7.** 3   **9.** $\log_p xy = \dfrac{\ln xy}{\ln p} = \dfrac{\ln x + \ln y}{\ln p} = \dfrac{\ln x}{\ln p} + \dfrac{\ln y}{\ln p} = \log_p x + \log_p y$

11. $\log_p x^y = \dfrac{\ln x^y}{\ln p} = y\dfrac{\ln x}{\ln p} = y\log_p x$ **13.** 0   **15.** 2   **17.** $t_1 < \ln a < t_2$ **19.** $f'(x) = 2(\ln 3)3^{2x}$

21. $f'(x) = \left(5\ln 2 + \dfrac{\ln 3}{x}\right)2^{5x}3^{\ln x}$ **23.** $g'(x) = \dfrac{1}{2\ln 3} \cdot \dfrac{1}{x\sqrt{\log_3 x}}$ **25.** $f'(x) = \dfrac{\sec^2(\log_5 x)}{x\ln 5}$

27. $F'(x) = \ln 2(2^{-x} - 2^x)\sin(2^x + 2^{-x})$ **29.** $\dfrac{3^x}{\ln 3} + C$ **31.** $\frac{1}{4}x^4 - \dfrac{3^{-x}}{\ln 3} + C$ **33.** $\log_5 |x| + C$ **35.** $\dfrac{3}{\ln 4}(\ln x)^2 + C$ **37.** $\dfrac{1}{e\ln 3}$

39. $\dfrac{1}{e}$ **41.** $f'(x) = p^x \ln p$ **43.** $(x+1)^x\left[\dfrac{x}{x+1} + \ln(x+1)\right]$ **45.** $(\ln x)^{\ln x}\left[\dfrac{1 + \ln(\ln x)}{x}\right]$ **47.** $x^{\sin x}\left(\cos x \ln x + \dfrac{\sin x}{x}\right)$

49. $(\sin x)^{\cos x}\left[\dfrac{\cos^2 x}{\sin x} - \sin x \ln(\sin x)\right]$ **51.** $x^{2^x}\left[\dfrac{2^x}{x} + 2^x(\ln x)(\ln 2)\right]$

55.

57.

59. $\dfrac{1}{4\ln 2}$ **61.** 2   **63.** $\dfrac{45}{\ln 10}$ **65.** $\dfrac{1}{3} + \dfrac{1}{\ln 2}$ **67.** approx. 16.999999; $5^{(\ln 7)/(\ln 5)} = (e^{\ln 5})^{(\ln 17)/(\ln 5)} = e^{\ln 17} = 17$

69. (a) the x-coordinates of the points of intersection are: $x \cong -1.198$, $x = 3$ and $x \cong 3.408$.

(b) for the interval $[-1.198, 3]$, $A \cong 5.5376$; for the interval $[3, 3.408]$, $A \cong 0.1373$

SECTION 7.6

1. (a) \$411.06   (b) \$612.77   (c) \$859.14   **3.** about $5\frac{1}{2}\%$: $(\ln 3)/20 \cong 0.0549$

7. (a) $P(t) = 10,000e^{t\ln 2} = 10,000(2)^t$ (b) $P(26) = 10,000(2)^{26}$, $P(52) = 10,000(2)^{52}$ **9.** (a) $e^{0.35}$ (b) $k = \dfrac{\ln 2}{15}$

11. $P(20) \cong 317.1$ million; $P(11) \cong 284.4$ million   **13.** in the year 2112   **15.** $200\left(\frac{4}{5}\right)^{t/5}$ liters   **17.** $5\left(\frac{4}{5}\right)^{5/2} \cong 2.86$ gms

19. $100[1 - \left(\frac{1}{2}\right)^{1/n}]\%$ **21.** 80.7%, 3240 yrs

23. (a) $x_1(t) = 10^6 t$, $x_2(t) = e^t - 1$

(b) $\dfrac{d}{dt}[x_1(t) - x_2(t)] = \dfrac{d}{dt}[10^6 t - (e^t - 1)] = 10^6 - e^t$.

This derivative is zero at $t = 6\ln 10 \cong 13.8$. After that the derivative is negative

(c) $x_2(15) < e^{15} = (e^3)^5 \cong 20^5 = 2^5(10^5) = 3.2(10^6) < 15(10^6) = x_1(15)$

$x_2(18) = e^{18} - 1 = (e^3)^6 - 1 \cong 20^6 - 1 = 64(10^6) - 1 > 18(10^6) = x_1(18)$

$x_2(18) - x_1(18) \cong 64(10^6) - 1 - 18(10^6) \cong 46(10^6)$

(d) If by time t_1 EXP has passed LIN, then $t_1 > 6\ln 10$. For all $t \geq t_1$ the speed of EXP is greater than the speed of LIN: for $t \geq t_1 > 6\ln 10$, $v_2(t) = e^t > 10^6 = v_1(t)$.

25. (a) $15(\frac{2}{3})^{1/2} \cong 12.25$ lb/in.^{2} (b) $15(\frac{2}{3})^{3/2} \cong 8.16$ lb/in.^{2} **27.** 6.4%   **29.** (a) \$29,045.86   (b) \$31,781.23   (c) \$35,833.24

31. $176/\ln 2 \cong 254$ ft   **33.** 11,400 years   **35.** $f(t) = Ce^{t^2/2}$ **37.** $f(t) = Ce^{\sin t}$

SECTION 7.7

1. (a) 0   (b) $-\frac{1}{3}\pi$ **3.** (a) $\frac{2}{3}\pi$ (b) $\frac{3}{4}\pi$ **5.** (a) $\frac{1}{2}$ (b) $\frac{1}{4}\pi$ **7.** (a) does not exist   (b) does not exist   **9.** (a) $\dfrac{\sqrt{3}}{2}$ (b) $-\dfrac{7}{25}$

11. $\dfrac{1}{x^2 + 2x + 2}$ **13.** $\dfrac{2}{x\sqrt{4x^4 - 1}}$ **15.** $\dfrac{2x}{\sqrt{1 - 4x^2}} + \arcsin 2x$ **17.** $\dfrac{2\arcsin x}{\sqrt{1 - x^2}}$ **19.** $\dfrac{x - (1 + x^2)\arctan x}{x^2(1 + x^2)}$

21. $\dfrac{1}{(1 + 4x^2)\sqrt{\arctan 2x}}$ **23.** $\dfrac{1}{x[1 + (\ln x)^2]}$ **25.** $-\dfrac{r}{|r|\sqrt{1 - r^2}}$ **27.** $2x \operatorname{arcsec}\left(\dfrac{1}{x}\right) - \dfrac{x^2}{\sqrt{1 - x^2}}$

29. $\cos\left[\operatorname{arcsec}(\ln x)\right] \cdot \dfrac{1}{x|\ln x|\sqrt{(\ln x)^2 - 1}}$ **31.** $\sqrt{\dfrac{c - x}{c + x}}$

33. (a) x (b) $\sqrt{1 - x^2}$ (c) $\dfrac{x}{\sqrt{1 - x^2}}$ (d) $\dfrac{\sqrt{1 - x^2}}{x}$ (e) $\dfrac{1}{\sqrt{1 - x^2}}$ (f) $\dfrac{1}{x}$ **35.** $\arcsin\left(\dfrac{x + b}{a}\right) + C$

39. $\frac{1}{4}\pi$ **41.** $\frac{1}{4}\pi$ **43.** $\frac{1}{20}\pi$ **45.** $\frac{1}{24}\pi$ **47.** $\dfrac{1}{3}\operatorname{arcsec} 4 - \dfrac{\pi}{9}$ **49.** $\frac{1}{6}\pi$ **51.** $\arctan 2 - \frac{1}{4}\pi \cong 0.322$

53. $\frac{1}{2}\arcsin x^2 + C$ **55.** $\frac{1}{2}\arctan x^2 + C$ **57.** $\frac{1}{3}\arctan\left(\frac{1}{3}\tan x\right) + C$ **59.** $\frac{1}{2}(\arcsin x)^2 + C$ **61.** $\arcsin(\ln x) + C$

63. $\dfrac{1}{3}\pi$ **65.** $2\pi - \frac{4}{3}$ **67.** $4\pi(\sqrt{2} - 1)$

69. $\sqrt{s^2 + sk}$ feet from the point where the line of the sign intersects the road.

71. (b) $\frac{1}{2}\pi a^2$; area of semicircle of radius a **75.** $\dfrac{1}{\sqrt{1 - x^2}}$ is not defined for $x \geq 1$.

77. estimate $\cong 0.523$, $\sin 0.523 \cong 0.499$ explanation: the integral $= \arcsin 0.5$; therefore $\sin(\text{integral}) = 0.5$

SECTION 7.8

1. $2x \cosh x^2$ **3.** $\dfrac{a \sinh ax}{2\sqrt{\cosh ax}}$ **5.** $\dfrac{1}{1 - \cosh x}$ **7.** $ab(\cosh bx - \sinh ax)$ **9.** $\dfrac{a \cosh ax}{\sinh ax}$ **11.** $2e^{2x}\cosh(e^{2x})$

13. $-e^{-x}\cosh 2x + 2e^{-x}\sinh 2x$ **15.** $\tanh x$ **17.** $(\sinh x)^x [\ln(\sinh x) + x \coth x]$ **27.** absolute max -3

31. $A = 2, B = \frac{1}{3}, c = 3$ **33.** $\dfrac{1}{a}\sinh ax + C$ **35.** $\dfrac{1}{3a}\sinh^3 ax + C$ **37.** $\dfrac{1}{a}\ln(\cosh ax) + C$ **39.** $-\dfrac{1}{a \cosh ax} + C$

41. $\frac{1}{2}(\sinh x \cosh x + x) + C$ **43.** $2\cosh\sqrt{x} + C$ **45.** $\sinh 1 \cong 1.175$ **47.** $\frac{81}{20}$ **49.** π

51. $\pi[\ln 5 + \frac{1}{4}\sinh(4\ln 5)] \cong 250.492$ **53.** (a) $(0.69315, 1.25)$
(b) $A \cong 0.38629$

SECTION 7.9

1. $2\tanh x \operatorname{sech}^2 x$ **3.** $\operatorname{sech} x \operatorname{csch} x$ **5.** $\dfrac{2e^{2x}\cosh(\arctan e^{2x})}{1 + e^{4x}}$ **7.** $\dfrac{-x \operatorname{csch}^2\sqrt{x^2 + 1}}{\sqrt{x^2 + 1}}$ **9.** $\dfrac{-\operatorname{sech} x(\tanh x + 2\sinh x)}{(1 + \cosh x)^2}$

15. (a) $\frac{3}{5}$ (b) $\frac{5}{3}$ (c) $\frac{4}{3}$ (d) $\frac{5}{4}$ (e) $\frac{3}{4}$

25. (a) absolute max $f(0) = 1$
(b) points of inflection at $x = \ln(1 + \sqrt{2}) \cong 0.881$, $x = -\ln(1 + \sqrt{2}) \cong -0.881$
(c) concave up on $(-\infty, -\ln(1 + \sqrt{2})) \cup (\ln(1 + \sqrt{2}), \infty)$; concave down on $(-\ln(1 + \sqrt{2}), \ln(1 + \sqrt{2}))$

(d)
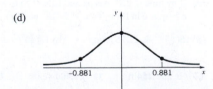

27. $(0, 0)$ is a point of inflection for both graphs   **31.** $\ln(\cosh x) + C$ **33.** $2\arctan(e^x) + C$ **35.** $-\frac{1}{3}\operatorname{sech}^3 x + C$

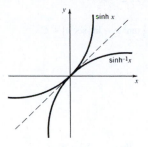

37. $\frac{1}{2}[\ln(\cosh x)]^2 + C$ **39.** $\ln|1 + \tanh x| + C$

Chapter 7. Review Exercises

1. $f^{-1}(x) = (x - 2)^3$ **3.** $f^{-1}(x) = \dfrac{x+1}{x-1}$ **5.** $f^{-1}(x) = \dfrac{1}{\ln x}$ **7.** not one-to-one   **9.** -4 **11.** $\frac{1}{2}$ **13.** $\dfrac{24(\ln x)^2}{x}$

15. $\dfrac{e^x - e^{3x}}{(1 + e^{2x})^2}$ **17.** $\dfrac{3x^2 + 3^x \ln x}{x^3 + 3^x}$ **19.** $(\cosh x)^{1/x}\left[\dfrac{\tanh x}{x} - \dfrac{\ln\cosh x}{x^2}\right]$ **21.** $\dfrac{2}{\ln 3\,(1 - x^2)}$ **23.** $\arcsin e^x + C$

25. $\dfrac{1}{2}\arctan\left(\dfrac{\sin x}{2}\right) + C$ **27.** $2\ln|\sec\sqrt{x} + \tan\sqrt{x}| + C$ **29.** $\dfrac{5^{\ln x}}{\ln 5} + C$ **31.** $\frac{3}{4}\ln\frac{17}{2}$ **33.** $\dfrac{\cosh 2^x}{\ln 2} + C$ **35.** $\frac{1}{12}\pi$

37. $2\tanh 1$ **39.** $\frac{1}{2}\ln 2$ **41.** $\frac{1}{6}\pi$ **45.** $a^2 \ln 2$ **47.** (a) $\frac{1}{3}\pi^2$ (b) 2π

49. (a) increasing on $(0, e]$, decreasing on $[e, \infty)$

    (b) absolute max $f(e) = \frac{1}{e}$

    (c) concave down on $(0, e^{3/2})$, concave up on $(e^{3/2}, \infty)$, point of inflection $(e^{3/2}, f(e^{3/2}))$

    (d) the x-axis is a horizontal asymptote

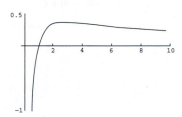

51. $b = a$ **53.** (a) 160 grams   (b) approx. 6.64 hours   **55.** (a) $A(t) = 100\,e^{-\frac{t}{140}\ln 2} \cong 100\,e^{-0.00495t}$ (b) approx. 58.12 days

57. (a) 6250   (b) approx. 5.36 years

CHAPTER 8

SECTION 8.1

1. $-e^{2-x} + C$ **3.** $2/\pi$ **5.** $-\tan(1 - x) + C$ **7.** $\frac{1}{2}\ln 3$ **9.** $-\sqrt{1 - x^2} + C$ **11.** 0 **13.** $e - \sqrt{e}$ **15.** $\pi/4c$

17. $\frac{2}{3}\sqrt{3\tan\theta + 1} + C$ **19.** $(1/a)\ln|a\,e^x - b| + C$ **21.** $\frac{1}{2}\ln[(x + 1)^2 + 4] - \frac{1}{2}\arctan(\frac{1}{2}[x + 1]) + C$ **23.** $\frac{1}{2}\arcsin x^2 + C$

25. $\arctan(x + 3) + C$ **27.** $-\frac{1}{2}\cos x^2 + C$ **29.** $\tan x - x + C$ **31.** $3/2$ **33.** $\frac{1}{2}(\arcsin x)^2 + C$ **35.** $\ln|\ln x| + C$ **37.** $\sqrt{2}$

39. (Formula 99) $\dfrac{x}{2}\sqrt{x^2 - 4} - 2\ln|x + \sqrt{x^2 - 4}| + C$ **41.** (Formula 18) $\frac{1}{2}[\sin 2t - \frac{1}{3}\sin^3 2t] + C$ **43.** (Formula 108) $\frac{1}{3}\ln\left|\dfrac{x}{2x + 3}\right| + C$

45. (Formula 81) $-\dfrac{\sqrt{x^2+9}}{x} + \ln|x + \sqrt{x^2+9}| + C$ **47.** (Formula 11) $x^4\left[\dfrac{\ln x}{4} - \dfrac{1}{16}\right] + C$ **49.** $2\sqrt{2}$

53. (a) $\frac{1}{2}\tan^2 x - \ln|\sec x| + C$ (b) $\frac{1}{4}\tan^4 x - \frac{1}{2}\tan^2 x + \ln|\sec x| + C$ (c) $\frac{1}{6}\tan^6 x - \frac{1}{4}\tan^4 x + \frac{1}{2}\tan^2 x - \ln|\sec x| + C$

(d) $\displaystyle\int \tan^{2k+1} x \, dx = \dfrac{1}{2k}\tan^{2k} x - \int \tan^{2k-1} x \, dx$

55. (b) $A = \sqrt{2}, \quad B = \dfrac{\pi}{4}$ (c) $\dfrac{\sqrt{2}}{2}\ln\left(\dfrac{\sqrt{2}+1}{\sqrt{2}-1}\right)$ **57.** (b) $-0.80, 5.80$ (c) 27.60

SECTION 8.2

1. $-xe^{-x} - e^{-x} + C$ **3.** $-\frac{1}{3}e^{-x^3} + C$ **5.** $2 - 5e^{-1}$ **7.** $-2x^2(1-x)^{1/2} - \frac{8}{3}x(1-x)^{3/2} - \frac{16}{15}(1-x)^{5/2} + C$

9. $\frac{3}{8}e^4 + \frac{1}{8}$ **11.** $2\sqrt{x+1}\ln(x+1) - 4\sqrt{x+1} + C$ **13.** $x(\ln x)^2 - 2x\ln x + 2x + C$

15. $3^x\left(\dfrac{x^3}{\ln 3} - \dfrac{3x^2}{(\ln 3)^2} + \dfrac{6x}{(\ln 3)^3} - \dfrac{6}{(\ln 3)^4}\right) + C$ **17.** $\frac{1}{15}x(x+5)^{15} - \frac{1}{240}(x+5)^{16} + C$ **19.** $\dfrac{1}{2\pi} - \dfrac{1}{\pi^2}$

21. $\frac{1}{10}x^2(x+1)^{10} - \frac{1}{55}x(x+1)^{11} + \frac{1}{660}(x+1)^{12} + C$ **23.** $\frac{1}{2}e^x(\sin x - \cos x) + C$ **25.** $\ln 2 + \dfrac{\pi}{2} - 2$

27. $\dfrac{x^{n+1}}{n+1}\ln x - \dfrac{x^{n+1}}{(n+1)^2} + C$ **29.** $-\frac{1}{2}x^2\cos x^2 + \frac{1}{2}\sin x^2 + C$ **31.** $\dfrac{\pi}{24} + \dfrac{\sqrt{3}-2}{4}$ **33.** $\dfrac{\pi}{8} - \dfrac{1}{4}\ln 2$

35. $\frac{1}{2}x^2\sinh 2x - \frac{1}{2}x\cosh 2x + \frac{1}{4}\sinh 2x + C$ **37.** $\ln x \arcsin(\ln x) + \sqrt{1 - (\ln x)^2} + C$ **39.** $\dfrac{x}{2}[\sin(\ln x) - \cos(\ln x)] + C$

47. π **49.** $\dfrac{\pi}{12} + \dfrac{\sqrt{3}-2}{2}$ **51.** (a) 1 (b) $\bar{x} = \dfrac{e^2}{4} + \dfrac{1}{4}, \bar{y} = \dfrac{e}{2} - 1$ (c) x-axis; $\pi(e-2)$, y-axis; $\dfrac{\pi}{2}(e^2+1)$

53. $\bar{x} = 1/(e-1), \bar{y} = (e+1)/4$ **55.** $\bar{x} = \frac{1}{2}\pi, \bar{y} = \frac{1}{8}\pi$ **57.** (a) $M = (e^k - 1)/k$ (b) $x_M = [(k-1)e^k + 1]/[k(e^k - 1)]$

59. $V = 4 - 8/\pi$ **61.** $V = 2\pi(e-2)$ **63.** $\bar{x} = (e^2+1)/[2(e^2-1)]$ **65.** area $= \sinh 1 = \dfrac{e^2-1}{2e}; \bar{x} = \dfrac{2}{e+1}, \bar{y} = \dfrac{e^4 + 4e^2 - 1}{8e(e^2 - 1)}$

69. $(\frac{1}{2}x^3 - \frac{3}{4}x^2 + \frac{3}{4}x - \frac{3}{8})e^{2x} + C$ **71.** $x[(\ln x)^3 - 3(\ln x)^2 + 6\ln x - 6] + C$ **73.** $e^x[x^3 - 3x^2 + 6x - 6] + C$

75. (a) $(x^2 - 5x + 6)e^x + C$ (b) $(x^3 - 3x^2 + 4x - 4)e^x + C$

79. (a) π
(b) 3π
(c) 5π
(d) $(2n+1)\pi, n = 0, 1, 2, \ldots$

81. (a) $\pi - 2 \cong 1.1416$
(b) $\pi^3 - 2\pi^2 \cong 11.2671$
(c) $(\frac{1}{2}\pi, 0.31202)$

SECTION 8.3

1. $\frac{1}{3}\cos^3 x - \cos x + C$ **3.** $\dfrac{\pi}{12}$ **5.** $-\frac{1}{5}\cos^5 x + \frac{1}{7}\cos^7 x + C$ **7.** $\frac{1}{4}\sin^4 x - \frac{1}{6}\sin^6 x + C$ **9.** $(1/\pi)\tan \pi x + C$

11. $\frac{1}{2}\tan^2 x + \ln|\cos x| + C$ **13.** $\frac{3}{8}\pi$ **15.** $\frac{1}{2}\cos x - \frac{1}{10}\cos 5x + C$ **17.** $\frac{1}{3}\tan^3 x + C$

19. $\frac{1}{2}\sin^4 x + C$ **21.** $\frac{5}{16}x - \frac{1}{4}\sin 2x + \frac{3}{64}\sin 4x + \frac{1}{48}\sin^3 2x + C$ **23.** $\sqrt{3} - \dfrac{\pi}{3}$ **25.** $-\frac{1}{5}\csc^5 x + \frac{1}{3}\csc^3 x + C$

27. $\frac{1}{6}\sin 3x - \frac{1}{14}\sin 7x + C$ **29.** $\frac{2}{7}\sin^{7/2} x - \frac{2}{11}\sin^{11/2} x + C$ **31.** $\frac{1}{12}\tan^4 3x - \frac{1}{6}\tan^2 3x + \frac{1}{3}\ln|\sec 3x| + C$ **33.** $\dfrac{2}{105\pi}$

35. $-1/6$ **37.** $\frac{1}{7}\tan^7 x + \frac{1}{5}\tan^5 x + C$ **39.** $\frac{1}{3}\cos(\frac{3}{2}x) - \frac{1}{5}\cos(\frac{5}{2}x) + C$ **41.** $1/4$

43. $\dfrac{\sqrt{3}}{2} - \dfrac{\pi}{6}$ **45.** $\pi/2$ **47.** $\dfrac{3\pi^2}{8}$ **49.** $\dfrac{\pi^2}{2} - \pi$ **51.** $\pi\left[1 - \dfrac{\pi}{4} + \ln 2\right]$

55. (a) $\frac{16}{35}$ (b) $\frac{5}{32}\pi$ **57.** $\frac{1}{2}\pi^2 \cong 4.9348$

SECTION 8.4

1. $\arcsin\left(\dfrac{x}{a}\right) + C$ **3.** $\frac{1}{2}x\sqrt{x^2-1} - \frac{1}{2}\ln|x + \sqrt{x^2-1}| + C$ **5.** $2\arcsin\left(\dfrac{x}{2}\right) - \frac{1}{2}x\sqrt{4-x^2} + C$

7. $\dfrac{1}{\sqrt{1-x^2}} + C$ **9.** $\dfrac{2\sqrt{3}-\pi}{6}$ **11.** $-\frac{1}{3}(4-x^2)^{3/2} + C$ **13.** $\dfrac{625\pi}{16}$ **15.** $\ln(\sqrt{8+x^2}+x) - \dfrac{x}{\sqrt{8+x^2}} + C$

17. $\dfrac{1}{a}\ln\left|\dfrac{a-\sqrt{a^2-x^2}}{x}\right| + C$ **19.** $18 - 9\sqrt{2}$ **21.** $-\dfrac{1}{a^2 x}\sqrt{a^2+x^2} + C$ **23.** $\frac{1}{10}$ **25.** $\dfrac{1}{a^2 x}\sqrt{x^2-a^2} + C$

27. $\frac{1}{9}e^{-x}\sqrt{e^{2x}-9} + C$ **29.** $\begin{cases} -\dfrac{1}{2(x-2)^2} + C, & x > 2 \\ \dfrac{1}{2(2-x)^2} + C, & x < 2 \end{cases}$ **31.** $-\frac{1}{3}(6x - x^2 - 8)^{3/2} + \frac{3}{2}\arcsin(x-3) + \frac{3}{2}(x-3)\sqrt{6x-x^2-8} + C$

33. $\dfrac{x^2+x}{8(x^2+2x+5)} - \dfrac{1}{16}\arctan\left(\dfrac{x+1}{2}\right) + C$ **39.** $\dfrac{3}{8}\arctan x + \dfrac{3x}{8(x^2+1)} + \dfrac{x}{4(x^2+1)^2} + C$

41. $\frac{1}{4}(2x^2-1)\arcsin x + \dfrac{x}{4}\sqrt{1-x^2} + C$ **43.** $\dfrac{\pi^2}{8} + \dfrac{\pi}{4}$ **45.** $A = \frac{1}{2}r^2\sin\theta\cos\theta + \displaystyle\int_{r\cos\theta}^{r}\sqrt{r^2-x^2}\,dx = \frac{1}{2}r^2\theta$

47. $\frac{8}{3}[10 - \frac{9}{2}\ln 3]$ **49.** $M = \ln(1+\sqrt{2}),\ x_M = \dfrac{(\sqrt{2}-1)a}{\ln(1+\sqrt{2})}$

51. $A = \dfrac{1}{2}a^2[\sqrt{2} - \ln(\sqrt{2}+1)];\quad \bar{x} = \dfrac{2a}{3[\sqrt{2} - \ln(\sqrt{2}+1)]},\ \bar{y} = \dfrac{(2-\sqrt{2})a}{3[\sqrt{2} - \ln(\sqrt{2}+1)]}$ **53.** $V_y = \frac{2}{3}\pi a^3,\ \bar{y} = \frac{3}{8}a$

57. (b) $\ln(2+\sqrt{3}) - \dfrac{\sqrt{3}}{2}$ (c) $\bar{x} = \dfrac{2(3\sqrt{3}-\pi)}{2\ln(2+\sqrt{3}) - \sqrt{3}},\ \bar{y} = \dfrac{5}{72[2\ln(2+\sqrt{3}) - \sqrt{3}]}$

SECTION 8.5

1. $\dfrac{1/5}{x+1} - \dfrac{1/5}{x+6}$ **3.** $\dfrac{1/4}{x-1} + \dfrac{1/4}{x+1} - \dfrac{x/2}{x^2+1}$ **5.** $\dfrac{1/2}{x} + \dfrac{3/2}{x+2} - \dfrac{1}{x-1}$ **7.** $\dfrac{3/2}{x-1} - \dfrac{9}{x-2} + \dfrac{19/2}{x-3}$ **9.** $\ln\left|\dfrac{x-2}{x+5}\right| + C$

11. $x^2 - 2x + \dfrac{3}{x} + 5\ln|x-1| - 3\ln|x| + C$ **13.** $\dfrac{1}{4}x^4 + \dfrac{4}{3}x^3 + 6x^2 + 32x - \dfrac{32}{x-2} + 80\ln|x-2| + C$ **15.** $5\ln|x-2| - 4\ln|x-1| + C$

17. $\dfrac{-1}{2(x-1)^2} + C$ **19.** $\dfrac{3}{4}\ln|x-1| - \dfrac{1}{2(x-1)} + \dfrac{1}{4}\ln|x+1| + C$ **21.** $\dfrac{1}{32}\ln\left|\dfrac{x-2}{x+2}\right| - \dfrac{1}{16}\arctan\dfrac{x}{2} + C$

23. $\dfrac{1}{2}\ln(x^2+1) + \dfrac{3}{2}\arctan x + \dfrac{5(1-x)}{2(x^2+1)} + C$ **25.** $\dfrac{1}{16}\ln\left[\dfrac{x^2+2x+2}{x^2-2x+2}\right] + \dfrac{1}{8}\arctan(x+1) + \dfrac{1}{8}\arctan(x-1) + C$

27. $\dfrac{3}{x} + 4\ln\left|\dfrac{x}{x+1}\right| + C$ **29.** $-\frac{1}{6}\ln|x| + \frac{3}{10}\ln|x-2| - \frac{2}{15}\ln|x+3| + C$ **31.** $\ln\left(\frac{125}{108}\right)$ **33.** $\ln\left(\frac{27}{4}\right) - 2$

35. $\dfrac{1}{6}\ln\left|\dfrac{\sin\theta - 4}{\sin\theta + 2}\right| + C$ **37.** $\dfrac{1}{4}\ln\left|\dfrac{\ln t - 2}{\ln t + 2}\right| + C$ **47.** $\dfrac{1}{ad-bc}\ln\left|\dfrac{c+du}{a+bu}\right| + C$

49. (a) $\frac{1}{4}\pi\ln 7$ (b) $\pi(4 - \sqrt{7})$ **51.** $\bar{x} = (2\ln 2)/\pi,\quad \bar{y} = (\pi + 2)/2\pi$

53. (a) $\dfrac{1}{x} - \dfrac{2}{x^2} + \dfrac{5}{x+1} - \dfrac{4}{(x+1)^3}$ **55.** $\dfrac{1}{3}x^3 + 2x - \dfrac{1}{2(4+x^2)} + \dfrac{3}{2}\arctan\left(\dfrac{x}{2}\right) + 2\ln|x-1| + \ln|x+3|$

(b) $\dfrac{1}{x^2+4} + \dfrac{3}{x+3} - \dfrac{4}{x-3}$

(c) $\dfrac{2x-1}{x^2+2x+4} - \dfrac{3}{x}$

57. (b) $3\ln 7 - 5\ln 3$ **59.** (b) $11 - \ln 12$

SECTION 8.6

1. $-2(\sqrt{x} + \ln|1 - \sqrt{x}|) + C$ 3. $2\ln(\sqrt{1+e^x} - 1) - x + 2\sqrt{1+e^x} + C$ 5. $\frac{2}{5}(1+x)^{5/2} - \frac{2}{3}(1+x)^{3/2} + C$

7. $\frac{2}{5}(x-1)^{5/2} + 2(x-1)^{3/2} + C$ 9. $-\dfrac{1+2x^2}{4(1+x^2)^2} + C$ 11. $x + 2\sqrt{x} + 2\ln|\sqrt{x} - 1| + C$ 13. $x + 4\sqrt{x-1} + 4\ln|\sqrt{x-1} - 1| + C$

15. $2\ln(\sqrt{1+e^x} - 1) - x + C$ 17. $\frac{2}{3}(x-8)\sqrt{x+4} + C$ 19. $\frac{1}{16}(4x+1)^{1/2} + \frac{1}{8}(4x+1)^{-1/2} - \frac{1}{48}(4x+1)^{-3/2} + C$

21. $\dfrac{4b + 2ax}{a^2\sqrt{ax+b}} + C$ 23. $-\ln\left|1 - \tan\dfrac{x}{2}\right| + C$ 25. $\dfrac{2}{\sqrt{3}}\arctan\left[\dfrac{1}{\sqrt{3}}(2\tan\dfrac{x}{2} + 1)\right] + C$ 27. $\frac{1}{2}\ln\left|\tan\dfrac{x}{2}\right| - \frac{1}{4}\tan^2\dfrac{x}{2} + C$

29. $\ln\left|\dfrac{1}{1+\sin x}\right| - \dfrac{2}{1+\tan(x/2)} + C$ 31. $\frac{4}{3} + 2\arctan 2$ 33. $2 + 4\ln\frac{2}{3}$ 35. $\ln\left(\dfrac{\sqrt{3}-1}{\sqrt{3}}\right)$

41. $2\arctan\left(\tanh\dfrac{x}{2}\right) + C$ 43. $\dfrac{-2}{1+\tanh(x/2)} + C$

SECTION 8.7

1. (a) 506   (b) 650   (c) 572   (d) 578   (e) 576   3. (a) 1.394   (b) 0.9122   (c) 1.1776   (d) 1.1533   (e) 1.1614

5. (a) $\pi \cong 3.1312$ (b) $\pi \cong 3.1416$ 7. (a) 1.8440   (b) 1.7915   (c) 1.8090   9. (a) 0.8818   (b) 0.8821

13. (a) $n \geq 8$ (b) $n \geq 2$ 15. (a) $n \geq 238$ (b) $n \geq 10$ 17. (a) $n \geq 51$ (b) $n \geq 4$ 19. (a) $n \geq 37$ (b) $n \geq 3$

21. (a) $n \geq 78$ (b) $n \geq 7$ 27. (a) $M_n \leq \displaystyle\int_a^b f(x)\,dx \leq T_n$ (b) $T_n \leq \displaystyle\int_a^b f(x)\,dx \leq M_n$

29. (a) 49.4578   (b) 1280.56   31. error $\leq 4.01 \times 10^{-7}$

Chapter 6. Review Exercises

1. $\frac{1}{2}\arctan\left(\dfrac{\sin x}{2}\right) + C$ 3. $2x\cosh x - 2\sinh x + C$ 5. $\dfrac{3}{x} + 4\ln|x| - 4\ln|x+1| + C$

7. $-\frac{1}{2}\cos x - \frac{1}{6}\cos 3x + C = -\frac{2}{3}\cos^3 x + C$ 9. $-\frac{3}{2} + \ln 16$ 11. $\ln|\sec x| - \frac{1}{2}\sin^2 x + C$ 13. $\frac{3}{4} - \frac{1}{4}e^{-2}$

15. $\frac{1}{4}\ln\left|\dfrac{e^x - 2}{e^x + 2}\right| + C$ 17. $\dfrac{x2^x}{\ln 2} - \dfrac{2^x}{(\ln 2)^2} + C$ 19. $-\dfrac{\sqrt{a^2 - x^2}}{x^2} - \arcsin\dfrac{x}{a} + C$ 21. $\frac{1}{2}x^2 e^{x^2} - \frac{1}{2}e^{x^2} + C$

23. $\frac{1}{6}\tan^6 x + C$ 25. $\frac{1}{2}\ln(2+\sqrt{3}) - \frac{1}{4}\ln 3$ 27. $\frac{1}{2}\left(\arcsin x - (x+2)\sqrt{1-x^2}\right) + C$ 29. $-\frac{3}{2}x - \cot x - \frac{1}{4}\sin 2x + C$

31. $x + \frac{1}{2}\sin^2 2x + C$ 33. $\ln\left|\dfrac{x-1}{x+2}\right| - \dfrac{2}{x-1} + C$ 35. $\frac{2}{3}\left[x^{3/2} + (x+1)\sqrt{x+1}\right] + C$ 37. $\frac{1}{2}x^2\sin 2x + \frac{1}{2}x\cos 2x - \frac{1}{4}\sin 2x + C$

39. $\tan 2x - \sec 2x - x + C$ 45. $\left(\dfrac{6-3\sqrt{3}}{\pi}, \dfrac{3\ln 3}{2\pi}\right)$ 47. (a) $\frac{1}{2}\pi\ln 3$ (b) $\pi(2 - \sqrt{3})$

49. $M_4 = 3.0270$, $T_4 = 2.7993$, $S_4 = 2.9975$ 51. (a) 123   (b) 10

CHAPTER 9

SECTION 9.1

1. y_1 is; y_2 is not   3. y_1 and y_2 are solutions   5. y_1 and y_2 are solutions   7. $y = -\frac{1}{2} + Ce^{2x}$ 9. $y = \frac{2}{5} + Ce^{-5x/2}$

11. $y = x + Ce^{2x}$ 13. $y = \frac{2}{3}nx + Cx^4$ 15. $y = Ce^{e^x}$ 17. $y = 1 + C(e^{-x} + 1)$ 19. $y = e^{-x^2}(\frac{1}{2}x^2 + C)$

21. $y = C(x+1)^{-2}$ 23. $y = 2e^{-x} + x - 1$ 25. $y = e^{-x}[\ln(1+e^x) + e - \ln 2]$ 27. $y = x^2(e^x - e)$ 29. $y = C_1 e^x + C_2 x e^x$

35. $T(1) \cong 40.10°; 1.62\,\text{min}$ **37.** (a) $v(t) = \dfrac{32}{k}(1 - e^{-kt})$

(b) $1 - e^{-kt} < 1; e^{-kt} \to 0$ as $t \to \infty$

39. (a) $i(t) = \dfrac{E}{R}[1 - e^{-(R/L)t}]$

(b) $i(t) \to \dfrac{E}{R}$ (amps) as $t \to \infty$

(c) $t = \dfrac{L}{R}\ln 10$ seconds

41. (a) $200\left(\frac{4}{5}\right)^{t/5}$ (b) $200\left(\frac{4}{5}\right)^{t^2/25}$ liters

43. (a) $\dfrac{dp}{dt} = k(M - P)$ (b) $P(t) = M(1 - e^{-0.0357t})$ (c) 65 days

45. (a) $P(t) = 1000e^{(\sin 2\pi t/\pi)}$ (b) $P(t) = 2000e^{(\sin 2\pi t)/\pi} - 1000$

SECTION 9.2

1. $y = Ce^{-(1/2)\cos(2x+3)}$ **3.** $x^4 + \dfrac{2}{y^2} = C$ **5.** $y\sin y + \cos y = \cos\left(\dfrac{1}{x}\right) + C$ **7.** $e^{-y} = e^x - xe^x + C$

9. $\ln|y + 1| + \dfrac{1}{y+1} = \ln|\ln x| + C$ **11.** $y^2 = C(\ln x)^2 - 1$ **13.** $\arcsin y = 1 - \sqrt{1 - x^2}$ **15.** $y + \ln|y| = \dfrac{x^3}{3} - x - 5$

17. $\dfrac{x^2}{2} + x + \dfrac{1}{2}\ln(y^2 + 1) - \arctan y = 4$ **19.** $y = \ln[3e^{2x} - 2]$ **21.** (a) $C(t) = \dfrac{kA_0^2 t}{1 + kA_0 t}$ (b) $C(t) = \dfrac{A_0 B_0(e^{kA_0 t} - e^{kB_0 t})}{A_0 e^{kA_0 t} - B_0 e^{kB_0 t}}$

23. (a) $v(t) = \dfrac{\alpha}{C\,e^{(\alpha/m)t} - \beta}$, where C is an arbitrary constant.   (b) $v(t) = \dfrac{\alpha v_0}{(\alpha + \beta v_0)e^{(\alpha/m)t} - \beta v_0} = \dfrac{\alpha v_0 e^{-(\alpha/m)t}}{\alpha + \beta v_0 - \beta v_0 e^{-(\alpha/m)t}}$,

(c) $\lim\limits_{t\to\infty} v(t) = 0$

25. (a) $y(t) = \dfrac{25,000}{1 + 249\,e^{-0.1398t}}, y(20) \cong 1544$ (b) 40 days   **27.** (a) 88.82 m/sec   (b) $v(t) = \dfrac{15.65(1 + 0.70\,e^{-1.25t})}{1 - 0.70\,e^{-1.25t}}$ (c) 15.65 m/sec

SECTION 9.3

1. $y = C_1 e^{-4x} + C_2 e^{2x}$ **3.** $y = C_1 e^{-4x} + C_2 xe^{-4x}$ **5.** $y = e^{-x}(C_1\cos 2x + C_2\sin 2x)$ **7.** $y = C_1 e^{(1/2)x} + C_2 e^{-3x}$

9. $y = C_1\cos 2\sqrt{3}x + C_2\sin 2\sqrt{3}x$ **11.** $y = C_1 e^{(1/5)x} + C_2 e^{-(3/4)x}$ **13.** $y = C_1\cos 3x + C_2\sin 3x$

15. $y = e^{-(1/2)x}(C_1\cos\frac{1}{2}x + C_2\sin\frac{1}{2}x)$ **17.** $y = C_1 e^{(1/4)x} + C_2 e^{-(1/2)x}$ **19.** $y = 2e^{2x} - e^{3x}$ **21.** $y = 2\cos\dfrac{x}{2} + \sin\dfrac{x}{2}$

23. $y = 7e^{-2(x+1)} + 5x\,e^{-2(x-1)}$ **25.** (a) $y = Ce^{2x} + (1 - C)e^{-x}$ (b) $y = C\,e^{2x} + (2C - 1)e^{-x}$ (c) $y = \frac{2}{3}e^{2x} + \frac{1}{3}e^{-x}$

31. (a) $y'' + 2y' - 8y = 0$ (b) $y'' - 4y' - 5y = 0$ (c) $y'' - 6y' + 9y = 0$

39. $y = C_1 x^4 + C_2 x^{-2}$ **41.** $y = C_1 x^2 + C_2 x^2 \ln x$

Chapter 9. Review Exercises

1. $y + Ce^{-x} - 2e^{-2x}$ **3.** $\ln(y^2 + 1) = x + \frac{1}{2}\sin 2x + C$ **5.** $y = -\dfrac{\cos 2x}{2x^2} - \dfrac{\sin 2x}{4x^3} + \dfrac{C}{x^2}$ **7.** $\arctan y = x + \frac{1}{3}x^3 + C$

9. $y = \dfrac{x^3}{5} + \dfrac{C}{x^2}$ **11.** $y = \dfrac{2}{x}\ln x + \dfrac{x}{2} + \dfrac{3}{2x}$ **13.** $y = \frac{1}{2} + e^{-e^{2x}}$ **15.** $y = e^x[C_1\cos x + C_2\sin x]$ **17.** $y = C_1 e^{2x} + C_2 e^{-x}$

19. $y = C_1 e^{3x} + C_2 xe^{3x}$ **21.** $y = e^{-2x}[C_1\cos 3x + C_2\sin 3x]$ **23.** $y = 1$ **25.** $y = e^{3x}[2\cos 2x - 2\sin 2x]$ **27.** $y^2 = C - x$

29. $r = -2, -1$ **31.** one year from now: \$3 million; one and one-half years from now: \$6 million; two years from now: ∞

33. (a) $y = \frac{a}{b}[1 - e^{-bt}]$ (b) $\frac{a}{b}$ (c) $\frac{1}{b}\ln 10$ **35.** (a) $T(t) = 70 - \frac{500}{7}\left(\frac{7}{10}\right)^{t/10}$ (b) $T(0) = -\frac{10}{7}$

37. (a) 20 minutes   (b) $P(t) = 4(20 - t) - \frac{7}{40}(20 - t)^2$ (c) 22.5 pounds   **39.** (a) $\cong 2742$ (b) $\cong 40$ days

CHAPTER 10

SECTION 10.1

1. vertex $(0, 0)$
focus $(0, \frac{1}{2})$
axis $x = 0$
directrix $y = -\frac{1}{2}$

3. vertex $(1, 0)$
focus $(1, \frac{1}{2})$
axis $x = 1$
directrix $y = -\frac{1}{2}$

5. vertex $(2, -2)$
focus $(2, -1)$
axis $x = 2$
directrix $y = -3$

7. vertex $(2, -4)$
focus $(2, -\frac{15}{4})$
axis $x = 2$
directrix $y = -\frac{17}{4}$

9. center $(0, 0)$
foci $(\pm\sqrt{5}, 0)$
length of major axis 6
length of minor axis 4

11. center $(0, 0)$
foci $(0, \pm\sqrt{2})$
length of major axis $2\sqrt{6}$
length of minor axis 4

13. center $(0, 1)$
foci $(\pm\sqrt{5}, 1)$
length of major axis 6
length of minor axis 4

15. center $(1, 0)$
foci $(1, \pm 4\sqrt{3})$
length of major axis 16
length of minor axis 8

17. center $(0, 0)$
transverse axis 2
vertices $(\pm 1, 0)$
foci $(\pm\sqrt{2}, 0)$
asymptotes $y = \pm x$

19. center $(0, 0)$
transverse axis 6
vertices $(\pm 3, 0)$
foci $(\pm 5, 0)$
asymptotes $y = \pm\frac{4}{3}x$

21. center $(0, 0)$
transverse axis 8
vertices $(0, \pm 4)$
foci $(0, \pm 5)$
asymptotes $y = \pm\frac{4}{3}x$

23. center $(1, 3)$
transverse axis 6
vertices $(4, 3)$ and $(-2, 3)$
foci $(6, 3)$ and $(-4, 3)$
asymptotes $y = \pm\frac{4}{3}(x - 1) + 3$

25. center $(1, 3)$
transverse axis 4
vertices $(1, 5)$ and $(1, 1)$
foci $(1, 3 \pm \sqrt{5})$
asymptotes $y = 2x + 1$, $y = -2x + 5$

31. center $(0, 0)$, vertices $(1, 1)$ and $(-1, -1)$, foci $(\sqrt{2}, \sqrt{2})$ and $(\sqrt{2}, -\sqrt{2})$, asymptotes $x = 0$ and $y = 0$, transverse axis $2\sqrt{2}$

33. $2\sqrt{\pi^2 a^4 - A^2}/\pi a$ **35.** $4c$ **37.** $A = \frac{8}{3}c^3$; $\bar{x} = 0$, $\bar{y} = \frac{3}{5}c$ **43.** $[2\sqrt{3} - \ln(2 + \sqrt{3})]ab$ **45.** $\frac{3}{5}$ **47.** $\frac{4}{5}$.

49. E_1 is fatter than E_2, more like a circle   **51.** The ellipse tends to a line segment of length $2a$.

53. $x^2/9 + y^2 = 1$ **55.** $\frac{5}{3}$ **57.** $\sqrt{2}$

59. The branches of H_1 open up less quickly than the branches of H_2.

61. The hyperbola tends to a pair of parallel lines separated by the transverse axis.

SECTION 10.2

1–7. See figure to the right.　**9.** $(0, 3)$　**11.** $(1, 0)$　**13.** $(-\frac{3}{2}, \frac{3}{2}\sqrt{3})$　**15.** $(0, -3)$

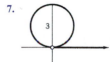

17. $[1, \frac{1}{2}\pi + 2n\pi], [-1, \frac{3}{2}\pi + 2n\pi]$　**19.** $[3, \pi + 2n\pi], [-3, 2n\pi]$

21. $[2\sqrt{2}, \frac{7}{4}\pi + 2n\pi], [-2\sqrt{2}, \frac{3}{4}\pi + 2n\pi]$　**23.** $[8, \frac{1}{6}\pi + 2n\pi], [-8, \frac{7}{6}\pi + 2n\pi]$

25. $\sqrt{r_1^2 + r_2^2 - 2r_1r_2 \cos(\theta_1 - \theta_1)}$　**27.** (a) $[\frac{1}{2}, \frac{11}{6}\pi]$　(b) $[\frac{1}{2}, \frac{5}{6}\pi]$　(c) $[\frac{1}{2}, \frac{7}{6}\pi]$

29. (a) $[2, \frac{2}{3}\pi]$　(b) $[2, \frac{5}{3}\pi]$　(c) $[2, \frac{1}{3}\pi]$　**31.** symmetry about the x-axis

33. no symmetry about the coordinate axes; no symmetry about the origin

35. symmetry about the origin　**37.** $r \cos\theta = 2$　**39.** $r^2 \sin 2\theta = 1$　**41.** $r = 4\sin\theta$

43. $\theta = \pi/4$　**45.** $r = 1 - \cos\theta$　**47.** $r^2 = \sin 2\theta$　**49.** the horizontal line $y = 4$

51. the line $y = \sqrt{3}x$　**53.** the parabola $y^2 = 4(x + 1)$　**55.** the circle $x^2 + y^2 = 6x$

57. the line $y = 2x$　**59.** $3x^2 + 4y^2 - 8x = 16$, ellipse　**61.** $y^2 = 8x + 16$, parabola

63. $(x - b)^2 + (y - a)^2 = a^2 + b^2$, center: (b, a), radius: $\sqrt{a^2 + b^2}$　**65.** $r = \dfrac{d}{2 - \cos\theta}$

SECTION 10.3

1. 　**3.** 　**5.** 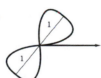　**7.**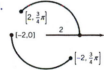

9. 　**11.** 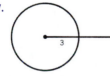　**13.** 　**15.**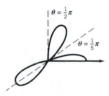

17. 　**19.** 　**21.** 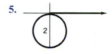　**23.**

25. 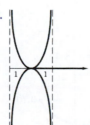　**27.** 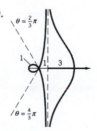　**29.** 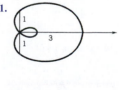　**31.**

33. yes; $[1, \pi] = [-1, 0]$ and the pair $r = -1, \theta = 0$ satisfies the equation　**35.** yes: the pair $r = \frac{1}{2}, \theta = \frac{1}{2}\pi$ satisfies the equation

37. $[2, \pi] = [-2, 0]$. The coordinates of $[-2, 0]$ satisfy the equation $r^2 = 4\cos\theta$, and the coordinates of $[2, \pi]$ satisfy the equation $r = 3 + \cos\theta$.

39. $(0, 0), (-\frac{1}{2}, \frac{1}{2})$　**41.** $(-1, 0), (1, 0)$　**43.** $(0, 0), (\frac{1}{4}, \pm\frac{1}{4}\sqrt{3})$　**45.** $(0, 0), (\pm\frac{\sqrt{3}}{4}, \frac{3}{4})$

47. $r = f(\theta - \alpha)$ is the curve $r = f(\theta)$ rotated counterclockwise α radians.

49.

The rectangular coordinates of the points of intersection are: $(0, 0)$, $(2, 0)$, $(-1, 0)$, $(-0.25, \pm 0.4330)$

51. (b) The curves intersect at the pole and at:

$r = 1 - 3\cos\theta$	$r = 2 - 5\sin\theta$
$[-2, 0]$	$[2, \pi]$
$[3.800, 3.510]$	$[3.800, 3.510]$
$[2.412, 4.223]$	$[-2.412, 1.081]$
$[-1.267, 0.713]$	$[-1.267, 0.713]$

53. butterfly **55.** a petal curve with $2m$ petals

SECTION 10.4

1. $\frac{1}{4}\pi a^2$ **3.** $\frac{1}{2}a^2$ **5.** $\frac{1}{2}\pi a^2$ **7.** $\frac{1}{4} - \frac{1}{16}\pi$ **9.** $\frac{3}{16}\pi + \frac{3}{8}$ **11.** $\frac{5}{2}a^2$ **13.** $\frac{1}{12}(3e^{2\pi} - 3 - 2\pi^3)$

15. $\frac{1}{4}(e^{2\pi} + 1 - 2e^{\pi})$ **17.** $\int_{\pi/6}^{5\pi/6} \frac{1}{2}([4\sin\theta]^2 - [2]^2)\, d\theta$ **19.** $\int_{-\pi/3}^{\pi/3} \frac{1}{2}([4]^2 - [2\sec\theta]^2)\, d\theta$

21. $2\left[\int_0^{\pi/3} \frac{1}{2}(2\sec\theta)^2 d\theta + \int_{\pi/3}^{\pi/2} \frac{1}{2}(4)^2 d\theta\right]$ **23.** $\int_0^{\pi/3} \frac{1}{2}(2\sin 3\theta)^2 d\theta$ **25.** $2\left[\int_0^{\pi/6} \frac{1}{2}(\sin\theta)^2 d\theta + \int_{\pi/6}^{\pi/2} \frac{1}{2}(1 - \sin\theta)^2 d\theta\right]$

27. $\pi - 8\int_0^{\pi/4} \frac{1}{2}(\cos 2\theta)^2 d\theta$ **29.** $\frac{5}{12}\pi - \frac{1}{2}\sqrt{3}$ **31.** $\dfrac{\pi a^2}{2}$ **35.** $(5/6, 0)$ **37.** $\dfrac{9\pi}{2}$ **39.** $\dfrac{4\pi}{3} + 2\sqrt{3}$ **41.** (c) $8 - 2\pi$

SECTION 10.5

1. $4x = (y - 1)^2$ **3.** $y = 4x^2 + 1, x \geq 0$ **5.** $9x^2 + 4y^2 = 36$ **7.** $1 + x^2 = y^2$ **9.** $y = 2 - x^2, -1 \leq x \leq 1$

11. $2y - 6 = x, -4 \leq x \leq 4$

13. $y = x - 1$

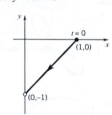

15. $xy = 1$

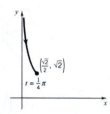

17. $y + 2x = 11$

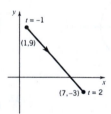

19. $x = \sin\frac{1}{2}\pi y$

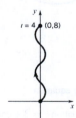

21. $y^2 = x^2 + 1$

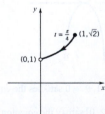

23. (a) $x(t) = -\sin 2\pi t, y(t) = \cos 2\pi t$ (b) $x(t) = \sin 4\pi t, y(t) = \cos 4\pi t$ (c) $x(t) = \cos\frac{1}{2}\pi t, y(t) = \sin\frac{1}{2}\pi t$

(d) $x(t) = \cos\frac{3}{2}\pi t, y(t) = -\sin\frac{3}{2}\pi t$

25. $x(t) = \tan\frac{1}{2}\pi t$, $\quad y(t) = 2$ **27.** $x(t) = 3 + 5t$, $\quad y(t) = 7 - 2t$ **29.** $x(t) = \sin^2\pi t$, $\quad y(t) = -\cos\pi t$

31. $x(t) = (2 - t)^2$, $y(t) = (2 - t)^3$ **33.** $\displaystyle\int_c^d y(t)x'(t)\,dt = \int_c^d f(x(t))x'(t)\,dt = \int_a^b f(x)\,dx = \text{area below } C$

35. $\displaystyle\int_c^d \pi[y(t)]^2 x'(t)\,dt = \int_c^d \pi[f(x(t))]^2 x'(t)\,dt = \int_a^b \pi[f(x)]^2 dx = V_x$;

$\displaystyle\int_c^d 2\pi x(t)y(t)x'(t)\,dt = \int_c^d 2\pi x(t)f(x(t))x'(t)\,dt = \int_a^b 2\pi x f(x)\,dx = V_y$

37. $A = 2\pi a^2$

39. (a) $V_x = 3\pi^2 a^3$ (b) $V_y = 4\pi^3 a^3$

41. $x(t) = -a\cos t$, $y(t) = b\sin t$; $t \in [0, \pi]$ **43.** (a) paths intersect at $(6, 5)$ and $(8, 1)$ (b) particles collide at $(8, 1)$

45. curve intersects itself at $(0, 0)$ **47.** curve intersects itself at $(0, 0)$ and $(0, \frac{3}{4})$

49. The particle moves along the parabola $y = 2x - \dfrac{x^2}{4}$ from $(0, 0)$ to $(12, -12)$.

51. The particle moves around the unit circle in the counterclockwise direction starting from the point $(1, 0)$.

53. (d) The curve has a loop if $a < b$; no loop if $a > b$ (e) $r = a - b\sin\theta$

SECTION 10.6

1. $3x - y - 3 = 0$ **3.** $y = 1$ **5.** $3x + y - 3 = 0$ **7.** $2x + 2y - \sqrt{2} = 0$ **9.** $2x + y - 8 = 0$ **11.** $x - 5y + 4 = 0$

13. $x + 2y + 1 = 0$ **15.** $x(t) = t, y(t) = t^3$; tangent line $y = 0$ **17.** $x(t) = t^{5/3}, y(t) = t$; tangent line $x = 0$ **19.** (a) none; (b) at $(2,2)$ and $(-2, 0)$

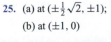

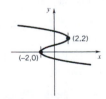

21. (a) at $(3, 7)$ and $(3, 1)$; (b) at $(-1, 4)$ and $(7, 4)$ **23.** (a) at $\left(-\frac{2}{3}, \pm\frac{2}{9}\sqrt{3}\right)$; (b) at $(-1, 0)$ **25.** (a) at $\left(\pm\frac{1}{2}\sqrt{2}, \pm 1\right)$; (b) at $(\pm 1, 0)$

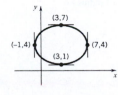   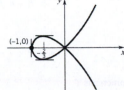   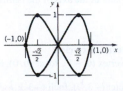

27. $y = 0$, $(\pi - 2)y + 32x - 64 = 0$

31.   **33.**   **35.**

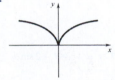      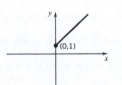

37. -8 **39.** 2 **41.** $\cot^3 t$ **43.** $y - 2 = -\frac{16}{3}\left(x - \frac{1}{8}\right)$

SECTION 10.7

1. $\sqrt{5}$ 3. 7   5. $2\sqrt{3}$ 7. $\frac{4}{3}$ 9. $6 + \frac{1}{2}\ln 5$ 11. $\frac{63}{8}$ 13. $\ln(1 + \sqrt{2})$ 15. $\frac{3}{2}$ 17. $\frac{1}{3}\pi + \frac{1}{2}\sqrt{3}$

19. initial speed 2, terminal speed 4; $s = 2\sqrt{3} + \ln(2 + \sqrt{3})$ 21. initial speed 0, terminal speed $\sqrt{13}$; $x = \frac{1}{27}(13\sqrt{13} - 8)$

23. initial speed $\sqrt{2}$, terminal speed $\sqrt{2}\,e^{\pi}$; $s = \sqrt{2}(e^{\pi} - 1)$ 25. 8a

27. (a) 24a   (b) use the identities $\cos 3\theta = 4\cos^3\theta - 3\cos\theta$, $\sin 3\theta = 3\sin\theta - 4\sin^3\theta$ 29. 2π 31. $\sqrt{2}(e^{4\pi} - 1)$

33. $\frac{1}{2}\sqrt{5}(e^{4\pi} - 1)$ 35. $4 - 2\sqrt{2}$ 37. $\ln(1 + \sqrt{2})$ 39. $c = 1$ 41. (a) $\left(\frac{1}{2}, -\frac{7}{2}\right)$

45. $L \cong 4.6984$ 47. (a)   (b) 2.7156

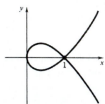

49. 4   51. (b) 28.3617   53. $\sqrt{1 + [f'(x)]^2} = \sqrt{1 + \tan^2[\alpha(x)]} = |\sec[\alpha(x)]|$

SECTION 10.8

1. $L = 1$, $(\bar{x}, \bar{y}) = \left(\frac{1}{2}, 4\right)$, $A_x = 8\pi$ 3. $L = 5$, $(\bar{x}, \bar{y}) = \left(\frac{3}{2}, 2\right)$, $A_x = 20\pi$ 5. $L = 10$, $(\bar{x}, \bar{y}) = (3, 4)$, $A_x = 80\pi$

7. $L = \frac{1}{3}\pi$, $\bar{x} = 6/\pi$, $\bar{y} = 6(2 - \sqrt{3})/\pi$, $A_x = 4\pi(2 - \sqrt{3})$ 9. $L = \frac{1}{3}\pi a$; $\bar{x} = 0$, $\bar{y} = 3a/\pi$, $A_x = 2\pi a^2$

11. $\frac{1}{9}\pi(17\sqrt{17} - 1)$ 13. $\frac{61}{432}\pi$ 15. $\pi[\sqrt{2} + \ln(1 + \sqrt{2})]$ 17. $\frac{2}{5}\sqrt{2}\pi(2e^{\pi} + 1)$ 19. (a) $3\pi a^2$ (b) $\dfrac{64\pi a^2}{3}$

23. (a) the 3, 4, 5 sides have centroids   25. $4\pi^2 ab$
   $\left(\frac{3}{2}, 0\right), (4, 2), \left(\frac{3}{2}, 2\right)$
   (b) $\bar{x} = 2$, $\bar{y} = \frac{3}{2}$
   (c) $\bar{x} = 2$, $\bar{y} = \frac{4}{3}$
   (d) $\bar{x} = \frac{13}{6}$, $\bar{y} = 2$
   (e) $A = 20\pi$

29. (a) $2\pi b^2 + \dfrac{2\pi ab}{e}\arcsin e$ (b) $2\pi a^2 + \dfrac{\pi b^2}{e}\ln\left|\dfrac{1 + e}{1 - e}\right|$, where e is the eccentricity $c/a = \sqrt{a^2 - b^2}/a$

31. at the midpoint of the axis of the hemisphere   33. on the axis of the cone $\left(\dfrac{2R + r}{R + r}\right)\dfrac{h}{3}$ units from the base of radius r

Chapter 10. Review Exercises

1. parabola; vertex $(0, -1)$, focus $(0, 0)$, axis $x = 0$, directrix $y = -2$

3. ellipse; center $(-3, 0)$, focii $\left(-3 \pm \frac{1}{2}\sqrt{3}, 0\right)$, major axis 2, minor axis 1

5. hyperbola; center $(1, -1)$, vertices $(1, -3)$, $(1, 1)$, focii $(1, -1 \pm \sqrt{13})$, asymptotes $y + 1 = \pm\frac{2}{3}(x - 1)$, transverse axis 4

7. ellipse; center $(1, -2)$, focii $(5, -2)$, $(-3, -2)$, major axis 10, minor axis 6   9. $(-2, -2\sqrt{3})$

11. $[4, \frac{3}{2}\pi + 2n\pi]$, $n = 0, \pm 1, \pm 2, \cdots$; $[-4, \frac{1}{2}\pi + 2n\pi]$, $n = 0, \pm 1, \pm 2, \cdots$

13. $[4, -\frac{1}{6}\pi + 2n\pi]$, $n = 0, \pm 1, \pm 2, \cdots$; $[-4, \frac{5}{6}\pi + 2n\pi]$, $n = 0, \pm 1, \pm 2, \cdots$ 15. $r = \sec\theta\tan\theta$ 17. $r = 4\cos\theta - 2\sin\theta$

19. $x = 5$ 21. $x^2 + y^2 - 3x - 4y = 0$

23. 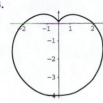   **25.**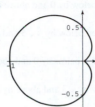

27. $\left[\frac{1}{2}\sqrt{2}, \frac{1}{6}\pi\right], \left[\frac{1}{2}\sqrt{2}, \frac{5}{6}\pi\right]$, the pole   **29.** 6π **31.** $\frac{1}{4}\pi$ **33.** $\frac{1}{2}\pi - \frac{1}{2}$

35. $y = 1 - 2x^2, 0 \le x \le 1$ **37.** $y = (x-1)^2, x \ge 1$

39. $x = 1 + 4t, y = 4 + 2t$ **41.** $x = 3\sin t, y = 2\cos t$ **43.** $5x + 3y = 30$

45. horizontal tangents at $(0, 10), (3, -\frac{7}{2})$; vertical tangent at $(-1, \frac{13}{2})$ **47.** $dy/dx = 2t, d^2y/dx^2 = 1/3t$ **49.** $\frac{19}{27}$

51. $\frac{1}{2}\sqrt{5} + \frac{1}{4}\ln(2 + \sqrt{5})$ **53.** 8   **55.** $\frac{992}{3}\pi$

57. $\frac{208}{9}\pi$ **61.**   **63.** $(\frac{2}{5}a. \frac{2}{5}a)$

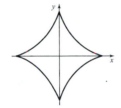

CHAPTER 11

SECTION 11.1

1. lub = 2; glb = 0   **3.** no lub; glb = 0   **5.** lub = 2; glb = −2   **7.** no lub; glb = 2   **9.** lub = $2\frac{1}{2}$; glb = 2

11. lub = 1; glb = 0.9   **13.** lub = e; glb = 0   **15.** lub = $\frac{1}{2}(-1 + \sqrt{5})$; glb = $\frac{1}{2}(-1 - \sqrt{5})$ **17.** no lub; no glb

19. no lub; no glb   **21.** glb $S = 0$, $0 \le (\frac{1}{11})^3 < 0 + 0.001$ **23.** glb $S = 0$, $0 \le (\frac{1}{10})^{2n-1} < 0 + (\frac{1}{10})^k$, $n > \frac{1}{2}(k+1)$

35. (a) 2.48832, 2.59374, 2.70481, 2.71692, 2.71815
(b) lub = e, glb = 2

37. (a)

a_1	a_2	a_3	a_4	a_5	a_6	a_7	a_8	a_9	a_{10}
1.4142	1.6818	1.8340	1.9152	1.9571	1.9785	1.9892	1.9946	1.9973	1.9986

(c) 2
(d) c

SECTION 11.2

1. $a_n = 2 + 3(n-1)$, $n = 1, 2, 3, \ldots$ **3.** $a_n = \dfrac{(-1)^{n-1}}{2n-1}$, $n = 1, 2, 3, \ldots$ **5.** $a_n = \dfrac{n^2 + 1}{n}$, $n = 1, 2, 3, \ldots$

7. $a_n = \begin{cases} n & \text{if } n = 2k - 1, \\ 1/n & \text{if } n = 2k, \end{cases}$ where $k = 1, 2, 3, \ldots$ **9.** decreasing; bounded below by 0 and above by 2

11. not monotonic; bounded below by 0 and above by $\frac{3}{2}$ **13.** decreasing; bounded below by 0 and above by 0.9

15. increasing; bounded below by $\frac{1}{2}$ but not bounded above   **17.** increasing; bounded below by $\frac{4}{5}\sqrt{5}$ and above by 2

19. increasing; bounded below by $\frac{2}{51}$ but not bounded above   **21.** increasing; bounded below by 0 and above by $\ln 2$

23. decreasing; bounded below by 1 and above by 4   **25.** increasing; bounded below by $\sqrt{3}$ and above by 2

27. decreasing; bounded above by -1 but not bounded below   **29.** increasing; bounded below by $\frac{1}{2}$ and above by 1

31. decreasing; bounded below by 0 and above by 1   **33.** decreasing; bounded below by 0 and above by $\frac{5}{6}$

35. decreasing; bounded below by 0 and above by $\frac{1}{2}$ **37.** decreasing; bounded below by 0 and above by $\frac{1}{3}\ln 3$

39. increasing; bounded below by $\frac{3}{4}$ but not bounded above

45. $a_1 = 1, a_2 = \frac{1}{2}, a_3 = \frac{1}{6}, a_4 = \frac{1}{24}, a_5 = \frac{1}{120}, a_6 = \frac{1}{720}$; $a_n = 1/n!$

47. $a_1 = a_2 = a_3 = a_4 = a_5 = a_6 = 1$; $a_n = 1$

49. $a_1 = 1, a_2 = 3, a_3 = 5, a_4 = 7, a_5 = 9, a_6 = 11$; $a_n = 2n - 1$

51. $a_1 = 1, a_2 = 4, a_3 = 9, a_4 = 16, a_5 = 25, a_6 = 36$; $a_n = n^2$

53. $a_1 = 1, a_2 = 1, a_3 = 2, a_4 = 4, a_5 = 8, a_6 = 16$; $a_n = 2^{n-2}$ for $n \geq 3$

55. $a_1 = 1, a_2 = 3, a_3 = 5, a_4 = 7, a_5 = 9, a_6 = 11$; $a_n = 2n - 1$

61. (a) n (b) $\dfrac{1 - r^n}{1 - r}$ **63.** (a) $150\left(\frac{3}{4}\right)^{n-1}$ (b) $\dfrac{5\sqrt{3}}{2}\left(\frac{3}{4}\right)^{(n-1)/2}$ **65.** increasing; limit $\frac{1}{2}$

67. (c) $2, 2.4142, 2.5538, 2.6118, \cdots, 2.6180$;   lub $= \frac{1}{2}(3 + \sqrt{5}) \cong 2.6180$

SECTION 11.3

1. diverges   **3.** converges to 0   **5.** converges to 1   **7.** converges to 0   **9.** converges to 0   **11.** diverges

13. converges to 0   **15.** converges to 1   **17.** converges to $\frac{4}{9}$ **19.** converges to $\frac{1}{2}\sqrt{2}$ **21.** diverges   **23.** converges to 1

25. converges to 0   **27.** converges to $\frac{1}{2}$ **29.** converges to e^2 **31.** diverges   **33.** 0   **35.** $\frac{1}{2}$

37. $\pi/2$ **39.** $-\pi/2$ **41.** (a) 1   (b) 0   **61.** converges to 0   **63.** converges to 0   **65.** diverges

67. $L = 0, n = 32$ **69.** $L = 0, n = 4$ **71.** $L = 0, n = 7$ **73.** $L = 0, n = 65$ **75.** (a) $\dfrac{3 + \sqrt{5}}{2}$ (b) 3

77. (a)

a_2	a_3	a_4	a_5	a_6	a_7	a_8	a_9	a_{10}
0.5403	0.8576	0.6543	0.7935	0.7014	0.7640	0.7221	0.7504	0.7314

(b) 0.7391; it is the fixed point of $f(x) = \cos x$.

SECTION 11.4

1. converges to 1   **3.** converges to 0   **5.** converges to 0   **7.** converges to 0   **9.** converges to 1   **11.** converges to 0

13. converges to 1   **15.** converges to 1   **17.** converges to π **19.** converges to 1   **21.** converges to 0   **23.** diverges

25. converges to 0   **27.** converges to e^{-1} **29.** converges to 0   **31.** converges to 0   **33.** converges to e^x **35.** converges to 0

37. 2   **41.** (b) $2\pi r$. As $n \to \infty$, the perimeter of the polygon tends to the circumference of the circle.   **43.** $\frac{1}{2}$ **45.** $\frac{1}{8}$ **51.** $L = 1$

53. (a)

a_3	a_4	a_5	a_6	a_7	a_8	a_9	a_{10}
2	3	5	8	13	21	34	55

(b)

r_1	r_2	r_3	r_4	r_5	r_6
1	2	1.5	1.6667	1.6000	1.625

(c) $L = \dfrac{1 + \sqrt{5}}{2} \cong 1.618033989$

SECTION 11.5

1. 0   **3.** 1   **5.** $\frac{1}{2}$ **7.** $\ln 2$ **9.** $\frac{1}{4}$ **11.** 2   **13.** $\dfrac{1+\pi}{1-\pi}$ **15.** $\frac{1}{2}$ **17.** π **19.** $-\frac{1}{2}$ **21.** -2

23. $\frac{1}{3}\sqrt{6}$ **25.** $-\frac{1}{8}$ **27.** 4   **29.** $\frac{1}{2}$ **31.** $\frac{1}{2}$ **33.** 1   **35.** 1   **37.** 0   **39.** 1   **41.** 1

43. $\lim\limits_{x \to 0}(2 + x + \sin x) \neq 0,\ \lim\limits_{x \to 0}(x^3 + x - \cos x) \neq 0$ **45.** $a = \pm 4, b = 1$ **47.** $-\dfrac{e}{2}$

49. $f(0)$ **51.** (a) 1   (b) $-\frac{1}{3}$ **53.** $\frac{3}{4}$ **55.** (a) $f(x) \to \infty$ as $x \to \pm\infty$ (b) 10   **57.** (b) $\ln 2 \cong 0.6931$

SECTION 11.6

1. ∞ **3.** -1 **5.** ∞ **7.** $\frac{1}{5}$ **9.** 1   **11.** 0   **13.** ∞ **15.** $\frac{1}{3}$ **17.** e **19.** 1   **21.** $\frac{1}{2}$ **23.** 0

25. 1   **27.** e^3 **29.** e **31.** 0   **33.** $-\frac{1}{2}$ **35.** 0   **37.** 1   **39.** 1   **41.** 0   **43.** 1   **45.** 0

47. y-axis vertical asymptote   **49.** x-axis horizontal asymptote   **51.** x-axis horizontal asymptote

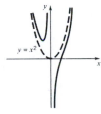

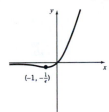

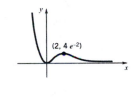

55. example: $f(x) = x^2 + \dfrac{(x-1)(x-2)}{x^3}$ **57.** $\lim\limits_{x \to 0^+} \cos x \neq 0$

61. (a) $A_b = 1 - (1 + b)e^{-b}$

   (b) $\bar{x}_b = \dfrac{2 - (2 + 2b + b^2)e^{-b}}{1 - (1 + b)e^{-b}}, \quad \bar{y}_b = \dfrac{\frac{1}{4} - \frac{1}{4}(1 + 2b + 2b^2)e^{-2b}}{2[1 - (1 + b)e^{-b}]}$

   (c) $\lim\limits_{b \to \infty} A_b = 1; \quad \lim\limits_{b \to \infty} \bar{x}_b = 2, \quad \lim\limits_{x \to \infty} \bar{y}_b = \frac{1}{8}$

63. $\lim\limits_{x \to 0^+}(1 + x^2)^{1/x} = 1.$ **65.** $\lim\limits_{x \to \infty} g(x) = -5/3.$

SECTION 11.7

1. 1   **3.** $\frac{1}{4}\pi$ **5.** diverges   **7.** 6   **9.** $\frac{1}{2}\pi$ **11.** 2   **13.** diverges   **15.** $-\frac{1}{4}$ **17.** π **19.** diverges   **21.** $\ln 2$

23. 4   **25.** diverges   **27.** diverges   **29.** diverges   **31.** $\frac{1}{2}$ **33.** $2e - 2$ **35.** (a) converges: $\frac{1}{32}$ **37.** $\frac{1}{2}\pi - 1$ **39.** π

                                               (b) converges: $\frac{\pi}{16}$

                                               (c) converges: $\frac{\pi}{16}$

                                               (d) diverges

43. (a)   (b) 2   (c) $V = \displaystyle\int_0^1 \pi\left(\dfrac{1}{\sqrt{x}}\right)^2 dx = \pi \int_0^1 \dfrac{1}{x}\,dx,$ diverges   **45.** (a)   (b) 1

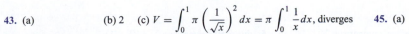

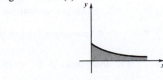

                                                                (c) $\frac{1}{2}\pi$

                                                                (d) 2π

                                                                 (e) $\pi[\sqrt{2} + \ln(1 + \sqrt{2})]$

47. (b) $V_y = \int_0^\infty 2\pi x\, e^{-x^2} dx = \pi$

49. (a)   (b) $\frac{4}{3}$ **51.** converges by comparison with $\int_0^\infty \frac{dx}{x^{3/2}}$

(c) 2π

(d) $\frac{8}{7}\pi$

53. diverges since for x large the integrand is greater than $\frac{1}{x}$ and $\int_1^\infty \frac{1}{x}\,dx$ diverges   **55.** converges by comparison with $\int_1^\infty \frac{dx}{x^{3/2}}$

59. $L = (a\sqrt{1+c^2}/c)\,e^{c\theta_1}$ **61.** $\frac{1}{s}$; dom $(F) = (0, \infty)$ **63.** $\frac{s}{s^2+4}$; dom $(F) = (0, \infty)$ **67.** $\frac{1}{k}$

Chapter 11. Review Exercises

1. lub $= 5$, glb $= -1$ **3.** lub $= 2$, glb $= -1$ **5.** no lub, no glb   **7.** increasing; bounded below by $\frac{1}{2}$, bounded above by $\frac{2}{3}$

9. not monotonic; bounded below by 0, bounded above by $\frac{3}{2}$ **11.** increasing from a_3 on; bounded below by $\frac{8}{9}$, not bounded above

13. diverges   **15.** converges to 1   **17.** converges to 0   **19.** converges to 0   **21.** converges to $\frac{3}{2}$ **23.** converges to 0

25. converges to 0   **29.** $L \cong 0.7391$ **31.** 5   **33.** $-\frac{1}{2}$ **35.** e^8 **37.** 0   **39.** $-\frac{1}{4}$ **41.** $\frac{1}{2}$ **43.** $2e^{-1}$

45. diverges   **47.** $2/\pi$ **49.** $\frac{1}{2}\pi$ **51.** $a \ln(1/a) + a$ **55.** (b) $\int_{-\infty}^\infty x\,dx$

CHAPTER 12

SECTION 12.1

1. 11   **3.** 15   **5.** -5 **7.** $\frac{13}{27}$ **9.** $\frac{85}{64}$ **11.** $\sum_{k=1}^{11}(2k-1)$ **13.** $\sum_{k=1}^{35} k(k+1)$ **15.** $\sum_{k=1}^{n} M_k \Delta x_k$ **17.** $\sum_{k=3}^{10} \frac{1}{2^k}$, $\sum_{i=0}^{7} \frac{1}{2i+3}$

19. $\sum_{k=3}^{10}(-1)^{k-1}\frac{k}{k+1}$, $\sum_{i=0}^{7}(-1)^i \frac{i+3}{i+4}$ **21.** set $k = n+3$ **23.** set $k = n-3$ **25.** $\sum_{k=1}^{\infty} \frac{a_k}{10^k}$ **27.** $1.33333\cdots$ **29.** $2.71828\cdots$

SECTION 12.2

1. $\frac{1}{2}$ **3.** $\frac{11}{18}$ **5.** $\frac{10}{3}$ **7.** $-\frac{3}{2}$ **9.** 24   **15.** $\sum_{i=0}^{\infty} x^{k+1}$ **17.** $\sum_{k=0}^{\infty}(-1)^k x^{2k+1}$

19. $\sum_{k=0}^{\infty}\left(\frac{3}{2}\right)^k$; geometric series with $r = \frac{3}{2} > 1$ **21.** $\lim_{k\to\infty}\left(\frac{k+1}{k}\right)^k = e \neq 0$ **23.** 18 ft   **25.** $\sum_{k=0}^{\infty} n_k\left(1+\frac{r}{100}\right)^{-k}$ **27.** \$9   **29.** 32

35. (b) (i) $\sum_{k=1}^{\infty} \frac{\sqrt{k+1}-\sqrt{k}}{\sqrt{k(k+1)}} = \sum_{k=1}^{\infty}\left(\frac{1}{\sqrt{k}} - \frac{1}{\sqrt{k+1}}\right) = 1$ (ii) $\sum_{k=1}^{\infty} \frac{2k+1}{2k^2(k+1)^2} = \sum_{k=1}^{\infty} \frac{1}{2}\left(\frac{1}{k^2} - \frac{1}{(k+1)^2}\right) = \frac{1}{2}$ **37.** $N = 6$

39. $N = 9999$ **41.** $N = \left[\left[\frac{\ln(\epsilon[1-x])}{\ln|x|}\right]\right]$, where $[[\ ]]$ denotes the greatest integer function.

SECTION 12.3

1. converges; comparison $\sum 1/k^2$ **3.** converges; comparison $\sum 1/k^2$ **5.** diverges; comparison $\sum 1/(k+1)$

7. diverges; limit comparison $\sum 1/k$ **9.** converges; integral test   **11.** diverges; p-series with $p = \frac{2}{3} \leq 1$ **13.** diverges; $a_k \nrightarrow 0$

15. diverges; comparison $\sum 1/k$ **17.** diverges; $a_k \nrightarrow 0$ **19.** converges; limit comparison $\sum 1/k^2$ **21.** diverges; integral test

23. converges; limit comparison with $\sum \dfrac{2^k}{5^k}$ 25. diverges; limit comparison $\sum 1/k$ 27. converges; limit comparison $\sum 1/k^{3/2}$

29. converges; integral test   31. converges; comparison $\sum 3/k^2$ 33. converges; comparison $\sum 2/k^2$

35. converges; comparison $\sum 1/k^2$ 37. $p > 1$

41. (a) 1.1777   (b) $0.02 < R_4 < 0.0313$ (c) $1.1977 < \sum\limits_{k=1}^{\infty} \dfrac{1}{k^3} < 1.209$ 43. (a) $1/101 < R_{100} < 1/100$ (b) 10,001

45. (a) 15   (b) 7   (c) 1.082   49. (b) $\sum a_k$ may either converge or diverge.

$\sum 1/k^4$ converges and $\sum 1/k^2$ converges; $\sum 1/k^2$ converges and $\sum 1/k$ diverges.

53. $N = 3$

SECTION 12.4

1. converges; ratio test   3. converges; root test   5. diverges; ratio test   7. diverges; limit comparison $\sum 1/k$

9. converges; root test   11. diverges; limit comparison $\sum 1/\sqrt{k}$ 13. diverges; ratio test   15. converges; comparison $\sum 1/k^{3/2}$

17. converges; comparison $\sum 1/k^2$ 19. diverges; integral test   21. diverges; $a_k \to e^{-100} \neq 0$ 23. diverges; limit comparison $\sum 1/k$

25. converges; ratio test   27. converges; comparison $\sum 1/k^{3/2}$ 29. converges; ratio test   31. converges; ratio test: $a_{k+1}/a_k \to \dfrac{4}{27}$

33. converges; ratio test   35. converges; root test   37. converges; root test   39. converges; ratio test   41. $\dfrac{10}{81}$

45. $p \geq 2$

SECTION 12.5

1. diverges; $a_k \not\to 0$ 3. diverges; $a_k \not\to 0$

5. (a) does not converge absolutely; integral test   (b) converges conditionally; Theorem 12.5.3   7. diverges; limit comparison $\sum 1/k$

9. (a) does not converge absolutely; limit comparison $\sum 1/k$ (b) converges conditionally; Theorem 12.5.3   11. diverges; $a_k \not\to 0$

13. (a) does not converge absolutely; comparison $\sum 1/\sqrt{k+1}$ (b) converges conditionally; Theorem 12.5.3

15. converges absolutely (terms already positive); $\sum \sin\left(\dfrac{\pi}{4k^2}\right) \leq \sum \dfrac{\pi}{4k^2} = \dfrac{\pi}{4} \sum \dfrac{1}{k^2}$ ($|\sin x| \leq |x|$)   17. converges absolutely; ratio test

19. (a) does not converge absolutely; limit comparison $\sum 1/k$ (b) converges conditionally; Theorem 12.5.3   21. diverges; $a_k \not\to 0$

23. diverges; $a_k \not\to 0$ 25. converges absolutely; ratio test   27. diverges; $a_k = \dfrac{1}{k}$ for all k 29. converges absolutely; comparison $\sum 1/k^2$

31. diverges; $a_k \not\to 0$ 33. 0.1104   35. 0.001   37. $\dfrac{10}{11}$ 39. $N = 39{,}998$

41. (a) 4   (b) 6   43. No.   For instance, set $a_{2k} = 2/k$ and $a_{2k+1} = 1/k$.

45. (b) $\sum 1/k^2$ is convergent, $\sum \dfrac{(-1)^k}{k}$ is not absolutely convergent.

SECTION 12.6

1. $-1 + x + \dfrac{1}{2}x^2 - \dfrac{1}{24}x^4$ 3. $-\dfrac{1}{2}x^2 - \dfrac{1}{12}x^4$ 5. $1 - x + x^2 - x^3 + x^4 - x^5$ 7. $x + \dfrac{1}{3}x^3 + \dfrac{2}{15}x^5$

9. $P_0(x) = 1$, $P_1(x) = 1 - x$, $P_2(x) = 1 - x + 3x^2$, $P_3(x) = 1 - x + 3x^2 + 5x^3$ 11. $\sum\limits_{k=0}^{n} (-1)^k \dfrac{x^k}{k!}$

13. $\sum_{k=0}^{m} \dfrac{x^{2k}}{(2k!)}$ where $m = \dfrac{n}{2}$ and n is even   **15.** $\sum_{k=0}^{n} \dfrac{r^k}{k!} x^k$ **17.** 0.00002 **19.** $n = 9$ **21.** $n = 6$ **23.** $|x| < 1.513$

25. $79/48$ $(79/48 \cong 1.646)$ **27.** $5/6$ $(5/6 \cong 0.833)$ **29.** $13/24$ $(13/24 \cong 0.542)$ **31.** 0.1745 **33.** $\dfrac{4e^{2c}}{15} x^5$, $|c| < |x|$

35. $\dfrac{-4\sin 2c}{15} x^5$, $|c| < |x|$ **37.** $\dfrac{3\sec^4 c - 2\sec^2 c}{3} x^3$, $|c| < |x|$ **39.** $\dfrac{3c^2 - 1}{3(1 + c^2)^3} x^3$, $|c| < |x|$ **41.** $\dfrac{(-1)^{n+1} e^{-c}}{(n+1)!} x^{n+1}$, $|c| < |x|$

43. $\dfrac{1}{(1-c)^{n+2}} x^{n+1}$, $|c| < |x|$ **45.** (a) 4   (b) 2   (c) 999   **47.** (a) 1.649   (b) 0.368

53. $\sum_{k=0}^{\infty} \dfrac{a^k}{k!} x^k$, $(-\infty, \infty)$ **55.** $\sum_{k=0}^{\infty} \dfrac{(-1)^k a^{2k}}{(2k)!} x^{2k}$, $(-\infty, \infty)$ **57.** $\ln a + \sum_{k=1}^{\infty} \dfrac{(-1)^{k-1}}{ka^k} x^k$, $(-a, a]$

59. $\ln 2 = \ln\left(\dfrac{1 + \frac{1}{3}}{1 - \frac{1}{3}}\right) \cong 2\left[\dfrac{1}{3} + \dfrac{1}{3}\left(\dfrac{1}{3}\right)^3 + \dfrac{1}{5}\left(\dfrac{1}{3}\right)^5\right] = \dfrac{842}{1215}$ $\left(\dfrac{842}{1215} \cong 0.693\right)$

63. (d) 0   (e) $x = 0$

65. $P_2(x) = x - \frac{1}{2}x^2$, $P_3(x) = x - \frac{1}{2}x^2 + \frac{1}{3}x^3$, $P_4(x) = x - \frac{1}{2}x^2 + \frac{1}{3}x^3 - \frac{1}{4}x^4$, $P_5(x) = x - \frac{1}{2}x^2 + \frac{1}{3}x^3 - \frac{1}{4}x^4 + \frac{1}{5}x^5$

SECTION 12.7

1. $P_3(x) = 2 + \frac{1}{4}(x - 4) - \frac{1}{64}(x - 4)^2 + \frac{1}{512}(x - 4)^3$ **3.** $P_4(x) = \dfrac{\sqrt{2}}{2} + \dfrac{\sqrt{2}}{2}\left(x - \dfrac{\pi}{4}\right) - \dfrac{\sqrt{2}}{4}\left(x - \dfrac{\pi}{4}\right)^2 - \dfrac{\sqrt{2}}{12}\left(x - \dfrac{\pi}{4}\right) + \dfrac{\sqrt{2}}{48}\left(x - \dfrac{\pi}{4}\right)^4$

$R_3(x) = \dfrac{-5}{128c^{7/2}}(x - 4)^4$, $|c - 4| < |x - 4|$ $R_4(x) = \dfrac{\cos c}{120}\left(x - \dfrac{\pi}{4}\right)^5$, $\left|c - \dfrac{\pi}{4}\right| < \left|x - \dfrac{\pi}{4}\right|$

5. $P_3(x) = \dfrac{\pi}{4} + \dfrac{1}{2}(x - 1) - \dfrac{1}{4}(x - 1)^2 + \dfrac{1}{12}(x - 1)^3$ $R_3(x) = \dfrac{c(1 - c^2)}{(1 + c^2)^4}(x - 1)^4$, $|c - 1| < |x - 1|$

7. $6 + 9(x - 1) + 7(x - 1)^2 + 3(x - 1)^3$, $(-\infty, \infty)$ **9.** $-3 + 5(x + 1) - 19(x + 1)^2 + 20(x + 1)^3 - 10(x + 1)^4 + 2(x + 1)^5$, $(-\infty, \infty)$

11. $\sum_{k=0}^{\infty} (-1)^k \left(\dfrac{1}{2}\right)^{k+1}(x - 1)^k$, $(-1, 3)$ **13.** $\dfrac{1}{5}\sum_{k=0}^{\infty} \left(\dfrac{2}{5}\right)^k (x + 2)^k$, $\left(-\dfrac{9}{2}, \dfrac{1}{2}\right)$ **15.** $\sum_{k=0}^{\infty} \dfrac{(-1)^{k+1}}{(2k + 1)!}(x - \pi)^{2k+1}$, $(-\infty, \infty)$

17. $\sum_{k=0}^{\infty} \dfrac{(-1)^{k+1}}{(2k)!}(x - \pi)^{2k}$, $(-\infty, \infty)$ **19.** $\sum_{k=0}^{\infty} \dfrac{(-1)^k}{(2k)!}\left(\dfrac{\pi}{2}\right)^{2k}(x - 1)^{2k}$, $(-\infty, \infty)$ **21.** $\ln 3 + \sum_{k=1}^{\infty} \dfrac{(-1)^{k+1}}{k}\left(\dfrac{2}{3}\right)^k (x - 1)^k$, $\left(-\dfrac{1}{2}, \dfrac{5}{2}\right]$

23. $2\ln 2 + (1 + \ln 2)(x - 2) + \sum_{k=2}^{\infty} \dfrac{(-1)^k}{k(k-1)2^{k-1}}(x - 2)^k$ **25.** $\sum_{k=0}^{\infty} \dfrac{(-1)^k}{(2k + 1)!} x^{2k+2}$ **27.** $\sum_{k=0}^{\infty} (k + 2)(k + 1)\dfrac{2^{k-1}}{5^{k+3}}(x + 2)^k$

29. $1 + \sum_{k=1}^{\infty} \dfrac{(-1)^k 2^{2k-1}}{(2k)!}(x - \pi)^{2k}$ **31.** $\sum_{k=0}^{n} \dfrac{n!}{(n - k)!k!}(x - 1)^k$

33. (a) $\dfrac{e^x}{e^a} = e^{x-a} = \sum_{k=0}^{\infty} \dfrac{(x - a)^k}{k!}$, $e^x = e^a \sum_{k=0}^{\infty} \dfrac{(x - a)^k}{k!}$ (c) $e^{-a} \sum_{k=0}^{\infty} (-1)^k \dfrac{(x - a)^k}{k!}$

35. $P_6(x) = \dfrac{\pi}{4} + \dfrac{1}{2}(x - 1) - \dfrac{1}{4}(x - 1)^2 + \dfrac{1}{12}(x - 1)^3 - \dfrac{1}{40}(x - 1)^5 + \dfrac{1}{48}(x - 1)^6$

SECTION 12.8

1. (a) absolutely convergent   (b) absolutely convergent   (c) ?   (d) ?   **3.** $(-1, 1)$ **5.** $(-\infty, \infty)$ **7.** $\{0\}$

9. $[-2, 2)$ **11.** $\{0\}$ **13.** $\left[-\frac{1}{2}, \frac{1}{2}\right)$ **15.** $(-1, 1)$ **17.** $(-10, 10)$ **19.** $(-\infty, \infty)$ **21.** $(-\infty, \infty)$ **23.** $(-3/2, 3/2)$

25. converges only at $x = 1$ **27.** $(-4, 0)$ **29.** $(-\infty, \infty)$ **31.** $(-1, 1)$ **33.** $(0, 4)$ **35.** $\left(-\frac{5}{2}, \frac{1}{2}\right)$ **37.** $(-2, 2)$

39. $\left[-\dfrac{1}{\sqrt{3}}, \dfrac{1}{\sqrt{3}}\right]$ 41. (a) absolutely convergent   (b) absolutely convergent   (c) ?

43. (a) $\sum |a_k r^k| = \sum |a_k(-r)^k|$ 45. (a) 1

   (b) If $\sum |a_k(-r)^k|$ converges, then $\sum a_k(-r)^k$ converges.   (b) $\sum a_k x^k = \dfrac{a_0 + a_1 x + a_2 x^2}{1 - x^3}$

SECTION 12.9

1. $1 + 2x + 3x^2 + \cdots + nx^{n-1} + \cdots$ 3. $1 + kx + \dfrac{(k+1)^k}{2!}x^2 + \cdots + \dfrac{(n+k-1)!}{n!(k-1)!}x^n + \cdots$

5. $\ln(1 - x^2) = -x^2 - \dfrac{1}{2}x^4 - \dfrac{1}{3}x^6 - \cdots - \dfrac{1}{n+1}x^{2n+2} - \cdots$ 7. $1 + x^2 + \dfrac{2}{3}x^4 + \dfrac{17}{45}x^6 + \cdots$ 9. $-72.$ 11. $\sum\limits_{k=0}^{\infty} \dfrac{(-1)^k}{(2k+1)!}x^{4k+2}$

13. $\sum\limits_{k=0}^{\infty} \dfrac{3^k}{k!}x^{3k}$ 15. $2\sum\limits_{k=0}^{\infty} x^{2k+1}$ 17. $\sum\limits_{k=0}^{\infty} \dfrac{(k!+1)}{k!}x^k$ 19. $\sum\limits_{k=1}^{\infty} \dfrac{(-1)^{k+1}}{k}x^{3k+1}$ 21. $\sum\limits_{k=0}^{\infty} \dfrac{(-1)^k}{k!}x^{3k+3}$ 23. $\dfrac{1}{2}$ 25. $-\dfrac{1}{2}$

27. $\sum\limits_{k=1}^{\infty} \dfrac{(-1)^{k-1}}{k^2}x^k,\ -1 \le x \le 1$ 29. $\sum\limits_{k=0}^{\infty} \dfrac{(-1)^k}{(2k+1)^2}x^{2k+1},\ -1 \le x \le 1$ 31. $0.804 \le I \le 0.808$ 33. $0.600 \le I \le 0.603$

35. $0.294 \le I \le 0.304$ 37. 0.9461 39. 0.4485 41. e^{x^2} 43. $3x^2 e^{x^3}$ 45. (a) $\sum\limits_{k=0}^{\infty} \dfrac{1}{k!}x^{k+1}$

51. $f(x) = x - \dfrac{2}{3!}x^3 + \dfrac{4}{5!}x^5 - \dfrac{8}{7!}x^7 + \cdots = \sum\limits_{k=0}^{\infty} \dfrac{(-1)^k 2^k}{(2k+1)!}x^{2k+1};\ \dfrac{1}{\sqrt{2}}\sin(x\sqrt{2})$ 53. $0.0352 \le 1 \le 0.0359;\ I = \dfrac{3}{16} - \dfrac{3}{8}\ln 1.5 \cong 0.0354505$

55. $0.2640 \le I \le 0.2643;\ I = 1 - 2/e \cong 0.2642411$

Chapter 12. Review Exercises

1. 4   3. 2   5. diverges; limit comparison with $\sum 1/k$ 7. absolutely convergent

9. absolutely convergent   11. diverges   13. conditionally convergent

15. converges   17. diverges   19. converges

21. $\sum\limits_{k=0}^{\infty} \dfrac{2^k}{k!}x^{2k+1}$ 23. $\sum\limits_{k=0}^{\infty} \dfrac{(-1)^k}{2k+1}x^{k+1}$ 25. $2\sum\limits_{k=0}^{\infty} \dfrac{1}{2k+1}x^{4k+3}$ 27. $1 - \dfrac{1}{3}x - \dfrac{1}{9}x^2 - \dfrac{5}{81}x^3$ 29. $\left[-\dfrac{1}{5}, \dfrac{1}{5}\right)$

31. $(-\infty, \infty)$ 33. $(-9, 9)$ 35. $(-4, -2]$ 37. $e^2 \sum\limits_{k=0}^{\infty} \dfrac{(-1)^k 2^k}{k!}(x+1)^k,\ r = \infty$ 39. $\sum\limits_{k=1}^{\infty} \dfrac{(-1)^{k+1}}{k}(x-1)^k,\ r = 1$

41. 0.494 43. 4.0816 45. 0.74316 49. $2/e$

Appendix A.2. Answers

1. -2 3. 0   5. 5   7. -6

9. Calculate the determinant of A^T and compare it with the determinant of A.

11. $\begin{vmatrix} 1 & 2 & 3 \\ 4 & 5 & 6 \\ 7 & 8 & 9 \end{vmatrix} = -\begin{vmatrix} 4 & 5 & 6 \\ 1 & 2 & 3 \\ 7 & 8 & 9 \end{vmatrix} = \begin{vmatrix} 4 & 5 & 6 \\ 7 & 8 & 9 \\ 1 & 2 & 3 \end{vmatrix}$ 13. $\dfrac{1}{2}\begin{vmatrix} 1 & 0 & 7 \\ 3 & 4 & 5 \\ 2 & 4 & 6 \end{vmatrix} = \dfrac{1}{2}(2)\begin{vmatrix} 1 & 0 & 7 \\ 3 & 4 & 5 \\ 1 & 2 & 3 \end{vmatrix} = \begin{vmatrix} 1 & 0 & 7 \\ 3 & 4 & 5 \\ 1 & 2 & 3 \end{vmatrix}$

15. (c) Let $D = \begin{vmatrix} a_1 & a_2 & a_3 \\ b_1 & b_2 & b_3 \\ c_1 & c_2 & c_3 \end{vmatrix}$. Then $x = \dfrac{\begin{vmatrix} \alpha & a_2 & a_3 \\ \beta & b_2 & b_3 \\ \gamma & c_2 & c_3 \end{vmatrix}}{D}$, $y = \dfrac{\begin{vmatrix} a_1 & \alpha & a_3 \\ b_1 & \beta & b_3 \\ c_1 & \gamma & c_3 \end{vmatrix}}{D}$, $z = \dfrac{\begin{vmatrix} a_1 & a_2 & \alpha \\ b_1 & b_2 & \beta \\ c_1 & c_2 & \gamma \end{vmatrix}}{D}$

■ INDEX

(*continued from the front*)

INVERSE TRIGONOMETRIC FUNCTIONS

64. $\displaystyle\int \arcsin u \, du = u \arcsin u + \sqrt{1 - u^2} + C$

65. $\displaystyle\int \arccos u \, du = u \arccos u - \sqrt{1 - u^2} + C$

66. $\displaystyle\int \arctan u \, du = u \arctan u - \tfrac{1}{2} \ln(1 + u^2) + C$

67. $\displaystyle\int \text{arccot}\, u \, du = u \,\text{arccot}\, u + \tfrac{1}{2} \ln(1 + u^2) + C$

68. $\displaystyle\int \text{arcsec}\, u \, du = u \,\text{arcsec}\, u - \ln |u + \sqrt{u^2 - 1}| + C$

69. $\displaystyle\int \text{arccsc}\, u \, du = u \,\text{arccsc}\, u + \ln \left|u + \sqrt{u^2 - 1}\right| + C$

70. $\displaystyle\int u \arcsin u \, du = \tfrac{1}{4}\left[(2u^2 - 1)\arcsin u + u\sqrt{1 - u^2} + C\right]$

71. $\displaystyle\int u \arctan u \, du = \tfrac{1}{2}(u^2 + 1)\arctan u - \tfrac{1}{2}u + C$

72. $\displaystyle\int u \arccos u \, du = \tfrac{1}{4}\left[(2u^2 - 1)\arccos u - u\sqrt{1 - u^2} + C\right]$

73. $\displaystyle\int u^n \arcsin u \, du = \frac{1}{n+1}\left[u^{n+1} \arcsin u - \int \frac{u^{n+1}}{\sqrt{1 - u^2}}\, du\right], n \neq -1$

74. $\displaystyle\int u^n \arccos u \, du = \frac{1}{n+1}\left[u^{n+1} \arccos u + \int \frac{u^{n+1}}{\sqrt{1 - u^2}}\, du\right], n \neq -1$

75. $\displaystyle\int u^n \arctan u \, du = \frac{1}{n+1}\left[u^{n+1} \arctan u - \int \frac{u^{n+1}}{1 + u^2}\, du\right], n \neq -1$

$$\sqrt{a^2 + u^2},\ a > 0$$

76. $\displaystyle\int \frac{du}{a^2 + u^2} = \frac{1}{a} \arctan \frac{u}{a} + C$

77. $\displaystyle\int \frac{du}{\sqrt{a^2 + u^2}} = \ln\left(u + \sqrt{a^2 + u^2}\right) + C$

78. $\displaystyle\int \sqrt{a^2 + u^2}\, du = \frac{u}{2}\sqrt{a^2 + u^2} + \frac{a^2}{2} \ln\left(u + \sqrt{a^2 + u^2}\right) + C$

79. $\displaystyle\int u^2\sqrt{a^2 + u^2}\, du = \frac{u}{8}(a^2 + 2u^2)\sqrt{a^2 + u^2} - \frac{a^4}{8} \ln\left(u + \sqrt{a^2 + u^2}\right) + C$

80. $\displaystyle\int \frac{\sqrt{a^2 + u^2}}{u}\, du = \sqrt{a^2 + u^2} - a \ln\left|\frac{a + \sqrt{a^2 + u^2}}{u}\right| + C$

81. $\displaystyle\int \frac{\sqrt{a^2 + u^2}}{u^2}\, du = -\frac{\sqrt{a^2 + u^2}}{u} + \ln\left(u + \sqrt{a^2 + u^2}\right) + C$

82. $\displaystyle\int \frac{u^2 \, du}{\sqrt{a^2 + u^2}} = \frac{u}{2}\sqrt{a^2 + u^2} - \frac{a^2}{2} \ln\left(u + \sqrt{a^2 + u^2}\right) + C$

83. $\displaystyle\int \frac{du}{u\sqrt{a^2 + u^2}} = -\frac{1}{a} \ln\left|\frac{a + \sqrt{a^2 + u^2}}{u}\right| + C$

84. $\displaystyle\int \frac{du}{u^2\sqrt{a^2 + u^2}} = -\frac{\sqrt{a^2 + u^2}}{a^2 u} + C$

85. $\displaystyle\int \frac{du}{(a^2 + u^2)^{3/2}} = \frac{u}{a^2\sqrt{a^2 + u^2}} + C$

$$\sqrt{a^2 - u^2},\ a > 0$$

86. $\displaystyle\int \frac{du}{\sqrt{a^2 - u^2}} = \arcsin \frac{u}{a} + C$

87. $\displaystyle\int \sqrt{a^2 - u^2}\, du = \frac{u}{2}\sqrt{a^2 - u^2} + \frac{a^2}{2} \arcsin \frac{u}{a} + C$

88. $\displaystyle\int u^2\sqrt{a^2 - u^2}\, du = \frac{u}{8}(2u^2 - a^2)\sqrt{a^2 - u^2} + \frac{a^4}{8} \arcsin \frac{u}{a} + C$